THE
ASTRONOMY
ENCYCLOPAEDIA

THE
ASTRONOMY
ENCYCLOPAEDIA

General Editor
PATRICK MOORE

MITCHELL BEAZLEY

THE INTERNATIONAL ENCYCLOPEDIA OF ASTRONOMY was edited and designed by Mitchell Beazley International Limited, Artists House, 14–15 Manette Street, London W1V 5LB

THE ASTRONOMY ENCYCLOPAEDIA was edited and designed by Mitchell Beazley International Limited, Artists House, 14–15 Manette Street, London W1V 5LB

Project Manager and Co-Editor	Frank Wallis
Art Editor	Mike Brown
Picture researchers	Brigitte Arora
	Anne Hobart
Production	Philip Collyer
Project Secretary	Janet Taylor

British Library Cataloguing-in-Publication Data

The Astronomy encyclopaedia.
　1. Astronomy——Dictionaries
　I. Moore, Patrick
　520′.3′21　　QB14

　ISBN 0–85533–604–8

The publishers will be grateful for any information which will assist them in keeping future editions up to date.

Photoset in Great Britain by Hourds Typographica Limited, Stafford
Reproduction by Nova Colour Limited, Birmingham
Printed in Spain by Graficas Estella, S.A., Navarra, Spain

Contributors

Contributors to the encyclopaedia, together with the initials used to identify them, are listed below in alphabetical order of initials. All except short entries are signed.

ADA Dr. A.D.Andrews, Armagh Observatory, Co Antrim

DAA Dr. David A.Allen, Anglo-Australian Observatory, Epping, New South Wales

DJA Dr. D.J.Adams, PhD, Astronomy Department, University of Leicester

HJPA H.J.P.Arnold, MA(Oxon), Dip CAM, FRAS, FBIS, Managing Director, Space Frontiers Ltd, Havant, Hants

RWA[1] R.W.Arbour, Pennell Observatory, South Wonston, Hants

RWA[2] R.W.Argyle, BSc, MSc, FRAS, Royal Greenwich Observatory (Tenerife, Canary Islands)

WIA Professor W.I.Axford, FRS, Max-Planck-Institut fur Aeronomie, Katlenburg-Lindau, FRG

AB Professor A.Boksenberg, FRS, Director, Royal Greenwich Observatory (Hailsham, Sussex)

DLB Dr. D.L.Block, FRAS, Senior Lecturer, Department of Applied Mathematics, University of Witwatersrand, Johannesburg

DPB D.P.Bian, Beijing Planetarium, Beijing, People's Republic of China

EB[1] Dr. E.Bowell, BSc, D del'u, FRAS, Lowell Observatory, Flagstaff, Arizona

EB[2] Dr. E.Budding, PhD, FRAS, Carter Observatory, Wellington, NZ

FMB Dr. F.M.Bateson, OBE, Managing Director, Astronomical Research Ltd, Greerton, Tauranga, NZ

GLB G.L.Blow, MSc(Hons), FRAS, Carter Observatory, Wellington, NZ

RFB Dr. Reta Beebe, PhD, Professor of Astronomy, New Mexico State University

RMB R.M.Baum, FRAS, Chester, Cheshire

SJBB Dr. S.J.Bell Burnell, Royal Observatory, Edinburgh

VB Professor V.Barocas, FInstP, FRAS, Emeritus Professor of Astronomy Preston Polytechnic, Former Director of the Preston Observatories & Past President of the British Astronomical Association.

DHC Dr. David H.Clark, PhD, Science & Engineering Research Council

DPC Dr. Dale P.Cruikshank, Professor of Astronomy, University of Hawaii

JLC Professor J.L.Culhane, FRS, Director, Mullard Space Science Laboratory, Dorking, Surrey

MC Dr. M.Cohen, Radio Astronomy Laboratory, University of California

PGEC P.G.E.Corvan, Armagh Observatory, Co Antrim

PJC Dr. P.J.Cattermole, Department of Geology, Sheffield University

VDC Von Del Chamberlain, MS in Astronomy, Director, Hansen Planetarium, Saltlake City, Utah & President of the International Planetarium Society

ACD Professor Audouin Dollfus, Observatorie de Paris, Meudon Principal Cedex

DD D.W.Dewhirst, MA, PhD, Senior Assistant Observer & Librarian, Institute of Astronomy, Cambridge, England

JKD Dr. J.K.Davies, Department of Space Research, University of Birmingham

LMD Commander L.M.Dougherty, MM, BSc(Hons), MSc, PhD, MRIN, FRAS, Halifax

MED M.E.Davies, The Rand Corporation, Santa Monica, California

RD Professor R.Davis, Jr, PhD Yale University 1942,

Research Professor, University of Pennsylvania

JPE Dr. J.P.Emerson, Department of Physics, Queen Mary College, London

GF Dr. G.Fielder, BSc, PhD, Reader Emeritus University of Lancaster

MWF Professor M.W.Feast, Director, South African Astronomical Observatory, Cape

NEF Norman Fisher, London

ACG A.C.Gilmore, Mt John Observatory, University of Canterbury, Lake Tekapo, NZ

KWG K.W.Gatland, FRAS, FBIS, Epsom, Surrey

MdG Dr. Mart de Groot, Director, Armagh Observatory, Co Antrim

OG Professor Owen Gingerich, Harvard-Smithsonian Center for Astrophysics, Cambridge, Mass

RHG Professor R.H.Garstang, ScD, PhD, Professor of Astrophysical, Planetary & Atmospheric Sciences, University of Colorado

AAH Dr. A.A.Hoag, Former Director, Lowell Observatory, Flagstaff, Arizona

AJH A.J.Hollis, Cheshire

DWH Dr. D.W.Hughes, Department of Physics, University of Sheffield

FH Professor Sir F.Hoyle, FRS, Penrith, Cumbria

GEH Dr. G.E.Hunt, DSc, PhD, FIMA, FRAS, FRMETS, FBIS, MBCS, London

HDH Commander H.D.Howse, FSA, FRIN, Sevenoaks

LH L.Helander, Stockholm

MAH Dr. M.A.Hoskin, Churchill College, Cambridge, England

MJH Michael J.Hendrie, FRAS, FBIS, Director of the Comet Section the British Astronomical Association

RFJ Dr. R.J.Jameson, Department of Astronomy, University of Leicester

RH Dr. R. Hutchison, FMS, Mineralogy Department, British Museum (Natural History), London

RMJ R.M.Jenkins, BSc, C Eng, MRAeS, FBIS, Dynamics Group, Space Communications Division, Bristol

ARK Dr. A.R.King, Department of Astronomy, University of Leicester

MRK Dr. M.R.Kidger, MSc, FRAS, Instituto de Astrofisicas de Canarias, Universidad de La Laguna, Tenerife

PvdK Dr P. van de Kamp, Sterrenkundig Instituut, "Anton Pannekoek", Universiteit van Amsterdam

WJK Professor W.J.Kaufmann, III, BA, MS, Adjunct Professor of Physics, San Diego State University

YK Dr. Y.Kozai, Director, Tokyo Astronomical Observatory, University of Tokyo

BL Sir Bernard Lovell, OBE, LLD, DSc, FRS, Nuffield Radio Astronomy Laboratories, Jodrell Bank, Cheshire

RJL R.J.Livesey, C Eng, FICE, FIWES, FRAS, Director, The Aurora Section the British Astronomical Association, London

SMcKL Professor Dr. S.McKenna-Lawlor, Experimental Physics Department, St Patrick's College, Co Kildare

AJM Professor A.J.Meadows, Department of Astronomy & History of Science, University of Leicester

CAM C.A.Murray, Royal Greenwich Observatory (Hailsham)

DFM David Malin, Anglo-Australian Observatory, Epping, New South Wales

HGM Howard Miles, Director, Artificial Satellite Section, British Astronomical Association

JCDM J.C.D.Marsh, MSc, C Eng, MIERE, FRAS, Director, The Hatfield Polytechnic Observatory, Herts

JWM Dr. J.Mason, BSc, PhD, John Mason Associates, Sussex

LVM L.V.Morrison, MA, MSc, FRAS, Royal Greenwich Observatory (Hailsham, Sussex)

PGM Dr. P.G.Murdin, Royal Greenwich Observatory (Hailsham, Sussex)

PM Patrick Moore, OBE, DSc(Hon), FRAS, Selsey, Sussex

RCM Dr. Ron Maddison, Senior Lecturer in Physics, University of Keele & Founder & Director of the Observatory

TJCAM T.J.C.A.Moseley, BA, FRAS, Co Antrim, N Ireland

IKN I.K.Nicolson, MSc, Senior Lecturer in Astronomy, Hatfield Polytechnic, Hants

JEO J.E.Oberg, Dickinson, Texas

WO Dr. Wayne Orchiston, Astronomy Unit, Victoria College, Victoria

JDHP Dr. J.D.H.Pilkington, Royal Greenwich Observatory (Hailsham, Sussex)

JLP J.L.Perdrix, FRAS, Wembley, Australia

MVP Dr. M.V.Penston, Royal Greenwich Observatory (Hailsham, Sussex)

CAR C.A.Ronan, MSc, FRAS, Project Co-ordinator, East Asian History of Science Trust, Cambridge, England

DJR Dr. D.J.Raine, Department of Astronomy, University of Leicester

HBR H.B.Ridley, FRAS, Somerset

RR Dr. R.Reinhard, Giotto Project Scientist, European Space Agency, Noordwijk, Netherlands

SKR Professor S.K.Runcorn, FRS, Head of School of Physics, University of Newcastle upon Tyne

ATS Dr. A.T.Sinclair, Royal Greenwich Observatory (Hailsham, Sussex)

EHS E.H.Strach, FRCS, MCh.Oeth, Liverpool

FRS Dr. F.R.Stephenson, MSc, DSc, PhD, FRAS, Department of Physics, University of Durham

RWS Dr. R.W.Smith, Department of the History of Science, The Johns Hopkins University, Baltimore

SS Dr. S.Saito, DSc in Astrophysics from the University of Kyoto, Kwasan & Hida Observatories, University of Kyoto

CWT Professor Clyde W.Tombaugh, Department of Astronomy, New Mexico State University

RFT R.F.Turner, FRAS, West Worthing, Sussex

AEW[1] A.E.Wells, Birmingham

AEW[2] Dr. A.E.Wright, PhD, Australian National Radio Astronomy Observatory, Parkes, New South Wales

APW Dr. A.P.Willmore, BSc, PhD, Department of Space Science, University of Birmingham

AWW Professor A.W.Wolfendale, FRS, Department of Physics, University of Durham

BW Professor B.Warner, Department of Astronomy, University of Cape Town

EAW E.A.Whitaker, FRAS, Member of IAU, AAS, BAA, Walter Goodacre Medal; Associate Research Scientist in Lunar & Planetary Laboratory, University of Arizona

ENW E.N.Walker, Royal Greenwich Observatory (Hailsham)

FGW Dr. F.G.Watson, UK Schmidt Telescope, Coonabarabran, New South Wales

GW Dr. G.Welin, PhD (Sweden), University Lecturer in Astronomy, Astronomiska Observatoriet, Uppsala University

JVW Dr. J.V.Wall, BSc, MASc, PhD, Royal Greenwich Observatory (Hailsham, Sussex)

KPW K.Wood, MSc, Department of Physics, Queen Mary College, London

LJCW Les Woolliscroft, Department of Physics, University of Sheffield

LW Dr. Lionel Wilson, BSc, PhD, FRAS, FGS, Department of Environmental Science, University of Lancaster

PAW Professor P.A.Wayman, MA, PhD, Dunsink Observatory, Dublin

SCW Dr. Sidney C.Wolff, Director, Kitt Peak National Observatory, Arizona

How to use this book

The encyclopaedia is arranged alphabetically according to these rules:

1. Entries follow the normal order of words a reader might look up, thus — **Lunar librations,** not **Librations, lunar.** There is one exception to this. Our Galaxy, the Milky Way, appears under **Galaxy, The** to differentiate it from other galaxies.

2. The seven colour essays in the book are, for manufacturing reasons, placed slightly out of strict alphabetical sequence. For example, the essay on the **Big Bang** falls slightly later in the Bs than it would otherwise do.

3. Entries beginning with numerals are treated as if the numerals were spelled out. Thus **3C147** follows **Three-body problem** and precedes **3C295.**

4. Biographies are entered by surname, with Christian names following the comma.

5. Entries with the same headword are listed in the order of people, places and things.

Cross-references SMALL CAPITALS in an article indicate a separate entry that defines and explains the word or subject capitalized. *See also* at the end of an article refers the reader to entries that contain additional and relevant information. There are no cross-references in the colour essays, but readers who encounter unfamiliar terms will find them defined in the A–Z section. There are no cross-references to the Sun or the planets because it is obvious that entries on them exist.

Foreign languages The system of transliteration used for Chinese is the Pin-yin. Greek letters are used in tables (see below).

Measurements Imperial measurements are given, but the metric equivalent (which follows in parentheses) is the more accurate. The exception is in historical articles where, for example, the diameter of an early English telescope will be given accurately in inches with a close metric approximation. Some figures are not converted, for example, negative powers of 10 and large positive powers of 10 (10^{-43} or 10^{39} for example). Densities, given in grammes per cubic centimetre, are usually not converted; nor are tonnes. Astronomical distances are most often given in astronomical units or light-years, but parsecs are also used. A billion is rendered as 1,000 million. For definitions of units of measurement, and conversion factors, and the handling of separation, period, co-ordinates and angles, see below.

Nomenclature Common practice is followed in naming lunar and planetary features. "Lakes", "seas", "marshes" and "bays" of the Moon will be found under "Lacus", "Mare", "Palus" and "Sinus". All other features — valleys, mountains and so on — will be found under their specific names. Thus **Mare Marginis** and **Sinus Iridum,** but **Tir Planitia** and **Caloris Montes.**

Finally, a note on consistency. No attempt has been made to settle differences of opinion between expert contributors when to do so would be to fly in the face of common sense. For example, the diameter of the asteroid Ceres is given variously as 933km and 1,023km. As astronomers are constantly revising data about it (and, indeed, about many other things in the universe), the differences have been allowed to stand; and the temptation to print "about 1,000km" resisted.

Abbreviations and conversions

Angle

°	degree	
′	minute	(set as 2°3′4″;
″	second	arc seconds are set as 1.″02)

Co-ordinates

E	east
N	north
S	south
W	west
R.A.	Right Ascension
Dec.	Declination

Dimensions

cm	centimetre (0.3937 inches)
ft	foot (0.3048 metres)
in	inch (2.54cm)
m	metre (3.2808ft)
mm	millimetre (0.0394in)
μm	micrometre (one-millionth of a metre)

Distance

AU	Astronomical unit (see entry – 92,955,630 miles)
Light-yr	Light-year (see entry – 63,240AU)
pc	parsec (3.26 light-years)
kpc	kiloparsec (1,000 parsecs)
Mpc	megaparsec (1 million parsecs)
km	kilometre (0.62137 miles)
mile	1.60934km

Force and pressure

gal	Galileo (1cm/sec/sec)
mgal	milligal (one-thousandth of a gal)
Newton	1kg accelerated at 1m/sec/sec
Pa	Pascal (1 Newton over an area of 1 sq m)
GPa	gigaPascal (1,000 million Pascals)
MPa	megaPascal (1 million Pascals)

Magnitude

m	(set as $3.^{m}4$)

Mass

gm	gramme (0.0022 pounds)
kg	kilogramme (1,000gm)
lb	pound (2.2046kg)
tonne	1,000kg

Period

P	(set as $0.^{p}1$)

Time

d	day
m	minute
s	second (set as, eg, R.A. $5^{h}38^{m}.3$)

Other abbreviations

AD	Anno Domini
BC	Before Christ
c.	*circa* (about)
cf	compare

Other abbreviations *continued*

eg	for example
EVA	extra-vehicular activity
f	focal ratio
fl	flourished
ie	that is
n	nano (10^{-9})
p	proton
$\bar{p}$	anti-proton
ρ	Rho (density)
rms	root-mean-square
T	tesla (magnetic flux density)
UV	ultraviolet
z	redshift factor
ZHR	Zenithal hourly rate

The Greek alphabet

Name	Sign		Name	Sign	
Alpha	A	α	Nu	N	ν
Beta	B	β	Xi	Ξ	ξ
Gamma	Γ	γ	Omicron	O	o
Delta	Δ	δ	Pi	Π	π
Epsilon	E	ε	Rho	P	ρ
Zeta	Z	ζ	Sigma	Σ(C)σ	
Eta	H	η	Tau	T	τ
Theta	Θ	θ	Upsilon	Y	υ
Iota	I	ι	Phi	Φ	φ
Kappa	K	λ	Chi	X	χ
Lambda	Λ	λ	Psi	Ψ	ψ
Mu	M	μ	Omega	Ω	ω

Contents

The seven essays listed below are in alphabetical order and evenly distributed throughout the encyclopaedia. A–Z entries fall appropriately either side of the essays.

A Family Named Universe

Leif J. Robinson
Editor
Sky and Telescope

The universe is 10 to 20 thousand million years old
and resulted from an explosion, unimaginably powerful,
called the Big Bang. The radiation from quasars, recognized
as the most energetic inhabitants of the universe,
derives from black holes — enormous gravity sinks in the
cores of ancient galaxies. A galaxy formed whenever the stuff
in space became a little more compacted than that nearby.
Galaxy shapes and compositions are straightforward

The dependence of humans on our star is symbolized by this photograph of a man silhouetted against the setting Sun. Indeed, were it not for previous generations of stars that exploded and cast their manufactured heavy elements into space, we would not be here at all. In February 1987, the brightest supernova in 383 years appeared in the Large Magellanic Cloud, a companion galaxy to ours 170,000 light-years away. Astronomers had long awaited such an event, to give them the opportunity to examine a supernova with modern techniques. Studies of SN 1987A, as it is called, will surely dramatically alter our understanding of these important stars.

consequences of internal motions, encounters with other galaxies, and the evolution of stars. Our Sun was born five thousand million years ago, and the Solar System's planets, asteroids, and comets condensed from its left-over debris.

Most astronomers would, in general, agree with all of those statements. But, I suspect, many of their darkest dreams are haunted by bogeymen who point to maverick evidence that doesn't fit such simple schemes. Let me cite three contemporary examples.

One very popular theory says that the primordial universe contained only three chemical elements: mostly hydrogen, some helium, and a tiny bit of lithium. But no star, among tens of thousands that have had their compositions examined critically, has been found to contain such a chemical composition. Where are these stars, the first-generation products of the Big Bang?

Since the 1930s evidence has mounted that there is "unseen" or "dark" matter in the universe. Yet, only after the case for such stuff became overwhelming a few years ago did astronomers begin a serious search for explanations. We still have no idea what this matter might be. Exotic elementary particles, black holes, dust, failed stars, and other bodies have all been proposed.

The basic process by which stars create energy, through the conversion of hydrogen to heavier elements, is presented as writ in astronomy textbooks. Isn't it odd,

however, that this process seems to produce only about a third as many neutrinos from the Sun as theory predicts?

That trilogy was *not* designed to throw cold water on the state of our astronomical knowledge. Rather, it attempts to focus on a problem — how to assimilate ever growing tidal waves of disparate information? Consider this: today 6,000 or so professional astronomers worldwide announce their discoveries in over 100 technical journals. For fun, I counted the number of pages published in 1985 by the two largest. Together, my issues of *The Astrophysical Journal* and *Astronomy and Astrophysics* contained some 15,000 pages and weighed 90 pounds!

All of this, of course, has come about thanks to the technological explosion that followed World War II. Suddenly, astronomers found themselves able to explore the universe in ways unparalled in human history. For example, by coupling newly developed radiation detectors to telescopes on the ground and in space they could sample the electromagnetic spectrum from gamma rays to radio waves, an energy range of more than a million, million times. Also, spacecraft have taken close-up photographs of all the planets from Mercury to Uranus (and many of their satellites). These robots have even dug samples from the surfaces of the Moon and the planets Venus and Mars.

Radical new ideas and knowledge are usually not

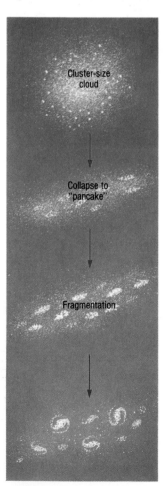
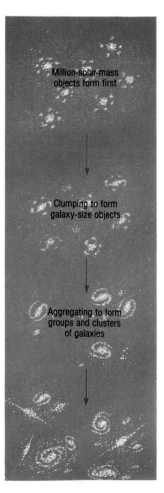

Cluster-size cloud

Collapse to "pancake"

Fragmentation

Million-solar-mass objects form first

Clumping to form galaxy-size objects

Aggregating to form groups and clusters of galaxies

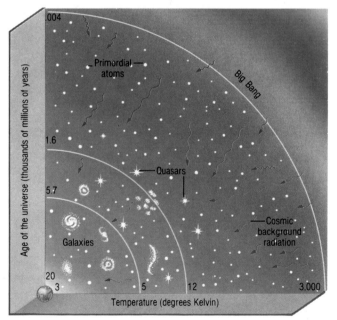

Two proposed methods for forming galaxies, *left.* In the first, enormous "pancakes" form and then fragment into galaxies. In the second, relatively small structures form first and then cluster together to produce galaxy-sized bodies.

As we look deeper into the universe, *above,* we look farther back in time. Since the universe

was opaque until it cooled to about 3,000K, about half a million years after the Big Bang, we cannot see what took place earlier than that. Our limit of "vision" is what we see as the omnipresent cosmic microwave background radiation. Quasars evolved when the universe was perhaps only a thousand million years old: galaxies formed much later.

Two images of the giant elliptical galaxy M87 in Virgo, *below*. At left, a computer-processed image reveals the galaxy's famous jet whose light comes from electrons travelling at relativistic speeds. The galaxy itself is coloured red. At right is a false-colour image in X-rays, revealing the galaxy's extensive halo of 50 million K gas. To prevent this gas from escaping into intergalactic space, M87 must have a mass 20 trillion (10^{12}) times the Sun's. Luminous matter can account for only about 10 per cent of that mass; the rest is dark matter, though astronomers do not know its composition.

One of the most useful tools in modern astrophysics is a colour-magnitude diagram, *opposite*. In one of its many forms the intrinsic brightnesses of stars are plotted against their colours (equivalent to temperature). Especially for a star cluster, which formed all its stars about the same time, successive stages of stellar evolution become evident. Stars like our Sun spend most of their lives on the main sequence. They arrive there when their core temperatures become high enough to allow hydrogen to be converted into helium; they leave after a substantial fraction of the core has been converted into helium. The star's surface then cools and expands — the star becomes a red giant. When the inner regions of the star become hotter than 100 million K, the helium fuses into carbon, accompanied by a shrinking of the star's atmosphere. The star then moves along the so-called horizontal branch and eventually becomes a white dwarf.

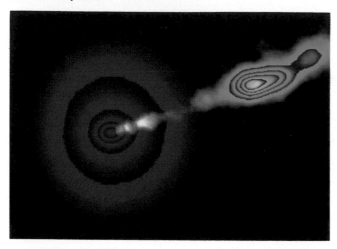

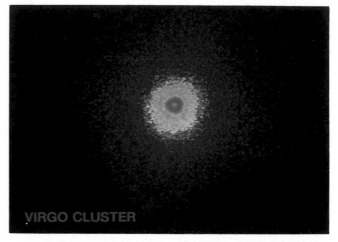

VIRGO CLUSTER

readily accepted. As a case in point, it took 1,800 years to overthrow Aristotle's model of the universe, in which the Earth lay at its centre. The Sun, Moon, planets, and stars were imagined to revolve around the Earth on crystalline spheres. But that simple and attractive scheme began to shatter in 1543 when Copernicus published his idea that the Sun, in fact, was at the centre of the Solar System. The nail was really driven into the geocentric coffin when Newton announced his universal law of gravity, which provided a theory that described how such a Sun-centred system could work. (Incidentally, that revelation occurred in 1687, exactly 300 years ago, and forever changed our perception of the cosmos.)

The Earth as the pivot-point of the universe took its final blow during the first half of the 20th century. Harlow Shapley and others proved that our Milky Way is a galaxy, like billions of others, and that our Sun is a very ordinary star situated on its outskirts. Ultimately, Edwin Hubble told us that the universe is expanding, with every galaxy flying away from every other one. Thus it turned out that even our Galaxy's location in the universe wasn't special, which is another way of saying that neither is our importance in Nature's grand design.

The fabric of the universe

Almost everyone agrees that the universe is expanding and that it got going with a bang some 10 to 20 thousand million years ago. We cannot pinpoint the time of this Big Bang more specifically because of uncertainties in the values of the so-called Hubble parameter (H_0) and the deceleration parameter (q_0). The first tells how fast the universe is expanding at present and the second tells how rapidly the expansion is slowing down. We can, however, readily determine how fast galaxies and quasars are moving away from us (and each other) by measuring how much the lines in their spectra are shifted toward red wavelengths (that is, transposed by the Doppler effect). The bugaboo is to calibrate the redshifts so they yield distances.

Despite enormous efforts over the past 60 years, astronomers still don't know the value of H_0 within a factor of two. Because of this uncertainty, we cannot determine the "critical density" of the universe, the quantity that tells whether the expansion will continue forever (the so-called open universe) or whether it will eventually halt and allow contraction to begin (the closed universe). Though of no practical consequence to us, the answer is definitely of interest to posterity.

In the first case the universe will continue to cool. Some 10 to 100 thousand million years from now even the longest-lived stars will wink out, and after a million, million, million years more have passed even their parent galaxies will be extinct. After another 10^{60} to 10^{70} years have elapsed (a time so long that it defies comprehension), gigantic black holes, the remnant cores of once brilliant galaxies, will have evaporated. The temperature of the entire universe will inexorably drift toward absolute zero, the point when all motion ceases.

In the second case, once the universe's expansion is reversed, matter will begin to heat up as it is ever more rapidly drawn together. By the time the universe has shrunk to a hundredth its present size, its temperature

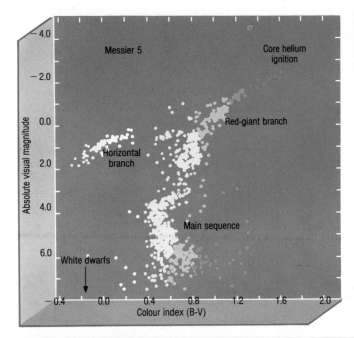

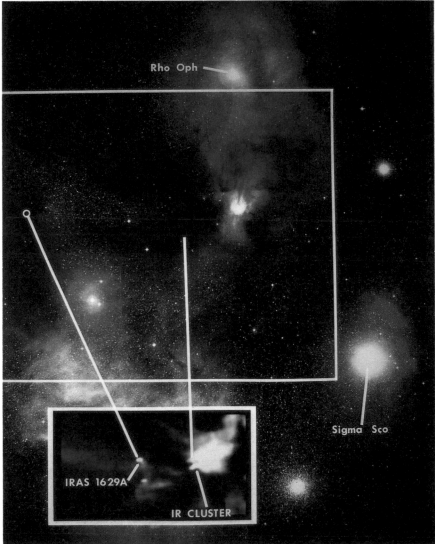

Rho Oph

IRAS 1629A

IR CLUSTER

Sigma Sco

When the centre of our Galaxy stands directly overhead and the Milky Way stretches from horizon to horizon, it seems as if we are viewing our Galaxy from intergalactic space, *above*. Along its entire length, the plane of the Galaxy is mottled with dark clouds of dust and gas; the brilliant star clouds around the galactic centre form a distinct bulge. To appreciate celestial distances, note that the meteor which made the streak just above the galactic centre was perhaps 100 miles (160km) from the camera; Halley's Comet (the small fuzzy object at right centre) was 39 million miles (62 million km) away; and the galactic centre was 160,000 million miles (260,000) million km) distant.

The best evidence yet for witnessing the actual birth of a star was found deep inside a gas and dust cloud known as the Ophiuchus dark nebula, *left*. Radio observations of carbon monosulphide molecules revealed that a cloud of gas, about 6,000AU in diameter and containing about 10 per cent of the Sun's mass, is falling inward in a manner consistent with contemporary theories of star formation. At the centre of this collapsing cloud there seems to be an embryonic star, perhaps only 30,000 years old, known as IRAS 1629A.

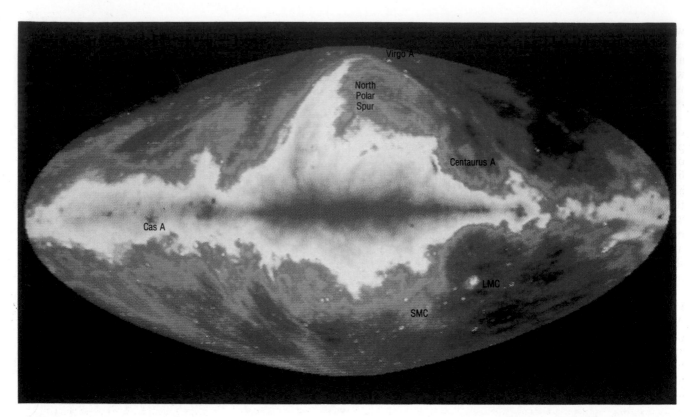

The radio sky observed at a wavelength of 73cm (slightly longer than UHF television, *above*; red represents the most intense emission. The galactic plane appears very strong, resulting from ionized hydrogen gas and supernova remnants such as Cassiopeia A. The North Polar Spur is believed to be the remnant of a nearby supernova explosion. The Large and Small Magellanic Clouds (marked LMC and SMC), Virgo A, and Centaurus A are galaxies.

When the Milky Way is observed at different wavelengths it takes on a different appearance. This all-sky infrared image, *below*, was obtained by the Infrared Astronomical Satellite and has the galactic centre in the middle. Red and yellow regions show dust concentrations along the plane of the Milky Way at a temperature of roughly 40K. Material some five times hotter, as indicated by a blue tint, follows the plane of the ecliptic (the Sun's path in the sky).

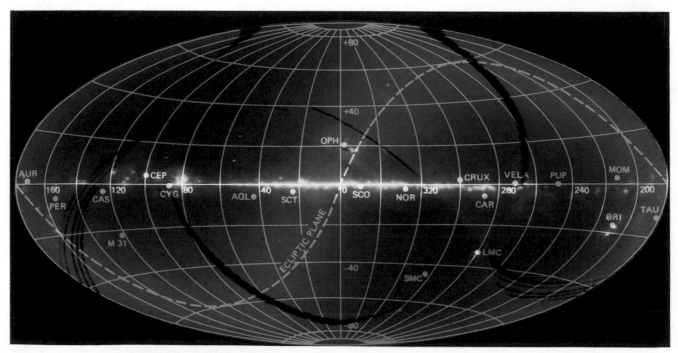

will have risen to that of boiling water. Eventually it will reach the same temperature it had at the time of creation; all matter will disolve into the sea of radiation from which it sprang. Maybe our progeny shouldn't care which way the universe goes; after all, neither alternative is very pleasant!

We may not know much about the beginning and end of the universe, but we do know something about its present structure. Galaxies, it seems, form on the surface of "bubbles", some of which are hundreds of millions of light-years across. Inside the bubbles are voids where galaxies are practically nonexistent. Nobody knows how the bubbles were created, though blast waves from exploding primordial stars or galaxies have been proposed. (An alternative interpretation to the bubble model is that the galaxies lie on a continuous convoluted skin akin to that of a sponge. If this is true, the voids are merely interstices in the sponge-stuff.)

How galaxies form and evolve

About 300,000 years after the universe was born, its temperature had dropped to about 3,000K, which was cool enough for hydrogen atoms to form. Tiny density variations in the very early universe could then begin to grow into protogalaxies. (How these variations began and were sustained is still a mystery, but no one can argue with the consequences.) The most popular view today envisions the initial formation of pancake-like structures, which ultimately fragmented into galaxies. Another hypothesis imagines that gobs of matter, each a million times more massive than the Sun, gathered themselves together to form clusters of galaxies.

Today we see an amazing variety of galaxies. Ellipticals are boringly similar and resemble pussy willow catkins. On the other hand, spirals like our Milky Way are each personalized by their curving arms. And then there are the odd-balls whose twisted and tortured shapes more closely resemble micro-organisms than star systems. How did all this diversity come about? One common thread seems to derive from collisions between galaxies or near misses.

Computer simulations have shown that the most bizarre configurations almost certainly arose from collisions. During such celestial accidents, particularly between spirals containing lots of gas and dust, gravity-induced tidal effects can tear out plumes of material or contort an otherwise normal disk. But the most important effect of a collision is to compress the gas enormously, which prompts the formation of hot young stars. These newborns then heat the dust, which releases a tremendous amount of infrared radiation. The galaxy's total luminosity soars to perhaps 10 trillion (10^{12}) times the Sun's and approaches that of the quasars. The recognition of such "starburst" galaxies came about largely due to observations made by the Infrared Astronomical Satellite in 1983.

So much energy is produced during a starburst that it could drive into intergalactic space all the matter that didn't form into stars. Since elliptical galaxies are known to be gas-poor, collision and merger seem like effective mechanisms for forming them. Thus, the dull-looking face of an elliptical may actually mask an extraordinary past. Furthermore, ellipticals are dominated by stars that are old, glowing orange like voltage-starved lamps, so the galaxies themselves must be ancient. Evidently, ellipticals are the products of a young universe, a time when everything was much more closely packed together and collisions must have been frequent.

Even the beautiful patterns evident in photographs of spiral galaxies appear to result from tidal interactions, caused by a bar in the centre of the galaxy or by a nearby companion. Either can cause a density wave to sweep

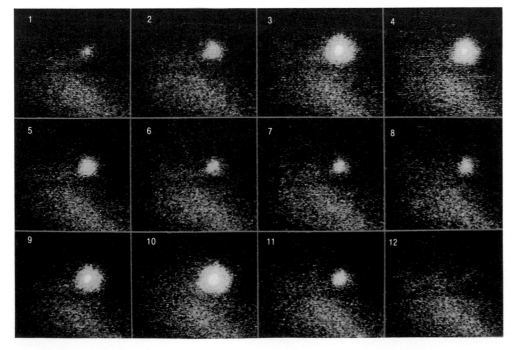

The pulsar that supplies energy to the Crab nebula is a neutron star, the remains of a supernova explosion that occurred some 7,000 years ago. This remarkable object spins around once every 0.033 seconds and each time sweeps its beam of radiation (from radio waves to gamma rays) over the Earth. These false-colour images, *left*, were obtained with the Einstein X-ray Observatory and show the two intensity peaks (frames 3 and 10 that occur every cycle.

During the past few years it has become evident that violent events take place at the centre of our Galaxy, which seems likely to harbour a million-solar-mass black hole. (Our Galaxy, therefore, would be directly related to the so-called active galaxies, which include the Seyferts and the quasars.) One of the strangest features found near the galactic centre is the arc shown in this image, *below*, from the Very Large Array. Situated about 700 light-years away from the hub of our

Galaxy, each of its filaments is about 3 light-years wide and 130 light-years long. They probably result from a magnetic field and radiate by the dynamo effect as gas is drawn across the field lines. The bright dot in the red area marks an object known as Sagittarius A West, which is believed to be the very hub of our Galaxy.

By suppressing the light from the bright star Beta Pictoris, here hidden behind the mask at centre, a faint disk of dust that surrounds it was made visible, *below*. The dust extends some 60,000 million miles (100,000 million km), 17 times the diameter of the Solar System. One popular interpretation is that this dust suggests the presence of another solar system. Some evidence indicates that the disk is tilted 5° to 10° to our line of sight.

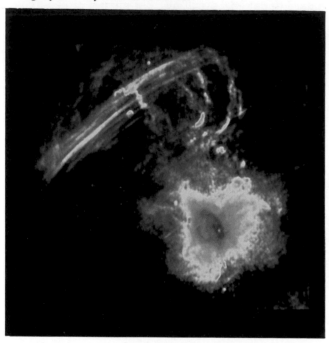

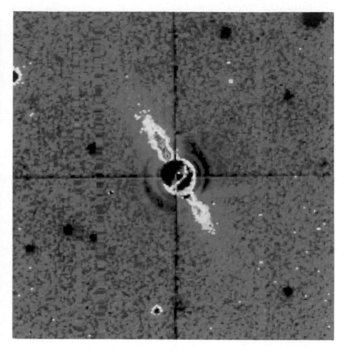

around the galaxy, thereby triggering the collapse of interstellar clouds and the subsequent formation of stars. The spiral arms, therefore, mark where a density wave passed and where young, hot stars were born.

The first hint that some galaxies are fundamentally different than others came in 1943, when Carl Seyfert noticed that certain systems had extraordinarily bright nuclei. Twenty years later Maarten Schmidt recognized that a starlike object called 3C 273 was farther away than anything then known in the universe. Because it appeared quite bright in the sky it had to be enormously powerful — the discovery of quasars literally made front-page news. Yet two more decades were to pass before it became clear that Seyfert's galaxies and Schmidt's quasars were related. We now know that the latter are bizarre manifestations of active galactic nuclei, outshining by at least 100 to 1,000 times the brightest normal spirals.

Such a statement can now be made since we have seen "fuzz" around many quasar images, in other words, indications of an underlying galaxy. And in at least the case of 3C 48 the fuzz has been shown to consist of stars. The power sources for quasars are less easily explained. From direct diameter measurements and from variations in quasar brightnesses, we know these beacons have to be small, a few light-years or less in diameter. The only object that seems capable of supplying enough energy

from such a small volume is a supermassive black hole, one having the mass of perhaps a thousand million suns. (Keep in mind that no one can say black holes truly exist; they remain merely products of theoreticians' pencils.)

The lives of stars

For us, one of the most important events in the history of the universe took place some five thousand million years ago. A very ordinary star was born, and we call it the Sun. It alone, through the fusion of four hydrogen atoms into one of helium, produces the energy necessary to sustain life on Earth.

For decades astronomers have had fairly concrete ideas about star birth. The actual process, however, has been literally shrouded from our eyes. The reason, of course, is that everything goes on deep within cold, dark, collapsing clouds of dust and molecules. Because light cannot escape from these clouds, only since the advent of advanced radio and infrared telescopes have we been able to peek inside. There we find gravity gathering protostar material together. Heating occurs until the temperature rises to some 10 million degrees, when the fusion process begins and a star comes to life. Intense winds, travelling at hundreds of kilometres a second, flow from such young stars and sometimes blow holes in the parent cloud. By such happy accidents we obtain our first glimpse of the newborn.

We have detected atmospheres around other stars and have mapped their surfaces. But the Sun, because of its nearness, remains the only star we can examine in detail. The photograph, *left,* showing many fine streamers in the Sun's atmosphere (corona), was taken during the total solar eclipse of February 16 1980. The bluish image in the centre is not the Sun but the Moon. Total solar eclipses occur only when the Moon completely covers the Sun's disk, thereby suppressing its intense light and allowing the faint atmosphere to be seen.

All stars are not born equal, and the amount of matter they start with is decisive in determining how they will live their lives. (Often this evolutionary process is depicted in a so-called Hertzsprung-Russell diagram, which was introduced during the second decade of this century.) The rate at which a star ages depends on its initial mass: a heavy one, say 15 times more massive than our Sun, will burn its hydrogen and helium in only 10 million years or so; in our Sun, however, these reserves should last 10 thousand million years. After the star exhausts its hydrogen and helium, it becomes redder and expands, becoming perhaps 10 to 100 times its original size.

Astronomers are sure that what happens next depends on the star's initial mass, but they aren't so confident of the details. A star like our Sun, once it becomes a giant, will probably eject a planetary nebula, then grow hotter and smaller, and eventually end up as an Earth-size white dwarf. Stars having masses between about 1.4 and 3 times the Sun's complete their lives as neutron stars, tiny bodies only 18 miles (30km) or so in diameter. The fate awaiting the heaviest stars is much more spectacular. They first become supergiants and then undergo a titanic supernova explosion in which a large fraction of their mass is blown into interstellar space. Depending on what's left, the shattered remnant will end up as a neutron star or a black hole.

Stars with less than about half the Sun's mass lead the most conservative lives of all, which seems proper for they are perhaps the most abundant type of star in our Galaxy. Due to their low mass, these dim objects cannot heat up enough to go beyond hydrogen burning, so they merely fade away as white dwarfs. But what about even lesser bodies? To "catch fire" the smallest star needs to have at least eight per cent of the Sun's mass (80 times that of Jupiter). Searches for these failed stars or "brown dwarfs" have been inconclusive. Yet, if it could be proved that many of these dim stellar embers exist, glowing at a temperature of about 1,000K, the missing mass problem for galaxies might be made to go away.

All of the above evolutionary scenarios assume that the stars are alone in space. But this may be true for less than 20 per cent of them; the rest are members of binary or multiple systems. We are fortunate that so many such systems exist, for binaries reveal abundant information about the masses, diameters, and other properties of stars. Life is especially interesting when two stars are so near each other that they physically interact. By exchanging material, for example, the lifestyles of the components may be drastically affected, and it is even possible to produce kinds of stars that otherwise wouldn't exist — explosive stars, high-energy X-ray sources, and bizarre objects like SS 433.

When single stars form (or very close or very wide

binaries), something else might happen. It now seems likely that planets are common by-products of star birth. Yet, despite many recent efforts to detect planets circling other stars, our Solar System remains the only known example. The basic process seems inescapable: a nascent star becomes centred in a flattened disk of dust and gas, perhaps enriched by heavy elements from a nearby supernova explosion, rotating around a well-defined axis. Especially dense pockets in the disk cause even more of the material to accumulate until the cores of planets are formed.

Particularly from observations by the Infrared Astronomical Satellite, dust has been found around at least a couple of dozen stars. But even so, there is still no indisputable evidence for planets. More conventional astronomical techniques, such as precisely measuring changes in a star's position on the sky or its speed toward or away from us, have also yielded negative results.

Things seen and unseen

In this review, I've had to leave out a lot of recent discoveries, in some cases very provocative ones. Glossed over are the pulsars, rapidly spinning neutron stars whose radiation is channelled into narrow beans by intense magnetic fields. Nor have I cited examples of naturally occurring maser action, where prodigious amounts of energy can be emitted at a single wavelength — from clouds of molecules in interstellar space, stellar atmosphere, comets, and even entire galaxies. Similarly

omitted is the phenomenon of jets, beans of matter and energy emitted from objects as diverse as nascent stars and quasars. And I've not mentioned the giant double-lobed radio galaxies that are the largest entities known in the universe; one of these, called 3C 236, spans nearly 20 million light-years.

I've also ignored the as-yet unsuccessful searches for extraterrestrial life, either from direct analysis of material from the Moon and Mars, or from passive attempts to listen for cosmic radio transmissions from intelligent beings. Equally unsuccessful has been the quest for gravitational waves, which Einstein's General Theory of Relativity says should appear when matter is rapidly accelerated, such as during a supernova explosion. And do so-called cosmic strings exist, as predicted by some theories of elementary particles? Are these infinitesimally thin, astoundingly massive, enormously large threads and loops of energy responsible for the formation of galaxies and their distribution in space?

A staggering amount of knowledge, much of it totally unexpected, has been uncovered about the universe since the middle of this century. Nevertheless, the old saw that each discovery brings fresh questions remains intact. It is almost frightening to contemplate that even more profound and enigmatic revelations await the next generation of astronomical instruments, such as the Hubble Space Telescope. The last half of the 20th century will surely be known as astronomy's "Golden Age", and the best is yet to come.

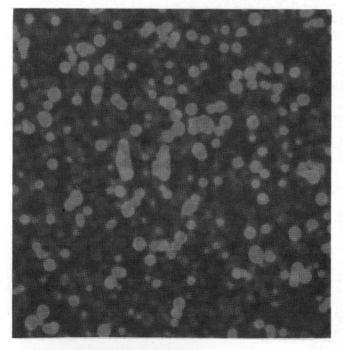 

One of the Hubble Space Telescope's biggest attributes is that it will be able to image details as small as 0.1 arc-second, equivalent to a metre stick seen at a distance of 36km. Such resolution is many times greater than that achievable with contemporary ground-based telescopes. The computer simulation, *above*, compares the appearance of the famous globular cluster Messier 13 at 0.8 arc-second resolution (left) as it would be seen from the surface of the Earth under the most favourable circumstances and the 0.1 arc-second resolution (right) attainable with the Hubble Space Telescope.

AAO *See* ANGLO-AUSTRALIAN OBSERVATORY.

Abbot, Charles Greely (1872–1961). American astronomer who specialized in studies of the Sun. In 1895 he joined the staff of the Smithsonian Astrophysical Observatory, becoming Director in 1907. His main work was a very accurate determination of the solar constant. His authoritative book *The Sun* was published in 1911.

Abenezra Lunar crater, 27 miles (43km) in diameter; 21°S, 12°E.

Aberration The small apparent displacement of a star from its true position in the sky caused by the velocity of the Earth in its orbit. Aberration was discovered by James Bradley in 1728. He

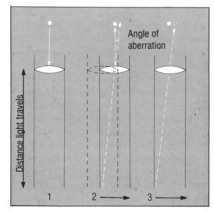

Aberration: apparent displacement of a star

was trying to measure the PARALLAX, a minute annual shift of nearby stars against the more distant background as the orbital motion of the Earth changed the observer's position in space. Instead, careful observations revealed that the apparent position of all objects shifted back and forth annually by up to 20 arc seconds in a way that was not connected to the expected parallax effect. Bradley's observations demonstrated both the motion of the Earth in space and the finite speed of light and have influenced arguments about the meaning of space and time to the present day. *DFM.*

Abetti, Giorgio (1882-1982). Italian solar physicist. As director of the ARCETRI

OBSERVATORY (1921–1952) he was instrumental in changing the activities of the observatory from positional astronomy to astrophysics. As a young postgraduate he went to Berlin, Heidelberg, Yerkes and Mount Wilson. The time spent at Mount Wilson and his friendship with G.E.Hale influenced his future work. With the Arcetri solar tower, added by him to the equipment of the observatory, he carried out his research into the structure of the chromosphere and into the motion of gases in sunspots (Evershed-Abetti effect). Under Abetti, Arcetri became a centre of solar studies. *VB.*

Ablation The process by which the surface layers of an object entering the atmosphere (eg, a spacecraft or a meteoroid) are removed through the rapid intense heating caused by frictional contact with the air. The heat shields of space vehicles have outer layers which ablate, preventing overheating of the spacecraft's interior.

Able Liquid-propellant rockets which formed the upper stages of the Thor and Atlas vehicles in early experiments to probe the Moon and interplanetary space.

Absolute magnitude (M). The visual magnitude which an astronomical body would have at a standard distance of 10 PARSECS. If m = apparent magnitude and r = distance in parsecs, M = m + 5 − 5 log r.

Absolute zero The lowest theoretically attainable temperature. Equivalent to − 273.15° Celsius or zero Kelvin (*see* K), it is the temperature at which the motion of atoms and molecules ceases.

Absorption As a beam of light, or other electro-magnetic radiation, travels through any material medium the intensity of the beam in the direction of travel gradually diminishes. This is partly due to reflective scattering by particles of the medium and partly due to absorption within the medium. Energy which is absorbed in this way may subsequently be re-radiated at the same or longer wavelength and may cause a rise in temperature of the medium.

The absorption process may be general or selective in the way that it affects different wavelengths. Examples can be seen in the colours of various substances. Lamp black, or amorphous carbon, absorbs all wavelengths equally and reflects very little, whereas

paints and pigments absorb all but the few wavelengths that give them their characteristic colours.

Spectral analysis of starlight reveals the selective absorption processes that tell us so much about the chemical and physical conditions involved. The core of a star is a hot, incandescent, high-pressure gas which produces a continuous spectrum. The atoms of stellar material are excited by the high-temperature environment but they are so close together that their electrons move easily from atom to atom, emitting energy and then being re-excited and so on. This gives rise to energy changes of all possible levels releasing all possible colours in the continuous spectrum.

The cooler low-pressure material comprising the atmospheres of both star and Earth, and the interstellar medium that lies between them, can be excited by constituents of this continuous radiation from the star core and such selective absorption produces the dark lines that are so typical of stellar spectra. These lines are not totally black, they are merely fainter than the continuum. This is because only a fraction of the absorbed energy is re-radiated in the original direction. *RCM. See also* MORGAN-KEENAN CLASSIFICATION.

Absorption lines *See* ABSORPTION.

Abulfeda Lunar crater, 40 miles (64km) in diameter, making up a pair with Almanon; 14°S, 14°E.

Abu'l-Wafa, al-Buzjani, Muhammad (994–997/998). Astronomer and mathematician, who helped develop trigonometry. His *Complete Book* (on astronomy) and his astronomical tables were used by many later astronomers.

Acamar The star Theta Eridani; magnitude 2.92. It is 17 parsecs (55 light-years) away, and is a fine double; both components are of spectral type A.

Accretion Collisions are commonplace in the universe and the outcome is, in the main, dependent on the energy of the two bodies involved. If this is large the result is fragmentation, the colliding bodies breaking up into a multitude of smaller particles, the number and masses of these particles depending on the density, composition, strength, phase (ie, gas, liquid and solid) and mass of the colliding bodies and the geometry of the collision. If the energy is small, accretion occurs and the two particles "stick" together. Usually both processes are occurring, one marginally dominating the other.

Consider first the early stages of the Solar System. The dust particles in the nebula surrounding the protosun have collapsed to form a disk, which is approximately in the equatorial plane of the Sun. This disk contains a range of particle sizes from submillimetre up to kilometre. The orbits deviate only slightly from circular. The particles' velocity with respect to the central body is large, about 18 miles (30km)/s near the Earth's orbit and 8 miles (13km)/s near Jupiter's. This velocity, however, decreases gently as the inverse square root of the distance from the Sun, so that particles in adjacent orbits can have a low relative velocity. The key parameter in accretion is the escape velocity of the largest particle. This is proportional to the square root of mass/radius and is, for example, 7 miles (11.2km)/s for Earth, 38 miles (61km)/s for Jupiter and 1.5 miles (2.4km)/s for the Moon. A small object falling from infinity under the influence of gravity hits the surface of a planet at the escape velocity. Imagine a collision at a much lower speed. The two objects bump together and become gravitationally interlocked. Added to this might be some physical or chemical bonding. If the objects are snowballs at a temperature just below their melting point the collision produces partial melting and considerable adhesion. The objects could be interstellar dust particles and have a layer of absorbed molecules on or near the surface. The effect of stellar ultraviolet and X-radiation plus cosmic rays on these simple H_2O, CO_2, HN_3, CH_4 molecules can convert them into complex long-chain organic molecules rather similar to sticky tars. These aid low-velocity accretion by providing a "glue".

If the impacting velocity is about half the escape velocity of the largest particle, considerable fragmentation can occur. The large majority of the resulting bits will not have velocities greater than the escape velocity and will fall back to the common centre of gravity forming a hot, ellipsoidal, rubble pile. In the case of the Earth, if *all* the potential energy of the total accreted mass was converted into thermal energy, we would have sufficient energy to raise the temperature by about 45,000K. Obviously only a small fraction of this energy is sufficient to make the "rubble" viscous and thus sticky. It is easy to see how a large accreting body can partially melt, differentiate and become nearly spheri-

cal. The temperature rise is proportional to the square of the radius of the body, multiplied by its density. So in the case of a 62-mile (100km) radius asteroid it is a mere 11°C, the rock remains solid and the shape of the final collection is irregular.

When the impacting velocity is equal to the escape velocity of the largest particle not only are both objects fragmented but also a considerable fraction of the mass is lost and we are approaching the stage where the final mass of the largest rubble pile is very similar to the mass of the original largest particle. A further increase in velocity reduces the size of the largest particle in a collection, and we have a fragmentation-dominated size distribution.

Evidence for both accretion and fragmentation can be found by measuring the mass distribution of specific collections of astronomical objects. This is usually represented by an index, s, defined so that the number of objects more massive than a specific mass, m, is proportional to $m^{(1-s)}$. If s is less than 2.0, the majority of the mass is contained in the large objects in the collection. If s is greater than 2.0, the majority of mass is now in the form of the smaller objects. Accretion-dominated systems have low s values, in the 1.5 to 1.8 range, whereas fragmentation-dominated systems tend to have s values of 2.0 and above.

Asteroids have two mass distribution indices. The big ones, of diameter greater than 160 miles (260km), have an index of 1.65, a value that clearly indicates that they have been produced by accretion. Smaller asteroids of diameter less than, say, 250km, have an index of 2.02 and this is strong evidence for them being fragments of larger, now disrupted, parents. Very small asteroid-type bodies such as Phobos and Deimos, the two which are now moons of Mars, are probably loose rubble piles. These asteroids are clearly made of silicate material which has a density of around 3.4g/cm³. Their actual density has been measured (by Mariner and Viking orbiters) to be of the order of 1.9 and 1.5g/cm³ respectively, so a considerable volume of their interior is empty.

Comets have a mass distribution index of 1.81 and are clearly the result of accretion processes. These "dirty snowballs" have been identified as planetesimals, the basic building blocks of the outer gas giant planets of our Solar System. Present-day observations indicate that there are at least a million

million comets with masses in excess of 10^{16}gm, ample to produce a steady state mass distribution index.

The picture is slightly more complicated in the case of stars. Accretion dominates the first stage of their evolution because their parent molecular cloud has to satisfy the Jeans Criterion before it can collapse. This criterion concerns the balance between gravitational energy (with its tendency toward contraction) and thermal energy (which favours expansion). A cloud will collapse only if its mass exceeds $8.5 \times 10^{22} T^{3/2} \rho^{-\frac{1}{2}}$ where T is its temperature in degrees Kelvin and ρ is its density in g/cm³. Clouds have to grow to this mass before collapse is possible and this growth is achieved by accreting smaller clouds by colliding with them at relative velocities below the escape velocity. When the Jeans mass is reached the subsequent collapse is usually followed by a break-up into a small (a few hundred) number of fragments, the resulting collection of stars having masses that follow a Weibull distribution.

On an even larger scale, accretion is obviously a dominant process. A big bang cosmology has a singularity starting point from which the mass of the universe expanded at a range of velocities up to that of light. Clearly the average density of the material was ever decreasing. But, far from this material dispersing into an ever more rarefied gas, it seemed to be governed by physical processes that led to accretion, firstly into galaxies of mass about 10^{46} gm which are now separated from each other by typically ten times their own dimensions. Then the galaxies broke up and their mass was accreted into 10^{33} gm stars, most stars being orbited by a handful of 10^{28}gm planets. So the universe today has around 90 per cent of its mass in the form of high-density objects, all of these being the result of accretion at some stage in their lives. *DWH.*

Achernar (Alpha Eridani) Brightest star in the southern constellation Eridanus. It has apparent magnitude 0.46, spectral type B3V and is at a distance of 38 parsecs.

Achilles (Asteroid 588). First Trojan, discovered February 22 1906 by Max Wolf in Heidelberg. It has a similar orbit to Jupiter (60° preceding); 44 miles (70km) in diameter.

Achondrites Stony meteorites formed by reprocessing on planetary bodies; may be grouped into one major "clan" and four minor ones, each probably originates from a single planet or asteroid. They are essentially metal-free.

The eucrite, diogenite, howardite clan is the product of ancient basaltic vulcanism, represented by 55, 15 and 24 meteorites, respectively, of each type. Eucrites are composed of the minerals plagioclase and pyroxene, the former (a calcium, sodium, alumino-silicate) almost sodium-free, the latter having roughly equal atomic amounts of iron, magnesium and calcium, combined with silica. Texturally the meteorites are intergrowths of needle- or blade-like plagioclase crystals largely enclosed by squat pyroxenes, like their terrestrial counterparts, basalt or diabase (dolerite). In some, one or other mineral became concentrated by crystal settling, "accumulating", from pools of parental liquid, some 4,500 million years ago. This process produced the diogenites by the concentration of calcium-poor pyroxene (hypersthene) from related liquids. Most eucrites and diogenites are brecciated — fragmented and re-welded by impacts. Howardites are regoliths ("soils") formed from this, plus minor chondritic, material. On the surface of their parent asteroid the howardites suffered radiation damage by cosmic rays. This was recognized in 1969, months before similar damage was discovered in lunar soils returned by Apollo 11. The eucrite, diogenite, howardite clan may be related to mesosiderites, pallasites and iron meteorites of Group IIIAB. Spectral observations indicate that their source is the asteroid Vesta, but there are dynamical objections.

Eight SNC meteorites (from *S*hergotty, *N*akhla and *C*hassigny meteorites) may be volcanic rocks from Mars. Nakhla crystallized only 1,200 million years ago, testifying to vulcanism on a planet after it had ceased on the Moon. There are four lunar meteorites. Seventeen ureilites are olivine-pyroxene rocks related to CO carbonaceous chondrites, and eleven aubrites are magnesium-pyroxene rocks related to the enstatite (E) chondrites. *RH*.

Achromatic The name given to a LENS which has been corrected for CHROMATIC ABERRATION, so that false colour is reduced. When a ray of light passes through a lens, the direction in which the light was travelling is changed as it passes from air into glass and back into the air again. This is called refraction. Ordinary white light is made up of a mixture of seven different colours—red, orange, yellow, green, blue, indigo and violet. Each colour of light is travelling as an electromagnetic wave which has a different wavelength. When a ray of white light passes through a lens, each colour is bent or refracted. The colour with the longest wavelength (red) is bent the least. That with the shortest wavelength (violet) is bent the most. This is called dispersion.

When a beam of white light passes through a simple lens, the white light is split up into the various colours so that each colour comes to a slightly different focus. The result is that the image of any object produced by the lens is surrounded by coloured fringes or false colour. This is called chromatic aberration. The effect can be very much reduced by using an achromatic lens. Such a lens is made up of several elements manufactured from glasses having different dispersive powers. Traditionally crown glass and flint glass are used. In this way the errors of each element tend to cancel each other out and the false colour is reduced. In practice a given lens can be correctly balanced only for a few wavelengths. Achromatic lenses having two elements are called doublets; those with three elements are called triplets. *JWM*.

Acidalia Planitia The main dark area in the northern hemisphere of Mars; 14–55° N, 0–60°W.

Acrux (Alpha Crucis). Brightest star in the Southern Cross, found in 1685 to be a visual double star. The brightest member, α^1Cru, is magnitude 1.6, and spectral type B0.5IV and is itself a spectroscopic binary with a period of 76 days. The companion, α^2Cru, of magnitude 2.1, spectral type BIV, is 4″.4 from α^1Cru. Their combined magnitude is 0.79. They are 125 parsecs from the Sun and orbit each other with a period of about 10,000 years. Another member of this multiple star system is a 4.9 magnitude B4IV star 90″ away. A line drawn from Gamma Crucis through Acrux passes close to the south celestial pole. *BW*.

Actinometer An instrument used for measuring at any instant the direct heating power of the solar radiation. Sir William HERSCHEL first noted, in 1800, that the heating effect of the Sun's rays was greatest beyond the red end of the spectrum. These infrared rays were further investigated by his son Sir John Herschel, who invented the actinometer in about 1825.

Adams, John Couch (1819–92). English mathematician and astronomer, born near Launceston, Cornwall.

He is best known for his analysis of the orbit of URANUS, which led to the prediction of the existence and position of NEPTUNE. Appointed Lowndes Professor of Astronomy, Cambridge, 1858, Adams succeeded Challis as Director of Cambridge Observatory in 1860.

Adhara The star Epsilon Canis Majoris; magnitude 1.50; type B2.

Adiabatic A process in thermodynamics in which a change in a system occurs without transfer of heat to or from the environment. An example is the vertical flow of air in the atmosphere. As it rises it expands and cools, and as it descends it contracts and grows warmer.

Adonis (Asteroid 2101). The second Apollo asteroid to be discovered. Adonis (diameter 0.6 miles/1km) missed Earth by 1,553,000 miles (2,500,000km) in 1936, but was not seen again until 1977.

Adrastea Jupiter's 15th satellite; magnitude 18.9.

Ænarium Promontorium Lunar cape on the eastern border of the Mare Nubium; 19°S, 9°W.

Æneas Trojan asteroid; diameter 81 miles (130km); period 11.7 years. Also, a crater on Saturn's satellite Dione.

Aeria Bright region on Mars, bordering the Syrtis Major.

Aerolite Old name for stony meteorite.

Aeronomy The study of the physics and chemistry of the upper atmosphere of the Earth and other planets. Because the region is rather inaccessible, being generally above the height that meteorological balloons can reach, research techniques rely heavily on the use of rockets and satellites together with remote sensing by radio waves and optical techniques.

The primary source of energy for the processes of aeronomy is the absorption of the incident solar energy before it reaches the surface of the planet. This energy can ionize the upper atmosphere to form the ionospheric plasma or can cause chemical changes such as the photo-dissociation of molecules to form atoms or the production of exotic molecules such as ozone and nitrous oxide. Some minor constituents have an important catalytic role in the chemistry of the upper atmosphere, hence, for example, the importance of chlorine compounds on the ozonosphere.

Consideration of the atmosphere as a fluid leads to an understanding of the various winds and circulation patterns. Fluid oscillations include tides and internal gravity waves, disturbances which propagate due to the buoyancy forces. Under certain conditions the atmosphere becomes turbulent, which leads to mixing and enhanced heat transport.

Optical phenomena include AIRGLOW, in which photoemission may be caused by a range of physical and chemical processes, and the AURORÆ, where the visible emissions are due to charged particles from the MAGNETOSPHERE. Associated with the auroræ are electric current systems, which cause perturbations in the magnetic field. There are other currents in the upper atmosphere due to tidal driven dynamos. *LJCW. See also* SOLAR TERRESTRIAL PHYSICS.

Agamemnon (Asteroid 911). A Trojan asteroid, discovered in 1919.

Agena *See* BETA CENTAURI

Agena One of the most successful American rockets, used extensively for launching satellites and as a second stage for lunar and planetary probes.

Ahnighito meteorite The largest, 31 tonne, mass of the Cape York, Greenland, iron meteorite shower, recovered by Peary, now in the Museum of Natural History, New York.

Airglow The faint ever-present glow of the night sky arising from various processes taking place in the Earth's upper atmosphere. On the clearest of nights, when there is no moon, the sky between the stars is not absolutely dark. All the starlight added together is insufficient to produce this glow. By the early 20th century, instruments had become available for analyzing the radiation produced. Oxygen and nitrogen were involved, as well as a number of less common elements.

We now know that the airglow is due to processes occurring in the ionosphere. This is a wide, diffuse band of the atmosphere where intense solar radiation falling upon the thin gases causes reactions that turn neutral atoms and molecules into ions (par-

ticles that carry a positive or negative electric charge). When these ionized atoms and molecules recombine, light is given off, and the emission continues for long after the Sun has set. In the daytime, the light emitted by these reactions (dayglow) is stronger, but our eyes are so dazzled by the direct sunlight that we cannot see it. From satellites orbiting in darkness outside the Earth's atmosphere, the airglow can be photographed. It appears like a luminous halo extending from a height of about 55 miles (88km) up to about 375 miles (600km). The airglow, often termed nightglow when seen at night, can interfere with the observation of faint celestial objects. *JWM.*

Airy disk The central spot in the diffraction pattern of the image of a star at the focus of a telescope. In theory 84 per cent of the star's light is concentrated into this disk. *See also* DIFFRACTION.

Airy, Sir George Biddell (1801–92). English astronomer, appointed seventh ASTRONOMER ROYAL in 1835.

A determined and ambitious young man, Airy entered Trinity College at Cambridge in 1819; here he showed great mathematical ability, graduating with the coveted position of Senior Wrangler in 1823. In 1826 he became Lucasian professor of mathematics, and two years later the Plumian professor of astronomy and Director of the Observatories. In 1835 he was appointed Astronomer Royal, and only then moved from Cambridge to Greenwich, where he remained until his retirement in 1881.

Airy was a punctilious man of regular habits, with a strong sense of order as his ruling passion. From early on he recorded all his personal transactions by double-entry book-keeping, never threw away any correspondence, cheque book stubs, or notes for tradesmen, filing everything chronologically.

In his day GREENWICH OBSERVATORY was concerned entirely with accurate measurement and when Airy arrived he found its standards had deteriorated seriously. He therefore reorganized the whole establishment, ensuring that all observations were systematically made in a standard way and with outstanding precision. Under his regime it became so thoroughly efficient an institution, that it was copied by other State observatories. But Airy would tolerate no independence of thought; his staff required method not judgement, and "very little science". He himself, however,

designed novel and effective observing equipment, and there is no doubt that his regime at Greenwich profoundly improved the standards of positional astronomy as a whole. *CAR.*

Aitken, Robert Grant (1864–1951). Leading American double star observer and Director of LICK OBSERVATORY (1930–35). His principal work was the New General Catalogue of Double Stars, published in 1932. It contains more than 17,000 double stars.

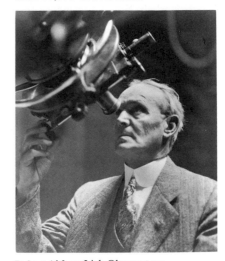

Robert Aitken: Lick Observatory

AI Velorum variables Dwarf Cepheids of RR LYRÆ type with periods shorter than 0^d25. They belong to disk population and are not to be found in star clusters.

Ajax Trojan asteroid; magnitude 16.6, period 11.89 years.

Akna Montes Mountains on Venus; part of the ISHTAR highland.

Alba Patera Low-profile shield volcano on Mars; 40°N, 110°W.

Albategnius (Al-Battani, *c.*858–929). Outstanding Arab observational astronomer who demonstrated that the Sun's distance varies, and whose *Astronomical Tables* were most valuable to later generations.

Albategnius Lunar walled plain, 80 miles (129km) in diameter, south of Hipparchus; 12°S, 4°E. The walls are fairly high, with terraces, though broken in the south-west by a large crater, Klein.

Albedo The fraction of incident light reflected from the surface of a body. Values range from 0 (perfectly black) to 1 (perfect reflector).

Albireo The star Beta Cygni. It is perhaps the most beautiful double in the sky, with a K5-type yellow primary (magnitude 3.2) and a blue companion of magnitude 5.4; separation 34″.6.

Alcor The star 80 Ursæ Majoris. It is the naked-eye companion to Mizar or Zeta Ursæ Majoris.

Alcyone The star Eta Tauri, magnitude 2.87; it is the brightest member of the Pleiades cluster. Its type is B7.

Aldebaran The star Alpha Tauri; magnitude 0.85, distance 68 light-years, luminosity 100 times that of the Sun. It is of type K5, and clearly orange in colour. It appears to lie in the Hyades cluster, but is not a true member of the group, and is in the foreground.

Aldrin, Edwin E. "Buzz" (1930–). American astronaut, born in New Jersey. Aldrin was assigned to Apollo 11 as Lunar Module pilot, and on July 20 1969 became the second man to walk on the Moon.

Alfvén waves Transverse magnetohydrodynamic waves which can propagate in a fluid with high electrical conductivity and containing a magnetic field. Found, for example, in the SOLAR WIND and the Earth's MAGNETOSPHERE (micropulsations). They are named after the Swedish-born astronomer Hannes Olof Gösta Alfvén, winner of the Nobel Prize in 1970.

Algenib The star Gamma Pegasi, in the Square; magnitude 2.8; type B2.

Algieba The star Gamma Leonis. A fine double, magnitudes 2.3 and 3.5; separation 4″.4. The primary is orange (type K).

Algol One of the most famous variable stars in the sky and a prototype of a class of ECLIPSING BINARIES.

Although it is conceivable that the ancients knew of the varying brightness of Algol (the name derives from the Arabic Ras-Al-Ghul – the Head of the Demon), the first recorded observation was made by Geminiano Montanari of Bologna in 1669. Just over a century later, in 1782, a young Englishman named John Goodricke established that the variability was periodic with a sudden fading occurring every 2.867 days. Goodricke explained that this change in light was due either to spots on the body of the star or the passage of a giant planet in front of the star.

Neither hypothesis was taken seriously at first but gradually the concept of an eclipsing companion became accepted and was finally confirmed by H.C. Vogel in 1889, when he showed that the radial velocity of Algol varied with the same period as that of the eclipses.

By this time, Algol was known to be a triple system. In 1855 ARGELANDER had observed that the period between primary minima had shortened by six seconds since Goodricke's observations. Fourteen years later he noted that the period between the times of minima varied in a regular fashion with a period of about 680 days. This was attributed to the variation in the distance which the light from the system had to travel because of orbital motion around the common centre of gravity with a third star (Algol C). This effect is known as the light time and was first used by Ole RØMER in 1675 to measure the velocity of light from timings of the eclipses of Jupiter's satellites.

In 1906 the Russian astronomer Belopolsky confirmed the existence of Algol C by showing that radial velocity variations in the spectral lines of Algol also had a period of 1.862 years superimposed on the period of 2.867 days for Algol AB. Several years later, Joel STEBBINS, a pioneer in stellar photo-

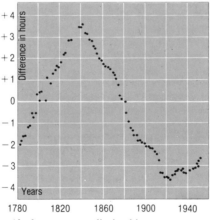

Algol: prototype eclipsing binary

metry, was observing Algol with a selenium cell. He found that there was a secondary minimum of much smaller amplitude occurring exactly halfway between the primary eclipses. This showed for the first time that the companion was not dark at all, but merely much fainter than A. Stebbins observed Algol again in 1920 with a photoelectric photometer. He observed the depth of the secondary minimum to be 0.06 magnitude and also found that the light from star A increased as secondary minimum approached. This

was interpreted as a reflection of light from the body of star B.

Since the early work by Stebbins, many photoelectric observations of Algol have been made and a number of periodic phenomena in the times of primary minimum have been well established. From all the evidence which is at our disposal today the current model of the Algol system would appear to be as follows:

There are two stars, of spectral types B8 and G, rotating about each other in a period of 2.867 days. The B8 star is a dwarf and is the visible component. The fainter star (whose spectrum was only observed directly for the first time in 1978) is a subgiant. The orbit is inclined to the line of sight by 82° which results in mutual eclipses corresponding to a drop in light of 1.3 magnitudes when A is eclipsed by B and 0.06 magnitudes when A passes in front of B. A recent measurement of the parallax of Algol from a series of more than 1,000 plates taken at Sproul Observatory shows that the distance is 30 parsecs with an error of 10 per cent. This corresponds to luminosities of 100 and 3 for A and B respectively. From the length and depth of the eclipses sizes of 2.89 solar radii and 3.53 solar radii have been derived for A and B respectively. The corresponding masses are 3.6 solar masses and 2.89 solar masses and this apparent anomaly gives rise to what is known as the "Algol paradox".

In current theories of STELLAR EVOLUTION, stars, as they evolve, advance in spectral type and the rate at which they do so is a function of their initial masses. Thus, if two stars form together from interstellar material, the more massive of the two should evolve more quickly. In Algol it is the more evolved star which is the less massive and the cause of this seems to be the transfer of mass from B to A. A stream of material between the two stars has been detected in the radio spectrum, and the current transfer rate is thought to be at least 10^{-7} solar masses per year. Optical spectra have shown very faint lines, thought to be emitted by a faint ring of material surrounding star A.

Algol C was first resolved by SPECKLE INTERFEROMETRY in 1974 and on several occasions since but the angular separation has never exceeded 0.1 arc seconds, which explains why the star has never been seen by visual observers.

Although many of the phenomena in the Algol system have been explained, the variations in the time of minima are still not clearly understood. An obvious explanation of the 32- and 180-year periods would be the existence of two further components in the system (Algol D and E) with the periods being explained as light-time effects from the corresponding orbits of the two stars around the system, AB-C. However, in order to produce the 180-year period, a star would have to have a mass of 3 solar masses and would be separated from AB by up to three arc seconds. If it existed it would be of naked eye brightness. In any case, comprehensive measurements of the proper motion of Algol since 1880 show that it is rectilinear in nature. The hypothetical, massive fifth component would have produced an observable perturbation. Another explanation of the jump around 1860 is that it corresponds to a particularly strong outburst of mass transfer.

The 32-year period can be explained by a magnetic cycle, similar to the 22-year cycle of the Sun. The magnetic field would affect the rate at which mass transfer took place and this in turn would affect the orbital period of AB and hence the time of primary eclipse. Another possible cause could be a rotation of the orbit of AB (apsidal motion) caused by the gravitational pull of star C. However, astrometric studies published in 1984 show that the orbital plane of C is either co-planar with, or at right angles to, the orbital plane of AB. If the former case is true then the explanation of the 32-year period as being due to apsidal motion is no longer tenable.

The real nature of the Algol system is still far from clear. Even after 200 years of continuous observation it still continues to evoke considerable interest from astronomers. *RWA*².

Algonquin A radio observatory at Lake Traverse in Ontario, Canada, run by the National Research Council. Its principal instrument is a 150ft (46m) fully-steerable radio dish in operation since 1967.

Alhena The star Gamma Geminorum; magnitude 1.93; type A3.

Aliacensis Regular lunar crater, 52 miles (84km) in diameter; 31°S, 5°E. It forms a notable pair with the crater Werner.

Alinda (Asteroid 887). Amor type. It is only a few kilometres in diameter.

Alioth Epsilon Ursae Majoris, the brightest star in the "Plough" or Dipper; magnitude 1.77; type A0.

Alkaid Eta Ursae Majoris, in the "Plough" or Dipper; magnitude 1.86; type· B3. Its alternative proper name is Benetnasch.

Allegheny Observatory Observatory of the University of Pennsylvania in Pittsburgh, founded in 1860. Most notable for its reliable determinations of trigonometric parallaxes of stars, using the Brashear 30in (76cm) refractor.

Almagest The Western corruption of the Arabic title of the great compendium of astronomy prepared in the second century AD by PTOLEMY. It is in essence

Almagest: Ptolemy's great compendium

a basic textbook: the reader need know only fundamental Greek geometry and some astronomical terms.

Al-Ma'mun (9th cent.). Arab astronomer, who built an observatory in 829 equipped with instruments better than those of the Greeks.

Almanon Lunar crater, 30 miles (48km) in diameter, forming a pair with its larger companion ABULFEDA; both lie west of the Mare Nectaris. Almanon's walls rise to 10,000ft (3,000m).

Alnair The star Alpha Gruis; magnitude 1.74; type B5; luminosity 230 times that of the Sun.

Al Nath The star Beta Tauri, formerly included in Auriga as Gamma Aurigæ.

Alnilam Epsilon Orionis, in Orion's belt; magnitude 1.70; type B0.

Alpha Capricornids A minor meteor stream, active from mid-July till mid-August, best seen from lower latitudes. Although weak – about six meteors per hour at most – the stream produces a high proportion of bright flaring meteors with long paths. Three distinct

maxima have been detected during the period of activity; this and the large radiant area indicate that the stream has a complex structure. Analysis of trails recorded photographically shows two sets of orbits, the longer-period group quite closely corresponding to the orbit of Comet P/Honda-Mrkos-Pajdusaková, a rather faint object with a period of 5½ years, last seen in 1985.

Alpha Centauri Rigil Kent, a −0.1 magnitude star close to the constellation of Crux (the Southern Cross) in the southern sky. It is a visual binary star with a period of 80 years and a separation of 1″.76. Associated with it is the much fainter star Proxima Centauri.

Alpha Centauri is notable in being the nearest of the bright stars: only Proxima Centauri lies closer to the Sun. Its distance was first measured by PARALLAX in 1839, one year after the first successful stellar parallax measurement (on 61 Cygni) by Bessel. Modern determinations of the parallax of Alpha Centauri yield $\pi = 0″.745$, giving a distance of 1.34 parsecs.

The brighter component (Alpha Cen A) of the visual binary is a G2 main sequence star, and the fainter component (B) is a K5 main sequence star. Analysis of the binary motions yields 1.1 and 0.9 solar masses respectively for A and B, an orbital inclination of 79° and an eccentricity of 0.52.

Proxima Centauri has an apparent magnitude of 15. It has a slightly larger parallax than Alpha Cen, with a corresponding distance of 1.31pc from the Sun. Although it is 2.2 degrees away from Alpha Cen, it is believed to be associated with the binary because it has a similar motion through space. *DJA*.

Alpha particles Helium nucleus consisting of two protons and two neutrons, positively charged. As helium is the second most abundant element, alpha particles are found in most regions of PLASMA such as inside stars, in diffuse gas around hot stars, and in cosmic rays. Alpha particles are also produced in radioactive decay of some elements. In the PROTON-PROTON chain of nuclear fusion reactions inside stars, 4 protons (hydrogen nuclei) are converted to one alpha particle (helium nucleus) with release of fusion energy, which powers stars. In the "triple alpha process", which is the dominant energy source in red giant stars, three alpha particles fuse to form a carbon nucleus with release of energy.

Alphard The star Alpha Hydræ; magnitude 1.98; distance 85 light-years; luminosity 115 times that of the Sun; type K3. It is reddish, and often called "the Solitary One". It has been suspected of variability.

Alpha Regio Plateau region of Venus centred at 25°S, 5°E. The circular central area has a mean elevation of 1,600ft (0.5km), a discontinuous rim and numerous NE/SW parallel lineaments.

Alphekka The star Alpha Coronae Borealis; magnitude 2.23; type A0.

Alpheratz The star Alpha Andromedae; magnitude 2.06. It is in the Square of Pegasus, and was formerly called Delta Pegasi.

Alphonso X of Castile (1221–84). Known as "Alphonso the Wise", he was a patron of learning, and especially astronomy, who is now primarily remembered for commissioning a new edition of the highly successful Toledan astronomical tables. Setting out the motions of the Sun, Moon and five naked-eye planets, they had been prepared originally by the Arab astronomer al-Zarqali in Toledo a century before. The new *Alphonsine Tables* incorporated more recent observations and so were of improved accuracy. Completed in 1252, the year in which Alphonso ascended the throne, they were not superseded for almost four hundred years. *CAR.*

Alphonsus Centred at 13.5°S, 3°W, this lunar crater, some 72 miles (117km) across, has fault-dissected walls which rise above the floor to over 10,000ft (3km). The floor is split in two by a ridge system, running nearly N to S, which is 9 miles (15km) wide and about 3,000ft (1km) high where it forms a prominent central peak. The origin of Alphonsus is unknown, some assuming it to be an ancient impact crater while others consider it to be a volcanic structure.

Within Alphonsus, small, elliptical craters, having their longer axes parallel to the run of the central ridge system, occur individually or in chains and provide evidence for the volcanic origin of the ridge which seems to lie in a right-lateral strike-slip fault. N.A. Kozyrev noted a reddish obscuration and secured spectra which indicated (in 1958) a cloud containing molecular carbon (mainly C_2) and (in 1959) red-hot material. These spectra related to

patches close to the central peak of Alphonsus. Carbonic gases emitted from lunar rocks could raise dust from the regolith and show characteristic spectral features when ionized by solar ultraviolet light.

Several of the kilometre-sized elliptical craters which populate the floor of Alphonsus have their longer axes set centrally in rilles; and some of these craters are surrounded by dark-halo material, which, like pyroclastic deposits, blankets local depressions and eminences. The dark-halo craters might represent volcanic vents from which dark, fragmental material was ejected from depth. The elliptical craters might have assumed their final forms as a result of subsidence, particularly since the rilles in which they lie seem to have been formed when regolith drained into underlying fractures. *GF.*

Alpine Valley (Vallis Alpes) This conspicuous, 124-mile (200km)-long valley cuts through the lunar Alps which border MARE IMBRIUM on its NE side. Varying in width from some 11 to 4 miles (18 to 7km) it tends to taper away from Mare Imbrium, but this narrowing is far from regular. Sharply contrasting with the hilly terrain of the Alps, the floor of the valley is relatively smooth, although detailed views show that it is crossed by faults and rilles, and that it contains some craters. The Alpine Valley is a trough that developed as a result of faulting accompanying the tectonic adjustment of Mare Imbrium. *GF.*

Alps (Montes Alpes) Forming the NE flanks of MARE IMBRIUM, these cross-faulted lunar mountains rise to 3,250ft to 9,750ft (1 to 3km) above the mare margins but degenerate to hills farther from the mare.

Al-Sūfi Abu'l-Husayn (903-986). Arab astronomer famous for his *Book on the Constellations of the Fixed Stars,* which was essentially a detailed revision of the star catalogue drawn up some eight hundred years earlier by PTOLEMY.

In this work Al-Sûfi discussed each constellation critically, and identified those stars with Arab names. He also provided a table of revised magnitudes and positions as well as drawings of each constellation, the latter becoming well known in the West. His work is valuable because it records the results of actual observation. He also prepared texts on the astrolabe and the celestial globe. *CAR.*

Altair The star Alpha Aquilae; magnitude 0.77; distance 16.6 light-years; type A7. It is ten times as luminous as the Sun, and one of the closest of the bright stars.

Altai Scarp A lunar scarp south-west of Mare Nectaris, formerly called the Altai Mountains; 25°S, 22°E. It has relatively few craters.

Altazimuth mounting A telescope mounting in which the motion to point the telescope and track stars is by movements in altitude and azimuth. These movements are like those of a gun, or theodolite. Altitude corresponds to the elevation angle about a horizontal axis, or motion up and down from the horizon to the zenith; "azimuth" corresponds to the bearing angle around a vertical axis, or motion from side to side parallel to the horizon. Altazimuth mountings contrast with EQUATORIAL MOUNTINGS. An equatorially-mounted telescope tracks stars by rotating only about the polar axis, with no rotation about the declination axis. An altazimuth telescope must move in both axes.

Historically, the first sighting tubes and early astronomical telescopes were mounted on stands which could be turned to any azimuth, with simple joints to give an adjustment in altitude. William HERSCHEL's large reflecting

A

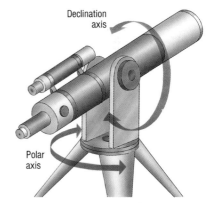

Altazimuth mounting: moves in both axes

telescopes were of this sort. Today's inexpensive small astronomical telescopes for amateur astronomers are also usually altazimuth mounted. On somewhat larger hand-driven telescopes it gives a smoother operation to rotate in one axis only, and an equatorial mounting is better. If the telescope mounting is to be driven by a simple constant-speed motor, rotating the axis at one revolution per sidereal day to compensate for the Earth's motion, then the telescope must be equatorially mounted.

If the astronomer is attempting photography, the easiest telescope mount to use is the equatorial, because the image of the sky remains fixed relative to the telescope. By contrast, in an altazimuth telescope, the image of the sky rotates about the optical axis, relative to the telescope structure. Thus, during a time exposure the image blurs while being photographed.

The image rotation can be visualized by imagining what happens as an altazimuth telescope tracks a planet whose path rises in the east, passes through the zenith, and sinks in the west. As the planet rises, the telescope azimuth is to the east, its altitude increases and the eastern or following side of the planet lies to the underside of the direction that the telescope is pointing. As the planet passes through the zenith, the telescope azimuth rotates to the west and its altitude decreases. The eastern, or following, side is now above the telescope and the planetary image has rotated through 180°, relative to the telescope structure and any camera which is photographing it. To compensate for this an altazimuth telescope must feature either an optical device to compensate for this rotation and hold the image steady (an image rotator) or a mechanical device to which the camera may be attached and rotated to follow the image (an instrument rotator).

Thus, while equatorial telescopes track stars by motion around a single axis, altazimuth telescopes need motion around three axes. Moreover, the single motion of the equatorial telescope is of constant speed, while the three motions of an altazimuth telescope are of varying speeds and must be coordinated. This is successful to a high accuracy only in a computer-controlled telescope.

It is obvious that altazimuth telescopes are more difficult and therefore more expensive to control than equatorial telescopes. The degree of sophistication required means that virtually all moderate-size telescopes are equatorially mounted. However, because the altazimuth telescope mounting is compact and symmetric, with the vertical axis lying along the direction of gravity, a large altazimuth telescope is cheaper and easier to make than an equatorial telescope. In particular, the observatory dome which encloses an altazimuth telescope is smaller than the one surrounding an equatorial telescope of the same size.

These benefits in the mechanical and civil engineering areas outweigh the additional cost of control devices, good bearings, etc.

RADIO TELESCOPES sample the sky point by point, so image rotation is not a critical problem, and radio telescopes do not need to be controlled to the same accuracy. Therefore all large fully-steerable radio telescopes are on altazimuth mountings, including the 250ft (76m) JODRELL BANK telescope, the 210ft (64m) PARKES radio telescope and the 210ft (64m) GOLDSTONE telescope at the Jet Propulsion Laboratory. These structures are so large that it is not practical to think of tilting them to a large angle, to line up with the Earth's axis — the practical limit was reached in 1959 with the 140ft (43m) radio telescope of the NATIONAL RADIO ASTRONOMY OBSERVATORY and problems with its equatorial mounting were not solved until 1965. Recent large altazimuth optical telescopes include the 20ft (6m) telescope of the Special Astrophysical Observatory in ZELENCHUKSKAYA, USSR, the 14ft (4.2m) WILLIAM HERSCHEL TELESCOPE in La Palma, the MULTI-MIRROR TELESCOPE in Arizona and the Advanced Technology Telescope at Siding Spring. The list includes the three most modern of the four largest optical telescopes in the world. The last two telescopes are housed in observatory buildings which co-rotate with the telescope around the azimuth axis. These buildings are even more compact than usual, with a minimum of empty space into which the telescope might need to sweep.

The altazimuth mounting makes possible an optical layout of particular convenience. It was invented by James Nasmyth. Based on a conventional reflecting telescope with concave primary and convex secondary mirrors, the Nasmyth telescope contains a third flat mirror at 45° to the optical axis. It is located at the intersection of the vertical and horizontal axes of the altazimuth mounting and directs the telescope beam along the horizontal axis through the altitude bearing. A platform at the bearing gives access to the focus: the platform is always horizontal, and slowly rotates around the vertical axis. In a large telescope the Nasmyth platform amounts to an optical laboratory. Nasmyth himself sat on the platform of his altazimuth telescope with his eye to an eyepiece, bringing the stars to him, rather than climbing a ladder to the stars as he would with an equatorial telescope of the same size. The same convenience applies to the design of modern optical instruments for the Nasmyth platform — they are not slung at all attitudes as the telescope moves over the sky and do not have to be so rigidly made and so compact as instruments for a Cassegrain or Newtonian focus. *PGM. See also* TELESCOPES, TELESCOPIC DRIVES.

Altitude The vertical angle between the horizon and a celestial body. The altitude of a particular star depends on the location of the observer and the time of the observation.

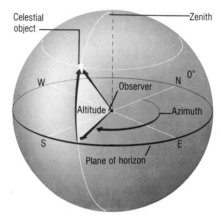

Altitude: not height, but an angle

Aluminizing The highly reflective coating used on the optics of reflecting telescopes consisting of evaporated aluminium evenly distributed and usually only a few microns thick. The process is carried out inside a vacuum chamber. The optical component is first thoroughly cleaned before being placed in the chamber together with some pure aluminium wire. After removing the air, the aluminium is vaporized by switching on the tungsten heating elements to which it is attached. The resulting gas condenses onto the clean surface of the optical component to form the aluminium coating. The process was pioneered by Dr. J. Strong, of the University of California. *NEF.*

Alya The star Theta Serpentis. It is a wide double; magnitudes 4.1 and 5.0; separation 22″.6. The combined magnitude is 3.4.

Amalthea The largest of the four innermost satellites of JUPITER, discovered by Edward BARNARD in 1892 at LICK OBSERVATORY during a systematic search for additional satellites of Jupiter. Because of its small radius, remote distance from the Sun, and low reflectivity, Amalthea, a faint object

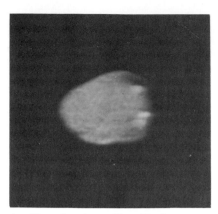

Amalthea: close-in satellite of Jupiter

nested in the glare of the scattered light from Jupiter, was a difficult object to observe from ground-based facilities. It was the last of the Jovian satellites to be discovered by Earth-based observers. Although the remaining satellites of Jupiter were not detected until the VOYAGER flybys, the orbit of Amalthea had been mapped with sufficient accuracy to allow pre-planning to be carried out to program the Voyager spacecraft and obtain observations.

The Voyager observations indicate that Amalthea is a non-spherical body with an irregular surface that is roughly an ellipsoidal shape. The longest axis of the ellipsoid is radially aligned with Jupiter, indicating that the satellite has resided in its present orbit for a period of time that has been sufficient for it to be tidally perturbed into its lowest energy configuration. In this orientation, the satellite rotates on its axis at the same rate that it revolves around the planet; hence one location on the body is always facing Jupiter, resulting in a minimum variation of gravitational force on the body.

The satellite has a low reflectivity and is a dark red colour that may be due to sulphur deposits ejected from the surface of IO by volcanic activity and swept up by Amalthea as the ejected molecules and atoms spiral in towards Jupiter. The surface of the satellite is heavily cratered, with the largest crater, Pan, spanning a distance of 56 miles (90km). In addition, there are grooves and ridges that extend for tens of miles across the surface of this small body whose total dimensions range from 170 to 90 miles (270 to 150km). *RFB.*

AM Canum Venaticorum Unique blue variable with changing period of about two minutes. It is probably a semi-detached binary of two white dwarfs, an accretion disk and a hot spot.

Ames Research Center NASA station located near San Francisco with broad range of programme responsibilities including computer science and applications, aeronautics and airborne/space sciences. In 1981 it was merged with the Dryden Flight Research Facility at Edwards in the Mojave Desert north of Los Angeles — the preferred landing site for the SPACE SHUTTLE.

AM Herculis stars Alternative name for POLARS.

Amirani Active volcano on Jupiter's satellite IO; 27°N, 119°W. The illustration below offers two theories of Io's volcanic activity: (1) an Earth-type mechanism; (2) liquid SO_2 exploding on contact with molten sulphur.

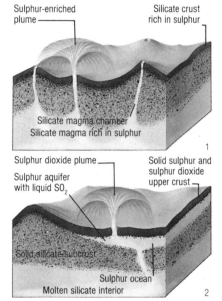

Amirani: structure of an Io volcano

Ammonia (NH_3). One of the first interstellar molecules discovered by radio astronomy and a constituent of the atmospheres of the giant planets.

Amor asteroids A type of Mars-crossing asteroid having, according to Eugene M. Shoemaker and colleagues, perihelion distances between 1.017AU (the aphelion distance of Earth) and 1.3AU. Due to close approaches to Mars and Earth, many Amors can become Earth-crossers (APOLLOS) for a time, and vice versa. Like all planet-approaching asteroids, Amors have finite lifetimes (about 10^7 years, typically) and most will be destroyed by collision with Mars or Earth.

Currently, about 40 Amors are known, though it is thought that there are about 500 larger than 6,500ft (2km)

in diameter. The first discovered, largest, and most studied is 433 EROS, a highly-elongated S-type asteroid measuring $8 \times 9 \times 22$ miles ($13 \times 15 \times 36$km) and rotating in 5.3 hours. For some time the Amor asteroid 1580 Betulia held the record orbital inclination of 52°. To judge by the broad variety of compositional types among the Amors, they have originated from several sources, among which are the 3:1 and 5:2 commensurabilities with Jupiter and perturbation from the inner main belt by Mars; some may be extinct cometary nuclei. *EB[1].*

Amphitrite (Asteroid 29). A main-belt non-family asteroid with a diameter of 120 miles (195km) and an orbital semi-major axis of 2.55AU. The reflectance in the infrared indicates that the surface is made up of iron and titanium silicates. The surface is inhomogeneous and this is thought to be due to differing physical depths of soil and not differing chemical composition. Amphitrite was discovered by A.Marth in 1854, from London. It has a spin period of 5.4 hours. *DWH.*

Analemma The analemma is the elongated figure-of-eight shape obtained by plotting (or photographing) the position of the Sun on the sky at the same time of day at regular intervals throughout the year. The figure graphically illustrates the effect of the Earth's orbital inclination on the declination of the Sun on the sky and underlines the seasonal variations in the Sun's position. A considerable degree of patience and technical skill is required to record the analemma photographically. *DFM.*

Ananke Jupiter's 12th satellite; diameter 18.6 miles (30km). It has retrograde motion, and may be a captured asteroid.

Anaxagoras (?500–?428BC). Greek philosopher, who sought to reconcile the idea of one single imperishable existence with evidence for change and the presence of different substances. This he achieved by claiming that several basic elements existed, and that material things could be infinitely divided.

Anaxagoras also believed that rarefied, dry and hot material formed the sky while the Earth, which he thought of as a disk, was composed of dense, wet, cold and dark matter. The Sun, Moon, stars and planets had, he believed, been torn from the Earth. They shone because their motion ignited them. *CAR.*

Anaxagoras Crater in the far north of the Moon; well-formed, with a diameter of 32 miles (51km); 75°N, 10°W. It is an important ray-centre.

Anaximander of Miletus (?610–?546BC). Greek philosopher who believed the Earth to be one of many existing worlds, and the Sun and Moon rings of fire.

Anaximenes of Miletus (mid-6th cent. BC). Greek philosopher who thought air to be the primeval substance, and the Sun and Moon to be made of fire.

Ancient Beijing Observatory Built during the reign of Zhengtong (1436–49) of the Ming Dynasty, the Ancient Beijing Observatory is situated at the southeast corner of the old city wall. The wall platform is 46ft (14m) high, with eight large astronomical instruments in bronze made in the Qing Dynasty. Six of the eight were made in 1673 under the leadership of the Belgian missionary Père Ferdinand Verbiest; they are the Celestial Globe, the Equatorial Armilla, the Ecliptic Armilla, the Altazimuth, the Quadrant and the Sextant. The other two are the Azimuth Theodolite (1715) and the New Armilla (1744). The Armillary Sphere and the Abridged Armilla, both built in the fourth year of the reign of Zhengtong (1439), and the GNOMON of the Ming Dynasty, originally kept here, were moved to the PURPLE MOUNTAIN OBSERVATORY in 1931. *DPB.*

Ancient Beijing Observatory: 15th century

Anderson, John August (1876–1959). American astronomer who, with Pease, used the Michelson interferometer on the 100in (254cm) Hooker reflector to measure the diameter of Betelgeux. He played a leading role in setting up the 200in (508cm) telescope.

Andromeda A large and important constellation of the northern hemisphere. It adjoins the Square of Pegasus, and indeed one of the stars in the Square,

Alpha Andromedæ (Alpheratz), was formerly known as Belta Pegasi.

Brightest stars

Name	Visual Mag.	Abs. Mag.	Spec.	Distance (light-yrs)
β (Mirach)	2.06	−0.4	M0	88
α (Alpheratz)	206	−0.1	A0p	72
γ (Almaak)	2.14	−0.1	K2,A0	121
δ	3.27	−0.2	K3	160

Other stars above the fourth magnitude are 51 (3.57), Tau (3.6, slightly variable) and Mu (3.87). The main stars form a line leading away from Pegasus.

Gamma (Almaak) is a fine double; magnitudes 2.2 and 5.0, separation 10″, P.A. 063°. The primary is orange; the blue companion is itself double, with a separation of 0″.6.

The only variable star in Andromeda with a maximum above magnitude 6.5 is the Mira-type star R, with a large range (mag. 5.9 to 14.9) and a period of 409 days; the spectral type is S.

Of course the most famous object in the constellation is the Great Spiral, M31. It has two companions, M32 and NGC205. The open cluster NGC752 is also an easy binocular object.

In mythology, Andromeda was the princess of the Perseus legend. *PM.*

Andromeda Galaxy The LOCAL GROUP OF GALAXIES is dominated by two giant spirals: our Milky Way Galaxy, and the Andromeda galaxy. Andromeda is the nearest spiral, some 700kpc away.

By virtue of its proximity the Andromeda galaxy has drawn the attention of astronomers through the last 100 years. Fundamental advances have come from these studies in such diverse fields as star formation, stellar evolution,

Andromeda Galaxy

Other designations	Andromeda Nebula, M31, NGC224
Apparent size	1.2 x 4.1 deg
Apparent (integrated) magnitude m_v	3.5 mag
Absolute magnitude M_v	−21.1 mag
Type (G de Vaucouleurs)	SA(s)b
Angle between plane of galaxy and line-of-sight	13 deg
Distance	700 Kpc, 2.2 MLy
Number of stars	$4 \times 10''$
Total mass	$3.2 \times 10'' M_0$
Diameter (optical)	50 Kpc
Dimensions of optical nucleus	5 x 8 Kpc
Satellite galaxies	M32, NGC147, NGC205

stellar nucleosynthesis, dark matter in the universe, and the distance scale and

evolution of the universe.

Andromeda is the faintest of the three galaxies to be seen by the naked eye, the other two being the irregular satellite galaxies of the Milky Way, the MAGELLANIC CLOUDS. To the eye, Andromeda is a faint smudge, even on a moonless night at a dark site. Through binoculars and small telescopes it is seen to have elongated structure some 1 × 4° in extent, with a nuclear concentration of light. Large telescopes are needed to reveal the spiral structure.

The "Great Debate" amongst astronomers in the 1920s concerned the nature of the nebulæ. Were these "island universes", complete star systems outside our own, as the philosopher Kant had suggested in the 18th century? Or were they gas clouds within our Galaxy, collapsing to form stars? The spiral nature of the Andromeda nebula had been observed by Isaac Roberts with a 20in (51cm) telescope in the 1880s, but it was in the 1920s that Andromeda provided the crucial clue to the nature of nebulæ, and to the nature of the the universe itself. Edwin HUBBLE resolved the outer parts of the nebula into stars. Even more essential was his discovery of blue stars in Andromeda with periods similar to a set of bright blue stars discovered in the globular clusters of our Galaxy. These stars were known to have a period-luminosity relation (*see* CEPHEID VARIABLES). Assuming that the same relation applied to the variable stars of Andromeda, Hubble was able to estimate their luminosities, and hence the distance to Andromeda, a distance which placed it clearly beyond the limits of the Milky Way. It was an "island universe"; our view of the Universe of Galaxies was established.

The next vital phase in the study of Andromeda was 1940-1955, resulting from the painstaking observations of Walter BAADE, first with the Mount Wilson telescope during wartime blackout conditions in California and then from Palomar, using the 200in (508cm) reflector. Baade began by trying to resolve stars in the central bulge of Andromeda, and he eventually found a combination of red-sensitive plates and filters which succeeded. The nucleus appeared to be filled with old red stars, substantially fainter than the bright blue stars of the outer regions and apparently similar to the red stars comprising globular clusters well known in our Galaxy. Baade referred to these as POPULATION II, labelling the hot bright stars of the disk of the galaxy as POPULATION I. The distinction remains in use, and is an essential feature of currently accepted theories of star formation, STELLAR EVOLUTION, and galaxy evolution.

The discovery of the two populations led in turn to a crucial discovery for cosmology. The CEPHEID VARIABLES turned out to be of two subsets, one belonging to each population. These obeyed two different period-luminosity relations, and since the Cepheids discovered by Hubble in Andromeda were in the disk, they were of Population I, while those of our Galaxy were amongst the old globular cluster stars of Population II. The result was to "increase" the distance to Andromeda by a factor of 2, and because Andromeda was a stepping stone in establishing distance indicators to galaxies yet farther away, the entire universal distance scale had to be doubled.

Radio continuum observations of Andromeda show it to have very weak nuclear activity in its nucleus, some 20 times below that of our Galaxy. However, the radio spectral-line information obtained from Andromeda is a gold-mine. The neutral hydrogen (HI) distribution in Andromeda has been extensively studied by radio astronomers observing the 21cm hyperfine-transition. This neutral hydrogen is a constituent of the gas of the galaxy, and as such is distributed like other Population I components. In fact the detailed studies of Emerson and collaborators (Cambridge) and Bajaja and collaborators (Westerbork) have shown a dramatic zone-of-avoidance at the centre of Andromeda, where the Population II stars dominated. The gas is distributed in doughnut form. Moreover, it does not rotate with perfectly circular motions about the nucleus, as the Population I stars seem to do; the innermost regions of the "doughnut" look to be falling towards the nucleus. The reasons for (and consequences of) this infall remain to be understood.

The dynamics of Andromeda have also been studied by Courtes and colleagues mapping the radial velocities of hot gas clouds across the galaxy. From these measurements and the HI observations a "rotation curve", a plot of rotation speed versus radius, can be constructed which in turn yields an estimate of the amount of mass within each radius. The mass given in the table is from such an estimate. However, the HI measurements in particular suggest that the outer regions of Andromeda contain substantial amounts of additional mass. Such unseen haloes, or "dark matter" associated with galaxies,

are crucial to current theories of galaxy formation and clustering, and indeed to modern theories of COSMOLOGY.

With regard to the outer regions of Andromeda, Hartwick and collaborators found a halo of GLOBULAR CLUSTERS some three times more extensive than that surrounding our Galaxy. The stars in these globular clusters show a generally higher abundance in heavy elements than those in the globular clusters of our Galaxy, and indeed with a much greater spread of abundances than expected. Globular clusters are deemed to be extreme Population II objects, objects in which star formation was complete at an early epoch in the history of a galaxy, and in

Andromeda: the nearest spiral galaxy

which subsequent enrichment of the heavy element proportion by nucleosynthesis in successive generations of stars was soon halted, keeping them poor in heavy elements. The great spread in element abundances of the globular clusters in Andromeda speaks for a much slower and much more irregular evolution than that of the Milky Way Galaxy.

It may be anticipated that the proximity of Andromeda would mean a substantial contribution to the theory of development of spiral structure. It has contributed controversy instead, and part of this is because the galaxy is so close to edge-on to us that the details of the spiral structure are hard to delineate. In fact it is not even known how many spiral arms there are. Arp has proposed two trailing spiral arms, with some disturbance of these due to the gravitational pull of M32. Kalnajs proposes instead a single *leading* spiral arm, set up via a gravitational "resonance" with M32. The dust clouds do not help the decision between these two models, but it is clear that resolution of the debate will ultimately advance our understanding of the mechanism gener-

ating spiral structure. (*See* DENSITY WAVE THEORY).

Andromeda is our sister galaxy, the nearest spiral, similar in most attributes to the Milky Way Galaxy. Much of the Milky Way is clouded out for us by massive dust aggregates; we rely on Andromeda for understanding our own Galaxy as well as the rest of the universe. *JVW*.

Andromedid meteors These meteors are also known as the Bielids, from their association with BIELA'S COMET. The comet, famous in its own right for having split into two comets in 1845, has not been definitely seen since 1852 and must now be considered lost or defunct, but it produced swarms of small particles which later, on encounter with the Earth, gave rise to some spectacular meteoric displays. Their name derives from the fact that the radiant is in the constellation of ANDROMEDA.

The first recorded appearance of the meteors was in 1741, when a modest shower was observed. Further small displays were seen in 1798, 1830, 1838, and 1847, in each case during the first week of December. When seen in 1867, however, they appeared on the last day of November, and this progressively earlier date of occurrence has been maintained. To become visible in our atmosphere, the meteor particles must be moving in orbits that intersect that of the Earth; the intersection point is known as the node and its location determines the date on which the encounter occurs. However, the orbits are disturbed by the attractions of the planets, and this causes the position of the node, and hence the date, to change. In the case of the Andromedids the node moves back (regresses) rather rapidly, making the meeting-date two to three weeks earlier per century.

In 1867 the association between a meteor shower and a comet had been demonstrated by Schiaparelli in the case of the Perseids, and other such connections were sought. It was known that the orbit of Biela's comet approached that of the Earth very closely, so that the comet could conceivably give rise to a meteor shower, and when the radiant was calculated it was found to agree closely with that of the meteor showers previously seen to emanate from Andromeda.

Biela's comet, if it still existed, would have been in the vicinity of the Earth in 1867, and since the meteor swarm would not be far displaced from its progenitor, a display could be expected

A

towards the end of the year. A good, though not spectacular, shower of Andromedids was seen on November 30, confirming the prediction. Since the orbital period was about $6\frac{1}{2}$ years,

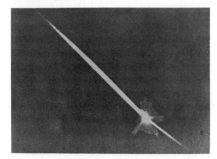

Andromedid meteor: photographed in 1895

A

Weiss, D'Arrest and Galle, who had made the first calculations, predicted another display for November 28 1872.

Soon after sunset on November 27 1872, a day earlier than expected, western Europe was treated to an awesome spectacle; meteors rained from the sky at the rate of several thousand per hour, emanating from near the star Gamma Andromedae. The event caused less alarm and terror among the general population than it might have done, for only six years previously, in 1866, there had been an equally dramatic display of the Leonids, and no harm had come of it. Although the Andromedids were about as numerous as the Leonids, they were not so brilliant because their speed through the atmosphere was very much less. The Leonid swarm travels around the Sun in the opposite direction (retrograde) to that of the Earth, so we encounter the meteors head-on at a relative velocity of 44 miles (70km)/sec. On the other hand, the Andromedids move in the same direction as we do and therefore overtake us from behind, at a relative velocity of 12 miles (19km)/sec.

The shower of 1872 led, incidentally, to one of the mysteries in astronomical history. Klinkerfues, of Göttingen, reasoned that since the comet was presumably ahead of the meteors it should be visible in the opposite direction to that from which the meteors came. He accordingly cabled to Pogson, an astronomer at Madras (the comet would not be visible from high northern latitudes): "Biela touched Earth November 27 — search near Theta Centauri". On December 2, the first clear night, Pogson made his search and sure enough, near the indicated position, he found a comet!

The object was observed again the following night, but then clouds inter-

vened and when a clearance finally came, there was no sign of it. The observations were inadequate for the calculation of an orbit and the prediction of future positions, so the comet, if such it was, was lost. If Biela's comet still existed and was pursuing its original orbit, it would have passed the position indicated by Klinkerfues some months previously. We must conclude that if Pogson, who was an experienced observer, did see a comet, it was not Biela's, and was either a laggard fragment of that object or another comet that just happened to be there at the time. Both alternatives are hard to believe, and the question remains open.

The next encounter was badly timed; no shower was seen, but two revolutions after the 1872 event, in 1885, Europe was, on November 27, delighted and thrilled by an even greater meteor storm — a stupendous display in which the meteors were estimated (counting was virtually impossible) to appear at the rate of 75,000 per hour. This compares with the rate of 100,000 per hour of the great storm of Leonids seen from the United States in 1966. Like that event, and other similar meteor storms, the intense activity was over in about six hours, though greatly diminished rates were seen some days before and after the maximum. This means that the core of the swarm was about 100,000 miles (160,000km) across, with weaker outlying shoals extending to a few millions of miles.

The average meteor particle is little more than a grain of sand, but if a meteor body were big enough — quite a large lump, in fact — it might survive its passage through the atmosphere and land on the ground, becoming a METEORITE. One might expect that in the course of a major meteor display at least a few of the meteor bodies would be big enough to produce meteorites, but only one instance is known of such an event. This happened at Mazapil in Mexico in 1885 while the Andromedid display was in progress. However, we know enough about comets to say that an iron meteorite is the last thing we would expect one to produce, and the fall in this case must be put down to pure coincidence.

Since 1885 the Andromedids have been quite undistinguished, and are now to all intents and purposes defunct. It appears that planetary perturbations have shifted the orbit of the swarm so that it no longer meets that of the Earth. A fairly strong display occurred on November 23 1892, and on November 24 1899 about

200 per hour were seen. Denning logged 20 per hour in 1904 and the last visual sightings were in 1940, when about 30 per hour were recorded, on November 15. A few individual meteors from the shower were caught by the Super-Schmidt meteor cameras in the United States in 1952 and 1953, but since then there have been no further reports. A prediction from an up-dated orbit for Biela's comet was recently (1986) made, but it did not lead to the recovery of that object. If the Andromedids appeared now they would do so about November 1, or late in October. It is too soon to say that we shall never see an Andromedid storm again — the Leonids gave us a miss for a century before returning, as strong as ever, in 1966. Maybe history will repeat itself with the remains of Biela's comet. *HBR. See also* METEORS.

Anglo-Australian Observatory Jointly funded by the UK and Australian governments, AAO operates the 13ft (3.9m) Anglo-Australian Telescope at SIDING SPRING OBSERVATORY.

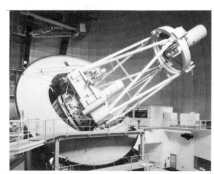

Anglo-Australian Observatory: the AAT

Ångström, Anders Jonas (1814–74). Swedish physicist who studied the solar spectrum with a diffraction grating.

Ångström unit A unit of length equal to 10^{-10} metres and sometimes used to describe wavelengths of light.

Angular measure The apparent diameter subtended by an object; or the difference in angle between two objects, measured in degrees or radians.

Angular momentum A quantity used in the analysis of the behaviour of rotating or orbiting bodies.

A planet orbiting a star has angular momentum relative to the star which is the product of its mass, its velocity and the radius of its orbit. In reality, however, orbits are seldom circular and nearly always elliptical, in which case the distance of the planet from the star varies as does the orbital velocity of the planet. In the absence of any external disturbance the angular momentum of the entire system is conserved and has a constant value. Since the mass of a planet remains constant, so must the product velocity (km/sec) × distance from star (km) also remain constant and it is clear that if the distance reduces, ie, towards PERIASTRON, the velocity must increase. If the distance increases, towards APASTRON, the velocity must decrease. From the definition of velocity (distance/time) it can be seen that for fixed intervals of time the product of the remaining dimensions (distance × distance) is an area that must also remain constant and this is in accord with Kepler's second law of planetary motion (*See* KEPLER'S LAWS).

The best known example of rotation on an axis is the behaviour of an ice skater pirouetting on the tip of one skate. She spins slowly when her arms are outstretched, but as she draws her arms in she spins faster. She is re-distributing the mass within her system and her angular velocity changes to compensate. For such axial rotation, where many particles spin as one object around a centre of gravity, we can calculate a quantity known as the moment of inertia (I) which is an effective mass × radius for the whole system. Here all the constituent particles have the same angular velocity (ω), measured in radians per second, and the angular momentum is expressed as $I\omega$.

The kinetic energy of a spinning body is $\frac{1}{2}I\omega^2$ which compares with the expression for the kinetic energy of a mass M in linear motion with velocity V, which is $\frac{1}{2}MV^2$. *RCM*.

Ann Arbor Observatory Observatory of the University of Michigan, founded in 1853. It possesses a meridian circle, refractors of 9.75 and 12in (25 and 31cm) aperture, and reflectors of 15 and 37.8in (38 and 96cm) aperture.

Annular eclipses When the Moon aligns between Earth and Sun (*see* SOLAR ECLIPSE) at or near APOGEE, due to eccentricity of the Earth and Moon orbits, annularity occurs because the apparent diameter of the Sun is greater than that of the Moon.

Anomaly An angular measurement used for determining the position of a body in an elliptical orbit. Can be defined as True, Eccentric or Mean Anomaly.

Anorthosite Type of basaltic rock found in the lunar highland crust. Highland

basalts are richer in aluminium and calcium and poorer in iron, magnesium and titanium than mare basalts.

Ansae Term applied by the early telescopists to define the opposite extremities of Saturn's ring system which, viewed foreshortened from Earth, resembles handles to the planet.

Antapex The point diametrically opposite the solar apex which is defined as the point on the CELESTIAL SPHERE towards which the motion of the Sun is directed.

Antares (Alpha Scorpii) is the brightest star in the constellation of Scorpius. It is a red giant star of the first magnitude lying at a distance of about 400 light-years from the Earth. Its reddish hue in the night sky is rather similar to Mars hence its name, which means "rival of Mars". It has a sixth magnitude, blue companion star, Antares B, lying about 3 arc seconds away which was discovered, in 1971, to be a source of weak, variable radio emission. There may also be a still-closer unseen companion star. *AW*.

Antimatter The idea of antimatter arose from the inadequacies of the quantum theory in describing the behaviour of electrons in terms of waves. The problem was that the most accurate and effective model for the electron supposes that it has a property which can best be described as "spin" that can exist in two possible directions. The quantum theory represents this by a quantum number "s" that has two possible values $+\frac{1}{2}$ and $-\frac{1}{2}$. The idea provides for the possibility that previously imagined energy levels could be resolved into two closely similar levels that could each hold an electron. This is a necessary supposition when combining the Bohr Theory of the atom with the Pauli Exclusion Principle, which states that no two electrons can have the same set of quantum numbers. The picture of the atom becomes clearer and its properties can be explained more simply.

The wave theory originated by Schroedinger in the early 1920s could not readily accommodate this idea of spin because it did not conform to the provisions of the Special Theory of RELATIVITY. However, in 1929 P.A.M. Dirac succeeded in setting up a proper relativistic wave equation for the electron with the most unexpected results. It appeared that the equation had four solutions, two corresponding to the

normal electron with its two directions of spin, and two corresponding to a pair of electron states having negative energy. The notion of negative energy has no physical meaning and was difficult to interpret. The Dirac equations describe the energy of the electron as being in two parts, the energy of motion or momentum and the mass-equivalent energy of the particle arising from the relativistic approach. The theory unambiguously requires the existence of electrons in states of negative energy. An inevitable consequence seems to be that if such a "negative" electron and an ordinary one were to collide they would annihilate one another with the release of all the mass-equivalent energy of both particles.

Dirac's interpretation was that there must be a full range of energy states for electrons but all the negative states must normally be occupied. If the negative states were empty then positive-energy electrons would immediately jump into them with the release of the annihilation energy.

Electrons in negative energy states are fundamentally unobservable, but Dirac pointed out that it would still be possible for such an electron to absorb a sufficiently large quantity of energy from a photon of radiation, for example an energetic gamma ray, to make it jump into a positive state, where it could be detected. The necessary energy would be the mass-equivalent energy of two electrons. As a result the vacancy in the "sea" of negative energy states would itself behave like a particle. The "hole" would be the absence of negative energy and negative charge, which would behave like an ordinary particle, the same mass as an electron but having a positive charge. Astonishingly such "positrons" were discovered in 1934 by C.D.Anderson during his studies of COSMIC RAYS. The particles showed as tracks in photographic emulsions curving under the influence of a magnetic field in the direction expected for positive charges. Moreover the cosmic ray was seen to convert into two particles, an electron and a positron, moving and curving in opposite directions. This process is called "pair production" and represents the direct conversion of energy into matter.

Positrons do not exist for long before they meet ordinary electrons and it is observed that they mutually annihilate on contact with the production of two gamma rays that move away in opposite directions, thus conserving momentum. This process is the direct

A

conversion of matter back into energy. In more recent years it has been found that many nuclear particles have their own anti-particles. There is the anti-proton and the anti-neutron and the thought readily occurs that there may be another part of the universe where whole anti-atoms exist. Maybe at this place the relative abundances of negative and positive particle energy states, as we would recognize them, are the other way around. From our point of view such places would be made of antimatter and the consequence of our meeting it would be an enormous outburst of annihilation radiation. *RCM. See also* ATOMIC STRUCTURE, WAVE MECHANICS.

Antilia (the Air-Pump). A small southern constellation, added to the sky by Lacaille in 1752; its original name was Antlia Pneumatica. It has no star brighter than magnitude 4.4; it adjoins Vela and Pyxis. Delta is a slow binary; magnitudes 5.6 and 9.7, separation 11″. The N-type semi-regular variable U Antilæ has a range of from 5.7 to 6.8, and a period of about a year.

Antinoüs A constellation formed by Tycho BRAHE; now included in Aquila.

Antoniadi Dorsa Scarp on Mercury; 28°N, 30°W.

Antoniadi, Eugenios (1870–1944). French astronomer, born in Turkey. A linguist, humanist and painter, he published in 1907 an atlas on Mosque Sainte Sophie of Constantinople and in 1940 *L'Astronomie Egyptienne*. In 1894 he met Camille FLAMMARION and devoted the rest of his life to telescopic observation of planetary surfaces. He moved, in 1909, from the Observatoire de Juvisy to the Observatoire de Meudon and for years used the large 33in (83cm) reflector for outstanding planetary observations. He settled the controversy about canals on Mars and his books *La Planète Mars* and *La Planète Mercure* (1934) are classics. They were translated into English by Patrick Moore in 1974. Antoniadi published more than 40 papers in the French journal *L'Astronomie* and many other papers. He received the JANSSEN award of Société Astronomique de France (1925), the Lacaille (1937) and the Gusinan (1940) awards of the French Academy of Sciences. *ACD.*

Antoniadi scale The recommended scale of seeing, used to record observations: (1) Perfect seeing, without a quiver; (2) Slight undulations, with moments of calm lasting several seconds; (3) Moderate seeing, with larger air tremors; (4) Poor seeing, with constant troublesome undulations; (5) Very bad seeing, scarcely allowing the making of a rough sketch.

Apastron The point in any orbit around a star that is farthest from that star. Mainly used in connection with BINARY STARS.

Apennines (Montes Apenninus) Much altered by strike-slip faulting, these lunar mountains rise to 2½ miles (4km) above the lava shores of MARE IMBRIUM and dwindle to hills 870 miles (1,400km) to the SE.

Aperture The clear diameter of the light collector (lens or mirror) of a telescope.

Aperture synthesis The history of radio telescopes has been a search for ever-improved sensitivity and higher resolution. Telescopes of even quite modest size have adequate sensitivity for many investigations. But resolution approaching that of Earth-bound optical telescopes — say one second of arc — would require a single radio dish several kilometres in diameter. To construct such a dish, and make it steerable to observe anywhere in the sky, has been impossible in the past.

The RADIO INTERFEROMETER provides one answer to this problem. Where only accurate positions and/or a simple angular size measurement for a radio source are required, the interferometer is adequate. Often, however, a full picture or "map" of the source is needed, perhaps for comparison with pictures produced by an optical telescope. It is to meet such demands that Earth-rotation aperture synthesis was developed. An aperture synthesis telescope is an array of small radio dishes linked electronically and which uses the rotation of the Earth to synthesize the effect of a much larger telescope.

In essence, all parabolic reflecting telescopes work in the same way. Photons from the particular part of the astronomical object being studied are reflected from different parts of the telescope surface and combine in the focal plane. In order to produce a good image, this combination must take place properly. The photons from the same part of the source must be focused to the same part of the focal plane and they must arrive in phase.

The parabolic-shaped reflecting telescope meets both these requirements at the same time. Other shapes of surface do not. But it is still possible to produce the same effect if signals arriving at different parts of the surface can be processed independently.

This is the basic principle of aperture synthesis telescopes. Here, individual radio dishes play the role of different parts of the surface of a hypothetical

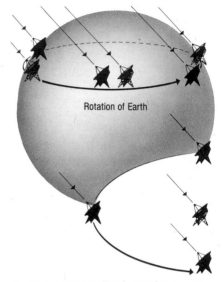

Aperture synthesis: Earth-rotation system

"super-telescope" which is synthesized by correctly combining their outputs.

In Earth-rotation aperture synthesis, relatively few separate dishes need be used since the Earth's daily rotation causes the missing parts of the super-telescope to be filled in. Consider, for example, a synthesis telescope located near the South Pole studying a source at declination −90° (the South Celestial Pole). Say the telescope comprises three dishes close together and arranged along a north-south line. Then, in 24 hours, the second and third dishes will have rotated once around the first to synthesize the super-telescope. In fact only 12 hours of observation is required, since the resolution information obtained in the second 12-hour period is symetrical with and redundant to that obtained in the first 12 hours. (The sensitivity, however, will be greater with the longer observation.)

If we were really observing with a single parabolic telescope as big as our super-telescope, then photons arriving at different parts of its surface would be combined in the focal plane after travelling different path lengths. In our synthesized telescope these path differences are reproduced by inserting electronic phase delays after the signals are received at the separate dishes.

A set of dishes working together in a synthesis telescope is frequently called an array. And the individual dishes are called elements of the array. In a typical synthesis array the separate dishes are often about 82ft (25m) in diameter. They must be small enough so that it is economic to have several of them and yet large enough so that they each have adequate sensitivity.

Cost, however, invariably dictates that our super-telescope is far from being "fully filled". That is, some parts of the synthesized surface are missing — even after Earth rotation. But this need not matter: the essential information needed to produce a good picture of a radio source comes from different parts of the super-telescope that are equally represented as regards separation from each other. Each pair of dishes in the array forms an interferometer. And we need as many different spacings as possible. For our purposes, a large, single-dish telescope has too many parts of the surface that have small separations from each other and not enough that have large separations.

The problem of how best to arrange a small number of dish elements in an array so as to get the optimum number of different spacings is a fascinating one. It is known as the minimum redundancy problem. For example, say we have only two dishes, A and B. We put them one unit of distance apart. Now introduce a third dish in line with the first two. We do *not* put it one unit beyond the second since then we would have two spacings of one unit and one of two units. Instead we put it 2 units from B, to get one spacing of one unit (A to B), one spacing of two units (B to C) and one spacing of three units (A to C). The second arrangement is preferred since there is no redundancy of spacings. A "perfect" solution to this problem exists for four dishes, but not for more. Instead we must accept a minimum redundancy of spacings.

Aperture synthesis techniques were developed simultaneously in Australia and England in the mid-1950s. But only after the completion of the Cambridge one-mile telescope were they widely used and became an important radio astronomical tool. At the present, the most productive existing radio telescopes — such as the VERY LARGE ARRAY — are of the aperture synthesis type. And this seems likely to be the trend for the future with the construction of the AUSTRALIA TELESCOPE and the Long Baseline Array (LBA) in the United States.

Why synthesis telescopes should be so popular is easy to understand: a synthesis array can be almost "all things to all people". It can provide good sensitivity and measure accurate positions as well as playing its more usual role as a mapping instrument. Its main defect is that it usually provides insufficient information about extended structure in the objects it maps, a fault resulting from a lack of very close dish element spacings. This deficiency can be made good to a large extent by complementary observations with the large, single-dish telescopes.

Several problems confront the user of an aperture synthesis radio telescope before his observations are visible as high-quality pictures. For example, it is rare for the same object to have been observed for a full 24 or even 12 hours. Because of this, parts of the super-telescope may not have been synthesized. This will cause defects in the resulting map. These defects take the form of grating-rings or bands of spurious emission across the area of interest: they are produced by strong, close-by sources.

If we know where these unwanted sources are in the map, their effects can, in principle, be removed. To do this several sophisticated computer image processing techniques have been devised. Examples are the Clean Algorithm and the Maximum Entropy Methods. Such techniques can be very valuable indeed, allowing faint detail to be seen in the object of interest, even though none was obvious in the original data. *AEW².*

Apex The point on the celestial sphere towards which a body, such as the Sun or the Earth, is moving. *See also* SOLAR APEX.

Aphelion The point at which a body moving in an elliptical orbit is at its greatest distance from the Sun.

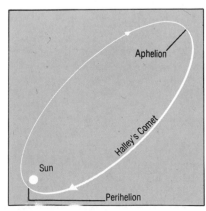

Aphelion: greatest distance from the Sun

Aphrodite Terra One of the major highland regions of Venus. Situated close to the planet's equator, it has three parts, a western mountainous region which rises $3\frac{1}{2}$ miles (5.5km) above datum, a central mountain complex $2\frac{1}{2}$ miles (4km) high and an eastern section (ATLA REGIO) which is in the form of a curved mountainous belt. The central and eastern zones are separated by a broad saddle which contains complex troughs (eg Diana Chasma) and ridges which show up sharply on radar images. The lack of crateriform structures on the imagery has been taken to indicate a relatively young age for the region. *PJC.*

Apochromat The CHROMATIC ABERRATION of a single lens may be greatly reduced by replacing it with a doublet combination of different glasses. Their dispersive powers are arranged to cancel while retaining the desired focusing effect. Here true achromatism can only be obtained for two individual wavelengths for which the focal length of the lens is precisely the same. Remaining colours have slightly different focal positions. To produce exact achromatism for three colours it is usually necessary to use three lenses of different materials. Such "three-colour achromats" are called "apochromats" — a term introduced by their inventor, the German physicist Abbe. *RCM.*

Apogee The point at which a body moving round the Earth in an elliptical orbit is at its greatest distance from the Earth.

Apollo asteroids have Earth-crossing orbits with semi-major axes greater than 1.0AU and perihelion distances less than 1.017AU (the aphelion distance of Earth's orbit). K.Reinmuth discovered Apollo (asteroid 1862), the first of this group, in 1932. To date more than 30 are known. Originally, they were thought to be extinct cometary nuclei but more recent orbital analysis indicates that Jupiter could perturb asteroids from near the 3:1 Kirkwood gap in the asteroid belt into Apollo-type orbits. They provide an adequate source of CHONDRITE meteorites and obviously have a chance of hitting the Earth, the large impacts resulting in cratering.

Apollo programme One of mankind's greatest adventures and triumphs — the programme that, employing some 500,000 people over more than a decade, and at a cost of $20 thousand million, landed men on the Moon and returned them safely to Earth. The first public announcement of the United States' intention to implement this feat was made by President Kennedy in 1961, but on the assumption that such a mission would eventually take place, NASA had already made at least preliminary plans for three different types of unmanned lunar missions — the Ranger, Surveyor, and Lunar Orbiter programmes — to investigate the feasibility of manned missions and to search for suitable landing sites.

Originally, 20 numbered Apollo test flights and missions were planned but funding cuts reduced this to 17. During this phase the disastrous fire in a test model Command Module that killed astronauts Chaffee, Grissom and White prompted a total change of cabin atmosphere composition. Nos. 1–6 were all unmanned flights to test the SATURN ROCKET assemblies. Nos. 7–17 were all manned flights with three astronauts aboard — the Commander (CDR), the Lunar Module Pilot (LMP), and the Command Module Pilot (CMP). The spacecraft consisted of four units in two pairs — the Command and Service Module (CSM), and the descent and ascent stages (DS and AS) of the lunar module. The CM housed the three astronauts during the

Apollo 11: Aldrin on the Moon

journeys to the Moon and back; it remained attached to the SM, which contained rocket engines, fuel, electrical supply, etc., until shortly before re-entering the Earth's atmosphere.

After achieving lunar orbit, the LM, containing two astronauts, undocked from the CSM, decelerated, and landed at the pre-selected location on the

Apollo missions

Date of mission () = landing	Astronauts (CDR, LMP, CMP)	Landing site and coordinates	Revs.	Days on Moon	Hrs. of EVA*	Wt (kg) of samples	Remarks
Apollo 7 11–22 Oct 1968	Schirra Eisele Cunningham	Earth orbit only	–	–	–	–	Only CSM put into Earth orbit; 163 revolutions
Apollo 8 21–27 Dec 1968	Borman Lovell Anders	Orbited Moon, no landing	10	–	–	–	Only CSM used; first manned circumnavigation of Moon
Apollo 9 3–13 Mar 1969	McDivitt Scott Schweickart	Earth orbit only	–	–	–	–	Both LM and CSM put into Earth orbit; undocking and docking. 151 revolutions
Apollo 10 18–26 May 1969	Stafford Young Cernan	Orbited Moon, no landing	31	–	–	–	First full test of complete spacecraft (except LM landing and lift-off). LM, 4 solo orbits
Apollo 11 16 (20)–24 Jly 1969	Armstrong Aldrin Collins	SW. part of Mare Tranquillitatis 0°.80N, 23°.46E	30	0.9	2.5	22	First lunar landing, the actual spot now officially named "Statio Tranquillitatis"
Apollo 12 14 (19)–24 Nov 1969	Conrad Bean Gordon	SE. part of Oceanus Procellarum 3°.04S, 23°.42W	45	1.3	7.7	34	Landed close to Surveyor 3, parts of which were removed and returned to Earth
Apollo 13 11–17 Apr 1970	Lovell Swigert Haise	no landing made	½	0	0	0	Mission aborted due to oxygen tank explosion in SM
Apollo 14 31 Jan (5) –9 Feb 1971	Shepard Mitchell Roosa	N. of Fra Mauro (highlands area) 3°.65S, 17°.48W	34	1.4	9.4	44	First landing in lunar highlands. First use of "push-cart" to carry equipment
Apollo 15 26 (30) Jly –7 Aug 1971	Scott Irwin Worden	Hadley Rille and Apennine Mtns. 26°.08N, 3°.66E	74	2.8	18.5	77	First use of "Moon Buggy", and of semi-automatic metric and panoramic cameras in SM
Apollo 16 16 (21)–27 Apr 1972	Young Duke Mattingly	N. of Descartes, Cayley Plains 8°.97S, 15°.51E	64	3.0	20.3	97	Generally similar to previous mission
Apollo 17 7 (11)–19 Dec 1972	Cernan Schmitt Evans	Taurus-Littrow, E. Mare Serenitatis 20°.17N, 30°.77E	75	3.1	22.1	110	First mission with fully trained geologist (Schmitt)

*EVA = extra-vehicular activity (on lunar surface)

Moon's surface. The third astronaut remained in the CSM. At the completion of extra-vehicular activities (EVAs) on the surface, the LMDS served as a launch platform for the LMAS, which re-docked with the orbiting CSM. Following the transfer of the two astronauts, film cassettes, Moon rocks and so on from the LMAS to the CM, the former was jettisoned, and usually programmed to impact the lunar surface, thereby providing a seismic signal of known energy to calibrate seismometers deployed on the surface. The following table provides some data relevant to the 11 manned Apollo missions.

As confidence was gained, each mission gathered more scientific data than its predecessors. Durations of EVSs and distances covered increased, with correspondingly more comprehensive selections of rock and soil samples. More experiments were deployed or conducted on the lunar surface, and the later orbiting CSMs obtained hundreds of high-grade stereophotographs of the surface. Within the constraints dictated by spacecraft design and landing safety, the landing sites were chosen to give a reasonably representative selection of different surface types, as determined from earlier ground-based or LUNAR ORBITER studies, or from preceding Apollo missions. Thus for

Apollo 11, a near-equatorial site was required for simplest landing and re-docking conditions, with a relatively smooth and level descent path and landing point. For these and other considerations, a point in Mare Tranquillitatis was chosen, not too far from the impact point of Ranger 8 and the landing point of Surveyor 5. Apollo 12 was again targeted for an equatorial mare site, this time on a surface of a browner colour; a point very close to the previously landed (2½ years earlier) Surveyor 3 was chosen, to determine the effects of a space environment on mechanical items over that period.

The ill-fated Apollo 13 was targeted for a near-equatorial site in the lunar highlands — specifically the so-called Fra Mauro Formation, thought to be ejecta from the cratering event that produced the Imbrium Basin. Unfortunately, an oxygen tank in the SM exploded soon after take-off, disabling most of the spacecraft systems. The astronauts returned safely. Apollo 14 was targeted for the same landing location and was completely successful. The Apollo 15 landing site, the first to be situated well away from the equator, combined a mare surface (Palus Putredinis), a nearby sinuous rille (Hadley Rille), and very mountainous highlands (the Apennines). Apollo 16 landed in an exclusively highlands area, on the so-called Cayley Formation — the level material that fills in craters and the intervening terrain here — but very close to the knobbly Descartes Formation, which was surmised to represent possible highlands vulcanism.

The choice of the Taurus-Littrow site for the final mission resulted largely from the Apollo 15 observations of small, dark-haloed craters, identified as possible volcanic vents, in the area. Additionally, the valley floor was one of the darkest mare surfaces on the Moon; a "splash" of bright material from the highlands covered part of the floor, and a cluster of secondary craters lay nearby. These plus the mountainous border of Mare Serenitatis were all within travelling distance.

The quantity of data resulting from the Apollo programme is overwhelming. Rock and soil samples have provided the lion's share of these data, having been subjected to every kind of test and analysis imaginable. These tests have been carried out and the results reported by internationally constituted teams of experts in the fields of geology, geophysics, geochemistry, petrology, petrography, and other relevant areas of expertise.

A major archive of data is the collection of many thousands of photographs taken during the transit and orbital phases of the missions. In particular, the metric and panoramic cameras used on the last three missions produced high-resolution, wide-angle, stereoscopic coverage of about 20 per cent of the lunar surface. Of the 16 or so other experiments conducted from orbit, ten were concerned with "remote sensing", allowing a tie-in between data from the landing sites and extended areas of the surface.

About 25 different types of experiment were carried out or deployed on the surface; several of these were simply using the Moon as a stable space-platform, but others, such as the seismometry, magnetometry, and heat-flow experiments were directed more towards the sub-surface structure and properties.

Although the cream has been skimmed from the returned data and samples, most of the rocks have never been analyzed and the photographs will remain an invaluable archive for the foreseeable future. *EAW*.

Apollonius of Perga (*c.* 260–early 2nd cent. BC). Greek mathematician and astronomer, whose two most important contributions were his *Conics* and his mathematical "model" to describe planetary motions.

Conics dealt with conic sections, the geometry of those curves — the circle, ellipse, parabola and hyperbola — obtained by slicing through a cone. All were important in describing the motions of celestial bodies. The methods Apollonius devised allowed their properites to be determined more readily and precisely than ever before.

Apollonius also introduced the powerful method of using epicycles and eccentrics to account for planetary motions. *CAR*.

Apparent magnitude The apparent brightness of a star (a measure of the light received at Earth) measured by the stellar magnitude system. Because stars are at different distances from us, apparent magnitude is not a reliable key to a star's real luminosity.

Appley Bridge Meteorite A chondrite, which fell in Lancashire, England, in 1914.

Appulse A close approach in apparent position, ie, line of sight, between two astronomical bodies; eg an ASTEROID apparently passing close to a star.

Apsides The points on an elliptical orbit closest to and farthest from the focus of gravitational attraction.

Ap stars Hot stars of spectral type A, ie, with surface temperatures near 10,000K, which have peculiar ("p") chemical compositions. In speech, both letters A and p are pronounced. The chemical peculiarities take the form of considerable enhancements of unusual elements such as manganese, mercury and even some of the rare earth elements. These have been manufactured by nuclear processes deep within the stars and brought to the surface by convection. Although convection is common inside most stars, it operates very effectively almost to the surface of stars of this temperature. *DAA*.

Apus (the Bee; originally Avis Indica, the Bird of Paradise). A far-southern constellation adjoining Octans, which contains the South Celestial Pole; Apus was introduced to the sky in Bayer's maps of 1603. It has two stars above the fourth magnitude, Alpha (3.8) and Gamma (3.9). It contains no objects of particular note.

Aquarius (the Water-bearer). A Zodiacal constellation; there are no well-defined legends attached to it.

Brightest stars

Name	Visual Mag.	Abs. Mag.	Spec.	Distance (light-yrs)
β (Sadalsuud)	2.91	−4.5	G0	978
α (Sadalmelik)	2.96	−4.5	G2	945
δ (Scheat)	3.27	−0.2	A2	98

Also above the fourth magnitude are Zeta (3.6), c² (3.66), Lambda (3.74), Epsilon (3.77) and Gamma (3.84).

Aquarius has no striking pattern; it lies between the Square of Pegasus to the north and Fomalhaut to the south. Zeta is a fine slow binary, with components of magnitudes 4.5 and 4.3 and a separation of 2 seconds of arc. R Aquarii is a symbiotic variable, with a range of from 5.8 to 11.5 and a mean period of 387 days. There are two important planetary nebulae: the "Saturn Nebula", NGC7009, and the so-called Helix Nebula, NGC7293, which is actually the brightest of all planetaries. Aquarius also includes two globular clusters, M2 and M72.

Aquarius contains several spiral galaxies above 13th magnitude. *PM*.

Aquila (the Eagle). A distinctive constellation, led by the first-magnitude Altair; most of it lies in the northern hemisphere of the sky, though it is crossed by the celestial equator.

Brightest stars

Name	Visual Mag.	Abs. Mag.	Spec.	Distance (light-yrs)
α (Altair)	0.77	2.2	A7	16.6
γ (Tarazed)	2.72	−2.3	K3	284
ζ (Dheneb)	2.99	0.2	B9	104
θ	3.23	−0.8	B9	199
δ	3.36	2.1	F0	52
λ	3.44	0.0	B8	98

Also above the fourth magnitude are the Cepheid variable Eta (maximum 3.7) and Beta or Alshain (3.7). Tarazed and Alshain lie to either side of Altair.

Eta has a period of 7.177 days, and a range of from 3.4 to 4.7; it lies between Theta and Delta, which make good comparison stars. The Mira variable R Aquilæ has a range of from 5.7 to 12.0, and a period of 300 days.

The Milky Way is very rich in Aquila, and runs through the constellation; there are also two open clusters visible with binoculars, NGC6709 and NGC6755.

Aquila is easy to identify; mythologically it is said to represent an eagle which was sent to collect a shepherd-boy, Ganymede, to become cup-bearer of the gods. *PM*.

Ara (the Altar). A fairly distinctive constellation, between Scorpius and Triangulum Australe; it is too far south to be seen from Britain.

Brightest stars

Name	Visual Mag.	Abs. Mag.	Spec.	Distance (light-yrs)
β	2.85	−4.4	K3	782
α	2.95	−1.7	B3	189
ζ	3.13	−0.3	K5	137
γ	3.34	−4.4	B1	1076

Next come Delta (3.62), Theta (3.66) and Eta (3.76).

The pattern of Ara is easy to make out. The brightest variable star in the constellation is R Aræ (5.9 to 6.91; period 4.4 days, Algol type; it has a distant third component). There are also some fairly prominent open clusters, NGC6167 (integrated magnitude about 6.4), NGC6193 (mag.5) and IC4651 (mag.8.), and three globulars, NGC6397 (mag. 5), NGC6362 (7) and NGC6352 (8).

Arago Lunar crater in Mare Tranquillitatis, diameter 18 miles (29km). Several prominent domes lie nearby; 6°N, 21°E.

Arago, François Jean Dominique (1786–1853). French astronomer, Director of the Paris Observatory from 1830.

Aratus (?310BC–?240BC). Greek poet, author of the famous poem *Phaenomena* dealing with astronomy and meteorology. It was very popular until the 16th century.

Arc (Galactic Centre). Radio emission arc NE of SAGITTARIUS A, with narrow filaments magnetically aligned perpendicular to the galactic plane. This unique structure may be associated with galactic centre activity.

Arcetri Observatory The observatory of the University of Florence (Italy) built by DONATI in 1872 at Arcetri not far from GALILEO's house, on the hills south of Florence. It replaced the observatory founded in 1807 which was in the town, where the sky was not suitable for spectroscopic work. In 1921 it became the first astrophysical observatory in Italy and, with the addition in 1924 of a solar tower, became a centre of solar research in Europe contributing much to the studies of the solar rotation, of the structure of sunspots and chromosphere, solar flares and solar radioastronomy. *VB.*

Archimedes (?287–212BC). Greek mathematician and physicist. An inventor of various mechanical devices, of which the Archimedean Screw for raising water is still with us, he developed the study of theoretical mechanics.

As a mathematician Archimedes computed the limits between which pi (π) should lie, discussed various geometrical figures, the centre of gravity of flat surfaces, and buoyancy. He also developed ways of calculating the areas and volumes of geometrical figures.

In considering the number of grains of sand which could be contained in a sphere as large as the universe, Archimedes expounded a new method of expressing very large numbers. *CAR.*

Archimedes This lava-floored lunar crater, at 30°N, 4°W, in MARE IMBRIUM, is 50 miles (82km) in diameter and 6,500ft (2km) deep measured below the wall tops.

Archytas Well-formed lunar crater, 21 miles (34km) in diameter, on the north border of the Mare Frigoris; 50°N, 5°E.

Arcturus Alpha Boötis, the brightest star in the constellation of Boötes, the Herdsman; Apparent magnitude −0.04; spectral class K2; distance 36 light-years.

Arecibo Observatory The radio telescope constructed near Arecibo in Puerto Rico is the largest single-bowl type of instrument in operation. Whereas the more familiar type of radio telescope is a paraboloidal bowl which can be steered in azimuth and elevation, the telescope at Arecibo is fixed to the ground, being built into a natural bowl in the karsh (limestone) terrain of the Montanas Guarionex of north-western Puerto Rico. The site is about 8 miles (12km) south of the city of Arecibo and 40 miles (64km) west of San Juan. The diameter of the telescope is 1,000ft (304m) formed as the segment of a sphere of 870ft (265m) radius. The area occupied by the bowl is 20 acres (8 hectares) and it collects a million gallons (4½ million litres) of water for 1in (2.5cm) of rainfall. Any excess which is not absorbed by the ground is pumped into the neighbouring chasm of the Tanama river.

Since the instrument is fixed to the ground it would normally transmit and receive only from the vertical direction. However, an ingenious feed system enables the beam to be swung up to 20°

from the zenith. A triangular steel girder platform with sides of 216ft (65m) is suspended over the centre of the bowl. This is supported by cables from three towers outside the perimeter of the bowl of height 385, 260 and 260ft (117, 79 and 79m). A 700ft (213m) long catwalk, and a cable car, enables this platform to be reached for adjustment of the primary-feed aerial and associated equipment. Underneath the platform a steel girder system holds the primary feed and this system can be tilted and rotated in azimuth to place the feed in the required position.

This motion is accomplished by mounting the feed on a steel arm 304ft (93m) in length. The surface of the arm on which the feed moves is curved to have the same centre of curvature as that of the spherical reflecting surface. The arm is carried on an azimuth ring 132ft (40m) in diameter. Thus, traverse of the feed along the arm and rotation of the azimuth ring enables the beam of the telescope to be directed to any point within the 20° region of the zenith, and computer control enables a celestial object to be tracked within this zenithal region. The weight of this feed support structure is 525 tons (more than 533 tonnes) and since movement in wind would produce an unwanted shift of the beam direction it has to be maintained rigidly in position. This is achieved by anchoring each corner of the steel triangle to the three towers by four 3in (7.6cm) cables and further bracing is produced by pairs of 1½in (3.8cm) guy wires to the ground. In a

Arecibo: world's largest radio "dish"

30 knot (55km/h) wind the line feed does not shift by more than 6in (15.2cm).

Whereas a paraboloid focuses the incoming radiation to a point, the spherical surface focuses the collected energy along a line and at Arecibo the primary feed is a long-phased waveguide. On a frequency of 430MHz on which the telescope was originally used this line feed was a 96ft (29m) tube of square cross section slotted along its length so that the correct phase was maintained. This feed projects downwards from the curved steel arm along which it can be traversed. The latitude of the site is 18°N and so the ability to move this line feed so that the beam can be directed up to 20° from the zenith provides a coverage of the northern half of the ecliptic, which includes all the planets and many galactic and extragalactic objects of interest in the accessible declination range of −2° to +38°.

The Arecibo telescope, dedicated on November 1 1963, was built by Cornell University in association with the United States Department of Defense at a cost of $8,300,000 (1963). Initially the instrument was equipped with a powerful radar transmitter giving 2½ million watts of peak power (150,000 watts cw) on a frequency of 430MHz when the width of the beam to half power was 10 arc minutes. The original intention was to devote half of the time of the telescope to radar observations of the Earth's ionosphere and the planets and, as a receiver, to radio-astronomical measurements. The remainder of the time was to be allocated for defence department requirements, such as military communications and missile detection and tracking. The prime mover in the development of the Arecibo facility was W.E.Gordon, the professor of electrical engineering at Cornell, and he was the first director. On October 1 1969 the National Science Foundation took over the facility from the Department of Defense and the Arecibo Observatory became part of the National Astronomy and Ionospheric Center (NAIC) operated by the Cornell University Center of Radiophysics and Space Research under contract with the National Science Foundation.

The original telescope had an excellent performance on the relatively long wavelengths used for the radar investigations of the ionosphere and for the planetary and other work for which it was intended. Immediate success was achieved in obtaining radar echoes from the planets. For example, on April 7 and 8 1964 radar echoes were obtained from the planet Mercury at a distance of 80 million miles (128 million km). However, with the change in emphasis of the researches after the transfer of responsibility to the National Astronomy and Ionospheric

Center in 1969 plans were made to improve the shape of the reflecting surface so that the telescope could operate on higher frequencies of particular interest to radio astronomical investigations. In the early 1970s the wire mesh surface was replaced by 38,778 adjustable aluminium panels perforated with $\frac{3}{16}$in (4.7mm) holes on $\frac{1}{4}$in (6.3mm) centres, so that the telescope could operate in the centimetre waveband. The surface accuracy of the reflector was improved by a factor of 10 to ±0.13in (3.3mm) by adding 29 cables to the ten already supporting the mesh from the rim. The weight of the surface increased from 101,000lb (220,200kg) to 616,000lb (1,355,200kg). Additional transmitters for operation in the centimetre waveband were installed and the supporting arrangements for the platform carrying the feed system were re-organized in order to achieve the necessary additional stability.

The modified instrument was rededicated on November 16 1974, when a coded message was transmitted towards the globular cluster M13 containing information about terrestrial civilization in the hope that it might be received by intelligent extraterrestrial beings around the year AD 30,000.

In 1985–86 plans for further modifications were announced. These call for a 60ft (18m) high mesh screen around the periphery to reduce the effect of ground radiation and for replacement of the line feeds by an 80ft (25m) diameter Gregorian reflector which will make possible a more efficient arrangement of the receiving equipment on the elevation track.

Both in the radar work on the planets and in passive radio astronomy the Arecibo radio telescope has been a major research instrument. For descriptions of its work see RADAR ASTRONOMY and RADIO ASTRONOMY; of particular importance has been the radar studies of the planets, especially the production of a detailed map of the planet Venus. In radio astronomy the telescope has been used with great success in the study of PULSARS and as an element in international VERY LONG BASELINE interferometric measurements of distant radio sources in the Universe. *BL.*

Arecibo Vallis Valley on Mercury: 27°S, 29°W.

Arend-Roland Comet Found at Uccle, Belgium; became a bright naked eye object by early 1957, having a 25° tail

and unusual anti-tail pointing sunwards.

Arend-Roland Comet: an unusual anti-tail

Arethusa (Asteroid 95). Diameter 143 miles (230km). It is exceptionally dark, with an albedo of only 0.019.

Argelander, Friedrich Wilhelm August (1799-1875). German astronomer, born in Memel, East Prussia. At the age of eighteen, he entered the University of Königsberg, where he came under the influence of the famed Friedrich BESSEL.

In 1823 Argelander was invited to become Director of the Finnish observatory at Åbo, where he remained until his appointment as Director of the Bonn Observatory in 1836.

Argelander's greatest title to fame undoubtedly is the Bonner Dürchmusterung (BD). With the aid of a modest 3in (7.5cm) refractor of short focal length, he and his assistants charted the positions of 324,198 stars to the ninth magnitude with greater precision than ever before.

As a result of an attack of typhoid fever Argelander died in Bonn on February 17 1875. *PGEC.*

Argo Navis (the Ship Argo). Old southern constellation, now broken up into Carina, Vela and Puppis.

Argument of the perihelion The angle measured along the orbital plane from the ascending node to perihelion (sometimes called "longitude of perihelion").

Argyre I A classical circular feature on Mars, 51°S, 43°W, formerly regarded as a plateau because it is often bright white when on the limb of Mars in the chill of night. Mariner spacecraft

A

identified it as a basin with a featureless floor covered by dust deposits: the floor is a light ochre colour when it is not covered by ice-fogs in autumn or frosts in winter. The basin was probably formed by a meteoroid impact that created its 9,800ft (3km) high mountainous rims. It is encircled by Nereidum Montes, on the north, and Charitum Montes, on the south.

Ariadæus Rille (Rima Ariadæus). Perhaps the best known lunar rille, this *en echelon* system of linear graben is more than 150 miles (250km) long and 3 miles (5km) wide — but less than 3,300ft (1km) deep — in its middle stretches; 7°N, 13°E.

A

Ariane launcher The Ariane family of launch vehicles has been developed by the EUROPEAN SPACE AGENCY (ESA) to

Ariane launcher: ESA's challenge to NASA

provide an economically competitive European capability for placing satellites in orbit.

Development started in 1973 and was funded by ten Western European countries with France contributing 64 per cent and West Germany 20 per cent.

The system became operational at the end of 1981 after four development flights. After a further four promotional flights conducted under the auspices of ESA, the responsibility for further launches was transferred to Arianespace.

Arianespace — the first commercially operational space transport company — thus assumed responsibility for the production, marketing and launching of operational Ariane vehicles.

Ariane was specifically designed to satisfy the requirement to launch geosynchronous orbit satellites but is

capable of undertaking other types of missions ranging from low Earth orbit to deep space injection.

A number of versions have been developed designated Ariane 1 to 4 providing payload capabilities into Geostationary Transfer Orbit (GTO) — an elliptical Earth orbit, perigee 124 miles (200km), apogee 22,245 miles (35,800km) — of 3,968 to 9,480lb (1,800 to 4,300kg).

The Ariane launcher is a three-stage vehicle. The first and second stages use the hypergolic liquid propellants ultra dymethyle hydrazine (UDMH) and nitrogen tetroxide (N_2O_4). The third stage is cryogenic utilizing liquid oxygen and liquid hydrogen.

Ariane 3 utilizes a pair of strap-on boosters to augment the thrust of the first stage and Ariane 4 allows a number of strap-on options to provide a high degree of operational adaptability. All versions are capable of launching dual payloads.

Ariane launches all take place from the Guiana Space Centre (CSG) situated near Kourou in French Guiana, South America. This site is close to the equator (5°N) which makes it most favourable for launches into GTO on account of the rotation speed of the Earth at that point.

Two launch complexes are available at CSG allowing a high throughput of vehicles. The second complex has separate vehicle preparation and launch pad zones. This arrangement allows work to commence on stacking a second launch vehicle when the first is undergoing final check-out and payload installation on the pad.

Launches are planned at the rate of about eight per year and the order book extends years ahead. However, the programme has not been without its problems and the rocket was grounded in 1986 after the fourth failure in 18 launches. Three of the four failures were concerned with the third stage where ignition proved to be a problem.

In 1985 CNES (Centre National d'Etudes Spatiales), the overall prime contractor for the Ariane development, was given the go-ahead for initial work on Ariane 5. Ariane 5 is a heavyweight successor to Ariane 4 and could give Europe in the 1990s an independent capability for manned missions. *RMJ. See also* ARTIFICIAL SATELLITES.

Arids A weak meteor stream, formerly called the Corona Australids, active in March. The radiant in in the constellation ARA.

Ariel A satellite of Uranus discovered in 1851 by Lassell. It has a radius of 360 miles (580 ± 5km), a density of 1.26 ± 0.39g/cm³ and an albedo of 0.40 ± 0.02. The surface consists of numerous craters, indicating an old surface bombarded by meteoroids; linear grooves suggestive of tectonic activity; smooth patches indicative of deposition of material. The satellite was observed during the Voyager 2 fly-by in January 1986 and was seen to have the brightest and geologically youngest surface in the Uranian system.

Ariel A series of six United Kingdom scientific satellites launched 1962–79 by the United States under cooperative agreement and used mainly for the study of the Earth's ionosphere, UV, X-rays, radio astronomy and cosmic rays.

Aries (the Ram). The first constellation of the Zodiac, though since precession has now shifted the vernal equinox into Pisces, Aries has technically lost this distinction. In mythology it represents the ram with the golden fleece.

Brightest stars

Name	Visual Mag.	Abs. Mag.	Spec.	Distance (light-yrs)
α (Hamal)	2.00	−0.1	K2	85
β (Sheratan)	2.64	2.1	A5	46

Next come c (3.63) and Gamma or Mesartim (3.9). Mesartim is a fine, easy double; magnitudes 4.6 and 4.7, separation 8″.2. The brightest variable star is U; 6.4 to 15.2, period 371 days, Mira type. There are no nebular objects above magnitude 10.

Aries lies south of the line of stars marking Andromeda. The little group of Alpha, Beta and Gamma is easy to recognize; Alpha (Hamal) is orange, and detectable with the naked eye. *PM.*

Aristarchus This 25-mile (40km) diameter lunar crater, at 23.6°N, 47.4°W, is known for its brightness and its bright rays: the crater reflects 20 per cent of the sunlight whereas the Moon as a whole is about half as reflective. The crater walls are terraced through a drop of 10,000ft (3km) to the floor, which, itself, occupies only half the diameter of the crater and supports a small, central peak. The floor consists of lava flows, which are considerably younger than the crater rim. Sightings (1963) of reddish glows on the crests of ridges in the region provide evidence for recent volcanic activity; although most consider the crater to be of impact origin. *GF.*

Aristarchus of Samos (?310–230BC). Greek astronomer and mathematician. He attempted to measure the distances of the Sun and Moon, and placed the Sun at the centre of the universe, thus being the first astronomer to propound a heliocentric theory.

The attempt Aristarchus made to measure the Sun's distance and that of the Moon was based on the simple expedient of observing the Moon at first quarter and, at the same time, measuring the apparent angle between the Moon and the Sun. Geometrically, this gave a right-angled triangle between observer, Moon and Sun, from which calculation provided both distances.

Though his method was excellent theoretically, the angle between the Sun and Moon was difficult to measure in practice, with the added disadvantage that any slight error in its value would introduce gross errors in the final result. In fact, Aristarchus measured the angle as 87°, whereas it

Aristarchus: measuring the Sun's distance

should have been 89° 50′. Thus, not only were his distances for the Sun and Moon severely in error — that of the Moon was four times too small, and of the Sun 21 times — but so too were his subsequent calculations of their sizes.

The idea of a moving Earth seems to have originated with the followers of Pythagoras, though they had it orbiting about a central fire, not the Sun. But it was Aristarchus who put the Sun at the centre. However, his hypothesis

appeared to present no advantages in calculating future planetary paths — the mainstay of Greek astronomy — while a moving Earth seemed to be contrary to reason and went against the laws of physics as then understood. In consequence, his hypothesis was not widely taken up and, indeed, created little interest in his own day. *CAR.*

Aristillus (about 300BC). Greek astronomer who worked at the Library and Museum at Alexandria. He measured and catalogued the positions of the principal stars.

Aristotle (384–322BC). Greek philosopher and polymath. Not only a philosopher and writer on ethics, he was also noted as a biologist and writer on physics.

Aristotle developed a grand self-consistent scheme of the universe which incorporated his teaching on the four elements — earth, air, fire and water — and the four qualities — hot, cold, dry and wet — as well as his laws of motion. These led him to reject the idea of a vacuum and, therefore, any atomic theory, because this demanded the presence of particles existing in empty space. Aristotle also considered and rejected any idea of a moving Earth; at the time his arguments appeared incontrovertible. In consequence, the Aristotelian universe followed the Earth-centred or geocentric theory accepted by almost all his predecessors.

Aristotle taught that the universe was spherical in shape, with the Sun, Moon and planets carried round on spheres nesting inside one another. All these celestial bodies were eternal and unchanging, and he thought of them being composed of a fifth element or essence. Change was confined to the terrestrial world, the "sublunary" region lying inside the Moon's sphere. As a result meteors, comets and other transitory events were thought to occur in the upper air; they were classified as meteorological phenomena.

In this universe everything had its natural place. "Earthy" materials fell downwards because they sought their natural place at the centre of the spherical Earth, which was itself at the centre of the universe. Water lay in a sphere covering the Earth, hence its tendency to "find its own level". Outside the sphere of water lay that of the air, and beyond this was the sphere of fire; flames always burned upwards because they were seeking their natural place above the air. *CAR. See also* GEOCENTRIC THEORY.

Arizona Crater The first crater on the Earth's surface to be recognized as being caused by meteoritic impact. It is situated on a flat plateau lying between Flagstaff and Winslow, Arizona, USA. The crater is a basin-shaped depression some 3,940ft (1,200m) in diameter. The height of the walls surrounding the crater rise 120 to 160ft (37 to 50m) above the surrounding plain. The outer

Arizona Crater: an impact crater of 3,900ft

slopes are quite gentle but the inner slopes are steep, being as much as 80° in the southern sector. A whole 2,000ft (600m) section of the pre-existing sedimentary rocks has been lifted 100ft (30m) to form the south wall of the crater.

The feature first created interest in 1891 when large quantities of meteoritic iron were discovered in the surrounding plain. In 1905 boreholes and shafts were sunk in the centre of the crater in an attempt to find the main mass of the meteorite. After passing through crushed sandstone and rock flour, undisturbed layers were found at a depth of 607ft (185m). In 1920 attention was concentrated on the southern rim without success. It is now known that at times of such impacts, the meteorite is either vaporized or shattered to extents that depend on the characteristics of the particular event. It is therefore concluded that no large mass exists.

The meteorites consist mainly of iron with just over 7 per cent nickel and 0.5 per cent cobalt. In addition to the irons, oxidized iron shale balls were found intermingled with the local rock debris. Silica glass and very finely divided white sand known as rock flour, together with forms of quartz known as coesite and stishovite, all point to the structure having been formed by meteoritic impact.

Studies of the distribution of the meteoritic material around the crater have led to the conclusion that the meteoroid responsible for the crater was travelling from NNW to SSE. This

A

is consistent with the evidence gained from studies of the tilt of the rock layers forming the rim.

Many attempts have been made to ascertain the age of the crater. Early attempts suggested 2,000 to 3,000 years; current estimates give an age of about 50,000 years. *HGM*.

Armagh Observatory Founded in 1790 by Archbishop Robinson who, from private funds, erected a fine Georgian building with a dome. The original telescope, an equatorial 2½in (6.3cm) refractor by Troughton, was restored in 1986.

The Armagh Catalogue of Stars, giving accurate positions of 5,345 stars, observed by Dr. Romney Robinson, Director from 1823 to 1882, and the New General Catalogue of Nebulae and Clusters of Stars (NGC) by Dr. J.L.E.Dreyer, Director from 1882 to 1916, are among the outstanding achievements during the 19th century.

In the 20th century the research interest at the observatory moved to the field of variable stars. Cepheids and flare stars are among those most intensively studied today, often in collaboration with astronomers abroad. In order to have access to the rich southern skies, Dr. E.M.Lindsay, Director from 1937 to 1974, collaborated with Harvard College Observatory in the United States and DUNSINK OBSERVATORY, Ireland, in the erection and operation from 1950 to 1976 of a 36-32 inch (91-82cm) Baker-Schmidt telescope, the so-called ADH, at the Boyden Observatory in South Africa.

Perhaps Armagh's most famous astronomer was Professor Ernst J.Öpik (1893–1985), an Estonian who came to Armagh in 1948 and did pioneering work in many fields of astronomy, especially meteoritics, planetary science, comets, stellar convection and the structure of red-giant stars.

While still the proud owners of a 10in (25cm) refractor, built in 1885, today's astronomers at Armagh use telescopes abroad and various astronomical satellites for their observations.

The observatory is operated on a budget provided by the Department of Education in Northern Ireland. In order to popularize astronomy, Dr. Lindsay conceived the Armagh Planetarium, which was opened in 1968 with Patrick Moore as its first Director. *MdG*.

Armanty meteorite A 20 tonne Chinese iron meteorite.

Armillary sphere The oldest known astronomical device, originally intended for observation, and later used for teaching.

In essence, it is a skeleton CELESTIAL SPHERE, originally geocentric, with a series of rings showing great circles in the heavens — equator, ecliptic, and usually tropics, Arctic and Antarctic circles, and colures — revolving on an axis inside the graduated horizon and meridian rings. It can be "rectified" to conform to any desired observer's latitude and the sphere can be rotated to represent the time of day.

ERATOSTHENES, HIPPARCHUS, PTOLEMY and Tycho BRAHE all used armillary spheres for observation. They were also used in China. *HDH*.

Armillary sphere: oldest astronomical device

Armstrong, Neil Alden (1930–). American astronaut, born in Ohio, and the first man to walk on the Moon. He graduated from Purdue engineering school, served in the Navy as a fighter pilot (78 combat missions in Korea), and then as a civilian with NASA as X-15 pilot (seven flights). In 1962 he was selected as an astronaut, commanded the first docking mission (Gemini 8), was backup commander of Apollo 8, and commanded the Apollo 11 lunar landing mission. He left NASA to teach engineering at the University of Cincinnati until 1980, when he became Chairman of the Board of Cardwell International. Armstrong remains an extremely private person, but serves on special space commissions (such as the "Challenger" investigation panel in 1986). *JEO. See also* APOLLO PROGRAMME.

Neil Armstrong: first man on the Moon

Arrhenius, Svante (1859–1927). Swedish chemist and Nobel Prize winner. In astronomy he proposed the PANSPERMIA THEORY, according to which life was brought to Earth by way of a meteorite.

Svante Arrhenius: life from space

Arsia Mons Shield volcano, 500 miles (800km) in diameter, situated in the Tharsis region of Mars; 9°S, 121°W. It rises to 15 miles (25km), has a large summit caldera and a plethora of radiating volcanic flows.

Artificial satellites The space age started on October 4 1957, when the Soviet Union launched the first artificial satellite into orbit round the Earth. By the end of 1985 there had been 2,766 successful launches and the current annual rate of launches is slightly more than 120. Many of these launches have placed more than one object in orbit, the instrumented payload itself, the rocket that put it into orbit, and sundry items such as protective nose cones.

All satellites move in orbits governed by Newton's Laws of Motion, an unperturbed orbit being an ellipse with the Earth's centre at one of the foci. However, the orbit will be modified continuously by external causes. Some of these can be predicted quite accurately, such as those caused by the fact that the Earth is not a perfectly homogeneous sphere but flattened at the poles. The main modifications are the changes in the direction of the orbital plane (like the wobble of a spinning top) and changes in the direction of the major axis of the ellipse within the plane.

The rates of change depend on the height of the satellite and the inclination of the orbital plane to the equator. Because of changes in the density of the atmosphere at a particular height above the Earth's surface, due principally to solar activity, the satellite will experience a variable drag, resulting in complex changes to the shape of the orbit. The overall effect is to change the eccentricity, with the orbit becoming more and more circular. Once a roughly circular orbit has been achieved, the satellite will spiral inwards and in doing so will experience increasing drag until finally it will plunge into the denser regions of the atmosphere, producing a spectacular fireball as it burns up.

Unfortunately, changes in atmospheric density cannot be predicted accurately and so predicted values for the lifetime of a satellite can be in error by 10 per cent or more. There are many other factors affecting an orbit such as atmospheric tides and winds, and whether the perigee of the orbit occurs in the northern or southern hemisphere and the local times at which this occurs. Solar radiation pressure has a major effect on satellites with a large area/mass ratio, such as balloon satellites. The drag produced by the atmosphere is such that no satellite can have a stable orbit with a height lower than about 100 miles (160km). At this height the period is about 88 minutes.

The precise purpose for the launching of a particular satellite is sometimes difficult to ascertain. Launches can be divided roughly into two main groups, military and civil, although there is considerable overlap. About 60 per cent of the launches are for military purposes, such as photo-reconnaissance, communications, electronic listening and navigation. Separating American military satellites from civil

ones is fairly straightforward, but Soviet sources rarely divulge the purposes of, for example, their Cosmos series. The title Cosmos covers all experimental, scientific and military launches. By the end of 1985 there had been 1,714 under that heading.

Common names are used extensively by launching organizations and those responsible for the onboard instruments. Such names as Skylab, Landsat, Nimbus and Salyut are well known, but confusion has arisen through two organizations using the same or similar names. For example, GEOS was used by the Americans for geodetic studies, but the EUROPEAN SPACE AGENCY (ESA) also used that name for the investigation of charged particles in the upper atmosphere.

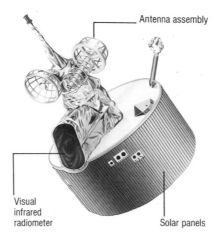
Antenna assembly
Visual infrared radiometer
Solar panels

Artificial satellite: US weather satellite

The use of acronyms has been used widely, adding to the confusion. Fortunately COSPAR (International Committee for Space Research) has provided a standard system for identifying all satellites and related fragments, based on the year of launch and the chronological order of successful launches in that year. For example, MIR, the Soviet space station, launched on February 19 1986, is known as 1986-17A, it being the 17th launch in that year. The A refers to the satellite. The rocket that put it into orbit was designated 1986-17B. If several satellites are launched by a single rocket, they are usually given the letters A, B, C, etc. Where, after an explosion and disintegration, more than 24 fragments exist, the letter Z is followed by AA, AB, . . . BA, BB, . . . etc. Before 1963 a system involving Greek letters was used. For example, Sputnik 3 was designated 1958 δ 2 whilst the associated rocket was known as 1958 δ 1. A modification of the COSPAR system is

often used in computer printouts and involves an all-figure number. For example, the MIR space station is known as 8601701, the first two digits giving the year, the next three the launch of that year and the last two identifying the component.

Tracking satellites is important scientifically and provides many useful data. The most accurate method of tracking involves laser techniques, but currently this is being carried out on a very limited scale. At the moment photographic and visual techniques are providing excellent data for geodetic and upper atmosphere studies. Apart from limitations due to meteorological conditions, satellites can be seen only when the observer's sky is dark but the satellite is still illuminated by the Sun. Such conditions exist after dusk and just before dawn. The lengths of these visibility periods depends on the time of the year, the latitude of the observer and the height of the satellite. The actual brightness of the object during these critical periods depends on the nature and curvature of the reflecting surface, the phase angle (the angle between the Sun and the observer as seen from the satellite) and the distance and altitude of the satellite from the observer. The brightest objects rival those of the planets, but others can only be seen through medium-sized telescopes.

The rate at which satellites cross the sky depends on their height above the surface of the Earth. Low satellites move quite quickly, crossing the sky in about two minutes, but those at heights of about 1,240 miles (2,000km) may take as much as half an hour. Even higher satellites, those at heights of about 22,370 miles (36,000km), take 24 hours to complete one revolution. As this is the rotational period of the Earth, if the orbital inclination is 0°, the net effect is that the satellite will appear motionless in the sky. Many such satellites have been launched and are used principally for meteorological and communication purposes. These orbits are known as geostationary orbits.

Roughly two-thirds of all the satellites launched have been from the Soviet Union and most of these have been from the launch complex at Plesetsk, just south of Archangel. About 300 have been launched from Tyuratam (Baikonur), east of the Aral Sea. A relatively small number have been launched from Kapustin Yar, near Volgograd. The Americans have also launched from three sites: Cape Canaveral in Florida, Vandenberg Air

Force Base in California and Wallops Island in Virginia. Of increasing importance is the French launch site of Kourou in French Guiana, from where the more recent ESA satellites have been launched. Japan uses two sites, Uchinoura and Tanegashima, both in southern Kyushu, and China also uses two, at Jiuquan and Xichang, in north and south-west China respectively. *HGM. See also* GEOSTATIONARY SATELLITES, IRAS, IUE NAVIGATIONAL SATELLITES, and individual satellites and series.

Arzachel Lunar walled plain, 60 miles (97km) in diameter; a member of the Ptolemaeus chain; 18°S, 2°W.

Arzachel Latinized name of the Arab astronomer al-Zarqali (*d.* 1100). He is mainly remembered for his preparation of the famous Toledo Tables of planetary motions.

Ascending node The point at which an orbit crosses from south to north of a reference plane; eg, the point at which the Moon's orbit crosses the CELESTIAL EQUATOR from south to north.

Ascraeus Mons Tharsis shield volcano on Mars, with complex caldera; 12°N, 104°W.

Aselli The "Asses": the stars Delta and Gamma Cancri, to either side of the open cluster Praesepe.

Ashen Light From the French *lumière cendrée*, literally the ash-coloured light. An expression loosely given to earthshine on the Moon, but more precisely the dim coppery glow infrequently seen on the dark side of Venus when the planet is visible as a thin crescent near inferior conjunction. Sometimes the entire night hemisphere is affected, at others the glow is patchy and localized, but has no preferred position.

The fact most observations are reported when Venus is a slender crescent may be attributed to the reduction of glare from the planet's sunlit part.

However, the Ashen Light is rare, fugitive and suspect. It is only seen, and then with extreme difficulty, if the planet is observed on a dark sky, and has its bright part hidden by an occulting device located inside the eyepiece of the telescope. Even this precaution is insufficient to dispel thoughts of illusion, and in the absence of photographic confirmation there is a natural tendency towards scepticism. Still, too many experienced observers have claimed sightings completely to discount the phenomenon.

Volcanic activity, phosphorescence of the surface, self-luminosity, accidental combustion and other illumining processes have all been proposed by way of explanation. The only conceivable physical mechanism that would be dependent on the phase or position of Venus as viewed from Earth would in fact be earthlight, but calculation clearly demonstrates that theoretical earthlight falls well below the threshold of visibility. Possibly the cause is to be found in an electrical phenomenon in the upper atmosphere of Venus. *RMB.*

A stars The stars of spectral type A are characterized by the great strength of the hydrogen absorption lines in their spectra. On the MAIN SEQUENCE they range in surface temperature from 7500K at type A9 up to 9900K at type A0. Their corresponding masses and radii, in terms of those of the Sun, are 1.8 and 1.4 respectively at A9, increasing to 3.2 and 2.5 at A0.

Whereas in the other principal spectral types spectral peculiarities are generally uncommon, among the A stars abnormalities are the rule rather than the exception. This is a result of the almost complete absence of convection near the surfaces of A stars: minor chemical composition anomalies that are mixed into the interiors of other stars are able to remain on the surfaces of A stars.

In stars of other spectral types EMISSION LINES from their chromospheres can be observed in their far ultraviolet spectra, but in the A stars such chromospheres, which derive their high temperatures from energy in the convective zones, are absent and emission line A stars are consequently extremely rare.

Roughly 10 per cent of the A stars have strong magnetic fields (from a few hundred to more than 20,000 gauss). In these, certain chemical elements become concentrated in their atmospheres and produce unusually strong spectrum lines. Such stars were classified as PECULIAR A (Ap) stars long before it was realized that they are magnetic. There are two principal kinds of spectral peculiarity in the Ap stars: the strontium-chromium-europium stars and the silicon stars. As the overabundances of these elements are concentrated in patches on the surface of the stars (often at the magnetic poles) the strengths of the anomalous spectrum lines in the Ap stars change as the stars rotate. Therefore most Ap stars are SPECTRUM VARIABLES.

Another kind of spectrum anomaly produces the metallic line (Am) stars. In these, the hydrogen line strengths and temperatures are those of normal A stars but the spectrum lines of the metals are usually strong — more characteristic of those seen in the F stars. The brightest star visible at night, Sirius, is an Am star.

The Ap and Am stars lie on or near the main sequence, but the metallic line phenomenon is also seen among subgiant and giant A stars. These are called, after the prototype, Delta Delphini stars.

There are other, rarer kinds of anomalous spectra among the A stars, including the Lambda Boötis stars, which have very weak metal lines without belonging to Population II.

The main peculiarities in the Ap and Am stars are currently thought to be a result of diffusion in their atmospheres. In an atmosphere that is not mixed by convection, separation of the chemical elements can occur because of imbalance between gravitational force and pressure among the minor constituents of the atmosphere. An element that absorbs light efficiently experiences more radiation pressure than one that does not. Thus, some elements are concentrated in the upper atmosphere of a star and others sink out of sight.

Even without convection, currents set up by rapid rotation can mix the outer regions of a star. Normal A stars rotate quite rapidly, preventing diffusive separation of the elements, but the Ap and Am stars are all found to be relatively slow rotators, in accord with the diffusion explanation of their peculiarities. The slow rotation of Ap stars probably results from magnetic braking (*see* ROTATING STARS), but the Am stars (which do not have magnetic fields) rotate slowly because they generally are the A stars in binary systems with orbital periods less than 100 days.

Four kinds of pulsational variability appear among the A stars: (i) Delta Scuti stars, which are Population I stars near the main sequence; (ii) Dwarf Cepheids, which resemble the Delta Scutis but with larger amplitude; (iii) the RR Lyrae stars of Population II, and (iv) the cooler Ap stars, which oscillate with periods of five to 15 minutes. No variability is known among the Am stars.

Bright examples (A stars appear white to the naked eye): Sirius A1m V; Vega A0 V; Altair A7 V; Deneb A2Ia. *BW.*

Asteroids Asteroids (also known as minor planets) are rocky bodies that, for the most part, orbit the Sun between Mars and Jupiter. Guiseppe PIAZZI'S discovery in 1801 of the largest asteroid, which he named CERES, confirmed the prediction of BODE'S LAW for a planet near 2.8AU and led the young Carl Friedrich Gauss to formulate a method of orbit calculation still in use today. Until the application of photography by Max Wolf in 1891, asteroids were discovered visually by laboriously checking telescope fields against star charts. Nowadays, SCHMIDT TELESCOPES are capable of recording many hundreds of asteroid images, which show up as trails due to their motion against the stellar background, on a single hour-long exposure. CHARGED-COUPLED DEVICE cameras are also being used to discover and recover asteroids. After an asteroid has been observed at several apparitions and an accurate orbit determined, it is assigned a permanent number and the discoverer has the right to name it. Thus 4 Vesta (after the Greek goddess) and 2138 Swissair (after the discoverer's favourite airline). About 3,500 asteroids are now numbered, though many thousands more have been glimpsed and lost.

The known population ranges in diameter from 579 miles (933km) (1 Ceres) to about 700ft (200m) (6344P-L) and in brightness from 6th to 22nd visual magnitude at mean opposition. There are thought to be a million asteroids larger than 3,300ft (1km) in diameter, the number increasing by a factor of 100 for each tenfold decrease in diameter, perhaps down to grapefruit-size METEOROIDS. The total mass of the asteroids is about 4×10^{24}g or 5 per cent that of the Moon. In the sky, all but the largest asteroids appear as starlike points of light, although fragments of them exist in terrestrial museums in the form of METEORITES.

Along with COMETS and meteoroids, asteroids constitute a remnant population of planetesimals left over from the time of formation of the Solar System. Most likely, the powerful gravitational influence of proto-Jupiter inhibited the formation of a major planet from asteroidal material, causing some of the material to be perturbed into planet-crossing orbits (which resulted in intense planetary and satellite cratering early in their history) and some to be ejected altogether from the Solar System. Shortly after formation, the largest asteroids evidently underwent heating, mainly due to radioactive

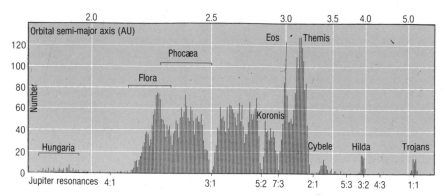

Asteroids: structure of the main belt produced by Jupiter's gravity

decay of the isotope aluminium-26, to the point where they melted and differentiated, acquiring metallic cores and layered mantles.

Certainly, Jupiter is largely responsible for the distribution of the bulk of the asteroids into a so-called main belt between 2.15 and 3.3AU, and for some of the belt's detailed structure (see illustration). Several prominent depletions, known as KIRKWOOD GAPS, are due to resonances with Jupiter's orbital revolution. At the 2:1 commensurability, for example, an asteroid would pass Jupiter in the same part of its orbit every second revolution and would thereby be subject to a cumulative gravitational perturbation. Other Jupiter resonances (e.g., 3:2, 4:3) have evidently resulted in enhancements by retaining asteroids in stable orbits. The most notable of these is the 1:1 grouping, where two clouds of asteroids, known as TROJANS, librate around the L_4 and L_5 LAGRANGIAN points 60° fore and aft of Jupiter in its orbit.

Many asteroids share orbits, not only of similar semi-major axis, but also inclination and eccentricity (at least, when the effects of planetary perturbations are averaged out by the calculation of so-called proper elements). These groupings are known as Hirayama families. Some, such as the Hungarias and Trojans, are dynamical groupings, but others (Eos, Koronis, Themis families, each containing more than 100 known members) comprise fragments that resulted from ancient catastrophic collisions among asteroids. It is possible that the solar system dust bands (see ZODIACAL BAND, ZODIACAL DUST), discovered by the INFRARED ASTRONOMICAL SATELLITE, originate from some Hirayama families. Altogether, about half of all asteroids are thought to belong to Hirayama families.

Viewed in three dimensions, the main belt is a somewhat wedge-shaped torus, increasing in thickness from its inner to outer edge. The median orbital inclination increases from about 5 to 9 degrees; whereas the median eccentricity is constant at about 0.15. Most asteroids therefore pursue orbits that are only somewhat more inclined and less circular than those of the planets. It follows that the majority of main-belt asteroids move among the stars just the way the planets do: retrograding through opposition and not straying too far from the ecliptic.

There are vivid exceptions, however. About one per cent of the known asteroids have orbits that cross those of one or more planets. A type known as AMOR ASTEROIDS cross the orbit of Mars; APOLLOS and ATENS cross that of the Earth; and a few Jupiter-, Venus-, and Mercury-crossers are known. One exceptionally distant asteroid, 2060 CHIRON, moves in a dynamically chaotic orbit between Saturn and Uranus, crossing Saturn's orbit near perihelion. The Earth-approachers can move very rapidly across the sky, sometimes at rates of several degrees per day. Because their orbits are subject to continual planetary perturbation, the fate of most planet-crossers is to collide with a planet on a time scale of about 10^7 years. It is estimated that there are about 1,000 Earth-crossers larger than 3,300ft (1km) in diameter, most of which are candidates for eventual collision with Earth. The impact of a 6-mile (10km) asteroid may have caused the widespread destruction of animal life (the Cretaceous-Tertiary extinction) about 65 million years ago. Removal of planet-crossers by planetary impact implies continuous replenishment, most likely by gravitational perturbation from regions near the Kirkwood gaps, although some planet-crossers, such as 3200 PHÆTHON, are possibly extinct cometary cores.

Since the 1970s, a number of remote-sensing techniques, such as optical polarimetry, infrared radiometry, spectrophotometry, broadband photometry, radar, and speckle interferometry, have been applied to asteroids to probe their physical characteristics. The diameters, surface albedos, rotation rates, and compositional types are now known for several hundred asteroids. The most accurate size and shape measurements of asteroids have been made by observing the duration of OCCULTATIONS of stars. Such observations have also raised the possibility that some asteroids possess satellites, though the evidence is far from conclusive. Asteroids have been divided into a number of classes according to surface composition. The most abundant type, called C (for "carbonaceous") asteroids, are of low albedo (typically 5 per cent — darker than coal) and predominate in the outer belt. They are thought to contain the same material as CARBONACEOUS CHONDRITE meteorites. S asteroids ("silicaceous") dominate the inner belt and are mixed with C-types in the central belt. Of intermediate albedo (15 to 25 per cent), they are probably analogous to the metal-bearing stony meteorites known as ordinary CHONDRITES. The third most populous class, M-types ("metallic"), also have moderate albedos, and may be the metal-rich cores of large differentiated parent asteroids that have been exposed by collisional breakup. A much rarer class, E-types ("enstatite"), have albedos of 40 per cent or more. The distributions of the various asteroid types in the belt, as well as the homogeneous compositions of many Hirayama families, suggest that most asteroids still orbit close to where they were formed in the solar nebula.

Asteroids change in brightness as they rotate, generally exhibiting two maxima and two minima per rotation. In most, the variation can be ascribed to non-spherical shape rather than albedo spots. The average rotational lightcurve has a peak-to-peak amplitude of 0.2 magnitude. Only the largest asteroids are near-spherical due to self-gravitation. Exceptionally, as with the Amor asteroid 433 Eros and the Trojan 624 Hektor, light variations exceed 1 magnitude, indicating elongated bodies two or three times longer than broad. The median rotation period is 10 hours, though there is a broad dispersion. The Apollo asteroid 1566 Icarus rotates in 2.27 hours, almost as fast as an asteroid can spin without shedding loose surface material centrifugally (about 2 hours). In contrast, 288 Glauke appears to have a period of about 47 days, which has led to the hypothesis that its rotation has been tidally slowed by a satellite. The M asteroids spin, on average, more quickly than S and C asteroids, presumably because of their higher density. The directions of the poles of rotation are known for some asteroids. There is no apparent preferential alignment.

Most asteroids have undergone considerable evolution by collisions with other asteroids. In the main belt, typical relative velocities of a few kilometres per second can cause catastrophic breakup of two colliding asteroids – this is how some Hirayama families were formed. Most collisions probably serve to fracture the surviving bodies, however, so that repeated impact events result in fragmented, gravitationally bound "rubble piles." In the process, asteroids are spun up to faster rotation rates, become less spherical, and gradually acquire mature, and perhaps deep, regoliths of dust and boulders. Close up, they must look battered and cratered, like PHOBOS and DEIMOS.

American, Soviet, and European space agencies are planning space missions to asteroids. Certainly, a flyby mission to several asteroids in the main belt (and perhaps a comet or two as well) will answer many questions about these primitive bodies. However, lander missions to some Earth-approaching asteroids require less energy than to any other celestial object, including the Moon. And since these bodies must contain huge stores of materials useful to future space colonists (abundant supplies of metals and even volatiles, such as water), they may well be perceived as an extremely valuable natural resource. *EB*[1].

Asterope Star in the Pleiades cluster; it is on the fringe of naked-eye visibility.

Asthenosphere The layer below the Earth's finite strength shell, or lithosphere, in which solid-state creep first plays a dominant role in the mantle due to the temperature reaching the threshold at which minute stress differences cause flow. Seismological evidence, a decrease of P wave velocities between about 62-155 miles (100-250km) depth and attenuation of S (shear) waves is taken to define the asthenosphere and has usually been interpreted in terms of a shell in which partial melting has occurred. The inference is drawn that its viscosity is some orders of magnitude less than that of the underlying mantle, which is assumed by some to have the effect of "decoupling" the latter from convection in the mantle. These conclusions are not securely based in the absence of a solid-state physics theory of seismic dissipation. Other planets and some satellites — including the Moon — may also have asthenospheres. *SKR.*

Astigmatism In order to possess the required focusing properties, a lens or mirror must be, to great accuracy, the correct shape. If the lens or mirror is large, these stringent requirements become even more of a problem. Furthermore, if an optical system is to give good definition for a wide field of view there are particular unwanted effects (known as aberrations) that may occur if care is not taken in the design of the optical components.

Astigmatism is an aberration which distorts the image of objects from which radiation is incident upon a lens at an angle to the optic axis of that lens. The focal length of an astigmatic lens varies across the lens' surface. The image is thus never a point, but at various distances from the lens appears elliptical, circular, or at two specific distances, a straight line. The best (or least distorted) result occurs when the image is circular, and this is known as the circle of least confusion.

The effect of astigmatism increases as the distance between the image and the optic axis increases. It can be partially corrected by using a stop to restrict the passage of light through the edge regions of the lens. *KPW. See also* LENS.

Astraea (Asteroid 5). Discovered by Hencke in 1845: diameter 73 miles (117km); period 4.2 years.

Astrograph A telescope specially designed for taking wide-angle photographs of the sky. Astrographs, traditionally, are refracting telescopes characterized by their OBJECTIVES, which have relatively fast FOCAL RATIOS (sometimes as fast as f4) and fields of view up to 6° wide—large by astronomical standards. The objectives may have three or four component lenses to achieve this performance, and the biggest astrographs have lenses 20in (50cm) or so in diameter. Astrographs are always placed on EQUATORIAL MOUNTINGS, which are motor-driven to allow the telescope to follow the apparent motion of the sky.

The early development of these telescopes was spurred by the realization, in the latter half of the 19th century, that detailed charts of the stars could be produced by photography rather than with the painstaking visual methods that had been used before. Although wide-angle astronomical photographs had been taken previously using portrait lenses, the first recognizable "modern" astrograph was built in 1886 at the Paris Observatory. It had an aperture of 13in (33cm) and its objective was designed to give the best images in blue light, early photographic emulsions being insensitive to other colours. The success of this telescope led, in 1887, to a major international conference of astronomers agreeing to embark on a photographic survey of the whole sky—the "Carte du Ciel". The Paris astrograph was adopted as a standard model for this ambitious project, and several similar telescopes were built, including one at Greenwich in 1890.

Astrographic photographs are always taken on glass plates rather than film, which lacks the stability of glass. They are principally used for ASTROMETRY, the accurate measurement of star positions. In recent years, the work of astrographs has largely been taken over by SCHMIDT TELESCOPES, which have the principal advantage that they can be made much bigger. *FGW.*

Astrolabe A two-dimensional working model of the heavens, with sights for observations. Invented in classical Greece and developed in Islam, it reached the Christian West about the 10th century.

The planispheric astrolabe consists essentially of two concentric flat disks, generally of brass, one fixed (the *mater* or *plate*) representing the observer on Earth, the other moving (the *rete* or *net*), which can be rotated to represent the appearance of the CELESTIAL SPHERE at a given moment, the altitudes and azimuths of bodies being read off on the graticule engraved on the plate beneath. It incorporates an alidade for measuring altitudes when the instrument is suspended from the hand.

Given latitude, date and time, the observer can read off the altitude and azimuth of the Sun, the brightest stars, and the planets. Conversely, having measured the altitude of a particular body, he can find the time. The astrolabe can also be used for many other problems in spherical trigonometry, such as finding times of rising, setting, twilight, etc., the astrological "houses" of bodies (important for horoscopes),

even the height of a church tower or the depth of a well, using angles measured with the instrument itself.

The mariner's astrolabe is essentially a planispheric astrolabe with the calculating parts removed, and specially designed for observing altitudes of Sun and stars at sea. It was developed by the Portuguese, superseding the simple wooden mariner's quadrant in the late 15th century and falling out of use in the 17th.

The prismatic astrolabe is a modern instrument for measuring local time

Astrolabe: working model of the heavens

and latitude by determining very precisely when a star reaches the almucantar (line of equal zenith distance) of 30°, the time being found by the equal-altitude method invented by GAUSS in 1808. Developed by Chandler and Claude in the 1880s, the modern version invented by André Danjon was first used in 1951. *HDH. See also* NAVIGATIONAL ASTRONOMY.

Astrology A proto-science which claims to be able to assess individual personality and behaviour and to foretell future trends and events from the aspects of the heavens.

In earliest times, when there was a general belief that deities ruled mankind, and the universe seemed of comparatively small extent, it seemed logical enough to suppose that there were strong links between the heavens and human behaviour. Indeed, in ancient China, where the whole of creation was viewed as a vast organism with every part interlinked, signs in the sky as well as natural disasters were thought to mirror the deeds and mis-

deeds of the Emperor and his government. Thus it was that celestial records were of singular importance, since using their evidence an emperor gained insight into the behaviour of his officials and could take appropriate action. But it is important to notice that traditional Chinese astrology was concerned primarily with the emperor and the bureaucracy, and with the mass of the population. Personal horoscopes were a later introduction.

In Western culture astrology developed along somewhat different lines. This was particularly so in Mesopotamia, that area which contained part of present-day Syria and the whole of Iraq. Here astrology was concerned at first only with celestial omens and grew slowly, taking an inferior place to other methods of divining the future. Like Chinese astrology, the omens originally dealt with events of national significance. Later on — probably about the 6th century BC — the Chaldæans in the south of Iraq are thought to have introduced the personal horoscope. This was concerned with the influences exerted at the time of birth by the fixed stars, as well as the Sun, Moon and five planets.

The Chaldæan horoscope exerted a strong influence on the Hellenistic culture, centred in Alexandria, where the concept of the microcosm and macrocosm — the "small world" of men and the large world of the universe or cosmos — was conceived. This idea led to a strong connection between astrology and medical treatment, which was later to permeate Western medieval thought. Though the medical connection is not now believed, astrology in the 20th century is still largely based on that of Hellenistic times.

In casting a horoscope, an astrologer first notes an individual's time and date of birth, then works out the sign of the Zodiac which is ascending or culminating, and the positions of the planets at that time. He next prepares a chart of this information and establishes elaborate geometrical relationahips which allow him to pronounce on the general characteristics and future of the person to whom the horoscope refers.

The procedures of astrology and the interpretation of horoscopes rely on ideas astronomers and most other scientists find subjective and unacceptable. In the first place the characteristics of the Zodiacal signs and of the seven planets are essentially those of pre-Hellenistic Greece, and appear purely subjective, depending on the particular legends of the civilization in

A

which they originated. Completely different legends and ideas permeate the myths of other nations and other civilizations all over the world.

This subjective nature of astrology is also underlined by the example of the influence attributed to Mars and Venus. In Western mythology Mars is the name of the god of war. With its red blood-coloured appearance, it is considered a warlike and belligerent influence on the human character. Venus, on the other hand is thought of as white and beautiful, the very epitome of the goddess of beauty after whom it is named. In contrast its astrological effects are peaceable and beautiful, concerned with love and similar attributes.

Things were very different in traditional Chinese culture. To the Chinese red is the most beautiful colour, so astrologically Mars has the benign influence of Venus in Western mind. White, on the other hand, is considered the colour of death, decay and destruction; Venus was therefore referred to as the "gloomy planet of war".

The modern astrologer has new problems to face. One is that due to PRECESSION, the Earth's axis no longer points in the same direction as it did in Hellenistic times. This causes the Sun to enter the Zodiacal signs at different times from those traditionally accepted. Thus it now enters Pisces rather than Aries at the time of the vernal equinox. Astrological prognostications in the popular press totally ignore precession, though serious astrologers now take it into account.

Work by the psychologists Françoise and Michel Gauquelin showing correlations between the appearance of certain planets and famous people in various professions is sometimes quoted by astrologers as a complete independent vindication of astrology. However, the Gauquelins themselves have made it clear that they can find no relationship between either the signs of the Zodiac or the aspects of the planets relative to each other at the time of birth, both tenets of traditional astrology. Moreover, even the link they claim to have discovered has not received unequivocal support from independent investigators, nor is their explanation of it accepted by many scientists as the only possible interpretation.

An independent attempt to assess the validity of horoscopes was made in California during the early 1980s. Astrologers as well as astronomers were concerned with "double-blind"

tests made on subjects whom they did not know and with whom they did not communicate directly. Independent selection by both subjects and astrologers of a horoscope which appeared correct gave results which, in every case, were the same as those expected by pure chance. However, the traits of the subjects involved were assessed on the basis of the California Personality Inventory, about which some psychologists have reservations, and this together with the fact that no psychologists were intimately involved in the testing, has led them to raise doubts over the outcome though many other scientists accept its conclusions.

A scientific approach to the natural world leads to the formulation of theories which have consequences which can be tested by observation and experiment. In this connection, it is significant that astrology never led its practitioners to question the possible existence of planets other than those known in antiquity. Yet the influences of Uranus, Neptune, Pluto and even the asteroids are now agreed by astrologers to exert effects which they take into account in the preparation of horoscopes. One would therefore expect that such effects would have become so obvious over the years that the presence of the planets would have been predicted by astrologers long before they had been discovered by astronomers.

In spite of its proto-scientific nature, astrology in early times played a useful part in promoting astronomical observation and providing funds to carry it out. It also stimulated the development of mathematical techniques for calculating the future positions of the planets. Moreover, the records kept by early astronomers and astrologers, especially in China, have proved recently to be of immense use in modern research into ancient eclipses, the appearance of novae and comets, as well as other long periods or unexpected celestial events.

In addition to predictive or "judicial" astrology, there was always the study of "natural" astrology, concerned with relationships between the heavens and inanimate nature. It led the Chinese, for example, to be the first to detect a relationship between the Moon and the tides, and in other cultures to the discovery of phenomena that were later taken over into the realm of astronomy proper. *CAR.*

Astrometric binaries Stars whose binary nature is recognized by their wavelike

proper motion, caused by their orbital motion with an unseen companion.

Astrometry The branch of astronomy concerned with the measurement of the apparent angular position of celestial objects by producing a fundamental reference frame on the CELESTIAL SPHERE. This frame, with the Sun at its centre, consists of the plane of the Earth's equator from which declinations are reckoned and the vernal equinox as the point from which right ascension is measured.

The position of the equator and the vernal equinox for a given date (called the epoch) is defined by making observations of the Sun and planets using a meridian circle. This instrument is similar to the transit circle, but it can also measure the declination of an object as it crosses the meridian. Once the equator and equinox are established, the positions of the stars can be measured relative to these by further observations with the meridian circle. These observations as such do not constitute an inertial or fixed frame of reference since the Earth moves through space in a complex fashion and the stars themselves possess their own individual motions. The components of these motions across the line of sight are called proper motions (*see* STELLAR PROPER MOTION) and were first discovered in 1718 by Edmond HALLEY. A FUNDAMENTAL STAR catalogue consists of stars whose positions have been corrected for the motion of the Earth, ie, they are referred to the centre of the Sun (heliocentric coordinates) with the effect of proper motions removed. So what needs to be done to convert the raw data from the meridian circle into the inertial frame of reference?

The first procedure is to correct the instrumental errors present so that the star positions are given with respect to the vertical line and plane of the true meridian. These coordinates are then referred to the centre of the Earth by allowing for the effect of REFRACTION, diurnal PARALLAX and diurnal ABERRATION. Further by making corrections for PRECESSION, NUTATION and polar drift the coordinates are then referred to the fundamental plane of the equator. Heliocentric coordinates are then obtained by correcting for the orbital motion of the Earth by allowing for the effects of annual parallax and aberration. It is then possible to go one step further and make a correction for the motion of the Sun itself which appears "reflected" in the proper motions of nearby stars. However, as this is a statistical procedure, it cannot be determined with great precision. The ultimate accuracy of the coordinates of the stars depends therefore on our knowledge of the astronomical constants of precession, aberration, nutation and the stellar proper motions.

The first of these was discovered by HIPPARCHUS as early as 125BC, but significant increases in positional accuracy did not emerge until the work of Tycho BRAHE and HEVELIUS in the late 16th and early 17th centuries. The invention of the telescope led eventually to the transit circle and the work of James BRADLEY at Greenwich. In 1726 he discovered aberration and two years later followed this with the discovery of nutation. These paved the way to the first catalogue of stellar positions which is still used with confidence today. Between 1750 and 1762 Bradley compiled a star catalogue which was to form the basis of the first fundamental star catalogue ever made, by BESSEL, in 1830. By comparing Bradley's work with that of PIAZZI in Palermo in about 1800, Bessel was able to calculate values of proper motions for certain bright stars and also to derive the constants of precession for the epochs involved. Bessel's name is also inextricably linked with two other aspects of astrometry. He was the first, in 1838, to measure the TRIGONOMETRICAL PARALLAX of a star and derived a distance of 9.3 light-years (the modern value is 11.2 light-years) for the nearby binary star 61 CYGNI. From his work on the fundamental catalogue he also found that the proper motion of SIRIUS was not constant and he correctly attributed this "wobble" to the presence of an unseen companion pulling Sirius from its path. The companion was fully discovered in 1862, and Sirus remains the first example of an ASTROMETRIC BINARY.

Further fundamental catalogues followed with those of Newcomb (1872, 1898) and Auwers (1879) being especially important. The latter was the first in a series which were compiled in Germany, the latest of which is the FK4 (FK standing for Fundamental Katalog). This is based on observations made around 1930 and the mean position error for that epoch was 0.04 arc seconds. Since that date accumulating errors in the proper motions mean that the current errors are now of the order of 0.10 arc seconds. Another catalogue, FK5, is currently being prepared and when finished will contain three times as many stars as FK4, ie about 4,500,

and the positions will have a mean error of 0.03 arc seconds at 1980. It will use a new theory of nutation and an updated value for precession.

Photographic astrometry is normally done with long-focus telescopes and most of our knowledge of positions, parallaxes and proper motions is a result of many years of research with these instruments. Combined with modern measuring machines, long-focus refractors have also revealed the existence of more than 20 astrometric binaries. In some cases the maximum displacement on the photographic plate amounts to no more than two microns. Such painstaking work usually requires up to 50 years of observations, amounting to thousands of plates, with the same telescope in order to reduce systematic errors. Visual binary stars can also be observed by photography, but there are few pairs that are both widely separated and of short period.

With the advent of telescopes and astronomical satellites in space astrometrists are looking forward to the next step in angular accuracy. A specially dedicated astrometric satellite called HIPPARCOS will be launched in 1988 to measure positions, proper motions and parallaxes to a mean accuracy of 0.002 arc seconds. In conjunction with the HUBBLE SPACE TELESCOPE it is also planned to measure the position of quasars as the basis of a new fundamental reference frame. Quasars are so distant that they have negligible proper motions, so using them would remove the uncertainty about the present reference frame, which depends on the value of the solar motion. *RWA*[2].

Astronautics The science of space flight. *See* colour essay on EXPLORING SPACE and related articles such as APOLLO PROGRAMME, ROCKET ASTRONOMY.

Astronomers Royal is the term now given to those British astronomers who hold their position by a specific royal warrant, though the first holder of the office did not immediately receive this title. Until comparatively recently the Astronomer Royal was always director of the Royal Greenwich Observatory. As the observatory at Greenwich was founded by King Charles II for the accurate measurement of positions of the Moon and stars for navigational purposes, the original holders of the office spent the greater part of their time making such observations.

Those Astronomers Royal who were

directors of the observatory and the dates during which they held office were: John FLAMSTEED (1675–1719); HALLEY (1720–42); James BRADLEY (1742–62); Nathaniel BLISS (1762–4); Nevil MASKELYNE (1765–1811); John POND (1811–35); George AIRY (1835–81); William CHRISTIE (1881–1910); Frank DYSON (1910–33); Harold Spencer JONES (1933–55); and Richard WOOLLEY (1956–71). So far only two Astronomers Royal have not been directors of the observatory while occupying the position of Astronomer Royal: Martin RYLE (1972–81); and Francis Graham-Smith (1982–).

It is sometimes claimed that the famous astronomer William HERSCHEL was an Astronomer Royal. He was never appointed to that position, though George III conferred on him the private appointment of court astronomer. For the whole time William Herschel pursued his astronomical career, Nevil Maskelyne was Astronomer Royal. *C.A R.*

Astronomical almanac An almanac is an annual calender listing astronomical events and phenomena such as phases of the Moon, times of sunrise and sunset, tides, eclipses, etc. By far the most authoritative of these, at least for astronomical use, is the *Astronomical Ephemeris* (*The American Ephemeris* in the United States) issued jointly by the Royal Greenwich Observatory and the United States Naval Observatory.

Astronomical clocks Clocks are important in astronomy because space is empty. Many objects of interest to astronomers exist in systems that are essentially isolated: they are affected to only a small extent by perturbations from outside, and relevant internal forces can be identified and accurately modelled. Examples are the rotation of the Earth, the more rapid spin of pulsars, and the orbital motion of the Earth, planets, satellites and space probes within the Solar System. The reading of a clock can be used to predict the results of observing such systems, and disagreement with the predictions, if found, offers the prospect of interpretation in terms of some physical cause — variation of rate of rotation as a result of changes of diameter or internal motion, the radiation of energy, acceleration caused by some otherwise invisible object, or the deformation of space by gravitational waves.

Clocks were developed originally not to test the predictions of scientific theories, but to mark the time for the various activities in monastic life. Later clocks actually used by working astronomers were often outstanding examples of contemporary technology, incorporating subtle features of design intended to improve the uniformity of their timekeeping and hence the accuracy of the observations in which they were used. Marine chronometers were triumphs of technical ingenuity applied to meet a real astronomically related need — that of providing an accurate and immediately accessible model of the rotation of the Earth for use in interpreting the astronomical "sights" taken during shipboard navigation. Their constancy of rate had to be maintained under conditions much more challenging than those faced by the "regulator" clocks in observatories on shore, whose readings could be compared as regularly as the weather allowed with the definitive time established by observations of the Earth's rotation.

Despite steadily improving technology, it was not until the late 1930s that groups of individually free-running clocks would agree among themselves better than they agreed with Earth rotation-based time. Before this stage was reached, the clocks were called upon only to cover the gaps between the observations that defined the time; afterwards the way was clear for the changeover to today's situation in which the clocks provide the standard against which the uniformity of all other processes can be compared.

For both land-based clocks and marine chronometers the developments relating to timekeeping stability before the 1930s were concerned with minimizing the effects of environmental changes and friction on the oscillations of the pendulum or balance-wheel whose repetitive movements were counted to measure elapsed time. Today the pendulum clock has gone from the most precise timekeeping for ever. The force of gravity on which it depends varies significantly at a particular site through tidal effects, through the changing distributions of air and ground-water, and even through the small changes of "centrifugal force" which arise because polar motion produces almost annual changes of a few metres in the distance to the Earth's axis of rotation. The clocks used in precise timekeeping now depend instead upon the intrinsic properties of crystalline quartz or of isolated but indistinguishable atoms of particular species which can be harnessed by electronics and can be made

almost immune from influence by uncontrollable external factors.

The quartz clocks that were introduced in and after the mid-1930s were originally developed as standards of frequency for use in telecommunications, and the quartz oscillator is still an essential component of the atomic clocks which have transformed scientific timekeeping since the mid-1950s. Quartz is used because it is a hard, elastic material with consistent mechanical properties which exhibits the property of piezo-electricity — in other words, deforming a crystal of quartz produces an electric field across it, and application of an electric field produces deformation. In a high-quality quartz oscillator a defect-free piece of quartz is

Astronomical clock: Strasbourg Cathedral

cut, ground and polished to form a "resonator" whose shape and orientation are precisely related to the molecular arrangement within the crystal. The resonator is mounted, rather like a bar in a xylophone, in a way that allows it to vibrate freely, and electrodes on or near its surface make the resonator a link between the output and input of an electronic amplifier which supplies energy to keep the crystal in controlled vibration in the desired mode but allows vibrations at other resonant frequencies to die away. The oscillatory voltage produced by this arrangement drives a chain of frequency dividers controlling a display or producing timing pulses at regular intervals.

The frequency produced by a quartz oscillator is mainly determined by the

size and shape of the resonator, and so can be controlled by the manufacturer; in this sense the resonator is like a pendulum, whose length and frequency of vibration can be varied at will. Research in the 1930s and 1940s demonstrated the possibility of relating radio frequencies in the microwave region to fundamental quantum-mechanical properties of some kinds of atom. The properties are related to those which produce the lines in the optical spectrum that are characteristic of each element, but the energy levels involved are those responsible for the "hyperfine structure" in the optical spectrum, and the energy change in each atom is only around a millionth of the changes that yield optical radiation. The correspondingly lower frequency could be synthesized and controlled by microwave techniques available in the 1950s. The first evaluation of an atomic resonance frequency (that of caesium 133) in terms of an astronomically-established timescale was published by Essen and Parry, working at the National Physical Laboratory near London, in 1955. The timescale used was the "Provisional Uniform Time" established by the Royal Greenwich Observatory from observations of Earth rotation. Later, in 1958, the same frequency was evaluated in terms of the second of "Ephemeris Time", which was based on observations and a theoretical model of the Moon's motion and was thought to be less variable in the long term than the second based on Earth rotation. The value so obtained was used in 1967 to define the SI (Système International) second as 9192 631 770 periods of the radiation corresponding to the hyperfine transition of the ground state of the Cs133 atom.

Atomic time — that is, time based on the uninterrupted accumulation of units of time interval defined by an atomic frequency standard — can be traced back to 1955. Corrections for the variable rate of rotation of the Earth before 1955, if they are required, must be derived from astronomical observations of other phenomena, such as occultations of stars by the Moon, and are necessarily much less precise than those available for current work.

Since 1972 the time signals provided for general use have been offset by a whole number of seconds from International Atomic Time (TAI), the offset being changed by one second on dates chosen by the Bureau International de l'Heure (BIH) according to an agreed code of practice in order to keep the scale actually disseminated within a tolerance of less than one second of rotational time (UT1). UT1, directly derived from observations of Earth rotation, is also known as mean solar time on the meridian of Greenwich. The internationally adopted name for the stepped atomic timescale given by the time signals is UTC (a language-independent abbreviation for Coordinated Universal Time), but the details of the procedures by which this scale is established are of concern to only a tiny minority of the population and the continuing general use of the familiar GMT is still entirely appropriate.

The fundamental timescale for use in dating astronomical events is now determined, in principle, from the uninterrupted operation of a single clock reproducing and counting SI seconds on the Earth's surface. There is no other source of "correct" time by which the clock can be reset. In fact TAI is based on the readings of more than 100 atomic clocks operated in laboratories and institutes around the world, although its long-term accuracy is controlled by a much smaller number of devices constructed and operated so as to provide not just replication of the SI second but an estimate of the accuracy with which this has been done. All contributing devices are intercompared, increasingly by the use of satellite techniques yielding uncertainties of around a hundredth of a millionth of a second, and computations performed in arrears by the BIH give the offset between the readings of real clocks and the international scale. Present indications are that TAI diverges from a perfect realization of its definition by less than 3 millionths of a second per year: for most astronomical purposes the "astronomical clock" provided by the international time services is essentially perfect.

The accuracy with which observations of astronomical events can be related to a suitable idealized timescale now almost always depends more on uncertainties in the links to TAI or UTC than on any inadequacies of these reference scales. The timing accuracy that is available to the user in fact, rather than in principle, is now determined largely by the depth of his pocket, since timing signals from navigation satellites can now be received almost anywhere in the world for several hours per day, and coverage is expected to become almost continuous as the Navstar-Global Positioning System (GPS) is brought into full operation towards the end of the 1980s. These signals can be decoded to

provide both the user's position, with an uncertainty of a few tens of metres, and "GPS time", directly linked to TAI, with an uncertainty of less than one microsecond. The cost of receivers is already subtantially less than that of caesium-beam clocks, and for many users the receiver will be a much more sensible buy.

Other sources of precise time are the Loran-C radio-navigation system, which has been used by the time services for almost 20 years but is now being largely supplanted by GPS, and other ground-based dedicated time-and-frequency services. These services cover a restricted area, and signal propagation times vary with iono-spheric conditions, but the time-coded 60kHz MSF signals, which are almost continuously available, meet the needs of the vast majority of users, who can tolerate timing uncertainties ranging upward from around one ten-thou-sandth of a second.

All the sources of time and clocks considered so far have produced time-scales and time intervals based on the SI second, which is a close approximation, of constant duration, to the slightly variable mean solar second; but of course astronomers also require access to local sidereal time in order to point their telescopes. Since UTC provides a good approximation to UT1, and sidereal time and UT1 are related by a formula, there is no difficulty of principle in determining sidereal time from UTC. It is, however, often convenient to have a clock that shows a reasonable approximation to sidereal time; the gaining rate of about 3m 56s per day relative to UTC may be obtained by shortening the pendulum or by gearing in a mechanical clock, or by their electronic equivalents. *JDHP.*

Astronomical mirrors Concave reflectors are the devices most often used by astronomers to collect and concentrate electromagnetic radiation from distant objects. Mirrors are usually taken to be reflectors working in the optical part of the spectrum, which includes the visible region. This extends from the far infrared, with wavelengths less than about 1mm, to the soft X-ray region with wavelengths about one million times shorter, though not all of this range penetrates the Earth's atmosphere. When a practical device for viewing distant objects incorporates a mirror, it is a reflecting telescope.

The first telescopes were refractors and employed lenses rather than mirrors to form an image. These early

instruments suffered from a variety of optical deficiencies, the most intractable of which was chromatic aberration, which gives bright objects such as stars a blue- or orange-coloured halo. It was these colours, produced by passing sunlight through pieces of glass, which intrigued Isaac NEWTON in the summer of 1666. He came to the firm but incorrect conclusion that all optical instruments incorporating lenses were destined to suffer from colour fringes. Never a man to be defeated by a technical problem, Newton devised a telescope that magnified by reflection from a curved metal mirror rather than by refraction through polished glass lenses. Although Newton's telescope was free from chromatic aberration the idea remained little more than a scientific curiosity. In the main this was because metal mirrors had low reflectivity, tarnished quickly and were difficult to make. Improvements in optical glasses and techniques, especially Dolland's re-discovery of lens systems free from chromatic aberration in 1758, gave little encouragement to the development of catoptric (mirror) telescopes, until in 1781 William HERSCHEL discovered the planet Uranus with a 6in (150mm) reflector he had made himself. This discovery enabled Herschel to devote himself to astronomical pursuits and he greatly advanced the art of making mirrors from "speculum metal", an alloy of 71 per cent copper and 29 per cent tin. This dense, brittle material was polished, initially by hand, and mounted in elaborate telescopes to explore "the structure of the heavens". The largest speculum mirror ever made was cast by Lord ROSSE in 1842. It was 6ft (1.8m) in diameter and weighed four tons (4,064kg). The last major telescope with a metal mirror was the ill-fated Great Melbourne Telescope, which was built in 1862 and had a 48in (1.2m) speculum. It was the problem of tarnishing, requiring frequent removal of the mirror from the telescope for re-polishing, which made all the telescopes awkward to maintain and defeated the efforts of the astronomers in Melbourne.

Many of these difficulties disappeared when a simple way of depositing a highly reflective film of silver on a polished glass surface was discovered. Although silver tarnished, it was easily replaced without refiguring the mirror and glass was lighter and easier to polish than speculum. However, the failure of the Great Melbourne Telescope held back the general acceptance

A

of reflectors until the outstanding success of the silver-on-glass mirrors of the 60in (1.5m) and 100in (2.5m) telescopes on Mount Wilson in the early years of this century. In the 1930s the silver-on-glass process was superseded by a vacuum deposition technique which gave a durable non-tarnishing aluminium reflective layer. The technical problems that had plagued Newton, Herschel and the astronomers in Melbourne were at last overcome and the way was clear for the new challenges posed by mirrors for the 200in (5m) Hale telescope and the 236in (6m) giant for the Soviet Special Astrophysical Observatory in the Caucasus. Although the Russian instrument has yet to reveal its full potential, even larger mirrors are planned.

The use of mirrors rather than transmission optics for astronomical telescopes offers numerous advantages. Lenses must be made from slabs of optical glass of excellent quality throughout. These requirements are difficult to meet in the large sizes demanded by astronomers. As the diameter of the lens increases so does the thickness, light absorption and weight, and heavy lenses supported only at the edge tend to distort. Mirrors can be fully supported both in use and during figuring and, of course, only one surface needs to be polished. The optical quality of the material is not important — indeed the low-expansion glass-ceramics (Cer-vit, Zero-dur), which have replaced quartz and Pyrex for large mirrors, are quite opaque though they polish as well as glass. These low-expansion materials can be cast in large moulds like glass but often require an elaborate cooling schedule to ensure their special thermal properties.

The shaping of the optical surface is as much an art as a science and in very few institutions in the world are the necessary skills available for figuring mirrors for big telescopes. The first stage of the operation involves polishing the blank to conform to a sphere. If the telescope is to be a Schmidt design a spherically formed mirror is all that is required. "Conventional" reflectors, however, have mirrors in the form of a paraboloid, or, in the case of modern instruments (the so-called Ritchey-Chrétien design), a hyperboloid. In both cases this is created by careful "deepening" of the spherical mirror with regular checks of the figure with the Foucault knife-edge test. This delicate operation can take many months and may require the removal of minute amounts of material from localized areas of the mirror by hand. The final process, which might await delivery of the mirror to the telescope site, involves coating the optical surface with an extremely thin layer of aluminium in a specially made high-vacuum chamber.

The problems inherent in polishing large mirrors have led to the development of a spin-casting process where the molten material is spun in a rotating annealing furnace, creating and maintaining its optical figure as it cools. Mirrors at least 315in (8m) in diameter are planned. Even greater collecting areas may be possible by building a mirror from individual segments. Already a MULTIPLE MIRROR TELESCOPE with six 71in (1.8m) paraboloids has been built. These individually supported and controlled mirrors direct light to a single focus and are equivalent to a 177in (4.5m) single reflector. A multiple-mirror system with an aperture equivalent to an 82ft (25m) telescope is also under consideration. *DFM. See also* LENS.

Astronomical spectroscopy Practice of obtaining and study of the spectra of astronomical objects. The spectra are obtained with a SPECTROSCOPE, and are records of the distribution of light over wavelength for the object concerned. From this distribution of light, astrophysicists interpret the properties of the object. The first gross property is whether the light is concentrated at a few isolated wavelengths (*see* EMISSION SPECTRUM) or forms a relatively broad distribution over a wide range (CONTINUOUS SPECTRUM), or is a mixture of the two. An emission-line spectrum comes from a tenuous gas (nebula). A continuous spectrum comes from a dense hot gas (a star's surface) or energetic electrons which are spiralling in a magnetic field (synchrotron source). On detailed examination, some wavelengths of the continuous spectrum will be discovered to be weakened, absorbed from the continuous spectrum on the light's passage through cool gas in space or in a star's atmosphere (*see* ABSORPTION). Mixed emission-line and continuous spectra come from complex objects, for example BINARY STARS in which tenuous gas is orbiting the surface of the stars, galaxies consisting of both stars and nebulae, or QUASARS consisting of gas and stars orbiting a synchrotron-emitting BLACK HOLE.

The wavelengths at which spectral lines occur identify the composition of the gas, and its temperature and density. The wavelengths may systematically depart from the laboratory values and can be interpreted by the DOPPLER EFFECT.

Astronomical spectroscopy is, in short, the key technique by which the secrets of astrophysics can be unlocked. Its power and versatility are indicated by the fact that modern large optical telescopes are used for astronomical spectroscopy for three-quarters of the time. *PGM.*

Astronomical twilight. The dawn and dusk periods when the Sun is less than 18° below the horizon. For nautical and civil twilight, the figures are 12° and 6° respectively.

Astronomical unit (AU). The mean distance between the Sun and the Earth: 92,955,630 miles (149,597,870km).

Astronomy, history of *See* HISTORY OF ASTRONOMY.

Astrophotography Today it is hard to imagine astronomy without photography. But while a few visionaries stressed the potential of the new invention in astronomy within a short time of its introduction in 1839, it took 50 years for photography to be accepted as a legitimate research tool by professional astronomers and another quarter of a century before continued improvements to equipment and plates raised photography to a position where it was vital to the progress of research in most areas of astronomy.

The human eye is a versatile organ but the advantages that a photograph has over it for the astronomer are readily apparent. A scene may be recorded for subsequent (more comfortable) study or measurement; that record will in most cases be much more accurate than any observer's sketch; and, since a film or plate stores light, the resulting image will show objects much fainter than those that the eye can discriminate. These advantages, however, are those of evolved photographic systems. For much of the 19th century after its invention photography suffered from insufficiently accurate TELESCOPE DRIVES, lack of sensitivity ("speed") in film and also film's inability to record light with equal facility across the whole visible spectrum (for many years it was "blind" to red). Despite this, the new technology was put to immediate use, mainly by enthusiastic amateurs.

The Moon was an obvious (because bright) target and the first successful

Daguerreotype plate of it is attributed to the American doctor J.W.Draper in March 1840. The exposure was 20 minutes. The French scientists H.Fizeau and L.Foucault obtained a successful Daguerreotype plate of the Sun in April 1845 — and the German photographer Berkowski an image of the eclipse on July 28 1851. Before the Daguerreotype system was replaced by the faster and somewhat more practical wet plate collodion technique in the early 1850s, W.C.BOND — director of the Harvard College Observatory in the United States — and a professional photographer, J.A.Whipple, succeeded in obtaining excellent images of the Moon in 1850 and (in the same year) the first star images — of Vega and Castor.

With the introduction of wet plate collodion came signs of a more systematic approach to astrophotography. Sir John HERSCHEL proposed that photography be used on a daily basis to record sunspot activity and the direction of this "solar patrol" (as well as responsibility for having a specially designed photoheliograph built) was placed in the hands of London businessman and scientist/photographer Warren de la Rue. This trailblazing work, which continued until 1872, was followed up and further developed by E.W.Maunder at the Royal Greenwich Observatory and J.JANSSEN — first director of the Meudon Observatory in France. Warren de la Rue (to whom Agnes Clerke, the historian of astronomy in the 19th century, attributed "the honour of having obtained the earliest results of substantial value in celestial photography") subsequently obtained excellent images of the July 18 1860 eclipse of the Sun which (together with the work of others such as Father A.Secchi) demonstrated reliably that the "red flames" (prominences) were solar and not lunar phenomena. Meanwhile, in the United States G.P.BOND — having succeeded his father as director of the Harvard College Observatory — recommended stellar photography with improved equipment and considerable success. He strongly argued the role of photography in astrometry which in his view could only become more extensive and valuable as plates and instrumentation improved.

Unfortunately, Bond died at an early age in 1865 but his opinions were to be proved well founded. An improvement came in the 1870s with the introduction of the more convenient dry plates and talented individuals such as L.M.RUTHERFURD and Henry Draper in the United States and subsequently A.A.COMMON and Isaac Roberts in Britain pushed the frontiers of astrophotography outwards. In 1880 Draper succeeded in obtaining a photograph of the ORION NEBULA, though both Common and Roberts were to improve on this within a few years. More significant, perhaps, were the advances in photographic spectroscopy. Draper obtained a spectrogram of Vega in 1872, but major work in this area over the next three decades was undertaken by W.HUGGINS, who in 1881 obtained a spectrogram of Tebbutt's Comet and in 1882 the first photographic record of the spectra of a gaseous nebula (Orion). Extensive and valuable work in photo-spectroscopy was conducted in the United States by E.C.PICKERING at Harvard College Observatory.

The 1880s and onwards marked photography's breakthrough in astronomy. This was not due to any single event or development but to a number — to improvements in instrumentation and plates and to the arrival of new minds (concerned with asking new questions) as well as the conversion of older minds to photography's undeniable case. In 1882 the Scot David GILL — an influential scientist and director of the Royal Observatory at the Cape of Good Hope — obtained the first completely successful image of a comet. He was impressed by the manner in which his plates showed star images despite exposures lasting up to 100 minutes. He set out to argue photography's case and in 1885 received a grant from the Royal Society which he directed toward constructing a photographic star map of the southern sky. In this he was joined by J.C.KAPTEYN of the Netherlands; and when the task was completed (despite the withdrawal of Royal Society support) the catalogue contained the magnitudes and positions of over 450,000 stars. As a professional astronomer Gill's enthusiasm for photography was mirrored by the Henry brothers at the Paris Observatory — and by their director, Admiral E.B.Mouchez. A meeting of minds developed on the possibility of international cooperation in the preparation of a major sky chart which eventually became known as the *Carte du Ciel*. The Carte was not completed until 1964, but it did mark (in the words of its official historian) "the systematic introduction of photography into astronomy".

As important as were astrometry and sky surveys, the "new" astronomy which was developing as the new century progressed was concerned with many other issues — which may be broadly described as the "what", the "how" and the "why" and not so much the "where". Photo-spectroscopy was an early indication of this and photography's role in that and other directions continued to expand. No area of research was more intriguing (or dramatic) than the greater and greater detail which came to be resolved — by such workers as J.E.KEELER and G.W.RITCHEY in the United States — in spiral nebulae, which E.HUBBLE and his co-workers using the 100in (254cm) Hooker reflector at Mount Wilson Observatory ultimately showed to be outside the Milky Way and indeed other galaxies. Spectrographic study of these led Hubble in time to propose the concept of the expanding universe.

As the 20th century has progressed, photography — defined here as a recording medium based on silver halide chemistry — has faced a challenge from electronic systems, of which the CHARGE-COUPLED DEVICE (CCD) has made the most impact on public awareness. In fact, the respective advantages and disadvantages of photography and electronic systems are such that they may be regarded as complementary. The major advantages of CCDs are their efficiency as light gatherers, their dynamic range, their linear response to light, and the ease with which their output can be digitized for analysis in a computer. On the other hand, CCDs make special demands in that for reasonable efficiency they must be kept at very low temperatures, they are expensive, they offer low resolution and are extremely small in size — a matter of a centimetre or two at present compared with the more than 350mm^2 of a typical SCHMIDT photo plate.

These differences suggest clearly demarcated roles for photographic and electronic systems in the near term at least. Photography is well suited for sky surveys and for search programmes to locate bodies such as comets, asteroids and quasars. CCD and similar systems are tailormade for imaging already located bodies (or those whose position has been accurately predicted) — particularly when such objects are very faint and the data obtained are to be processed by a digital computer.

Photographic techniques Astronomical photography is in most ways unlike normal picture-taking. Exposure times may be long, often in excess of an hour

A

(though greatly shortened by prior baking of the special emulsions), and the camera or telescope must follow the objects as they are carried across the sky by the Earth's rotation. The multitudes of objects that appear on any photograph are mostly self-luminous and exhibit an enormous range of brightness, perhaps 1,000,000:1 in a single exposure. In terrestrial photography, most images are made with reflected sunlight and have a brightness range usually less than 100:1. The astronomical photographer cannot rearrange the stars and galaxies as he pleases, nor can he alter the lighting to a more revealing angle. Similarly there is no correct exposure; in general the longer it is, the fainter are the objects recorded, though the practical limit is usually reached when the signal from the distant object is lost in the "noise" (photographic fog) from the ever-present airglow.

The introduction of the terms "signal" and "noise" alerts us to the true nature of astronomical photography. To the astronomer, a photograph may contain information about the location, brightness and colour of thousands of stars or the type, orientation and morphology of hundreds of galaxies or nebulæ. The photographic plate combines the function of a sensitive detector with excellent display and storage facilities. The thin layer of emulsion is as compact and efficient an information storage medium as a compact disk or magnetic tape—and can be read by eye.

Photography at a large observatory can be quite a complex operation. The special emulsions, coated on glass plates, are packed in dry ice and shipped by air from the Eastman Kodak Company in Rochester, New York. They are stored at very low temperatures until required. Even an hour or two at the daytime temperature of places like inland Australia (104°F/ 40°C) will ruin some materials. The emulsions are specifically made for the detection of faint light and are designed to be most efficient where exposures are long, perhaps an hour or more. Normal films work best under bright light where exposures are typically small fractions of a second. The long-exposure speeds of astronomical emulsions can be greatly improved by "hypersensitizing" just before use. This often involves carefully controlled baking in nitrogen or nitrogen/hydrogen mixtures (not to be attempted without professional help). Subsequent storage and exposure at the telescope is in an inert atmosphere. This can give speed gains of 10-30 times, essential when time on large telescopes is very difficult to obtain. Even bigger speed gains, perhaps 100-400 times, are obtained with some infrared-sensitive plates by bathing them in a silver nitrate solution.

Astronomical emulsions are available in a variety of spectral sensitivities,

Astrophotography: an 1882 comet picture

specially chosen to isolate well-defined parts of the spectrum so that measurements of star colours, for example, can be compared between observatories. The astronomer can also choose between several kinds of emulsions on the basis of contrast, resolution and granularity. Rather surprisingly, the most modern emulsions have low photographic speeds and very high contrast and are designed for the modern generation of fast (low f/ratio) Schmidt and 157in (4m) class telescopes where light from the night sky is recorded along with the faint images of distant stars and galaxies. The high contrast of the emulsion ensures that objects that are fainter than the night sky will be visible on the plate, while the fine, uniform grain structure gives good resolution with low granularity.

The photographs lend themselves to a variety of methods of post-processing image enhancement which are revealing new aspects of familiar objects and ever fainter stars and galaxies to challenge the imagination of astronomers. Quite simple contact copying techniques can reveal details so faint that less than one per cent of the light recorded by the plate comes from the object of interest, the rest being the unwanted, uniform glow of the night sky. This process of photographic amplification, and a related technique known as unsharp masking, has led to many new astronomical discoveries, perhaps the most unexpected and exciting being the existence of very faint shells of stars associated with otherwise undistinguished elliptical galaxies. These discoveries may yet reveal how this most common type of galaxy formed. Another recent advance from the photographic laboratories of the Anglo-Australian Telescope is the creation of colour photographs that are astronomically useful, revealing the colours of stars and nebulæ in a realistic manner. Many of these photographs appear in this book.

While the role of photography as the principal imaging system in astronomy has diminished in recent years, mainly as a result of the development of the charge-coupled device, the studies which require a wide field, and an extremely uniform, sensitive and compact detector, will still turn to the well-tried technology of photography.

With the development of new instruments and the coming of spaceflight, the study of astrophysics has extended well beyond the visible spectrum into ultraviolet, infrared, X-ray and radio wavelengths — with many new discoveries. Frequently data are obtained in non-imaging and numerical form, which is handled with the aid of a computer. Where images are obtained, photography has faced a challenge from electronic systems. In fact, photography and electronic systems are complementary, even if their roles are clearly demarcated in the near term at least. *HJPA/DFM.*

Astrophysics A relatively recent development of astronomy, astrophysics is the application of atomic and nuclear physics, particle physics, plasma physics, solid-state physics, thermodynamics and general relativity — indeed all known fields of physics — in a manner that attempts to interpret the observations and reveal the processes of nature obtaining in all bodies, systems and regions in the universe. These include the Moon, Sun, planets, interplanetary matter, stars, the Galaxy, interstellar matter, external galaxies, active galaxies and quasars, intergalactic matter, and also the universe as a whole, particularly in regard to its origin and evolution. Theoretical astrophysics applies mathematical methods and known physical laws (frequently through the use of simplifying models) to describe the observed or inferred physical conditions. The term "astronomy" is frequently used in place of, or to contain, astrophysics.

After their introduction in the 19th century, two basic observational techniques, photography and spectroscopy, led to new and quantitative methods for measuring the light received by telescopes from astronomical objects and enabled physical studies to be made of brightnesses, temperatures and chemical and physical natures of stars and nebulæ leading to the theoretical analysis and interpretation of their properties. In the 20th century, and particularly in the past few decades, such studies have been enormously increased and extended through the introduction of large telescopes and many new methods of radiation analysis, coupled with the introduction of photo-electronic detection techniques, photographic measuring machines, and digital handling of accumulated data using advanced image and data processing methods. Furthermore, new observational regimes have opened up in near-infrared and extended radio bands of the electromagnetic spectrum from ground-based observatories, and in far-infrared, ultraviolet, X-ray and gamma ray bands from high-altitude aircraft, balloons, rockets, artificial satellites and space probes where the obscuring effects of the Earth's atmosphere are more or less transcended. Primary cosmic rays and solar particle streams are now studied from space-based platforms and experiments are in hand to try to detect gravitational radiation.

Spectroscopy, ie, the dispersion of radiation into a spectrum of frequencies (or, more easily visualized, wavelengths), is the most powerful technique of analysis available in astronomy. Individual atoms, ions or molecules emit or absorb light at characteristic frequencies (or "lines": from their appearance in a spectroscopic instrument when the light to be analyzed passes into it through a narrow defining slit) and can thus be identified and their quantity assessed. Information on the physical and chemical state of an object can be obtained from an analysis of the way its spectrum is modified by the collective environmental conditions within the object, and line-of-sight velocities of the whole object or regions within it are indicated by the Doppler shift in frequencies. The most intensively studied spectral region is the ground-accessible "optical" region (which is defined by the window in atmospheric absorption extending from the near-ultraviolet to the near-infrared regions), but much vital information has come also from observations in the radio, far-ultraviolet and X-ray regions. The power of spectroscopy as an analytical tool in astrophysics has been evident from the earliest time of its use. The solar spectrum was studied in much detail by the German

physicist Joseph von Fraunhofer in 1814. He was able to map many hundreds of absorption lines, the most prominent of which still bear his name, and noted differences between the spectrum of the Sun and those of several bright stars. The astrophysical interpretation of the observed lines needed further developments in physics, augmented by further laboratory work, and in 1859 they were explained by the German physicists Gustav Robert Kirchhoff and Robert Bunsen as being due to absorption by familiar elements. In 1855 the English astronomer William Huggins began the visual study of spectra of the Sun, stars, nebulæ, comets and planets, and made a host of astrophysical discoveries. From the character of the emission spectrum of the Orion Nebula he showed that it is gaseous, while he found that the Andromeda "Nebula" had the appearance of a stellar spectrum, which led him to conclude that it was made up of stars. He made some of the first determinations of line-of-sight velocities of stars. Since then a great deal has been learned about the physical conditions of stars and the interstellar medium and about the internal structure and dynamics of galaxies.

In modern times much detailed theoretical work has gone into explaining how the distribution of radiation over the spectrum of a star depends on the structure of its outermost layers, or "atmosphere" (defined as the region from which radiation can escape into space and thus be detected), and more broadly, the processes occurring during the life cycles of stars: their formation from collapsed regions of the interstellar medium, their active burning phases, and their deaths, which can range from being an event in which a star mildly fades away to one exhibiting a violent explosion which at its maximum brightness can appear as brilliant as a whole galaxy (in the region of 10^{10} times the Sun's luminosity). The understanding of these can be counted a major success of astrophysics and will serve as a good example here.

Stellar spectra characteristically show a continuous bright background which by its colour distribution approximately indicates the "surface" temperature (but is made up of radiation contributed in some measure from throughout the whole of the stellar atmosphere). These continua are formed by certain processes in which radiation is emitted and absorbed in

stellar atmospheres: in cooler stars (like the Sun) the interaction chiefly involves electrons and hydrogen atoms, in hotter stars, electrons and protons, and in the hottest stars, electrons alone. The dark spectral lines superimposed on the continuum, which are due to atoms, ions and molecules, strictly do not arise from absorption by a cooler outer layer as a simple visualization suggests, but in a complex depth-dependent manner in which radiation dominantly escapes from outer and therefore cooler (darker) regions of the atmosphere near the centre of an absorption line, or deeper, hotter (brighter) regions for the continuum lines. The fact that the lines appear dark indicates that the outer layers are the cooler, and indeed the temperature generally is expected to decrease outward from the centre of any star. For the Sun and certain other stars some bright lines also are seen, particularly in the far-ultraviolet region, indicating that in these the temperature is rising again in extended, extremely tenuous layers of the outer atmosphere. The strengths, widths and polarization of spectral lines give information on the chemical composition, ionization state, temperature structure, density structure, surface gravity, magnetic field strength, amount of turbulent motion within the atmosphere of the star and stellar rotation.

A striking result of astronomical spectroscopy is the evident uniformity of chemical composition throughout the universe, extending even to the most distant galaxies known. Although the atomic constituents of the universe are everywhere the same, the proportions of these elements are not identical in all astronomical objects. Measurements of these differences give information on the nuclear processes responsible for the formation of the elements and the evolutionary processes at work. A star is kept in a steady state for a long time by the balance of the inward pull of gravity and the opposing force arising from the increase of pressure with depth which is maintained by a very high central temperature. For a star like the Sun (mass of 2×10^{33} grams, radius 435,000 miles/700,000km) the central temperature is about 15,000,000°C. The energy from such hot matter works its way tortuously outward through the star until it is radiated away from the atmospheric region near its surface. This process yields a relationship between the masses and luminosities of stars, with the luminosity increasing very dramatically with the mass; the luminosities of stars encompassed within the relatively small range of masses between 0.1 and 10 times that of the Sun, which represents most stars, range over more than a million. The high rate of loss of energy requires a powerful energy-producing process within the star, identified as thermonuclear fusion of light atomic nuclei (mainly hydrogen nuclei) into heavier ones (mainly helium nuclei). The transformation of hydrogen into helium in the case of the Sun — a small star — makes enough energy available to enable it to shine as it does today for a total of about 10^{10} years. In contrast, the lifetime of massive, very luminous stars, which burn up their hydrogen very rapidly, can be as short as a million years. During this burning process the core contracts, and it eventually becomes hot and dense enough for helium itself to ignite and to form carbon and oxygen. Although the contraction of the core is thus halted, the star then has a rather complicated structure, with two nuclear burning zones, and may become unstable. Later stages in the evolutionary cycle present many possibilities depending principally on the initial mass of the star. It may become a white dwarf, such as the companion of Sirius, which is a very dense, hot star with about the mass of the Sun but a radius nearer that of the Earth, then slowly cool; the immense pressures built up in such stars result in a special kind of behaviour called electron degeneracy, explained in physics by quantum theory. More massive stars progress to still more advanced stages of nuclear evolution, experiencing ignition of successively heavier nuclei at their centres. In some rare cases much of the outer layers of the star may be violently ejected in a supernova outburst in which heavy elements (some synthesized in the explosion) are scattered widely into the general interstellar medium. Out of this enriched material later generations of stars may be formed (as was the Sun); carbon and heavier elements in the universe are known to be produced in stars and dispersed in this way, while only the lighter elements existed before stars were formed. Supernova events are probably also the main source of cosmic rays. The residues of these events may be neutron stars, which can be observed as pulsars, and represent a still higher state of compression than white dwarfs, in which now electrons and protons have been crushed together to form a degenerate neutron gas. The Crab Nebula and the pulsar

within it are remnants of a supernova eruption seen in AD 1054. On the other hand, a star may go on contracting until it becomes a "black hole", of immense density, which can influence its surroundings by gravitational action only, and is intrinsically invisible, but whose violent accretion of nearby material makes its influence observable by strong emission in the X-ray region.

Such applications of physics to astronomical problems in general are progressing strongly on most fronts in the subject and much success has been achieved. Currently a great deal of attention is being directed toward the processes that occurred during the earliest phases of the formation of the universe, very close to the time of the Big Bang. But here astronomers and physicists are venturing out hand in hand, for not only is our knowledge of the behaviour of the universe incomplete, but so is our knowledge of physics itself. New physical theories unifying all forces of nature are needed before we can begin to understand the most fundamental properties of the evolving universe and, indeed, the origin of matter itself. *AB*.

Ataxites "structureless", nickel-rich iron meteorites.

Aten asteroids Asteroids with Earth-crossing orbits with semi-major axes less than 1.0AU and aphelion distances greater than 0.983AU. Aten (2062), Ra-Shalom (2100), Hathor (2340), and 1984 QA are the only known examples.

Atla Regio The eastern section of APHRODITE TERRA (Venus). This takes the form of a curved mountain belt superimposed upon which are linear features prominent on radar images. Prominent conical mountains, eg, Maat Mons and Ozza Mons, rise to 16,000ft (5km) above datum and may be volcanic in origin.

Atlas Lunar crater, 55 miles (88km) in diameter, east of Mare Frigoris; 47°N, 44°E. It forms a notable pair with Hercules.

Atlas The innermost satellite of Saturn; dimensions 24.8 × 12.4 miles (40 × 20km). It is the "shepherd satellite" of the A-ring.

Atlas The star 27 Tauri, in the Hyades; magnitude 3.63.

Atlas United States ICBM, later modified to launch Mercury spacecraft,

unmanned satellites and space probes; some with upper stages ABLE, AGENA or Centaur.

Atmosphere The gaseous envelope which surrounds a planet, satellite or star. The characteristics of a body which determine its ability to maintain an atmosphere are the temperature of the outer layers and the escape velocity. Small bodies, such as the Moon, Mercury and the satellites of the planets in the Solar System (apart from Titan and possibly Triton), do not have any appreciable atmosphere. The escape velocity for these bodies is sufficiently low so that it is easily exceeded by gas molecules travelling with the appropriate thermal speeds for their masses and temperature. The speed of a gas molecule increases with temperature and decreases with the molecular weight. Consequently, the lighter molecules, such as hydrogen, helium, methane and ammonia escape more readily to space than the heavier species such as nitrogen, oxygen and carbon dioxide. In the cold outer reaches of the Solar System, it is possible that Pluto and even Triton may have their potential atmospheric gases frozen out on their surfaces. The primary atmospheres of the bodies in the Solar System originated from the gaseous material in the solar nebula. The lighter gases were lost from many of these objects, particularly the terrestrial planets, Venus, Earth and Mars, which now have secondary atmospheres formed from internal processes such as volcanic eruptions. The main constituent of the atmospheres of Venus and Mars is carbon dioxide (CO_2); the main constituents of the Earth's atmosphere are nitrogen and oxygen. The only other nitrogen atmosphere found so far in the Solar System is on Titan, the satellite of Saturn. Only the giant outer planets, Jupiter, Saturn, Uranus and Neptune, still retain their primordial atmospheres of mainly hydrogen and helium. However, these major planets are also surrounded by envelopes of escaped hydrogen. Titan, too, is associated with a torus of neutral hydrogen atoms which have escaped from the satellite's upper atmosphere and populate the satellite's orbit. It is also possible for a body to gain a temporary atmosphere through a collision with an object containing frozen gases, such as a cometary nucleus.

The gaseous layers of an atmosphere are divided into a series of regions organized on the basis of the variation of temperature with altitude. For worlds with definite solid surfaces, such as the Earth, Venus, Mars and Titan, the levels start from the ground and relate to existing observations. The outer planets are huge gaseous envelopes with no solid surface and whose atmospheres have been investigated to just beneath the cloud tops. Consequently, the characteristics of the lowest layers are based upon theoretical models and the levels usually start from a reference position of the 1 atmosphere pressure surface. The names given to the various structured layers of the terrestrial atmosphere are also applied to all other planetary atmospheres. The precise variation of the atmospheric temperature with altitude will depend on the composition of the atmosphere and the subsequent solar heating and long-wave cooling at the various levels.

The troposphere is the lowest layer; for the Earth it extends from sea level, where the pressure is 1013mb, to the tropopause at an altitude of between 5 miles (8km) at the poles and 11 miles (18km) in the tropics. This region contains three-quarters of the mass of the atmosphere and is the meteorological layer containing the cloud and weather systems. The layer is heated from the ground and the temperature decreases with height to the tropopause, which marks the minimum temperature of $-67°F$ $(-55°C)$ in mid-latitudes and about $-112°F$ $(-80°C)$ in the equatorial region. Above the tropopause, the temperature increases with height through the stratosphere because the heating from the absorbed ozone (O_3) dominates the long-wave cooling by CO_2. The temperature increases to about $32°F$ $(0°C)$ at the 31-mile (50km) altitude which marks the stratopause. In the Earth's atmosphere, the region between 15 and 25 miles (25 and 40km) where the ozone is situated, is often called the ozonosphere. The temperature then decreases with altitude through the mesosphere to a minimum of about $-130°F$ $(-90°C)$ at a level of 53 miles (85km). This minimum level is called the mesopause. The atmosphere at these levels is very tenuous and is therefore sensitive to the heating from the Sun. Consequently, this region and the further layers of the atmosphere have a significant diurnal temperature cycle. Beyond the mesopause, we enter the thermosphere, where the temperature increases with height throughout the remainder of the atmosphere. This structure is created by the heating from the absorption of the far ultraviolet

solar radiation by oxygen and nitrogen. Solar X-rays, too, penetrate through to the mesosphere and the upper atmospheric layers and are therefore sensitive to changes in the solar radiation and the chemistry of the atmosphere. The solar UV radiation photo-ionizes the atmospheric constituents in these outer layers producing ionized atoms and molecules, which results in the formation of the ionosphere at an altitude of about 37 miles (60km). This is also the domain of meteors and aurorae found regularly in the upper atmosphere. Above 310 miles (500km) is the exosphere, from where the atmospheric molecules escape into space.

The individual planetary atmospheres each possess these basic structures but with specific variations as a consequence of their size, position in the Solar System relative to the Sun and atmospheric chemistry. The massive CO_2 atmosphere of Venus has a deep troposphere which extends from the surface to an altitude of about 62 miles (100km). About 90 per cent of the volume of the atmosphere is contained in the region between the surface and 17 miles (28km). There is no stratosphere or mesosphere but only a thermosphere, which makes the planetary atmosphere quite different from the Earth. By terrestrial standards, the Martian atmosphere is very thin, since the surface pressure is only 7mb, but the atmosphere does have some similarity in the vertical layering with the presence of a troposphere, stratosphere and thermosphere. The structure of the troposphere can change dramatically to an isothermal layer or even possess an inversion, during those periods of the global dust storms. However, there is much less ozone in the Martian atmosphere than on Earth, so that there is no reversal in the temperature gradient which is a characteristic of the terrestrial stratosphere. Lower temperatures are found in the Martian thermosphere, too, as a consequence of the CO_2 cooling effect. The basic structure of the atmospheres of the outer planets resembles that of the Earth. There is a troposphere that extends to unknown depths beneath the clouds and well-defined stratospheres, created by the heating from the solar absorption of methane (CH_4). The knowledge of the higher levels is still incomplete. Titan, too, has a well-defined atmospheric structure of a troposphere, stratosphere and thermosphere created more by the complex aerosol chemistry rather than by the gaseous composition alone.

These atmospheric structures will vary on a local scale as a consequence of the planetary weather systems. The atmosphere of the Earth receives its energy from the Sun. The atmosphere reflects 30 per cent of the incident radiation back to space and absorbs the remainder, which becomes redistributed by the weather patterns so that 50 per cent of the incident radiation actually reaches the surface. The heated atmosphere re-radiates some of the long-wave radiation to the surface while a portion is ultimately lost to space. The atmosphere behaves like a giant heat engine, trying to balance the absorbed solar energy with the long-wave radiation emitted to space: the small imbalances give rise to the changes in climate. In the Earth's atmosphere, clouds, which cover 50 per cent of the surface, are a key factor in weather and climate since a small change in their geographical and vertical distribution will have a profound effect on the environment.

Exploration of the Solar System has now provided a unique opportunity to examine the weather systems of all the planetary atmospheres. They have very different properties, ranging from Venus (slowly rotating and totally covered with cloud) to Mars (a thin atmosphere strongly affected by the local topography and annual global dust storms) and the outer planets, Jupiter, Saturn, Uranus and Neptune, with their huge rapidly rotating gaseous envelopes and additional internal heating effects. These planetary atmospheres of the Solar System are natural laboratories for investigating geophysical fluid dynamics. *GEH. See also* entries on individual planets.

Atomic structure The structure of an atom is extremely difficult to visualize because the dimensions of the component parts are so small that they begin to lose the properties of solid particles and behave more as if they are groups of waves. It also becomes impossible to specify energy, or even position, with unambiguous precision.

For best results we must use "models" to represent some aspects of the behaviour of atoms, but we need different models for different situations and the pictures we construct are often apparently incompatible. It should be remembered that all atomic theories are analogies and that they should not be interpreted too rigidly.

One effective model suggests that all atoms contain small heavy nuclei at their centres that are made of combinations of two fundamental particles — protons and neutrons. A proton has a mass of 1.6×10^{-24}gm and carries one unit of positive electrical charge, whereas a neutron of about the same mass carries no electrical charge.

The rest of the atom is mainly empty space, but to maintain electrical neutrality there are electrons in several spherical shells around the nucleus. Each one of these carries a unit of negative electrical charge and has a mass of 9.108×10^{-28}gm, which is about $\frac{1}{1836}$ of the mass of a proton.

The chemical nature of a nucleus, or what element it represents, is determined by the number of protons it has, but its chemical properties and reactivity are determined largely by the cloud of electrons surrounding it.

Altogether we know of a full range of 106 different elements and each one can be placed in an ordered list according to the number of protons it contains. The order number, which must, of course, be an integer, is known as the atomic number and the full table forms the basis of the system of classification of the elements (*see* PERIODIC TABLE). There are no gaps in the table although there may be more elements to be added at the end if they can be discovered.

The simplest atom, hydrogen, has one proton and one electron and it has the atomic number 1. This, however, is not the only form of hydrogen that can exist because it is possible for the single proton nucleus to be combined with one or two neutrons. Under such circumstances the nucleus makes a heavier variety of hydrogen which still has the atomic number 1. Because it occupies the same position in the table as normal hydrogen it is called an isotope but it has a different mass number, that is the total number of nucleons in its nucleus. In this case it would be two or three.

The second natural element in the table is helium, which has two protons and two neutrons in its nucleus and two orbiting electrons. It is given the description $_2He^4$, the subscript being its atomic number and the superscript being the mass number. Helium also has an isotope, $_2He^3$, but this is very rare and is formed transiently in nuclear reactions. Towards the heavier end of the table we find, for example, the common form of uranium, $_{92}U^{238}$, having an atomic number of 92 and 146 neutrons.

It is the outer part of an atom that determines its chemical activity. The distribution of the electrons around nuclei is complex but can be described by a set of rules first proposed by Niels Bohr in 1913.

Bohr's model views the electrons as orbiting around nuclei rather like the planets in a miniature solar system, but they are constrained to follow certain rules of behaviour that are in direct conflict with our classical mechanical experience. The first of Bohr's rules is that electrons may move only in certain permissible orbits, that is, not all orbits are allowed for the electrons. Each different nucleus or element has its own characteristic set of discrete possible orbits or energy levels.

Classical theory predicts that an orbiting electrical charge should radiate energy continuously but Bohr's theory denies this and postulates that as long as the electron revolves in one of its allowed orbits, then it will be stable and will not lose energy. It was further pointed out that orbiting electrons are still able to radiate or absorb energy provided that they do so by jumping from one allowed orbit to another. If each orbit is regarded as a particular energy level then an electron can jump from a level E_1 to a lower level E_2. The energy change is $E_1 - E_2 = hc/\lambda$. Here h is PLANCK'S CONSTANT, c is the velocity of light and λ the wavelength of a single photon of radiation emitted during the change. The reverse is also possible and an electron can absorb energy to become excited into a higher level. The energy required is given by the same expression and is a single photon of radiation having the necessary wavelength. These processes are the fundamental properties required in the production of EMISSION and ABSORPTION spectra.

It should be noted that the allowed orbits further from the nucleus have higher energies than those nearer to it and the nearest available orbit to the nucleus is the lowest possible energy level referred to as the "ground state".

This picture of an energy "jump" between orbits is somewhat misleading since the model insists that the electron can be found in one level or the other but never anywhere in between.

Very detailed spectroscopic observations reveal that many spectral lines are in fact multiple. The D line of sodium, for example, is an easily resolved doublet, but the simple Bohr theory is unable to explain this "fine" and other observed "hyperfine" line structure. Several improvements were necessary, which led to the introduction of a set of quantum numbers.

The principal quantum number "n" was already in use, being the number used to specify allowed orbits in order outwards from the nucleus. It was realized that the Solar System analogy could be taken a step further and that some electrons move in elliptical rather than circular orbits having differing eccentricities. Such orbits would have differing energies. The speeds of electrons would vary in their elliptical tracks and this in turn would cause the orbits to precess at certain frequencies. The need for a further quantum number was apparent. This number is usually designated "l" and is called the azimuthal quantum number. It serves to specify the shape of orbit involved; it must be an integer and the rule states that it can take all values up to $n-1$.

Another problem arose. If some electron orbits are elongated, then they must be orientated in particular directions because of the magnetic moment that must exist. The effect shows when atoms are placed in a magnetic field and the observed splitting of certain spectral lines is known as the Zeeman effect. A magnetic quantum number "m" was introduced to allow for the effect and its value is restricted to integers ranging from $-l$ to $+l$.

A final quantum number "s", specifying the two directions of axial spin of an electron, completes the set and allows the configuration of electron orbits around various nuclei to be described very accurately. The Pauli Exclusion Principle points out that no two electrons can have the same set of four quantum numbers and it can be seen that the electron energy level structures of all atoms can be worked out from these rules. For each value of the principal number "n" there are $2n^2$ sublevels of energy. This means that the shell closest to the nucleus may contain up to two electrons, the second shell can contain up to eight electrons, the third 18 and so on.

More detailed study of these shells shows how the chemical reactivity of various atoms can be worked out. *RCM.*

Atomic time *See* INTERNATIONAL ATOMIC TIME.

ATS Applications Technology Satellites, designed for developing applications such as weather, navigation, communications, etc. Used also for relaying signals from other satellites back to Earth.

AU *See* ASTRONOMICAL UNIT.

Auriga (the Charioteer). A large and important northern constellation; mythologically it represents Erechthonius, son of Vulcan, who became King of Athens and invented the four-horse chariot

Brightest stars

Name	Visual Mag.	Abs. Mag.	Spec.	Distance (light-yrs)
α (Capella)	0.08	0.3	G8	42
β (Menkarlina)	1.90	0.6	A2	72
θ	2.62	0.1	A0p	85
ι (Hassaleh)	2.69	−2.3	K3	267
ε	2.99v	−8.5	F0	4,560
η	3.17	−1.7	B3	199

Then follow Delta (3.72) and Zeta or Sadatoni (3.75, variable). The star Al Nath, formerly included in Auriga as Gamma Aurigæ, is now included in Taurus, as Beta Tauri.

The main pattern of Auriga is easy to make out; the brightest star, Capella, is almost overhead in northern latitudes in winter evenings. The triangle made up of Epsilon, Eta and Zeta is often called the Hædi or Kids; both Epsilon and Zeta are remarkable eclipsing binaries. It is pure coincidence that they lie close together in the sky, because they are in no way associated. Zeta is much the closer and less luminous of the two. The Milky Way flows through Auriga, and the whole constellation is very rich; there are several prominent open clusters, including M36, 37 and 38. *PM*.

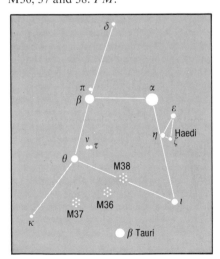

Aūriga: forms the shape of a kite

Aurora An illumination of the night sky known popularly as the northern and southern lights. When electrons and protons from space collide with atoms and molecules of the upper atmosphere, some particle energy converts

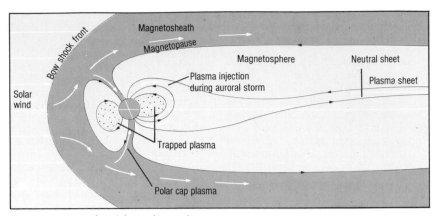

Aurora: magnetosphere/plasma interaction

to radiation in the manner of a neon sign. The resulting auroral light appears as diffuse glows, discrete arcs, bands, rays and veils or a mixture thereof. Between 50 to 150 miles (80 to 240km) above the Earth's surface the colour is mostly green, but may be white, yellow or blue. Below 60 to 50 miles (95km down to 80km) the lower border of a display may have a red fringe. From 150 to 370 miles (240 to 600km) above the Earth the aurora's hue is red.

The polar aurora comprises two oval haloes around the magnetic poles. The day side is about 20° of latitude from the magnetic pole. On the night side it is some 30°. The Earth turns under the ovals. The auroral zones are the regions where the aurora is most commonly seen — the areas swept by the night sides of the ovals.

The aurora normally remains within the oval. The steady state is upset by arrival of clouds of high-speed solar wind particles and their associated magnetic fields. The aurora may be driven polewards and equatorwards, to be seen in mid-latitudes. The transient aurora is linked with solar flares and appears first as a glow on the horizon in the direction of the magnetic pole. A homogenous arc then develops and may duplicate itself or twist into bands. Rays resembling searchlight beams rise from the arc and if the aurora forms above the observer then the rays converge in a pattern called a corona, the centre lying at the magnetic zenith. The storm may then dissolve into nebulous patches of light. During the rayed phase, and later, there may be movement of forms, flaming and flickering effects. Rays may be seen to travel eastward or westward. After the storm has died down it may recur an hour or two later. Sometimes observers may see only the upper part of a storm beyond their horizon.

The recurrent aurora derives its source from the high-speed particle streams emanating from persistent sources on the Sun such as coronal holes. As the Sun rotates, the flow of particles re-encounters the Earth at about 27-day interval and produces auroral activity. This activity is quieter than with the transient events, it does not expand so far south in latitude, and it may occur for several nights in succession.

Ions form at the base of auroral forms. In the E layer at a height of (70 miles) 112km VHF signals are scattered to increase the range between two stations from 50 to 1,000 miles (80 up to 1600km). Two stations at similar geomagnetic latitudes can talk to each other in Morse by pointing their aerials towards the pole, using the aurora like a mirror. Scattered Morse signals take on a low-pitched rasping sound. HF radio signals can be absorbed by the ionization and communications cut. The ion clouds enable the aurora to be studied by research radar and drift velocities determined by twin directional Doppler speed measures.

The particles causing the aurora follow the interplanetary magnetic field lines which connect either with the Earth's field lines in polar regions or with the magnetotail, when the particles become trapped in the plasma sheet. In an auroral storm magnetic energy is converted into kinetic energy and particles in the plasma sphere are injected into the atmosphere to form the aurora. Minor disturbances of the Earth's magnetosphere can cause particles to dribble downward and cause quiet auroral arcs. Discrete auroræ are thought to mark the boundary between the Earth's field lines connecting directly with the interplanetary magnetic field and the rest of the Earth, which remains a closed magnetic circuit. The diffuse aurora of mid-latitudes, which is red, forms a faint diffuse band at about latitude 40° within the closed field. It is not readily visible.

Other transient aurorae relate to varying sources of solar wind not as violent as flare sources. The frequency of the flare aurorae rises and falls with the sunspot cycle. The frequency of coronal hole aurorae peaks in the last years of the sunspot cycle.

Auroral light contains only distinct EMISSION LINES, brightest at the red end of the spectrum. It is the variable sensitivity of the human eye with colour that causes the green emission line of oxygen to appear the brightest.

A

Principal emission lines

Wavelength (Ångstroms)	Relative Light strength	Origin
3914	47.4	Molecular N
4278	24.4	Molecular N
5577	100.0	Atomic O
6300	10 to 600.0	Atomic O
6364	3 to 200.0	Atomic N
6611	–	Atomic N
6768	–	Atomic N

As auroral particles move in a magnetic field, electrical currents are caused to flow above the Earth, thereby inducing temporary fluctuations in the Earth's magnetic field. These are detected by magnetometers on the ground. These field changes induce electrical currents on the surface of the Earth, which can affect power transmission and telephone lines.

References to the aurora are found in pre-Christian, Greek, Chinese, Japanese and Korean texts. It is described in medieval church records and Viking chronicles. In 1621 Gassendi first used the name Aurora Borealis, the northern dawn, the auroral glow resembling twilight. HALLEY observed the great aurora in 1716 at London. In 1733 Mairan sensed the probable solar origin of the aurora. The Aurora Australis, or southern dawn, was first recorded by Captain Cook in 1773. Celsius and Hjørter measured magnetic perturbations coincident with auroral light. Cavendish tried to make height measurements in 1784 while Biot's polarimetry at Shetland in 1817 showed the aurora not to be reflected sunlight. Fritz published his catalogue of aurorae observed between 503BC and AD1872 in 1873 followed by a map of contours that joined places on the Earth where the aurora was visible with equal frequency.

In 1901 Birkeland studied the paths of auroral particles in a vacuum tank,

which STØRMER analysed mathematically in 1901. In 1910 Størmer began exposing and analysing thousands of pairs of photographs taken simultaneously from two observatories to determine the heights and positions of auroral forms. Since then many researchers such as Harang, Vegard, Feldstein, Akasofu, Eather and Legrand have contributed much to auroral theory.

Space vehicles now monitor the aurora from above the atmosphere, mapping the solar wind and the Earth's and interplanetary magnetic fields. They discovered the plasma belts and the theta polarcap aurora, so-called after the Greek letter theta (θ) by the resulting shape of the oval. Aircraft have monitored particle precipitation and have shown that mirror image aurorae form at conjugate points on the Earth's magnetic field lines. *RJL.*

Australia Telescope A large aperture-synthesis radio telescope being constructed in new South Wales. It will be the most powerful radio instrument in the southern hemisphere.

Auwers, Georg Friedrich Julius Arthur (1838-1915). German astronomer; specialist in fundamental astronomy.

Auzout, Adrien (1622–91). French astronomer; independent inventor of the micrometer. He was concerned in the founding of the PARIS OBSERVATORY.

Azimuth One coordinate of a body in the horizontal (or horizon) coordinate system. The azimuth of an object in the sky is its angular distance measured eastwards along the horizon from the north point to the intersection of the horizon with a vertical circle (meridian) running through the object.

Baade, Walter (1893–1960). German-born astronomer who worked mostly in the United States. Observing from Mount Wilson during the Los Angeles blackout in 1943, he was able to resolve for the first time a wide range of stars in the ANDROMEDA GALAXY. From their colour-magnitude diagrams he divided them into the bluish POPULATION I

stars of the disk, and the reddish POPULATION II stars of the nucleus and halo. These populations reflect different evolutionary phases of the GALAXY. After the Hale telescope came into use, he applied this new under-

Walter Baade: star populations

standing to improve the use of CEPHEID VARIABLES as distance indicators. With Minkowski, he was responsible also for the identification of the optical counterpart to the CYGNUS-A radio-galaxy. *APW.*

Baade's Window A region in one of the most crowded parts of the Milky Way where we have a relatively unobscured view towards the stars that comprise the bulge of our Galaxy. The region is

Baade's Window: a view of ancient stars

named after the German-American astronomer who distinguished between the hot, young population of stars found in the arms of spiral galaxies and the older, cooler stars which make up the nucleus.

Babcock, Harold Delos (1882–1968). American astronomer who was on the staff of Mount Wilson Observatory from 1908 to 1948. It was in 1908 that G.E.Hale announced the discovery of the ZEEMAN EFFECT in the spectrum of sunspots, postulating the existence of strong magnetic fields in sunspots, and in 1913 he detected the existence of an extremely weak general magnetic field in the Sun. The problem was how to detect with accuracy such weak magnetic fields.

Babcock became interested in the question of solar magnetic fields and in particular in the design of an instrument which could detect and record weak magnetic fields.

By 1948 he was able to confirm the existence of a weak general magnetic field on the Sun and also that this field

Harold Babcock: magnetic fields

showed variations with time. At the same time, in collaboration with his son Horace, he developed the magnetograph, which made the recording and measurement of magnetic fields possible. In 1958 Babcock and his son were able to establish the presence of magnetic fields also in several stars. Unlike the Sun, however, these fields were very strong, of the order of several thousand gauss. The discovery of the existence of these magnetic fields in stars and their variability opened a new field of research in astrophysics. *VB*

Background radiation *See* MICROWAVE BACKGROUND.

Backscatter The reflection of light back toward the direction of the light source. Light may be scattered from its direction of travel by fine particles of matter. For particles significantly larger than the wavelength of the incident light, reflection or backscattering of the light occurs.

Baikonur Also known as Tyuratam, this is the name of the important Russian space centre which is situated in a semi-arid desert, north-east of the Aral Sea in the Kazakh SSR.

Baily Vast lunar enclosure, described as a "field of ruins"; 185 miles (298km) in diameter; 50°N, 31°E.

Baily, Francis (1774–1844). English amateur astronomer, who was the first to describe the phenomenon during annular and total solar eclipses now known as BAILY'S BEADS.

Baily's Beads Francis Baily, an English amateur astronomer observed the annular eclipse of May 15 1836, in Scotland, and described "a row of lucid points, like a string of bright beads" around the Moon's limb. They are caused by the mountains of the Moon breaking up the very thin ring of the Sun, leaving isolated rays of sunlight shining through the lunar valleys. They can also be seen at the second and third contacts of a total solar eclipse.

Baker-Schmidt camera A fast meteor camera, typically f0.67, employing two spherical mirrors and a correcting plate. Largely achromatic, aplanatic, anastigmatic, and distortion-free with a wide, flat field.

Ball, Sir Robert Stawell (1840–1913). Irish astronomer (Astronomer Royal of Ireland from 1874) and author of many popular books.

Balloon astronomy Excluding ballooning for geophysics and COSMIC RAY physics, the first noticeable use of balloons for astronomical research was the memorable but tragic high-altitude flight of the balloon "Le Zenith" in France, on April 15 1875. Astronomer Jules JANSSEN proposed to search for the presence of water vapour in the SOLAR PHOTOSPHERE by analyzing, with a small SPECTROSCOPE, the decrease of the H_2O spectral bands with altitude. Due to a failure in the newly-invented-oxygen breathing system, two of the aeronauts, Sivel and Croce-Spinelli, lost their lives at an altitude of 26,000ft (8,000m); the third, Gaston Tissandier, survived.

Twelve years later the Russian chemist Mendeleiev flew alone in a balloon above the clouds over Tver near Moscow to observe the 1887 SOLAR ECLIPSE. In 1893 and 1899, the Société Française de Navigation Aérienne, again under the leadership of Jules Janssen, organized night flights above the clouds for visual observation of the Leonids shooting stars shower. Russian astronomer Hansky flew on November 13-14 1898 from Paris. Tikhov, another Russian astronomer

working in France, flew on November 15-16 1899. The following night a young astronomer from the PARIS OBSERVATORY, Dorothea Klumpke, also made a flight.

The first use of an astronomical telescope on a balloon was by Audouin Dollfus on May 30 1954. The purpose was to search for water vapour in the atmosphere of Mars. The 12in (30cm) telescope, operated manually from the open gondola, fed a wide-field birefringent filter selecting the H_2O band at 0.825μ m, then a spectral modulator and two photo-multipliers. Launched from MEUDON OBSERVATORY at night, Audouin Dollfus and his father Charles reached an altitude of 22,900ft (7,000m). This pioneering flight was the precurser of the 1959 stratospheric flight discussed below. Next, with the same balloon, Audouin Dollfus and the British astronomer Donald Black-

B

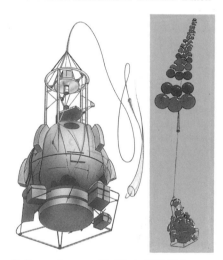

Balloon astronomy: Dollfus's 1959 flight

well took the first astronomical photographs from above the surface of the Earth. To discriminate between convection and turbulence acting in the solar photosphere, they analyzed the SOLAR GRANULATIONS with more sharpness than yet achieved from the ground, using a 12in (30cm) diameter telescope at an altitude of 19,600ft (6,000m). Images taken during the flights of November 22 1956 and April 1957 proved that convection was the force shaping the solar granules.

A few months later, the Princeton astronomer Martin Schwarzschild pioneered unmanned telescopic observation from balloon. The automatic Sun-pointing telescope was designated "Stratoscope". It was carried at an altitude of 79,000ft (24,000m) by a large plastic balloon. The flight of August 11 1959 recorded still sharper images of

B

the granulation of the photosphere.

Meanwhile, to improve the sensitivity of his method for the detection of water vapour on planets, Audouin Dollfus designed a new Cassegrain telescope, 20in (50cm) in diameter, selecting the strong H_2O band at 1.4μm and using a PbS detector. The instrument was combined with a sealed spherical cabin, 6ft (1.8m) in diameter, and the air was climatized to permit long duration flights at stratospheric altitude. This small observatory was lifted up to an altitude of 46,000ft (14,000m) by a train of 104 extensible rubber balloons, assembled in 34 clusters along a nylon rope of 1,476ft (450m) length. During the ascent, the balloons increased in diameter and, near ceiling altitude, successive bursts reduced the lift until stabilization was reached. The solo flight of Dollfus on April 22-23 1959, complemented by observations with the same telescope at Jungfraujoch, revealed the first signs of water vapour in the atmosphere of planets Mars and Venus (1963).

Another attempt to estimate the water vapour content of the atmosphere of Venus followed in the United States when the Johns Hopkins University astrophysicist John Strong organized with the United States Navy the project designated "Stratolab", with the assistance of Otto Winzen. The telescopic image of the planet was servo-controlled to remain on the slit of a spectrograph, despite the swings of the gondola. The flight occurred on November 28 1959, with Commander Malcolm Ross and C.B.Moore on board. The outcome determined the next step toward an unmanned flight, with an automatic pointing system for a telescope, 16in (41cm) in diameter, combined with an infrared spectrograph. The project was renamed "Balast". The spectra recorded during the flights of February 21 and October 28 1964 indicated ice crystals as a component of the Venus clouds.

All the flights in the United States used single large plastic balloons, a technique then newly developed at the University of Minnesota. In 1935 and 1938 E.Regener in Stuttgart, and the Belgian physicist Max Cosyns, had experimented with open-necked balloons made of cellophane. After World War II, Jean Piccard, James Ryan and Otto Winzen initiated the use of plastic material. Then Winzen collaborated with the United States Navy on project "Skyhook", which used large-capacity plastic balloons for high-altitude flights. The first success-ful Skyhook flight carried 70lb (32kg) of instrumentation at 98,000ft (30,000m) on September 1947. The plastic balloons used for astronomical observations have typically a capacity of 300,000m³, and are able to carry 5 tonnes at 92,000ft (28,000m). However, the safety for the Skyhook-type balloons having been considered marginal for manned flights, all the subsequent flights for astronomy were conducted unmanned.

Major balloon astronomy projects so far include: photopolarimetry of stars and planets in ultraviolet by Tom Gehrels at University of Arizona (project Polariscope); long-term ultra-violet photometric survey of stars organized by the Observatoire de Genève, the French Space Agency (CNES) and the United States organization NCAR; the detection and study of the outer SOLAR CORONA simultaneously by G. Newkirk at the University of Colorado (project Coronascope) and Audouin Dollfus with CNES in France (project Astrolabe, 1967); and the accurate spectroscopy of the solar PHOTO-SPHERE by the German K.O.Kiepenheuer (project Spectrostratoscope).

The largest balloon telescope in the United States was a 36in (91cm) Cassegrain instrument on project "Stratoscope II", by M.Schwarzschild and R.E.Danielson. The first flights occurred in 1963 on March 1 and November 26. Major results were infrared spectra of Jupiter, an estimate of the CO_2 content in the atmosphere of Mars, and sharp images of the planet Uranus. In the Soviet Union, the Leningrad astrophysicist V.Krat also directed a telescopic observation project, with large telescopes carried by plastic balloons, which included solar surface imaging and infrared spectroscopy. *ACD.*

Balmer lines EMISSION or ABSORPTION lines in the spectrum of hydrogen resulting from electron transitions down to or up from the second energy level of that atom.

Barium stars G and K giant stars with excesses of barium and other heavy elements in their atmospheres. *See also* K STARS.

Barlow lens Ancillary lens which increases the magnification of a telescope by typically 2× or 3×.

Barnard, Edward Emerson (1857–1923). A great observational astronomer who discovered numerous comets and Jupiter's moon AMALTHEA. BARNARD'S STAR is named after him and he established the nature of DARK NEBULÆ.

Barnard Regio Plain on Jupiter's satellite GANYMEDE; 22°N, 10°W.

Barnard's Loop A large, diffuse and very faint arc of nebulosity centred on the

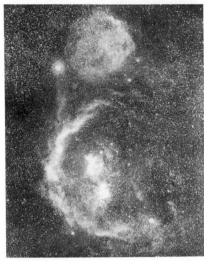

Barnard's Loop: cosmic bubble in Orion

Orion OB association, the group of hot stars south of Orion's Belt. The nebula is visible only on the eastern and southern sides of the constellation. The loop was named after the pioneer American astronomical photographer who discovered it in 1894 and if it were complete it would almost completely fill the region between Betelgeux and Rigel, covering almost 20° of sky. Its semi circular form suggests that it is a cosmic bubble of material blown in the interstellar medium by the burst of star formation which gave rise to the groups of hot stars in Orion's Sword.

Barnard's Star The star with the largest PROPER MOTION known. It was discovered by the American astronomer E. E. BARNARD in 1916, and crosses the sky so fast that it moves the Moon's apparent diameter relative to background stars in only two centuries. The rapid motion is caused in large part by the star's proximity, six light-years, yet the star is too faint to be seen without optical aid. It is intrinsically one of the faintest stars known, a RED DWARF, only one-hundredth as luminous as the Sun. Barnard's Star is the fourth nearest to the Sun. *DAA.*

Barred spirals Spiral galaxies in which the arms extend from a "bar" through the main plane of the system.

Barred spiral: an AAT picture of NGC1365

Barwell meteorite The biggest observed British meteorite fall. An L-group CHRONDRITE shower, it fell near Barwell, Leicestershire, on December 24 1965. Over 100lb (45kg) of fragments were collected.

Barycentre The centre of mass of the Earth-Moon system.

Baryon A heavy nuclear particle, such as a proton or neutron, which is acted on by the strong interaction.

Bayer, Johann (1572–1625). German amateur astronomer, who in 1603 published his *Uranometria* in which he allotted Greek letters to the stars in each constellation; his system is still used.

Becklin-Neugebauer object A very young star in the ORION NEBULA. It was found in 1969 by Eric Becklin and Gerry Neugebauer of the California Institute

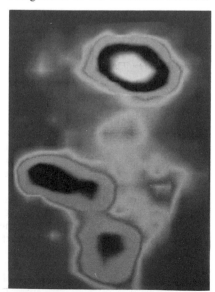

Becklin-Neugebauer object (top): in Orion

of Technology using the then new technique of INFRARED ASTRONOMY. The Becklin-Neugebauer object is invisible, and was originally thought to be a PROTOSTAR, but subsequent research has shown it to be a B STAR with a surface temperature of about 20,000K, which formed a few tens of thousands of years ago. It is hidden from view by Orion's extensive dust clouds. *DAA.*

Beer, Wilhelm (1797–1850). German amateur astronomer, co-author with J.H.MÄDLER of an important map of the Moon.

Beethoven Crater on Mercury; diameter 388 miles (625km); 20°S, 124°W.

Beijing Observatory A comprehensive research institute of The Academy of Sciences, China, giving priority to astrophysics, and founded in 1958. It has six departments: stellar physics; galactic physics; solar physics; radio astronomy; astrometry and latitude; and three special groups for satellites, the history of Chinese astronomy, and photoelectric image formation.

Attached to the observatory are five observation stations. Shahe, situated at 40°06′N, 116°20′E, and 130ft (40m) above sea level, deals mainly with the Sun, time and artificial satellites. It is equipped with a photoelectric transit instrument, a photoelectric astrolabe, an optical astrolabe, a set of atomic clocks, a 24in (60cm) solar telescope, a photospheric-chromospheric telescope, a laser satellite ranging system and so on. Xinglong, 40°24′N, 117°30′E, 3,083ft (940m) above sea level and 93 miles (150km) NE of Beijing, concentrates mainly on astrophysical observations and offers more than 200 fine nights yearly. The principal instruments are a 24/36in (60/90cm) Schmidt telescope, a 16in (40cm) double refractor and a 24in (60cm) reflector. Miyun at 40°33′N, 116°45′E, is devoted chiefly to radio astronomy and will be equipped with an APERTURE SYNTHESIS radio telescope consisting of 28 paraboloid antennæ at 30ft (9m) diameter each on a line of 1,094yd (1km). Huairou, the solar observation base, is equipped with a solar magnetograph and a photospheric-chromospheric telescope. Tianjin Latitude Station, at 39°08′N, 117°03′E, has a 7in (180mm) zenith telescope and a vacuum photographic zenith tube.

The observatory regularly publishes *Solar and Geophysical Data* and also *Acta Astrophysica Sinica,* a journal issued by the Chinese Astronomical

Society. *DPB. See also* ANCIENT BEIJING OBSERVATORY.

Bellatrix The star Gamma Orionis; magnitude 1.64; type B2.

Bennett's Comet Found by amateur J.Bennett at Pretoria; moved north to become a bright naked-eye object with a 30° tail during spring 1970.

Bessel Largest crater on the Moon's Mare Serenitatis; diameter 12 miles (19km); 22°N, 18°E.

Bessel, Friedrich Wilhelm (1784–1846). German mathematician and astronomer, born Minden, Westphalia. While still a shipping clerk, he was recommended by OLBERS to SCHRÖTER

Friedrich Bessel: star positions

as assistant at Lilienthal. In 1809, he was appointed the director of a new observatory at Königsberg (now Kaliningrad).

His reductions of Bradley's observations of 1750–62, and his publication of over 63,000 star positions based on his own and Greenwich observations, are generally considered as the beginning of modern ASTROMETRY.

In 1838 he was the first to make a successful stellar parallax measurement, of 61 Cygni. In 1844 he predicted the dark companions of Sirius and Procyon.

Bessel systematized the mathematical functions that bear his name. *HDH.*

Be stars Stars showing a characteristic B-type spectrum with the addition of hydrogen emission lines.

Beta Centauri (Agena). Second brightest star in the southern constellation

CENTAURUS. Alpha and Beta Centauri together form the Southern Pointers, which indicate the direction of the Southern Cross. Beta Centauri is a B1 giant with apparent magnitude 0.61, which, at 130 parsecs, is ten times further away than the nearest bright star, Alpha Centauri. It is actually the brightest in a multiple system: one member is visible in modest-sized telescopes as a 3.8 magnitude companion 1.′′3 away. Two other members have been detected spectroscopically and orbit with periods of 0.157 days and 352 days respectively. *BW*.

Beta Cephei variables Short-period pulsators. Light and radial velocity periods 0^d1 to 0^d6, amplitudes 0^m01 to 0^m3V. Also known as Beta Canis Majoris stars.

Beta Crucis The second star of the Southern Cross; magnitude 1.25, type B0.

Beta Hydri A 2.8-magnitude star; the nearest fairly bright star to the South Pole; declination $-77°\ 15'$.

Beta Lyræ variables A type of ECLIPSING BINARY in which gas escaping from a bloated primary star is falling onto an ACCRETION disk that surrounds the secondary star. Beta Lyræ, the third brightest star in the constellation of Lyra (the Harp), is the prototype.

The variability of Beta Lyræ was discovered in 1784 by John Goodricke, who followed up on a suggestion by William HERSCHEL that either Beta or Gamma Lyræ might be variable. The first light curve, obtained in 1859 by F.W.A.ARGELANDER from eye estimates, demonstrated that Beta Lyræ is an eclipsing binary. Spectrograms taken by A.A.Belopolsky in the 1890s showed that Beta Lyræ also is a single-line spectroscopic binary. This fortuitous combination of both eclipse and radial velocity data should normally be enough with which to compute many details of a double star system. Yet for many years Beta Lyræ remained one of the most enigmatic and controversial stars in the sky and in the 1970s was even proposed as a good candidate to contain a BLACK HOLE.

All Beta Lyræ variables show continuous magnitude variation between minima, resulting in a light curve that vaguely resembles the humps of a camel. In 1898, G.W.Meyers at the University of Illinois argued that the double-humped light curve of Beta Lyræ shows that the stars are deformed by their mutual gravitational interaction. The period of Beta Lyræ is only 12.9 days, and thus the two stars must be so close to each other that significant tidal distortion in their shapes seemed certain. Observed magnitude variation was then explained by the changing total amount of stellar surface area exposed to Earth-based view as the elongated stars revolve about each other. Assuming the stars to be ellipsoids, Meyers was able to reproduce Argelander's light curve.

This simple interpretation of Beta Lyræ survived until the 1950s, when improved observations revealed the system to be more complicated than originally believed. Spectroscopic observations had established that the primary star is a late-type B giant (B8.5II), but classification of the secondary was confounded by peculiar spectral features, some of which are caused by gas flowing between the stars and around the system as a whole. To study this system more thoroughly, the International Astronomical Union established the "1959 International Beta Lyræ Campaign", a month in late summer when extensive simultaneous photometric and spectroscopic observations of Beta Lyræ were made at observatories around the world.

One of the most important results of the campaign was the discovery that the secondary star in Beta Lyræ is roughly four to five times more massive than the primary star. According to the mass-luminosity relation, however, the more massive a star is, the brighter it is. Thus, astronomers were puzzled because the massive secondary is so much dimmer than the primary. An unknown process was causing the secondary to be significantly underluminous.

A major breakthrough in understanding this mysterious binary occurred in 1963 when Su-Shu Huang of Northwestern University published his interpretation that the secondary star in Beta Lyræ is enveloped in a rotating disk of gas captured from the primary. Material surrounding the secondary so severely subdues its luminosity that the star itself is impossible to observe.

Observations since the 1960s largely confirm Huang's model, although the accretion disk is believed to be much thicker than Huang had supposed. The primary star is overflowing its ROCHE LOBE and gases stream across the inner LAGRANGIAN POINT onto the disk at the rate of 10^{-5} solar masses per year. Ultraviolet spectra from OAO-3 (Copernicus) and Skylab revealed details of gas flow between and round the two stars. There are apparently clouds of gas at the Lagrangian triangular points L_4 and L_5. Some gas is constantly escaping from the system.

Astronomers now regard Beta Lyræ variables as members of a broad class of double stars called semidetached binaries. They are said to be "semidetached" because only one of the two stars fills its Roche lobe.

Closely related to Beta Lyræ variables are the ALGOL-type binaries, which are also semidetached systems. In Algol-type binaries, the detached component is a main sequence star and its less-massive companion is a red subgiant that fills its Roche lobe. In Beta Lyræ variables, the less-massive star also fills its Roche lobe but the detached component is shrouded in an accretion disk.

During the 1960s, astronomers laboured to understand semidetached

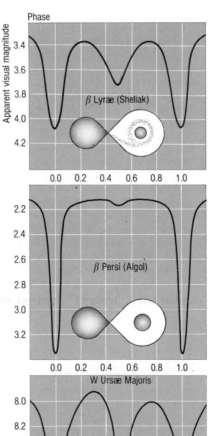

Phase

Apparent visual magnitude

β Lyræ (Sheliak)

β Persi (Algol)

W Ursæ Majoris

Beta Lyræ variables: three related systems

systems like Algol (Beta Persei). According to stellar evolution theory, the more massive a star is, the more rapidly it evolves. But in Algol, the more-massive primary is still on the main sequence while the less-massive secondary has evolved to become a red subgiant. The apparent contradiction that the less-massive star is also the more-evolved star was called the "Algol paradox".

The paradox was resolved by the work of Zdenek Kopal and others, who proposed that the red subgiant in Algol-type binaries was originally the more massive star. As it left the main sequence to become a red giant the star overflowed its Roche lobe, thereby dumping gas onto the originally less massive companion. Because of the resulting mass transfer, that companion is now the more massive star. David L.Crawford further argued that the initial phase of mass transfer is very rapid and slows down only after the originally more massive star has lost so much gas that it has become the less massive star.

It is possible that Beta Lyrae variables are binary stars near the end of the rapid phase of mass transfer. Thus some Beta Lyrae variables may evolve into Algol-type binaries. For example, as the primary star in a Beta Lyrae variable continues to evolve away from the main sequence, it will begin to resemble a red subgiant. As the rate of mass transfer subsides, the accretion disk shrouding the secondary may become transparent and a massive unevolved star will shine forth. The resulting system clearly would resemble Algol.

The fate of an Algol system depends on many factors, most notably on the stars' masses, which determine their rates of evolution. If the main sequence star in an Algol system is comparatively massive, it will evolve rapidly and expand to fill its Roche lobe while the companion star is still filling its own Roche lobe. The result is a "contact binary" in which both stars share the same photosphere. These binaries are often called W URSAE MAJORIS stars after the prototype of this class.

If the main sequence star in an Algol system evolves slowly, then its companion may become a white dwarf before the primary swells to fill its Roche lobe. When the primary finally does expand to become a red giant, gas flows across the inner Lagrangian point and goes into orbit about the white dwarf forming an accretion disk. These systems are called SS Cygni or U GEMINORIUM STARS or dwarf novae

because they exhibit rapid irregular flickering from the turbulent hot spot where the mass-transfer stream strikes the accretion disk. However, most SS Cygni stars are not eclipsing pairs (though a few of them are). It is important to note that eclipsing variables only appear to fluctuate in light because of the angle from which we are seeing them. If, for instance, we were observing Algol or Beta Lyrae "face-on", there would be no eclipses, and therefore no apparent variations.

Beta Lyrae variables, which account for about 20 per cent of all eclipsing variables, are an important but temporary stage in the early evolution of close binary systems. *WJK. See also* STELLAR EVOLUTION.

Beta Pictoris A5 dwarf, visual magnitude 3.9, distance 24pc. IRAS found a possible protoplanetary system around it (*see* VEGA), later supported by optical images of an edge-on disk.

Beta Regio One of the three main uplands on Venus (30°S, 75°W) discovered by the Pioneer Venus orbiter's radar. Satellite tracking shows that gravity is high over Beta Regio: the correlation between topography and gravity occurs over the other "continents", assumed to be blocks of less dense rocks, the product of early differentiation. Isostatic models, ie, in which the continent "floats" in the Venus mantle, cannot explain the magnitude of the gravity anomaly (80-100mgal) unless their "roots" extend down to more than 125 miles (200km). But the high surface temperature of Venus, 550°C, and the presumed similarity of geothermal gradient to the Earth, suggest a lithosphere thickness of only 6-18 miles (10–30km). The gravity anomalies probably arise deeper from downwelling (colder) convection currents, which control the position of the "continents". *SKR.*

Betelgeux (Alpha Orionis). The bright red star which forms the shoulder (top left as seen from the northern hemisphere) of Orion. It is a red supergiant star of spectral type M2 Iab at a distance of about 200 parsecs. Betelgeux is a variable; its V magnitude varies between +0.1 and +1.2.

Betelgeux is more luminous than most red variables. It also contrasts with other red variables in being a young star that has recently evolved from a bright blue star; its mass is about 20 times that of the Sun. Betelgeux is one of the few stars whose

angular diameter can be directly measured by speckle interferometry, and has a radius about 800 times that of the Sun. Speckle observations show it to have a faceted appearance, which probably results from convection in its deep atmosphere. *DJA.*

Bettinus Lunar crater, near Blancanus and Scheiner; it is aligned with several other large walled plains; 63°S, 45°W.

Biela's Comet Discovered by Montaigne at Limoges on March 8 1772 and then by Pons on November 10 1805. It was noted again by Wilhelm von Biela, an

Biela's Comet: split into two comets

Austrian army officer and amateur astronomer, on February 27 1826. He realized that it was the same as that which had already been observed in 1772 and 1805. It was subsequently named Biela's Comet and was shown to be periodic with a period of 6.6 years. It was seen again in 1832 and 1845-46. On December 19 1845 the comet appeared elongated and by the end of the year had split in two. By March 3 1846 the two parts were over 150,000 miles (240,000km) apart. At its next return in 1852, the separation had increased to $1\frac{1}{4}$ million miles (2 million km). The comet was never seen again, but in November 1872 and 1885 two tremendous showers of ANDROMEDID METEORS occurred on exactly the date when the Earth passed close to the comet's orbit — caused by the disintegration of the parent comet. *JWM.*

Big Bang theory *See* colour essay.

Big Dipper American nickname for URSA MAJOR (the Great Bear).

Billy Dark-floored lunar crater, 32 miles (51km) in diameter, south of Oceanus Procellarum; 14°S, 50°W; a pair with Hansteen.

B

B

Binary stars In 1802 Sir William HERS-CHEL defined a binary star as "a real double star — the union of two stars that are formed in one system by the laws of attraction".

Binary stars form three main groups in order of decreasing actual separation between the components. Visual binaries are the most common with 50,000 systems being known. SPECTROSCOPIC and ECLIPSING BINARIES are less common because they are more difficult to detect and observe.

The first visual binary star was discovered by RICCIOLI in 1650, when he observed Zeta Ursæ Majoris and saw that the star was actually double, with a separation of 14 arc seconds (this was distinct from the known companion Alcor: see MIZAR and ALCOR). Small numbers of binaries were accidentally found by telescopic observers in the course of the next century, but it was not until 1767 that a paper was presented to the Royal Society by the Rev. John Michell in which he argued that stars which appeared close together really were connected "under the influence of some general law". The observational proof of his theory did not appear until 1803 when Herschel produced his famous paper on the results of 25 years of double star observations. He had originally set out to determine the parallactic shift of the brighter (and, therefore, closer) member of an unequal pair of stars with respect to the fainter component. But he found that the relative motion which he measured, particularly in the case of CASTOR, could not be explained by parallactic shift. The motion of Castor B with respect to the primary star could be explained if the two stars were regarded as physically connected and rotating around a common centre of gravity. Later observations showed that the apparent motion of the companion described an arc of an ellipse.

In 1827 Savary was the first to calculate an orbit which would predict the motion of the companion star with respect to the primary. Another, and far more important, consequence was that the masses of the two stars were directly obtainable from elements of the true orbit. What is actually observed, the apparent orbit, is the projection of the true orbit onto the plane of the sky. The seven elements needed to define the size, shape and orientation of the true orbit are calculated from observations of position angle and separation. These are, respectively, the angle which the line between the stars makes with the north

point and the angular distance between the two stars in arc seconds and are usually made with a FILAR MICRO-METER. By plotting position angle and separation from measurements made at different times the apparent orbit can be drawn. The points are usually scattered since all such measurements depend on many factors — the closeness of the pair, the state of the atmosphere (or seeing), the size of the telescope and the skill of the observer being the most important ones. From the elements of the true orbit which follow, the size of the semi-major axis (a) and the period of revolution (P years) will allow the sum of the masses of the two stars to be calculated provided that the parallax of the binary (π arc seconds) is known. The relation connecting these quantities is:

$$M_1 + M_2 = \frac{a^3}{\pi^3 P^2}$$

where M_1 and M_2 are in terms of the Sun's mass.

In order to derive individual masses, however, we need another relation between M_1 and M_2. This can be obtained by plotting the position of each star against the background of fainter, relatively fixed stars in the same field. The two stars will both appear to describe ellipses, which vary only in size depending on the mass of each star, thus:

$$\frac{M_1}{M_2} = \frac{a_2}{a_1}$$

where a_1 and a_2 are the semi-major axes of the apparent ellipses described by the two stars. This process is a very time-consuming job, involving as it does hundreds of photographic plates taken over many years. Not surprisingly it has only been carried out for a few systems. A table is given below of

Masses of selected binaries

Name	Masses	Parallax (")	P (yrs)
η Cas	0.9,0.6	0.170	480
Castor	2.1,2.1	0.070	420
ξ Uma	1.1,0.9	0.130	60
γ Vir	1.2,1.2	0.090	171
α Cen	1.1,0.9	0.759	80
44 Boo	1.0,1.1	0.080	225
η CrB	1.1,1.0	0.059	41
70 Oph	0.8,0.6	0.500	88
61 Cyg	0.6,0.6	0.293	720
Sirius	2.3,0.9	0.375	50

the masses (given here in terms of the mass of the Sun) of ten well-known visual pairs.

A new category of double star, the spectroscopic binary, was discovered in 1889 when the spectral lines of Mizar

were shown to display doubling every 20.5 days. This is due to the DOPPLER shifts in the spectrum of each component as they rotate, with the star approaching the Earth displaying a

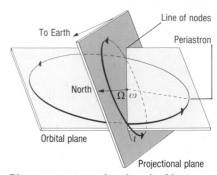

Binary stars: true and projected orbits

blue shift and the lines of the receding star being red-shifted. Equal intensity of the two spectra indicate that the two stars are of similar brightness. Most spectroscopic binaries, however, show only one set of spectral lines and are consequently termed single-lined. The determination of masses is more difficult than for the visual binaries because it is not possible to determine the inclination of the apparent orbit unless eclipses occur. The best that can be done is to assume that the inclination is 90° and to calculate from this the minimum values that the masses are likely to take. The following table lists minimum masses and periods for five bright spectroscopic binaries. Two spectral types indicate that the pair in question

Minimum masses and periods

Name	Masses	Spectra	P (days)
Capella	2.7, 2.5	G5,G0	104.0
δ Ori	20.5, 7.9	B1	5.73
Spica	7.5, 4.5	B2,B3	4.01
2 Lac	0.8, 0.6	B5,B6	2.62
ω Leo	1.3, 1.1	F5,A3	14.50

is a double-lined system. More than 2,000 systems are known with periods mostly between one and 100 days.

In photometric or eclipsing binaries duplicity is deduced from periodic variations in light due to mutual eclipses by stars whose orbital plane lies close to our line of sight. ALGOL was the first such binary discovered and today more than 4,000 are known. Most of the orbital periods lie between a few hours and ten days, shorter on the whole than the spectroscopic binaries because these are essentially closer pairs. In some cases, such as the stars known as the W. URSÆ MAJORIS binaries, the two components may even be in contact. If this happens, the strong mutual gravitational pull between the two components distorts their shapes until they become ellipsoidal. *RWA².*

Binoculars Twin telescopes, one for each eye, mounted on a single hinged frame giving binocular vision and some magnification. Interpupillary distance adjustment and independent eyepiece focusing are usual.

In the Galilean type each telescope comprises an objective lens followed by a strongly diverging eye lens. In the prism type the length is reduced and an erect image obtained by folding the light path through 360° by means of two prisms.

Binoculars are frequently designated by two numbers, eg, 7 × 50. The first number gives the linear magnification and the second the diameter of the object lens in millimetres.

Biot, Jean Baptiste (1774–1862). French physicist who first demonstrated the extraterrestrial origin of meteorites, after investigating the meteorite shower at L'Aigle on April 26 1803. He also accompanied Gay-Lussac on a balloon flight to collect scientific data about the Earth's upper atmosphere in 1804.

Bipolar flows Non-spherical flows of material from a star. Occasionally in the life of a star it loses matter at copious rates — at the end of protostellar life, and during the red giant phase. If these flows are organized into two oppositely-directed streams, they are called "bipolar". This exciting phenomenon is responsible for the creation of extremely rapid HERBIG-HARO OBJECTS, flying away from the precursors of the T Tauri stars. Very narrow, highly confined radio-emitting JETS characterize this youthful phase and these disturb the surrounding dark clouds and sweep up larger volumes of much slower molecular gas, also bipolar in pattern.

Usually the stars that generate bipolar flows are surrounded by extensive, flattened, dusty envelopes — huge toroids — orientated perpendicular to the star's rotation axis. The interplay of stellar rotation and mass outflow is believed to create bipolar outflows. Binarity of the driving star is implicated in at least some flows, and magnetic fields may be also. *MC.*

Birkeland Large crater on the Moon's far side; 30°S, 174°E. It is adjacent to the most interesting crater in the south-east region, Van de Graaff.

Birkhoff Large crater on the Moon's far side; 59°N, 148°W.

Birr Castle astronomy Birr Castle, in central Ireland, was for many years the site of the world's largest and most powerful telescope. It was a 72in (183cm) reflector and its maker, the third Earl of Rosse, used it to discover the spiral forms of the objects we now know to be galaxies.

The Earl of Rosse, originally known as Lord Oxmantown, was born in 1800. His interest in astronomy dated from early in his life, and he determined to make a telescope more powerful than any previously constructed. He built a 36in (91cm) reflector, and it worked well, so he decided upon something

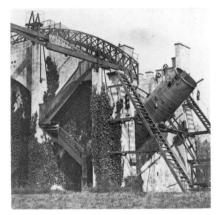

Birr Castle astronomy: 72in reflector

even larger. At that stage it was impossible to make glass mirrors of such size; Lord Rosse used speculum metal—an alloy of copper and tin—and even made his own forge to cast the mirror. Everything was done at his home at Birr Castle, and his only helpers were workmen from his estate whom he trained. During the process he also made many improvements in practical techniques.

Wisely, he decided against trying to make the telescope fully steerable. This would have been beyond the engineering of the time. Instead, the huge tube — it was 58ft (17.7m) long — was mounted between two massive stone walls, each 70ft (21m) long and 50ft (15m) high, and pivoted at the lower end, so that it could swing only for a very limited range to either side of the north-south line. One had to wait for the Earth's rotation to bring the required object into view.

By 1845 the telescope was ready, and Lord Rosse turned it toward the sky. He examined the objects then known as spiral nebulæ and found that many of them were spiral in form, like Cather-

ine-wheels. For many years only the so-called "Leviathan" was powerful enough to show the spirals, and Birr became an astronomical centre. There were various assistants, including Robert Stawell Ball (from 1865 to 1867), and Lord Rosse was himself a skilful observer. Moreover, he made his knowledge freely available to all.

For the decades following 1845 many discoveries were made with the great Birr telescope. When Lord Rosse died, in 1867, he was succeeded by his son, who became the fourth Earl, and was also an excellent astronomer, though he was concerned mainly with measuring the tiny quantity of heat sent to us by the Moon. (His results were much the most accurate of the time.) Between 1874 and 1878 J.L.E. Dreyer was assistant at Birr; it was during this period that he accumulated much of the information used later in his classic *New General Catalogue of Clusters and Nebulæ.*

Undoubtedly the most profitable period in the career of the 72in was in its early years. It could not be effectively driven, and could not be used for photography; its mirror, of metal, was by no means the equal of a glass mirror—though the observations show that it was still remarkably good. From 1900 activity gradually decreased; the fourth Earl died in 1908, and the 72in was dismantled. In 1916 the last Birr astronomer, Otto Boeddicker, departed, and work there came to an end.

However, in recent times the mounting of the 72in has been repaired, and so has the tube; there are plans for bringing the telescope back into operation, though admittedly for historical reasons only. In any case, there can be no doubt that the researches carried out at Birr between 1845 and 1900 were invaluable. Visitors to the castle can now see the tube of the telescope, the stone walls, a museum and an exhibition; it is a memorial to the two great astronomers, and the 72in should in the future be turned skyward once more. The story of Birr astronomy is unique. Nothing of the sort has ever happened before, and nothing comparable can ever happen again. *PM.*

Black body radiation Radiation emitted by an idealized perfect radiator. It has a continuous spectrum which depends only on the temperature of the source. *See also* WEIN'S LAW.

Black drop Optical effect visible during the early and late stages of a transit of

B

B

Mercury or Venus; apparent lengthening of the planet towards the Sun's limb.

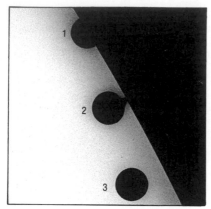

Black Drop: seen during transit of Venus

Black dwarfs The coldest stars. It has long been known that stars too cool to give out visible light might exist. These were generally known as black dwarfs. More recently the term BROWN DWARF has been used to describe the coolest stars on the MAIN SEQUENCE, and black dwarfs describes the ultimate state of WHITE DWARFS.

A white dwarf no longer has a mechanism to maintain its heat, and therefore cools steadily. Given long enough it cools to invisibility, and is then called a black dwarf. Such objects are hypothetical: none has been observed, and it is doubtful whether our Galaxy is old enough for any to have cooled sufficiently. *DAA*

Black holes One of the most exciting predictions of Einstein's general theory of RELATIVITY is the existence of black holes, where gravitational forces become so intense that they prevent the escape even of particles moving with the velocity of light.

There are two major classes of black hole: stellar, and primordial.

A stellar black hole is a region of space into which a star (or collection of stars or other bodies) has collapsed and from which no light, matter or signal of any kind can escape. This can happen after a sufficiently massive star whose mass exceeds both the Chandrasekhar and Oppenheimer-Volkoff limits (*see* COLLAPSARS) reaches the end of its thermonuclear life and collapses to a critical size, whereupon gravity overwhelms all other forces.

Stellar black holes may be either rotating, or non-rotating. In nature, it is expected that black holes do in fact rotate. Non-rotating black holes are called Schwarzschild black holes,

whereas rotating (and possibly charged) black holes form the Kerr-Newman variety.

The radius of a non-rotating Schwarzschild black hole may be computed by multiplying the mass M of the collapsing body by twice the universal constant of gravitation G, and dividing the result by the square of the velocity c of light in vacuum. In other words:

Radius of a non-rotating black hole

$$= \frac{2\,G\,M}{c^2}$$

When a star becomes smaller than this radius, gravity completely dominates all other forces: this radius determines the location of the surface of the black hole, called the event horizon. Only the region on and outside the event horizon is relevant to the external observer; events inside the horizon can never influence the exterior.

The Sun, for example, would have to collapse to a radius of 1.8 miles (2.95km) to form a black hole. The Earth would have to be squeezed to a radius of almost a third of an inch (0.89cm) if it were to form a black hole.

There is no lower limit to the radius of a black hole. Some of the primordial black holes (see below) could be truly microscopic.

When a stellar black hole first forms, its event horizon may have a grotesque

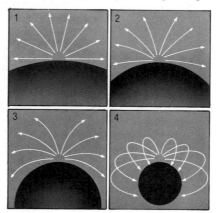

Black holes: stages in formation

shape and may be rapidly vibrating. Within a fraction of a second, however, the horizon settles down to a unique smooth shape. A Kerr-Newman black hole has an event horizon which is not circular, but rather is flattened at the poles (just as rotation flattens the Earth at its poles).

What happens to matter after it crosses the event horizon depends on whether or not the star is rotating. In the case of a collapsing but non-rotating star that is spherically symmetric, the matter is crushed to zero volume

and infinite density at the singularity, located at the centre of the hole. Infinitely strong gravitational forces deform and squeeze matter out of existence at the singularity, which is a region where physical theory breaks down.

In the case of a rotating Kerr black hole, however, the singularity need *not* be encountered. Rotating black holes have fascinating implications for space travel to other universes.

The density of matter in a star as it crosses the critical event horizon need not necessarily be very high at all: its density could even be less than that of water! This is because the density of any body is proportional to its mass divided by its radius cubed; the radius of a black hole is, as we have seen, proportional to its mass. These two facts combined imply that the density at which a black hole is formed is *inversely* proportional to the square of the mass.

Take a supermassive black hole with a mass ranging from 10,000 to 100 million solar masses—the mass of a black hole that might be found at the centres of certain active galaxies. Such a collapsing mass would reach the black hole stage when its average density was roughly that of water! If the mass of the collapsing sphere were that of an entire galaxy, the average density of matter crossing the event horizon would be less than that of air!

Attempts to discover stellar black holes must rely on the influence of their gravitational fields on nearby matter, and/or their influences on the propagation of radiation in the vicinity of the hole.

Black holes may appear as X-ray sources in BINARY STAR systems. In such a system, the black hole itself is of course optically invisible, but due to ACCRETION of gas from the companion star might emit X-rays, and can even be converted into a "glowing white body". When one searches the position of the X-ray sources, one is looking for a spectroscopic binary (that is, a star whose spectrum shows a Doppler shift indicating the presence of an invisible companion). Observational evidence is then needed to show that the unseen object is indeed compact, and not, for example, a bloated red star whose light is hidden by the brighter companion. Furthermore, the derived mass for the unseen component must be large enough for it not to have formed a WHITE DWARF or a NEUTRON STAR.

The most promising candidate where one companion might very well be a *(Continued on page 81)*

Big Bang

Martin Rees, FRS
Plumian Professor of Astronomy and Experimental Philosophy
University of Cambridge

Cosmology is a science that boasts few firm facts,
but each has great ramifications. The first key fact
emerged in 1929, when Edwin Hubble claimed that galaxies
recede from us with speeds proportional to their distance.
Later work has borne this out:
we seem to inhabit a homogeneous universe
where the distances between any widely separated pair
of galaxies stretch uniformly with time. Hubble's work

Studies of large samples of galaxies reveal that they are receding from us at a rate closely proportional to distance (in this figure, higher speeds are represented by longer arrows). The speeds are readily determined from spectra; the actual distances of galaxies, however, can be established only by a complex chain of observations and inferences. Consequently, although there is a well-established relationship between redshift and inferred distance (Hubble's law), the actual constant of proportionality, known as Hubble's constant, is uncertain by 50 per cent. The remotest observed galaxies recede at more than half the speed of light. The effects of relativity then become important, and there are further complications: because of the finite speed of light, such distant galaxies are being seen as they were in the remote past when the expansion rate might have been different from what it is today. The "age" of the universe, defined as the elapsed time since the big bang, is consequently uncertain by a factor of 2, being in the range 10–20 thousand million years.

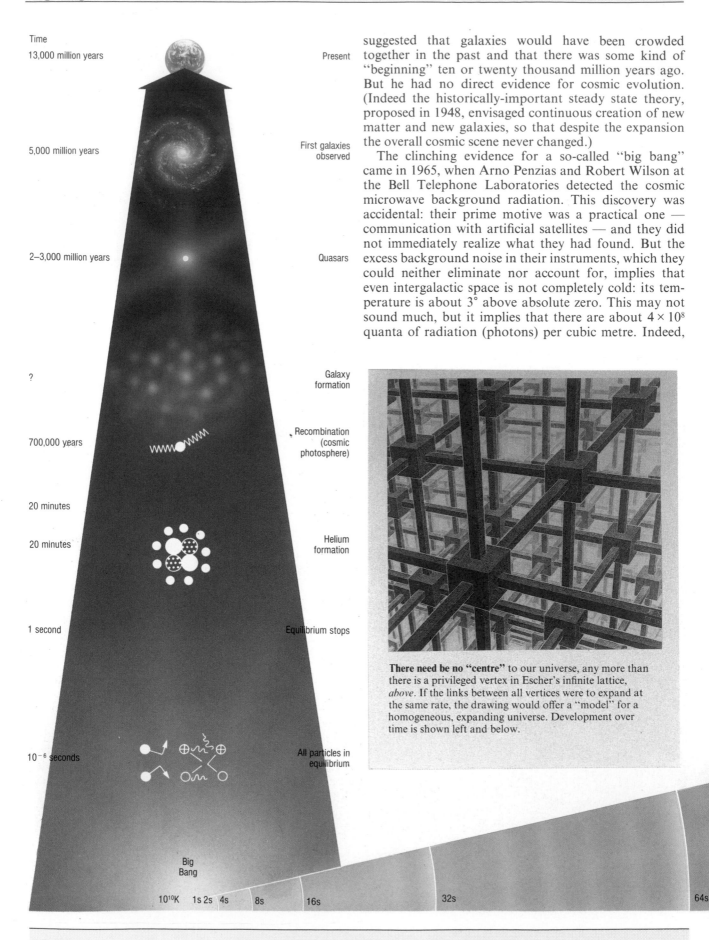

Time
13,000 million years

Present

5,000 million years

First galaxies
observed

2–3,000 million years

Quasars

?

Galaxy
formation

700,000 years

Recombination
(cosmic
photosphere)

20 minutes

20 minutes

Helium
formation

1 second

Equilibrium stops

10⁻⁶ seconds

All particles in
equilibrium

Big
Bang

10^{10}K   1s 2s 4s   8s   16s   32s   64s

suggested that galaxies would have been crowded together in the past and that there was some kind of "beginning" ten or twenty thousand million years ago. But he had no direct evidence for cosmic evolution. (Indeed the historically-important steady state theory, proposed in 1948, envisaged continuous creation of new matter and new galaxies, so that despite the expansion the overall cosmic scene never changed.)

The clinching evidence for a so-called "big bang" came in 1965, when Arno Penzias and Robert Wilson at the Bell Telephone Laboratories detected the cosmic microwave background radiation. This discovery was accidental: their prime motive was a practical one — communication with artificial satellites — and they did not immediately realize what they had found. But the excess background noise in their instruments, which they could neither eliminate nor account for, implies that even intergalactic space is not completely cold: its temperature is about 3° above absolute zero. This may not sound much, but it implies that there are about 4×10^8 quanta of radiation (photons) per cubic metre. Indeed,

There need be no "centre" to our universe, any more than there is a privileged vertex in Escher's infinite lattice, *above*. If the links between all vertices were to expand at the same rate, the drawing would offer a "model" for a homogeneous, expanding universe. Development over time is shown left and below.

there are about 10^9 photons for each atom in the universe.

There is no way of accounting for this radiation, and its spectrum and isotropy, except on the hypothesis that it is a relic of a phase when the entire universe was hot, dense and opaque. It seems that everything in the universe once constituted an exceedingly compressed and hot gas, hotter than the centres of stars. The intense radiation in this compressed fireball, though cooled and diluted by the subsequent expansion (the wavelengths being stretched and redshifted), would still be around — it fills the universe and has nowhere else to go!

The universal "thermal soup" from which galaxies evolved had a temperature exceeding ten thousand million degrees and initially expanded on a timescale of less than one second — that is, in about a second its temperature halved and any two parts of it doubled their separation. When the universe was only a few minutes old, its temperature would have been a thousand million degrees. Nuclear reactions would have occurred as it cooled through this temperature range. These can be cal-

culated — material emerging from the fireball would be about 75 per cent hydrogen and 25 per cent helium. This is specially gratifying, because the theory of synthesis of elements in stars and supernovae, which works so well for carbon, iron, etc., was always hard pressed to explain why there was so much helium, and why the helium was so uniform in its abundance. Attributing helium formation to the big bang thus solved a long-standing problem in nucleogenesis. It also bolstered cosmologists' confidence in extrapolating right back to the first few seconds of the universe's history, and assuming that the laws of microphysics were the same as now. (Without this firm link with local physics, cosmology risks degenerating into *ad hoc* explanations on the level of "just so" stories.)

More detailed calculations, combined with better observations of background radiation and of element abundances, have, over the last 20 years, strengthened the consensus among cosmologists that the hot big bang model is basically valid. Conditions were initially almost smooth and featureless, but not quite: there were fluctuations from place to place in the density or expansion

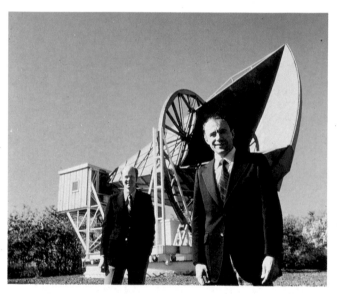

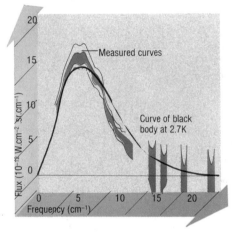

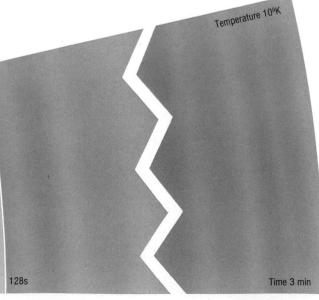

The microwave background intensity has now been measured at many wavelengths; it is well fitted by a thermal or "black body" spectrum, at a temperature 2.7 degrees above absolute zero, *above right*. Wilson and Penzias, *above*, made the first ground-based measurement at Holmdel, New Jersey, at about 7cm wavelength (4GHz). The temperature is intrinsically uniform across the sky: this is the best evidence that the overall cosmic expansion is isotropic, and that the universe is indeed highly homogeneous.

rate. Embryonic galaxies — slightly overdense regions whose expansion rate lagged behind the mean value — evolved into disjoint clouds whose internal expansion eventually halted. These protogalactic clouds collapsed to form galaxies when the universe was maybe 10 per cent of its present age.

The long-range forecast for our universe

The universe has been expanding for ten to twenty thousand million years. One may wonder, therefore, if the expansion will continue for ever. This depends on how much deceleration is occurring. In principle, this deceleration is directly measurable: the redshift of a distant galaxy tells us its speed when the light set out on its journey towards us; so in principle one can compare the expansion of the universe at early times with the present rate, and thereby infer how much it is decelerating from the way the redshift-distance relation deviates from linearity for remote objects. In practice, this type of work is bedevilled by observational difficulties and uncertain corrections, and can't yet give a reliable answer.

There is a wide spread in the properties of galaxies, just as there is for individual stars. One needs, therefore, to find some class of galaxies which serve as "standard candles", so that their distance can be inferred from their apparent brightness. Otherwise there will be a big scatter in the magnitude-redshift plot. The brightest galaxies in clusters are the best candidates, but even they display a 30 per cent scatter. Another problem is that galaxies become almost undetectably faint, even when observed with the largest telescopes, when they are far enough away (and receding so fast) that the effects of the deceleration should really show up.

But the worst problems stem from galactic evolution. The galaxies seen at large distances are systematically younger than nearby ones. Even if a certain class of galaxy provided precise standard candles at the present, one needs to know how each candle changes as it burns. The galaxies that are crucial for cosmological tests are those whose light has been journeying for about five thousand million years towards us, which are therefore being seen at roughly half their present age.

There are two aspects to the evolutionary correction. In a younger elliptical galaxy many stars would be shining which by now have died, and the present stars would be seen at an earlier stage in their evolution. This alters the brightness and the colour of the galaxy, the trend being that a younger galaxy would appear somewhat brighter than its present-day counterpart. But there is a secondary evolutionary correction stemming from the fact that a galaxy is not a self-contained isolated system. We can see many instances where galaxies seem to be colliding and merging with each other; and in rich clusters the large central galaxies may be cannibalizing their smaller neighbours. [In a few thousand million years this may, incidentally, happen in our own local group. The Andromeda galaxy is falling towards our own Milky Way, and there may be a collision between these two large disk galaxies, the likely remnant being a bloated amorphous "star pile" resembling an elliptical

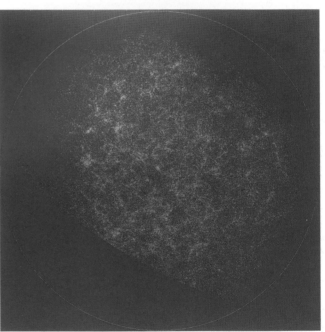

The brightest million galaxies in the northern sky, *above*, show an isotropic distribution, though many clusters are discernible. The clustering probably developed during the last 90 per cent of the universe's history. Regions with an above-average concentration of material would suffer more deceleration (due to the gravitational attraction of each galaxy towards its neighbours) and eventually evolve into conspicuous clusters. On the right is part of a "model" universe at three stages in its expansion. In this computer simulation (where the overall expansion has been scaled out) the emergence of clusters is clearly apparent.

Because of the finite speed of light, we observe remote parts of the expanding universe as they were at an early time when everything was more closely packed together. The "big bang" occurred 10–20 thousand million years ago. The universe as we actually see it resembles the Escher picture on the left — objects seem more and more crowded together towards our observational "horizon".

The building blocks of the universe are galaxies — basically self-gravitating assemblages of stars and gas. Extra power comes from the centres of some galaxies. The southern object Centaurus A, *below*, an intense radio source, is the nearest galaxy displaying conspicuous activity. It probably harbours a massive black hole at its centre, and may, earlier in its life, have rivalled the quasars in luminosity.

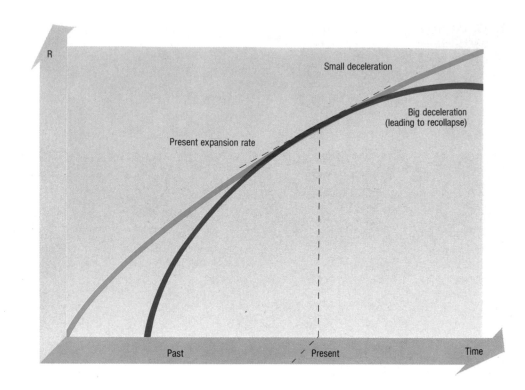

In an expanding universe the "scale factor" R, a measure of the distance between two typical galaxies, increases with time. If R is plotted against time (*t*), the slope of the resulting curve defines the expansion rate. The Hubble constant is, by definition, the expansion rate at the present time. However, the expansion rate in the past, relative to its present value, depends on the degree of deceleration, measured by the curvature of the R − *t* graph: if there were no deceleration, this would be a straight line. In the simplest models, deceleration depends on the mean density of material. For a given Hubble constant, the universe would be younger (ie, less time would have elapsed since R = 0) if the deceleration were large. If the mean density exceeds a "critical" value, the deceleration may be so severe that the expansion is eventually halted, and the universe collapses back to a "big crunch".

galaxy.] Many big galaxies, particularly those in the centres of clusters, may result from such mergers, traces of disturbance having by now been erased. This process would obviously result in big galaxies having been, on average, fainter in the past.

So we have two evolutionary corrections of opposite sign. The stars in a galaxy burnt more brightly in the past, but big galaxies may be gaining more stars as they age, by cannibalizing their neighbours. These corrections are both substantial — indeed as big as the effect we are looking for — and we must understand galactic evolution better before we can use them for probing the cosmic deceleration.

One might have hoped, and many astronomers initially did, that quasars would greatly aid this task. They are much more luminous than galaxies, and therefore detectable to vastly greater distances, thereby enabling us to probe much farther back into the past. But there is again the stumbling block that the evolution is not understood; indeed, everything about quasars is far more mysterious than ordinary stars and galaxies. Moreover, these evolutionary effects are *very* dramatic. A hypothetical astronomer observing the universe only two thousand million years after the big bang would have perceived a vastly more active and dramatic celestial environment. Whereas our nearest bright quasar, 3C273, is about two thousand million years distant, he would find a similar object fifty times closer, and appearing as bright as a fourth magnitude star. Galactic nuclei were much more prone to indulge in active outbursts when they were young, perhaps because there was then more uncondensed gas available to form and fuel central black holes.

The study of distant quasars, and attempts to detect

galaxies at even larger redshifts, are thus crucially important for the study of galactic evolution. All the problems are so interrelated that we can't use galaxies to probe the kinematics of the cosmos until we have a clearer perception of galactic evolution, and of what happens in active nuclei. Observations must be pursued on a broad front, in the hope that all issues will gradually clarify concurrently. This direct method can't yet tell us how much the universe is decelerating.

There is another line of attack on the "deceleration" problem. Everything in the universe exerts a gravitational pull on everything else, and it is this which slows down the universal expansion. One can calculate how much material would be needed to halt the expansion: it works out at about three atoms per cubic metre. The luminous gas and stars in all galaxies contribute no more than one per cent of the critical density.

But things are not as straightforward as this. There's another way of estimating the masses of galaxies; to look at pairs of galaxies orbiting around each other, or small groups which seem gravitationally bound. In such systems one can measure the orbital velocity. Then Newton's law tells us what mass is needed for gravity to balance the centrifugal force. The masses inferred in this way turn out, quite consistently, to be about ten times higher than those implied by the internal dynamics. This may mean that the luminous part of every galaxy is embedded in a much larger amount of diffuse dark material (a "halo"), and that clusters of galaxies contain a lot of dark material between the galaxies we see.

Supporting evidence comes from radio studies of edge-on, disk-like galaxies. Neutral hydrogen gas can be traced much farther out than the optical extent of the galaxy, and the rotation speed of this gas can be

Disk-like galaxies, of which our own Milky Way is an example, are held in equilibrium by a balance between gravity (which tends to pull everything towards the centre) and rotational motions (which, if gravity did not act, would cause the system to fly apart). In some galaxies viewed edge-on, such as NGC4565, *below*, a disk of tenuous gas can be detected which extends out to much larger radii than the visible stellar disk. The orbital speed of this outlying gas is surprisingly high — much higher than would be expected if it were "feeling" just the gravitational pull of the visible disk. Galaxies must therefore contain more material than we see; indeed, the entire visible content (stars and gas) of a galaxy like NGC4565 may be little more than a trace of "sediment" in a far larger and more massive dark halo, *left*. Dark matter" seems to be the main constituent of the universe; but its nature remains a mystery.

measured by radio astronomers from the 21cm line. The rotation of this gas indicates that it must be feeling an extra gravitating mass far less concentrated than the stars we see, and extending way beyond the visible image of the galaxy. The inference that 90 per cent of the gravitating stuff in the universe is in some unobserved and unknown form really shouldn't amaze us — there is no reason why everything in the universe should shine conspicuously, any more than it does on Earth. What we have already observed may be a small and atypical fraction of what actually exists.

What else could there be out there? The "unseen" mass could be in very faint low mass stars. Alternatively, it could be in the remnants of massive or supermassive stars, which were bright in early phases of galactic history but have now all died out.

These are the "conservative" options. But alternatively the hidden mass could be some relics from the big bang. Microwave radiation is left over from the fireball: 10^9 photons for every particle. But these photons have zero rest mass and do not contribute much to the present density of the universe. However, there are other particles, neutrinos, which would be produced in the big

bang about as prolifically as the photons. These used to be thought to have zero rest mass, like photons. But there is a possibility that they have a small mass. Because there are so many of them, even a mass 10^{-8} that of a proton would be very important. Attempts now being made to measure neutrino masses may however rule out even such a tiny value. But physicists have many other particles in reserve which could be important. There is real hope of using detailed observations to narrow down or decide among these options. It would be of special interest to particle physicists if astronomers could discover something fundamentally new about neutrinos, ghostly and illusive particles which hardly interact at all with ordinary matter, or about some still more bizarre particle. Any such particles, pervading intergalactic space (even the huge "voids" between clusters), could readily contribute the entire "critical" density.

If the universe recollapsed, the redshifts of distant galaxies would be replaced by blueshifts, and galaxies would crowd together again. Space is already becoming more and more punctured as isolated regions — dead stars and galactic nuclei — collapse to black holes. But this would then be just a precursor of a universal squeeze

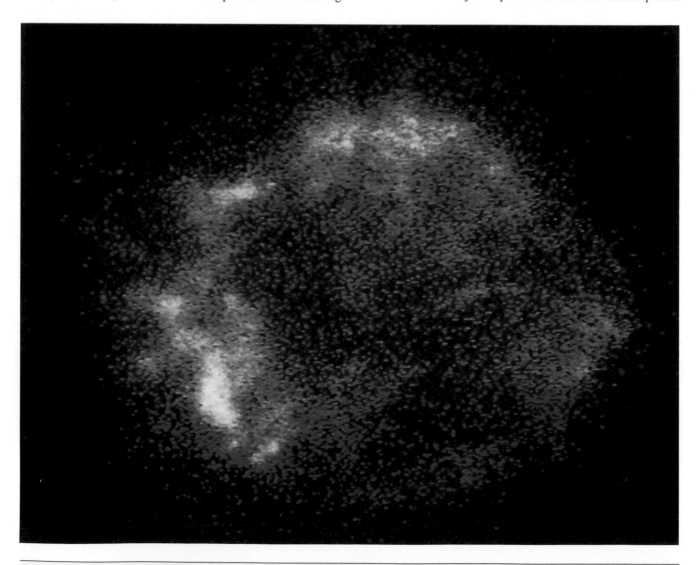

to a big crunch that engulfs everything. Galaxies merge; stars move faster, just as atoms of a gas move faster, as the universe is compressed; stars are eventually destroyed not by colliding with each other, but because the night sky becomes hotter than their centres. The final outcome would be a fireball like that which initiated the universe's expansion, though it would be somewhat more lumpy and unsynchronized. When will this happen? The earliest would be fifty thousand million years from now — at least ten times the remaining lifetime of the Sun.

But what about the other case, when there is not enough gravitating stuff ever to halt the universe's expansion? The facts that galaxies aren't exclusively composed of dead remnant stars, that there's still some uncondensed gas, and that not everything has been transmuted into heavier elements, could in itself have told us that our Galaxy hadn't existed for much longer than ten thousand million years, even if we knew nothing about the cosmic expansion. But, given enough time, the universe will undergo a "heat death": there *will* be enough time for everything to attain a terminal equilibrium. Various timescales are shown in the accompanying table. Stars all die (even the most slow-burning ones),

The far future of an ever-expanding universe

10^{14} years	Ordinary stellar activity completed
10^{17} years	Significant dynamical relaxation in galaxies
10^{26} years	Gravitational radiation effects in galaxies
[10^{32}–10^{36} years	Proton decay]
$10^{64}(\frac{M}{M\odot})^3$ years	Quantum evaporation of black holes
10^{1600} years	White dwarfs → neutron stars*
$10^{10^{26}}$–$10^{10^{76}}$ years	Neutron stars undergo quantum tunnelling to black holes which then "quickly" evaporate

*If protons do not decay

Possible forms of "hidden mass"

Black holes	10^{34}–10^{39} gm
(remnants of an early generation of massive stars)	
Faint low-mass stars	10^{32} gm
Neutrinos	10^{-32} gm
(if their rest mass is ~10ev)	
[or photinos, "...ijos", axions....]	

Any of the above could contribute 10 times as much gravitating mass as the stars and galaxies we see.

Cassiopeia A, *left,* is the debris from a supernova explosion that occurred within our Galaxy about 300 years. Star formation is still continuing, and this debris will eventually be recycled into new stars. When the Sun was born (4,500 million years ago), our Galaxy may have been about 5,000 million years old. The gas cloud from which our Solar System condensed would therefore have already been contaminated by effects from earlier generations of short-lived massive stars. We owe all the carbon, oxygen, iron and other heavy elements on Earth to nuclear transmutations in stars that formed early in galactic history and had already died by the time the Solar System formed.

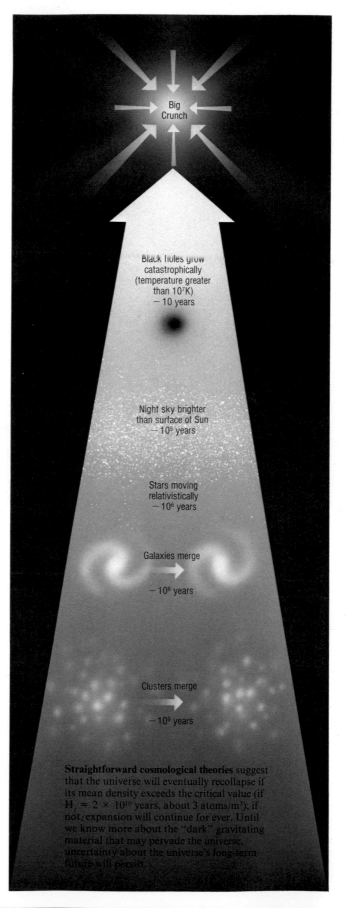

Big Crunch

Black holes grow catastrophically (temperature greater than 10^7K) — 10 years

Night sky brighter than surface of Sun — 10^5 years

Stars moving relativistically — 10^6 years

Galaxies merge — 10^8 years

Clusters merge — 10^9 years

Straightforward cosmological theories suggest that the universe will eventually recollapse if its mean density exceeds the critical value (if $H_i = 2 \times 10^{10}$ years, about 3 atoms/m³); if not, expansion will continue for ever. Until we know more about the "dark" gravitating material that may pervade the universe, uncertainty about the universe's long-term future will persist.

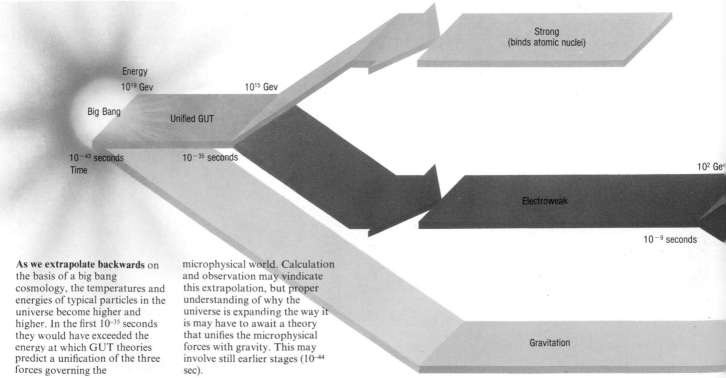

Energy

10^{19} Gev 10^{15} Gev

Big Bang   Unified GUT

10^{-43} seconds 10^{-35} seconds
Time

Strong
(binds atomic nuclei)

10^2 Ge

Electroweak

10^{-9} seconds

Gravitation

As we extrapolate backwards on the basis of a big bang cosmology, the temperatures and energies of typical particles in the universe become higher and higher. In the first 10^{-35} seconds they would have exceeded the energy at which GUT theories predict a unification of the three forces governing the microphysical world. Calculation and observation may vindicate this extrapolation, but proper understanding of why the universe is expanding the way it is may have to await a theory that unifies the microphysical forces with gravity. This may involve still earlier stages (10^{-44} sec).

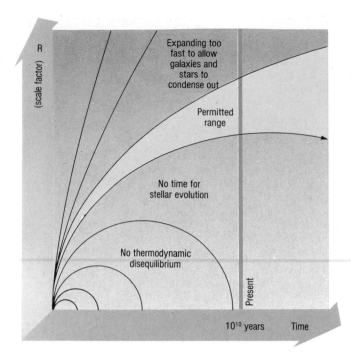

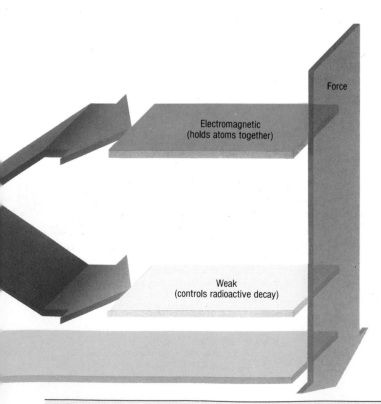

The largest particle accelerators, such as that a Fermilab, Illinois, *left,* can generate energies high enough to test the ideas of Salam, Weinberg and Glashow on the unification of the electromagnetic and the so-called weak forces, but not nearly enough to test a grand unified theory (GUT) that would relate the weak forces to the nuclear (or "strong") force.

"Fine tuning" of the dynamics of the early universe is demonstrated by the diagram, *above,* which plots the fate of hypothetical universes emerging from "bangs" with identically high initial density, but expanding at slightly different initial rates. Those that expand too slowly recollapse; those that expand too fast do not allow stars and galaxies to form.

galaxies experience dynamical evolution, black holes grow. If protons do not live for ever, then all ordinary stars will eventually decay, leaving only black holes. These too eventually decay by evaporation, the energy of all black holes (and everything they have swallowed) being recycled back into radiation. If protons did last for ever, then the final heat death would be spun out over a much longer period on which neutron stars can tunnel into black holes. (The longest timescale in the table is so enormous that, if written out in full, it corresponds to one followed by a number of zeros equal to the number of atoms in the observable universe!).

The ultra-early universe

The alternative long-range forecasts seem very different, but the initial conditions that could have led to anything like our present universe are actually very restrictive, compared to the range of possibilities that might have been set up. Our universe is still expanding after 10^{10} years. Had it recollapsed sooner, there would have been no time for stars to evolve or even form. (If it collapsed after less than a million years, it would have remained opaque, precluding thermodynamic disequilibrium.) The initial expansion cannot, on the other hand, have been too fast, otherwise kinetic energy would have overwhelmed gravity, and gravitationally-bound galaxies would never have been able to condense out. (This is the equivalent to saying that the present density is not orders of magnitude below the critical density.) The dynamics of the early universe must therefore have been "finely tuned". In Newtonian terms the fractional difference between the initial potential and kinetic energies of any spherical region must have been very small.

Why was the universe "set up" in this special way? And there are other issues that similarly baffle us. Why does the universe contain the fluctuations that constitute the "seeds" for galaxy formation, while being so homogeneous overall? Why are there 10^9 photons for each particle?

The primordial helium abundance allows us, with some confidence, to push our extrapolation back to t = 1 second; but maybe key features of the universe were imprinted at still earlier stages. The farther back one extrapolates, the less confidence one has in the adequacy or applicability of known physics (the material would exceed nuclear densities for the first microsecond). Particle physicists are interested in the possibility that the early universe may once have been at *colossally* high densities and temperatures. To motivate this interest, I shall digress to discuss the basic physical forces: electromagnetism, the weak force (important for radiative decay and neutrinos), the strong or nuclear force, and gravity. Physicists would like to discover some interrelation between these four forces — to interpret them as different manifestations of a single "primeval" force. The first modern step towards this unification was the Salam-Weinberg theory, relating the electromagnetic and the weak forces. The basic idea is that at high energies these two forces are the same. They acquire distinctive identities only below some critical energy. Energies

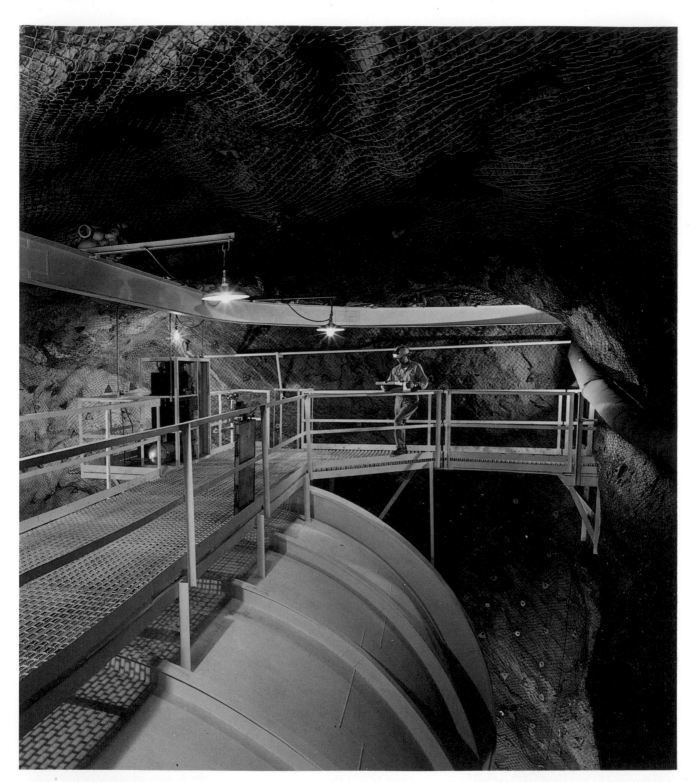

In the Homestake Mine in the United States, *above,* Dr. Raymond Davis has for many years been searching for neutrinos from the Sun, using a huge tank of carbon tetrachloride. A tiny fraction of the neutrinos passing through the tank interact with chlorine atoms, converting them to radio-active argon. The lower-energy primordial neutrinos that are expected to pervade the universe would be even harder to detect.

The Earth's atmosphere is opaque to some kinds of radiation, which therefore does not penetrate down to ground level. Studies of cosmic X-rays, and of ultraviolet and far infrared radiation, therefore had to await the development of space techniques. Detectors carried in satellites (like the IUE, *right*) now offer windows on the universe complementary to those provided by optical and radio telescopes on the ground.

in particle physics are measured in giga (10^9) electron volts, Gev for short, and the critical energy for this unification is about 100 Gev. This energy can just be reached by big particle accelerators, and the Salam-Weinberg theory has been vindicated by experiments at the European Centre for Nuclear Research (CERN) in Geneva. This development may prove as important in its way as Clark Maxwell's achievement a hundred years ago in showing electrical and magnetic effects to be manifestations of a single underlying force.

The next goal is to unify the electro-weak force with the strong or nuclear forces — to develop a so-called grand unified theory (GUT) of all the forces governing the microphysical world. But a stumbling block here is that the critical energy at which the so-called symmetry breaking occurs (the energy which is 100 Gev for the Salam-Weinberg theory) is thought to be 10^{15} Gev for the grand unification. This is a *million million* times higher than any feasible experiments can reach. It is hard therefore to test these theories on Earth. Only tiny effects are predicted in our low-energy world: for instance, protons would decay very slowly. (The lifetime is at least 10^{32} years, implying that only a few atoms, at most, would decay per year in a tank containing hundreds of tonnes of water. Experiments of this curious kind are now being carried out.) But if we are emboldened to extrapolate the big bang theory far enough we find that in the first 10^{-36} seconds, but only then, the particles would be so energetic that they would all be colliding at 10^{15} Gev. So perhaps the early universe was the only accelerator where the requisite energy for unifying the forces could ever be reached. However, this accelerator "shut down" ten thousand million years ago, so one can learn nothing from its activities unless the 10^{-36} second era left some fossils behind, just as most of the helium in the universe is left behind from the first few minutes. Physicists would seize enthusiastically at even the most trifling vestige surviving from that phase. But it has left very conspicuous traces indeed: it may be that all the atoms in the universe are essentially a fossil from 10^{-36} second.

If you were setting up a universe in the simplest way, you might make it symmetrical between matter and anti-matter, preparing it with equal numbers of protons and antiprotons. But the particles and antiparticles would then all annihilate as the universe expanded and cooled. We would end up with radiation but no matter, no atoms, and no galaxies. However, grand unified theories evade this problem in a way first outlined in a prescient paper by Andrei Sakharov, written in 1967. Although these theories predict that proton decay is now incredibly slow, at 10^{-36} seconds protons could readily be created or destroyed. As Sakharov first realized, the expansion and symmetry breaking, according to these theories, introduce a slight but calculable favouritism for creation of particles rather than their antiparticles, so that for every 10^9 proton-antiproton pairs, there is one extra proton. As the universe cools, antiprotons all annihilate with protons, giving photons. But for every 10^9 photons thereby produced, there is one proton that survives,

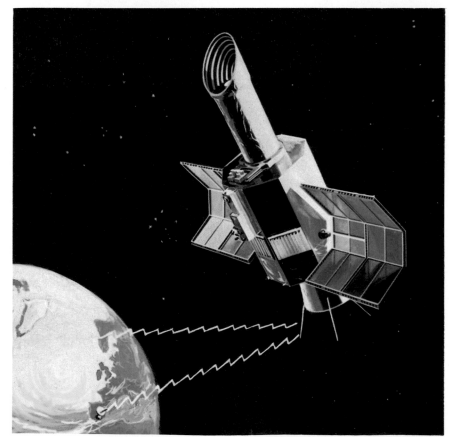

The material emerging from the Big Bang must contain more matter than antimatter: otherwise all protons would annihilate with antiprotons as the universe cooled, and we would be left with only radiation. In 1967, Andrei Sakharov, *above,* spelt out the conditions under which this excess might naturally arise. Recent progress in grand unified theories puts this suggestion on a firmer footing.

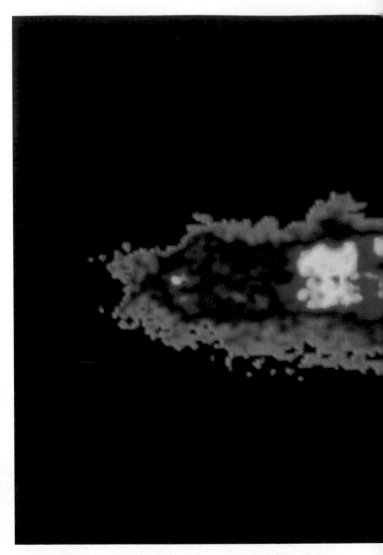

Although there was no real empirical support for the hot big bang theory until the 1960s, some theorists had explored the idea in earlier decades. Starting in the late 1940s, George Gamow, with his collaborators Ralph Alpher and Robert Herman, considered what nuclear reactions might have occurred in an early hot dense phase, and suggested that thermal background radiation might still pervade the universe as a relic of that era. This work was overlooked by experimenters, and the scientists at Bell Telephone Laboratories and at Princeton University who detected this radiation were at that time unaware of Gamow's "prediction". The idea of a "big bang" (or "primordial atom") was proposed by the Belgian astronomer Georges Lemaitre in the 1930s; he learnt of Penzias and Wilson's discovery just before his death in 1966.

George Gamow   Georges Lemaitre

because it cannot find a "mate" to annihilate with. The photons, now cooled to very low energies, constitute the 3° background. There are indeed about 10^9 of them for every atom. So the entire matter content of the universe could result from a small fractional bias in favour of matter over antimatter, imposed as the universe first cooled below 10^{15} Gev.

Grand unified theories are still tentative, but they at least bring a new set of questions — the origin of matter, for instance — in the scope of serious discussion. The realization that protons are not strictly conserved suggests, moreover, that the universe may possess no conserved quantities other than those, eg total electric charge, which are strictly zero. This, combined with the concept of a so-called "inflationary" phase whereby our entire universe could have burst forth from even a single quantum fluctuation, opens the way to envisaging "ex nihilo" creation of the universe.

The limitations of scientific cosmology

One theme that runs through astrophysics and cosmology is the inter-dependence of different phenomena. The everyday world is determined by atomic structure; stars are determined by the physics of atomic nuclei; and the much larger structures, galaxies and clusters, may be gravitationally bound only because they are embedded in particles which are relics of the high-energy initial instants of the big bang.

But in considering the early big bang, or the end of the universe, we are confronted by conditions so extreme that we know for sure that we *don't* know enough physics. In particular, we know that physics is incomplete and conceptually unsatisfactory in that there is still no adequate theory of quantum gravity. The two great foundations of 20th-century physics are the quantum principle and Einstein's general relativity. The theoretical superstructures erected on these foundations are still disjoint: there is generally no overlap between their respective domains of relevance. Quantum effects are crucial on the microscopic level of the single elementary particle; but gravitational forces between individual particles are negligible, weaker by almost forty powers of ten than electromagnetic forces. Gravitational effects are manifested only on the scale of planets, stars and galaxies, where quantum effects and the uncertainty principle can be ignored.

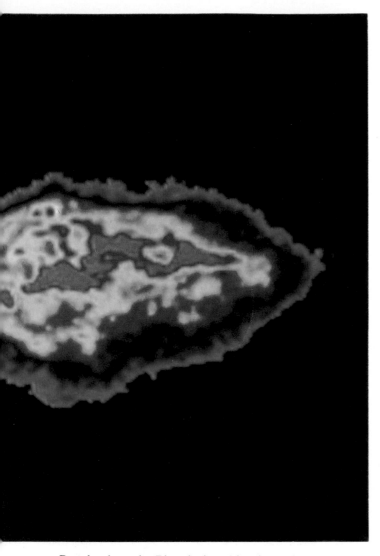

Gamow originally hoped that all chemical elements could be produced in the very early universe, but later calculations revealed that the material emerging from the big bang would be almost exclusively hydrogen and helium. Other elements of the periodic table result from nuclear transmutations in stellar interiors. "Processed" material is ejected in the supernova explosions whereby massive stars die; the gas in supernova remnants such as 3C 58, *left,* can then condense into new generations of stars. The concept of stellar nucleosynthesis was formulated by Fred Hoyle and William A.Fowler, and by Alistair Cameron and Geoffrey and Margaret Burbidge. This theory — which can explain why carbon and iron are common but gold and uranium are rare, and how all these elements came to be in our Solar System — ranks as one of the triumphs of astrophysics over the last 40 years.

Fred Hoyle   William Fowler

But, back at the Planck time (the time when quantum uncertainty effects became important in cosmology), the entire universe was squeezed to such densities that gravity is important even across the scale of a single elementary particle, and the effects of quantum gravity then cannot be neglected. Until we have such a theory, we cannot seriously address such really fundamental questions as "Why is the universe so large?" "Could a collapsing universe rebound phoenix-like into another cycle?" "Why are there three dimensions of space and one of time?" "Why are the constants of nature the same everywhere?" and above all, "Why does the universe have the overall symmetry and simplicity that is a prerequisite for any progress in cosmology?"

Cosmologists do not spend all their time pondering these really deep fundamental questions: it is sensible methodology to work on bite-sized topics where there is hope of progress, rather than to worry excessively about problems which may not be timely or tractable. But some bold theorists believe it is no longer premature to explore what physical laws prevailed at the Planck time, and have already come up with fascinating ideas; there is no consensus, though, about which concepts might

really "fly". We must certainly jettison "commonsense" notions of space and time: space-time on this tiny scale may have a chaotic foam-like structure, with no well-defined arrow of time; there may be no time-like dimension at all. Indeed, on the tiniest scale space may have extra dimensions, which aren't manifest in the everyday world because they are "compactified" — rather as a sheet of paper, a two-dimensional surface, might look like a one-dimensional line if rolled up very tightly.

Only when (or if) we have such a theory will we stand a chance of bringing the initial conditions (at the Planck time) within the scope of serious science. This may show us that our universe could not have been otherwise — that in some sense it is uniquely self-consistent the way it is — or, on the contrary, we may then be able to conceptualize all kinds of different universes, some more propitious than others as arenas for complex evolution.

Some cosmologists have expressed surprise that laboratory physics and its extensions suffice for interpreting the entire cosmos. It would not seem inherently implausible that phenomena on the extragalactic scale might involve some basically new law of nature. After all, a physicist whose laboratory was floating freely in space would probably never have discovered gravity because this force is very weak unless a large mass such as the Earth is involved. So there could be other effects, insignificant even on the scale of the Solar System, which were nonetheless crucial in galactic nuclei or cosmology.

Some astronomers have indeed argued for many years that the quasar phenomenon involves new physics. It is true that not much about quasars is yet explained, but my personal view is that we should think longer and harder before throwing in the sponge. Many phenomena in physics have eluded our understanding for decades, superconductivity and the solar cycle being just two examples. But these did not require new physics. Only in the ultra-early universe do we encounter phenomena that transcend the physics we learn on Earth.

The proverbial rational man who loses his key at night searches only under the street lamps, not because that is necessarily where he dropped it, but because his quest is otherwise quite certain to fail. Cosmologists approach their subject in a similar way. They start by using the physics that is validated locally, and making simplifying assumptions about symmetry, homogeneity, etc. There seems no reason why the universe *should* be so ordered that this permits any real progress — why the physics studied in the laboratory should apply in quasars thousands of millions of light years away, and in the early stages of the big bang. What is so surprising is that questions about how the universe began can at least be addressed scientifically, and not just in our unprofessional moments.

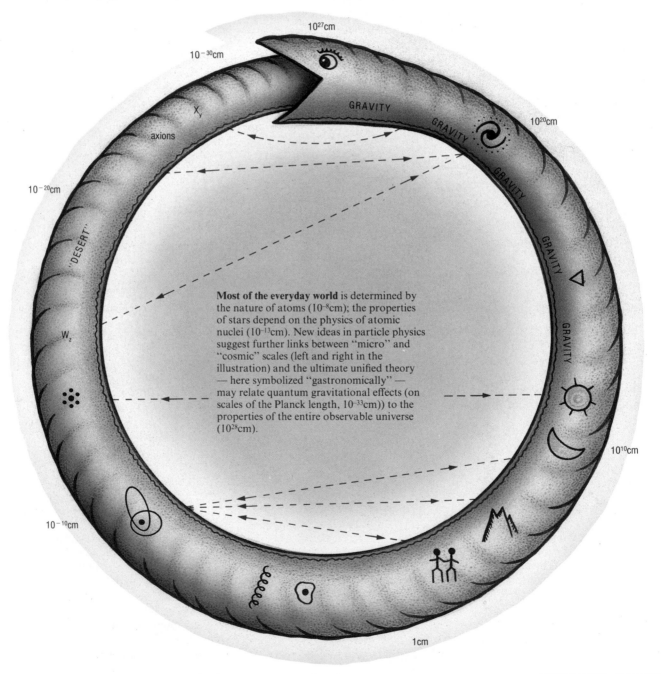

Most of the everyday world is determined by the nature of atoms (10^{-8}cm); the properties of stars depend on the physics of atomic nuclei (10^{-13}cm). New ideas in particle physics suggest further links between "micro" and "cosmic" scales (left and right in the illustration) and the ultimate unified theory — here symbolized "gastronomically" — may relate quantum gravitational effects (on scales of the Planck length, 10^{-33}cm)) to the properties of the entire observable universe (10^{28}cm).

black hole is the X-ray source called Cygnus X-1. At the position of this X-ray source lies a spectroscopic binary star HDE 226868, whose period is 5.6 days. It has been conjectured that the only model which readily fits the observational data is that of a distorted star from which matter flows into a black hole, having a mass of some eight times that of the sun. Hydrogen and ionized helium emission lines detected in the optical spectrum would originate from the "bridge" between the two companions, and the X-rays would originate from gases circling around and then disappearing into the black hole. Observed variations in light intensity would then be due to the gravitational distortion of the bright blue star HDE 226868 by the black hole as they both rotated about a common centre of mass.

Another promising candidate is LMC X-3: the third X-ray source to be discovered in the nearby Large MAGELLANIC CLOUD.

Interest in gravitational collapse was revived by the discovery of the QUASARS, with their apparently enormous energy requirements. The hypothesis has been advanced that accretion of matter onto a large central black hole might produce quasars. There is also the problem of the MISSING MASS—the density of the observable matter in our universe is much less than the theoretically computed value needed to "close" the universe, and it may be that at least some is in the form of black holes.

Not all black holes result from stellar collapse; in the early history of our expanding BIG BANG universe, some regions might have got so compressed that they underwent gravitational collapse to form so-called primordial black holes. Quantum mechanical effects become very important for very small primordial black holes. It is possible to show that such a black hole is *not* completely black, but that radiation can "tunnel out" of the event horizon at a steady rate. This then could lead to the evaporation of the hole! (Quantum mechanical effects are not important for conventional larger black holes.) Primordial black holes could thus be very hot, and from the outside could look like white holes, the time-reversals of black holes. Summed up in the words of one of the world's greatest relativists, Professor Stephen Hawking: "[Quantum mechanically] black holes behave in a completely random and time symmetric way and are indistinguishable, for an external observer, from white holes." *DLB.*

Blancanus Lunar walled plain, diameter 57 miles (92km); 64°S, 21°W; a pair with Scheiner.

Blashko effect A periodic change in the light curves and periods of some RR Lyræ variable stars discovered by Sergei Nikolaevich Blashko (1870–1956).

Blazar The term "blazar" is compounded from "BL LACERTÆ OBJECT" and "QUASAR"; it refers to a specific kind of extragalactic object. The blazars are the most active of the galaxies with active nuclei, that is, galaxies whose central regions are undergoing energetic processes which turn them into SEYFERT GALAXIES, BL LAC objects or quasars. The blazars show variable optical brightness, strong and variable optical polarization and strong radio emission. The variations in the optical region can be on timescales as short as days.

Much of the activity in active galactic nuclei is related to jets of gas expelled from their central regions with relativistic velocities. One explanation of the exceptional activity in blazars is that in these galaxies we are viewing the jets "end on". *BW.*

Blaze star The recurrent nova T Coronæ Borealis (outbursts 1866, 1945)

Blink comparator The blink comparator, or blink microscope, is an instrument that enables two photographs of a part of the sky, taken at different times, to

Blink Comparator: used to discover Pluto

be compared in a way that draws attention to any differences between them. The comparator has two optical paths so that the images of the two photographic plates can be seen together in one viewing eyepiece. By careful adjustment of the position of the plates the separate images are brought into exact coincidence. The plates are then alternately illuminated, changing from one to the other about once a second. All features that are identical on both plates appear unchanging in the alter-

nating illumination, but any object that is on only one of the plates is seen to blink on and off. Also, an object that changed its position between the times when the plates were exposed is seen to jump to and fro and an object that changed its brightness appears to pulsate in size at the illumination frequency.

The eye is very effective at detecting the few varying objects among what can be tens of thousands of star images on a long-exposure photograph. This simple technique makes possible the discovery of stars of large proper motion, minor planets, comets or variable stars without the need to compare individually every star image on two photographs.

Some examples of the product of the blink comparator are (i) the discovery of PLUTO by C.W.Tombaugh, (ii) the catalogue of over 100,000 stars brighter than magnitude 14.5 with detectable proper motion, produced by W.J.Luyten and (iii) the majority of the nearly 30,000 known variable stars discovered at various observatories around the world.

Other types of comparators also exist. In the stereocomparator binocular vision is used and discordant objects appear to stand out of the plane of the picture. In another type, discordant objects appear of different colour to the unchanging stars. *BW.*

Bliss, Nathaniel (1700–64). ASTRONOMER ROYAL from 1762 to 1764, born in Bisley, Gloucestershire. A graduate of Oxford, he later succeeded Edmond Halley as Savilian Professor of Geometry there. Bliss was an able observer who gave some assistance at Greenwich to James Bradley and succeeded him as Astronomer Royal. Unfortunately, he made little impression on the observatory, dying only two years after his appointment.

BL Lacertæ objects Rare, peculiar, extragalactic objects located in the nuclei of some galaxies.

In 1941, H. van Schewick discovered a variable star in the constellation of Lacerta. As is the case for all such objects, this one was given a conventional variable star designation, BL. It was noted that the light variations showed no repeatable pattern with time, and the variable was classified as irregular. No attention was paid the star for a considerable time.

That BL Lacertæ is no ordinary star became apparent in 1968, when it turned up as a prominent radio source, named VRO 42.22.01, in a survey made at the University of Illinois' Vermillion River Observatory. The radio source, like the visible counterpart, was found to vary.

Further research quickly revealed the peculiar nature of BL Lacertæ (or BL Lac as it is usually called). Most extraordinary was its optical spectrum. Normally, spectroscopy (*see* ASTRONOMICAL SPECTROSCOPY) is astronomy's most valuable tool, revealing the physics and chemistry of the object under study. This is accomplished by examination of the individual absorption lines at specific wavelengths, or in some cases the EMISSION lines of hot gas. The spectrum of BL Lac, however, contained neither ABSORPTION nor emission lines. It was completely featureless and yielded no information whatsoever.

Photography revealed a faint, hazy patch surrounding BL Lac. As astronomical equipment grew more sophisticated, spectroscopy of this haze became possible, and it was shown to be a giant galaxy at what was then thought of as a high REDSHIFT of 0.07, corresponding to a distance of some 1,000 million light-years.

Clearly, then, the variable star-like object BL Lac lies within a galaxy and totally outshines it. In this respect it resembles the QUASARS, but differs in not showing the strong emission features that characterize quasars' spectra.

The history of BL Lac is not unique. An almost identical story could be told of the southern variable "star" AP Librae. Other BL Lac objects have been found which were never accidentally included in catalogues of variable stars. In total about ten objects resembling BL Lac are known for which underlying galaxies have been observed. For these BL Lac objects we know the distance, and hence the luminosity.

The catalogue of BL Lac objects numbers over 50 members, however, and for most of them no underlying galaxy is seen. These objects are certainly closely similar to BL Lac itself, but even more luminous, so that the underlying galaxies are completely dominated by their light.

For the majority we do not know the redshift, though in some cases a lower limit to the redshift can be deduced. The limit arises because the light from very distant BL Lac objects, like that from quasars, can pass through intervening galaxies. By so doing it acquires recognizable spectral absorption features which give a redshift, and

hence a distance, not of the BL Lac object but of the intervening galaxy. We then know only that the BL Lac object lies at a greater distance. The highest absorption redshift for a BL Lac object is that of the Parkes radio source PKS 0525-250 at 2.77. This object lies well into the regime of the quasars in terms of distance and luminosity. Another which certainly competes with the quasars is PKS 0215-015, which has an absorption redshift of

BL Lacertæ: peculiar extragalactic objects

1.65; in the mid-1980s this object brightened to 14th magnitude and became the most luminous known object in the universe.

The characteristics of BL Lac objects are variability, wavelengths, optical spectra and polarization.

Variability Variable in output at all wavelengths. Appreciable change can occur in hours.

Wavelengths Radiates at all wavelengths from X-ray to radio. This is not a prerequisite, but as yet no BL Lac objects have been found which are not radio-emitting, whilst X-rays have been detected whenever sufficient sensitivity is available.

Optical spectra Always featureless except for absorption by intervening galaxies or gas clouds.

Polarization The light from BL Lac objects is usually highly polarized.

What are we to infer from these characteristics? We know that the luminosities can be as high as those of quasars, measured in terms of hundreds of normal galaxies. Like the quasars, this luminosity probably arises from an object at the centre of a galaxy, in a very compact source. Indeed, if BL Lac objects can vary in a few hours, then their dimensions cannot be much greater than the distance that light

travels in that time, or about the dimension of the Solar System.

The high polarization, which is itself variable, is a clue. Light can be polarized for only two reasons. Either it is formed polarized, or it acquires polarization, for example by bouncing off grains of dust. Various arguments favour the former in the case of BL Lac objects, and this in turn implies that special mechanisms are at work to create the radiation, quite distinct from starlight. The most likely is SYNCHROTRON RADIATION, produced when energetic electrons are released into a magnetic field. Most radio sources, and probably some X-ray sources, radiate by the synchrotron mechanism. The variability is also more easily explained by synchrotron radiation.

Moreover, synchrotron emission is thought to underlie quasars. In their case, however, it is masked by an additional component due to gas. A natural conclusion is that BL Lac objects are quasars in a region free from gas. This might, for instance, be the case if quasars are found in spiral galaxies, which generally have vast reservoirs of gas near their centres, and if BL Lac objects are restricted to the gas-poor elliptical galaxies. At present the evidence is not conclusive, but BL Lac objects are thought to lie only in ellipticals.

If this line of thought is correct, then it is particularly important to understand the BL Lac objects, for they are examples of quasars unencumbered by the confusing effects of gas. But the very absence of gas, and the consequently bland optical spectra, renders their study unusually difficult; progress is slow.

An interesting class related to the BL Lac objects is the optically violent variable (OVV) quasars. Whilst having variability and polarization properties characteristic of BL Lac objects, the OVV quasars show quite strong emission lines due to gas. Perhaps these are quasars around which only a minimal amount of gas survives, so that the "engine" in the nucleus of the galaxy is not fully camouflaged. Study of OVV quasars is, of course, important.

Currently the best interpretation of BL Lac objects and quasars assumes that a BLACK HOLE has formed in the nucleus of a galaxy, and has grown to a size measured in tens of millions of solar masses by absorbing many of the stars of the galaxy's central regions. Such a black hole continues to swallow stars and gas, liberating energy as it does so. When an object of mass *m* falls into a black hole, the energy released is most of mc^2, where c is the velocity of light. Compare this to the energy $0.001mc^2$ released when hydrogen burns to make helium in a star. By this means a single star being swallowed by a black hole liberates in a short time as much energy as a thousand would in their entire lives.

If the black hole weighs no more than a few million times that of the Sun, an infalling star settles first into a disk of gas, which may generate the quasar spectrum. If the hole weighs more than about 100 million Suns, stars can be swallowed almost intact, and gas features may not be seen in the spectrum. This is one possible explanation of the difference between quasars and BL Lac objects. *DAA*

Blue Moon Occasional blue colour of the Moon, due to effects in the Earth's atmosphere. The phenomenon was widely observed on September 26 1950, when it was due to Canadian forest fires, which had scattered high-altitude dust.

Bode, Johann Elert (1747-1826). German astronomer (Director of Berlin Observatory from 1772). He published a star catalogue and did much to popularize astronomy; he also drew attention to the relationship between planatary distances known as BODE'S LAW, though he did not in fact discover it.

Bode's Law A simple numerical relationship first noticed by Johann Titius of Wittenberg, but popularized by Johann Elert Bode in 1772, which was found to hold for the distances of the then known planets from the Sun. The formula is more correctly called the Titius-Bode Law. Bode started with the sequence of numbers 0, 3, 6, 12, 24, 48, 96, 192 . . . and added 4 to each number, giving 4, 7, 10, 16, 28, 52, 100, 196 . . . Then, if the distance of the Earth from the Sun was taken to be 10 units, it was found that Mercury fell into place at 4, Venus at about 7, Mars at 16, Jupiter at 52 and Saturn at 100. The discovery of Uranus at about 196, in 1781 by Sir William Herschel, initiated a hunt for the missing planet at 28, between Mars and Jupiter. This was supplied by Ceres, the first minor planet, in 1801. Neptune and Pluto fail to satisfy Bode's Law, and it has now declined in importance. *JWM*.

Bok, Bart (1906–1983). Dutch-born American astronomer best known for his study of compact dusty interstellar clouds (BOK GLOBULES) in the Milky Way.

Bok globules Globules (the modern name) are roundish dark clouds of relatively small dimensions which are likely precursors to the formation of protostars. Small globules (pointed out by Bart BOK) may have diameters as small as 0.04 parsecs (8,000AU) and can only be seen when they lie in front of a bright emission nebula. Larger globules are seen as dark patches against the stellar background of the Milky Way, and have dimensions up to 1pc across. Well- known small globules can be seen in photographs of Messier 8 (M8), NGC2244 and IC2944. Well known large globules lie near to the star Rho Ophiuchi, in Taurus, and in Crux (the famous COAL SACK nebula).

The mass of dust in a globule may be estimated from the extinction of background light, but it is certain that the dust makes up only a small fraction of the total mass of a globule. Being well shielded from ultraviolet stellar radiation, and at a temperature of 10K, most of the gas in a globule will be in

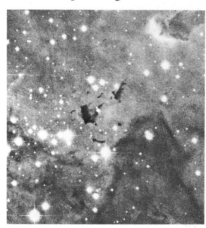

Bok globules: precursors of protostars

molecular form. The expected main constituent, molecular hydrogen, is very difficult to observe, but the much rarer molecules of carbon monoxide, formaldehyde and the hydroxyl radical have been observed at radio wavelengths. Total estimated globule masses range from about 0.1 solar masses for the smallest, to about 2,000 solar masses for the Rho Ophiuchi dark cloud.

Such a large mass of gas and dust in the small radius of a globule at a low temperature will suffer gravitational contraction to form firstly protostars, and eventually normal main sequence stars. This view is confirmed by the close positional association of globules with T-TAURI STARS and HERBIG-HARO OBJECTS, which are phenomena associated with star formation. *DJA*.

Bolides A term often used to describe a major fireball that produces a sonic boom. Such events are frequently associated with the deposit of meteorites.

Bolometer Instrument to measure the heat received in a telescope from a celestial body by its effect upon the balance of an electrical circuit.

Bolometric magnitude a measure of the total radiation of all wavelengths emitted by or received from a star expressed on the stellar magnitude scale.

Bond, George Phillips (1825–65). American astronomer; he succeeded his father, W.C.Bond, at Harvard.

Bond, William Cranch (1789–1859). American astronomer who became Director of the new Harvard College Observatory. He discovered Saturn's satellite Hyperion in 1848, and the Crêpe Ring in 1850.

Bonner Dürchmusterung Major catalogue of the stars, down to magnitude 9; it was mainly due to F.W. ARGELANDER. It contains 324,198 stars, and is still of great value. Usual abbreviation: *BD*

Boötes (the Herdsman). A prominent northern constellation containing Arcturus, the brightest star north of the celestial equator. In mythology Boötes is said to represent a herdsman who invented the two-oxen plough.

Brightest stars

Name	Visual Mag.	Abs. Mag.	Spec.	Distance (light-yrs)
α (Arcturus)	−0.04	−0.2	K2	36
ε (Izar)	2.37	−0.9	K0	150
η	2.68	2.7	G0	32
γ (Seginus)	3.03	0.5	A7	104
δ	3.47	0.3	G8	140
β (Nekkar)	3.50	0.3	G8	140

Also above the fourth magnitude are Rho (3.58) and Zeta (3.78).

Epsilon (Izar) is a double star; magnitudes 2.7 and 5.1, separation 2".9. The primary is yellowish or orange, the companion bluish. This is a fine pair, by no means difficult with a small telescope. Mu is a wide double; 4.5 and 6.7, separation 108".8. The only fairly bright cluster is NGC5466, a globular with an integrated magnitude of 8.5.

Boötes is fairly distinctive, and is

B

B

easy to locate because of the presence of Arcturus, which, with its declination of over 19°N, is visible from every inhabited continent. The Y arrangement made up of Arcturus, Epsilon and Gamma Boötis. *PM*.

Borealis Planitia The "Northern Plain" on Mercury; 75°N, 85°W.

Boscovich Very dark lunar formation on the border of Mare Vaporum; 10°N, 11°E; diameter 27 miles (43km).

Boss, Lewis (1846–1912) American astronomer who directed an extensive programme of precise positional measurements which were published as a series of star catalogues.

Boulder Location of the main campus of the University of Colorado and of several research organizations. *See also* HIGH ALTITUDE OBSERVATORY, JOINT INSTITUTE FOR LABORATORY ASTROPHYSICS, LABORATORY FOR ATMOSPHERIC AND SPACE PHYSICS.

Bovedy Meteorite Meteorite which fell in Northern Ireland on April 25 1969.

Bow shock The interplanetary space is filled with the tenuous ionized gas of the SOLAR WIND, which steams from

Bow shock: the "wave" set up by a planet

the Sun at supersonic speeds. The interaction of the individual planets with the interplantary medium creates a shock front called the bow shock, which is similar to the effect set up by an ocean liner moving through the sea.

Bowen, Ira Sprague (1898–1973). American astrophysicist who in 1927 showed that previously unidentified lines in the spectra of nebulae were due not to a new element "nebulium" but to so-called "FORBIDDEN LINES" of ionized oxygen and nitrogen.

Boxhole crater An impact crater 574ft (175m) across, discovered in 1937 in the Northern Territory of Australia. Produced by an iron meteorite; masses up to 180lb (82kg) were found.

Boyden Observatory In 1887 Harvard College Observatory received funds from the estate of U.A. Boyden, to be used for construction of an observatory at high altitude. The observatory, known as the Boyden Station, was erected in Arequipa, Peru, in 1890. Because of poor observing and living conditions the observatory was moved to Bloemfontein in South Africa in 1926. It was then known as Boyden Observatory and was equipped with a 60in (152cm) reflector as well as several small instruments used in surveys for variable stars. A 32/36in (81–91cm) Schmidt telescope was added in 1950.

The Boyden Observatory continued to be run by Harvard until 1954, when a consortium of observatories from Harvard, Dunsink, Armagh and others in Belgium, West Germany and Sweden took over. Since 1976 it has been operated by the University of Orange Free State. *BW*.

Bradley, James (1693–1762). Third Astronomer Royal. By meticulous measurement Bradley discovered the ABERRATION of starlight in 1728; this was the first observational evidence of the Earth's orbital motion. In 1732 he found a nodding or "NUTATION" of the Earth's axis and at Greenwich catalogued the positions of some 3,000 stars.

Brahe, Tycho (1546–1601). Danish astronomer. In 1572, Brahe discovered a supernova in Cassiopeia, and his report, *De Stella Nova* (1573), made him famous. The Danish king Frederik II gave him the island of Ven, and here Brahe built two observatories, URANIENBORG and Stjerneborg, where he carried out very accurate work. Brahe was not a Copernican but created a "Tychonian world system" with the planets revolving around the Sun, and the Sun and the Moon revolving around a fixed Earth. Brahe ended his days as imperial astronomer under Rudolf II in Prague. Johannes KEPLER was one of his assistants there. *LH*.

Brans-Dicke Theory An alternative theory of gravity to Einstein's General Theory of RELATIVITY, much discussed in the 1960s.

It was proposed by Princeton physicists C.H.Brans and R.H.Dicke in

1961. Its motivation was to give expression to an idea of the 19th century philosopher Ernst Mach, who held that there is a connection between unaccelerated "inertial" frames in which Newton's laws hold and the distribution of distant matter. This led Brans and Dicke to modify General Relativity, introducing a new "scalar field" which would make such a connection explicit. The strength of this extra field is governed by an arbitrary constant *w*.

The Brans-Dicke theory made one important prediction, different from General Relativity. The value of the rate of precession of the perihelion of the orbit of the planet Mercury should be some 10 per cent smaller according to Brans-Dicke. While the correct prediction of this rate had been a triumph of General Relativity, Dicke noted that a similar effect would be produced if the Sun were oblate (flattened at the poles). Then, in 1967, after a difficult and precise experiment, he set the cat among the scientific pigeons by measuring a solar oblateness just sufficient to account for the difference between the observed rate and that predicted by Brans-Dicke theory.

However, it soon became apparent something was wrong. The Brans-Dicke theory predicted that the constant of gravitation decreased with time as the universe evolved. This was seen to be inconsistent with theories of solar evolution since the Sun would have been smaller and brighter in the past. Gradually, as knowledge of solar system dynamics improved with the tracking of interplanetary spacecraft, the deviations allowed from General Relativity became smaller. This limited the Brans-Dicke parameter *w* to values closer and closer to 1, at which the Brans-Dicke theory becomes indistinguishable from General Relativity. Dicke's solar oblateness measurement has been attributed to the effects of SOLAR FACULÆ on his data. *MVP*.

Breccia A rock comprising angular fragments set in a fine-grained matrix. Lunar breccias, the most abundant rocks returned by the Apollo astronauts, may also contain impact glass.

Bremsstrahlung *See* FREE-FREE TRANSITION.

Bright Star Catalogue Modern extension of the HR (Harvard Revised photometry) Catalogue, containing information on positions, magnitudes, colours, spectra and motions of the

9,110 brightest stars. Published by Yale University Observatory.

British Interplanetary Society A London-based but international society of professionals and non-professionals founded in 1933 and dedicated to the promotion of space research and technology, and to manned/unmanned exploration of interplanetary and deep space.

Brooks' Comet (1893 IV). Though not a bright comet, it was one of the first to show on photographs a complex and rapidly-changing plasma tail.

Brorsen's Comet A faint periodical comet (period 5.5 years). It was seen at five returns, the last being in 1879, but has not been recovered and has almost certainly disintegrated.

Brorsen, Theodor (1819–95). Danish astronomer who discovered several comets and was one of the first to describe the GEGENSCHEIN.

Brown dwarfs Temperatures in the cores of stars that have masses less than 0.08 solar mass do not rise high enough to start thermonuclear reactions. Such stars are, however, luminous as they slowly shrink in size and radiate away their gravitational energy. As their surface temperatures are below the 2,500K lower limit of RED DWARFS, they are known as brown dwarfs.

Several such low-mass stars have been reported, either in ASTROMETRIC BINARIES or as isolated stars of very low luminosity. In the latter category the star LHS 2924 is the intrinsically faintest known single star, with a surface temperature near 2,000K and an absolute visual magnitude of +19, which makes its visual luminosity only a millionth of that of the Sun.

Among the nearby stars are several astrometric binaries that contain low, mass components. The low temperatures of these unseen companions puts most of their emitted radiation in the infrared, where they are difficult to observe. Nevertheless, the astrometric companion of the star van Biesbroeck 8 was successfully observed in 1985 by astronomers in Arizona using an infrared SPECKLE INTERFEROMETER. The companion, known as VB 8B, was detected about one second of arc from VB 8A, which is equivalent to an actual separation of six astronomical units, or a little greater than the distance of Jupiter from the Sun.

VB 8B has a surface temperature of only 1,400K and a total luminosity (mostly in the infrared) 30 millionths of that of the Sun. Its mass is about five times greater than Jupiter. Although it is the coolest and intrinsically faintest star yet observed, the failure of the infrared interferometer to detect the known companion to VB 10 implies that VB 10B is even less luminous and may have a mass similar to that of Jupiter. However, these observations have recently been called into question and there may be serious errors in them.

Bruno, Giordano (1548–1600). Italian philosopher and monk. An early supporter of Copernicus' heliocentric theory, he was burned at the stake in Rome for theological heresy and suspicion of magical practices.

B stars The spectral characteristics that define the B-type stars are the considerable strength of neutral helium absorption lines and the presence of hydrogen lines (although not as strong as in the A star spectra). On the MAIN SEQUENCE, the temperatures of B stars increase from 10,500K at B9 to 28,000K at B0 and their masses and radii increase from 3.2 solar masses and 2.5 solar radii to 17 and 10 times solar respectively. At their hottest, the B stars radiate 20,000 times the luminosity of the Sun.

Radiating at such high rates the B stars use up their nuclear fuel relatively quickly: a B9 star stays on the main sequence about 500 million years but a B0 star remains only five million years. The B stars that we see today must therefore have been formed comparatively recently in the life of the Galaxy—they are young stars and rare compared with cooler stars because of their shorter lifetimes. Because of their recent formation, the B stars are associated with regions of dust and gas which signify active areas of star formation in the Galaxy. The hottest B stars (and the even hotter O stars) form loose groupings in the sky known as OB ASSOCIATIONS.

The high surface temperatures of B stars result in most of their radiation being emitted in the far ultraviolet part of the spectrum. This energetic radiation is able to ionize gas in the vicinity of a B star, forming an HII REGION. The B stars and HII regions occur where there is dust and gas and as a result act as tracers of spiral arms in our Galaxy and in distant galaxies. This and their high luminosities, which enable them to be seen at great distances, make B stars

valuable aids in the study of galactic structure.

O and B stars evolving from the main sequence become SUPERGIANTS with radii up to 75 times that of the Sun. With luminosities up to 100,000 times solar, radiation pressure in these supergiants is sufficient to cause mass loss of about one-millionth of a solar mass per year. This is sufficient to reduce the mass of B stars significantly during their lifetime.

Most B stars rotate rapidly. The fastest have rotation periods of a few hours and are surrounded by disks of gas which extent to several stellar radii out from their equators. Such a disk, excited by radiation from the central star, emits an EMISSION LINE spectrum which gives the star a Be classification. The disks are larger and cooler than the stars and produce an observable excess of infrared radiation in Be stars relative to that in B stars.

Three types of pulsational variability are known among the B stars: the Beta Cephei Stars, the 53 Persei spectrum variables, and the Be variables. All arise from a mixture of radial and non-radial oscillations.

There are several kinds of peculiar spectra among the B stars. The Bp stars are a higher, temperature extension of the PECULIAR A STARS (Ap) into the B8 and B9 spectral types. Among the hotter B stars a few have exceptionally strong helium lines and are generally slow rotators. The Ap, Bp and Helium-strong stars all have magnetic fields and probably all have similar compositions, caused by diffusion in their atmospheres. Other anomalous spectra occur in the Mercury-Manganese (Hg-Mn) Stars, which are slow rotators and constitute a substantial fraction of the B dwarfs with temperatures between 11,000 and 16,000K, and the Helium-weak stars, Helium stars and OB Subdwarfs also have spectral peculiarities.

Bright examples (B stars are white or blue-white in colour): Achernar B5 IV-V, Regulus B7 V, Rigel B8Ia, Spica B1 V. *BW.*

Budh Planitia Plain on Mercury; 18°N, 150°W.

Bullialdus Magnificent lunar crater on Mare Nubium, with high walls and central peak; 21°S, 22°W; diameter 39 miles (63km).

Bürg Lunar crater, 28 miles (45km) in diameter, near Lacus Mortis; 45°N, 28°E; associated with a fine system of rilles.

B

Bursters Sources of X-ray bursts were discovered in 1975 by Grindlay (Harvard) and Heise (Utrecht) with the Astronomical Netherlands Satellite. They found that bursts were emitted from the GLOBULAR CLUSTER NGC 6624.

More than 30 X-ray bursters have been found in our Galaxy. Their distribution is concentrated towards the galactic centre. Several exist in globular clusters. They are thus associated with halo or old POPULATION II objects. The related steady X-ray emission shows no pulsations and no evidence for eclipses. Optical counterparts in the cases where they have been identified are faint blue objects with EMISSION LINE-dominated spectra. These steady X-ray emitters are known as galactic bulge sources. Thus X-ray bursters form a subset of this class which has quite different properties from X-RAY BINARIES.

The X-ray bursts have rise times of about 1s, fall times of about 60s, peak luminosities about 10^{38} erg/s and total luminosities of about 10^{39} erg, which is equivalent to one week's radiant energy output of the Sun. The intervals between burst are irregular, ranging from hours to days, while many sources undergo burst-inactive phases that can last for weeks to months.

The highly luminous pulsating X-ray binaries found in the Galaxy exist due to the ACCRETION of material from massive young stars onto NEUTRON STARS with strong magnetic fields. Studies of burst spectra by Swank (Goddard) indicate the source radius to be about 6 miles (10 km), which suggests the presence of a neutron star. An old neutron star would have no magnetic field, which explains the absence of pulsations. Delayed optical bursts (McClintock, MIT), optical spectra and periodic X-ray absorption dips (White and Swank) indicate that the neutron star is paired with an old low-mass star.

The individual bursts are due to thermonuclear explosions after accreted material has exceeded a critical mass on the neutron star surface. *JLC. See also* RAPID BURSTER.

Butterfly diagram At the beginning of the SOLAR CYCLE sunspots appear in latitudes about 35° north and south of the equator. Their number increases and their latitude distribution moves towards the equator throughout the cycle. At the maximum the mean latitude is approximately ±15° and at the minimum a few spots appear near the

equator at the same time as the first high-latitude spots of the next cycle. This is illustrated graphically by a plot of spot latitudes against dates of observation. It has, for each cycle, an outline suggestive of a butterfly. The diagram was originated by E.W.Maunder at Greenwich in 1904. *LMD.*

Byrgius Obscure lunar crater west of Mare Humorum; 25°S, 65°W; the small crater on its crest, Byrgius A, is a ray-centre.

Byurakan Observatory More properly, the Astronomical Observatory of the USSR Academy of Sciences. It is in Armenia (40° 20′N, 44° 17′5E) at a height of more than 4,900ft (1,500m). The largest telescope is a 102in (260cm) reflector.

Byurakan Observatory: founded in 1946

Cabæus Deep crater near the Moon's south pole; 85°S, 45°W.

Cælum (the Graving Tool). Originally Caela Sculptoris, the Sculptor's Tools. It was introduced by Lacaille in 1752, but for no good reason, because it has no star above magnitude 4.4; one variable, R, has a range of from 6.7 to 13.7 (period 391 days, Mira type). Cælum adjoins Columba, and is always very low from British Latitudes. *PM.*

Calendar A system of measuring time. The ancient Egyptians used a calendar based on a solar year, while the Babylonians (and modern Hebrews and Moslems) used a lunar year of 12 months, which is 11 days shorter than a solar year, so it has an extra month every third year.

Our present calendar is based on that of the Romans which originally had only 10 months. Then Julius Caesar, on

the advice of Sosigenes, introduced the Julian Calendar in 44BC. It had 365.25 days, ie, a Leap Year of 366 days every fourth year. The year commenced on March 25 (the Feast of the Annunciation, and approximately the VERNAL EQUINOX); this calendar is now known as "Old Style". As a year actually has 365.24219 days, an error of almost 8 days accumulates per 1,000 years.

Pope Gregory XIII introduced the Gregorian calendar in 1582: this shortened that year by 10 days, so that October 15 followed October 4, and decreed that only century years divisible by 400 (eg, 1600, 2000) are leap years; this is accurate to one day in 3,300 years: the start of the year was also changed to January 1.

Britain adopted the Gregorian calendar in 1752, by which time the error was 11 days, so September 14 followed September 2. This is why the British Financial Year ends on April 5: it is Old New Year's Day (March 25) plus the 11 days lost in 1752! *TJCAM.*

Calippus Well-formed lunar crater, diameter 19 miles (31km), at the north end of the Caucasus Mountains; 39°N, 11°E.

Callisto The outermost of the four large moons of Jupiter. It has the lowest average density, about 1.3gm/cm³, of the four satellites. This fact, coupled with its large radius, 1,500 miles (2,400km), indicates that the material from which it was formed must have contained large amounts of ices.

Callisto revolves about Jupiter in a nearly circular orbit in the equatorial plane of the planet. The satellite is

Callisto: icy Jovian satellite

locked in synchronous rotation with one face continuously directed toward Jupiter. In general, the rate of impact of debris due to the focusing effect of the gravitational field of Jupiter will decrease with distance from the planet. On the other hand, the frictional heating due to tidal torques is a steeper function of distance than the gravitational focusing; hence, the effect of impacts should dominate with distance. When Callisto is compared to the other Galilean satellites, it is the darkest and most heavily-cratered body. This implies that the visible surface of Callisto has preserved the record of impacting debris and has not undergone extensive modification as a result of tidal heating.

The surface of Callisto is saturated with craters; however, there is little vertical structure associated with these features. In addition to the dense cratering, there are at least nine large impact features on the surface of Callisto. These structures are surrounded by a series of concentric rings. The largest of these features, Valhalla, has a bright central, lightly cratered region about 370 miles (600km) in diameter with surrounding concentric ring structures extending outward for 1,250 miles (2,000km). The ring spacing and the crater counts increase outward. A possible explanation is that there has been an inward flow of underlying regions to refill the initial crater caused by the impact. The low density, lack of vertical structure and recoil of large features suggest a weak outer crust, overlying a water mantle. The remoteness of Callisto from Jupiter would have resulted in lower impact velocities and less tidal heating than the other satellites, allowing a crust to form that is thick enough to preserve a long cratering record. *RFB.*

Caloris Basin The largest (800 miles/ 1,300km in diameter) multi-ring structure on the illuminated hemisphere of

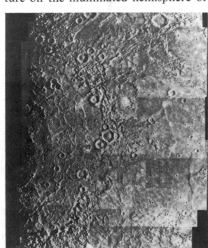

Caloris Basin: asteroid impact on Mercury

Mercury as imaged by Mariner 10. It is centred on 30°N, 190°W. The sunlit portion of the basin displays an outer ring of irregular mountain masses, highest towards the inner side and grading away with increasing distance from the basin centre. The mountain-free surface within the ring is marked by quasi-concentric and other ridges, which are transected by younger crack-like graben. The whole structure is undoubtedly the modified scar left by the impact of an asteroid-sized body, the floor being the end result of refilling of the crater by the crusting-over, semi-molten asthenosphere. *EAW.*

Caloris Montes The mountainous rim of the CALORIS BASIN.

Caloris Planitia The international name for the CALORIS BASIN.

Calypso Small satellite of Saturn; dimensions 21 × 13.5 × 13.5 miles (34 × 22 × 22km). It was discovered in 1980, and is co-orbital with Tethys and the small Telesto.

Cambridge Radio Observatory *See* MULLARD RADIO ASTRONOMY OBSERVATORY.

Camelopardalis (the Giraffe, alternatively known as Camelopardus). A barren northern constellation, introduced by Hevelius in 1690. It covers an area of over 750 square degrees, but it has no star brighter than magnitude 4, and there are no objects of particular interest. It adjoins Cassiopeia to the one side and Ursa Major to the other.

Campbell, William W. (1862–1938). Director of the Lick Observatory 1900–1930. His work included the discovery of 339 spectroscopic binaries.

Campo del Cielo Meteorite Argentinian shower of large iron meteorites, 1933.

Cancer (the Crab). A dim, barren Zodiacal constellation, between Castor and Pollux to the one side and the Sickle of Leo to the other. In legend it represents a crab which Juno, queen of the gods, sent to the rescue of the multiheaded hydra which was battling against Hercules. Not surprisingly, Hercules disposed of the crab by treading on it! There are only two stars above the fourth magnitude: Beta (3.52) and Delta (3.94).

The outline of Cancer recalls a very faint and distorted Orion. Zeta is a multiple system; the main components

are of magnitudes 5.1 and 5, and each is again a very close double. R Cancri is a Mira variable; range 6.2 to 11.8, period 362 days. RS Cancri, of type M, is a semi-regular variable (5.3 to 6.4; 120 days) and so is X (5.0 to 7.3, period 170 days, type N).

The open cluster Praesepe (M44) is one of the most famous in the sky, and is an easy naked-eye object; it is flanked by Delta (3.94) and Gamma (4.66) — the "Asses"; Praesepe has been nicknamed "the Manger" or "the Beehive". Also in Cancer is M67, a very old open cluster of integrated magnitude 6.1; it is thus on the fringe of naked-eye visibility. *PM.*

Canes Venatici (the Hunting Dogs). One of Hevelius' constellations, drawn in his map of 1690; it represents the dogs Asterion and Chara, held by the herdsman Boötes. The only star above the fourth magnitude is Cor Caroli; magnitude 2.90, absolute magnitude 0.1, type A0p, distance 65 light-years. The name was given to it by Edmond Halley, in honour of King Charles I; the star is the prototype magnetic variable. It is a double; magnitudes 2.9 and 5.4, separation 19″.7 — an excellent object for a small telescope.

Canes Venatici adjoins Ursa Major. Y is a semi-regular variable; 5.1 to 6.6, period 158 days, type N.

The famous Whirlpool Galaxy, M51, lies in Canes Venatici; this was the first spiral to be recognized as such, by Lord Rosse in 1845. Other spirals are M106 (integrated magnitude 8.6) and M63 (9.5). There is also the 6.3-magnitude globular cluster M3, which is quite distinct with binoculars. *PM.*

Canis Major (the Great Dog). One of Orion's hunting dogs; it is distinguished by the presence of Sirius, the brightest star in the sky, which lies in line with Orion's Belt. The leading stars of Canis Major are:

Brightest stars

Name	Visual Mag.	Abs. Mag.	Spec.	Distance (light-yrs)
α (Sirius)	−1.46	1.4	A1	8.6
ε (Adhara)	1.50	−4.4	B2	490
δ (Wezea)	1.86	−8.0	F8	3060
β (Mirzam)	1.98v	−4.8	B1	710
η (Aludra)	2.44	−7.0	B5	2500
ζ (Phurad)	3.02	−1.7	B3	287
o²	3.03	−6.8	B3	2800

Also above the fourth magnitude are Omicron¹ (3.86), Omega (also 3.86), Tau² (3.95) and Kappa (3.96). Mirzam is variable over a very small range, and is the prototype of the class of variables known either as Beta Canis Majoris or Beta Cephei stars. Sirius is much the least luminous of the chief stars.

There are three fairly bright variables; R (6.2 to 6.8, 1.14 days, Algol type), W (6.9 to 7.5, type N, irregular) and UW (4.7 to 5.3, 4.4 days, Beta Lyræ type). There is one very prominent open cluster, M41, which lies near the fourth-magnitude star Tau², not far from Sirius; it is an easy naked-eye object even from British latitudes, from which it is rather low down. *PM*.

Canis Minor (the Little Dog). The second of Orion's dogs.

Brightest stars

Name	Visual Mag.	Abs. Mag.	Spec.	Distance (light-yrs)
α (Procyon)	0.38	2.6	F5	11.4
β (Gomeisa)	2.90	−0.2	B8	137

Procyon, the eighth brightest star, makes the Little Dog very easy to identify. There are no other objects of note, though three Mira variables (V, R and S) can rise to above magnitude 8 at maximum brightness. *PM*.

Cannon, Annie J. (1863–1941). American astronomer who worked at the Harvard College Observatory and made an outstanding contribution to the classification of stellar spectra.

Canopus (Alpha Carinæ). Brightest star in the constellation Carina and, with apparent magnitude − 0.72, the second brightest star (after Sirius) in the sky.

Cape Canaveral *see* KENNEDY SPACE CENTER.

Capella Alpha Aurigæ, the brightest star in Auriga. Apparent magnitude 0.08, spectral class G8 + F (spectroscopic binary), distance 45 light-years.

Cape Observatory *See* ROYAL OBSERVATORY CAPE OF GOOD HOPE.

Capricornus (the Sea-Goat). One of the less brilliant of the zodiacal constellations; it has been identified with Pan.

Brightest stars

Name	Visual Mag.	Abs. Mag.	Spec.	Distance (light-yrs)
δ (Deneb al Giedi)	2.87	2.0	A5	49
β (Dabin)	3.08	4.0	F8	104

Also above the fourth magnitude are Delta² or A1 Giedi (3.57), Gamma or Nashira (3.68) and Zeta (3.74).

Alpha² forms a wide optical double with Alpha¹ (magnitude 4.5); the two are separated by 376″. The semi-regular variable RT has a range of from 6.5 to 8.1, and a period of 395 days; the spectral type is N. The most important nebular object is the globular cluster M30, with an integrated magnitude of 8.4.

Capricornus adjoins Aquarius, Piscis Austrinus and Sagittarius. There is no really distinctive pattern, but the group is not hard to identify. *PM*.

Captured rotation The case where a satellite's period of rotation is the same as its period of revolution around its primary, thus always presenting the same side to the primary: due to tidal effects. Also termed "synchronous rotation".

Carbonaceous chondrites Chondrite meteorites with atomic magnesium to silicon ratios greater than 1.02, further divided on trace element or textural grounds into four groups. The CI group (Carbonaceous. *I*vuna type) has five members, none with chondrules, less than 20 per cent water and about three per cent carbon as graphite, carbonate and organic compounds; the chemical composition is like the Sun, without volatiles. CM chondrites (33 known) are similar to CI, but with some chondrules; related to micrometeorites. The CV group (14 known) has large chondrules; rich in refractories (aluminium, iridium) with little water or carbon. The CO group (ten known) has small chondrules, poorer than CV in refractories and with little carbon or water. *RH*.

Carbon dioxide (CO_2). A gas which is the dominant constituent of the atmospheres of Venus and Mars.

Carbon-nitrogen cycle A cycle of nuclear reactions first described by Hans BETHE in 1938 which accounts for the energy production inside MAIN SEQUENCE stars of mass greater than the Sun.

The reactions (*See* NEUTRINOS) involve the fusion of four hydrogen nuclei into one helium nucleus at temperatures in excess of four million degrees. Some matter is converted into energy and the mass of one helium nucleus (4.0027 atomic mass units) is consequently less than the total mass of four protons (4.0304 units).

The presence of carbon is essential as one of the reactants, but it behaves like a catalyst and is reproduced in its original form after fusion. Isotopes of carbon, oxygen and nitrogen occur as transient intermediate products during the reactions. *RCM*.

Carbon stars Cool stars whose surface composition contains more carbon than oxygen. Many stars, when they age, expand to become RED GIANTS, and their surfaces cool. The two important elements carbon and oxygen may then combine to form carbon monoxide. Early in this ageing process most stars have more oxygen than carbon; excess oxygen combines with metals to give a characteristic spectrum. Some stars have more carbon, or else dredge up carbon that they have made within. The spectrum is then quite different, featuring many carbon-based molecules. Carbon stars are rare in our Galaxy, but common in the MAGELLANIC CLOUDS. *DAA*.

Carina (the Keel). This is part of the dismembered constellation Argo Navis, the Ship Argo. It is the brightest and richest part of Argo, and contains Canopus, the brightest star in the sky apart from Sirius.

Brightest stars

Name	Visual Mag.	Abs. Mag.	Spec.	Distance (light-yrs)
α (Canopus)	−0.72	−8.5	F0	1200
β (Miaplacidus)	1.68	−0.6	A0	85
ε (Avior)	1.86	−2.1	K0	200
ι (Tureis)	2.25	−4.7	F0	815
θ	2.76	−4.1	B0	750
ν	2.97	−2.0	A7	323
p	3.32	−1.7	B3	313
ω	3.32	−1.0	B7	228
q	3.40	−4.4	K5	913
a	3.44	−3.0	B2	619
χ	3.47	−3.0	B2	587

Next come u (3.78), R (maximum 3.9), c (3.84) and x (3.91).

Quite apart from Canopus — which is so remote that estimates of its distance vary quite widely — there are many interesting objects in Carina. Upsilon is a double; 3.1 and 6.0, separation 4″.6. Of the variable stars, special mention must be made of the unique Eta, which at one time in the last century was the brightest star in the sky apart from Sirius, but is now just below naked-eye visibility; it is associated with the magnificent Keyhole Nebula, NGC3372. Other variables with maximum above magnitude 6.5 are:

Variable stars

Name	Range	Period (days)	Spec.	Type
U	6.4-8.4	38.8	F-G	Cepheid
R	3.0-10	381	M	Mira
I	3.4-4.8	35.2	F-G	Cepheid
s	4.5-9.9	150	M	Mira

There are several naked-eye open clusters; NGC2516, NGC3114 and IC2581. The cluster round Theta is very spectacular when viewed with binoculars. There is also a sixth-magnitude globular cluster, NGC2808.

Carina adjoins Vela; the so-called False Cross is made up of Iota and Epsilon Carinæ and Kappa and Delta Velorum. The False Cross has been confused with the Southern Cross, but is larger and less brilliant, though its four stars are more equal. The constellation of Volans intrudes into Carina between Miaplacidus and Canopus. *PM.*

Carme Jupiter's 11th satellite; diameter 25 miles (40km). It has retrograde motion.

Carpathian Mountains Range forming part of the border of the lunar MARE IMBRIUM; 15°N, 24°W. Peaks rise to 7,000ft (2,100m).

Carrington, Richard C. (1826–75). English amateur astronomer who, in 1859, made the first observation of a SOLAR FLARE (a white light flare).

Carter Observatory The national observatory of New Zealand. It was established in Wellington city in 1941 from an 1896 bequest by a prominent local pioneer. The observatory is the primary source of astronomical information in New Zealand and is involved in education programmes from school to university level. Research is carried out on the morphology of galaxies, eclipsing binaries and occultations. Instruments include 9in (23cm) and 6in (15cm) refractors on the Wellington site. In 1978 an outstation with a 16in (41cm) Boller and Chivens Cassegrain reflector (equipped with astrograph and photoelectric photometer) was opened on Black Birch Mountain, on New Zealand's South Island. *DLB.*

Casatus Lunar crater, diameter 65 miles (105km), in the far south; 75°S, 35°W. It overlaps its neighbour Klaproth.

Cascade image tube A type of IMAGE INTENSIFIER, an electronic device used to enhance faint optical images. In a simple image tube, the incident light beam falls on a photocathode and a stream of electrons is liberated by the photoelectric effect. These electrons are accelerated by an electric field and are focused onto a phosphor screen, so that an image is produced. The image from one tube can be made to fall onto the photocathode of another image tube, and the intensification process then repeated. Several image tubes can be cascaded together in this way to give great image enhancement.

Cassegrain telescope Reflecting telescope with a concave paraboloidal primary mirror and convex hyperboloidal secondary mirror. Invented by M. Cassegrain in 1672, this was the second successful mirror-based telescope (after Isaac NEWTON's). The concave primary mirror gathers light from the sky, reflects it up to the convex secondary mirror; the second reflection passes back down through a hole in the primary mirror to the so-called "Cassegrain focus" below. The optical design with concave and convex mirrors tends to cancel the aberrations such as coma of the separate mirrors. The arrangement of mirrors gives an accessible on-axis position for the focus, which is conveniently near the ground and symmetrically located. This is good for installing equipment. It is the most common arrangement for reflecting telescopes of substantial size (24in/60cm aperture or bigger). *PGM.*

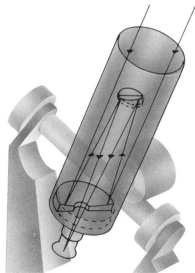

Cassegrain telescope: accessible focus

Cassini division The main division in Saturn's ring system, separating the two bright rings A and B. It is not empty; Voyager results show that it contains several narrow rings.

Cassini, Giovanni Domenico (1625–1712). Italian astronomer, also known as Jean Dominique Cassini, who became the first Director of the Paris Observatory. He discovered the rotation of Jupiter and Mars, published improved tables of Jupiter's satellites, discovered four new satellites of Saturn, and detected the main, CASSINI DIVISION in Saturn's ring system. He combined observations of Mars made at Paris and in Cayenne, French Guiana (by Jean Richer), to obtain the parallax and distance of Mars and hence deduced a value for the distance of the Sun (86,202,600 miles/138,730,000km), far superior to previous estimates. *IKN.*

Cassini, Jaques J. (1677–1756) Astronomer who succeeded his father, G.D.CASSINI, as Director of the Paris Observatory and attempted to measure the figure of the Earth.

Cassiopeia A familiar northern constellation, representing the proud queen of the Perseus legend. The five leading stars make up a W or M pattern, so far north that from British latitudes it is circumpolar; it and the main stars of Ursa Major lie on opposite sides of the Pole Star, and at roughly the same distance from it.

Brightest stars

Name	Visual Mag.	Abs. Mag.	Spec.	Distance (light-yrs)
γ	2.2v	−4.6	B0p	782
α (Shedir)	2.23v?	−0.9	K0	121
β (Chaph)	2.27	1.9	F2	42
δ (Ruchbah)	2.68	2.1	A5	62
ε (Segin)	3.38	−2.9	B3	522
η (Achird)	3.44	4.6	G0	19

Next come Zeta (3.67) and L (3.98). Gamma is an unstable star which is irregularly variable; Shedir may also be variable over a small range. Near Beta is Rho, which is a variable of uncertain type; usually it is of around magnitude 5, but on rare occasions falls to below 6. Other variables are R (5.5 to 13.0, 431 days, Mira type); RZ (6.4 to 7.8, 1.2 days, Algol type) and SU (5.9 to 6.3, type 1.9 days, Cepheid). Eta is a fine star binary, magnitudes 3.6 and 7.5, separation 10″. The period is certainly over 500 years. This is a good pair for viewing with a small telescope.

Cassiopeia also contains several open clusters within binocular range: M52, NGC457, NGC559, NGC663, and M103 (NGC581).

The constellation is crossed by the Milky Way; the area is very rich. *PM.*

C

Cassiopeia A 2321 + 58. Very powerful radio supernova remnant.

Castor Although known as Alpha Geminorum, this magnitude 1.6 star is in fact fainter than Beta Geminorum (Pollux).

Castor is one of the finest visual binary systems in the sky for the small telescope. It is a multiple system with each of the stars being in turn a spectroscopic binary. The period of the main pair is about 500 years and the separation is currently increasing (2.7 arc seconds in 1986).

A distant companion Castor C (*see* YY GEMINORUM) orbits the system and is itself an eclipsing binary. *RWA²*.

Cataclysmic variables A term given to a diverse group of stars that erupt irrespective of the cause of the outburst. Collectively they are referred to as CVs but include SUPERNOVÆ, NOVÆ, RECURRENT NOVÆ, nova-like stars, FLARE STARS, some X-ray objects, dwarf novæ and other erupting stars. CVs are, more correctly, very close binary systems whose outbursts are caused by interaction between the two components. A typical system of this type has a low mass secondary which fills its ROCHE LOBE, so that material is transferred through its LAGRANGIAN POINT onto the primary, which is usually a WHITE DWARF, either non-magnetic or only weakly so. The transferred material has too much angular momentum to fall directly on to the primary, but forms an ACCRETION disk, on which a hot spot is formed where the infalling material impacts on its outer edge. For any particular star outbursts occur at irregular intervals from about ten days to weeks, months or many years.

The foregoing general outline fits most CVs such as dwarf novæ, of which more than 200 are known. Their general characteristics are now outlined before other aspects are discussed.

Dwarf novæ are divided into three main subclasses in accordance with their optical behaviour. Each class is designated by the name of a type star. The first of these are SS Cygni or U GEMINORUM stars, named after the first two dwarf novæ discovered. Rising from minimum to maximum in one or two days, they return to minimum in several days. Their amplitudes are from two to six magnitudes. The average intervals, in days, between consecutive maxima is called the mean cycle. These vary from star to star but each star has its own characteristic mean cycle, which often varies within fairly wide limits. The cycles for stars of this class range from ten to several hundred days. Outbursts are either termed normal (or short) or wide (or long). The first have durations of a few days while the second last for up to about 20 days. Occasionally a star will take several days to rise to maximum. These maxima are called anomalous bursts.

Stars of the SU Ursæ Majoris type have two kinds of outbursts, which are either normal, lasting from one to four days, or super maxima, when the star becomes from half to one magnitude brighter than at normal, outbursts remaining at maximum for ten to 20 days. For almost all maxima, stars of this type rise to near maximum brightness in 24 hours or less. Some have a short pause at an intermediate magnitude on the rise for an hour or two. They also have periodic oscillations, termed superhumps, superimposed on the light curve with amplitudes $0^m.2$ to $0^m.5$. Their period is about 3 per cent longer than the orbital period.

The third class of dwarf novæ are called Z CAMELOPARDALIS stars and differ from other classes in remaining at an intermediate magnitude between maximum and minimum for several cycles. These stars take several days to rise to maximum at times. Their mean cycles are shorter, ranging from ten to 40 days. RECURRENT NOVÆ can be regarded as ordinary novæ that have had more than one outburst or as dwarf novæ with very long intervals between outbursts. They are close binary systems with a cool red star and a less massive dwarf. Typical recurrent novæ are RS Ophiuchi, which had outbursts in 1898, 1933, 1958 and 1985, and T Pyxidis which flared up in 1890, 1902, 1920 and 1944. The amplitudes are seven to nine magnitudes while reoccurence cycles are from ten to 80 years. These stars are brighter at minimum than common novæ, but not because of the cycle length; the stars of longer cycles tend to faster development and brighter maximal luminosity.

Most of the objects discussed so far show a relationship between maximal brightness and length of the mean cycle in the sense that the fainter the maximum magnitude is, the shorter the cycle. Nova-like variables are a less homogeneous group and also less well studied. Some have bursts of limited amplitude, whilst others have had no outburst but have spectra resembling old novæ. It can be concluded that some nova-like stars are true CVs because they exhibit some of the phenomena seen in CVs. There are many other objects that show some if not all of the characteristics of CVs. These include old novæ, some X-ray objects, AM HERCULIS stars and others.

Many models have been suggested to explain the cause of the observed outbursts. The two most probable theories are that they are caused either by variations in the rate of mass transfer or by instabilities in the accretion disk. Both models require the transfer of mass from the red secondary. It is fortunate that a few systems are so aligned that we see them undergoing eclipses, so that enables the main light source to be studied in detail. Typical examples of such systems are Z Chamaeleontis and OY Carinae; these stars can be examined both at outburst and minimum. It appears that during outbursts the disk increases in brightness. Both these systems have an inclination of 79°. The intervals between consecutive outbursts of the same type vary fairly widely as they do for most CVs. Z Chamaeleontis has a mean cycle for normal outbursts of 82 days; for super-outbursts 287 days. For OY Carinae the respective values are about 50 and 318 days. Their semi-periodic oscillations, timed in seconds, are small.

The advantage of eclipsing systems is that it is possible to see both primary and hot spot eclipsed. This gives the probable sizes of these components. Models can then be proposed and tested against observations. Certain details have to be assumed, such as probable masses of two components, and these have been given various values. Despite this, it is generally agreed that a variable mass rate can account for many observations. This poses a question. Why does the mass flow rate vary? There must be an instability in the red star, but no generally accepted theory has been advanced. Obviously the red star must relax after a burst until another surge again overflows, discharging another burst of gas, but exactly what causes this ebb and flow is a mystery. The angular momentum of the mass flow prevents it from falling directly on to the primary. Instead a disk is formed around it on which a hot spot is formed at the point of impact. Some of this matter must be carried away. Is this lost to the system or does it splash back on to the disk? Another unsolved problem! When an eclipsing system is at minimum, the main light source at primary eclipse comes from the red star. The disk is then in a steady state.

Theories that stipulate that the cause of the outburst is instabilities in the disk, while accepting variable mass transfer as the origin of the subsequent observed phenomena, contend that it is what happens on the disk that gives rise to outbursts. A number of disk models have been proposed, and depend on the data set assumed. None appears to fit the observed facts. For example, the disk increases in brightness during outbursts, at least in the eclipsing systems, and presumably in other dwarf novæ. Disk instability models differ on how this released energy spreads through the disk, which mainly radiates in the ultraviolet. This does not behave in the same way as the visual and appears to be against these models. If the disk dumps energy onto the primary, nuclear burning would be expected. Evidence on this is unclear. *FMB.*

Catadioptric system A closed tube telescope or astronomical camera employing a combination of lenses and mirrors to form the image. The SCHMIDT-Cassegrain and MAKSUTOV are two common types.

The primary mirror may be spherical or parabolic depending on the system. The secondary mirror has a matching curvature and the lens element is a full aperture correcting plate placed in front of both mirrors. As a telescope, the system combines the advantages of both refractors and reflectors as well as allowing a compact design. As a camera it is extremely fast and offers a wide flat field of view. *JCDM.*

Catharina Lunar crater, 55 miles (88km) in diameter; 18°S, 24°E; a member of the Theophilus chain. It has no central peak.

Caucacus Mountains An important lunar mountain range, 36°N, 8°E, forming part of the border between the MARE IMBRIUM and the Mare Serenitatis. Some of the peaks rise to 12,000ft (3,650m).

Celæno One of the brighter stars in the Pleiades cluster.

Celestial equator *See* CELESTIAL SPHERE.

Celestial latitude *See* CELESTIAL SPHERE.

Celestial longitude *See* CELESTIAL SPHERE.

Celestial maps Constellation figures appear in Egyptian royal tomb paintings from the second millennium BC,

and, though none has survived, star maps and globes were known to have been used in classical Greece. The earliest star chart (excluding those on globes and ASTROLABES) known to have survived is in a Chinese manuscript of c.AD940 (British Library MS Stein 3326), while the earliest Islamic map is preserved in the Bodleian Library in Oxford in a 1010 manuscript copy of AL-SÛFI's book (MS Marsh 144). The earliest Western maps, of both hemispheres, are in manuscripts

Celestial maps: Durer's star chart

dating from about 1440, preserved in Vienna and probably based on now-lost maps drawn about 1425, originally owned by REGIOMONTANUS.

The earliest printed star charts were the planispheres of Albrecht Dürer of 1515, which follow the Vienna planispheres, while the earliest printed book with star charts was Alessandro Piccolomini's *De le Stella Fisse* (1540), based on PTOLEMY's star positions.

The first serious star atlas was BAYER's *Uranometria* (1603), based on BRAHE's star positions. Important star atlases in the pre-photographic era include: HEVELIUS's *Uranographia* (1687), FLAMSTEED's *Atlas Cœlestis* (1729), BODE's *Uranographia* (1801), ARGELANDER's *Uranographia Nova* (1843), and GOULD's *Uranometria Argentina* (1877), mostly based on their authors' STAR CATALOGUES. For the southern skies, the planispheres of HALLEY (1678) and LACAILLE (1756-63) deserve mention.

The *Franklin-Adams Charts* of 1914 constitute the earliest photographic atlas of the whole sky. The magisterial *National Geographic Society Palomar Observatory Sky Survey* of 1954-8 is still in active use.

For "star gazing", *Norton's Star Atlas*, first published in 1910 with new editions to the present day, is a classic,

followed by the *Skalnaté Pleso Atlas of the Heavens 1950.0* and Vehrenberg's *Photographischer Stern-Atlas* (1962-4). The latest in this genre is Tirion's *Sky Atlas 2000.0* (1982). *HDH.*

Celestial mechanics Celestial mechanics (or dynamical astronomy) is concerned with using the laws of physics to explain and predict the orbits of the planets, satellites and other celestial bodies. The subject can be said to have started with the publication of NEWTON's *Principia* in 1687, in which are stated his law of gravitation, which describes the forces acting on the bodies, and his three laws of motion, which describe how these forces cause accelerations of the motions of the bodies. Then the techniques of celestial mechanics are used to determine the orbits of the bodies resulting from these accelerations.

One of the first results achieved by Newton was to give an explanation of KEPLER's laws. These laws are descriptions deduced from observations of the motions of the planets as being elliptical orbits around the Sun, but until the work of Newton no satisfactory explanation of these empirical laws had been given. However, Kepler's laws are true only for an isolated system of two bodies; in the real Solar System the attractions of the other planets and satellites cause the orbits to depart significantly from elliptic motion and, as observational accuracy improved, these perturbations became apparent. The greatest mathematicians of the 18th and 19th centuries were involved in the effort of calculating and predicting the perturbations of the orbits, in order to match the ever-increasing accuracy and time span of the observations. The orbit of the Moon was the major problem, partly because the Moon is nearby, and so the accuracy of observation is high, but also because its orbit around the Earth is very highly perturbed by the Sun. Various techniques were developed for calculating the perturbations of an orbit. An important technique is the method of the "variation of arbitrary constants", developed by Euler and Lagrange. An unperturbed orbit can be described by the six elements, or arbitrary constants, of the ellipse. The effect of perturbations can be described by allowing these "constants" to vary with time. Thus, for example, the eccentricity of the orbit may be described as a constant plus a number of periodic terms. The resulting expressions are called the "theory of the motion of the body",

C

C

and can be very lengthy in order to achieve the desired accuracy, perhaps hundreds of periodic terms. Around the middle of the 19th century an alternative method was developed in which the perturbations of the three coordinates of the body (eg longitude, latitude and distance from the Sun) are calculated instead of the perturbations of the six elements. Variants of this method have been used ever since for lunar and planetary theories, but the variation of constants is still more suitable for many of the satellites. The latest theories used to calculate the positions of the Moon and planets in the almanacs used by navigators and astronomers are those derived by Newcomb for the five inner planets, Uranus and Neptune; by Brown for the Moon; and by Hill for Jupiter and Saturn.

Overall, Newton's four laws and the techniques of celestial mechanics have proved successful at explaining the motions of the planets and satellites; problems with the orbit of the Moon were eventually resolved by improved theories, and anomalous perturbations of the orbit of Uranus led ADAMS and LE VERRIER to suspect the existence of a further planet, which resulted in the discovery of Neptune in 1846 close to the predicted position. Problems with the orbit of Mercury, however, resisted solution. In 1859 Le Verrier announced that the perihelion of Mercury was advancing at a rate that could not be explained completely by perturbations from the known planets. The existence of a ring of asteroids, or of a small planet inside the orbit of Mercury, was suspected but, despite many false alarms, none was discovered. The problem was eventually solved in 1915 with the publication of Einstein's General Theory of RELATIVITY. This is a more accurate representation of the laws of motion under the action of gravitation, but the differences from using Newton's laws are small, and only become noticeable in strong gravity fields.

With the advent of computers, methods other than the lengthy algebraic expansions by hand for calculating the positions of the planets have become feasible. One method is to use the same techniques as before, but to do the vast amounts of algebraic manipulation involved on a computer. Excellent theories of the Moon and some of the planets have been produced in this way. However, the method that has proved most effective is the less elegant but much simpler and more accurate method of integrating the equations of motion by numerical techniques, and the positions of the Moon and planets in the almanacs are now all computed in this way. However, the old algebraic methods are still best for most of the other satellites.

In recent years the main emphasis of celestial mechanics has changed somewhat, and many theoretical problems and problems of origin, evolution and stability are being studied. The three-body problem has been extensively studied. This problem has no general solution, but families of a particular type of orbit that repeats itself periodically can be found, and the study of these orbits gives an insight into all possible orbits. The stability of the Solar System has long been of considerable interest, and recent numerical calculations suggest that it is stable for at least 100 million years (but this is still only about one-fiftieth of its age). Another problem that has defied complete solution for more than a century is to explain the origin of the gaps in the asteroid belt (first noticed by KIRKWOOD), which occur at certain distances from the Sun that correspond to commensurabilities of period with Jupiter. A lot of progress has been made, and the mechanism that causes the gap at the 3:1 commensurability has been fairly well explained, but the cause of the widest gap at the 2:1 commensurability is still not completely clear. The many occurrences of commensurabilities in the Solar System, particularly among the satellites, presents a challenge, as there are many more than would be expected by chance. It has now been shown that some of these could have been caused by orbital evolution due to tidal action, which would continue until a commensurability was encountered, whereupon the satellites would become trapped. Following the discovery by the Voyager spacecraft of rings that are trapped by the repulsion of shepherding satellites, it is clear that another mechanism that can cause evolution of satellite orbits is the repulsion from a nearby ring. Calculations suggest that these effects are very significant, particularly in the SATURN satellite system. It is a major problem to understand how the tiny satellite Atlas, which is in orbit only 560 miles (900km) outside the main ring, can remain so close under the action of the large repulsive force that must exist.

Many other interesting problems of dynamics have arisen following the Voyager observations, such as the cause of the intricate structure of Saturn's rings, which consist of hundreds of individual ringlets. The causes of these have not yet been fully explained, but a major factor seems to be the increased perturbations at distances corresponding to commensurabilities with the satellites, with some other features possibly caused by small unseen satellites orbiting within the rings. *ATS.*

Celestial meridian *See* CELESTIAL SPHERE.

Celestial poles *See* CELESTIAL SPHERE.

Celestial Police Nickname for the team of astronomers who assembled in 1800 to search for a planet between the orbits of Mars and Jupiter. They discovered three asteroids (Pallas, Juno and Vesta) between 1802 and 1807.

Celestial sphere. The stars and planets are so far away from us that our limited human stereoscopic vision is totally unable to resolve the real three-dimensional structure of the sky and there is, as a result, an unavoidable impression that everything we see in the sky is projected onto an enormous screen extending all around us as if we were inside a gigantic planetarium. This is the illusion of the celestial sphere on which we base our charts of the sky and against which we make our measurements. The illusion is so strong that the early astronomers postulated the existence of a crystal sphere to which the stars are fixed and this was supposed to be of very great radius.

Our vantage point, on the surface of a spinning and orbiting planet, impresses on the sky an apparent cyclical motion which we have learned to understand and which has become the basis of our present system of position measurement using the coordinates of right ascension and declination.

The most obvious behaviour of the celestial sphere is its apparent diurnal rotation due to the axial spin of the Earth. There is a daily east to west rotation of the sky as if it were on the same extended terrestrial axis. In the northern hemisphere we see some stars that never set — the circumpolar stars — that appear to turn about the polar point near to the bright star Polaris. There is, of course, a similar polar point in the southern hemisphere. Over a short period of time the direction of the polar axis seems fixed in space, but it is in fact slowly drifting because of the precessional effect of the Earth's axis. It is actually describing a circle some $23\frac{1}{2}°$ in radius, but the period of the precession is over 26,000 years and the effect is only noticeable over a few decades.

Having recognized one easily observed direction within the celestial sphere it is possible to define an equatorial circle that is equidistant from the poles and touches the horizon due east and due west. This equatorial circle — the celestial equator — is a projection of the Earth's equator onto the celestial sphere and is our view of the plane passing through the centre of the Earth and dividing the sky into two hemispheres. This enables us to visualize a set of great circles of celestial "latitude", usually called declination, each being parallel with the equator and allowing us to specify the position of an object in terms of angle above or below the equator — usually referred to as positive or negative declination.

Right ascension, the celestial equivalent of longitude, is also an extension of the terrestrial system. On Earth the zero of longitude was chosen quite arbitrarily by someone scribing a line on the ground at Greenwich and defining it as that line. From this line longitude is measured in degrees east or west of the Greenwich meridian, the great circle passing through Greenwich and the two poles.

The celestial zero of right ascension has to be chosen in a less arbitrary fashion so that observers anywhere on Earth can find it — after all it is rather difficult to scribe a line on the sky! It is the apparent motion of the Sun around the sky during a year that is used to define this point. The motion is naturally the result of the Earth's orbital motion around the Sun and the track traced by the Sun through the constellations, called the ecliptic, is therefore an indication of the position of the plane of the Earth's orbit, the observer being in the centre of this plane. The ecliptic is not the same as the equatorial plane because the Earth's axis is tilted by about $23\frac{1}{2}°$ to the orbital plane and it is this tilt that is responsible for our cycle of seasonal changes.

For half of each year the Sun is in the northern hemisphere of the sky and for the other half it is in the southern hemisphere. On two dates of each year the Sun crosses the equator; it rises due east and sets due west and we have equal lengths of day and night. These are known as the equinoxes. It is one of these, the spring equinox, when the Sun crosses into the northern hemisphere, that is chosen as the zero of right ascen-

sion. It is known as the first point of Aries, which was an accurate description originally but precession has now carried the equinox into the constellation Pisces.

Right ascension is measured eastwards from the first point of Aries, but unlike terrestrial longitude it is measured in time units rather than degrees, minutes and seconds of arc. The reason for this lies in how we measure the axial rotation of the Earth. We have two slightly different time scales. The most familiar one, civil time, is related to the apparent position of the Sun in the sky. The time interval between successive noons is 24 hours of civil or Greenwich Mean Time. The Earth, however, makes almost one degree more than a complete revolution in this time because is has moved almost one degree in its orbit around the Sun — a circle being 360° and a year being $365\frac{1}{4}$ days.

The civil timescale arranges for us to have our working day in the hours of sunlight and it is the obvious timescale for running our daily affairs. The

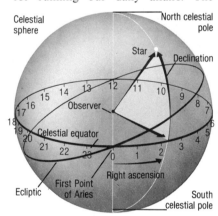

Celestial sphere: locating heavenly bodies

astronomer measures time relative to the first point of Aries rather than the Sun and his "sidereal" timescale has a day of 24 "sidereal" hours, which measure the rotation of the Earth relative to the stars, that is exactly 360° of axial rotation. In terms of civil time a sidereal day is 23 hours, 56 minutes and 4 seconds long, and the calendar year contains one more sidereal day then solar days because of the added orbital circuit.

The beginning of the sidereal day is when the first point of Aries lies on the meridian, the great circle linking the north and south points and passing directly overhead. After 24 sidereal hours it will be in that position again. The right ascension figures work like the face of a clock so that when the first

point of Aries is on the meridian the sidereal time is zero hours. At one o'clock sidereal time the sky will have rotated a further 15° and stars this angle east of the first point of Aries will then line on the meridian. A sidereal clock is an essential feature of an observatory since it tells exactly how the sky is orientated and makes identification of parts of the sky very much easier.

Extended observation of the celestial sphere shows that whereas the stars seem relatively fixed in position, and have only very small proper motions, there are a few objects that move quite quickly. These are the planets, which travel through the constellations in what appear to be rather convoluted paths. Early observers had identified Mercury, Venus, Mars, Jupiter and Saturn and had realized that these are restricted to a band of sky 7° wide on either side of the ecliptic. This belt of sky became known as the Zodiac. More recently it has been realized that the convoluted paths reflect the orbital motion of our own observing platform. The Zodiac is also seen to contain the tracks of Uranus and Neptune, but Pluto, discovered in 1930, strays as much as 17° from the ecliptic. This indicates the general planar structure of the Solar System, which becomes clear when distance measurements are taken into account to give a complete stereoscopic view. *RCM. See also* ASTROMETRY.

Celsius, Anders (1701–44). Swedish astronomer, professor in Uppsala, where he founded an observatory. Took part in Maupertuis's meridian arc measuring expedition to northern Sweden. Pioneer of stellar photometry and inventor of the (originally inverted) Centigrade or CELSIUS thermometer scale.

Celsius scale The temperature scale on which the freezing point of water is zero and the boiling point of water is 100°. Named after Anders CELSIUS (1701–44). Also known as Centigrade scale.

Censorinus Small but brilliant lunar crater near the south extension of the Mare Tranquillitatis; 0°, 32°E.

Centaurus (the Centaur). A brilliant southern constellation, unfortunately virtually inaccessible from European latitudes. There are no less than 13 stars above magnitude 3.5 (see table on next page).

C

Brightest stars

Name	Visual Mag.	Abs. Mag.	Spec.	Distance (light-yrs)
α	−0.27	4.7, 5.4	G2 K1	4.3
β (Agena)	0.61	−5.1	B1	460
θ	2.06	1.7	K0	46
γ	2.17	−0.6	A0	111
ε	2.30	−3.5	B1	489
η	2.31	−2.9	B3	359
ζ	2.55	−3.0	B2	359
δ	2.60	−2.5	B2	326
ι	2.75	1.4	A2	52
μ	3.04	−1.7	B3	290
κ	3.13	−2.5	B2	424
λ	3.13	−0.8	B9	186
ν	3.41	−2.5	B2	489

C

Next come Phi (3.83), Tau (3.86), Upsilon (3.87), d (3.88), Pi (3.89) and Sigma (3.91).

Alpha and Beta are known as the Pointers, because they indicate the direction of the Southern Cross; since Alpha is the nearest of all the bright stars, and Beta is very remote, there is no true connection between them. Alpha, which has no official proper name (though it has been called "Rigel Kent" or "Toliman") is a splendid binary; the components are of magnitudes 0.0 and 1.4, and the period is 80 years, with a separation ranging from 2″ to 22″. Gamma is a close binary with an 85-year period; the components are almost equal, but the separation is never greater than 1″.7.

R is a Mira-type variable; range 5.4 to 11.8, and the period is unusually long (547 days). Since the star lies between Alpha and Beta, it is very easy to locate. T is a semi-regular variable (5.5 to 9.0, 91 days, type M), and there are several other Mira and semi-regular variables that can become brighter than the eighth magnitude.

The globular cluster Omega Centauri is much the finest in the entire sky. It is an easy naked-eye object, and even a small telescope will resolve the outer regions into stars; it lies in line with Beta and Epsilon. There are also two bright open clusters (NGC5460 and NGC3766), and a fine planetary, NGC3918.

Centaurus is one of Ptolemy's original 48 constellations, and is one of the most splendid groups in the sky, but no well-defined mythological legends seem to be attached to it. *PM*.

Centaurus A Long known as NGC5128, this galaxy was discovered to be a powerful source of radio emission by the pioneer radio astronomers in Australia. Optically it appears to be a normal elliptical galaxy but crossed by a dark and very prominent dust lane. Very deep photographs reveal it to be more than 1° across but the radio image is much bigger. This enigmatic object was initially thought to be two galaxies in collision, a theory recently given more credence by the discovery of low-contrast internal "shells", which seem to indicate that what we see is a broad band of material left behind after the encounter of a massive elliptical galaxy with a smaller dusty spiral. The obscured nucleus of the galaxy is still embracing the interloper and is emitting a huge amount of energy at X-ray, optical and radio wavelengths. Centaurus A is at a distance of about 3 Mpc (about 10 million light-years). A supernova was observed in it in 1986—discovered by an Australian amateur, the Rev. Robert Evans. *DFM*.

Centaurus A: powerful radio source

Cepheid variables Stars so called because the first variable of this type was named Delta Cephei. They pulsate in a regular manner. These stars have left the MAIN SEQUENCE and occupy, on the HERTZSPRUNG-RUSSELL DIAGRAM, a position to the right of the upper main sequence and to the left of the red giants. This region is termed the Cepheid Instability Strip. Cepheids are passing through the first Instability Transition after leaving the main sequence.

During this brief period in their lives these stars oscillate, alternately expanding and contracting so that in each cycle a star may change in size by as much as 30 per cent. These regular, rhythmic changes in size are accompanied by changes in luminosity. The surface temperature also changes in the course of each cycle of variations in brightness, being at its lowest when the star is at minimum and at its highest when the star is brightest. This temperature change can equal 1,500K for a typical Cepheid. A change in temperature also means a change in spectral type, so that this can be F2 at maximum, becoming the latter type, G2, at minimum, changing in a regular manner as the temperature falls or rises. A Cepheid may continue to pulsate in this manner for a million years, which compared to the life span of a star is a comparatively short time.

Most stars spend at least some time as Cepheid variables. The stars like Delta Cephei have amplitudes of around 0ᵐ5 and periods usually not longer than seven days; there are, however, Cepheids with larger amplitudes and longer periods forming a separate sub-type. These include the naked eye stars, l Carinae, Beta Doradûs and Kappa Pavonis. The period of light changes is related to the average luminosity of the star. This means that the absolute magnitude of a Cepheid variable can be found by measuring the period of the light cycle. The apparent magnitude can be obtained directly. Once these three values are known—period, apparent and absolute magnitude — then it becomes possible to determine the distance to the star. It is this unique property that makes Cepheid variables so important as distance indicators.

Cepheids are visible in external galaxies because their absolute magnitudes are brighter than the RR LYRÆ VARIABLES by a factor of a thousand for the brightest stars. Their value as distance indicators is compounded by the fact that there are two types of Cepheids. Both types follow a period-luminosity relationship, but their curves are different. First there are the classical Cepheids, such as Delta Cephei itself, which are yellow supergiants of Population I. The second type are Population II stars and are found in globular clusters and in the centre of the Galaxy. Thus, in using Cepheids to determine distances it is necessary to know which type of Cepheid is being observed. At the time Cepheids were first used to determine distances it was not known that they were Cepheids with different period-luminosity values. This resulted in applying the value for type II Cepheids to the classical Cepheids in error, which affected the distances to external galaxies. When this error was found, in 1952, the result was to double the size of the universe.

The period-luminosity relationship means that the longer the period, the brighter the visual absolute magnitude. A comparison of the curves show that the classical Cepheids are about one magnitude brighter than type II

Cepheids. The light curves can be arranged in groups, according to their shapes, which progressively become more pronounced in each group as the period lengthens. Most Cepheid light curves fall into one of about 15 such divisions, each with a longer average period. They all follow the period-luminosity relationship, which commences with the RR Lyræ stars of very short period and after a break is continued by the Miras. This regular progression—the longer the period, the later the spectral type—is called the Great Sequence. A typical Cepheid would have a surface temperature varying between 6,000 and 7,500K and an absolute luminosity that is ten thousand times that of the Sun.

Since Cepheids are in a part of the H-R diagram where changes occur, observations are directed towards detecting changes in periods. Such changes are small but give information as to how stars progress through the instability strip. The method used to detect such changes in the long-period Cepheids, like 1 Carinæ, Beta Doradûs and Kappa Pavonis, is to observe many cycles in a particular year and determine a mean period. These observations are then repeated a few years later and compared with the previous results. *FMB*.

Cepheus In mythology, the king in the Perseus legend — husband of Cassiopeia, father of Andromeda. The constellation adjoins Ursa Minor, and is therefore in the far north, but is not very distinctive.

Brightest stars

Name	Visual Mag.	Abs. Mag.	Spec.	Distance (light-yrs)
α (Alderamin)	2.44	1.9	A7	46
γ (Alrai)	3.21	2.2	K1	52
β (Alphirk)	3.23v	−3.6	B2	750
ζ	3.35	−4.4	K1	717
η	3.43	3.2	K0	46

Next come Iota (just below 3.5) and two variables, Delta and Mu, which can exceed the fourth magnitude when at maximum.

Delta is the prototype Cepheid, with a range of from 3.5 to 4.4 and a period of 5.37 days; Mu, Herschel's "Garnet Star", is irregular, with a range of from 3.6 to below 5. Beta is variable over a small range, and is the prototype of the class known as either Beta Cephei or Beta Canis Majoris stars.

T Cephei is a Mira variable (5.4 to 11.0, 389 days). Another Mira variable is S (7.4 to 12.9, 488 days), which has an N-type spectrum and is very red. VV Cephei is a giant eclipsing variable; 4.9 to 5.2, period 7,430 days.

Delta has a wide optical companion of magnitude 7.5 (separation 41″); Beta has an 8.2-magnitude companion at 13″.7. There are no bright nebular objects in Cepheus. *PM*.

Čerenkov radiation Electromagnetic radiation emitted when a charged particle passes through a transparent medium at a speed greater than the local speed of light in that medium (for example, the speed of light in air or water is slightly less than in a vacuum). Radiation is emitted in a cone along the track of the particle. The Čerenkov radiation produced by high-energy charged particles can be detected by means of photomultipliers. COSMIC RAYS ploughing into the Earth's atmosphere produce Čerenkov radiation, which can be detected at ground level. This type of radiation was first discovered in 1934 by the Russian physicist Pavel Čerenkov. *IKN*.

Ceres The first asteroid discovered — on January 1 1801 by PIAZZI, at Palermo. It is named after the goddess of fertility, the patron of Sicily.

The possibility of a planet between Mars and Jupiter was first suggested by KEPLER, who had used his laws to find the distances of the planets. The Titius-Bode law gave 2.5AU as its distance from the Sun but attempts to calculate its position from this were unsuccessful. In 1800 six astronomers met at Lilienthal and agreed to form a society of 24 members (the so-called CELESTIAL POLICE) to carry out a search for it but before they began Ceres had been discovered. Piazzi was engaged in checking Lacaille's catalogue of Zodiacal stars when he found an extra one preceding the 87th which, over the next few nights, appeared to move. He followed it until prevented by illness. From Piazzi's measurements Gauss calculated the orbit, which allowed it to be recovered in December 1801.

Was Ceres a comet or a planet? After much study William HERSCHEL deduced that it was a new type of body which he called an ASTEROID.

Ceres lies in the middle of the main asteroid belt (2.77AU from the Sun). It is the largest minor planet, 636 miles (1,023km) in diameter, and has a dark (albedo 5.7 per cent) carbonaceous-type body. The axial rotation period is 9.075 hours with a light variation of only 0.04 magnitude. Ceres is nearly spheroidal in shape, but the polar flattening is unknown. Observations by radar show a loose powdery surface, which, from the study of polarized light, is uniform around the planet; no large isolated regions of a different composition (such as the Moon's dark maria) appear to exist.

The gravitational effects of Ceres, Vesta and Pallas have been studied and the mass of Ceres determined at $5.9 \pm 0.3 \times 10^{-10}$ that of the Sun. This is four times more than the next (Vesta and Pallas) and is about 30 per cent of the mass of the main asteroid belt. Its density is 2.3 g/cm³. *AJH*.

Cerro Tololo Inter-American Observatory The younger sister of KITT PEAK NATIONAL OBSERVATORY which, together with the National Solar Observatory at Sacramento Peak, the Space Telescope Science Institute at Baltimore (HUBBLE SPACE TELESCOPE),

Cerro Tololo Inter-American Observatory

and the Advanced Telescope Program (for the National New Technology Telescope), constitute the National Optical Astronomy Observatories (NOAO), itself managed by the Association of Observatories for Research in Astronomy and funded by the National Science Foundation.

Its foundation stemmed from a desire by the University of Chile to set up a new observatory at a superior site, coupled with a desire by G.P.KUIPER to find the best site on Earth (ie, clear, dark skies, good seeing, low humidity, and reasonable accessibility) to set up a major telescope for lunar and planetary research. Preliminary tests carried out by Jurgen Stock showed that the mountainous area east of La Serena, a town situated at 30° south on the Chilean coast, had great potential. Eventually, the 7,200ft (2,200m) Cerro Tololo peak was selected. It lies some 35 miles (56km) by air southeast of La Serena. Construction began in 1973, and the site and first instruments were dedicated in 1976. The observatory is

C

now home for eight instruments. The largest, the 158in (4m) reflector, is a twin of the largest Kitt Peak instrument, and is currently the largest in the southern hemisphere. The f2.6 primary mirror is made of CerVit and is thus unaffected by temperature changes. The Cassegrain focus is 125in (320cm). Other instruments are a 60in (150cm) Ritchey-Chrétien reflector, a 36in (92cm) reflector, two 16in (41cm) reflectors, the 24-36in (61-92cm) f3.5 Curtis Schmidt telescope of the University of Michigan, the Yale 40in (100cm) reflector, and a 24in (61cm) reflector from the Lowell Observatory. *EAW.*

C

Cetus The Whale or Sea-monster. A very large constellation, crossed by the celestial equator, but not very conspicuous.

Brightest stars

Name	Visual Mag.	Abs. Mag.	Spec.	Distance (light-yrs.)
β (Diphda)	2.04	0.2	K0	68
α (Menkar)	2.53	−0.5	M2	130
η	3.45	−0.1	K2	117
γ	3.47	1.4	A2	75
τ	3.50	5.7	G8	11.7
Next come ι (3.56), θ (3.60) and ζ (3.73)				

The most famous star in the constellation is Mira (Omicron Ceti) the prototype long-period variable, which attains naked-eye visibility near maximum. Tau Ceti is one of the two closest stars to bear any real resemblance to the Sun (the other is Epsilon Eridani). Part of Cetus extends into the Zodiacal region, so that planets can pass through it, but the pattern of the constellation is not well marked. *PM.*

Chaldæan astronomy Developed by the Chaldæans, who occupied an area north of the Persian Gulf in what is now Iraq. They are first mentioned by name in the 9th century BC, though there are earlier references to occupants of these "sea-lands". In the 7th century BC a Chaldæan dynasty was established in Babylon, to the north of the area, and in due course "Chaldæan" became synonymous with "Babylonian". Nevertheless many ancient authors used the term to refer to those skilled in the astrological and magical traditions of Babylonian culture.

Though heirs to a well-developed mathematical astronomy (*see* MESOPOTAMIAN ASTRONOMY), it was the Chaldæans who introduced the personal horoscope (*see* ASTROLOGY). *CAR.*

Challis, James (1803–1882). British astronomer who had unconventional ideas about the basic laws of physics, and is now chiefly remembered for his part in the search for Neptune.

Challis was born at Braintree, Essex, entered Trinity College (1821), and was ordained in 1830. He succeeded Airy as Plumian Professor (1836), and served until 1861 as Director of Cambridge Observatory. There in July 1846 he initiated a rigorous search for Neptune, and actually found it on August 4, though he failed to recognize its character. When he did, Galle had already effected the discovery.

Challis improved the collimating eyepiece and published numerous scientific works. *RMB.*

Chamæleon (the Chameleon). A small constellation near the south pole; it adjoins Octans, Carina, Musca and Apus. The brightest star, Alpha, is only of magnitude 4.07.

Chandler period The time, about 430 days, of the movement of the Earth's rotation axis about its axis of figure. This causes small variations in geographical latitude.

Chandrasekhar limit The maximum possible mass for a WHITE DWARF, first computed in 1931 by the Indian astrophysicist S. Chandrasekhar. The value computed by Chandrasekhar applies to a slowly rotating star and is about 1.4 solar masses. More recent computations suggest that a higher value is permissible for a rapidly and differentially rotating (rotating at different rates at different latitudes) white dwarf. According to R.H.Durisen (1975) such a white dwarf may have a mass as high as three solar masses.

A white dwarf is supported against its own gravitational attraction by electron degeneracy pressure (*see* DEGENERATE MATTER). The Pauli exclusion principle of QUANTUM MECHANICS states that no two electrons can occupy exactly the same state so that when all the low energy states have been filled, electrons are forced to take up higher energy states. With white dwarfs of progressively higher mass, as gravity attempts to squeeze the star into a smaller volume, so the electrons are forced into higher and higher energy states. They therefore move around with progressively higher speeds so exerting progressively higher pressures.

The greater the mass, the smaller the radius and the higher the density

attained by a white dwarf before electron degeneracy pressure stabilizes it against gravity. As the mass approaches the Chandrasekhar limit, electrons eventually are forced to acquire velocities close to the speed of light (ie, they become "relativistic"). As the limit is reached, the pressure exerted by relativistic electrons in a shrinking star cannot increase fast enough to counteract gravity. Gravity overwhelms electron degeneracy pressure, and a star which exceeds the Chandrasekhar limit collapses to a much denser state. Electrons combine with protons to form neutrons, and the collapse is eventually halted by neutron degeneracy pressure, by which time the star has become a NEUTRON STAR. *IKN.*

Chao Meng-Fu The south polar crater of Mercury; 87.5°S, 132°W; diameter 93 miles (150km).

Charge-coupled device (CCD). Complex, solid state, electronic devices which can be used as memories, delay lines and imaging detectors. CCD imaging detectors are now widely used in astronomy both for direct imaging applications and as spectroscopic detectors. A CCD is a silicon integrated circuit consisting of an oxide-covered silicon substate upon which is formed an array of closely spaced electrodes. The most common type has "frame transfer" architecture and this also is the type with the highest efficiency. Signal information is conveyed in the form of a quantity of electric charge (electrons). Potentials are applied to the electrodes in a certain pattern and the charge is attracted to the regions beneath the electrodes with the greatest applied potentials. It is therefore common to say that the electrons are being stored in "potential wells", suggested by the shape of the potential distribution beneath the electrodes. The charges also are confined to a system of closely spaced "channels", crossing the electrodes, which are bounded by electrically inactive p-type "channel stop" regions. Thus charge is localized to small elements (called pixels) arranged in a two-dimensional array. Arrays with elements typically of dimension 15–30 μm with several hundred on a side are readily available commercially. "Charge coupling" is the technique by which charge is transferred from under one electrode to the next, and is achieved by sequentially pulsing the voltages on the electrodes between high and low levels. By this means charge packets can be made to pass down the array of electrodes, confined within the channels, with hardly any loss and very little noise. Thus, the entire charge image can be shifted out bodily and is then caused progressively to enter a further line read-out section at the edge of the array. This accepts each line of charge packets one by one and is used to transfer them to a charge detection amplifier (also formed on the silicon layer) to give a serial video output signal. The noise in this amplifier represents a limit to the minimum detectable signal but in present devices can be as low as six electrons rms per image element. Cosmic ray events represent another limitation, as silicon is an efficient detector of high-energy charged particles.

The CCD operates as an imaging detector in the following way. Light imaged onto the surface penetrates the electrode structure and enters the substrate where hole-electron pairs are generated in numbers precisely proportional to the numbers of incident photons. The holes are conveniently lost by diffusion down into the depths of this substrate, while the electrons migrate rapidly to the nearest biased electrode, where they collect as a single charge packet. More efficient devices, which are particularly sensitive in the blue region, are made by thinning the substrate and illuminating the device from the rear, thus avoiding the initial hurdle of the electrode structure. For use in astronomy, inevitably for the detection of faint images, the devices must be cooled to avoid the build-up of thermally-induced charge, which otherwise would mask the signal. The general method of cooling is by thermal contact in a liquid nitrogen cryostat. Exposures of several hours can then be achieved without excessive interference due to dark current.

Broadly, a CCD can be regarded as the electronic equivalent of a photographic plate, but it is vastly more sensitive and more reproducible than photographic emulsion, responds over a wide spectral range from the near-ultraviolet to the near-infrared, and has a linear response over a very wide range of input light levels. It is, however, limited in size; the most common commercially available CCDs have an active area about a centimetre on a side, while some specially developed devices may be as large as 4 to 5 centimetres, although these are not generally available at the time of writing. In certain astronomical applications, for example, SCHMIDT TELESCOPE

wide field imaging, photographic plates still are more advantageous because of the very large sizes available. On the other hand, photoelectric imaging detectors have comparable sensitive areas to CCDs. While these devices, too, have much higher peak sensitivity than do photographic emusions, CCDs are greatly more sensitive in the red and near-infrared regions. Photoelectric devices, however, can respond to changes in light level more rapidly than CCDs, and those with internal intensifying stages have lower intrinsic noise and for very low signal levels can be more sensitive than CCDs. Thus in some respects CCDs and photoelectric detectors are complimentary in their applications in astronomy. *AB.*

Charlier, Charles Vilhelm Ludwig (1862–1934). Swedish astronomer; he advanced the theory that the Milky Way rotated, and proposed a hierarchical grouping of galaxies in a universe which extended to infinity.

Charon The only known satellite of the ninth planet, Pluto, appropriately named after the boatman who ferried the souls of the dead across the river Styx to Pluto's lower world (from ancient Greek mythology).

Charon was discovered by Dr. James W. Christy at the United States Naval Observatory in June 1978. While measuring positions of Pluto, Christy noticed that some of the Pluto images had a "bump" on one side as if deformed. On other plates, he noticed a bump on the opposite side. Christy realized that he was observing a very close satellite which was visible only at greatest "elongations". He found Charon to have an orbital period of revolution of 6.4 Earth-days at a distance of only 12,000 miles (19,400km) from the centre of Pluto.

Under the best observing circumstances, the image of Charon is not cleanly separated from the image of Pluto even on the large-scale plates taken with the powerful astrographic reflector at the Naval Flagstaff station. It is not surprising then that the discovery of Charon was delayed until Pluto had come in sunward almost to the orbit of Neptune.

The discovery of Charon was most fortunate, in that it provided the first reliable determination of the mass of Pluto, ending several decades of controversy about Percival Lowell's PLANET X.

Charon is 1.7 magnitudes fainter than Pluto, or a factor of only fivefold.

Since these two bodies are approximately similar in size, one might expect their compositions to be similar, which in turn would indicate that Charon's mass is about 10 per cent that of Pluto. Thus, Charon is the most massive satellite relative to its primary in the Solar System (greatly exceeding the Earth-Moon system ratio of 81 to 1).

In March 1976, Cruikshank, Pilcher and Morrison of the University of Hawaii discovered methane frost or ice in the near infrared spectrum of Pluto. The estimate of reflective power or albedo had to be revised upward; which in turn reduced the probable size of Pluto to about that of our Moon. The probable diameter of Charon comes out at about 750 miles (1,200km). Remarkably low mean densities are indicated, between 0.9 and 0.7 grams per cubic centimetre. Thus, both Pluto and Charon are large icebergs, composed mainly of water ice, with crusts of frozen methane.

Charon's orbital plane is now turned nearly edgewise toward Earth, and transits and eclipses were observed photo-electrically to have started in 1985. This series of events will continue every 6.4 Earth-days for the next five years. By measuring the drop in magnitude and the duration, these observations will provide accurate values for the shape, size, mean densities, and albedo surface differences, revising present estimates. *CWT. See also* colour essay on MOONS.

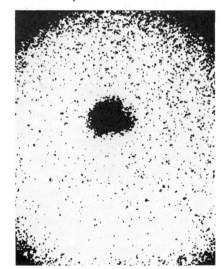

Charon: Pluto's only known satellite

Chebyshev A crater on the far side of the Moon, situated at 34°S, 133°W.

Chekhov Crater on Mercury 112 miles (180km) in diameter; 36°S, 62°W. It has an inner mountain ring.

C

Chi Cygni S7 type long period MIRA VARI-ABLE (407 days) star. It has the largest visual range of variation of any Mira and is very red at minimum, corresponding to a temperature of 1,700°C. Maximum magnitude 4.2, distance 70 pc. There is strong infrared excess and circumstellar emission from molecules (eg, CO and SiO). The gas lost from the star surface cools to form molecules and silicate dust grains in the circumstellar envelope. The surrounding dust absorbs some of the star's radiation, is itself heated, and radiates in the infrared, producing the infrared excess.

Chinese astronomy In recent years there has been renewed scientific interest in the astronomy of ancient China. This has largely arisen on account of the recognition of the value of early Chinese observations in modern astronomical research. Many of these observations are without equal elsewhere in the world before the introduction of the telescope (1609). In particular we may mention observations of sunspots and aurorae (useful in the study of long-term solar variability), Halley's Comet, solar eclipses (important in the study of changes in the length of the day) and supernovae.

Astronomy as practised in China was remarkably different from that in the West. Thus whereas in ancient and medieval Europe only 48 constellations were recognized, in China the corresponding figure was between about 250 and 300. The average number of stars in an oriental star group was only about five. Surprisingly few star patterns are common to both oriental and occidental astronomy; two notable exceptions are Shen ("The Triad"), equivalent to the main portion of Orion, and Beidou ("The Northern Dipper"), identical with the Plough.

The Zodiac, the apparent path of the Sun, Moon and planets across the CELESTIAL SPHERE, has never held a special place in Oriental astronomy. Equatorial co-ordinates were exclusively used for star positions. In place of the 12 signs of the Zodiac we find the 28 "lunar mansions", star groups lying fairly close to the celestial equator. Each of these lunar mansions was used to define a range of right ascension.

At least as early as 200BC — and possibly long before — the rulers of China employed official astronomers to maintain a regular watch of the sky. These observers were as much astrologers as astronomers; it was their duty to report the occurrence of any unusual celestial portent to the throne and provide an interpretation. Because of this deeply held belief in astrology, vast numbers of astronomical observations were recorded — and are still preserved today. The official histories of China down to the end of the last dynasty (1911) contain many thousands of reports of such diverse phenomena as eclipses of both Sun and Moon, sunspots, conjunctions of the Moon with planets and stars, planetary conjunctions, comets, "guest stars" (usually novae and supernovae, but sometimes comets), meteors and the aurora borealis. Many of these are accompanied by astrological prognostications.

Another reason for astronomical observation was the maintenance of a reliable calendar. From ancient times, a luni-solar calendar was adopted in China. At most periods the beginning of the year corresponded to January or February in the Julian or Gregorian Calendar. A typical year contained 12 months each of 29 or 30 days, totalling about 354 days. Hence in order to keep the calendar in step with the seasons, it was necessary to insert an occasional intercalary month, every 3 years or so.

Owing to the low survival rate of the more ancient texts, little is known regarding the beginnings of astronomy in China. Before the 8th century BC, only a few scattered references to eclipses and certain stars have survived. These are mainly found on the "oracle bones" — inscribed turtle shells and the shoulder blades of cattle and sheep — of the Shang Dynasty (c. 1500 to 1050BC). An impressive series of 37 solar eclipse observations made during the succeeding Zhou Dynasty indicates a high level of astronomy by then. The recorded dates of most of these eclipses, which lie between 720 and 480BC, agree exactly with modern calculations.

It is unfortunate that few other astronomical observations are extant down to the 3rd century BC. This circumstance is probably due to the infamous "Burning of the Books" at the command of Emperor Qin Shih Huang in 213BC and the subsequent sacking of the Chinese capital of Xianyang only seven years later.

Political astrology, the abiding motive behind celestial observation throughout most of Chinese history, seems to have had its origins in the Warring States Period (480 to 221BC). This period of turbulence was nevertheless marked by much philosophical activity and we also find the beginnings of cosmological speculation at this time. Although it seems that the Chinese never independently envisaged the form of planetary orbits — whether around the Earth or the Sun — and had no idea of the scale of the Solar System, one of the rival cosmological theories came remarkably close to 20th-century concepts. This was the Xuanye ("Infinite Empty Space") theory, in which the various celestial bodies were believed to be each following their own course in a void of infinite extent.

It was also during the Warring States Period that the first known star catalogues and charts were probably produced. These ancient works are no longer extant, but later sources tell us that in total 1,464 stars were mapped in 283 groups. Only medieval copies of a star catalogue which is reputed to originate from this period now survive. However, from the measurements which this work (the Xingjing) contains, the date may be calculated as around 70BC — two or three centuries later than the traditional date. The typical accuracy of measurements of star positions in this catalogue is fairly high — approximately one degree. Hence we may conclude that by this period the means were certainly available to produce a good star map. However, we have no detailed Chinese star maps made before 1000.

The official history of the Former Han Dynasty (206BC to AD9) contains the earliest known compilation of Chinese astronomical observations of various kinds, including the first detailed account of the motion of Halley's Comet through the constellations (12BC). Many of these observations are contained in a special section devoted to astronomy. A similar practice was followed in nearly all subsequent dynasties. Further, towards the end of the first millennium AD, observations in the Chinese style began to be made in both Korea and Japan. Hence in later centuries it is not uncommon to find two or even three reports of the same event — notably comets and supernovae. In all probability, as many as seven supernovae are recorded in Chinese history — in 185, 393, 1006, 1054, 1181, 1572 and 1604.

The earliest surviving original star maps which give a good representation of the night sky date from the Song Dynasty (960–1277). The most accurate of these maps is engraved on stone and dates from 1247. This is located in a Confucian temple at Suzhou in Jiangsu province. In all, 1,440 stars are charted in 313 groups. Modern measurements indicate a typical positional accuracy of about one degree for the Suzhou chart.

Although Chinese astronomy reached a high level of attainment by the beginning of the Christian Era, further progress was slow. The zenith was probably reached during the Song and Yuan dynasties (960–1367), when Chinese astronomy rivalled that of any other civilization. However, in the succeeding Ming Dynasty (1368–1644) there was a gradual decline and traditional Chinese astronomy was ultimately eclipsed by new developments in Renaissance Europe. When the first Jesuit astronomers such as Matteo Ricci and Adam Schall von Bell arrived toward the end of the Ming, they were soon able to achieve high positions in the imperial astronomical bureau. Although Chinese astronomy and astrology continued to exist until the beginning of the present century, it had now little to offer the rest of the world. *FRS. See also* ASTROLOGY.

Chiron An outer asteroid (No. 2060) discovered by Charles T. Kowal on photographic plates taken on October 18 and 19 1977. It is named after the Centaur Chiron, son of Saturn and grandson of Uranus. Its orbit lies almost entirely between those of Saturn and Uranus. A search of prediscovery photographic plates revealed its image on plates taken as early as 1895. Its orbital period is 49 years, inclination 6.92°, perihelion distance 8.51AU and aphelion distance 18.88AU. The orbit is unstable and a close approach to Saturn is inevitable on a timescale of 10,000 years.

Chiron is large, its diameter being somewhere between 60 and 200 miles (100 and 320km) dependent on what reflectivity you assume.

Its origin is a matter of considerable speculation. The simplest theory is that Chiron is a comet. Unfortunately it is much larger than any known comet (it is 250 times brighter than Halley's Comet at the same distance).

Another possibility is that it could have been perturbed out of the asteroid belt into its present orbit. But more asteroids (other than just Hidalgo) would be expected between Jupiter and Saturn if this were so.

It has been suggested that Chiron (and Pluto) were once satellites of Neptune and that both were tidally wrenched from that system by the close passage of a massive object. Similar theories have it escaping from Saturn or Uranus or the LAGRANGIAN POINTS of Saturn. It may just be the brightest

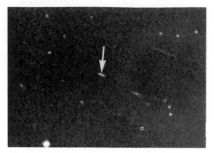
Chiron: the 1977 photo that revealed it

member of a new asteroid belt, members of which are difficult to discover because they move so slowly against the sky background.

A detailed investigation of its reflection spectrometry is needed together with a study of the polarization of its light. This should lead to its reflectivity and thus its size. Preliminary results indicate that the surface is darkish. The Palomar Solar System Survey is continuing its search for similar objects but to date Chiron is unique. *DWH.*

Chondrites Stony meteorites can be divided roughly into two main classes, chondrites and achondrites, according to whether or not the stones contain chondrules, spherical inclusions in the main mass of the meteorite. However, some carbonaceous chondrites do not contain the spheres. The chemical composition is, however, similar to the chondrule-containing stones and hence are classed with the chondrites. Modern classifications, however, tend to be based on chemical compositions rather than on the presence or absence of chondrules.

Chondrites can be divided into three main groups, though only two are relatively common. They are the enstatite (E); the ordinary (H or high-iron group, L or low-iron group, and LL or low-iron, low-metal group); and the carbonaceous (C). The classification depends to some extent on the ratio of MgO/FeO within the silicate mineral and the ratio of metallic Fe/Ni. The olivine-bronzite (H) chondrites contain rather more olivine than bronzite and about 20 per cent nickel iron. The most common type, the olivine-hypersthene chrondrites (L), are predominantly olivine, but the nickel-iron content is only 5 to 10 per cent. These two classes and a third class, the LL group, are known as ordinary chondrites. The important carbonaceous chondrites contain about 3 per cent carbon.

Chondrules are spherical bodies ranging in size from less than 1mm to more than 1cm (less than half an inch)

and normally consist of olivine and/or pyroxene. The latter are often made up of components radiating from a point offset from the centre of the sphere. The olivine chondrules may consist of a single or several crystals. Most are completely crystalline and tough.

The origin of chondrules is still not known with certainty. At one time they were thought to be fused drops of "fiery rain" from the Sun, or fragments of pre-existing meteorites which had been rounded by oscillation and attrition. Other theories include the cooling of droplets produced by collisions between asteroids and direct condensation from a hot gas. A more recent theory proposes the coalescence of micro-chondrules in an incandescent fog surrounded by a cooler dust envelope. *HGM.*

Christie, Sir William Henry Mahoney (1845–1922). Eighth ASTRONOMER ROYAL, who increased substantially the observing equipment at Greenwich Observatory, not least with a 28in (71.1cm) refractor. He also extended the observatory's programme to include physical as well as positional astronomy. Daily sunspot observations and spectroscopic observations of the stars were initiated too.

W.H.M.Christie: Astronomer Royal

Chromatic aberration Deficiencies in an optical system which are the result of the spread of wavelengths in light. The chromatically aberrated image shows coloured haloes or shifts. Light rays of different colours, and therefore wavelengths, are refracted differently as they pass through a lens, because the refractive index of glass varies with wavelength. A simple convex lens more strongly converges blue light than red, so the blue light focuses nearer the lens than the red. An eyepiece focused on

C

the blue image sees it surrounded by an out-of-focus red image, and vice versa.

Achromatic lenses are multiple lenses made of several different glass materials with different refractive indices; the combinations are designed to compensate for chromatic aberration. Although chromatic aberration can be controlled and minimized by careful optical design it can never be eliminated except in purely reflective optical systems operated in a vacuum. *PGM.*

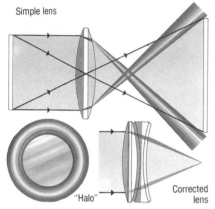

Chromatic aberration: coloured haloes

Chromosphere *See* SOLAR CHROMOSPHERE.

Chronometer *See* MARINE CHRONOMETER.

Chryse Planitia Extensive plains region on Mars which was the site of the VIKING 1 Lander probe. Situated just

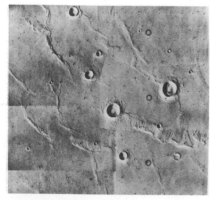

Chryse Planitia: from a height of 1,800km

north of the equator and centred on 40°W, it was shown by the Lander to consist of loose reddish material upon which were distributed large numbers of blocks of basaltic lava. Chryse occupies a large basin-like embayment into the cratered terrain of Mars, probably the infilled site of an ancient impact basin. In many respects it looks like a

lunar mare region. A large number of prominent channels converge on the region. These have their origin in the eastern end of MARINERIS VALLES and are of presumed fluvial origin. *PJC.*

Circe (Asteroid 34), diameter 69 miles (111km); period 4.4 years.

Circinus (the Compasses). A small constellation in the area of Alpha and Beta Centauri. Its only fairly bright star is Alpha; mag.3.19, absolute mag. 2.6, type F0, distance 46 light-years.

Circular velocity The velocity of a body in a circular orbit around a massive primary. Its value is given by $\sqrt{GM/R}$ where M is the mass of the primary, R is the radius of the orbit and G is the gravitational constant.

Circumpolar stars Stars which never set at the observer's latitude. *See also* CELESTIAL SPHERE.

Circumpolar stars: southern hemisphere

Cislunar Between the Earth and the Moon.

CK Vulpeculæ Slow NOVA of 1670. Magnitude 2.7 to <17.0.

Clark, Alvan (1804–87). American astronomer who made objectives for very large refractors. His son, Alvan Graham Clark (1832–97) continued the family business and discovered the companion star of Sirius.

Classification of galaxies First carried out by HUBBLE in 1925 on the basis of shape. Despite the wealth of subsequent types of galaxy observation, the term has retained its original meaning as morphological (shape) classification, and galaxy typing in this way

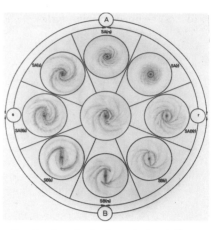

Classification of galaxies: spiral varieties

remains of central importance in modern astronomy.

Hubble recognized the four basic types of galaxies known today: ELLIPTICAL GALAXIES, LENTICULAR GALAXIES, SPIRAL GALAXIES, and irregular galaxies. He placed those of regular shape in his famous "tuning fork" diagram, in which the two prongs correspond to barred and non-barred spirals. An early hypothesis was that the classification represented an evolution sequence with ellipticals as progenitors, spiral structure developing in their smooth forms later. The hypothesis, long rejected, has left its mark in that galaxies at the left of Hubble's sequence: ellipticals are still known as "early-type", and those to the right, the looser spirals and indeed the irregulars, as "late-type".

The tuning fork has given way to more comprehensive schemes, developed principally by A. Sandage and G. de Vaucouleurs from studies of the several thousand brightest galaxies. One such revised classification is shown in the accompanying diagram. Such schemes remain based on "Hubble types", which run as follows.

E0 to E7: elliptical galaxies, the number indicating the ellipticity (flattening); the number is 10 (1–b/a), where a and b are the lengths of the major and minor axes respectively;

S0: lenticular galaxies, having a disk and a central bulge similar to spiral galaxies, but with no apparent spiral arms;

SAa, SAb, SAc, SAd: spiral galaxies, the A indicating simple spiral form, and "a" to "d" describing both the prominence of the bulge and the tightness of the arm-winding, which varies from a pitch angle of about 10° for "a" to about 25° for "d";

SBa, SBb, SBc, SBd: spiral galaxies, the B indicating a central bar, and "a" to "d" again describing bulge prominence and spiral tightness;

Irr I and Irr II: Irregular galaxies, the former intrinsically irregular (eg, the MAGELLANIC CLOUDS), the latter irregular due to disruption by an explosive nucleus or the gravitational attraction of a nearby galaxy.

The classification of galaxies remains essential because of the close correlation between morphology and three other properties of galaxies, namely rotation, the principal constituents, and the proclivity to cluster (*see* EXTERNAL GALAXIES). *JVW.*

Clavius One of the largest lunar craters, diameter 140 miles (225km), at 58°S, 14°W (highlands). Parts of the rim can be found within the 31-mile (50km) diameter Clavius B.

Cleomedes Lunar walled plain, diameter 78 miles (126km), at 27°N, 55°E, north of the Mare Crisium; the wall is interrupted by the deeper crater Tralles.

Clocks *See* ASTRONOMICAL CLOCKS.

Coal Sack. A dark, obscuring cloud in the southern constellation CRUX. The Coal Sack, like all such clouds of dust-laden gas, is seen only in silhouette, because behind it lies a bright background. In this case the background is the Milky Way. On a dark night the Coal Sack is very prominent, and appears to be the darkest spot in the entire sky, though that is purely an optical illusion. Near the centre of the Coal Sack is a faint star which can be seen only with good visual acuity from a very dark site. *DAA.*

Coal Sack: dark cloud in Crux

Coblentz, William Webber (1873–1962). American astronomer who worked at the Lowell Observatory. He was the first to verify PLANCK's law and was noted for planetary researches.

Cocoon Nebula H II REGION in the constellation Cygnus catalogued by DREYER as IC5146. The nebula resembles a cocoon around its embedded stars.

Cœlostat A plane mirror arranged to rotate east to west about a diameter parallel to the Earth's axis and at half the Earth's rotation rate. This counteracts the diurnal movement of the sky so that rays from a celestial object are reflected, by the cœlostat, into a fixed direction. Apparatus too heavy to be pointed skywards may then be positioned to receive these reflected rays. The primary and unique characteristic of the cœlostat is that the patch of sky presented to it is non-rotating.

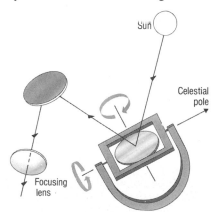

Cœlostat: rays in a fixed direction

For an observed object at DECLINATION δ the reflected ray appears to be coming from declination $-\delta$ and, as the angle between the incoming ray and the perpendicular to the mirror surface increases, the effective aperture decreases. Additionally, at these greater angles, the optical effects of surface inaccuracies on the mirror become more pronounced. The combination of these factors makes it desirable to limit observation to objects with declinations numerically less than 30° and to keep the angle between the incident and reflected beams as small as possible; preferably never exceeding 90°. These operating conditions so limit the directions available to the emergent beam that a second plane mirror is commonly used to reflect the cœlostat beam into the required final direction. This mirror is not driven, so that it is mechanically convenient to fit it with manual controls for manual correction of minor errors of the cœlostat drive.

Provided that the total reflection angle is kept below 90° at each mirror, there is no restriction on the direction of the final beam. The commonly held notion that this beam must lie horizontally and in the meridian is erroneous.

The second mirror must be sited on the axis of the observing apparatus and the cœlostat so positioned as to be clear of its shadow-zone. *LMD.*

Coggia's Comet Bright comet seen in 1874; period about 13,700 years.

Collapsars A star such as our Sun shines by converting hydrogen into helium in its core. Eventually the hydrogen becomes exhausted in the central region of the star, but hydrogen burning continues in a thin interior shell. When this happens, the star undergoes a drastic change in structure; the star can expand to such an extent that the outer layers drift off into space, while the central core collapses under the force of gravity.

Atoms in the core are squeezed closer and closer together, and electrons become dislodged from the nuclei about which they are orbiting. If the mass of the initial star is less than 1.4 times the mass of the Sun — called the CHANDRASEKHAR LIMIT – then at some point the pressure from the negatively charged electrons in the collapsing core becomes sufficient to halt the contraction. Such a star is a WHITE DWARF.

The death of stars considerably more massive than our Sun can be dramatic; the star, after first expanding outward, collapses under the force of gravity and then explodes, leaving behind a SUPERNOVA (exploding star) with a remnant core so dense that electrons are squeezed into the nuclei; they react with the positively charged protons there to produce neutrons (and neutrinos). An example of one such system is the Crab Nebula in Taurus, at the heart of which lies a NEUTRON STAR.

There is a mass limit called the Oppenheimer-Volkoff limit (about 3.2 solar masses), above which even the pressure of neutrons cannot stop the collapse. Very massive stars, after undergoing supernova explosions, leave behind cores so massive that they collapse in upon themselves under the force of gravity (with neither electron-electron pressure nor neutron-neutron pressure able to halt the contraction). Such objects, from which not even light can escape, are called collapsars, sometimes used instead of BLACK HOLES. *DLB.*

Colour index The difference in brightness of a star as measured at different wavelengths, and used as a measure of the colour of the star. The different wavelengths are isolated by optical filters of coloured glass (for example blue glass

C

and red) and light passing through each is expressed in magnitudes (B and R). The colour index B-R is zero for white stars (spectral type A0). It ranges from about −0.5 for the bluest stars to more than +2.0 for the reddest.

Colour index correlates well with the naked-eye perception of the colour of the brighter stars (the colour of fainter stars is hard to perceive for reasons to do with the physiology of the eye and the psychology of perception).

Colour index is principally a measurement of the temperature of stars. The bluest stars are hotter than 30,000K, white stars have surfaces at temperatures of about 10,000K. The reddest stars are very cool, say 3,000K: cooler stars exist but emit so little light that they may best be studied by infrared techniques.

The colour index of light from a star may be changed as the light passes through interstellar space. Dust in space interacts more easily with the blue light from the star and disturbs the blue light from its straight path (scattering). Red light passes relatively freely around the dust and carries straight on to the observer's telescope. Thus the starlight is reddened (it would be more accurate to say "de-blued"). The consequent increase of the star's colour index caused by the INTERSTELLAR DUST is called a colour excess due to INTERSTELLAR REDDENING.

The colour index of stars is also modified by the presence of atoms in their atmospheres. The light in different wavebands is affected to different degrees. It is possible to isolate the various effects by measuring the colour index between different pairs of wavelengths. A colour index formed with ultraviolet and with blue light (U-B) may be compared with (B-R) in a colour-colour plot. The position of a star in the plot gives clues about its chemical composition, temperature and interstellar reddening. *PGM.*

Colour-magnitude diagram A plot of the MAGNITUDE of a collection of stars versus their COLOUR INDEX, used as a diagnostic tool to study star clusters.

Columba (the Dove). A southern constellation adjoining Canis Major and Carina. It contains little of interest.

Brightest stars

Name	Visual Mag.	Abs. Mag.	Spec.	Distance (light-yrs.)
α (Phakt)	2.64	−0.2	B8	121
β (Wazn)	3.12	−0.1	K2	143

Colures Great circles on the CELESTIAL SPHERE which pass through the two celestial poles. The equinoctial colure passes through the celestial poles and the vernal and autumnal equinoxes. The solsticial colure passes through the celestial poles and the winter and summer solstices.

Coma An off-axis lens aberration typically resulting in points of light appearing comet- or fan-shaped.

Coma Berenices (Berenice's hair). A northern constellation adjoining Boötes. It has no star above magnitude 4.2, but is rich in faint stars and in galaxies; with the naked eye it gives the impression of a very wide star-cluster.

Coma cluster The Coma (Berenices) cluster is a very large group of galaxies, gravitationally bound to each other. Such groups have long intrigued astronomers but this one is particularly interesting because in 1971 it was found to emit X-rays, the first evidence that clusters of galaxies contained extremely tenuous gas heated to temperatures of a few million degrees by the orbital motion of the galaxies themselves. The Coma cluster is at a distance of about 100 Mpc (350 million light-years).

Cometary coma A spherical cloud of gas and dust around a comet's nucleus.

Cometary nebulæ Glowing gas clouds with specific morphology, somewhat cometlike in appearance but unrelated to these. A cometary nebula is always associated with a star that lies in a geometrically significant location. Five shapes are recognized: an arc, sometimes a complete ring, with a star at the centre; a fan-shaped nebula with star at the apex; a biconical (hourglass) nebula with star at the "waist"; a linear wisp protruding from a star; a complete or partial ring with the star on its rim. Most cometaries shine by reflecting the light of their allied star, though some are ionized by the ultraviolet radiation of hot central stars. Defined by these four shapes, all cometaries are associated with young, usually low-mass, stars and occur after accretion but during the pre-main-sequence phase. Therefore many T TAURI stars illuminate cometaries.

Recently, the biconical type has been generalized to include any bipolar system consisting of two separate nebulae with a star lying between these. Enlarging the class to incorporate the bipolar nebulae makes cometaries evol-

utionarily less homogeneous. The subclass of bipolar nebulae are not all indications of stellar youth. As red giants swell they throw off planetary nebular shells, and the evolved central stars heat up rapidly, eventually becoming white dwarfs. During this "proto-planetary nebula" phase some red giants lose mass by a BIPOLAR FLOW, thereby generating bipolar nebulæ

These nebulæ can be dramatically symmetric and some carry fanciful but descriptive names like "The Red Rectangle" and "The Egg Nebula".

The stars in cometaries are often embedded within dusty equatorial disks, some thick enough to render the stars optically invisible, but bright infrared sources as equatorial starlight degrades to cool dust emission. All cometary shapes may be explained by viewing from different directions the basic model — a star and dust disk that constrains its light to shine into a bicone. *MC.*

Comets The name derives from the Greek *kometes*, a long-haired star. At discovery a comet is usually just a round fuzzy spot of light, often brighter towards the centre. This coma, as it is called, gets brighter and larger as the comet approaches the Sun, and may show a distinct central point of light, the nucleus. A tail or tails may form, straight and narrow or broader and gently curved; perhaps both forms will appear. The brightest comets may show curved jets emanating from the nucleus, often forming expanding hoods or shells. The whole comet may be enveloped in a vast tenuous cloud of hydrogen, detectable only from spacecraft. More than 700 comets have been observed so far, and about six new ones are discovered each year.

The dimensions of comets can be enormous; the coma may be hundreds of thousands of kilometres across, and the tail may extend for a hundred million kilometres. The solid nucleus is small, that of HALLEY's COMET being only about 9 × 6 miles (15km × 10km). The material of the coma and tail is extremely rarefied and the total mass of all the comets in the Solar System is only a few times that of the Earth.

The nucleus is made of ices of various gases, mainly water, interspersed with solid grains and may possess a dark crust through which the jets of evaporated gas emerge. As a comet approaches the Sun, evaporation begins and the coma is formed, the

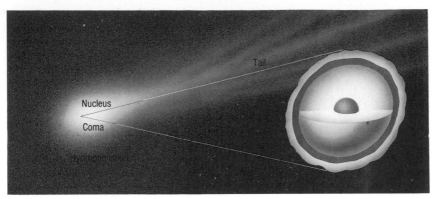

Comets: a small nucleus of ice and dust, a huge coma and an enormous tail

escaping gases containing entrained dust particles. Later, the gas and dust are swept away to form the tail. The tail gases are positively ionized and moving at high velocity, forming a straight narrow structure known as a plasma tail of Type I. The dust moves more slowly, and forms a curved tail of Type II. Spectroscopic study shows the light from the nucleus to be reflected sunlight; that from the coma and tail is a mixture of reflected and emitted light given by the dust and gas respectively. The emitted light is from various atoms, molecules, radicals and ions fluorescing under the influence of solar ultraviolet radiation.

The dust particles appear to be silicate grains, and there may be ice grains too. The gaseous phase consists of the simpler combinations of hydrogen, carbon, nitrogen, oxygen and sulphur, such as water, carbon monoxide and dioxide, cyanogen, hydroxyl, etc. Methane and ammonia are certainly present, but they are difficult to detect. When a comet is very close to the Sun, metallic emission, particularly from sodium, may occur.

Comets belong to the Solar System and are designated as short-period or long-period according to the size of their eccentric elliptical orbits and the corresponding orbital periods, the dividing line being set at 200 years. Most long-period comets take many thousands or even millions of years to complete a circuit of the Sun, and the short-period comets are believed to be derived from them as a result of planetary perturbations, mainly by Jupiter. About 130 short-period comets are known, and new ones are discovered almost every year; the total number is probably about a thousand, of which Halley's Comet, with a period of 76 years, is the brightest. ENCKE'S COMET has the shortest period of all, only 3.3 years.

Comets are normally named for those who discovered them, a maximum of three names being allowed, but the two comets just mentioned were awarded their names in recognition of the work done on their orbits by their namesakes. Comets are also given numerical designations: initially the year of discovery followed by a small letter according to the order of observation, but later these are rearranged in order of perihelion date, and Roman numerals are used. Thus Comet Kohoutek was originally 1973f, but is now 1973 XII. Short-period comets are indicated by the prefix P/, eg, P/Halley, 1982i.

The brightness of a comet is expressed as the equivalent stellar magnitude, as in the case of nebulae. Most comets show fadings and outbursts caused mainly by variations of jet activity and of the solar stimuli. It is difficult to describe the brightness by any fixed formula, and forecasts of comet magnitudes are notoriously subject to error. Closeness to the Sun has more effect than proximity to the Earth, and the famous Sun-grazing group has produced some of the most spectacular comets ever seen.

The tails of comets point away from the Sun, the inference being that something from the Sun drives them outward. Radiation pressure readily accounts for the motion of the dust particles, but it was not until the SOLAR WIND was discovered that the plasma tails could be explained. The outgoing electrons and protons of the wind, with their associated magnetic fields, accelerate the ionized plasma tail material to velocities of several hundreds of kilometres per second. Occasionally a plasma tail is disconnected and cast off when a comet crosses a sector boundary in the interplanetary magnetic field.

Rarely a comet is seen to have an "anti-tail" pointing towards the Sun. This is a purely geometrical effect owing to our line of sight projecting

trailing material on to the sky in a sunward direction. Comet AREND-ROLAND of 1957 is the best-known example of this effect.

Comets lose material permanently at every perihelion passage and must eventually become defunct; the lifetimes of short-period comets are probably of the order of about ten thousand years. The death-rate of these comets implies that the population must be diminishing — unless there is an influx of new objects to maintain it. Oort suggested that there exists a shell or halo of comets (the OORT CLOUD) extending from beyond the outermost planets to about halfway to the nearest stars. This is believed to be the reservoir from which comets are diverted towards the Sun by the action of passing stars; a small proportion of these newcomers would be captured into smaller orbits by the major planets, mainly by Jupiter, thus replenishing the numbers of short-period objects. This means that many of the long-period comets that we observe are making their first visit to the Sun, and must

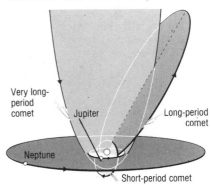

Comets: orbits of the three main classes

consist of matter that has remained virtually unchanged since the original formation of the Solar System. Hence the great scientific interest and importance of these bodies.

The planets all revolve around the Sun in the same direction (anticlockwise as seen from above the Earth's north pole); this is known as direct motion. Most of the short-period comets have direct motion, though a few (including P/Halley) revolve in the opposite or retrograde sense. The long-period comets, however, are just as likely to have retrograde motion as direct. These facts are consistent with an origin in the Oort cloud and subsequent capture in a few cases by planetary action.

The ultimate fate of a short-period comet appears to vary: it may disperse entirely into diffuse dust and gas, it

C

may break up into a number of smaller fragments, or it may become inert, leaving a dark solid core indistinguishable from some small minor planets (asteroids). Examples of all these ends are known, but the reasons for the differences are not fully understood, like so many other aspects of these fascinating and mysterious objects. *HBR.*

Comet wine Portuguese wine bottled in 1811; the growers believed that its excellent quality was due to the Great Comet of that year.

Commensurabilities There are many cases of pairs of planets or satellites whose orbital periods are close to the ratio of two small integers, and these are called "commensurabilities". They cause greatly increased PERTURBATIONS between the two bodies by a resonance effect, and it is a major problem of CELESTIAL MECHANICS to calculate these perturbations on the orbits. The principal examples among the planets are the 5:2 commensurability between Jupiter and Saturn, 3:1 between Saturn and Uranus, 2:1 between Uranus and Neptune, and 3:2 between Neptune and Pluto. The commensurabilities are strongest if they are close to the exact ratio and if the differences of the two integers is small. For Neptune and Pluto the commensurability is very close, and a special type of motion called a libration occurs. This prevents the two planets ever coming near to each other, even though their orbits cross, because they are never at that part of their orbits at the same time.

A similar effect occurs at the outer part of the asteroid belt. The Hilda group of asteroids at the 3:2 commensurability with Jupiter, and the asteroid Thule at 4:3, can only remain in orbits so close to Jupiter because their commensurabilities protect them from close approaches. The Trojan asteroids are an extreme example of this effect, as they move in the same orbit as Jupiter, but their 1:1 commensurabilities with Jupiter ensure that they remain far from Jupiter, close to 60° ahead or behind.

The many commensurabilities among the satellites provide a good means of determining their masses, due to the magnified perturbations that occur, but the effects are mostly not large enough to cause major problems in calculating their orbits. An exception is the 4:3 commensurability between Titan and Hyperion which causes very large perturbations of Hyperion's orbit. In the Jupiter system there is a complex commensurability involving the three satellites Io, Europa and Ganymede, which is a combination of two 2:1 commensurabilities. It is likely that this commensurability, and possibly some of those in the Saturn system, were formed by orbital evolution due to the tidal action. *ATS.*

Common, Andrew A. (1841–1903). English amateur who commissioned several large reflectors with which he took some of the finest astronomical photographs of his time.

Compton, Arthur H. (1892–1962). American physicist who discovered the COMPTON EFFECT in 1923.

Compton effect The loss of energy by, and consequent increase in wavelength of, a photon that collides with an electron and imparts some of its energy to it.

Conic section A curve that can be obtained by taking sections across a right circular cone at different angles. Examples are the circle, ellipse, parabola and hyperbola. In astronomy these curves represent the paths of bodies in a gravitational field.

Conjunction *See* INFERIOR CONJUNCTION, PLANETARY CONJUNCTION, SUPERIOR CONJUNCTION.

Constellations Arbitrary groupings of stars. Many systems have been proposed over the ages; we happen to follow that which has come down to us from the Greeks, though its origin is uncertain (it could be Chaldæan or possibly Cretan). The Chinese and Egyptian systems were completely different. It is clear that a constellation has no real significance, and is due to a line-of-sight effect, because the stars are at very different distances from us. Thus with the two "pointers" to the Southern Cross, which lie side by side in the sky and give a misleading impression of being associated. Alpha Centauri is just over 4 light-years away, Beta Centauri well over 450!

PTOLEMY gave a list of 48 constellations in his ALMAGEST. He could not, of course, include the far southern constellations, which never rise from Alexandria, where he lived, and he did not even cover the whole of the available sky; but all the constellations which he listed are retained on modern maps, though generally with altered outlines. Many of the named are drawn from ancient mythology; for example the Perseus legend is well represented. There are also a few animals and birds, and a few everyday objects.

Mapping the far-southern constellations involved some modern-sounding names, such as the Octant and the Telescope. There was, however, a period when almost every astronomer felt obliged to introduce new constellations, often with cumbersome names; thus Hevelius, in his maps of 1690, proposed Reticulus Rhomboidalis, now shortened merely to Reticulum. In 1775 J.E.Bode created various new groups such as Sceptrum Brandenburgicum (the Sceptre of Brandenburg) and Officina Typographica (the Printing Press) which have been rejected — though the name of one of them, Quadrans Muralis (the Mural Quadrant), is remembered because of the January meteor shower which we still call the Quadrantids. The nearest naked-eye star to the Quadrantid radiant is Beta Boötis.

Finally, in 1933, the International Astronomical Union decided to put matters on a more systematic footing. They reduced the accepted number of constellations to 88, also dividing the huge unwieldy Argo Navis (the Ship Argo) into its component parts of which Carina (the Keel) is the most important. The accompanying table gives the official list: Z indicates Zodiacal.

It is clear that these constellations are very unequal in size and in importance; they range in size from the vast Hydra down to the tiny but brilliant Crux. It is also true that there are some constellations which seem to have no justification for a separate identity; Horologium and Leo Minor are good examples of this. The 19th-century astronomer Sir John HERSCHEL commented that the constellations appeared to have been designed so as to cause as much confusion and inconvenience as possible, and though the IAU ruling has improved matters somewhat it is true that the overall result is still chaotic. However, things might have been worse. There have been periodical attempts to revise the entire system, even replacing the accepted figures with Biblical ones or politicians!

Very few constellations have more than one first-magnitude star (taking the "limit" for a first-magnitude star as the conventional 1.4). Note also that although Ophiuchus is not reckoned as a Zodiacal constellation, it does cross the Zodiac between Scorpius and Sagittarius. *PM.*

Constellations		Area (sq. degrees)	First magnitude stars
*Andromeda	Andromeda	722	—
Antlia	The Airpump	239	
Apus	The Bee	206	
*Aquarius	The Water-bearer	980 Z	
*Aquila	The Eagle	652	Altair
*Ara	The Altar	237	
*Aries	The Ram	441 Z	
*Auriga	The Charioteer	657	Capella
*Boötes	The Herdsman	907	Arcturus
Caelum	The Graving Tool	125	
Camelopardalis	The Giraffe	757	
*Cancer	The Crab	506 Z	
Canes Venatici	The Hunting Dogs	465	
*Canis Major	The Great Dog	380	Sirius
*Canis Minor	The Little Dog	183	Procyon
*Capricornus	The Sea-Goat	414 Z	
Carina	The Keel	494	Canopus
*Cassiopeia	Cassiopeia	598	
*Centaurus	The Centaur	1,060	Alpha Centauri, Agena
*Cepheus	Cepheus	588	
*Cetus	The Whale	1,232	
Chamaeleon	The Chameleon	132	
Circinus	The Compasses	93	
Columba	The Dove	270	
Coma Berenices	Berenice's Hair	386	
*Corona Australis	The Southern Crown	128	
*Corona Borealis	The Northern Crown	179	
*Corvus	The Crow	184	
*Crater	The Cup	282	
Crux Australis	The Southern Cross	68	Acrux, Beta Crucis
*Cygnus	The Swan	804	Deneb
*Delphinus	The Dolphin	189	
Dorado	The Swordfish	179	
*Draco	The Dragon	1,083	
*Equuleus	The Foal	72	
*Eridanus	The River	1,138	Achernar
Fornax	The Furnace	398	
*Gemini	The Twins	514	Pollux
Grus	The Crane	366	
*Hercules	Hercules	1,225	
Horologium	The Clock	249	
*Hydra	The Watersnake	1,303	
Hydrus	The Little Snake	243	
Indus	The Indian	294	
Lacerta	The Lizard	201	
*Leo	The Lion	947 Z	Regulus
Leo Minor	The Little Lion	232	
*Lepus	The Hare	290	
*Libra	The Balance	538 Z	
*Lupus	The Wolf	334	
Lynx	The Lynx	545	
*Lyra	The Lyre	286	Vega
Mensa	The Table	153	
Microscopium	The Microscope	210	
Monoceros	The Unicorn	482	
Musca Australis	The Southern Fly	138	
Norma	The Rule	165	
Octans	The Octant	291	
*Ophiuchus	The Serpent-bearer	948	
*Orion	Orion	594	Rigel, Betelgeux
Pavo	The Peacock	378	
*Pegasus	The Flying Horse	1,121	
*Perseus	Perseus	615	
Phoenix	The Phoenix	469	
Pictor	The Painter	247	
*Pisces	The Fishes	889 Z	
*Piscis Australis	The Southern Fish	245	Fomalhaut
Puppis	The Poop	673	
Pyxis	The Compass	221	
Reticulum	The Net	114	
*Sagitta	The Arrow	80	
*Sagittarius	The Archer	867 Z	
*Scorpius	The Scorpion	497 Z	Antares
Sculptor	The Sculptor	475	
Scutum	The Shield	109	
*Serpens	The Serpent	637	
Sextans	The Sextant	314	
*Taurus	The Bull	797 Z	Aldebaran
Telescopium	The Telescope	252	
*Triangulum	The Triangle	132	
Triangulum Australe	The Southern Triangle	110	
Tucana	The Toucan	295	
*Ursa Major	The Great Bear	1,280	
*Ursa Minor	The Little Bear	256	
Vela	The Sails	500	
*Virgo	The Virgin	1,294 Z	Spica
Volans	The Flying Fish	141	
Vulpecula	The Fox	268	

*Ptolemy's original 48 constellations. The 48th, Argo Navis, was later divided into four.

Continuous creation An adaptation of the physical laws according to normal mathematical practice which requires matter to be created as an ongoing process, instead of all matter appearing spontaneously at a supposed origin of the universe.

Continuous spectrum A continuous distribution of wavelengths emitted, eg, by an incandescent solid, liquid or dense gas. Also called a "continuum".

Convection A process of heat transfer in which heat energy is transported from one region of a gas or liquid to another by the flow of hot matter in bulk into cooler regions.

Cooke, Thomas (1807–1868). A leading English telescope maker.

Co-ordinate systems *See* CELESTIAL SPHERE and GALACTIC CO-ORDINATES.

Copenhagen Observatory "Det Astronomiske Observatarium"; established 1642 in Rundetårn Tower in the city centre; in 1861 moved to its present site at Copenhagen University.

Copernicus A lunar crater, striking for its bright rays, Copernicus (10°N, 20°W) is 57 miles (92km) across and the recent lava floor, with its multiple central peaks, is nearly $2\frac{1}{2}$ miles (4km) below the rim tops.

Copernicus: close-up of a lunar crater

Copernicus United States astronomy satellite launched August 1972 which studied stellar ultraviolet and X-rays. Observed Cygnus X-1, a BLACK HOLE candidate.

Copernicus, Nicholas (1473–1543) Renaissance astronomer who "stopped the Sun and threw the Earth into motion" with his radical proposal of a Sun-centred plan for the planetary system. Born in Toruń in Polish territory, he entered the University of Cracow, and then in Italy studied

C

C

Church law at Bologna and medicine in Padua. Returning home in 1503, he began his duties as a canon in the cathedral chapter of Warmia, the northernmost diocese of Poland, serving also as personal secretary and physician to the bishop, his uncle Lucas Watzenrode. Although technically a dependency of Poland, for most purposes Warmia was an ecclesiastical state under the rule of the bishop, so Copernicus held a position of considerable administrative responsibility in this small territory. Meanwhile, he maintained a lively side interest in astronomy.

In Copernicus' day it was universally accepted that the Earth was solidly fixed in the centre of the cosmos. The principal motions of the planets had been explained by PTOLEMY in terms of the compound effect of epicycles turning on larger circles around the Earth. Copernicus began to explore the consequences of fixing the Sun in the centre and placing the Earth in orbit around the Sun. He found that for the superior planets (Mars, Jupiter, and Saturn) the Earth's orbit could substitute in each case for the planetary epicycle, and that for the inferior planets (Mercury and Venus) he could get the same sightlines by centring their epicycles on the Sun instead of using separate carrying circles for each one. This resulted in a compact system in which all of the distances were linked together by the

Copernicus: "stopped the Sun"

common measure of the Earth's orbit. Furthermore, he noticed that Mercury, the swiftest planet, naturally circled closest to the Sun, whereas Saturn, the slowest, came at the outer bound of the planetary system, and all the others fell into place in the order of their periods of revolution. Precisely what motivated Copernicus' radical departure from

traditional astronomy is not known, but after he saw the elegant coherence of this pattern, he became persuaded that any astronomical theory "pleasing to the mind" had to take this linkage into account.

A Sun-centred arrangement had been proposed in antiquity by ARISTARCHUS, but only the barest hint of his scheme was known in the early 1500s, and definitely without any mention of the harmonious linkages that Copernicus found so compelling. Eventually Copernicus learned about his Greek predecessor, but only after he had independently hit upon the heliocentric system. It is not known precisely when Copernicus began to think about the new cosmology, but by 1514 he had written out a preliminary tract, the *Commentariolus* or "Little Commentary", and had distributed manuscript copies to a few friends.

At no time did Copernicus have any "proof" of his new cosmology. No available observation could have distinguished between the ancient geocentric scheme and the new heliocentric pattern. As GALILEO remarked in the next century, he could not admire enough those who accepted it *despite* the evidence of their senses.

After the publication in 1515 of Ptolemy's ALMAGEST, Copernicus realized that to compete with Ptolemy, he would have to present a complete description including tables for predicting planetary positions. He had already begun making observations to establish the characteristics of the planetary orbits, but some years had to pass before he could obtain full data for the slower-moving planets. Hence he continued with the administrative affairs of the cathedral chapter, biding his time, but he declined the opportunity to became a bishop in order to guarantee enough leisure to work on his astronomical treatise.

Besides rearranging the place of the Sun, Copernicus wished to fulfil the ancient axiom that celestial motions should be generated by combinations of uniform circular movements. As was well known, each planet runs faster (on the average) on one side of its orbit compared to the other. To achieve this Ptolemy adopted a geometrical device called an equant, which, in Copernicus' opinion, violated the principle of uniform motion, and his astronomy introduced small auxiliary circles or epicycles as an alternative. Although almost forgotten today because they have been replaced by Keplerian ellipses, this feature of his work was

one of its most admired aspects in the 16th century.

Copernicus was still perfecting his calculations when, in 1539, Georg Joachim Rhaeticus, a young Lutheran astronomer from Wittenberg, arrived to become his first and only disciple. Rhaeticus' enthusiasm persuaded

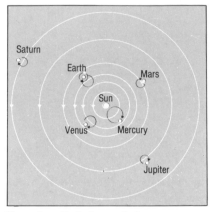

Copernicus: the heliocentric theory

Copernicus to allow him to publish a "first report" (*Narratio prima,* 1540) outlining the new ideas, and then to take the full treatise, DE REVOLUTIONIBUS ("The Revolutions") to Germany for publication. Copernicus finally received a complete printed copy of his work on the day he died, in 1543.

The 400-page treatise was quickly seen as the most important astronomical work since Ptolemy's *Almagest*. In the opening chapters Copernicus attempted to refute the ancient arguments for the immobility of the Earth, and he argued for the advantages of the new Sun-centred arrangement, which not only linked the system together but also provided an explanation for the so-called RETROGRADE MOTION of the planets. This apparent backward shift, caused by the swifter Earth bypassing the slower superior planets, was naturally smaller the more distant the planet. In addition, he correctly claimed that the annual backward motion of the stars expected from the postulated motion of the Earth could not be seen simply because the stars were so far away – "so vast, without any question, is the divine handiwork of the Almighty Creator".

Nevertheless, most astronomers considered his plan to be a hypothetical scheme for predicting planetary positions, but without any claim as a believable physical reality. Nearly everyone still followed Aristotle's physics, in which the Earth was viewed as a sluggish body unfit for motion. Not until the discoveries of KEPLER and Galileo did the system begin to make physical as well as geometrical sense. As a consequence of Galileo's writings, *De revolutionibus* was in 1616 placed on the *Index* of prohibited books by the Catholic Church. The book was not actually banned, however, but small corrections were introduced to make it appear entirely hypothetical. But by then Copernicus' book was already being made obsolete by Kepler's *Epitome of Copernican Astronomy*, which swept away the remaining geocentric vestiges that still remained in *De revolutionibus*, and by his far more accurate *Rudolphine Tables* (1627). *OG.*

Coprates Catena Crater-chain on Mars, 14°–16°S, 67°–58°W.

Cor Caroli The star Alpha Canum Venaticorum — a famous magnetic variable.

Cordoba Observatory National Observatory of Argentina, founded in 1871. Noted for its determinations of stellar positions, published in *Cordoba Dürchmusterung* and the *Astrographic Catalogue, Zone −24° to −31°.*

Coriolis effect To an Earthbound observer, anything that moves freely across the globe, such as an artillery shell or the wind, appears to curve slightly — to its right in the northern hemisphere and to its left in the southern hemisphere. This is the Coriolis effect. In 1835, Gaspard de Coriolis (1792–1843) first explained that this apparent curvature was not caused by some mysterious force. It simply shows that the observer is on a rotating frame of reference, namely the spinning Earth. This effect accounts for the circulation of air around cyclones.

Corona Australids *See* ARIDS.

Corona Austrinus (the Southern Crown). A small southern constellation adjoining Alpha Sagittarii. It contains no star above magnitude 4, but the little line of curved stars makes it easy to identify. It is sometimes called Corona Australis.

Corona Borealis (the Northern Crown). A constellation adjoining Boötes.

Brightest stars

Name	Visual Mag.	Abs. Mag.	Spec.	Distance (light-yrs.)
α (Alphekka)	2.23	0.6	A0	78
β (Nusakan)	3.68	1.2	F0	59

Though Alphekka is the only bright star, the curve made up of the two leaders plus Epsilon, Gamma and Theta makes the constellation easy to find. In the "bowl" of the crown is the famous variable R Coronae.

Corona *See* SOLAR CORONA.

Coronagraph *See* SOLAR CORONA.

Coronal holes Relatively cool low-density regions of the SOLAR CORONA which have been studied, especially by the SKYLAB mission, since the early 1970s. They appear as regions of low emission in those parts of the spectrum that are formed in the quiet corona (especially EUV and X-rays), and also as dark regions around the limb during SOLAR ECLIPSES. The polar coronal holes are dominant at the minimum of the SOLAR CYCLE and almost absent at SOLAR MAXIMUM. Coronal holes have an open magnetic structure leading to their association with the high-speed SOLAR WIND streams and are thus important to the discipline of solar terrestrial physics. *LJCW.*

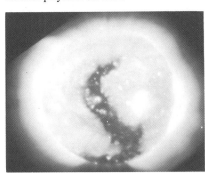

Coronal holes: relatively cool regions

Coronium A hypothetical element once thought to exist in the SOLAR CORONA in order to explain an unidentified line in the green part of the coronal spectrum. This line was shown in 1940 to be due to thirteen-times ionized iron.

Corvus (the Crow). A constellation adjoining Hydra. Its four leading stars form a quadrilateral which is very easy to find.

Brightest stars

Name	Visual Mag.	Abs. Mag.	Spec.	Distance (light-yrs.)
γ (Minkar)	2.59	−1.2	B8	186
β (Kraz)	2.65	−2.1	G5	290
δ (Algorel)	2.95	0.2	B9	117
ε	3.00	−0.1	K2	104

Corvus contains little of note.

Cos B *See* GAMMA RAY ASTRONOMY.

Cosmic rays These so-called "rays" are in fact almost entirely charged particles which arrive at the top of the atmosphere and come from cosmic space. Their energies extend up to 10^{20} electron volts per particle and thus represent the highest individual particle energies known to mankind.

Their discovery is accredited to the Austrian physicist Victor Hess, who, in 1912, carried electrometers aloft in a balloon (from Aussig in the then Austro-Hungarian empire) in an attempt to discover why it had proved impossible to eliminate completely a small residual background reading in the electrometers at ground level. Hess found that, after first falling, the reading started to increase as the balloon ascended and he made the remarkable claim that there was need "to have recourse to a new hypothesis; either invoking the assumption of the presence at great altitudes of unknown matter or the assumption of an extraterrestrial source of penetrating radiation". After arguments lasting a number of years the extraterrestrial origin idea finally won; Hess was awarded the Nobel Prize in 1936.

Initially it was thought that cosmic rays comprised some form of ultrapenetrating gamma radiation — and thus the term cosmic *rays* — but later work showed that particles, mainly PROTONS, were responsible.

The energy content of the cosmic radiation is not negligible. Near the Earth it is roughly one electron volt per cm^3 and is virtually the same as that in starlight, in the tangled magnetic field in the Galaxy and in the turbulent gas motions. It is presumed that the equality with starlight is fortuitous but the equality with the magnetic field energy density may possibly be because the field constrains the cosmic ray pressure — when the cosmic ray pressure is high, particles leak out of the Galaxy until the magnetic field regains control and creates an equilibrium situation. The GALACTIC MAGNETIC FIELD plays an important role in both trapping particles in the Galaxy and in deflecting the individual particle trajectories so that arrival directions at the Earth give virtually no information about the direction to the source of the particles.

The question of the origin of the cosmic rays is still an open one. Below 10^9eV a few come from time to time from the Sun, but the bulk are generated much farther afield. Most workers now think that the vast majority below

C

C

about 10^{19}eV come from sources within our Galaxy, with those above 10^{19}eV probably being extragalactic in origin. There are a number of contenders for the Galactic particles: SUPERNOVA explosions, shocks in supernova remnants (SNR) and PULSARS, to name the more likely. Of these, shocks in SNR have many supporters, the idea being that a significant fraction of the initial explosion energy in the supernova (about 10^{51}ergs) is converted over a timescale of several 100,000 years to ordered motion of cosmic rays. When the SNR merges into the general interstellar medium the cosmic rays then join the "Galactic pool". An efficiency of about 10 per cent for conversion of shock energy into cosmic ray energy plus a rate of one supernova of the required variety ("Type II") per century is sufficient to explain the observed fluxes of cosmic rays, at least up to about 10^{14} electron volts.

At higher energies other sources appear to be necessary and pulsars are a possibility. The dramatic object Cygnus X-3, situated at least 11 kiloparsecs from the Earth and which manifests itself by producing great fluxes of gamma-rays of energy up to 10^{16}eV, is the best example of this.

At the very highest energies, above 10^{19}eV, dramatic anisotropies (differences from a uniform distribution of flux from every direction) would be expected if cosmic rays were produced by Galactic sources. This is because the magnetic field in the Galaxy is too weak to deflect the cosmic rays much and we should be able to trace them back to their sources, which are no doubt concentrated in the general direction of the GALACTIC CENTRE. Quantitatively a proton of momentum p eV/c is deflected through an arc of a circle of radius R cm in a field of strength H gauss where pc = 300 HR; with pc = 10^{19}eV and H = 3×10^{-6} gauss (a typical Galactic field), R = 10^{22}cm = 3 kiloparsec. Such a distance is about one-third of that to the Galactic Centre and anisotropies should commence for Galactic sources. These are not seen, instead there is an excess of particles from high Galactic latitudes at the highest energies, specifically in the direction of the VIRGO CLUSTER of galaxies. There is the well-known giant jet coming from M87, a very big galaxy in the Virgo cluster, and this, and other similarly energetic extragalactic phenomena, are strong contenders.

An interesting feature at the very high energies is the expectation of an interaction between cosmic ray particles and the MICROWAVE BACKGROUND radiation (the "3K radiation"). The result should be a reduction in intensity of extragalactic cosmic rays coming from distances greater than a few per cent of the radius of the universe.

Turning to the different types of particles in the primary radiation, the nuclei represent roughly the relative abundances of the elements in the universe in general although there are significant and potentially informative differences. Protons predominate, α-particles (helium nuclei) form about one-seventh of the number of protons above the same rigidity (rigidity = momentum times the speed of light divided by charge) and heavier nuclei right up to the actinides comprise about one per cent of the protons. One of the unusual features is the presence of comparatively large fluxes of the elements lithium, beryllium and boron and these are identified as fragments of heavier cosmic ray nuclei which have struck nuclei of the interstellar medium (ISM). The presence of a very small fraction of radioactive beryllium-10, taken with other data, leads to an estimated lifetime of low-energy cosmic rays of at least 20×10^6 years.

About 3 per cent of the primaries are electrons. Electrons of energy of order 10^9eV generate SYNCHROTRON RADIATION when they are deflected by the magnetic fields in the ISM. The same process happens in supernova remnants and at other sites, not least in other galaxies.

A small flux of ANTIMATTER has also been seen, comprising positrons and anti-protons. Both are produced as the secondaries of cosmic ray interactions in the ISM although there is an excess of anti-protons which may be from extragalactic sources or exploding BLACK HOLES.

Very interesting phenomena occur when cosmic rays enter the atmosphere. Secondary particles, largely pions, are generated in the collisions with air nuclei and indeed a number of the so-called elementary particles (positron, pion, muon, strange-particles) were discovered in the cosmic radiation. Primaries of energy above about 10^{10}eV give rise to secondaries, which can reach ground level, the most important being the muon — the product of pion decay. About five muons pass through our heads every second at the latitude of the United Kingdom.

Both the magnetic field associated with the SOLAR WIND and the Earth's magnetic field shield the Earth from low-energy particles. The solar wind effect has ceased by about 10^{11}eV. The lowest-energy particles can arrive at the Earth's magnetic poles along the lines of force (as can particles from the Sun itself, for example during SOLAR FLARES). The spectrum is "hardest" at the magnetic equator. The Earth's field is responsible for the "latitude effect" and the "east-west effect".

Low-energy cosmic rays cause the production of radioactive carbon (C-14) and beryllium (Be-10) in the atmosphere and this leads to the possibility of studying intensity variations over 10^3 to 10^7 year periods from studies of C-14 in tree rings and Be-10 in deep-sea sediments.

The highest-energy particles produce gigantic showers of secondaries. At ground level these showers contain mainly electrons with about 5 per cent of muons and a few nucleons. The particles spread out to occupy an area of at least 10^5m². The rates of the big showers are very low, though — one per m² per year above 10^{16}eV, 100 per km² per year above 10^{18}eV and one per km² per century above 10^{20}eV, the highest energy detected so far. A 10^{20}eV shower contains over 10^{10} particles at ground level. *AWW.*

Cosmic ray satellites Satellites are necessary for precise measurements on COSMIC RAYS before they enter the atmosphere and interact with the nuclei of the air. The Soviet PROTON satellite results, reported by Grigorov in 1971, gave the proton- and all-particle spectra to modest accuracy. Contemporary American work using the SPACE SHUTTLE is measuring the elemental composition and the energy spectra of individual cosmic ray nuclei up to several 10^{12}eV per nucleon. Searches are being made for features which give information about the sources of the cosmic rays and the manner in which the cosmic rays propagate through the interstellar medium. *AWW.*

Cosmic ray telescopes Most COSMIC RAYS are charged particles moving at nearly the speed of light and their detection is often by way of the electrical signals caused when they pass through a gas or by flashes of light from atoms struck by them. A telescope comprises an array of detectors through which the particles pass; the direction of the particle can then be found. Telescopes have been operated in many different environments, from satellites down to the bottom of deep mines; directional telescopes in mines have shown upward-moving cosmic rays caused by NEUTRINOS which have penetrated the whole Earth before interacting to cause detectable secondaries. *AWW*.

Cosmic year The time taken for one complete revolution of the Sun around the centre of the GALAXY; about 225 million years.

Cosmogony The study of how the universe began. In antiquity the commonest theme was the emergence of the world from a watery chaos, little attention being paid to the distinction between the physical and animate world or between religion and reasoned argument. The early importance of cosmogony lay in the belief that the nature of a thing could be traced to the way it came into existence (as in Kipling's *Just-So Stories*). The opposite point of view, that the origin of things could be studied by the use of physical laws, has only been practicable in recent times.

Originally cosmogony referred to the creation of the Solar System and the formation of stars, subsequently to the formation of galaxies and the creation of the elements out of hydrogen. Nowadays each of these is a subject in its own right; cosmogony is taken to refer to the origin of the initial constituents of the big bang.

The discovery in the 1920s of the expansion of the universe and in the 1960s of its evolution led to the development of the big bang theory, and to the demise of steady-state and cyclical pictures. We can trace the big bang universe back at least to an age of 10^{-5} seconds when it contained matter, ANTIMATTER and RADIATION at a temperature around 10^{12}K. In the standard big bang theory we *cannot* extrapolate back to the moment of creation because the theory itself predicts its invalidity (at a "singularity").

Recently suggestions for an origin have arisen from progress in the unification of the fundamental forces in physics — the GRAND UNIFIED THEORIES (GUTs). These may lead to the inflationary cosmology: a finite region, of which the visible universe is part, starts off as a small, hot, expanding bubble of a phase of matter which can exist only at a temperature above around 10^{27}K. The bubble cools and expands rapidly as it changes into normal matter. The inflationary model would take the universe back to 10^{-35} or 10^{-45} seconds, depending on whether gravity is included in the unification of forces, but it may be more correct to

imagine time itself to begin at this point. *DJR. See also* essay on the BIG BANG.

Cosmology Cosmology is the science which studies the structure of the universe on the largest scale. Contained within it is COSMOGONY, dealing with the origins and evolution of the universe.

Man's early view of the universe was prejudiced by his belief that he occupied a special place within it — at the centre. Only in the present century have we realized that the Earth is but a small planet of a dim star, located in the outer suburbs of a galaxy typical of many others.

Perhaps the most important astronomical discovery of the early 20th century was HUBBLE's realization that the dim nebulae he observed were in fact enormous systems of thousands of millions of stars lying far outside our own galaxy.

Soon after came the evidence that these galaxies were receding from the Earth. Hubble obtained optical spectra of many galaxies and found that their spectral lines were shifted towards the red (longer wavelengths). He interpreted these REDSHIFTS as being caused by the DOPPLER EFFECT. Furthermore, the speed he inferred was found to be proportional to the galaxy's distance.

We now know that both the shift and the speed-distance proportionality follow naturally from an overall expansion of the scale size of the universe. Galaxies are redshifted because the universe has a different scale size now to that which it had when light was emitted from the galaxies.

Nevertheless, time has shown that Hubble was correct in his interpretation that the recession speed is proportional to distance. Today, the constant of proportionality bears his name (*See* HUBBLE CONSTANT).

Close to the Sun, the distance of galaxies can be determined from the properties of some of the VARIABLE STARS they contain — such as the Cepheids — or from the size of H II REGIONS. As we move farther out into the universe, however, these methods become increasingly inaccurate. Eventually distances can only be estimated by measuring the redshift and relying on the accuracy of the Hubble relation.

Unfortunately, for some of the most-distant and newly-found objects — such as the QUASARS — we do not have adequate confirmation that this procedure is valid. Some astronomers believe that at least part of the quasar

redshifts may originate from unknown, "non-cosmological" causes. Attempts to determine whether or not this is so have occupied much time on large telescopes in the past. Today the question is still unresolved.

However, perhaps the most important cosmological problems of the late 20th century are to determine the rate at which the universal expansion is taking place (determining the Hubble constant), how it has expanded in the past and how it will continue to behave in the future. To these must be added the question as to whether the overall geometry of the universe is "closed" or "open". In an open universe, the total volume of space is infinite, the universe has no boundary and will expand for ever. Closed universes contain a finite amount of space, may or may not have boundaries and will eventually collapse back on themselves.

Attempts to obtain a grand view of the universe have led to the construction of cosmological models.

A starting point for many cosmologists has been the finding that the universe appears much the same in all directions (the so-called isotropy) and at all distances (homogeneity). However, the expansion of the universe would at first seem to suggest that the overall density of the distribution of galaxies must decrease so that they become more sparsely distributed as time goes on.

Cosmological models have included the STEADY STATE models of BONDI, GOLD and HOYLE in which the universe is not only the same in all places but also at all times. It therefore had no beginning, will have no end and never changes at all when viewed on the large scale. This theory required matter to be created as the universe expanded in order that the overall density of galaxies should not decrease. For this reason it is also referred to as the CONTINUOUS CREATION model.

On the other hand, according to supporters of the big bang models — originally proposed by GAMOW, Alpher and Hermann — the whole universe was created in a single instant about 20 thousand million years ago and is presently expanding. In the future it may continue to expand or possibly collapse back on itself depending on the total amount of matter and energy in the universe, ie, whether or not the universe is open or closed. An important cosmological problem is the question of the MISSING MASS: the amount of matter we see in the universe is far smaller (by a factor of about 100)

C

C

than the amount we infer from the motions of the galaxies.

Definitive observations to discriminate between cosmological models are hard to make. The most informative parts of the universe are those farthest away. Unfortunately the objects we observe in such regions are faint and their nature is unknown. It is extremely difficult to tell to what extent the quasars, for example, are similar to the nearer — and more familiar — objects. And, if we cannot make comparisons, we cannot use them as standards to test cosmological models. Perhaps worst of all, we do not know if our Earth-derived physical laws are applicable in the universe at large. To assume they are is an item of faith.

Despite these problems, cosmologists have made several important discoveries: the Hubble constant is presently known to about 30 per cent accuracy. The deceleration parameter — which determines whether the universe will expand for ever or eventually collapse back on itself — has also been estimated. Surprisingly, it seems to have a value close to that expected for a universe which is marginally "closed" and will *just* continue to expand for ever. But why this should be so — if indeed it is true — is one of the most interesting coincidences in the whole of science.

Yet again, the discovery of the cosmic MICROWAVE BACKGROUND provided strong evidence against the steady state theory — but brought with it problems of its own: we don't understand how this background radiation can be so very uniform in all directions when it comes from different parts of the universe which have never been in communication with each other. Attempts have been made to invoke a very rapid period of expansion in the universe's history in order to remove this difficulty (the so-called inflationary models). But they appear, to many cosmologists, to be less than convincing.

At the present time, the best observational evidence favours an inflationary, big bang model for the universe with Hubble constant of around 75km per sec per megaparsec and deceleration parameter close to plus one-half. Its overall geometry is (surprisingly) close to the simplest possible flat, Euclidean model.

Cosmology is perhaps the most fascinating of the sciences. Nevertheless, it presently is based on a considerable amount of speculation fuelled by relatively little observational material. Future generations of cosmologists will

be presented with plenty of problems – and opportunities. And even if we *were* to obtain a good understanding of the present and future behaviour of the universe, we would still be far from comprehending what happened before the big bang. What form did matter and energy take then? Do such questions even have meaning at all? *AEW*[2]. *See also* essay on the BIG BANG.

Cosmos satellites A blanket name used by the Soviet Union for most of their scientific and military satellites. By the end of 1985 over 1,700 had been launched.

Cosmos satellites: more than 1,700 launched

Coudé system A coudé system is an arrangement of auxiliary mirrors which transfers the converging beam of light from a telescope to a focal point which is fixed in position. Often the coudé focus is at the entrance of a very massive SPECTROGRAPH, mounted on the floor of the observatory or even filling an entire room. Such a spectrograph could not be mounted at the CASSEGRAIN or NEWTONIAN focus of even the largest telescope, where it would have to be carried by the moving telescope.

The most efficient coudé system introduces only one additional small mirror, which must be moved at an appropriate speed with its own motor drive. This mirror reflects light coming from the secondary mirror and directs it across the observatory dome to the entrance slit of the spectrograph. To avoid the use of an independently driven mirror, most coudé systems use two or three additional mirrors which eventually lead the light down the hollow polar axle of the telescope to a spectrograph at the base.

As each extra reflection implies some loss of light, five-mirror coudé systems used to make very inefficient use of the light collected by the telescope. However, as these auxiliary mirrors are relatively small (1–2ft/30–60cm in diameter) they can be specially treated with thin layer coatings which give them essentially 100 per cent reflectivity over a selected wavelength range.

This may entail installation of several mirrors at each position, from which the one appropriate to the wavelength under study may be chosen.

Because of the ability to mount a spectrograph of almost unlimited size at the coudé focus it is here that the highest dispersion (resolution) spectra have traditionally been obtained. Detailed studies of the chemical compositions of atmospheres of stars and planets usually are made with the facilities at the coudé focus. For planetary work, where it may be important to obtain spectra for different regions of the surface, allowance must be made for rotation of the image at the coudé focus — an effect that does not occur at the other foci. This can be compensated by a motor-driven arrangement of prisms or mirrors called an image rotator. *BW*.

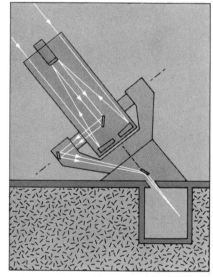

Coudé system: auxiliary mirrors

Counterglow Another name for the GEGENSCHEIN.

Crab Nebula Remnant of a supernova which was observed 900 years ago.

On July 4 1054 Chinese astronomers recorded a "guest star" in what is now the constellation Taurus. Many such stars have been recorded before: NOVAE in which for a brief period a WHITE DWARF rekindles and burns some hydrogen it has garnered from another source. The Chinese astronomers were not to realize that the 1054 guest star differed appreciably from its predecessors, though they did rate it brighter than most. It was no minor flicker of a white dwarf, but the total detonation of a star, a SUPERNOVA.

The star shone about as brightly as the planet Venus, being visible even in daylight. Surprisingly few records of so prominent an object have been found elsewhere in the world, though diligent searches are turning up some, including two rock engravings in the south-western United States which may depict the supernova near the crescent moon.

Today we can still see the remnant of that explosion as a small, bright nebula, a cloud of gas that originated in the star itself and now has expanded more than a millionfold in diameter. That nebula bears various names: Messier 1, NGC1952, Taurus A (a name given when it turned up as one of the most powerful radio sources in the sky). The name by which it is best known, however, is the Crab Nebula, a misnomer introduced because it was once sketched to resemble that creature.

The Crab Nebula is a supernova remnant — the nebula, radio and/or X-ray source left over from a super-nova. Supernova remnants usually take the form of an expanding, hollow shell. The star's material is ejected at a speed of typically 10,000km/s, so that it rams violently into the ambient interstellar gas. The effect of this continuous colli-sion is to keep the gas hot long after the supernova is forgotten. The gas is warmed to several million degrees, emitting light and X-rays. In so doing, the gas is slowed down, so that even-tually the remnant ceases to glow. Depending on its environment, a supernova remnant may shine for tens of thousands of years. The Crab Nebula is, therefore, a particularly young specimen.

The nature of the explosion is such that the gas forms filamentary struc-tures. The filaments give old supernova remnants a wispy appearance. In the Crab, the filaments are indeed present, on colour photographs glowing the characteristic red of hot hydrogen, like strands of red cotton wrapped around a soft yellow glow.

It is the yellow glow that is distinc-tive in the case of the Crab Nebula. Within the filaments is an ionized gas in which electrons, freed from their parent atoms, are spiralling in an intense magnetic field. When this occurs, a par-ticular kind of energy is emitted: SYNCHROTRON RADIATION. It is by the synchrotron mechanism that many objects, particularly supernova rem-nants, emit in the radio domain. Again, however, the radio radiation generally comes from a hollow shell of gas. The Crab Nebula, however, emits from its centre outwards, and is a member of a

very rare group of remnants known as filled or plerionic supernova remnants. Just why some remnants are filled is not clear. It may, but does not certainly, depend on their age.

The Crab Nebula is probably unique amongst filled remnants because the synchrotron radiation extends from the radio domain into the visible. It is the only synchrotron nebula that can be seen in a small telescope. In order to

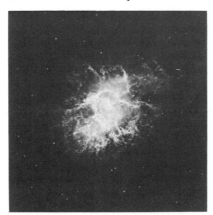

Crab Nebula: supernova remnant

emit at such short wavelengths, the electrons must be very energetic indeed. The very act of producing synchrotron radiation removes energy from the electrons, so in the Crab Nebula there has to be a continuing supply of ener-getic electrons now, 900 years after the star was seen to explode.

The source of electrons is known. When a massive star explodes, it does so from a region slightly outside its very centre. The inner region has, in effect, already burnt out. The explosion that drives off the outer layers also crushes the inner core to form a NEUTRON STAR or, in extreme cases, a BLACK HOLE.

Neutron stars spin very fast indeed, and the energy of their rotation is gradually converted to beams of ener-getic particles and radio waves. We recognize this activity as a PULSAR when the beam swings into view. The Crab Nebula contains a pulsar, spin-ning at the impressive rate of 30 times per second, and spraying out both radiation and electrons to replenish the synchrotron radiation.

The Crab Pulsar is the youngest known. A property of very young pul-sars is that they emit not only radio radiation but optical, ultraviolet and X-radiation too. If they are older than a few thousand years, the optical fades rapidly away. The Crab pulsar flashes at visible wavelengths, a discovery made by William Cocke and Michael

Disney in 1969 using one of the teles-copes of the KITT PEAK NATIONAL OBSERVATORY. They made the disco-very after many nights of trying to find optical flashes from many older radio pulsars. The Crab seemed to them a last resort. It happens that they worked with a tape recorder running; the tape of the discovery is justifiably famous.

Although the pulsar flashes too quickly for the human eye easily to resolve the pulses, stroboscopic tech-niques have been used to monitor its flash cycle in slow motion.

The Crab Nebula holds other sur-prises yet. Only recently a very faint extension was noted from its northern edge — a broad, parallel-sided jet of gas. The origin of this jet is still obscure. Various theories have been proposed, but none seems quite con-vincing at present. The Crab Nebula certainly will retain a vital place in the study of supernovae and supernova remnants for a considerable time. *DAA.*

Crab pulsar The pulsar in M1. It is one of a few to be detected optically.

Crater (the Cup). A small constellation adjoining Hydra and Corvus. Its brightest star, Delta, is of magnitude 3.56; it forms a triangle with Alpha (4.08) and Gamma (also 4.08). Crater contains no objects of note.

Crêpe Ring Common name for Saturn's Ring C.

Cretan astronomy Astronomy of the Minoan civilization, which was des-troyed around 1500BC, probably by the volcanic outburst at Thera. It is possible that the constellations listed by Ptolemy are of Cretan origin.

Crimean Astrophysical Observatory More properly known as the Sternberg Astronomical Institute Observatory. Its position is 44°43′42″ N, 34°01′ E; at an altitude above sea-level of 1,800ft (500m).

Of the various instruments at the observatory, the largest is the 104in (264cm) reflector, completed in 1960. There is also one of the world's largest MAKSUTOV TELESCOPES, a 26in (65cm), which was completed in 1958.

Conditions in the Crimea are much more favourable than those at the older Russian observatories, such as Pul-kova, and this has been shown by the excellent results from the Crimea. The observatory is also noted for its work in the field of solar research. *PM.*

Crimean Observatory: the radio telescope

C

Crommelin, Andrew Claude de la Cherois (1865–1939). Irish astronomer who, with P.H.Cowell, made the most accurate predictions of the return of Halley's Comet in 1910. A comet is named after him.

Crommelin's Comet Found by PONS in 1818. Rediscovered by Coggia and by Winnecke in 1873, it was again seen for only a few days. Forbes found a comet in 1928, shown by Crommelin to be identical with those of 1818 and 1873. Called Pons-Coggia-Winnecke-Forbes, it was later renamed Comet Crommelin. Predictions enabled the comet to be found in 1956 only 10° from the expected position by L.Pajdusáková and M.Hendrie. With a well-established orbit, recovery in 1983 was early and Crommelin was used to test the programme for the Halley's Comet return in 1985/1986. *MJH.*

Crossed lens Lens with convex surfaces so designed that SPHERICAL ABERRATION is reduced to a minimum for a parallel beam of light.

Cross staff Ancient instrument for measuring angular distances.

Cross staff: measuring angular distances

Crumlin meteorite Chondrite; fell Antrim, Northern Ireland, 1902.

Crux Australis (the Southern Cross). With Ursa Major, probably the most famous constellation in the sky. It is almost surrounded by Centaurus.

Brightest stars

Name	Visual Mag.	Abs. Mag.	Spec.	Distance (light-yrs.)
α (Acrux)	0.83	−3.9.−3.4	B1 + B3	360
β	1.25	−5.0	B0	425
γ	1.63	−0.5	M3	88
δ	2.80	−3.0	B2	257

These four make up a pattern which is more like a kite than a cross — and the symmetry is disturbed further by the presence of Epsilon (3.59), the only other star above magnitude 4. With an area of only 68 square degrees, Crux is the smallest constellation in the sky, but also one of the richest; interesting objects include the "Jewel Box" cluster Kappa Crucis and the dark nebula of the Coal Sack. *PM.*

Crux Australis: the smallest constellation

Culgoora radioheliograph A ring-array of 96, 46ft (14m) radio telescopes, 1.8 miles (3km) in diameter designed by J.P.Wild to study the Sun at low frequencies. It was operational in New South Wales from 1967 to 1984.

Culmination The maximum altitude of a celestial body above the horizon. This point is reached as the body crosses the observer's meridian. *See also* TRANSIT.

Curtis, Heber Doust (1872–1942). American astronomer; used novæ to measure distances to the galaxies. Excellent spectroscopist. Argued against SHAPLEY in the so-called "great debate" of 1920, correctly proposing that the spiral nebulæ were external galaxies.

Curvature of space The concept that space is curved in the presence of matter. According to the general theory of RELATIVITY, a massive body distorts space in its vicinity so that rays of light and particles of matter follow curved paths (known as geodesics).

In cosmological models the net overall curvature of space may be one of the following: zero — a flat, infinite (Euclidean) space in which the shortest distance between two points is a straight line; positive — a closed space like the surface of a sphere (Riemannain geometry); or negative — a hyperbolic open infinite space with the geometry of Lobachevsky. *IKN.*

Cusp The "horn" of an inferior planet in the crescent phase.

CV Serpentis An eclipsing star connected with the bright hydrogen nebula E41. Period of 29^{d}64, range 9^{m}86 to 10^{m}81B. It has a Wolf-Rayet and B0 stellar components.

Cydonia Region on Mars, adjoining ACIDALIA, considered as a landing-site for Viking 2 but then rejected.

Cygnus (the Swan). One of the most important constellations — often nicknamed the Northern Cross.

Brightest stars

Name	Visual Mag.	Abs. Mag.	Spec.	Distance (light-yrs.)
α (Deneb)	1.25	−7.5	A2	1800
γ (Sadr)	2.20	−4.6	F8	750
ε (Gienah)	2.46	0.2	K0	82
δ	2.87	−0.6	A0	160
β (Albireo)	3.08	−2.3	K5	390
ζ	3.20	−2.1	G8	390

Next come Xi and Tau (each 3.72). Kappa (3.77), Iota (3.79), Omicron¹ (3.79), Eta (3.89), Nu (3.94) and Omicron² (3.98). To these must be added the red Mira variable Chi, which can exceed the fourth magnitude.

Deneb is a very luminous supergiant. Albireo is a glorious double with a golden-yellow primary and a blue companion, separable with a very small telescope. There are many interesting objects of all kinds in Cygnus, and the constellation is crossed by the Milky Way, which is exceptionally bright in this region. *PM.*

Cygnus A Extragalactic radio source with the highest apparent flux of all such sources. It has two radio lobes centred on the parent galaxy and is also an X-ray source.

Cygnus loop The remnant of a Type II supernova, also known as the Veil

Nebula. It is a strong radio and soft X-ray source 6° in diameter and about 20,000 years old.

Cygnus X-1 Binary X-ray source. It probably contains a BLACK HOLE.

Cygnus X-3 Complex radio, X-ray and gamma-ray source. *See also* COSMIC RAYS.

Cyrillus Lunar crater 60 miles (97km) in diameter; one of the Theophilus chain.

Cytherean Adjective for "Venus" (often preferred to "Venusian").

Dædalus Large crater on the far side of the Moon; 6°S, 180°W.

D'Alembert Lunar crater on the far side; 52°N, 164°E.

Dall-Kirkham telescope A type of CASSE-GRAIN reflecting telescope in which, however, the concave primary mirror is ellipsoidal and the convex secondary is spherical. It is easier to make such mirrors than the paraboloids and hyperboloids of a conventional Casse-grain telescope and the telescope is insensitive to misalignment. Its field of view is only a third of that of a Casse-grain telescope.

Dark nebulæ Clouds of dust and gas are visible only because they block off

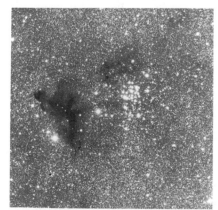

Dark nebulæ: Barnard 86

the light of stars and nebulae beyond them. They range in size from minute, more or less spherical BOK GLOBULES,

usually seen in photographs against the bright glow of emission nebulæ, to the naked-eye clouds of the southern COAL SACK and the gigantic RHO OPHIUCHI dark cloud, which affects 1,000 square degrees — two per cent — of the sky.

The clouds consist of a mixture of dusty particles and gas, the whole having a composition similar to the standard cosmic abundance of about 75 per cent hydrogen and 23 per cent helium with the rest heavier elements. The dust itself comprises about only 0.1 per cent of the mass of a cloud but it is believed to play an important part in the formation of molecules in space. The surface of the dust particles provides a surface to which atoms within the cloud can adhere and perhaps combine with others to form simple molecules (ie, H_2) or much more complicated compounds such as formaldehyde or even amino acids. These relatively fragile compounds survive because they are protected from energetic ultraviolet radiation from stars by the cloud itself. The interiors of these molecular clouds are thus very cold, typically only 10K, which allows them to gradually collapse under their own gravity, eventually to begin the process of star formation. *DFM.*

D'Arrest, Heinrich Louis (1822–1875). German astronomer. He assisted Galle to find Neptune (1846), investigated asteroids, comets and nebulæ and discovered three comets and one asteroid. He published measurements of nebulæ in 1858 and 1867.

D'Arrest's Comet Discovered by H.L. D'ARREST at Leipzig in 1851. It is a faint comet with a period of 6.4 years that has been seen at 14 returns, including that of 1982.

Darwin Low-walled lunar crater, 80 miles (130km) in diameter, in the Grimaldi area; 20°S, 70°W.

Date line The arbitrary line, mainly at 180° longitude, where the date changes.

Davida (Asteroid 511). With its diameter of 200 miles (323km) it is the seventh largest of the regular asteroid swarm.

David Dunlap Observatory The observatory of the University of Toronto, Canada. It was opened in 1935, its principal instrument being a 74in (188cm) reflector. Especially important has been its work on radial velocities of stars and studies of variable stars in globular clusters.

Dawes limit An empirical measure of the resolving power of an optical telescope. For a telescope of $d$cm aperture the resolution of a double star is:

$$r(\text{arc sec}) = \frac{11.58}{d\text{cm}} = \frac{4.56}{d\text{in}}$$

Dawes, William Rutter (1799–1868). A famous English double star observer who is chiefly remembered for the DAWES LIMIT. He was called the 'eagle-eyed" because of his acute vision.

William Dawes: "eagle-eyed"

Day Conventionally, the time taken for the Earth to complete one rotation on its axis. There are, however, various definitions.

A **sidereal day** is the time taken between successive meridian passages of the same star (23h 56m 4.091s). This is therefore the true rotation period; since the stars, which are used as reference points, are so far away that in this context they may be regarded as infinitely remote.

A **solar day** is the interval between two successive noons. Since the Sun is moving across the sky background at a rate of about 1° per day in an easterly direction, the solar day is slightly longer than the sidereal day; also, the Sun's rate of motion varies, and is greatest at perihelion. For convenience, use is made of a "mean sun", which is an imaginary body moving round the CELESTIAL EQUATOR at a constant speed equal to the average rate of motion of the true Sun along the ECLIPTIC. The mean solar day is equal to 24h 3m 36.555s.

In a **civil day** there are two 12-hour periods, am and pm, but these are never used astronomically and the 24-hour clock is always given. *PM.*

D

Daylight meteor stream A meteor shower whose radiant lies too close to the Sun for observation in a dark sky. The activity of these streams is monitored using radio techniques.

Decametric radiation (1). Radio waves having wavelengths of a few decametres or a few tens of metres, ie, from 10 up to 100m. It includes most HF radio frequencies and the upper part of the MF band.

Decametric radiation (2). Bursts of radio emission at wavelengths of tens of metres that correlate with the orbital position of the moon Io around JUPITER. Probably caused by charged particles moving in the current loop known to link the two bodies.

Deceleration parameter A measure of the rate at which the expansion of the universe is slowing down. *See also* COSMOLOGY.

De Chéseaux' Comet Brilliant multi-tailed comet of 1744, discovered independently by Klinkenberg (Dec. 9) and de Chéseaux (Dec. 13). It reached apparent magnitude − 7.

Decimetric radiation Radio waves having wavelengths of a few decimetres or a few tenths of a metre, ie, from 0.1 up to 1m. It includes most UHF radio frequencies and the upper part of the VHF band.

Declination (δ) Angular distance in degrees north (+) or south (−) of the CELESTIAL EQUATOR to a maximum of 90° at the poles.

Deferent In the PTOLEMAIC SYSTEM, the centre of the smaller circle, the EPICYCLE, on which a planet moved, itself moved in a circle, called the deferent, around the Earth.

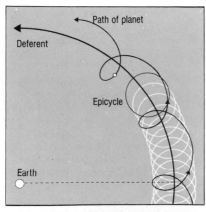

Deferent: aspect of Ptolemy's system

Degenerate matter At the enormous temperatures and pressures that exist at the centres of stars atoms are stripped of almost all of their electrons, leaving a gas of nuclei and electrons. Because these particles are much smaller than atoms, such a gas can be compressed to much higher densities than an ordinary gas. However, a quantum mechanical law known as the PAULI EXCLUSION PRINCIPLE prevents particles having similar velocities from getting closer together than a certain distance. When the particles reach that situation the gas becomes less compressible and is said to be degenerate.

In WHITE DWARFS, or the centres of giant stars and low-mass dwarf stars, the electrons are degenerate but the atomic nuclei, with higher masses and lower gas velocities, are not. Because of the high density the electrons are relatively confined in position and the action of the exclusion principle is to determine the velocities (more correctly, the momenta) of most of them. As a result, the electrons do not collide with themselves or other particles: any such interaction would require a change of momentum. Consequently the few electrons that are not degenerate can move without interacting with the majority, which makes degenerate matter an almost perfect conductor of heat. In degenerate matter the motions of the constituent particles are not determined by temperature, unlike the case for a normal gas, and therefore a degenerate gas does not obey the usual gas laws.

If energy is put into a degenerate gas the temperature rises, this being achieved by a fraction of the electrons with previously restricted velocities moving to higher and unrestricted states. At a sufficiently high temperature all of the electrons are removed from their degenerate condition. Degeneracy therefore can only exist at temperatures below what is called the Fermi temperature. *BW.*

Degree of arc $\frac{1}{360}$ of a circle.

Deimos The outermost of the two Martian moons has an irregular shape with dimensions of 9 × 7½ × 7 miles (15 × 12 × 11km). The moon is locked in SYNCHRONOUS ROTATION with Mars with its largest axis radially aligned with the planet and its smallest axis oriented perpendicular to the satellite's orbital plane. The surface of Deimos is dark, reflecting only 5 per cent of the incident sunlight, indicating that the body is composed of carbonaceous chondritic material similar to some asteroids.

The satellite orbits at a distance of 14,600 miles (23,500km) with a period of 1.262 days. Mars has a period of 1.026 days. As a result, Deimos would appear almost stationary to an observer on the Martian surface. The satellite would appear to move westward about 10° within a six-hour period. *RFB.*

Delambre Regular lunar crater, 32 miles (52km) in diameter, in the Mare Tranquillitatis area.

de la Rue, Warren (1815–1889). British businessman, scientist and astrophotographer. Notwithstanding early successful lunar images, his most lasting work was devoted to the study of the Sun — particularly sunspots and prominences.

Delavan's Comet Found at La Plata in 1913, then moved north to become a naked-eye object with a 10° tail in September 1914; last seen 1915.

Delphinus (the Dolphin). A small but very prominent constellation in the Aquila area. Its leading stars, between magnitudes 3.5 and 4, make up what looks at first sight like a compact cluster.

Delta NASA's most frequently used and successful rocket for launching intermediate-size spacecraft into Earth orbit. The first launch was in 1960 and as at mid-1986 there had been only 12 failures in 178 launches.

Delta: NASA's most successful rocket

Delta Aquarids An annual meteor shower, the maximum of 30-40 meteors per hour being on July 27. The orbit is small and highly eccentric, with perihelion very close to the Sun. The radiant is in Aquarius.

Delta Cephei Prototype of the CEPHEID VARIABLES. The magnitude range is from 3.7 to 4.4, and the period 5.366 days. It was discovered in 1784 by John Goodericke.

Delta Scuti variables Variable stars which pulsate in periods of 0^d01 to 0^d2. The range of light amplitudes is 0^m003 to 0^m9V, but usually only several hundreds of a magnitude. Spectral types A0-F6. Light curve shape, amplitude and period vary, with some stars having only sporadic variations which on occasions cease entirely as a result of strong amplitude modulations. They have radial and non-radial pulsations. The maximum expansion of surface layers does not lag behind maximum light for more than 0^d1. These stars are found in the lower part of the Cepheid INSTABILITY STRIP. They are also sometimes referred to as ultra-short-period Cepheids, due to their extremely short periods, or as dwarf Cepheids. *FMB.*

Dembowski, Ercole (1912–1881). Italian pioneer double-star observer.

Democritus (late 5th century BC). Greek physicist and mathematician who substantially developed the atomic theory of the material universe devised by his teacher Leucippus. He anticipated John Locke by distinguishing between primary and secondary qualities.

Democritus: atomic theory

Deneb The star Alpha Cygni, principal star of the Northern Cross. Although it has at a distance of 1,600 light-years, Deneb is the nineteenth brightest star in our night sky because it is highly luminous SUPERGIANT.

Deneb (upper left): 19th brightest star

Denebola The star Beta Leonis; magnitude 1.7, type A3.

Density The ratio of mass to volume for a given material or object. The average density of a body is its total mass divided by its total volume. Substances that are light for their size have a low density and vice versa. Water has a density of $1gm/cm^3$ under standard conditions. A wide variation in density is found throughout the universe, ranging from about $10^{-20}kg/m^3$ for the incredibly rarefied interstellar gas to over $10^{17}kg/m^3$ for the unbelievably dense neutron stars. The average density of matter in the entire universe is thought to be about $10^{-28}kg/m^3$. A typical white dwarf star has a density of between 10^7 and $10^{11}kg/m^3$. A large family suitcase of white dwarf material would have a mass of more than 125,000 tonnes. *JWM.*

Density wave theory Theory that describes the formation and persistence of the structure in spiral galaxies.

Spiral galaxies rotate differentially, not as solid bodies; the outer parts of spirals lag behind the inner parts. Moreover, the prevalence of spiral galaxies means that the spiral patterns must live for a substantial fraction of the life of the galaxies, and in particular the pattern must persist for perhaps 100 rotations of the galaxies. It follows that fixed spiral structure in the form of particular stars and gas clouds packed along trailing arms cannot be permanent and frozen features — if so, such patterns would be obliterated by the differential rotation after a few rotations. The solution to this problem is for the observed spiral structure to be a wave phenomenon, the spiral arms defining the regions of maximum density as density waves travel through the material of the galaxies. A good analogy is the expanding ripple pattern as a pebble is dropped into a still pool; the rings move outward with a pattern speed, while the water oscillates up and down locally without moving away from the spot at which the pebble entered. Similarly, the spiral-arm pattern travels through the galaxy as a fleeting enhancement, bunching the stars and gas locally before moving on to the adjacent stars at a rate set by the pattern speed. The controlling force is gravity, driving the stars in elliptical orbits around the galactic nucleus, and locking the orientation of the ellipses so the stars concentrate along lines of density enhancement which define the trailing spiral arms.

Lin and Shu in 1969 are generally deemed to have formalized density-wave theory, but there are many versions, and many aspects remain to be understood. For instance in its simplest form, the theory produces stable bars in galaxies. Why then are only a few galaxies (the "barred spirals"; *see* CLASSIFICATION OF GALAXIES) seen to have bars? Moreover, how is it that the gas does not damp out the waves in a few rotations? Perhaps vital roles in pattern longevity are played by massive dark haloes (*see* GALACTIC HALO) or the replenishment of the gas.

The effect of the density waves in the interstellar gas is particularly interesting. The pattern speed exceeds the sound speed, and shock waves, cosmic sonic booms, must result. Such severe compressions undoubtedly produce star formation (*see* STELLAR EVOLUTION); this and other observational evidence show that the arms of spiral galaxies are certainly the birthplace of hot young stars.

Dust (*see* INTERSTELLAR DUST) is swept along with the interstellar gas, and the dust lanes in spiral arms thus provide indirect evidence of density enhancements which correspond to density waves. However, detailed comparison of observation with theory is difficult. One approach is to make precise radio APERTURE SYNTHESIS maps of HI (neutral hydrogen) in nearby spiral galaxies. Another approach is to compare the widths and positions of spiral arms in different colour bands to see if the population balance of young (blue) and old (red) stars varies across the arms in accordance with predictions. Finally, developments in computing power mean that the N-body simulations and hydrodynamic calculations offer substantial hope of verifying or eliminating competing forms of density-wave theory. *JVW.*

D

De revolutionibus orbium coelestium "On the Revolutions of the Heavenly Spheres", COPERNICUS' masterwork setting forth the heliocentric arrangement of the Sun and planets, thereby establishing for the first time a Solar System. Published in 1543, the book quickly became a scientific classic, yet nearly a century passed before its cosmology was widely accepted. The opening pages defend the new Sun-centred system, but most of this technical treatise contains geometrical derivations and tables for predicting the positions of the Sun, Moon and planets. *OG.*

D

De Revolutionibus: Copernicus' masterwork

Descartes A concentric lunar crater, 31 miles (50km) in diameter; 11.5°S, 15.5°E.

Descending node The point at which an orbit crosses from north to south of a reference plane such as the CELESTIAL EQUATOR or ECLIPTIC.

De Sitter, Willem (1872–1934). Dutch astronomer, remembered for work on RELATIVITY. Double star observations showed that the speed of light did not depend on source velocity. The de Sitter model is a COSMOLOGY containing no matter. The Einstein-de Sitter model is a BIG BANG universe which expands forever.

Deslandres, Henri (1853–1948). French astrophysicist director of the Paris Observatory. An experimental astrophysicist who in 1894 designed the SPECTROHELIOGRAPH, invented independently at the same time by HALE.

Deuterium A heavy isotope of hydrogen. A deuterium nucleus contains one proton and one neutron and is denoted either by $^2_1$H or $^2_1$D.

Dew cap A tube fitted to the front cell of a telescope or element of a lens to prevent scattered light entering the system and to help prevent dew forming.

D-galaxy A galaxy with an elliptical nucleus surrounded by an extended envelope; or a giant radio galaxy of similar appearance.

Dialogues The greatest book written by GALILEO in defence of the Copernican system (*see* COPERNICUS). It led to his condemnation by the Inquisition in 1633.

Diamond ring effect The spectacle just before or after the total phase of a SOLAR ECLIPSE as part of the Sun shines like a diamond on the ring of the corona.

Diamond ring effect: from the 1966 eclipse

Diana Chasma The deepest fracture on Venus; 15°S, 150°E. It lies 6,500ft (2km) below the mean radius, and 13,000ft (4km) below the adjacent ridges.

Dichotomy The exact half-phase of the Moon or an inferior planet.

Diffraction The bending of light around a sharp edge or aperture which arises because each point on the wavefront, as it reaches the edge or aperture, behaves like a tiny source of wavefronts. Diffraction modifies the appearance of images produced by telescopes so that, for example, star images do not appear as points. *See also* AIRY DISK.

Diffraction grating A plate ruled with, typically, from 100 to 1,000 parallel grooves per millimetre which, by diffraction, disperses light into its constituent wavelengths and produces a spectrum.

Diffuse nebula Any luminous cloud of gas in the sky. The term "diffuse" refers to the fact that they cannot, like star clusters and galaxies, be resolved into individual stars; *nebula* is Latin for cloud.

Diffuse nebulæ come in two varieties. H II REGIONS, or emission nebulæ, shine by fluorescence, much as a street lamp does, the ultraviolet radiation of hot stars replacing the electricity used in terrestrial fluorescent lamps. The second type does not shine by itself; the motes of dust within the gas are lit by nearby stars, similar to smoke in sunlight. Such clouds are called REFLECTION NEBULÆ. *DAA.*

Digges, Leonard (?1520–1559?). Mathematician, surveyor, navigator and ballistics expert. He seems to have experimented with optical devices and may have devised some sort of telescope.

Dione A satellite of Saturn discovered in 1684 by CASSINI. It has a diameter of 696 miles (1,120km) and density of 1.4g/cm³, which is greater than the densities of all the other Saturn satellites apart from TITAN and PHŒBE. An important characteristic of Dione is its non-uniform surface brightness. The trailing hemisphere is dark with an albedo of 0.3 while the leading hemisphere has an albedo of 0.6. Only IAPETUS displays a greater range of surface brightness.

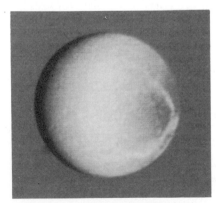

Dione: Saturn's satellite from Voyager 1

The satellite was observed during the VOYAGER flybys in 1980 and 1981 and the surface shows evidence of craters in the range of 18–25 miles (30–40km) in diameter and some large craters 100 miles (165km) across. The most striking feature is Amata, a crater 150 miles (240km) in diameter in the centre of a system of bright, wispy features that divide up the trailing hemisphere. There are also some broad ridges in the heavily cratered plains of the southern hemisphere and a valley more than 300 miles (500km) long near the south pole. Dione is accompanied by a small co-orbiting satellite Helene, discovered by Synnott in 1982. It measures 22 × 21 × 17 miles (36 × 34 × 28km). *GEH*.

Dionysius Brilliant lunar crater, 12 miles (19km) in diameter, on the edge of the Mare Tranquillitatis at 3°N, 17°E.

Dipole A pair of equal and opposite electric charges or magnetic poles separated by a finite distance. An ordinary bar magnet is a dipole.

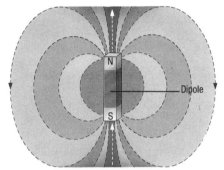

Dipole: a simple bar magnet

Direct motion Orbital motion in the same direction as the Earth's motion round the Sun. Also, west to east motion of Solar System bodies in the sky relative to background stars.

Discoverer Series of unmanned satellites with which the United States Air Force demonstrated the use of Thor-Agena rockets to orbit research payloads. Many had recoverable capsules.

Discovery Rupes Ridge on Mercury; 54°S, 38°W.

Dispersion The separation of light into its constituent wavelengths by refraction or diffraction; produced by a lens, prism or grating.

Distance modulus The difference between APPARENT MAGNITUDE (corrected for interstellar absorption) and ABSOLUTE MAGNITUDE of a star or galaxy. It is a measure of the distance to an object.

Diurnal arc *See* DIURNAL MOTION.

Diurnal motion Apparent daily motion of a heavenly body across the sky, from

east to west, caused by the Earth's rotation. At the poles, this is a circle parallel to the horizon; at other latitudes an arc from horizon to horizon, the length of which varies with latitude and declination, the arc's angle to the horizon varying with latitude alone.

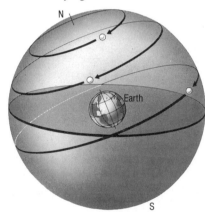

Diurnal motion: heavenly circles and arcs

Di Vico, Francisco (1805–1848). Italian astronomer; noted for his planetary work. In 1844 he discovered the periodical comet that bears his name.

Di Vico-Swift Comet A short-period comet and a member of the Jupiter family of comets, discovered by Francisco DI VICO. It was "lost" for 68 years after 1897, but was recovered in 1965 after calculations by Brian Marsden. The comet has a period of 6.3 years, orbital eccentricity of 0.52, and a perihelion distance of 1.62AU.

Dolland, John 1706–61). English optician who re-invented the achromatic LENS by combining crown and flint glasses. He was thus able to make refracting telescopes with greatly reduced colour aberration. Dolland, a silkweaver by trade, worked with his son Peter, an optician, whom he joined in 1752.

Dominion Astrophysical Observatory: major contributions to stellar spectroscopy

Dominion Astrophysical Observatory Located in Victoria, BC, Canada, it opened in 1917 and has 72in (183cm) and 48in (122cm) reflectors. The observatory has made major contributions to stellar spectroscopy, including the determination of spectroscopic binary orbits.

Donati, Giovanni Battista (1826–1873). Italian astronomer and spectroscopist, the first to make spectroscopic observations of COMETS. In 1864 he discovered that the tails of comets consisted of gases.

Donati's Comet Discovered by DONATI on June 2 1858. Four months later it developed a long curved tail and a bright nucleus. The head showed very remarkable changes.

Doppelmayer Lunar "bay" on the border of Mare Humorum; diameter 42 miles (68km); 29°S, 41°W.

Doppler, Christiaan (1803–1853). Austrian physicist, director of the Vienna

Christiaan Doppler: physicist

D

D

Physical Institute and professor at the University of Vienna. In 1842 he described the principle now called the DOPPLER EFFECT.

Doppler effect Principle in physics whereby the pitch of sound or the wavelength of light is altered by the velocity of the emitting object.

The Doppler effect can be visualized by imagining a transmitter emitting pulses regularly. The information from each pulse travels out in all directions, producing a pattern of circles, the wave crests. If the transmitter is moving, each circle has a different centre. In the

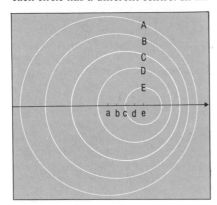

Doppler effect: detecting motion

accompanying diagram, the circles labelled A, B, C, D were emitted when the source was at a, b, c, d. In the forward direction the circles lie close together, whilst behind the transmitter they are stretched apart.

Consider the case of sound. The compression of the wave crests increases the pitch; the expansion of the pattern lowers it. We hear the effect when a speeding car passes. As it approaches, the pitch of the motor is high; the pitch falls abruptly as the car passes and recedes.

In the case of light, it is more convenient to think in terms of the wavelength, and hence colour. If an object approaches, the light it emits has shorter wave separations and so becomes bluer; if it recedes, the light is reddened. The latter is the much referred-to REDSHIFT.

If we examine a feature in a spectrum normally found at wavelength λ, then it will appear at the new wavelength $\lambda + \Delta\lambda$ if it recedes from us at velocity v, where:

$$\Delta\lambda/\lambda = v/c.$$

Here c is the velocity of light. Hence we can determine the velocity of an object by a sufficiently accurate measurement of the wavelength of known features. In this way all velocities of objects

along the line of sight are measured. Of course, all objects have some motion across the line of sight too, but the Doppler effect cannot measure these.

Distant galaxies and quasars recede at speeds approaching that of light itself. Under these circumstances Doppler's original mathematical formulation must be modified by EINSTEINS's theory of RELATIVITY. *DDA.*

Dorado (the Swordfish). A southern constellation, between Pictor and Reticulum. Its only bright star is Alpha (mag. 3.27, absolute mag. −0.6, type A0, distance 192 light-years), but it contains part of the Large Cloud of Magellan, including the magnificent TARANTULA NEBULA round 30 Doradûs.

Double stars Two stars which appear close together on the sky. BINARY STARS are connected by gravity, while optical double stars are chance alignments.

DQ Herculis A 14th-magnitude variable star, the remnant of Nova Herculis which rose to first magnitude in 1934. Modern photographs show the expanding shell of gas, which was ejected with a velocity of about 620 miles a second (1,000km/s).

DQ Herculis is an eclipsing BINARY STAR with an orbital period of 4^h39^m. The primary component is a WHITE DWARF star and the secondary is a normal K or M dwarf (MAIN SEQUENCE star). In addition to eclipses and flickering, DQ Herculis shows regular variations in brightness with a range of a few per cent and a period of 71.0745 seconds. These are thought to be caused by the effect of rotation of the white dwarf. *BW.*

DQ Herculis: remnant of a nova

Draco (the Dragon). A long, winding northern constellation, extending between Ursa Major and Ursa Minor, with the Dragon's "head" near Vega in Lyra.

Brightest stars

Name	Visual Mag.	Abs. Mag.	Spec.	Distance (light-yrs.)
γ (Eltamin)	2.23	−0.3	K5	101
η (Aldhibain)	2.74	0.3	G8	82
β (Alwaid)	2.79	−2.1	G2	270
δ (Taïs)	3.07	0.2	G9	117
ζ (Aldibah)	3.17	−1.9	B6	316
ι (Edasich)	3.29	−0.1	K2	156

Next come Alpha or Thuban (3.65), Xi (3.75), Epsilon (3.83), Lambda (3.84) and Kappa (3.87). Thuban used to be the north pole star in ancient times.

On the whole Draco is a barren group, but one of the stars in the "head", Nu, is a wide double, easily separable with binoculars; each component is of magnitude 4.9, and the separation is 62 seconds of arc. *PM.*

Draconid meteors *See* GIACOBINID METEORS.

Draper catalogue HENRY DRAPER'S widow supported continuing research in spectroscopy and this led in 1924 to the completion of a catalogue classifying stars by their spectra.

Draper, Henry (1837–1882). The son of J.W.Draper. A talented equipment maker and astro-photographer of the Moon, planets, comets and nebulæ who ultimately concentrated with distinction on photographic spectroscopy.

Draper, John William (1811–1882). Photographer who took the first Daguerreotype of the Moon in 1840.

Dreyer, John Ludwig Emil (1852–1926). Danish astronomer, mainly known for his historical works, including a major biography of TYCHO BRAHE.

D ring Innermost part of Saturn's ring system, but not a well-defined ring, and not observable from Earth.

Dubhe The star Alpha Ursæ Majoris; the brighter of the "Pointers" to the Pole.

Dumb-bell Nebula (M27). The nebula was so named by Lord Rosse, who observed its hourglass-like shape in his 72in (183cm) reflector. Messier 27 is a fine, nearby example of a planetary nebula, a low-mass star which has consumed most of its hydrogen fuel. The internal instabilities which appear as a consequence cause it to eject its outer layers in a spectacular nebula which signals its demise.

Dumb-bell Nebula: named by Lord Rosse

Dwarf galaxies: an elliptical variety

Dundrum meteorite 4lb (1.8kg) meteorite which fell in Ireland on August 12 1865.

Dunér, Nils Christoffer (1839–1914). Swedish astronomer. He was the first to measure the rotation of the Sun from Doppler shifts, showing it to be differential.

Dunham, Theodore (1897–1984). American astronomer who in 1932 made the first spectroscopic detections of CO_2 in the atmosphere of Venus and methane and ammonia in Saturn's atmosphere.

Dunsink Observatory Founded in 1783, and situated 5 miles (8km) from the centre of Dublin. The observatory was closed in 1937 and re-opened in 1947 as an integral part of the Dublin Institute of Advanced Studies.

Observational work with a transit instrument and a large Ramsden Circle in the early years included a diligent but unsuccessful search for stellar parallax by John Brinkley. The observatory attracted huge renown in the 1840s by the mathematical brilliance of its then director, Sir William Rowan Hamilton, and Sir Edmund Whittaker (director 1906–1912) was also an influential mathematician.

From 1947, under Hermann Brück, there was a new start in modern astronomy, with emphasis on photoelectric and solar work. Observing work has been done in South Africa and now in the Canary Islands. *PAW.*

Dust clouds Clouds of INTERSTELLAR DUST typically about eight parsecs in diameter. *See also* DARK NEBULÆ.

Dwarf galaxies Unusually small galaxies. Examples are the MAGELLANIC CLOUDS, which are dwarf irregular galaxies, and Messier 32, a dwarf elliptical galaxy.

Dwarf stars The most common type of star in the GALAXY. They constitute some 90 per cent of the approximately 10^{11} stars in the Galaxy and contribute about 60 per cent of its total mass. Dwarfs are thus the principal stellar building blocks of the universe.

The term "dwarf" derives from the early history of the HERTZSPRUNG-RUSSELL DIAGRAM, when it was realized that stars can be divided into two main classes: those falling on the MAIN SEQUENCE and those of great luminosity. The terms dwarf and giant were used respectively for these two types, referring to their luminosities rather than size (although giants are of course much larger than dwarfs of the same temperature). Thus "dwarfs", also known as main sequence stars, should be thought of as normal rather than diminutive. The Sun is a typical dwarf, being roughly midway in the range of properties of dwarf stars.

From a theoretical point of view dwarfs are common because stars spend most of their evolutionary lives on the main sequence. The main sequence in the Hertzsprung-Russell Diagram is the region which stars occupy while they are converting hydrogen into helium in their cores. A star spends longer in this core hydrogen-burning phase of its life than in its previous star formation stages or in its subsequent high-luminosity (giant) phases.

The internal structure of a dwarf star consists of a core more or less depleted in hydrogen (according to how long it has been on the main sequence — ie, since NUCLEAR REACTIONS first started in its interior), surrounded by a hydrogen-rich envelope. In stars heavier than the Sun the principal nuclear reaction network is the carbon-nitrogen-oxygen cycle and the core is mixed by convection. In lower-mass stars the nuclear energy generation comes from the PRO-TON-PROTON CHAIN and there is no convection in the core. An outer convection zone, extending in from the surface of the star, increases in depth as we consider lower-mass stars. For stars of spectral type A0 (about 3 solar masses) there is little or no convective envelope; at the mass of the Sun the convection zone reaches in about a quarter of the way to the centre; at about 0.3 solar masses the zone reaches the centre; stars of even lower mass are fully convective.

While on the main sequence all of the gross properties of a star — its surface temperature, radius, luminosity, lifetime, etc. — are determined by its mass. The accompanying table illustrates the mass-luminosity and mass-radius relationships for dwarf stars of "normal" composition, ie, with element abundances similar to those of the Sun. There is some dependence of the structure of the star on its composition, but most dwarfs have the same abundances of hydrogen and helium as the Sun, and abundances of the heavier elements which are within a factor of two of those in the Sun.

The relatively short lifetimes of the hottest dwarfs result in their being very rare in the Galaxy, whereas the lifetimes of the lowest mass (K and M dwarfs) are so long that none has evolved from the main sequence since the Galaxy was first formed. This strong lifetime dependance, acting with the fact that the processes of star formation produce far more low-mass than high-mass stars, results in the lower mass dwarfs being much more populous.

For masses less than 0.08 times that of the Sun the core temperatures of stars do not become high enough to start hydrogen-burning nuclear reactions; as a result the main sequence terminates at stars with surface temperatures of about 2,500K. However, lower-mass stars are formed, probably in even greater numbers than the M dwarfs, and become what are termed BROWN DWARFS.

An upper limit to the masses of dwarf stars is determined by the increase of radiation pressure in their interiors: above about 60 solar masses the outward force exerted by radiation exceeds the attractive force of gravity, preventing such a star from being formed, or making it unstable with a short lifetime if it does form.

The average rotation speed of dwarf stars varies along the main sequence, with rapid rotation for high-mass stars and very slow rotation for stars of one

D

Properties of dwarf stars

Spectral type	Effective temperature (K)	Mass	Absolute magnitude	Luminosity	Radius	Surface gravity	Lifetime yrs	Velocity dispersion km/s
O	50,000	50	−6	10^6	20	0.13	5×10^5	12
B	28,500	17	−4.3	2×10^4	7.5	0.30	1×10^7	15
A	10,000	3.2	0.6	100	2.5	0.50	4.5×10^8	20
F	7,450	1.8	+2.5	8	1.35	0.99	3×10^9	29
G	6,050	1.1	+4.3	1.3	1.05	1.00	8×10^9	40
K	5,000	0.8	+5.8	0.4	0.85	1.11	1.7×10^{10}	45
M	3,500	0.5	+8.8	0.06	0.65	1.18	5.6×10^{10}	50
	2,500	0.08	+18	0.001	0.15	3.6	6×10^{11}	55

The masses, luminosities, gravities and radii are in solar values.

D

solar mass or less. (*See* ROTATING STARS).

At various places along the main sequence variability of luminosity is found: thus the Beta Cephei, Delta Scuti, Oscillating Ap stars and the FLARE STARS are included among dwarf variable stars. (*See* individual spectral types and VARIABLE STARS.) Magnetic fields and X-ray emissions are detected in some dwarf stars.

As well as the intrinsic properties of dwarf stars, their space velocities show some interesting correlations. The dispersion of their velocities increases toward lower masses. This is a consequence of the increase in average age of the dwarf stars at lower masses: stars of all masses are probably formed with approximately the same dispersion in space velocities, but the mechanisms that act to increase the dispersion have longer to work on those stars that remain as dwarfs for the longest time. Thus O and B stars evolve off the main sequence before their velocity dispersions have changed much from their initial values, but among the M dwarfs there is a spread of ages from that of the Galaxy itself down to the most recent ones formed. Over long periods of time, encounters with massive interstellar clouds (and, to a lesser extent, other stars) change the orbits of stars in the Galaxy, increasing their velocity dispersion. Thus the oldest M dwarfs have greatly increased dispersions and the average over all M dwarfs is significantly increased.

About three per cent of F, G, K and M dwarfs show very high space velocities (greater than 40 miles/s (65km/s) relative to the Sun). These high velocities have not been acquired through increase in dispersion, but rather are a relict of the motions of the stars produced in the earliest stages of formation of the Galaxy. These dwarf stars are deficient (relative to solar composition) in elements heavier than helium and are termed subdwarfs. *BW.*

Dynamo effect Convective motions in the fluid electrically conducting core of a planet, generating its magnetic field by electromagnetic induction. The power sources must be sufficient to overcome the natural decay of electric currents in a conductor of finite conductivity, ie, the magnetic Reynolds number must exceed a critical value. The motions must have helicity, an asymmetry — just as the field coil in a self-exciting dynamo is wound in one sense — attributed to the Coriolis force arising from the planet's rotation. Cores in the terrestrial planets are iron-rich: in the major planets, Jupiter and Saturn, metallic hydrogen; in Uranus probably an electrolyte. The fields produced are strongly dipolar: roughly aligned along the axes of the planets, except for Uranus. Reversals of polarity seems an inherent property of the core dynamo. *SKR.*

Dyson, Sir Frank Watson (1868–1939). Ninth ASTRONOMER ROYAL. He made a lengthy study of STELLAR PROPER MOTION and notable contributions to the values of fundamental measurements vital to astronomy's progress.

Frank Dyson: Astronomer Royal

Eagle Nebula (M16). The Eagle Nebula is an emission nebula (H II REGION) which surrounds NGC6611, a bright cluster of newly formed stars. The nebula and its associated stars are at a distance of about 2,000 parsecs from us in the plane of the Milky Way. The association of stars and nebula is no coincidence for these stars have formed recently from the gas and dust which they so effectively illuminate.

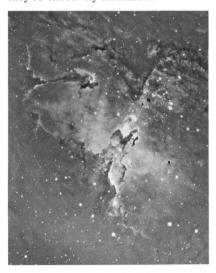

Eagle Nebula: about 2,000 parsecs away

Early-type stars High-temperature stars of spectral types O, B or A. The term was devised when astronomers believed the sequence of spectral types represented an evolutionary sequence. This is no longer accepted and the term, therefore, is misleading.

Earth The Earth is the largest of the inner group of "Terrestrial Planets" in the Solar System and has an equatorial diameter of 7,926 miles (12,756km). It is the third in order of distance from the Sun, which it orbits at a mean distance of 92,955,630 miles (149,597,870km) and it has a rotation period of 23h 56m 4s. When imaged from space it is dominated by blue, brown and green colours, with a continually changing pattern of whitish clouds. The solid surface is mantled by an ATMOSPHERE dominated by nitrogen and oxygen, the presence of which has allowed the development of animal and plant life.

The internal structure of the Earth has been derived by geophysical means and is known to consist of a partially liquid nickel-iron core, surrounded by a thick mantle dominated by magnesium and iron silicates. The thin outermost skin, or crust, is in part composed of silicic rocks, eg, granite, and in part by denser basaltic materials. Temperatures at the centre of the planet are believed to be of the order of 4,000°C, with pressures of around 3,850kbars. Much of the heat residing in the Earth is a legacy from the initial ACCRETION process, which must have been completed by about 4,750 million years ago — the age of the Earth as derived by radiometric dating techniques. In addition the decay of long-lived radioisotopes of elements such as uranium, thorium and rubidium contributed to the energy budget. The gradual escape of both heat and volatiles from the interior has been the driving force for the various dynamic processes which typify the planet.

Modern plate tectonics theory explains how the structure of the crustal layer has its origins in the creation of new crust along linear mobile zones manifested in oceanic ridges and rises such as the Mid-Atlantic Ridge and the East Pacific Rise. Hot basaltic magma rising from the Earth's mantle regions escapes along these zones and spreads out laterally to form new sea-floor at rates which vary between 0.5 and 10cm (4in) a year. The creation of such new crust implies that, since the terrestrial surface area remains the same, crust must be being destroyed to maintain the status quo. This occurs at SUBDUCTION ZONES — steeply inclined belts where one slab or crust is underthrust beneath that adjacent. The processes which operate at subduction zones are complex but include the downward plunging and eventual melting of the subducted slab, together with a portion of the veneer of sedimentary debris that typically overlies the sea-floor basalts. Another portion of the sedimentary veneer is scraped off the plunging slab and becomes highly deformed and may eventually give rise to new mountain ranges, such as the cordilleran chains of North and South America.

The activity at subduction zones expresses itself in a high incidence of earthquakes and active volcanicity, magmas being generated along the active belts by both frictional heat and by the melting of downward-plunging crust as it reaches deeper levels. Once magma is formed in this way, it is less dense than the overlying rocks and

emerges at the surface along arcs of volcanically-active islands (eg, Japan, Indonesia and New Zealand). Plate tectonics thus explains how the Earth's uppermost layer is made up from a number of huge plates (lithospheric plates) which are constantly in motion relative to one another. These plates are rigid and move slowly over a more mobile layer known as the ASTHENOSPHERE. Plate boundaries are of three kinds: (i) divergent — where adjacent plates are moving away from one another; (ii) convergent — where they are colliding; and (iii) conservative — where adjacent slabs are sliding past one another. The boundaries between plates are marked by dynamic activities such as seismic disturbance, volcanicity, major faulting and, over long periods of time, by the creation of new mountain ranges.

The modern methods now available for the absolute dating of geological events, based on the decay of long-lived radioisotopes, allow geologists to establish how the pattern of Earth's continents and oceans has changed with time. The concept of "continental drift" was first mooted by an Austrian meteorologist, Alfred Wegener, who first lectured on the subject in 1912, but

Earth: an Apollo II view

did not gain widespread approval. Subsequently, the notion that Earth's continents constantly change their position with respect to the rotational axis, has become a consensus view and explains how climatic changes witnessed in the rock successions of the continental regions have come about.

The presence of the Earth's mobile molten core is responsible for a strong MAGNETIC FIELD which interacts with the solar wind and forms an extensive series of radiation belts around the planet — VAN ALLEN BELTS. This field takes the form of a magnetic DIPOLE. At the equator the field strength

measures about 0.3 gauss, while at the poles it is roughly double this value. The study of the palæomagnetism fossilized in the crustal rocks indicates there have been relatively regular reversals in the polarity of the field. The imprint of these reversals in the basaltic rocks of the ocean floor has allowed geophysicists to assess the rate and pattern of sea-floor spreading.

The Earth's climate is a function of a number of factors. Thus differing intensities of incoming solar radiation experienced at different latitudes produce temperature differences which, in turn, generate pressure gradients. Once these are established winds are generated and the possibility of atmospheric circulation becomes a reality. The actual circulatory pattern of the atmosphere is very complex, and is a function not only of pressure differences, but also of the topographical configuration of the Earth and of interactions between the atmosphere and the vast areas of ocean which are a unique feature of the Earth. The day-to-day weather is largely a function of the activity of cyclonic and anticyclonic systems.

Geological evidence, in the form of faunal and floral remains locked up in the crust, of lithological differences exhibited by these strata and of isotopic characteristics, indicates that during the long span of time the Earth's climate has changed. Thus as little as 10,000 years ago, there was a great Ice Age; similar ice ages have left their imprints amongst older rocks on all the continents, even those now enjoying tropical climates. While continental drift can explain how any individual continent may shift into a different latitude and, consequently, into a different climatic zone, it cannot account for every kind of major change. Some of these may be explained by variations in the eccentricity of the Earth's orbit; others may occur in response to the changing chemical composition of the atmosphere. Then, of course, the Sun's energy output may have fluctuated through time. *PJC.*

Earthshine Changing clouds and snow lead to our planet's reflecting differing amounts of sunlight, but the eye can usually detect this Earthshine illuminating the night hemisphere of the Moon.

Eccentricity (e). The degree of elongation of an ELLIPSE such as an orbit. It is obtained by dividing the distance

E

E

between the foci by the length of the major axis. A circle has e = 0; a PARABOLA is the extreme case, with e = 1.

Eclipse *See* LUNAR ECLIPSE, SOLAR ECLIPSE.

Eclipsing binaries Eclipsing binaries may be regarded as essentially a subset of the close binary stellar systems (*see* BINARY STARS) — pairs of stars whose separation tends to be of the order of ten times, or less, the average size of the component stars. Occasionally, it might happen that a star is observed to undergo eclipse from a companion located at a relatively large distance (eg, Alpha Coronae Borealis), but clearly the likelihood to observe eclipses in binary systems will increase when the mean separation of components is but a few times greater than the mean radius of the stars themselves.

Notice also that for most of the classical eclipsing pairs, seen as VARIABLE STARS at optical wavelengths of light, the component stars have radii which are not very dissimilar from each other. The light from a system in which one star is very much larger than the other (say, more than 10 times in radius), would tend to come preponderantly from that large star, so that the eclipse of the small star (OCCULTATION), or its passage in front of the large star (TRANSIT), would tend to have an effect sufficiently small as not to be readily discovered, unless that small star were relatively very bright (per unit area of its surface) compared to the large star.

From this picture then we may understand eclipsing binaries to be undergoing a regular cycle of variation in apparent brightness, occasioned by the circumstance of their orbital motions under gravity taking place in a plane which is oriented at a relatively low angle to the line of sight. The term "orbital inclination" refers to the angle between the axis of the orbital motion and the line of sight. Hence, for eclipsing binaries, the inclination is usually not far from 90°, and, in any case, greater than about 58°. There will be a succession of two alternating photometric minima — one a transit, the other an occultation (unless the stars are *exactly* equal in size). Eclipses are said to be "complete" when the disk of the smaller star at mid-minimum projects to an area *within* that of the larger star: a "total" eclipse at occultation, and an "annular" eclipse at the transit. If the eclipses are not complete they are described as "partial".

As there are two stars in the system, two eclipses may be expected during the course of an orbital cycle, though it sometimes happens that one of these events causes so slight a loss of overall light from the system as not to be noticeable. The deeper eclipse minimum, associated with the eclipse of the star of greater brightness per unit surface area, is usually called the primary minimum. In photometric contexts the star so eclipsed (which may be either the greater or smaller in size) would tend to be called the primary star, and, in practice, this star is very often, though not always, the more massive star. The shallower minimum is called the secondary one, and in the same way is associated with the eclipse of the secondary component.

The components of most eclipsing binary systems revolve in orbits that are sensibly circular. Elementary geometry will then show that the same effective area of stellar surface is eclipsed at both minima — the minima are also of equal overall duration (for circular orbits). Since the eclipsed areas are equal at corresponding phases of either minimum, the ratio of depths in the minima at these phases is just proportional to the brightness per unit area. The primary star is thus the star of greater surface temperature; though it is possible for the secondary, though cooler, to be bigger in size (the primary minimum would then be of occultation type), and actually to put out more light overall than the primary. This situation, though not too common, can be confusing, particularly, for example, where spectroscopy is being combined with photometry. The spectroscopic primary (source of greater fraction of overall light) would then differ from the photometric primary (the hotter star).

Eclipsing binary systems attract attention because it is possible to analyse their characteristic pattern of photometric variability ("light curves") to determine key parameters which are of great, often, in some sense, *basic*, astrophysical significance. They thus form a means whereby information on the sizes, luminosities, surface temperatures, masses, and, to some extent, composition and structure of the component stars can be derived. Actually, it would not normally be possible to derive such absolute parameters as have been mentioned directly from a single monochromatic light curve (though surface temperatures can be inferred from two such curves at different wavelengths). But if spectro-

scopic evidence is available on the system, eg, the spectral type and spectral luminosity class of the brighter, or perhaps both components, then values of the absolute quantities can begin to be inferred. More powerful arguments are available on these when the radial velocity variations of one, or, especially favourably, both stars through the course of the orbit are given.

As the use of the word "eclipsing" suggests, however, it is the photometric evidence on these stars which tends to take phenomenological precedence in the study of the objects as a class. For this reason, such binaries are classified into three basic types on the basis of their light curves. Historically, these were described as the ALGOL type (EA), the BETA LYRÆ type (EB), and the W URSÆ MAJORIS type (EW). The distinction is this, that EA types have a sensibly constant (to within 0.1 magnitude) light level outside of eclipses, which are thus clearly defined (at least for the primary eclipse). The EB-type light curves show a continuous pattern of variability. The abrupt variations which mark out the eclipse regions are still distinct, but proximity effects render the regions outside the minima (sometimes called "shoulders") to be markedly curved. In EW light curves these proximity effects show such a scale of variation as to merge smoothly into the eclipse minima, which can thus no longer be clearly discerned. It is a fact, about which there is not a fully established understanding, that the EW systems also show more or less equal depths of primary and secondary minima, and this condition seems to be part of the EW-type definition. Hence a light curve showing pronounced proximity effects with minima depths differing by more than about 0.1 magnitude, would tend to be assigned an EB classification, even if the eclipses were not so well demarcated, though examples of this type are seldom encountered.

The proximity effects are generally explained in terms of two basic types of interaction, which increase at reduced separation of the components, namely the radiative ("reflection effect") and gravitational ("ellipticity effect") interactions. The full, detailed discussion of these interactions can become quite complicated, both as physical processes and as mathematical representations: for the purpose of convenient light curve analysis a number of simplifying assumptions are usually made; they are probably valid so long as the proximity effects are not too marked.

Besides the classification based on light curve morphology, schemes have been proposed which may be more influenced by our ideas of the physics or evolution of the stellar components. A common terminology is the "detached", "semi-detached" and "contact" system, which is based on the relationship of the member stars to a certain surface, often called the Roche surface, lobe or limit. If a binary component were to expand beyond this limiting surface, a dynamical instability of some kind would ensue. The detached systems are generally seen as unevolved, each star lying quite within its own ROCHE LOBE. Probably the most frequently observed type is, however, the semi-detached kind, of which Algol (Beta Persei) the first eclipsing binary to be recognized as such, is the prototype. In these systems the (present) secondary, invariably — as it seems — is in or very close to "contact" with its surrounding Roche lobe. Physical Algol-type close binary systems do not, regrettably for the terminology, always correspond with the EA-type light curves — the physical Algols show light curves that are actually a subset of the EA kind. The contact binaries are characterized by both stars filling their Roche lobes and would be thus in contact with each other. In fact, the general opinion nowadays about these binaries, which, as a class, concur very strongly with systems showing the EW-type light curve, is that they overflow beyond their common (inner) Roche limiting volume and into a surrounding common envelope.

Apart from these major groupings of eclipsing binaries there are various smaller subgroups to which attention has been drawn, usually with regard to some attributes of deeper physical significance. The RS CVn binaries, which though not necessarily eclipsing, are, for historical/selection effect reasons, mostly known as eclipsers, have been especially studied in recent years within the context of extending the notion of solar activity to stellar examples. Some dwarf novae, eg, U. GEMINORUM, exhibit eclipses, and the special geometrical circumstance can be particularly revealing on physical properties. In a more extreme physical condition are the X-RAY BINARIES, for which the eclipse effect can sometimes be detected affecting the X-ray radiation itself; again providing insight into the geometrical arrangement of the components.

Though there is no definitely estab-lished case of an eclipsing binary member of a globular cluster, eclipsing systems have been observed in the MAGELLANIC CLOUDS, and, in principle, offer the means of comparing stellar properties, in a direct way, over very large reaches of space.

A little more than 4,000 examples are known at the present time, references to some 3,500 of which are contained in the "Finding List" of Wood *et al*, which was published by the University of Pennsylvania in 1980. *EB²*. *See also* ASTRONOMICAL SPECTROSCOPY, PHOTO-METER.

Ecliptic The apparent yearly path of the Sun against the background stars, passing through the patterns of the Zodiac. It is really the projection of the Earth's orbit around the Sun onto the CELESTIAL SPHERE. The angle between the ecliptic and the Celestial Equator is about $23\frac{1}{2}°$, due to the tilt of the Earth's axis.

Eddington, Sir Arthur Stanley (1881–1944). English theoretical astronomer. One of the leading proponents of RELATIVITY theory, he made observations of star positions during the total solar eclipse of 1919 which confirmed EINSTEIN's prediction that light rays would be deflected in a gravitational field. He was a pioneer of the theory of stellar interiors and evolution, and correctly suggested that nuclear energy causes the Sun and stars to shine. He was also a leading exponent of the theory of the expanding universe. He was, too, an outstanding popularizer of the subject. Among his most notable books were *The Internal Constitution of the Stars* (1926, which became one of the classics of astronomy, and *The Expanding Universe* (1933). *IKN*.

Arthur Eddington: relativity

Effective temperature For a star, the temperature which a black body (an ideal radiator) would have if it had the same radius and luminosity as the star.

Effelsberg Radio Observatory The 328ft (100m) diameter steerable radio telescope 25 miles (40km) south-west of Bonn of the Max-Planck-Institut für Radioastronomie. Completed in 1970, the comparatively lightweight instrument is designed on the homology principle so that the bowl always deflects into a true paraboloid whose focal length varies with the elevation. Extensively used for radio astronomical measurements in the centimetre band.

Effelsberg Radio Observatory: 100m dish

Egyptian astronomy There is no doubt that the Egyptians were some of the very earliest astronomers, even though they had no concept of the real nature of the universe. They were capable of carrying out very accurate measurements, but they made the initial mistake of assuming that the universe took the form of a rectangular box with a flat ceiling supported by pillars at the four cardinal points. In some areas it was held that Egypt lay in the centre of the flat Earth, and was surrounded on all sides by a vast ocean. Of course astronomy and religion were merged; in 1367BC the young Pharaoh Akhenaten founded a new religion based upon sun-worship, though it did not last for long.

Egyptian constellations differed from those now in use, but there are a few similarities; thus Orion is shown as a man in the famous Dendereh Zodiac, an Egyptian star-map which was discovered in 1798 by Napoleon's General Desaix. The Zodiac is 5ft (1.55m) in diameter. Our Cassiopeia is the Egyptian Hippopotamus, while our Lyra is the Egyptian Hawk.

The Egyptians needed a good calendar, as their economy depended upon the annual flooding of the Nile,

E

and for this they used to measure the heliacal rising of Sirius. They finally derived a calendar of $365\frac{1}{4}$ days.

Days and nights were each divided into 24 hours; time measurements were made by sundials and clepsydræ (water-clocks) and seem to have been fairly accurate.

Of course the Egyptian astronomers are best remembered because of the Pyramids, which are undoubtedly astronomically aligned with remarkable precision.

Egyptian astronomy declined with the end of the country's political power. The last great astronomer of ancient times, PTOLEMY, lived in Alexandria, and was probably Greek, though he has an Egyptian name.

Today there is considerable astronomical activity in Egypt and there is one major observatory (Helwan). *PM*.

Einstein, Albert (1879–1955). German theoretical physicist, noted above all as the originator of the theories of Special and General RELATIVITY.

He was born in Ulm in 1879 and graduated in Zürich in 1900. He became a Swiss citizen in 1901 and entered the Patent Office in 1902. In one great year, 1905, he published papers expounding the Special Theory of Relativity, on the theory of the Brownian motion, introducing the motion of the quantization of energy, and on the equivalence of inertial mass and energy ($E = mc^2$). The Special Theory was a response to earlier work, especially by Lorentz, and the experiments of Michelson and Morley. It was based on two fundamental postulates, the invariance of the equations of mechanics in inertial reference frames and the invariance of the velocity of light. The resulting concepts of the dilation of time and space, of four-dimensional space (due to Minkowski) and the relation of mass and energy (confirmed experimentally by Cockcroft and Walton in 1932) were widely influential in physics and astronomy and created remarkable popular interest. In 1909 he was appointed to a chair at Zürich and was by that time recognized internationally as a leading theoretical physicist.

The General Theory of Relativity, published in 1916, introduced, inter alia, the idea that the geometry of space is determined by the presence of matter. A quantitative prediction of the deflection of starlight passing close to the Sun was confirmed in 1919 by eclipse expeditions led by Eddington. The effects of the curvature of space are most evident on a universal scale, and Einstein then applied the theory in cosmology, introducing the notion of a finite but unbounded universe. He was awarded the 1921 Nobel Prize for Physics mainly, oddly enough, for his contribution to understanding the photoelectric effect, not for his theories of relativity. From this time to the end of his life, his main scientific interest was the quest for a unified field theory of gravitation and electromagnetism. A Jew, he found conditions in Germany intolerable after Hitler's accession in 1933 and became the first professor at the Institute for Advanced Study in Princeton. He retired in 1945. *APW*.

Einstein Observatory Name given after its launch in November 1978 to NASA's second High Energy Astronomical Observatory satellite, which contained four nested grazing incidence X-RAY TELESCOPES of maximum diameter 2ft (0.6m) and focal length 11ft (3.4m).

Einstein Observatory: X-ray sources

With an angular resolution of four arc sec, this instrument revolutionized X-ray astronomy. Operational for two and a half years, it detected thousands of new X-ray sources. Many of these were ordinary stars, thus establishing that stars of almost all types emit X-rays. Our knowledge of the X-ray properties of QUASARS has been greatly enhanced by the Einstein mission. Observations of very many GALACTIC CLUSTERS have transformed our view of these extended X-ray sources. *JLC*.

Einstein shift The REDSHIFT (ie, increase in wavelength) of spectral lines in radiation emitted from within a strong gravitational field.

Elara The seventh satellite of Jupiter; diameter 50 miles (80km).

E-layer A distinct layer within the Earth's IONOSPHERE where the density of free electrons is higher than average. It is situated at a height of about 75 miles (120km) above the Earth's surface. Radio waves of sufficiently low frequency are reflected by the E-layer. Sometimes, a phenomenon known as Sporadic-E occurs in the months of May to August. This effect causes VHF radio waves with frequencies up to 200 MHz to be reflected. It is still not properly understood.

Electra 17 Tauri, in the Pleiades; magnitude 3.7. It is an easy naked-eye object.

Electromagnetic radiation Radiation consisting of periodically varying electric and magnetic fields which vibrate perpendicularly to each other and travel through space at the speed of light. Light, radio waves and X-rays are all examples of electromagnetic radiation (*see* ELECTROMAGNETIC SPECTRUM).

The theory of electromagnetic waves was first developed by the Scottish physicist James Clerk MAXWELL, who deduced that an oscillating charged particle would emit a disturbance which would travel through the electromagnetic field at the speed of light. He identified light as a form of electromagnetic radiation.

Electromagnetic radiation is usually considered to be a transverse wave motion, vibrating perpendicularly to the direction of propagation, just as a water wave vibrates up and down as it moves along. The distance between successive wavecrests is the wavelength, and wavelengths range from less than 10^{-14} metres (for the shortest gamma rays) to kilometres or more (for the longest radio waves). The number of wavecrests per second passing a fixed point is the frequency. The relationships between wavelength (λ), frequency (f) and the speed of light (c) are:

$$c = f\lambda; \quad f = c/\lambda; \quad \lambda = c/f$$

High frequency corresponds to short wavelength and low frequency to long wavelength. The unit of frequency is the hertz (Hz) where $1\text{Hz} = 1$ wavecrest per second. Thus a wavelength of 1 metre is equivalent to a frequency of 3×10^8Hz; a wavelength of 1 nanometre (10^{-9}m) is equivalent to 3×10^{17}Hz, and so on.

Electromagnetic radiation may be polarized. The state of polarization describes the orientations of the vibrations. An unpolarized beam consists of waves vibrating in all possible directions perpendicular to the direction of propagation, while a plane-polarized beam consists of waves vibrating in one plane only. In circularly polarized radiation the electric vector (which specifies the magnitude and direction of the electric field of the wave) has a constant magnitude and rotates as the wave moves along, while in elliptically polarized radiation, the direction and magnitude changes so that the vector traces out an ellipse as the wave moves along.

Electromagnetic radiation may also be described as a stream of massless particles, called PHOTONS, each of which is a little packet, or "quantum" of energy. The energy of a photon (E) is related to wavelength (*l*) and frequency (f) by the following relationships:

$$E = hf = hc/l$$

where *h* is a constant known as Planck's constant ($h = 6.63 \times 10^{-34}$ Joule-seconds). The energy of a photon can be described in ordinary energy units (Joules) but more often is measured in electron volts (eV), where 1 electron volt is the energy acquired by an electron when accelerated by a potential difference of 1 volt ($1eV = 1.62 \times 10^{-19}$ Joules). The shorter the wavelength (or the higher the frequency), the greater the energy of the photon. Thus gamma rays and X-rays have the highest energies (and are damaging to living cells) while radio waves have the lowest energies.

Light and other forms of electromagnetic radiation exhibit wave properties in some circumstances (REFRACTION, INTERFERENCE, DIFFRACTION, POLARIZATION) and particle properties in others, notably the photoelectric effect (*see* PHOTOELECTRIC AIDS).

In general, electromagnetic radiation is emitted when a charged particle is accelerated, or when it loses energy by some other mechanism.

Electromagnetic radiation can be emitted as a continuous spectrum (or continuum) — an unbroken distribution of radiation spanning a wide range of wavelengths — or in lines or bands at specific wavelengths only (*see* EMISSION SPECTRUM). Line emission occurs when an electron in an atom or molecule drops down from a high energy level to a lower one (the reverse process absorbs radiation and produces ABSORPTION lines). A CONTINUOUS SPECTRUM is produced by electrons involved in free-free and free-bound transitions. In a free-free transition,

radiation is emitted when an electron is accelerated in the field of an ion or atomic nucleus but is not captured (this radiation is sometimes called "bremsstrahlung", or "deceleration radiation"). A free-bound transition involves the capture of an electron by an ion or nucleus with the release of the lost kinetic energy of the electron in the form of radiation. Since free electrons can have a wide range of energies, free-free and free-bound radiation likewise spans a broad range of wavelengths.

Sources may be thermal — where the radiation emitted depends on the temperature of the source (eg, a star or hot gas cloud) — or non-thermal — where radiation is emitted by charged particles for some other reason. With thermal sources, the higher the temperature the shorter the wavelength at which the greatest amount of radiation is emitted (eg, X-rays are emitted predominantly by gas at temperatures in excess of 1,000,000K while infrared sources are cooler than about 3,000K). Black body radiation is a particular case of thermal radiation. SYNCHROTRON RADIATION (radiation emitted by charged particles moving at a large fraction of the speed of light in a magnetic field) is a common form of non-thermal radiation (*see also* ČERENKOV RADIATION). Non-thermal processes produce continuous spectra which differ in "shape" from thermal spectra (the intensity varies with wavelength in a different way). Synchrotron radiation is polarized whereas thermal radiation is not. *IKN. See also* COSMIC RAYS, GAMMA RAY ASTRONOMY, RADIATION, X-RAY ASTRONOMY.

Electromagnetic spectrum The complete range of electromagnetic radiations from the shortest to the longest wavelength, or from the highest to the lowest energy. Although there is no fundamental difference in nature between, say, X-rays and radio waves, the complete range of wavelengths has been divided into a number of sections.

From the shortest to the longest wavelengths, the principal divisions are:
gamma rays (less than 0.01 nanometres [nm])
X-rays: (0.01–10nm)
ultraviolet: (10nm–400nm)
visible: (400 nm–700nm)
infrared: 700nm–1mm)
microwaves: (1mm–0.3m)
radio: (greater than 0.3m)
Further subdivisions which may be encountered include:
hard X-rays: (0.01nm–0.1nm)
soft X-rays: (0.1nm–10nm)
extreme ultraviolet (EUV): 10nm–120nm)
near infrared: (1–4μm)
middle infrared: (4–10μm)
far infrared: (10μm–1mm)
submillimetre: (0.3–1mm)
millimetre wave: (1mm–10mm)
The Earth's atmosphere prevents most wavelengths from penetrating to ground level. Gamma-rays, X-rays, and the great bulk of ultraviolet radiation are absorbed at altitudes of tens or even hundreds of kilometres, where they excite atoms and molecules, dissociate molecules (break them apart) or ionize atoms or molecules (expel electrons from them).

Radiation of between about 300nm and 750nm can penetrate readily to ground level and this range of wavelengths is termed the "optical window". Water vapour and carbon dioxide are the major absorbers of infrared radiation, but there are a few narrow wavebands at which infrared can penetrate to ground level or, at least, to mountain-top observatories, notably at around 2.2, 3.5 and 10 micrometres (*see* INFRARED ASTRONOMY). In the far infrared there are no atmospheric windows, but a small proportion of submillimetre and millimetre wave radiation can penetrate to high-altitude sites. Radiation in the wavelength range from about 20mm to nearly 30m can penetrate to sea level, and this spectral band is referred to as the radio window. Waves longer than 30m are

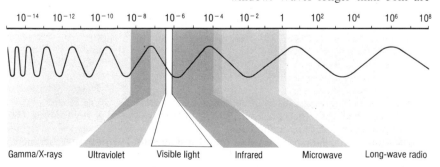

Electromagnetic spectrum: from gamma rays to long-wave radio

reflected back into space by the ionosphere.

Following early balloon-borne and rocket-borne experiments, instrumentation carried on orbiting satellites, high above the obscuring effects of the atmosphere, has opened up the entire electromagnetic spectrum and revolutionized our knowledge of the gamma-ray, X-ray, ultraviolet and infrared universe. *IKN. See also* ELECTROMAGNETIC RADIATION.

Electron A stable elementary particle; one of the family of leptons. Electrons are the constituents of all atoms, moving around the central, far more massive, nucleus in a series of layers or electron shells. Electrons can also exist independently as free electrons. The electron has a mass of 9.1×10^{-31}kg and a negative charge of 1.6×10^{-19} Coulomb. The antiparticle of the electron is called the positron.

Electronic devices In still comparatively recent times the control of telescopes was done manually or mechanically, observational assessment relied on naked-eye or photographic measurements, and the readings were recorded by hand. Now the rapidly increasing use of electronic devices of all kinds has revolutionized the practice of astronomy. Such devices pervade the whole observational activity from the complete control of telescopes to the final processing of the data. While this may be obvious for the relatively new radio observatories or for the space-borne observatory satellites, it is no less true of the observatories working in the traditional optical regime.

Modern optical telescopes are controlled electronically and operated through an accurately-geared TELESCOPIC DRIVE system. The position of the telescope is sensed by electronic encoders which give feedback to the drive system. It is now conventional to use a computer — itself a highly sophisticated electronic device — to receive the astronomers' requirements through a keyboard, and the information from the encoders, then output the appropriate instructions to the drive system, both to set the telescope in the desired direction and then to guide it at the sidereal rate. The computer allows very precise setting of the telescope by taking account of the inevitable flexure of the telescope structure, which can be very complex, and the refraction effects of the atmosphere. Telescopes operated in this way now can be reliably set in any direction

on the sky within a few seconds of arc.

To give the astronomer confirmation that the object of interest has indeed been acquired, a television camera is employed to avoid the need to sight through an optical eyepiece on the telescope. Image integration (allowing "time exposures") within the camera, and storage in an external digital memory from which the image is displayed on a screen, all assist in the rapid observation of faint objects in the telescope field. The same system is then used to set the telescope more precisely still, so that the light from the object to be studied passes through the entrance aperture of the instrument mounted on the telescope at that time, and finally to show what progressive corrective movements of the telescope are needed to maintain precise tracking of the object while the exposure proceeds, perhaps for several hours. Such corrections generally are very fine and stem, for example, from inadequate knowledge of the air temperature and pressure throughout the atmosphere. A more satisfactory fine guiding technique is to use an autoguider head, which is an electronic device capable of sensing the true position of the telescope by reference to a suitable guide star in the field of the required object. The autoguider head can be thought of as a crude television camera whose function is optimized to give precise positional information. The detecting device generally is a special form of PHOTOMULTIPLIER tube having an electronic scanning stage between the photocathode and the electron multiplying structure, but in recent developments this is replaced by a CHARGE-COUPLED DEVICE (CDD). The head unit is mounted on a carriageway which can be moved in the telescope focal plane to acquire a selected guide star. The error signals from the autoguider are used by the telescope control system in a closed-loop servo to maintain it tracking accurately on target.

The instruments attached to a telescope to enable the light from target objects to be analysed (for an example, *see* ASTRONOMICAL SPECTROSCOPY) have now become highly complex and rely upon the application of a variety of electronic devices. Most instruments are operated remotely from a control room near the telescope. In the most modern systems a computer is used to deliver instructions to the instrument which cause it to set itself into a specified mode, to check the state of the instrument and its electrical parameters, giving alarms if necessary, to

time the exposure, to record the data in an associated digital memory and display it, and finally to transfer the data to magnetic tape. In the most sophisticated instruments an individual microprocessor is often incorporated into its electronic substructure to organise the detailed functioning of the various motorized mechanisms and keep track of the data from encoders and sensors, while ensuring no conflicts occur in its operation. Communication with the control computer consequently is simplified and the computer then can more easily handle all the instrumental and auxiliary functions at once, including the various telescope functions (through its own computer), and so become central to the whole system. It is a relatively small step to operate the instruments and even the telescope, sited on a distant mountain top chosen for its good astronomical properties, from a control room in one's own home institution.

The electronic devices which are the most fundamentally important in astronomy today, without which the instruments could not function efficiently, are the detectors and their associated electronic systems. Photomultipliers are used mainly in PHOTOMETERS for standard photometric measurements within defined spectral passbands or for measurements on rapidly varying objects such as PULSARS. The signal pulses due to detected photons are electronically shaped and then accumulated in digital registers for a prescribed time after which the count is passed to a data recording system and the next accumulation begins. CCDs, although a comparatively recent development, are now commonly used at the major observatories for direct imaging or spectroscopic applications. These devices are intrinsically highly intricate, but relatively simple to operate. The related electronic circuitry provides the controlling and drive voltages which enable the device to accumulate and read out the image information in the form of electronic charge packets, then to amplify, digitize and store it in an external memory, and finally to display the result. Another type of detecting device which is widely applied in astronomy is the IMAGE INTENSIFIER; like a CCD this is used for direct imaging or spectroscopic measurements. Several forms of this device exist, but all produce a more or less intensified version of the incident image. Each is contained in a vacuum envelope which has a photocathode deposited inside an input window to convert the incident faint optical image into an electron image, then a means of amplifying the electron image and, finally, a phosphor conversion screen deposited inside an output window at which the enhanced electron image is returned to an optical image. In some cases intensification by 10^7 times is achieved. Such high gains allow scintillations due to single photon detections to be observed individually. The output image can be recorded photographically but this has now given way to electronic recording as in the Image Photon Counting System (IPCS). Here the highly intensified image is received by a fast-scanning television camera which continually records the photon scintillations from a faint input image and passes the video signal to a digital processing unit. This unit registers the central positions of the photon events in each frame and accumulates the counts in corresponding locations in a large digital memory, in which the detected image then is gradually formed, while the progression can be watched on a monitor screen. *AB.*

Elements All substances found on Earth are composed of elements either occurring discretely or combined into compounds. Elements are substances which cannot be decomposed by simple chemical processes into two or more different substances. So far 107 elements have been identified and 90 of the 92 so-called *naturally occurring* elements have been found in nature, either free (eg, Cu, Fe, Au S) or combined with other elements (eg, SiO_2, $CaCO_3$); the missing two are technetium, identified in the cyclotron transmutation products of molybdenum in 1937, and promethium, isolated in 1945 from nuclear fission products. The remainder have been synthesized by nuclear bombardment or fission.

The atoms of each element have a characteristic number of protons in their nuclei (*see* ATOMIC STRUCTURE); this number is called the atomic number (Z). Isotopes are atoms of the same elements which differ in the number of neutrons in their nuclei. There are two isotopes of helium; atoms of each have two protons and either one or two neutrons. The number of protons plus neutrons is called the mass number (A), thus isotopes may be distinguished in symbol form, eg, $_2He^3$ and $_2He^4$, where the atomic number is the subscript and the mass number is the superscript. Isotopes of the same element have the

same chemical properties. Because some plants can selectively absorb one isotope of an element (eg, potassium) rather than another, our definition therefore should be that an element is a substance, all of whose atoms have the same atomic number.

Some elements exist in two or more molecular or crystalline forms differing in physical properties and some chemical ones also. This is known as allotrophy. Phosphorus, for example, has four allotropes — two white, one violet, and one black. With some allotropic elements, temperature determines which form exists, eg, *a* and *b* sulphur, whereas others undergo irreversible transformations as with explosive antimony.

The first two elements (H, He) were

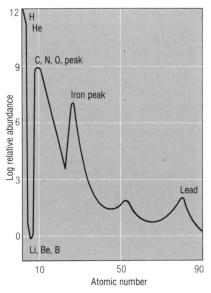

Elements: relative cosmic abundances

formed in the BIG BANG, elements up to $Z = 26$, iron, are produced in stars, and the heavier elements are the result of nucleosynthesis in SUPERNOVÆ. Cosmically, hydrogen and helium are the most abundant, approximately 97 per cent, but lithium, beryllium, and boron are very rare compared with their neighbours. These are destroyed, not created, in stellar nuclear processes. On Earth, elements with even atomic numbers are about ten times more abundant than those with odd atomic numbers; however, the ratio is nearer to 70 in meteorites. The accompanying diagram shows the relative cosmic abundances of the elements with the odd-even fluctuation smoothed out to illustrate general trends. In Earth's crust, which includes oceans and atmosphere, the approximate relative abundances of the four commonest

elements are: oxygen 49.2 per cent, silicon 25.7 per cent, aluminium 7.5 per cent, and iron 4.7 per cent.

The elements fall into three classes according to their physical properties: metals, non-metals, and metalloids. Metals are the largest class and are physically distinguished as malleable, ductile, good conductors of heat and electricity, have a high lustre, a close-packed arrangement of atoms, and are usually strong. These properties are explained by metallic bonding in which the atoms cast their valence electrons into a sea of delocalized electrons surrounding a lattice of positively charged ions. The ions remain stationary in the highly mobile electrons.

Non-metals are abundant in Earth's crust and are important biologically; included in them are the noble gases. Generally they are poor conductors of heat and electricity, brittle, hard, lack lustre, and have very high melting points. Solid non-metals have a continuous lattice of atoms in which the atoms are bound to their nearest neighbours by covalent bonds. Graphite (carbon) is an exception having covalent bonds within discrete layers.

Elements included in the metalloids are boron, silicon, germanium, arsenic, antimony, and tellurium with both metallic and non-metallic properties.

The chemical properties of an element may be related to its position in the PERIODIC TABLE. The inert gases became the noble gases in 1962 when it was shown experimentally that they could form compounds with other elements, particularly the halides, eg, krypton fluoride (KrF_2). The ability to form compounds of the noble gases shattered one of chemistry's most vigorously held principles, the octet theory, whereby the electrons of an atom are arranged in shells which are most stable when each occupied shell is complete, usually with eight electrons in the outer (valence) shell. The noble gases possess such completed shells, which were supposed to neither gain nor lose any electrons, that is, to possess no chemical reactivity. Metals act as electron donors; they have up to four electrons in an outer (valence) shell (in excess of a noble gas configuration), which are shed easily to leave positive ions (cations). Non-metals accept electrons to become negative ions (anions). Salts are combinations of cations and anions, eg, sodium chloride (NaCl). Polyatomic ions are bound internally by covalent bonds, eg, sulphate SO_4^{2-}, NB_4+. The anions need not consist of non-metals, but may also

E

contain permanganate ion (Mn_4^-), this corresponds to a higher oxide, which is acidic, Mn_2O_7.

The most reactive metals occur in groups IA and IIA (the alkali and alkaline earth metals) whose oxides and hydroxides are extremely basic (alkaline). None of groups IA or IIA metals is found in nature because of their extreme reactivity; sodium, potassium, and calcium react with air or water; however, their salts are very stable. The number of metals is swelled by the *transition* metals (Z 21 to 30, 39 to 48, and 57 to 80), by the lanthanides (58 to 71), and by the actinides (90 to 103).

The elements of group VIII, part of the transition metals, differ from the other groups in that there are three triads of metals. Iron, cobalt, and nickel (not rare metals, particularly iron) have similar atomic weights as have the other rare platinum metals of each of the other two triads. They form complex radicals, and such salts as potassium ferrocyanide, $K_4[Fe(CN)_6].3H_2O$; the sulphates of iron, cobalt, ruthenium, and iridium form *alums*, $(NH_4)_2SO_4.Fe_2(SO_4)_3.24H_2O$. The oxidation states (valency) of group members range from $+1$ to $+8$, the brightest shown by any element.

Although fewer in number, the nonmetals are more abundant, usually have nearly completed octets, and generally gain electrons in chemical reactions. The most reactive nonmetals are the halides, group VIIB in the periodic table, which have seven electrons in the outer (valence) shell. In sodium chloride (NaCl, common salt), the sodium atom donates its lone valence electron to the chlorine atom whose valence shell is then filled with eight electrons. When two non-metals combine, there is no opportunity for one to accept an electron from the other and so a compromise is reached with the two atoms sharing one or more electrons pairs, each of which constitutes a covalent bond between the atoms. This allows each atom to reach the stable configuration of a noble gas. In water (H_2O), one atom of oxygen uses two of its original six valence electrons to share with the two electrons from the two hydrogen atoms.

The oxides of the non-metals are generally acidic; they react with water to give oxyacids, eg, HNO_3, used to form salts with metallic oxides and hydroxides mentioned above.

Carbon is a special non-metal, the vast majority of whose compounds are called organics. There are over three million known organics. This is possible because carbon atoms with four valence electrons can share pairs of electrons with each other forming chain and ring molecules, and at the same time can share electrons with atoms of other elements.

A radioactive element is one which breaks down spontaneously into an element or elements of lower atomic number with the emission of particles and/or rays. Naturally occurring radioactive isotopes are many; however, better known ones are radon, radium, thorium and uranium.

Although identification of new elements has slowed down over the two decades, there is no other reason to believe that we have exhausted the range of the elements. *JLP.*

Ellipse One of a family of curves known as conic sections. The longest line that can be drawn through the centre of the ellipse is the major axis; the shortest line is the minor axis. These two axes are perpendicular to each other. Inside the ellipse are two foci, on the major axis. The total distance from one focus to any point on the ellipse and then to the other focus is constant, and equals the length of the major axis. The orbits of bodies around the Sun are ellipses. The Sun lies at one focus; the other focus is empty.

Ellipsoid A three-dimensional geometrical figure whose intersection with any plane is an ellipse.

Ellipsoidal variables Eclipsing binaries with components almost in contact, with the result that they are distorted to non-spherical shapes. Periods less than one day; range less than $0^m.8$; spectra F to G.

Elliptical galaxies Galaxies that have optical images which appear as smooth ellipsoidal shapes; in contrast to SPIRAL GALAXIES they are composed of old (red) stars, contain little gas or dust, and show very little rotation. They predominate in clusters of galaxies; are the central dominant galaxies in very rich clusters; and are the host galaxies for powerful radio sources. *JVW.* See also CLASSIFICATION OF GALAXIES, EXTERNAL GALAXIES, RADIO GALAXIES.

Elongation The angular distance between the Sun and a planet, or the Sun and the Moon. An elongation of 0° is called conjunction, one of 90° is called quadrature, and one of 180° is called opposition. When Mercury and Venus are at eastern elongation they may be visible in the evening sky; when at western elongation they may be seen in the morning. They never reach quadrature. The greatest elongation east or west varies between 18° and 28° for Mercury, and 47° to 48° for Venus. All planets are at superior conjunction when on the far side of the Sun as seen from the Earth. Mercury and Venus are at inferior conjunction when they lie between the Earth and the Sun. *JWM.*

Elysium Fossæ Aligned valleys crossing the flanks of Elysium Mons on Mars; 21°–30°N, 217°–224°W.

Elysium Planitia Elevated Martian plains region centred at 25°N, 210°W.

Emden, Robert (1862–1940). Swiss theoretical astronomer whose main contribution was on the thermodynamics of gaseous spheres. He introduced in astrophysics the concept of radiation pressure and photon statistics.

Emission lines *See* EMISSION SPECTRUM.

Emission nebulæ Clouds of gas such as the ORION NEBULA which are perceived because they emit light (contrast DARK NEBULÆ, which absorb the light of stars behind, and reflection nebulæ which reflect the light of stars nearby). Planetary nebulæ are emission nebulæ associated with individual stars, which have the telescopic appearance of a planet.

The mechanism by which the light is created in emission nebulæ is basically that the gas is heated, causing its constituent atoms to break up. Ultraviolet radiation from a nearby star is the usual cause of the heating, and ultraviolet photons which interact with the atoms in the gas cause the atoms to eject one or more electrons. The electrons gain energy from the ultraviolet radiation. The electrons eventually recombine with the atoms and re-emit the energy, some of it in the form of light. It is this light which enables the nebula to be discovered optically. The free electrons in the gas also lose energy in the form of radio waves, so radio telescopes can also detect emission nebulae.

The most abundant element in interstellar gas is hydrogen, which is very susceptible to ionization by ultraviolet light: nearby to a hot star which emits ultraviolet light essentially all the hydrogen is split into its components (*Continued on page 145*)

Exploring Space

David Morrison
Professor of Astronomy
University of Hawaii

Space has frequently been called the last frontier. Certainly this is true in the sense of physical frontiers, since humans have by now explored the entire land area of the Earth, and our sway is rapidly being extended to the oceans and the air as well. Our species has thus reached a fundamental turning point in history, imposed by the finite size of the Earth itself. Either we are to remain an Earth-bound civilization, working out our future within the

The United States Space Shuttle provides access to low Earth orbit for astronauts and cargoes of up to 25 tonnes. The Shuttle, which began operations in 1981, was the first space vehicle capable of returning to Earth for repeated flights. This photograph shows the cargo bay, which is about the size of a large railway goods wagon, on Shuttle flight 41-C. European Space Agency astronaut James D. van Hoften is using the manned manoeuvring unit; the arm of the Canadian-built remote manipulator system can be seen on the right. Docked at the rear of the cargo bay is the Solar Maximum Mission satellite, brought back on board the Shuttle for repair.

confines of this single planet, or else we will expand into the Solar System and perhaps some day even beyond. This is a fundamental choice that we or our children must make.

Those who argue for a policy of space exploration frequently point to the history of exploration of our own planet. The "western" civilization that has come to dominate the world economically and politically has strong roots in exploration and the exploitation of new lands, as well as in the constantly expanding spheres of science and technology. The very concept of progress has been linked to new discoveries, new markets, and an increasing mastery over nature. It is the scientific and exploring branch of humanity that has come to dominate our planet and to shape our expectations.

Space represents an opportunity, but also a barrier. Space, like the oceans that separated the New World from the Old on our planet, is a void that must be crossed to reach the fabled lands on the other side. Near-Earth space plays much the role of coastal waters — primarily of concern for defence, communications, and observations of the home country. Here, in Earth orbit, we establish our first outposts and learn to function in this new environment. Meanwhile, we send a few scouting missions to reconnoitre the distant worlds far beyond. Only after we have a firm foothold in these coastal waters can we contemplate large-scale expansion into the rest of the Solar System.

Today we are still at the threshold of space. Fewer than a hundred humans have left the Earth, and none of them has stayed more than a few months in the space environment. The first small space stations are just now making continuous human presence in orbit possible, and a few dozen robotic spacecraft have begun the scientific exploration of the other planets. But most of our interest in space remains firmly bound to our own

A possible lunar meteorite, *above,* designated ALPHA 81105, collected in the Antarctic in 1981. This may be a sample of material from the lunar highlands, ejected from the Moon in a crater-forming impact.

The Very Large Array (VLA) radio telescope, *right,* in Soccoro, New Mexico, built and operated by the United States National Radio Astronomy Observatory. This spectacular installation is one of the most powerful astronomical instruments in the world. It consists of 27 telescopes, each of aperture 82ft (25m), that can be moved along a "Y" configuration with a total span of 22 miles (36km). The VLA normally operates at three wavelengths, 3cm, 6cm and 21cm, and it can record "pictures" of the sky with resolutions of one arc-sec or better.

planet. Even from orbit, we still look primarily down at our own Earth, rather than upward toward the stars.

Pioneers in space exploration

The dream of space is only about a century old, dating from the pioneering writings of Konstantin Tsiolkovsky in Russia and Hermann Oberth in Germany, and from the inspired backyard engineering carried out by men like Robert Goddard in the United States, and by Wernher von Braun and others in the private space and rocket organizations that briefly flourished between the world wars in Germany, England, and Russia. These men defined the objectives of space exploration, worked out the basic principles, and began to design the necessary rocket hardware to make space flight possible.

The first rocket capable of rising to the upper fringes of the atmosphere was the German A-4 (the military V2), which was adapted for research purposes in the United States following World War II. By the early 1950s a number of nations had developed high-altitude research rockets. But the real spur for the development of vehicles that were ultimately to reach Earth orbit came from the imperatives of the nuclear arms race.

In an era where a single bomb could devastate a city, the means of delivery for such bombs assumed a new importance. For the first time it was justifiable to design and build large expendable rockets capable of spanning intercontinental distances. During the 1950s both the United States and the Soviet Union built such ICBMs, rockets that still play a major role in space 30 years later. But the same military demands that led to the development of the intercontinental nuclear missile also created pressures to limit and control such arms. The key to arms control became verification, which in turn demands surveillance — the "open skies" proposed by President Eisenhower. No nation wished to give up sovereignty

The Earth from space, *above.* This global view was obtained by Apollo 17 on its return from the final manned flight to the Moon in 1973. Such views of our planet, with its beautiful blue oceans and swirling clouds, have had a major impact on our increasing realization of the unique and fragile nature of the Earth, and they have helped spur humanity to try to curb the increasing overpopulation and pollution that threatens the ecosystem. Much remains to be done, however, and it seems likely that half of the total species present on Earth in 1950 will be extinct by early in the 21st century.

over its airspace, but observation carried out from orbit offered a possible solution. Thus the same rockets that had been built to carry the bomb were also modified to launch satellites that would ultimately make arms control possible by establishing networks for global surveillance from orbit.

The result of these efforts were the first artificial Earth satellites: the Soviet Sputniks 1 and 2 launched in 1957 and the United States Explorer 1 in 1958. In the wake of these pioneering satellites, the United States and the Soviet Union each rapidly developed a series of orbiting spacecraft for both scientific and military missions. In the Soviet Union the military and civil programmes were combined, but in 1958 the United States placed its civil and scientific space efforts within a new civilian agency, NASA (the National Aeronautics and Space Administration). Not until 1985 did the Soviets establish Glavcosmos, which serves some of the same functions as NASA.

The first satellites carried instruments to study the Earth, especially its newly discovered magnetosphere. Soon efforts were extended deeper into space, however. In January 1958 the Soviet Luna 1 became the first craft to escape from Earth entirely, and in September Luna 2 crash-landed on the Moon itself, followed three weeks later by Luna 3, which transmitted to Earth the first photos of the lunar farside. In 1960 the United States launched the first weather satellite (Tiros 1), the first navigation satellite (Transit 1B), the first communications satellite (Echo 1), and the first successful surveillance satellite (Discoverer 13). By the end of 1961 the United States had launched 63 spacecraft, 33 of them military, while the Soviet Union had launched just 15.

Both the Soviet Union and the United States developed programmes aimed at manned spaceflight. The new era began in April 1961 with the orbital flight of Yuri Gagarin in Vostok 1, followed by Gherman Titov in Vostok 2 in August. The Soviets continued with a series of additional "firsts": first dual spacecraft (Vostok 3 and 4); first woman in space (Valentina Tereshkova in Vostok 6): first three-man crew (Voskhod 1 in 1964); and first spacewalk (by Alexei Leonov in Voskhod 2). The United States meanwhile pursued its Mercury and Gemini programmes with equal success, but always seemingly one step behind the Soviets, up to the Apollo Moon shots of the late 1960s.

There are many reasons for the exploration of space, ranging from the practical to the visionary, and from the love of adventure to the lure of profits. Some of these are summarized below.

The military in space
One of the prime motivations for space exploration has been the potential military advantages of "the high ground". It is military rockets (ICBMs) that have powered most of our flights into space. Even vehicles developed primarily for civil purposes, such as the United States Space Shuttle, have major military applications. In addition, almost all Soviet cosmonauts and about half of the American astronauts have been military pilots.

Traditionally, three primary military uses for space have been identified, in addition to the ICBMs that form the basis for the nuclear deterrence that underlies modern global strategy. The first and most important is surveillance. Current spy satellites can photograph features as small as 4in (10cm) across, and these are supplemented by radar satellites and those with infrared and other sensors. Orbital surveillance is essential for mili-

A scientific satellite in space, *above*. This is the Solar Maximum Mission (SMM) spacecraft, photographed from the Shuttle after its repair in orbit. The ability to retrieve and repair satellites is one of the unique capabilities of the Shuttle.

Another view of the first Mariner, *below*, being prepared for launch to Venus. Mariner 4 was the first probe to reach Mars; Mariners 6 and 7 extended the mapping of Mars; and Mariner 10 was targeted for Mercury.

The Space Shuttle "Challenger", *opposite*, photographed by free-flying astronaut Bruce McCandless, with the Earth in the background. The Shuttle's cargo bay will carry a payload of 65,000lb (29,500kg) of deployables or rock-mounted instruments.

The first Mariner interplanetary probe, *above*, on an Atlas-Agena rocket, awaiting launch to Venus in 1962 from the United States spaceport at Cape Canaveral, Florida.

tary planning, and without it arms control in the modern world would be impossible. Communications represent the second major military requirement for space. Most military long-distance communications (just like most civil ones) are carried by satellites. Third, military navigation satellites are essential for precise location of ships (especially missile submarines, whose locations must be known if their weapons are to reach their targets), and also have increasing use in land- and air-based navigation. None of these military uses for space involves weapons in orbit, and all have become fully integrated into the major armed forces of the United States, Soviet Union, NATO and Warsaw Pact nations.

Of much greater concern to many people is current renewed interest in the development of space weapons systems. Both the United States and the Soviet Union have built and tested anti-satellite weapons (ASATs), which are destabilizing because of their potential to blind the surveillance and communications capabilities of an enemy. In addition, the acceleration of research and testing of exotic space weapons (such as the laser and particle-beam weapons proposed under the United States' "star wars" Strategic Defense Initiative) is a matter of worldwide interest. Today military space expenditures substantially exceed those of civil and scientific programmes, and the trend is toward increased militarization of near-Earth space.

Science in space

The space frontier is perhaps most of all a scientific frontier. Many research disciplines can gain by using the space environment, and in areas such as astronomy, planetary science, and Earth science access to space is essential. Global-scale study of our own planet has revolutionized our concepts of atmospheric and ocean science, and we are still discovering the major outlines of the magnetosphere that surrounds the Earth and of the complex interrelationships among the Sun, the magnetosphere, and the atmosphere of the Earth. Before 1960 study of the other bodies in the Solar System was a minor branch of astronomy; now planetary science is a major discipline, with emphasis on comparative studies of the planets (including Earth) and on efforts to understand the origin and early evolution of the solar system. Astronomy gains tremendously by lifting telescopes above the atmosphere and probing new regions of the electromagnetic spectrum. And we have just begun to investigate the effects on humans of weightlessness, or the insights into fundamental physics and chemistry that might be obtained from experimentation carried out on a space station.

Today NASA spends nearly $2,000 million, or about 20 per cent of its budget, on space science. ESA (the European Space Agency) spends substantially less, although a larger fraction of its space budget is devoted to science. The space science effort in the Soviet Union is probably similar to that in the United States. Although science is often quoted as the major reason for going into space, less than 10 per cent of the world's space activities are actually science-related.

Retrieval of a satellite in low Earth orbit, *left*. Astronaut Joseph P. Allen is holding on to the Palapa B-2 communications satellite, with his colleague Dale A. Gardner assisting (at lower right). While a great deal of publicity has been given to the ability of the Shuttle to carry out such operations, they have not proved to be cost-effective, and it is usually cheaper to replace faulty spacecraft than to retrieve them for repair. Until access to space is cheaper, most spacecraft will continue to be launched with unmanned expendable rockets.

A scientific satellite being deployed from the Shuttle, *above*. This 1984 photograph shows the Earth Radiation Budget Satellite (ERBS) being held by the Canadian-built remote manipulator system.

The Infrared Astronomical Satellite (IRAS), *below*, carried out an all-sky survey in 1983. This NASA diagram shows the optical system of the satellite enclosed within a cryogenic system that maintained the entire telescope at a temperature only about 10° above absolute zero.

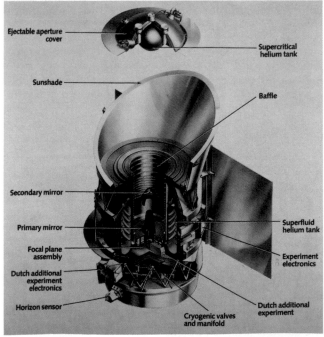

Ejectable aperture cover

Supercritical helium tank

Sunshade

Baffle

Secondary mirror

Superfluid helium tank

Primary mirror

Focal plane assembly

Experiment electronics

Dutch additional experiment electronics

Horizon sensor

Dutch additional experiment

Cryogenic valves and manifold

The lunar surface, *below,* photographed by the Apollo 15 astronauts. Shown is Mt Hadley, about 13,000ft (4km) high, which makes up a part of the rim of the Imbrium basin on the Moon. Like other lunar mountains, it was raised by the force of the basin-forming impact. Because there is little erosion on the Moon, mountains remain smooth and rounded, with no steep slopes or sharp peaks such as are carved by ice and water on the Earth.

An artist's impression of the United States Viking spacecraft, *right,* in orbit around Mars. In this view, the Viking Lander in its aerodynamic entry shield is just separating from the Orbiter, in preparation for the descent to the Martian surface. The first Viking successfully touched down on the plains of Chryse on July 20 1976.

The commercialization of space

Space has had one great commercial success: communications satellites, which now represent a multibillion-dollar business. The potential of satellites in 24-hour geosynchronous orbit as communications nodes was first publicized more than 40 years ago in Britain by Arthur C. Clarke, and the development of such systems has progressed rapidly in support of the worldwide demand for video and data transmission circuits as well as voice (telephone) lines. The first geosynchronous communications satellite, Syncom 1, was launched in 1963. Today the largest orbital communication system is managed by Intelsat (the International Telecommunications Satellite Organization), with more than 300 Earth stations in nearly 150 countries, and more than 40,000 voice circuits alone. From plane hijackings to Olympic Games, these systems have brought the whole world into our living rooms and created a "global village" of communications.

Other commercial successes remain for the future. These range from the bizarre (orbiting human last remains as an alternative to burial) to the practical (manufacturing pharmaceuticals in zero-gravity) to the visionary (solar power stations in orbit to solve the Earth's energy crisis). Most will depend on the development of a better space infrastructure, with orbiting platforms accessible to private industry and a new generation of rockets to provide cheaper access to space.

Humans in space

For many space supporters the ultimate objective is a major human presence in space. Certainly we all identify with the astronauts and cosmonauts whose courage and accomplishments have fascinated the world for a quarter of a century. While there is a long-standing debate between proponents of human and robotic missions, the fact is that both have a role in the exploration of space. Europe is joining the United States and the Soviet Union in developing a manned space programme, and it seems clear that the basic human desire to explore will continue to be a major motivation for space exploration.

Apollo: the greatest adventure

The greatest triumph of the early years of space exploration was the conquest of the Moon. Project Apollo represented a major national effort in the United States, costing about $100,000 million in current (1987) dollars and requiring the development of the Saturn V rocket, still the largest such vehicle ever flown.

On July 20 1969, this effort reached its culmination when NASA astronaut Neil Armstrong stepped from his Apollo craft onto the lunar surface. Between 1969 and

The first man on the Moon, *above*. NASA astronaut Neil Armstrong is working next to the Apollo 11 lunar landing module. Note the flat plains of Tranquillity Base, with the soft lunar soil imprinted by the astronauts' boots. Between 1969 and 1973 a total of 12 Americans visited the lunar surface, but today no nation has the capability for manned flights to the Moon.

Launch of one of the Saturn V rockets, *below*, developed for the Apollo Moon programme. These giants, with first-stage thrust of 7.5 million pounds (3.4 million kg), are the largest rockets ever flown at the time of writing (1987). The last Saturn V flights took place in the mid-1970s and since then the largest payloads that can be launched by either the United States or the Soviet Union into low Earth orbit have been limited to between 20 and 30 tonnes.

1973 12 Americans walked on the Moon, traversing a hundred kilometres of the surface and returning 842lb (382kg) of samples to the Earth. Both scientifically and as an expression of human accomplishment this unique exploration of another planet was a great success. However, America turned back from this heroic outward thrust, and the Moon was abandoned. The Saturn V rockets lie rusting on the lawns of NASA centres at Houston and Cape Canaveral, and spacecraft built to land on the Moon instead ended in museums. Now, 15 years later, no nation has either the capability or the will to return to the Moon.

Exploring the planets

In only one area do we continue to probe outward, and that is in space science. Scientific missions, carried out by small robotic spacecraft, have not stopped at the Moon but have continued into the rest of the Solar System. Some notable milestones in this exploration have been the first interplanetary flight by Mariner 2 (to Venus in 1962); first orbit of another planet (Mariner 9 at Mars in 1971); first spacecraft to the outer planets (Pioneer 10 launched in 1972); first photos from the surfaces of Venus (Venera 9 and 10 in 1975) and Mars (Vikings 1 and 2 in 1976); and the remarkable "grand tour" of Jupiter, Saturn, Uranus, and Neptune carried out by Voyager 2.

The Martian soil, *opposite,* has a dark orange
hue, similar to that of many terrestrial
deserts, while suspended dust gives the
atmosphere of the planet a pink, rather than
blue, colour. For reasons of safety, both
Viking landers were directed to smooth
lowland basins far from the most interesting
Martian topography: the volcanoes, canyons,
channels and layered terrain of the polar
caps. With improved navigation and landing
techniques, future spacecraft should be able
to land in more interesting regions of the
planet and to explore their surroundings with
mobile rovers.

The Soviet Mir space station, *below,* and its
Flight Control Centre, *above,* at Kaleningrad,
near Moscow. Mir, launched on February 20
1986, is a successor to the highly successful
series of Salyut orbital stations, with which
the Soviet Union has accumulated far more
experience with humans in zero gravity than
has any other nation. Though the basic Mir
module has a mass of only about 20 tonnes
and is smaller that the United States Skylab
of the mid-1970s, it is equipped with six
docking ports for attachment of additional
modules. The first of these should be added in
1987, leading toward a large continuously-
manned station before the end of this decade.
The United States space station, in contrast,
will not be operational before 1995.

Although it represents only a small part (currently less
than five per cent) of the NASA programme, planetary
exploration has always had wide appeal. In addition to
its scientific impact, sending spacecraft to other planets is
responsive to our basic urge to expand our horizons and
explore new worlds. While in the United States the
planetary programme has been cut back, other nations
have moved ahead. Four spacecraft from ESA, the
Soviet Union, and Japan flew close by Halley's Comet in
1986, and the Soviet planetary programme (with
considerable European involvement) has recently
accelerated with a new focus on Mars.

Humans in orbit

Manned space programmes have continued to play a
major role in both the Soviet Union and the United
States, with increasing involvement from other countries.
Both the Soviet agency Intercosmos and NASA have
encouraged international participation in their programmes, and both agencies are committed to establishing international space stations in Earth orbit. By the
end of 1985, astronauts from 18 nations had flown in
space.

The Soviet Union has concentrated its efforts since
1970 on long-term human flight leading toward the
development of a permanently manned space station. Its
primary vehicle for transport to and from Earth orbit is
the three-man spacecraft called Soyuz, which also participated in a joint 1975 United States/Soviet mission
called Apollo/Soyuz. In 1977 the first true space station,
Salyut 6, was placed in orbit, and several records for
continuous manned space flight were set in it and its successor Salyut 7. In 1984 cosmonauts Oleg Atkov, Leonid
Kizim, and Vladimir Solovyev completed a flight of 237
days, while Valeri Ryumin logged 361 days in orbit,

counting multiple flights. A further major step took place in 1986, with the launch of a modified Salyut called Mir, the first modular space station capable of expansion into a full-scale facility to be manned permanently by a crew of six or more cosmonauts. This entire programme, with its many manned flights each year, has been carried out using the same basic rockets, only slightly modified from the time they were introduced in the late 1950s.

During the same period the United States developed the Space Shuttle, the first reusable spacecraft designed to fly repeatedly to and from orbit with a crew of up to seven astronauts. The Shuttle first flew in 1981, and during the next five years a total of four vehicles made 24 highly successful flights, demonstrating a wide range of activities in space, including the retrieval and repair of satellites. Among those flying on the Shuttle were a United States Senator and Congressman, and a public school teacher was on the final flight of the "Challenger" in January 1986. Unfortunately, this exciting period was terminated on the 25th Shuttle flight when "Challenger" exploded during launch killing its entire crew. While this was not the first fatality in space — the Soviets had lost one cosmonaut in 1967 and three in 1971 — it shocked an overconfident NASA and set back the American space programme by several years. This accident also reopened the debate in the United States concerning the role of humans in space flight.

As the world's space programmes have matured, the limits of what can be done are set less by technological capabilities than by financial resources. We know how to do much more than governments feel they can afford. Thus the future will be shaped less by science and engineering than by politics and economics. In this environment, strategic planning has become an important activity on both sides of the Atlantic.

Halley's Comet, *above,* in 1986. Comets are the most primitive material in the Solar System and hence most likely to reveal the circumstances of the origin of the planets. Following the successful European, Soviet and Japanese missions to the comet, it is expected that considerable future effort will be expended on continued exploration of comets.

The great observatories in space

One area of space science that has defined its hopes very clearly is astronomy, primarily through recommendations made by the Astronomy Survey Committee of the United States National Academy of Sciences. During the 1960s and 1970s a number of small telescopes were flown in space, the most successful of which (like the X-ray satellites Uhuru and Einstein and the infrared satellite IRAS) opened up major new portions of the electromagnetic spectrum. The next step in space astronomy will be marked by the launch, probably in 1988, of the Hubble Space Telescope (HST), a 2.2m optical telescope of unprecendented accuracy and sensitivity, designed to be serviced by the Shuttle and to remain operational for at least 15 years. The HST will be the first real observatory in space.

Following the HST, NASA hopes to launch three other long-lived space observatories to cover other parts of the electromagnetic spectrum. These are the Gamma Ray Observatory (GRO), the Advanced X-ray Facility (AXAF), and the Space Infrared Telescope Facility (SIRTF). Together, these four instruments, each of

which costs close to a thousand million dollars, are called the "four great observatories in space". NASA studies show that each of these telescopes will be tens of times more sensitive than any instruments that now exist, with a promise of a major revolution in our understanding of the universe.

Europe also is participating in the new astronomy from space, with several optical and X-ray telescopes in orbit or under development. ESA's ISO (Infrared Space Observatory) will be launched about 1990, and the ESA astronomy cornerstone mission for the late 1990s is an advanced X-ray telescope complementary to AXAF.

A return to the planets

In the area of planetary exploration, the United States strategy has been defined by the Solar System Exploration Committee of the NASA Advisory Council. The plan recommended by this committee calls for three types of missions. Least expensive are the Planetary Observers, which will explore the inner planets using the same commercial technology developed for Earth orbiters. The first of these missions is a Mars Observer,

to be followed perhaps by missions to the Moon and several near-Earth asteroids. The second mission type is to be based on a new generation of Mariner spacecraft, capable of reaching the outer planets and comets. If approved, the initial mission in this class will be a comet rendezvous, followed later by a joint NASA/ESA mission to Saturn and its satellite Titan. The final, and most technologically challenging, planetary missions will involve sample return from Mars and comets, but no moves have been made by NASA to initiate such ambitious missions.

While the United States programme lags, other nations are moving into the field of planetary exploration. The most ambitious efforts are expected from the Soviet Union, which is directing its planetary programme toward Mars and the small bodies (comets and asteroids) after more than a decade of successful Venus missions. The first of these new missions, Phobos, will be launched to Mars and its satellite Phobos in 1988, to be followed by a Mars/asteroid mission called Vesta in the early 1990s. Some people predict a Soviet Mars sample return mission before the year 2000.

Two Voyager images of Jupiter's satellite Io, *opposite,* taken in 1979, show (top) two active volcanoes and (below) a closeup of one erupting more than 125 miles (200km) above the surface. Io's surface is variegated soft tans, yellows, and white, due primarily to sulphur compounds erupted from the active volcanoes. All four of the large Galilean satellites of Jupiter, revealed to us for the first time by Voyager, are planet-sized objects of great interest to both geologists and astronomers.

Jupiter's Great Red Spot, *above,* from Voyager 2 in July 1979. The Great Red Spot is a high-pressure storm system large enough to engulf two Earths: it was first seen nearly 300 years ago, but could not be studied in detail before the Voyager flybys.

Saturn photographed from Voyager in 1980, *above*. After the Jupiter flybys of 1979, both Voyager 1 and Voyager 2 continued on to Saturn. Voyager 1 was targeted for a very close flyby of Saturn's large cloud-covered satellite Titan, while Voyager 2 swung by Saturn on a trajectory that would carry it on to Uranus. The view here, looking back at the crescent planet and its shadow cast on the rings, is one that could never be obtained from the Earth. Because Saturn is twice as far from Earth as Jupiter, the incremental addition to our knowledge provided by the Voyager data was even greater than it had been for Jupiter.

Space stations and other advanced hardware

The exciting scientific exploration programmes outlined above, as well as others more directed toward the study of the Earth, all should be viewed in the context of a broader international effort at space exploration. The key elements of such a programme include less expensive means of getting payloads into orbit and the development of space platforms and space stations. The Soviet Union, ESA, and the United States all have plans for re-usable shuttle vehicles and orbiting modules of various kinds. The Soviet Mir space station is clearly the most advanced, with similar United States and European capabilities not anticipated before the second half of the 1990s.

The largest existing launch vehicles — the American Shuttle and Titan-III and the Soviet Proton — all have a capability of lifting between 20 and 30 tonnes to low Earth orbit, or 2–3 tonnes to high geosynchronous orbit. Thus the individual space station modules are all limited to about 20 tonnes on both sides, and even the most ambitious interplanetary spacecraft have masses of no more than 2,200lb (1,000kg) or so. So-called Heavy Launch Vehicles are needed to surpass these payload limitations.

The possible shape of space activity beyond the mid-1990s has been addressed by the United States National Commission on Space in its 1986 report "Pioneering the

Phobos, *below,* the larger satellite of Mars, photographed by the Viking Orbiter. Phobos is a dark object of chemically primitive composition (*see* the essay on MOONS). It is the primary target of the 1988 Soviet Mars missions, and it has been suggested as the ideal staging point for intensive exploration of Mars — a kind of natural space station for our use.

Uranus, *right,* photographed by Voyager in 1986. Using the gravity of Saturn to modify its trajectory, Voyager 2 continued its mission of exploration to Uranus, more than 1,250 million miles (2,000 million km) from the Earth. The spacecraft is *en route* to Neptune, which it will reach in 1989, the final target in its remarkably successful "grand tour" of the outer Solar System.

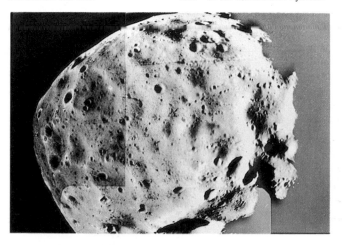

Space Frontier". The commission emphasizes the need to build up a space infrastructure and to reduce the cost of access to orbit if the plans of scientists or commercial interests are to become reality. It calls for a new generation of passenger and cargo vehicles to replace the Shuttle by the year 2000. One possibility considers an aerospace plane capable of flying directly from a runway into orbit and back. Unfortunately, in the wake of the "Challenger" accident such calls for new technology directed toward major goals in the next century seem beyond the perspective of most of the managers employed by NASA.

The future of the Soviet programme is also unclear, although for different reasons. A new generation of heavy-lift vehicles has been expected for some years, and a Soviet version of the Shuttle may fly in 1987. Beyond that, and the promised expansion of the Mir space station, little can be predicted with certainty.

Focusing on Mars

One common element of United States and Soviet programme plans is the exploration of Mars. Both the Solar System Exploration Committee and the National Commission on Space place Mars missions, including eventual visits by astronauts, near the top of their recommendations. And possible cooperative Mars missions have been on the agenda of several recent United States/ Soviet Union summit meetings.

Mars is the planet that is most like the Earth, and the one most accessible (after the Moon). It has always fascinated scientists and public alike, and with increasing knowledge our enthusiasm has grown. Besides, where else can we go? Venus is too hot, the Moon and Mercury lack air and water, the asteroids are too small, and everything else is too far away. If there is ever to be a practical extraterrestrial focus to space exploration, it will have to be Mars.

Mars missions also offer the potential for meaningful international collaboration. Joint space efforts in Earth orbit are complicated by military rivalry, and in any case the major space powers are already committed to their own space stations. In contrast, Mars is distant enough in space and time to present a more acceptable target, while the expense of major Mars missions is great enough to motivate nations to pool their resources toward a common goal.

One further advantage of Mars is the existence of two natural space stations in the form of its small satellites Phobos and Deimos. These two objects, each about 12 miles (20km) across, are ideally located to serve as jumping-off places for the Martian surface, and in addition they can be readily mined to yield water, oxygen and rocket fuels. Their presence makes an already fascinating planet that much more accessible.

Steps toward the red planet

The first steps toward the international exploration of Mars are likely to come when we consider sending instrumented rover vehicles and returning samples of Martian material to the Earth. Such a mission has been recommended to NASA by the Solar System Exploration Committee for 1996 or 1998. If carried out as a joint United States-Soviet project, one partner could supply the rover, which would traverse up to 60 miles (100km) of the Martian surface exploring and collecting samples. The samples would then be returned to Earth (probably initially to a space station) by a vehicle supplied by the other partner. Such cooperation would maximize the scientific results with a minimum of technology transfer between nations that are otherwise competing in space.

Following successful joint missions of the sort described above, the National Commission on Space proposes a sequence of visits by humans to Mars in the 21st century. The initial trips might be to Phobos or Deimos, with descent to the surface coming later. The transit times from Earth to Mars are only a little longer than the current Soviet records for continuous human habitation in space — a fact that has often triggered speculation as to the ultimate target of Soviet manned space flight. And in terms of energy, Phobos and Deimos are approximately as easy to reach as the Moon.

Conclusions

The eventual colonization and even terraforming of Mars are possibilities far beyond the scope of this essay. Nor would I want to suggest that Mars should be the sole objective of space exploration in the next century. There are many other worlds to investigate in the Solar System, not least of which is our own planet Earth, as well as an entire universe to be observed with orbiting telescopes. A human return to the Moon also has much to commend it. But Mars looms as a clear and obvious target if humanity chooses to leave the confines of the Earth (and low Earth orbit) and seek a future among the planets of our own system.

A possible Martian base is currently being tested by NASA in the Arizona Desert. This artist's impression shows it as it might look when assembled on the planet. The central building encloses a self-supporting ecological system that provides sustenance for the year or 18 months the base will be manned. The space vehicle, buried in soil to reduce solar radiation to a minimum, provides living quarters for the astronauts. Mars is the one planet besides Earth where humans could establish a self-supporting base with relative ease. Air and water are available, and rocket fuel could be manufactured from the Martian soil. Both the United States and the USSR have shown interest in manned landings, perhaps in the first decade of the 21st century.

E

— a free proton nucleus and a free electron. The symbol for this is H II (pronounced "H-two") and emission nebulæ produced this way are therefore also called H II REGIONS.

Hot stars that emit ultraviolet light are usually young stars, recently formed from gas and dust by the collapse of part of the gas cloud; usually many stars are formed simultaneously and form a star cluster. Thus hot stars in clusters are associated with emission nebulæ — the association between stars and gas cloud is not by chance (as is usually the case with reflection nebulae) but is generic. The connection between the Orion Nebula and the TRAPEZIUM stars is of this kind.

Emission nebulæ can derive their energy from sources other than hot stars. Gas clouds that collide, such as the remnants of SUPERNOVÆ, re-radiate the kinetic energy of the collision. They are called collisionally-excited emission nebulæ. *PGM. See also* NEBULÆ PLANETARY NEBULÆ.

Emission spectrum Any hot, solid body will emit visible light which, when split by a PRISM, or by a SPECTROGRAPH, into its component colours, shows the familiar rainbow effect. A typical example of this is an electric light bulb.

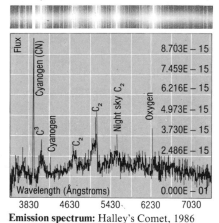

Visual spectrum

Emission spectrum: Halley's Comet, 1986

In contrast, when the light from a gaseous body, for example, a sodium light, is examined, it does not split into the familiar rainbow, but instead shows distinct lines, or bands of colour. That is, instead of emitting all colours, it emits certain ones preferentially. Each line in an emission spectrum is due to a particular element, or molecule, so the wavelengths of the individual lines act like a fingerprint, allowing us to identify, without question, what has caused them. Not only that but, since each element can have many different lines and the relative strength of these different lines depends on conditions within the region that has emitted them, astronomers have a powerful tool for investigating any object that has an emission spectrum. A typical example is that of PLANETARY NEBULÆ. These emit almost all their light as emission lines. From a study of line ratios (that is, how powerful certain lines are compared to others), an astronomer can calculate not only the temperature but also the density and composition of the planetary nebula.

Certain spectral lines, called FORBIDDEN LINES, because they are normally impossible and occur only in very thin gases, were first observed in nebulae. As they could not be identified with any known element, they were, until the true explanation was discovered, identified as a new element, called NEBULIUM.

Other objects, such as COMETS, SUPERNOVÆ, galaxies and QUASARS, also have emission lines in their spectra. The emission lines in quasars are often very broad; while certain other effects can cause them to appear wider than normal; the broadening is often due to the material that causes the lines moving at a wide range of velocities. The width of the lines gives astronomers an idea of the speed of motion within the object. *MRK.*

Empedocles (?490–432?BC). Greek philosopher who devised the four elements — earth, fire, air and water — to explain the material substances in a universe he thought to be spherical and made of crystal. He held that the atmosphere was substance, not a void.

Empedocles: crystal universe

Enceladus Moving outward from Saturn, Enceladus is the second of the larger icy satellites and was discovered by William HERSCHEL in 1789. Although Enceladus is similar in size to Mimas, with a diameter of 310 miles (500km), and is located at a distance of 147,800 miles (237,900km) from the planet, it displays evidence of extensive resurfacing. The surface is divided into types of terrain: an old cratered terrain with flattened craters; smooth planes that are lightly cratered; and ridged planes containing parallel grooves up to 3,300ft (1km) in elevation.

Enceladus: a Voyager 2 view in 1981

These features are probably the result of extrusion of water from the interior. The fact that the period of revolution is one-half that of DIONE indicates that tidal forces involving larger orbital eccentricities than those currently observed may have been the source of the energy necessary to melt the satellite. *RFB.*

Encke, Johann Franz (1791–1865). German astronomer who compiled star charts and calculated the path of the comet that bears his name.

Encke's Comet In 1786 MÉCHAIN found a comet just visible to the naked eye. New comets were found in 1795 by C. Herschel, in 1805 by PONS, Huth and Bouvard, and in 1819 by Pons. The work of ENCKE linked the orbits of these four comets and the year 1822 saw the first predicted return of Encke's Comet. To 1984 it had been seen at 53 returns, having a period of 3.3 years, the shortest known: modern instruments make it possible to observe the comet all round the orbit. Changes in the period were thought due to a resisting medium in the Solar System but are now attributed to the "rocket effect" of material expelled from the comet's nucleus. *MJH.*

Encke's division The main division in Saturn's Ring A, discovered by J.F. Encke. It is not a difficult object telescopically when the rings are suitably placed. Its mean distance from the centre of Saturn is 82,900 miles (133,500km).

Endymion Dark-floored lunar crater, 73-miles (117km) in diameter; Humboldtianum area: 55°N, 55°E.

Entropy Entropy is a measure of the available energy, or extractable work (not the same as the total energy content), of a body or system. This is closely related to the amount of disorder in a system (for example the molecules in water vapour are in a state of disorder, but the molecules in ice are in a well-ordered lattice). Entropy, like temperature, pressure or volume, is a physical parameter that defines the state of the system.

Eötvös, Roland Baron von (1848–1919). Hungarian physicist who used a torsion balance to show that gravitational and inertial mass are equal to within five parts in one thousand million.

Ephemeris A book published regularly, usually every year, giving tables of the daily predicted positions of the Sun, Moon and planets, over the period covered. It also contains other information such as the positions of certain bright stars, times of eclipses, times of sunrise and sunset, moonrise and moonset, times of beginning and end of twilight, etc. It is used in astronomical observation and as an aid to navigation. The *Astronomical Ephemeris* has been published annually by HM Nautical Almanac Office at the Royal Greenwich Observatory. In the United States, the *American Ephemeris and Nautical Almanac* is produced by the US Naval Observatory. *JWM.*

Ephemeris time (ET). Timescale used since 1960 for ephemerides of Sun, Moon and planets, based on the average length of the mean solar day during the 19th century. Ephemerides are a series of predicted positions.

Epicycle Once used to explain planetary motion. *See also* PTOLEMY.

Epimetheus Small satellite of Saturn; dimensions 87 × 75 × 62 miles (140 × 120 × 100km). It was discovered in 1978, and is co-orbital with JANUS.

Epoch A particular instant of time selected for reference purposes, so that data may be compared. For example, the positions of stars on star charts

may be shown for the epoch 2000.00.

Epsilon Aurigæ Epsilon Aurigae is one of the most interesting eclipsing BINARY STARS. More than 150 years of observation and scientific endeavour have so far failed to produce a wholly satisfactory understanding of it. O. STRUVE, one of the leading investigators of binary stars, was once moved to comment that the history of Epsilon Aurigæ "is in many respects the history of astrophysics since the beginning of the 20th century".

Binary stars are by no means rare. Surveys show that many stars in the Sun's neighbourhood are bound to at least one companion by mutual gravitational attraction. In the special case of an ECLIPSING BINARY, the orbital inclination is such that the component stars periodically pass in front of each other as viewed from Earth, thereby causing variations in apparent brightness.

The variability of Epsilon Aurigae was first observed as long ago as 1821 by a German pastor, J.Fritsch, but it was observations by H.LUDENDORFF at the turn of the century that produced the first significant data. Using spectroscopic methods Ludendorff determined the period of revolution of Epsilon Aurigae to be 27.1 years. This period is now accepted as approximately 9,885 days, and is the longest known period of an eclipsing binary system.

The figure accompanying this article illustrates the results one might expect from photometric observations of an ordinary eclipsing binary star. In practise the components will differ in some respects, and in particular the more luminous component is known as the primary: the darker component is known as the secondary.

Note that the light curve exhibits two minima per period of revolution. During the primary eclipse (ie, when the primary component is obscured) the total brightness of the system diminishes by a greater proportion than during the secondary eclipse. It is clear therefore that the relative size and shape of the minima give some indication of the relative surface brightness, size, and possibly structure of the two components.

The spectrum observed will under normal circumstances exhibit contributions from both components of a binary system. It is therefore possible to gauge with some accuracy the relative motion of the two stars by observing the Doppler shifting of selected spectral lines. As one component of the binary

system moves to some degree towards the observer, lines in its spectrum will be shifted to shorter wavelength (blueshifted) while as the other component recedes from the observer, lines from its spectrum will be shifted slightly to longer wavelength. Thus spectral lines common to both stars will be seen to split and recombine twice within each period of revolution of the system.

It was in attempting to apply this kind of analysis to observations of Epsilon Aurigae that astronomers first realized that here was no ordinary eclipsing binary system.

The primary component of Epsilon Aurigae is a luminous supergiant star which exhibits an F0-type spectrum. It is a conspicuous naked-eye object with apparent magnitude +3.0 at maximum brightness and +4.1 during primary eclipse, (4h 58.4m, +43° 44'). Recent reports associate semi-regular pulsing with the primary, but it is on the whole not a particularly remarkable star.

All the interest in Epsilon Aurigae is associated with the mysterious companion star. The primary F0 spectrum is observed at all phases of the system's revolution (even at primary minimum) and, more surprisingly, no single feature has yet been observed that can be assigned to the secondary!

The light curve of Epsilon Aurigae is also somewhat unusual. Primary eclipse has been observed in 1928–30, 1955–57, and 1982–84. This minimum is "flat-bottomed" such that the total phase lasts approximately 400 days within an entire eclipse phase of approximately 700 days.

The flat-bottomed nature of the primary minimum normally implies that the primary component is eclipsed by a larger companion. At primary eclipse the apparent magnitude of Epsilon Aurigae decreases by 0.8 magnitudes. This is a considerable reduction in brightness from which one would normally conclude that the two components have similar surface brightness. We might therefore expect to observe a similar loss in brightness as the secondary component is eclipsed by the primary and therefore a secondary minimum of comparable depth to the primary minimum. In the case of Epsilon Aurigae, however, no secondary minimum is observed at all!

So it seems that scientists obtain inconclusive and in some respects apparently contradictory evidence concerning the nature of the secondary component of Epsilon Aurigae. Faced with this unique challenge, more sophisticated observations have been

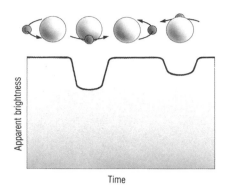

Epsilon Aurigæ: a missing minimum

made and various models proposed in order to explain them. It is perhaps not surprising that some of these models are somewhat exotic.

Early models concentrated on the possibility of a secondary component larger than the primary. One idea in particular proposes an object some 3,000 times larger than the Sun. Further, it is suggested that the object is cool (less than 1,000K perhaps), therefore emitting strongly at infrared wavelengths but remaining sufficiently dim at optical wavelengths to explain observations. Furthermore, this object would be tenuous and therefore semi-transparent. This would then explain why the primary component spectrum is observed unchanged even during primary eclipse.

The major problem facing models of this kind is that of justifying the depth of the observed primary minimum. Such large tenuous bodies are not capable of blocking enough radiation from the primary component during eclipse to reduce the apparent brightness by the observed amount.

It was not until the 1960s that techniques in INFRARED ASTRONOMY became sophisticated enough to provide assistance in matters such as these. Infrared photometry of Epsilon Aurigæ in 1964 by F.Low and R.Mitchell showed no evidence for a cool object of the type envisaged in these models.

It was long before the observations of Low and Mitchell, however, that the Italian astronomer Margareta Hack proposed an alternative model for Epsilon Aurigæ. It was thought that a small secondary component might explain the complete absence of a secondary minimum in the light curve. If this object were hot and luminous, however, it would be difficult to understand the absence of a secondary contribution to the observed spectrum. Moreover, with a small secondary object there is difficulty once again in

explaining the extent to which brightness is reduced during primary eclipse. How could the secondary object block so much radiation from the primary and yet not be noticeable as it, in turn, is eclipsed?

From detailed analysis of the spectrum of Epsilon Aurigæ it was Struve who first pointed out that the primary component seemed to possess an extended atmosphere. This naturally led to interpretations based on the idea of a small dark secondary star and a region of dusty matter situated between the two binary components. A disk-shaped region of dust viewed edge-on, for example could conceivably mask radiation from the primary star to the observed degree.

Attention was once again lavished upon Epsilon Aurigæ during the most recent eclipse of 1982–84. Results of many detailed observations have been summarized in a special report dedicated to this one binary star. Yet there is still much to learn about it and much work remains to be done. *KPW.*

Epsilon Boötis (Izar). The second star of Boötes; magnitude 2.37, type K0.

Epsilon Eridani K2 dwarf, visual magnitude 3.7, distance 3.3 pc. In 1983 IRAS found a possible protoplanetary system of radius 3.5AU around it. *See also* VEGA.

Epsilon Indi The least luminous star visible with the naked eye; magnitude 4.7, absolute magnitude 7.0, distance 10.7 light-years, type K5.

Epsilon Lyræ The famous "double-double" near VEGA. This consists of two fourth-magnitude stars 208 arc seconds apart, each of which can be split again into two stars by a small telescope.

The four components are physically associated, but the two main pairs are widely separated.

Equation of time The difference between mean (clock) time and apparent (sun-dial) time at any moment.

Because of the tilt of the Earth's axis and the fact that the orbit is not circular, solar days vary in length through the year. With the coming of accurate clocks, it became necessary to introduce a fictitious Mean Sun which moves along the CELESTIAL EQUATOR at a constant rate. From this is derived mean (or average) time.

On November 4, for example, the real Sun crosses the meridian some 16

minutes before mean noon, while on September 2 mean and apparent times are the same. *HDH.*

Equatorial mounting A telescope mounting in which the two mutually perpendicular axes, required to facilitate telescope pointing over the whole of the visible sky hemisphere, are arranged with one axis — the polar axis — parallel to the Earth's rotation axis. The other then lies in the plane of the equator with the telescope perpendicular to it and, because rotation about it sweeps the telescope pointing in DECLINATION, it is called the declination axis. The primary design characteristic of this mounting is that, for any telescope pointing, the diurnal motion of a celestial object can be followed by rotation of the telescope about the polar axis only.

Instruments with a polar axis were used by Scheiner in 1627, but it was not until 1749 that the first design which could properly be called an equatorial mounting was presented, to the Royal Society, by James Short. It was a further seventy years before the design characteristic was brought to full practical realization by FRAUNHOFER with the type which became known as the German mounting.

To ensure a smooth and uniform driving motion, which gives an accurate following of the observed object, the required torque must be constant for all pointings of the telescope. The necessary conditions for this are that the centre of mass of the telescope and its immediate fittings is at the intersection of the telescope and the declination axis and, secondly, that the centre of mass of all the moving parts lies at the intersection of the declination and polar axes. This is the balanced condition. Additionally the bearings must be of high quality and well maintained.

In the German mounting the centre of mass of the telescope and its immediate attachments lies near to one end of the declination axis. Balancing of the telescope is achieved by large and small counterweights fitted at the opposite end of the declination axis together with small auxiliary weights on the telescope. This axis, with the telescope and counterweight, is effectively a beam loaded at both ends and supported in the middle by the polar axis. Even with much care in its engineering design to ensure mechanical rigidity it forms an easily excited vibrating system. Similarly the telescope vibrates about its attachment to the declination axis and, lastly, the whole weight of the

E

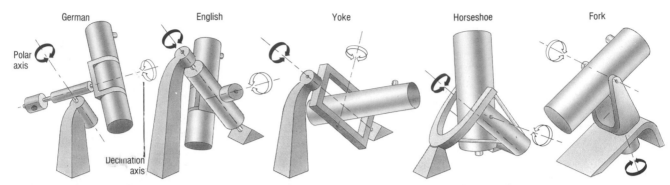

German   English   Yoke   Horseshoe   Fork

Polar axis

Declination axis

Equatorial mounting: mechanical solutions to the problems of tracking celestial bodies smoothly and accurately

E

moving parts at the upper end of the polar axis overhangs outside its bearings. In small instruments the vibrations arising from these design limitations are nearly always significant and they are often severe. With increasing size the greater inertia reduces the likelihood of vibrations, but if they do occur they may well take longer to die out. The increased weight also aggravates the design problem of overhang outside the bearings.

At any specific latitude there is a range of declinations and HOUR ANGLES over which the telescope can be correctly pointed when set to the east or to the west of the pier, but not in both positions because of obstruction to the telescope movement by the pier or the driving mechanism. Within this range it is not normally possible to follow an object across the meridian for much more than half an hour or so. To avoid obstruction the telescope must then be rotated through 180° about the declination axis and 12 hours (180°) about the polar axis. This brings it to the correct pointing on the other side of the pier but rotates the field of view through 180°. To the visual observer this is an inconvenience which is best avoided during a critical observation, but the 180° rotation of the image on the film is unacceptable to the photographer. Despite its limitations the German mounting continues in use for small portable and transportable instruments. Commercially the mechanical parts can be assembled, adjusted for the observer's latitude, and supplied as one unit — the head — which can be set upon a single pier or tripod by the purchaser, who need only set the polar axis in the meridian and pointing toward the pole to bring it into use.

The overhang of the telescope and declination axis beyond the polar axis bearings may be avoided by extending this axis poleward and supporting its upper bearing upon a second pier. Provided the distance between the bearings

is sufficiently great the telescope is free of mechanical obstruction; it can be pointed toward the pole and be driven across the meridian except in the region below the pole. Confusingly this type is called the English mounting, or the cross-axis, or the modified English mounting.

To be rid of the counterweight, and the overhangs outside the declination axis bearings, the extended polar axis of the English mounting is developed into a frame, or yoke, within which the telescope swings, hung between declination bearings fitted one on each side of the yoke. This yoke mounting (sometimes called the English mounting!) is the engineering acme of equatorial design. The counterweight is eliminated and the moving parts all lie within their bearings. Most of the sky is conveniently accessible and the telescope can be driven through the meridian except for the region below the pole. This is obscured by the poleward pier. There is mechanical obstruction to pointing to the polar region and this may be significant for some observing programmes.

By developing the upper polar axis bearing in the form of a horseshoe with a gap sufficiently large to admit the telescope, the polar region becomes accessible. Such a bearing is intrinsically weak so the horseshoe, or open yoke, mounting is invariably of massive construction to ensure the required rigidity. For this reason it calls for very careful design.

To access the meridional areas of sky below the pole the extended yoke with its upper bearing and poleward pier must be abandoned. The lower part of the yoke, including the declination bearings, is retained and this part, together with the polar axis, is massively redesigned to attain the rigidity required to provide against the considerable overhang that is introduced. This is the fork mounting, which can access all parts of the visible sky and

drive through the meridian both above and below the pole.

In general, refracting telescopes are long and slender whereas reflecting telescopes are short and stubby. For this reason reflectors are suitable for design into any of the mountings discussed above, but for most refractors the German mounting, despite its limitations, is the only suitable type. Some refractors of large aperture and very short focal length, intended for wide-field photography, are to be found mounted similarly to reflectors.

Many large modern reflecting telescopes are on yoke mountings but the most recent are altazimuth types. The development of computer control systems during the last 25 years has made the electronic solution of the problem of driving the telescope simultaneously about two axes much simpler and cheaper than the mechanical solution of constructing an equatorial mounting to enable single-axis drive.

In the balanced condition, with well-maintained modern bearings, the polar axis can be driven at the required sidereal rate by a small electric motor. This is typically a stepper or synchronous type supplied by a variable frequency pulse generator or oscillator. The observer then guides the telescope by manual correction of minor random drifts. *LMD. See also* ALTAZIMUTH MOUNTING.

Equinoctial colure *See* EQUINOX.

Equinox The two days each year when the Sun crosses the CELESTIAL EQUATOR, and day and night are of equal length all over the Earth. When the Sun crosses from south to north, about March 21, it is the spring or vernal equinox in the northern hemisphere; when it crosses from north to south, about September 23, it is the autumn equinox, or First Point of Libra: in the southern hemisphere, the seasons are, of course, reversed.

The vernal equinox, or First Point of Aries, is the zero of celestial longitude, the point on the celestial sphere, at any moment, where the ECLIPTIC and celestial equator meet at the ASCENDING NODE, where the Sun crosses the Equator from south to north. This is the true or apparent equinox, or equinox of any date. Precession of the equinoxes causes it to move westwards along the ecliptic by about one-seventh of a second of arc daily, but the Sun's position in the sky relative to it remains almost the same on the same day each year for thousands of years because of the refinements of the calendar. The stars, of course, change their positions relative to it because of the precession. When used in connection with measurements, the vernal equinox always means this moving true equinox. The literal equinox is the actual moment when the centre of the Sun crosses the celestial equator. The mean equinox, used for measurement of positions in star charts and catalogues, is the true equinox corrected for NUTATION in right ascension, an irregularity with a maximum of $1\frac{1}{4}$ seconds.

The First Point of Libra is the autumnal equinox, the point of the Sun's descending node. Due to PRECESSION, the First Points are no longer in the constellations of their names — that of Aries is now in Pisces and that of Libra is now in Virgo.

The equinoctial colure is the HOUR CIRCLE of the vernal and autumnal equinoxes, ie, of 0^h and 12^h R.A. The equation of the equinoxes is the nutation in longitude, or nutation measured along the ecliptic. *TJCAM.*

Equuleus (the Foal). A very small and obscure constellation in the area of Aquila. Its brightest star, Alpha, is only of magnitude 3.92. It has no features of particular interest.

Eratosthenes Well-formed lunar crater 38 miles (61km) in diameter, 15°N, 11°W, near the end of the Apennine chain. It has high terraced walls and a central elevation.

Eratosthenes of Cyrene (?276–195?BC). Greek geographer, mathematician and polymath, noted for his measurement of the size of the Earth. He spent most of his adult life at the Library and Museum at Alexandria, where he later became Chief Librarian.

In addition to his scientific work, Eratosthenes was also a poet, grammarian and chronologist, being the first Greek known to have made a scientific attempt to date historical events.

His mathematical work led him to discuss the essential basic concepts of geometry, and to invent a mechanical device for solving the fundamental problem in geometrical proportions known as "duplicating the cube". In arithmetic he invented his famous "sieve", a method for finding prime numbers, and in astronomy he determined the obliquity of the ECLIPTIC.

In geography Eratosthenes divided the Earth's globe into five zones — a torrid one bordered by two temperate zones, which in turn were bordered by two frigid regions. And though his *Geography* included the usual descriptive sections, it broke new ground by putting the whole subject on a mathematical basis.

Eratosthenes' measurement of the size of the Earth rose from the fact that at the summer solstice, the Sun cast no shadow down a vertical well at Syene, south of Alexandria. By setting up a GNOMON (a vertical stake) at Alexandria to obtain the Sun's altitude there at noon on the same day, and then having the distance between Syene and Alexandria carefully measured, Eratosthenes found that the two places were separated by one-fiftieth of a circle

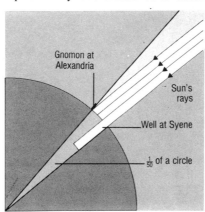

Gnomon at Alexandria

Sun's rays

Well at Syene

$\frac{1}{50}$ of a circle

Eratosthenes: measuring the Earth's size

(7°2). Though there were errors in the scheme — Syene, for instance, was not precisely due south of Alexandria as he assumed — he obtained a value for the Earth's circumference which was about 29,000 miles (46,661km), and very close to the present value of 25,000 miles (40,225km). *CAR.*

Eridanus (the River). One of the largest constellations in the sky; it extends from near Rigel in Orion to the south polar area, so that its brightest star, Achernar, is inaccessible from Europe.

Brightest stars

Name	Visual Mag.	Abs. Mag.	Spec.	Distance (light-yrs.)
α Achernar	0.46	−1.6	B5	85
β Kursa	2.79	0.0	A3	91
θ Acamar	2.92	0.6. 1.7	A3 + A2	55
γ Zaurak	2.95	−0.4	M0	143

Other stars above the fourth magnitude are Delta (3.54), Upsilon⁴ (3.56), Phi (3.56), Tau⁴ (3.69), Chi (3.70), the nearby Epsilon (3.73), Upsilon² (3.82), F1.53 (3.87), Eta (3.89), Nu (3.93) and Upsilon³ (3.96).

Acamar is a fine double, separable with a small telescope. It was ranked as of the first magnitude in ancient times, but is now only of the third — though it is probable that the discrepancy is due to an error in observation or translation. Eridanus contains surprisingly few interesting objects. *PM.*

Eros (Asteroid 433). Discovered in 1898 by Witt of Berlin. It was the first asteroid known to come within the orbit of Mars. The distance from the Sun ranges between 1.133 and 1.783AU; period 1.76 years; orbital eccentricity 0.223, inclination 10°83. It is small, and elongated in form; the length is about 8 miles (12km), the width no more than about 9 miles (14km). At the close approach of 1931, when it came within 15,000,000 miles (23,000,000km) of the Earth, worldwide measurements of its position were made, in order to determine its distance and, hence, the length of the astronomical unit — though the final result derived was rather too high. The last close approach was that of 1975. At its best Eros may reach magnitude 8.3; slight variations yield a rotation period of 5.3 hours. *PM.*

Eruptive variables Variable stars can be broadly classified in three groups: (1) ECLIPSING BINARIES, including ellipsoidal variables; (2) Pulsating variables (eg, CEPHEIDS); (3) Eruptive variables. This last group is a large heterogeneous class which contains many little-understood objects. Members of the class range from SUPERNOVÆ, where violent processes occur involving the whole star, through the outbursts seen from a variety of interacting binary systems, to flare activity on single stars and the ejection of shells of matter by a variety of objects. This article deals with the physical mechanisms in these stars.

Cataclysmic variables, symbiotic stars and novæ are interacting binary systems in which matter flows from one

E

component to the other, which is generally a WHITE DWARF (but may be a NEUTRON STAR). The complex phenomenon in these systems (rapid variations, large outbursts, X-ray emission, etc) are primarily due to the effects of orbital motion, to varying rates of mass exchange and to thermonuclear burning of hydrogen that the white dwarf has acquired from its companion.

Related objects are X-ray transients such as A0538-66 in which a neutron star is in a highly eccentric orbit about a hot (B-type) giant. When the outer envelope of the hot star is sufficiently distended, the neutron star passes through it, giving rise to a burst of optical and X-ray radiation. Supernovae owe their outbursts to the rapid release of thermonuclear energy leading to the ejection of large amounts of mass and the formation of a neutron star. In the case of type I supernovae, mass transfer from a companion to a white dwarf in a binary system may initiate its explosion.

UV Ceti stars, of which PROXIMA CENTAURI, the nearest star, is an example, are red dwarf stars showing flaring activity (increases of brightness ranging from a few tenths of a magnitude to six magnitudes, the flare lasting a few minutes — *see* FLARE STARS).

Some red dwarfs (the BY Draconis stars) show light variations which are attributed to large star spots (roughly analogous to SUNSPOTS or groups of sunspots). The rotation of the star carries the spots periodically in and out of view causing the observed light variations. The spots are not only larger than those on the Sun (covering up to 20 per cent of the stellar surface) but are different in having little or no penumbra (the outer, brighter, parts of sunspots). As with sunspots this is also a magnetic phenomenon. Magnetic fields constrain the flow of matter and energy near the stellar surface and lower temperature regions (the spots) form where the magnetic tubes of force cluster together. As might be expected from the fact that both flares and spots have a magnetic origin, some red dwarfs are simultaneously BY Dra and UV Ceti stars. Star spots are also found on the giant and subgiant components in some close binary systems (the RS CVn stars). Most (though probably not all) BY Dra stars are in binary systems as well. The concentration of spotted stars to binary systems is simply due to the need to produce a high magnetic field. In a binary system the high rate of rotation can be maintained, despite

magnetic braking forces, by tidal interaction of the two stars.

T TAURI STARS are very young (preMAIN SEQUENCE) objects and are generally found in regions of recent star formation. They show flaring activity and a strong, variable, chromospheric-type emission spectrum. It seems likely that the basic physical mechanism at work is the same as that in UV Ceti stars. That is a very strong magnetic field produced in a rapidly rotating star.

S Doradûs variables (also called Hubble-Sandage variables) are hot, very luminous, supergiants (spectral type B-F, luminosity about a million times that of the Sun) with dense expanding envelopes. The mass-loss rate changes with time and the corresponding changes in the envelope (which redistributes the energy coming from the star) leads to the observed irregular visual brightness changes, although the energy output of the underlying star remains sensibly constant. The ejection of matter is probably connected with the rapid evolution of a star whose luminosity is near the upper limit for stable stars. The unusual object ETA CARINÆ, which underwent a major brightening in 1843, is sometimes included in this class though without any strong reason. At a somewhat fainter level the Be stars (hot, rapidly rotating stars with emission lines in their spectra) often show irregular variability due to outflow of matter in their equatorial planes forming disks and rings. The most famous star in this class is Gamma Cassiopeiæ.

R CORONÆ Borealis variables are solar mass stars which have evolved past the red giant stage and have ejected their hydrogen-rich envelopes exposing a carbon-rich, hydrogen-poor core. Their irregular declines in brightness (of up to about 9 magnitudes) are due to obscuration by the clouds of carbon particles they "puff" off. Each one of these clouds covers only a limited area of the stellar surface (perhaps 3 per cent). Declines occur when a cloud is ejected in our line of sight. Clouds ejected in other directions may be detected spectroscopically and in the infrared. The process of mass ejection may be related to the pulsations which at least some stars of this group undergo. *MWF.*

ESA *See* EUROPEAN SPACE AGENCY.

Escape velocity The minimum velocity with which a body must be projected from the surface of a massive body in order not to fall back onto it, or, the

minimum velocity needed to escape from a closed orbit round a massive body. It is, in effect, the velocity required to achieve a parabolic trajectory relative to the massive body. The escape velocity at the surface of the Earth is 6.9 miles/s (11.2 km/s).

ESRO European Space Research Organization, set up in 1964. It launched 184 rockets and built 7 satellites, all launched by the United States. it was absorbed into EUROPEAN SPACE AGENCY in 1975.

ESSA Second generation of United States' meteorological satellites, launched by the Environmental Science Services Administration. They were placed in near polar orbits so as to cover the entire Earth each day.

Eta Aquarid meteors One of two meteor showers associated with HALLEY'S COMET, and seen when the Earth passes within 0.065AU of the comet's orbit on about May 8 every year, near the descending NODE. The Eta Aquarids were first recorded in 74BC, and they may be seen between about April 24 and May 20 every year. There are four or five peaks in shower activity. Highest rates are obtained between May 4 and May 8; the zenithal hourly rate is about 35 meteors per hour. The meteors are swift-moving with long trails. There are several radiants centred around R.A. 22h 20m, Dec. −01°, quite close to the star Gamma Aquarii. *JWM.*

Eta Carinae Peculiar, bright star in the southern Milky Way, possibly the heaviest and most luminous in our Galaxy.

When the southern stars were catalogued, Eta Carinae was a thirdmagnitude star, but from 1833 it began to vary irregularly, becoming at its brightest second only to SIRIUS. What makes this performance remarkable is the great distance of the star, then unknown but now recognized to be about 8,000 light-years. At its peak brightness, Eta Carinae shone some four million times as brightly as the Sun.

The star then faded to just below naked-eye visibility, where it remains to this day. At the same time, a nebulous patch formed and expanded around it. This nebula, milky-white and irregular, is called the HOMUNCULUS because its shape resembles a manikin. It continues to grow at a speed of about 300 miles (500km) per second.

In 1968, Gerry Neugebauer and

James Westphal, from the California Institute of Technology, discovered that Eta Carinae is the brightest INFRARED source in the sky (apart from Solar System objects). They proved that Eta Carinae continues to this day to shine about as brightly as it did last century. Most of the radiation is absorbed by grains of dust in the Homunculus, which then re-emit it at infrared wavelengths.

The exact nature of Eta Carinae remains mysterious. Its mass is at least 100 times that of the Sun (it is the most massive star known in our Galaxy). It also is known to have consumed most of its hydrogen. Kris Davidson, of the University of Minnesota, has made the plausible suggestion that it has begun to evolve away from the MAIN SEQUENCE, to become a yellow supergiant star, and has reached an unstable

Eta Carinæ: exact nature mysterious

state where it needs to shed material in a series of eruptions. There certainly is evidence that Eta Carinae has produced at least one nebula before the Homunculus. If Davidson is right, it is highly likely that Eta Carinae's eventual fate, perhaps only a few centuries from now, is to become a SUPERNOVA. *DAA.*

Eta Geminorum A bright semiregular and eclipsing variable with a range of 3^m15 to 3^m9V in a period of 232^d9. The eclipses re-occur at intervals of 2,984 days. Both components are red giants with M3 spectra.

Euclides Lunar crater 7.5 miles (12km) in diameter, 7°S, 29°W, near the Riphæan Mountains; it is very bright.

Eucrites Stony meteorites belonging to the achondrite group. Eucrites are calcium-rich and, with the Howardites, are termed basaltic achondrites. They are like some lunar samples.

Eudoxus of Cnidus (?400–347?BC). Greek astronomer, mathematician, and polymath of immense reputation.

In geometry Eudoxus provided much of the basic thinking lying behind Euclid's later work, he investigated the subjects of mathematical proportion and the theory of numbers, as well as other problems.

The achievement of Eudoxus which exerted most influence was his explanation of the movement of the Sun, Moon and planets. He suggested that they were fixed to the surfaces of rotating spheres, all centred on the Earth. Though the spheres were probably regarded by Eudoxus purely as a mathematical model, ARISTOTLE imagined that they had a physical existence. *CAR.*

Euphrosyne (Asteroid 31). Diameter 230 miles (370km); the fifth largest of the regular asteroids.

Europa The second of Jupiter's Galilean satellites. It is the smallest of the four, with a diameter of 1,942 miles (3,126km) and is the only one of the Galileans which is smaller than our Moon. It moves around Jupiter at a mean distance of 417,000 miles (670,900km) in a period of 3.551 days. The orbital eccentricity and inclination are both very low.

Europa was surveyed by both the Voyager probes, with surprising results. The surface is icy, with virtually no craters and little vertical relief; new terms have had to be introduced to describe the surface features — "linea" (a dark or bright elongated marking, either straight or curved); "flexus" (a very low curvilinear ridge with a scalloped pattern); and "macula" (a dark, sometimes irregular spot). Only three craters have been identified with reasonable certainty, and all are shallow and obscure.

It has been said that Europa is as smooth as a billiard ball; the general aspect has also been likened to a cracked eggshell. One region looks much like another, and no particular object on the surface stands out. If craters ever existed, they have been obliterated, perhaps by water seeping out from below and then freezing; certainly the general aspect gives the impression of being relatively young. Presumably there is a silicate core; the ice crust may overlie a thick layer of what may be termed "slush". There are suggestions that an ocean of liquid water may exist beneath the outer ice crust. In many ways Europa is quite unlike the other Galileans; the reasons for its exceptional appearance are by no means fully understood. *PM.*

European Southern Observatory (ESO). A major European observatory at La Silla, Chile. Includes a 150in (3.8m) reflector.

European Space Agency The convention creating the European Space Agency, signed in May 1975, brought together under a single agency the activities of the two earlier European space organizations – ESRO, the European Space Research Organization, and ELDO, the European Organization for the Development and Construction of Space Vehicle Launchers.

The agency is financed by its 13 member states, who all contribute to the general and scientific budgets but choose their contributions to the application programmes, which include telecommunications, Earth observations, space transportation systems and microgravity research.

ESA has not only been responsible for a series of highly successful scientific and application spacecraft but has also developed SPACELAB and the ARIANE LAUNCHER. *RMJ.*

Evection Irregularity in the Moon's motion caused by perturbation by the Sun and planets, displacing it from its orbit by a maximum of $\pm 1° 16'$ in a period of 32 days.

Event horizon *See* BLACK HOLES.

Evershed, John (1864–1956). English astronomer who discovered radial motions of gases in SUNSPOTS (Evershed effect). Designer of a high-dispersion solar spectrograph using large liquid prisms as the dispersive element.

Excite An atom or molecule possessing extra internal energy is excited. It may return to its normal state by emiting radiation of specific wavelengths.

Exobiology The study of the possibilities of, and nature of, life beyond the Earth. *See also* LIFE IN THE UNIVERSE.

Exosat Name given to the European Space Agency's X-ray astronomy satellite launched in May 1983. Smaller than the EINSTEIN OBSERVATORY it contained two X-RAY TELESCOPES

E

whose total area was about 30 per cent of the Einstein telescope. It included other large-area, non-imaging X-ray detectors. Placed in a highly elliptical orbit, observations of up to 90 hours' duration were possible. Einstein observations were interrupted by Earth occulation for 30 minutes in every 90. The long observing times made possible by Exosat permitted major advances in studies of X-RAY BINARIES, STELLAR CORONÆ, quasi-periodic oscillations in X-ray sources and the properties of QUASARS and SEYFERT GALAXIES. The mission ended in 1986. *JLC*.

E

EXOSAT: under test in Holland

Exosphere The pressure of an atmospheric layer decreases with increasing altitude so that, eventually, the atmosphere is so thin that the collisions between the gaseous species become very infrequent. Ultimately, there is a level when there are no collisions; the fast-moving atmospheric constituents may then escape the gravitational influence of the planet altogether or perhaps go into orbit around it. The region where this situation occurs is the exosphere, which for the Earth's atmosphere is situated at an altitude of 250–300 miles (400–500km). This layer is therefore beyond the thermosphere, but includes the radiation belts extending to the magnetopause, where it meets the interplanetary medium. *GEH*.

Expanding universe The theory that the universe is expanding, ie, that each galaxy, or cluster of galaxies is moving away from every other one. That the universe is expanding was first demonstrated observationally in 1929 by Edwin HUBBLE, although the possibility had been suggested earlier on theoretical grounds by W.DE SITTER

(1917), A.FRIEDMANN (1922) and G.LEMAÎTRE (1927).

Explorer Series of United States scientific satellites used to explore orbital space. Explorer 1, instrumented by James A. Van Allen, discovered the radiation belts girdling the Earth.

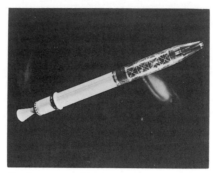

Explorer: a model of Explorer 1

Extensive air showers Extensive air showers (EAS) are generated by COSMIC RAY particles of the highest energy known to man, the particles probably being hydrogen nuclei, ie, protons. Individual particles have energy as high as 10^{20}eV but the manner in which their energy is achieved is not understood. The shower starts when the primary particle hits an oxygen or nitrogen nucleus in the upper levels of the atmosphere. The initial secondary particles are largely pions but these decay to the more stable muons; at ground level the showers contain mainly electrons with an admixture of perhaps 5 per cent of muons. *AWW*.

External galaxies Complete star systems located beyond our own Milky Way Galaxy. Optical telescopes can now detect them down to magnitudes fainter than 25; in principle more than 3,000 million galaxies are accessible to study by modern astronomers. Only the brightest few thousand have been catalogued completely and studied systematically.

Recognition that there were external galaxies came with the resolution of the "Great Debate" over the nature of nebulæ in the early 1920s. At this time some of the catalogued nebulæ ("clouds"; *see* MESSIER CATALOGUE) were resolved into stars, some of which were observed to vary in a manner identical to the CEPHEID VARIABLES of the Milky Way Galaxy. The external or "extragalactic" nature of these nebulæ was thus established, and it was soon realized that most nebulous objects observed away from the plane of the Milky Way were extragalactic. The

term "nebula" for this type of object has been supplanted by "galaxy".

There were three fundamental discoveries about galaxies during the 1920s and 1930s. Firstly, galaxies were found to be receding, and the greater the distance from us, the faster the velocity of recession. This established our modern view of the EXPANDING UNIVERSE (*see also* HUBBLE'S CONSTANT, DISTANCE MODULUS); faint galaxies can now be detected with RED-SHIFTS greater than 1, corresponding to a distance of perhaps 2,000Mpc, and a light travel time of 90 per cent of the age of the universe. Secondly, galaxies were observed to cluster. Thirdly, galaxies were seen to have very different forms or morphologies, which are closely related to clustering and to other physical properties.

Galaxies are ensembles of stars, perhaps a few hundred million, immersed in an interstellar medium of tenuous gas and dust. The two main types are elliptical and spiral (*see* CLASSIFICATION OF GALAXIES). ELLIPTICAL GALAXIES contain 10^8 to 10^{13} stars, and the most massive galaxies are of this kind. Old, cool red stars predominate (RED GIANTS of Population II; *see* STELLAR POPULATIONS), and little interstellar dust or gas is apparent. Star formation does not appear to be taking place. The light distribution is in the form of a smooth flattened ellipsoid, but rotation is generally observed to be so slow that the flattening is not simply the result of this rotation. SPIRAL GALAXIES are less massive but more luminous, and have central regions ("bulges") which resemble small elliptical galaxies. The distinctive arms are formed (*see* DENSITY WAVE THEORY) in thin, flattened disks which rotate rapidly, and which are dominated by hot, young blue stars and clouds of interstellar gas and dust (Population I objects). Star formation is clearly under way in such regions. LENTICULAR GALAXIES, intermediate between ellipticals and spirals, show a bulge and a disk, but no spiral-arm structure, no young stars, and little evidence of gas and dust.

The hierarchy of galaxy clustering extends uniformly from single or "field" galaxies, to groups, to rich clusters of thousands of galaxies, to SUPERCLUSTERS (*see* colour essay) containing many clusters. There is no preferred size of concentration, except that the tendency to cluster drops rapidly beyond a scale of about 20Mpc — superclusters this size are the largest agglomerates of matter in the universe.

The LOCAL GROUP, of radius 1.3Mpc, contains some 30 galaxies, with its mass dominated by two giant spirals — our Galaxy and M31, the ANDROMEDA GALAXY. Some 50 such groups of galaxies with several tens of members exist out to 16Mpc. The nearest rich clusters of galaxies are the VIRGO CLUSTER at 20Mpc (on which the LOCAL SUPERCLUSTER is centred), and the COMA CLUSTER at 100Mpc. Groups of galaxies contain spirals and ellipticals, clusters contain mostly ellipticals, and the richest clusters contain solely ellipticals or lenticulars. The poor clusters are usually irregular in shape, but rich clusters tend to be spherically symmetric and centrally condensed, with a giant elliptical galaxy often dominating the central region. The richest clusters may contain 10,000 galaxies, and the central regions may have a density of galaxies one million times greater than the mean galaxy density.

The theory of galaxy/cluster formation remains one of science's most difficult and challenging problems. GRAVITY is the universal force which achieves the creation of stable structures; for success, self-gravity, the self-attraction between different portions of the structure, must dominate forces which tend to disrupt, such as internal pressures generated in several astrophysical processes. Current theories suggest that all galaxies were formed from growing density enhancements by self-gravity at roughly the same time, some 15 thousand million years ago, soon after the universe expanded from its hot, dense big-bang phase. In elliptical galaxies, star formation took place rapidly with perhaps a time-scale of only several hundred million years, quickly using up all the interstellar gas in the process. The subsequent mature shape is the result of star orbits within the galaxy formed by the exchange of energy in mutual interactive pulls by gravity between all the individual stars. Spirals, however, are the result of a two-stage formation process in which gas was left over from the initial and interrupted star formation which produced the elliptical-like bulge component at the centre. The subsequent evolution was drastically modified by the excess gas rapidly settling into a disk, in which shock waves and density waves proceeded to form spiral arms. The predominance of ellipticals in rich clusters suggests that the gravitational attraction of the extra matter within clusters may assist in the initial rapid and uninterrupted star formation.

After they formed, the dominant D GALAXIES in the richest clusters may have grown by cannibalism, the overwhelming gravitational force of such objects dragging in the smaller galaxies.

Clusters themselves appear in all stages of evolution: mature like the largest and richest, in which the gravitational redistribution of energies has left the galaxy orbits so that the overall shape is regular; and immature or evolving, such as the smaller, irregular clusters in which this redistribution is still in progress. In some such cases the structure is unstable, and the self-gravity will never be adequate to form a single bound entity. The spread of velocities of the individual galaxies in clusters shows that large amounts of dark matter are present, the nature of which remains the subject of intense speculation.

Most galaxies in clusters have a good chance of one near-collision in their lifetimes. The results of such encounters are severe distortions by gravitational tides which produce filaments, bridges, tails, and deformed disks. Interactions of gas-rich galaxies such as spirals can result in shock waves producing bursts of star formation, and such interactions are possible trigger mechanisms for generating active galactic nuclei, of which SEYFERT GALAXIES, N-GALAXIES and QUASARS are examples. These objects are generally believed to represent normal galaxies with nuclei in extreme states of activity. *JVW.*

Extinction Loss of light from the line of sight as it passes through a medium. The loss may be by SCATTERING, in which case the light energy is taken up by the medium and promptly re-emitted at the same energy but redirected out of the line of sight. Or the loss may be by ABSORPTION, in which case the light energy is eventually re-emitted in a different form altogether, eg, as heat.

Typical examples of extinction in astronomy are the loss of starlight as it passes through the interstellar medium and is scattered and absorbed by interstellar grains, and the loss of starlight as it passes through the Earth's atmosphere and is scattered by air molecules. *PGM.*

Extragalactic gamma rays There is a tiny flux of cosmic gamma rays which appear to be coming from outside our Galaxy. Some of these are generated by discrete objects such as SEYFERT GALAXIES and QUASARS, but it is likely that there is also a diffuse flux which may have a cosmological origin. There is a "bump" in the energy spectrum of the gamma rays at about two million electron volts — a result first detected by the Apollo astronauts — and an exotic possibility is that it arises from the annihilation of matter and antimatter in the early universe. *AWW.*

Extra-solar planets For obvious reasons, extra-solar planets are very difficult to detect. One method is to study perturbations in the PROPER MOTIONS of nearby stars. This was initiated in 1937 at the Sproul Observatory with the 24in (61cm) refractor. The attainable photographic positional accuracy is about one micron, or $0''.02$ per exposure. The threshold value of the total amplitude of a measurable perturbation is about $0''.005$, and over a dozen cases have been discovered. Since 1964 the astrometric reflector at the United States Naval Observatory has also been engaged in this work.

Most of the stars concerned are RED DWARFS; half a dozen perturbations point to companions with masses well below 6 per cent of the Sun's mass, the critical value above which a "star" can be a real star. Five of these companions, or "dark dwarfs", are substellar objects with masses ranging from 0.005 to 0.020 times the Sun's mass.

However, in one case (BARNARD'S STAR) the perturbation is best represented by assuming that there are two planetary companions, with masses well below that of Jupiter.

Planetary systems and binary or multiple star systems are completely different dynamically. Binary and multiple systems do not show the characteristics of having come out of a rotating disk. Their orbits are eccentric (averaging 0.6) and the mass ratios range from 1 to 0.1. The differences are therefore genetic.

At present it is believed that there are five established cases of perturbations which indicate the existence of small but not planetary companions with less than 0.03 times the mass of the Sun.

Substellar objects

Name	Mass of companion (Sun = 1)
Van Biesbroeck 10	0.005
BD + 68°946	0.008
BD + 43°4305	0.009
Stein 2051 A	0.018
CC 1228	0.022

The orbital eccentricities range from 0.3 to 0.9, averaging 0.56, as compared with 0.6 for visual or spectroscopic binaries. Note the absence of small eccentricities. Therefore these objects are comparable with visual or spectroscopic binaries so far as orbital shapes are concerned, and all in all it does not seem that the companions are planetary in nature. Infrared studies by Don McCarthy have not shown companions to any of these stars, but infrared SPECKLE INTERFEROMETRY has suspected a faint dwarf companion for Van Biesbroeck 8 with an estimated mass five times that of Jupiter.

Barnard's Star (visual magnitude 9.5, spectrum M5, distance 6.0 light-years) has been photographed on 1,200 nights with the Sproul telescope over the period 1938–1982. A distinct pattern is shown in the perturbations, with a period of 12 years, but since 1975 it has become evident that a single perturbation is not the complete answer to what the residuals show; a second period with a longer perturbation is indicated. The latest analysis supports 12 and 20 years for the periods of the two perturbations. One interesting thing is that circular orbits are entirely satisfactory in the analysis; there is no indication of even slight eccentricity. The radii of the orbits are $0\rlap{.}''0070$ and $0\rlap{.}''0065$ or, in astronomical units, 0.013 and 0.012 respectively. Adopting a mass of 0.14 for Barnard's Star (Sun = 1), the orbital radii work out at 2.7 and 3.9AU, with masses of 0.7 and 0.5 times that of Jupiter. The mass ratios relative to Barnard's Star are 1:210 and 1:330.

The possibility, therefore, is that we may here be dealing with a planetary system of which we have found the two most massive Jovian-type members, a situation rather analagous to that of our own Solar System, in which Jupiter and Saturn would be the first members to be found by the same technique.

What about other nearby red dwarfs, such as PROXIMA CENTAURI, at 4.2 light-years? Unfortunately it has not yet been sufficiently observed, but many more planetary systems may be ready for discovery, and we are simply awaiting patient astrometric studies by conventional photographic or other techniques. But we should be ready, able and willing to rescue cosmic phenomena from the sea of errors.

There is, of course, no reason to assume that our Sun is in any way exceptional; it is a normal G-type dwarf star, and there is no reason to believe that planetary systems are unusual. Quite apart from the astrometric method, we now have evidence

of quite a different kind.

The highly successful IRAS satellite, launched in 1983, operated for much of that year. When testing the infrared telescope carried on the satellite, two American astronomers working at the Rutherford-Appleton Laboratory in England made a remarkable discovery. They found that VEGA (Alpha Lyrae) had "a huge infrared excess" which was presumably due to cool material associated with the star. The investigators came to the conclusion that the material must contain bodies much larger than dust-grains, because very small particles would already have been drawn back into the star. This would leave intermediate and large-scale debris in orbit, and debris could have developed into planets. The temperature of the material was estimated as $-184°C$, about the same as the temperature of the icy particles making up the rings of Saturn.

Before long it was found that Vega was not unique in its infrared excess. Another case was that of FOMALHAUT (Alpha Piscis Austrini). Even more significant was BETA PICTORIS, a southern-hemisphere star of visual magnitude 3.8 and spectral type A5 — 40 times as luminous as the Sun, and 78 light-years away. The infrared excess was so marked that Bradford Smith and Richard Terrile, using the Las Campanas reflector in Chile, decided to make a careful study in ordinary light, using the most sensitive electronic devices together with the telescope. They found obvious signs of the material which is responsible for the infrared radiation. It extends out to either side of the star, to a distance of 400AU, with its thickest part 100AU wide. The appearance seems to be due to a disk of material which we are viewing almost edgewise-on. The star itself is dimmed by only about half a magnitude, so that presumably the disk does not extend all the way to the star's surface — and it is in this region that planets might possibly be formed.

All in all, some 40 stars have now been shown to have infrared excesses. Of course this is not to claim that any of these — even Beta Pictoris — are attended by planets, but it does make the idea much more plausible. Whether any of these hypothetical planets could support life-forms is quite another matter, though here too it would be dangerous to claim that the Earth is unique.

Further research will depend on astrometric studies, and on infrared and visual observation. *PvdK/PM*.

Eyepieces An astronomical telescope is made up of an objective lens or mirror together with an eyepiece. The eyepiece is the lens nearest to the eye and is frequently detachable and mounted in its own tube. Its purpose is to magnify the image formed by the objective or mirror without degradation.

It may be a single lens but is more likely to be a compound set of lenses, either cemented together or airspaced. In general there will be a field lens, which accepts the rays of light from the telescope objective or mirror, and the eye lens which directs these rays of light into the eye.

There are many types of eyepiece; some for general viewing but some for specific purposes. The main parameters which are under the designer's control are focal length, angular field of view or acceptance angle, eye relief and exit pupil, but optimizing all of these is not possible and some imperfections will remain.

The focal length of a typical eyepiece lies in the range 3mm to 40mm (about $\frac{1}{8}$in to $1\frac{1}{2}$in) with the most common values 40mm, 25mm, 12mm, 6mm. It is the focal length in conjunction with the focal length of the telescope objective or mirror which determines the magnification of the overall system: for example, an 8in (20cm) f10 telescope mirror has a focal length of 2,000mm (almost 6ft 6in). The magnification in conjunction with a 25mm eyepiece would be 2,000 divided by 25, which equals an 80 times magnification.

The observed field of view of the telescope is largely determined by the acceptance angle of the eyepiece, which is typically around 40°. The telescopic field of view is then the acceptance angle of the eyepiece divided by the magnification. Wide-angle eyepieces can be designed to have acceptance angles up to 80°, but this additional width is paid for by increased distortion, particularly towards the edges of the field.

The exit pupil is the image of the telescope objective or mirror produced by the eyepiece. All the rays pass through the exit pupil so that if all the light gathered by the objective is to be passed to the eye without loss, the diameter of the exit pupil must be less than the diameter of the pupil of the eye at maximum dark adaptation, ie, about 7mm. Clearly if the diameter of exit pupil exceeds the diameter of the pupil of the eye some light will be lost: if it is less, the resolving ability of the eye is not being fully utilized.

The eye relief is the distance from the

eye lens of the eyepiece to the exit pupil. For comfortable viewing this should be around 7 to 10mm (between $\frac{1}{4}$ and $\frac{3}{8}$in).

The principal defects of eyepieces are spherical aberration, chromatic aberration, distortion, field curvature, and astigmatism.

Spherical aberration results in a slightly out-of-focus image. It is caused by the annuli of the lens at different radii having different focal lengths. It cannot be eliminated from a simple lens without employing an aspheric surface, but for a given focal length it may be minimized.

Chromatic aberration is caused by the refractive index varying with different wavelengths of the incoming light. The effect is to cause the image to be surrounded by a blurred band of colour. Chromatic aberration may be minimized by a compound lens comprising two or more glasses of different refractive indices — the original combination was crown and flint glass.

Distortion is a defect caused by unequal magnification at the centre and the edge of the field of view. Depending on which is the greater, the result is either barrel or pincushion distortion; the sides of a square figure would be made either convex or concave.

Field curvature causes the image to become defocused as it is moved away from the optical axis. A sharp image may be obtained even at the edge of the field by refocusing, but the defect is particularly serious where a wide, flat field is required as, for example, in ASTROPHOTOGRAPHY.

Astigmatism occurs when the lenses in the eyepiece do not have exactly the same curve across each of their diameters. It affects mainly the outer part of the field of view and is most pronounced in wide-angle eyepieces.

Apart from the single lens eyepiece, which is rarely used, the simplest type of eyepiece is the Ramsden, which has two plano-convex lenses with the convex sides facing each other. The lenses are of equal focal length and may be identical. The Ramsden has a reasonably wide angular field of view, 40° to 50°, but rather poor eye relief. Being a positive lens the focal plane is external to the eyepiece, allowing crosswires to be attached for guiding purposes.

The Huygenian eyepiece is also made up of two plano-convex lenses, but they have different focal lengths and the convex surfaces are turned away from the eye. It is a negative eyepiece with

the focal plane between the lenses, making crosswires difficult to incorporate. The eye relief is good, the angular field of view is large and the eyepiece is free from distortion. The main defect is rather severe spherical aberration. The Kellner eyepiece is a form of achromatic Ramsden using for the eyelens a positive meniscus achromat, which results in reduced chromatic aberration. Like the Ramsden, the Kellner is prone to producing ghost images, which result from internal reflections within the eyepiece. It is, however, popular with many observers.

Solid as opposed to airspaced eyepieces may be constructed from either a solid piece of glass or several pieces of glass cemented together. These types are characterized by high transmission and freedom from ghosts. However, they suffer from a small field of view and very poor eye relief. The best-known solid type is the Tolles, whilst the Steinheil monocentric is the most frequently encountered cemented triplet. These eyepieces are useful in particular applications such as the study of double stars and planetary observations.

For special purposes such as observations of stellar clusters, a wide-field eyepiece is essential. There are several types, all of which make use of a large number of elements, up to 6 or more, to produce the wide field. Most of them employ aspheric lenses and can offer fields of view up to 80°. The most common type is the Erfle, which consists of three doublets and has an angular field of view of about 68°. All wide-field eyepieces suffer from astigmatism distortion near the edge of the field of view, but this is unimportant where a broad picture is required.

The Plossl or Dialsight eyepiece comprises two nearly symmetrical achromatic doublets with their crown surfaces facing inward and nearly touching. They have large eye relief and a flat 40° to 50° angular field of view.

Probably the best general-purpose eyepiece is the orthoscopic. Typically, this uses a plano-convex eye lens and a field lens in the form of a cemented triple achromat, which supplies aberrations opposite to those of the eye lens. The angular field of view can be up to 50° and the eye relief is large, making for comfortable viewing. There is some loss of light due to the large number of lenses, ie, the transmission is less than average, but this can be improved to up to 80 per cent by coating the lenses. *JCDM.*

Fabry-Perot interferometer OPTICAL INTERFEROMETER in which the two parts of the incoming beam are recombined after multiple partial reflections between parallel lightly silvered glass plates in a Fabry-Perot etalon. It is capable of making extremely accurate measurements of the wavelengths of light emissions in atomic spectra.

Fahrenheit scale *See* TEMPERATURE SCALES.

Fallows, Fearon (1789–1831). Mathematical Fellow of St. John's College, Cambridge. He was appointed (in 1821) the first Director of the ROYAL OBSERVATORY, CAPE OF GOOD HOPE.

False Cross A pattern of stars made up of Iota and Epsilon Carinæ, and Kappa and Delta Velorum. It is often mistaken for the Southern Cross, but is larger, less brilliant and more symmetrical.

Faye's Comet Found by H. Faye, French amateur in 1843, recovered in 1850, from a prediction by Leverrier. Period 7.3 years, last seen in 1984.

F-Corona The outer part of the SOLAR CORONA. It is responsible for the greatest part of the brightness of the corona beyond about two solar radii. It consists of slow-moving particles of interplanetary dust. It extends for several million kilometres, and eventually merges with the interplanetary medium and with the particles causing the ZODIACAL LIGHT.

Ferguson, James (1710–1776). Scottish astronomer, lecturer and maker of clocks and orreries. Author of *Astronomy Explained on Sir Isaac Newton's Principles,* 1756, from which William HERSCHEL learned the subject. George III granted him a pension of £50 a year for astronomy.

Fermi Large crater on the Moon's far side; 20°S, 122°E.

F, G and K subdwarfs POPULATION II F, G and K dwarf stars with low metal abundances appearing below the normal main sequence in the colour-magnitude diagram. In fact they are not subluminous — they are merely more blue than dwarfs with normal metal content.

Field equations The equations of EIN-STEIN'S General Theory of RELATIVITY which describe the gravitational field in terms of the curvature of space.

Field stars Background or foreground stars in the same area of sky as, but not physically associated with, the object of interest.

Filar micrometer The filar micrometer is used for accurate visual measurement of the separation and position angle of two neighbouring astronomical objects, such as the components of a visual double star or a planet and its satellite. It is attached between the eyepiece and the end of the telescope tube. The view through the eyepiece shows at least three fine wires or filaments as well as the double star, all sharply in focus. The wires include one that may be termed "horizontal", with two "vertical" wires accurately perpendicular to the first, one of which is fixed and the other movable with a finely threaded screw of the highest quality. The position of the movable wire is read from a drum or vernier turned by the screw. The entire micrometer assemblage can be rotated around the optical axis of the telescope, its position and hence that of the "horizontal" wire, being read from a circular scale divided at least as finely as $0°.1$.

An observation consists of moving the telescope until one object is precisely at the intersection of the horizontal and fixed wire, rotating the micrometer assemblage until the companion star lies along the "horizontal" wire, and finally adjusting the micrometer screw until the movable wire bisects the companion. After reading the screw and circle readings the movable wire is run back to the fixed one and a further reading taken. The difference between the two screw readings gives the separation in terms of revolutions of the screw. This may be calibrated by observation of wide double stars of accurately known separation.

The zero point of the position angle circle is found by switching off the telescope drive and adjusting the horizontal wire until the star trails exactly along it, thus defining the east-west direction. In practice, spider webs are often used for the wires and they are faintly illuminated with a source of red light. *BW*.

Filters An optical filter is a piece of material that will transmit light of certain wavelengths while absorbing others. If a red object is viewed through a red filter it will appear bright and objects of other colours will appear darker.

Filters are used in planetary astronomy to good effect. If a light blue filter is used to observe JUPITER, the Red Spot will appear much darker than normal and show more contrast against the body of the planet. Using a red filter to view MARS will make the dark markings appear much darker than the body of the planet, which reflects red light.

The effect of filters when used in combination with photographic emulsions can be even more dramatic. A red-sensitive emulsion and red filter will record only red light. In this way the astronomer can completely isolate various colours; in this example, the light of hydrogen. Colour emulsions are not generally available in large photographic plates. By making separate exposures using red, green and blue filters with the appropriate emulsions, the astronomer can obtain a colour picture by combining the three black and white images in the darkroom.

A more complex form of filter is a monochromatic filter which will transmit only a single spectral line. Another form of filter is the interference filter, which can be designed to block out light pollution. *RWA[1]*.

Finlay's Comet Faint comet found by W.H.Finlay at Cape Town in 1886. He recovered it in 1893. Seen at 10 returns to 1981, period 7.0 years.

Fireballs When a solid fragment enters the atmosphere it is ablated, producing an intense point of light known as a meteor or shooting star. If the intensity of the light is brighter than the planets, it is known as a fireball. Major fireballs can be brighter than the full Moon and occasionally as bright as the Sun. The objects producing these brilliant fireballs are of two types, most being thought to consist of cometary material, a loose agglomeration of small particles held together by ice. The remainder consist of solid interplanetary material, possibly from the asteroidal belt. The fragments remaining after ablation reach the ground as meteorites. *HGM*.

First Point of Aries *See* CELESTIAL SPHERE, EQUINOX.

First Point of Libra *See* EQUINOX.

Fitzgerald contraction A shrinkage in the lengths of moving objects which was suggested in 1889 by the Irish physicist George Francis Fitzgerald (1851–1901) and, independently, by LORENTZ, to explain away the null result of the MICHELSON-MORLEY EXPERIMENT. Generally known as the "Lorentz-Fitzgerald contraction".

Fizeau, H. (1819–96). French physicist who succeeded in measuring the velocity of light with apparatus on Earth for the first time in 1849. He used a rapidly rotating toothed wheel which interrupted a fine beam of light before and after reflection from a distant mirror more than 5 miles (8km) away.

Flammarion, Camille (1842–1925). French astronomer, who specialized in observations of the planets and was a noted popularizer of astronomy; he was also the founder of the periodical *L'Astronomie*. In 1883 he established a private observatory at Juvisy-sur-Orge, where he worked until his death.

Flamsteed, John (1646–1719). First ASTRONOMER ROYAL, noted for his painstaking and accurate observations of the positions of the Sun, Moon and stars.

A sickly youth who, because of his health, never went to a university, he became nevertheless an accomplished amateur astronomer, so that his opinion was sought when the important question of finding longitude at sea by observing the position of the Moon among the fixed stars was considered by the British government. Flamsteed's recommendation that the positions of the Moon and stars needed redetermining with greater precision, led to the establishment of GREENWICH OBSERVATORY by King Charles II.

Appointed "our astronomical observator" in 1674, Flamsteed moved into the observatory as soon as it was completed in July 1675, but had to supply his own instruments and pay for any skilled assistance he needed.

At Greenwich Flamsteed continued to study the Moon's motion and that of the Sun, as well as to measure star positions using equipment fitted with sighting telescopes instead of the then customary "open" sights; this innovation provided substantially greater accuracy than had hitherto been obtainable.

Because of his desire to perfect his observations, Flamsteed was slow to publish his results. With an irascible disposition, partly due to continued ill-health, and an inability to get on with other people, it is not surprising that after continued and protracted delays, the Royal Society finally decided to publish Flamsteed's observations deposited with them under seal. Bitterly opposed to this, in later years Flamsteed bought up and destroyed most of the copies. His corrected observations were published posthumously,

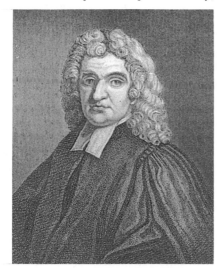

John Flamsteed: Astronomer Royal

appearing in 1725 as *Historia Cœlestis Britannica*. A companion *Atlas Cœlestis* appeared four years later. Both are monuments to Flamsteed's conscientious work and helped set the seal on Greenwich as a centre of excellence in observational astronomy. *CAR.*

Flare stars Among the intrinsically faintest stars in the solar neighbourhood, up to a distance of 25 parsecs, are found at least 444 RED DWARF stars of spectral class dM, with absolute visual magnitude 8 to 16 and age about 100 million years. Of these, 67 show EMISSION LINES of hydrogen and ionized calcium, apart from the characteristic dM-type titanium oxide bands, and are distinguished as the so-called dMe stars. About 34 of the dMe stars are presently known to exhibit flares. A stellar flare may be defined as a transient phenomenon involving the release of large quantities of energy in a non-periodic manner. During flares stellar emission lines become enhanced and a strong blue-ultraviolet continuum appears. The timescales of these energetic outbursts are typically between one minute and one hour with rise times usually of only a few seconds. The initial decline is usually equally rapid and followed

by a still-stand and a quasi-exponential decay. Small precursors or pre-flares are sometimes observed and occasionally slow rising flares are also seen. The stellar flare light is seen against the relatively faint light from the normal quiescent star, which has a cool photosphere at 2,800–3,600K, and at maximum optical brightness the larger flares may contribute more than 10 to 100 times the total output of the whole visible stellar hemisphere.

Since the photographic discovery by HERTZSPRUNG in 1924 of DH Carinae, the first flare star, the number of such objects found has been small. The earlier discoveries were usually by chance, for example, during routine photographic PARALLAX measurements or were recorded unexpectedly on spectral plates being exposed for an entirely different purpose. Good examples of this class of object, and their apparent visual magnitudes, are UV CETI (13.0), AD Leonis (9.4), EV Lacertae (10.1), PROXIMA CENTAURI (11.0) and BY Draconis (8.2). Binarity appears to be a common feature, notably of the BY Draconis variety, which for several years were studied separately on account of their quasi-periodic variability of very small visual amplitude (0.05 to 0.3 magnitude) attributed to starspots passing across the visible hemisphere as the star rotates. These rotational periods for the few well-observed objects are from 0.5 to 10 days. An extremely large flare, such as that observed by Andrews and Perrott at Armagh, Lovell at Jodrell Bank and Kunkel at Cerro Tololo in YZ Canis Minoris on January 19 1969, was

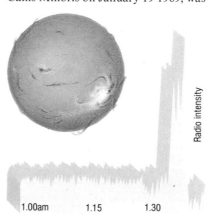

Radio intensity

1.00am   1.15   1.30

Flare stars: an impression of UV Ceti

of amplitude 8.5 magnitudes in the ultraviolet and 1.7 magnitudes in the visual and lasted more than 4 hours with an estimated energy output of 7×10^{27} Joules in the optical region of the electromagnetic spectrum alone.

Classical photographic spectroscopy has shown that during flares the Balmer emission lines of hydrogen become greatly enhanced so do many other lines, for example of helium, calcium and other metallic elements. Modern techniques using fast pulse-counting photometry, spectroscopy with sensitive CCD (CHARGE-COUPLED DEVICE) detectors, as well as observations from above the Earth's atmosphere, such as with the INTERNATIONAL ULTRAVIOLET EXPLORER (IUE) and EXOSAT X-ray satellites, have recently revealed that flare stars are amongst the most active stellar objects in our Galaxy, and that flare energies are often 1,000 times their solar counterparts.

Collaborative international observations are an essential part of present-day flare star research where large numbers of hours are devoted to a few selected stars, and where it is necessary to collect simultaneous spectral, optical, radio and X-ray data. In the far-ultraviolet the IUE satellite registered strong emission enhancements of doubly-ionized magnesium, the carbon and silicon ionic species up to CIV and SiIV and also the high-temperature nitrogen line, NV, reminiscent of the solar outer atmosphere. Furthermore, micro-flaring in a few dMe stars has been observed in both optical spectra and, simultaneously, in coronal X-rays which suggests that some stars are frequently active at a timescale of a few seconds to great heights above the stellar surface. One school of thought, that pursued by the Armagh-Catania-JILA (Boulder) group, is of a solar-type scenario for flare stars, that of a spotted star with overlying active atmospheric layers corresponding to an extremely wide range of observed temperatures, from 3,000° in the spots, through 10,000° to 200,000° in the chromosphere and transition region to more than one million degrees in the corona. The flare mechanism, however, is poorly understood, although magnetic fields associated with differential stellar rotation (*see* ROTATING STARS) are probably the cause. The low mass of the nearby dMe stars, many less than one tenth of a solar mass, and their relatively young age, means that following gravitational collapse from the primordial material the steady thermonuclear burning in the interior which maintains normal MAIN-SEQUENCE stars with a radiative core may not have established itself. Wholly convective interiors may occur in the youngest stars of the lowest mass, and

F

F

flaring appears to be a consequence of these instabilities, linked with complex magnetic fields. Difficult theoretical problems need to be solved as to the storage of energy prior to flares and their role in the heating of the intense stellar coronæ, and also as to the source of energy producing the flares.

Closely similar to the nearby flare stars are the Orion variables, RW Aurigæ and T Tauri-related type stars in the region of the ORION NEBULA (distance 460–500 parsecs), the PLEIADES (130 parsecs) and most young galactic clusters (age a few million years and less). They abound in regions rich in dust and gas where star formation is still in progress, where young stellar aggregates and GIANT MOLECULAR CLOUDS with stars still in their dusty cocoons are found. Due to their generally being very faint (apparent visual magnitude 12 to 20) these flare stars have almost invariably been discovered with large SCHMIDT TELESCOPES using multiple-exposure techniques, and those known today required several thousand hours of patient observations and searching of photographic plates. Haro at Tonantzintla and Rosino at Asiago pioneered these techniques. Flares of amplitude 4 to 6 magnitudes in the ultraviolet are fairly common amongst the fainter of these stars, although they are not often seen to flare repeatedly, a characteristic perhaps not wholly due to the difficulties of observation. There are more than 400 such flare stars in and around the Orion Nebula and approximately the same number in the Pleiades, although a small fraction of these may be foreground dMe stars. Flares have also been observed in Praesepe, the Hyades, and the Cygnus and Taurus dark clouds, NGC2264 and 7000. These abundant objects occur at earlier spectral types, dK0 to dM6, and were consequently differentiated by the name "flash stars" by earlier astronomers. Although little is known still about these stars, there is good evidence that flares occur in quite a variety of objects including, for example, the more massive RS Canes Venaticorum variables, which are binaries with G- or K- type subgiant components. The term "flare star" is therefore becoming outmoded by its simplicity, and the word "flare" itself will alsobecome more meaningful in the wider context of active regions on stars of differing ages and with differing rotational velocities and over a broader range of mass, spectral type and cosmic environment. *ADA.*

Flash spectrum A spectrum of the Sun is produced by concentrating its light onto the slit of a spectrograph, (*see* SPECTROSCOPE), collimating the emerging rays into a parallel beam, dispersing the beam using a prism or diffraction grating and then focusing it onto a photographic film. Each colour appears as an image of the slit and forms a spectral "line".

For stellar spectrography it is not always necessary to have the slit and collimator. Stars are so distant that their light is already collimated and their images are so small that they can be drawn into narrow spectral lines simply by moving the telescope during the exposure in the appropriate direction.

A similar technique is used in the study of the SOLAR CHROMOSPHERE. Just before and after the total phase of a SOLAR ECLIPSE the Sun appears as a slim crescent that can be used as if it were a narrow curved slit. After the dispersion the spectrum appears as a series of crescents, but it lasts for just a few seconds and is known as the "flash spectrum".

Chromospheric light is very faint compared with the brilliance of the photosphere and is usually seen in absorption, but during an eclipse the photosphere is covered and we see bright curved emission arcs. At the beginning of an eclipse the crescents become narrower and shorter as the Moon covers the Sun, but, at any single instant, the crescents are of different lengths because the colour-emitting processes occur at differing altitudes above the photosphere. Colours produced low down in the chromosphere have shorter arcs because this zone is deeper into the eclipse. Higher altitude processes give rise to longer arcs which have yet to be covered by the Moon. The length of arc is related to height above the photosphere and thus the physical nature of this region can be investigated. *RCM.*

Fleming, Wilhelmina (1857–1911). Scottish astronomer who spent much of her career in America. She joined the Harvard Observatory staff, and was in charge of the famous DRAPER CATALOGUE; she discovered 222 variable stars and 10 novae.

Flexus Low curved ridge with scalloped pattern; several are recorded on Jupiter's satellite Europa.

Flickering In many close binary stars, especially novae and dwarf novae, rapid variations in brightness over times of minutes, or even as short as seconds, occur, and are called flickering.

Flora (Asteroid 8). Diameter 94 miles (151km), magnitude 8.7.

Focal length The distance between the centre of a lens or mirror and its focal point or focus.

Focal plane A plane through the focal point at right angles to the optical axis in which the image of a distant object will be formed.

Focal ratio The ratio between the focal length of a lens or mirror and its effective aperture. The focal ratio is often called the f-number.

Focus The point at which parallel rays near the optical axis are refracted by a LENS or lenses, or reflected by a mirror or mirrors, or a radio dish. In practice, the focus is a plane or curved surface. *See also* FOCAL LENGTH.

Fomalhaut Alpha Piscis Austrini. A3 dwarf, visual magnitude 1.2, distance 7pc. IRAS found a possible proto-planetary system around it. *See also* VEGA.

Fontana, Francisco (1585–1656). Italian amateur astronomer; one of the earliest telescopic observers.

Forbidden lines Emission lines found in the spectra of gaseous nebulae which are absent in terrestrial spectra because the physical conditions cannot be satisfied in the terrestrial laboratory.

Emission lines result from transitions of an atom from a high energy level to a lower level, often as a result of collisions, emitting a photon of characteristic wavelength in the process.

If gas pressure and density are low as in gaseous nebulae, collisions are rare and weak. Only atoms with an easily accessible ionized state will be excited, typically nitrogen, oxygen and neon. These atoms then decay radiatively, emitting forbidden lines.

Forbush effect The reduction in cosmic ray intensity at the Earth due to a magnetic storm. First observed and interpreted by Scott Forbush in 1937.

Fornax (the Furnace). A southern constellation, adjoining Eridanus. The only star above magnitude 4 is Alpha (3.87), but the constellation includes an important cluster of galaxies.

47 Tucanæ (NGC104). The second brightest globular cluster at magnitude 4.0. 30′ in diameter with a sharp concentration toward the centre. It is 14,000 light-years away and about 120 light-years across. Appears beside the Small MAGELLANIC CLOUD: Epoch 2000 coordinates 0^h24^m0, $-72°04′$.

Foucault, Jean Bernard Léon (1819–68). French physicist; demonstrated with a long pendulum the Earth's rotation in

Jean Foucault: Earth's rotation

1851, and invented the Foucault knife-edge test for mirrors and lenses. He developed the heliostat and silver coatings for mirrors, measured the speed of light, and demonstrated that it travels more slowly in water than in air.

Fracastorius (or Fracastor). Lunar bay 60 miles (97km) across, 21°S, 33°E, leading off the Mare Nectaris.

Fractional method A way of visually estimating a whole number to represent the difference of two stars of constant brightness and of estimating by what fraction a variable differs from each comparison star.

Fra Mauro Low-walled lunar crater 50 miles (81km) in diameter, 6°S, 17°N, in the Mare Nubium; it makes a group with Bonpland and Parry. Apollo 14 landed in this area.

Franklin-Adams camera A wide-angle (15°×15°, 10in/25cm objective) telescope used by J Franklin-Adams (1843–1912) and assistants to make a photographic survey of the whole sky. From its publication (1913–14) this was the standard photographic atlas for more than 50 years. Now located near Johannesburg, the camera was used to

study HALLEY'S COMET in 1910 and again in 1985-86.

Fraunhofer, Joseph von (1787–1826). German optician and physicist, son of a glazier. In 1806 he joined the Munich optical institute of Reichenbach, Utzschneider and Leibherr. He took charge of their new institute at Benediktbeuern in 1809 and became a partner in the firm.

His research into what came to be known as FRAUNHOFER LINES led to the development of ASTRONOMICAL SPECTROSCOPY, a crucial factor in the development of astrophysics.

Inventor of the "German" EQUATORIAL MOUNTING, he produced many superb telescopes, particularly the Dorpat (now Tartu) 9.5in (24cm) refractor for Wilhelm Struve and the Königsberg 6.25in (15.8cm) heliometer with which BESSEL measured the first stellar parallax. *HDH.*

Fraunhofer lines Dark lines in the spectrum of sunlight. Caused by the atomic absorption of specific wavelengths in the upper, cooler, parts of the Sun's atmosphere.

Fraunhofer lines: dark lines in the spectrum

Free fall The state of falling freely in a gravitational field, ie, experiencing no resistance to the acceleration due to gravity. A freely-falling body follows a path determined by the combination of its velocity and the acceleration due to gravity; this path may be a straight line, circle, ellipse, parabola or hyperbola. A satellite in a circular orbit is subject to a constant acceleration toward the central body, but its transverse motion ensures that it gets no closer to that body. The gravitational acceleration continuously alters the direction of the satellite's motion, but not its speed. A freely-falling body experiences no sensation of weight. *IKN.*

Free-free transition Radiation emitted when a free electron is decelerated in the vicinity of an ion without being captured. Also known as Bremsstrahlung ("deceleration radiation").

Frequency The number of oscillations of a vibrating system per unit of time. For an electromagnetic wave, the frequency is the number of wave crests passing a given point per unit time.

Freyja Montes Mountain block on the north flank of Ishtar Terra (Venus) and rising to 4 miles (6.5km) above datum; 72°N, 310°W. An area of high radar backscatter and linear structure.

Friedmann, Alexsandr A. (1888–1925). Russian mathematician who in 1922 worked out the basic details of the open and closed models of the EXPANDING UNIVERSE.

Friendship 7 First United States orbital manned spacecraft.

Friendship 7: with escape tower in place

F ring Ring in Saturn's system outside Ring A; mean distance from Saturn's centre, 87,360 miles (140,600km). Its "shepherd satellites" are Prometheus and Pandora.

Fringes Patterns of light and dark produced by the interference of light waves which are alternately in and out of phase with each other. *See also* INTERFERENCE PRINCIPLE.

F stars Stars of spectral type F have surface temperatures between 6,100K and 7,400K on the MAIN SEQUENCE, but the giants and supergiants are about 300K cooler. F dwarfs range in mass from 1.2 to 1.7 times that of the Sun, but most F giants and supergiants have evolved from considerably higher mass stars. The great majority of F stars have chemical compositions very similar to the Sun's. The cool end of the Peculiar A stars, however, just reaches into the

F

F

F star range, producing Fp stars, and the oldest (POPULATION II) F dwarf stars have low metal abundances, which give them an F subdwarf classification. The coolest F stars are slow rotators, but the hottest F dwarfs rotate as fast as the A stars.

Bright examples: Canopus F0Ib, Procyon F5 IV. *BW.*

Fundamental stars A measurement of the position of a star (ie, its RIGHT ASCENSION and DECLINATION) is called "absolute" if it does not make use of the already known positions of other stars. Absolute positions are obtained from a transit circle or prism astrolabe and an accurate clock. Many observatories have specialized in the determination of absolute positions, eg, the Royal Greenwich Observatory, the US Naval Observatory, the Royal Observatory, Cape of Good Hope and Pulkova Observatory. A comparison of the independent measurements made at these observatories shows the presence of small systematic errors (as well as the inevitable small random errors). From a careful analysis, and elimination as far as possible, of the systematic differences the best possible star positions are derived. The stars for which this has been done are called fundamental stars and they are published in a fundamental catalogue.

The first fundamental catalogue was prepared by A.Auwers and published in two parts in 1879 and 1883. Since that time a series of fundamental catalogues have been produced in Germany, the Neuer Fundamentalkatalog in 1907, a third one (FK3) in 1938 and the FK4 in 1963, which is considered to be the best available. In the United States two important fundamental catalogues have been published — the General Catalogue by B.Boss in 1937 and the N30 in 1952.

The system of fundamental stars provides the best available approximation to an inertial frame of reference. Positions of all other celestial objects can be measured relative to this system. However, uncertainties in the proper motions of the fundamental stars limit the accuracy with which the method can be applied. Consequently, there is underway a major programme for determination of absolute positions of extragalactic objects which have negligible proper motions, using both optical and radio observations. These will define an improved system of fundamental positions. *BW. See also* related entries on stellar positions, eg, CELESTIAL SPHERE.

FU Orionis Type star for a small group of young stars which have been seen to flare by a large factor to a permanent bright state. Many young stars may pass through this phase.

Furnerius Lunar crater 80 miles (129km) in diameter, 36°S, 60°E, in the Petavius chain; it has irregular walls.

Fusion Nuclear fusion is a process in which two light atomic nuclei join together to form the nucleus of a heavier atom. In the very high temperatures found in the core of the Sun (about 15 million K) two hydrogen nuclei are fused together to give a deuterium nucleus, plus a positron and a neutrino. Later stages in this fusion process eventually produce the nucleus of the gas helium. Such processes are called thermonuclear reactions. With light elements, nuclear fusion releases vast amounts of energy. The fusion of elements up to the iron-peak results in the production of energy. The fusion of elements heavier than iron requires an input of energy to keep the reaction going. Fusion is the energy-producing process in stars. The formation of elements up to iron by fusion is called nucleosynthesis. *JWM. See also* NUCLEAR REACTIONS.

Gagarin, Yuri (1934-1968). Russian cosmonaut who became the first man in space, when he made a full orbit of the

Yuri Gagarin: first man in space

Earth in Vostok 1 (April 12 1961). He was killed during a test flight of a jet-aircraft in 1968.

Galactic centre *See* SAGITTARIUS A.

Galactic clusters Open star-clusters in our Galaxy.

Galactic co-ordinates REFERENCE SYSTEM based on the plane of the GALAXY, using galactic latitude and longitude. The north galactic pole is at R.A. 12h 49m, Dec + 27° 24m (1950.0). The zero of galactic longitude is the direction of the galactic centre, R.A. 17h 42.4m, Dec. − 28° 55′ (1950.0). The galactic longitude of the north celestial pole is 123°.

Galactic halo The spheroidal distribution of old stars and globular clusters which surround the disk and nucleus of the GALAXY. There appears also to be a much more extensive halo of "dark matter" which extends out to a radius of at least 200,000 light-years.

Galactic magnetic field An interesting and important property of the Galaxy is a tangled magnetic field of average strength locally about 3×10^{-6} gauss (ie, about 10^{-5} of the Earth's field at the surface). The direction is roughly along the spiral arms and the energy content is about the same as that of starlight and gas motions. The field plays an important role in the condensation of gas clouds to form stars, although the exact details are not understood; similarly, the origin of the field is a matter of some conjecture, although dynamo action associated with galactic rotation is a strong candidate. *AWW.*

Galaxies, classification *See* CLASSIFICATION OF GALAXIES.

Galaxies, external *See* EXTERNAL GALAXIES.

Galaxy, The Those who have gazed upon the sky from the northern hemisphere know the faint, luminous band that meanders from the deep south in Sagittarius, glides through Aquila and Cygnus, embraces the W of Cassiopeia, then fades gradually past Perseus and Orion to dip once more below the southern horizon in Puppis. It has been known since time immemorial. In every mythology it has a place — a river of light, a sacred road, pathway of the spirits. To us it is the Milky Way.

Those who have viewed it only from the northern latitudes have seen but a pale version of the Milky Way. In the southern hemisphere it is far brighter, far more spectacular. From Puppis it flows into Carina, and there it is richly textured, and bright enough even to be glimpsed from within large cities. Continuing bright it sweeps past the stars of Crux, where a black scoop has been removed from it by an intervening cloud of dusty gas, the COAL SACK. From there a narrow dark streak bisects the Milky Way as it passes through the giant constellation of Centaurus, swells and brightens through the tail of Scorpius, and enters Sagittarius. When seen against a dark sky, and when it rides nearly overhead, the Milky Way in Sagittarius reaches its widest, more than 30° in breadth, and still split by a dark central band. That band continues and the Milky Way narrows northwards through Aquila to Cygnus. Northern hemisphere dwellers have termed this dark band the Great Rift, but its full extent, one-third of the entire sky, can be appreciated only from the south.

It is well known that at the eyepiece of a telescope the Milky Way is revealed to be not a luminous band but a vast huddle of stars, so numerous that a human lifetime would be insufficient to count them all. Stars in such profusion exist in only one type of object — a galaxy. The Milky Way is a galaxy, and the very fact that it completely circles the sky clearly indicates that we live within it. It is our Galaxy, and to distinguish it from all others it is honoured with a capital letter.

What type of galaxy is ours? External galaxies basically have three forms — spiral, elliptical and irregular. The Milky Way is really quite symmetrical about the constellation Sagittarius, so we can eliminate an irregular classification. Elliptical galaxies are almost spherical blobs of stars, so that if we lived inside one we would see its constituent stars in every direction. The narrowness of the band of the Milky Way shows that our Galaxy is basically flat; as such it must be a spiral galaxy.

In fact, when viewed from the southern hemisphere, a spiral classification is readily appreciated. We lie among the spiral arms, so that at least one wraps round outside us like a cloak, while others curl between us and the centre. And, like other spiral galaxies, ours contains bands of opaque dust clouds along the spiral arms, and a central bulge like a miniature elliptical galaxy within the orbit of the arms. That bulge is in Sagittarius, which is why the Milky Way widens there. In front of

the bulge passes a prominent spiral arm that produces the great brightness of the Centaurus-Crux-Carina region. This arm curls away from us in Carina, where, seen effectively end-on, a great accumulation of stars and nebulæ causes the extreme brightness and complexity. Immediately outside this arm,

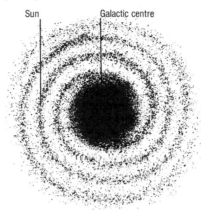

Galaxy: plan view of the Milky Way

and therefore superimposed in front as we view it, is the band of gas and dust, a region where stars have yet to form, that causes the dark rift from Cygnus to Crux. Beyond Carina we see the fainter arm that wraps itself round outside the Sun. It crosses the northern sky, gaining brightness as we follow it gradually inward toward Cygnus.

Although the trained eye easily sees these features, it took a long time for the Galaxy's structure to be deduced. Even today, the details remain contentious, for instance of how many spiral arms it has. Radio astronomy has been of paramount importance. In 1931 Karl Jansky discovered radio emission from Sagittarius, and before long the first maps of the radio sky were made. Apart from a few individual spots of radio emission, which were later to prove of immense importance, these maps showed a band exactly following the Milky Way. The mechanisms producing visible light and radio emission are grossly different, and very few objects liberate both in comparable amounts. Nonetheless, the Galaxy contains sources of both in a well-mixed soup.

In the northern sky the radio emission was found weak, but it rose to a crescendo in Sagittarius. The difference between the radio and the visible Milky Way was, quite simply, that the radio waves passed unimpeded through the dark clouds. They showed what the Galaxy would be like if we could see all the stars. In particular, they focused our attention on the very centre of the

Galaxy. This theme is enlarged on under the heading SAGITTARIUS A.

As radio astronomy matured it began to make more sophisticated observations. One of the most important of these was the study of radiation by hydrogen atoms at a wavelength of 21cm. Observations at this wavelength could pick out the gas clouds associated with spiral arms, revealing them irrespective of whether optically opaque clouds intervened. The radio astronomers put together a map of the Galaxy revealing its spiral arms. In practice the Galaxy appeared highly complex, and spiral arms could be sketched through the map only with the eye of faith. More recently, however, similar observations made from Australia, of the less ubiquitous gas carbon monoxide, have painted a much clearer canvas which suggests that there are four arms to our Galaxy.

Our view from within the disk of the spiral arms can never be adequate to disentangle the full structure of the Galaxy. What follows, therefore, is a description based in part on studies of our own, and in part on what we know of other spiral galaxies.

About 100,000 million stars populate the Galaxy, which is somewhat above average in size. Put in perspective, that is the number of grains of rice that could be packed into an average cathedral. The Galaxy is, however, mostly empty space: if we were to build a model in which stars were scaled down to the size of rice grains, then to match the Sun's environment we should have to scatter them at the rate of a handful across an area the size of the British Isles. The diameter of our scale model would be 240,000 miles (400,000km), the separation of the Earth and Moon. In astronomical terms our Galaxy is nearly 100,000 light-years across, though its outer boundary is not well defined, and we live 25,000 light-years from the centre.

The bulge of stars in the central part comprises those which formed relatively early in its history, although stars are still forming in some regions right in to the centre. These stars have evolved away from the MAIN SEQUENCE, many becoming cool giants, while those which remain main sequence stars are small and also cool. As a result their average colour is yellow, and they are collectively referred to as POPULATION II stars. By contrast the disk that spreads even beyond the Sun's orbit, and contains the spiral arms, is a site of star formation, as seen in the ORION NEBULA and

G

G

similar locations. The integrated light of young stars is dominated by the hottest, brightest specimens, and is therefore blue. We call these POPULATION I stars. Interspersed among the stars, and especially between the curving spiral arms, are dense clouds which will continue to fuel new stars for thousands of millions of years.

Just like planets round the Sun, the stars of our Galaxy orbit the central bulge. Each star does so at a speed dictated by the mass of material (stars, gas, and any unseen objects such as BLACK HOLES) within its orbit. The stellar orbits are not quite circular, so stars shuffle relative to one another. In the bulge some orbits are quite elliptical while others swing up and down in complex gyrations which may be visualized by drawing a sine wave around the surface of a slightly squashed cylinder. Our Sun circles the centre at about 160 miles (250km) per second, taking some 200 million years per orbit.

If we measure the motions of the outermost stars of the Galaxy, we can calculate its total mass. This measurement is only partially feasible, because we can determine velocities only along the line of sight (using the DOPPLER EFFECT) and not across it. A further complication is our own motion, itself not very accurately known. Various statistical analyses of the measured motions seem to suggest that the Galaxy weighs nearly one million million times as much as the Sun, which is nearly ten times the mass of the visible constituents. Whether the extra mass really exists or is a statistical peculiarity, and what material is responsible for it if it is real, remain contentious research topics. *DAA. See also* CLASSIFICATION OF GALAXIES and related entries.

Galilei, Galileo (1564–1642). Italian physicist and astronomer, noted especially for his research in physics, his pioneering telescopic observations of the heavens, and his championing the Sun-centred or heliocentric theory of COPERNICUS.

Eldest of seven children and educated first at a Jesuit school, Galileo later went on to Pisa university, where, at his father's insistence, he was entered as a medical student. But Galileo's interest in medicine was marginal; he preferred physics and mathematics, and though opposed by his father, studied both at Pisa and afterwards privately. He showed a flair for the subjects, making great progress in

studies of Euclid and ARCHIMEDES, and discovering the isochronism of the pendulum by noting that, whether long or short, its swings always occupy the same interval of time.

Galileo's father, Vincenzio, was a professional musician of advanced and independent views, and his eldest son seems to have inherited much of his father's scorn for outdated authority. Thus when in 1589 he became professor of mathematics at Pisa, Galileo soon set about attacking the teaching of physics at the university because this was still based on old-fashioned Aristotelian ideas. Whether or not he actually dropped bodies of various materials from the Leaning Tower is uncertain,

Galilei Galileo: tried in Rome

but if he did his aim would have been to demonstrate the error of Aristotle's view that bodies of different weights fell to the ground at different speeds. He also wrote a tract, *De Motu,* attacking Aristotle's laws of motion.

In 1591 Vincenzio Galilei died and support of the family then became Galileo's responsibility. Poorly paid at Pisa, with little likelihood of the renewal of his post at the end of a three-year period because of his anti-Aristotelian opinions, his radical views about academic behaviour, and his severe criticism of Giovanni de' Medici's scheme for dredging the nearby harbour at Leghorn, he managed through patronage to obtain a similar chair at the university of Padua. Here the intellectual atmosphere was more to Galileo's taste, and his mathematical approach to physics developed. On receipt of *Mysterium Cosmographicum* from KEPLER, Galileo expressed his private belief in the heliocentric universe, though he did not

then follow Kepler's advice to publicize his support.

At Padua Galileo began supplementing his stipend by manufacturing a calculating device. Then in 1604 the appearance of a "new star" (a supernova) led him to take part in attacks on Aristotle's view that the heavens were unchangeable.

In October 1608 Hans Lippershey, a Dutch spectacle-maker, applied to the Netherlands government for a patent for a telescope, and news of the device reached Galileo in July 1609. From descriptions of such an instrument which had appeared in Padua for a short time, Galileo worked out its principle and constructed a sample himself. He soon improved the design, and by late August, brought to Venice one which magnified nine times. For this he was given life tenure of his chair at Padua and a large increase of salary.

However, the document confirming this did not do all he had understood it would, and he began seeking a position in Tuscany, the region in which he had been born. By the summer of 1610 he was with the Tuscan court in Florence.

By the time he got there, Galileo had already been busy observing the heavens using a still further improved telescope magnifying 30 times. His results were revolutionary and appeared in his *Sidereus Nuncius* ("The Starry Messenger"), which was published in March 1610, only two months after he had begun his observing programme. They showed the Milky Way to be composed of separate stars, the presence of valleys, mountains and craters on the Moon, and Jupiter to be accompanied by four orbiting satellites. The first demonstrated that there was more to be discovered about the universe than was known in ancient Greece, a result at odds with received opinion. The second that Aristotle was wrong in claiming that celestial bodies were unblemished globes different from any earthly body. The third seemed utterly to disprove the argument that the Moon would be left behind if the Earth orbited the Sun. Then late in the summer of 1610 Galileo discovered that Venus presents phases similar to those of the Moon. This again was evidence against ancient teaching.

The observations convinced him that the Copernican theory was correct, though in fact they did allow of a different interpretation. For instance, the scheme proposed by Tycho BRAHE in which the planets orbited the Sun, but with the Sun itself orbiting a stationary Earth at the centre of the universe, would also fit Galileo's results.

A stern critic and a past master at heaping ridicule on his opponents, Galileo soon made enemies by his support of Copernican views. Though contrary to biblical teaching, they had been accepted by many churchmen as a purely mathematical way of accounting for observed planetary motion; Galileo, however, believed firmly in their physical reality. He summed up his attitude in the dictum of Cardinal Baronio that the Bible teaches the way to go to heaven, not the way the heavens go. Nevertheless, the Church was concerned to ensure that the Bible, as the divine word of God, should not be found untrue in any respect, and it was with this in mind that it placed the Copernican text *De Revolutionibus* on the *Index of Prohibited Books* and admonished Galileo not to promote such ideas publicly.

When his friend Maffeo Barberini, a trained mathematician, became Pope in 1623 Galileo hoped for a more liberal outlook. He did get permission to write about heliocentric ideas, but only with the proviso that his arguments in their favour should not be conclusive. Unfortunately his book, *Dialogue on the Two Chief World Systems*, put the Aristotelian case and the Pope's views about the interpretation of observation, into the mouth of Simplicius, an Aristotelian commentator from the 6th century. Galileo's enemies persuaded Barberini that he had been purposely ridiculed, with the result that Galileo was summoned to Rome to be tried by the Inquisition. Some exaggerations of Galileo's original admonition were presented to the court, but he was able to defend himself by producing the original letter of instruction. His sentence was still severe, though not as drastic as it might otherwise have been. He was forced to recant publicly his belief in the Copernican theory, and forbidden to concern himself any more with astronomical problems. Then he was sent home to live under permanent house arrest.

This condemnation effectively stifled the study of the heliocentric theory in Italy, and research on it passed to those in Protestant countries.

In spite of his sentence, Galileo continued his scientific work, though now he concentrated on physics. Welding together the subjects of statics — the study of bodies at rest — and dynamics — a study of bodies in motion — he also showed how to create a vacuum experimentally. So although prevented from doing astronomical research, he still managed to attack the Aristotelian outlook by demolishing its basic physics. However, since no books by Galileo could be published in Italy, the manuscript reporting his results was smuggled to the Netherlands and issued at Leyden in 1638. The book, *Discourses and Mathematical Demonstrations Concerning Two New Sciences* laid the foundations of mathematical physics. *CAR.*

Galileo probe The Galileo project is a NASA mission to orbit the planet Jupiter and to send an instrumented probe into the giant planet's atmosphere. The mission is named after the Italian astronomer who discovered the four main satellites of Jupiter, Io, Europa, Ganymede and Callisto. The 2¼-tonne (2,250kg) spacecraft will be launched by shuttle and use the inertial

Deceleration module
aft cover

Descent module

Deceleration module
aeroshell

Galileo probe: mission to Jupiter

upper stage to provide the additional thrust necessary for interplanetary space missions.

The probe will separate from the orbiter about 150 days before the arrival at Jupiter and it will then follow an independent path to the planet to provide the first direct sampling of the planetary atmosphere. As the probe strikes, at great speed, the upper layers of the atmosphere, it will have to be rapidly decelerated before commencing the *in situ* measurements of the physical and chemical properties of the atmosphere. A parachute is used to slow the probe in the final portion of the descent when the measurements are made and related to the orbiter for transmission to the Earth. The probe is expected to make measurements for about 60 minutes, when it will have reached a pressure of about 20 atmospheres.

The orbiter incorporates a new dual-spin design. This enables some of the instruments, such as imaging, to operate on a three-axis stablized platform while others, such as magnetometers can be mounted on a section that will spin regularly in a circular motion. Galileo uses a new 16ft (4.8m) furlable antenna and radioisotope thermoelectric generators to provide the power for the spacecraft and instruments in the outer reaches of the Solar System. A few hours before the probe entry, the orbiter will pass within 620 miles (1,000km) of Io. This will further slow the spacecraft, which will then be in an elliptical orbit around Jupiter. During at least one orbit, the spacecraft will fly through the Jovian magnetotail, which has not been observed by the Pioneer 10, 11 or Voyager spacecraft and cannot be observed from the Earth. The orbiter will complete 11 different orbits of Jupiter in 20 months while observing the satellites, rings and the planet. *GEH.*

Galle, Johann Gottfried (1812–1910). German astronomer. First to recognize Neptune (1846). Studied comets and meteors and detected Saturn's C-ring (1838). Determined solar parallax by observation of asteroid Flora (1873).

Gamma ray astronomy The latest, and essentially the last, part of the electromagnetic spectrum to be studied by astronomers comprises that of the gamma radiation. Gamma rays can be regarded as ultra-penetrating X-rays and conventionally they have quantum energies greater than the rest-mass of the electron, 0.51MeV (510keV). Their extreme penetration makes gamma-rays of considerable value in probing regions of the Galaxy and beyond from which other radiations are absorbed.

Gamma rays are produced by a variety of processes, all associated with energetic mechanisms in the cosmos. Indeed, the association with COSMIC RAYS is very close — cosmic rays were first thought to comprise a form of gamma radiation, they have to be detected against a background due to

G

cosmic rays, and many are produced by the interactions of cosmic rays with the gas in the interstellar medium.

Balloon experiments gave tantalizing evidence for the tiny flux of cosmic gamma rays (about 10^{-6} of the flux of cosmic ray particles), but the subject did not really start until March 1967 with the launch of the American satellite OSO III. Measurements were made which showed a broad peak towards the Galactic Centre at energies above 100MeV.

The bulk of the data available at present has come from the gamma ray satellites SAS II and COS B. SAS II was launched by NASA on 15th November 1972 but failed after only six months in orbit because of faults in a power supply. COS B — a European collaborative venture — had a longer life, 1975–82, but suffered from rather restricting background problems. Both satellites covered the "middle" energy region from some tens of MeV to several GeV (approximately $3 \times 10^7 - 3 \times 10^9$eV).

The gamma ray view of the heavens in this energy range is not too dissimilar from that of starlight in that there is a general concentration of emission in the galactic plane and an increase in the general direction of the galactic centre. However, the poor angular resolution of the gamma ray telescopes (several degrees uncertainty in individual arrival directions) has led to many problems, particularly with the nature of the "hot spots" of emission. Some of these are genuine discrete sources in that they can be identified by their time profile with pulsars, specifically the CRAB (PSR 0531+21) and VELA (PSR 0833−45). Another ("Geminga") is so strong and away from confusing alternatives that it too is certainly discrete — perhaps a pulsar. A small signal has been seen from the QUASAR 3C-273. Of the other 25 in total sources claimed by COS B some are probably due to pulsars undetected in radio, but others may well be due to dense clouds of molecular gas irradiated by cosmic rays, the gamma rays being produced by the processes outlined in the accompanying diagram.

It is generally agreed that most of the galactic radiation is produced by cosmic rays and attempts have been made by using information about the way in which the gas is distributed in the Galaxy to tell how the cosmic ray intensity varies over the Galaxy. It is generally agreed that the lower energy gamma rays (below about 100MeV) are produced by electrons and that the electron intensity is higher in the inner Galaxy than locally and lower in the outer Galaxy. The manner in which the important proton component, which gives most of the higher energy gamma rays, varies is the subject of heated

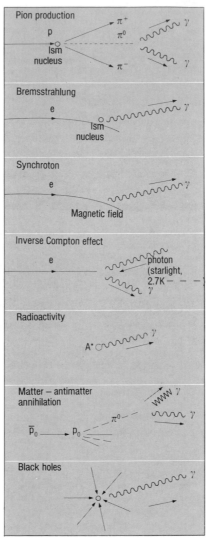

Gamma ray astronomy: gamma ray sources

debate. The answer may come from the forthcoming (1988 +) "Gamma Ray Observatory".

SAS II appeared to find a weak extragalactic gamma ray flux which had an interesting bump in its energy spectrum at several MeV. Explanations abound; most likely it owes much to the sum of contributions from individual galaxies, particularly Seyferts and quasars, which are not resolved by the detectors.

Turning to lower energies, there is considerable interest in gamma ray lines and gamma ray bursts. The electron-positron annihilation at 511keV has been observed from the galactic centre region. Apart from its considerable strength it is variable on the scale of a year; a possible explanation is that there is a BLACK HOLE at the galactic centre which has somewhat irregular "meals" of interstellar gas. A line at 1,809keV, identified as due to aluminium-26, has been observed and appears to be coming from a wide region of the Galaxy. Supernova ejecta should contain this radioactive nucleus but the observed line seems to be too strong; novæ may also be contributing.

Lines have also been detected from specific objects, notably the Crab Nebula and the remarkable object SS433.

Gamma ray bursts were discovered, quite surprisingly, by satellite-borne detectors meant to monitor violations of the nuclear explosion test ban treaty. The discovery itself was made in 1967. The bulk of the energy detected is usually in the hard X-ray region (above about 20keV), but the bursts are nevertheless usually called "gamma ray bursts". Several hundred GRB have been detected and it is often remarked that "no two are alike". Most bursts have rise-times in the range 10-1,000 milliseconds and durations of a few tens of ms to a few hundred seconds.

It is generally thought that GRB come from highly magnetized neutron stars but the exact mechanism is unclear; perhaps they are due to interstellar material falling on to the neutron star. Satellite studies are continuing and searches are being made for the association of GRB with known celestial objects. A few associations have been claimed already; it is possible that GRB 790305 is associated with the supernova remnant N49 in the Large Magellanic Cloud. A few other GRB may be associated with optical transients seen in archival photographic plates and with known X-ray sources, but the bulk has not yet been identified.

Moving now to the highest energy we come to a very lively area, that of VHE (very high energy: $10^{11} - 10^{12}$eV) and UHE (ultra-high energy: $10^{14} - 10^{16}$eV) gamma rays.

VHE gamma rays are currently detected by way of the electromagnetic cascades they generate in the upper levels of the atmosphere. Electrons are generated which are travelling faster than the speed of light *in air* and these emit ČERENKOV RADIATION in the visible region which can be detected by mirrors and photomultipliers at ground level. The technique is difficult because of the high background from conventional particle-initiated showers, but variable sources have been detected, notably Cygnus X-3, Crab and Vela pulsars. There are also some other candidates.

UHE gamma rays are recognized by the cosmic ray shower secondaries that actually reach ground level. This is a most exciting field. Cygnus X-3, with its 4.8-hour period, is the best candidate and its energy emission is so great that if the initial particles accelerated by the object (pulsar?) are protons then only a few such sources need be "on" at any one time to explain the bulk of the high-energy cosmic radiation. *AWW.*
See also colour essays on INTERSTELLAR MATTER and PULSARS.

Gamma Velorum The brightest of the WC-type stars; it is 3,800 times as luminous as the Sun, with a visual magnitude of 1.78. Like all stars of this type, it is decidedly unstable. It is 520 light-years away.

Gamma Virginis (Arich). Beautiful double star, magnitude 2.76; components 3.6 and 3.6, both yellow (spectrum F0). Period 180 years; separation about 4"; closest in 2016; distance 10 parsecs.

Gamow, George (1904–68). Russian nuclear physicist and cosmologist. Born in Odessa in 1904, Gamow emigrated in 1933 and became professor at George Washington University and the University of Colorado. He was a very successful popularizer of science and created a cartoon character, Mr Tomkins, the common man, by whose adventures Gamow explained relativity, atomic structure, and biology. His career began in nuclear and atomic theory, and he applied this training to ASTROPHYSICS, particularly stellar structure and the study of the creation of the elements. One early problem solved by him was how the relatively low collision velocities of nuclei in the interior of the Sun could trigger NUCLEAR REACTIONS. Gamow showed that there was a low but non-zero probability (in the Sun, 1 chance in 10,000 million million million) that two colliding nuclei could overcome their mutual repulsion, tunnelling through the potential barrier which keeps them apart, and sticking together.

Gamow was an early proponent of the BIG BANG theory of the universe, visualizing the early stage as a hot dense matter called ylem. In the ylem Gamow proposed that the elements were created, proton being added to proton to build up the elements in

sequence like the letters of the alphabet. He wrote up this theory principally with his colleague R.A.Alpher, inviting a friend H.Bethe to join the list of authors so that the theory could become known as the Alpher, Bethe, Gamow Theory (abbreviated to $\alpha\beta\gamma$-theory).

Gamow's greatest prediction was that the cosmic fireball of radiation left over from this ylem stage should have cooled to about 10K in the subsequent expansion of the universe. The 3K MICROWAVE BACKGROUND radiation discovered by Penzias and Wilson in 1965 realized the prediction. *PGM.*

George Gamow: predicted 3K background

Gan De Chinese astronomer who, in 365BC, recorded a naked-eye observation of Ganymede, Jupiter's brightest moon.

Ganges Catena Crater-chain on Mars: 2 to 3°S, 71 to 67°W. "Ganges" was also the name of one of the "canals".

Ganymed Amor asteroid 1036, discovered in 1924. It has a diameter of 22 miles (35km) and is probably the largest "close-approach" asteroid.

Ganymede The largest and second most distant from Jupiter of the Galilean satellites. It has a low density, 1.9 gm/cm³, indicating that it has a mantle of water, consisting of as much as half of the volume of the satellite. Voyager images of Ganymede revealed that this mantle is overlain by an extremely modified crust. Large isolated regions of low reflectivity are surrounded by grooved lighter material. Greater crater densities in the dark areas indicate that they are of greater age.

The satellite orbits Jupiter in a nearly circular orbit in the equatorial plane

and synchronously rotates so that one hemisphere faces into the direction in which the planet is revolving, while the other trails. Large, ancient, dark, heavily cratered regions are found on both the leading and trailing hemispheres, with Galileo Regio, the largest and nearly circular region, located north of the equator in the leading hemisphere.

The lighter, grooved terrain covers about 60 per cent of the surface mapped by Voyager 1 and 2. The grooves occur in sets, with the spacing between the grooves ranging from 1.8 to 6 miles (3 to 10km) and with a range in altitude of 2,300ft (700m). A set of grooves may extend for several hundred kilometres; however, there is complex cross cutting of the grooves. The complex patterns of the grooves and the nature of their interface with the ancient cratered terrain suggest that the grooved terrain forms by fluid or slushy material breaking through a weakened crust. The source of energy for this process could be derived from tidal heating and energy released from isotopic decay or phase changes in the interior of the planet.

Craters on Ganymede's surface deform with age, with large craters appearing more flattened than smaller craters. Some features are so flattened that they appear as mere stains. New craters display bright rays extending outward from the centres. Greater brightness of rays from craters located in the grooved terrain than those

Ganymede: largest Galilean satellite

located in the ancient terrain suggests that the material from which the grooved terrain is formed contains a larger percentage of water ice than the disturbed surface layer of the ancient terrain. *RFB.*

Garnet Star Nickname for the very red irregular variable Mu Cephei.

G

G

Gas exchange experiment Viking Mars experiment utilizing gas chromatography technique to study samples given liquid nutrient treatment. The experiment was used to test for organic life on Mars.

Gassendi Lunar crater 18°S, 40°W, diameter 55 miles (89km). It borders the Mare Humorum, and its "seaward" wall has been greatly reduced. It has a central mountain structure and an impressive system of clefts on its floor.

Gassendi, Pierre (1592–1655). French mathematician and astronomer. In 1631 he made the first recorded observation of a TRANSIT of the planet Mercury across the face of the Sun.

Gauss, Carl Friedrich (1777–1855). Outstanding German mathematician and physicist who, in 1801, computed the orbit of the asteroid Ceres. He made contributions to many aspects of mathematics and physics, ranging from probability theory to magnetism.

Gay-Lussac Irregular lunar enclosure north of Copernicus; 14°N, 21°W; diameter 15 miles (24km).

Gegenschein A very faint patch of light sometimes visible on very clear moonless nights near to the ecliptic at the anti-solar point. It is oval in shape, the major axis in the ecliptic plane extending for some 20° with the minor axis for some 10°. The size can be somewhat larger in the tropics. In the northern hemisphere at high latitudes it is best seen in the months of December and January when the ecliptic lies at its greatest altitude above the horizon. In the southern hemisphere June and July are the best months. The gegenschein, or counterglow as it is also called, may be joined around the ecliptic by means of a faint band of light, called the ZODIACAL BAND, to connect with the ZODIACAL LIGHT.

The gegenschein is caused by the scatter of sunlight backwards towards the Earth from tiny dust particles in the interplanetary medium. The effect is similar to the glow of light surrounding the shadow of a person projected onto a distant surface due to ice or dust particles in the air.

The presence of the gegenschein tends to indicate that the Earth lies within a cloud of interplanetary particles which are surrounding the Sun and forming an extension of the Sun's corona. One theory suggests that the Earth may possess a faint dust tail stretching away from the Sun rather like that of a comet. Such a tenuous structure would, like an interplanetary dust cloud, scatter light backward from the night side toward the Earth. The very existence of a counterglow can be theoretically accounted for if the particles have a rough surface. The brightness at the anti-solar point is similar to the intensification of moonlight at full moon due to direct specular reflection and the elimination of shadow on the reflecting bodies. *RJL.*

Gemini (the Twins). A large and important Zodiacal constellation.

Brightest stars

Name	Visual Mag.	Abs. Mag.	Spec.	Distance (light-yrs)
β (Pollux)	1.14	0.2	K0	36
α (Castor)	1.58	1.2	A0	46
γ (Alhena)	1.93	0.0	A0	85
μ (Tejat)	2.88v	−0.5	M3	230
ε (Mebsuta)	2.98	−4.5	G8	690
η (Propus)	3.1 max	−0.5v	M3	186
ξ (Alzirr)	3.36	0.7	F5	75

Also above the fourth magnitude are Delta (3.53), Kappa (3.57), Lambda (3.58), Theta (3.60), Zeta (variable. 3.7 at maximum) and Iota (3.79).

Castor, though lettered Alpha, is fainter than Pollux, though in ancient catalogues it was given as the brighter of the two. Eta is a semi-regular variable, 3.1 to 3.9, with a very rough period of 233 days; Zeta is a Cepheid, range 3.7 to 4.3, period 10.15 days.

Gemini is crossed by the Milky Way, and is very rich. The open cluster M35, near Eta, is easily visible with the naked eye. *PM.*

Gemini Series of two-man United States spacecraft launched by Titan II rockets between 1965 and 1966. Experiments included orbital rendezvous and EVA. The longest mission (Gemini 7) lasted for 330hr 35m 17s.

Geminid meteors An important and active meteor shower that peaks on about December 13/14 every year. Geminid meteors are visible between about December 7 and 16. Regular displays of the shower have been observed since 1838. It is currently one of the richest annual showers, and at peak the zenithal hourly rate reaches 60 meteors per hour or more. The rise to peak activity is more gradual than the fall form the peak. At maximum the radiant is situated at R.A. 7h 28m, Dec. + 32°, near to the star Castor in Gemini. The meteoroids hit the atmosphere with a velocity of about 22 miles/s (36km/s). Visual observations have shown that the peak for bright Geminids is significantly later than that for the fainter Geminids. The meteor stream has an unusual orbit, with a very small perihelion distance of 0.14AU, and a period of only 1.57 years. There is no known parent comet, but an Earth-crossing asteroid named PHAETHON orbits within the meteor stream and is probably connected with it. The Geminids have an unusually high density compared to the other major annual meteor streams. *JWM.*

Geminus Lunar crater, 56 miles (90km) in diameter; 35°N, 57°E, in the Crisium area. It has broad, terraced walls and a central elevation.

Genesis rock Clast in lunar anorthositic breccia, mistakenly thought to represent the original highland crust.

Geocentric theory In the astronomy of all early civilizations the universe is conceived of as geocentric with the central Earth at rest while the Sun and stars complete daily rotations. Slightly more detailed observation reveals that the risings and settings of the Sun and stars change with the seasons while the planets ("wanderers") regularly appear to execute complex retrograde (looping) motions. By the 5th century BC the early Greeks were aware that the apparent motion of the Sun could be represented as a sum of a daily rotation about the Earth and a yearly motion round the Earth along the ECLIPTIC, the axes of the two rotations being inclined by some 24°. The complex solar motion could thereby be resolved into the perfect uniform circular motion deemed to be the only appropriate motion for "heavenly bodies". This is said to have led the great Greek philosopher PLATO to set astronomers the problem of accounting for the observed movements of the Moon and planets by means of similiar compositions of circular motion.

The first geocentric theory of planetary motion is due to EUDOXUS, who was a young man when Plato died. Inspired perhaps by the then recent discovery of the sphericity of the Earth, Eudoxus proposed a heavenly arrangement of 26 revolving spheres upon which the Sun, Moon and planets were borne through space. By attaching the axes of each of the planetary spheres to other nested spheres with relatively inclined axes of rotation it was possible to produce retrograde motion, albeit largely only in qualitative agreement with the observed movements, particularly with regard to the frequency of occurrence. This scheme was elaborated not long afterwards by Callippus, who added nine further spheres. With these additional spheres some of the grosser discrepancies of the planetary retrogressions were removed and in addition account was taken of the non-uniformity of the Sun's apparent motion as revealed in the inequality of the seasons (between equinoxes and solstices). Later still, the Greek philosopher ARISTOTLE added counter-rotating spheres between the groups for each planet in order to enable the complete systems to be joined together while ensuring that the spheres for different planets had no effect on each other. Aristotle appears to have been uncertain as to the number of spheres required for this, a total of either 55 or 47 spheres is suggested, but it is in the Aristotelian form that the theory was known in the later Middle Ages. Perhaps the most obvious problem with this version of the geocentric theory, of which Eudoxus himself may well have been aware, was that it failed to explain the variations in distance from the Earth of Venus and Mars in particular, as judged by their changes in brightness.

An alternative theory was proposed by APOLLONIUS around 200BC. According to this the Earth remains at rest at the centre of the Universe, but the Moon, Sun and planets move in circular motion about centres displaced from the Earth (*see* accompaning diagram 1). Not only does this allow varying Earth-planet distances, but it also accounts, qualitatively at least, for the non-uniformity of orbital motion, since viewed from the Earth the planet appears to move through a larger angle in a given time when it is closer. Apollonius showed that this eccentric orbit is equivalent to one in which the planet moves on an epicycle, ie, on a circle the centre of which itself moves uniformly on a circle (the DEFERENT). Provided the speeds are chosen so that P and E remain at the vertices of a parallelogram the planet will move on an eccentric circle (2). So far the two interpretations are equivalent, but whereas the eccentric theory does not suggest a modification to incorporate retrograde motion, such a change is simple in the epicyclic version. It is necessary only to give up the restriction on the speed of P round the epicycle and allow the speeds of the two circular

motions to be chosen independently. This modified theory is equivalent to an eccentric circle with a moving centre.

It is important to point out that at this level the geocentric theory is *entirely equivalent* to a heliocentric version in which the planets move in circles around the Sun with uniform velocity. In this regard we might note that reduced to the scale of the figures here the departure of actual planetary orbits from uniform circular motion would not be detectable.

Even in antiquity the geocentric theory did not go unchallenged. Heraclides, one of Plato's pupils, was the author of a hybrid theory according to which Mercury and Venus were taken to be satellites of the Sun, while the Sun

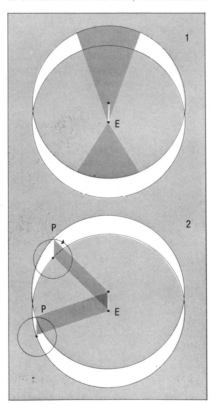

Geocentric theory: Apollonius' view

and planets revolved round the spinning Earth at the fixed centre. This arrangement was also championed by Tycho BRAHE, even after COPERNICUS had proposed his heliocentric theory. ARISTARCHUS, who is usually credited as the first to suggest that the Earth rotated daily on its axis, postulated a completely heliocentric theory in the 3rd century BC. But generally speaking, such ideas were not taken up. The scientific reasons for this were at least twofold. First the motion of the Earth did not appear to be detectable by any

observations independent of planetary theory, for example by stellar PARALLAX. This is a good argument, but we now know it fails because the stars are too far away to give significant parallaxes. The second obstacle was the authority of Aristotle's science of motion. According to this theory, which was really the only logically coherent theory of motion before NEWTON, heavenly substance is endowed with natural circular motion, whereas earthly matter moves naturally in straight lines toward or from a fixed centre. For this reason the stony material of the Earth agglomerates at the centre of the universe. For this reason too it is possible for the heavens to whirl round in accordance with their nature without flying apart, but not possible for the Earth so to move. Aristotle's science of motion supports a geocentric view because it was formulated to do so.

The geocentric theory reached its acme in the 2nd century AD in Alexandria, where PTOLEMY wrote what has become known as the ALMAGEST ("the greatest"). Since the time of HIPPARCHUS two and a half centuries earlier it had been known that the old epicyclic theory was inconsistent with the ever-growing accumulation of observational data. In order to achieve consistency Ptolemy found it necessary to increase greatly the number and complexity of the epicyclic motions, even, in the view of some of his critics, to the point of violating the principle of uniform circular motion. Ptolemy's aim was quite clearly not only to provide a numerically accurate account of planetary motion — to "save the appearances" — as in the Babylonian astronomy from which he freely borrowed, but to discover the real geocentric motions of the universe according to the principles of geometry and uniform circular motion. Possibly with some slight adjusting, Ptolemy was able to obtain agreement with observations from both longitudes and latitudes of solar system bodies, the latter by suitable inclinations of the planes of the epicycles. On the other hand, distances were quite clearly wrong: the size of the Moon should change periodically by a factor of two according to the theory. Nevertheless, the results in the *Almagest* were not improved on for a thousand years.

The demise of the geocentric hypothesis resulted from two parallel developments. Firstly, in contrast to the decline of learning in the western Roman empire, science in the East continued

G

G

and Aristotelian dynamics was subjected to searching criticism. While these criticisms were largely based on philosophical arguments rather than experiment, they were sufficient to undermine Aristotle as the sole and ultimate authority. Secondly, with increasing accuracy of astronomical observation following the 13th-century revival of learning in the West, additions to the Ptolemaic epicycles became increasingly elaborate and unappealing. Nevertheless, the reign of geocentrism did not end with the advent of the Copernican heliocentric system in 1543. For many years the even more complex system of epicycles employed by Copernicus to describe planetary motion in terms of circular orbits in a Sun-centred picture was taught in parallel with the Ptolemaic system as an alternative hypothesis to aid calculation. In Copernicus's native Poland the heliocentric theory was not taught in Church schools until well into the 18th century. But the Copernican system, which allows the incorporation of planetary distances as well as positions in the sky, paved the way for KEPLER's abandonment of circular orbits and the final heliocentric explanation of the Solar System in terms of motion under gravity in Newtonian theory. *DJR*.

Geodetic satellites The orbital behaviour of many satellites has been used to study the shape of the Earth, but several have been launched solely for this purpose, among them the 98ft (30m) diameter balloon satellite

Geodetic satellites: studying the Earth

PAGEOS, launched into a circular orbit about 2,600 miles (4,200km) above the Earth's surface. its purpose was to provide a point source of light which could be photographed from widely separated points on the Earth's surface,

giving a positional accuracy of about 32ft (10m). More recently LAGEOS, a sphere 2ft (60cm) in diameter and covered with corner reflectors, has been tracked by lasar beams, giving ranges to an accuracy of about an inch (2.5cm). Future developments using lasars will provide accurate values for tectonic plate movements. *HGM*.

Geographos (Asteroid 1620). An APOLLO, so that it can cross the Earth's orbit. It was discovered in 1951, recovered in 1969. Its orbit is now well known. Its diameter is only a few kilometres.

Georgian planet Obsolete name for URANUS.

GEOS An acronym used by the United States for three satellites designed for geodetic studies. The name was later used by ESA for two satellites investigating charged particles in space.

Gerard Lunar crater in the Sinus Roris area, 54 miles (87km) in diameter; 44°N, 75°W.

Giacobinid meteors Also known as the Draconid meteors, this shower is associated with the GIACOBINI-ZINNER COMET. Showers of Giacobinids occur quite rarely and last only a few hours. They seem to take place only when the Earth closely follows the parent comet to the descending node of the comet's orbit, and passes very close to, but just inside, the comet's orbit. Giacobinid showers occurred on October 9 1926 (17 meteors per hour), October 9 1933 and October 10 1946 (5,000 meteors per hour), October 9 1952 (180 meteors per hour) and recently on October 8 1985 (about 400 meteors per hour). The displays of 1933 and 1946 were certainly very spectacular. The shower seen in 1985 was a surprise because at the time the Earth was some 0.033AU from the comet's orbit. This may indicate that the stream is wider in places than previously thought. The radiant of the Giacobinids is at R.A. 17h 23m, Dec. +57°, near the Head of Draco. *JWM*.

Giacobini-Zinner Comet This short-period comet was discovered by Giacobini at Nice Observatory on December 20 1900, and independently by Zinner at the Remeis Observatory on October 23 1913. With a period of 6.5 years, the comet is easily observed near perihelion about every 13 years. Its perihelion distance is 1.028AU. It was poorly seen in

1940, 1966 and 1979, and missed altogether in 1907, 1920 and 1953. On September 11 1985, the American ICE spacecraft flew through the tail of the comet. It is possible that the spin axis of the comet's nucleus was severely perturbed around 1959.

Giant molecular clouds Radio astronomy measurements since the early 1970s have shown the existence of remarkably large amounts of molecular gas in the Galaxy, most of which is in the form of molecular hydrogen. The total mass of this gas is probably very similar to that in atomic hydrogen, in the inner galaxy. The molecular gas is very clumpy and mixed with dust. There is a wide range of masses of the clumps, the biggest — the so-called giant molecular clouds — having masses up to several million solar masses. It is the condensation of gas within these clouds which forms most of the stars seen in the heavens. *AWW*.

Giant molecular clouds: Omega Nebula

Giant stars Stars that are placed well above the main sequence in the HERTZSPRUNG-RUSSELL DIAGRAM are called giant stars. Although a giant star has a larger radius than a main sequence (dwarf) star of the same temperature, giant stars of intermediate temperatures are smaller in size than the hottest dwarfs. Thus the term "giant" really refers to the relative luminosity of a star.

Giant stars result from evolution of dwarf stars after the latter have exhausted hydrogen throughout their central regions. They thus represent later states in the lives of stars. Because of their high intrinsic luminosity, giants are quite common among naked-eye stars (eg, Arcturus, Aldeberan,

Capella) but are in fact relatively rare in space. *BW*.

Gibbous phase The phase of an illuminated body when between half and full.

Gibbous phase: a gibbous Moon

Gill, David (1843–1914). Fifth Director of the Cape Observatory (1879–1907). He made the most accurate pre-photographic measurements of stellar parallaxes. His photograph of the Great Comet of 1882 led him to propose photography as the best method of preparing star catalogues, thereby founding the Astrographic Catalogue.

Gioja The north polar crater of the Moon; diameter 22 miles (35km). It is fairly regular in form.

Giotto The Giotto spacecraft of the EUROPEAN SPACE AGENCY (ESA) was launched on July 2 1985 by an ARIANE-1 rocket from French Guiana, in South America. After an interplanctary journey of eight months it encountered HALLEY'S COMET. The closest distance to the nucleus of 376 miles (605km) was reached at 00: 03: 02UT on March 14 1986. Giotto passed the nucleus on the sunward side at a speed of 42.48 miles/s (68.373km/s): during the encounter the comet was 0.89AU from the centre of the Sun and 0.98AU from the centre of the Earth.

The mission was named Giotto after the Italian painter Giotto di Bondone, (*c*.1267–1337), who in 1304 depicted Halley's Comet as the "Star of Bethlehem" in a fresco in the Scrovegni chapel in Padua, Italy.

The Giotto spacecraft has a cylindrical shape with a diameter of 6ft (1.86m) and an overall height of 9.3ft (2.85m). It is spin-stabilized with a nominal spin period of 4s. At launch Giotto weighed 2,072lb (960kg). A solid-propellant motor was mounted centrally. It carried 825lb (374kg) of propellant, giving a velocity increment of 4,593ft/s (1,400m/s) needed to inject it into heliocentric orbit.

Giotto's scientific payload comprised 10 hardware experiments with a total mass of just less than 132lb (60kg); a camera for imaging the comet nucleus and inner coma; three mass spectrometers for analyzing the elemental and isotopic composition of cometary neutrals, ions and dust particles; various dust impact detectors; a photopolarimeter for measuring the coma brightness; and a set of plasma experiments for studying the interaction processes between the solar wind plasma and the cometary ionosphere. Almost all the experiments are mounted on the "experiment platform" of the spacecraft, looking down into the stream of cometary particles (*see* accompanying illustration). During the encounter the spacecraft spin axis was aligned with the relative velocity vector ("relative" in the comet's frame of reference), so that cometary particle (dust and gas) streaming was from below.

Giotto had no onboard data storage capability and all data were transmitted back to Earth in real time at a rate of 40 kilobits a second. This high data rate over a distance of 1AU could be achieved by using a high-gain parabolic dish antenna on Giotto and a 210ft (64m) ground station at Parkes, in Australia. The antenna could be operated in either the S-band (receiving at 2.1GHz, transmitting at 2.3GHz) or the X-band (transmitting at 8.4GHz). The antenna beam was inclined 44.3° with respect to the spacecraft spin axis and the antenna itself was despun so that the beam could point continuously at Earth during the encounter.

At its other end Giotto carried a dual-sheet bumper shield to protect it from high-velocity dust impacts. A dust particle striking the thin aluminium front sheet was completely vaporized; the vapour cloud then expanded into the empty space (9in/23cm) between the two sheets, and impacted on the rear, Kevlar-sandwich sheet, where its energy was dissipated by being distributed over a large area.

The small nucleus of a comet cannot be seen from Earth and it is masked by the dust and gas in the coma: its position can only be estimated, which presented a problem for the targeting of Giotto. The cameras on the earlier-arriving VEGA 1 and 2 spacecraft located the nucleus and passed the information onto the Giotto Project, enabling the craft to be targeted with high precision.

During the encounter, just seconds before closest approach, a dust particle of about 1gm impacted on Giotto and the spacecraft started to nutate around an axis which was 0.9° off from the relative velocity vector. As a consequence, the high-gain antenna beam

G

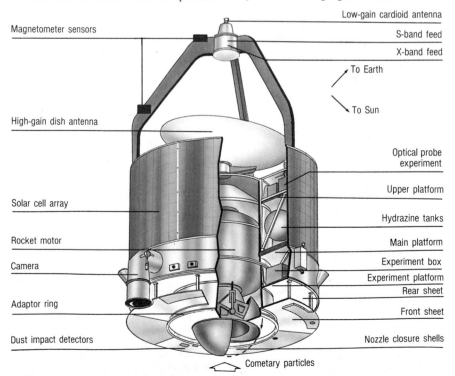

Magnetometer sensors

High-gain dish antenna

Solar cell array

Rocket motor

Camera

Adaptor ring

Dust impact detectors

Low-gain cardioid antenna

S-band feed

X-band feed

To Earth

To Sun

Optical probe experiment

Upper platform

Hydrazine tanks

Main platform

Experiment box

Experiment platform

Rear sheet

Front sheet

Nozzle closure shells

Cometary particles

Giotto probe: examining Halley's Comet

was not pointed at the Earth and the telemetry stream was interrupted intermittently. Within the next 30 minutes the nutation was damped, however, and continuous data were again received. Up to closest approach all instruments performed well and returned a wealth of scientific data.

After completing the flyby, the spacecraft and half of the experiments were damaged. A series of orbit correction manœuvres was, however, successfully carried out and Giotto will return to Earth on July 2 1990 and may possibly be re-targeted using Earth gravity to encounter GRIGG-SKJELLERUP COMET on July 14 1992. *RR. See also* COMETS.

G

GK Persei Nova, 1901. Magnitude 13.5 to 0.2. Notable for the asymmetrical expanding shell now visible.

Glacial cosmology A strange theory, proposed in the 1930s by H.Hörbiger of Austria, that most of the bodies in the universe are made of ice. The theory became popular in pre-war Germany and almost became a part of Nazi creed.

Glenn, John Herschel (1921–). The first American to orbit the Earth, in Mercury 6 (February 20 1962). He retired from NASA in 1964, and was elected Senator (Democrat) for Ohio in 1974.

Glitches Jumps in the period of a PULSAR. The pulsation period of a pulsar is gradually increasing (the pulsar is slowing) typically at a rate of a few hundredths of a second in a million years. A few pulsars, particularly PSR 0833-45 (in the Vela nebula, called the VELA PULSAR) and PSR 0531+21 (in the Crab nebula, called the CRAB PULSAR) occasionally show an abrupt and major speeding up — the pulsation period shortens. The largest of these glitches have been seen in the Vela pulsar; they occur every few years and typically decrease the period by a few millionths of the period (or "undo" the slowing of the pulsar that occurred during the previous three weeks). No glitch has been directly observed; all have occurred between successive observations of the pulsar, separated by about a week. It is suspected that the glitch takes only a matter of minutes. The glitch appears to overshoot and for the next few weeks the period increases more rapidly than normal, settling back to the rate that pertained before the glitch. This settling back provides information on the internal structure of

the pulsar, and is strong evidence for there being several layers within the star, some of which must be superfluid. The connection between the layers is not tight and after a change in spin rate it takes several weeks for all the layers to come to the same spin rate. As the pulsar slows it loses angular momentum. In the superfluid layers there are vortices (whirlpools) which move outward. Because they are pinned at top and bottom, the vortices are not able to move outward freely, but move jerkily when the pinning breaks. It is thought that these jerks cause the glitches. *SJBB.*

Globular clusters Almost spherical clusters of stars of considerable antiquity.

When our Galaxy first condensed from a huge cloud of gas it was roughly

Globular clusters: M13 in Hercules

spherical. Only as the collapse proceeded did the rotation of the cloud begin to take effect, eventually causing it to become a flattened disk rather than a ball. However, in the earliest stages of the collapse stars formed in some locally denser sections. These stars, the oldest in the Galaxy, occupy a more nearly spherical distribution collectively known as the Galactic spheroid or halo. They are not very numerous, except in a couple of hundred specific locations where vast numbers of stars formed from a condensation in the original gas cloud. The resulting congregations are known as globular clusters because each is almost spherical.

A globular cluster is an impressive sight. Through a small telescope it is merely a fuzzy blob, and indeed when Comet Halley passed near to the brightest globular cluster OMEGA CENTAURI in 1986 many people were misled into believing that they had found a second comet. Through a large

telescope the hazy patch is resolved into many thousands of stars, concentrated so tightly near the centre that they cannot be fully separated by any ground-based telescope. Large globular clusters contain several million stars.

Globular clusters can be seen around other galaxies, but merely as extremely faint spots of light which can be distinguished from foreground stars only by careful observation. There are always far more globular clusters around elliptical than around spiral galaxies, perhaps running into thousands of clusters for the largest elliptical galaxies. Our Galaxy is typical of spirals having fewer than 150 globular clusters. Although they occupy the spherical halo, most globular clusters actually lie no farther from the centre of our Galaxy than does the Sun; as a result they congregate in the general area of sky containing the centre of our Galaxy — in particular in Sagittarius, Ophiuchus and Scorpius. Globular clusters are therefore better studied from the southern hemisphere, though a few specimens do lie conveniently northwards, and the best known of these is M13 in Hercules. Some globular clusters occupy the flattened disk of our Galaxy; because they must orbit the galactic centre, inevitably they make periodic passages through the disk. The obscuration by dusty gas clouds in the Milky Way renders globular clusters lying near the plane difficult to recognize, and a few probably remain to be discovered.

What makes globular clusters particularly important is that they contain a mix of stars of various sizes which all lie at the same distance from us, and which all evolved from the same cloud of gas at the same time. For this reason, the globular clusters have played a vital role in understanding the life history of stars. Without them we would have required much longer to appreciate the HERTZSPRUNG-RUSSELL DIAGRAM, and in particular the MAIN SEQUENCE and RED GIANTS.

Because the stars of a globular cluster lie at the same distance from us, we can plot a Hertzsprung-Russell diagram for them simply by measuring their colour and their apparent magnitude (brightness), knowing that there is a constant difference between the apparent and the absolute magnitude according to the cluster's distance. It is then possible to deduce that distance by comparison with a standard Hertzsprung-Russell diagram, because the main sequence will appear in the correct place only if the distance is right. In the early days of trying to measure the size of the Galaxy, this technique was vitally important.

The Hertzsprung-Russell diagram of a globular cluster tells more than its distance, however. As time passes, the brightest stars in a cluster exhaust their hydrogen and become red giants. This happens progressively to fainter and fainter stars. The Hertzsprung-Russell diagram clearly shows which stars have begun their transformation to red giants, and leads directly to the age of the cluster. Globular clusters have thus been dated about 10,000 million years old; they must have formed before the universe attained half its present age.

In those remote times the chemical mix of our Galaxy was quite distinct from the present. Stars, especially those which turn into SUPERNOVÆ, create elements such as carbon, oxygen, silicon and iron out of their hydrogen. Over the years, the proportion of these elements, which astronomers call metals, has increased, so we expect to see more in young stars than in the old globular clusters. And indeed we do in most cases. It turns out, however, that the globular clusters neatly divide into two groups. The larger group is metal poor, as expected; the smaller group has far more of the heavier elements. The origin of the metal-rich globular clusters remains something of a mystery. One suggestion is that they once belonged to a small galaxy with more metals than our own, which was swallowed up by ours in the remote past.

When we talk of the chemical composition of a globular cluster, we really mean that of its constituent stars, for they are what we measure. Of course, because all the stars formed from the same gas cloud, they all have pretty much the same chemical mix. There is, however, an exception to this obvious rule. In the large globular cluster Omega Centauri the chemical mix differs from centre to edge, the inner part containing more of the "metals". This too remains something of a mystery.

Perhaps the most important role of the globular clusters has been to give us a means of estimating distances. Globular clusters contain a particular type of variable, the RR LYRÆ stars. We can determine the intrinsic brightness of RR Lyrae stars in the globular clusters, and can recognize them in other locations, so allowing the distance to be derived for different objects. Currently the RR Lyrae stars near the centre of our Galaxy are providing a more accurate estimate of our distance from there than was possible from the globular clusters alone. *DAA.*

Globules Dense compact dust clouds. *See also* BOK GLOBULES.

Gluons Some scientists believe that many of the more than one hundred known sub-atomic particles may be made up of just a few even tinier ones called quarks. These quarks have to be bound together by some force. Scientists have proposed that this force occurs by the interchange of particles called gluons. However, it is not at all certain how many different types of quarks or gluons exist, or even if they do exist!

Gnomon Ancient instrument for measuring the altitude of the Sun. Also the pointer on a sundial.

Goclenius Lava-flooded lunar crater on the edge of the Mare Fœcunditatis; 10°S, 45°E. It is 32 miles (52km) in diameter, and has a low central elevation. It lies east of the rather larger formation Gutenberg.

Goddard, Robert Hutchings (1882–1945). American rocket pioneer, who flew the first liquid-propellant rocket in 1926, and had earlier speculated on the possibilities of sending vehicles to the Moon.

Robert Goddard: American rocket pioneer

Goddard Space Flight Center A NASA facility in Greenbelt, Maryland, USA, which is named in honour of the American rocket pioneer Robert H. GODDARD. The centre also operates the Goddard Institute for Space Studies in New York City, the Wallops Island facility and a global network of tracking stations. Goddard operates satellites and sounding rockets for a wide range of research purposes including astronomy, solar-terrestrial relations and environmental monitoring. It also houses the National Space Science Data Center.

The centre also manages the Delta launch vehicle programme and plays a major role in the international Search and Rescue Satellite Aided Tracking Project (SARSAT). *IKN.*

Goddard Space Flight Center: NASA base

Godin Regular Lunar crater 27 miles (43km) in diameter; 9°N, 10°E; south of Mare Vaporum. It has a central peak. It is a pair with Agrippa.

Goethe Crater on Mercury; 80°N; 044°W; diameter 210 miles (340km). It lies near the edge of the Planitia Borealis.

Goldschmidt, Hermann (1802–66). German astronomer living for much of his life in Paris. Using a small telescope, he discovered 14 asteroids between 1852 and 1861.

Goldstone radio dish Goldstone was the first 210ft (64m) radio antenna in the NASA/JPL Deep Space Tracking Network. It is located in the Mojave desert in Southern California and became operational in 1966. Other 64m

Goldstone radio dish: desert location

antennæ in the network are sited in Australia (Tidbinbilia) and Spain.

The antenna was originally built to support NASA's missions to the planets. However, it has since been used regularly for radioastronomical studies. Perhaps most notable amongst these has been its role in Very Long Baseline (recording) Interferometery, working with other telescopes to produce accurate positions and structural information for compact radio sources, such as active galaxies and quasars. *AEW².*

Göttingen Observatory The observatory at the University of Göttingen in West Germany, and the site of important work by Brandes and Benzenberg on meteor astronomy between September and November 1798. They were the first people to carry out triangulation and determine the heights of meteors in the atmosphere. Out of a sample of 402 meteors seen, they deduced that the meteors were occurring at an average height of about 60 miles (98km), very close to the accepted modern value. The present Göttingen Observatory lies at an altitude of 530ft (161m) at 51.53°N, 0.94°E.

Gould, Benjamin Apthorp (1824–96). American astronomer who catalogued the southern stars from the CORDOBA OBSERVATORY, Argentina. He founded the *Astronomical Journal* in 1849.

Gould's Belt A belt of stars and gas tilted to the galactic plane by about 20° which contains many O- and B-type stars and which probably represents the local spiral arm of which the Sun is a member. It is named after Benjamin Gould.

"Grand Tour" In the late 1960s planners and scientists in the United States and elsewhere concerned with interplanetary exploration evolved concepts in which spacecraft could travel great distances at high speed by means of gravity assists provided by the planets themselves. At the same time, it was realized that in the late 1970s/1980s the four major outer planets — Jupiter, Saturn, Uranus and Neptune — would be in an approximate alignment repeated only once every two centuries. Ambitious plans were made by NASA and the scientific community in the United States for dual gravity-assist launches to Jupiter/Saturn/Pluto in 1976-77 and similar launches to Jupiter/Uranus/Neptune in 1979 — the so-called *Grand Tour Missions.* These

plans fell foul of the political and economic domestic United States climate of the late 1960s and early 1970s and were reduced to a two-spacecraft mission

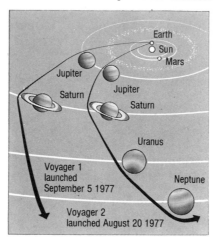

"Grand tour": paths of the Voyager probes

targetted (initially) to only Jupiter and Saturn. Begun in 1972 as Mariner-Jupiter-Saturn (MJS), the project's name was changed in 1977 to VOYAGER. Subsequently a Grand Tour was in fact accomplished almost by the back door as the Voyager 2 mission was extended on from Jupiter and Saturn to Uranus and then to Neptune (in August 1989). *HJPA.*

Grand unified theory (GUT). A theory that attempts to "unify" three of the four forces of nature, ie, to demonstrate that they are different manifestations of a single force.

In descending order of strength, the four forces are: the strong nuclear interaction (which binds atomic nuclei); the electromagnetic force (which holds atoms together); the weak nuclear interaction (which controls the radioactive decay of atomic nuclei); and gravitation. The weak and strong interactions operate over tiny distances within atomic nuclei while the other two are infinite in range. With the exception, so far, of gravitation, all the forces can be described by quantum field theories, which imply that forces are communicated by means of particles (*see* VIRTUAL PARTICLES); the more massive the force-carrying particle, the shorter its life and the shorter the range of the force. GUT theories attempt to unify the weak, strong and electromagnetic forces. Many physicists believe that a more elaborate theory will eventually be found which will unify all four forces.

The first step toward unification was the Weinberg-Salam-Glashow theory

of the "electroweak" force, published in 1967, which implied that at particle energies greater than about 10^{12}eV, the electromagnetic and weak forces should be equal in strength and behave in the same way. The theory was confirmed in 1983 when the predicted force-carrying particles — the massive W and Z bosons — were detected.

Grand unified theories imply that at energies greater than about 10^{24}eV the electroweak and strong nuclear interactions merge into one common force. Such energies far exceed the capabilities of man-made particle accelerators, but would have existed everywhere in the Big Bang universe during the first 10^{-35} seconds after the initial event. Thereafter, as the universe expanded and cooled, the unified force would have split into the strong force and the electroweak force which, in turn, split into the electromagnetic and weak forces.

The simplest GUT theory, proposed in 1973 by Glashow and Georgi, requires the existence of a set of supermassive "X-particles" which can transform quarks into leptons and vice versa. Since quarks are the basic constituents of PROTONS, the theory predicts that protons themselves must eventually decay into lighter particles. The half-life of a proton is believed to be at least 10^{32} years, but given enough protons it should be possible at any time to detect the sporadic decay of a few. Experiments are under way to test this prediction which, if confirmed, implies the eventual disintegration of all the familiar chemical elements. *IKN.*

Gravitational lenses The curvature of space which is caused by a heavy mass, and which can deflect light from a distant object, distorting its image.

A ray of light (or a radio wave) which passes near to a massive body is deflected: this is a prediction of Einstein's General Theory of RELATIVITY. It was first successfully tested during the solar eclipse of 1917 when light from distant stars was photographed beyond the darkened Sun. The images of the stars were shifted outward from the Sun, which acted as a magnifying gravitational lens.

Another manifestation of the same effect occurs if a massive galaxy should happen to lie almost on the line of sight between us and a quasar. The image of the quasar may appear doubled in the gravitational lens associated with the massive galaxy. This consequence of Einstein's theory, championed by Fritz ZWICKY in 1937, became a reality in

1979 (Einstein's centenary year) with the discovery by D.Walsh, R.Carswell and R.J.Weymann of the first of the known "double quasars". In these objects two quasars, with identical spectra and REDSHIFTS, are found just a few arc seconds apart. Rather than accept the astonishing and unlikely coincidence that separate but identical quasars could be found so close together by chance, astronomers attribute the doubled images to gravitational lensing of single quasars by very large galaxies of mass in the region of 10 million million Suns. In one of the double quasars, 0957+561, the galaxy is bright enough that it can be identified between the pair of quasar images — it is the brightest, and presumably most massive, in a cluster. In the two other definite cases, (1115+080, 2016+112), the lensing galaxy is too distant to be detected and its existence is only surmized.

A fourth example of a gravitational lens was discovered in 1984. Right in the centre of an unremarkable, relatively near spiral galaxy sits the bright image of a distant, high redshift quasar, 2237+0305. Its image has been intensified by the gravitational lens made by the galaxy. It is probably a double image, but the separation of the pair of images is too small to detect.

There are half a dozen further possible gravitational lenses, of which the widest separated is 1146+111, with the two images 146 arc seconds apart. The two images in this double quasar are not quite identical, and this is attributed to the wide separation and therefore difference in path length from the quasar to each image as viewed from Earth. This difference in path length amounts to many light-years, so each image represents the original quasar at a different time. Since quasars vary, the two images might not look identical. In general, if the quasar itself brightens, then first one image will brighten and the other will follow. The time lag of one image behind the other could give interesting information on how gravitational lenses work. *PGM.*

Gravitational waves Disturbances in a gravitational field which, according to Einstein's General Theory of RELATIVITY, propagate through space at the speed of light.

Any changes in a gravitational field are expected to travel at the speed of light so that, for example, if the Sun were to be annihilated, 8.3 minutes would elapse before the Earth ceased to "feel" the gravitational influence of the Sun. Just as an accelerating or oscillating charged particle emits electromagnetic waves, so an accelerating, oscillating, or violently-disturbed mass is expected to radiate wave-like gravitational disturbances — gravitational waves.

Because gravitation is by far the weakest of the forces of nature, gravity waves are expected to be weak and difficult to detect. The most likely sources are rapidly spinning bodies distorted from perfect sphericity, close binary systems — particularly those involving collapsed objects like NEUTRON STARS or BLACK HOLES — SUPERNOVÆ, and events involving massive black holes. *IKN.*

Gravity The fourth physical force exerting a mutual attraction between two masses.

In his *Principia* in 1687, NEWTON announced that the force of gravity between any two masses is proportional to the product of their masses and inversely proportional to the square of the distance between them or algebraically:

$$F = Gm_1 m_2/r^2$$

Thus, he was able to link the acceleration with which objects, such as the apocryphal apple, fall to the Earth's surface with the speed of the Moon in its orbit.

Although, compared to electromagnetic and nuclear forces, gravity is very weak at close range, it becomes vitally important on large scales because it has a longer range than the nuclear forces and because, unlike electric charges, all mass is positive and attracts every other mass. The resulting absence of any possibility of cancellation means that gravitation is the dominant force on large scales and it is astronomical observations that have provided the tests of physical theories of gravitation such as Newton's original theory, Einstein's General RELATIVITY and the BRANS-DICKE THEORY.

Astronomers often determine the masses of astronomical objects such as the Sun, planets, stars, galaxies or quasars using their orbital dynamics. They actually find the product of the mass and the constant of gravitation, *G*. This constant can only be determined by measuring the attraction between known masses on Earth in the laboratory. In such experiments gravity is already extremely weak and very careful measurements are needed. Because all masses attract, there is no way to screen out forces from masses outside the laboratory and yet the experimenter must measure a disturbance which is a tiny fraction, perhaps one-tenth of a thousand-millionth of the Earth's attraction, on the object being tested. *MVP.*

Gravity gradient The rate of change with distance of the strength of a gravitational field. Tidal forces arise because of gravity gradients (*see* TIDES). The magnitude of the gravity gradient at a distance R from a mass M is proportional to M/R^3. It follows that the greatest gravity gradients (and, therefore, the strongest tidal forces) are to be found close to highly compressed massive bodies such as NEUTRON STARS and BLACK HOLES. In the absence of other forces an elongated satellite will align itself along the gravity gradient of its parent planet so that its long axis points toward the centre of the planet. This effect can be used to stabilize the orientation of an artificial satellite. *IKN.*

G

Gravity wave detectors Devices intended for the detection of gravitational waves.

A passing gravitational wave is expected to stretch and squeeze a solid body or to cause small displacements in the separation between two masses. In principle, these effects could be measured. In practice, few, if any, events are likely to produce displacements larger than 1 part in 10^{17}, so that two masses separated by a metre would be displaced by an amount smaller than a nuclear particle.

Current detectors include solid bars (cooled, to minimize noise) and interferometer systems, which attempt to measure the relative displacement of two or more "freely" suspended mirrors (multiple reflections are used to increase the effective separation between the test masses). Present-day systems are still too insensitive to have any reasonable expectation of detecting gravity waves, but they are capable of further development. Possibilities for the future include measuring the relative displacements between the Earth and satellites or spacecraft. *IKN.*

Great Bear The constellation of Ursa Major. Its seven main stars make up the PLOUGH or Dipper pattern and are circumpolar from Britain.

Great circle A circle (on the surface of a sphere) whose plane passes through the centre of a sphere. Examples are the equator, or a meridian (a great circle passing through the poles).

Great Red Spot of Jupiter A huge anti-cyclonically-rotating meteorological eddy, which currently is 14,900 miles (24,000km) long and 6,800 miles (11,000km) wide and centred at a latitude of 22°S. *See also* JUPITER.

Great Red Spot of Jupiter: huge eddy

Green Bank Observatory The major facility of the National Radio Astronomy Observatory in West Virginia, USA. It comprises a range of instruments including a 298ft (91m) dish and an interferometer.

Green flash This is purely an atmospheric phenomenon seen normally for a few seconds as the last remnant of the setting Sun vanishes below the horizon. It can also be seen as the Sun rises over the horizon. During the last seconds of visibility, the strong red colour of the Sun suddenly changes to a vivid green. The duration of the phenomenon increases with increasing latitude, ie, as the angle of descent of the Sun decreases, and it can last for minutes in the polar regions. It is seen most frequently during the summer months when the sky is very clear and the horizon well defined over the sea. Cool weather and the absence of a red sky tend to favour production of the flash. *HGM.*

Greenhouse effect A phenomenon whereby the surface of a planet is heated by the trapping of infrared radiation by the atmosphere. This mechanism is important in the atmospheres of the Earth, Venus and Titan. The increased burning of fossil fuels is increasing the amount of carbon dioxide in the Earth's atmosphere. Sunlight is transparent to the CO_2 in an atmosphere, which absorbs and re-emits the infrared radiation, increasing the surface temperature. On Venus, this mechanism has created the currently hostile surface temperature of 737K.

Greenwich mean time (GMT). Mean solar time on the Greenwich meridian, which was chosen as the world's prime meridian by the International Meridian Conference in 1884 when they proposed "the adoption of the meridian passing through the centre of the transit instrument at the Observatory at Greenwich as the initial meridian for longitude."

That instrument still survives in working order, though it went out of use in 1954.

The conference then went on to define what they proposed as the fundamental unit of time: "That this universal day is to be a mean solar day; is to begin for all the world at the moment of mean midnight of the initial meridian, coinciding with the beginning of the civil day and date of that meridian; and is to be counted from zero up to twenty-four hours."

Over the next 40 years, the international TIME ZONE system was adopted country by country for civil use, the standard time in each zone differing from GMT by an integral number of hours.

In 1928, the INTERNATIONAL ASTRONOMICAL UNION recommended that GMT should in future be known as UNIVERSAL TIME for scientific purposes. Today, that usage has extended to legal and other purposes, though the term GMT is still used in many contexts.

In 1766, MASKELYNE published the annual *Nautical Almanac and Astronomical Ephemeris* based on the Greenwich meridian, giving the navigator for the first time the means of finding longitude at sea. That publication was so successful that, by 1880, 72 per cent of

the world's shipping tonnage had the Greenwich meridian on their charts. That, and the fact that the vast North American rail network already used today's time-zone system, was the main reason why Greenwich was chosen in 1884 to be the world's prime meridian, the basis of the world's time. *HDH.*

Greenwich Observatory It was the need to establish a reliable system of navigation, more particularly a practical way of finding the longitude from the rolling deck of a ship at sea, which led to the founding of an astronomical observatory overlooking London's River Thames at Greenwich. In the last months of 1674, a mysterious Frenchman by the name of Le Sieur de St Pierre announced that he had discovered such a method. Since the first nation "to solve the problem of the longtitude" would gain an enormous naval advantage, a Royal Commission was appointed to examine St Pierre's claim and to report to King Charles II "upon how farre it may be practicable and usefull to the Publick". The Commission engaged the services of an up-and-coming young astronomer, John FLAMSTEED, who told them that while St Pierre's idea might work in principle, the basic astronomical data for use at sea was simply not available. When this was relayed to the King he immediately appointed Flamsteed his "astronomical observator" with instructions to "apply himself with the most exact care and diligence to the rectifying the tables of the motions of the heavens .... for the perfecting the art of navigation". Flamsteed was to be paid a meagre £100 a year and a site at Greenwich was chosen for the observatory,

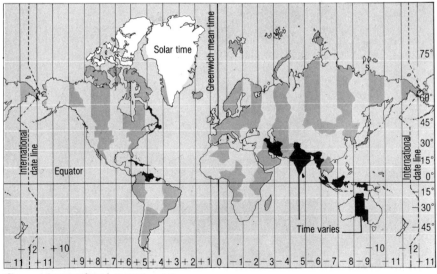

Greenwich mean time: international time

mainly on the advice of Sir Christopher Wren, designer of the first buildings.

Flamsteed's 43 years as ASTRO-NOMER ROYAL were filled with an enormous variety of activities, though toward the end of his life his dispute with NEWTON and HALLEY obscured his achievements. Halley succeeded Flamsteed as second Astronomer Royal in 1720 and devoted himself largely to measuring the apparent place of the Moon against the background stars, the information that was lacking in St. Pierre's proposed method of finding the longitude. Subsequent Astronomers Royal, especially James

Greenwich Observatory: designed by Wren

BRADLEY, Halley's successor, greatly increased both the quality and quantity of observations, though a long drawn-out dispute about the ownership and publication of results interfered with their dissemination for over 100 years after the founding of the observatory. However, with the first publication of the *Nautical Almanac* (1766) under the guidance of Nevil MASKELYNE, the fifth Astronomer Royal, the requisite astronomical information to enable mariners to determine their position at sea was at last available and King Charles' 1675 warrant had been satisfied. Apart from astronomical tables, successful navigation also depended upon the availability of accurate marine chronometers and sextants and the observatory played an important part in developing new and convenient forms of these for use at sea.

Until the beginning of this century the work of the Royal Observatory was directed more to determining the movement and precise location of astronomical objects rather than understanding their physical nature. However under W.H.M.CHRISTIE, the eighth Astronomer Royal, several major new instruments appeared at Greenwich and the work of the observatory took on a recognizably modern

character. As photography began to replace the eye as a detector, it became increasingly apparent that the polluted skies of London were unsuitable for astronomical research, but it was not until 1948 that the observatory was moved to the darker but often cloudy skies of Herstmonceux Castle in Sussex, leaving the buildings of the original Royal Observatory and the site of the world's prime meridian to be developed as a museum and historical monument.

At Herstmonceux the construction of the 98in (2.3m) Isaac Newton Telescope began immediately and the observatory grew rapidly in complement and capability to accommodate its new role as provider of major national facilities for British astronomy, a task shared with the Royal Observatory in Edinburgh. From the beginning it was obvious that Herstmonceux was a poor place for a large telescope and it was clear that if Britain was to be a significant force in world astronomy an overseas site was needed. The British half-share in the 153in (3.9m) ANGLO-AUSTRALIAN TELESCOPE, while it provided an enormous impetus to United Kingdom optical astronomy, merely underlined the need for major facilities in the northern hemisphere. A vital step in this direction was taken in 1979 with an agreement with the Spanish government, in conjunction with Denmark and Sweden, to build a major facility on La Palma in the Canary Islands. Since then Germany, the Netherlands and Eire have joined in.

Now, on what appears to be an ideal site for astronomy, major facilities are appearing. The rebuilt Isaac Newton telescope, complete with a new, 100in (2.5m) mirror has been moved from Herstmonceux and is operational. It has been joined by the 168in (4.2m) WILLIAM HERSCHEL TELESCOPE, completed in 1987. The Danish automatic meridian circle and a Dutch 39in (1m) telescope are also functional. This major new venture for the Royal Greenwich Observatory has been undertaken at a time of severe financial constraint and reduction in staff numbers which has delayed the completion of both telescopes and instruments. Despite these difficulties, Britain has at last one of the finest observatories in the world beneath the clear dark skies of the Canary Islands.

There are plans to transfer the RGO to Cambridge, but at the time of writing (early 1987) Herstmonceux is still operating. The time service, which

made Greenwich a household name, is no longer as important as it once was, but the preparation and publication of the *Astronomical Almanac,* carried on jointly with the United States Naval Observatory, is continued, together with other publications related to navigation. These enduring and important links with the origins of the Royal Greenwich Observatory provide an essential reference frame for the work of modern astronomers throughout the world. *DFM.*

Gregorian telescope Reflector designed in 1663 by Scottish astronomer James GREGORY. A parabolic primary mirror with a central hole reflects light to a concave elliptical secondary placed outside the focus, and thence back through the hole to an eyepiece behind the primary where it gives an erect image. It was popular in the 17th and 18th centuries with makers such as SHORT, as the secondary was easier to make than the convex secondary required for the CASSEGRAIN, and the tube length was shorter than that of a NEWTONIAN. It is not much used today, as the Cassegrain is even more compact. *TJCAM.*

G

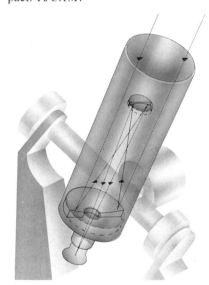

Gregorian telescope: designed in 1663

Gregory, James (1638–75). Scottish mathematician. In 1663 he described the principle of a reflecting telescope, though he had no practical skill and did not attempt to make a telescope himself.

Grigg, John (1838–1920). New Zealand amateur astronomer who discovered comets Encke in 1898; 1902II; 1903III; and 1907II. He supplied one of the first time services in New Zealand.

G

Grigg-Mellish Comet 1907II long-period comet; once thought to have a period of 164 years, but now believed to be much longer.

Grigg-Skjellerup Comet Found in New Zealand in 1902 by J.GRIGG and rediscovered in 1922 by J.Skjellerup. Period 5.1 years; last seen 1982.

Grimaldi Lunar walled plain 6°S, 68°W, 120 miles (193km) in diameter, west of Oceanus Procellarum and never well seen from Earth. It has no central peak, and a floor which is regarded as the darkest point on the Moon.

G ring A ring in Saturn's system, lying between the orbits of Janus and Mimas.

Gruithuisen, Franz von Paula (1774–1852). German astronomer. He was an energetic observer of the Moon and planets, but had a lively imagination — he even announced the discovery of a "lunar city". He originated the impact theory of lunar crater formation.

Grus (the Crane). A southern constellation.

Brightest stars

Name	Visual Mag.	Abs. Mag.	Spec.	Distance (light-yrs)
α (Alnair)	1.74	−1.1	B5	69
β (Al Dhanab)	2.11	−2.4	M3	173
γ	3.01	−1.2	B8	230
ε	3.49	1.4	A2	80

Then follow Iota (3.90) and Delta[1] (3.97). Grus is the most distinctive of the "Southern Birds"; the whiteness of Alnair contrasts with the orange of Al Dhanab. In the line of fainter stars, Mu and Delta are optical pairs.

G stars Stars of spectral type G have surface temperatures between 5,000K and 6,050K on the main sequence, and between 4,600K and 5,600K among giants. The coolest G-type supergiants are about 4,200K. G dwarfs range in mass from 0.8 to 1.1 times that of the Sun; the giants and supergiants have generally evolved from more massive stars farther up the main sequence.

Almost all G stars are dwarfs with compositions similar to the Sun. A few are subdwarfs with lower metal abundances and a small fraction of the G giants have peculiar abundances that give them the name barium stars. The G stars are all slow rotators (unless in a short-period binary system).

Bright examples (G stars look yellowish in colour): the Sun and Alpha Centauri G2 V, Capella G8 III. *BW*.

Guericke Lunar walled plain, 33 miles (53km) in diameter; 12°S, 14°W, making up a group with Fra Mauro, Bonpland and Parry in the Mare Nubium. It is irregular, with broken walls. Apollo 14 landed in this region.

Guide star A star on which the guide telescope is tracked to ensure that the main telescope is accurately aligned on its subject, for the duration of an astronomical exposure.

Gum, Colin (1924–60) Australian astronomer who made a photographic survey of the southern Milky Way to look for new nebulæ and after whom the GUM NEBULA was named.

Gum Nebula An extensive nebula in the constellations of Puppis and Vela which spans 30°–40° of sky and includes the Vela supernova remnant. It is named after the late Colin GUM.

Guo Shoujing (1231–1316). Chinese engineer, mathematician and astronomer. In mathematics he developed the equivalent of spherical trigonometry; he put this to great astronomical use.

A very accurate observer, Guo Shoujing was also a noted designer of precision astronomical instruments. He constructed a giant tower for measuring the Sun's altitude, a large armillary sphere for determining celestial positions, and a vast clockwork-driven celestial globe. Guo Shoujing is best remembered, however, for devising the first equatorial mounting of an astronomical instrument. *CAR*

Guzman Prize A prize offered in France in 1901 to be awarded to the first man to contact beings from another planet — Mars being excluded as being too easy! The prize has yet to be claimed.

Hadriaca Patera One of the oldest recognizable volcanoes of Mars, located at 31°S, 267°W on the northern rim of the Hellas basin. It has a 37-mile (60km) central caldera, but little relief.

Hædi Nickname for the three stars Eta, Zeta and Epsilon Aurigae, near Capella. It means "Kids" (since Capella was known as the She-Goat).

Hæmus (Montes Hæmus). Bounding Mare Serenitatis to the south-west, this intensively striated, fault-controlled lunar range, some 218 miles (350km) long, has rounded summits attaining 6,560ft (2km) in height.

Hagecius Lunar crater in the Vlacq group; 60°S, 46°E; diameter 50 miles (81km). Its walls are broken by smaller craters.

Halation ring The bright ring around a point image on a photograph caused by reflection and scattering within the emulsion.

Hale, George Ellery (1868–1938). American astronomer; originally a solar observer, who invented the SPECTROHELIOGRAPH and discovered the magnetic fields of SUNSPOTS; but who is best remembered for his part in setting up very large telescopes — the 40in (101cm) refractor at Yerkes, and the 60in (152cm) and 100in (254cm) reflectors at Mount Wilson — and persuading millionaires to finance them. He also masterminded the 200in (508cm) Palomar reflector, though unfortunately it was not completed until well after his death. *PM*.

George Hale: finding funds for telescopes

Hale Observatories For a time the observatories of MOUNT WILSON, near Pasadena, and MOUNT PALOMAR, near San Diego, were combined as the "Hale Observatories". However, the arrangement did not prove to be satisfactory, and was discontinued.

Hale reflector the 200in (508cm) reflector at PALOMAR OBSERVATORY.

Hall Crater on Phobos, the larger of the two satellites of Mars.

Hall, Asaph (1829–1907). The discoverer of Phobos and Deimos, the two tiny satellites of Mars. He made the discovery from the United States Naval Observatory, Washington DC, in 1877.

Halley, Edmond (1656–1743). Second ASTRONOMER ROYAL, astronomer and polymath.

In astronomy Halley devised a highly successful method of determining the Sun's distance from transits of Venus. He also showed that comets travel in elliptical orbits about the Sun,

Edmond Halley: father of geophysics

and made the first scientific prediction of a comet's reappearance — the 1758 return of HALLEY'S COMET.

He opened up stellar astronomy, discovered that stars have motions of their own, first suggested that nebulae were clouds of gas, and discussed both novae and the infinite nature of the universe. Halley also completed a long series of observations of the Moon and stars when appointed Astronomer Royal in 1720, at the age of 64. His other astronomical work included studies of meteors, planetary orbits, the aurora, and eclipses of both Sun and Moon.

Halley was the father of modern geophysics, developing the first truly scientific theory to account for the changing variation between true and magnetic north. He tried to use variation to determine longitude at sea, voyaging twice across the Atlantic for this purpose. His resulting charts introduced isogones (lines joining places with

equal variation). He wrote about tidal phenomena, and broke new ground by calculating the age of the Earth from physical principles.

A pioneer meteorologist, Halley wrote on trade winds and monsoons, the evaporation of sea water, and the relationship between barometric pressure and height above sea level. He wrote also on thermometry and optics.

A noted mathematician and linguist, publishing papers on historical and modern mathematics, Halley also laid the foundations of modern life assurance. A practical man, he invented new diving equipment and floated a company to salvage wrecks. With diplomatic gifts he used for promoting science, Halley was much loved by his contemporaries, who thought of him as the "greatest astronomer of his age". *CAR.*

Halley's Comet Brightest of the comets whose paths we can predict, and the only one which has been seen regularly with the naked eye. Halley's comet has been observed and recorded for more than 3,000 years; it was first noted by the Chinese in the winter of 1059/1058BC, and there is a chance that it is the comet referred to in a report from 2467BC. Halley's comet travels around the Sun in a highly elliptical orbit having an eccentricity of 0.967. When farthest away it lies below the orbit of the planet Neptune, at a distance of 35.295AU. It returns to perihelion about every 76 years; the last occasion was on February 9 1986. At closest it is only 0.587AU from the Sun, between the paths of Mercury and Venus.

The comet is named after the second English Astronomer Royal, Edmond Halley, not because he discovered it but because he was the first person to calculate its path around the Sun. Halley realized that the comets seen in 1531 and 1607 were the same as the one he saw in 1682. He predicted that the comet would return in 1759, and it did. Since then the comet has been named in Halley's honour. We now have observations of Halley's Comet at every return since that of 240BC. Its appearance in 1066 shortly preceded the Norman conquest of England, and the comet is depicted in the Bayeux Tapestry with the Saxon courtiers looking on horrified and King Harold tottering on his throne. In 1301, the comet was seen by the Florentine painter Giotto di Bondone, and he used it as a model for the Star of Bethlehem in his painting "Adoration of the Magi" in the Arena Chapel in Padua. In 1456 there is a story that Pope Calix-

tus III may have condemned the comet as an agent of the Devil.

Unfortunately, at some returns of the comet it does not pass close to the Earth and so we do not get a very good view of it. This was so during the latest return in 1985/86, and the next, that of 2061, will be no better. At closest the Earth was still 39 million miles (63 million km) from the comet in April 1986, and this is far from a good close approach. In AD837 the two bodies

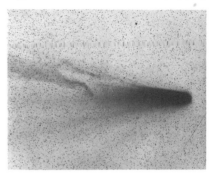

Halley's Comet: brightest periodical comet

passed only 3,700,000 miles (6 million km) apart, and the comet was a breathtaking spectacle, with a head as bright as Venus and a tail stretching for 90° across the sky.

The recent return of Halley's Comet in 1985/86 has been the most important. In early March 1986 a flotilla of spacecraft passed close to the comet, sending back to Earth an immense quantity of data on its structure and composition. The central part of Halley's Comet is a tiny elongated nucleus, shaped a bit like a peanut, about 9 miles (15km) long by 5 miles (8km) wide. This nucleus is made up primarily of dirty ice, which is protected from the tremendous heat of the Sun by an insulating layer of very black dust which surrounds it. The nucleus is one of the darkest objects in the Solar System. Its surface has been likened to a piece of black velvet. The dust layer cracks in places when violently heated by the Sun, forming deep fissures which expose the underlying ice. Jets or streamers of gas and dust shoot out from these fissures, giving rise to all the activity of the Comet. Every time Halley's comet passes around the Sun it loses about 250 million tonnes of material. It is estimated that if this rate of loss is maintained then the comet will survive for about another 170,000 years, before all the ices have been dissipated. *JWM.*

Hamal The star Alpha Arietis, of the second magnitude.

H

H

Hamburg Observatory Also known as the Bergedorf Observatory, Hamburg, was the site of the world's first Schmidt camera, constructed in 1930. Its main mirror was 17in (44cm) in diameter, but unfortunately it was destroyed in World War II. The site has been important for cometary discoveries; comet Schwassmann-Wachmann I was found there in 1928. More recently, Dr. Lubos Kohoutek, a Czech astronomer working at Hamburg, has found several comets, including the famous one discovered in March 1973, Kohoutek 1973 XII, which never became as brilliant as expected. The Hamburg Observatory lies at an altitude of 135ft (41m) at 53.48°N, 10.24°E. *JWM*.

Hansen Lunar crater in the Crisium area; 44°S, 83°E, diameter 22 miles (36km).

Harbinger Mountains A formation in the Imbrium area of the Moon; not a true range, but "clumps" of highlands.

Harding, Karl Ludwig (1765–1834). German astronomer, for some years assistant to J.H.SCHRÖTER and subsequently professor at Göttingen. He discovered the asteroid Juno, observed variable stars and nebulæ, and was the author of an excellent star atlas.

Harpalus Deep lunar crater, 32 miles (52km) across, near the boundary of Mare Frigoris; 53°N, 43°W.

Harriot, Thomas (*c* 1560–1621). English mathematician, astronomer and surveyor. He started telescopic observations the same year as Galileo. He drew the earliest Moon map, studied Jupiter's satellites, and also derived the Sun's rotation period.

Hartmann, Johann (1865–1936). German astronomer; discovered the presence of interstellar matter by studying the spectrum of the SPECTROSCOPIC BINARY Delta Orionis in 1904. He developed a comparison photometer, and a spectrocomparator to measure radial velocity.

Harvard College Observatory The first major observatory in the United States.

Harvest moon The full moon nearest the autumn EQUINOX, in the northern hemisphere. Because the Moon's daily motion along the ECLIPTIC is then bringing it northward, the RETARDATION is then at a minimum, and the Moon may therefore rise no more than 15 minutes later each night.

Hathor (Asteroid 2340). Discovered in 1976; it has a period of only 0.76 year, and is only a few kilometres across.

Hayashi track The evolution of a pre-main sequence star may be pictured in the HERTZSPRUNG-RUSSELL DIAGRAM as three distinct phases. An initial rapid collapse of the protostar, which is poorly understood, moves the star leftward in the H-R diagram out of the so-called Hayashi forbidden zone. This is followed by a large decrease in luminosity with only a small increase in effective temperature, called the Hayashi track. During this phase the star is almost entirely convective with only a small radiative core. Finally, the star becomes radiative and approaches the main sequence along the HENYEY TRACK *RFJ*.

Haystack Vallis Valley on Mercury; 4°N, 46°W.

HDE 226868 A blue supergiant star which is the optical component of the X-ray source Cygnus X-1. It has a mass of about 20 solar masses, and is a member of a binary system of period 5.6 days which may well contain a BLACK HOLE.

Heaviside Formation on the Moon's far side; 10°S, 167°E.

Heaviside layer More correctly known as the Kennelly-Heaviside layer, this alternative name for the Earth's ionosphere is no longer used. After Marconi's successful transmission of radio signals across the Atlantic, it was A.E.Kennelly and O.Heaviside who first suggested that an electrically conducting region existed in the atmosphere at high altitudes capable of reflecting radio waves back to Earth.

Hebe (Asteroid 6). Discovered by Hencke in 1837. Its diameter is 121 miles (195km).

Hecates Tholus Prominent shield volcano on Mars; 32°N, 210°W.

Hector A preceding Trojan asteroid having a diameter of 118 miles (190km), twice the size of any other Trojan. It is dumb-bell shaped, the light from it varying by a factor of 3.1.

Heemskerck Rupes Ridge on Mercury: 27°N, 125°W; adjoining Planitia Budh.

Heis, Eduard (1806–77). German expert on variable stars.

Heisenberg, Werner Karl (1901–76). A German physicist who devised the Uncertainty Principle of quantum mechanics. He was awarded the Nobel Prize for physics in 1932.

Helene. Small satellite of Saturn, co-orbital with DIONE. Its dimensions are 22 × 20 × 18 miles (36 × 32 × 30km). It was formerly known as 1980 S6; it was discovered in that year by Lacques and Lecacheux.

Heliacal rising Strictly, the rising of a celestial body simultaneously with sunrise. More commonly, it refers to the date when the body first becomes visible in the dawn sky. The heliacal rising of SIRIUS was used by the ancient Egyptians to predict the Nile flooding.

Heliocentric theory Theory that the planets revolve around the Sun. Proposed by COPERNICUS in 1543 to replace the GEOCENTRIC THEORY where the Earth is the centre of the universe.

Heliometer A double-image micrometer in the form of a telescope with a split objective lens. Used historically for accurate measurement of very small angles.

Helios probes Two probes launched in 1974 and 1976 designed to study the Sun and interplanetary space. They both approached the Sun to within 26.7 million miles (43 million km).

Heliosphere The region of space within which the interplanetary magnetic field and solar wind dominate over the interstellar medium. Its boundary probably lies at a distance of some 50–100AU.

Heliostat Mirror equatorially-mounted and clock-driven to follow the apparent motion of the Sun. The reflected ray is directed to feed a fixed telescope.

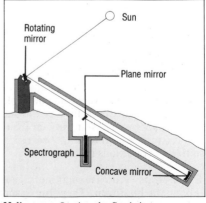

Heliostat: reflecting the Sun's image

Helium The second lightest chemical element, symbol He, helium is also the second most abundant element in the universe. Helium has an atomic number of 2 and an atomic mass of 4.0026 (denoted by $_2^4$He); ie, a helium nucleus comprises two protons and two neutrons.

Of the identifiable mass content of the universe, about 72 per cent is hydrogen and 27 per cent is helium. The ratio of the number of hydrogen nuclei (protons) to helium nuclei is about 11:1.

Helium is believed to be generated by nucleosynthesis in stars. The basic reaction, which is thought to power all MAIN SEQUENCE stars, involves, essentially, the FUSION of hydrogen to form helium. The two principal reactions are the PROTON-PROTON CHAIN and the CARBON-NITROGEN CYCLE.

The present abundance of helium in the universe cannot be explained by nucleosynthesis in stars (the "helium problem") but, according to the Big Bang theory of cosmology (see colour essay), most of the helium observed today was synthesized during the first few minutes of the history of the universe.

Helium was first identified in the solar spectrum by Janssen and Lockyer in 1868 as a set of lines which did not correspond to any known element on Earth. It was named "Helios", the Greek for "Sun". Helium was first identified on Earth in 1895 as a gas released by radioactivity in the mineral clevite. Helium is the lightest of the "noble" gases — those elements which are virtually inert chemically. Its boiling point is 4.2K (-268.9°C) and its melting point (at standard pressure) is the lowest of any element (0.9K, -272.2°C). It is used to cool certain astronomical detectors for infrared work in order to minimize thermal noise. *IKN.*

Helium flash A theoretically predicted event in the evolution of a lower mass star. After the hydrogen in the core of the star has been exhausted, and the star has become a red giant, the central temperature will rise to the point at which helium FUSION commences suddenly. It is unlikely that the explosion will reach the surface of the star. In a more massive star, helium fusion will commence gradually because of different conditions in its core.

Helium stars Stars lacking the usually abundant hydrogen. They are formed when a star ejects its hydrogen-rich

envelope (producing, for example, a planetary nebula), exposing its helium core. The R Coronæ Borealis stars are examples of helium stars. The term "helium star" was originally used for B-type stars.

Helix Nebula (NGC7293). A bright planetary nebula in AQUARIUS. It has the largest angular diameter of any known planetary nebula (about 0.5°) and lies at a distance of some 450 light-years (140 parsecs).

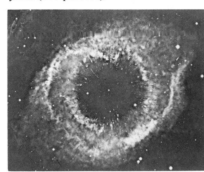

Helix Nebula: 450 light-years away

Hellas Hellas appears on early maps of Mars as a roughly circular region of ochrous hue. Images obtained by Mariner 9 and Viking Orbiter spacecraft reveal it to be the largest of Mars' impact basins, with a diameter of 1,100 miles (1,800km). It is situated south of the Martian equator, in a hemisphere predominantly well above datum. Hellas itself lies well below this level and at its deepest point — near to the western rim — descends to $2\frac{1}{2}$ miles (4km).

Large sections of the rim are either missing or difficult to trace due either to erosion or later burial by younger deposits. The most continuous section lies on the western side where it passes into the blocky mountains of Hellespontus Montes. Viking photographs reveal there to be several annular, inward-facing scarps outside of the main rim, together with a small number of large ancient volcanic structures, such as Amphitrites and Hadriaca Pateræ. The floor of Hellas is occupied by rather complex plains units upon the surface of which are small impact craters and some short channel systems. *PJC.*

Helmholtz, Hermann von (1821–94). German scientist, who was the first to propose that the Sun derives its energy by slow contraction.

Henbury craters. Set of at least 12 meteorite craters near Alice Springs, Central Australia, formed about 4,700

years ago by a body which fragmented just before impact.

Henderson, Thomas (1798–1844). Second Director of the Cape Observatory (1831) and first Astronomer Royal for Scotland (1834). His observations in 1832 provided the first parallax of Alpha Centauri.

Henry, Paul & Prosper (1848–1905) and (1849–1903). Optically and mechanically talented brothers employed at the Paris Observatory who photographically discovered nebulosities in the PLEIADES and whose successes helped to stimulate the *Carte du Ciel* concept — an international attempt, begun in 1896, to chart the entire sky.

Henyey track The evolutionary path in the HERTZSPRUNG-RUSSELL DIAGRAM of a star from the base of the HAYASHI TRACK to the main sequence. Along the Henyey track both effective temperature and luminosity increase such that luminosity, L, is proportional to temperature, T ($L \propto T_e^{4/5}$). During this phase energy transfer within the star is largely radiative rather than convective, in contrast to the Hayashi track. The length of the Henyey track, increase in T_e, is greater for high-mass stars than low-mass stars although the time spent on this track is shorter for the high-mass stars. Finally, before reaching the main sequence, the star's luminosity dips slightly due to the expansion of the star caused by the onset of nuclear burning. *RFJ.*

Hephaistos (Asteroid 2212). An Apollo object, discovered in 1978. Its period is 3.17 years.

Heraclides Promontorium One of the lunar capes bordering Sinus Iridum. The other cape is Laplace. Between them there are faint indications of a former wall to the Sinus, which leads off the Mare Imbrium.

Herbig-Haro objects A class of small, faintly luminous nebulæ independently discovered by G.Herbig and G.Haro in the 1950s. HH objects are irregular in outline, contain bright knots, and are found in regions rich in interstellar material. Their spectra reveal a weak continuum dominated by EMISSION LINES from hydrogen, oxygen, nitrogen and iron.

The association of HH objects with star-forming regions initially hinted that they were recently formed stars, but the relatively rapid changes in some

H

H

of them, with knots vanishing or changing shape, ruled this out. Then the discovery of a faint nebulosity similar to an HH object near to the star T Tauri suggested that HH objects were produced when the powerful STELLAR WIND from a T TAURI star collided with nearby gas and dust. The energy released during this collision would excite the gas, producing the emission

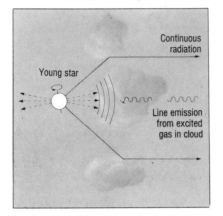

Herbig-Haro objects: current theory

lines, but could not explain the weak continuum radiation observed.

An alternative idea was put forward when infrared sources, which proved to be young stars deeply embedded in dust, were discovered near to HH objects. Since such stars have many emission lines in their spectra, it was

Herbig-Haro objects: hidden stars

proposed that HH objects were simply dust clouds which reflected the light from hidden stars. Unfortunately this did not explain why the continuum radiation detected from the HH objects was so weak.

These two ideas have now been reconciled and the current picture of an HH object is of a very young star hid-

den from Earth by an interstellar cloud. The powerful and asymmetric stellar wind of this star crashes into parts of the cloud, producing the emission lines. The light from the star is reflected by other parts of the cloud, which are unaffected by the stellar wind, allowing the continuum radiation from the star to be detected, albeit weakly. *JKD*.

Hercules Though named after the great hero of mythology, Hercules is not very conspicuous.

Brightest stars

Name	Visual Mag.	Abs. Mag.	Spec.	Distance (light-yrs)
β (Kornephoros)	2.77	0.3	G8	100
ζ (Rutilicus)	2.81	3.0	G0	31
α (Rasalgethi)	3 – 4	–2.3v	M5	220
δ (Sarin)	3.14	0.9	A3	90
π	3.16	–2.3	K3	390
μ	3.42	3.9	G5	26

Next come Eta (3.53), Xi (3.70), Gamma (3.75), Iota (3.80), Omicron (3.83), 109 (3.84), Theta (3.86), Tau (3.89) and Epilson (3.92).

Rasalgethi is a semi-regular red star, with a rough period of about 100 days. Zeta is a fine close binary, with a period of 34 years; the separation is below 2″. The most interesting telescopic objects are the two globular clusters, M13 (visible with the naked eye) and M92 (easy in binoculars). *PM*.

Hercules Lunar crater, 47°N, 39°E. Hercules is 42 miles (67km) in diameter, approximately 9,840ft (3km) deep, and has a dark lava floor sporting one 8-mile (13km) crater.

Hercules X-1 A binary X-ray PULSAR comprising a visible star, HZ Herculis, and what is believed to be an accreting neutron star. The pulsar period is 1.24 seconds and the orbital period, 1.7 days. There is also a 35-day modulation of the X-ray emission possibly due to precession of the accretion disk.

Herculina (Asteroid 532). On June 7 1978 Herculina occulted a star, the data indicating that it had a diameter of 135 miles (217km) and was a binary asteroid with a 31-mile (50km) diameter companion about 620 miles (1,000km) away.

Hermes (Asteroid 1937UB). An Apollo-type asteroid discovered October 28 1938 by K.Reinmuth at Königstuhl Observatory near Heidelberg, Hermes

passed within 500,000 miles (800,000km) of the Earth, closer than any other asteroidal object on record. Only a few kilometres in diameter, it was lost after only a few days.

Herodotus Lunar crater, 23 miles (37km) in diameter, adjoining Aristarchus; 23°N, 50°W. It marks the start of the great winding Schröter Valley. Unlike Aristarchus, Herodotus does not have bright walls or a central peak.

Herschel, Caroline Lucretia (1750–1848). German astronomer born in Hanover. She came to England in 1772 and assisted her much older brother William in both his musical and astronomical life. When they moved to Datchet and Slough she started sweeping the sky herself, picking up a good many nebulae (including the Sculptor Galaxy, NGC253, in 1783) and eight comets (P/Encke, in 1795, 1786II, 1788II, 1790I, 1790III, 1792I and 1797). She returned to Hanover in 1822, after the death of Sir William. Caroline produced a revised version of FLAMSTEED's catalogue (in 1798) and was awarded the Gold Medal of the Royal Astronomical Society in 1828. *DWH*.

Herschelian reflector A type of telescope developed by Sir William HERSCHEL in the late 18th century. It uses a paraboloidal primary as in the NEWTONIAN, but the primary is tilted so that the image is formed to the side of the incoming light, obviating the need for a secondary mirror: there is thus no central obstruction. The eyepiece looks down the tube, with the observer facing the main mirror, but to one side out of

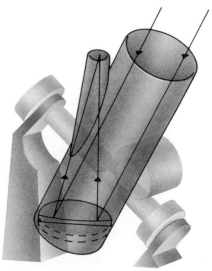

Herschelian reflector: tilted mirror

the incoming light. The optical arrangement is also known as the front-view system.

Herschel made many telescopes that were the best of their day, culminating in ones of 20ft (6m) focal length, 18in (48cm) aperture; and 40ft (12m) focal length, 48in (122cm) aperture, then the world's largest.

The tilted mirror introduced distortions in the image, and the Herschelian was soon superseded by the achromatic refractor and the silver-on-glass Newtonian. *TJCAM*.

Herschel, John (1782–1871). Son of Sir William Herschel. He read mathematics at Cambridge, and in 1813 became a Fellow of the Royal Society. He continued his father's astronomical work, and in his early years paid special attention to double stars. Then, in 1834, he went to the Cape of Good Hope, taking with him an excellent telescope with a focal length of 20ft (6m). Before returning home, in 1838, he produced the first really detailed survey of the southern stars; he catalogued 1,707 clusters and nebulæ and well over 2,000 binary stars. His results were published in 1847, and two years later he wrote a popular book, *Outlines of Astronomy*, which for many years was regarded as the standard text. He was also a pioneer of photography. *PM*.

Herschel-Rigollet Comet Found 1788 and seen again in 1939, period now 155 years.

Herschel, William (1738–1822). Pioneer of stellar astronomy and perhaps the greatest observer of all time. Herschel was born in Hanover and trained as a musician. In 1757 he came to England and in 1766 became organist to a chapel in Bath. Attracted to astronomy by his bedtime reading, he experimented in the 1770s in making reflectors for observation of distant objects, and became highly skilled in grinding and polishing large mirrors and in making eyepieces of high magnification. In 1776 he built a 20ft (6m) reflector with 12in (31cm) mirrors, but with a primitive mounting. In 1783, however, he built another 20ft reflector, this time with 18in (46cm) mirrors and a stable mounting that supported the observer in safety. Although in 1789 Herschel completed a monster 40ft reflector with 48in mirrors, he had now exceeded the scale appropriate to his mounting and the 40ft was not a success. The "large"

20ft, however, served Herschel for one of the greatest observing campaigns in the history of astronomy, and in the last years of his life it was refurbished under his supervision by his son John, who later took it to the Cape of Good Hope and extended his father's observations to the southern hemisphere.

With his home-made telescopes Herschel became a skilled observer. His first serious move was to examine all the brighter stars and to list those that were double. Following a suggestion of GALILEO, he hoped to determine stellar distances by observing the apparent ("parallactic") motion of the nearer

William Herschel: discovered Uranus

component of a double star relative to the more distant component which would then serve as a virtually fixed reference point. In all Herschel published catalogues of some 848 double stars, but John Michell had already used a probability argument to show that most doubles must be physically connected binary systems rather than the chance or optical doubles needed for the measurement of parallax. Herschel confirmed this visually when he re-examined some of his doubles in 1802, providing the first direct evidence that attractive forces (presumably gravity) operated beyond the Solar System.

On March 13 1781 Herschel was using a 7ft (214cm) reflector in his search for double stars when he came across an object that immediately struck his experienced eye as "curious". A few days later he again observed it and found that it had moved, and so was a member of the Solar System. Herschel assumed it was a comet, but it proved to be the planet now known as Uranus, the first to be discovered in recorded history. It had

been observed many times previously, but Herschel was the first observer able to detect at a glance its non-stellar nature.

The discovery brought him immediate fame, and his allies in the scientific establishment successfully lobbied the king to grant Herschel a royal pension that would enable him to devote himself completely to astronomy. Herschel thereupon moved to the neighbourhood of Windsor Castle, his only duties being to show the royal family the heavens when they so requested. His pension he supplemented by making reflectors for sale, even after marriage in 1788 to a rich widow brought him financial security.

An early (1783) essay in stellar astronomy, and one that owed nothing to Herschel's prowess as an observer, consisted of an attempt to determine the direction in which the Solar System is moving: the "solar apex". Herschel argued that the known proper motions fell into a pattern that implied the solar apex is near Lambda Herculis, a result very close to the modern value, though this closeness was partly a matter of luck.

In 1783 Herschel decided to advance the understanding of nebulae, first by acting the natural historian and systematically sweeping the sky for specimens of nebulae. In twenty years he "collected" some 2,500 to add to the hundred or so previously known. In the 1770s he had, he believed, detected changes in the Orion Nebula, which therefore must be small and close at hand, and not a vast star system disguised by distance. Such nebulae consisted of "true nebulosity". They could be distinguished from distant star systems by their milkiness, in contrast to the mottled appearance of the star systems. But observations of the Omega Nebula and the Dumbbell Nebula in 1784, in which both kinds of nebulosity were present in the same object, persuaded him that he had been mistaken, and that in fact all nebulae were star systems, the milky being the more distant. Some such systems, including the Orion Nebula and Andromeda Nebula, "may well outvie our milky-way in grandeur" and be complete galaxies. In papers on "the construction of the heavens" in the 1780s, he discussed the development of star systems in time, under the action of gravity. He also used star counts to estimate the distance to the borders of our Galaxy in any given direction: assuming that within the borders the stars are distributed uniformly, he

H

H

counted the number of stars visible in a given direction and converted the number into the (relative) distance to the border in that direction.

He had, however, been puzzled by "planetary nebulæ", which looked nebulous and yet seemed to have a planetary-like disk. In 1790 he came across what was in fact another planetary nebula, but one that was large enough to appear to him as a "nebulous star", consisting of a central star surrounded by a halo. Herschel decided the star was condensing out of the surrounding nebulosity. "True nebulosity" therefore existed, and in subsequent papers he extended his existing cosmogony backward in time, deploying nebulae from his catalogues to illustrate how diffuse clouds of nebulosity might collect into smaller, more concentrated clouds, which in turn would condense into scattered stars and thence into clusters of stars. Meanwhile, light being given out from stars would eventually collect into diffuse clouds to begin the cycle over again.

The study of variable stars had revived in the 1780s, and Herschel decided to facilitate their discovery by compiling elaborate catalogues of "the comparative brightness of stars", in which the stars of a constellation were arranged in delicately descending sequences of apparent brightness, the intention being that variation in a particular star would disturb the order of the sequence and so be readily detected.

Herschel was a maverick astronomer who presented a threat to the received wisdom. No one could deny his achievements as an observer and a telescope builder, or as the first man in history to discover a planet. But he was lacking in the basic skills of the professional, resistant to advice, and committed to speculation. He imported the methods of natural history into astronomy, and he devoted himself to the study of "the construction of the heavens", which hitherto had been the province of philosophical speculators of questionable credentials. But Herschel had free access to the pages of *Philosophical Transactions*, and so new questions, new methods and new answers came to the attention of the next generation of astronomers. *MAH.*

Hertzsprung, Ejnar (1873–1967). Danish astronomer who in 1911 discovered the giant and dwarf subdivisions of late-type stars.

Hertzsprung-Russell diagram A plot of the absolute magnitude of stars against

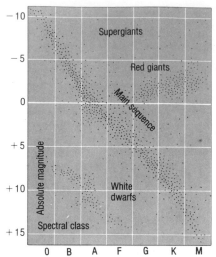

Hertzsprung-Russell diagram (A): see text

their spectral·class. Equivalently, a plot of the luminosity of stars against their surface temperature.

In the early 1900s, Ejnar Hertzsprung in Denmark and Henry Norris Russell in the United States independently began to consider how the brightnesses of stars might be related to their spectra. In 1911, Hertzsprung plotted the apparent magnitudes of stars in several clusters against their spectral class. In 1913, Russell plotted the absolute magnitudes of stars in the solar neighbourhood against their spectral class. Both astronomers were quick to emphasize the non-random patterns they saw in the arrangement of the data on these graphs. Such graphs have become extremely important in modern astronomy and are today called Hertzsprung-Russell diagrams.

A Hertzsprung-Russell diagram, or H-R diagram, is any graph on which a parameter measuring stellar brightness is plotted against a parameter related to a star's surface temperature. For example, the ordinate of the graph may be apparent magnitude, absolute magnitude, absolute bolometric magnitude, or luminosity. The abscissa may be spectral class, colour index, or surface temperature.

Two equivalent H-R diagrams, A and B, are shown with this article. A resembles Russell's original graph. Each dot represents a star whose absolute magnitude and spectral class have been determined from observations. Most of the data lie along a broad line called the main sequence that extends diagonally across the diagram. The Sun (absolute magnitude +5, spectral type G2) is a typical main sequence star. A second prominent grouping of data points represents very large, cool stars

called red giants. In the lower left corner, a third grouping identifies compact, hot stars called WHITE DWARFS.

An H-R diagram can also be drawn by plotting the luminosity of stars against their surface temperature, as shown in·B. Notice that temperature increases toward the left. The spectral classes OBAFGKM constitute a temperature sequence and the first H-R diagrams were drawn with the hot O stars on the left and the cool M stars on the right.

These H-R diagrams show that the total range in stellar brightness is 27 magnitudes, corresponding to a factor of 10^{11} in luminosity, and the range in the surface temperature of stars is from 2,200K to 50,000K. The size of a star is related to both its luminosity and surface temperature as indicated by the dashed lines. Most main sequence stars are roughly the same size as the Sun. White dwarfs are about the same size as the Earth while red giants and supergiants can be as big as the Earth's orbit.

The statistical distribution of data on an H-R diagram demonstrates that stars fainter than the Sun are far more numerous than those brighter than the Sun. In a given volume in the solar neighbourhood, about 90 per cent of the stars are main sequence stars, about 10 per cent are white dwarfs and about one per cent are red giants or supergiants.

The existence of three main groupings on the H-R diagram means that a typical star passes through three very different stages during its life. Calculations performed on computers reveal how the luminosity and surface temperature of a star changes as it evolves. With this information, the path followed by an evolving star can be plotted on an H-R diagram.

The evolutionary tracks of contracting PROTOSTARS show that they change rapidly as gravitational contraction compresses and heats their cores. When a protostar's central temperature reaches several million degrees, hydrogen burning begins and the evolutionary track stops at the main sequence. Consequently, main sequence stars are comparatively young and have hydrogen only recently ignited burning at their cores.

Pre-main sequence evolutionary tracks deposit new-born stars along the main sequence in agreement with the MASS-LUMINOSITY RELATION, so that the more massive a star is, the brighter it is. Thus the main sequence is a sequence in mass, as well as temperature and luminosity. These tracks also show that the more massive a star is, the more rapidly it evolves.

As a star consumes hydrogen, it gradually becomes slightly brighter and cooler. When the hydrogen supply is exhausted, the star's core contracts, its atmosphere expands, and it becomes a red giant. Ageing stars spend their final years as bloated red giants.

Stellar evolution calculations indicate that massive stars can end their lives with supernova explosions that leave behind NEUTRON STARS and BLACK HOLES. These are detected at non-visible wavelengths so the H-R diagram has no relevance to the post red giant evolution of massive stars.

A star less massive than about three solar masses can eject as much as half its mass at the end of its life, producing

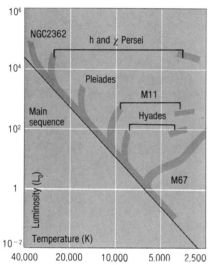

Hertzsprung-Russell diagram (B): see text

a planetary nebula. The exposed, burned-out core of the star then contracts to become a white dwarf. Detailed calculations of this process do indeed produce evolutionary tracks that rapidly take low-mass stars from the red giant region in the upper right of the H-R diagram to the white dwarf region in the lower left.

Stellar evolution calculations relate the H-R diagram to time, and thus these graphs can provide information about the ages of stars. Such age dating is especially applicable to star clusters.

A very young cluster, such as NGC2362, consists primarily of main sequence stars. However, the more massive a star is, the more rapidly it consumes its core supply of hydrogen and become a red giant. Since the main sequence is a progression in stellar mass, the stars at the upper end of the main sequence are the first to leave. As the cluster ages, its main sequence

"burns down like a candle", as stars of progressively decreasing mass and luminosity become red giants. M67, for example, is an old cluster because only those stars less massive and less luminous than the Sun still remain on the main sequence.

Horizontal branch stars are believed to be low-mass stars that have recently ignited helium in their cores via the HELIUM FLASH. In an asymptotic red giant, helium burning is occurring in a thin shell surrounding the star's centre.

Between the horizontal branch stars and the red giant is a gap. When the evolutionary track of a low-mass star takes it across this region, the star pulsates as an RR LYRÆ VARIABLE. This gap is at the lower end of the so-called instability strip, a vertical region in the middle of the H-R diagram where stars are unstable against radial oscillations. When a massive star passes through this region, it pulsates and is called a CEPHEID VARIABLE.

The H-R diagram is a valuable tool wherein observational and theoretical information are blended to produce a deep understanding of the stars. *WJK. See also* related articles on magnitudes, star types and STELLAR EVOLUTION.

Hesiodus Lunar crater south of the Mare Nubium; 29°S, 16°W; 28 miles (45km) in diameter. It is the companion of the larger Pitatus. A prominent cleft runs south-west from it.

Hevel Lunar walled plain, 76 miles (122km) across; 2°N, 67°W. It is a member of the Grimaldi chain. Hevel has a modest central elevation; a convex floor and rather low walls; it contains a system of clefts.

Hevelius, Johannes (1611–87). Polish astronomer and instrument maker.

A rich man, he built at Danzig (Gdańsk) what, for a time, was the world's leading observatory. It was destroyed by fire in 1679.

Hevelius is famous for his *Selenographia* with its lunar map, and his books about comets and observing equipment.

Hexahedrites Iron meteorites poor in nickel.

Hidalgo (Asteroid 944). An unusual asteroid, travelling on a comet-like orbit which takes it from the inner edge of the asteroid belt (2.0AU) out to 9.7AU, beyond the orbit of Saturn. It is 18 miles (29km) in diameter. The period is 14.2 years.

High Altitude Observatory A division of the National Center for Atmospheric Research, with headquarters in Boulder, Colorado. It operated a solar observing station with coronagraphs at Climax, high in the Colorado mountains, from 1940 to 1972, when the observing facilities were moved to Hawaii. The staff also made many successful eclipse expeditions.

Himalia Jupiter's sixth satellite, discovered by Perrine in 1904. It has direct motion but may well be a captured asteroid. Its diameter is about 115 miles (185km), so that it is much the largest of Jupiter's satellites apart from the Galileans.

Hippalus Lunar bay, 38 miles (61km) across, leading out of Mare Humorum; 25°S, 30°W. It is associated with clefts.

Hipparchus An ancient, flat-floored crater, 84 mile (135km) in diameter, located near the centre of the Moon's Earthward-facing hemisphere. It is somewhat irregular in outline with a degraded rim.

Hipparchus of Nicæa (first quarter of 2nd cent.–c 127BC). Greek astronomer who probably spent most of his later life at Rhodes; he is sometimes known as Hipparchus of Rhodes.

Little is known of Hipparchus' life and the only book of his to survive is a commentary he made on the work of ARATUS. Most information about him comes from the *Almagest* of PTOLEMY; from this it is clear that Hipparchus was probably the greatest observational astronomer of antiquity.

Hipparchus was also a noted geographer and a fine mathematician. He seems to have been the first to construct a table of chords — a mathematical quantity approximately equivalent to the sine in modern trigonometry. This allowed astronomical calculations to be made more readily.

Astronomically, Hipparchus determined the lengths of the four seasons, measured the length of the year to within 6.5 minutes, and derived mathematical theories to express the apparent motion of the Sun and Moon, basing his work on observations from Babylonia as well as those he made.

Hipparchus also determined the sizes and distances of the Sun and Moon, computing his results from observations of eclipses of the Sun and Moon. Finding the Sun's parallax too small to measure, he assumed a value based on the accuracy of his observing

H

equipment. This was almost five times too large, giving him a value for the Sun's distance almost five times too small, and a figure more than 16 times too small for the Sun's diameter. Nevertheless, his results were a vast improvement on previous estimates.

Hipparchus' values for the Moon were far more successful. He found the Moon's average distance as 63 times the Earth's radius (modern value 60.4), and its diameter 0.33 times that of the Earth (modern value 0.27). He is most noted for his discovery of the precession of the equinoxes. *CAR.*

Hipparcos The satellite Hipparcos, which is due to be launched by the European Space Agency in 1988, will be the first-ever satellite to be designed for ASTRO-METRY. By systematically scanning the sky over a 2.5-year mission, the relative positions, annual PROPER MOTIONS and TRIGONOMETRIC PARALLAXES of some 120,000 selected stars will be measured. The average accuracy, which depends on magnitude, for each of these astro-metric parameters, will be about 0″.002. The effective limiting magnitude will be about 12, but the majority of stars in the programme will be brighter than magnitude 11.

The principle of measurement is direct triangulation using a telescope which can view simultaneously two directions separated by a constant angle of 58°. The satellite rotates at approximately 11 revolutions per day about an axis perpendicular to the two fields of view. The direction of this axis of rotation is changed slowly so that the whole sky is scanned many times during the mission.

The measuring device in the focal plane of the telescope is a finely ruled grid of 2,688 parallel lines, with effective spacing of about 1″.2, perpendicular to the direction of scan. As a star from either field of view crosses the grid, the light is modulated, and the exact position in the field is given by the phase of the modulation. The whole field of view is 0°.9 square. In order to avoid confusion between the many stars which could be on the grid at the same time, an image dissector tube is used to count photons received in a very small area, known as the "instantaneous field of view", which can be switched rapidly from star to star, thus enabling the relative phases of signals from selected stars to be measured.

In order to derive celestial co-ordinates from the measured grid phases, it is important to know the direction of

the instantaneous axis of rotation of the satellite to within a small fraction of a second of arc. This is achieved by measuring the transits of stars with accurately known positions across an auxilliary grid system known as a "star mapper", which consists of four slits parallel to those of the main modulating grid and four chevron-shaped slits with arms inclined at 45°. Photons from the complete star mapper are counted by two photo-multipliers, one sensitive to blue light and the other to yellow. In deriving the satellite attitude, the combined output from the two photo-multipliers is used.

Although only selected stars are used for attitude determination, photons from the star mapper will be counted continuously throughout the mission. This, known as Project Tycho, will provide astrometric positions and two-colour photometry for all stars brighter than about magnitude 10. *CAM.*

Hirayama families *See* ASTEROIDS.

History of astronomy The science of astronomy is normally understood to start with the Greek civilization, especially the "golden age" of the 6th and 5th centuries BC. This is so because the Greeks were the first to understand that general principles could be derived by understanding particular occur-rences. Because astronomical objects are remote, this process has been regarded as the essence of understand-ing astronomy. The body of astronomi-cal knowledge thought for a long time to be most of the knowledge acquired in Grecian times was encapsulated in the work *Megistæ Syntaxis*, commonly called the *Almagest*, of the philosopher PTOLEMY of Alexandria.

Nevertheless, much knowledge existed previously and much existed elsewhere. Chinese and Indian civiliza-tions, mostly seeking astrological sig-nificance to what was seen in the sky, were able to establish concepts of the solar and lunar motions, the distinction between planets (wanderers) and the fixed stars, and to determine the obli-quity of the ecliptic. It is therefore still possible, especially by newly discovered artefacts, or new reading of records long known to exist, to make new dis-coveries in the early history of astron-omy. The main historical trend, however, follows on from the discover-ies of the Greek philosophers.

ARISTARCHUS (*c.* 280BC) and HIP-PARCHUS, one of the greatest of the ancient astronomers, who began work around 160BC, were able to draw on

much that went before, especially the work of the School of Pythagoras (*c.* 580–500BC), of Plato (*c.* 427–347BC) and ARISTOTLE (384–322BC), and on earlier and contemporary Baby-lonian astronomy. Aristarchus undoubtedly embraced the concept of a moving and rotating Earth, although unable to convince his hearers. Hippar-chus discovered PRECESSION by the apparent movement of the EQUINOX position along the ecliptic. Hipparchus was important also because he recog-nized that observation, when suffi-ciently refined, provides criteria for judging theory.

Ptolemy was able to incorporate extensive numerical detail in his *Alma-gest*, but some of this detail suffered from lack of appreciation of the influence of observational error. Occurrence of error was understood and indeed led to preference for philo-sophical interpretation of what was observed and development of complex mathematical abstractions. Thus circu-lar motions, superimposed one on the other, provided the technical material that had to be mastered in medieval scholarship for over a thousand years following Ptolemy. The subject matter of astronomy became intricate, inac-cessible and fossilized until a new spirit of enquiry dawned in the Renaissance of the 15th century.

European scholarship, in spite of pressures to conform to accepted ideas, gradually produced individuals who were able, using especially the contri-butions of the great Arabian observers of 800–1250, to question the correct-ness of views that had held sway for far too long. Nicolaus COPERNICUS (1473–1543) when quite young developed in his own mind the concept of the moving and rotating Earth, although he did not publish his great work DE REVOLUTIONIBUS until the year of his death. By this work the artificialities of the Ptolemaic-Aristotelian system became understood, although the prac-tical application of Copernicus' work was minimal because he retained the hypothesis of uniform and circular motion. Gradually the old ideas were overturned. Within one hundred years Tycho BRAHE (1546–1601), the great Danish observer, GALILEO Galilei (1564–1642), the Italian savant, and Tycho's pupil Johannes KEPLER (1571–1630), in Germany, had produced the material and ideas that were sufficient for modern astronomy to emerge.

While Galileo's observations with a telescope and experiments emphasizing the importance of scientific principles made the Copernican hypothesis plausible, or even compelling, Kepler's interpretation of Tycho's systematic naked-eye measurements of positions of celestial objects provided definite and eventually incontrovertible evidence. Kepler used elliptical orbits to represent planetary motions and his empirical laws concerning the time taken in those orbits led to the truly epoch-making work of Isaac NEWTON (1642–1727). In this work the physical cause of Kepler's empirical laws was ascribed to the force of gravity familiar at the Earth's surface but diminished by distance according to the law of inverse squares. To deal mathematically with the implications of this law and to demonstrate the elliptic, parabolic or hyperbolic form of planetary orbits, Newton arrived at the processes of differential and integral calculus. In this way astronomy at this time had a major impact on the rest of science.

Newton's success was particularly appreciated and stimulated by Edmond HALLEY (1656–1741), one of the first three ASTRONOMERS ROYAL at Greenwich. FLAMSTEED (1646–1711), Halley and BRADLEY (1693–1762) sought to use the best knowledge of astronomy for the purpose of improving the accuracy of navigation at sea and at the same time sought to understand better from the scientific viewpoint the more subtle effects shown up in their systematic observations. Halley's discovery of periodic orbits for comets and Bradley's discovery of the ABERRATION of light, matching RØMER's measurements of the VELOCITY OF LIGHT, and of NUTATION as the essential component of precession, were discoveries in pure science that had influence on practical procedures as well as establishing credence for the laws of gravity and the laws of motion.

There followed, during the latter part of the 18th century and through the whole of the 19th century, two main streams in the development of astronomy. On the one hand Newton's dynamical ideas and the associated mathematical analyses were developed, particularly in France and Germany, to an extent that permitted very detailed comparison with observation and merited greatly increased accuracy of observation, revealing small elements in the motion of the Earth that still require active investigation. This work was principally carried out in the great national observatories at Greenwich, Paris, Poulkova, Berlin and Washington. Eventually the theory of gravitation was found to be able to account

for the stable Solar System as regards all the main features but, with the emergence of the theory of RELATIVITY in the 20th century, has needed to be supplanted.

On the other hand, new types of enterprise were initiated by William HERSCHEL, a German musician working in England in the last quarter of the 18th century. He devised means for constructing large reflecting telescopes for himself rather than relying on the relatively small lenses produced by opticians. He proved the existence of binary star systems, observed nebulous objects, and in 1781 startled the intellectual world by discovering a new major planet subsequently named Uranus. Objects unknown previously such as diffuse nebulae, planetary nebulae, new satellites, and spiral nebulae were observed and catalogued by the 1840s, finally making obsolete the astronomy of the ancients. New cosmologies became possible and by the middle of the 19th century even speculation as to the plurality of inhabited worlds became fashionable.

A special advance occurred near the middle of the 19th century when it was found by KIRCHHOFF that the dark lines in the solar spectrum could be used to discuss the chemical elements present in the Sun and the physical conditions affecting the outer layers. The science of ASTROPHYSICS was created thereby and special astrophysical observatories were founded. It became realistic to study the stars as similar to, or different from, the Sun and the theme of STELLAR EVOLUTION has continued to be one of the most important features of astronomy. The emergence of the HERTZSPRUNG-RUSSELL DIAGRAM, which has remained a prominent tool for over fifty years, has epitomized this interest at a popular level.

With the rise of astronomy as a systematic activity in the United States in the second half of the 19th century the idea became established that the physical nature of celestial bodies as well as their mere motions could be understood. This required application of the laws of physics and chemistry discovered in the laboratory and verified by experimental methods to be applied to the conditions existing far from the Earth's surface. The advent of quantum theory was an essential part of this process. Also, quantitative appreciation of stars' properties was needed and the accurate determination of distances and brightness of astronomical objects became and remains of

crucial importance. This work stimulated the construction of large telescopes, especially in the United States, from the 1890s up to the present time, and the frontiers of the observed universe were extended outward in a relentless and spectacular manner.

In the period since 1945, the dominant contribution to the advance of astronomical knowledge and understanding has been the application of new technologies. The most important contributions have been through the application of electronics to new types of detectors of electromagnetic radiation at all wavelengths, especially in the radio bands, to measuring instruments, to computing methods, to control mechanisms, and to the achievement of observation from Earth satellites. Engineering advances have always gone hand in hand with advances in astronomy and, in an era when new design "philosophies" are appearing with increasing frequency, astronomy will advance at an increasing pace. The most notable achievements of the past 40 years include an understanding of most aspects of stellar evolution, recognition of essentially new types of object such as QUASARS and PULSARS, detection of the 3° blackbody background radiation, discovery of infrared sources, ability to probe the vicinity of planets and planetary satellites with space vehicles, and, perhaps above all in grandeur, the building up of plausible cosmologies dating back to the hypothetical "origin" of the universe. Some of these developments have been entirely unexpected, while others have been foreshadowed by early speculation. Borne on the shelves of every major astronomy library are the published records of the work of thousands of astronomers of the past who have contributed in small or large ways to the ever-expanding history of astronomy. *PAW.*

History of observatories *See* OBSERVATORIES.

Hoba Meteorite The largest single meteorite known, estimated weight, 60 tonnes. An ataxite, found 1920, near Grootfontein, Namibia, where it still lies in the ground.

Hodr Crater on Jupiter's satellite CALLISTO: 69°N, 87°W.

Hoffmeister, Cuno (1892–1968) Founder of the Sonneberg Observatory in the modern German Democratic Republic. Cuno Hoffmeister devoted most of his

H

H

life to studies of meteor phenomena and variable stars. He made many fundamental discoveries. He also started important projects to monitor the night sky photographically to identify VARI-ABLE STARS and establish their variations. He published two important books on meteors, namely *Die Meteore* in 1937 and *Meteorströme* in 1948. His classic work on variable stars, *Veränderliche Sterne,* was published in 1970, after his death. *JWM.*

Holmes' Comet Periodical comet (period 7 years) discovered in 1892, when it was a naked-eye object. It was seen again in 1899 and 1906, but then lost until 1964, when it was below magnitude 18. It has since been recovered at each return, but is very faint.

Homer Crater on Mercury, 200 miles (320km) in diameter; 1°S, 37°W.

Homestake Mine The Homestake Gold Mine is an operating mine, the largest producer of gold in the United States. This mine is the site of an underground laboratory devoted to astronomical observations. The facility is located 4,850ft (1,500m) below the surface in a space 30 × 60 × 32ft (9 × 18 × 10m) with auxiliary chambers. It was initially excavated for Brookhaven National Laboratory in 1965, to house an experiment to observe NEUTRINOS from the Sun. Solar neutrinos are detected by a radiochemical method that depends on the neutrino capture reaction, $v_e + {}^{37}Cl \rightarrow {}^{37}Ar + e^-$.

Since 1985, the Homestake underground laboratory has been operated

Homestake Mine: neutrino detector

by the University of Pennsylvania, and now houses the following experimental facilities: a scintillation counter telescope for observing energetic neutrinos, cosmic ray muons, and magnetic monopoles; an air shower detector on

the surface, operated in coincidence with the under-ground scintillation telescope (University of Pennsylvania); a radiochemical detector for studying electromagnetic interactions of energetic muons (Smithsonian Astrophysical Observatory); a Ge double beta-decay experiment (University of South Carolina-Battelle Northwest Laboratory); and the chlorine solar neutrino detector (University of Pennsylvania). *RD.*

Homunculus Nebula Cloud of gas and dust thrown out by the star ETA CARINÆ during the 19th century.

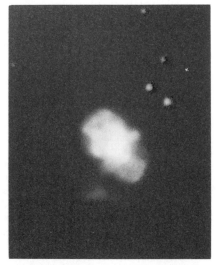

Homunculus Nebula: from Eta Carinæ

Hooke, Robert (1635–1703). English experimental philosopher and architect. Royal Society curator 1662, secretary 1677–82, Gresham professor of Geometry 1665. Many inventions in optics, horology, mechanics, etc.

HI clouds Zones of neutral hydrogen gas. Most of the universe by mass consists of hydrogen. Several forms are common: neutral atomic ("HI"); ionized ("HII"); and molecular (the hydrogen molecule is like a tiny dumbbell of two hydrogen atoms). The void between the stars is called "the interstellar medium" and several different regimes of hydrogen exist, distinguished by density and temperature. Threads and bubbles of very hot (millions of degrees Kelvin) but extremely rarefied "coronal" (like the solar corona) gas permeate the diffuse medium of warm (about 80K), neutral hydrogen embedded in which are much denser (perhaps 10,000–100,000 hydrogen molecules per cubic centimetre), cooler (only 10–30K) molecular cores.

Neutral hydrogen is detected by its

radio line emission at 21cm wavelength (1,420 Megacycles/s, in frequency). Large radio telescopes can "see" HI even in extremely distant galaxies. Different kinds of galaxy contain different proportions of HI. Our own Galaxy and other spiral or irregular galaxies contain an appreciable fraction of their total mass in HI. Ellipticals contain no (or extremely little) detectable HI.

In our Galaxy HI clouds are heavily concentrated to a flattened disk in which a spiral pattern occurs. Spiral arms are made conspicuous, even in very distant galaxies, by the groups of hot, high-mass, blue stars lying along the arms, like pearls on a string. These represent recent star-forming activity from the HI clouds in the arms triggered by global "spiral density waves". Not all HI clouds are capable of forming stars; only the densest ones that also contain molecular (carbon monoxide is a good tracer of the hydrogen) cores. A number of HI clouds lie out of the galactic plane and are very diffuse. Even dense hydrogen clouds are much more tenuous than the best laboratory vacuum achievable on Earth! *MC.*

Horizon The great circle on the CELESTIAL SPHERE 90° away from the observer's zenith. The term is also used to describe a boundary beyond which the observer cannot see (eg, "event horizon") or beyond which particles cannot yet have travelled (particle horizon).

Horologium (the Clock). An obscure constellation bordering Eridanus. Alpha (3.86) is the only star above magnitude 4.

Horrocks, Jeremiah (1619–41). English clergyman and amateur astronomer, who predicted and observed the TRANSIT of Venus in 1639. He carried out important mathematical work.

Horse's Head Nebula A dark absorption nebula in Orion, shaped like the head and mane of a horse and measuring some three light-years across, which is seen in silhouette against an emission nebula, IC434. It is located about half a degree south of Zeta Orionis, the most easterly star in Orion's belt, and a similar distance to the east of Sigma Orionis. IC434 is colliding with an extensive dusty molecular cloud on the east side of Orion, within which star formation is taking place. The Horse's Head is the most distinctive intrusion of that dark cloud into IC434. *IKN.*

Horse's Head Nebula: 3 light-years across

Hour angle The angle which the hour circle of a heavenly body makes with the meridian. *See also* CELESTIAL SPHERE.

Hour circle The great circle that passes through the celestial equator, a heavenly object, and the celestial pole. It measures declination. *See also* CELESTIAL SPHERE.

HII regions Volumes of space in which hydrogen is ionized. Most HII regions contain hot, high-mass, blue stars whose energetic outpouring of ultraviolet radiation includes many photons with sufficient energy to ionize atomic hydrogen (ie, rob it of its solitary electron). Spiral arms are traced by HII complexes, entire groups of recently-formed luminous blue stars, that lie like beads on a wire along the arms. Ionized hydrogen also produces radio emission, either continuous ("free-free" radiation as liberated electrons curve around hydrogen nuclei, namely protons) or at specific frequencies ("recombination line" radiation as some electrons recombine with protons to form new hydrogen atoms). High-mass (hot) stars form in dense clouds of gas and dust; consequently they are often optically invisible — obscured by the dust. Both radio and infrared (detecting thermal radiation from the dust mixed with the gas and heated by the embedded hot stars) techniques can render even these HII regions "visible".

The simplest HII regions are those that arise around single hot stars embedded in HI clouds. The radius out to which the gas is ionized is calculable given the spectral type and luminosity (yielding the surface temperature and radius) of the star. Such a "radiatively ionized" zone is termed a "Stromgren sphere". When substantial dust contaminates the atomic hydrogen, the Stromgren radius shrinks since dust competes with hydrogen atoms for the energetic stellar ultraviolet photons. Other HII regions can be created simply by the impact of a high-speed gas stream or blob on the surrounding medium. This process of "collisional ionization" may account for the presence of HII regions around cool, low-luminosity objects without the ultraviolet photons to cause radiative ionization but possessing appreciable STELLAR WINDS.

Abundant and conspicuous in the optical, radio and infrared, HII complexes characterize "starburst galaxies" in which high-mass star-forming activity was recently initiated. *MC.*

Hubble constant The parameter that gives the relative rate at which the scale size, R, of the universe is changing with time. It is also the rate at which distances between all "co-moving" objects — ie, those objects which have no individual motion relative to the fabric of space — are increasing. The units for the Hubble constant are such that its inverse is a time, which can very loosely be thought of as an "age" for the universe. More usually, however, the units employed are kilometres per second per megaparsec since, for small REDSHIFTS, galaxy recession velocities increase by this amount for each megaparsec of distance.

The Hubble constant is normally designated H_0, the zero subscript specifying that it is the expansion rate at the present epoch that is meant. The relative expansion rate changes with time and the Hubble parameter and its rate of change are related to the DECELERATION PARAMETER, q_0, by
$$H_0 = \dot{R}/R$$
$$q_0 = \ddot{R}/(R.H^2_0)$$
both values being calculated for the present epoch.

The search for good values for H_0 and q_0 has been the driving force behind much of observational COSMOLOGY in the last 50 years. Many different methods of obtaining values for H_0 have been devised. But they all have in common redshift and distance measurements of a set of similar objects carefully chosen to eliminate selection and measurement bias to the greatest extent possible. The best values of H_0 available at present lie in the range from about 50 to 80km per second per megaparsec. *AEW².*

Hubble, Edwin Powell (1889–1953). American astronomer, who carried out vitally important work in studies of the objects we now know to be galaxies, and was the first to give definite proof that they lie far beyond the Milky Way.

Hubble was born in Missouri. For a while he practised law, but in 1914 went to the Yerkes Observatory as a research assistant. Following military service in the United States Infantry from 1917 to 1919, he went to MOUNT WILSON OBSERVATORY, where he remained for the rest of his career. He paid special attention to the so-called spiral nebulæ, and decided to do his best to find out whether or not they were contained in our Galaxy. This involved using CEPHEID VARIABLES, and he was able to take advantage of the powerful 100in (254cm) Hooker reflector. By 1924 he was able to announce that the Cepheids, and hence the systems which contained them, were immensely remote — though in fact the distances which he gave for them were too low, because of an error in the Cepheid scale which was not discovered until many years later.

Together with his colleague Milton HUMASON, Hubble produced a CLASSIFICATION OF GALAXIES, from loose spirals and barred spirals through to elliptical systems; with modifications, his system is still accepted. He

Edwin Hubble: classifed galaxies

measured the RED SHIFTS of the spectra of galaxies, and established the all-important fact that the speed of recession increased with distance; this led on to what is called "Hubble's Law", which links distance with recessional velocity, though its precise value is still somewhat uncertain (*see* HUBBLE CONSTANT). The time that has elapsed since the beginning of the universe in its present form (ie, since the BIG BANG) is often referred to as Hubble Time, and the Space Telescope has been named the HUBBLE TELESCOPE in his honour.

Hubble was involved in the completion of the Hale reflector on MOUNT PALOMAR, and remained active in research until shortly before his death. *PM.*

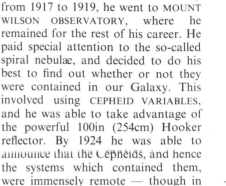

H

Hubble-Sandage variables *See* ERUPTIVE VARIABLES.

Hubble Space Telescope Probably the most sophisticated scientific satellite ever built, the Hubble Space Telescope (HST) is due to be launched into orbit 378 nautical miles (512km) above the Earth by the SPACE SHUTTLE in the late 1980s. It is the product of more than two decades of detailed planning, and by the time it flies will have cost around $1,400 million.

By operating above the Earth's atmosphere, HST has two big advantages over ground-based telescopes. First, the HST will avoid completely the problem of seeing. The excellence of its images will then be fixed only by the laws of optics, the quality of its mirrors, and how accurately and steadily the HST can be directed toward its targets, not the state of the atmosphere. HST should, in consequence, provide as a matter of course star images of only 0.1 arc seconds in diameter. In addition, even the Earth's upper atmosphere glows faintly. Second, from its orbit high above the Earth, HST will be able to detect radiation from the ultraviolet through the visible to the infrared part of the spectrum, a much broader band than equivalent ground-based telescopes.

The HST is essentially a very precisely engineered CASSEGRAIN telescope. Its launch weight will be approximately 11.6 tonnes, and it will be 43ft (13m) long and 14ft (4.3m) in diameter. It has four basic elements: the Optical Telescope Assembly (OTA); the Support Systems Module (SSM); the scientific instruments; and the solar arrays.

The SSM has been built by Lockheed Missiles and Space Company. Its main sections are the aft shroud enclosing the scientific instruments, the SSM equipment section (housing, for example, the spacecraft's computers and devices for directing the HST to different points of the sky), and the forward light shield, the main attachment point for the solar arrays.

The idea of putting a large telescope in space was discussed by some of the space pioneers of the 1920s. However, in 1946 Lyman Spitzer Jr. became the first astronomer to consider in detail the astronomical advantages such a telescope would possess over ground-based telescopes. Spitzer became the leading advocate of building a large telescope in space, but it was not until the late 1950s and 1960s that he began to win much support from his colleagues. The advent of NASA in 1958, and a growing awareness among astronomers of the potential of observations from space, convinced an initially small but soon growing group that a large space telescope could and should be built.

The initial plans focused on a 10ft (3m) telescope. But when NASA's plans to build such a telescope were presented to the United States Congress in 1974 they were judged to be too expensive. There then ensued a three-year struggle to win the political approval and support necessary to build the

Hubble Space Telescope: $1,400 million

telescope. This was not forthcoming until 1977 and required several economy measures. One was to reduce the size of the primary mirror from 10ft to 8ft (2.4m). The overall design too was substantially changed. In particular, the SSM's equipment section now surrounded the main ring of the primary mirror instead of sitting behind the mirror.

The explosion of the Space Shuttle "Challenger", and the subsequent grounding of the Shuttle fleet, has delayed the launch. Nevertheless, the HST should be such a big advance over existing optical/ultra-violet telescopes that it promises to have a profound effect on astronomy. *RWS.*

Huggins, Sir William (1824–1910). English amateur astrophysicist, born Stoke Newington, London. Pioneer in the application of spectroscopy and photography to astrophysics.

Huggins was educated at City of London School and privately. After a few years in business, he came into a modest inheritance and in 1854 built an observatory at his house in North London, with a 5in (13cm) equatorial by Dolland. In 1858, he acquired an 8in (20cm) objective by Alvan CLARK.

He found conventional positional and planetary astronomy not to his lik-

ing and turned toward the "new astronomy" in 1862 after reading of KIRCHHOFF's spectroscopic researches of the Sun. With his Tulse Hill neighbour, William Allen Miller, professor of chemistry at King's College, he set about applying Kirchhoff's methods to the stars. Together they devised a new instrument, a star SPECTROSCOPE, which was attached to Huggins's 8in Clark refractor.

In 1864 he made the discovery that certain nebulae were gaseous whereas spiral nebulae showed stellar-type spectra, concluding that, contrary to what was thought at the time, not all nebulae were aggregations of stars too faint to be resolved. In 1866–68, he was the first to examine a nova and a comet spectroscopically, and, with Miller in 1868, made the first stellar DOPPLER EFFECT measurement.

In 1870, the Royal Society placed a 15in (38cm) refractor and an 18in (46cm) Cassegrain reflector, both on the same mounting, at his disposal. With this he continued his work on stars and nebulae and made many advances in techniques for observing solar spectra. As early as 1863, he had attempted to photograph spectra but was frustrated by the limitations of the wet collodion process. In 1875, however, he photographed the spectrum of Vega on a gelatine dry plate, the first such application by an astronomer.

Huggins was knighted in 1897. He was President of the Royal Society from 1900 to 1905, and was awarded the Order of Merit in 1902. *HDH.*

Humason, Milton (1891–1972). American astronomer. He had no formal training, and was at first a mule driver at MOUNT WILSON OBSERVATORY, but he became an expert observer, and was Edwin HUBBLE's main assistant during the pioneer studies of the nature and distribution of galaxies.

Hun Kal Crater on Mercury, 4,900ft (1.5km) in diameter; it has been agreed that the 20° meridian passes through its centre.

Hunter's moon The full moon following the HARVEST MOON; it usually occurs in October in the northern hemisphere.

Huygens, Christiaan (1629–95) Dutch physicist and astronomer, born and died at The Hague. He studied maths and law at Leiden and Breda. In 1666 he commenced 15 years, work at the Bibliothèque Royale in Paris, before returning to the Hague.

Working with his brother Constantijn he became a skilful telescope maker, and developed the eyepiece which bears his name and is still used today. With his first telescope he discovered TITAN in 1655, and with a better instrument of 23ft (7m) focal length, he correctly interpreted in 1666 the nature of Saturn's rings. He established the wave nature of light, and explained refraction and reflection. He studied centrifugal force and gravity, discovered the laws of elastic collision, and invented the pendulum clock. He was elected FRS in 1663. *TJCAM.*

Hyades Recognized as a configuration of its own by the early Greeks — it is, eg, mentioned in Homer's *Iliad* — the Hyades derived their name from a group of daughters of Atlas and Æthra, half-sisters of the PLEIADES. Rising at dusk in autumn they were considered as foretelling the season's rains, "the rainy Hyades". Originally encompassing various parts of Taurus, the name was used later on for the third-magnitude stars lying to the west of bright Aldebaran, forming the V of the Bull's head; unlike in the Pleiades, the individual names of the sisters are not used for the Hyades stars.

From studies of STELLAR PROPER MOTIONS it was realized in the 19th century that the Hyades move together with numerous other stars in the surrounding parts of the sky in a general southeasterly direction, the Taurus stream. Refined proper motion measurements subsequently revealed that the Hyades share a common motion to within very narrow limits with an appreciable number of stars in their vicinity, and it was thus established that all these stars form an open STELLAR CLUSTER. Earlier this cluster was designated the Taurus moving cluster; the name "Hyades" was often used to denote the central denser clustering including the original stars. Today that name is generally used for the whole cluster; individual members are sometimes termed "Hyads".

The cluster covers a sky area several degrees in diameter, elongated in a direction roughly parallel to the galactic equator. Its extension in depth is also observable. Situated at a distance of about 46 parsecs or 150 light-years from the Sun, the Hyades form the nearest star cluster but one. It has been extensively studied by innumerable astronomers over the past 100 years.

The Hyades cluster is believed to contain a few hundred stars, most of them within a radius of 6pc from the

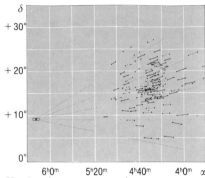

Hyades: the convergent point

centre. Although the cluster stars show a marked concentration toward the centre, the density is no larger than about 0.2 solar masses per cubic parsec. Therefore the cluster is susceptible to disruption in a comparatively short time, by tidal forces from the Galaxy, and especially by encounters with interstellar gas and dust clouds — in fact its estimated age of about 650 million years is some ten times longer than the expected time for disruption by cloud encounters! It seems that the cluster has so far managed to avoid disastrous cloud encounters altogether, despite its being situated close to the galactic plane.

As they are so close to us, there is no apparent extinction of light by cosmic dust in front of the Hyades stars, and also no dust has been found within the cluster.

The majority of the cluster member stars belong to the MAIN SEQUENCE, the brightest of these being of spectral type A2. It is from this fact that the age estimate is derived, since more massive stars have had time to evolve away from the main sequence. A few such stars are known in the Hyades; they are now yellow GIANT STARS. Also belonging to the cluster are a score of WHITE DWARF stars. A considerable share of the cluster members is believed to consist of BINARY STAR systems. The abundance of chemical elements heavier than helium ("metals") is enhanced relative to the Sun, on average by a factor of 1.3.

The central concentration, primarily of the more massive stars, indicates that the cluster is nearly relaxed, ie, the stars have had time to reach a smoothed distribution of velocities with near equipartition of kinetic energies. The relaxation is, however, not complete, as is seen from the somewhat ragged distribution of stars in the outskirts of the cluster.

The distance to the Hyades has been determined in several ways, generally

with rather concordant results. The cluster is near enough to permit decent measurements of TRIGONOMETRIC PARALLAXES for its stars. The distance may also be obtained from the apparent magnitudes of the stars, provided their ABSOLUTE MAGNITUDES are known. These may be determined in different ways, eg, by comparison with other, similar stars or with theoretical stellar models, through the Wilson-Bappu effect in giant stars, or from the MASS-LUMINOSITY RELATION for those stars in binary systems that have had their masses determined by means of KEPLER's third law.

Perhaps the most important method for deriving the distance to the Hyades is based on the discovery by L.BOSS in 1908 that the proper motions of the Hyades stars all seem to be nearly directed toward the same place in the sky, the convergent point (*see* accompanying diagrams). The internal motions in the Hyades are known to be small — the velocity dispersion is less than 1km/s — and thus it may be assumed that the cluster is neither expanding nor contracting, and also that it has no appreciable systemic rotation. Hence the cluster stars may be considered to follow nearly parallel paths through space, in the direction of the convergent point. That the proper motions seem to converge may thus be interpreted as a perspective effect similar to the apparent running together of the rails of a long, straight railway.

In the moving cluster method for distance determinations — which has also been applied to other clusters besides the Hyades — the position of the convergent point must first be established. Then one proceeds by measuring the RADIAL VELOCITIES for a number of cluster stars. From this and the angle between the line of sight to the star and the direction to the convergent point it is easy to calculate the total space velocity of the star, and also the component of the velocity perpendicular to the line of sight, the transverse velocity. The transverse motion shows up in the

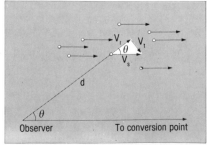

Hyades: determining distance

H

H

proper motion of the star across the sky. The size of the proper motion μ depends on both the transverse velocity v_t and the distance d to the star according to the formula $v_t = C\, d\, \mu$. Once μ and v_t are known, the distance may thus be readily obtained; C is a constant depending on the units used.

From the measured radial velocities the mean space velocity of the whole cluster is calculated, whereby the transverse velocity of any cluster star may be deduced. Its proper motion then yields its distance in the above way; the mean of all distances thus obtained is taken as the cluster distance.

This is then used to derive absolute magnitudes for cluster members of different kinds, and these are in turn used as calibrators for determinations of distances to other stellar aggregates, containing also intrinsically brighter objects. In this way the distance to the Hyades forms one crucial step on the distance ladder of the universe. It might thus be considered one of the most fundamental data in astronomy. *GW.*

Hydra The Watersnake, representing the monster killed by Hercules. It is the largest constellation in the sky.

Brightest stars

Name	Visual Mag.	Abs. Mag.	Spec.	Distance (light-yrs)
α (Alphard)	1.98	−0.2	K3	85
γ	3.00	0.3	G5	104
ζ	3.11	0.2	K0	124
ν	3.11	−0.1	K2	127
π	3.27	−0.1	K2	153
ε	3.38	0.6	G0	111

Then come XI (3.54); Lambda (3.61); Mu (3.81) and Theta (3.88). Hydra extends from near Gemini to the south of Virgo; it is recognizable because of its long line of stars. The MIRA VARIABLE R Hydræ can reach naked-eye visibility. There is also one bright open cluster, M48. *PM.*

Hydrocarbons Compounds of hydrogen and carbon of which thousands may be made because strong covalent bonds form between carbon atoms. Seven (CH, CH^+, CH_3C_2H, C_2H, C_2H_2, CH_4, C_4H) have been identified in interstellar material.

Hydrogen in space Hydrogen is by far the most abundant element in the universe, comprising about 75 per cent of its mass, with most of the remainder being helium. Paradoxically these light-weight, transparent elements and their transformation from one into another give rise to the most conspicuous features of the cosmos, the stars and galaxies that we strive to understand. All the other elements, including oxygen and iron, carbon, calcium, nitrogen and silicon which make our planet and our bodies are merely local concentrations of impurities in a universe dominated by the simplest element of all, a single proton with one electron in orbit around it.

Our most familiar contact with hydrogen is in the form of one of its two oxides, water (the other is hydrogen peroxide), or more rarely as the gas which appears when an acid and a metal react. It is widely used in industry, to make both metals and margarine and, in liquid form, as a convenient, if hazardous, rocket fuel. Beyond our immediate environment, the everyday chemistry of hydrogen is of little consequence. Within the Sun, for example, the immense pressures and temperatures of the interior turn hydrogen into a nuclear fuel, generating the energy that warms the Solar System.

More distant still, the Milky Way contains huge amounts of non-luminous material which is mostly hydrogen and a little helium mixed with various concentrations of tiny solid particles — dust. These clouds can vary in size from a few times the diameter of the Solar System to vast irregular aggregations hundreds of parsecs across, concentrated in a narrow band straddling the Milky Way. About half of the mass of our Galaxy is hydrogen; most of the rest is stars (again largely hydrogen) with the dispersed dust comprising less than one per cent of the total. Thus although hydrogen is itself invisible, it is associated with an impurity that actually hides most of the Galaxy and a substantial fraction of the universe from our view. The presence and location of hydrogen between the stars can be detected in several ways.

Emission nebulæ If the hydrogen gas is in the vicinity of hot stars, those with surface temperatures in excess of 30,000K (the Sun's surface is about 6,000K), the hydrogen gas is almost completely ionized. Ultraviolet radiation from these stars has sufficient energy to separate the hydrogen atom into a proton, the hydrogen nucleus, and an electron, which are then free to wander independently through the cloud. A very hot O-type star such as Sigma Orionis may ionize all the hydrogen within a sphere over 100pc in radius, transforming it from a cool dark cloud of neutral hydrogen (HI) into a softly-glowing patch of nebulosity known as an HII REGION. In such nebulae the gas is remarkably tenuous, with at most between 10^3 and 10^4 atoms (or ions) per cubic centimetre, an excellent vacuum by terrestrial standards.

The energy imparted by the exciting star or stars raises the temperature of the photoionized hydrogen to around 10,000K, but it is not its temperature which causes an HII region to glow; rather it is the energy released by the recombination of the electron with a proton to re-form an electrically-neutral atom of hydrogen. During such a recombination the electron may fall into any one of several precisely-defined energy levels around the proton, emitting a photon as it does so. The energy (ie, wavelength) of the emitted photon is restricted to one of several monochromatic EMISSION LINES characteristic of the element involved. In the visible part of the spectrum, hydrogen shows strong recombination lines at 434nm in the blue-violet (Hγ), 486nm in the blue (Hβ) and the very prominent red line of Hα emission at 656nm. Colour photographs show this mixture of monochromatic emission lines as a pinkish-red, typical of HII regions. Occasionally colours are tinged with emission lines from other elements, most often the green line of oxygen at about 500nm, which gives the red Hα emission a yellowish hue.

The complex interplay of photoionized gas, obscuring dust and bright stars gives rise to some of the most photogenic objects in astronomy. Well-known examples are the LAGOON NEBULA, M8 in Sagittarius, the magnificent nebula around ETA CARINÆ in the southern Milky Way and the famous ORION NEBULA, all of them visible to the unaided eye. These nebulae are the sites of star formation, with the young stars responsible for the nebulosity having recently been formed there. As they mature these stars will disperse the surrounding gas and the nebula will fade, leaving an open cluster of stars like NGC4577, the Jewel Box.

Stars create new kinds of nebulae as they move toward the end of their lives, returning hydrogen and other elements which have been processed in the nuclear furnace to the interstellar medium to take part in another round of star formation. A low mass star loses its outer layers to create a planetary nebula which is excited by the intense ultraviolet light of the stellar core left behind at its centre. More dramatic are the death throes of massive stars which explode as supernovæ at the end of their short lives. In both cases a hydrogen-rich emission nebula appears briefly and quickly fades as the star which created it is extinguished.

Radio detection of hydrogen In the hydrogen atom both the proton and its orbiting electron have axial spins, rather like the spin of the Sun and the daily rotation of the Earth in orbit around it. The spin on the proton may either add energy to, or subtract it from, the electron, depending on their relative orientations. Collisions between atoms can alter this condition and with the spins on electron and proton aligned the atom is in a slightly excited state. When the atom returns to its normal, ground state a photon of 21.11cm wavelength is emitted and can be detected by sensitive receivers attached to radio telescopes. Although this transition is an extremely rare event, there is so much hydrogen present that the 21cm radiation can be detected from quite small clouds.

Observations at 21cm show that the hydrogen in the Milky Way is distributed in a remarkably flat layer typically 100pc or less thick throughout the plane of the Galaxy. From our position within the plane we can detect the DOPPLER shift of the radiation from both the systematic circular motion due to rotation of the Galaxy and the smaller random motions within the clouds of gas. Since the propagation of 21cm radiation is hardly affected by intervening matter, much of the neutral hydrogen in the Galaxy is detectable. Rotation curves of external galaxies can be made in the same way and recently vast clouds of hydrogen, apparently devoid of luminous matter, have been detected in the space between the galaxies.

The hydrogen molecule (H_2) exists in space only where it is protected from energetic short-wavelength radiations from stars — in the dark clouds where dust filters out the ultraviolet light. Molecular hydrogen cannot easily be detected directly but it is usually associated with other, more massive molecules, which are detectable. In particular carbon monoxide (CO) is used as a tracer of molecular hydrogen and reveals the presence of massive molecular clouds throughout the Galaxy. Within these clouds the first stages of star formation occur, a process which eventually transforms the

cold, tenuous clouds of hydrogen into hot, dense stars. *DFM.*

Hydrus (The Little Snake). A far-southern constellation.

Brightest stars

Name	Visual Mag.	Abs. Mag.	Spec.	Distance (light-yrs)
β	2.80	3.8	G1	20.5
α	2.86	2.6	F0	36
γ	3.24	−0.4	M0	160

There are few interesting telescopic objects. Beta Hydri is the nearest fairly bright star to the south pole; Alpha lies close to Achernar.

Hygeia (Asteroid 10). Discovered in 1849 by de Gasparis in Naples. The fourth largest member of the main belt (254 miles/409km). Dark carbonaceous type. At closest approach (perihelion) Hygeia is 2.84AU from the Sun and at its farthest recession is 3.46AU away. Its albedo is 0.041.

Hyginus Rille (Rima Hyginus). Of the "linear crater-rille" type, with collapse-craters set along parts of it. Hyginus (6 miles/10km across) is the largest of these graben craters, located at a bend in the rille.

Hyperbola The conic section obtained when a right circular cone is cut by a plane which makes an angle with its base greater than that made by the side of the cone. It is an open curve (not closed like a circle or ellipse).

Hyperion The eighth satellite of Saturn to be discovered. It was initially viewed by W.C. and G.P.BOND in the United States and W.LASSELL in England at about the same time in 1848. The orbit of Hyperion lies outside of that of the large satellite, TITAN; they are in resonance with their orbital periods near a ratio of three to four. The mass of Hyperion is negligible compared with that of Titan.

In 1981, Voyager 2 took high-resolution pictures of Hyperion, revealing its irregular shape, large craters, and arcuate scarps. P.Thomas found it to be bi-axial, measuring 230 × 174 × 140 miles (370 × 280 × 225km). J. Wisdom, S. Peale, and F. Mignard predicted that Hyperion would be found to be tumbling in chaotic rotation due to its irregular shape, its eccentric orbit, and the gravitational effects of Saturn and Titan. This chaotic rotation has been confirmed by Voyager 2. *MED.*

Iapetus The second satellite of Saturn to be discovered. It was first observed by G.D.CASSINI in France in 1671. The orbit of Iapetus lies outside that of Hyperion. It is nearly circular and is highly inclined to Saturn's equator. It is the third largest satellite of Saturn, behind TITAN and RHEA, with a diameter of about 895 miles (1,440km). Telescopic observations since the time of Cassini showed that the leading hemisphere in orbital motion of Iapetus was dark, and the trailing hemisphere was bright. The difference in albedo is about two magnitudes.

Voyagers 1 and 2 encountered the Saturnian system in November 1980 and August 1981, returning a great deal of data regarding Iapetus, including many high-resolution pictures. The pictures confirmed that the leading hemisphere is covered with a very dark, reddish material. The images revealed a highly-cratered surface with many dark-floored craters in the bright region. The low density of Iapetus suggests that is composed primarily of ice. *MED.*

Iapetus: Saturn's third largest satellite

IAU *See* INTERNATIONAL ASTRONOMICAL UNION.

Icarus (Apollo asteroid 1566). Its orbit crosses Mercury's orbit. When last close to Earth it was detected by radar; its cross-section proved to be about 295ft sq (0.09km²). Its period of rotation is 2.27 hours.

ICE (*International Cometary Explorer*). Originally launched in 1978 as ISEE-3, this spinning spacecraft was renamed and diverted by means of lunar gravitational assist to fly through the tail of Comet Giacobini-Zinner on September 11 1985 — the first comet intercept in history. ICE passed HALLEY'S COMET in March 1986 on the sunward side at a distance of 17 million miles (28 million km).

IC2602 A large open cluster centred on the hot young O-type star, THETA CARINÆ, and faintly visible to the naked eye. Its distance is about 700 light-years.

Igaluk Crater on Jupiter's satellite CALLISTO: 5°N, 315°W.

Igneous rock A rock produced by the crystallization of a rock melt (magma). This may be extruded at the surface of a planet (extrusive rock) or may solidify at depth (intrusive rock).

Ikeya-Seki Comet Brilliant comet seen in 1965; among the brightest of the century.

Image intensifier An opto-electronic device mounted within a sealed glass tube, and operating in a vacuum, which uses electronic amplification to increase the brightness and contrast of the

incoming optical signal and its wavelength, but on the photo-emitting material from which the photocathode is formed. This is frequently a caesium compound which responds to wavelengths from less than 400 nanometres at the blue end of the spectrum to more than 700nm in the near infrared.

The electrons are accelerated down the intensifier tube by a dc potential of up to 40kV and are focused by an electric or magnetic field onto an anode, the phosphor coating of which forms the viewing screen. In addition to being accelerated the number of electrons may be increased by cascading successive stages giving an electronic amplification of several million times. The image formed at the anode is typically a pale, luminous green and is measurably brighter than the original optical image, giving perhaps an overall optical gain of up to five stellar magnitudes. The image so formed may be viewed directly, photographed or coupled to a television system.

The principal disadvantage of the image intensifier is that electronic focusing results in geometric distortion, which, while not more than a few per cent, does tend to reduce the resolution normally expected from the optical image. Other minor disadvantages are the fragility of the tube and the need

Inferior conjunction The passage of an INFERIOR PLANET (and a few asteroids and comets) between the Earth and the Sun, when they have the same RIGHT ASCENSION.

Inferior planets Mercury and Venus; which are closer to the Sun than the Earth is.

Inflationary universe Recent revision to the Big Bang cosmological model of the universe to explain the uniformity of the MICROWAVE BACKGROUND. *See also* colour essay on the BIG BANG.

Infrared astronomy The portion of the electromagnetic spectrum lying between the optical wavelengths to which our eyes are sensitive and the radio wavelengths is the infrared. It is normally considered to include wavelengths in the range 1 to 1,000 micrometers (μm). The Earth's atmosphere is opaque to much of this wavelength range, and astronomers wishing to study planets, stars, and galaxies in the infrared must confine their observations to the atmospheric "windows", or make their observations from high-altitude aircraft, balloons, or from space.

The prominent atmospheric windows in which telescopic observations can be made from ground-based observatories on high mountain tops (above about 10,000ft/3,000m) are as follows:

$1.0–2.5\mu$ $3.0–5\mu m$ $7.5–14.5\mu m$
$17–14\mu m$ $320–370\mu m$ $420–400\mu m$
$600–1,000\mu m$

The windows beyond $40\mu m$ are only partially transparent and can be used only at exceptional times when the water vapour content of the air above the observatory is less than about 1mm of precipitable water.

To avoid the Earth's atmosphere, infrared observations are made from high-flying aircraft, notably the KUIPER AIRBORNE OBSERVATORY (NASA), from unmanned balloons, and from earth-orbiting spacecraft. The INFRARED ASTRONOMICAL SATELLITE (IRAS) was the most productive of the latter projects. IRAS surveyed the entire sky at several infrared wavelengths for a 10-month period in 1983.

Astronomical sources of infrared radiation include the planets and other Solar System bodies, the stars and dusty regions in the Milky Way, other galaxies, and peculiar extragalactic objects of unknown nature. These sources emit continuous radiation (Continued on page 209

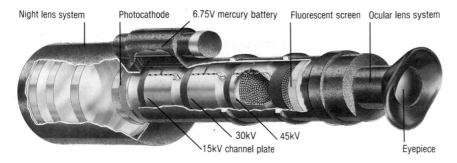

Night lens system   Photocathode   6.75V mercury battery   Fluorescent screen   Ocular lens system

30kV   45kV
15kV channel plate   Eyepiece

Image intensifier: a device for increasing brightness and contrast

optical image normally displayed by the telescope. It is a special category of photo-multiplier tube mounted at the focus of the telescope. The intensifier in the accompanying illustration would be adapted for astronomical purposes by omitting the night lens system (which the telescope's lens or mirror would replace) and by displaying the image on a television screen.

The optical image from the telescope is focused through a sealed window, usually of mica, onto a photocathode from which the incoming photons cause electrons to be liberated.

The number of electrons released depends not only on the intensity of the

for kilovolt supply voltages. *JCDM*.

Image photon-counting systems *See* ELECTRONIC DEVICES and PHOTOELECTRIC AIDS.

Inclination The angle between the plane of an orbit of a planet, asteroid or comet and that of the ECLIPTIC; for a satellite orbit, with the equator of its primary.

Indus (The Indian). A small constellation near Grus. The only star above magnitude 3.5 is Alpha (3.11; absolute magnitude 0.2, type K0, distance 124 light-years). Next comes Beta (3.65).

Interstellar Matter

Alexander Boksenberg, FRS
Director
Royal Greenwich Observatory

The universe at large seems to be mostly empty.
There are vast reaches of open space between the stars
in our Galaxy and it is clear that the same is true in
other galaxies; we term this *interstellar space*. There are enormous
distances between the galaxies. But it is a fact that
nowhere is the universe quite empty. At least a slight trace of
matter is present everywhere in space, although mostly
it is far more sparsely filled than the best laboratory vacuum.

The Orion Nebula, 450 parsecs from the Earth, is the nearest HII region and the best studied. Optically it is seen as a 4th magnitude fuzzy patch of light in the Sword region of the Orion constellation. On this long photographic exposure the nebula appears to extend about half a degree across. The beautiful series of loops and filaments are apparent mainly in the characteristic red light of hydrogen. The central part of the nebula is so bright that the four extremely hot stars of the Trapezium cluster, whose ultraviolet radiation ionizes the nebula and makes it glow, are hidden in this picture. These stars are probably less than one million years old. The dark region near the Trapezium is a thick neutral cloud which has not yet been ionized by the stars. Just behind the nebula is a huge molecular cloud, observed at radio wavelengths, showing concentrations associated with two strong infrared sources. One of these, called the Becklin-Neugebauer object, is a star in the process of formation; evidence for this lies in the presence of two masers, due to OH and H_2O. These regions are the densest parts of the molecular cloud and are contracting gravitationally.

Throughout the entire universe main constituents of interstellar matter are everywhere the same – dominantly hydrogen with some helium – but the proportions of the minor constituents differ widely between different locations. The total density and other properties of this matter also differ widely from place to place. Mostly the matter is gaseous but a small proportion is embodied in the form of minute solid dust particles. On average there is only about one atom per cubic centimetre and one dust grain per 100,000 cubic metres of space. Although the density of interstellar matter is extremely low, the volume of the space in a galaxy is so great that the total quantity of interstellar material is very considerable. Our own Galaxy contains about 10^{10} solar masses of material between the stars, making up about 10 per cent of its total mass. Most of this matter is distributed in the spiral arms and the disk of our Galaxy and is confined to a layer only a few hundred parsecs thick.

There is a continuing exchange of matter between the stars and the interstellar medium within the galactic disk. First, stars condense out of the primordial material which ultimately forms into galaxies, although it is not yet clear when in the early life of a galaxy this occurs. The less massive stars live on for many thousand millions of years and die relatively quietly. Other stars may shed some of their mass in more or less violent ways before they die. Some may produce, for example, visible shells such as the planetary nebulæ which can be seen to be expanding on photographs taken over several years. In the extreme, the very massive stars, many times more massive than the Sun, burn out after only a few million years and end their lives in a brilliant supernova explosion. In this way massive stars return to the interstellar medium much of the hydrogen and helium from which they were formed. But, what is more important, heavier elements than are present in the primordial gas, which have been produced in the centres of stars by nuclear fusion processes, also are shot out in large amounts during the explosion. And yet a further consequence of a supernova explosion is the formation of additional nuclei of iron and other elements, including very heavy elements even beyond uranium, which also participate in the event. All this material, ejected at enormous velocities, expands into interstellar space and eventually becomes more or less mixed with the gas already there.

The star cluster NGC 2264, *left,* is approximately two million years old and is one of the youngest open clusters known. A large number of stars is still in the process of being formed out of part of the cloud of gas and dust that is evident. The nebula, an HII region in a giant molecular cloud, is ionized by the ultraviolet emission from hot young stars inside it.

In the interior of the Trifid Nebulae, *right,* intense ultraviolet radiation from hot stars ionizes the hydrogen and causes the dominantly red emission that is manifest. A network of dark bands of interstellar dust is developing at the heart of this expanding HII region; globules of cold gas are present which are still not ionized by the stars and these may condense and give birth to new stars. The bright blue region to the north is a reflection nebula; not ionized, its bluish colour arises from the associated bright star whose light has been scattered back by dust.

The Ring Nebula in Lyra, *below left,* is a planetary nebula which has an apparent diameter of about 0.5 light years and is expanding at a velocity of almost 12 miles a second (19km/sec). It results from the ejection of the envelope of a red giant star which has now progressed to the white dwarf phase, evident as the central star. This very hot star ionizes the hydrogen and other gaseous constituents, and resultant fluorescence and electron collision effects gives rise to the observed colours.

The red glow of hydrogen dominates the emission in the expanding gaseous shell of the Aquila Planetary Nebula, *below right.*

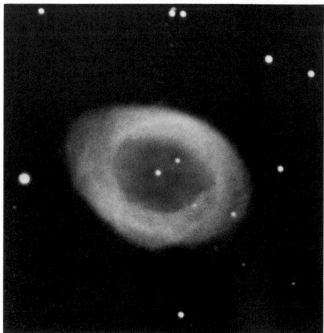

The brightest stars of the Pleiades Cluster, *right*, are each surrounded by blue nebulosities which are parts of the initial gas cloud out of which the cluster was formed. These are termed reflection nebulæ. The relatively cold gas around these stars is mixed with a substantial quantity of dust which diffuses the light from the neighbouring bright stars like a screen. The gas itself is too cold to become ionized so it does not glow as an HII region. The blue colour is partly due to the fact that the dust grains are of a size that gives them greater power for scattering blue light than for red and partly because the bright stars are bluish. The combed appearance shows the irregular distribution of dust grains within the nebulae. The regular circles around some of the stars, like the crosses through them, are optical effects of the telescope. The Pleiades Cluster actually contains more than 3,000 young stars and is situated at a distance of some 120 parsecs.

The Crab Nebula, *above*, is the remnant of a supernova explosion which occurred in 1054. It is about 2 kiloparsecs away and is now about 4 parsecs in diameter. A pulsar in rapid rotation (33 milliseconds period), near its centre, is the reservoir for the power radiated by the nebula. Streams of highly energetic electrons from the pulsar spiral in the magnetic field within the nebula and produce synchrotron radiation seen as the whitish emission which also shows a high degree of polarization. Radiation is emitted from this region at all wavelengths from radio to gamma rays. The orange filaments, which contain an amount of material close to one solar mass, are expanding at a velocity of 930 miles a second (1,500km/sec); these are the remains of the stellar envelope. They are heated and ionized not by the star but by the radiation from the rest of the nebula, which actually is of much lower mass than the filaments.

The supernova explosion that gave rise to the extensive Vela Supernova Remnant, *right*, probably dates back more than 10,000 years. The visible emission is at the edge of the shell separating the ambient interstellar medium from the expanding gas in the interior of the remnant, which has been heated by the shock wave from the explosion. Collisions between gas particles of the two components results in the observed emission dominated by the red glow of hydrogen. The remnant itself is detected in radio waves and X-rays. At the centre of the nebula is the rotating pulsar (about 90 milliseconds period) which is the stellar relic of the supernova. The gaseous remnant is cooling with time and finally will merge with a general interstellar medium, enriching it with heavy elements.

The new heavy elements may either exist in a gaseous state or somehow condense to make dust grains. Later, when the density of a cool cloud exceeds a critical value and it can be compressed by its own internal gravitational forces, new stars may form out of this enriched interstellar material and these will have proportionally more of the heavier elements than the initial population of stars. This process of star formation, synthesis of heavy elements, expulsion of enriched material back into interstellar space, followed again by star formation, is a constant cycle of events which occurs throughout the Galaxy. The stuff out of which the Sun and the planets condensed, and out of which we are made, has been recycled more than once in this way.

The most obvious manifestations of interstellar material are nebulæ. Several types have been classified: reflection nebulæ, ionized hydrogen (HII) regions, planetary nebulæ, supernova remnants and dark nebulæ. The differences between these stem mainly from the way the material is illuminated or the light from other sources is obscured, their density, or their recent history. In a reflection nebula the dust grains preferentially reflect and scatter light from nearby stars. Dark nebulæ, on the other hand, are dense clouds of gas and dust

which have no suitable stars to illuminate them and are so opaque that they prevent light from background stars or bright nebulæ from passing through them; they appear in silhouette as a "hole" in the sky. The light from HII regions and planetary nebulæ is initiated by the ionization of gas by ultraviolet photons from a very hot star. Visible light is produced dominantly when electrons subsequently are recaptured by hydrogen or helium ions or following the collision of electrons with one or more times ionized oxygen, neon or nitrogen atoms. Excitation processes of a more complicated nature occur in supernova remnants, including the production of synchrotron radiation by electrons emitted from the residual pulsars and collisional excitation of interstellar gas by energetic particles transported in the explosions. Other types of nebulæ not apparent on optical photographs are dense molecular clouds, which can be detected by their emission or absorption of radio, microwave or infrared radiation rather than of visible radiation. As we shall see, several of these nebular manifestations are related.

The existence of gas between the stars was discovered in 1904 by Johann Hartmann through the observation that a few of the absorption lines recorded in the spectrum of the binary star Delta Orionis did not change in wavelength (by the Doppler effect) as the star moved around its orbit; this followed the observation in 1874 by William Huggins that certain nebulæ had a spectrum characteristic of rarefied gas. Since that time interstellar absorption lines have been recorded in the spectra of a large number of stars used as background sources. In the optical region these lines are few in number and are usually much narrower than the stellar features themselves. Frequently they have multiple Doppler-shifted components, arising from clouds with different line-of-sight velocities. The strongest optical lines are due to neutral sodium and singly-ionized calcium atoms. Neutral potassium, calcium and iron atoms and singly-ionized titanium atoms also can be seen by their visible absorption lines and in addition a few simple molecules have been detected. However, a large number of atoms in various stages of ionization have their absorption lines at ultraviolet wavelengths and have been studied by

means of telescopes borne on balloons, rockets or satellites. The Lyman alpha transition of neutral hydrogen, falling at 121.6nm wavelength, is by far the strongest of all absorption lines observed. While most of the hydrogen is neutral, some elements exist in the interstellar medium primarily in an ionized state. The ionization of such elements arises largely from the presence of energetic photons in the all-pervading bath of galactic light due to stars. The low density of the general interstellar medium ensures that the time an atom in the ionized state has to wait before it can recombine with a free electron which has wandered close enough is quite long. In denser regions, hydrogen in the form of H_2 molecules is strongly detected at far ultraviolet wavelengths if a suitable hot star is available in the background. The total number of atoms or molecules of each kind between us and the background star can be calculated from the shape and strength of its characteristic absorption features. When this is done, significant differences in the

The Rho Ophiuchi Dark Cloud, *right*, is about 230 parsecs from the Earth and its apparent size on the sky extends over an area about 300 times larger than the Moon. The cloud contains a large amount of irregularly distributed dust, which blocks the visible light of the stars within it and behind it. The apparently empty dark areas represent the regions where the absorption of light is the greatest. Nevertheless, the cloud can be penetrated to the core at radio and infrared wavelengths and this has revealed evidence for the existence of more than 60 young stars buried in the cloud, and several compact HII regions. It is estimated that about 10 per cent of the mass of the cloud has already condensed to form new stars. The thick parts of the cloud are very cold (around 10K) and consist mainly of hydrogen in the molecular form (H_2) with the densest regions containing more than 10,000 molecules per cubic centimetre. The formation of molecules is facilitated both by the high concentration of gas, leading to a high frequency of atomic collisions, and by the high concentration of dust which catalytically assists in the formation of molecules and at the same time shields them from the destructive effect of ultraviolet radiation from hot stars. The molecular hydrogen can not be observed directly, but the presence of other molecules such as formaldehyde, OH or CO can be mapped through their absorption at radio wavelengths. The regions of strongest absorption correlate well with the visibly thick parts of the obscuring cloud, confirming that the dust and the molecules cohabit in the cloud. The cloud is remarkable because it contains so many reflection nebulæ, showing a range of colours from blue to red, and so few extended HII regions.

The Horsehead Nebula, *above,* is a dense dark cloud in which dust is very plentiful; it completely absorbs the light from the regions behind it and contrasts strongly with the fairly uniform background, which actually is a diffuse gaseous emission nebula. In fact, the Horsehead is only a protruberance of a highly extended dark cloud covering all the lower part of the picture. For this reason the density of stars visible there is much lower than in the upper part; only the few stars which are in front of the cloud can be seen, while those behind are completely obscured through extinction by the dust. The bright bluish-white nebula to the lower left of the Horsehead is a reflection nebula, produced by the illumination of the front part of the cloud by a bright star not seen because it is washed out in this overexposed photograph.

Small dark clouds called Bok Globules, *left,* of radius less than 1 parsec and mass usually between 100 and 200 solar masses, can be seen in silhouette against the stellar background of the Milky Way or bright areas of gaseous nebulae. They are opaque to visible light because of the large amount of dust they contain. Radio observations reveal their richness in molecules and also that some globules are in gravitational contraction, indicating that eventually these may become star forming regions.

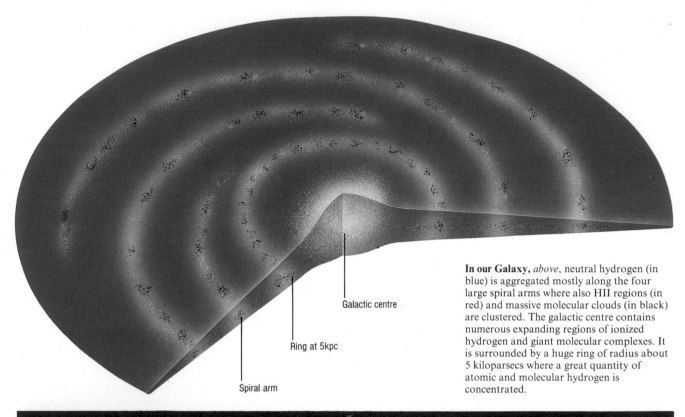

Galactic centre

Ring at 5kpc

Spiral arm

In our Galaxy, *above*, neutral hydrogen (in blue) is aggregated mostly along the four large spiral arms where also HII regions (in red) and massive molecular clouds (in black) are clustered. The galactic centre contains numerous expanding regions of ionized hydrogen and giant molecular complexes. It is surrounded by a huge ring of radius about 5 kiloparsecs where a great quantity of atomic and molecular hydrogen is concentrated.

A global view of the night sky, *above*, projected in galactic coordinates, shows the stars, dark dust clouds and bright HII regions extending along the disk of our Galaxy as seen from the inside. In the south are the Magellanic Clouds.

The galaxy NGC 4565, *right*, in the Virgo cluster, is a good example of a spiral galaxy viewed edge-on from the outside. Throughout the disk are dark regions indicating the presence of dust, and bright HII regions.

abundances of the heavier elements relative to hydrogen are found between the interstellar medium and the stars. For example, in some well-studied cases the interstellar gas seems to contain only a hundredth of the iron and a thousandth of the calcium that is commonly present in the atmospheres of stars. The answer to this is easy to see: the missing proportions are in the interstellar dust grains.

When light meets an interstellar grain it is partly absorbed and partly scattered back into space. At visible wavelengths approximately equal amounts are absorbed and scattered. In fact, about a quarter of the diffuse light we see from the Milky Way is starlight scattered from the pervading dust. Reflection nebulæ are seen in those cases when a bright star lies close enough to a dense cloud or patch of gas and dust for it to be directly illuminated. These nebulæ appear bluer than their associated stars since blue light is scattered more efficiently than red light by the grains in the clouds. Nevertheless, the general spectral characteristics of the light from such a nebula otherwise resemble the continuous spectrum with absorption lines characteristic of the nearby star, in contrast to the light from HII regions whose generally red colour comes predominantly from within the gas. The combined action of absorption and scattering dust also dims our direct view of stars, an effect known as inter-

stellar extinction. Extinction varies enormously from place to place but in the plane of the Galaxy the strength of starlight at visual wavelengths is reduced typically by two magnitudes for every kiloparsec it travels. In some very dense interstellar clouds the visual extinction can be as much as several hundred magnitudes per parsec. The amount of extinction is found to depend quite strongly on wavelength. Red light is dimmed less than blue light. This causes the apparent colour of a star to be altered by an amount related to the degree of extinction its light has undergone; this colour effect is known, subjectively, as interstellar reddening. A similar kind of reddening occurs in the Earth's atmosphere and explains the Sun's apparent colour changes as it rises and sets. Towards shorter wavelengths the extinction increases further and in the far ultraviolet region is several times larger than for visible light. Following the same trend the extinction rapidly reduces towards longer wavelengths and in the infrared at 2 microns is about a tenth of the value it has in the visible region. This large improvement in the transparency of the interstellar medium at long wavelengths allows astronomers to see deep into extremely dusty regions and to great distances in the plane of the Galaxy, which would be unreachable at shorter wavelengths. The curve describing the change in extinction from infrared to ultraviolet wavelengths is not entirely

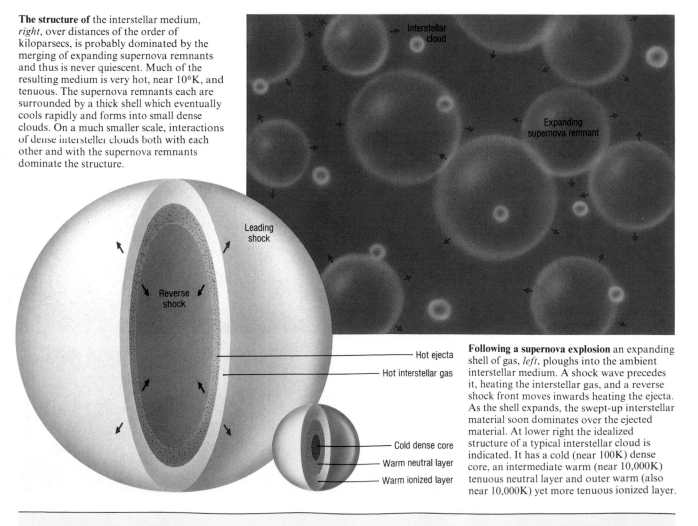

The structure of the interstellar medium, *right*, over distances of the order of kiloparsecs, is probably dominated by the merging of expanding supernova remnants and thus is never quiescent. Much of the resulting medium is very hot, near 10^6K, and tenuous. The supernova remnants each are surrounded by a thick shell which eventually cools rapidly and forms into small dense clouds. On a much smaller scale, interactions of dense interstellar clouds both with each other and with the supernova remnants dominate the structure.

Interstellar cloud

Expanding supernova remnant

Leading shock

Reverse shock

Hot ejecta

Hot interstellar gas

Cold dense core

Warm neutral layer

Warm ionized layer

Following a supernova explosion an expanding shell of gas, *left*, ploughs into the ambient interstellar medium. A shock wave precedes it, heating the interstellar gas, and a reverse shock front moves inwards heating the ejecta. As the shell expands, the swept-up interstellar material soon dominates over the ejected material. At lower right the idealized structure of a typical interstellar cloud is indicated. It has a cold (near 100K) dense core, an intermediate warm (near 10,000K) tenuous neutral layer and outer warm (also near 10,000K) yet more tenuous ionized layer.

smooth; several humps exist, one in the ultraviolet with peak wavelength near 220nm and two in the infrared near 3 microns and 10 microns. The extinction curve gives us much of the information we have about the nature of the dust grains. One conclusion is that most of the grains which cause extinction at visible wavelengths have dimensions near 0.1 micron, smaller than the wavelength of light, although very little is known about their shape from this. A clue about their shape, however, comes from the observation that the direct light from stars is mildly polarized; this points to at least some grains being elongated, so giving them a tendency to line up under the influence of the magnetic field which exists in interstellar space throughout the Galaxy. Even smaller particles are needed to explain the extinction at ultraviolet wavelengths. The hump in the ultraviolet portion of the extinction curve indicates that graphite particles may be present, and those in the infrared portion indicate water ice crystals and some kind of silicate material. Different materials may be present in the same dust grain. Other substances also have been proposed as possible constituents, including quite exotic organic molecules.

Where are the grains formed? There is not much doubt that some grains condense from the gas in the outer atmospheres of late-type stars; many such stars are seen to have a great deal of dust in their immediate vicinity, detected from the infrared emission reradiated by the grains, which must have been made in this way. These grains are eventually blown away into interstellar space and make way for more grains to be formed out of new material from the star. Grains may also condense out of gas clouds in the process of collapsing to form new stars. Once existing, grains may grow by steadily accreting more atoms and molecules from the interstellar medium. On the other hand, interstellar grains can be destroyed by being heated through collisions between gas clouds, by the effects of intense radiation, or by being incorporated into new stars. However, whatever the processes of formation or destruction, we do not expect that the proportion of interstellar matter which is in the form of grains is on average more than about one per cent by

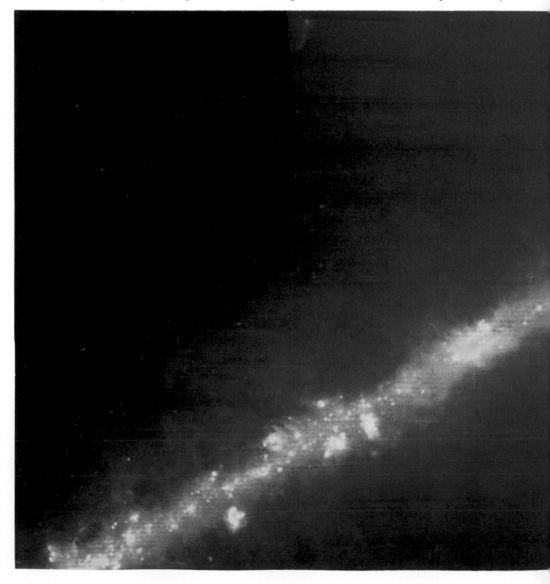

Far infrared measurements made by the Infrared Astronomical Satellite (IRAS) have produced this large colour image of the plane and centre of our Galaxy. The warmer material is shown as blue (around 300K) and colder material as red (30K). The infrared telescope and instrumentation carried by IRAS penetrates the dense veil of dust that obscures the view of optical telescopes. The yellow and green knots and patches distributed along the band are giant clouds of interstellar gas and dust heated by nearby stars. Some regions are heated by newly formed stars embedded in the cloud, others by nearby massive, hot stars which can be tens of thousands of times brighter than the Sun.

A particularly interesting pattern of emission in the overall background was observed by IRAS in its longest wavelength band at 100 microns. When the data in this band were spatially filtered to emphasise features that range in apparent size from half a degree to about 10 degrees, the celestial sphere was seen to be covered by wispy clouds; these have been termed "infrared cirrus". Most of the infrared cirrus is probably dust in the interstellar medium in the Sun's immediate neighbourhood. Some of the features are coincident with clouds of hydrogen gas observed at radio wavelengths which are known to be interstellar.

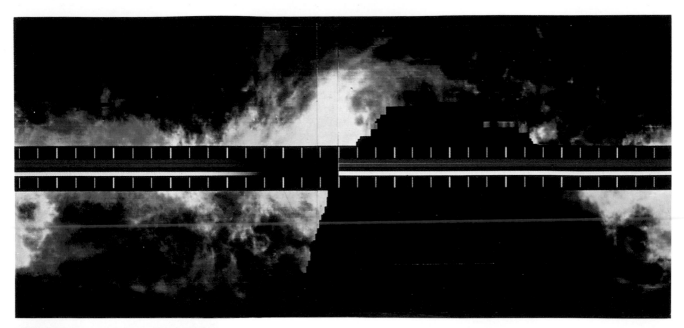

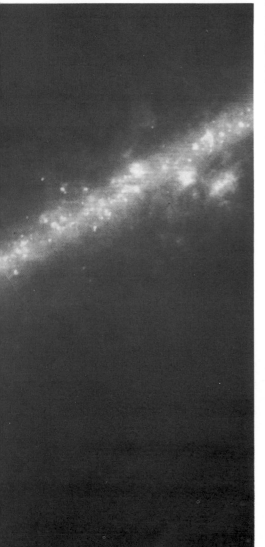

This 21cm radio map of the sky, *above*, shows the distribution of neutral hydrogen within a few hundred parsecs from us. The velocity of recession is indicated red and of approach, blue. The galactic plane is in the centre; regions near the plane and near the pole are excluded observationally.

Individual interstellar clouds impose their own spectral signatures on the light from a distant star through absorption at characteristic wavelengths, *below*. These lines are sharper than the stellar features and here are indicated at different Doppler shifts. The cloud structures can be probed by using different background stars.

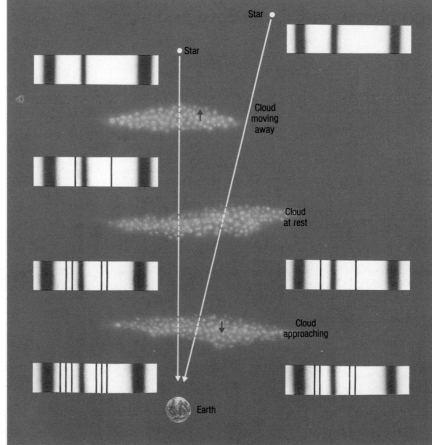

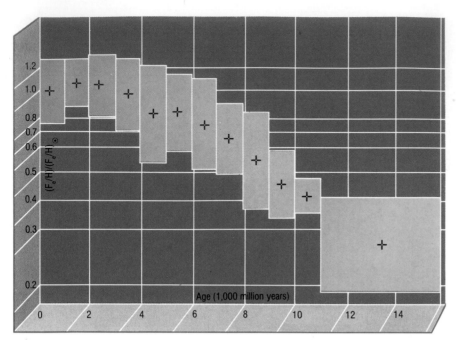

The stellar metallicity relative to the solar value is plotted for stars in the solar neighbourhood as a function of age, *left*. The size of the rectangle corresponding to each age band represents the variation in metallicity for stars of about the same age. Together with the distribution of the number of stars as a function of age, this diagram enables us to delineate the chemical evolution of the region of our Galaxy in the vicinity of the Sun. In the simplest conceivable model, the solar neighbourhood initially contained gas of zero metallicity. Then stars formed and variously enriched the interstellar medium with heavy elements which had been synthesized within them. However, this does not work well in predicting the observed metallicities. Many difficulties are removed by introducing a dynamical evolution of the stars progressing from the initial formation of halo stars to the subsequent formation of stars in an already enriched disk, combined with continuing dilution of disk gas by infalling halo gas.

mass. The simple reason for this is that grains are made predominantly out of the heavy elements, which altogether constitute less than one per cent of the material in the Galaxy.

Although the average intensity of the interstellar magnetic field is only a few microgauss, it strongly influences the flight of charged particles in the interstellar medium. As a result, synchroton radiation is seen to come from throughout the Galaxy, produced by high energy cosmic ray electrons whose paths are curved by the field. The other constituents of cosmic rays are atomic nuclei and elementary particles whose kinetic energy can reach 10^{21}eV. Cosmic rays have a negligible density compared to other interstellar particles, but a considerable total energy. It is thought that supernovæ play a dominant role in producing them.

Most of our knowledge about the large scale distribution of the interstellar gas in our Galaxy has come from the study of the spectral line emission and absorption of radiation detected at the 21cm radio wavelength transition of neutral atomic hydrogen. This transition arises from a change in the alignment of the spins of the proton and electron between the states when they are parallel and when they are opposed. In interstellar space about three-quarters of the hydrogen atoms are in the spins-parallel state, which is slightly higher in energy than the other state by an amount corresponding to a photon of 1420.4MHz frequency or 21cm wavelength. A collision with another particle induces a transition between the states, but given the low average density of matter in the interstellar medium this may happen to a particular atom only once in several hundred years. Even so, radiation in the 21cm line readily can be detected from nearly everywhere in the Galaxy because the number of interstellar hydrogen atoms is so large. From the Doppler shift in frequency of the line as received, clouds with different velocities along the same line of sight can be distinguished and this is extremely useful in

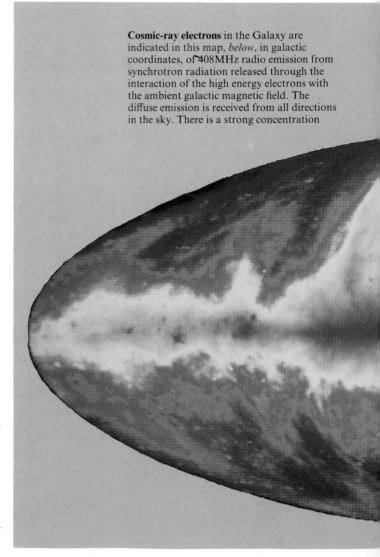

Cosmic-ray electrons in the Galaxy are indicated in this map, *below*, in galactic coordinates, of 408MHz radio emission from synchrotron radiation released through the interaction of the high energy electrons with the ambient galactic magnetic field. The diffuse emission is received from all directions in the sky. There is a strong concentration

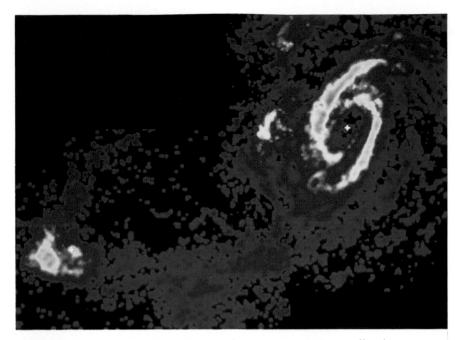

The structure of the spiral galaxy M81, as detected in the 21cm emission line of neutral hydrogen, is shown in this picture, *left*, obtained with the Westerbork radio telescope. At any point, the density of the gas is colour-coded, with the densest regions indicated green. The spiral structure of the galaxy is clearly seen, with the neutral gas being concentrated in the arms. Additional information is obtained from the observed Doppler shifts of the 21cm line, which show the rotation of the galaxy. Very little neutral hydrogen exists at the centre, contrary to what we see for the stars in the galaxy; in this region hydrogen exists in forms other than as neutral atoms.

towards the plane of the Galaxy but there are other prominent features as well. The largest of these is the North Polar Spur, which rises from the plane near galactic longitude 30°; it may be the result of a supernova explosion relatively near the Earth some 100,000 years ago.

allowing us to study the distribution and motion of the neutral gas in distant parts of the Galaxy. Such observations show that neutral hydrogen clouds are clustered mostly along the spiral arms.

With the discovery of interstellar molecules it became generally recognized that the spectacular HII regions actually are relatively small hot cavities formed at the edges of or inside much more massive molecular clouds out of which the hot stars were formed. Molecular clouds are the coldest form of matter in the Galaxy, with temperatures from a few K ranging to more than 100K in regions of active star formation. Most of the received radiation from these clouds is emitted thermally by dust in the far infrared region and through rotational transitions by molecules both in the far infrared and the radio regions. More than sixty types of molecule and radical have been detected. The most abundant, of course, is molecular hydrogen although it does not itself emit radiation unless its temperature exceeds about 500K. A frequently used tracer of molecular clouds is the CO molecule, which does radiate strongly. It is now clear that CO-emitting molecular gas is widespread in the Galaxy. The emission is received mostly from clouds confined to a layer only about 150 parsecs thick lying near the plane of the disk. From our viewpoint, seen in the light of such molecular emission, the Galaxy appears confined to a narrow band along the plane, only about 2° wide. Only the clouds closest at hand are seen in directions far off the plane of the Galaxy. This gas mostly is contained within the spiral arms, with the greatest concentration occurring in a ring at about 5 kiloparsecs radius, although some is more widely distributed throughout the disk and there is a strong presence at the galactic centre. Molecular clouds exist in a large variety of sizes, masses and types. Most of the molecular gas in our Galaxy is in immense clouds of up to a million solar masses with diameters around 40 parsecs and central densities which may be more than 10,000 particles per cubic centimetre. These giant clouds are the largest self-gravitating bodies in the disk of the Galaxy and are the sites where most of the stars at present are being formed. The smallest clouds, the so-called Bok Globules, have masses ranging from only a few to some hundreds of solar masses. Some of these have formed low-mass stars and others may be collapsing on the way to forming stars. There is also a wide range of intermediate mass clouds or cloud complexes. The total amount of molecular gas in the interstellar medium probably exceeds the amount in atomic form.

Even when taken together, molecular clouds and neutral hydrogen clouds very far from fill the volume of interstellar space. Although it is still a subject of speculation, much of the gas between the clouds is thought to be very hot and very tenuous, at a temperature around a million K and density a few times 10^{-3} particles per cubic centimetre, being the direct outcome of the expansion of supernova remnants which have been produced in abundance throughout the Galaxy. We know that in a galaxy like ours a supernova explosion occurs once in every few tens of years and the gaseous remnant will expand violently out to a radius of roughly 100 parsecs. Consequently, in a time as short as a few million years the gas at any point throughout the entire volume of the disk will be exposed at least once to a blast wave of hot gas from a supernova. The structure of the interstellar medium over a scale of kiloparsecs probably is dominated by a foam of such vast, hot, coalescing supernova remnants of various ages. The expanding front of a remnant progressing through the interstellar medium will envelope the denser clouds, and tend to sweep away any low-density material present and itself fill the space. The displaced material will be compressed into denser sheets

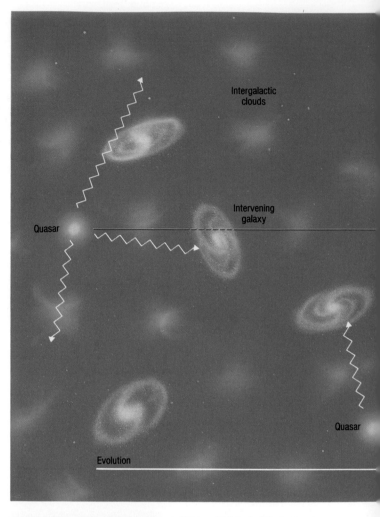

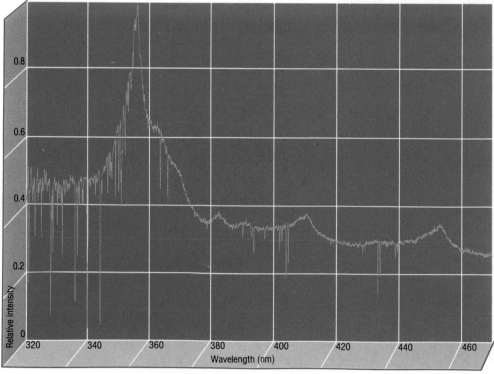

Using a high red-shift quasar as a source of background light, *above*, the distribution of intervening gaseous matter in the universe from early times can be deduced from the evidence in its spectrum. In this way, distant galaxies and primordial intergalactic gas have been identified. Quasars also provide a general intergalactic radiation field which ionizes the tenuous haloes of galaxies.

In a typical spectrum of a very distant quasar, *left*, the strongest emission feature is the Lyman alpha line of neutral hydrogen highly red-shifted from its rest wavelength at 121.6nm. The sharp absorption lines to the left of this line probably are mostly due to primordial intergalactic hydrogen gas clouds at various redshifts, again seen in the Lyman α line. The sharp absorption complexes to the right of the strong emission line indicate the presence of heavy elements in the interstellar gas of very distant galaxies.

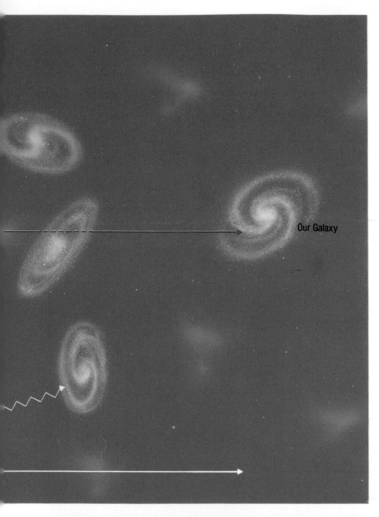

Our Galaxy

or filaments in a shell around the remnant. After a time, this cools and then fragments to form an additional population of cold dense clouds. The observational evidence for this ubiquitous hot gas comes from the widespread presence of five times ionized oxygen atoms (ie, stripped of five electrons), as indicated by ultraviolet interstellar absorption features, and from soft X-rays received from most regions of the sky. This hot gas must extend far from the galactic plane, forming what is termed a galactic corona by analogy with the Sun's corona.

In general, the clouds surviving within this hot corona are thought to have a composite structure. A relatively cold, inner core of radius up to a few parsecs, temperature less than 100K and density some tens of particles per cubic centimetre, is surrounded by a warm, neutral layer of a hundredth of the core density and temperature near 10,000K, which itself is enveloped by a warm, ionized shell at the same temperature as the intermediate layer but of somewhat lower density. The outer envelope actually is heated and ionized not by the corona but by the ultraviolet radiation of hot stars at large. However, the intermediate layer is heated although not much ionized by X-rays which do come from the corona. Generally, the core itself also is heated by stellar radiation, although relatively mildly. Very dense cores become self-shielding and turn into cold molecular clouds. The various clouds move at an average speed of about 6 miles a second (10km/sec); mutual collisions sometimes will lead to clouds coalescing, sometimes to the formation of several independent fragments. Clouds will also be disrupted by passing supernova remnants. Others will be reformed. New generations of stars arise, out of which in turn new supernovæ erupt. Thus the

The Coma Cluster, *left,* is a well-known cluster of galaxies which extends over several degrees of the sky. It contains more than a thousand bright galaxies and is one of the nearest rich clusters. Its mean recession velocity is 4,300 miles a second (6,900km/sec), indicating that the centre of the cluster is at about 100 megaparsecs distance. The cluster is spherically symmetric and exhibits a strong central concentration; it is therefore in a relaxed state with the galaxies in the core having undergone many mutual gravitational interactions. The total diameter of the cluster is estimated to be nearly 3 megaparsecs while the core is more than five times smaller than this. Practically all of the massive galaxies in the cluster lie within the core, to which they have gravitated. Overall, the core is populated with elliptical and spheroidal galaxies while spirals are distributed further out. Most of the round objects are foreground stars situated in our Galaxy.

interstellar medium is a far from quiet and empty place.

It is obvious from the appearance of other galaxies nearby that mostly they too have active interstellar media, although the gas content varies according to the type. Spiral galaxies contain the most gas and star formation in them generally is very evident, while elliptical galaxies contain much less gas and show fewer signs of star forming activity. The optical spectra of high redshift quasars show a large number of sharp absorption lines in addition to their characteristic emission lines. It is now generally accepted that these arise in absorbing matter located somewhere between the distant quasar and our Galaxy. Two such populations are observed. One of these appears as a forest of absorption lines shorter in wavelength than the quasar's redshifted atomic hydrogen Lyman alpha emission line. These lines are identified as absorption lines of Lyman alpha at various redshifts, and attributed to primordial intergalactic hydrogen clouds distributed throughout the universe. The other population is apparent at wavelengths longer than the emission lines. These lines have been clearly identified as being due to various atoms and ions of heavy elements, including carbon, oxygen, nitrogen, silicon, sulphur, iron and magnesium, grouped into several redshift batches. There is now considerable evidence that such groups represent sets of absorption lines arising from interstellar matter in the extended haloes of very distant galaxies which we see as they were at much earlier times. Most of these galaxies are too faint to be detected by any other means.

The spiral galaxy NGC253 is about three million parsecs distant. It appears to be elongated because it is highly inclined in our view. The galaxy contains a large quantity of interstellar material and is one of the dustiest known. The blue colour of the spiral arms comes from the presence of young, massive stars. The red spots are HII regions, where the youngest stars are sited. The yellow colour of the nucleus indicates an older population of stars containing relatively low mass main sequence stars and red giants. The nucleus emits most of its energy in the infrared wavelength region.

throughout the electromagnetic spectrum according to the black-body radiation laws of Stefan and Wien. The distribution of energy with wavelength from an emitting black body is described by the Planck's Law. In astronomical sources, bodies of relatively low temperature (less than 1,000K) are often of greatest interest in the infrared because in that region their maximum energy is emitted.

Other astronomical infrared sources emit radiation in one or more spectral lines or bands, depending upon the temperature and composition of the emitting gas (see EMISSION SPECTRUM). Line emission regions are typically gas envelopes surrounding stars, or are gas-rich regions in which stars are about to form. Emission in the infrared lines is sometimes the main mechanism for the cooling of such gas-rich regions in our Galaxy. Among the infrared atomic lines important in this context are the FORBIDDEN LINES of oxygen ([OIII] at $88.35\mu m$) and carbon ([CI] at $610\mu m$ and [CII] at $157\mu m$), and various other lines of nitrogen, silicon, neon, and argon.

Important molecular bands found in emission and in ABSORPTION in sources outside the Solar System include carbon monoxide (CO) and hydrogen (H_2) which are found in gas-rich regions in the galactic plane and in protostellar objects (see PROTOSTARS). These bands are not only indicative of chemical composition but of the physical processes in progress, such as the passage of shock waves through regions in which star formation occurs.

Interstellar molecular clouds have infrared sources imbedded within them, and from infrared observations the composition, mass, and dynamics of a molecular cloud can be determined. Most of the interstellar dust and gas are contained in Giant Molecular Clouds (GMCs). Maps of GMCs at infrared wavelengths show the warm or hot regions where star formation is already in progress or is about to begin as the cloud collapses locally. Infrared observations are the key to the composition and the energetics of star formation in GMCs.

Probably the most studied and famous molecular cloud is in Orion's sword, where star formation is observed in progress in the most dense parts of the Orion Molecular Cloud. Temperatures in this cloud reach about 2,000K in regions where molecular hydrogen is excited, but other regions range from a few tens to a few hundred degrees. Turbulence and large-scale motion of the gas within the cloud can be observed by the techniques of infrared astronomy, giving some idea of the dynamics with a collapsing interstellar cloud.

On a much smaller scale, highly compact and opaque BOK GLOBULES of interstellar dust and gas with just a few tens of solar masses are potential star-forming regions. Infrared observations show their temperatures to be very low, just 10–20K, indicating that the globules are in an early state of gravitational collapse.

Infrared spectroscopy of sources in the Galaxy, particularly those surrounded or obscured by large dust concentrations, shows numerous unidentified emission bands, and some spectral features that have been identified as ice and as hydrocarbons. Crystals of water ice have formed in some cold molecular clouds as evidenced by the spectral features detected. On grains of interstellar dust, other molecules of carbon-bearing compounds condense, undergo catalytic reactions, and are converted to complex hydrocarbons of presently undetermined composition.

In regions of our own Galaxy and other galaxies where clouds of hot hydrogen gas occur (HII REGIONS), infrared observations afford a means of studying the chemical composition of interstellar gas and of the physical conditions within the hot gas clouds, where most of the electrons are stripped from their atomic nuclei. These hot, turbulent clouds of atomic particles have densities up to 10^5 electrons per cm^3 or higher.

A great virtue of observations in the infrared is that many of the wavelengths available for study can penetrate the otherwise opaque clouds of dust in the plane of our Galaxy. Thus, it is possible to detect stars and other radiation sources toward the centre of the Galaxy which are completely obscured at optical wavelengths. Through infrared observations it has been possible to map the inner structure of our Galaxy to reveal the structure and dynamics of the innermost portions, located in the constellation Sagittarius. The observations show a high density of late-type stars and a large ring of molecular gas and dust, as well as many regions of active star formation. In the central parsec of the Galaxy, there are clusters of compact infrared sources, late-type stars, HII regions, and the possible existence of a BLACK HOLE at the centre has recently been discussed seriously.

Infrared observations give us fundamental information on the energetics of other galaxies. In "normal" galaxies, we learn about the distribution and motion of molecular clouds, HII regions, and the rate of star formation. For "active" galaxies, quasi-stellar objects, and other peculiar objects in extragalactic space, infrared observations are a key to understanding their overall output of energy and the physical processes occurring within them. Infrared measurement of active galaxies found originally by optical astronomers has moved us a small way toward understanding of the scope of physical phenomena in progress, and now a great wealth of additional extragalactic objects has been revealed by the IRAS survey mission from space. The follow-up detailed observations of these infrared active galaxies will be highly important in coming to some understanding of the basic physics of matter under extreme conditions of temperature and density.

Closer to home, Solar System scientists have used the techniques of infrared astronomy to work out the basic chemistry of the planets, their satellites, and the asteroids. In the near-infrared, molecules in planetary atmospheres exhibit a rich absorption spectrum, the analysis of which permits the composition and temperature of the atmospheres to be established. This work has been undertaken from ground-based telescopes, from aircraft, and from spacecraft flying past the planets. Spectral reflectance observations of asteroids, planetary satellites, and small planets without substantial atmospheres have been used to discover the mineralogical compositions of their surfaces. Specific igneous rock-forming minerals have diagnostic absorption bands in the region $1 - 5\mu m$, and several individual silicate minerals have been found. Ices of water, methane, and other volatile substances also have diagnostic absorption bands in the same spectral region, and infrared techniques have established the presence of water and carbon dioxide ices in the polar caps of Mars, water ice on many of the satellites of Jupiter, Saturn, and Uranus, and methane ice on Triton and Pluto.

Measurement of the bulk black-body emission of many small Solar System bodies has given us information on their sizes and surface albedos where no other techniques succeed. As a result, astronomers have begun to compile a detailed inventory of the myriad small bodies in the Solar System to

I

complete our picture of the nature of the Earth's neighbourhood. *DPC.*

Infrared Astronomical Satellite INFRA-RED ASTRONOMY began during the 1960s and 1970s with the development of semiconductor detectors sensitive to radiation between one and 20 microns. Using these, a number of pioneering infrared sky surveys were made which showed that there was a need for a sensitive infrared survey of the entire sky. Unfortunately, conventional infrared astronomy is limited by the absorption of extraterrestrial infrared radiation in the atmosphere, which prevents the study of wavelengths longer than approximately 20 microns, and by emission from the atmosphere and the telescope itself. To avoid these restrictions IRAS was conceived as an orbiting infrared telescope which could operate unhampered by the atmosphere. Since it would be operating in space it was also possible to cool the IRAS telescope to very low temperatures without condensing vapours onto the optical surfaces, enabling IRAS to make observations at wavelengths as long as 100 microns without being swamped by emission from the telescope structure.

The main objective of IRAS was to make a reliable all-sky survey at wavelengths of approximately 12, 25, 60 and 100 microns, wavelengths that correspond to objects with black-body temperatures between about ten and a few hundred degrees Kelvin. IRAS was also able to make "additional observations" of objects of special interest. Some additional observations were planned before the launch; others were chosen later to follow up on new discoveries made during the survey.

IRAS was an international project involving the United States, the Netherlands and the United Kingdom. The heart of the satellite was a 23in (60cm) reflecting telescope which focused the incoming radiation onto an array of 64 semiconductor detectors. These detectors operate most efficiently at low temperatures and so the entire telescope was surrounded by a cryostat, essentially a giant dewar flask, containing about 475 litres of superfluid helium. The helium boiled away slowly throughout the mission and in doing so maintained the detectors at 1.8K and the rest of the telescope below 10K, ensuring maximum sensitivity at the longer wavelengths. The supply of helium determined IRAS's useful lifetime since once it had boiled away the telescope

warmed up, saturating the detectors, and the mission was at an end. To reduce the rate at which heat could leak into the cryostat the dewar was surrounded by many layers of insulation and the telescope aperture was topped by a sunshade which prevented stray sunlight from being reflected down the telescope. The telescope, including its detectors and cryostat, and the launch were provided by the United States. The Netherlands was responsible for the development of the spacecraft, which provided IRAS with essential

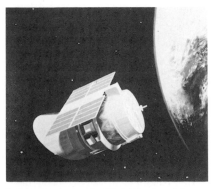

Infrared Astronomical Satellite: in orbit

services such as attitude control, communications and power. The Netherlands also provided a low-resolution SPECTROMETER, known as the Dutch Additional Experiment, which automatically measured the 7 to 23 micron spectrum of bright sources detected by the telescope. The United Kingdom was responsible for tracking and controlling IRAS from a "preliminary analysis facility" at the Rutherford Appleton Laboratory near Oxford, where a team, made up of scientists and engineers from all three nations, checked that the satellite was operating correctly and followed up new discoveries made during the survey. The task of producing the final infrared catalogues was carried out at the JET PROPULSION LABORATORY in Pasadena, California.

Since IRAS was making the first all-sky survey in the infrared it was important that the resulting catalogue be as reliable as possible. This was a difficult task since the infrared detectors could be triggered by COSMIC RAYS or by energetic electrons in the Earth's RADIATION BELTS. The satellite could also be confused by infrared emission from warm objects within the Solar System, such as other Earth satellites, comets and asteroids. To eliminate these false detections the layout of detectors in the focal plane, and the orbit of IRAS, was such that real

sources registered on two different detectors a few seconds apart (seconds confirmation) and were then detected at the same celestial position on the following orbit (hours confirmation) and detected again during a rescan of the same region of sky a few days later (weeks confirmation). This strategy improved the reliability of the final catalogues enormously and made possible some remarkable discoveries within the Solar System.

IRAS was launched by a Delta rocket from Vandenberg Air Force Base, California, on January 26 1983. It was placed in a Sun-synchronous, circular orbit, 560 miles (900km) high inclined at 99° to the equator. After a five-day engineering checkout a telescope aperture cover, which had protected the supercold optics from contamination before the launch, was jettisoned and IRAS detected its first infrared source. Next a number of additional observations were made to calibrate the infrared detectors, and a small survey of a strip of sky 5° wide was made. These data were then used to test and refine the data-processing techniques which would be used throughout the mission.

The main survey began in early February 1983 and continued until the end of August, by which time the entire sky, apart from a few areas missed due to engineering problems, had been observed twice. Due to the scan strategy this meant that most of the fixed infrared sources had been detected by the satellite a total of eight times and accidental detections of asteroids had been rejected. To improve the reliability of the final catalogues still further, and to increase the chances of detecting faint sources, a second sky survey was begun in September and this continued until the helium coolant ran out on November 21 1983. By then IRAS had covered 97 per cent of the sky at the required level of reliability and only a few small areas remained unfilled. This enormous amount of survey data was then processed by the Jet Propulsion Laboratory and a catalogue containing 245,000 infrared point sources, more than a hundred times the number known before the launch, was released in late 1984. Other catalogues, containing information on small extended sources and a catalogue of spectra from the Dutch additional experiment were released later. A catalogue of asteroid observations has also been produced using data which were rejected by the main survey because they referred to moving sources.

As well as producing the infrared catalogues, astronomers using IRAS made many exciting discoveries both during the mission and during subsequent analysis. Solar System discoveries included six new comets, including Comet Iras-Araki-Alcock, huge invisible tails on Comet TEMPEL-2 and other comets, several asteroids, including the unusual object 3200 PHÆTHON and unexpected bands in the ZODIACAL DUST. A dust shell, possibly related to planetary formation, was discovered around the star VEGA and subsequently similar shells were found around other stars. Star formation regions in DARK NEBULÆ were studied in great detail and many PROTOSTARS discovered. The star BETELGEUX was found to have ejected three huge dust shells, and clouds of dust named INFRARED CIRRUS were discovered all over the sky. IRAS also studied the Galactic centre in great detail. Beyond the Milky Way IRAS observed that many galaxies are powerful emitters of infrared radiation and some of these, the STARBURST GALAXIES, emit much more infrared than visible light. Some IRAS sources were found to have no optical counterparts, and their nature remains mysterious. *JKD.*

Infrared cirrus Faint infrared emission from cool dust in our Galaxy. When displayed on computer-produced maps it appears as wispy structures resembling cirrus clouds across the night sky.

The existence of clouds of gas and dust in interstellar space has been known for many years and the study of interstellar dust was a major objective of the INFRARED ASTRONOMICAL SATELLITE (IRAS). IRAS scientists discovered that as well as the relatively compact dust clouds associated with star formation, the sky was crisscrossed by bands of dust which shone faintly at wavelengths of 60 and 100 microns. This emission became known as the infrared cirrus.

The fact that the dust responsible for the cirrus emits mainly at long infrared wavelengths means that it must be very cool, probably only 30–40 degrees above absolute zero. Measurements of the PARALLAX of the cirrus show that it must lie at least 1,000AU away from the Sun. The cirrus is thus separate from the ZODIACAL DUST, which is much closer to the Earth (1–5AU) and considerably warmer.

It was originally thought possible that the cirrus might be due to the OORT CLOUD of comets. Although material in the Oort cloud would be at these very low temperatures, some of the cirrus has been identified with clouds of interstellar gas well beyond our immediate solar neighbourhood. This link to interstellar clouds implies that much of the cirrus is too far away to be part of the Oort Cloud and is probably due to emission from normal interstellar dust grains.

The cirrus probably consists of tiny grains of dust, no larger than particles of tobacco smoke and composed of carbon in the form of graphite. These grains may have condensed in the STELLAR WINDS of cool stars or been formed during SUPERNOVA explosions. *JKD.*

Infrared radiation That portion of the electromagnetic spectrum in the wavelength range one to 1,000 micrometres (μm). Objects at any temperature emit infrared radiation with a total energy given by Stefan's law, and distributed with wavelength according to WIEN'S LAW and the PLANCK CONSTANT. *See also* RADIATION.

Innes, Robert Thorburn Ayton (1861–1933). Director of the Union (later Republic) Observatory in Johannesburg (1903–27), double star observer and discoverer of PROXIMA CENTAURI.

Innes' Star Discovered by R.T.A. INNES. *See* PROXIMA CENTAURI.

Instability strip The narrow region of the HERTZSPRUNG-RUSSELL DIAGRAM where pulsating stars are located. It extends from the brightest Capheids down to the pulsating white dwarf (ZZ Ceti) stars and also includes RR Lyræ stars, Delta Scuti stars and Dwarf Cepheids.

Intelsat Satellites produced and operated by the multinational International Telecommunication Satellite Corp., set up in 1964. All are placed in geostationary orbits.

Intensity interferometer *See* OPTICAL INTERFEROMETER.

Inter-Agency Consultative Group (IAGC). Group that co-ordinated all space missions to HALLEY'S COMET and all Halley observations from space. The members are the EUROPEAN SPACE AGENCY, NASA, Intercosmos in the Soviet Union, and the Institute of Space and Astronautical Science in Japan. After 1986 the group continued to exist, co-ordinating a solar terrestrial physics programme involving eight spacecraft from the four agencies.

Interamnia (Asteroid 704). Diameter 210 miles (338 km). It has an unusual, possibly dust-free, surface.

Interference principle When two trains of waves of the same wavelength meet they interfere. If maxima (crests) of the waves arrive simultaneously at the same place, their displacements add together to produce a wave of larger amplitude. This is constructive interference. If the maxima of one train coincide with the minima of the other then they cancel totally if their amplitudes are identical, otherwise they cancel partially. This is destructive interference. Thus two wave trains crossing each other produce an interference pattern, with alternative lines of constructive interaction and of destructive interaction. The principle applies to any wave motion: electromagnetic radiation, sound waves or waves on the surface of liquids. *BW.*

Intergalactic matter Matter which exists in the space between galaxies.

The transparency of intergalactic space implies that there are no significant quantities of intergalactic dust, but there is ample evidence for the existence of intergalactic gas.

Numbers of radio galaxies have a "head-tail" appearance, rather like a tadpole, as if their radio-emitting clouds of electrons were ploughing through a resisting medium. Extended X-ray emission from clusters of galaxies can be explained as emission from a hot, tenuous intracluster gas with a temperature of 10 to 100 million K. For example, the X-radiation from the Perseus cluster, and the head-tail appearance of the member galaxy NGC1265, are consistent with that cluster containing about 5×10^{13} solar masses of intergalactic gas with a mean density of about 1,000 ions per m³.

Whether the intergalactic gas is predominantly primordial matter (ie, original hydrogen and helium from the BIG BANG) falling into the clusters, or matter expelled from galaxies, is uncertain. The identification of X-ray EMISSION LINES due to highly-ionized iron suggests that some of the intergalactic gas must have been processed inside stars and expelled by supernova explosions. This "processed" matter may then have been driven out of parent galaxies by supernova "winds" or tidal encounters between galaxies; alternatively, it may have been swept out by the intergalactic gas as the galaxies plunged through it.

Tidal interactions between galaxies

I

are common, and 21cm wavelength observations reveal streams of neutral hydrogen connecting and enveloping interacting pairs and groups of galaxies such as M81 and M82. A similar gas stream, the Magellanic stream, links the MAGELLANIC CLOUDS and the Milky Way system.

In order to hold itself together a cluster of galaxies must exert sufficient gravitational attraction on its members to prevent them escaping. In all cases, the mass required to bind a cluster is far greater than the value obtained by adding up the estimated masses of its visible member galaxies (*see* MISSING MASS PROBLEM). The intergalactic gas so far identified is insufficient to make up the deficit, and it has been suggested that additional mass may exist within clusters in forms ranging from old dead stars and BLACK HOLES to exotic nuclear particles. *IKN.*

International Astronomical Union (IAU). The controlling body of world astronomy.

International atomic time (TAI). The international reference scale of atomic time, based on a second as defined by the *Système International (SI)*: the duration of 9, 192, 631, 770 periods of radiation corresponding to the transition between two hyperfine levels of the ground state of cæsium 133. It is approximately equal to the ephemeris second.

Since 1971, TAI has been formed by the *Bureau international de l'heure* (BIH) in Paris from clock data supplied by co-operating establishments world-wide.

The first operational atomic clock system was developed in 1948–49 in the United States, using an absorption line of ammonia. Attention soon turned to the element cæsium and, though most of the early development took place in the United States, it was at England's National Physical Laboratory that a cæsium beam standard was first used on a regular basis, brought into use for the calibration of quartz clocks and as a frequency standard in June 1955. Over the next few years, further cæsium standards came into use at Boulder (Colorado), Ottawa, and Neuchâtel.

All clocks have to be adjusted so that they go at the correct rate and also have to be set to time. Atomic clocks are no exception and between 1955 and 1958 atomic clocks in the United Kingdom and the United States were calibrated against astronomical time-scales

at Herstmonceux and Washington. In 1959, the United States Naval Observatory's time-scale A1 (since renamed TAI) was brought into use world-wide, its point of origin being arbitrarily set so that Atomic Time and UT2 (*see* UNIVERSAL TIME) were the same at 1958 January 1d 0h. At the same time, the atomic second was defined in terms of the resonance of the cæsium atom, as above: in 1967, this was adopted instead of the ephemeris second as the fundamental unit of time in the SI system. *HDH.*

International Geophysical Year (IGY). An 18-month world-wide study of the Earth's physical nature, running from July 1 1957 to December 31 1958.

International Halley Watch (IHW). Group that co-ordinated all ground-based observations of HALLEY'S COMET from 1982 until 1989. More than 1,000 professional astronomers from 51 countries, and almost 1,200 amateur astronomers from 54 countries, participated in the IHW and provided observations that are included in the "Halley Archive". The IHW is organized in eight observing nets: astrometry; infrared spectroscopy and radiometry; large-scale phenomena; near-nucleus studies; photometry and polarimetry; radio studies; spectroscopy and spectrophotometry; and meteor streams studies. *RR.*

International Ultraviolet Explorer This satellite, widely known to astronomers by its acronym IUE, provides (until the launch of the HUBBLE SPACE TELESCOPE) the main instrument for ULTRAVIOLET ASTRONOMY and takes spectra of planets, stars and galaxies in that wavelength region. The satellite carries an 18in (45cm) telescope and SPECTROGRAPHS to provide both high and low dispersion spectra of astronomical objects. It was launched in 1978 as a joint project of the American National Aeronautics and Space Administration (NASA), the European Space Agency (ESA) and the British Science and Engineering Research Council (SERC).

IUE was the brainchild of Professor Bob Wilson, now of University College, London. He originally led the team which proposed a "Large Astronomical Satellite" to the then-European Space Research Organization (ESRO) in 1964. Various studies followed but the project was rejected on cost grounds by ESRO in 1968, even after it was simplified and renamed the

"Ultraviolet Astronomy Satellite". However, this work attracted the interest of NASA, who saw a gap in their programme between the flights of early ultraviolet satellites and that of the Hubble Space Telescope. As a result they made a collaborative agreement for the construction of the satellite in 1972 — initially only with the United Kingdom. A year later, ESRO (the forerunner of ESA) rejoined the project and a final division of the work emerged.

Under this agreement, Britain was to provide the ultraviolet detectors, which were modified television cameras, and data-analysis software. ESA provided the solar arrays, which are the power source for the satellite, and a European ground-station. Meanwhile NASA took the lion's share of the work building the telescope, spectrograph and spacecraft and providing the spacecraft operations and an American ground-station. After two more name changes, the mission finally became known as the International Ultraviolet Explorer, symbolizing its character as a highly successful and genuinely international scientific endeavour.

The observing time on IUE was initially shared out between the three collaborating agencies in approximate proportion to their contributions with NASA getting two-thirds of the time and Britain and ESA one-sixth each. Since 1981, however, the combined European share (British and ESA's) has been assigned by a single joint committee on scientific merit alone. At the time of writing, eight and a half years after its launch, IUE is still working well and providing excellent astronomical data for ultraviolet astronomers world-wide.

The scientific instrument on IUE covers the wavelength region from 1,150 to 3,200 Ångstroms with two spectrographs. One covers the "long wavelength region" from 1,900 to 3,200; the other the "short wavelength region" from 1,150 to 1,950. The division is necessitated by the difficulty of making a spectrograph which covers more than a factor two in wavelength; however, little observing efficiency is lost because for most astronomical objects two observations would in any case be needed to correctly expose both wavelength regions. The ranges overlap each other and, to a small extent, the wavelengths visible from the ground.

Both high- and low-resolution modes exist in the same spectrograph. For the high resolution an "echelle" spectrograph is used because this gives a compact layout of the spectrum consistent with the television-type detectors. The high resolution is matched to the solution of astronomical problems involving gas flows in and around stars such as STELLAR WINDS and mass transfer between components of BINARY STARS. Much of the evidence for these flows appears strongly in the ultraviolet spectra. Movement of a single mirror in the IUE spectrographs converts them to a lower spectral resolution suitable for work on fainter objects where there is insufficient light for the finer work. Particular targets studied at low resolution include GALAXIES and QUASARS.

IUE was launched into a geosynchronous (24-hour) orbit some 22,700 miles (36,000km) above the Earth. This was different to earlier astronomical spacecraft which were placed in near-Earth orbits and which pass over their ground stations once every 90 minutes. As a result IUE is above the horizon at the ground station for substantial periods. It can be continuously tracked from the European ground station, VILSPA, near Madrid, for more than 10 hours per day and is permanently visible from the United States ground station at Goddard Space Flight Center (GSFC) at Greenbelt, Maryland, just outside Washington.

The existence of continuous communications between satellite and ground station eases the problems of controlling the spacecraft. It no longer needs to be a self-contained robot able to act autonomously when out of contact and which needs to dump its data and be reprogrammed at every ground station "pass". Instead events on the satellite can be monitored and controlled in real-time from the ground and activities develop at a diurnal rate consistent with the lives of its human operators. In fact IUE is controlled eight hours a day from VILSPA and 16 hours a day (in two shifts) from GSFC.

To take advantage of this, the astronomer responsible for a programme of IUE observations goes to the appropriate ground station to direct operations. He is able to act just as a guest astronomer does at a modern optical observatory. The control room closely resembles that at such a ground observatory with television screens and monitors showing the field of view seen by the telescope, the data from the observations and the activities of the computers controlling the telescope and instrument. The only difference is that, instead of the telescope being at the end of a wire in the next-door room

(dome) as in a ground observatory, it is at the end of a radio communications link 36,000km up in the sky!

The visiting astronomer has at his disposal a team of technicians who understand the capabilities of IUE and who are responsible for spacecraft safety. The visitor, however, makes the real-time programme decisions, choosing which stars to observe, the wavelength region, resolution and exposure times for his observations. He is also responsible for recognizing the star field so that the correct star is observed. The data collected by IUE can be seen

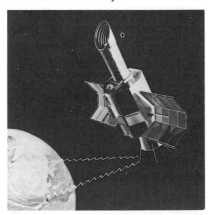

International Ultraviolet Explorer

within some 20 minutes of the completion of each observation to aid in programme decisions. Lastly the visiting astronomer collects his data fully reduced and processed and then takes them away to his home observatory or laboratory for further analysis.

It is this availability to visiting astronomers, modelled on the guest observer programmes of optical observatories, which has led to the widely recognized success of the IUE project. Many astronomers can be involved without a detailed knowledge of the IUE hardware. As a result of this style of operation, more than 700 different European scientists have used IUE. These have included theoreticians and astronomers active at other wavelengths and have brought the talents of the very best astronomers to bear on the problems posed by the IUE data with remarkably fruitful results.

All the data taken by IUE is kept in three archives, one operated by each of the collaborating agencies. These keep the reduced data private to the original observer for six months after his observation but then make them available to all astronomers. These archives have also made an important contribution to the widespread use of the IUE data and the success of the project. *MVP.*

Interplanetary dust A large tenuous cloud of dust and electrons surrounding the Sun. The solid particles may be of cosmic origin and include debris from the disintegration of comets. The presence of electrons in the cloud implies that it may be considered to be an extension of the SOLAR CORONA. Dust particles probably lie in the size range between one to 300 microns. In 1683 CASSINI discovered the ZODIACAL LIGHT, which he supposed was sunlight reflected from a cloud of matter surrounding the Sun. Robley (1980) has suggested that the dust may become fluorescent due to impact of the high speed solar wind particle streams that Legrand (1985) ties to recurrent AURORA. *RJL.*

I

Interstellar absorption The absorption or extinction of light caused by INTERSTELLAR DUST. On average, the extinction amounts to one stellar magnitude per kiloparsec of distance.

Interstellar chemistry An increasingly important research area concerns chemical reactions taking place in the interstellar medium, particularly in the clouds of gas which are such a feature of the view of the heavens in the infrared and millimetre wavelength regions. Many of the reactions are between ions (electrified atoms) and neutral atoms. COSMIC RAYS are an important agent for producing the ions. Studies in visible light of the spectral lines of objects such as the dramatic ORION NEBULA also provide useful information of the nature of the chemistry of the regions. *AWW.*

Interstellar dust Small grains of matter, typically about 100 nanometres in diameter, in interstellar space. Although, on average, there are only about 500 grains per cubic kilometre, they are very effective at absorbing and scattering visible and ultraviolet light, so making distant stars appear fainter (*see* INTERSTELLAR ABSORPTION) and redder. Because the grains produce polarization of starlight, they are believed to be elongated particles aligned by the galactic magnetic field. Most grains appear to be composed of silicates or graphite (a form of carbon); some have icy mantles. They may form by condensing from gas flowing out of the atmospheres of cool stars. *IKN.*

Interstellar matter *See* colour essay.

Interstellar molecules Molecules which exist in interstellar clouds. Most have been identified from their emission lines at mm and cm wavelengths.

Interstellar reddening The reddening of starlight passing through interstellar dust which arises because the dust is less effective at attenuating long-wave (red) light than short-wave (blue) light.

Invariable plane A plane of reference of the Solar System passing through its centre of gravity; inclined 1° 35′ to the ECLIPTIC.

Inverse square law A commonly-encountered relationship whereby the magnitude of a physical quantity (eg, the force of gravity or the apparent brightness of a light source) diminishes in proportion to the square of distance.

Invisible astronomy Astronomical research at wavelengths invisible to the human eye: GAMMA RAY, X-RAY, ULTRAVIOLET, INFRARED, RADAR, RADIO, and the study of particles such as COSMIC RAYS and NEUTRINOS, are the main areas.

Io The innermost Galilean satellite, with a volume and density similar to the Moon. Io is locked in a synchronous rotation and revolves around Jupiter with a period of 1.77 days. The satellite encounters EUROPA every 3.55 days. As Io catches up and passes Europa the interaction forces Io to oscillate about its synchronous position, creating internal stresses and generating heat. This heat is vented to the surface in the form of ongoing volcanic processes. The rate of deposition of volcanic material is large enough that Io, deep in Jupiter's gravitational well where the impact rate of debris should be large, has no impact craters visible on its surface.

A wide range of coloration and reflectivity is observed on the surface. Yellow, orange, red, brown, black and white are present. Many of these colours can be attributed to sulphur; however, impurities can drastically modify the colour of sulphur. There is considerable controversy concerning the amount of sulphur present in the crust of Io.

Voyagers 1 and 2 mapped about 35 per cent of the surface with a resolution better than 3 miles (5km). About 5 per cent of the total area consisted of volcanic vents. The vents are larger and occur more often in the equatorial regions, where tidal torques should have the largest effect. Nine active volcanoes were observed by the Voyager

cameras. Material is propelled to heights of 190 miles (300km) and deposits on areas with radii up to 370 miles (600km). SO_2, a heavy gas, requiring high energies to gain escape velocity, has been spectroscopically identified. Compression and heating of the gas could drive the volcanic mechanism, leading to a repaving of the surface at a rate of 1cm per century.

In addition to the plains areas containing the volcanic vents, there are mountains with elevations as high as 30,000ft (9km) above the nearby plains and layered structures with faults and scarps as high as 5,600ft (1.7km) that tend to be located near the south pole. These structures provide evidence that silicate is present in the crust. Sulphur would flow under the pressure generated by such a large overburden.

Io appears to be a body that has been constantly reworked. The satellite should be differentiated, with a core of heavy elements and mantle of sulphur and silicate compounds. Lighter gases, including water, have been lost. *RFB*.

Io: volcanic satellite of many colours

Ion An atom which has lost (or gained) one or more electrons compared to the normal, or "neutral", atom (in which the number of orbital electrons equals the number of nuclear protons so that the atom has zero net electrical charge). A positive ion has fewer electrons while a negative ion has more electrons than a neutral atom. For atoms which have lost electrons, the degree of ionization (ie, the number of electrons which have been lost) is denoted by a Roman numeral. For example, neutral iron is Fe I, singly ionized iron (one electron missing) is Fe II, 25-times ionized iron is Fe XXVI, and so on. *IKN*.

Ionization The name given to any process by which normally electrically neutral atoms or molecules are converted into ions, by the removal or addition of one

or more electrons. This gives them a positive or negative electrical charge. An ion can itself be ionized, such as by losing or gaining a second electron. The minimum energy required to remove an electron from an atom or molecule is called its ionization potential.

Ionosphere A region in the Earth's atmosphere that extends from a height of about 40 to 300 miles (60–500km) above the Earth's surface. Within this layer most of the atoms and molecules exist as electrically charged ions. This high degree of ionization is maintained by the continual absorption of radiation from the Sun. These free electrons and ions can disturb the transmission of radio waves through the ionosphere. There are several distinct ionized layers called the D, E, F_1, F_2 and G layers. These are rather variable with the D layer disappearing and the F_1 and F_2 layers merging at night. The D layer is situated at a height of between 40 to 56 miles (60–90km), the E layer at 75 miles (120km), the F_1 layer at 120 miles (200km) and the F_2 layer at 190 to 250 miles (300–400km). The free electrons in the E and F layers strongly reflect some radio waves. They are very important for long-distance radio communications, by successive reflections between it and the ground. In radio astronomy, it makes ground-based observations almost impossible below 10MHz. The D layer, where collisions between the molecules and ions are more frequent, is more an absorber of radio waves than a reflector. *JWM*.

Iota Aquarids A minor meteor stream, active in August.

IPCS (*Image Photon Counting System*). *See* ELECTRONIC DEVICES and PHOTO-ELECTRIC AIDS.

IQSY International Quiet Sun Year, 1964–65. A year in which astronomers of many nations collaborated in studies of the Sun and solar phenomena.

IRAS *See* INFRARED ASTRONOMICAL SATELLITE.

Iris (Asteroid 7). Discovered in 1847; diameter 130 miles (209km), magnitude 7.8. Of the regular asteroids only Vesta, Ceres and Pallas are brighter.

Iron A metal, chemical symbol Fe, which is the seventh most abundant element by numbers of atoms, and the fifth most abundant in terms of the mass content of the universe. Its properties include: atomic number, 26; atomic mass 55.847; melting point, 1,808K; boiling point, 3,023K.

Iron is the major constituent of the cores of the planets Mercury, Venus and Earth, and of certain types of meteorites. Iron grains are believed to make up a proportion of the interstellar dust population.

Elements up to and including iron can be built up by fusion reactions which release energy and which, therefore, can act as energy sources in stars (*see* STELLAR EVOLUTION). *See also* METALLIC CONTENT IN STARS.

Iron meteorites Meteorites composed essentially of iron-nickel metal, although some have up to 10 per cent by weight of stony material. Some 725 iron meteorites (irons) are known; 610 have been classified. Originally, irons were classified on a structural basis into three major groups, but now thirteen groups are recognized by their distinctive trace element chemistry. A single chemical group may have members from different structural classes. Fifteen per cent of irons do not belong in a chemical group.

It is thought that each chemical group derives from a parent asteroid. The structure of an iron was determined by its nickel content and the rate at which it cooled, the latter having been controlled by the size of, and its location in, the parent body. Iron-nickel metal with less than 6 per cent nickel at low temperature consists of the mineral kamacite; metal with nickel greater than 15 per cent consists of the mineral taenite. However, during cooling from above 800°C, metal with nickel between 6 and 15 per cent develops a textural intergrowth of kamacite and taenite; irons with this texture are called "octahedrites", and on polishing and etching with acid they display a Widmanstätten pattern.

The largest chemical group, IIIAB (eg, Cape York; Rowton), has members that grade chemically from low to high nickel contents, with concommitant increases or decreases in trace element (eg, gold) contents. They are therefore interpreted as having come from the once-molten core of an asteroid. Gradual solidification produced iron-rich, nickel-poor solids first, the most nickel-rich IIIAB irons representing the last liquid to solidify. Another major group, IAB, contains stony inclusions and although the nickel content has a wide range, trace elements do not. The IAB irons are

thought to be from pods of metal, near the surface of their parent asteroid, that were never molten. They are represented by the object that produced ARIZONA CRATER. Irons with less than 6 per cent nickel (hexahedrites) are represented by the North Chile shower; those with greater than 15 per cent nickel (ataxites) by the HOBA meteorite. *RH.*

Irradiation The process by which a region of space or material is subjected to radiation, whether by light, radio, infrared or other forms of electromagnetic waves, or by energetic particles such as protons and electrons.

Irregular variables Stars with no period in their light changes, due to flares in their outer layers; variable stellar winds. Others are slowly changing stars.

Ishtar Ishtar Terra is one of the major highland blocks of Venus and claims the most elevated point on the surface of the planet. It is Australia-sized and comprises Lakshmi Planum, a vast plateau containing two depressions (Colette and Sacajewa) and bordered on the west and north by Akna and Freyja Montes respectively. Immediately east are Maxwell Montes, which rise to 6.8 miles (11km), the highest point on Venus. Like the mountains surrounding Lakshmi Planum, they have a complex banded structure clearly seen on radar images. The eastern region takes the form of a rather hummocky plateau which rises at maximum to 8,200ft (2.5km) above datum. *PJC.*

Isidis Planitia Low-lying Martian plains region centred at 15°N, 170°W.

Isidorus, Bishop (*c.* 570–636). Bishop of Seville, who was one of the first to differentiate between astronomy and ASTROLOGY.

Islamic astronomy A flowering of celestial science in the Middle East, North Africa, and Moorish Spain that took place in the 8th to 14th centuries.

Two circumstances fostered the growth of astronomy in Islamic lands. One was geographic proximity to the world of ancient learning, coupled with a tolerance for scholars of other creeds. A second impetus came from Islamic religous observances, which presented a host of problems in mathematical astronomy, mostly related to time-keeping. In solving these problems the

Islamic astronomers went far beyond the Greek mathematical methods. These developments, notably in trigonometry, provided essential tools for the creation of Western Renaissance astronomy.

The foundations of Islamic science in general and of astronomy in particular were laid two centuries after the emigration of the prophet Muhammad from Mecca to Medina in AD 622. Toward the end of the 8th century in Baghdad, the capital of the new Abbasid dynasty, a massive effort produced in a few decades Arabic translations of all the major scientific texts of antiquity. The work was done by Christian and pagan scholars as well as Muslims. The most vigorous patron was Caliph al-Ma'mun, who came to power in 813. He founded an academy called the House of Wisdom, which became the centre for the translation project.

The academy's principal translator of mathematical and astronomical works was Thabit ibn Qurra, who wrote more than 100 scientific treatises, including a commentary on PTOLEMY'S *Almagest.* Another mathematical astronomer there was al-Khwarizmi, whose *Algebra* may well have been the first book on this topic in Arabic. Khwarizmi's name has given rise to the modern mathematical word "algorithm". Still another astronomer in 9th-century Baghdad was Ahmad al-Farghani, whose *Elements* helped spread the more elementary and nonmathematical parts of Ptolemy's geocentric astronomy into the West. By 900 the foundation had been laid for the development of an international science with one language — Arabic — as its vehicle.

From the 9th century onward spherical trigonometry was rapidly improved. Islamic astronomers discovered a variety of procedures that made the solution of problems involving the CELESTIAL SPHERE far simpler and less clumsy. Of the six modern trigonometric functions, five seem to be Arabic in origin, only the sine function having been introduced from India.

Among the conspicuous examples of modern astronomy's Islamic heritage are the names of stars. Betelgeux, Rigel, Vega, Aldebaran, and Fomalhaut are among those that are directly Arabic in origin or are Arabic translations of Ptolemy's Greek descriptions. In the *Almagest* Ptolemy had produced a catalogue of more than 1,000 stars. The first critical revision of the catalogue was compiled by 'Abd ar-Rahman as-Sufi, a 10th-century Persian

I

astronomer who worked both in Iran and Baghdad. His *Kitab suwar al-kawakib* ("Book on the Constellations of Fixed Stars") followed Ptolemy's often faulty list, but it did give improved magnitudes and Arabic identifications. His splendid pictorial representations became known in the Latin West.

Meanwhile the Arabic star nomenclature trickled into the West by another route: the making of ASTROLABES. The earliest dated astrolabe, in Arabic, comes from 927–8. Only a handful of 10th-century Arabic astrolabes exist, whereas nearly 40 survived from the 11th and 12th centuries, including several from Moorish Spain. It is primarily from Spain that astrolabe making, together with Arabic names for the stars, entered the West with England providing a gateway in the late 13th and 14th centuries.

Despite the preponderance of Arabic star names, Islamic astronomers did not make exhaustive observations of the sky. They restricted their sightings, or at least those they chose to record, primarily to measurements that could be used for rederiving key parameters of solar or planetary orbits. An impressive example of an Islamic astronomer working strictly within a Ptolemaic framework but establishing new values for Ptolemy's parameters was Muhammad al-Battani, a younger contemporary of Thabit ibn Qurra. Al-Battani's *Zij* ("Astronomical Tables") is still admired as one of the most important astronomical works between the time of Ptolemy and the Renaissance. COPERNICUS cites his 9th-century predecessor no fewer than 23 times.

In contrast, one of the greatest astronomers of medieval Islam, 'Ali ibn 'Abd ar-Rahman ibn Yunus, remained virtually unknown to European astronomers until around 1800. Working in Cairo a century after al-Battani, Ibn Yunus wrote a major astronomical handbook called the *Hakimi Zij*. Unlike other Arabic astronomers, he prefaced his *Zij* with a series of more than 100 observations, mostly of eclipses and planetary conjunctions. His handbook was widely used in Islam, and his timekeeping tables were used in Cairo into the last century.

Throughout the entire Islamic period astronomers stayed securely within the geocentric framework of Ptolemy and Aristotle. Nonetheless, criticism arose concerning particular technical details of Ptolemy's treatment, which seemed to violate the ancient injunction to use only uniform circular motions or combinations thereof for explaining the eternal celestial movements. One of the first critics was Ibn al-Haytham (Alhazen), a leading physicist of 11th-century Cairo, who went so far as to declare the planetary models of Ptolemy's *Almagest* false.

Meanwhile, in the 12th century in the western Islamic region of Andalusia, Ibn Rushd (Averroës) developed a more extreme criticism, saying that "to assert the existence of an eccentric sphere or an epicyclic sphere is contrary to nature". An Andalusian contemporary, Abu Ishaq al-Bitruji, actually tried to formulate a strictly geocentric astronomy, with disastrous results. Nevertheless, his work was translated into Latin and had considerable philosophical influence on Western astronomers.

At the other end of the Islamic world a fresh attack on the Ptolemaic mechanisms was undertaken in the 13th century by Nasir ad-Din at-Tusi. A prolific writer with 150 known titles to his credit, at-Tusi constructed a major observatory at Maragha (in present-day Iran). Other astronomers at the Maragha observatory also offered new arrangements of circles, but a fully acceptable alternative (from a philosophical point of view) did not come until the work of Ibn ash-Shatir at Damascus around 1350. Although Ibn ash-Shatir's solution, as well as the work of the Maragha observatory, remained generally unknown in the West, it is curious that Copernicus adopted a very similar arrangement of small circles or epicyclets in his constructions. Whether or not Copernicus received some technical hints from the Islamic tradition, the idea of criticizing Ptolemy was certainly part of the climate of opinion inherited by the Latin West from Islam. *OG.*

Ismenia Fossa Long, shallow depression on Mars: 37° to 42°S, 315° to 337°W. On old maps the area was called "Ismenius Lacus".

Isostasy Principle which recognizes there to be a state of balance between topographic masses and the underlying materials which support them. Owes much to the work of Sir George AIRY.

Isotherm A line drawn on a weather map, joining all places which are, at a given moment of time, experiencing the same temperature.

Isotopes Atomic nuclei which have the same atomic number but different atomic masses; ie, they contain the same number of protons but different numbers of neutrons.

ISEE (*International Sun-Earth Explorer*). A series of three scientific satellites launched in 1977-78 in an international project to study the near-Earth environment (particularly the magnetosphere) and its interaction with solar phenomena.

Ithaca Chasma A large terraced canyon system on the surface of Tethys. It extends more than 620 miles (1,000km) from 60°S, 0° to 50°N, 340°W. It is named for the homeland of Odysseus.

IUE *See* INTERNATIONAL ULTRAVIOLET EXPLORER.

Ius Chasma Prominent Martian canyon forming west part of Valles Marineris.

Ivar (Amor asteroid 1627). Diameter 4.3 miles (7km).

Jansky, Karl Guthe (1905–50). American radio engineer, who was the first to detect radio waves from the Milky Way (1931) during research into "static" for Bell Telephone Laboratories. In his honour the IAU in 1973 assigned the name "jansky" to the unit of strength of a radio wave emission.

Jansky unit The unit of flux density adopted by the International Astronomical Union in 1973. It is used in radio astronomy to describe the energy received from a source of radio noise per unit time per unit area. One jansky, denoted by the symbol Jy, is equal to 10^{-26} watts per square metre per hertz.

Janssen Lunar walled plain; 46°S, 40°E. Its walls are low, and broken by the deeper crater Fabricius. Janssen is 106 miles (170km) in diameter.

Janssen, Jules César (Pierre) (1824–1907). French astronomer, who founded MEUDON OBSERVATORY (1875) and the observatory at Mont Blanc (1895). He discovered the telluric line (1862), the spectral line of helium (1868), and the way to analyse SOLAR PROMINENCES without waiting for an eclipse (1868). He also recorded the SOLAR GRANULATION on photographs, and pioneered cinematography (1874).

Jules Janssen: studied solar phenomena

Janus Small satellite of Saturn; dimensions 137 × 124 × 99 miles (220 × 200 × 160km). It was discovered in 1978, and is co-orbital with EPIMETHEUS. The two may originally have been one body.

Jeans, Sir James Hopwood (1877–1946). English astronomer, who made notable contributions to ASTROPHYSICS and was also a noted popularizer of astronomy; he was famous for his broadcasts as well as his books.

James Jeans: astrophysicist and writer

Jet Propulsion Laboratory (JPL). Founded by Theodor von Kármán. It is situated near Pasadena, California, and is the centre of operations for all American unmanned space missions.

Jets Long, thin gaseous structures characterized by radio emission and/or optical nebulosity. Jets arise from two vastly different engines. In highly luminous, double-lobed radio sources and QUASARS, two oppositely-directed highly supersonic energetic beams of relativistic electrons bore into the ambient intergalactic medium. Jets remain collimated unless the surrounding gas causes wiggles and bends by refraction, or until they impact upon sufficiently dense material that they dissipate much of their energy of ordered motion in shocks, creating radio "hot spots" of synchrotron emission within the voluminous radio lobes. Intercontinental arrays of radio antennae reveal exceedingly well-collimated and thin radio analogues of the electron beams.

Sometimes galaxies even generate knotty optical nebulae along their jets. For periods of order 100 million years, the central engines maintain their memory, always pumping jets in the same two directions, perpendicular to any dusty lanes that might delineate the disks of these galaxies.

The second generator of jets is the low-luminosity PROTOSTAR. Fragments of dark cloud collapse to form dense

Jets: coma, dust jets and nucleus

cores in stellar nurseries. Infalling dusty envelopes surround the fledgling protostars that quietly accrete this material until their cores undergo nuclear fusion of hydrogen into deuterium. This drives convection throughout the young stars. The process of stellar readjustment reverses the infall, replacing this eventually by an outflow. However, this early wind, millions of times more vigorous than our present SOLAR WIND, cannot blow from the entire stellar surface because of the pressure of the still-accreting envelope. The initial outflow is bipolar, as thin jets puncture the circumstellar shell along the rotational axis of the star. The confined flow creates radio-emit-

ting jets and strings of optical blobs (HERBIG-HARO OBJECTS).

Rotation and magnetic fields are the parameters common to these two, apparently distinct, jet-forming mechanisms. *MC.*

Jet streams A narrow belt of high-speed wind at a level near the tropopause in the atmosphere. There are normally two westerly jet streams, subtropical and polar in each hemisphere, where the wind speeds are in the range of 100–200mph (160–320kph).

Jewel Box Nickname for the star cluster round Kappa Crucis (NGC4755).

Jewel Box: cluster in the Southern Cross

Jodrell Bank The radio astronomy observatory of the University of Manchester situated in Cheshire about 25 miles (40km) south of the City of Manchester. The main instrument is the 250ft (76.2m) aperture steerable radio telescope. Completed in 1957, this radio telescope attracted attention when it detected by radar the carrier rocket of the Soviet Sputnik I. For some years until large radio telescopes became available elsewhere, the telescope filled an important role in tracking the Soviet and American deep space probes. However, the telescope was built primarily to study the radio emissions from the universe and this has been its main use either as a single steerable paraboloid or linked with one or more remote radio telescopes in an interferometric combination. Since 1980 it has been possible to link a network of five remote telescopes to Jodrell Bank in the MERLIN array (Multi-Element Radio-Linked INterferometer). These remote telescopes are of 82ft (25m) diameter situated at distances of 11, 18,

J

J

24, 68 and 127km from Jodrell, distributed in direction to give the most effective sky coverage for producing detailed intensity contour maps of distant radio sources. The remote network is computer-controlled by landline from Jodrell Bank and the radio signals from each telescope are transmitted by microwave radio links to a correlator-computer system at Jodrell. When used on a frequency of 5GHz the resolving power of this array is 0.08 arc seconds. The system is also used in collaboration with a European network of radio telescopes (mainly those at EFFELSBERG, near Bonn, ONSALA in Sweden and WESTERBORK in Holland). In this VLBI (Very Long Baseline Interferometer) network the signals at each site are tape recorded and maps of QUASARS and other distant objects can be made with a resolution of 0.03 arc seconds on a frequency of 5GHz. In order to improve the details of the maps produced by the MERLIN array the linkage was extended to a sixth remote telescope at Cambridge in 1986.

The original 76.2m steerable paraboloid was known as the Mark I. During the years 1969–71 modifications were made to this telescope to improve the shape of the paraboloid and accuracy of control. The focal point of the original reflector was in the plane of the aperture. A new reflecting surface was built above this original structure of longer focal length (82ft/25m) from the apex instead of the 62ft (19m) of the original). The modified telescope is known as the Mark IA. In 1964 a smaller telescope of elliptical aperture 125ft × 83ft (38.1m × 25.4m) was completed at Jodrell (the Mark II). Like the Mark IA this is used as a single telescope or incorporated into the MERLIN network.

In December 1945 Bernard Lovell, returning as a lecturer in physics to the University of Manchester after wartime service, borrowed from the army three trailers of radar equipment with which he hoped to detect by radar the ionization from large cosmic ray air showers. On the university site electrical interference made observations impossible and he was given permission to take the trailers for a few weeks to the grounds of the botany department at Jodrell Bank. The original intention was never fulfilled. The transient radar echoes observed transpired to be from the ionized trails of meteors and not from cosmic ray ionization. New radar techniques for studying meteors were evolved leading to the discovery of hitherto unknown meteor showers

active in the daytime and to a method of measuring meteor velocities with which the controversy over the nature of the sporadic meteors was settled in favour of a Solar System — as distinct from interstellar — origin. The detection of echoes from the cosmic ray showers remained a prime target and to this end in 1946–47 a 218ft (66m) diameter telescope was built using scaffolding poles and steel cables on which was wound 16 miles (26km) of thin wire to form a paraboloidal reflector. This telescope was fixed to the ground, but by tilting the 126ft (38m) mast holding the primary feed at the focus

Jodrell Bank: the 250ft (76m) "dish"

the beam could be directed several degrees away from the zenith. This "transit telescope" was a great success, but not for the original intention. In the developing science of radio astronomy the transit telescope had the highest gain and narrowest beam available and the first accurate surveys of the radio emissions from the zenithal strip were soon made. In particular, in the early autumn of 1950, R. Hanbury Brown and his student C. Hazard produced the first conclusive evidence that an extragalactic system (M31 in Andromeda) was a radio source similar to the local Galaxy.

The success of the transit telescope provided the major arguments for the design of a similar instrument that could be steered to study the radio emission from any part of the sky. Construction began in September 1952 and after many difficulties this Mark I radio telescope was first used in October 1957 — as a radar telescope when the Soviets launched Sputnik I. For some years the Mark I was used in the radar mode as well as a receiving instrument. Radar studies of the Moon provided information about the nature of the scattering from the lunar surface

and systematic measurements were made of the electron density in the Earth–Moon space. In the early 1960s radar echoes from the planet Venus were obtained and this led to a definitive measurement of the solar parallax and of the rotation rate of Venus. At that time superior radar facilities became available in the Soviet Union and in the United States and thereafter the researches at Jodrell Bank were largely concentrated on the study of the radio emissions from the universe.

In the late 1950s and early 1960s a Jodrell group, led first by Hanbury Brown and then by H.P.Palmer, pioneered the development of the long baseline interferometers for the measurement of the angular diameters of radio sources. Until 1957 the 218ft transit telescope was used with a smaller remote aerial, and then the Mark I. Successive increases in the baseline led by 1960 to measurements over a baseline of 71 miles (115km) and to the discovery that out of nearly 400 sources at least seven were unresolved, showing that their diameter was less than 3 seconds of arc. This was a primary incentive that led to the optical searches with the 200in (5m) telescope on Palomar and to the discovery of the blue stellar-like objects in 1960 and subsequently to the realization in 1963 that there were remote objects in the universe — the quasars. In 1965 the first measurements were made over a baseline of one million wavelengths. This sequence of measurements eventually grew into the VLBI work and to the MERLIN network.

In 1972–73 a survey of nearly 800 radio sources was made using the Mark IA and II telescopes. The attempts to establish optical identifications for these sources culminated in 1979 when D.Walsh and his collaborators using the optical telescope on KITT PEAK in Arizona discovered the GRAVITATIONAL LENS. Since the discovery of PULSARS at Cambridge in 1968 a significant amount of the research effort has been directed to the study of these phenomena and many of the known pulsars in the northern hemisphere have been found by using the Mark IA telescope. A large research effort is also devoted to the study of the spectral line emissions from neutral hydrogen and the hydroxyl radical in the local Galaxy and in extragalactic systems. Other researches cover a wide field in planetary, galactic and extragalactic astronomy. In 1960–61 bursts of radio emission from red dwarf FLARE STARS were discovered and the study of stars both in their formative and eruptive stages continues.

As a teaching institution Jodrell Bank forms part of the physics department of the university and many hundreds of masters and doctors degrees have been awarded for research work carried out at Jodrell. Sir Bernard Lovell remained the director until his retirement in 1981, when he was succeeded by Sir Francis Graham-Smith, the Astronomer Royal. Since 1960 the Observatory has been known as the Nuffield Radio Astronomy Laboratories, Jodrell Bank, in recognition of the financial help given by Lord Nuffield and the Nuffield Foundation to the cost of the Mark I telescope. *BL. See also* RADAR ASTRONOMY, RADIO ASTRONOMY and RADIO TELESCOPES.

Johnson Space Center The operations headquarters for all United States manned space missions. Located on a $2\frac{1}{2}$ sq mile (650 hectares) site some 25 miles (40km) south-east of Houston, Texas, it commenced business in 1963. Originally known as the Manned Spaceflight Center (MSC), it was renamed on the death in 1973 of Lyndon B. Johnson, the United States President who engineered its establishment in his home State. The centre's responsibilities "include the design, development and testing of the spacecraft and associated systems for manned flight; selection and training of astronauts; planning and conducting

Johnson Space Center: space missions HQ

the manned missions; and extensive participation in the medical, engineering and scientific experiments that are helping Man to understand and improve his environment."

JSC received world-wide publicity during the APOLLO missions because of the extensive television coverage that originated from the centre. The Visitors' Center displays a wealth of relevant exhibits including a complete Saturn rocket system and descent and ascent modules as used on the Apollo missions. Most of the rock and soil samples from the Moon are kept in safe

storage at JSC, awaiting new avenues of testing and analysis. *EAW.*

Joint Institute for Laboratory Astrophysics Located in Boulder, Colorado, and founded in 1962, the institute is jointly operated by the University of Colorado and the National Bureau of Standards. Its staff work on problems mainly in atomic physics, chemical physics and astrophysics.

Jones, Sir Harold Spencer (1890–1960). Tenth Astronomer Royal. He arranged the move of Greenwich Observatory to Herstmonceux, and did notable research on determining the Sun's distance.

Joy, Alfred Harrison (1882–1973). American astronomer, famous for his work on variable stars, stellar distances and radial motions. He spent many years on the staff of the MOUNT WILSON OBSERVATORY in California.

Julian calendar The calendar that was devised by Julius Cæsar and Sosigenes of Alexandria and used in the Roman Empire from 46BC. It was in general use in the West until 1582, when the Gregorian calendar was instituted. In the Julian calendar, each year had 12 months, and there was an average of 365.25 days per year; three years of 365 days followed by a leap year of 366 days every four years. This simple four-year rule for leap years was modified in the Gregorian calendar.

Julian date The number of days which have passed since noon GMT on January 1 4713BC (the Julian day number), plus the decimal fraction of a day that has elapsed since the preceding noon. The starting point was chosen quite arbitrarily by the French mathematician Joseph Scaliger. The name "Julian" is in honour of his father Julius, and has nothing to do with Julius Cæsar. This method of numbering the days is independent of the length of the month or year. It is used to calculate how often different phenomena occur over long periods, and the intervals between them.

Julius Cæsar Very dark-floored, irregular lunar formation: diameter 44 miles (71km), 9°N, 15°E. It is in the Mare Vaporum area, not far from another dark-floored crater, Boscovich.

Juno (Asteroid 3). Discovered by Harding in 1804. Though numbered 3, it is not, however, the third largest of the

swarm; its diameter is only 179 miles (288km). Its rotation period is 7.2 hours, and its mass estimated at 2.0×10^{19}kg.

Jupiter The largest planet in the Solar System: it is 318 times as massive as the Earth and revolves about the Sun in an orbit with an average radius of 5.203AU. At opposition, when Jupiter appears on the observer's meridian at midnight, it subtends a diameter of about 47 arc seconds. Earth-based observers' ability to resolve the disk allowed the polar and equatorial radii to be established at an early date and, using the mass derived from Newton's law of gravity, the average density was determined. A volume equal to 1,335 times that of the Earth and an average density of 1.31 grams/cm^3, less than one-fourth that of the Earth, indicated that internally Jupiter was not earthlike.

Early spectroscopic studies revealed that Jupiter's spectrum was highly similar to that of the Sun; however, in 1934, R.Wildt identified absorption features of methane and ammonia in its spectrum. Because multi-atom molecules dissociate at high temperatures, this indicated that the observed spectrum was reflected sunlight from a cool planet with a molecular atmosphere. Hydrogen molecules and helium atoms do not possess readily observable absorption features in visible light; however, with access to infrared data, it has been ascertained that the composition of Jupiter, with respect to the relative abundance of hydrogen, helium, carbon, and nitrogen, is solarlike.

Because Jupiter revolves at an average distance of 5.2AU, the effective solar heating per unit area is reduced by a factor of 0.037 relative to the Earth. Utilizing the data from the PIONEER and VOYAGER missions, the total light scattered in all directions by the planet can be accurately determined; thus, the total absorbed energy is known. When this is compared to the non-solar infrared component of Jupiter's radiation, the ratio of emitted to absorbed solar radiation is 1.668 ± 0.085. Hence, heat generated internally from decay of unstable isotopes or slow gravitational contraction enters the atmosphere from the interior of the planet.

Even with its internal heat source, the outer regions of Jupiter's atmosphere are extremely cold, with the cloud-deck temperature near 150K (−120°C). The visible cloud deck

J

occurs at a depth of about one bar of pressure and is composed mainly of ammonia ice. Beneath this deck temperatures and pressures increase inward. At a depth of 5–10 bars the water cloud cycle should occur, moving energy from the deeper regions by convection and condensing and releasing the energy at higher levels. At still deeper levels the gaseous atmosphere becomes a liquid ocean which eventually changes into a region, composed chiefly of hydrogen and helium that, under the large pressures due to gravitational force on the overburden, exhibits a crystalline structure and is a good conductor of heat. This is referred to as the "metallic hydrogen" core. The heavy elements in the material from which the planet formed would migrate inward, forming a dense central core of about 15 earth masses.

The rotation rate of the observed DECIMETRIC and DECAMETRIC radio signal, presumed to be associated with the rotation of the conductive core, is approximately 9h 55.5m. Based on this data, a standard rotating co-ordinate system, System III, has been selected by the International Astronomical Union. The zero longitude has been chosen corresponding to the central meridian of the planet at 0 hours UT on January 1 1965. Longitude has been defined to increase with time at a rate of 870.536° per 24 hours. Within this system, the latitude and longitude of small eddies within the cloud deck can be measured as a function of time and prevailing winds can be derived. These measurements are interpreted as atmospheric motions relative to the interior of the planet and, if fully understood, could reveal a great deal about energy transport within the planet.

Average zonal winds are characterized by strong eastward winds within 10° of the equator and a series of alternating westward and eastward jets extending toward the poles. At latitudes equatorward of 40°, bright zones are bounded on the equatorward side by a westward jet and on the poleward side by an eastward jet. These regions have anticyclonic flow and correspond to upwelling regions in the Earth's atmosphere. The brown intervening belts have cyclonic shear and, by analogy, should be regions of downward flow where ammonia ice melts. Detailed photometric studies indicate that there is an aerosol haze extending from high altitudes to below the cloud deck. This haze contributes a yellowish-brown coloration to the planet and the size of the particles and the depth to

which the line-of-sight penetrates play a large role in determining the hue and shade of a specific cloud feature. The number of times the reflected photons have been scattered off particles, and the amount of ammonia ice located at high elevations, both play a role in the observed albedo.

Individual cloud features within the Jovian atmosphere are unique in that they are large and long-lived. A low effective temperature causes Jupiter to lose the excess energy that is brought up by convection cells more slowly than the Earth's atmosphere; however, this does not explain the extreme longevity of some cloud systems. The most well-known feature is the Great Red Spot, centred at about 20°S. Nested between a westward wind along the northern edge and an eastward wind on the south, the giant elliptical cloud system rotates in a counter-clockwise or anticyclonic sense. It has been observed by ground-based observers for more than 300 years. First reported by Robert Hooke in 1664, it was followed for several years by Cassini. Drawings by Heinrich Schwabe in 1831, William Dawes in 1851 and

Jupiter: the largest planet

Alfred Mayer and the fourth Earl of Rosse in the 1870s all show the Red Spot. The feature became quite visible in the 1890s and British Astronomical Association records contain extensive data from that time to the present. To a large extent, the visibility of the Red Spot is determined by the amount of turbulence that is present in the westward jet that is deflected around its northern perimeter. White eddies enter the feature during periods of active convection, rendering the contrast so low that detection is difficult. Actually, the Red Spot reflects no more red light than the surrounding white clouds. Its uniqueness lies in the fact that the circulation of the feature brings some

unknown constituent to the surface that is a strong ultraviolet and violet absorber; hence observing through a broad-band violet haze filter enhances the visibility of the feature. Small red spots, displaying the same ultraviolet absorber, appear at a similar latitude in the northern hemisphere; however, they do not grow to over-fill their windspace and become long-lived features. Many models have been proposed to explain the long-lived well-defined cloud features. These models deal with stable wave solutions and energy and momentum transport.

Jupiter's magnetic field is similar to that of the Earth. To a first order of approximation, it can be represented as a tilted dipole that is 10 times stronger than the Earth's field. Deviations from a dipole shape and failure of the magnetic axis to be aligned with the rotation axis or to pass through the centre of mass indicate that complex structure may occur deep in the interior. Interaction of the magnetic field with high-speed charged particles streaming outward from the Sun causes the particles to decelerate and become entrapped in the magnetic field, creating a complex region around the planet called the MAGNETOSPHERE. The high velocities of the entrapped particles create a hazard for exploring spacecraft. High-speed electron and hydrogen and helium nuclei penetrate the craft and cause electronic or radiation damage. The shape of the magnetic field has been sampled only near the equatorial plane by Pioneer and Voyager spacecraft. Advance planning for a Jovian polar orbiter has begun.

Currently ongoing research to obtain a model that is self-consistent involves several major regions of the planet. Within the deep core, problems concerning the nature of the contaminated hydrogen-helium alloy and the manner in which it conducts heat outward are challenging. How this region interfaces with a convective envelope, and the degree to which the rapid rotation introduces an organized cylindrical circulation about the axis of rotation within the envelope, is under investigation. The manner in which the envelope couples with the atmosphere, and the degree to which the dynamics of the atmosphere is driven from below, is an area of active research and controversy. As we continue to explore the Solar System, comparison of Jupiter with the other outer planets — Saturn, Uranus, Neptune and Pluto — will enhance our understanding of this giant planet. *RFB.*

Jura Mountains The surviving half of the rim of a 155-mile (250km) diameter crater that forms the north-west border of Sinus Iridum on the Moon. Spectacularly beautiful as the Sun rises on it.

K The symbol for the Kelvin, the unit of temperature on the Absolute, or Kelvin, temperature scale (which commences at ABSOLUTE ZERO). 1 Kelvin is equal in magnitude to 1° Celsius. The symbol °K is sometimes used, but is not strictly correct in the International (SI) System of units.

Kaiser, Frederick (1808–72). Dutch astronomer who became Director of Leyden Observatory. He specialized in double stars, and also produced a map of Mars.

Kant, Immanuel (1724–1804). German philosopher and cosmologist, born Königsberg.

His early work concerned the philosophy of science. In 1754, he suggested that the Moon's tide-raising forces must explain why the Moon always presents the same face to the Earth and must have the long-term effect of slowing down the Earth's rotation rate.

In 1755, he published a paper commenting on WRIGHT's cosmological theories in which he suggested that the Milky Way could be compared with the Saturn system of swarms of particles rotating around a central nucleus; the Milky Way, he said, was just another nebula, thus anticipating the ideas of HERSCHEL and LAPLACE. The term "Island Universe" is usually attributed to him. *HDH*.

Kappa Crucis (NGC4755). Open cluster of stars in the southern constellation CRUX, bright enough to see with the unaided eye. Also called the Jewel Box because it contains a single red supergiant star.

Kappa Cygnid meteors A minor meteor stream with a radiant at R.A. 19h 20m, Dec. +55° (near Kappa Cygni), that peaks on about August 20. The meteors are typically bright, slow-moving fireballs. The shower may be seen from about August 17 to 26.

Kapteyn, Jacobus Cornelius (1851–1922). Dutch astronomer renowned for his work in photographic astrometry and fundamental measurements of the parallaxes and proper motions of stars; his systematic plan of Selected Areas to study the structure of our Galaxy; and his discovery of star streams.

Kapteyn telescope The 39in (1m) Anglo-Dutch telescope at the ROQUE DE LOS MUCHACHOS OBSERVATORY on La Palma. Its novel Harmer-Wynne f8 optical design uses, as an alternative to its conventional f15 CASSEGRAIN design, a parabolic primary mirror, a spherical secondary and a doublet corrector lens. This design yields wide-angle photographs, with good images and minimum distortion, which are used to fix star positions accurately.

Karl Schwarzschild Observatory A major observatory 1,150ft (350m) above sea level near Tautenburg, in East Germany. The main telescope is a 78in (2m) reflector made by VEB Carl Zeiss, Jena; it can be used at the Schmidt, Cassegrain or coudé focus. The observatory is part of the Central Institute for Astrophysics of the East German Academy of Sciences; the group also includes the observatories at Babelsberg, Potsdam and Sonneberg. *PM*.

Kasei Vallis Major outflow channel entering the Chryse plains on Mars.

K-corona The inner part of the SOLAR CORONA. It is responsible for the greatest part of the brightness of the corona out to about two solar radii. It consists of rapidly moving free electrons. The K-corona reaches a temperature of about 2 million degrees Kelvin at a height of about 46,600 miles (75,000km) above the photosphere. There may be streamers (regions of higher density) and coronal holes (areas of very low density) within the inner corona.

Keeler, James Edward (1857–1900). American astronomer who became Director of the LICK OBSERVATORY. He specialized in spectroscopy, the study of nebulæ, and planetary research. In 1895 he made a study of Saturn and its rings which proved that they could not be solid.

Kelvin, William Thomson, Baron (1824–1907). Distinguished Scottish mathematician and physicist who gave his name to the Kelvin temperature scale, based on absolute zero. *See also* K.

Lord Kelvin: scale based on absolute zero

K

Kennedy Space Center NASA's manned spaceport, located on 138,000 acres (56,000 hectares) of Merritt Island, north of (and contrary to popular misconception, not on) Cape Canaveral, Florida. The complex was commissioned in 1961 to be the location of the Saturn-V launch site for APOLLO missions, and two pads, "39A" and 39B", were built. The facilities were modified to support SPACE SHUTTLE missions.

Kennedy Space Center: Apollo missions

Kepler Lunar crater: 8°N, 38°W on the Oceanus Procellarum. It is 22 miles (35km) in diameter, and is a major ray-centre.

Kepler, Johannes (1571–1630). German astronomer and mathematician, remembered particularly for his three laws of planetary motion.

Kepler was born prematurely at Weil der Stadt, west of Stuttgart. His father was a mercenary, his mother an innkeeper's daughter. A weak but intelligent child, he obtained scholarships and in 1587 went up to Tübingen University. Here he showed himself to be an able mathematician, and though he began studying Protestant theology after gaining his Master's degree, he was nominated provincial mathematician and teacher of mathematics at the Lutheran school in Graz. There he soon made his mark by issuing a calendar for 1595. In spite of his mixed views on ASTROLOGY — he thought of it as "the foolish little daughter of astronomy" — Kepler followed fashion and provided astrological predictions about the weather, peasant uprisings and Turkish invasions, all of which happened to be fulfilled.

While at Tübingen, Kepler had come under the influence of Michael Maestlin, the Professor of Astronomy, who was well versed in the HELIOCENTRIC THEORY of COPERNICUS. Kepler was fired with enthusiasm for the Copernican idea that the Earth is a planet and, with the rest of the planets, orbits the Sun. A staunch believer, too, in a divine plan of the universe, he suddenly came upon what seemed to him a proof of the idea. He found that if one took a spherical universe with the Sun at the centre, and then inserted the spheres on which the planets were then supposed to travel, the sizes of these spheres were such that one of the five regular solids of classical geometry could be inserted between them. Thus between the spheres of Saturn and Jupiter, a cube could be fitted almost exactly; between Jupiter and Mars a tetrahedron (solid with five faces), and so on for the other pairs of planets. This novel scheme worked so well with the planetary distances then accepted, that nowhere was there an error greater than five per cent.

Kepler's grand design was published in 1596 with a long title but is now usually known as *Mysterium cosmographicum* (The Mystery of the Universe). The work, which he circulated to noted scholars such as GALILEO and Tycho BRAHE, established his reputation as a brilliant mathematician.

Religious upheavals forced Kepler and his wife to consider moving from Graz, and he visited Tycho in Benatky, outside Prague. Here Kepler saw the superb precision of Tycho's observations and recognized the need for their mathematical analysis to improve planetary theory. For his part, Tycho did not accept the Copernican theory, but was so impressed with Kepler's abilities that in 1600 he invited him to Benatky. Kepler accepted and when Tycho died the next year, Kepler was appointed his successor as imperial mathematician.

In 1604 a SUPERNOVA appeared in the constellation of OPHIUCHUS and Kepler wrote his *Stella Nova . . .*, describing its astronomical and astrological aspects. Indeed, his post as imperial mathematician kept him busy providing the Holy Roman Emperor Rudolph with astrological interpretations of astronomical phenomena, such as the conjunction in 1603 of Saturn and Jupiter. All the same Kepler managed to continue with his analysis of Tycho's observations of Mars. It was fortuitous that it was on Mars that he worked first, because it has the greatest eccentricity of all the planets except Mercury and Pluto. This was to prove of the greatest significance.

From 1600 to 1609 Kepler worked at Tycho's observations. Their precision was such that he was able to find that they did not fit the Copernican theory, and nor did they give any support to Tycho's own planetary scheme. After

Johannes Kepler: brilliant mathematician

much calculation he was forced to the conclusion that Mars orbits the Sun, but in an ellipse not a circle. This was a fundamental break with tradition, for previously everyone had followed the Greeks and taken it for granted that the planets move in circular orbits. Moreover, they had also accepted the Greek belief that the planets move at an unvarying rate throughout their entire orbits. This Kepler showed to be untrue. The orbital velocity of Mars was greatest when closest to the Sun in its elliptical path, and slowest at its most distant reaches of its orbit. Kepler published his results in his *Astronomia nova* of 1609, and in the years that followed began examining other results of Tycho's with a view to discovering whether all the other planets moved in elliptical orbits. He found that they did, and, in expressing it, formulated what

we now know as Kepler's first law of planetary motion.

It was characteristic of Kepler that, having taken the important step at letting his theoretical interpretation be led by observation, he should next seek to adopt another scientific attitude, and try to find a physical reason to account for a planet's elliptical motion. Having recently read two books about magnets and magnetism, he speculated about a magnetic force emanating from the Sun, and sweeping the planets round as the Sun rotated. This was wrong, and led him to suggest that the force diminished directly with distance. This again was an error, but it nevertheless enabled him to formulate a law about orbital velocity which, though strictly only correct at the nearest and most distant points to the Sun (aphelion and perihelion) in a planet's orbit, he made a general rule. This brought him to his second law of planetary motion, that the line Sun to planet (the "radius vector"), sweeps out equal areas of the ellipse in equal times. He published his two laws in his *Epitome astronomiæ Copernicæ* (Epitome of Copernican astronomy).

With the idea of a divinely harmonious universe still in mind, Kepler turned to see whether he could now reframe the idea in the light of his recent research. By fitting the different orbital velocities at aphelion and perihelion to the musical scale to provide an insight into celestial harmony, he was able to derive what seemed to him an example of divine law — the relationship between a planet's distance from the Sun and the time it takes to complete an orbit. This is known now as Kepler's third law, and appeared in 1619 in his *Harmonices mundi* (Harmonies of the World).

Kepler wrote a number of other books. *Tabulæ Rudolphinæ* (Rudolphine Tables) a star catalogue and listing of future planetary motions based on his research, appeared in 1627; he also wrote another set of astronomical tables and fine books on optics, one of them a comment on *The Starry Messenger* by GALILEO, in which he proposed his own design of telescope. He wrote also on the geometrical but practical problem of determining the volumes of barrels; a text, the *Somnium . . .* (The Dream), which is a description of an imaginary trip to the Moon and a defence of the heliocentric outlook of Copernicus; and a book on astrology.

Kepler suffered from poor health all his life, but when his mother was tried for witchcraft, he went to help her, defending her with great success to his own advantage as well as hers, for her conviction would have damaged his career. Twice married, he lived at a time of religious upheaval in Europe, ending his days in Regensburg, chronically short of money due to delays in the payment of his stipend. *CAR. See also* KEPLER'S LAWS.

Kepler's Laws The three fundamental laws of planetary motion, announced by KEPLER in 1609 and 1618. (1) The planets move in elliptical orbits, the Sun being situated at one focus of the ellipse. (2) The radius vector, an imaginary line joining the centre of the planet to the centre of the Sun, sweeps out equal areas in equal times (thus, a planet moves fastest when closest to the Sun). (3) The squares of the SIDEREAL PERIODS of the planets are proportional to the cubes of their mean distances from the Sun.

These laws, which Kepler saw as of divine origin, disproved the PTOLEMAIC theory, and vindicated COPERNICUS. The laws apply to all bodies in closed orbits around the Sun, and to satellites orbiting planets. *TJCAM.*

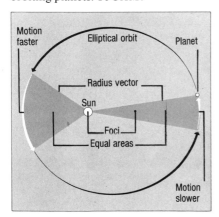

Kepler's Laws: vindicated Copernicus

Kepler's Star In 1604, only 32 years after TYCHO'S STAR, a second supernova appeared in OPHIUCHUS. No galactic supernova has definitely been seen since that time.

Reaching an apparent magnitude of −3, the 1604 star was extensively observed in Europe and the Far East. Both KEPLER and the official astronomers of Korea made fairly systematic brightness estimates for up to a year. The smooth light curve closely resembles that of a Type I supernova. Accurate measurements by both Kepler and David Fabricius fixed the position of the star to better than one arc minute. The remnant is a powerful source of electromagnetic waves. *FRS.*

Kew Observatory Historical observatory at Kew in Surrey, England, established at the order of King George III. It still stands.

Keyhole Nebula Dust cloud seen dark against the Carina Nebula. Named by Sir John Herschel, but has since changed its appearance, being illuminated by the variable ETA CARINÆ.

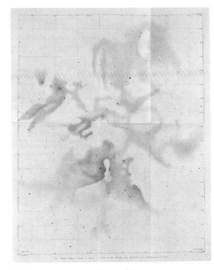

Keyhole Nebula: a change in appearance

Kiloparsec One thousand PARSECS.

Kirchhoff's Laws. Laws of spectroscopy discovered by Kirchhoff and Bunsen and published in 1856. In essence they state that an incandescent solid, liquid of high-density gas emits a continuous spectrum whereas a low-density gas emits or absorbs light at particular wavelengths only. Each element has its own characteristic pattern of lines.

Kirkwood, Daniel (1814–95). American mathematician. Theorized on Solar System evolution, its lesser bodies and structure in the asteroid belt.

Kirkwood Gaps Distances where few ASTEROIDS occur in the main belt, kept clear by Jupiter's gravity. Noticed by KIRKWOOD in 1857, explained in 1866.

Kitt Peak National Observatory The national research centre for ground-based optical astronomy in the United States. Located in the Quinlan Mountains 56 miles (90km) south-west of Tucson, Arizona, Kitt Peak is the site of one of the world's most advanced stellar telescopes, the 158in (4m) Mayall Telescope. Kitt Peak National Observatory operates six additional telescopes: an 84in (2.1m) reflector optimized for infrared as well as optical

observations; a 50in (1.3m) reflector, used primarily for infrared studies; two 36in (91cm) general-purpose reflecting telescopes; a 36in (91cm) instrument designed to feed light to the main spectrograph at the 2.1m telescope; and the Burrell-Schmidt telescope, which is operated jointly with Case Western Reserve University and is used for spectral and imaging surveys of large regions of the sky.

All of the telescopes operated by Kitt Peak National Observatory are available to qualified scientists from around the world. Allocations of observing time are based on evaluation of written proposals. Every year more than 300 astronomers use Kitt Peak facilities. Most of these researchers are staff members or graduate students at universities or government research laboratories, and they travel to Kitt Peak for observing runs that last typically only three to five nights.

The programmes carried out at Kitt Peak span the full range of modern astrophysics. The recent apparition of HALLEY'S COMET stimulated a great deal of work on the chemistry of COMETS and on the physical properties of comets and ASTEROIDS. Observations at Kitt Peak have provided the first empirical evidence for the existence of BROWN DWARFS, objects of such low mass that they cannot ignite nuclear reactions and become true stars. Kitt Peak has also pioneered in the study of distant galaxies and QUASARS in an effort to understand the development of structures during the early evolution of the universe.

Kitt Peak is also the site of the McMath Solar Telescope, which is the world's largest solar telescope and is operated by the National Solar Observatory. In addition, the mountain is home to facilities managed by the

Kitt Peak National Observatory

National Radio Astronomy Observatory, by the University of Arizona's Steward Observatory, and by the McGraw Hill Observatory, which

K

K

supports observers from the University of Michigan, Dartmouth, and Massachusetts Institute of Technology.

Kitt Peak National Observatory was established by a consortium of universities, which joined together in 1957 for the purpose of seeking funding from the recently formed National Science Foundation in order to create and manage national centres for research in optical astronomy. This consortium, which became known as AURA, the Association of Universities for Research in Astronomy, established Kitt Peak National Observatory in 1958. In 1963, CERRO TOLOLO INTER-AMERICAN OBSERVATORY was opened in Chile to provide access to the southern skies. AURA also now operates the National Solar Observatory, with facilities on Sacramento Peak in New Mexico, as well as on Kitt Peak. These three observatories together with the Advanced Development Program, which explores the application of new technologies to astronomy, form the National Optical Astronomy Observatories. *SCW.*

Kocab The star Beta Ursæ Minoris, one of the "Guardians of the Pole". It is noticeably orange; magnitude 2.06.

Kohoutek's Comet 1973 XII. A comet discovered by L. Kohoutek in 1973, when it lay near the orbit of Jupiter. It was expected to become brilliant, but as a spectacle proved a disappointment. It was, however, well observed by the Skylab astronauts. Its orbit is essentially parabolic.

Konkoly Observatory The leading observatory in Hungary. It is particularly noted for its work in connection with variable stars.

Kopff's Comet Found at Heidelberg in 1906, passed near Jupiter in 1942 and 1954 changing orbit; period now 6.4 years; last returned in 1983.

Korelev Concentric lunar crater, more than 250 miles (400km) wide, at 4°S, 157°W.

KREEP Type of lunar basaltic rock with distinctive content of potassium (symbol *K*), *R*are *E*arth *E*lements and *P*hosphorus. KREEP samples have been excavated from the Moon's interior by impact-cratering events.

Krüger 60 A twelfth-magnitude binary star at a distance of 12.8 light-years in Cepheus. Both stars are M-type RED DWARFS. The fainter component is a FLARE STAR of exceptionally low mass.

K stars The visible spectra of K stars are characterized by numerous absorption lines and the presence of molecular bands of CH and CN. On the MAIN SEQUENCE their surface temperatures range from 3,550 to 4,900K; giants are about 400K cooler and supergiants 300K cooler still. The dwarf K stars are 0.5 solar mass at their lowest point on the main sequence, rising to 0.8 solar mass at their highest. As K giants and supergiants may be either evolved old stars of about one solar mass, or evolved younger stars of higher masses, a wide range is represented. However, most of the K giants are stars of a few solar masses. Except where they are members of short-period binary systems, all K stars are slow rotators.

Only a small fraction of K stars is variable: the coolest CEPHEID VARIABLES have K-type spectra, as do the hottest of the MIRA VARIABLES.

Among the K giants a few per cent show spectral peculiarities which resemble those in the S and N stars, with over-abundance of carbon and certain heavy elements. The largest excesses of carbon produce spectra of R-type. Moderate carbon abundances combined with heavy element excesses are present in the BARIUM STARS. The same behaviour is seen in the POPULATION II K giants, in which these selective enhancements of chemical composition are superimposed on a general metal deficiency. Such stars have weak atomic lines but very strong molecular bands of CH and are known as CH stars.

Bright examples (K stars are yellow or orange in colour): Dubhe and Pollux KO III, Arcturus K2 III, Epsilon Pegasi K2Ib, Alpha² Cen K5 V, Epsilon Eridani K2 V. *BW.*

Kuiper Crater on Mercury: 11°S, 32°W, diameter 25 miles (40km). It intrudes into the 78-mile (125km) crater Murasaki. Kuiper is a ray-centre, and was the first crater to be recognized on Mercury during the approach of Mariner 10 in 1974.

Kuiper Airborne Observatory Currently the world's largest aircraft-borne observatory, comprising a C-141 Star-Lifter jet aircraft, a 36in (91cm) short-focus (f2 primary) Cassegrain reflecting telescope mounted in an airtight compartment and capable of motion in altitude, plus all the necessary supporting equipment for operating the telescope

and the various spectrometers and other instruments that may be attached to it. The KAO is operated by NASA from its Ames Research Center at Moffett Field, near San José in California, and stems from G.P.KUIPER'S desire to obtain infrared spectra of Solar System bodies with minimal interference from water vapour in the terrestrial atmosphere. Solar spectra from a high-flying Canberra jet aircraft from RAE Farnborough had been obtained in 1957,

Kuiper Airborne Observatory

but the comparative faintness of planets, etc, clearly required much more sophisticated apparatus. The first trials were made from an A-3B jet in 1965, the success of which led to the equipping of a Convair 900 jet with facilities for much larger spectrometers or interferometers. Experience gained from this, together with that from similar missions in a Lear jet, culminated with the creation of the KAO, which was put into service in 1975 and named later that year in honour of the man whose efforts had led to its becoming an entity.

Several discoveries and much new data have resulted from KAO operations including the co-discovery of Uranus' rings, of water vapour in Jupiter's atmosphere and HALLEY'S COMET, of condensed cores in BOK GLOBULES, and of more than 60 previously unobserved spectral features due to atomic, molecular and solid-state species in the interstellar medium. *EAW.*

Kuiper, Gerard Peter (Gerrit Pieter, 1905–73). An outstanding and dynamic Dutch-born astronomer who, more than any other person, was responsible for restoring Solar System astronomy to prominence in an era dominated by stellar and galactic research.

After making many noteworthy contributions to stellar astronomy following his move to the United States, he turned to his original intention of pursuing a vigorous programme of lunar and planetary research. Resumption of this work after World War II duties resulted in a string of discoveries — a methane atmosphere on Titan, a tenuous CO_2 atmosphere on Mars, a fifth satellite of Uranus (Miranda), and a second satellite of Neptune (Nereid). In 1960 he was finally able to fulfil an early ambition — to set up an institute devoted to Solar System studies. This was the Lunar and Planetary Laboratory at the University of Arizona in Tucson, the oldest and largest establishment of its type.

He was intimately involved in most of NASA's space programmes, and was Chief Experimenter for the pioneering Ranger series of Moon missions. His search for superior observatory sites led to the choice and development of CERRO TOLOLO in Chile and MAUNA KEA in Hawaii. He also initiated the use of telescopes in high-flying jet aircraft for infrared observations. For these and many other original contributions to Solar System astronomy, Kuiper's is the only name commemorated on three different Solar System bodies (a crater on each of Mercury, the Moon, and Mars). *EAW.*

Kwasan and Hida Observatories Affiliated with the Faculty of Science, University of Kyoto, the Kwasan and Hida Observatories are engaged in research in co-operation with the Department of Astronomy.

Kwasan Observatory was established in 1929 and is situated on Kwasan Hill (35°0′N 135°48′E), at an altitude of 768ft (234m), 4 miles (6km) south of the university. The instruments at that time were a Cooke 12in (30cm) refractor (now re-formed to a Zeiss 18in/45cm refractor), a Sartorius 7in (18cm) refractor, a Bamberg 3½in (9cm) meridian instrument and solar equipment with an Askania spectroheliograph.

In 1958 the observatory separated from the Department of Astronomy administratively and extended its field of research to solar, lunar and planetary physics; a 27in (70cm) coelostat with high-dispersion spectrographs was installed.

In the early 1960s viewing conditions began to deteriorate and after years of survey, Hida Observatory was built in 1968 at the top of a mountain (36°15′N 137°18′E), at an altitude of 4,180ft (1,276m), at the western end of the North Japan Alps in Gifu, 217 miles (350km) north-east from Kyoto.

Hida Observatory has a Zeiss 25in (65cm) refractor, a Tsugami 27in (60cm) reflector, and a Zeiss 27in (60cm) domeless solar telescope, with subsidiary instruments. Recently, a Perkin-Elmer μ-10 microdensitometer and DEC VAX 11/750 minicomputer were installed at Kwasan Observatory. The observatories co-operate in such fields of research as the physics of the Sun and Solar System. *SS.*

Labelled release experiment An experiment on the Viking lander spacecraft. A sample of Martian soil was moistened with a nutrient containing radioactive carbon 14. Any gas emitted by biological activity would have been revealed by its radio-carbon content. No clear evidence for living organisms was found.

Laboratory for Atmospheric and Space Physics An institute of the University of Colorado. It engages in the study of the planets, especially Venus, Mars and Saturn, and their atmospheres and rings, using instruments on spacecraft.

Lacaille, Nicolas-Louis de (1713–62). French astronomer. He was the "father of southern astronomy", observing southern skies at Cape of Good Hope 1750–55, and naming 14 new constellations. He pioneered the lunar-distance method of finding longitude.

Lacaille: 14 new constellations

Lacerta (The Lizard). A small, obscure constellation bordering Cepheus.

Lagoon Nebula (M8, NGC6523). An emission nebula in Sagittarius, surrounding the star cluster NGC6530 which includes the two naked-eye stars 7 and 9 Sagittarii. The cluster was discovered by John FLAMSTEED in 1680, and the fainter nebula was first noticed by Le Gentil in 1747. At a distance of 4,500 light-years, the nebula is cut by a dark cloud which gives it its name.

Lagrange, Joseph Louis (1736–1813). Great mathematician, born in Turin but who went to Paris in 1787 and spent the rest of his life there. He made many contributions to general dynamics and pure mathematics.

Lagrangian points Points in the orbital plane of two massive bodies, in circular orbits around their centre of mass, at which much less massive objects can remain in equilibrium.

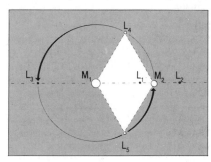

Lagrangian points: orbital equilibrium

Lakshmi Planum The western part of Ishtar Terra, one of the two main highlands on Venus. Its height is 9,800ft (3km) above the standard level.

Lalande, Joseph Jérome Le François (1732-1807). French astronomer, who became Director of the PARIS OBSERVATORY; a skilled mathematician and observer, and also a popularizer of astronomy.

Lallemand camera An electronic camera which converts the telescopic image into electrons that are accelerated and focused onto a nuclear emulsion to form an enhanced picture.

Lambda Scorpii Shaula, magnitude 1.63, the brightest star in the "sting" of Scorpius.

Lambda Tauri ECLIPSING BINARY in Taurus; range 3.3 to 4.2, period 3.95 days.

Landsat satellites A series of five United States satellites designed to monitor the Earth's resources. They contained

L

multispectral scanners scanning the Earth at four wavelengths and three television cameras operating on three wavelengths from which coloured images could be obtained. Because different surface features have different spectral responses, it is possible to monitor such features as forests, crops, pollution and marine biology. The first three were placed in polar orbits and were capable of scanning the whole Earth in 18 days. The last two were placed in a lower polar orbit, giving better resolution. *HGM.*

Langley Research Center A NASA facility at Hampton, Virginia. It concentrates on research and development of advanced aircraft and spacecraft systems. It has a major visitors' centre.

Langrenus A prominent lunar crater 82 miles (132km) in diameter, which has finely-terraced walls, a generally level floor, a prominent central peak, and a weak ray system.

Langrenus, Michael Florentius (Michiel Florenz van Langren, 1600–75). Compiler and publisher, in 1645, of the first map of the Moon. A member of a prominent Belgian family of map and globe makers, he was appointed "Mathematician and Cosmographer" to King Philip IV of Spain. After making drawings of 30 or more different phases in order to include the major (and some minor) topographical features as well as the surface shadings, he combined these into a single map 13in (34cm) in diameter. He portrayed the craters and mountains as if everywhere illuminated by a rising Sun, a technique currently used on the best lunar maps. He also introduced the scheme of naming features after famous persons, also the current scheme, although only about 65 of his original names have survived.. He later turned to the maintenance and improvement of Belgian ports and waterways. *EAW.*

La Palma Observatory *See* ROQUE DE LOS MUCHACHOS OBSERVATORY.

Laplace, Pierre-Simon de (1749–1827). French ▸ mathematician and astronomer. In its field, his *Méchanique céleste* (1799–1825) is second only to Newton's *Principia* in importance. He proposed the nebular hypothesis of the origin of the Solar System.

Las Campanas Observatory Major observatory near La Serena in Chile,

operated by the Carnegie Institute in Washington. The main telescopes are the 98in (2.5m) Irénée du Pont reflector and the 39in (1m) Swope reflector, completed in 1977 and 1972.

Laser A device that produces a pure beam of high-intensity coherent (all waves in phase) monochromatic (all of one wavelength) electromagnetic radiation at infrared, optical, or shorter wavelengths. The acronym derives from *l*ight *a*mplification by *s*timulated

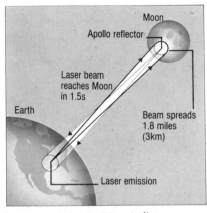

Laser: measuring the Moon's distance

*e*mission of *r*adiation. A similar device at microwave frequencies is known as the MASER.

The stimulated emission of a photon of particular wavelength occurs when an electron in a high-energy level of an atom is induced to drop to a lower level by an encounter with a photon of energy exactly equal to the difference in energy between the upper and lower levels. The emitted photon has exactly the same wavelength and direction of propagation as the stimulating photon, and the emitted radiation is said to be coherent. In a laser, large numbers of electrons are pumped up into a higher energy level by means of a suitable energy source (in a simple device, such as a ruby laser, a flashlight may be used) so that there are more electrons in the higher level than the lower level; this distribution of electrons is known as an "inverted population". The electrons are then stimulated to emit a beam of coherent radiation.

Laser action can be produced in solids, liquids or gases, and lasers may be pulsed or continuous in operation.

Lasers are used in various astronomical contexts. For example, very precise measurements of the lunar distance, and the Moon's slow rate of recession, have been made by bouncing laser beams off reflectors placed on the lunar surface by the APOLLO astro-

nauts, and laser beams reflected from orbiting satellites allow the rate of continental drift to be measured.

Lasers are also used to check and maintain the alignment of optical components, particularly in complex optical systems like the MULTIPLE MIRROR TELESCOPE. An important application of lasers is in microdensitometers and plate-measuring machines: the pure beam provides a sensitive means of measuring the brightness and position of each image, or part of an image, on a photographic plate. *IKN.*

La Silla Observatory Otherwise known as the European Southern Observatory (ESO). It is situated in Chile, where the seeing conditions are exceptionally good. The main telescope is the 150in (381cm) reflector, completed in 1975. La Silla is one of three important modern observatories set up in Chile, the others being CERRO TOLOLO and LAS CAMPANAS.

Lassell, William (1799–1880). English amateur astronomer who built two large reflecting telescopes. He discovered 600 nebulæ and several faint planetary satellites.

Late-type stars Conventionally, stars of types, K, M, R, N and S. The name was given when it was (wrongly) believed that the types represented an evolutionary sequence, hot stars being the youngest and cool stars the oldest.

Launch window The period of time during which it is possible to launch a particular space mission.

Launton meteorite Chondrite which fell Oxfordshire, England, in 1830.

Leavitt, Henrietta Swan (1868–1921). American astronomer, who spent

Henrietta Leavitt: studied Cepheids

much of her career at the Harvard College Observatory. She specialized in variable star work, and it was her studies of CEPHEID VARIABLES in the Small Cloud of Magellan which led on to the all-important Cepheid PERIOD-LUMINOSITY relationship.

Leda The 13th satellite of Jupiter, discovered by C.Kowal in 1974. It is only about 9 miles (15km) in diameter. It forms a group with Himalia, Lysithea and Elara. Though it has direct motion, it may well be a captured asteroid.

Leibnitz, Gottfried Wilhelm (1646–1716). German mathematician and philosopher who invented the calculus independently of Newton and supported the "relational" theory of space and time which contradicted Newton's view that space and time were "absolute" (ie, had independent existence).

Lemaître, Georges Édouard (1894–1966). Belgian Catholic priest who was professor of astronomy at Louvain from 1927. He began his work on the expansion of the universe, for which he is now most widely remembered, during an earlier visit to America. He considered how this expansion would vary in the time according to EINSTEIN's General Theory of RELATIVITY, including the effect of the so-called cosmological term and of the pressure of matter, and was the first to apply such solutions as models for the actual universe. His ideas on the origin of the universe from the explosive, radioactive decay of a primeval "atom", which later gave rise to the BIG-BANG terminology, gained a notable general popularity, but were not of much scientific interest. *DJR*

Lemonnier, Pierre Charles (1715–99). French astronomer and physicist. He surveyed 1° arc of meridian near Arctic Circle; studied planetary perturbation theory; edited *Atlas Céleste*; and made prediscovery observations of Uranus.

Lens Basic light-transmitting optical component, which changes the path of light rays by refraction at its surface, usually in such a way as to form, improve or modify an optical image. Lenses are familiar to everyone as simple pieces of glass with curved surfaces, but they often hide a sophistication that embodies the highest levels of modern technology. Even the lenses in a humble pair of spectacles are made of specialized glass types with precisely-known characteristics, and have surfaces whose shape has been carefully controlled so that the lenses compensate exactly for the defects in the wearer's vision. At the other extreme, lenses for special purposes usually incorporate several component lenses (or "elements") which may be made of a wide variety of glass types, or sometimes more exotic materials like sapphire. Their accurately-polished sur-

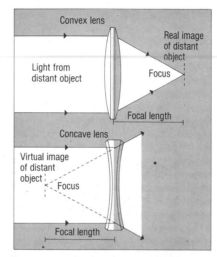

Lens: convex (top) and concave

faces are frequently coated with thin layers of other optical materials to minimize unwanted reflections.

Lenses have been known since ancient times and, indeed, occur in nature in the eyes of most living creatures. The word "lens" is simply the Latin for "lentil", an accurate description of the cross-section of a common form of lens. The cross-sectional shape is normally used to categorize different varieties of simple lenses, and there are two basic forms with fundamentally different properties: those that are thicker at the centre than the edge are called "convex lenses" (because at least one of the surfaces is convex toward its surroundings) and those that are thicker at the edge than in the centre are called "concave lenses". Within these two basic categories, further subdivisions are possible, and we find "bi-convex" lenses (both surfaces convex) or "plano-concave" lenses (one surface flat), for example. The curved surfaces of lenses are actually small segments of spheres, whose centres lie on an imaginary line called the "optical axis" of the lens. The optical axis nearly always passes through the centre of the lens itself, and at right angles to it.

How does a lens work? A ray of light passing through one of the surfaces of a lens is bent in accordance with the laws of refraction, so that its direction is changed. The amount of bending is determined by an intrinsic property of the glass — its refractive index — and the angle of incidence of the ray. It is the smooth change of the angle of refraction over a spherical surface which allows lenses to be used to form and manipulate optical images. Light rays from distant point objects (as from a star, for example) are effectively parallel, and when these are intercepted by a convex lens they are made to converge to pass through a single point, which is an image of the distant object. It is called a "real image" because it could be projected onto a screen placed at the convergent point. This is basically the configuration in which the lens of a camera is used. The OBJECTIVES of TELESCOPES or BINOCULARS are similarly used to form real images. A concave lens, on the other hand, causes parallel light rays to diverge as if coming from a single point, again an image of the distant object is formed, but as a "virtual image". This image can be seen through the lens, but cannot be projected onto a screen as the light rays do not actually pass through it.

The image formed by a lens from parallel light directed along the optical axis is important because it lies at the "focal point" of the lens. The distance of the focal point from the lens — the FOCAL LENGTH — is its single most important characteristic. Related to this is the "power" of the lens, defined as the reciprocal of its focal length, and expressed in units of "diopters" if the focal length is in metres. Lenses with high powers (ie, short focal lengths) have surfaces with relatively small radii of curvature.

The imaging properties of lenses are not confined to light from distant objects. For any object position along the optical axis of a lens there will be a corresponding real or virtual image position, also on the optical axis. These object and image points are said to be "conjugate". For example, if an object is placed at the focal point of a convex lens, the rays from the object will be made parallel, and the conjugate (virtual) image position will be at infinity. This is the way in which a simple magnifying glass is used and, in more complex form, the EYEPIECE of a telescope.

Images produced by a simple lens suffer from defects caused by "aberrations". The aberrations arise because of the slightly imperfect geometry imposed on light rays by refraction at a spherical surface so that, for

L

example, parallel rays passing through different zones of a convex lens will not converge to precisely the same point. CHROMATIC ABERRATION occurs because all types of glass disperse light into its component colours, and so is a consequence of the material from which the lens is made. The remaining five primary aberrations are due to the shape of the lens, and will occur even if light of only a single colour is used. These aberrations were investigated in detail by the 19th-century mathematician Ludwig von Seidel, and are usually referred to as the Seidel aberrations. All but the first, SPHERICAL ABERRATION, occur only when images are formed away from the optical axis, ie, when light is passing obliquely through the lens. They are "coma" (the elongation of a point image in a direction away from the optical axis), ASTIGMATISM, curvature of the image, and distortion of the image.

These aberrations occur together in greater or lesser degree, and it is usually the main problem facing lens designers to eliminate them, or reduce them as far as possible. This is done by combining lenses of different glass types, thicknesses and radii of curvature to produce a "corrected" lens; for example, an ACHROMATIC lens usually has two elements, and an APOCHROMAT three. Modern lenses normally contain many elements and are optimized for specific purposes using computer-aided design techniques.

Lenses are manufactured from blanks of the exact types of glass specified by the designer. Their surfaces are roughly shaped to the required radii of curvature with diamond machine tools, before being polished with fine abrasives. Highly sensitive test procedures, relying on the interference of light waves, are used to ensure that the surfaces are within the specified tolerances for curvature and uniformity.

Unconventional lenses are sometimes made for special purposes. For example, "cylindrical lenses" have surfaces that are portions of a cylinder rather than a sphere, and therefore focus light in one plane only. "Zoom lenses", familiar to photographers, have multiple elements whose relative positions can be changed to vary the overall focal length. But perhaps the most unusual of all are "grin" (or graded index) lenses, rods of glass with flat surfaces having a carefully designed refractive index gradient from the centre to the edge, so that they behave like ordinary convex lenses. Light actually follows curved paths inside these lenses. Not surprisingly, perhaps, even this miracle of modern technology has been anticipated in nature, for the eyes that are now reading these words contain lenses that have a similar refractive-index gradient within them. *FGW.*

Lenticular galaxies Lenticular galaxies appear lens-shaped when seen edge-on, hence their name. They are of a type intermediate in form between the much more common elliptical and spiral types and are classified S0, which indicates that they have the flattened form of spirals but not spiral arms.

Lenticular galaxies: NGC5102

Leo (the Lion). A large, bright Zodiacal constellation — in mythology the Nemæn lion killed by Hercules.

Brightest stars

Name	Visual Mag.	Abs. Mag.	Spec.	Distance (light-yrs)
α (Regulus)	1.35	−0.6	B7	85
γ (Algieba)	1.99	0.2	K0 + G7	91
β (Denebola)	2.14	1.7	A3	39
δ (Zosma)	2.56	1.9	A4	52
ε (Asad Australis)	2.98	−2.0	G0	310
θ (Chort)	3.34	1.4	A2	78
ζ (Adhafera)	3.44	0.6	F0	117

Then come Eta (3.52), Alpha (also 3.52), Rho (3.85), Mu (3.88) and Iota (3.94).

Regulus is the brightest of the curved arrangement of stars making up the famous "Sickle". Denebola is the brightest of a triangle of stars making up the rest of the main pattern. Algieba is a fine, wide binary; magnitudes 2.3 and 3.5, period 407 years; separation now 4".4. The MIRA VARIABLE R Leonis can reach magnitude 5.4 (minimum 10.5; period 313 days). There are four galaxies in Messier's list; M65, M66, M95 and M96. *PM.*

Leo Minor (the Little Lion). A small constellation adjoining Ursa Major. It has no star above magnitude 3.8.

Leonid meteors A very interesting meteor shower, which nowadays reaches its maximum on about November 17 every year. The radiant is situated at R.A. 10h 08m, Dec. + 22°, within the "Sickle" of the constellation Leo, the Lion. Although some Leonids can be seen every year (ZHR about 10), if the weather is fine, the best showers occur about every 33 years. Certainly the two exceptional meteor displays which took place on November 12 1799 and again on November 12 1833 did much to stimulate interest in the scientific study of meteors.

One astronomer, H.A.Newton at Yale College, Princeton, pondered on the origin of these spectacular Leonid meteor storms. He found historical records of the shower going back to AD902, including exceptional displays in 934, 1002, 1101, 1202, 1366, 1533, 1602, 1698, 1799, and 1833. Newton realized that these tremendous displays were caused by a dense cloud of meteoroid particles which the Earth encountered at intervals of roughly 33 years. He made a bold prediction, confirmed by Olbers, that there would be another meteor storm in 1866, and on the night of November 13/14 1866, the tremendous storm took place. At peak, Leonid rates attained over 120 meteors per minute (7,200 per hour).

No great storm occurred at the predicted return in 1899. The reason for the failure was the planet Jupiter. On its way toward the Sun, the swarm passed by Jupiter in 1898, and the gravitational pull of the planet deflected the particles away from the Earth by 1,250,000 miles (2 million km). The displays of 1932 and 1933 were not particularly exciting either, and many people felt that the Leonids were past their best. However, in 1966 the greatest meteor display in recorded history was seen by observers in the mid-western United States. At peak, rates reached 40 meteors per second (about 144,000 per hour). The peak was very brief, however, and observers in Europe saw only modest rates.

The Leonid meteor stream is associated with Comet P/Tempel-Tuttle 1866 I, first seen in 1865, which has a return period of 32.9 years. Recent studies have shown that most of the dust ejected from the parent comet evolves to a position lagging behind the comet and outside its orbit. Significant Leonid showers are possible only up to about 2,500 days before or after the parent comet reaches perihelion — but only if the comet passes closer than 0.025AU inside or 0.010AU outside the Earth's orbit. The meteoric dust is concentrated near the comet, and has not yet spread out all the way around the comet's orbit.

What are the chances of good showers when the comet next returns to perihelion in 1998? In 1997, 1998 and 1999 the Earth will lead the comet by 108 days, lag it by 257 days, and lag it by 622 days respectively. The miss distance will be about 0.008AU, with the Earth outside the comet's orbit. There is certainly a good chance of fine displays during the period 1997–99, but we cannot be certain. The best date is likely to be November 17 1998. *JWM.*

Leonov, Alexei Arkhipovich (1934–). Soviet cosmonaut who was the first man to walk in space (Voskhod 2 spacecraft, March 18 1965).

Lepaute, Madame (née Brière) (1723–88). French mathematician, who played a major role in the prediction of the 1758 return of HALLEY'S COMET.

Lepus (the Hare). A small constellation south of Orion — in legend placed there because hares were often hunted by Orion.

Brightest stars

Name	Visual Mag.	Abs. Mag.	Spec.	Distance (light-yrs)
α (Arneb)	2.58	−4.7	F0	945
β (Nihal)	2.84	−2.1	G2	316
ε	3.19	−0.3	K5	163
μ	3.31	−0.8	B9	215

Then come Zeta (3.55). Gamma (3.60). Eta (3.71) and Delta (3.81). Lepus is quite distinctive; there are various interesting objects, notably the very red MIRA VARIABLE R Leporis (5.9 to 10.5, 432 days, type N) and the globular cluster M79.

Leptons A small class of elementary particles, which are the simplest forms of matter. The lepton family consists of the electron, the muon, the tauon and their associated NEUTRINOS. Leptons do not take part in strong interactions. They are restricted to weak interactions and, if electrically charged, to electromagnetic interactions. The neutrinos take part in weak interactions only. Leptons do not join together to form other particles, and do not appear to be made up of QUARKS. *JWM.*

Le Verrier, Urbain Jean Joseph (1811–77). Experimental chemist who switched to astronomy (1837) and specialized in celestial mechanics.

He was born in St Lô, Normandy. He investigated the stability of the Solar System, undertook a comprehensive analysis of planetary theory, especially the intricate problem of mutual

Le Verrier: co-discoverer of Neptune

influences, and predicted the existence of Neptune from the irregular motion of Uranus, being honoured, in 1846, for his contribution to its discovery.

Le Verrier attributed the orbital anomaly of Mercury to an intramercurial planet, Vulcan, but it was never found. As director of the Paris Observatory (1854) he introduced major reforms, but was dismissed from office in 1870 after a staff walk-out. He was reappointed following the death of his successor, Delaunay.

Le Verrier was also involved in the formation of the meteorological network across the continent of Europe. *RMB.*

Lewis Research Center NASA station located in Cleveland, Ohio, specializing in aircraft/spacecraft propulsion, satellite communications, development of power systems for the space station and microgravity materials research.

Lexell, Anders (1740–84). Finnish astronomer, who specialized in cometary research and was the first to prove the planetary nature of Uranus.

Lexell's Comet The first short-period comet on record. Discovered by Charles Messier, it is one of the best-known cases where a cometary orbit was altered considerably after being perturbed by the gravitational pull of the massive planet Jupiter. In 1770, this comet in its approach toward peri-

helion, passed between the satellites of Jupiter. Later, it came within 1,500,000 miles (2½ million km) of the Earth on July 1 1770. As it passed by, the transient gravitational grasp of the Earth caused a decrease in the comet's period by almost three days — but the change in the period of the far more massive Earth was so small as to be immeasurable. The orbit of the comet was investigated by LEXELL, who found it to have a period of 5.6 years. He showed that the comet had been highly perturbed by Jupiter in May 1767, when its orbit had been changed from a much larger ellipse to its present shape, which explained why it had never been seen previously. The comet was never seen again after 1770. The reason for this was provided by LAPLACE. He showed that a second close approach to Jupiter had occurred in 1779. On this occasion it had passed so close to Jupiter that its orbit was altered dramatically, with large changes in its orbital period and perihelion distance. Lexell's Comet is another example of how the man who calculated the orbit gave his name to the comet. *JWM.*

Libra (the Scales or Balance). One of the most obscure of the Zodiacal constellations.

Brightest stars

Name	Visual Mag.	Abs. Mag.	Spec.	Distance (light-yrs)
β (Zubenelchemale)	2.61	−0.2	B8	121
α (Zubenelgenubi)	2.75	1.2	A3	72
σ (Zubenalgubi)	3.29	−0.5	M4	166

Then follow Upsilon (3.58), Tau (3.66) and Gamma (3.91). Sigma was formerly included in Scorpius, as Gamma Scorpii. The only object of real interest is the eclipsing binary Delta Libræ; 4.8 to 6.1, period 2.33 days. It is of the Algol type. Libra was originally known as Chelæ Scorpionis, the Scorpion's Claws.

Lick Observatory The first United States observatory on top of a mountain (Mt. Hamilton in northern California, elevation 4,200ft/1,283m). It was completed in 1888. The eccentric James Lick, who made millions during the gold rush by property speculation in San Francisco, funded the observatory and is buried beneath the world's second-largest, 36in (91cm) refractor, Lick's oldest telescope. Four reflectors and a 20in (51cm) astrograph maintain a strong and varied research programme despite nearby San José's

L

bright sky. The reflectors are the 120in (300cm) Shane (1959); the 40in (100cm) Nickel (1983). the Crossley (an 1890s' gift; and a 24in (61cm). *MC.*

Lick Observatory: a great observatory

Life in the universe One of the questions which has been asked throughout history is: "Can life exist beyond the Earth — and if so, what form is it likely to take?" Studies of this problem have led to an entirely new branch of science which has been termed exobiology, though as yet we cannot pretend to be able to give any firm answers.

We must first decide what is meant by "life". According to all the available evidence, living molecules depend upon one type of atom — that of carbon — which alone has the ability to build up the complicated molecules required; its only possible rival is silicon, but there is at present no evidence of silicon-based life. We may therefore assume that life, wherever it exists, must be based upon the same elements as that on Earth, which eliminates the fascinating science-fiction creatures known popularly as BEMs or Bug-Eyed Monsters! It is fair to say that if BEMs do exist, then most of our modern science is wrong; and this does not seem likely.

There is still considerable discussion about the origin of terrestrial life. In the early 20th century the Swedish scientist Svante ARRHENIUS proposed the PANSPERMIA theory, according to which life on Earth did not originate here, but was brought by way of a meteorite; this idea never met with much support, mainly because it appeared to raise more problems than it solved. Much more recently Sir Fred Hoyle and Professor Chandra Wickramasinghe have revived the idea, though in very modified form; on their hypothesis life originates in space, and is brought to Earth by way of a comet. They argue that the creation of living material involves a whole sequence of

events, each of which is itself inherently improbable, so that the whole "resources of space" are needed rather than a restricted area such as Earth. However, this theory, like that of Arrhenius, has met with scant support.

In carrying out practical searches for life in the Solar System, we depend upon space research methods. Worlds which are devoid of atmospheres, such as the Moon, may be ruled out at once; Venus is unsuitable in every way, as the probes have shown; the giant planets have no solid surfaces, and consist largely of hydrogen. Saturn's satellite TITAN does indeed have a relatively dense, nitrogen-rich atmosphere, and the ingredients for life must exist there, but the very low temperature seems to preclude it. There remains MARS, which has always been regarded as a possible abode of life; indeed, the MARTIAN CANALS were once regarded as possibly artificial. A detailed investigation was carried out from the landers of the VIKING probes in 1976. No evidence for Martian organisms was found, and it seems likely that at the present epoch, at least, Mars is lifeless.

We must therefore search beyond the Solar System, and there are obvious difficulties here. All we can do is to take what facts we have, and put the most reasonable possible interpretation on them.

Our Galaxy contains about 100,000 million stars, many of which are very like the Sun — and this is only a start; the number of separate galaxies is very great indeed. It therefore seems logical to assume that planetary systems are common, and there is considerable evidence that this is the case. What can happen to the Sun can presumably happen to other solar-type stars also. We must cast around for relatively close stars which are of the same spectrum and luminosity as the Sun, and the two nearest candidates are Tau Ceti and Epsilon Eridani, which are around 10 light-years away. There is, of course, no proof that either star has a planetary system, but there seems no valid reason why not, though both stars are considerably smaller and cooler than the Sun.

We must also bear in mind that even if we can locate a planet similar to the Earth moving round a star similar to the Sun, we still have no guarantee that life will appear there. However, in 1960 the first attempts were made to use radio as a means of interstellar communication. A "listening watch" was kept upon Tau Ceti and Epsilon Eridani at a wavelength of 21.1cm (the

wavelength of radiation emitted by clouds of cold hydrogen in the Galaxy) to see whether any signals could be received rhythmical enough to be regarded as artifical. Not surprisingly, the results were negative, and the OZMA PROJECT was soon discontinued, but further attempts along the same lines have since been made, and there is even a full-scale planned programme, using large radio telescopes, which has been named SETI: the Search for Extra-Terrestrial Intelligence.

Obviously, contact would be limited even if signals could be picked up. A message to a star which is, say, 11 light-years away would involve a delay time between transmission and answer of 22 years, and most suitable candidate stars are much farther away than that. Yet if we could establish the existence of extra-terrestrial life, the effects upon all our thinking — scientific, philosophical, even religious — would be profound.

If other civilizations do exist, they may be expected to be of various types; some less advanced than ours, some technologically comparable, some much more advanced. There is always the possibility that other beings will contact us before we are able to contact them, and the lack of "visitors from space" so far has been cited as an argument in favour of the idea that life on Earth is unique in the universe. We can certainly discount the wild stories of flying saucers and UFOs, but we must remember that advanced life on Earth is relatively new, and an alien craft which landed here only a few millions of years ago would have found no sign of civilization. It is wrong to claim that such visitations are impossible; all that can be said is that there is no evidence that they have taken place since the emergence of Man.

If we are ever to succeed in establishing contact, it can only be by means of radio in the context of our present technology. Once we launch forth into ideas of (say) telepathic communication, teleportation or thought-travel, we are straight into the realm of science fiction, though it is worth commenting that such exotic theories are no more far-fetched to us than television would have been to Julius Cæsar.

There have been suggestions that to establish contact with an alien civilization would be unwise, but it seems much more likely that any civilization sufficiently advanced will have overcome our own tendency to self-destruct — in which case we would have no cause for alarm, and could only benefit from contact. On the other hand, there is also a school of thought which maintains that life on Earth is really unique in the universe, and that there are no other civilizations anywhere. This is a matter for debate; in the future we may find out, though it is fair to say that all our previous attempts to classify ourselves as "important" have ended in humiliating failure. *PM*.

Light The nature of light is now thought to be fairly well understood, yet was the subject of controversy for more than 200 years, as experimenters uncovered conflicting evidence as to its true nature. Certain evidence suggested that light was composed of individual particles, other evidence suggested that it was composed of waves. It was even found necessary to invent a substance to carry these waves, the so-called luminiferous ether, which is now known not to exist. Isaac NEWTON, in one of his rare errors, inclined to the former point of view as to the nature of light, something that, due to his enormous reputation, held back the study of the subject for many years. Not until the pioneering experiments of the Scottish physicist Robert Young was the wave-like nature of light firmly established for the scientific community.

It is now known that light is *both* particle and wave. That is, that light comes in small packets, called photons. This is true whether we consider visible light, or radio waves, or X-rays (all of which are types of light). One photon is the smallest amount of light possible, just as one atom is the smallest amount possible of a substance. This packet, though, is composed of an electric field and a magnetic field, each of which "oscillates" at right angles to its direction of motion. The interaction of these fields was first described in mathematical form by James Clark MAXWELL. His basic laws are now known to all students of physics as "Maxwell's laws" and are the basis of the study of electromagnetism, the study of all phenomena related to the interactions of electric and magnetic fields. *MRK*.

Light curve A graph of the variation with time of the brightness of an astronomical body (usually a variable star).

Light pollution The effect of street and other lighting concentrated mainly in urban areas on astronomic observing and imaging. For amateurs, it limits the celestial objects (particularly extended objects) that can be seen and for professionals its effects can so

impede or distort scientific observations as to result in the eventual closure of observatories enveloped by urban sprawl in the years since their foundation.

Light, velocity of *See* VELOCITY OF LIGHT.

Light-year A unit of distance measurement equal to the distance travelled by a ray of light in a vacuum in one year (5.87×10^{12} miles, 9.46×10^{12} km, or 63,240 ASTRONOMICAL UNITS.)

Limb The apparent edge of the Sun, Moon or a planet. It could refer to any celestial body that shows a detectable disk.

Limb brightening Increase in brightness from the centre to the limb of an astronomical body. This is observed with the Sun at radio and X-ray wavelengths, because those radiations come predominantly from the corona.

Limb darkening Decrease in brightness of an astronomical body from the centre to the limb. This is observed at optical wavelengths on the solar disk because an observer "sees" vertically down to deeper hotter layers at the centre of the disk, compared to what can be seen looking obliquely into the PHOTOSPHERE near the limb.

Limiting magnitude The faintest MAGNITUDE visible or recordable by photographic or electronic means, dependent on APERTURE, sky transparency, seeing, exposure and sensitivity of the eye or recording apparatus.

Liner Galaxy in which gas is apparently heated by a very faint QUASAR-like nucleus, lacking the broad-line emission of Seyfert galaxies. The acronym derives from *Low Ionization Narrow Emission-line Region*.

Linné A famous lunar feature on the Mare Serenitatis: 28°N, 12°E. Some 19th-century observers, including Beer and Mädler, described it as a small but deep and well-formed crater, comparable with Bessel, the largest crater on the Mare Serenitatis. In 1866 J. Schmidt announced that the crater had disappeared, to be replaced by a white patch. Apollo photographs have shown that Linné is now a small, deep crater of regular form, with a bright area round it. The alleged change has caused great controversy, but it seems quite certain that there has been no

genuine alteration — if only because Mädler, in 1868, wrote that it looked just the same to him as it had done in 1832. *PM*.

Lippershey (or Lipperhey), Hans (*c.* 1570–*c.* 1619). Spectacle-maker of Middelburg in the Netherlands. A possible inventor of the telescope, he applied for a patent for his device in October 1608.

Lithosphere The outer semi-rigid shell of the Earth. This comprises the whole of the crust, together with the uppermost layer of the underlying mantle.

Lobes In practice an antenna (or aerial) neither radiates nor receives radio waves equally in all directions. This is shown by its antenna pattern, a graph of the sensitivity (or gain) of the antenna as a function of direction. A number of distinct lobes may often be identified in the antenna pattern. The lobe corresponding to the direction of best transmission or reception is the main lobe. All the others are called side lobes and are usually unwanted. A good design of antenna keeps the magnitude of these side lobes to a minimum in most cases.

Local group The small group of galaxies to which our own Milky Way belongs. It also contains the GREAT NEBULA in Andromeda, its satellites, the MAGELLANIC CLOUDS — satellites of the Milky Way — and various dwarf ellipticals.

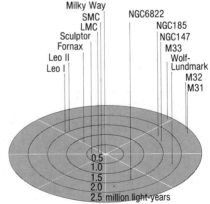

Local group: the Milky Way is a member

Local supercluster The collection of galaxies and clusters of galaxies to which our Milky Way galaxy, the LOCAL GROUP and the VIRGO and the COMA clusters of galaxies belong. *See also* colour essay on SUPERCLUSTERS.

Lockyer, Sir Joseph Norman (1836–1920). English astronomer. Founder

L

and editor of *Nature* for 50 years. Discoverer of helium in the Sun and founder of the Observatory at Sidmouth.

Loki Volcano on Jupiter's satellite IO, active during both VOYAGER passes: 19°N, 305°W. It was originally called Plume 2. The temperature was measured at 17°C. The plume rose to 60 miles (100km) as measured from Voyager 1 (March 1979) and 120 miles (200km) from Voyager 2 (July 1979).

Lomonosov, Mikhail (1711–65). The first great Russian astronomer. He was the son of a fisherman, and had little formal schooling, but in 1735 he gained an entrance to the University of St. Petersburg, subsequently studying in

L

Mikhail Lomonosov: atmosphere of Venus

Germany before returning to his homeland. By 1745 he was a full member of the Academy, and was appointed Professor of Chemistry at St. Petersburg. His interests were wide; he was a pioneer of the kinetic theory of gases, and made contributions to geology and meteorology. He was also a poet and grammarian. Astronomically, he was a strong supporter of COPERNICUS' theory (then unpopular in Russia). In 1761 he observed a transit of Venus, and correctly deduced that the planet has a dense atmosphere. *PM*.

Longomontanus Lunar walled plain, 90 miles (145km) in diameter, in the Clavius area, with complex floor-detail. 50°S, 21°W.

Loop I Measurement of the radio emission from the Galaxy show a

number of arches of which "Loop I" is the most prominent. It is almost certainly the remains of a SUPERNOVA which exploded several hundred thousand years ago. The supernova was about 500 light-years away and the nearest surface is now some 5 light-years distant. There is probably only about one supernova of this type exploding in the Galaxy every century. It is likely that COSMIC RAYS are accelerated by the shock waves inside these supernova remnants. *AWW*.

Loop Nebula *See* TARANTULA NEBULA.

Lorentz, Hendrick Antoon (1853–1928). Dutch physicist who suggested independently of Fitzgerald (*see* FITZGERALD CONTRACTION) that moving bodies contract. He developed the LORENTZ TRANSFORMATIONS.

Lorentz transformations. A set of equations, worked out by Hendrick LORENTZ, which relate space and time co-ordinates in frames of reference which are in uniform relative motion (*see* RELATIVITY).

Lost City Meteorite A CHONDITE which fell in Oklahoma in 1970. Its fall was photographed.

Lowell Observatory Founded by Percival LOWELL at Flagstaff, Arizona in 1894, this privately-endowed observatory is managed by a sole trustee who is a Lowell descendant. Now (1986) staffed by 30 persons, half of whom are research astronomers, the observatory operates eight telescopes. Four of these are used at the Anderson Mesa dark-sky site southeast of Flagstaff. The largest of these is the Ohio Wesleyan University's 72in (183cm) Perkins reflector operated jointly with the Ohio State University. The remaining telescopes are used at the original site at Flagstaff. A Planetary Research Center, primarily supported by the National Aeronautics and Space Administration, contains facilities for planning and analyzing Solar System explorations and includes an extensive archive of ground-based planetary photographs augmented in recent years by an International Planetary Patrol operated by Lowell Observatory.

The observatory is historically famous for work done by, or stimulated by, Lowell, including studies of Mars, the discovery of galaxy REDSHIFTS by V.M.SLIPHER, and the trans-Neptunian planet predictions that led to Tombaugh's discovery of PLUTO.

Subjects of current study include not only Solar System objects with emphasis on asteroids, comets, planetary atmospheres, and solar variability, but also interstellar matter, stellar variability, high-resolution observations of stars by lunar occultations and interferometry, galaxies and star formation, quasars, and participation in the Space Telescope Astrometry and Wide Field Planetary Camera teams. *AAH*.

Lowell, Percival (1855-1916). Businessman, Orientalist, diplomat, and author, who turned to fulltime astronomy at the age of 39. Though notorious for his conclusions about life on Mars, based on then generally widespread observations of "canals", his heritage includes a major observatory with outstanding archives of planetary and celestial photographs, the discovery of galaxy REDSHIFTS by V.M.SLIPHER (1914), done at Lowell's urging, and predictions that led to Tombaugh's discovery of PLUTO in 1930. *AAH*.

Percival Lowell: believed in life on Mars

L² Puppis Bright semi-regular variable in Puppis; range 3.4 to 6.2, period around 141 days.

Lucian of Samosata (*c*. AD180). Greek satirist, who wrote a book about a voyage to the Moon.

Ludendorff, Frederich Wilhelm Hans (1873–1941). German astrophysicist, born at Köslin (Koszalin) in Pomerania. In addition to scientific contributions such as the analysis of variable star periods, he is noted for outstanding work in chronology and the history of astronomy.

Luminosity function. The numerical distribution of stars (or galaxies) among different values of luminosity or ABSOLUTE MAGNITUDE. Values of the luminosity function are usually expressed as the number of stars per cubic parsec within a given range of luminosity or absolute magnitude.

The value of the luminosity function in the solar neighbourhood increases to a maximum at absolute magnitudes of 14-15, and decreases again at higher (fainter) magnitude values. This implies that faint M-type dwarfs of about one ten-thousandth of the solar luminosity are the most abundant type of star locally.

Luminosity functions can also be plotted for special categories of object such as stars of particular spectral classes, stellar clusters, galaxies, or clusters of galaxies. *IKN.*

Lunæ Planum Martian plain; formerly known as Lunæ Lacus; 5-23°N, 75-60°W. It lies to the west of the Tharsis ridge upon which stand the giant volcanoes.

Luna probes One of the two series of spacecraft used by Soviet Russia for exploration and research of the Moon, the other being the ZOND series.

There were 24 numbered missions in the series, which began in early 1959 and ended in the latter half of 1976. They encompassed various types of mission, including impact, flyby, orbital and soft-landing. *See* table below.

Lunar craters Roughly circular depression-like features seen on all parts of the lunar surface. It is difficult to give a universal definition of the term "crater" as applied to the Moon; all the features to which the word is applied involve a depression in the surface, but there is a wide range of other attributes. The largest craters are hundreds of kilometres in diameter

and it is conventional to refer to depressions larger than about 200 miles (300km) in width as basins, though there is no fundamental difference in shape. Craters as small as a few centimetres in size are recognizable in photographs of the surface taken by APOLLO astronauts, and microscopic examination of lunar rocks returned to Earth shows crater pits as small as a few microns on their surfaces.

The morphologies (ie, shapes and structures) of most craters show systematic variations with crater size. Large craters with diameters greater than about 6 miles (10km) and basins generally have pronounced rims raised above the level of the surrounding terrain and much of the central part of the depressed interior is nearly flat. In many cases a mountain complex rises from the centre of the flat floor, and in craters larger than about 75 miles (120km) this is replaced by a series of concentric rings of mountains — up to five rings in some large basins. The steep, inner walls of these craters show signs of modification by slumping and sliding of fragmented rocks to form characteristic terraces, and the more gently sloping flanks outside the rim crest consist of a hummocky terrain which grades back to the general surface level. Beyond the hummocky terrain region are sometimes found large numbers of small, irregular secondary craters, often occurring in groups and elongated along a line roughly radial to the main, or primary, crater. Even farther out from the primary crater rim, roughly radial patterns of brightening of the surface, called LUNAR RAYS, are visible. In plan view, the primary crater rims may range from nearly circular, through polygonal (with six-sided polygons commonest), to quite irregular.

Small craters, in the size range from 6 miles (10km) down to a few centimetres, generally have simpler shapes. The inner walls are generally smooth, with occasional outcrops of bedrock, and show only minor modification by slumping; there is no central mound of rock. In profile, many of these features are bowl-shaped, though some have flat floors and some stepped interiors occur; in plan view, a near-circular shape is generally seen. A small but significant fraction of craters in this size range, especially those between 300ft and 2 miles (100m and 3km) in diameter, have no raised rim or hummocky terrain beyond the edge of the depression. Many of these craters tend to be elongate rather than round, and

L

Luna probes

Luna No.	Launch date	Landing site and co-ordinates	Remarks
1*	Jan 2 1959	—	Missed Moon by 3,700 miles (6,000km)
2*	Sep 12 1959	30°N 0°W	First spacecraft to impact Moon (on Sep 13)
3*	Oct 4 1959	—	First to photograph Moon's farside (portion of)
4*	Apr 2 1963	—	Attemped soft landing; missed Moon by 5,300 miles (8,500km)
5	May 9 1965	1.6°S 25°W	Attempted soft landing; impacted
6	June 8 1965	—	Attempted soft landing; missed Moon by 100,000 miles (161,000km)
7	Oct 4 1965	9.8°N 47.8°W	Attempted soft landing; impacted
8	Dec 3 1965	9.6°N 62°W	Attempted soft landing; impacted
9	Jan 31 1966	7.13°N 64.37°W	First soft landing; 27 close-up panoramic photos
10	Mar 31 1966	—	First to achieve lunar orbit; physical data transmitted for almost 2 months
11	Aug 24 1966	—	Similar to Luna 10
12	Oct 22 1966	—	Similar to Luna 10; high-resolution photos also transmitted
13	Dec 21 1966	18.87°N 63.05°W	Soft landing, similar to Luna 9; also soil density data
14	Apr 7 1968	—	Gravity measurements from lunar orbit
15	Jly 13 1969	Mare Crisium	Orbiter; test of automatic navigation system; impacted. Unsuccessful attempt to obtain lunar samples
16	Sep 12 1970	0.68°S 56.30°E	Soft lander; returned a soil sample to Earth from Mare Foecunditatis
17	Nov 10 1970	38.28°N 35.00°W	Soft lander; deployed roving vehicle "Lunokhod I" which transmitted photos and physical data
18	Sep 2 1971	3.57°N 56.50°E	Similar to Luna 15, but with upgraded system; impacted
19	Sep 28 1971	—	Orbiter; photographic mission, successful
20	Feb 14 1972	3.53°N 56.55°E	Soft lander; returned a soil sample to Earth from highlands north of Mare Foecunditatis
21	Jan 8 1973	25.85°N 30.45°E	Soft lander; similar to Luna 17, but with "Lunokhod II"
22	May 29 1974	—	Orbiter; transmitted photos and physical data
23	Oct 28 1974	Mare Crisium	Soft lander; soil sample return not achieved
24	Aug 9 1976	12.75°N 62.20°E	Soft lander; returned a soil sample to Earth from Mare Crisium
*Originally named "Lunik"			

L

to be associated with LUNAR RILLES or to form parts of chains of similar features. Some of the smallest, so-called micro-craters (those less than a few centimetres in size) have smooth interiors, and their rims appear to consist of material from the central depression which has flowed and piled up in a plastic manner; others show fracture patterns in the walls as well as signs of plastic flow.

The morphological features of most craters with raised rims, slumped inner walls and hummocky exteriors are consistent with the idea that they were formed by some explosive process which shattered the lunar surface rocks in the vicinity of the crater, excavated them from the depression and deposited them in the vicinity to form the rim and the hummocky terrain. Large fragments of material thrown out at relatively high speeds would have formed the secondary, irregular craters nearby, while smaller, high-speed particles would have produced the bright rays. Materials forming the outer parts of the hummocky terrains of the largest craters must have been thrown at least 300 miles (500km), implying speeds greater than about 3,000ft/s (1,000 m/s). Such speeds could be reached by debris from volcanic explosions, caused by the build-up of pressure in trapped gases, only if pressures greater than about 300 million pascals were reached in a volume of the crust which is comparable with the volume of the crater depression. These pressures are 30 times greater than the strengths of the rocks involved, implying that they could never be reached in the first place — the rocks would fracture too soon.

An alternative type of explosive vulcanism — the same kind of nearly steady, high-speed discharge of gas and fragments that currently takes place on Jupiter's satellite IO — almost certainly occurred on the Moon to produce the extensive areas of dark deposits seen thinly mantling the surface in some places. But no volcanic explosion mechanism consistent with the detailed structures of the largest craters has yet been demonstrated. The maximum diameter of a crater with a rim and external debris blanket that could be produced by a discrete volcanic explosion is currently estimated to be only a few kilometres, and good examples of such features are the small craters, surrounded by dark, halo-like debris blankets, on the floor of the larger crater Alphonsus.

The high pressures distributed over large volumes needed to explain the largest craters and basins as explosion structures can easily be reached when a METEOROID, ASTEROID or COMET collides with the lunar surface. At the average impact speed, about 9 miles/s (15km/s), the conversion of kinetic energy to internal energy which takes place at the point of contact will produce pressures in excess of a thousand million pascals within a volume which is roughly equal to the size of the impacting object. These conditions can readily explain the basic morphologies of near-circular craters with rims and external debris blankets and are consistent with the mineralogical indications of transient high pressures and local melting seen in many of the rocks collected from the lunar surface. This, together with the near-randomness of their locations, has led to the majority opinion that most of the lunar craters of all sizes are of impact origin.

However, the common occurrence in large craters of flat floors produced by extensive fluid materials flooding the interior, and of significant non-circularity of the rim, indicate that many primary features were modified during

Lunar craters: Daedalus, on the far side

or after their formation by tectonic stresses in the crust or by volcanic processes. The most obvious example is the extensive flooding of the largest basins on the side of the Moon facing the Earth by lava flows resulting from the eruption of magma from the lunar interior, mainly between 4,000 and 3,000 million years ago. Various kinds of elongate or irregular craters up to a few kilometres in size are associated with the source vents of these lavas, and some rimless craters were formed by the subsidence of surface rocks into underground lava tubes or evacuated, near-surface magma reservoirs.

The relative numbers of craters of a given size seen on different parts of the surface has been used to obtain relative ages of terrains. Also, when debris from a crater is seen to lie on top of other features it provides a time marker for those features. By combining these relative ages with absolute ages measured on rocks returned to Earth, it is found that there was a major reduction in the rate of formation of large craters about 4,000 million years ago, possibly marking the end of the period when the planets were sweeping up the interplanetary material left over from their initial formation. *LW.*

Lunar domes Roughly circular, rounded volcanic hills up to 12 miles (20km) in diameter found in LUNAR MARIA areas. These rare features appear to consist of unusually viscous lavas.

Lunar eclipses A lunar eclipse occurs when the Moon passes through the shadow of the Earth. The shadow falls in two regions "behind" the night hemisphere of the Earth (that is, starting from the twilight zone and extending in the direction away from the Sun). The darker part of the shadow, called the UMBRA, is formed by the rays of light which form the "external" tangents joining the Sun and Earth (see accompanying diagram). Light refracted in, and emerging from, the Earth's atmosphere can enter the umbra, whereas direct sunlight cannot do so. The less dark parts of the Earth's shadow are called the PENUMBRA. It is limited by the "internal" tangents joining the Sun and Earth. In the penumbra, light levels rise as these internal tangents are approached by the shortest route from the umbra. An observer in the umbra would not see the Sun, although he would be aware that the Earth was obscuring it because of the scattering of light in the Earth's atmosphere. In the penumbra, the observer would see only that portion of the Sun not obscured by the Earth. Thus, when, in the accompanying diagram, the Moon is at M_1 it is illuminated by the whole of the Sun. At M_2, as the Moon moves toward the umbra through the penumbra, the moonlight fades gradually. The change is not marked to the eye, however, until the Moon enters the umbra. Then the limb of the Moon that first enters the umbra becomes invisible. The curve of the unsharp umbral boundary, seen in projection on the Moon, mirrors the curved surface of the Earth. An hour or so later, the Moon may be totally eclipsed. At M_3, near to the central phase of totality, the longest wavelengths of the sunlight illuminate the Moon faintly after having been refracted in the Earth's atmosphere. This is because the shortest wavelengths of the sunlight are preferentially scattered and absorbed by atmospheric particles and gases. Since these particles vary in concentration from place to place and from time to time (for example, our cloud cover is very variable) the dull, red-brown light passing through the atmosphere to reach the fully eclipsed Moon changes in intensity from one eclipse to the next.

The Moon's path, or orbit, is tilted at about 5° to the plane, called the ECLIPTIC, in which the Earth lies as it

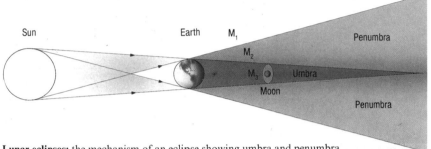

Lunar eclipses: the mechanism of an eclipse showing umbra and penumbra

orbits the Sun. So the path of the Moon must transgress the ecliptic when the Moon passes from the southern side of the ecliptic to its northern side. This occurs at a point in the Moon's orbit called the ASCENDING NODE; and there is a corresponding point in the orbit called the DESCENDING NODE. The line joining the ascending and descending nodes is not fixed in space relative to the stars: viewed from the northern star hemisphere this line moves slowly clockwise, completing one revolution in approximately $18\frac{2}{3}$ years. The nodes of the Moon are said to regress.

If the centres of the Sun, Earth and Moon lie in a line (the "line of opposition") and if, at the same time, the Moon is at one of its nodes, then the Moon will pass centrally through the Earth's shadow and totality might last for some two hours. The duration of the total phase of the eclipse will, however, decrease as the distance of the nearer node from the line of opposition increases. Commonly, the nodes are at points other than M_3 in the orbit. Totality may only just occur or, most frequently, there may be a normal, full Moon and no eclipse of the Moon at all. In the course of a year there may be up to three total, or partial, lunar eclipses in which the umbra wholly, or partly, covers the full Moon, respec-

tively. On the other hand, there may be none at all.

Because a given node of the Moon's orbit is regressing at a measured rate in the ecliptic, it can be shown that the node will have the same longitude as the Sun every 346.62 days. This period, shorter than the year, is known as the "eclipse year" because of its relevance to the problem of eclipse prediction. Taking this period together with the time between successive full moons and the time the Moon takes to move between its successive closest approaches to the Earth, it is possible to evaluate the time elapsing between (almost) identical geometrical arrangements of the Sun, Earth and Moon. That interval of time turns out to be 18 years $10\frac{1}{3}$ or $11\frac{1}{3}$ days, depending on the number (four or five, respectively) of intervening leap years, and is known as the SAROS. The saros has, for 24 centuries, proved to be useful in the prediction of eclipses for, after the eclipses of a particular saros are complete, every one of those eclipses (be it lunar or solar) will be closely reproduced (although not repeated exactly as to form) if viewed from some point on the Earth's surface.

How useful are lunar eclipses? Because they repeat in a systematic, predictable way, the times and details of past eclipses can be calculated. In this manner, attempts can be made to set accurate dates to historical records which incorporate descriptions of eclipses. The timing of lunar eclipses led astronomers to evaluate the principal inequalities (changes of velocity) in the motions of the Moon and Earth. These inequalities arise because the orbits of the Moon and Earth are ellipses rather than circles. The motion of the Moon's PERIGEE (that point, in the lunar orbit, which is closest to the Earth) and node may also be studied from eclipse observations; and the analysis of ancient eclipse timings lends substantial weight to the overall view that the Moon's orbit is slowly enlarg-

ing. Modern observations confirm that the Moon has such a "secular acceleration", principally as a consequence of tidal friction, the distance between the Earth and the Moon must have been less in the past than it is now. Although old lunar eclipse measurements are still invaluable for certain studies such as that of the secular acceleration, eclipse methods are now being superseded by measurements of the Moon's position and motion using timed laser pulses. In these new studies, experimenters make use of the four retro-reflectors established on the Moon since 1969.

The relative heat conductivity of different patches of lunar terrain can be studied, profitably, during an eclipse of the Moon. Certain craters emit more heat than their surroundings when the sunlight is cut off and the lunar rocks cool by radiating their warmth to space. In this way it was shown (R.W.Shorthill, H.C.Borough and J.M. Conley, 1960) that the crater Tycho, for example, was warmer than the surrounding country during a lunar eclipse. Rocks of higher heat conductivity would appear the warmer during an eclipse and the cooler at times around full moon. *GF.*

Lunar librations Real and apparent irregularities in the Moon's motion, revealing a total of 59 per cent of its surface as seen from Earth, in spite of its CAPTURED ROTATION. Geometrical librations include libration in longitude, caused by variations in the Moon's orbital velocity (whereas its rotation is uniform), revealing a maximum of about 8° of the eastern or western limbs; libration in latitude, caused by the $6\frac{1}{2}$° tilt of the Moon's equator to its orbit, so that we see more of the northern and southern limbs alternately; and diurnal libration, caused by the observer's position on a rotating Earth. There is also a small irregularity in the Moon's motion caused by the Earth's pull on the Moon's equatorial bulge. *TJCAM.*

Lunar maria Massive lava flows which infill some lunar impact basins and irregular depressions. Ancient astronomers believed these to be lunar oceans. They are arguably the most spectacular features on the Moon and were produced early in lunar history, as shown by Apollo radiometric dating.

Mare lavas are most abundant on the nearside, although the youngest and most perfectly preserved basin (MARE ORIENTALE) occurs on the farside and has been partially flooded with lava. It is at least 560 miles (900km) across. Mare materials also have flooded less regular depressions to give irregular maria, eg OCEANUS PROCELLARUM and MARE TRANQUILLITATIS. MARE IMBRIUM dominates much of the northern half of the nearside and has associated with it distinctive radiating furrows and other lineations. These were recognized by G.K.Gilbert in 1893 and called by him "Imbrium Sculpture"; he attributed them to impact processes. The mare surface exhibits a number of topographic features which indicate it owes its formation to the effusion of fluid lavas. Such features include flow lobes, sinuous rilles and domes.

It has been estimated that mare lavas cover 3,900,000 square miles (6.3×10^6km²), or about 17 per cent of the Moon's surface. Their thickness is rather difficult to measure, but may be as much as 2.4 miles (4km) in places. Analysis of returned samples by Apollo and Luna missions indicates that the mare lavas are like terrestrial basalts, but lack hydrated minerals. Radiometric dating shows that the lavas crystallized between 3,900 million and 3,100 million years ago. They represent a period of intense volcanic activity on the Moon which lasted 800 million years. *PJC. See also* entries on particular maria.

Lunar mascons Gravity anomalies (about 100–150 mgal) on the Moon discovered by P.M.Muller and W.L.Sjögren from early Apollo satellite tracking. Their close correlation with the circular LUNAR MARIA, Imbrium, Serenitatis, Crisium, Humorum and Nectaris, is attributed to the upper 5,000ft (1.5km) thick plates of lava flows covering these basins being out of isostatic equilibrium. The basins were probably originally formed by the impact of lunar satellites, whose orbits decayed by tidal friction between 4,200 million and 3,800 million years ago; the resulting basins acquired isostatic equilibrium. Lava flooding occurred 3,600–3,200 million years ago: cooling of the Moon had then strengthened the lithosphere so that the lava sheets have since been supported by it. *SKR.*

Lunar module The section of an Apollo spacecraft that took the astronauts onto the surface of the Moon; on departure, the lower part of the module was used as a launching base and was left on the lunar surface. The procedure was successful with all the landings (Apollos 11, 12, and 14 to 17);

previously there had been tests with the modules of Apollos 9 (in Earth orbit) and 10 (in orbit round the Moon). In the case of the crippled Apollo 13, it was the power provided by the lunar

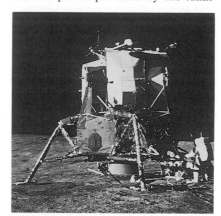
Lunar module: Apollo Moon lander

module that enabled the astronauts to return safely to Earth. Had the explosion in the command module happened on the return journey, when the lunar module would have been jettisoned, rescue would have been impossible. *PM. See also* APOLLO PROGRAMME.

Lunar Orbiter A series of five NASA spacecraft, launched at 3-monthly intervals from August 1966 onward, which were designed for lunar photography with emphasis on choosing suitable landing sites for the APOLLO missions. Images, which were recorded on fine grain photographic film that was processed on board, were transmitted back to Earth by radio through a very high-resolution scanning system. The highest resolution achieved was about 2m, from the 24in (61cm) camera in nominal (70 miles/112km high) orbit, but larger areas at lower resolution were secured by the 3in (7.6cm) camera, and also by both cameras from higher orbits. Almost all of the Moon's far side was imaged, and many scientifically interesting formations on the nearside were photographed. Orbiter 4 imaged most of the nearside under favourable solar illumination conditions with a resolution averaging 100m or better. *EAW.*

Lunar rays Diverging for typical distances of 10 diameters from well-formed, young craters, these systems of straight or curved streaks look like splash marks. Many rays traverse relief without deviation or interruption. Most rays are brighter — but some minor rays are darker — than the underlying rocks. The exceptional

Tycho, 54 miles (87km) in diameter, has rays extending to 23 diameters. Some rays form ovals; others seem to follow the crests of ridges. Many apparently continuous rays resolve into linear elements which point only roughly toward the centres of their parent craters.

The rays are widely thought to be surficial deposits of finely comminuted ballistic ejecta from craters. But rays also contain small craters and blocks,

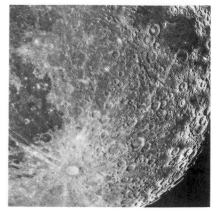
Lunar rays: deposits of ejected material

and ray-elements commonly fan out from these "secondary" craters. *G.F.*

Lunar rilles Channels or valley-like features found mainly near the edges of the mare basins. Linear rilles have straight, nearly parallel sides whereas arcuate rilles are gently curved in plan view. Both types appear to represent subsidence of the ground between two parallel, near-vertical fractures in the crust, caused by tensional forces. The meandering channels of sinuous rilles are of two main types. Some are dis-

Lunar rilles: subsidence on the Moon

continuous and trace out underground lava tubes whose roofs have partly collapsed. Others are continuous, with well-defined source depressions; these channels were cut into earlier terrain by voluminous, hot, turbulent lava flows which slowly melted and carried away the ground beneath them. *LW*.

Lunar rocks Knowledge about lunar rocks derives from photogeological analyses, remote sensing observations (physical measurements made from the Earth, from man-made satellites or space probes, and from geophysical observatories established on the Moon), and from examination of rock specimens either *in situ* or returned to the Earth via soft-landing spacecraft.

A particulate layer of broken and ground rocks covers the entire Moon. This "regolith" may be metres in thickness and has been produced, largely, by impact comminution of solid pieces of rock over the ages. The larger pieces of rock from the highlands' regolith are mostly aluminium-rich breccias produced by cementation of various regolith materials as a result of local pressure and/or temperature increases. By contrast, in the darker-toned maria, where extensive sheets of once-fluid lavas may be detected, most of the rocks are titanium-rich basalts.

The times at which the various mare rocks were last molten lie within the bracket 2,000–4,000 million years ago. This melting, and the differentiation between the Moon's major rock types, is relatively easy to explain in terms of the heat liberated during the decay of long-lived radioactive materials occurring naturally in the rocks. But the highlands are very much older: the igneous rocks generally underlying the breccias appear to have been differentiated and then solidified from a melt some 4,600 million years ago; and that date refers to the earliest period of formation of the Solar System. Various sources of heat for this very early mobility of at least the outer parts of the Moon have been suggested but none has been proved.

Both the mare basalts and the solid rocks called anorthosites from the highlands differ chemically from most terrestrial and meteoritic rocks. Following the Apollo missions, a few completely new minerals were discovered among the returned lunar samples. *GF*.

Lunar terminator The boundary between the day and night hemispheres of the Moon.

Lunar vehicles A total of five self-propelled vehicles has traversed different small areas of the lunar surface. Three of these were used on APOLLO missions 15, 16, and 17, to which they provided immeasurable assistance, enabling the astronauts to gather samples of rocks and soils and conduct various experiments from different surface geological units. Distances travelled were about 16½, 17, and 22 miles (27, 28, and 35km) respectively. Each of these vehicles was equipped with, amongst other things, a television camera, allowing worldwide live viewing of the astronauts at work. The other two vehicles were the Russian Lunokhods 1 and 2, which were unmanned and piloted, through television, from ground control in the USSR. Lunokhod 1 landed near Promontorium Heracclides, while the second traversed part of the floor of Le Monnier crater. *EAW*.

Lunar vehicles: equipped with television

Lunations The interval between successive new (or full) moons — the synodic month. The mean is equal to 29 days 12 hours 44 minutes 2.9 seconds (29.53059 days).

Lund Observatory Swedish observatory that traces its roots back to 1672. The present building in central Lund was built in 1867. The central tower originally housed a 9½in (24.5cm) refractor. In the western wing there is a meridian circle. Modern equipment includes computer-based systems for image processing and microphotometry, and a small planetarium for educational purposes.

Since 1966 the observatory has had a branch station at Jävan, 11 miles (18km) south-east of Lund. The main instrument is a 24in (61cm) Cassegrain-Nasmyth reflector. The old refractor was earlier located there, but has now been replaced by a 14in (36cm) Celestron. *GW*.

Lunokhod Automatic roving vehicle soft-landed by the USSR's Luna 17 in NW MARE IMBRIUM in November 1970: another by Luna 21 in eastern MARE SERENITATIS in January 1973.

Lupus (the Wolf). A constellation in the southern hemisphere, bordering Scorpius and Centaurus.

Brightest stars

Name	Visual Mag.	Abs. Mag.	Spec.	Distance (light-yrs)
α	2.30	−4.4	B1	685
β	2.68	−2.5	B2	360
γ	2.78	−1.7	B3	260
δ	3.22	−3.0	B2	590
ε	3.37	−2.3	B3	720
ζ	3.41	0.3	G8	137
η	3.41	−2.5	B2	490

Then come Theta[1] (3.56), Kappa (3.72), Pi (3.89) and Chi (3.95). Lupus has no distinctive shape, and no objects of immediate interest. *PM*.

L waves Long-period seismic waves that travel around Earth's periphery.

Lyman series The series of lines in the ultraviolet spectrum of hydrogen resulting from electronic transitions to and from the lowest energy level (the ground state).

Lynx (the Lynx). A dim northern constellation adjoining Ursa Major. Its leading star is Alpha (3.13; absolute magnitude −0.4, type M0, distance 166 light-years). Lynx covers a wide area, but the only other star above magnitude 4 is 38 (3.92).

Lyot, Bernard Ferdinand (1897–1952). French astrophysicist at MEUDON and

Bernard Lyot: astrophysicist and inventor

L

PIC-DU-MIDI Observatories. He invented the coronograph to observe the SOLAR CORONA in the absence of an eclipse (1930), the birefringent filter (1933–39), the visual fringe polarimeter (1922), and the photoelectric polarimeter (1923, 1958). He discovered 11 coronal lines, analyzed by cinematography the SOLAR PROMINENCES and CHROMOSPHERE, and introduced high resolution in planetary studies.

Lyot filter Monochromatic birefringent interference filter devised by French astronomer Bernard LYOT in 1933. It consists of alternating layers of polaroid and quartz (or calcite) plates. Polarized light suffers double refraction by the quartz plate and when realigned by a further polaroid interference occurs: some wavelengths are cancelled, some are reinforced. After several similar passages the wavebands become widely spaced and the desired wavelength can be isolated by a suitable coloured glass.

Lyra (the Lyre or Harp). A small but prominent constellation — in mythology, the harp given to Orpheus by Apollo.

Brightest stars

Name	Visual Mag.	Abs. Mag.	Spec.	Distance (light-yrs)
α (Vega)	0.03	0.5	A0	26
γ (Sulaphat)	3.24	−0.8	B9	190
β (Sheliak)	3.4 (max)	−0.6	B7	300

Lyra also contains the famous double-double or quadruple star Epsilon Lyræ, whose components are each of magnitude 5.1, giving a combined magnitude of 3.9; the two main stars can be separated with the naked eye, and each is again double. Beta Lyræ or Shaliak is the prototype eclipsing binary (see BETA LYRÆ VARIABLES); between it and Gamma (Sulaphat) even small telescopes will show the Ring Nebula, M57, a planetary nebula.

Vega is the brightest star in the northern hemisphere of the sky apart from Arcturus. As seen from latitudes of Britain and the northern United States it is almost overhead during summer evenings, and is easily identified because of both its brightness and its steely-blue colour. *PM.*

Lyrid meteors A meteor stream which peaks on about April 21 or 22 every year. Members of this shower can be seen from about April 19 to 25. Although at the present time the Lyrids

usually give only poor displays (ZHR about 15), there have been some spectacular showers in the past, notably on March 27 15BC, when "stars fell like rain". In 1803, 1922 and as recently as 1982, the Lyrids produced very brief but modest displays of over 100 meteors per hour. The Lyrids are a stream of considerable antiquity; they have been traced back for over 2,500 years. It is thought they are associated with Comet Thatcher 1861 I. The Lyrid radiant is at R.A. 18h 08m, Dec. +32°, close to the border between Lyra and Hercules. *JWM.*

Lysithea Jupiter's tenth satellite, discovered by Nicholson in 1938; diameter about 22 miles (35km). It forms a group with Leda, Himalia and Elara. It has direct motion, but may well be a captured asteroid.

Mach, Ernst (1838–1916). Austrian physicist and philosopher whose greatest influence was in philosophy, where he rejected from science all concepts which could not be proved by experience. This freed EINSTEIN from the absoluteness of Newtonian space-time and helped him toward his theory of relativity. Mach's Principle is an idea that all the inertial properties of a piece of matter are in some way attributable to the influence of all the other matter in the universe.

Maclear, Sir Thomas (1794–1879). Irish astronomer who become HM Astronomer at the Cape (South Africa) in 1833. He reorganized the observatory during his long régime, and also carried out major researches into stellar parallaxes.

Maculæ Dark sometimes irregularly-shaped patches on EUROPA.

Mädler, Johann Heinrich (1794–1874). German astronomer. With his colleague W.Beer, he produced the first really good map of the Moon (1837-38). Subsequently he became Director of the Dorpat Observatory in Estonia.

Mädler Land (or Mädler Continent). Old name for Chryse; the Martian plain on which Viking 1 landed in June 1976.

Maffei galaxies Two nearby galaxies so close on the sky to the plane of the Milky Way Galaxy as to be almost completely obscured optically by galactic dust. They were originally discovered in 1968 by the Italian astronomer Paolo Maffei as two infrared sources close (on the sky) to the bright galaxy IC1805. The two infrared sources were then identified with faint but very extended optical objects, with all the attributes of nearby galaxies, and with the exciting possibility that these were newly-discovered members of our LOCAL GROUP of galaxies.

Maffei I is now known to be an S0 galaxy which is indeed on the edge of the Local Group, at a distance of 1,000kpc and with an absolute magnitude of $Mv = -20$ mag. Neutral hydrogen (HI) observations by Wright and Seielstad have been shown that Maffei 2 is a rotating gas-rich galaxy, inclined at about 67° to the line of sight. These and other observations indicate that it is a nearby Sb spiral; but at a distance of 5 Mpc, it is clearly unrelated to Maffei 1, and indeed it lies beyond the generally-accepted radius of the Local Group. *JVW. See also* CLASSIFICATION OF GALAXIES

Magellan NASA's Venus Radar Mapper spacecraft, renamed and scheduled for launch in the late 1980s. It will map up to 90 per cent of the planet's surface at a resolution of 3,200ft (1km).

Magellanic Clouds The two Magellanic Clouds, the nearest galaxies outside our own, are of crucial importance in many branches of astronomy, including the establishment of the extragalactic distance scale. Their distances are only about one-tenth that of the ANDROMEDA GALAXY (M31): they are thus near enough that we can make detailed studies of individual objects in them using large modern telescopes. At the same time they are far enough away that we can assume all the objects in any one cloud are at (roughly) the same distance from us.

Both clouds are easily visible to the naked eye in a dark sky, appearing like isolated off-shoots of the Milky Way with apparent diameters of about 6° (Large Cloud) and 3° (Small Cloud). The LMC has an "axial bar", with a high concentration of faint stars, as one of its main features. At its north-east end lies the 30 Doradûs complex, one of the largest and brightest groupings of hot stars and bright nebulosity known in any galaxy. The irregular distribution of star clusters and H II REGIONS (diffuse nebulæ) in the LMC contrasts with the more uniform stellar distribution in the SMC. In the latter, however, there seems also to be a main axial bar broadening out at its south end. Although the initial impression of the clouds is one of irregularity, G. de Vaucouleurs, in particular, has stressed their barred nature and the faint outer structure that can be interpreted as rudimentary spiral arms. Optical and radio studies of the motions of stars and gas in the clouds have indeed shown that there is considerable regularity in their underlying structure. The Magellanic systems. As with the clouds themselves, members of this class fremagellanic systems. As with the clouds themselves, members of this class frequently come in pairs, suggesting gravitational interaction.

The two clouds lie south of −65° declination and the early navigators passing round the southern point of Africa knew them as the Cape Clouds, but they are now associated with the name of Magellan, who described them. The Abbé de Lacaille, who worked at the Cape (1751–53), named the constellation Mons Mensæ (= modern Mensa) in honour of Table Mountain and depicted the LMC as the "table cloth" that often covers it.

The importance of the clouds to astronomy began to emerge when Henrietta LEAVITT discovered (1912) that the CEPHEID VARIABLES in the SMC obeyed a period–luminosity relation. She showed that the brightness (expressed in magnitudes) increased linearly with the logarithm of the period of pulsation. It is on this foundation that our current views of a universe built of galaxies is based, since it was Edwin HUBBLE's application (1923) of the P–L relation to the Cepheids he had discovered in M31 (the Andromeda Galaxy) and M33 that established these spirals as galaxies in their own right at very large distances from our own Galaxy. Since Miss Leavitt's time the Cepheid P–L relation, and its more accurate descendant the period–luminosity–colour (P–L–C) relation have been continually extended and improved. The best current distances for the LMC and SMC are 160,000 and 185,000 light-years with an uncertainty of about 10 per cent. Young objects (hot stars, Cepheids, etc) in the LMC lie in a thin disk, which is seen nearly face-on (an angle to the plane of the sky of about 27°). Older objects such as planetary nebulæ form a somewhat thicker disk structure, but

there is no evidence at present for a spherical halo such as is formed by very old objects in our own Galaxy. Recent work on Cepheids in particular has shown that the SMC is greatly extended in the line of sight (a depth of 60,000 light-years). Its depth is five

Magellanic clouds: the Large Cloud

times its diameter on the plane of the sky. The curious, long, twisted structure that has been found (and which may in fact divide the cloud into two distinct portions) suggests that this galaxy has been distorted by strong tidal forces. One suggestion is that these forces operated during a close passage of the SMC and LMC some 200 million years ago.

A major revision of views on the Magellanic Clouds occurred in 1952. Until that time no RR LYRÆ VARIABLES had been found in either cloud despite searches for them. RR Lyræ stars in our Galaxy belong to a very old stellar population (they are found in galactic globular clusters) and their absence had been taken to indicate that at least the LMC contained no old stars. RR Lyræs are also relatively good distance indicators. In 1952 A.D. Thackeray and A.J. Wesselink used the new 74in (1.9m) Radcliffe Reflector (Pretoria) to discover RR Lyræs in both clouds. They were 1.5 magnitudes fainter than had been predicted. Both clouds were thus seen to contain objects of a wide range of ages (as does our Galaxy) and also to be twice as far away as had previously been thought. At the same time W. Baade was failing to find RR Lyræs in M31 with the new 200in (5m) telescope (Palomar), again suggesting an increase in the extragalactic distance scale by at least a factor of two. Later work on Cepheids in open clusters in our Galaxy showed that the zero point of the Cepheid P–L relation needed changing and this explained the earlier, erroneously small, extragalactic scale.

In our own Galaxy, globular clusters are all old objects. However, Thackeray showed that the cluster NGC1866

in the LMC, though globular in form, was a young system. Several such "blue globulars" exist in the clouds. No similar objects are known in our own Galaxy.

MIRA VARIABLES were first reported in the LMC by Lloyd Evans and Glass (1981). They follow a well-defined period–infrared luminosity relation. They are thus an important new distance indicator, especially as the effects of interstellar absorption are very small when working in the infrared.

Spectroscopic analyses of supergiant stars and (especially) of H II regions (diffuse nebulæ) have shown that these young objects are deficient in "metals" (ie, elements heavier than helium) in both clouds, compared with similar objects in the solar neighbourhood. In the mean the deficiency is about a factor of 1.4 in the LMC and 4.0 in the SMC. Metals are made by nuclear processing in the interior of stars. SUPERNOVA explosions then spread them into interstellar space, where they can be used to form a new generation of stars. The Magellanic Cloud results show therefore that the rate of processing of matter through stars there has been different from that in our Galaxy. This makes the clouds important testing grounds of STELLAR EVOLUTION theories since we can examine objects of the same age or mass in the LMC, SMC and our Galaxy and compare with predictions for different metallicities. This is an area of much current work.

Comparative studies of the Cepheid variables in the two clouds have been invaluable in showing the effects of metallicity on the P–L and P–L–C relations.

Studies of supergiants in our own Galaxy are greatly hampered by the uncertain distances of these objects. In the Magellanic Clouds we are presented with a large sample of such objects all at the same, known, distance. Work at the Radcliffe Observatory in the 1950s showed that there was a clear upper limit to the luminosities of stars in the clouds. This limit is about a million times the luminosity of the Sun, at least for stars evolved well away from the MAIN SEQUENCE. Theory says that such stars are about 100 times more massive than the Sun.

The Magellanic Clouds are under intensive study at all the major southern observatories and, as more and more sophisticated equipment is brought to bear on them, we can be sure that they will keep their central importance to astrophysics and cosmology. *MWF.*

M

Maginus Lunar walled plain 110 miles (177km) in diameter, in the Clavius area; 50°S, 6°W. It has irregular walls, and becomes very obscure near full moon.

Magma Any hot rock melt.

Magnetic fields One of the fundamental forces of nature. On Earth a weak magnetic field exists capable of swinging a well-balanced compass needle. In astronomical objects the field can be more than one million million times stronger, and it then can control the motion of gas and the shape of objects.

Magnetic stars Stars, first observed in 1948, that have strong magnetic fields ranging from a few hundred to several thousand gauss. They are also known as magnetic variables because their magnetic field is variable. The variability can be periodic or irregular accompanied sometimes by a reversal of polarity of the field and variations in their spectra. Magnetic stars do not appear to be restricted to any particular spectral class or part of the H-R diagram and present some still unsolved problems such as the reason for their periodic variability and their place in STELLAR EVOLUTION. *VB.*

Magnetic storms Sudden fluctuations in the Earth's magnetic field, due to events on the Sun — generally associated with SOLAR FLARES. They are often accompanied by AURORÆ.

Magnetohydrodynamics The study of the behaviour of PLASMA (a mixture of ions and electrons) in magnetic fields. This is relevant to a wide range of astrophysical phenomena such as the solar atmosphere or the MAGNETOSPHERES of planets. The science of magnetohydrodynamics was founded largely by the pioneering work of the Swedish-born astronomer Hannes Olof Gösta Alfvén (1908–).

Magnetopause The magnetopause marks the boundary between the MAGNETO-SPHERE of a planet and the external SOLAR WIND.

Magnetosphere A region surrounding a planet which contains ionized particles controlled by the planet's magnetic field. We may consider this region as the magnetic sphere of influence of the planet whose domain is limited by the interaction between its magnetic field and the solar wind. It is not a necessary condition for the planet to have a

magnetic dipole moment in order to possess a magnetosphere. The presence of surface magnetization or an ionosphere can also provide one.

The solar wind originates in the atmosphere of the Sun and consists of ionized hydrogen atoms and electrons, so that overall, it is composed of neutral particles. Consequently, the solar wind is an example of a PLASMA in motion. The outward speed of the solar wind is about 250mps (400km/s).

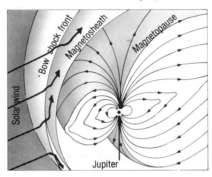

Magnetosphere: ionized particles

While this speed remains almost constant as the solar wind moves outwards, the density of the ions and electrons decreases in proportion with the square of the distance from the Sun. Consequently, the solar wind affects the entire Solar System and restricts the size of each planetary magnetosphere. However, the solar wind loses its identity in the interstellar medium at a distance of about 100AU from the Sun. When the solar wind encounters the magnetic field of a planet it compresses the upwind side so that the planetary magnetosphere has the visual appearance of a wind-sock. Indeed, if it were not for the presence of the solar wind, the terrestrial magnetosphere would extend for $100R_E$ (Earth radii). The downwind side is therefore called the magnetotail and the terrestrial feature has been detected as far as $1,000R_E$ from the planet. The boundary between the solar wind and the planetary magnetosphere is the BOW SHOCK, since it resembles the shape in the water in the neighbourhood of the bow of a moving ship. The precise position of the bow shock will vary with the solar activity, since when the Sun is active, such as during a solar storm, the increased solar wind pressure will compress further the magnetosphere. Between the bow shock and the MAGNETO-PAUSE, the solar wind is greatly decelerated, which creates a region of agitated motions and high temperatures. The magnetopause marks the inner bound-

ary of this region, which is called the magnetosheath. Then the region inside the magnetopause is the magnetosphere which contains the regions of trapped radiation such as the VAN ALLEN BELTS of the Earth. A small number of neutral atoms from the solar wind readily cross the magnetopause into the magnetosphere. However, most of the ions and electrons are deflected, while a very small number actually reach the magnetotail.

There is now direct evidence for magnetospheres associated with Mercury, Jupiter, Saturn and Uranus and it is expected that Neptune also will be found to possess a substantial magnetosphere. Although the magnetic field of Mercury is very weak and only one-hundredth of the terrestrial field, it still provides a definite interaction with the solar wind. There is a well-defined bowshock, but the field is too weak to generate the equivalent Van Allen Belts.

The Jovian magnetosphere is the largest in the Solar System and it has a cross-section equivalent to that of the Sun if seen visually from the Earth. It is extremely variable in size, extending to distances of between 50 and $100R_J$ in the direction of the Sun as a result of the changes in the solar wind. The high-energy particles in Jupiter's magnetic field also form Van Allen Belts, although they are 10,000 times more intense than those of the Earth. The magnetotail of Jupiter is so large that it extends over more than 5AU and interacts with Saturn. Jupiter also emits waves at radio frequencies, corresponding to DECAMETRIC and DECIMETRIC regions of the electromagnetic spectrum, which are greater in intensity than any other extraterrestrial source other than the Sun. The Saturnian magnetosphere is intermediate in size between those of Jupiter and the Earth and it is also very sensitive to the behaviour of the solar wind. The VOYAGER 1 spacecraft crossed the bow shock five times in all, at distances ranging from 26.1 to only 22.7 R_S. A major difference of the Saturnian region compared with Jupiter is the concentration of highly electrically-charged dust in the ring plane. It is possible that this material is the source of the high-energy discharges in the rings. Uranus has an unusual magnetosphere associated with the dipole magnetic field whose axis is offset 60° with respect to the planetary rotation axis. At the time of the Voyager encounter the Uranian magnetosphere extended $18R_U$ and the tail had a radius of $42R_U$ at a distance of $67R_U$. There are also radio emissions from Uranus, although weaker than those of Saturn signals. The measurements of the rotational period of the charged particles in these magnetospheres of the outer planets provides a direct observation of the true rotational period for each planet since the field lines terminate in the planetary interior. This measurement is essential for the meteorological investigations, since it provides the reference period for the observed atmospheric winds. There are interactions between the charged particles in each of these planetary atmospheres, producing auroral activity in the polar regions. While the phenomenon is well known on the Earth as the polar AURORA, on these planets it is much more intense and at Jupiter it is 60 times brighter than the terrestrial aurorae. However, there is also a further magnetospheric interaction with the atmosphere not known on Earth, called the electroglow, which is only seen on the dayside of Uranus, where the aurora is not observed.

Each of the major planets is surrounded by a system of rings and an extensive family of satellites. Many of these bodies reside inside the magnetospheres of the individual planet and have a unique interaction with the charged particle environment of the magnetosphere. In this aspect, the major planets differ markedly from the Earth, whose moon resides outside the main magnetosphere, passing through the magnetotail only as it orbits the planet. At Jupiter, seven of the innermost satellites and the rings are in this hostile region. The satellites are constantly bombarded by the high-energy particles, which erode the surfaces and alter the chemistry of the satellites. The interaction between Jupiter and IO is the most intense and the resulting "sputtering" action is responsible for the creation of the sodium, potassium and magnesium in the cloud stretching out along Io's orbit. Io and Jupiter are also connected by a flux tube carrying a massive current of 5 million amps with a potential difference of 400,000 volts. The power in this mass of electricity is more than 70 times the generating capabilities of all the nations on Earth. This electrical energy plays an important role in the local heating of the surface of Io and the generation of its volcanic activity.

At Saturn the massive system of rings and many of the satellites have the effect of sweeping a path through the charged particle environment. The rings and the satellites are very efficient in absorbing the protons in the magnetosphere. TITAN sits at the edge of this region. It is inside or outside the magnetosphere as a consequence of the solar activity, which will then have a variable effect on the upper atmosphere chemistry of the satellite. Similar effects occur in the Uranian system, where MIRANDA and ARIEL have major effects on the trapped radiation fluxes. It is possible that this additional energy is responsible for the creation of the unusually dark material found on the satellites and rings of Uranus. *GEH.*

Magnitude The brightness of an astronomical body. Apparent magnitude 1 is exactly 100 times brighter than magnitude 6, each magnitude being 2.512 times brighter than the next. Magnitudes brighter than 0 are minus figures, thus Sirius is -1.4, and the Sun is -26.8. The faintest objects yet photographed are about magnitude 26. *See also* ABSOLUTE, APPARENT, BOLOMETRIC, PHOTOGRAPHIC, PHOTOVISUAL and VISUAL MAGNITUDES.

Main Lunar crater, 81°N, 9°E, diameter 30 miles (48km). It intrudes into its 35-mile (56km) "twin", Challis.

Main Sequence A region of the HERTZSPRUNG-RUSSELL DIAGRAM where most stars are found. It links points running from hot blue stars to cool red stars (top left to bottom right). These points represent stars during the greater part of their lifetimes as they produce energy primarily from the hydrogen-to-helium fusion reaction. The original line formed by stars shortly after their ignition is called the zero-age main sequence. As compositional changes alter the internal structures of these stars, the points that represent them will move away from the main sequence and trace an evolutionary track on the diagram. *RCM.*

Maksutov, Dmitri Dmitrievich (1896–1964). A Soviet optician and telescope maker. After an eventful career as a soldier in World War I and the Russian Revolution, Maksutov became an optician in Odessa in 1921, beginning work in the theory of astronomical optics. He organized a laboratory of astronomical optics at the State Optical Institute in Moscow and it was there that he first invented and then built the first of several MAKSUTOV TELESCOPES. He also built more conventional large telescopes such as the 15in (38cm) SCHMIDT TELESCOPE for the Engelhardt Observatory, near Kazan, the Pulkovo solar telescope, and the 32in (82cm) Pulkovo refractor. In 1952 he transferred all his activities to Pulkovo, near Leningrad, where he headed a section which made all kinds of astronomical instruments. *PGM.*

Maksutov telescopes A modification of the basic SCHMIDT TELESCOPE construction, designed to remove the small but noticeable aberrations of the Schmidt optics. In the early 1940s D.D. MAKSUTOV at the State Optical Institute in Moscow, and A.A.Bouwers in Holland, independently discovered that instead of the aspherical corrector plate used in the Schmidt design, a thin negative meniscus lens with two spheri-

Maksutov telescopes: strong and compact

cal surfaces can be used. The convex side of the lens faces toward the primary mirror. The spherical aberration that such a lens introduces can be arranged to cancel the spherical aberration of the primary mirror. Fortunately, the weakly negative meniscus lens also can be designed to be almost free of CHROMATIC ABERRATION.

Instead of the meniscus lens being placed at the centre of curvature of the primary mirror, as is the corrector plate of the Schmidt telescope, the lens is slightly diverging and consequently can be mounted considerably nearer to the primary. As a result, a much shorter tube can be used, which reduces the overall weight and cost of the telescope and its dome. This and the comparative ease of producing the spherical surfaces of the meniscus lens, as opposed to the

M

M

aspheric surface of the Schmidt corrector, have made the Maksutov a popular choice for telescopes and cameras with wide fields of view and short FOCAL LENGTHS. The Maksutov produces better images than the Schmidt for f ratios down to about f1; below that the Schmidt is superior.

Used as an astronomical camera, the Maksutov construction shares with the Schmidt the problem of a curved FOCAL PLANE. That is, for the stars to be photographed in focus over the entire field, the photographic plate must be deformed to a curved shape. This severely limits the thickness of the glass and makes large plates very fragile to handle. The necessity of mounting the photographic plate in a holder supported in the middle of the telescope tube is an added problem. These difficulties are not serious for small Maksutov telescopes with correspondingly small plate sizes. For small and intermediate-size Maksutov telescopes a field-flattening lens mounted in front of the photographic plate (sometimes incorporated into the plate holder) solves the problem of field curvature.

For the large apertures required for professional use, the meniscus corrector lens must be quite thick and deep in curvature, making it difficult to manufacture. As a result, few large Maksutov telescopes have been made. The first, in 1950, was installed at the USSR Academy of Sciences Observatory at Alma-Ata. It has a 20in (51cm) mirror and a 16½in (42cm) diameter corrector lens. A 26in (65cm) instrument was added to the Crimean Astrophysical Observatory in 1952. It was followed by a 28in (71cm) instrument at Abastamani, in Georgia (1956), and then, in 1958, by a 28/20in (71/51cm) at the Sternberg Astronomical Institute Observatory in the Crimea. There is a 25in (63cm) instrument at Ondrejov Observatory, in Czechoslovakia.

As with the Schmidt design, the basic Maksutov has its focus within the telescope tube and cannot be used visually. However, it may be adapted by introduction of a flat diagonal mirror to produce a NEWTONIAN focus, or with secondary mirrors that are convex hyperboloidal or concave ellipsoidal to produce CASSEGRAIN and GREGORIAN constructions respectively. The compactness of the telescope is maintained if the secondary mirror is figured as a small area at the centre of the back side of the meniscus corrector. This is aluminized and reflects the light through a central hole in the primary mirror to a focus behind the primary. An advan-

tage of this arrangement is that the secondary mirror has no radial supporting arms (or "spider") so images produced by the Maksutov design are free of diffraction spikes.

The small aperture Maksutov telescope is very popular with amateur astronomers. The images it produces are free from coma and astigmatism and almost completely free of chromatic aberration. As with a refracting telescope, the closed tube eliminates air currents within the tube which, with the freedom from diffraction spikes, ensures excellent images. The telescope is, however, far more compact than a refractor: a Maksutov with a tube 18in (46cm) in length may have an effective focal length of 100in (254cm). Such compactness makes for a strong and rugged construction. In its Cassegranian form, the telescope has the additional benefit of a particularly flat field.

Despite the necessity of figuring the meniscus corrector lens, the Maksutov design has proved popular with amateur telescope makers. Modern optical manufacturing processes enable the production of small meniscus corrector lenses at relatively low cost, with the result that Maksutov telescopes in the 4-10in (10-25cm) diameter range are relatively inexpensive.

As with the Schmidt, the wide field of view and fast optics make the Maksutov construction a good choice for the cameras used in spectrographs that depend on photographic recording of the spectrum. *BW.*

Malmquist, Gunnar (1893–1982). Swedish astronomer. "Malmquist's relation" is important in stellar statistics and observational cosmology.

Maps *See* CELESTIAL MAPS.

Maraldi, Giovanni (1709–88). Italian-French astronomer who specialized in planetary observation.

Marduk Active volcano on Jupiter's satellite IO (originally known as Plume 7). During the Voyager 1 pass the plume rose to 75 miles (120km). Position 28°S, 210°W.

Mare Australe The Southern Sea, on the south-eastern limb of the Moon's near-side. It is not a major sea, but contains various lava-flooded craters. Under good conditions of libration it can be quite conspicuous.

Mare Crisium A hexagonal area, spanning roughly 310 miles (500km), of

dark grey lava surrounded by mountains in the eastern hemisphere of the Moon. It is crossed by mare ridges and rays (*see* LUNAR RAYS).

Mare Foecunditatis (the Sea of Fertility). One of the major lunar "seas" in the south-east quadrant. It is joined to the Mare Tranquillitatis, but is much less regular. The large craters Langrenus and Vendelinus lie on its borders; it contains the Messier craterlets.

Mare Humboldtianum (Humboldt's Sea). A minor lunar sea, on the north-east limb as seen from Earth and therefore very foreshortened. It is of the same basic type as the Mare Crisium.

Mare Humorum (the Sea of Humours). Lunar sea leading off the Mare Nubium; Gassendi, Hippalus, Vitello and Doppelmayer lie on its borders. There are no major craters upon its comparatively smooth floor.

Mare Imbrium A dark plain, 800 miles (1,300km) in diameter, north of the Moon's equator, basaltic flood lavas layered in sheets sloping down (generally toward the centre) at only about 1° from the horizontal. Basalts returned from the Apollo 15 mission to SE Mare Imbrium were dated at 3,300,000,000 years; but some Imbrium lavas were formed more recently. One of the latest flows is 15–30 miles (25–50km) wide, 80 miles (130km) long and has a thickness in excess of 80ft (25m). Mare Imbrium is associated with a mascon and forms the centre of the most pronounced lunar fault system of radial and transverse lineaments, in evidence in the surrounding highlands. While the mare is widely believed to be a vast impact crater flooded by later lavas from within the Moon, the associated faulting is a result of slowly acting tectonic forces. *GF.*

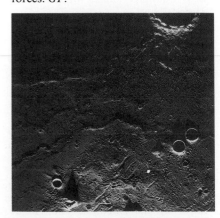

Mare Imbrium: largest well-marked "sea"

Mare Ingenii A sub-circular, lava-flooded crater roughly 200 miles (330km) wide centred at 34°S, 164°E. It contains several large, ill-defined craters.

Mare Marginis Spanning the longitude range 81° to 94°E at about 13°N, this is an irregular, lava-inundated area.

Mare Moscoviense The lava-flooded, inner parts of a much larger, sub-circular multi-ring structure. The mare itself is centred at 28°N, 148°E and averages 155 miles (250km) across.

Mare Nectaris The central, lava-flooded, polygonal part of a multi-ring structure south of the lunar equator. Associated with a LUNAR MASCON and with major, sub-radial faults.

Mare Orientale (the Eastern Sea). A lunar mare, discovered by Patrick Moore and H.P.Wilkins in 1948 and named by them; because of the reversal of east and west, it is now on the Moon's western limb as seen from Earth. It has proved to be a vast ringed structure, most of which lies on the Moon's farside.

Mare Serenitatis One of the major seas of the Moon, bounded in part by the Hæmus Mountains; it adjoins Mare Tranquillitatis to the east, while Mare Imbrium lies to the west. There are notable ridges on Mare Serenitatis, and some craters, of which the largest is Bessel; this is also the site of the controversial formation Linné.

Mare Smythii Centred close to the lunar equator, this irregular, lava-inundated area spans the longitude range 81° to 91°E and is associated with a mascon.

Mare Spumans (the Foaming Sea). Small darkish area on the Moon, south of Mare Crisium.

Mare Tranquillitatis An irregularly-bordered area of dark lavas, in the eastern lunar hemisphere, containing the igneous complex Lamont, the fault Rupes Cauchy and, nearby, a major rille.

Mare Undarum (the Sea of Waves). Small, patchy dark area on the Moon, south of the Mare Crisium.

Mare Vaporum A lava-inundated area, centred 13°N, 5°E, bordered on the north by the Apennine and Hæmus mountains.

Maria Michell Observatory Located on Nantucket Island, off the coast of Massachusetts, this small observatory was founded in 1902 to commemorate the first woman astronomer in the United States. It houses a 5in (12.5cm) Alvan Clark refractor and a 7.5in (19cm) photographic refractor.

Marine chronometer Precision timepiece specially designed for navigational use at sea, particularly for finding longitude.

To do this astronomically, the navigator has to measure his local time — a fairly simple procedure and compare that with the local time of the reference meridian at the same moment — much more difficult. In 1530 the Louvain mathematician Gemma Frisius suggested that the navigator should take with him one of the newly-invented portable clocks or watches, set at the beginning of the voyage to the time of the reference meridian. Easier said than done! Making a timepiece that would keep time to the required precision for months or years, in any climate in a moving ship, took another 250 years.

The first attempts to show promise were those of Christiaan HUYGENS, several of whose marine timepieces were tried at sea between 1662 and 1687. In 1714 the British government offered up to £20,000 — perhaps over £1 million today — to anyone discovering a method of finding longitude that was "practicable and useful at sea" within certain limits. The Yorkshire clockmaker John Harrison was eventually awarded £18,750 in 1772 for his large silver watch, still preserved at Greenwich.

Other early makers of chronometers (as they came to be called) from the 1770s were John Arnold and Thomas Earnshaw in England, and Pierre Le Roy and Ferdinand Berthoud in France.

By the middle of the 19th century, chronometers had become cheap enough for most foreign-going ships to have three or more, but demand fell off in the 1930s with the rise in numbers of radio time signals available worldwide. In recent years, quartz movements, often in the form of a wrist watch, have begun to replace the traditional spring-driven chronometer at sea. *HDH.*

Mariner spacecraft A series of American planetary probes. Mariner 2 was the first successful probe to reach Venus (1962). Mariner 4 was the first probe to

reach Mars, sending back pictures of a cratered surface (1964). Mariners 6 and 7 extended the mapping of Mars,

Mariner spacecraft: the Venus probe

including the south polar region (1969). Mariner 9 went into orbit round Mars, it being the first to be placed in orbit round another planet (1971). Mariner 10 flew past Venus on its way to Mercury, revealing that Mercury had a heavily-cratered surface.

Mariner Valley (Mars) A series of enormous, interconnected canyons situated between 30°W and 110°W, just south of the Martian equator. These extend from the crest of the Tharsis Rise to the vicinity of Chryse Planitia, a distance of 2,500 miles (4,000km). In the west the canyons system emerges from Noctis Labyrinthus then trends approximately west-east. The main section of the system comprises a number of parallel, straight-sided chasms, beginning in the west with Ius and Tithonium Chasmæ. The central section sees both the widest and deepest expression; here three vast troughs, each 124 miles (200km) wide and up to 4 miles (7km) deep, slice through the plains. At the eastern end the canyons pass into chaotic terrain. *PJC.*

Marius Lunar crater in Oceanus Procellarum; 12°N, 51°W. It is fairly well-formed, with a low central peak; its twin is Reiner. There are many ridges and hills in this area.

Marius, Simon (1570–1624). German astronomer. Claimed discovery and naming of Jupiter's satellites (1609) but his contemporaries credited GALILEO. First to observe ANDROMEDA GALAXY with telescope (1612).

M

Markarian galaxies Galaxies catalogued by the Armenian astronomer Markarian, distinguished by their brightness in blue light.

Mars The planet Mars has a diameter roughly half that of Earth (4,200 miles/ 6,794km) and orbits the Sun at a distance which varies between 128 million miles/206 million km (perihelion) and 155 million miles/249 million km (aphelion). Its orbital period is 686 days, while the Martian day has a length of 24hrs 37m 22.6s. It is popularly known as the "Red Planet" on account of the pervasive orange/red colour of the disk when viewed through Earth-based telescopes. This coloration is a function of widespread regions of reddish dust which on occasions may be raised high into the tenuous Martian atmosphere by winds. Such dust storms had been observed regularly by observers for many years, but one in particular, that of 1971, was observed at close quarters by the American spacecraft Mariner 9 as it approached the planet and eventually went into orbit around it.

The first telescopic observations of Mars were made by GALILEO in 1610; he noted the phase but no surface markings. Subsequently HUYGENS, CASSINI and MARALDI all observed Mars, noting the polar caps and some of the dark features apparent in telescopes. Most early workers believed the dark features to be seas, but when, in the 19th century, it became apparent that the atmosphere was too tenuous for this to be the case, the consensus view became that they were dried-up sea beds supporting lowly vegetation.

In 1877, SCHIAPARELLI drew a much more detailed chart of the planet and, for the first time, depicted numerous dark linear markings he called *canali* (*see* MARTIAN CANALS). By way of an explanation these canals were suggested to have been dug by intelligent beings, as irrigation channels across the Martian deserts. In the latter part of the 19th century, Percival LOWELL founded the famous LOWELL OBSERVATORY at Flagstaff in Arizona, with the express purpose of charting the Martian surface, particularly the canals. He mapped hundreds of these features and described them in his books and papers. Yet despite these observations, many contemporary workers saw either no sign of them, or depicted them as much broader, diffuse, markings. Only in 1971 was the matter of the canals finally resolved. In that year Mariner 9 approached the

planet, went into orbit and showed quite clearly that the linear features do not exist. Evidently what Schiaparelli, Lowell and others drew, was some kind of optical illusion.

The atmosphere of Mars is extremely tenuous, being only one-hundreth that of Earth. Ninety-five per cent of it is due to carbon dioxide, the rest being due to nitrogen (2.7 per cent), argon (1.6 per cent) and very small amounts of oxygen, carbon monoxide and water vapour. Because of the low temperatures experienced at the distance of Mars (equatorial temperature ranges from 26°C just after noon in summer, to −111°C just prior to sunrise), and despite there being but 0.03 per cent water vapour in the air, it is relatively moist. One major source of such vapour is the north polar cap which has been shown to be composed of water ice. Strangely, the southern cap is largely carbon dioxide ice ("dry ice"), the reasons for which are not really known.

Mars: the surface from Viking Lander 2

Because the atmosphere is thin, its movements are largely controlled by solar heating; there are no oceans or major mountain ranges to complicate the circulation. Clouds do form, however, particularly over the higher regions of Tharsis. Mists have also been observed in valley bottoms, such as that of the great Valles Marineris (*see* MARINER VALLEY).

Since the advent of Mariner 9 and the Viking spacecraft, the topography and geology of Mars has become known in some detail. There is a hemispheric asymmetry to the planet: much of the southern hemisphere is well above datum and is heavily cratered. The density of such craters is high, suggesting that it is as old as the highland regions of Earth's Moon. The two large impact basins, HELLAS and ARGYRE are located in this hemisphere. In several places this ancient cratered terrain is

cut by gullies and complex channel systems which appear to have been incised by water. Several tongues of the cratered surface extend into the northern hemisphere, the largest occurring on either side of the 330° meridian.

Most of the northern hemisphere is less heavily cratered and much is below datum. The surface looks altogether "smoother" on the large scale. The most impressive features of this hemisphere are the vast shield volcanoes of Tharsis and the extensive valley network known as Valles Marineris, both first revealed in Mariner 9 images. The Tharsis region is in the nature of a huge bulge in the Mars lithosphere; it has a size roughly the same as that of Africa south of the Congo river, 2,500 by 1,900 miles (4,000 × 3,000km). The entire Tharsis region is about 4 miles (6km) above datum but it rises to heights of between 12 and 16 miles (20 and 26km) in the large shield volcanoes, such as OLYMPUS MONS. A similar but smaller bulge resides in the region known as Elysium.

Superimposed on Tharsis (and Elysium) are several huge shield volcanoes which, in terms of profile, are analogous to Earth's Hawaiian volcanoes, such as Mauna Loa. They are, however, substantially larger and have vast summit depressions (calderas) crowning them. Very extensive lava flows may be traced radiating out from these constructs. Radiating out from the Tharsis (and Elysium) regions is a vast array of fractures which must have developed in response to the formation of the two crustal bulges.

To the east of Tharsis is an impressive canyon system which has been named Valles Marineris, after the Mariner 9 spacecraft which discovered it. It extends from near the summit of the bulge at 5°S,100°W, for 2,500 miles (4,000km) eastward, eventually merging into an immense area of "chaotic terrain". In most places the canyons are 2 miles (3km) deep and may be up to 125 miles (200km) across.

Perhaps the most puzzling of features are the Martian channels. So-called "outflow channels" may be several hundred kilometres long and tens of kilometres wide. Generally they start abruptly without tributaries. Most are located north of the great canyon system and converge on the plain known as CHRYSE PLANITIA, although others appear related to Elysium's northwest edge. They appear related to a period of flooding in the distant past, presumably when the climate was quite different from today and water could exist in quantity at the Martian surface.

The Viking Lander probes took panoramic photographs of the two landing sites, Chryse and Utopia — both plains areas. They recorded dune-like features and a reddish dusty surface crowded with dark blocks. Many of the blocks were vesicular and appeared similar in aspect to terrestrial basaltic rocks. Chemical analyses carried out on the surface revealed that the soil was probably the result of weathering of basaltic rocks, with much oxidation having occurred. The deep reddish coloration is believed due to the presence of smectite clays.

In addition to the volcanic shields and the surrounding lightly-cratered plains, the heavily-cratered terrain of the southern hemisphere and the polar caps, there are extensive regions adjacent to the poles which have been dissected to reveal strongly-laminated deposits. Much of the high latitudes also appears mantled in dust which, in places, is strongly etched by the winds. Scientists are still trying to understand how these deposits formed and so expand our knowledge of the Martian environment. *PJC.*

Mars crawler Descriptive term applied to a proposed remote-controlled rover vehicle to be taken to Mars on a future space mission.

Mars probes Unmanned vehicles to Mars. To date 16 have been launched by the United States and the USSR.

Mars probes

Name	Country	Record
Mars 1	USSR 1962	Failure
Mariner 3	USA 1964	Failure
Mariner 4	USA 1964	Passed at 6,200 miles (10,000km), July 1965
Zond 2	USSR 1964	Failure
Mariner 6	USA 1969	Passed at 2,100 miles (3,390km), July 1969
Mariner 7	USA 1969	Passed at 2,200 miles (3,500km), August 1969
Mariner 8	USA 1971	Failure
Mars 2	USSR 1971	Failure
Mars 3	USSR 1971	Failure
Mariner 9	USA 1971	Orbited; complete success
Mars 4	USSR 1973	Failure
Mars 5	USSR 1973	Failure
Mars 6	USSR 1973	Failure
Mars 7	USSR 1973	Failure
Viking 1	USA 1975	Orbiter and lander (1976); success
Viking 2	USA 1975	Orbiter and lander (1976); success

Martian basins Roughly circular basins of presumed impact origin which occur on Mars. The most prominent is the Hellas Basin, which is centred at 40°S, 295°W and which measures 1,000 × 1,250 miles (1,600 × 2,000km). Hellas has a rugged hilly rim which is between 30 and 250 miles (50 and 400km) in width and has associated with it a concentric pattern of fracturing that extends as far as 1,000 miles (1,600km) from the centre of the floor. The floor

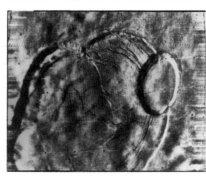

Martian basins: a Mariner 9 photograph

of the basin is occupied by plains materials which are believed to be volcanic in origin, and mantled by wind-blown debris. The deepest part of the basin, which lies on the western side, is 2.5 miles (4km) below Mars datum.

Also situated in the southern hemisphere is the ARGYRE basin. This is 560 miles (900km) in diameter and is also floored with plains units that are extensively mantled by aeolian materials. A third basin, the Isidis basin, lies in the northern hemisphere between SYRTIS MAJOR and the Elysium Rise. This is approximately 700 miles (1,100km) across but is poorly defined on its eastern side, where it merges onto the volcanic plains associated with the Elysium volcanoes.

A number of other basins have been recognized on VIKING and MARINER images. Generally these are less well-defined. For instance, CHRYSE PLANITIA is believed by some workers to occupy an ancient multi-ringed basin, although the surrounding ramparts are not clearly defined like those of Argyre and Hellas. All of the basins are ancient, indeed they represent the oldest recognizable features on the planet. *PJC.*

Martian canals Reported regular features on the surface of Mars, once believed by many astronomers to be due to artificial irrigation systems.

Though linear markings had been drawn by various observers from the mid-19th century, the canal system was

first described in detail by the Italian astronomer G.V.SCHIAPARELLI in 1877. Using a good refractor under the clear skies of Milan, Schiaparelli produced a map of Mars, revising the old nomenclature. Crossing the red "deserts" he showed linear features which he called *canali*; the English translation into "canals" was more or less inevitable! The canal system as shown by Schiaparelli was a comprehensive network; where two canals crossed each other there was a dark patch. Schiaparelli considered the possibility of artificial construction, but kept an open mind.

The canal question was taken up by Percival LOWELL, who founded the LOWELL OBSERVATORY at Flagstaff mainly to study Mars. He was convinced that the system was artificial, built to pump water from the polar ice-caps through to the arid equatorial regions. His ideas met with strong opposition, though at the time of his death in 1916 his theory was still taken seriously in many quarters.

However, other observers (notably E.E.BARNARD) could see no trace of the canal system, and attempts to photograph it were not successful. Gradually Lowell's concept of a canal as a narrow waterway (possibly piped) flanked to either side by irrigated land was discarded, but it was still thought possible that the canals existed in some form, as natural features. It was not until the Mariner flights of the 1960s that it became clear that the canals did not exist at all; they had been due to nothing more significant than tricks of the eye. There is no correlation between the Schiaparelli or Lowell canal network and features on Mars such as crater chains, valleys or ridges. *PM.*

Maser Acronym for "*m*icrowave *a*mplification by the *s*timulated *e*mission of *r*adiation". A device or celestial object in which radio emission from molecules stimulates further radio emission at the same energy from other molecules. *See also* LASER.

Maskelyne Lunar crater, 2°N, 30°E; diameter 15 miles (24km). It is in the Mare Tranquillitatis, not far from the original landing site for Apollo 11.

Maskelyne, Nevil (1732–1811). Fifth ASTRONOMER ROYAL. Having studied mathematics at Cambridge University, and though ordained a clergyman in 1755, he devoted his spare time to assisting James BRADLEY, the third Astronomer Royal.

Maskelyne is noted for producing in

M

1763 his *British Mariner's Guide*, explaining how to determine longitude at sea. In 1767, two years after his appointment as Astronomer Royal, he published the first issue of the annual *Nautial Almanac* for the same purpose. This is his greatest claim to fame.

Maskelyne was involved with assessing the accuracy of the newly invented chronometer, and also measured the density of the Earth. *CAR.*

Nevil Maskelyne: the *Nautical Almanac*

Mass-luminosity relation For stars on the MAIN SEQUENCE there exists a one-to-one relationship between luminosity and mass, expressed approximately as $L \propto M^\alpha$. The exponent α is about 3.5, but varies slightly with mass.

Mass number The number of nucleons (protons and neutrons) in the nucleus of an atom. For example, the mass number of hydrogen is 1 (1 proton), while that of carbon is 12 (6 protons and 6 neutrons).

Mauna Kea Observatory The summit of Mauna Kea, the "White Mountain" of the Island of Hawaii, was recognized in 1964 to have the potential for an excellent astronomical observatory site. At an elevation of 13,796ft (4,200m), the summit stands above the tropical clouds and moist maritime air of Hawaii, and the absence of major population centres ensures a dark nighttime sky. Airflow over the mountain is laminar, resulting in stable images ("seeing") of stars and planets.

The first major telescope put into use on Mauna Kea was the 88in (2.24m) reflector of the University of Hawaii, commissioned in 1970. The predicted qualities of the site were soon realized as the telescope was used for infrared

and optical studies of the planets, the stars, galaxies, and the interstellar medium. Many of the first infrared observations of Solar System bodies were made in the early 1970s with this telescope, and the exceptional quality of Mauna Kea for infrared measurements was established in those years.

Those who built the observatory, and the astronomers who first worked there, quickly recognized the difficulties of operating at nearly 14,000ft elevation, where the oxygen concentration in the air is only 60 per cent of that at sea level. Astronomers and technicians have learned to accommodate to their reduced mental and physical capacity at the summit, and to carry out the scientific work with relatively little impediment.

The 88in reflecting telescope was built according to the Ritchey-Chrétien optical design to give a wide field at the CASSEGRAIN focus (f10). Various spectrographs and imaging instruments, photometers, and cameras are used in this focal position just behind the primary mirror. A CHARGE-COUPLED DEVICE (CCD) is the most commonly used detector because of its high sensitivity, linearity, and ease of use. Images, either direct or spectroscopic, are recorded in digital form on magnetic tape for subsequent analysis. The telescope also has an infrared Cassegrain focus (f35) and uses a small secondary mirror, which is oscillated over a small angle to modulate the starlight falling on the detector, thereby making it possible to detect infrared radiation that is a million times weaker than the thermal radiation of the sky itself. This special configuration is used

Mauna Kea Observatory: exceptional site

on other telescopes at Mauna Kea as well. A COUDÉ focus position is also available, primarily to feed the large spectrograph, which has a system of five individual cameras for light analysis in which photographic emulsions as well as electronic detectors such as the CCD are used.

Recognizing the exceptional qualities of Mauna Kea ias a site for modern large telescopes, various other nations and scientific establishments have sought to locate telescopes there. In 1979, three major new telescopes were put into operation at the summit.

The United Kingdom Infrared Telescope (UKIRT) has a primary mirror of 150in (3.8m) diameter, and was built to a specialized design to facilitate infrared observations with the most sensitive modern detectors. The Cassegrain secondary mirror oscillates on a special mechanism to permit electronic cancellation of the background thermal radiation. A collection of specialized instruments consisting of cooled detectors and devices for spectrum analysis is available, and is upgraded as new techniques and detectors become available. At 6.6 tonnes, the telescope's main mirror is especially light in weight for its great diameter. The Cassegrain focal ratio is f35.4, and the telescope can also be used in a coudé mode at f20.1. The UKIRT dome is some 60ft (18.3m) in diameter, and the building includes shops, and a control room where the computers are kept and the observers and telescope operators work. The UKIRT is operated by the Royal Observatory, Edinburgh, on behalf of the United Kingdom Science and Engineering Research Council. It is used chiefly by scientists from the United Kingdom and Holland in a co-operative arrangement, and astronomers from the University of Hawaii are also awarded observing time. The telescope's main work is in the measurement of infrared radiation from other galaxies, our Milky Way Galaxy, star-forming regions, and bodies in the Solar System.

The Infrared Telescope Facility (IRTF) of the US National Aeronautics and Space Administration was also dedicated in 1979. Also built to a special design for infrared work, the IRTF has a primary mirror 120in (3m) in diameter, and an oscillating secondary mirror as described above. Its focal ratio is f35 at the Cassegrain focus, but it can also be used at the coudé focus of f120. The IRTF has an especially large and rigid mounting which permits very precise setting on celestial objects. The scientific programmes are very similar to those at the UKIRT, but the IRTF is operated as a national facility for United States and other astronomers by the University of Hawaii's Institute for Astronomy in Honolulu.

The 142in (3.6m) Canada-France-

Hawaii Telescope (CFHT) was also commissioned in 1979. It is a highly versatile optical and infrared telescope with an exceptionally wide field of view and high-quality optical system. It is the major national telescope for scientists from France and Canada, and is also used in part by astronomers of the University of Hawaii. The Cassegrain focus operates at f8 and is equipped with modern photometric and spectroscopic equipment. At the prime focus configuration, working at f3.8, exceedingly faint galaxies and stars can be studied with highly sensitive instrumentation. A coudé focal system (f20) can also be used to feed a large spectrograph and other instruments.

Other major telescopes at Mauna Kea include a new United Kingdom/Netherlands millimetre wave telescope named after James Clerk Maxwell (JCMT). It has an antenna of diameter 49ft (15m) to collect weak millimetre wave radiation from galaxies, stars, and clouds of gas and dust in our Galaxy. The JCMT will be operated remotely by astronomers in Europe using satellite links to Hawaii, but a technical staff at Mauna Kea will maintain the equipment and ready it for use by the astronomers, who can do much of their work from their home institutions halfway around the world.

For observations in the far infrared and submillimetre wavelengths, the California Institute of Technology has built a 33ft (10m) special telescope with a mirror made of epoxy resin.

The new telescope works in the spectral region between those of the infrared telescopes (UKIRT and IRTF) and the JCMT, and makes observations important in studies of emission from star-forming gas clouds and from other galaxies.

Contruction began in 1986 on the W.M.Keck telescope at Mauna Kea. This optical instrument will be the largest telescope in the world, with a mirror of 33ft (10m) effective diameter. The California Institute of Technology and the University of California have joined to design, build, and operate this facility, which is expected to become operational in 1992. The unique primary mirror will consist of a mosaic of 36 hexagonal segments, each 6ft (1.8m) wide and 3in (7.5cm) thick, and be controlled by a computer system to adjust each segment some 100 times per second to combine the light to form a single image of a distant star or galaxy.

DPC. *See also* INFRARED ASTRONOMY, TELESCOPES.

Maunder Minimum In 1890, E.W. Maunder, studying the number of sunspots recorded during nearly 300 years of observations, discovered that very few sunspots had been recorded during the period 1645-1715. This could have been the result of a scarcity of observations during those years, but in 1976 J.Eddy obtained confirmation of the low activity of the Sun over that period by studying old records of aurorae and the amount of ^{14}C in old tree rings. The period 1645-1715 is called the Maunder Minimum and coincides with years of sustained low temperatures in the terrestrial northern hemisphere known as the Little Ice Age. *VB.*

Maurolycus A lunar crater 65 miles (105km) in diameter, located at 42°S, 14°E in the southern highlands.

Maury, Antonia C. (1866–1952). American astronomer (Harvard), niece of Henry DRAPER. She developed a new stellar spectral classification for blue stars (1897), later explained by HERTZSPRUNG.

Max Planck Institute for Radio Astronomy To increase the scope of Bonn Observatory (West Germany), plans for a radio telescope were initiated in 1951. In 1956 an 82ft (25m) instrument was set up and provided the basis for the establishment of the present institute, sited close to Bonn. In 1972 came the 328ft (100m) "dish" at Effelsberg, 25 miles (40km) south-west of the city, which remains the largest fully-steerable instrument of its kind. It is now used in an inter-continental interferometer network, and the institute is also associated with institutes in Sweden, the USA, the USSR and Britain.

Maxwell, James Clerk (1831–79). Scottish physicist who unified the forces of electricity and magnetism and showed that light is a form of ELECTROMAGNETIC RADIATION. He also contributed to developing the kinetic theory of gases.

Maxwell was appointed the first professor of experimental physics at Cambridge in 1871 and directed the organization of the Cavendish Laboratory. One of his early papers (1859) demonstrated mathematically that Saturn's rings must consist of small solid, particles.

James Maxwell: electricity and magnetism

Maxwell Montes The highest mountains on Venus; 61°–67°N, 355°–10°W. They occupy the eastern end of Ishtar Terra and rise to 6.8 miles (11km) above the mean level of the planet (5 miles/9.2km above the adjacent plateau). They show no sign of active vulcanism.

Mayer, Christian (1719–83). Jesuit who became an astronomer and director of observatories at Schwetzingen and Mannheim. He was the pioneer observer of DOUBLE STARS.

Mayer, Johann Tobias (1723–62). Mapmaker and astronomer who produced important tables from his careful observations of the Moon and stars to help in determining longitudes.

Mean Sun A fictitious sun moving along the CELESTIAL EQUATOR at a constant speed for timekeeping purposes. *See also* EQUATION OF TIME.

Méchain, Pierre (1744–1805). French astronomer, best known for his work on comets. He discovered eight comets between 1781 and 1799, including the famous ENCKE'S COMET, which has the shortest known period (3.3 years), in 1786. He also discovered a comet in 1790, but was again unlucky not to have it named after him. The same comet was recovered by H.Tuttle in 1858, and it was thereafter known as TUTTLE'S COMET.

Megaparsec One million PARSECS. Abbreviation, Mpc.

Megrez The star Delta Ursae Majoris in the Plough or Dipper. It is now of mag-

M

M

nitude 3.3, but ancient astronomers ranked it as of magnitude 2. However, there is little chance that it has really faded; its spectral type is A3.

Melpomene (Asteroid 18) discovered in 1852; diameter 93 miles (150km). There have been suggestions that it is attended by a smaller body which could be classed as a satellite, but the evidence is not conclusive.

Mendeleev Lunar enclosure; 6°N, 141°E. It is on the far side of the Moon, and therefore invisible from Earth.

Menelaus Brilliant lunar crater, 20 miles (32km) in diameter, in the Haemus Mountains; 16°N, 16°E.

Menkar The star Alpha Ceti, in the Whale's "head"; magnitude 2.53. It is of type M2, and obviously orange-red.

Mensa (the Table). A far southern constellation, with no star above the fifth magnitude. A part of the Large Cloud of Magellan extends into it.

Menzel, Donald Howard (1901–76). American astronomer, who made notable contributions to ASTROPHYSICS and, in particular, to studies of the Sun.

Merak The star Beta Ursae Majoris, the southern, and fainter, of the Pointers to the Pole Star.

Mercator projection Map projection showing meridians as parallel lines, with latitude scale increasing as secant of the latitude. Used for navigation because compass directions appear as straight lines.

Mercury The innermost densest planet, orbiting the Sun in 87.969 days and revolving about its axis in two-thirds of the orbital period, 58.646 days. The PRECESSION of its APSIDES could not be entirely explained by Newtonian theory and the anomalous rotation of 43 sec of arc per century was explained

Mercury: innermost and densest planet

by Einstein's General Theory of RELATIVITY. Its radius is 1,515 miles (2,438km) and its density of 5.4gm/cm³ can be explained only by a fivefold fractionation of iron over silicate compared to the Earth. Its core is taken to be about 1,118 miles (1,800km) in radius. The magnetic field of 300nT is approximately a DIPOLE aligned along the axis of rotation and is attributed to dynamo action if the core is assumed fluid; but, if not, it could just arise, alone among the terrestrial planets, in the remanent magnetization of the crust acquired from primeval internal or external magnetic fields.

The slow rotation (*cf* Venus) is assumed to result from the slowing down by the tidal action of the Sun from a state of rotation more comparable with the other planets. The two-thirds resonance requires, however, the "trapping" of Mercury in this state and a non-hydrostatic bulge of perhaps 10^{-5} is inferred, probably caused by convection in its mantle occurring by solid state creep.

Only a part of the surface has been imaged, by the MARINER 10 flyby, but like the Moon and many other very different bodies in the Solar System, eg, the satellites of the major planets, Mercury's surface has preserved the bombardment by planetesimals and it must be inferred that the age of the surface is similar to the Moon's.

One basin, CALORIS, seems likely to be due to an impact by a body similar to those forming the multi-ring basins (circular maria) on the Moon: perhaps a satellite whose orbit decayed. No strike slip faults are seen on Mercury: no plate tectonics took place, but compressional features are seen which could be explained by a contraction due to cooling since primeval times. Estimates of the contractions required are about one part in 1,000 or 10,000 and would be consistent with temperature decreases of a few hundred degrees. Solidification of a once-fluid core would be very effective. *SKR.*

Mercury Project Series of one-man spacecraft in which Americans first gained experience of space flight. First flights by modified Redstone boosters were sub-orbital: Alan SHEPARD, May 5 1961, reached an altitude of 116 miles (186km) and landed 297 miles (478km) downrange; Virgil Grissom, July 21 1961, max. alt./distance 118 miles (196km)/303 miles (487km).

Orbital missions by Atlas D boosters: John GLENN, February 20 1962, 3 Earth revolutions, 4h 55m 23s; Scott

Carpenter, May 24 1962, 3 revolutions, 4th 56m 5s; Walter Schirra, Oct 3 1962, 6 revolutions, 9h 13m 11s; Gordon Cooper, May 15–16 1963, 22 revolutions, 34h 19m 49s. *KWG.*

Mercury Project: space flight experience

Meridiani Terra Dark Martian feature; 5°S, equatorial. It is easily seen with a small telescope when Mars is well placed.

Meroe Patera Volcanic feature on Mars; 7°N, 291°W.

Merope The star 23 Tauri, in the PLEIADES; magnitude 4.18.

Mesopotamian astronomy Unlike the Egyptians, the Babylonians developed methods of calculating lunar and planetary movements. From 700BC or thereabouts they were able to draw up good calendars; they left eclipse records, and made important advances in mathematics.

Mesosiderites Shock-produced stony-iron meteorites.

Mesosphere The layer in the Earth's atmosphere directly above the stratosphere, in which the temperature falls with height to reach a minimum of about −90°C at the mesopause, at an altitude of about 50 miles (85km). Thereafter, the temperature rises again. Within the mesosphere, the heat contributed by absorption of solar ultra-violet radiation by the ozone becomes less plentiful. The upper part of the mesosphere merges into the thermosphere, the lowest region of the IONOSPHERE.

Messier Catalogue Catalogue containing a list of galaxies, globular clusters, open clusters, planetary nebulae, and a supernova remnant compiled by Charles MESSIER in the latter part of the 18th century.

Messier, searching for comets, was irritated by the appearance of fuzzy objects that could be confused with them. These objects were usually galaxies or clusters, but there was no atlas or catalogue of objects likely to be mis-identified as comets. Messier decided to compile his own so that he, and others, would not waste time and effort trying to confirm them as comets.

The first entry, M1, is dated September 12 1758. All objects listed in the catalogue are prefixed by the letter "M", which denotes their compiler, Messier. His instruments were very poor by today's standards, the best probably as effective as a modern 4in (10cm) refractor; therefore all the objects are relatively bright, and can be observed with amateur equipment. Some objects, such as the ANDROMEDA GALAXY (M31), and the PLEIADES (M45), can be seen with the naked eye. There are other objects, of low surface brightness, that can challenge the visual observer: M74, a spiral galaxy in Pisces, is an example.

The first edition of the catalogue was published in 1774, with supplements in 1780 and 1781. Not all the objects in the catalogue were discovered by Messier, neither did he claim them to be. Included are some discovered by HEVELIUS, HALLEY, LACAILLE and his friend and rival comet hunter Pierre MÉCHAIN, who willingly contributed 32. Today Messier's catalogue of objects totals 110 but there are mistakes and omissions, reducing the number to 105. The incorrectly catalogued objects are M40, M47, M48, M91 and M102. *RWA*[1].

Messier, Charles (1730–1817). French astronomer noted for his catalogue of deep-sky objects.

Messier arrived in Paris at the age of 21 to take up an appointment as a draughtsman and recorder of astronomical observations. He was also trained as an observer and moved to the Marine Observatory, where the astronomer in charge, Delisle, employed him to recover HALLEY'S COMET. He was beaten by the amateur astronomer Palitzsch after searching for over a year. Messier discovered his first comet in 1764 and virtually all comets for the next 15 years, which earned him the title "ferret of the comets" by King Louis XV.

Although remembered for his catalogue of deep-sky objects, Messier should also be remembered as a very successful comet hunter. He discovered a total of 15 comets and was the co-discoverer of six more. *RWA*[1]. *See also* MESSIER CATALOGUE.

Me stars Stars basically of type M but with hydrogen EMISSION LINES in their spectra.

Metallic content in stars Astronomers use the word "metal" to describe all elements heavier than helium. Thus the metal content of a star, denoted by Z, is the mass fraction of all elements heavier than helium combined. Mass fractions of hydrogen and helium are denoted by X and Y respectively, thus a typical star may have X = 0.75, Y = 0.24 and Z = 0.01. While the values of X and Y are relatively constant from one unevolved star to another, Z changes dramatically over a range approximately 10^{-4} to 0.02. Indeed, stars are postulated (Pop III stars) with Z = 0, although these have not been discovered.

The relative abundances of the elements are usually quoted by the logarithm of their number density relative to hydrogen set arbitrarily at 12.0. The following are the figures for the solar abundances for the more common elements; these probably reflect the relative abundance of the heavy elements on a cosmic scale: H (12.00). He (10.93), O (8.82), C (8.52), N (7.96), Ne (7.92), Fe (7.60), Si (7.52), Mg (7.42) S (7.20).

The abundance of elements in a star is measured from the ABSORPTION LINES in its spectrum. These lines are formed above the photosphere so that the chemical composition deduced only refers to the outer atmosphere of the star, although this should reflect the original composition of the whole star before changes in its centre occur due to "nuclear burning". Elements may be recognized by the lines from both neutral atoms and singly or multiply ionized atoms. Clearly, the relative abundance of atoms to ions varies with the gas temperature. Furthermore, the strength of absorption lines depends on the population of the electron energy levels in the atom/ion, which again is temperature dependent. Also, the gas density must be known; thus to deduce the abundance of an element relative to hydrogen it is necessary to make a complete model of its outer atmosphere. A complete, or near complete, abundance analysis is an immense task which has been performed for only a few stars, most notably the Sun.

For stars other than the Sun the abundances of the elements are not generally so well known or easily measured. Thus [Fe/H], the iron to hydrogen ratio, is often used to measure the "metallicity" (total abundance of elements heavier than helium) of stars; it being assumed that the other heavy elements will be in the same proportion to iron as found in the Sun. Formally, [Fe/H] is $\log (N_{Fe}/N_H)(\text{star}) - \log (N_{Fe}/N_H)(\text{Sun})$.

The reason for the large variation in metal content, Z, of stars is that the heavy elements are not primordial. Only hydrogen and helium (and very small quantities of lithium) are formed in the BIG BANG. All other elements are synthesized in the nuclear reactions that take place in the centres of stars. This processed material may eventually be returned to the interstellar medium by stellar mass loss, of which the most dramatic example is the SUPERNOVA explosion. Thus the interstellar medium is continually being enriched with "metals". It also follows that young stars that have recently formed from the interstellar medium will have large Z. Thus POPULATION I stars, which are young, have large Z whereas POPULATION II stars, which formed first and are now old, have low Z values. The cosmic lithium abundance is an important indicator of the nature of the early universe. Unfortunately, there are many reasons why stellar lithium abundances will not accurately reflect primordial lithium abundances.

The metallicity Z has important consequences for a star. There are two basic effects. The most important effect is that a larger value of Z gives rise to a greater opacity of the stellar gas. The increase in opacity is due to the many possible lines, free-bound transitions and free electrons caused by ionized and partially-ionized heavier elements. This will radically effect the structure of a star at any stage and also its evolution. Probably the most striking example of this is that Population II MAIN SEQUENCE stars (low Z) are both hotter and more luminous than Population I stars of the same mass. Thus the Population II main sequence lies leftward of the Population I main sequence in the HERTZSPRUNG-RUSSELL DIAGRAM. Furthermore, Population II stars will spend a shorter time on the main sequence than their Population I counterparts. Thus a good knowledge of Z is essential for judging the age of star clusters from the main sequence. Likewise, Z must be well known if the main sequences of two clusters are to

M

M

be compared for the purpose of measuring cluster distances. A similar effect occurs on the horizontal branch in the H-R diagram. Thus low metallicity RR LYRÆ stars are thought to be intrinsically brighter than high-metallicity RR Lyrae stars. There is some controversy about this, but given the fundamental importance of RR Lyrae stars for distance measurement, metallicity is clearly important.

The original metallicity of a star can also affect its structure and evolution by modifying the nuclear burning in its centre. The most obvious example of this is that the abundance of carbon and nitrogen will affect the "carbon cycle" for hydrogen burning in the more massive main sequence stars.

It must also be mentioned that certain stars have particular elements or groups of elements that are relatively either over- or under-abundant. For example, Ap stars show over-abundances of manganese or europium, chromium and strontium. This is thought to be caused by some diffusion process which brings these elements up from the interior. Similarly the Am stars show a slight over-abundance of elements near iron and an under-abundance of calcium and scandium. Again, a diffusion mechanism is thought to operate.

Amongst the cool giant stars, C (carbon) stars show carbon to oxygen ratio four times that of normal stars. S stars show an over-abundance of zirconium, ythrium, barium and even technetium, which has a half-life of only 2×10^6 years. Some G and K giants also show over-abundance of barium, strontium and other elements produced by the S-process as well as over-abundant carbon. These stars are called BARIUM STARS. All these giant stars are long-period irregular variables and it is possible that their abundance anomalies are due to mixing with material in the stellar core.

Many late-type stars have atmospheres that are sufficiently cool to allow the metals to form molecules. The molecules commonly recognized in such stars are CO, H_2O, TiO and the CN radical. *RFJ. See also* entries on particular star types.

Metal-poor stars Stars with low abundances of elements heavier than helium. In general they are POPULATION II stars. The term is more meaningful when applied to a specific group of stars, such as RR Lyræ stars, whose average metallicity is known. *See also* METALLIC CONTENT IN STARS.

Metal-rich stars Stars that are the opposite to METAL-POOR STARS and are generally POPULATION I stars. *See also* METALLIC CONTENT IN STARS.

Metastable state A relatively long-lived excited state of an atom. In the rarefied interstellar medium an electron can remain in such a state for its natural lifetime, but under terrestrial conditions, atoms are very soon knocked out of metastable states by frequent collisons. *See also* EXCITE and FORBIDDEN LINES.

Meteor The brief streak of light seen in a clear night sky, when a small particle of interplanetary dust, called a meteoroid, burns itself out in the Earth's upper atmosphere. The meteoroid is drawn into the upper layers of the atmosphere under the influence of the Earth's gravitational pull. As the meteoroid collides with air molecules in the

Meteor: the Leonid shower of 1966

atmosphere at tremendous speed, a large quantity of heat energy is produced which normally vaporizes the particle completely by a process of ABLATION. The vaporized atoms from the ablating meteoroid make further collisions with air molecules, and first excitation (*see* EXCITE) and then IONIZATION occurs as electrons are stripped from the atoms and molecules by the energy of the collisions. After ten or so collisions, the energy of an ablated atom has dropped below the level at which it can cause further excitation or ionization.

The ablating meteoroid leaves behind it a trail of highly excited atoms that then de-excite to produce the streak of light, which we call a visual meteor. The ionization results in a long trail of positively-charged ions and negatively-charged free electrons (both meteoric and atmospheric) also being formed behind the meteoroid. This can scatter or reflect radio waves transmitted from ground-based radio equipment, causing a radio meteor.

The meteoroid starts to ablate at a height of about 70 miles (115km). It then dissipates its energy, distributes its ablated atoms and molecules in the atmosphere, and produces the visual and radio meteors in the meteor region between about 40 and 70 miles (70 and 115km) above the Earth's surface. A typical meteor reaches its maximum brightness and greatest electron density at an altitude of about 60 miles (95km). Usually, a visual meteor will persist for between one-tenth and eight-tenths of a second. Most of the visible light from a meteor comes from the region immediately surrounding the ablating meteoroid. Sometimes a brighter meteor leaves a faintly glowing trail of luminosity along its path. This is called a "wake" if it lasts for less than a second, and a "persistent train" if the duration is longer. Occasionally a meteor will display sudden bursts of light during its path; these are called flares.

Meteors may be seen at any time of the year, and it is often quoted that on a clear, moonless night a casual naked-eye observer may see between six and eight of these so-called sporadic meteors every hour. The eye will detect meteors down to about fifth visual magnitude, with the most probable magnitude being $+2.5$, and over three-quarters of the meteors observed having magnitudes in the range 3.75 to 0.75. However, the number of sporadic meteors visible per hour varies not only during the course of a single night, but also throughout the year. In the early evening, an observer is on the trailing side of the Earth as it moves along its orbit around the Sun at a velocity of about 18 miles (30km) per second. This means that the meteors seen are catching up the Earth from behind, so the velocity with which they enter the atmosphere is reduced, and less energy is given out. Later in the night, during the early morning hours, an observer is on the leading side of the Earth, and any meteors seen are meeting the Earth more or less "head-on". In this case their velocity of entry into the atmosphere will be correspondingly higher, and considerably more energy will be released. The consequence of the Earth's motion is that more meteors render themselves visible to the naked eye in the early morning than in the evening. Hence, a visual observer may see only six sporadic meteors per hour (6m/h) in the early evening, but this number rises steadily during the night, reaching a peak of about 14m/h at around 0400 hours local time. In addition to this diurnal variation in meteor rates, they also vary on an annual basis. Sporadic meteor rates are generally rather higher in August and September than they are in February and March, but there are short-term fluctuations superimposed on this annual cycle.

Meteroids enter the atmosphere at a velocity somewhere between 6.9 and 45 miles per second (11.2 and 72km/s). The lowest velocity is just the free-fall velocity of a particle hitting the Earth (assuming it started with zero velocity relative to the Earth at an infinite distance away). The greatest velocity of the incoming meteoroid is obtained by summing its maximum velocity relative to the Sun at a distance of 1AU from the Sun — namely, 26 miles (42km) per second — with the Earth's mean velocity around its orbit, which is 18 miles (30km) per second.

A large percentage of the meteoroid influx to the Earth's atmosphere (some 16,000 tonnes per year) is made up of particles in the size range that produces visual and radio meteors, with masses in the range from 10^{-6} to 10^{+6} grammes. Although sporadic meteors occur continuously over the entire Earth's surface (day and night), most people are aware that at certain times of the year meteors seem to be more numerous than normal. These are the occasions of the well-known meteor showers. They are caused as the Earth passes through a narrow region where the meteoroids are far more concentrated. This is called a meteor stream, and is due to the clouds of meteoric dust strewn along the orbit of a decaying periodic comet. Dust particles are pushed away from the cometary nucleus by gas pressure, when the nucleus is near perihelion. The ejected particles then have slightly different orbits to that of the parent comet. They form a dense cloud slightly ahead of, and also slightly behind, the comet's nucleus. This is the first stage of formation of a meteor stream, where the cloud of particles extends around only a very small fraction of the comet's orbit. If the Earth's orbit intersects that of the parent comet, or passes near to it at a distance of not more than about 0.1AU (usually near the ascending or descending node of the comet's orbit), then a very strong meteor shower will be seen, but only very infrequently. The particles are distributed only across a narrow region in space, and although meteor rates may be exceptionally high, the resulting meteor "storm" will be of very short duration — probably lasting only a few hours. Furthermore,

there may be many years, between the major storms, when few, if any, meteors are seen.

As a meteor stream gets older, various processes cause the meteoroid debris to spread out and extend all around the orbit of the parent comet. The time taken for this to occur is called the loop formation time, and depending upon the size of the particle and orbit of the comet, may take anything from only a few tens of years for a really short-period comet, up to several thousand years in the case of a parent comet of long period. The meteor stream will still be rather narrow, although it may be up to ten times broader at aphelion than it is at perihelion. The shower duration is still likely to be fairly short, but meteors will be visible every year, although rates may not be particularly constant from one year to the next. The older a meteor stream becomes, the more broadening occurs, due to collisions between the particles within the stream. This occurs mainly near perihelion, where the stream density is highest. Planetary perturbations will also tend to cause broadening of the stream, and in some cases particles may be lost from the stream altogether. There may also be some segregation of the different masses within a stream, as the smaller dust particles slowly spiral in toward the Sun by virtue of a phenomenon called the Poynting-Robertson effect. This effect is greatest for the very small dust grains and is negligible for the largest ones. The oldest meteor streams will be the widest, with the result that

the shower duration is a week or more. Meteor rates will tend to be reasonably constant from year to year.

All the meteors of a particular shower will seem to come from one particular region of the sky, known as the radiant. Although all the meteors within a stream are moving along parallel paths, it is an effect of perspective that makes them appear to radiate out in all directions from one point in the sky. Imagine you are standing on a bridge which crosses over a long, straight stretch of motorway. The parallel lanes of traffic seem to "radiate" from a distant point near the horizon. It is essentially the same with the meteors of a shower. The meteor showers are named according to the constellation which contains the radiant. The exception is the January shower, the Quadrantids, which come from an area of the now-rejected constellation Quadrans Muralis, near Boötes, the Herdsman. Nowadays, most of the major annual meteor streams are associated with specific periodic comets. The two exceptions are the Quadrantids, where it is believed that the parent comet no longer exists, and the Geminids (December), where no parent comet has been found, but an Earth-crossing asteroid, 3200 PHAETHON, appears to have an orbit that is identical with that of the Geminid meteor stream particles, and is almost certainly associated with it. The accompanying table lists some of the more important annual meteor showers, the listed Zenithal Hourly Rates (ZHR) are not constant,

M

Some important annual meteor showers and cometary associations

Stream	Maximum	Normal limits	ZHR	Radiant R.A. Hrs.	Radiant R.A. Mins.	Radiant Dec. (degrees)	Parent comet
Quadrantids	Jan 3/4	Jan 1 – Jan 6	80	15	28	+50	No known parent comet
April Lyrids	Apr 21/22	Apr 19 – Apr 25	15	18	08	+32	Thatcher 1861 I
*Pi-Puppids	Apr 24/25	Apr 21 – Apr 26	40?	07	18	−44	P/Grigg-Skjellerup
Eta Aquarids	May 5/6	Apr 24 – May 20	35	22	20	−01	P/Halley
Daytime Beta-Taurids	Jun 29/30	Jun 23 – Jly 05	20	03	40	+15	P/Encke
Alpha-Capricornids	Aug 2/3	Jly 15 – Aug 25	8	20	36	−10	P/Honda-Mrkos-Pajdusaková
Perseids	Aug 11/12	Jly 25 – Aug 21	75	03	04	+58	P/Swift-Tuttle
*October Draconids	Oct 8/9	Oct 7 – Oct 10	high?	17	23	+57	P/Giacobini-Zinner
Orionids	Oct 21/22	Oct 16 – Oct 30	25	06	24	+15	P/Halley
Taurids	Nov 4/5	Oct 20 – Nov 30	12	03	44	+14	
				03	44	+22	P/Encke
*Leonids	Nov 17/18	Nov 15 – Nov 20	10?	10	08	+22	P/Tempel-Tuttle
Geminids	Dec 13/14	Dec 8 – Dec 16	60	07	28	+32	No known parent comet; asteroid 3200 Phaethon?
Ursids	Dec 21/22	Dec 17 – Dec 25	5?	14	28	+78	P/Tuttle

Notes: In addition to the Daytime Beta-Taurids, there are two other permanent daytime streams; the Arietids, active from May 29 to June 17 and the Xi Perseids, active from June 1 to June 15.

*The Pi-Puppids, October Draconids and Leonids are periodic streams and only give good displays when the parent comets return to perihelion.

but act as a reasonable guide to activity as seen by a visual observer. *JWM*.

Meteor Crater Impact crater located between Flagstaff and Winslow, Arizona, USA. It is some 3,940ft (1,200m) in diameter. *See* ARIZONA CRATER.

Meteorite A natural object that survives its fall to Earth from space, and is named after the place where it was seen to fall, or was found.

When an object enters the atmosphere, its velocity is greater than the escape velocity, 7mps (11.2km/s), and unless it is very small (*see* MICRO-METEORITES) frictional heating produces a fireball. Fireballs may rival the Sun in brightness. For example, a brilliant fireball that occurred on June 25 1890, at 1pm, was visible over a large area of the mid-west of the United States; the CHONDRITE fall at Farmington, Kansas, was the result. If an object (meteoroid) enters the atmosphere at a low angle, deceleration in the thin upper atmosphere may take tens of seconds. The fireball of April 25 1969 travelled from south-east to north-west and was visible along its 310-mile (500km) trajectory from much of England, Wales and Ireland. As commonly occurs, towards the end of its path the fireball fragmented. Sonic booms were heard after its passage, and two meteoritic stones were recovered, some 37 miles (60km) apart, the larger at Bovedy, Northern Ireland, which gives its name to the fall. Fragmentation in the atmosphere may produce a meteorite shower; one that fell at Pultusk, Poland, in 1868, is estimated to have had a total weight of 2 tonnes among some 180,000 individual stones. Large meteoroids of over about 100 tonnes that do not break up in the atmosphere are not completely decelerated before impact. On striking the surface at hypersonic velocity, their kinetic energy is released, causing them to vaporize and produce explosion craters, such as the ARIZONA CRATER.

Photographic observations of fireballs indicate that more than 19,000 meteorites heavier than 3.5oz (100g) land annually, but of these most fall in the oceans or deserts and fewer than 10 become known to science.

Most meteoroids, however, are composed of low-density, friable material that does not survive to the surface; it may be of cometary origin. In contrast, photographic observations and visual sightings of meteorite-producing fireballs show that they have orbits similar to those of Earth-crossing (Apollo)

asteroids. It is certain, then, that most meteorites are pieces broken from asteroids, but of the 2,800 identified falls, four are from the Moon, eight may be from Mars, and some may be cometary.

Material from those 2,800 different meteorite falls is preserved in collections. Some 95 per cent of meteorites seen to fall are composed dominantly of stony minerals and hence are known as stony meteorites. Iron meteorites constitute the bulk of the remainder, while meteorites composed of 50-50 mixtures of iron-nickel metal and stony material — the stony-irons — are very rarely seen to fall. However, many

Meteorite: origin uncertain

more iron and stony-iron meteorites have been found than were observed to fall, which merely reflects their resistance to erosion and their distinctive appearance relative to terrestrial rocks rather than a change in the composition of the meteorite flux with time. Since 1969 meteorites have been found in large numbers on the surface of the ice in parts of Antarctica. The small number of Antarctic iron meteorites relative to stony types is similar to the ratio in observed falls.

Meteorites are classified on various chemical criteria. Stony meteorites are subdivided into the chondrites — metal-bearing, unreprocessed, with three major groups, further divided into nine (EH, EL; H, L, LL; CI, CM, CV, CO) — and the ACHONDRITES — metal-free, derived directly or indirectly from planetary volcanic melting. Iron meteorites comprise 13 chemical groups, plus other apparently unrelated ones, and there are two groups of stony-irons. From the spectrum of their chemical properties, meteorites in collections must be representative of more than 20 parent-bodies.

Although frictional heating during atmospheric flight causes the outside of a meteoroid to melt, the molten material is swept into the atmosphere as droplets. The bulk of the heat is

removed with the melt and the inside of the object stays cold. Only the melt at the last second of hypersonic flight solidifies on the surface as the potential meteorite falls to Earth under gravity. The solidified melt is known as fusion crust. On most stony meteorites it is dull black, but on many achondrites it is shiny black. Iron-nickel metal conducts heat more efficiently than stone, and some of the heat generated in atmospheric flight may penetrate to the interior of an iron meteorite. Stony meteorites, however, preserve a record of their history before their encounter with our planet. Most preserve evidence of conditions in the Solar System of 4,550 million years ago, and none has an age or isotopic signature consistent with an origin outside it.

Various "ages" can be measured.

Age of formation Most meteorites, or their components such as chondrules, went through a high-temperature event early in their history. The age of formation is the time, to the present, since a meteorite first cooled to become a closed chemical system. For example, uranium decays to lead at a fixed rate, and the uranium-lead age of a meteorite is the time that has elapsed since the uranium and lead were able to exchange freely, when the body was hot. The lead formed from the decay of uranium can be measured; the quantity is proportional to the uranium content, also measured, and to time, which usually is close to 4,550 million years.

Exposure age When an object is broken from its parent asteroid it continues to orbit around the Sun. Cosmic rays from the Sun, and beyond, bombard its surface. The exposure age is the time during which this bombardment takes place. Radiation damage may be measured in various ways, such as the content of a substance produced in nuclear reactions, perhaps a radioactive isotope of aluminium. (The levels of radioactivity in meteorites are so low that specially prepared, ultrasensitive counting equipment is required for their measurement.) Exposure ages range from a few hundred thousand years for some stony meteorites to 1,000 million years for a few irons. These reflect the susceptibility of stony types to erosion by impact in space, and the durability of iron-nickel metal.

Terrestrial age The time since a meteorite landed on Earth. The oldest meteorite known is a chondrite within a Swedish Ordovician limestone 460 million years of age; the second oldest may be an iron meteorite found in coal in the Soviet Union, and about 300 million years old. Apart from these, most meteorites are much younger. Some from Antarctica have lain in the ice for 700,000 years, but these are exceptional.

Meteorites often record shock or thermal events when they were part of their parent bodies. For example, many L-group chondrites were shock-reheated 500–1,000 million years ago; these include the Bovedy meteorite. The chemical elements heavier than hydrogen and helium are synthesized in stars, and many meteorites preserve a record of these processes. The chondrites often contain the decay products of plutonium, indicating that this element was present in the matter from which the Solar System formed. From the quantity of plutonium that must have been present relative to other elemental abundances the plutonium must have been formed within about 200 million years of the formation of the Solar System. This period between stellar processing and the incorporation of an element into a meteorite is often known as the formation interval. Meteorites, therefore, provide many important clues to the origin and history of the Solar System; they also contain records of conditions in interplanetary space. *RH.*

Meteoroids Potential meteorites; natural objects in Earth-crossing orbits around the Sun, that may land on Earth. Most are cometary and burn up; 100,000 tonnes fall annually.

Meteosat Operational geostationary meteorological/climatology satellite system managed by ESA on behalf of Eumetsat and located at a nominal 0° longitude. Meteosat 1 was launched in November 1977, Meteosat 2 in June 1981 — with three further satellites to be launched in the period to 1990.

Meteosat: Europe, January 4 1979

Methane (CH_4). A hydrocarbon which is a constituent of the ATMOSPHERES of Jupiter and Saturn. It was identified in interstellar space in 1977 by its emissions at a wavelength of 3.9mm.

Metis (Asteroid 9). Discovered in 1848; it is 93 miles (150km) in diameter. Magnitude 9.1.

Metonic cycle An interval of 19 years after which the phases of the Moon recur on the same days of the year. This occurs because 19 years (6939.60 days) is almost exactly equal to 235 lunar months (6939.69 days). Its discovery is often attributed to the Greek astronomer Meton around 433BC.

Meudon Observatory The ancient castle of Meudon, near Paris, partly destroyed, was converted by astronomer Jules JANSSEN to an astrophysical observatory in 1876. A large refractor of 32in (83cm) aperture and a 39in (100cm) reflector were commissioned in 1893. Emphasis was given to solar physics by Janssen — who discovered helium — and by Henri DESLANDRES, following as director in 1908, who invented the SPECTROHELIOGRAPH, independently of HALE, to analyse the SOLAR CHROMOSPHERE (1891), an instrument fully exploited during the following half-century by Lucien d'Azambuja. Then Bernard LYOT invented the coronograph for artificial eclipses (1930 and the birefringent filter for monochromatic imaging.

Comets and novæ (namely those of 1934 and 1936) were analyzed by Henri Deslandres and by Fernand Baldet. Accurate planetary surface observations were conducted with the large refractor from 1909 to 1939 by E.M. ANTONIADI, and Lyot analyzed the planets and the Moon by polarimetry.

After 1950, the observatory was expanded under André Danjon. The large 83cm refractor was modernized for a double-stars survey conducted by Paul Muller. Solar physics was developed by a group of scientists with Raymond Michard, and instrumented with a 23in (60cm) aperture and 118ft (36m) high solar tower (1969), a magnetograph (1962), a new birefringent filter and high-accuracy polarimeters. A radioheliograph was erected at Nançy. Solar research currently includes an active centres survey, flare theory, photospheric magnetic field analysis, chromospheric dynamics and coronal physics.

Planets and other Solar System bodies were analyzed by imaging and

polarimetry at Pic-du-Midi and with the modernized 39in (100cm) telescope at Meudon by Audouin Dollfus, than by spectroscopy and in the infrared. Use is made of spacecraft missions in the Solar System.

Radio astronomy was introduced in 1953 by J.F.Denisse, with J.E.Blum and J.L.Steinberg, for solar, stellar, galactic and extra-galactic work and now forms a large part of the observatory's activity. The observatory has a staff of more than 500 scientists, engineers and technicians; it also has a section for space research, a computing centre, and a teaching activity *ACD*

Michelson interferometer The optical interferometer invented by A.A. Michelson and used by him in 1920 for the measurement of diameters of stars.

Michelson-Morley experiment. An experiment carried out by Albert Michelson in 1881 and, with greater precision, by he and Edward Morley in 1887, to attempt to detect the motion of the Earth through the ETHER. Their apparatus split a beam from a common source into two parts, one travelling at right angles to the other. Both beams were reflected and recombined to produce an interference pattern. If one beam were to travel out and back along

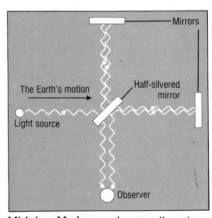

Michelson-Morley experiment: null result

the direction of the Earth's supposed motion through the ether then, if light moved through the ether at a constant speed, it should take a marginally longer time to cover the same distance as a beam travelling at right angles to the Earth's motion. Rotation of the two beams should have produced a readily detectable change in the observed interference pattern. No change was detected. The null result of this experiment was explained by the Special Theory of RELATIVITY. *IKN.* *See also* FITZGERALD CONTRACTION.

M

M

Micrometer Device used to measure the angular separation of celestial objects such as binary stars, the diameter of extended objects such as planets, or features on planets or comets.

In principle, early micrometers used some means of projecting against the background stars a wire, grid, bar or some similar device in the focal plane of the telescope so that it was viewed through the eyepiece together with the stars to be measured.

The magnification which the eyepiece fitted to the micrometer gives when used on any particular telescope must be known so that the angular separation as shown by the micrometer can be calculated in seconds of arc.

Most micrometers can be rotated against a divided circle so that the position angle of the object can be read off. There are several types.

Binocular micrometer This instrument produces an artificial image of a pair of stars using pinholes in front of a light source, a focusing lens, mirrors or prisms and a collimating lens. The image is directed into an eyepiece.

A separate eyepiece is used by the observer to view the real pair of stars, so that each eye sees a different image.

The separation of the artificial pair is adjusted by means of the focusing lens, which is next to the pinhole and light source, until they are seen to match the real star images. The position angle is adjusted by revolving the entire instrument and reading the angle directly.

Comparison image micrometer This uses an artificial pair of stars obtained from a pinhole and a polarizing prism. By altering the distance of the prism and the pinhole the separation of the artificial images can be adjusted until they match the real binary pair. A beam splitter combines the real and artificial stars in front of the eyepiece.

Cross wire micrometer Two webs or bars are crossed at about 90° in the focal plane of the telescope. The transit times of stars at 45° to the cross and including stars of known position will enable the right ascension and declination of an object to be determined.

Double image micrometer Basically, this type of micrometer uses either a split objective lens or alternatively a split BARLOW LENS (negative lens) placed inside the focal plane of the objective. This component, divided into two halves, is mounted so that one half is movable laterally, causing a doubling of the image.

Another type uses a Wollaston (polarizing) prism fitted into the optical axis of the telescope; movement along the axis also doubles the images. The orientation of the field may be read off from a position circle and adjusted to bring the line of the separated lenses into line with a pair of double stars to be measured, so determining their positon angle relative to the north point or zero.

Objective-grating micrometer This instrument uses a coarse grating or bars. The diffraction pattern produced forms a row of bright segments which elongate the stellar images. A diaphragm, having slits which are widely spaced, enhances the effect and produces a first order false image of the stars with more images farther away. All these are on a line perpendicular to the orientation of the slits.

By ensuring that the two rows of real and false images are at right angles to each other, the position angle is read off; 0° is parallel with the CELESTIAL EQUATOR. If the grating is rotated to make the pair of double stars and their first order images form a square, then the separation can also be determined.

Reticulated micrometer This consists of a grid of parallel lines which cross each other at 90°; some types also have concentric circles. These lines are etched onto glass. In use, it is necessary to know the separation of the webs. This is determined by letting a star of known declination trail a web, and timing the interval between squares. Any two stars in the field can be so lined up on the webs that their relative positions can be estimated. Alignment should be east-west and north-south.

Ring micrometer This consists of a ring of opaque material on glass placed in the focal plane of the telescope's objective or primary mirror. The ring, whose internal dimensions against the sky have been determined, is used to time any two stars close enough in the field of view to cross the ring so that their relevant angular separations can be calculated.

Transit micrometer The use of a fixed web to time the transit of stars across the meridian in order to fix their position in right ascension can be improved by the use of a moving web, so that a number of readings can be made and the average taken over the course of a few minutes.

Repsold, in 1889, first used a travelling web, moved by a micrometer screw that kept the travelling wire on top of the star as it transited. As the screw was turned it made an electrical contact at regular intervals which was recorded against time. *NEF. See also* FILAR MICROMETER.

Micrometeorites Natural objects from space small enough to escape destruction during atmospheric flight. Meteoroids of diameter less than 200 micrometres (*see* MICRON) radiate away much frictional heat because of their large surface-to-weight ratio. They survive to the surface and are recognizable as "cosmic spherules" in slowly accumulated deep-sea sediment, or Greenland or Antarctic ice. The spherules are probably related to CM2 CARBONACEOUS CHONDRITES. Even smaller, "Brownlee particles" are completely unscathed by atmospheric entry, are often friable and porous, but may be collected by high-flying aircraft. Such interplanetary dust is our only sample of primitive material. *RH.*

Micron A unit of measurement equal to one-millionth of a metre. Strictly speaking, in the SI system of units, the term micrometre (symbol μm) should be used.

Microscopium (the Microscope). A small southern constellation, adjoining Grus. It has no star brighter than 4.6.

Microwave background The remnant electromagnetic radiation from the formation of the universe in a hot BIG BANG about 15 thousand million years ago. It is also known as the cosmic background or the 3° background. The radiation now detected corresponds to BLACKBODY RADIATION at a temperature of 3K (3° above absolute zero, equivalent to $-270°$ Centigrade) and is the redshifted (*see* REDSHIFT) remnant of the very much hotter radiation field that existed in the early universe.

The discovery of the microwave background was serendipitous. In 1964 Penzias and Wilson, at the Bell Telephone Laboratories in Holmdel, New Jersey, USA, were calibrating a microwave (7.35cm) antenna and receiver constructed for satellite communications. The system seemed to produce more random noise fluctuations than would have been expected from its design or could be accounted for as arising from the Earth, or its atmosphere, or even local pigeons! Eventually by careful work Penzias and Wilson concluded that the noise arose from space, was uniform in all directions, and corresponded to the signal strength of radiation from a blackbody at a temperature of about 3K.

Puzzled by their result they discussed it with Burke of Massachussets Institute of Technology. Burke told them of a group (Dicke, Peebles, Roll and Wilkinson) at Princeton University who had been carrying out calculations to explain the present-day observed abundance of helium in the universe in terms of a hot big bang. A prediction of their calculation was that the redshifted remnant of the radiation associated with this big bang should be observable at the present in the form of blackbody radiation at about 10K and uniformly filling the universe. The Princeton group was in the process of building an instrument to try to detect this radiation at 3cm. The background detected by Penzias and Wilson was identified with that predicted by the Princeton group.

The two groups contacted each other and simultaneously published their respective experimental results and predictions. Subsequent measurements have confirmed the blackbody nature of the spectrum. In 1978 Penzias and Wilson were awarded the Nobel Prize in Physics for their discovery.

The theoretical prediction of a microwave remnant of the initial fireball of radiation had actually first been made in the early 1940s by Alpher, Herman and Gamow, but had not been followed up at the time and had been forgotten.

Up to 1965 there were two main competing theories of COSMOLOGY: the big bang and the steady state theories. The detection of the microwave background was very strong evidence in favour of the big bang, so much so that it has now become the standard model. The steady state theory offers no immediate natural explanation of the existence of such a background.

According to the standard model the universe came into being about 15 thousand million years ago in a highly compressed and hot state (a temperature of over a million million degrees K during the first 0.1 thousandths of a second) and immediately began to expand uniformly in all directions. As a consequence of the expansion of all lengths the wavelength of electromagnetic radiation also increased (became redshifted). The temperature of a blackbody that would give rise to such radiation decreases as the wavelength of the radiation increases. Thus as the universe expands, it cools.

As the universe expands, temperatures become sufficiently low that matter becomes increasingly stable, until after three minutes the universe has cooled sufficiently (to less than a few thousand million K) for the formation of the nuclei of the major elements hydrogen and helium to have been completed. The present-day observed abundances of these elements, and of deuterium and lithium, are mostly determined in the first three minutes, with some further modifications and synthesis of other elements occurring much later during nuclear FUSION reactions in the interior of stars, in SUPERNOVÆ and in COSMIC RAY interactions.

At this stage the universe still had so much energy density that positively charged nuclei and the oppositely charged electrons were not able to be held together by electrical forces. After some half a million years the universe had cooled to a few thousand K, at which point the thermal energy no longer overcame the electrical attraction between nuclei and electrons so that the electrons and the nuclei combined.

Up to this stage the universe had been opaque to radiation because it could not travel very far before being absorbed in interactions with other radiation or with matter. In particular, before electrons and nuclei combined, free electrons had been the main particles with which the radiation interacted, providing the major source of opacity. When the electrons became attached to the nuclei they were no longer available to interact with the radiation and the universe ceased to be opaque and became transparent to the background radiation. As the radiation henceforward hardly interacted with the matter, it is said to have decoupled from it.

The background radiation continued to increase in wavelength as the universe expanded until today it peaks in the millimetre region of the spectrum and corresponds to a blackbody temperature of 3K. It is known as the microwave background as it was first detected in the form of microwaves. Because this radiation is essentially unchanged, except for being redshifted, since it decoupled from the matter it can tell us about the universe at the time of decoupling and earlier.

The microwave background radiation is found to be very isotropic (to better than one part in a thousand). This isotropy provides strong empirical evidence for one of the most important assumptions of cosmology, the "Cosmological Principle", which states that at any instant of time the universe appears, on a large-scale average, homogeneous and isotropic to all observers. In other words, from one position it appears the same in all directions, and it would also appear the same from other positions. This means the Earth is not in any special position with respect to the rest of the universe.

Although the microwave background radiation is highly isotropic there is a small (0.2 per cent) dipole anisotropy with maximum intensity toward the VIRGO CLUSTER of galaxies and a minimum in the opposite direction. This is what would be expected, from the Doppler effect, if our Galaxy is moving relative to the microwave background at a velocity of 600km/sec in the direction of the Virgo cluster.

Anisotropy on other angular scales, and small deviations from a blackbody spectrum, are not yet clearly detected but could potentially give evidence about the state of and processes in the early universe. NASA's Cosmic Background Explorer satellite (COBE) should give more detailed information about the microwave background radiation when it is launched in the late 1980s. *JPE.*

M

Microwaves Electromagnetic waves of wavelength between 1mm and 30cm, longer than optical, infrared and submillimetre waves but shorter than radio waves. The corresponding oscillation frequencies are between 300 and 1 Gigahertz one Gigahertz is a frequency of 10^9 cycles per second. Microwaves are usually detected by radio-type receivers that directly measure the wave's varying electric field. The development of microwave receivers led to discoveries such as the MICROWAVE BACKGROUND radiation and interstellar molecules.

Midas Apollo asteroid (number 1981) discovered by Charles Kowal at Palomar. The orbital period is 2.37 years and the ARGUMENT OF THE PERIHELION librates in a comet-like fashion.

Midnight Sun The Sun seen above the horizon at midnight, visible from inside

Midnight Sun: a north Norwegian view

the Arctic and Antarctic circles at different times of the year. At the North Pole the Sun is visible all the time it is north of the CELESTIAL EQUATOR; conversely at this time it is below the horizon at the South Pole.

Mie Large Martian crater, situated 48°N, 220°W, to the east of Viking 2 Lander site.

Milankovič theory A theory proposed by the Yugoslav scientist M.Milankovič to explain the periodical Ice Ages which have affected the Earth throughout its history.

Over long periods, there are changes in both the Earth's orbit and its axial inclination. The period of precession is 26,000 years, and the period of the full range of axial inclination is just over 40,000 years, while there is a regular change in the orbital eccentricity over 90,000 to 100,000 years. All these produce slight changes in climate, and, according to Milankovič, when all these are "in phase" the result is a drop in the Earth's overall temperature, producing an Ice Age.

The theory has met with considerable support, particularly from analyses of deep-sea cores which seem to show periodicities in climate, though it is not universally accepted. *PM*.

Mills' Cross One of the earliest radio arrays, designed by B.Y.Mills of the CSIRO, Sydney, Australia. The aerials were arranged along two lines 1,500ft (457m) long at right angles.

Milne, Edward Arthur (1896–1950). British theoretical astrophysicist and cosmologist. Between 1920 and 1932 he attacked fundamental problems in stellar atmospheres and stellar structure, but in his later life became interested in the expansion of the universe. His cosmological ideas, which he called the kinematic theory of relativity, were based on attractive, general, philosophical ideas, such as the "cosmological principle" that all observers in the universe, no matter in which galaxy, see the same "world picture". He was a particular advocate of the idea that the "constants" of physics, like the gravitational constant G, actually changed over the life of the universe. These predictions proved unfounded, as did the theoretical basis of the kinematic theory of relativity. *PGM*.

Mimas The innermost of the large Saturnian satellites, discovered by William HERSCHEL in 1789. The satellite orbits Saturn in a slightly elliptical orbit at an average distance of 115,000 miles (185,500km) and is locked in a synchronous rotation with the planet. The low density, 1.2gm/cm³, indicates that the satellite contains a large fraction of ice. Although it is near Saturn, there is little evidence that resurfacing has occurred.

The surface is cratered, with craters larger than 12.4 miles (20km) having central peaks. The dominate feature is a large crater, Herschel, located in the

Mimas: photographed by Voyager 1, 1980

leading hemisphere at 0° latitude and 104° longitude. This crater is 6 miles (10km) deep and has a central peak 3.7 miles (6km) high. Grooves up to 6 miles (10km) wide and 6,600ft (2km) deep appear to have formed along with Herschel. The colliding object, capable of forming so large a crater, must have nearly shattered the satellite. *RFB*.

Minkowski, Hermann (1864–1909) Professor of physics at Zürich who, in 1908, suggested that space and time were intimately related and that events should be considered as points in a four-dimensional spacetime.

Minkowski, Rudolph (1895–1976) German-born astronomer who worked at Mount Wilson Observatory in California, and was also a pioneer of radio astronomy. He had a lifelong interest in supernovæ.

Mini-quasar A term sometimes used to describe the bright core of an N-GALAXY.

Mintaka Delta Orionis, the northern star of Orion's belt, only 0°17′ 57″ south of the equator. Its magnitude is 2.2, but it is an ECLIPSING BINARY with a very small range.

Minute of arc One-sixtieth of a degree.

Mir Basic long-duration space station facility launched by the Soviet Union on February 19 1986. The vehicle has six docking ports, at least four of which are "building block" modules dedicated to specific scientific disciplines.

Mira (Omicron Ceti). The first star to be discovered to vary in a periodic manner. David Fabricius, a Dutch amateur astronomer, and a disciple of Tycho BRAHE, first noticed Mira as of third magnitude on August 13 1596. He could not find it in star catalogues, atlases or globes. A few months later it was invisible, but he saw it again on February 15 1609 at third magnitude. Bayer, in 1603, lettered it Omicron and noted it as of fourth magnitude. It was observed from 1659 to 1682 and thought to be a new star, catalogued 68 of sixth magnitude. Holwarda, in 1638, noted that Mira became visible to the naked eye from time to time and invisible in between these times.

In 1660 the star was shown to vary in an approximate period of eleven months. Its period is $331^d.96$, but is subject to irregularities. Mean range in brightness is $2^m.0$ to $10^m.1$. The maxima have been observed as bright as $1^m.7$ and as faint as $4^m.9$. Minima show a similar variation of from $8^m.6$ and $10^m.1$. The magnitude that a future maximum or minimum will have is not predictable. Usually Mira has a protracted minimum phase followed by a steep rise and a slow decline.

The spectrum varies M5e to M9e, which is that of a cool star. There are strong bands of titanium oxide, and also lines of neutral metals. An EMISSION SPECTRUM becomes superimposed on the ABSORPTION spectrum when Mira has reached seventh magnitude on the rise, with the intensities of the lines increasing to well after maximum light. They then disappear. The emission spectrum is mainly hydrogen.

Mira Ceti lies at a distance of 196 light-years. Aitken discovered a faint companion to Mira in 1923. The separation is 0″.61; the companion is now known as VZ Ceti, a variable with a range of $9^m.5$ to 12^m. It shows small variations in times of several hours on which are superimposed 10- to 15-minute variations. It also has very rare flares which last for about two minutes. It is possible that VZ Ceti is a WHITE DWARF with an ACCRETION disk. It orbits Mira in 1,800 years. *FMB*.

Mira variables So called after the first star discovered to vary in an approxi-
(Continued on page 273)

Moons

Joseph Veverka
Professor of Astronomy
Cornell University

Fifty-four satellites are known in the Solar System,
the last ten having been discovered in January 1986 during
the Voyager flyby of Uranus. In size they range from
Jupiter's Ganymede, which with a diameter of 3,280 miles (5,270km),
is bigger than either of the two smallest planets
Pluto or Mercury, to tiny bodies no more than a few tens of
kilometres across. Deimos, one of the two moons of Mars,
with a diameter of only 7 miles (12km), is currently

During the past 15 years, 43 of the 54 known moons in our Solar System have been investigated by spacecraft — all but three of these by Voyagers 1 and 2. We have learned that far from being dull clumps of rock or ice, many of these objects preserve evidence of fascinating geologic histories on their surfaces. Interpreting these clues, using data such as this colour image of Jupiter's icy satellite Callisto, obtained by Voyager in 1979, is a dynamic area of planetary research which is contributing to a better understanding of how our Solar System formed and how it has evolved during the past 4,600 million years.

EARTH'S MOON

Moon

MAR'S MOONS

Phobos
Deimos

JUPITER'S MOONS

Metis
Adrastea
Amalthea
Thebe
Io
Europa

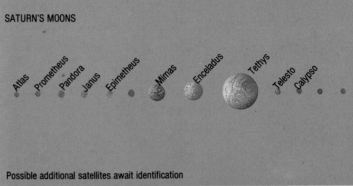

SATURN'S MOONS

Atlas
Prometheus
Pandora
Janus
Epimetheus
Mimas
Enceladus
Tethys
Telesto
Calypso

Possible additional satellites await identification

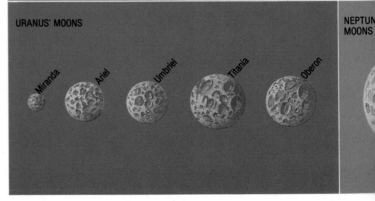

URANUS' MOONS

Miranda
Ariel
Umbriel
Titania
Oberon

NEPTUN
MOONS

the smallest known satellite. Our own Moon is one of the larger satellites in the Solar System (diameter 2,160 miles/3,476km), and is the only one whose surface features can be observed directly from Earth.

Compared to the planets, most satellites are very small bodies that have weak surface gravities. For the smaller ones, gravity is too weak to readjust the shape to an approximately spherical form if a sizeable portion is removed by a large impact. The transition between irregularly shaped and round satellites occurs somewhere in the range of 125–190 miles (200–300km) across. Saturn's Hyperion, with dimensions of 255 × 160 × 140 miles (410 × 260 ×220km), is the largest known irregularly shaped satellite. Due to their weak gravitational attraction, most satellites have not been able to retain atmospheres. Their surfaces are therefore exposed to the near-vacuum of space and are subject to direct bombardment by charged particles and ultraviolet light from the Sun, cosmic rays, and meteoroids of all sizes. If one discounts the very tenuous sulphur dioxide atmosphere produced by Io's volcanoes (see below), only Saturn's Titan and Neptune's Triton have atmospheres. Both of these satellites may also be partially covered by oceans – of liquid ethane in the case of Titan, and liquid nitrogen in the case of Triton.

Undeniably, most small satellites have experienced histories of limited geologic activity, but it is certainly not true that satellites are geologically duller than planets. Some of the larger satellites have experienced, and in a few cases continue to undergo, geologic processes as diverse and interesting as those of the Earthlike planets. Today Jupiter's Io is geologically much more alive than is the planet Mercury. Miranda (a small satellite of Uranus) has had a geologic history as complex, if not more so, than most planets.

Satellite surfaces are exposed to both external and internal processes. The dominant external process is cratering by impacting Solar System debris, which produces the characteristically pockmarked appearance of most satellite surfaces. Internal processes, such as rifting and faulting of the crust and the eruption of molten materials, are driven by internal heat. Important sources of heat include those initially released during the accretion of the body and subsequently perhaps by differentiation, as well as that due to the decay of radioactive elements. For most satellites, one expects that these sources of energy would have been important only during the earliest part of the object's history. A crucial recent realization has been that tidal heating (due to frictional dissipation of tides caused by a planet within one of its satellites) can be an important source of internal energy (see below).

Direct information on the composition of satellites comes from two major sources: remote sensing of the spectral characteristics of surface materials, and determinations of the satellite's mean density from mass and volume measurements. Indirect information comes from inferences based on chemical condensation models which attempt to predict the types of solid materials that would have formed in various places in the Solar System. The few satellites in the inner Solar System, like the Earthlike planets, are made of rocky material. On the other hand, many satellites in the outer Solar System have large admixtures of ices. Water ice is the dominant icy constituent, and there is evidence that other ices (those of ammonia and methane) become more important with increasing distance from the Sun. There

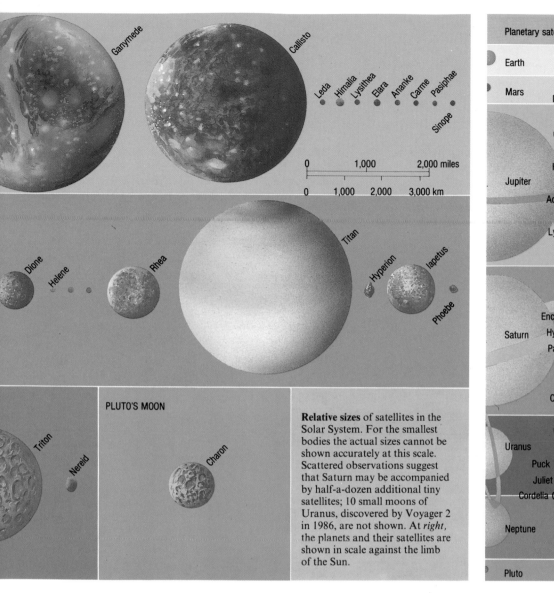

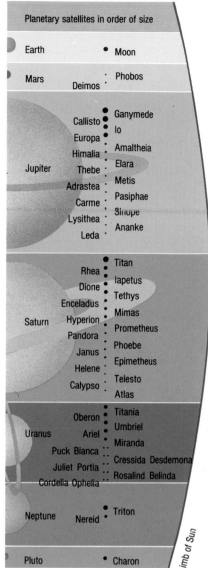

Planetary satellites in order of size

Earth	Moon
Mars	Phobos
	Deimos

Jupiter	Ganymede
Callisto	Io
Europa	Amalthea
Himalia	Elara
Thebe	Metis
Adrastea	Pasiphae
Carme	Sinope
Lysithea	Ananke
Leda	

Saturn	Titan
Rhea	Iapetus
Dione	Tethys
Enceladus	Mimas
Hyperion	Prometheus
Pandora	Phoebe
Janus	Epimetheus
Helene	Telesto
Calypso	Atlas

Uranus	Titania
Oberon	Umbriel
Ariel	Miranda
Puck Bianca	Cressida Desdemona
Juliet Portia	Rosalind Belinda
Cordelia Ophelia	

Neptune	Triton
Nereid	

Pluto	Charon

Limb of Sun

Relative sizes of satellites in the Solar System. For the smallest bodies the actual sizes cannot be shown accurately at this scale. Scattered observations suggest that Saturn may be accompanied by half-a-dozen additional tiny satellites; 10 small moons of Uranus, discovered by Voyager 2 in 1986, are not shown. At *right*, the planets and their satellites are shown in scale against the limb of the Sun.

is indirect evidence that some of the satellites of Saturn may have incorporated ammonia ice (probably in the form of ammonium hydrate) in addition to water ice, and those of Saturn methane ice as well. It is likely that Pluto's Charon contains methane or water ice.

The origin of any particular satellite is uncertain. Some satellites almost certainly formed during the formation of their parent planet by a process vaguely analogous (but not well understood) to that which was responsible for the formation of the planets around the Sun. Others probably were captured. Still others (especially the smallest ones) may be fragments of more ancient satellites that were fragmented by catastrophic impacts. It is likely that most of those satellites which occupy nearly circular, uninclined orbits (sometimes called regular satellites) formed somehow along with their parent planet. Satellites with distant, inclined, and eccentric orbits (sometimes referred to as irregular satellites) may have been captured (by processes which cannot be reconstructed with much certainty).

Most satellites have direct orbits, that is, they move around their planet in the same sense as the planet spins. Of the 54 known satellites, only six move in the opposite sense, or in retrograde orbits.

Most satellites in the Solar System are in a synchronous spin state – that is, like our own Moon, they spin once about their axis in one revolution around their planet. This situation is the result of tides exerted on the satellite by its planet. Since the strength of tides falls off rapidly with increasing distance, one would expect that only remote satellites might not spin synchronously. This expectation is confirmed in the Jupiter system, where Callisto and all the satellites inward of Callisto have synchronous spins, whereas those outside do not; and in the Saturn system, where satellites inside the orbit of Phœbe spin synchronously, but Phœbe does not. (Phœbe's spin period is 9.4 hours, considerably shorter than its orbital period of 550 days.)

The orbital evolution of close satellites can be affected markedly by tides. For direct satellites the general effect is for tides to pull a satellite inward if its orbit is within the so-called synchronous limit (distance from the centre of the planet at which the orbital period equals the spin period of the planet), but to push it out if it lies outside. Of the two satellites of Mars, Phobos is currently being pulled in rapidly by tides, while Deimos is being pushed slowly out. Our Moon is in the same category as Deimos; tides are pushing our Moon out at a measurable rate of 1.3in/year (3.3cm/yr). Retrograde close satellites are always pulled in by tides; hence their orbits are unstable. Neptune's Triton may be in such a predicament.

A satellite that gets too close to its planet may be pulled apart by tides. This is one possible origin for some planetary rings. Another is the disruption by impact of a small satellite that is within the Roche limit of its planet. The Roche limit is the distance within which the differential pull of the planet on two small particles exceeds their mutual gravitational attraction. Thus, within this distance "tides" prevent small particles from accumulating into larger bodies.

The Inner Solar System

Of the fifty-four known moons, only three are associated with inner, or Earthlike, planets. Earth has one large moon; Mars has two tiny ones; and both Venus and Mercury are devoid of satellites. In fact, Mercury and Venus are the only planets in the Solar System that have no moons. Whether this is an accident or a consequence of being close to the Sun is not known. A particularly intriguing speculation is that the backward spin of Venus is the result of the spiralling in of a large retrograde satellite in the planet's past. (Recall that tides tend to pull retrograde satellites ever closer to their planet.)

Our Moon is one of the larger satellites in the Solar System. Once styled as the largest moon in comparison with the size of its planet, that distinction is now known to belong to Pluto's Charon.

The Moon is the only satellite whose surface features can be studied in detail by telescopes on Earth, and also the only one from which samples have been obtained. Analysis of such samples provides much information, including data on the chemistry and the age of the rocks. Compared to the Earth, the Moon is enriched in elements that melt at high temperatures, and depleted in iron and in volatile substances such as water. One aim of lunar studies has been to use the detailed information that is becoming available to unravel the origin of our satellite. Three principal types of theories have been discussed: (a) capture, (b) fission from the Earth, and (c) co-origin with our planet. There is strong evidence that the Moon has been associated with the Earth for much of the Earth's history. A currently favoured mode of origin

The best-studied of all satellites, our own Moon, *left,* underscores the difficulty in unravelling origins. Is the Moon a captured object? Did it split off from a rapidly spinning Earth long ago? Did it perhaps form at the same time as our planet by a similar process of accretion? Deciding among these choices was a key motivation behind the Apollo programme, which produced a wealth of new data about our neighbour in space, including definite clues that the origin of our Moon may have been much more complex than we had guessed.

The Moon's orbit, *above opposite.* Since the Moon, a comparatively large satellite, is relatively close to Earth, tidal interactions are strong. Right now tides are pushing the Moon slowly outward, a clue that long ago the Moon was much closer to us than it is today.

A typical impact crater, *below opposite,* on the Moon — a reminder that like all bodies in the Solar System the Moon has been exposed to a continued flux of impacting objects. Studies prove that the cratering flux, far from being constant with time, was considerably more fierce during the first 500 million years of Solar System history.

is a highly complex one involving elements of all of the possibilities above. According to this view, very shortly after the beginning of the Solar System 4,600 million years ago, a Mars-sized object hit proto-Earth, leading to the ejection of a vast amount of material. Some of this material formed a ring of debris around the Earth, out of which the Moon accumulated.

The Moon has no atmosphere. Its surface is heavily cratered on all scales, a process that has produced a ubiquitous rubble layer, or regolith. Broadly speaking, the surface can be divided into two types of regions which differ in albedo, composition, elevations, and age: (1) darker, lower, relatively less cratered (therefore younger) maria made of basalt; and (2) brighter, higher, much more heavily cratered (therefore older) uplands (or highlands) made of anorthositic rocks. Dating of returned samples shows mare materials ranging in age from 3,500 million to 4,000 million years, whereas the uplands date from the earliest 500 million years of the Moon's history. Since the uplands are significantly more cratered than the maria, it is evident that the Moon, and presumably other planets and satellites, were subjected to a much more intense flux of impacting objects during the first 500 million years, than has been the case since 4,000 million years ago. It is likely that this period of very intense bombardment represents the tail end of the accretion process by which the sizable bodies of the Solar System are believed to have accumulated out of smaller planetesimals. The bombardment was probably intense enough to fracture the upland crust (producing a megaregolith) to depths of tens of kilometres. During an earlier phase, the influx may have been intense enough to melt the surface layers, producing a transient "magma ocean".

The geologic history of the Moon can be divided into three major episodes: (1) accretion beginning 4,600 million years ago, melting of the outer layers by impact, and formation of an anorthositic crust; (2) tail end of heavy bombardment about 4,000 million years ago and the eruption of basaltic lavas into low-lying areas. Some of the lowest areas were the floors of large craters (called "basins"), of which Imbrium, with a diameter of 780 miles (1,250km), is the largest. The basaltic lavas were produced by partial melting at depths of many hundred kilometres due to the accumulation of heat from the decay of radioactive elements, principally U, Th, and K. Such eruption peaked between about 3,900 million and 3,500 million years ago, and declined as the interior of the Moon cooled. Few, if any, took place later than 2,500 million to 2,000 million years ago. (3) For the past 2,000 million years or so, the Moon has been a "dead planet", its surface features modified predominantly by a continuing flux of impacting objects. It is estimated that at present one or two new craters 6 miles (10km) or larger in diameter form every ten million years. Such a crater involves the impact of a body about 1,600ft (500m) across. The impact that produced the Imbrium basin long ago must have involved an asteroid-sized body about 30 miles (50km) across.

Next to the Moon, Phobos and Deimos, the two tiny satellites of Mars, are the best studied moons in the Solar System. This is so thanks to thorough investigation carried out by the Mariner 9 and Viking spacecraft during the 1970s.

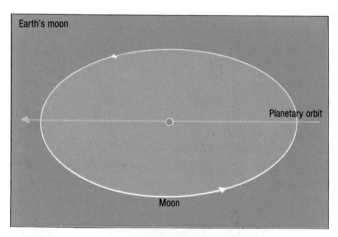

Earth's moon

Planetary orbit

Moon

Mars

Both satellites are small, heavily cratered, irregularly shaped objects. The inner of the two, Phobos, is about 17 × 13 × 11 miles (27 × 21 × 18km) across; the outer, Deimos, has dimensions of 9 × 7 × 6 miles (15 × 12 × 10km). Both are made of a very dark grey, rocky material. Additional information on composition comes from the low mean densities, about $2g/cm^3$. (The Viking spacecraft made close flybys of the satellites – within 19 miles/30km in the case of Deimos – to determine their masses, and hence, mean densities.) The dark grey colours and low mean densities are consistent with the properties of some water-rich carbonaceous chondrite meteorites believed to have originated in the outer portions of the asteroid belt. If so, Phobos and Deimos could be captured objects, but the capture mechanism is far from clear. Such capture could have been facilitated very early in the history of Mars, when the planet is presumed to have been surrounded by a distended atmosphere (such an extended atmosphere could provide the drag needed to effect capture).

The present orbits are such that Phobos is within the synchronous orbit distance of Mars (about $6R_M$), while Deimos is just outside. Tides are currently pulling Phobos toward the planet at a readily measurable rate, which implies that the orbit will decay in the next 100 million years.

The surfaces of both satellites are covered with regoliths, the texture of which ranges from that of ejecta blocks many metres across to fine dust. Prominent craters are abundant; the largest on Deimos is about 2 miles (3km) across. Stickney (diameter 7.5 miles/12km) and Hall (3 miles/5km) are the two largest craters on Phobos. Such craters are large enough to have signifi-

cantly affected the overall shapes of the satellites: the very irregular outlines of these bodies are clearly the result of a long history of cratering.

A remarkable aspect of Phobos is that its surface is covered by long, linear grooves some 300–600ft (100–200m) wide and 30–60ft (10–20m) deep, which seem to be associated with the largest crater, Stickney. The energy of the impact that formed Stickney was close to that needed to break up the satellite; one speculation is that the grooves are surface expressions of fractures produced by this nearly catastrophic impact. No grooves are observed on Deimos. For other small satellites, we as yet lack images of sufficient resolution to search for features.

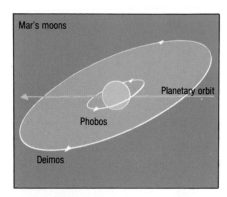

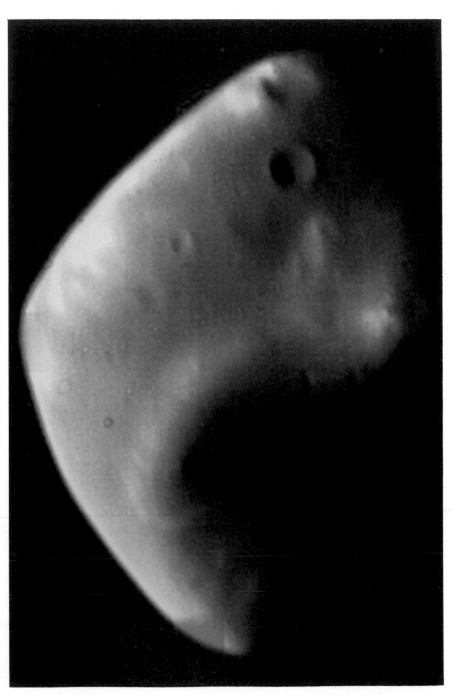

The two tiny satellites of Mars orbit, *above*, very close to the planet. At present, Mars tides are pulling the inner satellite (Phobos) toward the planet and pushing Deimos, *right*, slowly out. Attempts to calculate the orbits back in time to shed light on the satellites' origin have yet to yield conclusive results. Both satellites are made of dark, low-density rocky material and covered with rubble.

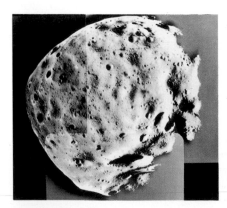

The irregular outline of Phobos, *above*, betrays a long history of cratering. The largest crater on this 20km body is Stickney, a 12km-wide scar produced by an ancient impact which almost shattered the small satellite. The grooves discovered in high-resolution Viking images may be fractures dating back to this nearly catastrophic event.

Satellites of Jupiter

Jupiter's 16 known satellites are divided conveniently into three groups – the four small inner satellites (Metis, Adrastea, Amalthea, and Thebe), the four large Galilean satellites (Io, Europa, Ganymede, and Callisto), and the eight small distant "irregular" satellites.

Prior to the Voyager flybys of Jupiter in 1979, only one inner satellite, Amalthea, was known. Voyager discovered Metis, Adrastea, and Thebe, all within the orbit of Io. All are small, very dark objects, ranging in diameter from about 50 miles (80km) (Thebe) to 16 miles (25km) (Adrastea). Metis and Adrastea orbit at the outer edge of Jupiter's ring. It is likely that ejecta from these two satellites resupply small particles to the Jovian ring.

Voyager images show Amalthea to be a very heavily cratered, irregular object with dimensions 170 × 100 × 90 miles (270 × 165 × 150km). The satellite is in a synchronous spin state, with its long axis pointing toward Jupiter. The surface is covered with a dark, red material that may owe its spectral peculiarities to sulphur ultimately derived from Io (see below). There is no determination of Amalthea's mass; hence the satellite's mean density and possible internal composition remain unknown.

The eight outer satellites are divided into two orbital families: the first in eccentric, retrograde orbits at about

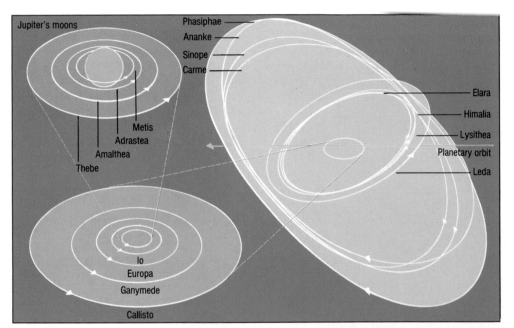

Jupiter's 16 known satellites can be divided into three groups, *left*. The four large satellites (Io, Europa, Ganymede and Callisto) have nearly circular orbits close to Jupiter's equatorial plane. The small distant satellites have very eccentric orbits that fall into two families: an inner group in direct (prograde) orbits and an outer group in retrograde motion. The orbits of four small inner moons within the orbit of Io have been omitted for clarity.

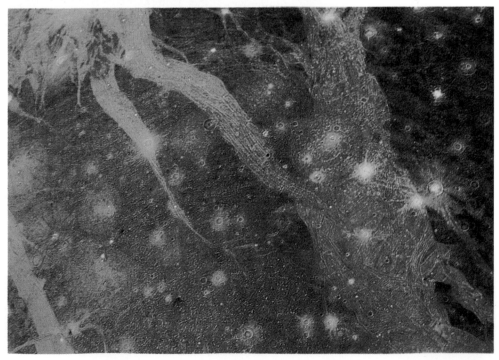

A close-up of Ganymede, *right,* the Solar System's largest satellite, shows that the very old, heavily cratered icy surface has been rifted and replaced in part by swaths of newer, cleaner ice forming the characteristic grooved terrain. It appears that the surfaces of icy satellites darken with time, but the processes, and the composition of the dark, non-icy materials involved, are uncertain. Note the numerous bright ray craters, evidence that impacts continue to punch through a relatively thin dirty crust to layers of cleaner, brighter ice below.

300 R$_J$; the second, in direct moderately inclined, and less eccentric orbits at about 160 R$_J$. Speculations exist that these may be captured bodies, and that the clusters result from the break-up of protosatellites due to drag forces during capture by Jupiter's extended very early atmosphere. Little is known about the physical characteristics of these objects. They range in diameter from some 105 miles (170km) (Himalia), to 12 miles (20km) or less (eg, Sinope). A lightcurve observed for the brightest (Himalia) shows that this satellite is irregular in shape and spins with a period much shorter than its orbital period. (The inferred spin period is 9.5 hours, compared with an orbital period of 251 days.) Spectrally, these dark, small satellites fall into two categories (not clearly connected with orbital families). One group is similar to the C-asteroids; the other to D-asteroids and the Trojans near Jupiter's orbit.

The four Galilean satellites of Jupiter – Io, Europa, Ganymede, and Callisto – are large, planet-sized objects. The smallest, Europa (diameter 1,950 miles/3,130km), is slightly smaller than our Moon; the largest, Ganymede (diameter 3,280 miles/5,270km), is the biggest satellite in the Solar System, and is larger than either Pluto or Mercury. With the exception of Io, which has an extremely tenuous atmosphere associated with current volcanism, none of the satellites has an atmosphere. Judging from the progression of mean densities (from 3.5 at Io to 1.8 at Callisto), there is a gradation in composition with increasing distance from Jupiter. Io is completely "rocky", whereas by the time Callisto is reached, the composition is about half rock, half ice. Models suggest that the original rocky material was similar to that in carbonaceous meteorites, and that the ice was predominantly water ice. In its very early history, Jupiter is

In this global view, *opposite*, of volcanic Io, the evident absence of impact craters is an unmistakable clue to the satellite's continuing geologic activity. The whitish areas are sulphur dioxide frost, while various forms of sulphur may predominate in the coloured regions. All are products of volcanic activity. There is evidence that the crust may be basaltic, with only a veneer of sulphur.

Volcanic areas appear as dark spots. One of the largest, Loki, seen at upper right, was active at the time of the Voyager flyby in 1979. Infrared monitoring from Earth indicates that this volcano continues to be active. Io's intense volcanic activity is powered by Jovian tides.

Evidence of a colossal ancient impact is preserved on Callisto's surface, *below*, in the form of Valhalla, a concentric structure some 1,250 miles (2,000km) across. The morphology is quite distinct from that of large impact basins (such as Imbrium) on our own Moon. The very subdued topography suggests that when this feature was formed Callisto's crust was far from rigid and probably close to the melting point at shallow depths. However, there is no evidence that Callisto's crust was ever ruptured by internal expansion as was that of Ganymede.

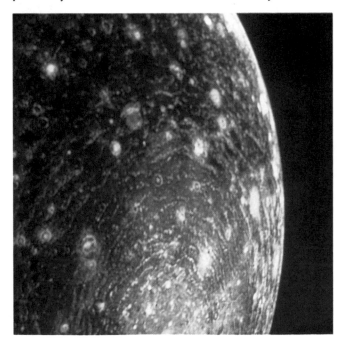

believed to have been more luminous than it is today, due to heat released by the contraction of the planet under gravity. The result was a temperature gradient in the region of formation of the satellites, which led to chemical differences somewhat mimicking those seen among the planets: only rocky objects close in, and volatile-rich bodies farther out.

The outermost Galilean satellite, Callisto, has a diameter of 3,000 miles (4,840km) and a mean density of 1.8, indicating that it is approximately half ice, half rock. The surface is heavily cratered on all scales, the largest impact feature being the basin Valhalla, 1,250 miles (2,000km) in diameter. Valhalla has a peculiar, subdued topography which suggests that at the time of its formation (3–4,000 million years ago?), the crust of Callisto was close to the melting point of water ice at fairly shallow depths. There is no evidence of crustal rupturing or

of the eruption of materials from the interior to the surface. Spectrally, the surface of Callisto consists of water ice heavily contaminated by a dark, supposedly rocky, material. Only in association with some of the fresher craters are brighter patches of cleaner water ice seen.

The biggest satellite of the system, Ganymede, has a diameter of 3,280 miles (5,270km). Its mean density, 1.9, slightly higher than that of Callisto, suggests a somewhat higher rock/ice ratio. Yet, the surface appearance of Ganymede is dramatically different from that of Callisto. The surface is divided into two types of areas: the first, reminiscent of the heavily cratered surface of Callisto; the second, made up of a brighter, less cratered terrain with a characteristic grooved topography. The grooved terrain occurs in swaths typically 60–120 miles (100–200km) wide, made up of sub-parallel grooves spaced 2–6 miles (3–10km) apart, and on average 1,000–1,300ft (300–400m) in relief. Models suggest that within the first 500 million to 1,000 million years of its history the interior of Ganymede got hot enough (due to energy released by radioactive elements in the rocky component) to melt. The melting resulted in differentiation – the process whereby the rocky materials sank toward the centre to form a dense core and the lighter ice rose to form a lower density crust. Differentiation would have led to an expansion of the crust, which ruptured and allowed molten material (water/ice slush?) to escape to the surface, forming the swaths of grooved terrain. While we lack an absolute chronology for events on outer planet satellites, the number of impact craters accumulated on the grooved terrain suggests that the groove-forming processes came to an end long ago (3,000 million years ago?). No evidence of similar processes is preserved on Callisto. Callisto may not have differentiated to the extent that Ganymede did, for reasons which are obscure but may be related to less radioactive heating due to a smaller rock/ice ratio.

Spectral observations show that water ice is the dominant constituent of Ganymede's surface. All areas of Ganymede are brighter (cleaner ice) than Callisto. Unlike Callisto, Ganymede has thin polar cap-like deposits of cleaner water ice at high latitudes and prominent rays made of bright clean ice associated with some of the more recent craters.

Europa, the smallest of the Galilean satellites (diameter 1,950 miles/3,130km) must be ranked as one of the most intriguing moons in the Solar System. With a mean density of 3.0, Europa is almost entirely made of rock, except for a thin mantle of water ice that cannot exceed some 60 miles (100km) in thickness. From spectrophotometry, the surface is known to be made of water ice contaminated with a reddish material, probably sulphur or sulphur compounds derived from Europa's neighbour Io (see below). The surface has a high albedo (about 70 per cent).

Europe's surface is remarkable in several ways. First, there is the almost complete lack of impact craters, indicating a geologically young surface. Second, the surface is rifted by globally extensive patterns of lineaments, many of which appear to be troughs and other ridges. In morphology they are somewhat similar (except in scale) to leads and pressure ridges observed on sea ice in the polar regions of Earth. It is possible that today Europa is a satellite whose icy mantle is partially molten, except for a thin surface crust of ice. Such a condition would explain the lack of preservation of impact craters, the observed global pattern of lineaments, as well as the amazingly low vertical relief on the satellite (less than 650ft/200m from Voyager measurements).

The crustal rifting that produced the lineaments has been attributed to a number of possible causes: tidal stresses from Jupiter, stresses due to a slight general expansion of the satellites due perhaps to a general warming of the satellite's interior, or to fracturing by impacts. All three may have played a role.

Calculations show that if melting and differentiation occurred during the first thousand million years of Europa's history (due to heat released by the decay of radioactive elements in Europa's rocky component) tidal heating by Jupiter (see below) would be sufficient to keep part of the satellite's mantle partially molten to the present day. Europa's surface is a prime target for ultra-high-resolution imaging during the Galileo Orbiter mission, which will take place in the 1990s.

Even before the flights of Voyagers 1 and 2, Io, the innermost of the Galilean satellites, was a puzzle to astronomers. The satellite was known to have an unusual yellow-red colour, bizarre thermal properties, and to move in an orbit surrounded by atoms of sodium, sulphur, etc. One idea was that these were atoms sputtered from the surface of Io, by high-energy charged particles (electrons and protons) trapped by Jupiter's magnetic field which embays Io.

In size (diameter 2,200 miles/3,640km) and in density (3.5), Io is similar to our Moon. Yet its surface, far from being covered by grey rocks, is coated with yellowish, orange, and reddish material suspected of being sulphur and sulphur compounds, and with bright whitish patches known to be sulphur dioxide frost. Unlike our Moon, Io has a surface devoid of impact craters, an observation that suggests that Io must remain geologically active to this day. In fact, the Voyager spacecraft imaged at least eight active volcanic eruptions during their brief flybys. In many respects the morphology of the volcanic features seen on Io is similar to those produced by basaltic volcanism on Earth, and an unresolved issue is the extent to which the putative sulphur deposits are only a surface patina.

The question of why sulphur is so abundant on Io's surface is answered readily. Volcanic activity on the satellite is driven by frictional heat produced by tides due to Jupiter, a possibility suggested theoretically shortly *before* Voyager 1 discovered intense volcanism on the satellite. It is likely that Jupiter has been heating Io in this way throughout most of the satellite's history. The result is that Io has been "cooked dry", most of its volatile substance having been driven to the surface by volcanic processes. Once at the surface, all but the heaviest substances, of which sulphur and sulphur dioxide are the most abundant, have escaped, leaving the

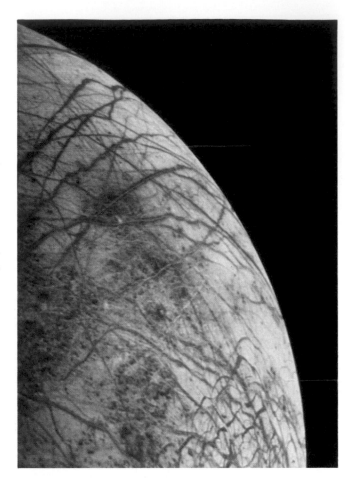

surface rich in sulphur and sulphur compounds.

The eruptions imaged by Voyager included the ejection of plumes of gas (sulphur dioxide) and particulate material (possibly sulphur and condensed sulphur dioxide) to heights of 125–185 miles (200–300km). Some of the gases erupted eventually condense in colder regions of the surface, some remain to form a very tenuous atmosphere. Material is also believed to escape from the tops of the plumes, especially if a molecule or atom is ionized. Such charged particles are readily swept up by Jupiter's magnetic field, which swings by Io. (Jupiter's field sweeps around with a period of about 10 hours, while Io orbits the planet once every 42.5 hours). Many escaping particles eventually populate the "Io torus", a crudely doughnut-shaped concentration of atoms (mostly sodium) and ions (mostly sulphur and oxygen). Some of the material around Io's orbit eventually diffuses throughout the Jovian magnetosphere. There is strong circumstantial evidence that Amalthea and Europa are being contaminated in this way.

Although surface features on Io cannot by resolved usefully by terrestrial telescopes, it is possible to monitor volcanic activity on the satellite through variations in the thermal emission detectable at infrared wavelengths. By the time of the next spacecraft visit to Io — by Galileo in the mid-1990s — one can expect that the appearance of many areas will have been modified substantially from that recorded by Voyager in 1979.

Europa, *opposite,* an enigmatic satellite, remains a prime target for future space missions. Available evidence suggests that its relatively thin mantle of water ice may be molten except for an outer rind. While predominantly of water ice, Europa's surface has a reddish tinge which may be due to sulphur contamination from nearby Io.

Amalthea, a small inner satellite of Jupiter, *above,* cannot be studied well from Earth. Voyager revealed it to be a very irregularly shaped body whose surface is covered by an unusual material of undetermined composition which is both very dark and very red (in this photograph the surface brightness has been exaggerated for clarity).

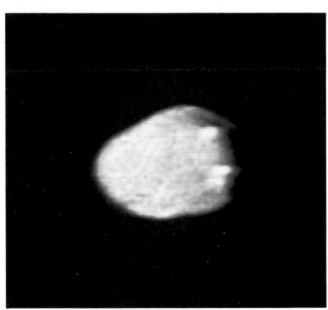

Satellites of Saturn

Following the Voyager exploration of the Saturn system, we know that this planet has at least 17 satellites. About half are tiny, presumably icy objects. The others, with the exception of Titan, arc small- to medium-sized bodies ranging from 250 to 930 miles (400 to 1,500km) in diameter. Titan, probably the Solar System's most unusual satellite, has a diameter of 3,200 miles (5,150km) and a thick, cloudy atmosphere whose visible extent is at least 125 miles (200km) above the surface. With the possible exception of Neptune's Triton (which Voyager 2 is scheduled to investigate in 1989), Titan is the only satellite with a substantial atmosphere. Spacecraft and Earth-based measurements reveal that the atmosphere is principally nitrogen (surface pressure about 1.5 atmospheres; temperature about 90K), with a variety of hydrocarbons among the secondary constituents. Of these, methane is the most abundant, but ethane, propane, acetylene, and ethylene have been identified. In addition, traces of hydrogen cyanide, cyanoacetylene, cyanogen, and carbon dioxide are also present. Many of tnese minor compounds are believed to be the products of photochemical processes occurring high in Titan's predominantly N_2/CH_4 atmosphere. These processes have also led to the production of reddish-orange particulate material of complex chemical composition which forms a planet-wide optically thick cloud layer and completely obscures the surface. This reddish-orange material has

been successfully duplicated in laboratory simulations. There is indirect evidence that closer to the surface, below the level of the visible orange clouds, is a layer of methane cloud. One interesting speculation is that the satellite's surface is covered partially by an ocean of liquid ethane, but direct tests of this possibility will probably have to await the next spacecraft mission to the Saturn system, scheduled to be the Cassini mission of the late 1990s.

Speculations about possible ethane oceans aside, we know almost nothing about what the surface of Titan is like. It may bear some resemblance to that of Ganymede, a satellite which Titan resembles closely in size and mean density. But unlike Ganymede, Titan must contain ices of ammonia and methane in addition to water ice to account for the make-up of its atmosphere. (The presence of nitrogen is best explained as a photodissociation product from the break-up of ammonia). Given the complex chemistry of Titan's atmosphere, hypotheses about biochemistry on the surface abound in the literature. There is no evidence for life on Titan; certainly the extremely low surface temperature would seem to make active biology unlikely.

Another remarkable satellite of Saturn is Iapetus (diameter 900 miles/1,460km), which circles the planet once every 79 days between the orbits of Hyperion and the outermost known moon, Phoebe. From the fact that Iapetus varies in brightness by about a factor of five during each orbit, astronomers deduced long ago that the satellite is in a synchronous spin state, and suspected that one of its hemispheres (the leading hemisphere facing the direction of motion) is almost black, while the other (the trailing hemisphere) is very bright. Voyager images and recent spectral measurements made by large telescopes on Earth confirm that the trailing hemisphere of the satellite is covered by bright water ice, whereas the leading side is covered by a very dark substance similar to that in some carbonaceous meteorites. The mean density of the satellite is very low (1.2). Therefore, Iapetus cannot be a rocky object with a thin hemispherical coating of water ice; rather, it is an icy body and the problem is to explain the origin of the dark material and its concentration on the leading face. Numerous hypotheses have been proposed: according to some, the dark material is external in origin and is picked up preferentially by the leading hemisphere as Iapetus orbits the planet; others explain the dark material as of internal origin. Due to limited resolution, Voyager images were inconclusive in resolving this debate. The spacecraft data did disprove one version of the external theory, according to which the dark material represents ejecta from the outer retrograde satellite Phoebe, which spiral into Iapetus. Voyager revealed Phoebe to be a 140-mile (220km) object made of a very dark material, but one whose colour is significantly less red than that found on Iapetus.

The quality of Voyager images of the dark hemisphere of Iapetus is inadequate to show surface detail well, but the bright hemisphere is seen to be a very heavily cratered, and therefore presumably very old, surface.

Mimas, *right,* **and Enceladus,** *below,* two contrasting satellites of Saturn. Both are small icy bodies: Mimas has a diameter of 390km, while Enceladus is 500km across. The profusely cratered surface of Mimas suggests that impact cratering has been the major geologic process during the moon's history. By contrast, Voyager 2 images of Enceladus reveal a remarkable variation in crater density (and therefore age). The youngest areas are crossed by groove-like sinuous features indicative of internal geologic processes. There is evidence that tidal heating by Saturn has kept Enceladus geologically active to this day. But why has Mimas, which orbits closer to Saturn than Enceladus, remained ostensibly immune to these effects?

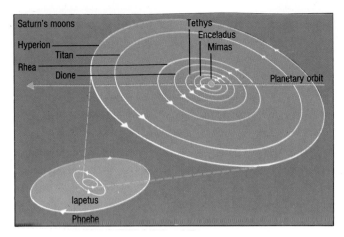

Orbits, *above,* of the nine larger satellites of Saturn. In addition, there are at least eight smaller moons, some in very unusual orbits. For instance, two tiny satellites move at the same distance as Dione, one about 60° ahead of, the other 60° behind Dione in its orbit. Some of these small moons could be debris from past collisions which destroyed larger bodies.

Saturn's Iapetus, *below,* a puzzle even after two Voyager flybys. One hemisphere is covered with water frost; the other with black, possibly carbonaceous, material of uncertain composition and unknown origin. Was it extruded from inside? Or is it a covering of external debris accumulated on the satellite's leading hemisphere?

Just inside the orbit of Iapetus is Hyperion, one of the largest irregularly shaped satellites in the Solar System. From Voyager data the dimensions are 255 × 160 × 140 miles (410 × 260 × 220km); the surface is cratered and covered by water ice contaminated by a darker, more opaque component of undetermined composition. Since the satellite's mass has never been measured, the mean density is unknown. Thus, we do not know whether Hyperion is primarily an icy or a rocky object. The satellite's most remarkable property is its chaotic spin state. Hyperion appears to be tumbling through space — a peculiar, unstable motion believed to be due to its very irregular shape and gravitational interaction with giant Titan. At the time of the Voyager 2 flyby, Hyperion's rotation axis was more or less in its orbital plane and the period was 14 days (compared to the orbital period of 21 days).

Incidentally, the outermost satellite, Phœbe, is known to be in a stable, but non-synchronous, spin state. Phœbe spins around once every 9.4 hours about an axis approximately perpendicular to its orbital plane. It orbits the planet once every 550 days.

Saturn's five remaining sizeable satellites, Mimas (diameter 240 miles/390km), Enceladus (310 miles/500km), Tethys (660 miles/1,060km), Dione (696 miles/1,120km), and Rhea (950 miles/1,530km) all have water ice surfaces, and mean densities ranging from 1.2 to 1.4, indicating that they are primarily icy, with somewhat varying admixtures of a rocky component. With the exception of Enceladus they have very heavily cratered surfaces, but in most cases (except Mimas), there is evidence of internal activity: rifting of the surface (possibly due to crustal expansion as a result of refreezing following melting and differentiation), and even extrusion of water/ice slush onto the surface.

Evidence of internal activity is most striking on Enceladus. The surface is divided into two major types of areas: an older, moderately cratered terrain, and much younger areas of swaths of groove-like terrain bearing some resemblance to the grooved areas of Ganymede. Compared to Ganymede, Enceladus is a tiny satellite, and some different mechanism of internal heating must be sought to explain its geologic activity. The best guess is tidal heating by Saturn, but there are difficulties. In its simplest form the tidal heating model predicts similar or even more intense heating of Mimas, but there is no evidence of internal activity on that satellite.

Enceladus has a remarkably very high albedo (0.9) and other photometric characteristics suggesting that its surface is uniformly covered by a layer of very clean (presumably, fresh) water frost. This observation, combined with the strong evidence that Enceladus is the source of the tiny water-ice particles that make up Saturn's E-ring, suggest that Enceladus remains geologically active to this day.

Saturn has at least eight other satellites, all tiny, irregularly-shaped bodies ranging in average diameter from about 6 to 60 miles (10 to 100km), and in albedo from 0.4 to 0.8. All are probably primarily made of water ice. Two of these tiny satellites, Prometheus and Pandora,

Voyager found that older, heavily cratered regions on Ariel, *right,* have been rifted and replaced locally by smoother, less cratered regions which preserve evidence of flow. It has been suggested that tides heated this icy satellite enough to expand and rupture its frozen crust, and to extrude new icy "lavas" to its surface. Although absolute timescales are unknown, the significant number of impact craters superimposed on even the smoothest areas suggest that all this happened long ago. Why was the tidal heating of Ariel so much more effective in the past than it appears to be today?

Due to the extreme tilt of the pole of Uranus, the orbits, *below,* of the five larger satellites (Miranda, Ariel, Umbriel, Titania and Oberon) describe a bullseye pattern. Voyager discovered 10 small moons, all within the orbit of Miranda, not shown in this diagram.

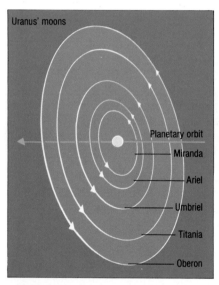

Uranus' moons

Planetary orbit
Miranda
Ariel
Umbriel
Titania
Oberon

orbit just inside and just outside, respectively, of Saturn's thin F-ring. It is believed that the gravitational effect of these "shepherding" satellites confines the ring particles and keeps them from spreading. Prometheus is about 90 × 60 × 50 miles (140 × 100 × 80km) across; its companion has dimensions of 70 × 55 × 45 miles (110 × 90 × 70km). The irregular shapes, and the unusual orbits of some (eg, Telesto and Calypso move approximately 60° ahead of and 60° behind Tethys in its orbit, respectively; the two "co-orbital" satellites Epimetheus and Janus circle Saturn at almost the same distance) have led to suggestions that these tiny bodies are collision fragments of larger objects. Judging from the cratering record preserved on satellites such as Iapetus and Rhea, one can expect very severe cratering to have occurred in the past, especially close to Saturn, since impacting objects will be accelerated significantly as they approach the planet. One model predicts that close satellites even as large as Dione had a significant probability of suffering a catastrophic impact during the age of the Solar System. The tiny irregular satellites, and the myriad that makes up the rings, may be debris from collisions.

Satellites of Uranus
In addition to providing detailed information on Uranus' five previously known satellites (Miranda, Ariel, Umbriel, Titania, and Oberon), Voyager 2 in January 1986 discovered ten smaller bodies. Two of these new satellites, Cordelia and Ophelia, orbit just inside and just outside the Epsilon Ring and probably play a shepherding role in confining the extent of this ring. They are among the smallest of the newly discovered moons, with diameters estimated between 25 and 30 miles (40 and 50km). Eight other new satellites — of which Puck, an irregularly shaped object about 100 miles (170km) across, is the largest — orbit between Miranda and the Epsilon Ring. Puck is known to be made of a very dark material (albedo about 0.07) that may be similar to that making up the ring particles. This material is definitely non-icy, unlike the surfaces of the five large satellites, and one suggestion is that it may be a carbon-rich residue left over from the irradiation of methane ice by electrons and protons trapped by the planet's magnetic field. Indications are that the other new moons also have very dark surfaces.

Miranda, *below,* is proof that even small icy satellites can have complex geologic histories. About half of the southern hemisphere of this 490km satellite consists of a rolling, heavily cratered terrain. The remainder is made up of three areas of younger banded and grooved topography, which were emplaced by fluid material from the interior. Evidence of pervasive and contained faulting is abundant. What is the energy source responsible? And when did all this take place?

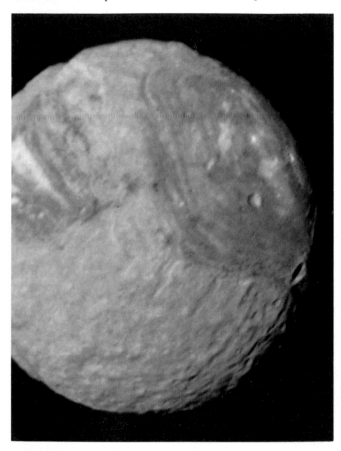

Of the previously known satellites, by far the most exciting turned out to be Miranda (diameter 300 miles/ 490km). The satellite's low density (1.3) indicates that it is primarily an icy body. Spectrally, only water ice has been detected on the surface, but suspicion is strong that other ices (ammonia, methane?) must be present, especially in the interior. Voyager images reveal that Miranda has had a remarkable geologic history. Its cratered terrain is broken by extensive patches of several distinct types of ridged terrain (some of which are vaguely similar to those found on Enceladus) and which clearly indicate the eruption of materials from the interior. The satellite's crust must be quite rigid today to support the large vertical relief (of at least 3–6 miles/ 5–10km) that is common. Some of this relief is associated with dramatic fault scarps, an indication of intense tectonic activity. Most of the complex geologic activity on Miranda may have taken place long ago, since most of the surface is heavily cratered. The energy source that produced this much activity on such a small icy satellite is a puzzle. By default, tidal heating has been postulated, but detailed models remain to be worked out.

The four remaining satellites come in two pairs, each pair consisting of objects that are closely similar in diameter and mean density, yet noticeably different in terms of their surface characteristics. This is especially true of the Ariel-Umbriel pair. Ariel (diameter 720 miles/ 1,160km) is the brightest of Uranus' satellites (albedo = 0.40), and like Miranda has had a geologically very active history. Much of the surface consists of a cratered terrain which has been pervasively rifted and locally replaced by smoother, less cratered areas which show some evidence of flow features. The density of craters on Ariel is the lowest of any of the Uranian satellites, but is still appreciable, indicating that much of the geologic activity evidenced by the surface took place long ago. Model calculations have shown that Ariel could be subject to significant tidal heating. Ariel's mean density (about 1.5) is within measurement uncertainties, identical to that of the other three larger satellites, Umbriel, Titania, and Oberon. Spectra of the surface show that a dominant constituent is water ice; the spectra may also contain some evidence of ammonia ice.

In spite of similarities in size and density, the appearance of Umbriel (diameter 740 miles/1,190km) is very different from that of Ariel. Umbriel's surface is much darker (albedo = 0.19), and shows little evidence of internal geologic activity. Impact craters form the ubiquitous surface features. The surface density of large craters (the only ones discernible at Voyager resolution) is similar to that seen in the lunar highlands. Umbriel's mean density is, within the uncertainty of the determination, the same as that of Ariel. Spectrally the surface also shows evidence of water ice. Therefore, it is likely that both satellites are closely similar in composition, and in the intensity of heat sources due to the decay of radioactive materials. An inescapable conclusion is that the vast difference in the geologic activity of Ariel and Umbriel reflects the fact that for Ariel tidal heating was important, whereas for Umbriel it was not.

Both Titania and Oberon are ice-rich satellites (density about 1.5), with similar albedos (0.28 and 0.24, respectively) and surfaces that spectrally show the presence of water ice. Titania (diameter 1,000 miles/ 1,610km) is Uranus' largest satellite. It has a heavily cratered surface, but one fractured by a planet-wide system of scarps and troughs. There is evidence of the emplacement of smoother material from the interior in association with some of these fractures, but the evidence is much more subtle than in the case of Ariel.

Oberon (diameter 960 miles/1,550km) was imaged less well and much less extensively by Voyager than Titania. Impact craters and their ejecta dominate the surface. There is some evidence of crustal faulting, but much less so than in the case of Titania.

Any discussion of surface features on Uranus' satellites must be qualified by the following caveat: due to very inclined axial orientation of the Uranus system, Voyager 2 imaged only one half (the southern hemisphere) of the satellite surfaces. Caution is in order in judging the current geologic state and past evolution of a body from such incomplete glimpses.

Satellites of Neptune

Two satellites of Neptune are known, although there is every reason to expect that additional ones will be discovered when Voyager 2 flies by the planet in August 1989. The smaller of the two, Nereid, orbits Neptune once every 365 days in a very eccentric orbit (e = 0.75). From Earth Nereid appears as a very faint + 19 magnitude object. Nothing is known about its size, nor about the albedo or composition of its surface. A likely guess is that the diameter lies in the range of 125–250 miles (200–400km).

The larger of the two satellites, Triton, is a most remarkable object. Earth-based spectroscopy indicates that it has at least a thin atmosphere, containing methane, and that there is methane ice on its surface. An absorption band in the spectrum near 2.1 microns has been identified tentatively with liquid nitrogen. At the very low temperatures of Triton's surface (50–60K), it is conceivable that the satellite is covered in part by an "ocean" of liquid nitrogen. Voyager 2 will make a close flyby of Triton to clear up this and other questions, including those concerning the size and mass of the satellite. There has been a long controversy over how big Triton is. Current estimates centre around a diameter close to that of our Moon (between 2,100 and 2,200 miles/3,400 and 3,600km), but the issue is far from resolved. Similar uncertainties surround estimates of the satellite's mass.

Triton's orbit is unusual in that it is close to its planet ($14.6\ R_N$) and retrograde. In principle, such an orbit is unstable, and tides should gradually bring Triton closer to Neptune. The timescale for the decay of Triton's orbit has been variously estimated to be as short as 10^8 years, and as long as 10^{10}, depending on very uncertain assumptions about how effectively tidal energy is dissipated within the planet.

Pluto-Charon

The discovery in 1978 that Pluto has a satellite came as a complete surprise. Charon orbits Pluto once every 6.4 days, at a distance of about 11,800 miles (19,000km). The planet itself is about the size of our Moon (diameter about 1,800 miles/3,000km); Charon's diameter is estimated at 800 miles (1,300km). Both Pluto and its satellite are believed to spin around their axes once every 6.4 days: that is, both spins are synchronous with the orbital period — a situation unique in the Solar System. The evidence for Charon's synchronous spin is indirect. In the case of the planet, it is based on observed periodic light variations due to spots on the planet's surface, rather than to mutual eclipses of Pluto and Charon, although these are also observed.

From the orbital motion of Charon one can determine the mass of the Pluto + Charon system. The result yields mean densities close to unity, suggesting that both bodies are predominantly icy. Evidence for methane ice has been found in Pluto's spectrum, as have indications of a thin atmosphere of methane gas. Comparable measurements cannot be carried out for Charon: it is about five times fainter than Pluto and very close to the planet, making it extremely difficult to separate its light from Pluto's in spectroscopic measurements.

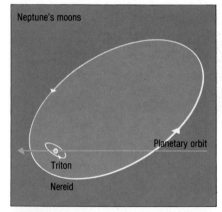

Neptune's moons

Triton

Nereid

Planetary orbit

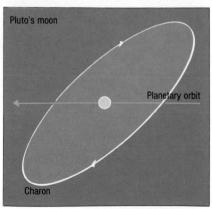

Pluto's moon

Charon

Planetary orbit

Neptune's two known moons have unusual orbits, *left above*. Nereid moves in a very elongated orbit, while the larger inner moon Triton circles Neptune in a retrograde sense: an inherently unstable situation. The discovery that Pluto has a moon, *left below*, came as a complete surprise.

An artist's impression, *above,* of what it might be like to stand on the surface of Triton. Due to its vast distance, it is difficult to learn much about Triton from Earth. Voyager 2 will flyby the satellite in 1989 and test current speculation about its nature.

mate period of several months. They are also called long-period and red variables. They are giants or super giants belonging to the disk population and with periods ranging from one hundred to one thousand days. The difference in their periods stems firm whether they are Population I or II. The former generally have periods longer than 200 days, while the periods of Population II stars tend to be shorter and less than 200 days.

The visual light variations of Mira variables range from 2^m5 to 11 magnitudes, but in the infrared their amplitudes are much smaller, and are less than 2^m5m. In fact, it is less than one magnitude for most of them. Each star has a mean cycle; that is, an average period during which it goes through a complete cycle from maximum and back to maximum again or from minimum to minimum. These periods for any one star can vary by about 15 per cent. Thus their periods are to some extent irregular.

Their amplitudes can also vary widely from one cycle to the next. If the changes in magnitude are plotted against time, a light curve can be drawn from the plotted points. They show a wide diversity. However, they fall into three main divisions. There are those stars that have rises steeper than the fall. These tend to have wide minima and sharp, short maxima. As the asymmetry becomes greater, the period lengthens. Stars with symmetrical curves have the shortest periods. A third group show humps on their curves, or have double maxima and have long and short periods.

Mira variables have late-type emission spectra. The longer the period, the later the spectral type. They show a spectral (or temperature) relationship which is a continuation of the same relationship found in CEPHEIDS. Both Mira variables and Cepheids are hotter at maximum, Mira variables having the smaller range in temperature. Pulsation is thus the underlying cause of their variability. Pulsations send running waves through the surface layers. *FMB.*

Miranda Satellite of Uranus, discovered in 1948. Though only 300 miles (490km) in diameter, Voyager 2 showed in 1986 that it has an amazingly varied landscape, with craters, grooved terrain, scarps, fractures and large enclosures. The Voyager pictures had a resolution of 2,000ft (600m), so that the views of Miranda are extremely detailed.

Mirzam The star Beta Canis Majoris. It is a pulsating variable with a very small range, and is the prototype of the variables known either as Beta Canis Majoris or BETA CEPHEI stars.

Missing mass problem The quandary whereby most of the matter in the universe seems to be either underluminous or else invisible. This dark matter surrounds individual galaxies and pervades clusters of galaxies.

The first evidence of dark matter came from the work of Jan Oort, who, in 1932, determined the thickness of our Galaxy in the solar neighbourhood from the locations of RED GIANTS in the sky. These bright stars are sufficiently numerous that a map of their distribution in space reveals that the galactic disk has a thickness of about 2,000 light-years. In addition, however, Oort used the distribution and speeds of the red giants in his study to calculate the vertical gravitational field of the Galaxy as well as the mass in our vicinity needed to produce it. A dilemma arose when Oort compared this calculated mass with the observed masses of all the stars and gas clouds in the solar neighbourhood. The observed mass is only about half that needed to confine the red giants to a disk 2,000 light-years thick. Thus half of the matter in our part of the Galaxy seems to be invisible.

Observations of the rotation of the Galaxy confirm the existence of this invisible matter. The rotation rate of the inner regions of our Galaxy is determined from bright stars and EMISSION NEBULAE. In the dim outer regions, radio observations of hydrogen and carbon monoxide in giant gas clouds provide the required data. In all cases, Doppler shift measurements (*see* DOPPLER EFFECT) are combined with information about the position of a source to deduce its circular velocity around the galactic centre. The results are best displayed on a plot of circular velocity against distance from the galactic centre.

Beyond the confines of most of the Galaxy's matter, the orbital velocity of outlying stars should decrease according to KEPLER's third law, just as the velocities of the planets decrease with increasing distance from the Sun. The dashed line in the accompanying diagram indicates such a Keplerian decline. However, the rotation curve of the Galaxy is nearly flat, even to a distance of 60,000 light-years from the galactic centre. This means that astronomers have still not detected the

edge of the Galaxy and thus a substantial quantity of invisible matter must exist beyond the observable stars, nebulae, and gas clouds.

The conventional picture of our Galaxy is a flattened disk 100,000 light-years in diameter surrounded by a spherical halo of old stars and GLOBULAR CLUSTERS roughly 130,000 light-years across. Recent analyses of rotation curve data strongly suggest that the galactic halo is embedded in an enormous "corona" that is roughly 600,000 light-years in diameter and contains at least 10^{12} solar masses of subluminous matter. Thus the corona

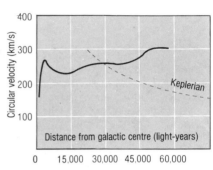

Missing mass problem: invisible matter

is at least five times more massive than the disk and halo combined. Most other spiral galaxies also exhibit flat rotation curves and thus they too must be surrounded by large, massive coronae.

The distribution of matter in such a galaxy can be compared with its luminosity by constructing the mass-to-light ratio (M/L). In essence, M/L tells us how much matter there is in a given region compared to the radiation that it generates. By definition, M/L for the Sun is 1.0 (that is, 1 solar mass produces 1 solar luminosity), and M/L is 0.8 for a typical globular cluster.

From a rotation curve, astronomers calculate the rate at which the matter density declines with distance from a galaxy's centre. However, the galaxy's surface brightness decreases much more rapidly with distance from the galaxy's centre. Since the brightness falls much more rapidly than the matter density, the mass-to-light ratio climbs to 50 or more in the outer regions of such a galaxy. This dramatic increase in mass-to-light ratio is the crux of the missing mass problem: one finds an increasing proportion of non-luminous matter as one moves outward from a galaxy's centre.

Apparently, the space between galaxies in a cluster is dominated by this non-luminous matter. In the 1930s,

M

Fritz ZWICKY and Sinclair Smith pointed out that the VIRGO CLUSTER must contain a substantial amount of non-luminous matter, otherwise there would not be enough gravity to hold the cluster together.

The mass of a cluster of galaxies can be determined from the VIRIAL THEOREM, which relates the average speed of the galaxies to the size of the cluster. For example, the COMA CLUSTER contains roughly 1,000 bright galaxies spread over a volume 10 million light years in diameter. Doppler shift measurements are available for 800 Coma galaxies and the average velocity relative to the cluster's centre is about 530 miles/s (860km/s). These data along with the virial theorem give a total mass of 5×10^{15} solar masses for the Coma cluster. If each of the 1,000 brightest galaxies have a mass of 10^{12} solar masses then we are able to account for only one-fifth of the cluster's mass.

Similar results are obtained for other rich clusters. Such clusters typically have mass-to-light ratios in the range of 200 to 350, clearly indicating a substantial presence of dark matter.

The issue of missing mass also arises on the largest cosmological scales. According to the General Theory of RELATIVITY, the expansion rate of the universe must be slowing down because of the mutual gravitational attraction of all the matter in the universe. Astronomers detect this cosmic deceleration by measuring the recessional velocities and distances of extremely remote galaxies. The purpose of such observations is to determine the so-called DECELERATION PARAMETER (q_0) which is directly related to the average density throughout space. Thus, by measuring the rate at which the cosmic expansion is slowing down, astronomers can deduce the average density of matter in the universe.

Observations indicate that the average density of matter in the universe is near the "critical density" of 5×10^{-30}g/cm³, which is equivalent to about three hydrogen atoms per cubic metre of space. This density is called "critical" because it ensures that the universe will just barely continue expanding forever, without ever collapsing back upon itself. However, the average density of matter that astronomers actually observe in space is about 3×10^{-31}g/cm³. Thus the observed matter is less than a tenth that needed to account for the behaviour of the universe.

A wide variety of proposals tries to account for the dark matter that apparently dominates the universe. A reasonable possibility involves low-mass stars, which are intrinsically very dim. Perhaps our Galaxy's corona and intergalactic space in rich clusters contain numerous extremely faint stars that have simply escaped detection.

A controversial proposal involves NEUTRINOS which are elusive particles created in great profusion during the BIG BANG. Until recently, these particles were assumed to be massless. However, some physicists now claim that the neutrino has a very tiny mass. If the mass of the neutrino is only a ten-thousandth of an electron mass, then there is an adequate abundance of neutrinos left over from the Big Bang to account for all the missing matter in the universe.

Speculative theories at the frontiers of theoretical physics suggest that particles such as "photinos" and "gravitinos" should be abundantly scattered around the universe. The existence of these particles is required by the latest attempts to devise a completely unified field theory that explains all the forces of nature. Like the neutrino, these hypothetical particles should be very difficult to detect.

A resolution of the perplexing issue of missing mass might involve a combination of both astronomical discoveries and major advances in theoretical physics. *WJK.*

Mitchell, Maria (1818–99) The first famous American woman astronomer, who founded the MARIA MITCHELL OBSERVATORY.

Maria Mitchell: founded an observatory

Mizar (Zeta Ursae Majoris) A second-magnitude star of spectral type A2 in the constellation of Ursa Major. With the naked-eye Mizar and its close neighbour, the fourth-magnitude ALCOR, can be seen as two distinct stars separated by 12 arc minutes. With a small telescope a fourth-magnitude companion can be seen 14 arc seconds from Mizar. These two stars form a very long-period binary system and the spectrograph shows that Mizar A and Alcor are close binaries and that Mizar B may even be a spectroscopic triple. *RWA².*

M numbers The numbers allotted to clusters and nebulae by Charles MESSIER in his famous catalogue published in 1781.

Molniya satellites Soviet communications satellites placed in highly eccentric orbits with apogee in the northern hemisphere.

Monoceros (the Unicorn) A constellation adjoining Orion; it is crossed by the celestial equator. It has no star brighter than magnitude 3.7, but it is crossed by the Milky Way, and contains some famous clusters and nebulae, including the ROSETTE NEBULA NGC2237-9 and the open cluster M50.

Monochromatic filter A filter that transmits light of a small range of wavelengths or colour. The smaller the range, the more monochromatic the filter. Most work by destructive interference of all wavelengths except that required. This is done by making the light suffer multiple reflections between parallel layers of accurately spaced material. These filters are used for observations at specific wavelengths, for example, the strong hydrogen emission line at 656.3nm, to give results unconfused by possibly unrelated emission at nearby wavelengths.

Montanari, Geminiano (1633–87). Italian astronomer who discovered the variability of ALGOL (1669) and invented a primitive form of micrometer.

Month A unit of time based on the synodic period of the Moon (the mean time interval between successive New Moons) which is equal to 29.5305882 days. Since there is not a whole number of months in a year, the month and the year cannot simply be reconciled into a common calendar. However, 235 lunar months is almost exactly equal to 19 years (*see* METONIC CYCLE). While the modern civil calendar is based on the year, the dates of religious festivals (such as Easter) are still set by reference to the lunar month.

Moon The Earth's natural satellite. By far the closest of the Earth's neighbours in space, the Moon has been observed from Earth since antiquity. It is the best-studied body in the Solar System in terms of Earth-based remote sensing techniques and investigations from various spacecraft missions. These include the ZOND flyby probes, the RANGER spacecraft, which impacted the Moon, the LUNA and SURVEYOR soft-landers, the LUNOK-HOD remote-controlled roving vehicles, the ORBITER spacecraft and APOLLO command/service modules, which orbited the Moon, the Apollo manned lander missions, which collected surface samples, and the later Luna missions, which included automated sample return.

The evolution of the Moon's surface and interior has been deduced from many lines of evidence. Radio-isotope methods can be applied to lunar rocks returned to Earth to find the time since they were most recently molten or subjected to violent stresses. The bulk chemistries and the mineral contents of these rocks are a guide to the temperatures and pressures at which they were formed. Since some rocks have been erupted from the interior as lavas and others have been excavated from considerable depths during the formation of impact craters and basins, they give information about the lunar interior as well as the surface.

The Apollo landing missions placed seismometers on the surface in several locations. The depths of the boundaries between various kinds of rock layers within the Moon, together with the densities and elastic properties of these rocks, have been deduced by observing the speeds and arrival patterns of seismic waves from natural and artificial "moonquakes". The orbiting parts of the Apollo spacecraft carried detectors for finding the amounts of elements such as aluminium and silicon in the lunar surface layers by measuring the X-rays being produced from the excitation of these elements by bombarding COSMIC RAYS.

Finally, other information has been obtained by geological mapping and by spectroscopic examination of images of the lunar surface made from Earth and from spacecraft at wavelengths in the visible and infrared parts of the spectrum. For example, the relative ages of different parts of the lunar surface can be estimated by counting the numbers of small impact craters seen on them, while the absorption, at particular wavelengths, of the solar energy reflected from the surface indicates the presence of particular minerals in the rocks.

Using a combination of all these sources of information, it has been shown that the Moon formed about 4,600 million years ago. It is probable that the outer 60 to 120 miles (100 to 200km) of the Moon melted and then cooled to form a crust of anorthositic rocks, these forming the bright part of the lunar surface as seen from Earth. During this early time (perhaps the first

Moon: taken by the Lick 36in refractor

10 million years) a small, iron-rich core also probably formed at the centre. Over the next 500 to 700 million years the crust was heavily bombarded by large and small meteoroids to produce craters and basins, the largest basins mainly forming between 4,100 million and 3,800 million years ago. The general bombardment continues down to the present time, but at a much slower rate.

Subsequent further cooling of the outer layers was overtaken by the heating of the deeper interior due to the decay of naturally radioactive elements. Partial melting of the lunar mantle took place and molten rocks of basaltic composition found their way to the surface, mainly during the period from 3,900 million to 3,300 million years ago. The easiest routes were those involving the fractures in the crust beneath and around the edges of large basins, and lava flows from the resulting volcanic activity progressively flooded the interiors of basins and low-lying areas to form the dark mare areas easily visible from the Earth. The word mare (Latin for sea; plural maria) has been used since the time it was thought that these areas were water ponded in depressions. The term terræ (singular terra; Latin for land) is now only rarely used for the highlands.

An overall pattern of net internal heating of the Moon with a consequent tendency toward slight global expansion, prior to about 3,500 million years ago, followed by net cooling with a tendency toward global compression, is supported by the analysis of lunar surface features which indicate the presence of stresses. These features include ridges (called "wrinkle ridges") which meander across the surfaces of the mare lava flows; linear and arcuate rilles (valley-like depressions with parallel, straight or gently curved sides); and the nearly straight segments of polygonal crater rims. The linear and arcuate rilles, apparently formed when blocks of the crust sank between parallel tensional fractures, are generally found on the older surfaces, whereas the ridges, assumed to be the products of compressional folding of the surface rock layers, are found mainly on the younger mare surfaces.

The orientations of most of the Moon's linear surface features (fractures, ridges and crater rim segments) form simple patterns related to stresses induced either by the formation of major impact basins or by the tidal distortion of the Moon by the Earth. The Earth causes tides in the body of the Moon in the same way that the Moon causes tides in the oceans and the body of the Earth. The consequent flexing of the rocks causes stresses that can lead to fractures. Currently, the Moon rotates in such a way as to keep essentially the same face pointed permanently toward the Earth, leading to a simple pattern of tidal fractures, but there is evidence from the ancient parts of the fracture systems, strongly supported by theoretical analyses of the tide-raising process, that the Moon was once closer to the Earth and rotated freely on its axis. There may then have been a significant tidal contribution to the Moon's internal heat sources.

Recently, Earth-based spectroscopic measurements on small, dark areas in the otherwise bright lunar highlands have indicated that volcanic activity may have begun earlier in lunar history than previously thought, with eruptions in the highlands forming deposits that have been masked by the crustal disruption from later impacts. It is clear that the complete story of the thermal history of the Moon, especially

M

in terms of how much of the interior was molten at a given stage in lunar history, is far from worked out.

Equally uncertain is the origin of the Moon. Theories of its formation include the splitting (fission) of a larger body in to the Earth and Moon, either as a result of its rapid spin rate or its collision with a separate body; the capture by the Earth of a Moon formed elsewhere in the Solar System; and the simultaneous accretion of the Earth and Moon from a common reservoir of primitive material. The overall chemical composition of the lunar interior can be estimated from the surface chemistry and the inferred internal structure, and it is found that the Moon is enriched, relative to the Earth, in refractory elements (those with high melting and boiling temperatures) and depleted in volatile elements and compounds, especially hydrogen, carbon, nitrogen and water. These differences are easily explainable if the two bodies formed in different parts of the Solar System, but can also currently be accommodated by suitable adjustments to the other theories, leaving the matter unresolved. *LW*. *See also* entries under LUNAR.

Moon hoax A joke due to a New York journalist, Richard Locke, in 1835. In the New York *Sun* Locke published articles in which it was claimed that Sir John HERSCHEL, using powerful telescopes at the Cape in South Africa, had discovered weird and wonderful creatures on the Moon. The articles were cleverly written, and were quite widely accepted, though the hoax was soon exposed. Herschel apparently took it all in good part! *PM*.

Moon illusion The false impression that the full moon looks larger when low down than when higher in the sky.

Moonquakes Slight ground tremors on the Moon, due to mild internal disturbances.

Moretus High-walled lunar crater; 70°S, 8°W. it is 65 miles (105km) in diameter, with a particularly lofty central elevation.

Morgen-Keenan classification The categorization of a stellar spectrum from the visual appearance of its ABSORPTION features. The astronomer's primary source of physical and chemical information about a star is the stellar spectrum. Using a prism or grating, starlight is broken into its different frequencies and its various messages decoded. Systematized in the early 1940s (by W.W. Morgan and P.C. Keenan), the "MK" classification established formal rules for describing a star by spectral type (sequence, from hot to cold surface temperature, O, B, A, F, G, K, M; other chemically peculiar cool stars N, C, R, S) and luminosity class (I = supergiant; III = giant; V = dwarf with interpolation to Ia, Iab, Ib, II, IV and VI as required). Early spectra were photographic with plates whose best response was in the blue. Consequently the MK scheme depends upon criteria at low dispersion at blue wavelengths. It compares nearby pairs of absorption lines in the spectrum and determines relative strengths, or absences. By eyeball alone, one can categorize a star in some detail following the MK procedure.

Absorption lines are due to atoms in the cooler outer layers of a star absorbing photons from the deeper hotter atmosphere. Their patterns of frequencies are uniquely diagnostic of specific atoms and molecules. Line shapes (sharp, shallow or broad, deep) also provide clues about the extent of the atmosphere and the luminosity class (supergiants are vastly extended with sharp shallow lines; the compact dwarves show broad deep lines).

Basic MK features are as follows: O, B, the hottest stars (11,000 – 50,000K) — a few weak lines of neutral and singly-ionized helium; A (typically 9,000K) — dominated by a series of deep hydrogen lines; F (7,000K) start to show a wealth of metallic lines (Fe, Mg, Ca); G (solar: 6,000K) reveal some molecular absorption features against the busy pattern of metal lines; K (4,500K) — molecules and atoms; M (3,000K) — heavily mutilated by broad absorptions of TiO molecules; N, C and R — carbon-bearing molecular bands; S — a mixture of M- and C-type characteristics, together with exotic rare earth oxides (ZrO, LaO). *MC*.

Morning Star (Phosphorus). Old name for Venus as a morning object before sunrise.

Mount Huygens High peak in the lunar Apennines; 20°N, 3°W.

Mount Palomar Following the completion in 1917 of the 100in (254cm) telescope at MOUNT WILSON, several designs for a larger telescope were put forward and by 1923 $6,000,000 had been gifted by the Rockefeller Foundation to finance it.

The main mirror was manufactured in Pyrex in 1934 by Corning Glass Works. It was a little more than 16½ft (5m) in diameter and had a honeycomb back. It thus weighed only 20 tons (20,320kg), about half the weight of a solid piece of Pyrex the same size.

The mirror, which took eight months to cool and anneal, was twice in danger of being flooded by the river on which the glass works were situated.

The telescope tube weighing 125 tons, together with the yoke of 300 tons

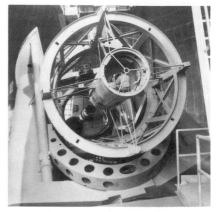

Mount Palomar: the 200in reflector

and the mirror, took six years to make. They were made with such accuracy that the whole telescope could be moved by a one-sixteenth horsepower motor.

The site chosen for the telescope's erection was long hogback in the San Jacinto Mountains an hour's drive to the east of San Diego, Southern California. Although many telescopes have been installed at the observatory, Mount Palomar is still recognized as the home of the 200in (508cm) telescope. When completed in 1946, it was the largest telescope in the world for decades and today is still in the forefront of astronomical research.

The first Director of the observatory, and its guiding light, was George Ellery HALE, who died in 1938.

Today, a modern highway, first laid by prisoners from the local Alimony Jail, leads to the group of telescopes. Dominated by the dome of the 200in, possibly the greatest of all telescopes, the site is still visited by thousands of people. *RFT*.

Mount Stromlo Observatory One of Australia's leading observatories, located on the south-western outskirts of Canberra, the Australian capital, at an altitude of about, 2,500ft (760m). It was founded in 1924 as the Commonwealth Solar Observatory, and during the 1920s and 1930s important research was done on upper atmospheric physics and the solar spectrum.

Dr R.v.d.R.Woolley took over as director in 1939, and after World War II succeeded in establishing a Department of Astronomy at the newly-founded Australian National University in Canberra, and had the observatory affiliated to it. He also shifted the observatory's research focus from solar to stellar work, and in the process introduced a name change.

The main instruments used at the time were 30 and 50in (76 and 127cm) reflectors, but in 1953 they were joined by a new 74in (188cm) telescope, which for 20 years was the largest telescope in the southern hemisphere. Many pioneering studies of the southern skies were carried out with these three instruments, particularly during the later directorship of Professor Bart BOK (1957–66).

One of Bok's major achievements was to establish a Mount Stromlo field station at SIDING SPRING in north-western New South Wales, where 16, 24 and 40in (40, 61 and 101cm) telescopes were set up away from the rapidly-deteriorating skies of Canberra

Although Siding Spring is now the main research facility for Stromlo staff and students (there is also a new 91in/230cm reflector there), the 30 and 74in reflectors at Mount Stromlo are still in regular use for photoelectric and spec-

Mount Stromlo: magnificent research

troscopic work, and the 50in telescope is currently being modified for infrared research. In addition, a 26in (66cm) refractor installed by Yale and Columbia universities in 1954 is used occasionally for double star work.

Over the years Mount Stromlo Observatory has built up a magnificent research record, and important contributions have been made to our understanding of stellar structure and

evolution, globular clusters, various types of variable stars, planetary nebulae, supernova remnants, the Magellanic Clouds, and extragalactic nebulae. In addition, there have been exciting new developments in astronomical instrumentation, while the reputation of the observatory, coupled with the promise of the southern skies, continues to attract top graduates from around the world as PhD students. *WO.*

Mount Wilson In 1902 Andrew Carnegie, as part of a $10,000,000 grant to science, provided sufficient funds for George Ellery HALE to set up a solar observatory in the clear air of California.

Situated in the San Gabriel range near Pasedena, Mount Wilson was selected and in 1904 the Snow telescope, a 60ft (18m) SPECTROGRAPH with an 18ft (5.48m) focus, was installed. Three times larger than anything yet produced, the new instrument soon produced research results on SUNSPOTS that led to the development of Mount Wilson as a major astronomical centre.

The French glass makers Saint Gobain provided a 5 ton (5,080kg), 60in (152cm) disk 8in (20cm) thick, which was also financed by Carnegie and took four years to turn into a telescope mirror. The designer and optician George Willis RITCHEY added many innovations to this new telescope, including the first sheet-steel dome and observatory, which broke with the convention of the massive brick building.

By 1908 the 60in telescope had been installed and was working well. A decision was taken to attempt to build a larger telescope, initially designed as an 84in (213cm) but later, due to the generosity of John D. Hooker, re-designed as 100in (254cm) in diameter.

Saint Gobain were again asked to cast the disk, but as it had to be poured from three ladles temperature differences created layers of bubbles in the glass. Four disks were cast, the best being sent to California. Ritchey ground the disc flat and, considering it unsuitable to make into a telescope, used it as an assembly surface for more than a year while the project made no progress.

In 1911, on the advice of Dr. Arthur Day, Ritchey, still sceptical, started to manufacture the mirror. That took the next six years. Unconvinced of its success, Ritchey refused to attend the first tests in 1917 and it was left to Hale

and two fellow astronomers to try the telescope. Not daring anything else, Hale first looked at Vega — and found

Mount Wilson: the 100in reflector

the telescope did not work. It was not until five hours later that it was realized that the dome had been open all day and that temperature changes had rendered the telescope useless. By holding the dome at night temperature, the telescope — the world's largest at that time — was a success and assured the prominence of Mount Wilson for the next three decades as a major centre for astronomy.

Now more than 70 years later, the lights of Los Angeles intrude on the night sky, severely curtailing the use of this famous observatory. *RFT.*

Moving groups Groups of stars, often widely scattered, that share the same motion through the Galaxy.

Although the birthplaces of stars are hidden from view by opaque clouds of dust and gas, we know something of the process entailed. Some stars form from small, isolated clouds known as BOK GLOBULES. Many stellar nurseries are, however, huge. The ORION NEBULA for example is a maternity ward for thousands of stars. Those stars have been born in a cluster, and many spend most of their lives therein. Gradually the natal shrouds of gas and dust will be burnt and blown away, passing through the phase we see in the PLEIADES, where some gas clings to the stars, and eventually ageing to a loose group of old red giants.

The individual stars must remain in the cluster if their speeds are lower than that needed to overcome the gravitational pull of all their neighbours. At birth this is inevitably so, but as the stars mill around some will be accelerated by a close passage to one or more

M

M

of their kin, gaining in the process enough of a fillip to escape. Clusters gradually expand in size and shrink in population, but cannot be totally disbanded unless they happen to collide with something bigger, like another cluster or a nebula.

This being so, we might expect to find clusters of all sizes and complexities persisting late into the lifetime of the member stars. Indeed we do. The moving groups, also sometimes known as moving clusters, are the sparsest and most diffuse of clusters. They can best be illustrated by an example. The Ursa Major group includes five of the stars of that well-known asterism the Plough or Big Dipper. The star at the tip of the tail, Eta UMa, and the pointer that lies nearer to the pole, Alpha UMa, are the non-members. Over long periods of time the motion of these two stars will distort the present star pattern. The other five stars share the same basic motion through the Galaxy and have about the same distance from us. They therefore resemble a very open cluster of a few stars. When we plot the motions of other bright stars through the Galaxy, however, we find more showing the same motion as the Ursa Major stars. Sirius is one such star, and others include Delta Leonis, Beta Aurigae, Beta Eridani and Alpha Coronæ Borealis. In total about 100 members of the group are known. Many are widely scattered across the sky from Ursa Major. Their separation in space is moderately large, cartainly larger than in a normal cluster. Our Sun is taking us through the outskirts of the group, which is why its member stars can appear on all sides of the sky.

Other moving groups are known, largely thanks to the work of the American astronomer Olin Eggen, currently based at CERRO TOLOLO INTERAMERICAN OBSERVATORY, in Chile. Many of these centre on well-known clusters such as the Pleiades and the HYADES in the constellation Taurus, and PRÆSEPE in Cancer. It appears that the Sun is moving through more than one moving group, which also means that groups pass through one another, each carrying a memory of the motion of its birthcloud.

As moving groups age, so they become sparser and harder to recognize. Eggen believes that he has identified one group as old as the Sun itself, roughly half the age of the Galaxy. If he is right, then this collection of stars has made many complete circuits of the Galaxy without being fully disbanded. *DAA*.

Mozart Crater on Mercury, 140 miles (225km) in diameter, at the very edge of the region surveyed from Mariner 10: 8°N, 191°W.

M stars The defining characteristic of an M-type spectrum is the presence of ABSORPTION bands of titanium oxide molecules.

The M dwarf stars, the hottest having surface temperatures of 3,500K, define the lower end of the MAIN SEQUENCE. According to theory, stars with masses less than 0.08 of the solar mass do not ignite hydrogen-burning nuclear reactions in their interiors; they simply contract and cool as BROWN DWARFS. As a result, the main sequence terminates at its low end with M stars having surface temperatures of about 2,500K. The M dwarfs cover the mass range 0.08 to 0.5 solar mass and have radii between 0.1 and 0.6 times that of the Sun. At their most massive their luminosities are nearly one-tenth that of the Sun, but this diminishes rapidly at lower masses, to become less than one-thousandth of the Sun's luminosity at the bottom of the main sequence. As these very low-mass stars are also very cool, most of their radiation is emitted in the infrared and their visual luminosity decreases even more precipitately: from one-twenty-fifth of the Sun for the hottest M dwarfs to one-ten-thousandth for the coolest.

The lifetimes of M stars on the main sequence are very long–greater than the age of the Galaxy. As a result, all of the M dwarfs formed during the lifetime of the Galaxy are still with us. This, and the fact that the process of star formation makes far more low-mass than a high-mass stars, means that they are the most common of dwarf stars. Because of their individual low luminosities, they do not contribute greatly to the total radiation from the Galaxy. However, they are so numerous that most of the luminous material in the Galaxy is in the M dwarfs.

All of the M dwarfs (with the exception of the small number in short-period binary systems) are slow rotators; their periods range from a few days to a few months. From computations of the structure of low-mass dwarfs it is found that convection in the envelope extends deeper at lower masses until at 0.3 solar mass it reaches the centre of the star. Consequently, M dwarfs cooler than about 3,000K have their interiors completely mixed.

M-type giants and supergiants are all evolved stars with masses much larger than those of the M dwarfs. Their surface temperatures are typically a few hundred degrees cooler than dwarfs of the same spectral type. The most luminous M supergiants rival the energy radiated by the luminous hot stars (up to a million times the luminosity of the Sun). As the temperatures of the M supergiants are so much lower it follows that they must have enormous surface areas. The largest have radii several thousand times that of the Sun — as large as the orbits of Saturn or Uranus around the Sun. Even the coolest M giants have radii as large as that of the Earth's orbit. Such distended stars are unstable so the M supergiants and most of the M giants are irregular or semi-regular VARIABLE STARS. For example, the bright star BETELGEUX (spectral type M2Iab) varies with a range of about half a magnitude on a timescale of years. Many M giants vary in brightness more regularly and are known as long-period variables, or MIRA VARIABLES, with periods ranging from 100 to 2,000 days.

Variability of a different kind is also common among the M dwarf stars. The surface activity visible on the Sun — spots and flares — associated with magnetic fields is common among all low-mass stars, but the very low luminosities of the M dwarfs makes such activity more readily observable. Most low-mass stars are FLARE STARS (UV Ceti variables) or spotted stars (BY Draconis variables) with dMe spectral types. The EMISSION LINES of ionized calcium, which on the Sun are associated with regions of magnetic activity, are strong in the dMe stars and observations made over many years show that intensities vary regularly, enabling both the rotation periods of the stars to be measured and also longer periods, probably analogous to the sunspot cycle in the Sun.

At the low temperatures of M star atmospheres molecules can form and remain intact. In the infrared spectra of M stars, molecular bands of carbon monoxide (CO) and OH are strong, and at temperatures below 3,500K bands of H_2O (water, or, more specifically, steam) become strong. Certain other bands, such as those of calcium hydride (CaH), have strengths that are sensitive to the pressure in the atmosphere of the star and therefore are useful as indicators of the luminosity of the star. The high luminosity and low surface gravity of M giants and supergiants causes a wind of gas in which the mass loss can be as high as one solar mass in 100,000 years. In the relatively cool environments surrounding the M stars, molecules and solid particles quickly condense from the expanding gas shells. At the highest rates of mass loss the stars themselves are hidden from view by the resulting dust shells, but can be detected by their far infrared radiation or OH microwave emission (OH/IR stars). At lower rates the M star is visible, but spectral bands of the dust particles, which are mostly composed of silicates, appear in the infrared. The shells themselves, whose temperatures are only a few hundred degrees Kelvin, are detectable from their far infrared radiation. They also can be observed from the light that they scatter: the shell around Betelgeux is in this way observed to have a diameter of 3 minutes of arc. At the low velocities of expansion of the shells (typically 6mps/10km/s) material at the outer edge of this shell left Betelgeux 10,000 years ago.

As with the K giants, the M giant stars can at some stage in their evolution dredge up nuclear-processed material from their interiors. This material is rich in carbon and heavy elements and when sufficient has mixed into the atmosphere of an M star its spectrum and hence classification changes. In this way, many M giants pass through the N-type or S-type spectral classes.

For all stars that reach the M giant or M supergiant phases, these represent their final period of life as an extended star, at the end of which their envelopes are ejected, leaving the hot core to settle down as a degenerate star. This process can be a spectacular SUPERNOVA explosion for the massive M supergiants, or a quieter event producing a PLANETARY NEBULA in the lower mass stars.

Bright examples (M stars appear red to the eye. There are no M dwarfs visible to the unaided eye): Betelgeux M2Iab, Gamma Crucis M3II, Antares M1Ib, Alpha Ceti M2III, Mu Cephei M2Ia. *BW.*

Mu Cephei An irregular VARIABLE STAR known as the "Garnet Star". It is an M2 supergiant, absolute magnitude −7, which fluctuates in apparent magnitude between 3.6 and 5.1. Its distance is about 1,600 light-years.

Mullard Radio Astronomy Observatory The MRAO is the radio astronomy observatory of the University of Cambridge and one of the research divisions of its department of physics (Cavendish Laboratory).

The first observations of natural radio emission of extra-terrestrial origin, made by Jansky and Reber in America in the 1930s, were extended by the discovery of radio emission from the Sun during World War II. The explosive development of radio astronomy in the post-war years stemmed in part from the return to civilian life of physicists who had worked on radar and similar radio technologies. In Cambridge there was an existing radio research group in the Cavendish Laboratory under the direction of J.A.Ratcliffe, which resumed peacetime research in 1945 when Martin RYLE and others joined the

Mullard Radio Astronomy Observatory

group. In the British Commonwealth similar studies developed, for example, in Manchester (Nuffield Radio Astronomy Laboratories) and in Australia.

In Cambridge the first radio telescopes were built on land in West Cambridge close to the university rugby football ground. Attention was soon directed to the study of "radio stars", of which only a few were then known, and which were recognized to be beyond the Solar System. This has long continued to be the major interest of the MRAO. The Manchester and Cambridge groups developed different types of telescope, the former concentrating on large single paraboloidal "dishes", and Cambridge on the design and construction of smaller but widely spaced aerials used as an interferometer to attain a better determination of the position of the sources in the sky. In 1951 accurate determinations of the positions of the bright sources in Cassiopeia and Cygnus by F.Graham-Smith (now Astronomer Royal) led to their optical identification respectively as a new type of supernova remnant, and a distant peculiar galaxy. A four-element interferometer was completed in 1953 and, operating at a frequency of 159 MHz. was used eventually to

compile the "Third Cambridge" catalogue of 470 radio sources, first published in 1959 and revised in 1961: modern designations of radio sources as, for example, 3C.273 (the serial number of the object in that catalogue) are the preferred ones for the more intense sources of the northern sky.

In 1956, by the generosity of Mullard Ltd., and with the support of the Department of Scientific and Industrial Research, it became possible to vacate the earlier site and build larger instruments on a tract of land at Lords Bridge 5 miles (8km) to the south-west of Cambridge, the present MRAO. The first instruments built here used the powerful new technique of APERTURE SYNTHESIS developed at the observatory by M.Ryle and A.Hewish, in which an array of both fixed and moving antennæ or dishes is used to attain very high angular resolving power. The method is now widely used in radio astronomy and was one of the several contributions to the subject that earned these two scientists the joint award of the Nobel Prize for physics in 1974. The major aperture synthesis telescopes with dates of completion are:

The 4C Telescope (1958). This consists of an east-west parabolic trough of stainless steel wires 1,558 × 69ft (475 × 21m) in area, in conjunction with a smaller aerial movable north-south. This was used for the observation of the 5,000 sources of the 4C catalogue.

The One-Mile Telescope (1964). In this and later instruments the rotation of the Earth is used to produce by aperture synthesis the equivalent (in this case) of a dish one mile (1.6km) in diameter. Along the east-west line are three dishes of 60ft (18m) diameter, one being movable along half a mile (0.8km) of very accurate rail track of 43ft (13m) gauge.

The Half-Mile Telescope (1968). This specialized instrument is fitted with a multi-channel 1420 MHz (21cm) receiver for the observation of the radio emission from neutral hydrogen gas in the Galaxy and nearby galaxies. The two movable dishes of the total of four 30ft (9m) paraboloids share the track of the One-Mile Telescope.

The 5km Telescope (1972). The acquisition of a narrow tract of land running in an east-west direction and some 3 miles (5km) long, for the construction of this powerful instrument, was fortuitously solved by the closure of the adjacent Cambridge to Bedford railway, including Lords Bridge station, which almost exactly met the requirement. There are four fixed and

M

four movable dishes of 43ft (13m) diameter, the latter movable on a specially constructed track of remarkable precision permitting the location of the dishes to better than 1mm. At the highest of the three operating frequencies the instrument has a resolving power of 0.3 arc sec, equivalent to that of the best optical telescopes.

In addition to these instruments there are other more specialized arrays. The 151 MHz synthesis array (1975) consists of 60 fixed aerials (like TV aerials) along the line of the 5km Telescope, and is used as a fast survey instrument giving an angular resolution of 1 minute of arc. The 3.6 hectare array is a large "forest" of 4,096 dipoles covering this area. It was designed for the study of the SCINTILLATION of small-diameter radio sources that is caused by the irregular density in the solar wind. It was in the course of the analysis of the records produced by an earlier version of this instrument that Jocelyn Bell in 1967 recognized the important new type of rapidly and regularly fluctuating radio sources now known as PULSARS.

The main thrust of the work of the MRAO from its early years has been the invention and construction of new types of radio telescopes, and their use in the discovery and investigation of the number, location and structure of the many thousands of discrete radio sources in the northern sky. The early accurate measurements of the positions of these objects enabled optical astronomers (whose large telescopes were then mostly also in the northern hemisphere) to identify hitherto unrecognized but remarkable classes of objects such as giant galaxies and QUASARS, and contributed significantly to the development of whole new areas of astronomy that are now being investigated not only by optical and radio astronomers, but also by studies from spacecraft at X-ray, ultraviolet, and infrared wavelengths.

The high angular resolution of the telescopes at the MRAO is used to make detailed maps of the radio emission from such objects as galaxies and the remnants of supernovæ, which aid the understanding of the origin of the radio energy they emit. The telescopes also give information about the large-scale structure and evolution of the universe, since the faintest extragalactic sources are the most distant known objects and show the universe as it was in the remote past.

In addition to these extragalactic and cosmological investigations the work of the observatory has diversified over the years. The MRAO has more recently become closely involved in the development of the joint UK-Netherlands millimetre-wave telescope, a 49ft (15m) diameter high-precision dish which has been installed at a high-altitude site on Hawaii. This telescope (the James Clerk Maxwell Telescope) is particularly important for studying the cool gas clouds from which stars are formed in the Galaxy. Other important work includes the study of disturbances in the solar wind, such as interplanetary shocks which affect spacecraft, and improved methods of image enhancement and analysis. A new type of telescope using a method of optical aperture synthesis, analogous to the radio method, is also under development.

The major instruments of the observatory are financed by government grants allocated by the Science and Engineering Research Council. The total complement of research and supporting staff is about 80; the Director for many years was Professor Sir Martin Ryle, who died in 1984 after some years of declining health and was succeeded by Professor Antony Hewish. *DWD.*

Multi-Mirror Telescope A reflecting telescope of entirely novel design that makes simultaneous use of six 72in (183cm) diameter, relatively lightweight mirrors, the six images being brought together at a common focus. The six mirror blanks were made available to the University of Arizona in 1969 by the United States Air Force, at which time Smithsonian Astrophysical Observatory astronomers were investigating multiple-mirror designs for their Mt. Hopkins Observatory (now the Whipple Observatory) located some 40 miles (65km) south of Tucson and the university. The two groups joined forces and resources, culminating with the existing design. The MMT occupies the very apex of Mt. Hopkins, at an altitude of 8,550ft (2,600m) and about 1,000ft (300m) higher than the rest of the Whipple Observatory.

The six component telescopes lie at the apices of a regular hexagon, adjacent optical axes being separated by about 8.5ft (2.6m). They are of traditional Cassegrain design except that the convergent beams from the secondary convex mirrors do not pass through holes in the primary mirrors. Instead, they are reflected radially inward toward the central axis of the hexagon by elliptical-plane mirrors located just above the primary mirrors. Before reaching the central axis, however, the beams are intercepted by six small plane mirrors which combine the beams to a common focus on the central axis. Because of the small tilts of the beams, the images are not exactly coplanar. The resulting combination has the light-gathering power of a 176in (4.5m) telescope, and is thus currently the world's third largest. Each primary has a focal length of just over 16ft (5m) (f2.7), the effective focal length being about 190ft

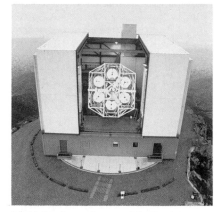

Multiple mirror telescope: Mt. Hopkins

(58m), giving an image scale of 3.5 arc sec/mm.

A major problem was to design a practical system that would maintain all six telescopes both in alignment to a fraction of a second of arc and also in optimum focus. This was solved by mounting a 30in (76cm) Cassegrain reflecting guide telescope on the central axis, and employing a laser as an artificial star. The 30in diameter parallel beam of laser light emerging skyward from this telescope is sampled by six corner cube-type optical systems that are located at the top of the main framework. These reflect the samples to servo systems via the edges of the six main mirrors. These servo systems make minuscule tilts to the six secondary mirrors by means of variable voltages applied to piezo-crystal stacks. The whole system is computer controlled.

The MMT was designed specifically to operate efficiently in the infrared portion of the spectrum as well as the visual and ultraviolet parts. Thus an IR sensor placed at the main focus "sees" an absolute minimum of mirror supports, none of the telescope framework, and no reflections of the observatory floor, walls, etc. For wavelengths of 10 microns (one-hundredth of a millimetre) or more, any opposite pair of mirrors can be used as an interferometer, giving much higher resolution when the incoming beams are maintained in phase with each other.

Because of the large overall lateral dimensions of the main framework of the MMT — about 25 by 30ft (7.6 × 9m) — and from various other considerations, the instrument is not mounted equatorially, but in an altazimuth configuration. Furthermore, the telescope itself does not rotate within a conventional building with the usual hemispherical dome — in this case the approximately rectangular building rotates along with the telescope. Each end of the building houses several rooms which include the control room, laboratories, storage rooms, living-dining area, etc. The basically simple, rectangular design of the building made construction both more straightforward and less expensive, factors that may equally be applied to the instrument and its mounting. The MMT was dedicated in May 1979 and has been in constant use since then. It has added significantly to the body of data in the fields of Solar System, galactic and extragalactic astronomy, and has been indispensable in the discovery and recognition of the existence of GRAVITATIONAL LENSES ("twin" quasars), of protogalactic clouds, and of a new type of WHITE DWARF star. *EAW*.

Multiple stars These are gravitationally connected groups of stars with a minimum of three components. The point at which highly multiple stars and star clusters overlap is not clear, but systems such as CASTOR and Alpha² Capricorni, with six components each, are still regarded very much as multiple stars. In these systems there is a recognized hierarchy or order. In Castor, for instance, the two main pairs rotate about a common centre of gravity whilst a pair of cool red dwarfs rotates about the same centre, but at a much larger distance. The proportion of double stars that are known to be triple may be as high as 30 per cent with the same proportion of triples known to be quadruple, and so on. It is estimated that systems such as Castor compose about 0.1 per cent of known double stars.

The similarity between multiple stars and star clusters suggests that they form in similar ways, ie, by condensation from interstellar clouds. A good example of this is the Trapezium in the ORION NEBULA. This consists of four very young stars enmeshed in nebulosity with a further five possible

members nearby. The Trapezium is the prototype of a special type of quadruple star in which very little relative motion is observed. It may be that the stars will spend most of their lifetimes in this configuration. Many other examples of Trapezium systems are known. In hierarchical quadruple systems the most common situation consists of two pairs revolving about each other, ie, EPSILON LYRÆ.

Formation of multiple stars by chance encounter, such as that postulated in the formation of binary systems cannot be taken seriously. If binary stars are formed by fission from a single PROTOSTAR then multiple systems born in the same way might be expected to be rather rare. *RWA²*.

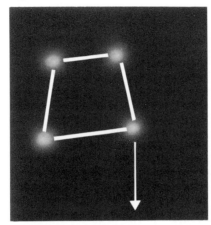

Multiple stars: the Trapezium

Mundrabilla meteorite Shower of large irons, Western Australia.

Mural Quadrant Early instrument used for measuring DECLINATION: a graduated circle with sighting arm and telescope. The "mural" indicated that the instrument was usually attached to a wall (Latin *murus*).

Murasaki Crater on Mercury; 12°S, 31°W. It is 78 miles (125km) in diameter. Its wall is broken by the ray-crater Kuiper.

Musca Australis (the Southern Fly; more usually known simply as Musca). A small but distinctive constellation adjoining Crux.

Brightest stars

Name	Visual Mag.	Abs. Mag.	Spec.	Distance (light-yrs)
α	1.94	−2.3	B3	230
β	3.42	1.2	A5	90

Also above the fourth magnitude are

Delta (3.62); Lambda (3.64) and Gamma (3.87). The globular cluster NGC4833 is visible with binoculars.

Naburiannu (*c.* 500BC). Babylonian astronomer. He determined the position of the equinox, and devised early tables of the Sun and Moon.

Nadir The point on the CELESTIAL SPHERE directly below the observer and directly opposite to the ZENITH.

Naked singularity A singularity which would be visible to the outside world. *See also* BLACK HOLE, SINGULARITY.

Nanjing Observatory *See* PURPLE MOUNTAIN OBSERVATORY.

Nanometre A fraction of 10^{-9} or 1/1,000,000,000 (one-thousand millionth) of one metre.

NASA The National Aeronautics and Space Agency, with headquarters in Washington DC. It was created in 1958 with responsibility for all non-military aspects of the United States space programme, though inevitably military considerations have intruded from time to time.

NASA, whose annual budget reached $5 thousand million, and at its peak employed 34,000 people with another 400,000 contract employees,

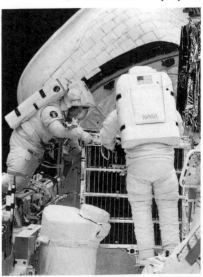

NASA: astronauts at work

organized and controlled all the lunar and planetary programmes, including APOLLO, as well as the first space-stations such as SKYLAB.

Its stations include AMES RESEARCH CENTER, Cape Canaveral, KENNEDY SPACE CENTER, LANGLEY RESEARCH CENTER, LEWIS RESEARCH CENTER, Lyndon B. Johnson Space Center, Hugh L.Dryden Flight Research Facility, GODDARD SPACE FLIGHT CENTER, the JET PROPULSION LABORATORY, Marshall Space Flight Center and WALLOPS FLIGHT FACILITY.

During the 1970s the idea of the recoverable SPACE SHUTTLE became all-important. The shuttle took much longer to develop than expected, and then, in 1986, came the disaster when the "Challenger" shuttle exploded after take-off, killing all seven astronauts on board. Quite apart from the human tragedy, the loss of the "Challenger" has had devastating effects upon a pro-gramme of planetary exploration which had in any case been cut back on financial grounds (for example, the mission to HALLEY'S COMET was can-celled as being too expensive).

It is fair to say that NASA's credi-bility has been badly damaged by the Shuttle disaster and the subsequent suggestions of mismanagement. At the moment (1986) no further planetary launchings have been confirmed, and many projects have been either can-celled or postponed indefinitely. Among those postponed is the launch of the HUBBLE SPACE TELESCOPE. However, NASA has announced plans for major space-stations before the end of the century. When these will come to fruition cannot yet be said. *PM. See also* essay on EXPLORING SPACE.

Nasmyth focus The focal point of a reflecting telescope in which the con-verging beam is reflected perpendicu-larly by means of a third mirror to a point outside the lower end of the tele-scope tube. In the altazimuth mounting as used by James Nasmyth, and also employed in the 236in (6m) reflector of the USSR Academy of Sciences, the beam exits through hollow elevation trunions.

Navigational astronomy Seamen have used the heavens to help them navigate since the earliest times. Homer tells how the goddess Calypso instructed Odysseus to keep the Greater Bear (then as close to the CELESTIAL POLE as the Lesser Bear) on his left hand to sail east, and an epic poem by Lucan of about AD74 gives the first reference to

the use of the stars for estimating latitude.

The next fundamental navigational developments — the appearance of the magnetic compass at sea from about 1180; and the appearance of the sea chart in about 1270 — were not astro-nomical, but in the 15th century, when the ships of Portugal, and later of Spain and England, began to make ocean voyages, there occurred a second navigational revolution which was.

Shipmasters learnt how to find their latitude at sea, first by measuring the ALTITUDE of POLARIS (by then only a degree or so from the North Celestial Pole), then, as the equator was approached and passed (by Portugal in 1474), by measuring the altitude of the Sun at noon, and applying its DECLINA-TION for the day. For observing, the simple hand-held quadrant with plumb-line was superseded about 1480 by the CROSS-STAFF for Polaris and the Mariner's ASTROLABE for the Sun. About 1590, Captain John Davis invented the backstaff, which, by using the sun's shadow, proved far easier to use in a seaway than the astrolabe.

Because latitude could be measured but not longitude, a highly successful navigational technique known as "running down the latitude" was used. To make Barbados, say, a navigator aimed a few hundred miles to the east and sailed until he reached its latitude (13°N), when he turned west, measur-ing latitude frequently so as to steer along the parallel of 13° until the island was sighted.

The two methods of finding longi-tude eventually used were proposed early in the 16th century. The lunar-distance method demanded an instru-ment accurately to measure the dis-tance of the Moon from the Sun or a zodiacal star, and the ability to predict the Moon's place with precision. The MARINE CHRONOMETER method demanded a timekeeper capable of keeping time to a few seconds a week, for months, perhaps years, on end, in a moving ship in any climate. But neither of these was then practicable.

The longitude-at-sea problem became so urgent that huge prizes were offered for its solution — by Spain in the 16th century, Holland, Portugal, Venice and France in the 17th, and — the largest of all, up to £20,000 — by Britain in 1714. Greenwich Observ-atory was founded in 1675 specifically to provide data for the lunar method.

In 1732, John Hadley produced his double-reflection quadrant, later devel-oped into the sextant we know today,

accurate enough for lunar distances. It immediately superseded all previous angle-measuring instruments for alti-tudes at sea also.

The first publication by Nevil MAS-KELYNE in 1766 of the British *Nautical Almanac,* based on the Greenwich meri-dian, finally made the lunar method practicable at sea. Meanwhile, the chronometer method became practi-cable also through the efforts of Harri-son and others. Over the next 50 years, progress in chronometer design and production was such that, by 1830, most foreign-going ships carried one or more chronometers, causing the lunar

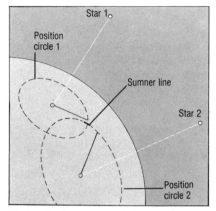

Navigational astronomy: fixing position

method to fall into disuse. Like any clock, a chronometer must first be set to time and frequently checked: during the 19th century, visual time signals were installed at ports the world over, one of the earliest being that at Green-wich in 1833.

Hitherto, finding latitude and longi-tude had been looked upon as two dis-tinct operations: latitude by meridian altitude; and longitude by chronometer (or lunar-distance), using an assumed latitude. In 1837, an American ship-master, Captain T.A.Sumner, sailing up St. George's Channel uncertain of his latitude, calculated his longitude using three different assumed latitudes. He discovered that the three positions lay on a straight line perpendicular to the Sun's azimuth. Actually, this Sumner line (as it came to be called) is part of a circle with centre at the sub-solar (or sub-stellar) point, radius the body's zenith distance. Because of the large radius, the arc of that circle can be considered as a straight line when plotted on a large-scale chart.

Sumner realized that a single obser-vation defined a *line* on which the ship must lie. Of course, if two or more posi-tion lines intersect at a reasonably large angle, one can define a position *point.*

This can be achieved by obtaining near-simultaneous observations of stars and planets when the horizon is visible at twilight; or near-simultaneous observations of the Sun and Moon in daytime when both are visible at suitable altitudes; or by taking two observations of the Sun in daytime, long enough apart in time to get a reasonable change of azimuth, allowance being made for the ship's run between observations (a "running fix").

A French naval captain, Marcq St Hilaire, greatly facilitated the use of position lines when he proposed in 1873 the intercept or cosine-haversine method, called the "New Navigation" in England, where it rapidly became popular, widely replacing the basic Sumner-line method. However, the combination of latitude-by-meridian-altitude and longitude-by-chronometer was never superseded.

Air navigators began to be specially catered for in the 1930s with the production of an integrating bubble sextant, an *Air Almanac*, and direct-entry tables for use with the intercept method, giving pre-calculated solutions to the necessary spherical triangles, to an accuracy of 1 arc minute. Similar *Tables of Computed Altitude and Azimuth* calculated to 0.1 arc minutes, suitable for surface navigation, were produced in the United States in the 1930s, the many thousands of spherical triangles being solved logarithmically as part of Roosevelt's New Deal. The methods used in the *Air Almanac* had been so successful that the *Nautical Almanac* (by now an international publication) was redesigned from 1952, tabulating Greenwich hour angle instead of RIGHT ASCENSION, with a day's tabulations to a page.

From 1905, radio time signals led to the final demise of the lunar-distance method. Starting with radio-direction-finding about 1915, radio then began to give the navigator non-astronomical ways of finding position out of sight of land — radar, Loran, Decca and Omega and, since 1964, satellite navigation systems. Soon, the NavStar Global Positioning System, with 18 geostationary satellites, will give virtually continuous world-wide cover with three-dimensional positional accuracy to a few metres.

Finally, two new pieces of equipment must be mentioned: the quartz crystal chronometer; and the dedicated navigational calculator, which can provide the answers to nearly all navigational calculations in a second or two; some have almanac quantities pre-programmed for many years ahead, some have built-in chronometers, most give the answer for a single observation in the form of an intercept and azimuth, but some will integrate several observations to give latitude and longitude. *HDH. See also* NAVIGATIONAL SATELLITES.

Navigational satellites Satellites designed specifically as a navigational aid for shipping and aircraft. They were developed originally for naval vessels, but are now widely used commercially. The first satellite (Transit 1B) was launched in 1960. The satellite transmits information about its orbit and position and the signals are picked up on board ship. Analysis of the DOPPLER shift in the signals enables a highly accurate position to be obtained. A later series called Timation, and containing an atomic clock, produces even more accurate positions. A further series called Nav-Star, developed by the US Navy and Air Force, commenced in 1978. Eleven launches in this series had been made by the end of 1985. *HGM*.

Nebulæ Celestial objects that appear larger and fuzzier than stars, and especially those comprising glowing dust or gas.

It is apparent on almost any dark night that the pinpoint stars are not the only objects in the sky. A faint, luminous band circles the entire sky, sometimes high overhead, sometimes girdling the horizon: this is the Milky Way, our GALAXY. We now know that the Milky Way actually comprises many millions of faint, distant stars, crowded together to give the impression of a luminous streak on the sky. But in the days before the telescope existed, before this was known, the Milky Way deserved a separate classification. The term adopted is nebula, the Latin word for cloud, and as fitting a description as ancient astronomy could hand down to us.

Other diffuse, faint nebulæ were known before the telescope's revolution. Compared to the Milky Way, however, all were tiny and most were faint. The telescope, offering a larger area to collect light and magnification to make the small easier to see, was to garner from the depths of the sky nebulæ by the thousand. Patient vigils at the eyepiece, particularly by the HERSCHEL family, Caroline, William and John, netted much of the material for a huge catalogue, the NGC, published late last century. A modern catalogue, if compiled along the lines of the NGC, would list some hundreds of millions of nebulæ, thanks to the impact of detectors vastly superior to eyeballs.

Many astronomers, among them the Herschels, believed that all nebulæ would break up into stars at the eyepiece of a sufficiently powerful telescope. This was an entirely natural way to think given that the Milky Way had done just that in the telescope they used. What put paid to the idea was the invention of the SPECTROSCOPE. In 1864 the English astronomer William Huggins examined the nebula NGC6543, and found to his surprise that it did not show the spectrum of starlight, but instead emitted its light in a few specific wavelengths. Huggins realized that here was an object made not from blended stars but from a cloud of gas.

Today, at least in professional parlance, objects which can potentially be resolved into stars are no longer termed nebulæ; they are galaxies or very compact clusters of stars. Nebulæ comprise the material between stars, congregated sufficiently densely to be obvious against the backdrop. But, despite this simple definition, nebulæ still have, like Proteus, a great many forms, and may shine by three completely different mechanisms.

Most of the brighter nebulæ emit their light in a series of EMISSION LINES. In these, the gas fluoresces, much as the mercury gas in our domestic fluorescent lamps. Instead of electricity, ultraviolet radiation pumps in the energy that comes out as light. In these emission nebulæ atoms lose their electrons (are ionized), and light is actually emitted as the electrons are recaptured. Hydrogen is the most abundant element in almost all nebulæ, and because it is ionized they are called HII REGIONS. However, the light actually comes from the tiny fraction of hydrogen atoms that have temporarily captured an electron to become neutral (HI). Hydrogen gives off its strongest light in the red at 656nm, where our eyes respond poorly but photographic emulsions are readily excited. HII regions photograph red, but often look green due to a prominent pair of lines at 496 and 501nm emitted by oxygen. If the star exciting an emission nebula is very hot, so that most of its radiation is in the ultraviolet, the nebula can appear much brighter than the star, since it processes the ultraviolet radiation to visible light. Although looking bright and almost solid, emission nebulæ are very tenuous. Typically every

N

N

gram of material is spread over a volume of one million cubic kilometres.

Another cause of a nebula is reflection of starlight. In this case we see not the gas itself but the myriad tiny motes of dust that normally are mixed with it. The dust is lit up like smoke in sunlight. Reflection nebulæ are always fainter than the star that illuminates them, unless the star is hidden from us by a denser portion of the nebula. However, some reflection nebulae, in particular the so-called cirrus that was brought to our attention by the INFRARED ASTRONOMICAL SATELLITE, are lit not by an individual star but by the combined

Nebulæ: NGC2070, in Dorado

light of the whole Galaxy, much like clouds above a distant city. Reflection nebulæ generally appear blue on photographs. A physical principle dictates that they must be bluer than the star that illuminates them.

Dark nebulæ are identical to reflection nebulæ except that they are not lit by any star. They can be seen only when silhouetted against something brighter. The best example is the COAL SACK.

One final type of nebula shines by the exotic mechanism SYNCHROTRON RADIATION. A single example is bright enough to merit mention: the CRAB NEBULA.

The subdivision of nebulæ by the origin of their light, as above, is not sufficient to describe all their varieties. A second system categorizes them according to their nature. Many of the types briefly mentioned below are expanded into separate entries elsewhere in this encyclopaedia.

HII Regions Techically any emission nebula is an HII region. The term is, however, also applied specifically to the clouds of gas in which stars are forming. Example: ORION NEBULA.

Diffuse nebula Amateur astronomers' term for either a reflection nebula or an HII region.

Planetary nebula Emission nebula formed when a MIRA VARIABLE sheds its outer layers, leaving a hot core to excite the gas. Example: RING NEBULA.

Supernova remnant Emission nebula formed when a star rips itself apart as a SUPERNOVA. Unlike other emission nebulae, the gas is made to glow not by the ultraviolet radiation of the star (which no longer exists) but by the frictional heating as it collides with ambient gas. Example: CYGNUS LOOP.

Bipolar (or **biconical**) **nebula** Light escapes to illuminate a reflection nebula, or more rarely to excite an emission nebula, only in two opposite directions, being constrained elsewhere by a doughnut-shaped ring of denser nebulosity.

Cometary nebula Bipolar nebula in which one of the two lobes is hidden, normally by a dark nebula. Example: Hubble's nebula attached to the star R Monocerotis.

Cometary globule Dense blob of gas squeezed by a passing shock wave, as from a supernova, and illuminated from outside, or else shining by the same emission mechanism as a supernova remnant. Usually has a dark centre and a long tail where material has been blown back.

Herbig-Haro object Tiny emission nebula with a distinctive spectrum, associated with young stars.

Zodiacal light The Sun's own reflection nebula. *DAA.*

Nebular hypothesis A theory that the Solar System was formed from nebulous material in space.

In 1755 the philosopher Immanuel KANT published his *Universal Natural History and Theory of the Heavens*, in which he suggested that the planets condensed out of diffuse "primordial" matter. This theory was elaborated — independently it seems — by Pierre-Simon LAPLACE in his *Exposition du système du monde* of 1796. Though known as the "nebular" hypothesis it is more atmospheric in nature, the very tentative proposal being that because of the motions and disposition of the planets in the Solar System, it was logical to suppose that they each condensed out of the Sun's atmosphere, which Laplace envisaged extending far into space. *CAR.*

Nebular variables Also known as T TAURI STARS, these are very young stars which have started to shine, but have not yet reached the MAIN SEQUENCE. They are believed to represent the last stage in the development of protostars, before they settle down on the Main Sequence. They vary irregularly in brightness. Nebular variables are F-, G- or K-type giant stars, many still surrounded by warm cocoons of dust and gas. There are a great number in the ORION NEBULA. While contracting, nebular variables probably lose large amounts of material in the form of a stellar or T Tauri wind. The rate of mass loss is about one-ten millionth of a solar mass per year. They are also rotating very rapidly, throwing off material at speeds of up to 180 miles (300km) per second. Sub-classes of the nebular variables are named after their prototype stars YY Orionis (extremely young), T ORIONIS and RW AURIGÆ. *JWM. See also* VARIABLE STARS.

Nebulium An "element" invented by 19th-century astronomers to explain certain unidentified lines in the spectra of nebulae. The lines were eventually shown to be lines of ionized oxygen and nitrogen, which are not found under normal terrestrial conditions. "Nebulium", therefore, does not exist. *See also* BOWEN, IRA SPRAGUE, and FORBIDDEN LINES.

Neison, Edmund (1851–1938) English astronomer whose real name was Nevill, who in 1897 published an important lunar map and text. From 1881 to 1910 he was Director of the now-defunct Natal Observatory in South Africa.

Neptune Neptune is the eighth planet from the Sun. It lies in a very nearly circular orbit, inclined 1.77° to the ecliptic, and orbits the Sun in 164.79 years. Its mean distance from the Sun is 30.110AU.

Neptune is too faint to be seen with the unaided eye, and its discovery awaited the invention of the telescope. The story of the discovery is quite unusual and merits a brief explanation. After HERSCHEL's discovery of Uranus in 1781, predictions were made that yet another planet lay farther out in the Solar System. From observed irregularities in the motion of Uranus, the mathematicians John Couch ADAMS working in England and U.J.J. LEVERRIER working independently in France both predicted the location of the unseen planet. A comedy of errors and ineptitude prevented the English astronomers from following up on Adams' predictions, and it was Leverrier who eventually pursuaded the German astronomer J.G.GALLE in

Berlin to search the region of the sky in which he had calculated that the new planet would lie. On September 23 1846, Galle and his assistants located the new planet very close to the predicted position.

Scholars have largely given joint credit for the discovery to Adams and Leverrier because their predictions were based on a sound analysis of data and a thorough knowledge of gravitation theory. At least, this was the situation for 134 years, but a surprise was in store. The astronomical world was astonished when in 1980 an astronomer and a GALILEO scholar announced their conclusion that Galileo himself had first seen Neptune with his telescope in January 1613, some 233 years before it was identified by Galle! It was uncharacteristic of Galileo not to recognize the nature of the moving object he sketched close to Jupiter just a few years after he first began to use his telescope, but he apparently did not, and as a result, Adams and Leverrier are given credit for the discovery.

Although Neptune is sometimes grouped with Jupiter, Saturn, and Uranus as a "jovian" planet, the resemblence to Jupiter and Saturn is superficial. Neptune is, however, quite similar to Uranus in many ways, including diameter (30,200 miles/ 48,600km), composition of the atmosphere, and mass (17.2 times the mass of the Earth). Seen through the telescope, both planets are nearly featureless, in strong contrast to the abundant cloud structures on Jupiter and the banded appearance of Saturn's clouds. While Uranus is a faint greenish colour, Neptune is weakly tinted blue.

The atmosphere of Neptune is composed mainly of molecular hydrogen, with an estimated 10–15 per cent helium. Mixed with these gases is a small amount of methane. The methane, though only a trace atmospheric gas, has a very active spectrum and it causes not only the planet's colour but controls its heat budget as well through absorption of sunlight. Another hydrocarbon, ethane, has also been detected in Neptune's atmosphere from spectroscopic observations. Though dense clouds appear to be rare in Neptune's hydrogen atmosphere, hazes are usually present, and they can be photographed with Earth-based telescopes. The hazes probably consist of crystals of frozen methane or some other atmospheric constituent, and they occur very high in the atmosphere, perhaps similar to the high-altitude NOCTILUCENT CLOUDS on Earth.

Clouds and hazes come and go on Neptune, sometimes as quickly as a few hours, and sometimes in the course of several weeks or months. The overall haziness of the atmosphere appears to be anti-correlated with the sunspot cycle in that at times of maximum solar activity Neptune appears in our telescopes to be a little fainter than average, while at sunspot minimum it is brighter. The Sun's activity thus appears to influence the atmosphere of Neptune some 2,798 million miles (4,504 million km) away by the formation of a global haze that affects the apparent brightness of the planet as seen from Earth.

Astronomers can follow the motions of clouds or hazy patches in Neptune's atmosphere either by pictures or by the effect they have on the infrared brightness of the planet, and thereby measure the period of the planet's rotation. Different astronomers have obtained various results from these techniques, but the rotation period of the upper atmosphere is about 17.7 hours. The atmosphere behaves as a fluid, however, and its rotation period lower down may well be different.

We do not know if Neptune has a solid surface, but it is unlikely that it resembles the Earth with a distinct boundary between solid planet and the atmosphere. The interior structures of Neptune and the other large planets of the outer Solar System are mostly unknown because we do not know in detail the properties of their component materials under conditions of high pressure and temperature that pertain to the internal regions. The fact that Neptune's mean density is 1.66 g/c^3, somewhat higher than the other outer planets, indicates that it contains a substantial quantity of rocky material in its interior. The nature of this material is unknown, but it is assumed for computational purposes to be the same as chondritic meteorites (*see* CHONDRITES). In addition. as Neptune formed, it must have incorporated a vast amount of water, methane, and ammonia, nearly all of which condensed to form an "icy" mantle surrounding the rocky core. Hydrogen and helium under the high pressure of the atmosphere must also be included as solids or liquids in the "icy" mantle. It is thought that this liquid or mushy mantle is in convective motion, carrying heat up from the core region of a few thousand degrees Kelvin.

As Neptune continues to contract from its own strong gravitational field and as the constituents of the icy

mantle gradually "unmix", with the heavier molecules settling to the bottom, energy is liberated from the planet. We see this energy emitted as excess heat. With Earth-based telescopes we measure the temperature of the upper atmosphere of Neptune to be 57K, though the temperature predicted on the basis of Neptune's distance from the Sun and other parameters is a few degrees cooler.

We do not know if Neptune has a magnetic field, auroræ, or radiation belts. Studies of the planet from Earth are complicated by its great distance, and as a result we know relatively little about it.

Neptune has two known satellites, Triton and Nereid. Searches for additional satellites in recent years have given unusual results. During several observations of stars being occulted by Neptune as the planet slowly passed in front of them, occultation events attributed to the presence of small, unseen satellites were reported. Further analysis suggests that Neptune is instead surrounded by one or more ring arcs, or partial rings, that do not fully encircle the planet as do the rings of Saturn and Uranus. The presence of these incomplete rings is possible only if tiny satellites (undetectable with Earth-based telescopes) in orbit around Neptune act to focus swarms of small particles or ice or rock into streams sharing the same orbits as the satellites.

The VOYAGER 2 spacecraft, launched from Earth on September 5 1977, has made its historic flybys of Jupiter (1979), Saturn (1981), and Uranus (1986), and is now *en route* to Neptune. It will make a close pass over Neptune in August 1989, and five hours later will encounter TRITON. A successful encounter with the most distant giant planet will vastly increase our knowledge of this least understood major planet of our Solar System, and will complete mankind's first reconnaissance of the Earth's neighbours. *DPC.*

Nereid Smaller and outermost of NEPTUNE's two known satellites, lying in a highly ellipical orbit with a period of 360.14 days. Thought to be a captured body of diameter between 220 and 600 miles (350–1,000km). Discovered by G.P. KUIPER in 1948.

Nestor (Asteroid 659). A Trojan, discovered in 1908. It never rises much above magnitude 15.8.

Neutral point The position between two bodies where their gravitational pulls

N

are equal in strength; the neutral point is closer to the more massive body. It was wrongly used by Jules Verne in his classic novel of 1865, in which his lunar voyagers became "weightless" when they reached the neutral point between the Earth and the Moon; in fact they would have been weightless from the moment of launching (in free fall). Moreover, in the Earth–Moon region the gravitational pull of the Sun is dominant in any case, and the "neutral point" has no real significance. *PM.*

Neutrinos In the last 20 years, detectors have been developed which have sufficient sensitivity to observe neutrinos from the Sun, collapsing stars (*see* COLLAPSARS), and COSMIC RAYS. Observing neutrinos from the cosmos is of great fundamental interest to astronomy. Neutrinos have the ability to penetrate enormous distances in matter. Because of this unique property, neutrinos are used as probes to study the energy-producing processes in the Sun, and have great potential applications in studying other astronomical phenomena.

Neutrinos are neutral particles with an almost immeasurably small mass. They travel essentially with the velocity of light. As particles, they are classified as leptons, a family of "light particles" which includes the electron (e^-, e^+), the muon (μ^-, μ^+), and the tauon (τ^-, τ^+). The negatively charged lepton is the particle, and the positively charged lepton is the anti-particle. to each of the charged leptons there is a corresponding neutrino, designated respectively (v_e, $\bar{v}_e$), (v_μ, $\bar{v}_\mu$) and (v_τ, $\bar{v}_\tau$), where the bars designate anti-neutrinos. All leptons carry one-half unit of spin. Neutrinos are polarized in the sense that their spin axes are oriented along the direction of motion; the direction of spin is left-handed for neutrinos and right-handed for anti-neutrinos. There have been many attempts to measure the masses of the various neutrino types. These experiments have only served to set upper limits to their masses; for example, the electron-neutrino has a mass less than 1/20,000 the mass of the electron. It is generally presumed that all neutrinos have a very small mass, and that the electron-neutrino is the lightest one.

Electron-neutrinos are created in beta-decay processes, and it is these processes that produce neutrinos in the interiors of stars. Based upon our understanding of these beta-decay processes, we are able to devise experimental methods of detecting neutrinos, and thereby use this particle as a probe to study the energy-producing processes in the interior of stars. Neutrinos have been detected by two processes, (1) the absorption of a neutrino in a nucleus with the emission of an electron, and (2) by the elastic scattering of a neutrino with an electron. By observing the electron produced from these processes one is able to infer the presence of the neutrino, and measure its energy. The specific processes will be discussed later in connection with a description of experimental neutrino observations.

Neutrinos from the Sun The Sun and all main sequence stars produce energy by thermally fusing hydrogen to form helium by the overall process, $4H \rightarrow {}^4He + 2e^+ + 2v_e + 25.6 Mev$ of energy. There are two sequences of exothermic nuclear reactions for hydrogen fusion, the PROTON-PROTON CHAIN and the CARBON-NITROGEN CYCLE. These are shown in the accompanying table.

The reactions shown are chosen from our knowledge of nuclear processes involving the light elements. Most of the reactions emit gamma (γ) rays or positrons which are absorbed in the Sun and converted into thermal energy. The neutrinos (v_e) produced pass through the solar matter.

The Sun generates most of its energy by the proton-proton chain because this chain will operate at lower temperatures. There are three competing branches in the proton-proton chain; each branch emits a characteristic energy spectrum of neutrinos. The carbon-nitrogen cycle is not particularly important in the Sun at present, but will become more so as the Sun grows older and its interior temperatures increase.

The Sun has been studied for many years, and we know quite precisely its luminosity, radius, chemical composition, mass and age. One can calculate its interior structure, temperature profile, rate of the nuclear reactions and the changes in these parameters with time. The calculation is based upon three general assumptions. The star is presumed to be in hydrostatic equilibrium, ie, the outward kinetic and radiation pressures are balanced by the inward gravitational forces. Various processes such as diffusion and turbulence are regarded as unimportant. It is assumed that the only sources of internal energy are the thermal fusion reactions listed in the table. Finally, the transport of energy throughout the inner regions of the Sun is through elementary atomic processes involving the interactions of electrons, photons and atoms in various states of ionization. In the outer regions, the Sun is fully convective, and a simple theory of this process is used. Detailed calculations have been carried out for the Sun, using the astronomical parameters: mass, age, luminosity, chemical composition, and radius. The laboratory measurements of the nuclear reaction cross-sections are used. The measured values are extrapolated to the low energies corresponding to the thermal conditions in the solar interior.

This complex theory is the basis of our understanding of the evolution of all stars and their lifetimes as MAIN SEQUENCE stars. Later in their evolution, other thermal fusion processes occur and the star will move away from the main sequence phase of hydrogen burning. It is important to test the theory quantitatively using the Sun as a model star, the only star sufficiently well understood to allow one to make a quantitative calculation of its structure. The only possible direct means of studying the thermal fusion processes is to observe the neutrino radiation, and of course the Sun is the only star close enough to allow these observations to be made.

A solar neutrino detector was built in the HOMESTAKE MINE at Lead, South Dakota, USA, by the Brookhaven National Laboratory during the period 1965-67. This detector depends upon the capture of neutrinos by the chlorine isotope ${}^{37}Cl$ to produce the radioactive isotope ${}^{37}Ar$ (half-life 35 days)

$$\text{capture}$$
$$v_e + {}^{37}Cl \rightleftarrows {}^{37}Ar + e^-.$$
$$\text{decay}$$

The capture of a neutrino by a ${}^{37}Cl$ nucleus depends on the neutrino energy. The cross-section for capture can be calculated accurately from our knowledge of beta-decay processes. The capture reaction has a threshold energy of 0.814 Mev, so all solar neutrinos above this energy would be observed. The capture rate for each source in the Sun depends on the energy of the neutrino. The detector observes a total neutrino capture rate corresponding to the fluxes (ϕ cm^{-2} sec^{-1}) multiplied by their respective neutrino capture cross-sections (σ in cm^2). Since the neutrino capture cross-sections are extremely small, in the range of 10^{-42} to 10^{-46} cm^2 for solar neutrinos, a special unit called a "solar neutrino unit", or "SNU", is used to express the rate. It is defined as the product, $\phi \sigma$, for each neutrino source in the Sun per target atom, multiplied by the factor 10^{36} ($SNU = \phi \ \sigma \times 10^{36}$ per second per atom). The neutrino flux and capture rate expected from the theoretical solar model is listed in the table for each neutrino source in the Sun.

The chlorine solar neutrino uses a radiochemical method that depends upon a chemical recovery of ^{37}Ar from a very large volume of a chlorine-containing liquid, and observing the concentrated ^{37}Ar radioactivity in a very small detector. A volume of 380,000 litres of tetrachloroethylene (C_2Cl_4) is used as the neutrino-capturing medium. Because of the low neutrino capture rate, less than one per day, the experiment must be carried out nearly a mile underground to shield it from cosmic rays. The ^{37}Ar is recovered by purging the liquid with a stream of helium gas, and circulating the helium through a charcoal adsorber at a low temperature ($-195°C$) to collect the argon gas. The argon recovered in this way is then purified and placed in a small proportional counter to measure the ^{37}Ar activity. Since the neutrino capture rate was found to be much lower than originally anticipated, it was necessary to make measurements over a longer period of time. Fifteen years of observations showed that the average neutrino capture rate was only 0.39 ± 0.05 per day, a rate corresponding to 2.1 ± 0.3 SNU. This rate is one-third the value expected from the most recent theoretical solar models.

There has been an effort to explain the lack of agreement between theoretical solar models and experimental results. Solar models have been suggested that introduce mixing, diffusion, a rapidly rotating core, or a chemical composition of the solar interior that is lower in the abundance of the heavier elements such as O, C, Fe, N, Si, etc. These models do give neutrino fluxes in the range of 1.5 to 2.5 SNU, in agreement with the results of the chlorine experiment. A basic feature of these models is that a mechanism is introduced which would diminish the internal temperature. This has the effect of reducing the flux of ^8B neutrinos. It may be noticed from the table that the neutrinos from ^8B decay are expected to contribute a major fraction of the neutrino capture.

Another possible explanation for the low neutrino capture rate in the chlorine experiment is that the neutrinos are affected in some way in their passage through the Sun or during their trip to the Earth. At present it is believed possible that all neutrinos have a small mass and have the property of changing into neutrinos of a different type. Solar neutrinos produced in the fusion processes are electron-type neutrinos, v_e, and they may change into neutrinos of the muon or tauon type; $v_e \rightleftarrows_\mu \rightleftarrows v_\tau$. It is thought that the effect most likely to cause the transformation of a solar neutrino is a resonance scattering process in the dense interior of the Sun. If neutrinos do indeed have masses and mixing parameters in the correct range, these properties would severely effect the solar neutrino measurements and other observations using neutrinos.

The chlorine solar neutrino experiment is the only one with sufficient sensitivity to measure the flux of neutrinos from the Sun. Clearly, additional solar neutrino experiments are needed to verify the results of the chlorine experiment and give information on the spectrum of neutrinos from the Sun. A measurement of the flux of low-energy neutrinos from the primary proton-proton reaction ($H + H \rightarrow D + e^+ + v_e$) would be of great importance. This reaction initiates and dominates the proton-proton chain of reactions. It operates over a wide temperature range and consequently is the most intense source of neutrinos from the Sun. These low-energy neutrinos would not be observed by the chlorine experiment. However, a radiochemical experiment based on the reaction, $v_e + ^{71}Ga \rightarrow ^{71}Ge + e^-$, has been developed, capable of observing the neutrinos from the proton-proton reaction. Because of the high cost of the gallium needed, about 25 tonnes, building a full-scale detector has been long delayed. However, there is a hope that two experiments will be built in 1988–90, one in the Soviet Union, and one in Western Europe.

In recent years very large-scale underground detector systems have been built to observe the decay of the proton. Some of these facilities use 3,000 tonnes or more of water as the detecting medium. This technique may also be applied to observing neutrinos. In water, the neutrino would be observed by its interaction with an electron in the water molecule. When a neutrino with sufficient energy scatters from an electron, the struck electron, in passing through the water, will emit a cone of light (ČERENKOV light), which can be observed by sensitive light detectors (photomultipliers) located on the walls of the water tank. There is now great interest in employing these large detector systems to observe neutrinos from the Sun and from other astronomical sources. The Kamiokande water Čerenkov detector at Kamioka, Japan (University of Tokyo), has been modified to allow observations of low-energy neutrinos and there is a great hope that the flux of ^8B decay neutrinos will be observed by

Energy production processes in the Sun, and solar neutrino fluxes

Reaction		Neutrino Flux (per cm^2 per sec.)	Neutrino Energy (Mev)	Capture Rate (^{37}Cl in SNU)
The Proton-Proton Chain				
PP-I	$H + H \rightarrow D + e^+ + v_e$	6.1×10^{10}	0 – 0.42 spectrum	0.0
	$H + H + e^- \rightarrow D + v_e$	1.5×10^8	1.02 line	0.24
	$H + D \rightarrow ^3He + \gamma$			
	$^3He + ^3He \rightarrow ^4He + 2H$			
PP-II	$^3He + ^4He \rightarrow ^7Be + \gamma$			
	$^7Be + e^- \rightarrow ^7Li + v_e$	4.0×10^9	0.86, 0.38 lines	1.0
	$^7Li + H \rightarrow ^8Be \rightarrow 2^4He$			
PP-III	$^7Be + H \rightarrow ^8B + \gamma$			
	$^8B \rightarrow ^8Be + e^+ + v_e$	4.0×10^6	0 – 15.0 spectrum	5.0
	$\longrightarrow 2^4He$			
The Carbon-Nitrogen Cycle				
	$H + ^{12}C \rightarrow ^{13}N + \gamma$			
	$^{13}N \rightarrow ^{13}C + e^+ + v_e$	5.0×10^8	0 – 1.20 spectrum	0.08
	$H + ^{13}C \rightarrow ^{14}N + \gamma$			
	$H + ^{14}N \rightarrow ^{15}O + \gamma$			
	$^{15}O \rightarrow ^{15}N + e^+ + v_e$	4.0×10^8	0 – 1.73 spectrum	0.24
	$H + ^{15}N \rightarrow ^{12}C + ^4He$			
				6.6 SNU

N

this advanced detector in the near future.

The field of observational neutrino astronomy has been confined to a single experimental attempt to observe neutrinos from the Sun. The results have raised questions about the structure and dynamics of the Sun and the properties of neutrinos. At the present stage of our understanding of neutrino physics, one must be cautious in interpreting a neutrino signal from astronomical sources. Astrophysicists tend to look toward neutrino astronomy as a means of studying the basic properties of neutrinos. The subtle, and often the most significant, properties of neutrinos are not easily revealed. Because of their feeble interaction with matter and with magnetic fields, these properties can only become known through astronomical observations.

Supernovæ After a massive star has exhausted its nuclear fuel, the collapse of its core results in a brilliant pulse of light followed by the formation of a NEUTRON STAR. It is believed that this event produces a very intense burst of neutrinos lasting a fraction of a second, followed by a longer pulse of neutrinos and anti-neutrinos, lasting a few minutes. The first pulse is attributed to the capture of electrons by the nuclei to form a neutron star, with the emission of neutrinos, a process that can be represented by the elementary equation, $e^- + p \rightarrow n + v_e$. The longer pulse is attributed to the rapid cooling of the neutron star by the emission of neutrino/anti-neutrino pairs. The energy in the form of neutrinos that is radiated during this event is believed to be in the order of 10^{53} ergs, a quantity that can be compared to the Sun's total luminosity of 2.3×10^{35} ergs per minute! These theoretical ideas were invoked to explain the sudden collapse of a Type II supernova, the presence of a fast-rotating radio pulsar as a residue, and the synthesis of the heavier elements.

Neutrino detectors currently operating are capable of observing the neutrinos produced by a supernova event if it occurs within 10 kiloparsecs of the Earth. The chlorine solar neutrino experiment would respond to the flux of electron neutrinos, and the Kamiokande water detector would respond principally to anti-neutrinos, through their capture in hydrogen by the reaction, $v_e + p \rightarrow n + e^+$. There are numbers of smaller neutrino detectors in the USSR, Italy and the USA, using a few hundred tonnes of liquid scintillator. A liquid scintillator is mineral oil, containing a small percentage of light-emitting molecules, which are excited when a charged particle passes through the liquid. The light is registered by photomultiplier tubes. Scintillation detectors also respond primarily to anti-neutrino capture in hydrogen. Unfortunately, supernova events are extremely rare. The rate of occurrence of these events can best be estimated by observations on neighbouring galaxies. The estimates range from one event every 10 to 30 years in our Galaxy, but one must bear in mind that the most recent visible supernovæ observed in our Galaxy were in 1572 (Tycho) and 1604 (Kepler). If one occurs nearby, the neutrino pulse will be detected at a few underground observatories.

Neutrinos in the cosmos Cosmic rays pervade the entire universe. These are extremely high-energy particles consisting of ionized hydrogen, helium, and heavier atoms, and gamma rays. When these very high-energy particles come through the Earth's atmosphere, a cascade of nuclear particles is produced, including very high-energy neutrinos. These are predominantly muon neutrinos and anti-neutrinos (v_μ, $\bar{v}_\mu$). Cosmic ray-produced muon neutrinos have been observed by the large-scale water Čerenkov and scintillation detectors discussed above. In this case, the neutrino is absorbed in one of the atoms in the rock surrounding the detector, to yield an energetic muon, for example, $v_\mu + {}^{16}O \rightarrow {}^{16}F + \mu^-$. The energetic muons ($\mu^+$ or μ^-) are observed passing through the detector, and because the energy is high the direction for the incoming neutrino can be determined. By observing an upward-going muon, one can distinguish the neutrino-produced muons from the normal downward-travelling cosmic ray muons. These upward-travelling muons result from muon-neutrinos that have travelled through the Earth. A total of about 1,000 neutrino-produced muons of this nature has been observed by detectors worldwide. The number observed, and their angular distribution, are consistent with the number expected from cosmic ray interactions in the atmosphere. These observations could be used to search for high-energy neutrinos from sources within our Galaxy by recording an excess of neutrinos coming from a particular direction. So far, no sources have been found, but this branch of neutrino astronomy is just beginning.

In recent years there has been great interest in relic neutrinos that could have been created by the presumed BIG BANG. Neutrinos would have decoupled from other matter a second or so after the event. At the present time, relic neutrinos could pervade the universe with a number density and momentum distribution similar to the 2.7K photon background. The kinetic energy of these neutrinos would be too low to observe by any presently feasible detection method, although a number of techniques for observing the neutrino background have been contemplated. If neutrinos have a small mass, they would contribute significantly to the mass of dark matter in the universe. There is evidence for the presence of dark matter in the universe to explain the velocity of stars in galaxies, the behaviour of clusters in galaxies, and the motion of stars in our Galaxy. If neutrinos have a mass within the range of present experimental limits, greater than or equal to 25 electron volts (1/20,000th the mass of the electron), this would be sufficient to provide mass to close the universe. These matters have been the subject of active discussion among theoretical astrophysicists during the last decade. Interest in this subject will continue until the question of the masses of all neutrino types is resolved. *RD.*

Neutrino telescope A detector used to observe the flux of NEUTRINOS from space. Since neutrinos produce a very feeble signal, it is necessary to carry out the observations deep underground with detectors weighing 100 to 3,000 tonnes. Existing large-scale neutrino detectors are based on a neutrino capture process, or an elastic scattering from atomic electrons. A radiochemical detector depends on the capture of a neutrino in a specific isotope, producing a radioisotope which may be removed from a large mass of the target element. The radioisotope, which accumulates over a period of time, is removed and measured in a specially designed radioactivity detector. Neutrino detectors based on elastic scattering processes depend on observing the energy of the recoil electron in a scintillating liquid, or on observing the ČERENKOV light emitted by the recoil electron in water. Large-scale neutrino detectors are located in Japan (water Čerenkov), the Soviet Union (scintillation) Italy (scintillation), and the United States (radiochemical, water Čerenkov, and scintillation). *RD.*

Neutron stars The densest and tiniest stars known, made largely of neutrons and sometimes likened to a giant nucleus, neutron stars are possibly the most peculiar things in the universe. They are observed as PULSARS, as one of the components in X-RAY BINARIES, and perhaps as gamma ray burst sources. (*see* GAMMA RAY ASTRONOMY).

The neutron, which along with the PROTON makes up the atomic nucleus, was discovered in 1932. It is said that within a few hours of the news of the discovery reaching him, Landau had suggested the possibility of cold, dense stars, comprised principally of neutrons. At the end of 1933 BAADE and ZWICKY tentatively suggested that in SUPERNOVA explosions ordinary stars are turned into stars that consist of extremely closely packed neutrons. They called these stars neutron stars. Seminal work on the structure of a neutron star was carried out by J.Robert Oppenheimer along with Volkoff and Tolman at the end of the 1930s.

Neutron stars were thought to be too faint to be detectable and little work was done on them until November 1967, when Pacini pointed out that if the neutron stars were spinning and had large magnetic fields, then electromagnetic waves would be emitted. Unbeknown to him, radio astronomers at Cambridge were shortly to detect radio pulses from stars (the pulsars) which are now believed to be highly magnetized, rapidly spinning neutron stars.

"Guest" star or "new" star was the name given by the ancient Chinese astronomers to the catastrophic explosion of an old star. It is today called a supernova. Some stars when they reach the end of their life go out with an enormous bang, spewing debris into space. In the explosion the core of the star is compressed and the ordinary material of which it is made converts largely to neutrons. The electrons, which originally surrounded the protons and neutrons of the atomic nuclei, merge with some of those protons to form yet more neutrons, and the atomic structure collapses. In its place is material of a density comparable to the nucleus of the atom. In the demise of a normal star a neutron star is born. In terms of stellar evolution, therefore, neutron stars represent life after death!

Comparison of the estimated number of pulsars in the GALAXY with the estimated supernova rate sometimes suggests that there are too many pulsars and too few supernovæ for this to be the only way neutron stars can be formed. Estimation of both quantities is difficult and liable to error. There has, however, been some investigation of whether (and how) a neutron star could be formed near the end of an ordinary star's life without a supernova explosion.

A neutron star has a mass comparable with that of the Sun, but is only about 6 miles (10km) in radius, so has an average density one thousand million million times that of water. (If the cap of a ball-point pen was filled with this material it would weigh a thousand million tonnes.) Such a large mass in such a small volume produces intense gravity — objects weigh one hundred thousand million times more on the surface of a neutron star than on the surface of the Earth. The intense gravitational field affects light and other electromagnetic radiation emitted by the star, producing significant REDSHIFT ($z \sim 0.2$). The strong gravitational attraction allows neutron stars to spin rapidly (hundreds of revolutions per second) without disintegrating. Such spin rates are expected if the core of the original star collapses without loss of ANGULAR MOMENTUM. If the original star has a magnetic field, then this too may be conserved and concentrated in the collapse to a neutron star. Pulsars, gamma ray burst sources, and the neutron stars in some X-ray binaries are believed to have magnetic fields of strength about 100 million Tesla (roughly a million million times the strength of the Earth's magnetic field).

When formed the neutron star has a high temperature, but initially cools very quickly through the radiation of NEUTRINOS. After roughly a thousand years the surface temperature has fallen to about a million degrees, and thereafter the cooling is slower. The rate of cooling then depends on the composition of the star and determination of the temperature of the surface of the star can place constraints on the internal structure of the neutron star.

There are four fundamental forces in nature — gravitational, electromagnetic, the weak nuclear force and the strong nuclear force. The physics of the interior of the neutron star involves all four. In addition quantization effects predominate, and these in turn force particles into high energy levels so that relativistic effects are also important. The physics of the surface of the neutron star is bizarre; the properties of the core are beyond modern physics.

The neutron star is like a raw egg — it has a hard thin shell on the outside, and some peculiar liquids inside. It is unlike an ordinary star in that it is not a sphere of ionized gas radiating energy that originates in nuclear reactions. The neutron star has no such source of energy.

The surface of the neutron star is made of iron. In the presence of a strong magnetic field the atoms of iron polymerize and the polymers pack to form a strong lattice with density about ten thousand times that of terrestrial iron. Its strength is a million times that of steel. It has excellent electrical conductivity along the direction of the magnetic field, but is a good insulator perpendicular to this direction.

Immediately beneath this surface the star is still solid, but its composition is changing. Larger nuclei, particularly rich in neutrons, are formed, and materials that on Earth would be radioactive are stable in this environment. With increasing depth the density rises. When it reaches 400 thousand million times that of water the nuclei can get no larger and neutrons start "dripping" out. As the density goes up further the nuclei dissolve in a sea of neutrons. The neutron fluid is a superfluid — it has no viscosity, no resistance to flow or movement.

Within a few kilometres of the surface the density has reached the density of the atomic nucleus. Up to this point the properties of matter are reasonably well understood, but beyond it understanding becomes increasingly sketchy. The composition of the core of the star is particularly uncertain. It may be liquid or may be solid; it may consist of other nuclear particles (pions, eg, or hyperons); there may be another phase change where QUARKS start "dripping" out of the neutrons, forming another liquid.

The discovery of neutron stars has led to considerable advances in our understanding of the behaviour of material at high densities. Greater undertanding of the internal structure of neutron stars is being pursued vigorously by physicists and astrophysicists. *SJBB. See also* ATOMIC STRUCTURE and essay on PULSARS.

New Moon The phase of the Moon when the dark side is turned toward us. The true New Moon can be seen only at a solar eclipse.

New Quebec Crater Canadian impact crater 9,840ft (3km) in diameter.

Newall telescope A 25in (63cm) aperture refractor built by Thomas Cooke for R.S.Newall of Gateshead, England. It took seven years to build and was completed in 1869, a year after Cooke's

N

death. It was, at that time, the largest refractor in the world.

Newcomb, Simon (1835–1909) American astronomer who carried out very important work on lunar and planetary theory and became head of the American Nautical Almanac Office. He is also remembered for his contention that no man-made heavier-than-air machine could ever fly!

Newton, Isaac (1642–1727) English mathematician, physicist and astronomer.

Born prematurely at Woolsthorpe, a hamlet near Grantham, Lincolnshire, three months after his father's death, Newton was left in the care of his grandmother, during which time his mother married again. Hating his stepfather, Newton suffered all his life from a sense of insecurity; later it was this which probably made him abnormally sensitive to criticism of his scientific work.

His stepfather having died, Newton's mother determined that he should become a farmer to manage her estates,

Isaac Newton: the laws of gravitation

but Newton showed no interest in such a career and, fortunately, she was persuaded to allow him to be prepared for university. In 1661 Newton entered Trinity College, Cambridge, as a subsizar — an improverished undergraduate who had menial tasks to perform to help pay for his keep.

Newton graduated in April 1665, having already invented his mathematical method of fluxions, though few people were aware of the fact. However, the university was then closed because of the plague and the scholars dispersed. Newton went back to Woolsthorpe, and here, except for a

return for a couple of months to Cambridge, he consolidated his mathematical work and his nascent optical theory of the nature of colour. He also recognized that each planet was held in its orbit by a force emanating from the Sun which diminished according to the square of its distance — an "inverse square' law. But he published nothing.

On his return to Cambridge in October 1667, Newton was elected to a Fellowship at Trinity, and two years later as Lucasian professor of mathematics; Isaac Barrow, the previous holder, is said to have resigned in his favour.

By 1671 Newton had built his first reflecting telescope (*see* NEWTONIAN TELESCOPE). At the time telescopes were refractors, forming their image by a lens system, not a mirror. However, in order to avoid displaying coloured fringes around bright objects and provide good sharp images they had to be extremely long, and so were cumbersome. Using his theory that white light is a mixture of the light of all colours, Newton mistakenly believed that the refractor could not be cured of this defect. He therefore designed a reflecting telescope — the first successful one ever to be made — and it aroused much interest; not only did it display no coloured fringes, but it was also very much shorter than any refractor. A duplicate instrument was sent to the Royal Society in London at their request, and Newton was forthwith elected a Fellow. At his own suggestion he then submitted a paper on his theory of light and colour. His epoch-making paper was, however, criticized and the ensuing controversy severely upset him, though in 1675 he did submit two other optical papers. One was about the colours seen in very thin films of oil and other materials, the other on his theory that light was composed of tiny particles or "corpuscles". These also caused controversy, and in the end, Newton shut himself away in Cambridge and avoided further publication.

Not until late in 1679 did correspondence start again with Robert Hooke, who was Honorary Secretary at the Royal Society and previously a severe critic of Newton's theories of light and colours. An exchange of letters led Newton to re-examine his work on planetary orbits, but again he published nothing; not until 1684 was he stimulated to consider making his ideas known. This was due to a visit from Edmond HALLEY.

In London Halley had been discussing the problem with Hooke and with

Christopher Wren, astronomer as well as architect. No one had been able to produce a mathematical proof that an inverse square law would result in elliptical planetary orbits, and Halley took it upon himself to visit Newton and enquire whether he knew the reason. Newton replied that he did and had a mathematical proof. Unable to locate it among his papers, he promised to send it on to Halley later. This he did, sending a small tract with the title *De Motu* (On Motion). When Halley received it, he was mathematician enough to realize that here was a document of the greatest significance. He therefore prevailed on the Royal Society to publish an entire book on the subject by Newton, and also persuaded Newton to continue expanding his ideas for publication.

There were difficulties, however. On the one hand the Royal Society was short of money and could not afford to publish; Halley himself had to edit the manuscript and pay the printer out of his own pocket. On the other hand, Hooke raised objections, claiming that Newton had stolen his results. Newton was furious, went through his text deleting nearly every reference to Hooke, and refused to complete the book. Only the exercise of all Halley's undoubted diplomatic gifts saved the situation, so that at last, in July 1687, the entire volume appeared.

With the title *Philosophiæ Naturalis Principia Mathematica* (The Mathematical Principals of Natural Philosophy) it covered the whole question of the motion of bodies and incorporated Newton's new theory of universal gravitation. In it Newton laid down new laws of motion, discussed the behaviour of bodies when they move through a resisting medium like air or water, and gave the theoretical reason why all bodies fall to the ground at an equal speed, a fact already studied in detail by GALILEO but still requiring an underlying theoretical explanation. Newton then used gravitation to explain the laws of planetary motion discovered by KEPLER. Cometary paths and their computation from a few observations only were also discussed, and led Halley to his famous discovery that comets moved in ellipses, and that their returns could be calculated precisely. The book was a *tour de force*, and has been recognized as one of the most important scientific works ever written, for it not only solved virtually all the questions and problems of traditional astronomy, but also stimulated a vast amount of further research.

Newton was exhausted after this vast intellectual effort, and seems to have suffered a nervous breakdown. He withheld publication of his *Opticks* and his acceptance of the presidency of the Royal Society until after Hooke was dead, and did no more major scientific work. Thereafter he spent his time on biblical research and on alchemy and chemistry, as well as seeking public preferment. In the event, he was appointed Warden of the Mint in 1696 to take charge of a great recoinage scheme; later he became Master. In both posts he carried out the work with outstanding ability, being knighted in 1705 in recognition of his exceptional service.

Newton's later years were darkened by two major controversies. One concerned the publication of the observations of FLAMSTEED which he persuaded Halley to edit, the other an argument over the invention of the calculus. Newton had expanded his fluxions into the technique now known as been discovered by Gottfried LEIBNITZ. Newton was encouraged to accuse Leibnitz of stealing his ideas, and the controversy became very bitter, Newton pursuing it even after Leibnitz' death. It is known that both discoveries were quite independent.

By the end of his life Newton was considered not only the doyen of British science but his reputation had become international. He died on March 20 1727 in London, and was buried in Westminster Abbey following a State Funeral. *CAR. See also* GRAVITY.

Newtonian telescope The reflecting telescope characterized by a paraboloidal mirror as primary and a diagonal plane

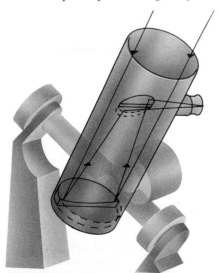

Newtonian telescope: first reflector

mirror, or in some small instruments a 45° prism, positioned to reflect the rays through 90° to a focus near the side and toward the pointing end of the telescope tube. Its development by NEWTON, and presentation to the Royal Society in 1671, as the first reflecting telescope, arose from his erroneous opinion that it was not practicable to make achromatic refracting telescopes. For apertures less than 6in (15cm), which includes Newton's original at 1in (2.5cm), simpler spheroidal primaries are used as there is no significant loss of optical performance. *LMD.*

N galaxies Galaxies whose light is dominated by a point-like stellar nucleus. This classification, due to Professor W.W.Morgan of Yerkes Observatory, is thus a morphological

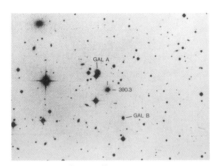

N galaxies: often radio sources

one based on the appearance of the galaxy on direct photographs.

They attracted great interest soon after the discovery of the QUASARS, which they resemble, but in which the extended part of the galaxy is not seen. Like the quasars, N galaxies are often radio sources and show the ultraviolet excess, broad EMISSION LINES and variability of quasars. Their REDSHIFTS are somewhat lower, however, since they are closer and less luminous, making it easier to see the stars in the outer parts of the galaxy.

They are now recognized as one of many groups of active galaxies between normal galaxies and quasars. *MVP.*

NGC (New General Catalogue). Catalogue of clusters and nebulae compiled by the Danish astronomer J.L.E. Dreyer, and published in 1888. Generally, professional astronomers use its numbers in preference to Messier's. The NGC includes over 13,000 objects.

Nice Observatory Erected in 1881 as a private observatory by a French sponsor, Raphael Bishoffsheim, Nice Observatory is equipped with a large

refractor, 30in (76cm) in diameter, a 15in (38cm) refractor, and a double astrograph of 16in (40cm). After a famous period when work was done on asteroids, comets, double stars and astrometrics, which was followed by a decline, the observatory was modernized from 1962 onward by J.C.Pecker and now specializes in theoretical and observational astrophysics.

Nicholson, Seth Barnes (1891–1963). American astronomer who discovered four faint moons of Jupiter and made thermocouple measurements of the temperatures of planets and the eclipsed Moon.

Nicol prism An optical polarizing prism, which transmits only the extraordinary ray, made up of two cut and polished pieces of doubly refracting calcite crystal cemented together with Canada balsam.

Nitrogen An element, chemical symbol N, which in order of cosmic abundance is fifth by numbers of atoms and seventh in terms of the mass content of the universe. Its properties include: atomic number, 7; atomic mass, 14.0067; melting point, 73.3K ($-209.9°C$); boiling point 77.4K ($-195.8°C$).

Nitrogen comprises 78.08 per cent by volume of dry terrestrial air (75.52 per cent by mass). Under normal terrestrial conditions, it exists as a diatomic molecule, N_2, and is chemically inert because the bond between each pair of atoms is hard to break. It is also contained in a wide range of compounds, including ammonia (NH_3), and is an important constituent of many organic molecules and foodstuffs for living organisms. Atomic and ionized nitrogen are produced in the upper atmosphere by the action of solar radiation, and react with oxygen and ozone to produce nitric oxide. Sudden enhancements of atomic and ionized nitrogen resulting from the impact of energetic particles from SOLAR FLARES can therefore result in significant temporary depletions of the terrestrial ozone layer.

Apart from the Earth, the only Solar System body to possess a nitrogen-rich atmosphere is TITAN, the giant satellite of Saturn. Nitrogen makes up 94 per cent by volume of Titan's atmosphere.

Nitrogen plays an important catalytic role in the CARBON-NITROGEN CYCLE, which is the major helium-producing nuclear reaction in stars more massive than the Sun. A variant of the

N

carbon-nitrogen cycle, the carbon-nitrogen-oxygen "bi-cycle", involves the capture of further PROTONS to produce nitrogen and helium as end products, and a large fraction of the cosmic abundance of nitrogen-14 may have been produced by this process.

Liquid nitrogen is widely employed to cool sensitive astronomical detectors in order to reduce unwanted background "noise". *IKN*.

Noctilucent clouds Pearly white or blue clouds at a mean height of 52 miles (83km) above the Earth's surface which are lit by the Sun lying some 6 to 16° below the observer's horizon. In the northern hemisphere they are visible, when present, from May till August between the latitudes of 50 to 70°. They have a marked ripple or wave structure with veils, bands, billows and whirls, sometimes associated with a directional drift. The brightest part of the display lies over the sun beneath the horizon.

Rocket and satellite studies suggest that the clouds consist of water ice of about 0.1 micron in size, frozen onto a dust core. *RJL*.

Noctilucent clouds: water ice on dust

Noctis Labrinthus A vast maze of interconnecting Martian canyons, 46,000 sq miles (120,000km²) in extent, located at 5–15°S, 90–110°W, at the summit of the THARSIS RIDGE. Individual canyons are relatively short, their trend being constrained by intersecting fractures.

Nodes The intersection between the orbit of a body and the ECLIPTIC.

Norma (the Rule)—formerly known as Quadra Eucladis, Euclid's Quadrant). A small constellation near Ara and Lupus. It contains no star above magnitude 4.

North America Nebula HII REGION in Cygnus crossed by a thick dust lane hiding the ionizing stars. The east portion, NGC7000, resembles North America; the west, IC5070, is the "Pelican" Nebula.

North polar distance Angular distance from the north celestial pole.

North polar sequence A list of 96 stars near the north celestial pole, ranging in magnitude from 2 to 20, whose PHOTOGRAPHIC MAGNITUDES have been accurately measured. Used for comparison purposes.

Novae The appearance of a bright star-like point of light in the sky over only a few nights of observation, at a position where previously no star had been known, was a matter of great interest in ancient times. The phenomenon tended to be ignored by European chroniclers during the Middle Ages because the prevailing culture belief was that the heavens represented a perfect and therefore unchangeable creation by God. On the imperfect Earth alone was change supposed to be permitted, and when change did in fact occur in the sky the matter was dealt with by ensuring that no written record of it was made, a practice not unknown in modern times. When astronomers today are concerned to know about what happened a thousand years ago they perforce must consult the records of cultures other than the European, notably of the Chinese, who held a very different but equally erroneous belief, that by noting changes in the heavens they could predict the course of events on the Earth.

Such flare-ups in the sky are known nowadays to be of two distinct kinds with quite separate explanations, novae and SUPERNOVAE. This article is concerned solely with novae.

Since astronomers nowadays have photographic plates of the whole sky on which even exceedingly faint stars can be seen, it is often possible by examining previously taken plates to find what was there at exactly the same position before the outburst of a nova. An even greater amount of information can be obtained by following the spectral features of a nova through its sudden outburst into a decline in brightness over a time interval of weeks and months. The outcome of such investigations is that novae are in the main, if not indeed wholly, WHITE DWARF stars that happen to be members of BINARY STAR systems.

Novae can be observed in galaxies other than our own, and attempts have been made to use such observations as an indicator of the distances of other nearby galaxies. But the great majority of observed cases are within our own Galaxy, where for the limited region of

the solar neighbourhood they are found to occur at a rate of about two per year. The solar neighbourhood is determined by the fogging effect of interstellar dust, which cuts down the range of effective observation to about 5 per cent of the whole galaxy, for which the rate of occurrence of novae is estimated to be about 40 per year.

Our subjective perception of a nova is confined to light in the visual and photographic ranges of wavelength, omitting the ultraviolet. Since the ultraviolet is of variable importance, probably being of greater importance before outburst than during it, our subjective perception tends to exaggerate the contrast between the prenova state and the maximum emission of light during the outburst. The observed contrast for visible light is usually about 10,000 to 1, but if all wavelengths were included the contrast would probably be about 100 to 1.

The emission of visible light in the prenova stars is of a similar order to the emission from the Sun, whereas the emission at maximum outburst is of a similar order to that of a supergiant star of type F8, of which Wezea, the fourth brightest star in the constellation of Canis Major (the Great Dog) is an example. A typical nova rises to its maximum in a few days, and declines thereafter in brightness by a factor of about 10 in 40 days, although cases of both slower and more rapid declines are known and studied.

Clouds of gas are ejected at high speeds during outbursts, speeds typically of about 930 miles/s (1,500km/s), which is more than sufficient for the expelled gases to become entirely lost into interstellar space, together with myriad fine dust particles that condense within the gases as they cool during their outward motion. The total amount of material thus lost is estimated to be about one part in ten thousand of the total mass of the parent white dwarf star, although the amount in especially violent cases is almost surely appreciably larger than this.

Much thought has been given in recent years to the cause of novae. The consensus of opinion is that the basic logic of the explosion of a nova is the same as for the explosion of a man-made nuclear weapon. To begin with, energy is produced at a comparatively gentle rate by NUCLEAR REACTIONS in material that is covered by a layer of relatively inactive other material, tamped as one says. Because of the tamping material the energy produced by nuclear reactions cannot escape and must therefore accumulate within the reacting material itself. Provided the tamping effect is strong enough, the rising temperature and pressure which thus occurs cannot be alleviated by expansion, and the rising temperature and pressure causes the nuclear reactions to become less and less gentle as the process proceeds. The same cycle of events is repeated and repeated again with the energy released from the nuclear reactions accelerating at an ever-increasing rate, until eventually the situation gets quite out of hand. Or at any rate until the tamping effect of overlying material fails at last and the

Novæ: stellar outbursts

material is blown entirely out of the strong gravitational field of the white dwarf star. Depending on the mass of the white dwarf star, its chemical composition, and the amounts of the reactions and the tamping materials, the details of explosion can vary in ways that are subject to mathematical calculation, and which have been found to agree with many of the observed features of novae.

Some stars are known to undergo explosions repeatedly, for example T Coronae appeared as a nova in 1866 and 1946, while T Pyxidis did so in 1890, 1902, 1922 and 1944. It is thought that all stars experiencing the nova phenomenon probably do so repeatedly and that it is the ones undergoing explosion most frequently that tend to be noted, like T Coronae and T Pyx.

In order to go through the explosion several times it is necessary for a white dwarf star to receive new material at its surface to replace what has been lost in previous explosions. It is important that there should be a binary companion, for the companion star serves as the source of material to the white dwarf star. The favourable case is when the companion serving as the

source is of giant type, which is to say a star of very large radius from which material escapes rather easily. So the favoured system for understanding the nova phenomenon is a binary system with components not very far apart, one component a giant star and the other a white dwarf. Material can then drain from the surface of the companion star where gravitation is weak to the surface of the small white dwarf star where gravitation is strong, and where only violent explosions can serve to blow the material eventually out into space.

The circumstances in which the nuclear reactions occur, especially if a supply of protons is available for mixing with carbon and oxygen, lead to the production of some nuclides that are not synthesized by nuclear processes occurring toward the centres of stars. Examples are ^{15}N and ^{26}Al. It is also possible that neutrons from reactions of alpha particles with ^{13}C lead to a form of r-process, a process in which very heavy neutron-rich nuclei are synthesized. Condensing solid grains in the gases expelled by novae may be expected to contain such unusual nuclides, and so would form a component of interstellar material with an unusual chemical composition.

When the Solar System formed from interstellar grains, some grains derived from novae would be present. It is an interesting and controversial question as to whether a fraction of early grains of unusual composition have been preserved to this day, for example in meteorites. There is evidence to show that this may be so.

The total energy of explosion of a typical nova has been estimated at 10^{45} ergs, which is to say about as much energy as the Sun emits in 10,000 years, about as much as would be realized in the explosion of 1,000,000,000,000,000,000,000 manmade nuclear weapons. *FH.*

N stars The N spectral class was part of the original Harvard system of SPECTRAL CLASSIFICATION and was introduced to describe very red stars which lack the titanium oxide molecular band absorptions of the M stars, but instead have very strong absorptions of molecular carbon and carbon compounds. The N class was divided into subdivisions N0 to N9.

The N stars, together with those of type R, comprise the carbon stars and are nowadays classified on the Keenan and Morgan system of C types (*see* R STARS). The later R stars and all of the

old N types now fall in the C4–C9 spectral classes. Many of the reddest N stars are thus not as cool as their colour suggests.

R and N stars are giants but the latter, being more luminous and having stronger spectrum band absorptions, are easier to discover in general objective prism surveys of cool stars. Large numbers of N stars have been found in the MAGELLANIC CLOUDS: in our own Galaxy the ratio of N stars to late M giants is about 1:100 but in the Small Magellanic Cloud it is 25:1.

Probably all the N stars are VARIABLE STARS, most are irregular variables but some are long period variables of the MIRA type.

As well as having excesses of carbon in their atmospheres, N stars have overabundances of many heavy elements. In this way they are similar to the S stars, but evidently have greater abundances of carbon. The extreme redness of many N stars arises from the weakening of the blue and ultraviolet light by strong sources of opacity in their atmospheres. The molecules responsible for their absorption are thought to be silicon carbide (SiC) and triatomic carbon (C$_3$). *BW.*

Nuclear reactions The interactions between atomic nuclei or between atomic nuclei and fundamental particles (*see* ATOMIC STRUCTURE). Nuclear reactions produce transformations from one ELEMENT or ISOTOPE to another. Such reactions are responsible for building up all the elements from the primordial "soup" of fundamental particles coming from the PRIMEVAL FIREBALL. Nuclear reactions are therefore responsible for the world of our experience, and indeed for the elements from which the molecules of life itself are assembled. They are also responsible for the maintenance of life — the energy output from the Sun and all stars is from nuclear reactions.

There are two basic types of reaction, fusion and fission, and two subclasses of each of these as summarized in the table overleaf.

Fusion is the binding together of different nuclear constituents to build up elements of heavier MASS NUMBER, while fission is the breaking of nuclei into lighter nuclei to form elements of smaller mass number. Fusion processes are exothermic, liberating energy, while fission processes are usually (but not exclusively) endothermic, requiring energy input to make them happen. The amounts of energy involved are vast, and are calculable from $E = mc^2$,

N

Types of nuclear reaction

Reaction	Where occurring	Elements involved (Mass number)	Elements produced (Mass number)
Fusion			
Thermonuclear	Stellar interiors Early universe Novæ	Hydrogen (2) to iron (56)	Helium (4) to iron (56)
Neutron absorption Proton absorption	Stellar interiors Supernovæ	Iron and heavier elements (>56)	Elements heavier than iron (>56)
Fission			
Photodisintegration	Novæ, supernovæ	Sulphur (32) to iron (56)	Some stable elements between neon (20) and iron (56)
Spallation	Interstellar medium	Carbon (12) nitrogen (24) oxygen (16)	Lithium (3) beryllium (5) boron (7)

N

Einstein's famous equation relating the change in mass (m) to the energy (E) released or absorbed; c is the velocity of light. The change in mass comes from the difference in total mass of the final products of the reaction and the initial products. For instance in the commonest nuclear reaction, the transmutation of HYDROGEN to HELIUM, one gram of hydrogen becomes 0.993 grams of helium, and the mass difference, 0.007 grams, releases 6×10^{11} joules or 175,000 kW/hours of energy.

The fusion reactions build elements of heavier mass number from the lighter. The first of these, thermonuclear fusion, is the process which created most of the helium in the universe in the first few minutes after the BIG BANG; it provides the main energy output of the stars, and indeed it governs STELLAR EVOLUTION; it provides the energy released in hydrogen bombs; and it may eventually be harnessed for power generation in fusion reactors (*see* THERMONUCLEAR REACTIONS).

The second type of "building" reaction is neutron absorption. Thermonuclear fusion reactions can build the elements up to iron (mass number 56), the most stable element, because the fusion reactions involving the lighter elements liberate energy, ie, the total mass of the final nuclei is slightly *less* than the mass of the initial nuclei. The reverse is the case for elements heavier than iron. The principal mechanism to build such elements becomes neutron absorption, in which one or more neutrons penetrate a nucleus and are captured, increasing the mass number and so creating a heavier element. It is the electrical neutrality of neutrons which enables them to penetrate the charged nucleus. In principle this could happen at low temperatures, and indeed the process is favoured by lower

temperatures, except for one crucial aspect: neutrons live only about 10 minutes before decaying into (positively charged) PROTONS. Thus a ready supply is required for the neutron capture process to be effective. This is available only in regions of very high temperature, where thermonuclear fusion reactions are occurring to liberate adequate numbers of free neutrons. Such conditions exist in stellar interiors, in the explosively-burning shells of NOVÆ atmosphere, and in the disintegration of stellar structure in SUPERNOVÆ explosions. There are two types of neutron capture process, known as slow (s-process) and rapid (r-process). In the r-process, two neutrons are captured almost simultaneously, and the result is a stable isotope. In the s-process, if an "extra" neutron is absorbed to form an unstable isotope, it has time to decay into a proton before a subsequent neutron penetrates and stabilizes the nucleus. The processes thus result in distinctive elements or isotopes: the r-process produces neutron-rich nuclei, while the s-process produces atomic nuclei with approximate balance between the numbers of neutrons and protons. In fact all elements heavier than lead need to form by the r-process. It obviously requires very large neutron supplies, and neutrons in appropriate numbers are liberated during explosive phases of stellar evolution (eg, the onset of supernovæ). With regard to the s-process, favourable conditions exist during the RED GIANT phase of stellar evolution, in which adequate supplies of neutrons are produced during the mixing of the zones burning hydrogen and helium.

A few heavy isotopes are very proton-rich and cannot be made by either s- or r-processes; these are probably produced by proton capture (the p-pro-

cess). Again it is likely that the late stages of stellar evolution produce the free protons in required quantity.

In photodisintegration reactions, very energetic PHOTONS (gamma rays) are absorbed by nuclei and split them into lighter nuclei. The energy of the photons is required to overcome the binding energy of the nuclei, and the reactions hence are endothermic. Temperatures of perhaps 10^9K are required to produce photons of adequate energy. Perversely, photodisintegration can result in the formation of heavier elements by creating a succession of semi-stable nuclear fragments which then recombine in different forms. Photodisintegration works for elements both heavier and lighter than iron. Because iron and the elements adjacent in mass number are the most stable, photodisintegration produces the "iron peak", the excessive abundance of iron in the universe compared to that of elements with higher and lower mass numbers.

Spallation fission reactions are from high-speed collisions between light particles such as hydrogen nuclei (protons) or helium-4 nuclei (ALPHA PARTICLES) and heavier nuclei. The result is the transmission of enough energy to some of the nucleons (protons or neutrons) for them to leave the heavier nuclei. The process allows formation of some of the lightest elements, lithium, beryllium and boron, from elements of intermediate mass such as carbon, nitrogen and oxygen, formed in thermonuclear fusion in stellar interiors. Spallation reactions happen in the interstellar medium which is bombarded by cosmic rays produced in supernova explosions; protons and alpha particles are major constituents of cosmic rays. Spallation reactions can also take place at the surface of some stars where streams of fast light particles can be created.

In summary, nuclear reactions began a few seconds after the big bang, the primeval fireball. At this time the thermonuclear fusion reactions produced most of the helium. Nucleosynthesis in the interiors of stars subsequently formed the heavier elements. Thermonuclear fusion created elements up to iron; neutron absorption and photon absorption processes created elements heavier than iron. Some of the lightest elements required the process of spallation to break up moderately heavy nuclei, while other heavy elements were destroyed by photodisintegration reactions, producing the iron peak. Explosive phases in stellar evolution produced complex mixtures of nuclear reactions: fusion reactions, and neutron absorption and photodistintegration. *JVW*.

Nucleon A proton or a neutron (ie, one of the two types of particle which comprise an atomic nucleus).

Nutation A small slow variation in PRECESSION, discovered by J. BRADLEY in 1747, caused by the 5° inclination of the Moon's orbit to the ECLIPTIC. This produces a slight "nodding" of the Earth's axis of 9″ arc on either side of the mean position, in a period of 18 years 220 days.

Nysa (Asteroid 44). It has the highest albedo (0.377) of any known asteroid. Its diameter is about 51 miles (82km).

O and B subdwarfs Hot stars with luminosities less than those of O and B dwarfs but greater than those of WHITE DWARFS. They are generally stars in the final stage of collapse to white dwarfs.

OAO NASA's Orbiting Astronomical Observatories. OAO-2 and OAO-3 (Copernicus) made ultraviolet observations of stars and interstellar matter from Earth orbit in the early 1970s.

O-association A loose grouping of young hot O STARS that appear to share a common origin, but which is likely to disperse in a time that is short compared to the rotation period of the GALAXY.

OB-associations Regions of space in which recent and ongoing star formation has led dominantly to visible high-mass (blue) O and B STARS. These are contrasted with "T-associations" whose dominant young population is low-mass stars (T TAURI STARS). The predominance of the very luminous O and B stars renders OB-associations recognizable at great distances, even through the dusty haze of interstellar space. Tens or hundreds of these massive stars can occur in a single association. Low-mass (faint) stars may exist in abundance too, but are far harder to see in distant OB complexes. The Great Nebula in ORION, with its core of hot stars, is part of an OB-association that

is relatively close to us, hence low-mass young stars are also detected.

Recent computer models suggest that OB-associations could be self-perpetuating: that is, an appreciable population of massive stars can lead to the preferential formation of new massive stars. The mechanism involves the combined and powerful winds from the surrounding O stars.

OB-associations are a feature of the gas-rich spiral and irregular galaxies, but not of elliptical galaxies in which no young stellar population is apparent. The frequency with which stars of different mass occur in a galaxy (or part of one) is called the "mass spectrum". In our Galaxy, low-mass stars (like our Sun, and smaller) are prodigiously more abundant than high-mass stars and contribute most of the mass, yet the tremendous luminosities of hot blue stars make them unrepresentatively obvious. Some consequences of these high luminosities are: the exceedingly short lifetime of high-mass stars (no more than 2–3 million years for stars above about 20 solar masses); their likely interactions with the ambient medium (vigorous OB stellar winds could disperse parental clouds, inhibiting the formation of lower mass stars); and the high incidence of SUPERNOVÆ within OB complexes. *MC*.

Oberon The outermost satellite of Uranus; diameter 942 miles (1,516km). Like Titania, which is almost the same size, Oberon has an icy, cratered surface, but inside some of the craters there is dark material of unknown origin which has presumably flowed out from below the visible surface.

Objective The principal light-collecting and imaging optical element of a telescope. The word derives from "object glass", used by early astronomers to mean the glass nearest to the object being imaged. Although sometimes used also for the main mirror of a reflecting telescope, "objective" nowadays almost always implies the lens of a refracting telescope.

The performance of a telescope objective depends on its ability to produce a sharp image, free of distortion or false colour. The earliest objectives, starting with those made by LIPPERSHEY and GALILEO at the beginning of the 17th century, were single biconvex lenses. A simple lens suffers from several kinds of aberration, the most important of which are spherical and chromatic (*see* CHROMATIC ABER-

RATION). In the former, light rays passing through the central portions and outer portions of the lens arrive respectively at foci at different distances from the lens; in the latter rays of light of different colours arrive at different foci. To reduce the effects of these aberrations early telescope makers produced objectives of small diameter but enormous focal length, such as the 150ft and 210ft telescopes of Johannes HEVELIUS.

In 1722 John Hadley produced the first successful reflecting telescope, which, being free of spherical and chromatic aberrations, terminated further use of long-focus objectives. In 1729 Chester Moor Hall discovered how to make achromatic objectives, but these did not come into general use until 1759 when the optician John Dollond took up their manufacture.

In an achromatic doublet lens a convex lens, usually of crown glass, is combined with a concave (diverging) lens of flint glass. Their radii of curvature are chosen so that the positive spherical aberration of the former counteracts the negative aberration of the latter. At the same time, the different dispersive powers (ie, variation of refractive index as a function of wavelength) of the two types of glass are used to reduce the amount of chromatic aberration. In a doublet, light at two chosen wavelengths can be focused at the same point; all other colours are out of focus at that point, but not as badly as in a single lens. The residual chromaticism (ie, the range of focus for light of different colours) is known as the secondary spectrum and in a doublet is typically one-twentieth of that of a single lens. It can be reduced further by introducing additional component lenses in the objective; such a compound lens is known as an APOCHROMAT. Triplet objectives were produced by Peter Dollond from about 1763. Doublet objectives have residual aberration which can be reduced by careful figuring of the lens surfaces, that is, by changing the surfaces to become aspherical.

For visual observations an achromatic doublet, designed to have minimal secondary spectrum in the green-yellow spectral region where the eye is most sensitive, is usually employed. As the secondary spectrum decreases with increasing FOCAL RATIO, visual objectives usually have FOCAL LENGTHS at least ten times their diameters. To achieve a wide field of view, however, shorter focal lengths are necessary, in which case apochromats must be used.

O

Some refracting telescopes are constructed purely for photographic use. Early photographic emulsions were sensitive only to the shorter wavelengths, so photographic objectives, for example those used in astrographs, were achromatized for blue light. Later photographic objectives were achromatized to allow the telescope to be used both in the blue and yellow — photography through a blue filter giving "photographic" response and through a yellow filter giving "photovisual" response. To make use of modern emulsions, which are sensitive over a wider range of wavelength, apochromatic triplets are often used.

Many visual objectives, being designed only for the use of an eyepiece close to the optical axis, have curved FOCAL PLANES, which complicates or precludes their use for photography over wide fields of view. Photographic objectives are designed to have flat focal planes, but the presence of residual aberrations produces a systematic variation of image quality over the field of view. In a well-designed objective such effects are kept to an amount below the usual size of star images (determined by atmospheric conditions). For very large (50–60°) fields of view, off-axis aberrations such as coma and ASTIGMATISM are important. Objectives which minimize these are known as anastigmats.

Almost all astronomical refracting telescopes have objectives in which one component is flint glass. This is a glass rich in lead oxide with high refractive index (familiar in its use for cut-glass ware), but in thicknesses greater than half an inch, opaque at wavelengths shorter than about 4000Å. These cannot therefore be used for observations of ultraviolet light from celestial objects.

The component lenses of an objective are mounted in a rigid metal cell to ensure that their optical surfaces are correctly aligned and that the precise spacing between the components is maintained. In order to prevent distortion of the lenses from stresses they are supported at only three points around their edges.

As a lens can only be supported around its perimeter its weight distorts the shape of its optical surfaces as the telescope changes direction. A lens must be made thick enough, and hence rigid enough, to keep the optical effects of such distortion to an imperceptible amount in the image. For objectives of large diameter the total thickness of glass results in absorption of substantial amounts of light. This produces a practical upper limit to the size of an objective, greater than which the additional light collected is more than offset by loss through absorption in the glass.

The largest objective in use, that of the 40in (101cm) refractor at Yerkes Observatory, weighs 500lb (227kg) and has a focal length of 62ft (19m). It was completed by Alvan Graham Clark in 1895; he had plans to construct an objective of 60in diameter, but did not live long enough to start work on it. A 41in refractor was built by Howard Grubb in 1919 for the Siemeis observing station in the Crimea. Although the optical blanks were cast, their glass was not considered to be homogeneous enough to warrant grinding them into lenses, so the telescope was never completed. The largest objectives actually used were the 49.2in doublets, one visual and the other photographic, made for a horizontally mounted refractor at the Paris Exhibition of 1900. The refractor was fed light by an 80in mirror but was later broken up as a failure.

Because of competition from reflecting telescopes, large refractors are no longer constructed. The 24in (61cm) refractor at Saltsjobsaden, Sweden, built in 1931, possesses the last really large general purpose astronomical objective to be made. However, the superior quality of images in refractors ensures that objectives up to 10 or 12in diameter will continue to be made. These are used for visual observation or for transit circles. Larger objectives, specially designed for solar telescopes, are also occasionally made. *BW. See also* LENS, TELESCOPES, and related entries.

Objective prism A narrow-angled prism placed in front of the aperture of a telescope to produce a low-resolution spectrum of every star in the field of view.

Oblateness The ratio of the difference between the equatorial and polar radii to the equatorial radius of a body.

Obliquity of ecliptic The angle between the ECLIPTIC and the CELESTIAL SPHERE: 23° 26′27″ (1988), decreasing by about 0″5 per year. After about 1,500 years it will increase again; the maximum range is 21° 55′ to 28° 18′, and the period is 40,000 years.

Observatories An observatory is perhaps best defined as "a structure specifically intended for making celestial observations and measurements". On this basis, it is debatable whether Stonehenge and other megalithic monuments in north-west Europe can be properly termed observatories; their true nature and purpose is still very much in doubt. The earliest extant structure which can be identified as an observatory with reasonable confidence dates from as late as AD640 and is located in south-east Korea. This bottle-shaped building, which is about 30ft (9m) high, is still well preserved.

We have little evidence for the existence of observatories anywhere in the world before about 750BC — a relatively recent date in ancient history. Among the great early civilizations of the Fertile Crescent, the Egyptians do not seem to have had much interest in astronomy apart from keeping a regular watch on the bright star Sothis (SIRIUS). Although the Babylonians were very active astronomers, our knowledge of astronomy in Babylon before about 750BC is very fragmentary. However, from this time onward down to about AD100 a fairly well-equipped observatory was maintained. This was under royal patronage and was staffed by professional astronomers whose main interest was astrology. Little is known about the instruments that these astronomers possessed, but from their surviving records it is apparent that they used water clocks to time eclipses and other events to within a very few minutes. They also measured the apparent distances between two celestial bodies to the nearest 0°1. According to the 1st century BC historian Diodorus of Sicily, the astronomers of Babylon used to observe from the top of the ziggurat in the city — a tower 300ft (90m) high.

Probably the first major observatory established by the ancient Greeks was that built by HIPPARCHUS at Rhodes around 150BC. From this he catalogued 850 stars, observed eclipses and determined the times of equinoxes. Little is known regarding the instruments which Hipparchus employed. PTOLEMY in his *Almagest* notes Hipparchus' use of an equatorial ring for determining when the Sun reached the equinox and a dioptra for observing changes in the Moon's apparent diameter. Hipparchus presumably also employed some sort of ARMILLARY SPHERE. Soon after AD100, Ptolemy established a great observatory at Alexandria. This was equipped with a wide variety of instruments such as the armillary sphere, GNOMON, quadrant and water clock. Using these instruments, Ptolemy measured the positions of more than 1,000 stars with a typical accuracy of about 0°2. After Ptolemy, astronomy fell into a decline in the western world until its revival by the Arabs some 700 years later.

Little is known about observatories in China until as late as about 200BC, the time of the unification of the various states into a single empire. In earlier centuries — from about 500BC – there are records of official astronomers appointed by rulers of the various states into which China was then divided. Also we have evidence that detailed star maps were produced around 300BC. It thus seems likely that observatories were in existence long before 200BC, but just when is impossible to say. An imperial observatory staffed by civil servants is attested in China from about 200BC. The main motive was astrological. Instruments available in this observatory at first included sundials, gnomons, sighting tubes and water clocks. Around 50BC the first fixed equatorial ring was introduced and by about AD100 an armillary sphere was designed. Observatories in the Chinese style were in later centuries — particularly after about AD600 — built in both Japan and Korea.

Perhaps the most famous Chinese observatory was that established by the astronomer Su Song around AD1090.

huge gnomon—some 40ft (12m) in height — a water clock and probably an armillary sphere. Observatories in China lost much of their traditional character with the arrival of the Jesuit astronomers in the 16th century. The imperial observatory was staffed by Jesuits and it was through them that the telescope was first introduced in China around 1630.

Scientific observation of the heavens by Arab astronomers commenced early in the 9th century AD and rapidly attained a high level. The first Islamic observatory of note was founded in 829 by Caliph AL-MA'MUN at Baghdad, then capital of the great Abbasid Empire. This was the centre of astronomical activity in the Arab world for about a century. The various instruments at Baghdad were in the main modelled on ancient Greek devices — as listed above. Other important observatories were established over the next two centuries, eg, by al-Battani (*see* ALBATEGNIUS) at Raqqa (AD887), AL-SUFI at Shiraz (*c.* 965), Ibn Yunus at Cairo (*c.* 990) and al-Biruni at Ghazna (*c.* 990). Gradually, new instruments were invented for specific purposes. In particular, the precision of measuring angles reached about 0°1 at this period, surpassing Ptolemy.

Of the many Islamic observatories, the earliest for which a detailed list of equipment has survived is that founded

quadrant of 33ft (5.3m) radius; this was graduated in degrees and minutes of arc. The last great Islamic observatory was that established by ULUGH BEIGH at Samarkand in 1420. This was noted for its huge meridian arc some 130ft (40m) in radius. Around the same time, interest in astronomy was being rekindled in Europe after the "Dark Ages".

Beginning in 1471, Johann Müller (REGIOMONTANUS) and his pupil Bernhard Walther made numerous fairly accurate observations of the position of celestial bodies from a room in Müller's house. However, it is doubtful whether this can be rated an "observatory". The earliest major European observatory after the Renaissance was that built by Tycho BRAHE at Uraniborg (Denmark) in 1576. Here Tycho, under the patronage of King Frederic of Denmark, designed and built a wide range of instruments – quadrants of various kinds, sextants and armillary spheres. Tycho went to considerable lengths to achieve the utmost precision in constructing his instruments and his accuracy of measurement far surpassed that of all earlier observers. One of his quadrants could be read to 10 seconds of arc and in his numerous measurements Tycho typically realized a precision of 1 arc minute or better. Only eight years after the death of Tycho, the telescope was first used in astronomy, but several decades elapsed before any advance on Tycho's accuracy of measurement was made.

There were, of course, many other naked-eye observatories, notably in India; those at Jaipur and Delhi are still intact. However, the modern-type observatory dates only from the invention of the telescope in the early 17th century. One of the earliest was set up at Leiden, in Holland, mainly to house a large quadrant. The "Round Tower" observatory at Copenhagen was founded in 1637, on the recommendation of the Danish astronomer Longomontanus. In 1641 the great observer HEVELIUS built an observatory on the roof of his house in Danzig (now Gdańsk), in Poland and equipped it with one of the small-aperture, long-focus refractors of the time.

The national observatory of France was built in Paris from 1667 to 1671, though the French King was so anxious to make it architecturally imposing that the building was more or less useless as an observatory; the first Director, G.D.CASSINI, had to take his telescopes into the grounds! In England, the Royal GREENWICH OBSERVATORY dates from 1675. It was

O

Observatories: Hamburg, site of the world's first Schmidt camera

This was equipped with a splendid water clock and a mechanized armillary sphere and celestial globe. No trace of this observatory now exists, but much of the Mongol observatory founded at Gaocheng (Henan) in 1276 still stands. This was equipped with a

at Maragha (Iran) in 1259. Some of the instruments of this observatory still survive in museums. Maragha had a staff of at least 15 astronomers, and contained a huge library of more than 400,000 volumes. The largest of the impressive list of instruments was a founded, by order of King Charles II, so that a new star catalogue could be drawn up for the use of British seamen. Subsequently it became a major centre for all branches of astronomical research. In the 1950s the main equipment was shifted to Herstmonceux, in Sussex, because of the deteriorating conditions round London. A current proposal to close Herstmonceux, and transfer the observatory elsewhere, is being strongly opposed by the astronomical community.

Observatories proliferated from the late 18th century. Some were amateur, such as those of Johann SCHRÖTER (destroyed by the invading French in 1814) and the third Earl of ROSSE (completed in 1845; the great reflector was mounted between two massive stone walls – surely the most remarkable "observatory" of all time). But with the appearance first of large refractors and then of large modern-type reflectors, major observatories became essential. Some, such as the LOWELL OBSERVATORY in Arizona, the LICK OBSERVATORY in California and the YERKES OBSERVATORY in Wisconsin, were complete before 1900, though naturally their equipment has been extended since. Then came the MOUNT WILSON Observatory, with its 60in (152cm) and 100in (254cm) reflectors. In 1948 the 200in (508cm) Hale reflector came into operation on MOUNT PALOMAR in California, and for many years remained the world's largest.

Just as photography replaced the human eye for most branches of research, a century ago, so today photography is itself being superseded by electronic devices, and this naturally causes modifications in observatory design and technique. There is considerable emphasis upon southern-hemisphere sites, and on the advantages of altitude; for example the summit of Mauna Kea in Hawaii is now festooned with large telescopes of all kinds — including UKIRT, the United Kingdom Infra-Red Telescope, which is therefore above much of the atmospheric water vapour, the infrared astronomer's worst enemy. There is, too, a flying observatory, the KAO or KUIPER AIRBORNE OBSERVATORY, which has the additional advantage of being manoeuvrable!

In the foreseeable future it is hoped to have a major telescope in space: the HUBBLE SPACE TELESCOPE, which will be placed in orbit round the Earth and will be virtually clear of any atmospheric interference.

Another development of vital importance is the remote control of telescopes. No longer does an observer have to sit in a cold dome, guiding his equipment; he need not be in the observatory, or even in the same country — thus the telescopes at La Palma, in the Canary Isles, can be operated by an observer who is situated in Britain. Times have changed; it is no longer far-fetched to suppose that before long we may even have an observatory on the surface of the Moon. *FRS./PM. See also* CHINESE ASTRONOMY, HISTORY OF ASTRONOMY, ISLAMIC ASTRONOMY.

Occam's Razor A principle formulated by William of Ockham (or Occam) in about 1300. This says that "the simplest theory which fits the facts corresponds most closely to reality". It has many applications in science.

Occultation An event in which one body is obscured by another: ie, a distant object is hidden by a nearer one. The most obvious example is a SOLAR ECLIPSE, which is an occultation of the Sun by the Moon.

The term can be applied to a range of celestial objects, but the most commonly observed are lunar occultations. During a lunar occultation a star is obscured by the passage of the Moon along the band of the ECLIPTIC. The most commonly observed occultation is the disappearance of a star (immersion) at the dark limb. An inexperienced observer is often surprised at the suddeness with which the star vanishes. This is due to the Moon having no atmosphere and the star being at such a great distance that it is virtually a point source of light in space. The lunar limb acts as a cutting edge against the background of stars so that the star instantly disappears from view.

After full moon a second type of lunar occultation may be seen: a re-appearance of a star from behind the dark limb (emersion). This is much harder to see as the star is not visible until the occultation takes place. To be successful an observer needs to know the time as well as the point on the limb where the star is to appear. When the star suddenly appears at emersion it may well catch the observer unawares.

Visual study of occulations has been undertaken on a regular basis for a century and still plays a major role in astronomy. However, with the development of fast, sensitive electronic photo detectors it is possible to increase the applications to which occultation study can be applied.

Recently the linking of a detector to a computer has enabled the light-curve of stars during immersion to be recorded. Identification of previously unknown close double stars, and the confirmation of known double star systems, has been possible using this technique. Attempts to measure the diameters of some of the nearer stars has also given results. Undoubtedly the most important aspect of such recording systems is the generation of light-curves that can be re-examined at a future date. This ability will mean that as new applications are found for occultation results the information can be analyzed.

Photoelectric measurement is not possible for amateurs unless they are well equipped, but visual study can be undertaken with anything from binoculars to telescopes. The main object of watching occultations is to obtain an accurate time for the point of occultation, but it is often impossible to obtain an observation due to local cloud conditions at the time. In the course of a year the Moon may well occult more than 4,000 stars; though this may seem a great many, a keen observer would be lucky to see a few hundred events during a year. Major observatories can only rarely afford to commit their instruments to this type of study. For this reason the professional astronomer encourages the amateur to do the work for him. Predictions of when stars are to be occulted are issued; the amateur attempts to time the event and records the information onto a report form. These reports are then sent to the International Lunar Occultation Centre for data reduction. The information thus obtained acts as a store for professionals to use.

The best time to see an occultation is just after new moon when only a thin crescent will be illuminated. Earthshine will often illuminate the dark limb, thus making it easier to judge when the star will be occulted. During most evenings only a few stars will lie in the path of the Moon, but occasionally an open cluster such as the PLEIADES will be suitably placed. On these occasions many occultations will occur during a few hours and given suitable conditions a great number of timings will be possible.

Stars are not the only bodies to be occulted by the Moon; on occasions a planet may also be occulted. In the case of Jupiter with its own satellite system, the satellites too will be occulted during the approach to the lunar limb of the planet. Such events, though much rarer, do produce spectacular views when seen through a telescope or binoculars. The planet, being much closer than the stars, exhibits a disk so that the occultation is a gradual affair lasting many seconds. Because of the brightness of a planet the event that takes place at the bright limb may also be seen, though this is rare in stars due to the glare of the Moon. In the predawn sky at a location in the middle of England on October 4 1980 an even rarer event was seen. The planet Venus grazed along the limb of the Moon, being partially obscured by it. Many lunar features, mountains and valleys were seen in profile against the bright disk of Venus. Occultations of this type are called "graze occultations" as the star or planet grazes along the edge of the lunar limb. When a star grazes the Moon then this event can be seen only along a track no more than 1,100yd (1km) wide on the Earth's surface. For this reason these events are rare, but if observed by a team can yield much useful information about the contours of the limb at the point of graze.

During a lunar eclipse the glare from the bright limb is much reduced. This enables stars or even deep sky objects to be seen and the number of occultations visible increases during the time that the Moon is eclipsed. Occultations that involve planets and stars permit the study of the planet's atmosphere. The star will become dimmer the more atmosphere that is present until it is totally obscured by the main body of the planet. These events are also rare, but are within the range of amateur equipment. The discovery of the rings of Uranus was made using occultation techniques as the star was temporarily dimmed as it passed behind each ring.

The ASTEROIDS can also occult stars and information about their shape and diameter can be obtained provided that several observatories are employed in timing the event. Recently the method has been employed to see if it is possible to detect any satellites of a minor planet before and after the main occultation. This is similar in principle to the method employed to discover the Uranus ring system. The chances of detection are much lower, as the satellite would have to be in the track of the star as seen from Earth and the observing teams must spend some hours before and after the event to ensure that the satellite occultation is not missed. *AEW.*[1]

Oceanus Procellarum Vast, lava-inundated area, in the western hemisphere of the Moon, containing the notable ray-craters Aristarchus and Kepler and the unusual features Rümker and Schröter's Valley.

Octagon Room Room in the Royal Observatory, Greenwich, London, designed by Sir Christopher Wren with many tall windows for use with the long telescopes of the day. It was originally called the Great Room. *See also* GREENWICH OBSERVATORY.

Octahedrites Iron meteorites; two metallic minerals intergrown.

Octans (The Octant). A faint constellation, notable only because it contains the south polar star, Sigma Octantis (mag. 5.46). The brightest star in Octans, Nu, is only of magnitude 3.76.

Odessa Crater Located in Texas, 525ft (160m) in diameter, 18ft (5.5m) deep, with two smaller ones; associated with meteoritic iron, mostly weathered, of the type that formed ARIZONA CRATER.

Odysseus Crater on Saturn's satellite TETHYS. It is 250 miles (400km) in diameter — though the diameter of Tethys itself is only 1,050km! The position of the crater's centre is 30°N, 130°W. Any theory about the formation of such an object is beset with difficulties. An impact of sufficient violence would probably have shattered Tethys, whose density is only about that of ice; the curve of the interior of Odysseus is the same as that of the globe, so that Tethys cannot have been fragmented and later re-formed; but neither is it easy to see how Odysseus can have been produced by internal forces — though, according to some authorities, this seems more probable than an impact origin. *PM.*

Œsel Crater (Kaalijarv). Six craters, in the Estonian Soviet Socialist Republic; iron meteorites have been found in the vicinity.

Off-axis guiding system Equipment used by serious amateur astro-photographers whereby light from a guide star at the edge of the camera's field of view is directed at right angles by a small prism up to an eyepiece with an illuminated guiding cross hair, thus enabling telescope tracking errors (revealed by movement of the guide star) to be corrected.

Officina Typographica (the Printing Machine). Rejected constellation,

shown on Bode's maps *c.* 1775.

Of stars Stars basically of type O, but with EMISSION LINES of doubly ionized nitrogen in their spectra.

Ogygis Rupes Martian ridge; 32°–35°S, 53°–55°W.

OH Symbol for the hydroxyl radical, the first interstellar molecule to be discovered at radio wavelengths (in 1963). Also found, for example, in comets as a result of the decomposition of water by solar ultraviolet radiation.

Olbers' Comet A bright long period (69.47 years) comet that has been seen three times, as 1815, 1887V and 1956IV. Its orbital parameters were first calculated by BESSEL.

Olbers, Heinrich (1758–1840). German doctor, who was as renowned in medicine as he was as an amateur astronomer. He was one of the CELESTIAL POLICE who searched for the missing planet between the orbits of Mars and Jupiter, and discovered two of the first four asteroids (PALLAS and VESTA). He made several cometary discoveries, and invented a new method of calculating their orbits; he was also the originator of OLBERS' PARADOX. He was tireless in encouraging others, and it was he who was responsible for the brilliant career of F.W. BESSEL. He wrote numerous papers on many astronomical subjects. *PM.*

Heinrich Olbers: Celestial Policeman

Olbers' Paradox A paradox discussed in 1826 by Heinrich OLBERS (although it had been raised earlier, eg, by HALLEY) when he posed the question "Why is the sky dark at night?" If space were infinite and uniformly filled with stars, then in whatever direction the observer were to look he or she would eventually end up looking at the surface of a star and so the entire sky should be as bright as the surface of the Sun.

The paradox can be resolved in various ways, for example: (i) if the universe is not sufficiently old, light from the more remote objects cannot yet have reached us; (ii) the expansion of the universe ensures that radiation emitted by galaxies is weakened by the RED-SHIFT and cannot be detected beyond a certain range. *IKN.*

Oljato (Asteroid 2201). An APOLLO ASTEROID, photographed in 1947, lost, and rediscovered in 1979. Oljato is of interest because, like 3200 PHÆTHON, it may be the burned-out remains of a cometary nucleus.

Olympus Mons Undoubtedly the most spectacular shield volcano in the Solar System. Situated on the north-western flank of the Tharsis Rise on Mars, it rises some 15 miles (25km) above the surrounding plains and is more than 430 miles (700km) in diameter. It comprises a central nested caldera 50 miles (80km) across from which the flanks slope away at angles of only 4°. The flanks are terraced and traversed by numerous lava flows and channels with a roughly radial arrangement. In many respects it is similar to terrestrial basaltic shields like those of Hawaii, but with a volume that is between 50 and 100 times greater.

Toward the north and south ends of the main shield, lava flows drape a prominent peripheral escarpment which, in places, forms a cliff 4 miles (6km) high. Numerous landslides degrade this and spread out onto the adjacent plains. The flows themselves are very long compared with terrestrial flows, a phenomenon which may be explained either by very high eruption rates, very large-volume eruptions, or a combination of the two.

Surrounding the shield is a wide and very complex region of lobate ridged terrain that is termed the "aureole". In places this extends 430 miles (700km) from the basal scarp. Associated with it is a positive free-air gravity anomaly. Much controversy attaches itself to the origin of this feature. Some workers believe it to have been formed by eruptions of ash (nuées ardentes), while others suggest it may have been the result of massive gravity slides off the main volcano. A further theory has the

summit of the surrounding cliff representing the top of a former ice-sheet surrounding the volcano. If so, the aureole material was the product of eruptions beneath an ice cover. *PJC.*

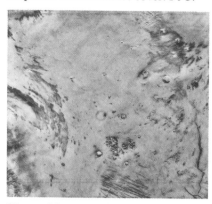

Olympus Mons: an artist's impression

Omega Centauri (NGC5139; 2000 coordinated R.A. 13ʰ 26ᵐ.8, Dec. − 47° 29′) The brightest globular cluster; it is visible to the naked eye as a fuzzy star of magnitude 3.7 and is about 40′ in diameter, with a broad, bright central region. Its brightest stars are 11th magnitude. It is 17,000 light-years away, and is hence one of the nearer globular clusters. It contains several hundred thousand stars in a volume 200 light-years across. Omega Centauri is intrinsically the brightest globular cluster in the Galaxy and is noted for wide variations in the heavy element ("metal") content of its stars, suggesting that they were formed at different times. Some 200 VARIABLE STARS have been found in it. *ACG.*

Omega Nebula Ionized gas cloud in Sagittarius. Also known as M17, this is the only visible part of an extensive complex in which stars are forming.

Onsala Space Observatory Radio observatory, outside Gothenburg, Sweden, run by Chalmers Institute of Technology. It has an 85ft (26m) dish for longer wavelengths — also used in VLBI observations — and a 65ft (20m) dish for millimetre-wave observations.

Oort Cloud According to a theory due to the Dutch astronomer Jan Oort, there is a "cloud" of comets orbiting the Sun at a distance of between 30,000 and 100,000AU. If a comet is perturbed for any reason (possibly by a passing star, or even a remote planetary body) it may swing in toward the Sun. If it is perturbed by a planet — usually Jupiter — it may be captured and put into a short-period orbit; alternatively

it may be expelled from the Solar System altogether, or else simply return to the Oort Cloud, not to return to the inner Solar System for an immensely long period. There is no positive proof of the existence of the Oort Cloud, but the theory does seem to fit the facts as we know them at present. *PM.*

Opacity A measure of the ability of a gas to absorb radiation.

Open clusters *See* STELLAR CLUSTERS.

Ophir Chasma Prominent canyon situated in the central part of the MARINERIS VALLES on Mars. Ophir Chasma is one of two rather spatulate canyons, the other being Candor Chasma, which are situated to the north of the central canyon complex. The west and east ends of the canyon are both rather blunted and the trend of the canyon walls is continued laterally by lines of

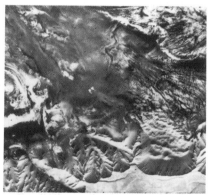

Ophir Chasma: evidence of water?

crater chains, indicating an underlying structural control.

Great interest attaches to a number of high-resolution VIKING images, since they reveal the existence not only of erosional debris but also of rhythmically-layered deposits which, by analogy with terrestrial rocks, would be expected to have been laid down in quiet conditions beneath a cover of water. Such deposits are seen generally to form flat-topped hills (mesas) or irregular blocks within the troughs but are seldom visible in the canyon walls. Nevertheless, in one locality, namely in the divide between Ophir and Candor Chasmæ, the layered deposits are seen in finely-etched mesas and quite clearly transect the major ridge-and-gully topography of the canyon dividing wall.

The origin of these desposits is uncertain. If they are seen as the deposits of lakes, the implication is that liquid water existed at the time of canyon formation (a relatively recent event, based on crater ages). There is little evidence that this has been the case in the recent past. If they are simply the erosional remains of canyon formation, then it is difficult to interpret the relationships displayed in the intercanyon divide.

As to the formation of the canyon system itself, there is no consensus view. Faulting must have played a role, but whether or not it was the major one is open to debate. Erosion also could have been extremely important, but attached to this process is the problem of where the eroded material has gone to. Certainly surface water could not have removed the debris, since many of the canyons are closed at either end. The solution must await further evidence. *PJC.*

Ophiuchid meteors A minor meteor stream which peaks on about June 20. The radiant is at R.A. 17h 20m, Dec. −20°, close to the star Xi Ophiuchi. Meteors from this shower may be seen from about June 17 to 26. Activity is low (ZHR about 6).

Ophiuchus (the Serpent-bearer). It commemorates Æsculapius, son of Apollo, who became so skilled in medicine that he even restored the dead to life!

Brightest stars

Name	Visual Mag.	Abs. Mag.	Spec.	Distance (light-yrs)
α (Rasalhague)	2.08	0.3	A5	62
η (Sabik)	2.43	1.4	A2	59
ζ (Han)	2.56	−4.4	09.5	550
δ (Yed Prior)	2.74	−0.5	M1	140
β (Cheleb)	2.77	−0.1	K2	120
κ	3.20	−0.1	K2	117
ε (Yed Post)	3.24	0.3	G8	104
θ	3.27	−3.0	B2	590

Next come 72 (3.73), Gamma (3.75), Lambda (3.82) and 67 (3.97).

Ophiuchus is not well defined; Rasalhague lies close to Rasalgethi or Alpha Herculis. However, the constellation contains several fairly bright globular clusters: M9, M10, M12, M14, M19, M62 and M107. *PM.*

Oppolzer, Theodor Egon Ritter von (1841–86). Austrian mathematician and astronomer who calculated the time and track of every lunar and solar eclipse from 1207BC to AD2163. This work was published as the *Canon of Eclipses* in 1887. He was elected President of the International Geodetic Association in 1886.

Theodor von Oppolzer: worked on eclipses

Opposition The position of a SUPERIOR PLANET at 180° celestial longitude from the Sun.

Optical interferometer The resolution of a telescope, that is, its ability to separate two stars (or two regions of the same star) into distinguishable images, is limited by the diameter of its lens or mirror. The largest optical telescopes currently have mirrors of 13–16ft (4–5m) aperture, which, given perfect atmospheric conditions, could resolve directly stars separated by about 0″02 (two-hundredths of a second of arc). In order to make full use of the resolution of a large telescope in normal atmospheric conditions, and as a means of effectively increasing the aperture of telescopes, various kinds of optical interferometer are used. These use the INTERFERENCE PRINCIPLE: light from a point source brought through one optical path will interfere with light from the same source brought through a different optical path, producing interference fringes. However, if the telescope or interferometer has sufficient aperture to resolve the source, ie, the source is no longer point-like, then interference does not take place.

The first optical interferometer to be applied to measuring the diameters of stars made use of this principle. In late 1920, A.A.MICHELSON and F.G.Pease observed with an interferometer constructed from a 20ft (6m) rigid steel beam mounted over the front of the Mount Wilson 100in (254cm) reflector. Two mirrors, whose separation could be varied along the steel beam, fed light from a star to two other mirrors, which

directed the light into the telescope. If the movable mirrors were set close together then interference fringes would be seen at the focus of the telescope, but as the separation of the mirrors increased, the fringes would gradually disappear if the star had a large enough angular size to be resolved at the full 20ft separation. The diameter of Betelgeux was measured to be 0″047 with this instrument. A 50ft (15m) interferometer was built in the 1930s, but did not work successfully.

Only recently have renewed attempts been made to revive the stellar optical interferometer in its original form (but with the use of modern photoelectric detectors rather than the eye at the telescope). However, several alternative forms of the instrument had been proposed and used routinely. One of these, pioneered by C.H.Townes, uses separate telescopes and a laser beam to communicate between them. In its essence, instead of the two separate beams of light interfering with each other, each is made to interfere (or, more correctly, beat with) the laser beam. The beats, which are an interference pattern that varies with time, occur at much lower than optical frequencies and can be recorded electronically. In principle this technique can be

Optical interferometer: at Mt Wilson

extended to employ telescopes separated by hundreds of kilometres, which will give angular resolutions of millionths of arc seconds and allow observation of details such as spots on stars with the largest angular diameters.

A completely different type of interferometer, the intensity interferometer, was developed by R.Hanbury Brown and R.Twiss, starting in 1949. Instead of using the wave nature of light and producing interference directly, the intensity interferometer makes use of the photon structure of light. The times of detection of individual photons are noted at two separate telescopes. If the star is unresolved, then the photon arrival times are correlated, that is, if a photon arrives at one telescope then there is a higher probability of a photon arriving at the other telescope than when the star is resolved. By increasing the spacing between the two telescopes and measuring electronically the correlation between photon detections, the spacing is found at which the correlation begins to decrease. This can be converted into a measurement of the angular size of the star.

An intensity interferometer was constructed at Narrabri, in northern New South Wales, in 1962 and completed observations of the 32 stars that it was capable of observing by 1972. It consisted of two reflecting arrays, each 20ft (6m) in diameter covered with a mosaic of 252 hexagonal glass mirrors. The two reflectors were separately steerable around a circular track about 600ft (180m) in diameter. The results from this interferometer provided the most accurate measurements of diameters of stars. Stellar diameters down to 0".001 were measured.

An entirely different technique, SPECKLE INTERFEROMETRY, was developed by A.Labeyrie in the 1970s. Using a large telescope and a highly magnified image of a star, enhanced in brightness with the aid of image tubes, photographic or electronic pictures are taken with exposures of about one-fiftieth of a second. This "freezes" the effects of atmospheric turbulence ("seeing") and shows that the instantaneous image of a star observed through the atmosphere consists of a large number of speckles. Each speckle is in some sense an image of the star resolved to the theoretical limit for the size of telescope used.

Using computer or other electronic processing, information is obtained about the separation and orientation of a double star or the diameter and intensity distribution across the surface of a resolvable star. Many hundreds of double stars with separations smaller than that measurable by direct observation at the telesocpe have been observed by speckle interferometry.

A quite different approach to achieving the high resolution necessary to measure angular diameters of stars and separations of close double stars is the use of lunar OCCULTATIONS. A point source on the edge of an opaque boundary produces an interference pattern. As the Moon moves in its orbit it occults stars and at the moment of occultation the star at the Moon's limb

generates such an interference pattern. The movement of the Moon sweeps this pattern across the surface of the Earth; so a telescope equipped with an intensity measuring device of high time resolution can record the interference fringes as they pass by. The interference pattern, containing a dozen or more fringes, sweeps past in about one-tenth of a second.

If the star is fully resolved then interference effects disappear and the occultation produces a simple gradual disappearance. For stars with angular diameters from 0".002 to about 0".05 the shape of the recorded fringe pattern reveals the diameter. Double stars produce two superimposed interference patterns. Over one hundred diameters of stars and several dozen double star separations have been measured in this way. An obvious restriction of the method is that only those stars capable of being occulted by the Moon may be studied.

Future developments in the field of optical interferometry include the simultaneous use of many telescopes, each linked by laser beam to use the technique of APERTURE SYNTHESIS as already employed for several decades by radio astronomers. A pilot project is underway in France. Larger Michelson interferometers and larger intensity interferometers are under consideration. Eventually, interferometers with very large separations of their component telescopes will be used in orbit, away from the difficulties produced by the Earth's turbulent atmosphere. *BW*. See also RADIO INTERFEROMETRY.

Orbit The path of one celestial body in the gravitational field of another.

Orbiter probe Any spacecraft that make at least one revolution around a Solar System body. If the probe is injected into a polar orbit, the whole surface of the body can, in general, be scanned by cameras; radar; gamma-ray, far-UV, and mass spectrometers; X-ray fluorescence, and alpha-particle detectors; IR radiometers; magnetometers; etc.

Origin of planets One of the major unsolved mysteries in astronomy. The difficulties are as follows: the initial conditions arc unknown; it took place (in the case of the Sun's planets) about 4,570,000,000 years ago; many of the chemical and physical processes that occurred in the interim were not time-reversible; and, worst of all, only one planetary system (the one we live in) can be studied in any detail. Even here

we are not sure which facts are of importance and which are inconsequential. Theories are usually divided into three classes: (a) planets formed as a direct consequence of the formation of the Sun; (b) planets formed from material dragged from a passing star *after* the formation of the Sun; and (c) planets formed after the Sun out of material captured from interstellar space. Theories in the first of these classes are currently in favour.

What are the important facts when it comes to the origin of our planetary system? Listed in order of decreasing importance they are:

(1) The planets and the Sun are of about the same age $(4.57 \pm 0.03) \times 10^9$ years.

(2) The Sun is 745 times more massive than all the other objects in the Solar System put together, but has only 0.5 per cent of the angular momentum.

(3) Planets divide into two groups, small, dense, rocky terrestrial planets and large, gas-giant Jovian ones. If they were made from the same solar-type material, the NEBULA from which they condensed would have a mass between 0.05 and 1.0 times that of the Sun. Compositional differences can be accounted for by postulating a variation in temperature throughout the initial condensing disk, it being hotter and more dense, close to the Sun.

(4) The Solar System is small, Neptune being 30AU from the Sun in comparison to the present-day closest star, which is 270,000AU away. It is reasonable to suppose that there is no major planet beyond Neptune. Not only do we run out of material beyond 30AU but also the condensation time becomes inordinately long.

(5) Of the 32 closest stars to the Sun, four out of the 16 single stars have planetary systems and one star in the eight binary systems seems to have planets. This conclusion comes from a careful study of the irregular celestial motion of the stars caused by the gravitational effects of their planets. So planets are a common, but not essential, adjunct to star formation.

(6) The orderliness of the system is probably due to a combination of turbulence in the initial condensing disk and 4.57×10^9 years of gravitational PERTURBATION, so that now the planets try to keep out of each other's way.

(7) As the Sun was most likely a member of an open cluster of a thousand or so stars at the time of its birth, a handful of these stars would have gone SUPERNOVA during the accretion time of the planets and some supernova

remnants would have been mixed in with the condensing nebula disk.

Planetary formation is closely associated with star formation, and both processes occurred at nearly the same time and out of the same material. A dense cloud of interstellar material, orbiting the galactic nucleus on a path that keeps it in the plane of the Milky Way, passes into an arm of our galaxy about every 200 million years. This gravitational shock decelerates the cloud and causes contraction. If this contraction forces the cloud into a condition where the Jeans criterion is satisfied, it will collapse. (Jeans used a balance of gravitational and thermal energy to calculate that a cloud of mass M and temperature T could collapse only if its density (in g/cm^3) became greater than 7.2×10^{45} $(T(K)^3/M(g)^2)$. As the cloud collapses, it loses heat because its molecules act as efficient radiators. Magnetic energy is transferred back to the galactic arm because the field lines are still connected. ANGULAR MOMENTUM is conserved and as the cloud fragments into "star mass" pieces, this momentum is converted (in the main) from spin angular momentum to orbital angular momentum. The final fragment continues to collapse and spin-up until the condensation mechanism changes from being ISOTHERMAL to ADIABATIC. This occurs when all the constituents, such as H_2, H_2O, NH_3, CO_2, CH_4, etc, have been dissociated and ionized. We are now confronted with a lens-shaped cloud of mass around two solar masses, radius around 30AU, and central temperature a few tens of thousands K. Even 5AU from the centre, at around Jupiter's present orbit, the temperature is above a few thousand K. This flattened lens of gas and dust then slowly separates into two discrete components, a central protosun (about the size of Mercury's orbit) and a surrounding, more tenuous, nebula of gas. As the protosun shrinks, the metallic elements in the inner regions of the nebula start to condense out. Moving farther away the temperature drops even lower and progressively refractory and earthy materials, and then volatile substances such as water, ammonia and methane, condense. The disk is spinning and the particles have orbits with a range of eccentricities and inclinations. Particles collide frequently, these collisions being non-elastic, resulting in energy loss. This has a tendency to reduce both the eccentricity and the inclination leading to the dust sinking into a flat equatorial disk made up of parti-

cles with circular orbits. This dense disk is inherently unstable and suffers from turbulence powered by the new sun's energy and by shock waves from nearby dying stars. ACCRETION and condensation occur leading to the dust disk breaking up into regions with sizes in the centimetre to tens of kilometre range. Subsequent collisions lead to both growth and fragmentation. But growth wins in the end — the planets are here to prove it. The accumulation was, however, an extremely wasteful process. From a nebula which originally had cosmic composition (ie, a preponderance if hydrogen and helium) and a mass similar to that of the Sun, all that remains is a planetary residue of a mere 0.0013 solar masses. The Sun itself was responsible for this loss. During its early evolution it went through a brief (a million years?) T TAURI stage, when its luminosity increased to 10,000 times its present value and the solar wind freshened into a gale sweeping clean the nascent Solar System and removing much of the gas and dust. The remaining planetesimals continued to orbit the Sun in rings, similar to Saturn's but of much higher spatial density.

More low-velocity encounters occur, leading to the growth of the largest planetesimals. The gravitational influence of these big objects increases and they eventually mop up the smaller objects until each ring of the nebula essentially accretes to form one planet. The process is slow. The ring that now only contains Uranus and Neptune took about 300 million years to accrete to that state from a time when it had about 10^{25} bodies of size up to 1,000kg. Inner planets formed faster (Earth took about 5 million years).

The gravitational potential energy brought to the growing planet is considerable and can lead to melting and differentiation. The resulting object is thus spherical and has a dense core and a lighter mantle. The end product is a set of planets, spinning in the same direction as their orbital motion and moving on nearly circular co-planar paths around their central star.

Our set of nine make up a beautiful example. Alas, we don't know how typical they are. *DWH. See also* STELLAR EVOLUTION.

Orion The celestial Hunter; perhaps the most splendid of all the constellations. Orion is so distinctive that it must have been noticed from very early times, and many famous legends about it are found in almost every country.

Brightest stars

Name	Visual Mag.	Abs. Mag.	Spec.	Distance (light-yrs)
β (Rigel)	0.12	−7.1	B8	900
α (Betelgeux)	var.	−5.6v	M2	310
γ (Bellatrix)	1.64	−3.6	B2	360
ε (Alnilam)	1.70	−6.2	B0	1200
ζ (Alnitak)	1.77	−5.9	O9.5	1100
κ (Saiph)	2.06	−6.9	B0.5	2100
δ (Mintaka)	2.23v	−6.1	O9.5	2350
ι (Hatysa)	2.76	−6.0	O9	1860
π^3	3.19	3.8	F6	25
η (Algjebbah)	3.36	−3.1	B1	750
λ (Heka)	3.39	−5.1	O8	1790

Then come Tau (3.60). Pi^4 (3.69). Pi^5 (3.72) and Sigma (3.73).

Orion's characteristic pattern makes it unmistakable. The celestial equator passes closely north of Mintaka, the northern star of the Hunter's Belt, so that Orion can be seen from every inhabited continent. It is a very rich constellation; the Great Nebula, M42, is an easy naked-eye object (*see* ORION NEBULA), and a small telescope will show the main stars of the Trapezium (THETA ORIONIS). The Nebula is actually the main part of a vast molecular cloud. *PM.*

Orion Arm The local spiral arm of the GALAXY.

Orionid meteors One of two meteor showers associated with HALLEY'S COMET, and seen when the Earth passes within 0.154AU of the comet's orbit on about October 25 each year, near the ascending node. The Orionids may be seen between about October 16 and 30 every year, with peak activity between October 20 and 24, when the ZHR is about 25 meteors per hour. There are four or five peaks in shower activity. Outstanding displays of the Orionids may have occurred in AD288 and 1651. Following a careful study by J.P.M.Prentice between 1928 and 1939, it was shown that the Orionids have a multiple radiant structure, centred around R.A. 6h 24m, Dec. + 15°, close to the star Gamma Geminorum. *JWM.*

Orion Nebula Gas cloud in the constellation ORION, visible to the unaided eye but illuminated by faint, young stars.

Almost the entire region of sky we call Orion is occupied by a vast cloud of gas and microscopic dust grains, lying at a distance of about 1,600 light-years. Although the cloud is extremely tenuous by the standards of terrestrial vacuums, its diameter is so great that no light ray trying to travel through

O

can make the passage without encountering a dust grain and being absorbed or deflected. Much of the Orion Nebula is, therefore, opaque to light, and can be probed only by infrared and radio techniques.

Within the densest portions, the collapsing cloud is forming stars: the Orion Nebula is one of the most important regions of star formation readily available for study. The most recent stars to be formed lie unseen within it, and include the BECKLIN-NEUGEBAUER OBJECT. The next youngest are the Trapezium stars, THETA ORIONIS. They are probably less than one million years old, and provide almost all the ionizing radiation that renders visible the small portion known as the Orion Nebula. It is the Trapezium stars' ultraviolet radiation that makes the nebula glow; because they are extremely hot — about 50,000K — most of their energy is emitted in the ultraviolet, and is processed by the gas to make visible light, which is why the nebula appears to our

Orion Nebula: star-forming region

eyes to outshine the stars. Such a nebula is an HII REGION.

The Trapezium stars have destroyed most of the volatile dust in the nebula, etching out a roughly spherical hollow that is still growing. That hollow has opened out on the near side of the dark cloud, allowing us to see in. The visible Orion Nebula is little more than the inside of an incomplete spherical hole in a much larger dark nebula.

Orrery Geared model of the Solar System. The first orrery, probably that preserved in the Adler Planetarium, Chicago, was made by George Graham

about 1708. The name is usually attributed to John Rowley, who made a similar instrument for Charles Boyle, 4th Earl of Cork and Orrery, in 1712.

Orrery: 17th century

Many "grand orreries" up to 3ft (1m) in diameter, showing the motions of the six known planets and their satellites, with clockwork movements, were made in England during the 18th century. From about 1775, hand-driven portable orreries came on the market, often with alternative fittings so that an orrery could be a "planetarium", "tellurium", or "lunarium" as desired. *HDH.*

Oscillating universe If the amount of mass in the universe is sufficiently large, its current expansion phase will eventually be halted by gravitational attraction and the motions of the galaxies will be reversed. At the end of the collapse phase, with the universe packed into a small volume of great density, it is possible that a "bounce" will occur which will result in a further expansion phase. The universe would thus oscillate between BIG BANG and "Big Crunch" episodes and could be infinite in age.

A major task of modern astronomy is to estimate whether there is sufficient matter in the universe to halt its expansion. *BW.*

O stars The defining characteristic of O-type spectra is the presence of ABSORPTION lines of ionized helium; neutral helium lines are also present. On the MAIN SEQUENCE the O stars represent an extreme of dwarf stars: they are the hottest, with surface temperatures in the range of 30,000 to 50,000K, the heaviest, with masses from 20 to 50 times solar, and intrinsically the brightest at luminosities from 50,000 up to one million times that of the Sun. At these luminosities the consequences are

the same as described for the hottest B STARS: the time that O stars remain on the main sequence is less than a million years and as a result they are a rare kind of star. This, however, gives quite the wrong impression of their importance in the evolution of the GALAXY. Although there are currently only about 300,000 O stars in the Galaxy, which is a very small fraction of the total of 100,000 million stars, there has been time for 10,000 generations of O stars in the lifetime of the Galaxy. Therefore, in all there have been about 3,000 million O stars formed from interstellar gas. As the O stars are more than 30 times heavier than the average star, and they return most of their material back to the interstellar gas, we deduce that a large fraction (perhaps greater than one-third) of all of the matter that currently makes up stars of all types has passed through earlier generations of O stars. The element-building NUCLEAR REACTIONS that occur in O stars are therefore an important source of many of the heavy elements in our Galaxy.

The O stars, being relatively young stars, are associated with the interstellar clouds of gas and dust in which star formation occurs. Together with the hotter B stars, they form OB ASSOCIATIONS, which show that high-mass stars tend to form in groups rather than in isolation. Most O and B stars form in binary systems, many with components of similar mass. The more massive star of such a double evolves fastest and is likely to explode as a SUPERNOVA, which can release its companion from orbit at high velocity. It is estimated that as many as 20 per cent of the O stars are such runaways.

As with the hottest B stars, the higher luminosity of the O stars drives a high rate of mass loss via a stellar wind — from two-millionths to 20-millionths of a solar mass per year in the O stars. Summed over all of the O stars in the Galaxy, this can account for one-third of the gas being returned to space from stars (RED GIANTS and PLANETARY NEBULÆ account for most of the rest).

The high rate of mass loss from O stars has another effect of galactic significance. If a binary system consisting of two O or B stars survives intact, the supernovæ explosion of one of its components, then the result can be a NEUTRON STAR orbiting through the stellar wind from the unevolved star. The gas falling onto the neutron star emits energetic X-rays. Although such systems are very rare — probably only

50 in the entire Galaxy — they account for a substantial fraction of the hard X-ray sources in the Galaxy.

Rotation among O stars is similar to that found in B stars — they are rapid rotators and the fastest possess circumstellar disks, which produce EMISSION LINE spectra leading to the Oe classification. These stars also have the highest rates of mass loss. Some of the Oe stars have, in addition to hydrogen and helium emission, strong emission lines of doubly ionized nitrogen. These are given the classification Of and probably represent O stars that were once very massive (greater than 50 solar masses) which, like the WOLF-RAYET STARS, have been stripped of the outer layers to reveal a nitrogen-rich interior. About 15 per cent of O and early B stars show emission lines.

There are no large amplitude pulsating variables among the O stars, but most vary by a few per cent in brightness over months or years. There is also evidence of small changes from night to night in some stars.

Many subluminous stars — hot white dwarfs, hot subdwarfs, nuclei of planetary nebulae, and cataclysmic variables at minimum brightness — have been given O or Oe spectral classifications. Such descriptions are on the old Harvard system of classification; these stars are not O stars in the standard MK system.

Bright examples (O stars appear bluish-white): Delta Orionis 09.5 II, Zeta Orionis 09.5, Ib, Zeta Puppis 05, Zeta Ophiuchi 09.5 V. *BW.*

Oterma's comet Period 7.88 years, seen as 1942VII, 1950III and 1958IV. A close approach (0.10AU) to Jupiter in 1963 changed the period to 19 years and the perihelion distance to 5.4AU and thus made further recoveries very doubtful.

Outgassing Terrestrial bodies, formed from primeval Solar System material, contained volatiles; some residual may still be coming out. This is termed "outgassing".

Owl Nebula A planetary nebula (M97) in Ursa Major of apparent magnitude 11 and apparent diameter 3 arc minutes. Its distance is about 2,000 light-years and its diameter about 1.5 light-years.

Oxygen A colourless, odourless gas which makes up 21 per cent by volume of the Earth's atmosphere and which is the third most abundant chemical element in the universe (about 3 oxygen

nuclei to every 10,000 hydrogen nuclei). Chemical symbol O, atomic number 8, atomic mass 15.9994.

Ozma Project An attempt to pick up intelligent messages from beyond the Solar System, initiated in 1960 by a group of radio astronomers at Green Bank, West Virginia. They used the large radio telescope there to "listen out" at a wavelength of 21cm, which is the wavelength emitted by clouds of cold hydrogen in the Galaxy — reasoning that radio astronomers in other systems would be likely to concentrate upon this wavelength also. The search was unsuccessful, and was abandoned after a few months, though similar experiments have been undertaken more recently. The name "Ozma" was drawn from the Wizard of Oz in Frank Baum's famous children's book, though the investigators tended to refer to it as Project Little Green Men! *PM.*

Ozone A highly reactive form of oxygen in which the molecule has three atoms (denoted by O_3). In the terrestrial atmosphere it is the major absorber of solar ultraviolet radiation in the 230–320nm wavelength range and has its greatest concentration at altitudes of between 7 and 31 miles (12–50km).

Pallas (Asteroid 2). Discovered in 1802 by H.OLBERS. With a mean opposition magnitude of 8.0, it is brighter than any other asteroid apart from VESTA and CERES. The diameter has been given as 378 miles (608km); the period is 4.6 years. The orbit has the very high inclination of 34°.85. The rotation period has been given as 10 hours. In type it has been classed as "peculiar carbon".

Pallasites Stony-iron METEORITES: olivine and metal.

Palomar *See* MOUNT PALOMAR.

Palus Putredinis (the Marsh of Decay). Part of the lunar MARE IMBRIUM, near Archimedes.

Palus Nebularum (the Marsh of Clouds). Part of the eastern MARE IMBRIUM. The name has been deleted from some maps, but lunar observers use it.

Pandora Third satellite from Saturn in order of distance. It measures 68 × 56 × 44 miles (110 × 90 × 70km) and is the outer SHEPHERD SATELLITE of the F ring. It was discovered in 1980 from VOYAGER photographs.

Panspermia Theory A theory proposed in 1906 by the Swedish chemist and Nobel Prize winner Svante ARRHENIUS, according to which life on Earth did not begin here, but was brought to the Earth by way of a METEORITE from outer space. The theory caused considerable interest but was never widely accepted, if only because it seemed to raise more difficulties than it solved. However, in modified form it has been revived by Sir Fred Hoyle and his colleague Chandra Wickramasinghe. Their ideas differ markedly from those of Arrhenius, but they too adopt the principle that life was brought to the Earth from beyond the Solar System — not by a meteorite, but by a COMET. Hoyle also maintains that interstellar "grains" are bacteria, and it is certainly true that in recent years many organic molecules have been identified between the stars.

The main argument used by Hoyle and Wickramasinghe is that the appearance of life involves many steps, each of which is in itself inherently unlikely, so that the Earth is much too small for such evolution to have occurred; instead it needs the whole "resources of space". Therefore, life will be carried around, mainly by comets, and will take root wherever conditions are suited to it — as on Earth. Hoyle and Wickramasinghe also believe that comets can deposit bacteria in the Earth's atmosphere, thereby causing epidemics. They were certainly right in claiming that the nucleus of HALLEY'S COMET would be dark, and covered with organic material, rather than bright and icy as had been generally believed until the GIOTTO mission of March 1986.

It is fair to say that the overall theory has met with little support from astronomers and virtually none at all from medical experts, but it has at least led to renewed discussion about the origin of life, about which there is still a great deal of uncertainty. We cannot, as yet, prove that there is life anywhere beyond the Earth, though most authorities believe that it must exist. *PM.*

Parabola An open curve, one of the conic sections, obtained by cutting a cone in a plane parallel to the opposite side of the cone. It can be regarded as

P

an ELLIPSE with only one focus, an infinite major axis, and eccentricity = 1.

Parallax *See* SPECTROSCOPIC PARALLAX and TRIGONOMETRICAL PARALLAX.

Paris Observatory The oldest observatory in the world still functioning, Observatoire de Paris was commissioned by King Louis XIV, on the recommendation of his Prime Minister Colbert, and built by the famous architect Claude Perrault in 1667. Early astronomers were the Frenchman Father Picard, who first measured the segment of the meridian; the Dane Ole RØMER, who demonstrated with Jupiter's satellites the limited speed of light (1676); the Dutchman Christiaan HUYGENS, inventor of the astronomical clock; and the Italian Giovanni Dominico CASSINI, discoverer of the red spot of Jupiter, of the gap in Saturn rings, of the Saturn satellites Iapetus (1671), Rhea (1672), Tethys and Dione (1684), and author of a detailed lunar map (1692) and of tables of Jupiter's satellites.

Successors were Jacques CASSINI, who extended the geodesic measurements of the meridian (1720), and then, in 1771, César-François Cassini III, author of a detailed map of France in 183 sheets. Cassini IV was the director from 1784 until 1793. The observatory sought to clear up the controversy about the shape of the Earth by using meridian measurements and accordingly sent Godin and Bouger to Peru from 1735 to 1744, and Maupertius and Clairant to Lapland in 1736. The meridian of Paris remained the origin of longitudes until 1884.

After the French Revolution in 1789, the observatory created the metric system and Jérôme de LALANDE published his stellar catalogue. Other directors were DELAMBRE, MÉCHAIN and Bouvard, who inaugurated the era of celestial mechanics. François ARAGO was elected director in 1834. His disciples were Fresnel, FIZEAN and FOUCAULT.

Urbain LE VERRIER, who discovered Neptune by computation in 1846, succeeded as director in 1854 and raised celestial mechanics to its highest level, when Delaunay worked out the theory of the Moon and Tisserand published in 1889 his masterpiece, the *Treatise of Celestial Mechanism*, which is still in use. Le Verrier also organized the first world meteorological network.

In 1880 Admiral Mouchez initiated the international photographic star survey "Carte du Ciel". Loewy and

Paris Observatory: oldest still in use

Puiseux supervised the photographic lunar atlas from 1896 to 1910. In 1911, General Ferrié set up the world time service "Bureau International de l'Heure". Nicolas Stoyko, in 1937, demonstrated the variation of Earth rotation.

Today, the observatory is co-ordinating the activities of four research institutes, with a total staff of 700, at Paris; MEUDON OBSERVATORY; the radio-astronomy Centre of Nançy; and at the astrometric, geodesic and geodynamic station of CERGA near Grasse. The observatory houses a collection of astronomical instruments from the 16th to the 19th centuries, and a major library on the history of science. *ACD.*

Parkes Radio Astronomy Observatory A major radio astronomy observatory in New South Wales and Australia's

Parkes Observatory: the radio "dish"

National Radio Astronomy Observatory; the main telescope, a 210ft (64m) "dish", was completed in 1961. The Parkes observatory works very closely with SIDING SPRING OBSERVATORY, also in NSW.

Parking orbit An initial orbit into which a spacecraft is placed before carrying out another manœuvre.

Parsec 3.26 light-years: the distance at which a star would subtend a PARALLAX of one second of arc.

Parthenope (Asteroid 11). Discovered in 1850; diameter 93 miles (150km), magnitude 9.9.

Paschen series The series of lines in the infrared spectrum associated with electronic transitions down to or up from the third energy level of the hydrogen atom.

Pasiphaë The eighth satellite of Jupiter discovered in 1908 by Melotte. It is 30 miles (50km) in diameter, and has retrograde motion, so that it may well be an ex-asteroid.

Patientia (Asteroid 451). Discovered in 1899. With an estimated diameter of 172 miles (276km) it is among the dozen largest known asteroids.

Patroclus (Asteroid 617). The second Trojan to be discovered (by Kopff in 1906). The estimated diameter is 91 miles (147km).

Pauli Exclusion Principle The quantum-mechanical principle that no two fermions (particles like electrons, neutrons or protons) may exist in the same state, ie, they may not have exactly the same set of quantum numbers (quantum numbers define a particle's energy, angular momentum, and spin). *See also* WAVE MECHANICS.

Pavo (the Peacock). One of the Southern Birds.

Brightest stars

Name	Visual Mag.	Abs. Mag.	Spec.	Distance (light-yrs)
α	1.94	−2.3	B3	230
β	3.42	1.2	A5	90

Then come Delta (3.56), Eta (3.62) and Epsilon (3.96). Kappa is a W Virginis star (a type II CEPHEID) with a range of from 3.9 to 4.8 and a period of 9.1 days. Its mean absolute magnitude is only 3.4, so that it is much less luminous than a classical Cepheid. *PM.*

Pavonis Mons A massive Martian shield volcano situated at 1°N, 113°W, along the crest of the Tharsis Rise. It has a large summit caldera surrounded by concentric graben and radiating volcanic flows.

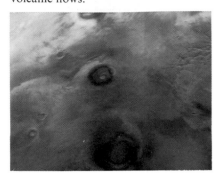

Pavonis Mons: massive Martian volcano

Payne-Gaposchkin, Cecilia (1900–79). British astronomer who spent much of her life in America, mainly at Harvard. She carried out important work on stellar evolution and galactic structure, and wrote a number of successful books on astronomy.

P Cygni The galactic B1 supergiant P Cygni (34 Cygni) is an important VARIABLE STAR easily identified about 2° south of the centre star (Gamma Cygni) of the Northern Cross. This highly luminous star has an apparent visual magnitude 5. P Cygni is 724,000 times more luminous than our Sun, with an effective surface temperature of 19,300K. Its distance is 1.8 kiloparsecs, radius 76 solar radii and mass 50 solar masses. P Cygni's past history is relatively well documented in the 17th century. It appeared as a reddish star of visual magnitude 3 in 1600–06, faded to magnitude 6 in 1620, rose again in 1655–59 and then fluctuated around minimum until 1715. In modern times P Cygni appears as a relatively constant, yellowish-white star. Its spectral signature, the "P Cygni profile", is seen particularly in O, B and A supergiants showing episodes of catastrophic mass loss.

The characteristic spectrum exhibits strong EMISSION LINES accompanied by equally strong variable blue-shifted ABSORPTION components. These P Cygni line profiles are seen in the strongest BALMER hydrogen lines, in ionized helium, nitrogen, silicon, sodium and iron. The episodic nature of the mass loss, witnessed in the 17th century although not then understood, is supported by the detection by Wendker of a remarkable radio arc interpreted as the remnant of an earlier gas shell.

P Cygni loosely resembles Zeta[1] Scorpii and Eta, AG and HR Carinæ, all within our Galaxy. It has extra-galactic counterparts in the Large MAGELLANIC CLOUD, eg, the somewhat redder variable, SS Doradûs, and in the ANDROMEDA GALAXY, eg, AE and AF Andromedæ.

Calculations suggest that P Cygni's evolutionary track across the HERTZSPRUNG-RUSSELL DIAGRAM as it expends its internal nuclear fuel, and also episodically loses mass, may be exceptional. P Cygni lies on the borderline below which STELLAR EVOLUTION brings other massive stars into the red supergiant phase. It is sometimes referred to as a "hypergiant". According to Lamers, de Groot and Cassatella it is not clear whether P Cygni has entered the red supergiant phase yet or is oscillating across an upper luminosity limit in the Hertzsprung-Russell diagram. *ADA.*

Peculiar A stars The Peculiar A, or Ap, stars are basically of spectral type A but they have some unusual spectral features. For example, a common peculiarity is the great strength of lines of silicon, strontium or europium. These anomalies occur in A stars with strong magnetic fields. The Ap stars are classified into types (eg, silicon stars) according to their spectral peculiarities.

Peenemünde German military rocket base on the Baltic, established in the 1930s. The German Government took over the old German Society for Space Travel (VfR) and most of its leading scientists, including Wernher von Braun, were taken to Peenemünde. It was here that the V2 weapon was developed, and used to bombard southern England during the final stages of World War II. Activities at Peenemünde were temporarily interrupted by a bombing raid by the Royal Air Force, in which one of the main scientists, Dr Thiel, was killed, but the V2s were produced until the launching sites were overrun by the Allies and research at Peenemünde ceased.

Most of the Peenemünde scientists, notably von Braun, went to the United States to continue their research into rocketry. In 1958 the team led by von Braun launched America's first artificial satellite, Explorer 1. *PM.*

Pegasus (the Flying Horse). A very conspicuous constellation of the northern hemisphere. Its four main stars make up the famous Square, though Alpheratz, formerly known as Delta Pegasi, has been transferred to Andromeda, and is now classified as Alpha Andromedæ.

Brightest stars

Name	Visual Mag.	Abs. Mag.	Spec.	Distance (light-yrs)
ε (Enif)	2.38	−4.4	K2	520
β (Scheat)	2.4 max	−1.4v	M2	176
α (Markab)	2.49	0.2	B9	100
γ (Algenib)	2.83	−3.0	B2	490
η (Matar)	2.94	−0.9	G2	173
ζ (Homan)	3.40	0.0	B8.5	156
μ (Sadalbari)	3.48	0.2	K0	147

Also above magnitude 4 are Theta (3.53) and Lambda (3.76). Pegasus is not a rich constellation, but it does contain a bright globular cluster, M15, which is an easy binocular object. Scheat is a semi-regular variable, with a rough period of 36 days and a small range (2.4 to 2.8). *PM.*

Peiresc, Nicholas (1580–1637). French amateur astronomer who made the first recorded telescopic observation of the ORION NEBULA.

Peking Observatory *See* BEIJING OBSERVATORY.

Penelope Large crater on Saturn's satellite TETHYS; 10°S, 252°W.

Penumbra The area of partial shadow surrounding the main cone of shadow cast by the Earth or other non-luminous body in sunlight. In that area, a partial eclipse is visible.

Perfect Cosmological Principle The hypothesis that on the large scale the universe looks the same everywhere, in all directions and at all times. It was postulated in 1948 by Hermann Bondi and Thomas Gold as the basis of their formulation of the STEADY-STATE THEORY (Hoyle approached the theory from a different standpoint).

Most modern cosmological theories assume the validity of the ordinary "cosmological principle", that on the large scale the universe is homogeneous (looks the same everywhere) and isotropic (looks the same in every direction). The perfect cosmological principle went further in asserting that the large-scale appearance of the universe remains the same at all times. Since the galaxies are moving apart, the perfect cosmological principle requires that new galaxies be formed at a rate just sufficient to maintain a constant average number of galaxies in each given volume of space. This implies the continuous creation of matter at a very slow but steady rate, in the region of 10^{-44}kg/m³/s, ie, about one proton per cubic metre of space every ten years.

The perfect cosmological principle also implies that the universe is infinite, with no beginning and no end, and that there never was a time when all matter was densely concentrated together (as the BIG BANG theory requires). It follows, too, that galaxies accelerate away from each other so that, despite being infinite, the universe has a "horizon" beyond which galaxies cannot be seen because their light is too severely red-shifted. The principle also asserts, by definition, that the laws of nature do not change with time (some cosmologists contend that this need not be valid in an evolving universe).

The "timeless" quality of the perfect cosmological principle made it attractive to many people, but it is now widely believed to be untenable in the face of the present observational data. *IKN.*

Periastron The point in the orbit of a member of a double or multiple star system nearest to the primary star.

Perigee The point in a satellite orbit nearest to the Earth.

Perihelion The point in the orbit of a planet or comet nearest to the Sun.

Period–Luminosity Law A relationship discovered by Henrietta LEAVITT in 1912 between the pulsation periods of CEPHEID VARIABLES and their median luminosities or absolute magnitudes. The brighter the Cepheid, the longer is its period. The importance of this relationship lies in the fact that observed periods are distance independent, whereas apparent magnitudes are dimmed by distance according to the inverse square law. Absolute magnitude is related to luminosity and is the apparent magnitude as measured from a standard distance. The measurement of period indicates absolute magnitude, which may then be compared with apparent magnitude to indicate the distance of the Cepheid. *RCM.*

Periodic table Classification of the elements arranged in ascending atomic number into vertical groups (0–VIII) and horizontal periods 1–7). Groups I–VII have A and B families, elements in each family have similar chemical properties because they have similar electron configurations in their atoms. Periods 1–6 terminate with a noble gas that has the valence electron shell full for that period.

Three-quarters of the elements are metals and occupy the left and centre of the table; non-metals (except hydrogen) occupy the right. Metalloids lie adjacent to a diagonal from boron to polonium. Position in the table is used in predicting the properties of an element and formation and formulae of its compounds. *JLP. See also* ELEMENTS, ATOMIC STRUCTURE.

Perrine, Charles Dillon (1878–1951). American astronomer, who worked at Lick and then went to direct the CORDOBA OBSERVATORY in Argentina. He published an important star catalogue, and discovered several comets as well as Jupiter's satellite Himalia.

Perseid Meteors One of the most reliable of the annual meteor showers. It reaches its maximum on August 11 or 12 every year, when the radiant is at R.A. 3h 04m, Dec. + 58°, not far from the star Eta Persei. Most people in the northern hemisphere will know that during the first three weeks of August, meteors seem to be exceptionally bright and numerous. The reason is that between August 1 and 21, the Earth intersects a very broad stream of debris, left behind by Comet P/Swift-Tuttle 1862 III. During its period of maximum activity, which occurs between about August 10 and 13, peak rates of over 50 meteors per hour are normal for this shower. Indeed, since the mid-1970s peak rates have exceeded 70 meteors per hour, and in 1980 and 1981 were over 100 meteors per hour.

The history of the Perseid stream goes back over 1,900 years, the first recorded display being noted by Chinese observers on July 17, AD 36. The next documented return occurred in late July 714, and there have been many Chinese, Japanese and Korean records since that time. The shower was first noted in Europe from 811 and records are abundant from 830. The existence of the Perseids as a major annual meteor shower was probably recognized in about 1835, by A. Quételet in Brussels, E.C.Herrick in the USA, and several others. The stream has been observed ever since.

For many years the Perseids were known traditionally as "the Tears of St. Lawrence", in memory of the Spanish martyr who was killed on August 10, 258, close to the date of maximum of the meteor shower. Between 1864 and 1866, the great astronomer Giovanni Schiaparelli computed the orbit of the Perseid stream, and showed that it was almost identical with the periodic comet Swift-Tuttle. This was the first time that mathematical proof had been found linking meteor showers with periodic comets. The Perseid meteoroids enter the atmosphere at about 135,000 miles per hour (60 km/s), and produce a large number of bright meteors, many with persistent trains. *JWM.*

Perseus A prominent northern constellation.

Brightest stars

Name	Visual Mag.	Abs. Mag.	Spec.	Distance (light-yrs)
α (Mirphak)	1.80	−4.6	F5	620
β (Algol)	2.12v	−0.2	B8	95
ζ (Atik)	2.85	−5.7	B1	1100
ε	2.89	−3.7	B0.5	680
γ	2.93	0.3	G8	110
δ	3.01	−2.2	B5	320
ρ	3.2 max	−0.5v	M4	190

Next come Lambda (3.76), Gamma (3.77), Kappa (3.80), Sigma (3.83) and Tau (3.95). ALGOL is the prototype eclipsing binary. Rho is a semi-regular variable, with a range of from 3.2 to 4.2 and a very rough period of from 33 to 55 days. Perseus — the hero of the famous legend of the Gorgon and Andromeda — is a rich constellation, crossed by the Milky Way; there are various open clusters visible with binoculars, including M34, which can be seen with the naked eye. *PM.*

Perseus Arm The spiral arm of our Galaxy, immediately exterior to the Orion arm, which contains the Sun. Spiral arms are seen by the 21cm emissions from hydrogen and the presence of young blue stars.

Perturbations According to KEPLER'S LAWS, the orbit of a planet around the Sun, or of a satellite around a planet, is an ellipse. In reality attractions by other planets and satellites, and by the non-uniform distribution of mass in the planets, cause a significant displacement of the position of a body from where it would be in pure elliptic motion. This displacement is the perturbation of the orbit, and has periodic and secular components. The periodic components consist of fairly small short-period perturbations with periods similar to the orbital periods, and much larger long-period perturbations which are mostly caused by close COMMENSURABILITIES of the orbital periods. The most notable example is the so-called great inequality of Jupiter and Saturn which is due to their 5:2 commensurability, and it causes variations of their longitudes with a period 880 years and amplitudes of 0.°3 and 0.°8 respectively. The secular perturbations are continual changes of an orbit in one direction, which are mostly such that the mean ellipse of the orbit rotates in the direct sense, and the normal to the orbit plane describes a retrograde conical motion. These are caused by the long-term average attractions of the other bodies, and, for the satellites, by the flattened shape of their planets also. For the planets there appear in addition to be steady changes of the eccentricity and inclination of the orbit. In reality these are very long-period perturbations, with periods ranging from 25,000 to 2,000,000 years, but over the few hundred years of observations it is convenient to represent them as secular terms.

Perturbations can cause very large changes in the orbits of COMETS. All the short-period comets (periods mostly less than 200 years) almost certainly originated as long-period comets (periods of thousands of years), and were perturbed into their present orbits following many close approaches to the planets, principally Jupiter. *ATS.*

Petavius Majestic lunar crater 106 miles (170km) in diameter; 25°S, 61°E. It has a complex central mountain group, and a major cleft crosses the floor. It is one of the great Langrenus chain of formations.

Petrarch Prominent crater on Mercury 100 miles (160km) in diameter; 30°S, 27°W.

Phaethon (Asteroid 3200). An APOLLO ASTEROID, discovered in 1983 by Simon Green and John Davies during the INFRARED ASTRONOMICAL SATELLITE mission. Originally known as 1983TB, Phaethon is only a few miles in diameter, has the smallest perihelion distance of any asteroid (0.14AU), and its orbit is virtually identical to that of the GEMINID METEORS. Since the Geminids are believed to be dust shed from a vanished comet, Phaethon may be the remains of a cometary nucleus that has lost all its volatile material. Infrared observations made in 1984 suggested that Phaethon has a rocky surface, not the dusty crust predicted for an extinct comet, but the link to the Geminids remains unexplained. *JKD.*

Phases The shape of the part of the Moon or a planet illuminated by the Sun as seen from Earth. The four main lunar phases are new (between Earth and Sun), full (opposite the Sun) and first and last quarter (90° from the Sun). Mercury and Venus (and any asteroids passing inside the Earth's orbit) can show all the phases, though in this case full occurs at SUPERIOR CONJUNCTION. Mars can appear markedly GIBBOUS at QUADRATURE, Jupiter barely so.

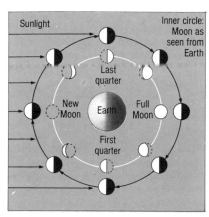

Phases: the main phases of the Moon

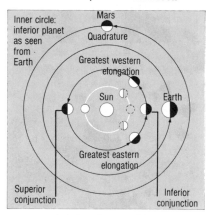

Phases: an inferior planet and Mars

Phekda (or Phecda). Alternative proper name for Phad (Gamma Ursæ Majoris).

Phillips Lunar crater west of W. Humboldt; 26°S, 78°E; diameter, 75 miles (120km). It has a central ridge.

Phobos The innermost of the two Martian moons. It has an irregular shape with dimensions of 17 × 13 × 12 miles (27 × 21 × 19km). It is in synchronous rotation, with its largest axis aligned along a Martian radius and its smallest axis perpendicular to its orbital plane.

The satellite orbits at a distance of 5,830 miles (9,380km), with a period of 0.319 days. This implies that Phobos revolves from west to east at a rate of 47°/h, while Mars rotates at 14.°6/h in the same sense. To an observer on the surface of Mars, Phobos would appear

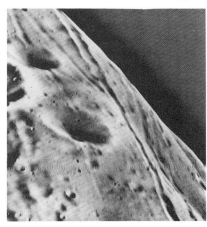

Phobos: from Viking 1, 1977

to rise in the west and set in the east 5.5 hours later.

Surface properties, a low reflectivity of 5 per cent, a dark grey colour, and a density of about 2gm/cm³, indicate that Phobos is composed of CARBONACEOUS CHONDRITIC material similar to some asteroids. *RFB.*

Phocylides Lunar crater in the Schickard group; diameter 60 miles (97km); 54°S, 58°W.

Phoebe The outermost satellite of Saturn, discovered in 1898. Its diameter is only 99 miles (160km) and as it has retrograde motion it may be a captured asteroid; its revolution period is 550.4 days, but its rotation period is no more than 9 hours. It was not well imaged from the VOYAGERS, but seems to have a darkish surface with some brighter patches. *PM.*

Phoebe Regio On Venus: 10°–20°N, 275°–300°W. It is not far from the volcanic Beta Regio area, and has been the landing site for several Russian unmanned probes.

Phoenicid Meteors A minor southern-hemisphere meteor shower which peaks on about December 4 or 5 each year. The radiant is located close to the star Zeta Phoenicis, at R.A. 1h 00m, Dec. −55°. It produces only low activity (ZHR about 5).

Phœnicus Lacus Region of Mars on the southern flank of the Tharsis Rise. It is crossed by extensive lava flows and large numbers of fractures.

Phœnix (the Phœnix). One of the Southern Birds. (*Table overleaf*)

P

Brightest stars

Name	Visual Mag.	Abs. Mag.	Spec.	Distance (light-yrs)
α (Ankaa)	2.39	0.2	K0	78
β	3.31	0.3	G8	130
γ	3.41	−4.4	K5	910

Phœnix lies not far from Achernar in Eridanus. Zeta Phœnicis is an eclipsing binary, with a range of from 3.6 to 4.1 and a period of 1.67 days.

Photoelectric aids Electronic devices used to convert the energy from a beam of ELECTROMAGNETIC RADIATION into an electrical current. Such devices are of great importance in modern astronomical photometry (see PHOTOMETER). Inside each of these photoelectric devices there is an electrode, called a photocathode, which emits electrons when a beam of electromagnetic radiation strikes its surface. By a suitable choice of photocathode material, a good response may be obtained from near infrared wavelengths up to low-energy X-rays. The percentage of the incident photons which are converted into electrons is known as the quantum efficiency of the photoelectric device. The earliest instruments were PHOTO-ELECTRIC CELLS. These were evacuated and worked by the photoelectric effect. At best their quantum efficiency was only about 30 per cent, although this percentage was usually lower when taken over a wide spectral region.

An important development of the simple photoelectric cell is the photomultiplier tube. This is an evacuated electronic device which is used to convert a low-intensity light signal into an electrical current, and then amplify this signal very considerably. Most astronomical light sources are extremely faint, so the ability to amplify the very small currents produced when only a few PHOTONS strike the photocathode is of great importance. Inside the photomultiplier tube there is a photocathode from which electrons are liberated by the incident photons of light. These negatively-charged electrons (called primary electrons) are accelerated down the tube by a series of positively-charged electrodes, called dynodes, to which an increasingly high potential is applied. As each of the primary electrons strikes the first dynode, it liberates a number of additional electrons (called secondary electrons). This process is repeated at each dynode, and so the initial number of primary electrons is magnified enormously in a tube where the number of

dynodes might be eight or nine or even more. The electrons are all collected at the final electrode in the dynode-chain (the anode), which is held at a very high positive potential relative to the photocathode. A large current pulse is thus produced in the external circuit connected to the anode, which can be further amplified or filtered, and fed to some recording apparatus.

Most modern photocells now consist of a layer of semiconductor material sandwiched between two electrical contacts. When electromagnetic radiation falls upon the sandwich, its electrical conductivity increases considerably as a result of photoconductivity. This is where the photons of radiation create electron-hole pairs within the semiconductor. The electrons are negatively-charged and the holes (where there is an absence of electrons) have a net positive charge. The electron-hole pairs are called charge carriers and they cause an electrical current to flow in the external circuit connected across the semiconductor sandwich. The current is again proportional to the amount of radiation falling on the photocell. This type of device is often called a photoconductive cell, or sometimes a photoresistive cell, because the resistance of the device to the passage of an electrial current through it falls with increased intensity of the incident light. Photoconductive detectors are used in all kinds of light sensors, light-sensitive switches, light meters, and in the vidicon television camera tube. Most employ photoconductive semiconductors such as lead telluride or cadmium sulphide.

There are two types of television camera tube, the image orthicon and the vidicon tube. The former has a target, on which the light beam is focused, coated with a photoemissive material. This allows an electric charge to build up on its surface according to how much light falls upon it. The image is divided into a large number of picture elements or pixels. Each pixel is scanned by an electron beam, so that the amount of charge at that point on the target is changed into an electrical signal. This varies according to the amount of light falling on the pixel which is being scanned. The vidicon tube has a target made of photoconductive semiconductor. The vidicon, and its development the plumbicon, are smaller, simpler and cheaper than the image orthicon. Such television cameras are widely used not only in conjunction with ground-based telescopes, but also in space probes and

satellites. However, in recent years the completely solid-state CHARGE-COUPLED DEVICE camera has tended to take over from the more conventional television camera tubes.

A photodiode is different from a photoconductive cell in that it consists of two different types of semiconductor sandwiched side by side between the two electrical connections of the device. A diode normally acts like a "one-way valve"; it allows a modest electrical current to pass in one direction, but allows virtually no current to flow in the other direction, which is known as the reverse-bias direction. A very small reverse current does flow, and it is called the dark current, because it exists when there is no illumination. If light is allowed to fall on the junction between the two types of semiconductor, the reverse current increases. An increase in the light intensity will result in a proportional increase in the reverse current. The photodiode is particularly useful as a light-sensitive switch. The quadrant photodiode may be used for telescope control. Its light-sensitive surface is divided into four equal quarters. Light from a star is focused at the exact centre of the photodiode, and as long as it remains there nothing happens. However, if the telescope is not tracking perfectly, the image wanders into one of the four quadrants of the photodiode. An electrical current is produced, and by knowing from which quadrant this emanated, it is possible to correct the telescope drive in exactly the right way for the image to be brought back again to the exact centre of the device. A large matrix of photodiodes can also be used as an image sensor, the magnitude of the reverse current from each diode being proportional to the intensity of the image at that point.

Another type of photoelectric aid is the image tube or IMAGE INTENSIFIER. This is an evacuated electronic device that is used to intensify a very faint optical image. The beam of visible light or near-ultraviolet radiation falls onto the photocathode. Electrons are liberated by the photoelectric effect, and they are accelerated by an electric field using electrodes at a high positive potential relative to the photocathode. Using magnetic field coils, the electron beam can be focused onto a positively-charged phosphor screen, so that an optical image is formed, many times brighter than that from the original beam. This image can be made to fall onto the photocathode of a second image tube, and the intensification process repeated. Several such stages are used in the cascade image tube. The image on the final phosphor screen can be photographed.

In the electronographic camera, the image is formed on a very fine-grain, high-resolution "nuclear" photographic emulsion in which the high-energy electron tracks are recorded directly. The image can be developed in a similar way to an ordinary photograph. The density at any point on the electronograph is proportional to the intensity at the corresponding point in the optical image over a very wide range of intensity. The device has a very high quantum efficiency. In the Image Photon-Counting System (IPCS), pioneered by Professor Alex Boksenberg, there is a high-gain multistage image tube, and a television camera records the two-dimensional image on the final phosphor screen. The gain of the device is such that every electron liberated from the photocathode can be detected as a burst of light by the television camera. The signal from the camera is fed into the memory of a computer for picture storage, pending later image analysis. Modern photoelectric aids used by astronomers are reaching such a level of sophistication that individual photons can now be detected and counted by such devices. *JWM.*

Photoelectric cell An electronic device that converts a light signal, or a signal in the near-infrared or ultraviolet regions of the spectrum, into an electric current. Light falls upon the photocathode and electrons are emitted by the photoelectric effect. This is where radiation energy is transferred in the form of photons to the electrons within the atoms of the photocathode. If the energy is above a certain threshold level, the most weakly-bound electrons can escape from their atoms. These electrons are attracted to a positively-charged anode so that a current flows in the external circuits. The current increases linearly with the intensity of radiation falling on the photocathode over a wide range. The photo-electric cell, which was evacuated, was the earliest form of photocell, one of a number of PHOTOELECTRIC AIDS. *JWM.*

Photographic magnitude The magnitude of a star as recorded on a traditional photographic emulsion which has its peak response towards the blue end of the spectrum and records very poorly at wavelengths longer than 500nm. Denoted by m_{pg}, the photographic magnitude differs significantly from the

VISUAL MAGNITUDE recorded by the eye (m_v).

Photography, astronomical   *See* ASTROPHOTOGRAPHY.

Photometer An instrument with which astronomers can measure the intensity (ie, the brightness) and various other properties of visible light, and also of infrared and ultraviolet radiation. The earliest instruments were visual equalization photometers and extinction photometers. In equalization photometers, the intensity of a star image is steadily reduced until it is judged to be equal in brightness to that of another star (natural or artificial) in the same field of view. The magnitude difference of the two stars is then 2.5 times the logarithm of the star's reduced brightness expressed as a fraction of its original brightness. If the magnitude difference is great, a reduction in the intensity of the brighter source, by a known amount, to near equalization, may be required. This can be achieved by a reduction of aperture, use of polarized filters, or an optical wedge, etc. The eye can detect fairly slight differences in intensity between the astronomical source and the comparison source, so that reasonably accurate measurements can be made. Examples of this type of instrument are the Zöllner photometer, where the brightness of the artificial source is changed using an optical wedge; the polarization photometer, where the reduction is carried out using polarized filters; and Danjon's cat's-eye photometer, where natural stars are used for comparison purposes, the brightness reduction being effected by using a combination of two prisms and a cat's-eye diaphragm. The flicker photometer is a true equalization photometer where the equality of the artificial and real light sources is assessed by the absence of flicker between the two images, when viewed in rapid succcession from the same position.

In extinction photometers, an optical wedge or polarized filter, marked with a scale, is used steadily to reduce the intensity of the light source until it just becomes invisible. The position on the scale at which this occurs is noted, and in this way the brightness of the source can be determined. Although visual photometers have been employed to obtain some useful results in the past, and the human eye has the ability to detect a wide range of light levels, it is comparatively insensitive to very small changes in brightness. In most cases, a

difference in brightness of about one-tenth of a magnitude between the known brightness of the reference source and the unknown brightness of the source under study is the best that can be achieved with a single measurement. Greater precision may sometimes be achieved by taking the average of a series of measurements. However, for the far greater accuracies now required in astronomical research, or in cases where the intensity is constantly changing or varying very rapidly, physical photoreceptors such as the photocell, photomultiplier tube or bolometer must be used. In all of these devices the incident electromagnetic radiation is converted into an electrical signal, whose magnitude can be determined very precisely.

In optical, infrared and ultraviolet astronomy, magnitudes are determined using photometers over a selection of wavelength bands. Present systems such as the internationally-accepted UBV system use a suitable combination of a special filter and a photocell or photomultiplier tube (plus an electronic amplifier) to select light or other radiation of a desired wavelength band and measure its intensity. The UBV system of stellar magnitudes is based on photoelectric photometry. The photoelectric magnitudes denoted by U, B and V are measured at three broad bands, U (ultraviolet), B (blue) and V (visual — ie, yellow-green) centred on wavelengths of 360nm, 420nm and 540nm respectively. The light intensity can be measured for a particular wavelength band as a greatly increased (amplified) electrical signal. Magnitudes can also be measured for various infrared bands. Narrower-wavelength bands are used in the UVBY system. Photoelectric photometry can be used to study the light output from an isolated star or small region of the sky.

In a typical photoelectric photometer, the light from the telescope is fed via a pinhole and a Fabry lens to the photocathode of the photomultiplier tube (PMT). The pinhole is located in the focal plane of the objective mirror or lens of the telescope. It effectively excludes all light except that from the object under examination. The Fabry lens ensures that the image is correctly positioned on the photocathode. Provision must also be made between the pinhole and the Fabry lens for a viewing eyepiece, to ensure that the object is centred in the pinhole, and for a set of standard colour filters. These, together with suitable circuitry

P

for the PMT, are enclosed in a light-proof box, termed the photometer head, which is placed at the eyepiece position of the telescope. The current from the PMT is fed to the measuring equipment via a suitable amplifier. In older photometer systems, a simple meter or a chart-recorder was used to measure the amplified output from the PMT. Nowadays, most photometer systems employ a microcomputer which, with peripherals, is used to receive, store, process and print out the data.

A wide variety of telescopes is used for carrying out photoelectric photometry. It is vital that the telescope and its mounting are sturdy enough to carry the photometer head, which can weigh a couple of kilogrammes. It is also essential that the object under study is kept in the centre of the pinhole during each observation, which will typically last for 30 seconds or more, so the telescope mounting must be accurately aligned and driven precisely at sidereal rate. Most astronomers use the technique of differential photometry, where the brightness of the object under study is compared with standard stars of constant, known brightness. Regular procedures are carried out whereby the object being studied, the comparison star(s), and the background sky alone are observed in a sequence. The contribution from the background sky is then deducted from the other readings, so that the true difference in brightness between the object and the comparison star(s) may be calculated. The time taken for each observation is typically a few tens of seconds, and usually a whole series of observations is made, and the average then taken. Of course, this may not be possible in the case of very rapidly varying sources. Where appropriate, the standard colour filters are used to obtain observations that can be reduced to one of the standard photometric systems (eg, UBV or UVBY), thus enabling data from one observatory to be accurately related to that from another. *JWM*.

Photomultiplier An extremely sensitive electronic device used to detect photon signals in many ground-based and space-borne instruments, including PHOTOMETERS, scanning spectrophotometers and polarimeters.

The components of the device are contained within a sealed, evacuated envelope, normally of glass, incorporating an input window and a multi-pin connector. Photons pass through the window to a thin photocathode layer

deposited on its inside surface, and electrons are emitted into the vacuum space by the photoelectric effect. The fraction of the photons which yield electrons (the quantum efficiency) depends on the materials chosen for the window (defining the short-wavelength cut-off) and the photocathode (giving the long-wavelength extension) and varies markedly with wavelength. Responses can span from the ultraviolet to the near infrared regions and generally peak between 10 and 30 per cent. The photoelectrons are electrostatically accelerated and focused onto the first electrode (dynode) of an electron multiplying structure. On average, each photoelectron liberates many secondary electrons which are, in turn, accelerated onto the next dynode yielding another, yet larger generation of secondaries, and so on. Finally, the electrons are collected at an anode from which the signal can be measured. The ratio of secondary to primary electrons at each impact depends on the type of surface chosen and the energy of the primaries. By using a variable high-voltage supply and a potential divider network to provide the inter-dynode voltages, the size of the output signal can be varied over a wide range. The more stages, the higher the gain at a specific overall voltage and the higher the maximum gain attainable; a gain of 10^8 can be readily achieved in a 14-stage device. High electron gain ensures that the signal from each photon detection is much larger than the noise in the recording electronics. *AB*.

Photon A discrete packet or "quantum" of electromagnetic energy. Radiation of wavelength λ and frequency f can be regarded as a stream of photons each of energy E given by $E = hf$ or hc/λ where c denotes the speed of light and h is the Planck constant. The higher the frequency (or the shorter the wavelength) of the radiation, the greater the energy of its photons. A photon can also be visualized as an elementary particle with zero rest-mass, zero charge, and spin 1, which travels at the speed of light. The photon theory explains several aspects of the behaviour of ELECTROMAGNETIC RADIATION including the spectrum of black body radiation (*see* PLANCK), the photoelectric effect (*see* PHOTOELECTRIC AIDS) and the COMPTON EFFECT. *IKN*.

Photosphere The luminous surface of the Sun that is visible as a sharp disk. Generally, the boundary between the interior and the exterior of a star.

Photovisual magnitude The magnitude of an object measured photographically using a combination of filters and photographic emulsions which mimic the spectral response of the eye.

Piazzi, Giuseppe (1746–1826). Italian astronomer-monk. Professor of mathematics, Palermo, where he directed the observatory. He discovered CERES, the first asteroid (1801), compiled star catalogues, and was Director of the Government Observatory, Naples.

Giuseppe Piazzi: discoverer of Ceres

Pickering, Edward Charles (1846–1919). American astronomer; from 1876 Director of Harvard Observatory. He was concerned mainly with stellar classification and spectra, and masterminded the famous *Harvard Photometry* and the *Draper Catalogue*.

Pickering, William Henry (1858–1938). Younger brother of E.C.PICKERING; assistant at Harvard. In 1891 he organized the Arequipa outstation. He concentrated mainly upon lunar and planetary observation.

Pico Prominent isolated lunar mountain in MARE IMBRIUM, south of Plato.

Pictor (the Painter). An unremarkable constellation in the southern sky, near Canopus. Its leading star is Alpha (magnitude 3.27, absolute magnitude 2.1, spectrum A5, distance 52 light-years). The only other star above the fourth magnitude is Beta (3.85). The bright nova of 1925, RR Pictoris, flared up here.

Pillar-and-claw mount An old-fashioned form of telescope mounting, sometimes used for small refractors. It consists of a central pillar, on which the telescope is pivoted, with three claw-shaped feet. It looks neat, but is quite useless for astronomical work. Almost all pillar-and-claw-mounted telescopes can be fitted to a better mounting.

Pioneer United States spacecraft used for a variety of purposes. The first of the series was launched toward the Moon in 1958; the last is still (1986) sending back data from the region of Venus.

The Pioneer series has been one of the most successful in the history of space research, even though it had its early failures. There were several notable "firsts". Pioneer 1 was the first American probe to go anywhere near the Moon; Pioneer 6 was the first to make direct investigation of a comet;

Pioneers 6 and 7 set a "longevity record" for probes in solar orbit. Pioneer 10 was the first probe to obtain close-range data from Jupiter, and Pioneer 11 was the first probe to Saturn. Finally, Pioneer Venus Orbiter (sometimes called Pioneer 12) was the first to obtain detailed radar maps of virtually the whole of the surface of Venus.

Pioneer 7 also set another record. Its transmitters had been deliberately turned off when it was badly placed to send signals, but in September 1972 it was located and re-activated — something which is easy today, but was far from easy in the early 1970s.

Pioneers 10 and 11 are, to date, the most remote vehicles still in contact; they will never return, and carry plaques just in case they are ever picked up by

some alien civilization. It is hoped that they will maintain contact until they reach the boundary of the HELLIO-SPHERE. They were also the first probes to pass through the asteroid belt, and it was with a considerable sense of relief that space scientists followed their progress; previously it had been believed that material in the asteroid belt might prove a hazard to space probes.

New versions of the spin-stabilized Pioneers are being proposed, with the aim of carrying out further studies of the Moon, Mars and Venus, though whether the name of "Pioneer" will be retained for them is not certain. *PM.*

Pisces (the Fishes). One of the Zodiacal constellations; but it is very obscure, consisting of a chain of rather faint stars south of Pegasus. Its brightest star, Eta, is only of magnitude 3.62.

Piscis Austrinus The Southern Fish, also known as Piscis Australis. It lies south of Pegasus and Aquarius, and has only one star above magnitude 4; this is Fomalhaut (magnitude 1.16; absolute magnitude 2.0, spectrum A3, distance 22 light-years).

PKS2000–330 A very distant quasar. The redshift equals 3.78.

Planck constant The fundamental constant, denoted by h, which relates the quantum of energy of a PHOTON to its frequency. Its value is 6.626×10^{-34} Joule second (Js).

Planck, Max Karl Ernst Ludwig (1858–1947). German physicist who, in 1900,

P

Pioneer missions

Vehicle	Launch date	Programme	Remarks
1	Oct 11 1958	Lunar probe	Failed to reach the Moon, but achieved 70,760 miles (113,854km). It was also used to study solar activity, and to investigate the radiation zones round the Earth.
2	Nov 8 1958	Lunar probe	Failure; 3rd stage failed to ignite.
3	Dec 6 1958	Lunar probe	Failed to reach the Moon, but achieved 63,600 miles (102,333km) and confirmed the findings of Pioneer 1.
4	Mar 3 1959	Lunar probe	Passed within 37,280 miles (59,983km) of the Moon, and sent back useful data. It then entered solar orbit (0.987 × 1.142AU).
5	Mar 11 1960	Solar data	Put into solar orbit (0.806 × 0.995AU) to send back data about solar flare activity and the solar wind. It remained operative until June 26 1960 when its distance from Earth was 23,000,000 miles (37,000,000km).
6	Dec 16 1985	Solar data	Put into solar orbit (0.814 × 0.985AU). It gave a good description of the solar atmosphere, and with Pioneers 7, 8 and 9 made up a network of solar probe "stations". It also measured the tail of Kohoutek's Comet of 1973. It was scheduled to operate for 6 months, but was still active two decades later.
7	Aug 17 1966	Solar and magnetic data	Put into solar orbit (1.010 × 1.125AU). It measured the Earth's magnetic tail out to over 11,000,000 miles (19,000,000km). In 1986 it was still capable of sending back data.
8	Dec 13 1967	Solar wind, magnetic data, cosmic rays	Put into solar orbit (1.0 × 1.1AU) to join Pioneers 6 and 7 on obtaining data about solar wind, the magnetic fields in space, cosmic rays, etc.
9	Nov 8 1968	Solar data	Put into solar orbit (0.75 × 1.0AU).
E	Aug 27 1969	Solar probe	Failure; did not enter orbit (it was intended to be Pioneeer 10).
10	Mar 3 1972	Jupiter probe	Passed by Jupiter on December 3 1973 at 82,000 miles (132,000km) and sent back extensive data. Now on its way out of the Solar System, but still in contact.
11	Apr 5 1973	Jupiter probe	Passed by Jupiter on December 2 1974 at 26,600 miles (42,800km), confirming and amplifying the results from Pioneer 10. It was then diverted on to a rendezvous with Saturn on September 1 1979 at 13,000 miles (20,800km), obtaining the first close-range results from Saturn — over a year before the rendezvous of Voyager 1.
Pioneer Venus 1	May 20 1978	Venus orbiter	In orbit round Venus, sending back data and also obtaining radar maps of the surface.
Pioneer Venus 2	Aug 8 1978	Venus probe	Multi-probe, consisting of a "bus" and five small landers. The "bus" burned up in Venus' atmosphere on December 9 1978; all five landers were successful. They were not designed to transmit after impact, though in fact one (the Sounder Probe) did so for 68 minutes.

Max Planck: originated the quantum theory

explained the distribution of intensity of black body radiation by asserting that radiation is emitted in discrete quantities called quanta (*see* PHOTON). He thus originated the quantum theory (*see* WAVE MECHANICS).

Planck time 10^{-43} seconds, the timescale within which the known laws of physics break down. A quantum theory of gravity is necessary to discuss the state of the BIG BANG universe at times earlier than 10^{-43}s after the initial event. Named after Max PLANCK.

Planetarium The word "planetarium" means, literally, of or pertaining to the planets. Thus, a planetarium is a model or representation of the Solar System. Over time there has been a variety of such models, ranging from simple hand-driven mechanisms to elaborate computer-controlled projection devices. Through usage, the word has come to refer to an optical device for projecting celestial images and effects onto a domed screen representing the sky, and even to a room or building for housing such a projector. It is interesting to note that most people associate a modern planetarium more with a representation of the starry sky than with planets, but a planetarium would not be properly named without its ability to illustrate the apparent behaviour of the planets, the wandering bodies of the sky.

We can assume that the motivation for a planetarium goes back to prehistoric times when people looked into the sky with desire to comprehend the objects they saw there. Right away, they created explanations for the existence of these objects and for their apparent movements. At first these interpretations were merely verbal expressions of belief, followed by written descriptions. The *Phenomena*, by Aratus, for example, is a poem that lists constellations in the starry vault as they were understood by the Greeks possibly about 2000BC. Eventually, people rendered models portraying what they imagined.

The first representational globes we know of were models of the sky, not Earth. An old example is the Farnese Atlas, dating to about 300BC. Now at the National Museum in Naples, this marble sphere of constellation figures is supported on the shoulder of the kneeling Atlas. About AD150 PTOLEMY made a celestial sphere which could be manipulated to simulate change of latitude and even precession as well as diurnal (daily) motion. These same changes in the sky are fundamental to modern planetaria.

The first real planetarium we know of from literature was made by Archimedes, possibly around 250BC. It was described by Cicero as showing the movements of the Sun, Moon and planets. Apparently it had a spherical frame to represent the starry sky and inside of this was an Earth-centred model which was capable of accurately portraying the apparent relationships of the Moon, Sun and planets, including phases of the Moon and solar and lunar eclipses.

Moving forward in time, we note that things such as ARMILLARY SPHERES, celestial globes, ORRERIES and ASTRONOMICAL CLOCKS are very much part of the chain of developments leading to modern planetaria. Indeed, some very large celestial spheres began to resemble modern planetaria. About 1650 Adam Olearius and Andreas Busch constructed the "Gottorp Globe", a 10ft (3.1m) hollow copper sphere into which several people could enter, sit upon a platform surrounding a small Earth globe, and watch constellations with gilted stars parade overhead as the giant sphere was rotated by a water-wheel. Glass balls could be moved in the Zodiac to simulate the planets. Variations of these large globes included one built in Cambridge by Roger Long, Pembroke College's Lowndean professor of astronomy and geometry. This 18ft (5.4m) globe, turned by a winch, contained holes of appropriate relative diameter to represent the stars. External lighting combined with a dark interior presented an impressive starry firmament to viewers inside.

Celestial spheres and orreries became very elaborate and instructionally more significant and these led eventually to the projection planetarium. In the early part of the current century, Oskar von Miller, founder of the Deutsches Museum in Munich, initiated a series of developments which finally involved the Carl Zeiss optical works in Jena. Moving from the idea of a large rotating globe with perforations for stars and with luminous mobile planetary disks, the Zeiss people settled upon the idea of an array of projectors at the centre of a stationary dome. This concept came from Walter Bauersfeld in 1919:

"The great sphere shall be fixed; its inner white surface shall serve as the projection surface for many small projectors which shall be placed in the centre of the sphere. The reciprocal positions and motions of the little projectors shall be interconnected by suitable driving gears in such a manner that the little images of the heavenly bodies, thrown upon the fixed hemisphere, shall represent the stars visible to the naked eye, in position and in motion, just as we are accustomed to see them in the natural clear sky."

With this concept, the modern projection planetarium was born.

"The wonder of Jena," as it was soon known, projected the first simulated starlight in August 1923 on a plaster surface inside a dome constructed on the roof of the Zeiss building. Then it was installed and publically opened on October 21 1923 in the Deutsches Museum, where it was used until 1960. This historic projector presented a reasonable simulation of the appearance of the stars, Sun, Moon and planets as seen from the latitude of Munich.

Soon, the Zeiss planetarium was improved in many ways, including the capability to simulate the sky from anywhere on Earth. Such instruments appeared in major population centres in Europe and Japan. Two Zeiss companies, Zeiss-Jena in East Germany and Zeiss-Oberkochen in West Germany, resulted from World War II.

A number of additional planetarium projectors joined those made by Zeiss. The leader for smaller, less expensive, instruments was Armand N. Spitz, founder of the Spitz planetarium. His educationally motivated efforts resulted in hundreds of planetaria located in schools and small museums as well as in larger institutions. Several planetarium instruments, small and large, with many innovations, have been developed by the companies resulting from the efforts of Armand N. Spitz and those who followed him. Other manufacturers, most notably Goto and Minolta, both in Japan, have produced planetaria, each having their own special optical and mechanical qualities.

A quantum leap in cosmic projectors was introduced in 1983 by Evans and Sutherland Computer Corporation of Salt Lake City, Utah. This computer graphic image device, known as Digistar, projects simulations consisting of points and lines programmed in computers. Thus it is possible to move at will within three-dimensional data fields, including simulated planetary and star fields. Simulated flight to stars within a few hundred light-years of the Sun, and proper motions forward and backward as much as one million years in time, can be shown with illustrative accuracy. An unlimited host of additional projections, such as three-dimensional models of atoms, molecules or galaxies, can be composed and shown with the same instrument.

Modern planetaria employ large arrays of additional audiovisual devices to work in concert with the central planetarium instruments. High-quality slide, motion picture, video and special-effect projection, blended with superb sound, leads to elaborately-illustrated programmes on astronomy, space science and many other topics. In recent years sound and light shows, typically employing lasers, have become popular and profitable alternative presentations under planetarium projection domes. Plays, concerts and other innovative programmes make use of the special features of planetarium theatres.

Typical well-established planetarium theatres currently consist of domed projection screens ranging from about 30 to 100ft (10 to 30m) in diameter, at the centres of which there are planetarium projectors, surrounded by dozens to hundreds of additional devices. Some projector banks are used to create panoramas, or even full-dome integrated images. All of these and more are likely to be automated for use in concert for presentation of precisely-structured, carefully-written and produced programmes involving hundreds to thousands of colourful images in addition to stars and planets.

Very large-image motion picture capability is a recent addition to some planetarium theatres. In such places, the motion picture projector has become at least equal in importance to the planetarium projector. Omnimax projectors employ 70mm film, transported horizontally, for ultra-high-quality, large-format images on either tilted or horizontal domes. In addition, 35mm projectors, most notably Cine 360, are used in planetaria. The tilted-dome facilities are often referred to as "space theatres", because they are designed to make viewers feel that they are in space with images filling almost all of their forward field of vision, including that somewhat below as well as above them.

All of this development, involving progress in technology as well as in the astronomically-related sciences, has resulted in planetarium professions. The International Planetarium Society, a worldwide affiliation of national and regional planetarium professional groups, was formed in 1970. Infor-

mation about this organization can be obtained by writing to: The International Planetarium Society, Hansen Planetarium, 15 South State Street, Salt Lake City, Utah 84111, USA. *VDC.*

Planetary conjunction A close alignment in the sky between a planet and another celestial body. *See also* INFERIOR CONJUNCTION, SUPERIOR CONJUNCTION.

Planetary nebula A nebula associated with and deriving from a star, in principle having a disk-like appearance, similar to the telescopic view of a planet. About 1,000 planetary nebulæ are catalogued, perhaps one-tenth of the number in our Galaxy. Their radii range from that of our own Solar System to about 0.5 to 1 light-year.

The name was coined in 1785 by William HERSCHEL, echoing remarks by Antoine Darquier about the RING NEBULA. Herschel thought at first that planetary nebulæ were unresolved groups of stars like distant globular clusters, but in 1790 he observed NGC1514, a planetary nebula which looked like a "star of about the eighth magnitude, with a faint luminous atmosphere, of a circular form . . . The star is perfectly at the centre . . ." The particularly simple form of the nebula, and the strikingly central position of the very bright star, so much brighter than others nearby, convinced Herschel that true nebulosity existed, which, no matter how powerful his telescopes, could never be resolved into stars.

Astronomers are attracted to the circularly shaped planetary nebulæ, for the very good reason that they have a simple symmetry for which it looks relatively easy to construct a mathematical model — how can one explain the complicated if one cannot explain the simple? However, only 10 per cent of planetary nebulæ have a circular shape. About 70 per cent have a bipolar structure and have two lobes. The shapes suggest a wide variety of names.

Planetary nebulæ

Name	Catalogue nos.	Constellation	Diameter
Butterfly	NGC6302	Scorpius	20″
Dumbbell	NGC6853, M27	Vulpecula	8′
Eskimo	NGC2932	Gemini	40″
Helix or Sunflower	NGC7293	Aquarius	30′
Owl	NGC3587, M97	Ursa Major	150″
Ring	NGC6720, M57	Lyra	80″
Saturn	NGC7009	Aquarius	25″

Even the simpler shapes are open to various interpretations. Is the Ring

Nebula a hollow ellipsoidal shell completely surrounding the star at its centre, or is it a torus (circular doughnut) seen at a slightly oblique angle? The strong contrast between the brightness of the ring and the hollow space within suggests that the hollow space is truly empty, and that there are no front and rear faces of a shell to contribute to the light from this line of sight. Minkowski and Osterbrock concluded therefore that the Ring Nebula was toroidal. If viewed exactly in the plane of the toroid the Ring Nebula would have a twin-lobed shape typical of many planetary nebula. Suppose the height of the toroid were exaggerated so that it became a squat cylinder centred on its star? If such a cylinder were viewed obliquely, then the open ends could give the hollow eye-like appearance of the Owl Nebula.

A planetary nebula shines because of ultraviolet light emitted by the central star. The atoms of the nebular gas can be ionized by ultraviolet photons of sufficient energy; hydrogen atoms, for instance, can be ionized by photons with wavelength 912Å or shorter. When the ion recombines with its electrons it emits photons of light from a structured ladder of energy steps. Each ultraviolet photon input into the nebula produces a spectrum of photons out: in this sense planetary nebulæ are photon counters. Their symmetry and the fact that the central stars are identifiable means that planetary nebulæ have been laboratories for atomic astrophysics.

Each central star has enough light output to ionize a certain volume of planetary nebula. There may be gas beyond the visible boundary of the nebula, but ultraviolet photons do not reach far enough to make it visible. In fact within each planetary nebula numerous sub-nebulæ exist, each corresponding to the ionization of a given type of atom or ion. Thus pictures taken in different optical spectral lines from different ions show stratification — the images of the planetary nebula are nested like a set of Russian matroshka dolls made of coloured glass. The Ring Nebula M57 is a good example of this: the blue, green and red images are successively larger because each is dominated by a particular spectral emission from a different atom or ion.

Because it is relatively simply structured, the celestial "laboratory" of M57 can be analyzed to determine what it contains. The nebula itself is in total about 0.2 times the mass of the

P

Sun. Its density is about 10,000 ions of hydrogen per cubic centimetre (compare with the density of air at 30 million million million atoms per cubic centimetre). Each cubic centimetre contains the common elements in the following proportion: hydrogen, 10,000 ions/cc; helium, 800; oxygen, 6; nitrogen, 3; neon, 1.

The input of energy from the central star to the nebula warms the gas: the electrons which are split from the constituent ions have temperatures of about 12,000K. Radio telescopes detect the thermal emission from planetary nebulæ, and in fact give a more complete view of the structure of some nebulæ that are obscured by dust.

All the planetary nebulæ are expanding. This can be determined from two techniques. The Doppler Shift of the optical spectral lines provides a measurement of the expansion velocity along the line of sight: values of 20km/s are typical. The increase in size of the image of a planetary nebula on photographs taken over the years shows its expansion across the line of sight. One of the closest planetary nebulæ, the Dumbbell, has been measured by Pulkovo astronomers to grow in angular size at 0.068 arc sec per year — the annual growth rate of the trunk of a tree at 100km whose rings grow at 1mm per year!

Optical techniques for measuring the growth of planetary nebulæ are likely quickly to be superseded by measurements of the radio emission from planetary nebulæ with the VERY LARGE ARRAY of radio telescopes in New Mexico. The amazing angular resolution of this telescope and its ability to measure angle absolutely makes it possible to measure the expansion of planetary nebulæ after only a few years. The expansion in the radial and tangential directions can be combined to yield for the first time reliable distance for planetary nebulæ.

The central stars of planetary nebulæ are very hot and blue (typical temperatures 30,000K to 400,000K), and form a range of stars somewhat fainter than the usual MAIN SEQUENCE of stars (absolute magnitude 0 to +10). They form a missing link between the GIANT STARS and the WHITE DWARFS, a transitional stage at the end of stars' evolution. This is also indicated by the fact that planetary nebulæ form a population in our Galaxy which lies neither in the galactic plane of young stars nor in the halo of very old stars: they form a so-called disk population which is of intermediate age. The

precise evolutionary status of only one planetary nebula is known: it lies within the globular cluster M15 in Pegasus.

It is clear that the planetary nebula stage in a star's life marks a change from advanced middle age to senility and is associated with a form of loss of material from the star. But it is unclear how the mass loss takes place. Some planetary nebulæ such as the Eskimo or Saturn nebulæ, which have a bright central nebula surrounded by fainter extensions, give the impression that a series of explosions has occured in the central star, ejecting successive shells

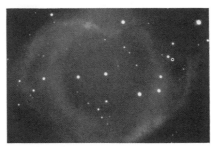

Planetary nebulæ: evolved stars

into space. This conception of planetary nebulæ is based an analogies with the ejection of shells in novæ, such as the one which was produced by Nova Persei 1901. It is an idea which is directly traceable back to William Herschel's papers of the 1780s which first recognized planetary nebulæ. However, the ejection velocities of novæ are thousands of times faster and, in novæ, the amount of mass ejected is hundreds of thousands of times smaller.

The modern view relates the planetary nebulæ to the stellar wind material blown out from a red giant star, identifying the central star as the red giant's core, as it evolves to become a white dwarf. The gravity at the surface of a red giant is very low, because the star is so big, and storms on its surface akin to solar prominences and other solar phenomena can throw off material easily. A star up to about 8 solar masses can lose three-quarters of its mass into space before it turns into a white dwarf of mass about 1.4 times solar mass. Such a process could readily provide enough material to make a planetary nebula. The visible nebula is simply the central region of a much larger object, the only part which the limited ultraviolet light from the central star is capable of illuminating.

The idea that the outflow is connected to stellar winds and magnetic storms leads to a connection between

the curious shapes of planetary nebulæ and the magnetic field and rotation properties of the star. The magnetic lines of force might control the outflow, channelling it into particular directions, and the rotation of the star could add spiral twists. The bipolar structure of most planetary nebulæ must be in general terms connected with the existence of a preferred direction, such as a spin or magnetic axis.

The fact that the nebulæ are expanding as the central stars fade means that planetary nebulæ are relatively transitory objects, with lifetimes measured in tens of thousands of years. Given the estimated number of planetary nebulae in our Galaxy (10,000), this means that several per year are forming. One possible event which astronomers believe may represent the birth of a planetary nebula has occurred on the star FG Sagittæ. Its brightness rose steadily from 13.6 in 1894 to 8.9 in 1970. In 1960 it was noticed that the star had acquired a nebula of radius 18 arc sec; there were indications that the old photographs of the star showed a fuzzy image as if the nebula had formed in the 19th century. The atmosphere of the star is expanding and it appears to be ejecting a second shell. This shell contains spectral lines of unusual chemical elements such as yttrium, zirconium, cerium, neodymium and samarium. The significance of these elements is that they are formed by one nuclear process called the S-process: the elements have risen to the star's surface and become visible in its ejected shell. *PGM.*

Planetary satellites The satellites of the planets of the Solar System are varied in size, shape, colour, mass, and density, indicating a diversity of origin. Since planetary rings are composed of large and small particles and chunks in orbit about the planet, the number of satellites in the Solar System is very large; in fact, uncountable. Thus, only the larger satellites have their orbits determined and are named.

The two planets closest to the Sun, Mercury and Venus, do not have any known satellites. The Earth's Moon is the best explored of the satellites and is large (1,080 miles/1,738km radius) compared with its primary. Mars has two small satellites, Phobos and Deimos. Jupiter has 16 named satellites including the four large Galilean satellites: Io, Europa, Ganymede, and Callisto. Galileo monitored the movement of these bodies in orbit around Jupiter and became convinced that the Copernican theory of planetary orbits was correct. Io and Europa are about the size of the Moon and Ganymede and Callisto are about the size of Mercury. Ganymede (1,635 miles/2,631km radius) is the largest satellite in the Solar System. The Voyager spacecraft discovered active volcanoes on Io.

Saturn has at least 17 satellites. Its largest satellite, Titan (1,600 miles/ 2,575km radius), is surrounded by a dense cloudy atmosphere. Other large satellites are Mimas, Enceladus, Tethys, Dione, Rhea, Hyperion, and Iapetus. Uranus has five large satellites: Miranda, Ariel, Umbriel, Titania, and Oberon, and ten small satellites discovered by the Voyager 2 spacecraft. Neptune has one large satellite, Triton, and one small satellite, Nereid. Pluto has a large satellite, Charon. *MED, See also* colour essay on MOONS AND MINOR PLANETS.

Planetary volcanism Volcanoes are recognized on the Earth, Mars, Moon and IO: some have been suspected on Mercury, Venus and Ariel. Active volcanoes have been observed unequivocally only on the Earth and Io.

In the volcanic process, materials heated at depth reach the surface of the planet where they are erupted. The heating process can result from natural radioactivity; from the conversion of gravitational energy into heat, as with the tidal heating of Io; or as in the case of isostatic adjustments of extensive loads on a planet's surface; from chemical reactions; or from combinations of these processes. Because planetary crusts consist of a variety of minerals, heating these materials does not necessarily lead to a unique melting temperature. Some minerals may melt while others, still solid, may be carried along in a fluidized medium. The volatile component, viscosity and yield strength of this mixture will vary between planets and between different rock types on a given planet: thus, extreme types of eruption are "quiescent" and "violent". Large volumes of basaltic lava are erupted relatively quiescently from calderas (eg, Hawaii) and fissures (eg, Iceland) on the Earth, and are also found on the Moon and Mars. Cones of silica-rich pyroclastic material are produced by repeated explosive eruptions in which the gas pressure shatters the containing rock into fountains of fragments.

The Earth sports a wide variety of volcanic landforms. The volcanic materials reach the surface through conduits and fissures associated with the active margins of tectonic plates and mid-oceanic ridges. Lunar lavas were erupted from fractures. The shield volcano OLYMPUS MONS, on Mars, may be the largest volcanic edifice known. Tidal stressing, by Jupiter, will have fractured and heated Io: and its eruption products — of both pyroclastic and flow form — are unique because of their high content of sulphur and its compounds. *GF.*

Planet X The designation that Percival LOWELL gave to the predicted hypothetical planet he believed existed beyond the orbit of Neptune. Lowell became interested in the possibility of such a planet in the first decade of the 20th century. He founded the Lowell Observatory in 1894, principally for an observational study of the planet Mars.

Lowell's mathematical analysis of the small "perturbation residuals" of the planet Uranus (not attributable to Neptune) seemed to indicate the existence of an unseen planet beyond Neptune's orbit.

From 1905 to 1907, using a 5in (13cm) aperture camera at his observatory, Lowell initiated a search along the "Invariable Plane" (the mean orbital plane of the planets, inclined 1°.6 to the ecliptic) without success.

In the following years, Lowell attempted roughly to pinpoint a particular region of the sky where the unseen planet might be found in order to lessen the task of observational searching. The favoured region was in

Planet X: beyond Neptune's orbit?

the constellation of Libra. Search plates were taken at the Newtonian focus of his 40in (101cm) reflector in 1911. What he gained in light power was offset by the small field, only 1° across each plate.

The Lowell Observatory obtained the loan of a more practical search camera, a 9in (23cm), from Swarthmore College for a renewed photographic search from 1914 to 1916. Also, Lowell had purchased a "Blink-Microscope-Comparator" (*see* BLINK COMPARATOR) from the Karl Zeiss Works in Germany. This device permits a much more thorough examination of pairs of plates.

It is ironic that images of Pluto were recorded on a pair of these plates taken in 1915, but were missed. Lowell expected his Planet X to be over two magnitudes brighter with a mass seven times that of Earth.

In his continued study of the Uranus residuals, Lowell drastically shifted his favoured position of the unseen planet to eastern Taurus.

Lowell died suddenly of a stroke on November 16 1916, at the age of 61, a discouraged and exhausted man. The search for Planet X ceased until 1929.

The more rigorous and extensive planet search by Clyde Tombaugh and the discovery of the small planet, Pluto, beyond the orbit of Neptune, in 1930, triggered a spirited controversy as to the reality of Lowell's Planet X. *CWT. See also* PLUTO.

Plaskett's Star HD47129, an exceptionally massive O-type supergiant in Monoceros. In 1922, J.S.Plaskett showed it to be a spectroscopic binary of period 14 days. Each component has an estimated mass of 55 solar masses.

Plasma An almost completely ionized gas, containing equal numbers of free electrons and positive ions, moving independently of each other. Plasmas such as those forming the atmospheres of stars, or regions in gas discharge tubes, are highly electrically conducting, but electrically neutral, since they contain equal numbers of positive and negative charges. The temperature of a plasma is usually very high. The electrical currents flowing within a plasma mean that it carries with it an intrinsic magnetic field. A plasma has been described as "the fourth state of matter"; the others are solid, liquid and gas. *JWM.*

Plate tectonics The theories of continental drift and sea floor spreading are complementary: rising lava at the mid-ocean ridges creates the new ocean floor, filling the gaps left by the continents, which are parting at rates of a few centimetres per year. Older, cooler and thicker ocean floor disappears at the ocean trenches. Plate tectonics is a unifying model describing these processes, in which the lithosphere is divided into seven main plates, in

P

which ocean floor and continental blocks move together: the creation and disappearance of the plate occurring at its edges, delineated by the shallow earthquakes at the ridges and the deep-focus earthquakes at the trenches. Collision of the plates, such as the African and European plates, and the Indian-Australian and Asiatic plates, accounts for Alpine and Himalayan orogenies respectively. Plate motions have been attributed to gravity sliding, away from the ridges or sucked downward at the trenches, or to mantle convection. *SKR.*

Plato (427–348/347BC). Greek philosopher, who with Socrates and ARIS-TOTLE formulated the basis of Western philosophy. He emphasized the importance of mathematics and claimed that the philosopher (and scientist) must study the essential reality behind things. Though experiment and observation were to him unimportant, astronomy was nevertheless considered a fundamental part of education. *CAR.*

Plato: regarded astronomy as fundamental

Plato Lunar walled plain, 60 miles (97km) in diameter, on the edge of MARE IMBRIUM; 51°N, 9°W. Its very dark, relatively smooth floor makes it easy to identify when it is in sunlight.

Platonic year The time required for the CELESTIAL POLE to describe a circle around the pole of the ECLIPTIC, as a result of PRECESSION: about 25,800 years.

Pleiades A young galactic or open cluster in the constellation of Taurus, the Bull. It is the most famous of all the open clusters and is popularly known as the "Seven Sisters". This is because seven

of the stars within the cluster are easily seen with the unaided eye on a clear night. These seven are named after Atlas and his daughters. The brightest is called Alcyone (Eta Tauri). The others are Maia, Atlas, Electra, Merope, Taygete and Pleione; Pleione is

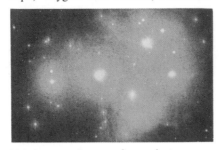

Pleiades: most famous of open clusters

somewhat variable, and may once have been brighter. It is thought to be unstable due to its fast rate of rotation, and is throwing off shells of gas.

Keen-sighted people will see more than seven Pleiades; the record is said to be 19. Binoculars will show further members of the cluster, and with a telescope it is best to use a low power in order to see the whole cluster in the field of view. It probably contains as many as 3,000 stars altogether.

The brightest stars in the Pleiades are blue-white (spectral type B or Be) and are highly luminous. The less brilliant ones are mainly spectral type A and F stars. Gas and dust surround the brighter stars and reflect the starlight, thus producing faint reflection nebulæ around the stars. The Pleiades are thought to be about 126 parsecs (410 light-years) away from us. *JWM.*

Plinius Prominent lunar crater, 30 miles (48km) in diameter, between Mare Serenitatis and Mare Tranquillitatis; 15°N, 24°E.

Plough Nickname for the seven main stars of Ursa Major (Alkaid, Mizar, Alioth, Megrez, Phad, Merak and Dubhe). An alternative nickname is the Big Dipper.

Plough: the stars of Ursa Major

Pluto The ninth planet from the Sun, with a mean distance of 39.44AU (9.38AU beyond that of Neptune). By definition, Earth's mean distance is 1.0AU. It is of interest that Pluto and Mercury are the smallest major planets (Pluto's diameter is only 1,000 miles/1,600km less than Mercury's). Also, that these nearest and farthest planets have the highest orbital inclinations and the most eccentric of all the nine planets. The best estimated diameter of Pluto is four times greater than that of the largest asteroid, Ceres.

The period of revolution of Pluto around the sun is 247.7 tropical Earth years. Pluto makes two revolutions around the Sun to Neptune's three, exactly commensurate.

At the present time, Pluto is slightly inside the orbit of Neptune, having crossed it in 1979. In 1989, Pluto will be at its perihelion and then will recede from the Sun until 2113, when it reaches its aphelion at 49.3AU. At aphelion, Pluto will receive 2,430 times less light from the Sun than the Earth receives. Even then, Pluto's landscape will be illuminated 164 times brighter than our full-moon light! If Pluto had been at aphelion during Clyde W. Tombaugh's Trans-Neptunian planet search, it would have been detected at photographic magnitude $16\frac{1}{2}$ (one full magnitude brighter than the plate limit).

Pluto was discovered on February 18 1930 at 23 hours Greenwich Mean Time by Tombaugh while making a "blink" examination (*see* BLINK COMPARATOR) of a pair of wide-angle search plates (12 by 14°) he had taken with the new 13in (33cm) telescope. With one-hour exposures, the instrument records stars to magnitude $17\frac{1}{2}$. The region was centred around a third magnitude star, Delta Geminorum. Pluto was singled out from 160,000 other star images by its shift in position of 3.5mm against the star background.

When the plates of a given region are taken at opposition (180° from the Sun), the Earth's orbital motion in a few days produces a parallax effect in which the amount of shift is inversely proportional to the distance of the exterior planet. It was just what Tombaugh was looking for, and he instantly realized that the object was far beyond the orbit of Neptune. The plates of the blinked pair were taken on January 23 and 29 1930. The planetary nature of the object was confirmed by its position on a third plate taken on January 21.

The new object was carefully observed on every possible night over the next several weeks. The discovery was announced to the world on March 13 (the 149th anniversary of the discovery of Uranus by William HERSCHEL in 1781 and Percival LOWELL's 75th birthday anniversary). The interest in the new planet was intense.

The first computed orbits were wild because the extremely short arc of only a few weeks' motion was not definitive. Astronomers dug back in their old plate files, and the images of Pluto were found on plates taken as far back as 1908. This much longer interval provided reliable new orbits to be computed.

The actual orbit had a remarkable similarity to the one calculated by Lowell (published in 1916, just before his death). Although the planet was about 2½ magnitudes or ten times fainter than Lowell predicted, the remarkable agreement of the orbital elements seemed to indicate that the new planet was Lowell's PLANET X — a fulfilment of his prediction. Pluto's position in the sky also coincided with William H. Pickering's Planet O.

A spirited controversy ensued, challenging the validity of Lowell's prediction. His solution indicated a mass nearly 7 times greater than the Earth. Pluto's photographic magnitude of 15.5 (14.7 visual) hinted that the planet was too small to have produced the perturbation residuals on Uranus. Was Pluto a less important, unexpected interloper and the real Planet X yet to be found?

in June 1930, V.M. SLIPHER instructed Tombaugh to continue his thorough searching over larger areas of the sky, because the effectiveness of the new 13in (33cm) telescope was so impressive.

Over the next 13 years, Tombaugh extended his planet search, covering 70 per cent of the entire sky to 17½ magnitude (six times fainter than Pluto). Tombaugh sat at the blink-comparator a total of 7,000 hours of tedious "blinking", seeing 90 million star images. He checked about 10,000 faint planet suspects with a third plate (all false). No more planets showed up. A planet like Neptune would have been picked up at seven times Neptune's distance (over 200AU) from the Sun. Certainly Neptune is the last of the giant planets. Tombaugh concluded from his thorough and extensive search that Lowell's Planet X does not exist.

In an effort to "save" the mass of Pluto, several investigators in the 1930s entertained the concept that the unexpected fainter magnitude was due to

specular reflection (in which the light of Pluto was merely a virtual image of the Sun, as would be seen in a convex mirror), and, therefore, its true disk would be imperceptible.

For more than a decade, several astronomers could not dismiss the remarkable agreement of Lowell's predicted orbital elements and its position within six degrees.

From a study of the perturbations on Neptune, Nicholson and others deduced the mass of Pluto to be 0.8 and 0.9 that of the Earth, considerably short of Lowell's required value.

An ideal opportunity to measure the angular diameter of Pluto came on the night of April 28 1965, when Pluto

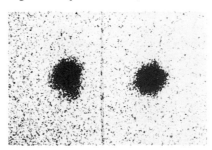

Pluto (left): the smallest planet

might occult a 15.3 magnitude star, from an ephemeris prepared by Ian Halliday of the Dominion Observatory at Ottawa, Canada. The event was widely observed, but no dimming was measured. This near miss indicated that the disk of Pluto could not exceed a diameter of 4,225 miles (6,800km) — the size of Mars. The value of the mass of Pluto was revised downward to 0.1 Earth mass or less.

In March 1976, Cruikshank, Pilcher, and Morrison of the University of Hawaii, using the 157in (4m) Mayall telescope on Kitt Peak, detected the presence of methane frost on Pluto. As a result, the albedo of Pluto was revised upward, which reduced the value of Pluto's diameter to about that of the Earth's Moon (2,000 miles/3,200km). Pluto is the only planet cold enough for methane to exist as a solid.

In June 1978, the greatest breakthrough in our knowledge of Pluto occurred. In measuring Pluto positions on plates taken with the large astrographic reflector at the Naval Observatory Flagstaff station, James W. Christy noticed a bump on several Pluto images. This indicated that Pluto possessed a relatively large satellite with a period of revolution equal to 6.4 Earth-days (coinciding with the length of Pluto's day). Using Kepler's Harmonic Law, a reliable determi-

nation of the mass of Pluto was made. The value was shockingly low, only $\frac{1}{400}$ that of Earth, and the mass of its satellite, Charon, only one-tenth of that (see CHARON).

Pluto's very low mass indicates a mean density of less than 1.0 (water), perhaps as low as 0.7. Thus, Pluto is a huge iceberg, consisting mainly of water ice and possessing a crust of methane ice. Since water ice is a very weak rock, geologically, gravitational isostasy would mould it to a sphere (or more precisely to a spheroid of small oblateness, in accordance with its slow rotation on its axis). Made of ice, the crust of Pluto could not support any high mountain.

The orbital plane of Charon is now turned edgewise to our view, and a series of transits and eclipses started early in 1985 and will continue for five years. Photo-electric observations of the mode and duration of dimming will permit precise determinations of the size and shape of both Pluto and Charon in the near future. *CWT*.

Pogson step method Assigns an arbitary figure to magnitude differences between constant stars. A variable is placed in fractions of a step between these stars.

Point source A celestial object whose angular extent is so small that it cannot be measured.

Polarimeter A device that measures the polarization of radiation. *See also* ELECTROMAGNETIC RADIATION, POLARIZATION.

Polaris Alpha Ursæ Minoris, which now lies within one degree of the polar point. It is a pulsating variable, with a very small range (1.95 to 2.05, period 3.9698 days) and has a 9th-magnitude optical companion at 18″.30.

Polarization In the wave description of ELECTRO-MAGNETIC RADIATION, light consists of vibrations at right angles to the direction of propagaton. The vibrations can be at any angle around the direction of motion; ordinary unpolarized light is composed of vibrations with the same amplitudes in all the directions. If the vibrations have greater amplitude in a preferred direction, then the light is said to be linearly polarized — ranging from partially to totally polarized. Some crystals transmit only one direction of vibration, thus converting unpolarized into linearly-polarized light (polaroid sun-

P

P

glasses use this property). Light reflected from a surface or scattered in a gas becomes partially polarized. Such polarization occurs in planetary atmospheres and through the action of interstellar dust grains.

If the plane of vibration rotates systematically in the direction of motion, then the light is circularly polarized. Combinations of linear and circular polarization produce elliptical polarization. Reflection from a metallic surface is a source of elliptical polarization.

In astronomy, the occurrence of circularly-polarized light is an indication of the presence of a strong magnetic field in the region where the light is emitted. Examples are found in the

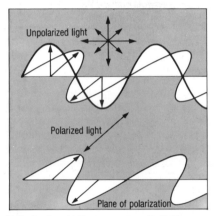

Polarization: indicates a magnetic field

radio and visible emissions from PULSARS and SUPERNOVA remnants (eg, the CRAB NEBULA) and also in MAGNETIC STARS. *BW.*

Polars Binary systems (also known as AM Herculis stars) with periods in the range 1 to 3 hours. They show strongly variable linear and circular polarization and also eclipses, the phase of which is colour dependent (red early, blue late, UV later). They are strongly variable X-ray sources and their light curves change from orbit to orbit. They also show changes in brightness and in variability with time scales of decades. They seem to be related to dwarf novæ, in that one component is a white dwarf, but differ in that the magnetic field of the white dwarf is sufficiently strong to dominate the mass flow and thus cause the effects observed. *ENW. See also* CATACLYSMIC VARIABLES, DWARF NOVÆ.

Pole star *See* POLARIS.

Pollux The star Beta Geminorum, the brighter of the two Twins in Gemini.

Pond, John (1767–1836). Sixth ASTRONOMER ROYAL. His term of office was notable for improvements to observing equipment at Greenwich Observatory, while the subsequent accuracy he obtained exceeded anything previously achieved there. He also instituted the first public time signals, using a time-ball descending a mast at 1pm every day.

John Pond: Astronomer Royal

Pons, Jean-Louis (1761–1831). Director of Florence Observatory. He began his career as the handyman at Marseilles Observatory, where he learned astronomy. He was one of the most successful comet hunters of all time, discovering 37 between 1803 and 1827.

Jean Louis Pons: began as a handyman

Pons-Brooks Comet Found by PONS in 1812, rediscovered by W.R.Brooks in 1883. Found as predicted in 1954; just naked eye visibility; period 70.9 years.

Pons-Winnecke Comet Found by PONS in 1819, rediscovered by Winnecke in 1858. A faint comet; period 6.4 years; with the 19th apparition in 1983.

Pontlyfni meteorite Unique, stony meteorite; fell Gwynedd, Wales, 1931.

Populations I and II *See* STELLAR POPULATIONS.

Porro prism A set of two right-angle prisms (as used in BINOCULARS and as a separate telescope accessory) which corrects an image so that it is both the correct way round and the right way up.

Posidonius Lunar walled plain, 60 miles (96km) in diameter, on the edge of M. Serenitatis; 32°N, 30°E. It has narrow walls, and much floor detail.

Position angle The direction in the sky of one celestial body with respect to another, measured from 0° to 360° eastwards from the north point. It can also indicate the angle at which the axis or some other line of a celestial body is inclined to the HOUR CIRCLE passing through the centre of the body. That angle is also measured eastwards from the north.

Positron The antiparticle of an electron. It has the same mass as an electron but the opposite (positive) charge. When an electron meets a positron, the particles annihilate each other and are converted to a pair of gamma-ray photons.

Pourquoi Pas Rupes Major ridge on Mercury; 58°S, 156°W.

Poynting-Robertson effect An effect whereby radiation pressure causes small orbiting particles to spiral slowly into the Sun. Because of the particle's orbital motion, photons tend to strike it on its leading edge, so exerting a small but finite drag sufficient to cause its orbit to shrink.

Præsepe (Messier 44, NGC2632). Noted already by HIPPARCHUS as a hazy patch at the centre of Cancer, this open STELLAR CLUSTER was first resolved into individual stars by GALILEO. It is a beautiful object for binocular observation. Præsepe also carries the old names the Beehive and the Crib.

Præsepe contains a few hundred stars, and, as the brightest member on the main sequence is of spectral class A0, the cluster is believed to be about *(Continued on page 337)*

Pulsars

Professor Antony Hewish, FRS
Nobel Laureate (Physics) 1974
University of Cambridge

The discovery of pulsars in 1967 was a lucky chance
which occurred because a new type of radio telescope was
brought into operation for a special purpose.
In most observations in radio astronomy high sensitivity
is achieved by averaging the incoming signals
for at least several seconds, while sharply defined images of
radio galaxies and quasars require the use
of centimetre wavelengths. My radio telescope was

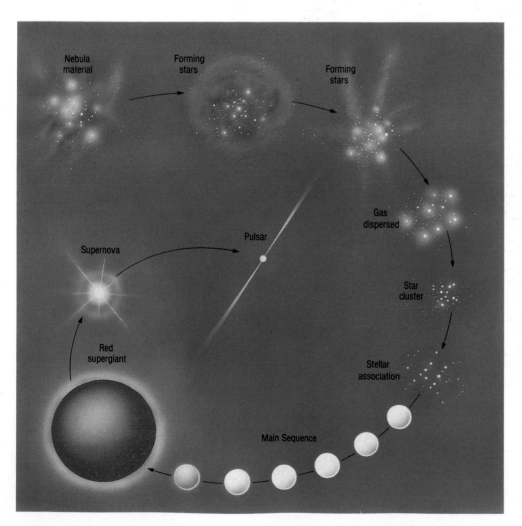

The life cycle of a star, *left,*
begins in a cool cloud of
interstellar gas and can end in an
explosion, a supernova, which
marks the birth of a neutron star.
Little is yet known about the
origin of stars because the
interstellar clouds are opaque to
light and the condensations
which will turn into stars cannot
be seen. As a large cloud
collapses under its own gravity
the denser portions will shrink
faster, generating heat until the
density and temperature are
suitable for hydrogen nuclei to
commence nuclear fusion. The
internal pressure then rises until
it balances gravity and a new star
is created. Young stars tend to be
bright, like the Pleiades, and
many similar groups can be
found which obviously
condensed more or less
simultaneously.

The middle-age of a star like
the Sun will last for 9,000 million
years before nuclear reactions
cease and it expands to become a
red giant, before finally shrinking
to a white dwarf. Stars heavier
than the Sun ultimately collapse
more violently, forming a rapidly
spinning neutron star, which can
be observed as a pulsar.

The radio telescope which detected the first pulsar (seen *left* with the author) was constructed at the Mullard Radio Astronomy Observatory of the University of Cambridge to operate with high sensitivity at a frequency of 81.5MHZ. The east-west rows of dipoles are connected so as to scan the sky in 16 strips simultaneously and each celestial radio source is recorded for a few minutes while in transit across the meridian. It was designed to study interplanetary scintillation and was particularly suitable for the first observations of pulsars.

different because it was designed to study interplanetary scintillation — a rapid fluctuation of intensity caused by clouds of ionized gas ejected continuously from the Sun to form the solar wind. These clouds cause radio galaxies to "twinkle", but the effect is pronounced only at metre wavelengths, for that is where the clouds refract more strongly, and scintillation is blurred out if the radiation is averaged for more than one second. A suitable instrument had to be large and the most economical design was an array of 2,048 dipole aerials covering an area of 4.4 acres (1.8 hectares). Quite fortuitously, the requirement of high sensitivity to fluctuating signals at metre wavelengths was exactly right for detecting the first pulsars.

A survey of the whole sky accessible to the telescope began in July 1967 with the aim of locating every radio galaxy which scintillated. I had already shown that quasars scintillated more strongly than other radio galaxies and the goal was to identify as many quasars as possible. Initially the survey was carried out by Jocelyn Bell, who joined my team as a graduate student in 1965. She was responsible for running the survey and analyz-

Jocelyn Bell Burnell, *right,* who joined Antony Hewish as a graduate student from Glasgow University in 1965. She took part in the construction of the array and was responsible for the analysis of the first routine survey which led to the pulsar discovery.

A survey record showing the first indication of signals from a pulsar on August 6 1967. The fluctuations appear similar to normal scintillation (or possibly man-made radio interference), but Jocelyn Bell noted that this source was observed near midnight when scintillation due to the solar wind is usually very weak. Regular pulses cannot be distinguished on this chart and it was not until late November, when a much faster chart speed could be employed, that the extraordinary regularity of the signals was first recognised.

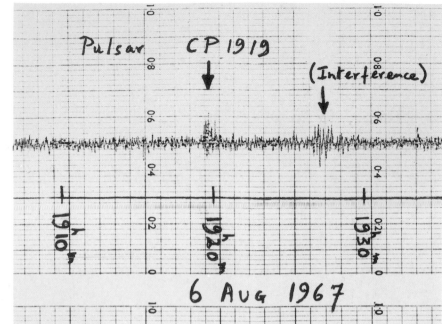

Pulsar CP 1919 (Interference)

19¹⁰ʰ 19²⁰ʰ 19³⁰ʰ

6 AUG 1967

ing the observations, which involved the careful scrutiny of every inch of the charts to check for fluctuating signals.

After a few weeks Jocelyn pointed out an unusual record which showed scintillation in a direction at nearly 180° to the Sun-Earth line. This was strange because the solar wind becomes less dense farther away from the Sun and only weak scintillation normally occurs in this part of the sky. At first we thought that the signals might have been simply local radio interference of some kind, but this became highly unlikely when repeated scans at intervals of one week showed similar fluctuations. On the other hand, on some occasions the source was not detectable and its position seemed to vary slightly in an irregular way.

One possibility was that we had discovered a flare star, rather like the Sun but much more distant. To check this idea we installed an improved recorder to study the signals with higher time resolution, but our efforts were initially frustrated. The source faded below the level of detection for several weeks! In late November it returned and good observations were made on the 28th. Jocelyn

The giant radio telescope at Arecibo, Puerto Rico, *left*, is the largest reflecting bowl so far constructed. Radio waves incident on the reflector, which is 1,000ft (305m) in diameter, are focused onto a receiving antenna suspended by cables at the centre. The bowl was excavated from the limestone hills and lined with a metallic mesh. Telescopes covering a very large area, such as this and the Cambridge array, are essential to achieve the high sensitivity required for observing pulsars.

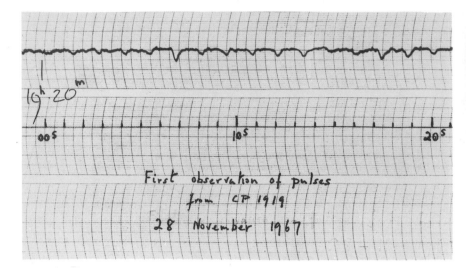

First observation of pulses from CP 1919 28 November 1967

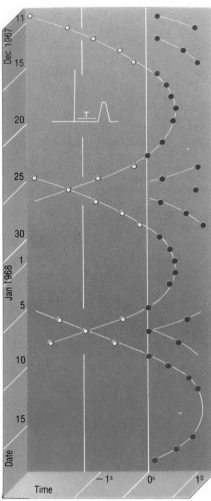

The first record of regular pulses, *above,* which marked the discovery of the pulsar PSR 1919 at Cambridge on November 28 1967. The amplitude varies rapidly from one pulse to the next, but time interval between pulses is precisely maintained.

Typical intensity variations on another pulsar over a somewhat longer time scale are shown below.

A plot of accurate timing measurements, *right,* which were carried out from day to day to see if the pulses were emitted from an unknown planet in orbit about a distant star. The remarkable pattern is caused by our own motion around the Sun. No other effect is present, which shows that the emitter cannot be an alien planet. These observations were crucial in establishing the stellar nature of the source.

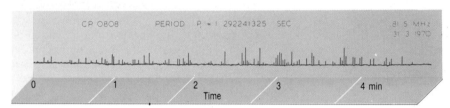

telephoned me with the astonishing news that the signals were a succession of sharp pulses, regularly spaced at intervals of just over one second. It was hard to believe and I was not convinced until I actually saw the pen tracing similar pulses on the following day.

As soon as possible I started to time the pulses more accurately and was amazed to find that they were in step to one-millionth of a second. Meanwhile other members of the team, John Pilkington and Paul Scott, found that two receivers tuned to slightly different frequencies registered the pulses at slightly different times. This provided a rough estimate of the distance to the source, because radio signals travel through the rarefied gas between the stars at slightly different speeds depending on the wavelength. We found the distance to be a few hundred light-years, beyond the nearest stars but well inside our Galaxy. The final surprise was the realization that the object emitting the pulses had to be very small. A radiating source, such as a star, cannot emit a pulse the duration of which is shorter than the time taken for radiation to travel a distance equal to the stellar radius. Shorter pulses would be smeared out by the different travel-time of radiation from different points on the star. In our case the emitter could not have been larger than a small planet such as the Earth.

At this stage, in early December, I contemplated the notion that some message from extra-terrestrial beings provided the simplest explanation, although we could see no intelligence in the pulse trains. Since life can only exist on planets, I began measurements to test whether the pulses were coming from an unknown planet in orbit about some distant star. Orbital motion would have shown up as a change in the pulse-rate caused by the Doppler effect. After carefully timing the pulses for several weeks I ruled out the possibility of a planetary source. When a substantial Doppler shift due to the Earth's motion was allowed for no evidence for planetary motion remained.

I finally concluded that the source must be some new kind of star, akin to the well-known Cepheid variables, which show regular changes in their emission of light on a timescale of days. The small radius excluded all possibilities except white dwarf stars or hypothetical neutron stars. Cepheids vary because of radial vibrations, and I assumed that similar effects could generate the pulses.

Publication of our work in February 1968 caused great excitement and within one year it became generally accepted that pulsars were neutron stars, not vibrating but spinning and radiating a well-defined beam like a lighthouse.

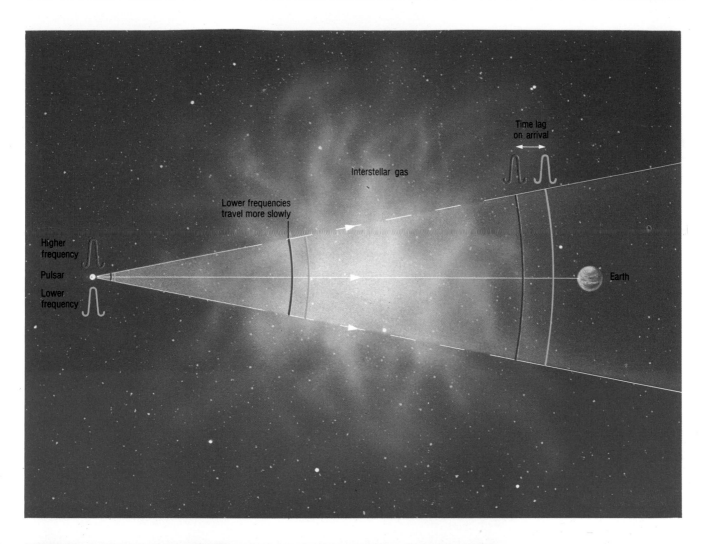

Time lag on arrival

Interstellar gas

Lower frequencies travel more slowly

Higher frequency

Pulsar

Lower frequency

Earth

The distances of pulsars can be estimated from the difference in the arrival time of pulses observed at two wavelengths, shown red and blue, *above*. A very low density gas fills space between the stars and some of this gas is ionized. Radio waves passing through ionized gas are slowed down and the effect is greater at longer wavelengths. The measured time delay when the same pulse is observed at two wavelengths can be used to calculate the distance if the density of the ionized gas is known. Distances up to 30,000 light-years have been obtained, but most of the pulsars lie in the disk of our Galaxy at a few hundred to several thousand light-years.

It is believed that pulsars emit radiation, *left*, in a narrow beam so that rotation of the star produces regular flashes like a lighthouse.

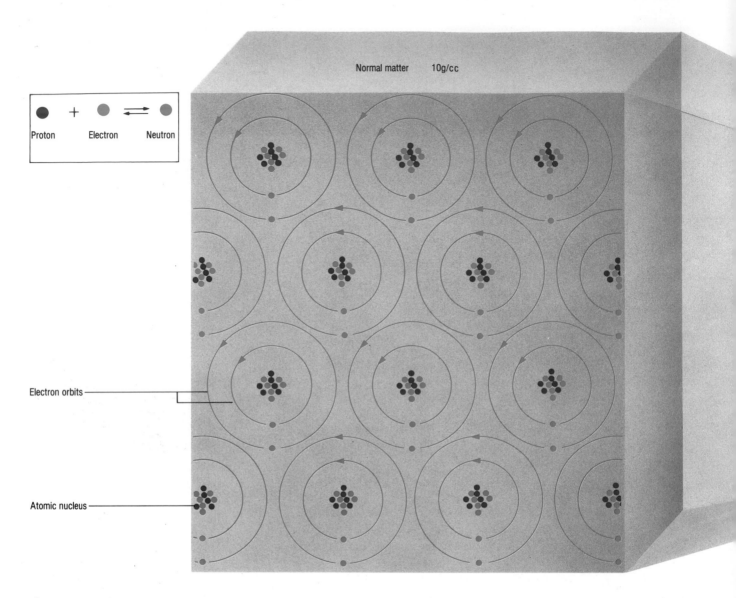

Proton   +   Electron   ⇌   Neutron

Normal matter   10g/cc

Electron orbits

Atomic nucleus

Neutron Stars

To understand the remarkable nature of neutron stars, and their place in the family of stellar types, we must first consider how matter behaves under extreme compression. Gravity is enormous inside bodies of large mass, like the Sun, and under certain conditions it can crush matter into new forms that do not exist on Earth.

Imagine an experiment in which we could subject ordinary matter to an ever-increasing pressure. Normal atoms can be squashed together until the orbiting electrons which surround the nucleus begin to overlap. Such a close-packed arrangement occurs in solids like a block of iron. Further compression demands that the electrons move faster, because quantum theory treats the particles like waves and the wavelengths must become shorter to fit into the smaller volume. Their speed is inversely proportional to the radius of the orbit. At sufficient compression the electrons move so fast that orbits bound to individual nuclei are disrupted and they move randomly between the nuclei. This new state is called degenerate matter; it occurs in white dwarf stars and a teaspoonful

would weigh more than one tonne! Under further compression the electrons move almost at the speed of light and begin to interact with protons in the nuclei to form neutrons. Ultimately nearly all the protons and electrons are converted and the result is a material composed of neutrons at a density of 100 million tonnes in a teaspoon! Yet more compression and gravitation becomes so powerful that the material collapses to become a black hole in space.

A common star like the Sun is powered by nuclear fusion in an active central core, and its size is determined by the balance between gravitational forces and the outward pressure due to heat generated in the reaction. When the nuclear fuel is finally spent, gravity becomes dominant and pulls the material into a smaller volume. The star's eventual fate then depends upon its total mass, since this governs the compression. When the Sun burns out, roughly 4,000 million years hence, it will ultimately become a sphere of degenerate matter about the size of the Earth. It will enter retirement as a white dwarf star, very hot but cooling steadily. A substantially heavier star

In solid materials, *left*, neighbouring atoms are placed so that the electron orbits are just touching. Under intense gravitational compression inside a star matter can take a different form, called degenerate matter, in which the electrons move randomly between the nuclei. A teaspoonful of degenerate matter would weigh one tonne. Inside very massive stars larger compression causes the electrons to combine with protons to form neutrons. Matter can then become so dense that 100 million tonnes would occupy the volume of a sugar cube.

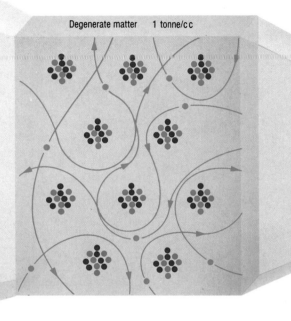

Degenerate matter   1 tonne/cc

Neutron matter
100 million
tonnes/cc

The Sun maintains its present size because gravitational forces are balanced by the internal pressure caused by heating from nuclear fusion, *below*. When fusion ceases, about 4,000 million years hence, gravity will compress the star until it is little larger than the Earth. It will then become a white dwarf, composed of degenerate matter, and will spend the rest of its life cooling down.

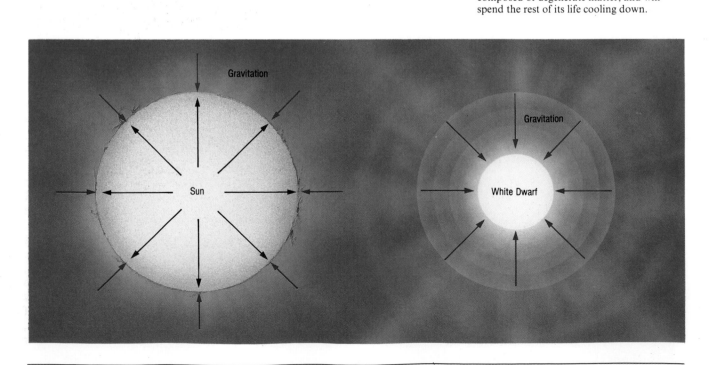

cannot do this because increased gravitational compression converts the material to a sphere of neutrons — a neutron star. In this state a star containing 1.5 times the mass of the Sun would have a diameter of roughly 12 miles (20km), a truly remarkable object.

In reality a neutron star must be more complicated than a sphere of pure neutrons owing to the great variation of pressure from the surface to the centre. The probable structure is expected to comprise a rigid crust, starting with a skin of iron-like material which soon becomes degenerate so that the electrons move randomly within a lattice of ^{56}Fe nuclei. At deeper levels neutrons replace the electrons and protons to produce a neutron fluid of immensely high density. The behaviour of neutrons in bulk cannot be studied on Earth because once outside the atomic nucleus these particles decay into electrons, protons and neutrinos within a few minutes. From the observed properties of the neutron particle it is predicted that the interior of a neutron star will be superfluid so that the liquid has zero viscosity. This strange condition has been achieved for liquid helium in the laboratory at temperatures close to absolute zero. Near

the centre of the neutron star, where there is maximum pressure, it is possible that there exists a core of quarks, or other exotic particles, but no theory is yet adequate to discuss this region.

The possibility that neutron stars might exist was realized in the 1930s, soon after the neutron particle was identified, but it was thought that their tiny size would render them almost undetectable. It was, however, foreseen that the final collapse to such a tiny volume would be a violent process, with material falling inward at enormous speed. The pile-up at the centre must be accompanied by a violent release of energy so powerful that it blasts outwards a substantial fraction of the stellar material. This was one theory to account for stellar explosions, known as supernovæ, which are occasionally seen in our own and other galaxies.

The Crab Nebula is a well-known remnant of a supernova. The explosion was witnessed by Chinese and other oriental astronomers in AD1054, who saw what they thought to be a new star suddenly appearing in the sky. About one year after the discovery of pulsars a careful search of the Crab Nebula revealed a pulsar near the

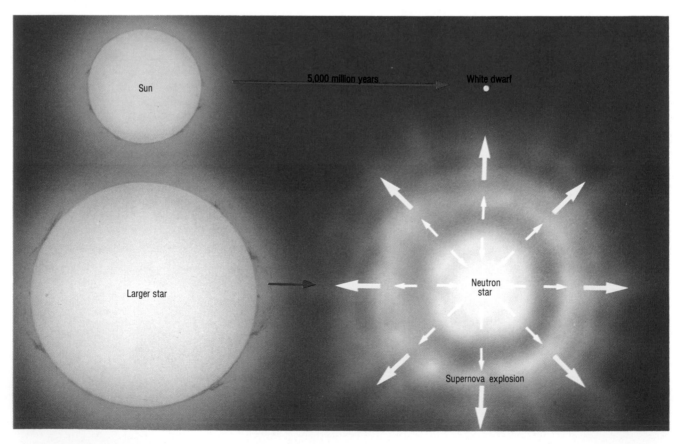

When nuclear fusion ceases, what happens to a star depends upon its mass, *above*. Stars like the Sun will ultimately collapse under self-gravitation to become white dwarfs. Heavier stars form neutrons in their cores, but their final collapse is violent as material falls inward at speeds approaching one-tenth the speed of light. Shock waves from the implosion explode the outer layers of the star, to form a supernova, while a neutron star condenses at the centre.

The Crab Nebula, *right*, is the remnant of a supernova that was seen to explode in AD1054 by oriental astronomers. Shreds of material blown outward from the explosion are still moving at speeds of several thousand kilometres per second. The central neutron star is the only pulsar that is easily seen with optical telescopes.

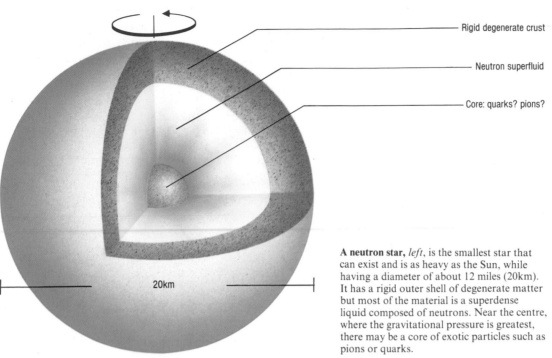

Rigid degenerate crust

Neutron superfluid

Core: quarks? pions?

20km

A neutron star, *left,* is the smallest star that can exist and is as heavy as the Sun, while having a diameter of about 12 miles (20km). It has a rigid outer shell of degenerate matter but most of the material is a superdense liquid composed of neutrons. Near the centre, where the gravitational pressure is greatest, there may be a core of exotic particles such as pions or quarks.

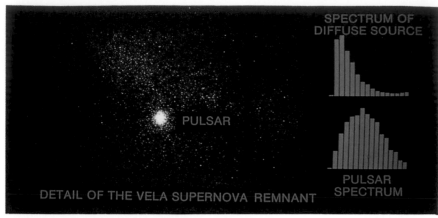

SPECTRUM OF
DIFFUSE SOURCE

PULSAR

PULSAR
SPECTRUM

DETAIL OF THE VELA SUPERNOVA REMNANT

The Vela supernova remnant contains a pulsar, *above*, which emits X-rays in addition to radio waves. Out of more than 400 pulsars so far discovered only three produce detectable radiation at frequencies outside the radio band. The reason for this is not known, but is probably related to the fact that these neutron stars were formed in quite recent supernova explosions.

Very weak pulses of light were just detectable when the Anglo Australian Telescope was turned towards the Vela pulsar in 1977. The first two frames, *below*, show the pulsar clearly just above the centre, but it is much weaker than the other stars nearby. The pulsar flashes 11 times a second, a rate which, occasionally, suddenly increases by about one part in a million.

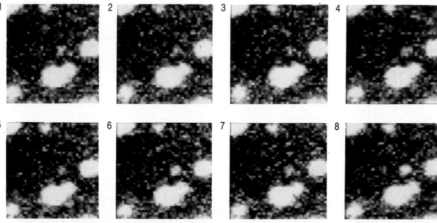

1   2   3   4

5   6   7   8

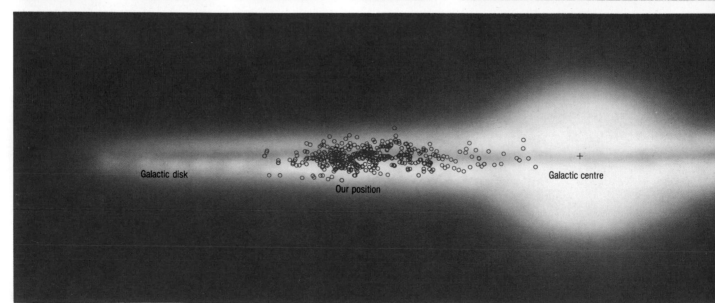

Galactic disk

Our position

Galactic centre

The shreds of hot gas that now constitute the Vela supernova remnant, *left*, indicate that this stellar explosion must have taken place about 11,000 years ago (as seen from the Earth). It is one of the few examples, like the Crab Nebula, where the exploded remains of the original star which suffered gravitational collapse can still be seen. The age of a neutron star can be estimated from the rate at which it is slowing down. Precise measurements of the steady decrease of the pulse rate indicate that most pulsars have existed for around one million years after the supernova occurred. During this time all traces of the original explosions would have dispersed.

If we could see our Galaxy, *below,* from the outside its shape would be a spherical bulge surrounded by a flat disk at a distance of roughly 30,000 light-years from the centre. The pulsars located to date (points) lie mainly within a few thousand light years from us, and beyond this distance they would be undetectable. Obviously the Galaxy contains a far greater number and there must be about 50,000 in the whole Galaxy. This distribution of pulsars is similar to that of the known remnants of supernovæ, in accordance with the creation of neutron stars in stellar explosions.

centre which emitted at a rate of 30 pulses per second. On the "lighthouse" theory, first suggested by Thomas Gold, only a neutron star could be responsible. Any other star, such as a white dwarf, would fly to pieces when rotated at this speed. Additional confirmation of the neutron star theory came from timing observations which showed that the pulse rate was slowing down. During the next 1,000 years the pulse rate will fall to half its present value. This agrees well with the predicted decrease, assuming that the rotational energy of the neutron star is being converted into other forms, such as radiation or particle emission. Long before this pulsar was discovered it was known that the Crab Nebula contained some energy source to sustain the strong emission of X-rays observed, but the origin of the power was a mystery.

Problems and theories

At the present time more than 400 pulsars have been located and the neutron star theory has become firmly established. The slowest one pulses at intervals of about four seconds and the fastest at 1.5 milliseconds, but the most common period is just under one second. Pulsars have been detected at distances up to 150,000 light-years but they are concentrated towards the flat disk of our Galaxy where supernovae most commonly occur. Only two other pulsars definitely lie inside supernova remnants. One of these is the Vela pulsar in the southern sky which has a period of 89 milliseconds. It is believed that most pulsars have ages exceeding one million years so that any remnants of the supernovæ which attended their birth would have dispersed and faded from view long ago. On the other hand efforts to find pulsars in other supernova remnants have not been successful and the lack of them is puzzling. To account for the number of pulsars found in our Galaxy requires the formation of one neutron star every 50–100 years, and this implies that most supernovæ should give birth to neutron stars. One possibility for not finding these pulsars is that they may not begin to radiate until after the remnant has dispersed.

After eighteen years of study it is still not known how pulsars radiate or why the radiation is beamed. Only the Crab and Vela pulsars emit pulses of light as well as radio waves; in these instances the spectrum extends into

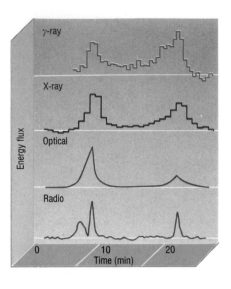

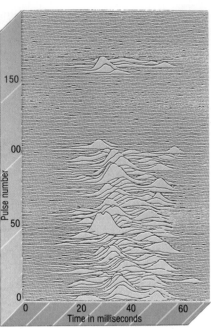

The pulsar in the Crab Nebula radiates over a wide range of the electromagnetic spectrum extending from radio to γ-rays. The average pulse shape, *left*, is similar across the spectrum and is double-peaked, indicating that the beam is almost perpendicular to the rotation axis, so that two pulses are obtained for each complete revolution of the neutron star.

Although the average pulse shape remains the same, successive pulses can vary greatly as shown, *right*, in this sequence from the pulsar PSR 1944 + 17. Some of the peaks drift sideways, as though the lighthouse beam is not sweeping at exactly the same rotation speed as the neutron star. At other times the pulses are not seen. No satisfactory explanation has yet been given for these rapid pulse-to-pulse variations.

The pulsar PSR 1237 + 25 has a complex pulse shape, *below left* showing five peaks, the strongest being located at the edges of the pulse. The two outer peaks on each side periodically wax and wane in strength. These effects suggest that the overall beam must contain a number of narrower beams which turn steadily about the magnetic axis.

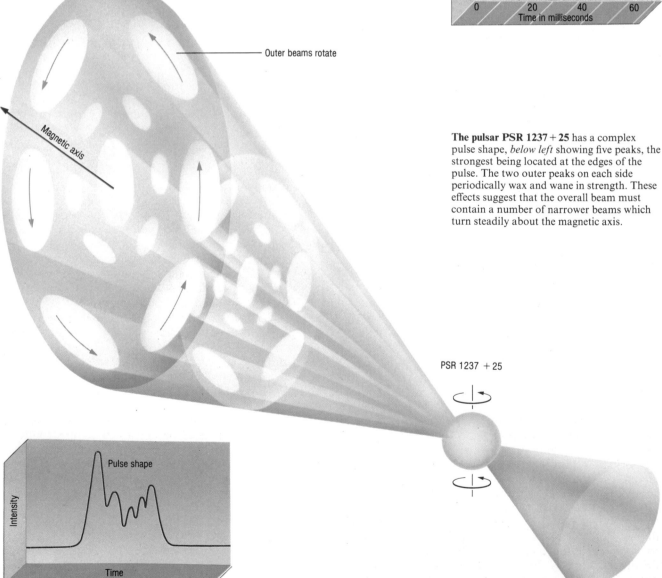

PSR 1237 + 25

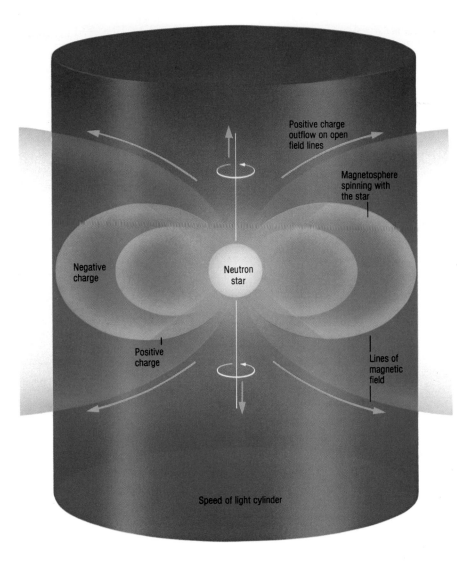

Positive charge outflow on open field lines

Magnetosphere spinning with the star

Negative charge

Neutron star

Positive charge

Lines of magnetic field

Speed of light cylinder

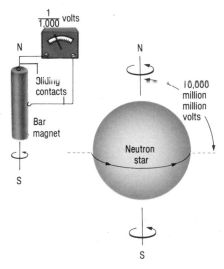

Electrical forces resulting from the dynamo-effect of a spinning neutron star, *left*, are extremely strong. Charged particles, such as ions and electrons, are torn from the surface and hurled into space. Guided by the magnetic field, these charges form a magnetosphere which extends far from the neutron star. Separate zones contain either positive or negative charge, unlike planetary magnetospheres where the charges are mixed.

$\frac{1}{1,000}$ volts

N

Sliding contacts

Bar magnet

S

N

10,000 million million volts

Neutron star

S

When a star collapses to form a neutron star the original magnetization is compressed into much smaller volume. The star becomes an immensely strong magnet, *above*. As it is also spinning rapidly, it acts as a dynamo and generates a huge voltage between the pole and the equator. In actual neutron stars the magnetic axis is not aligned with the rotation axis and the voltage is less symmetrically arranged.

the X-ray and γ-ray band. The remarkable similar waveform over the whole spectrum for the time-averaged pulse of the Crab pulsar points to a highly organized mechanism for generating the radiation. On a short timescale, however, some pulsars exhibit a bewildering complexity of behaviour. The pulse amplitude and shape can vary dramatically, and characteristic peaks in successive pulses sometimes drift steadily backward or forward within the overall pulse. This phenomenon of drifting sub-pulses may also be accompanied by a total fadeout of the radiation for varying intervals of time. Unfortunately, we have no means of observing the beam in three dimensions and its form must be deduced from the time variations produced as it sweeps rapidly across the Earth. One theory supposes that the radiation is emitted along the magnetic axis of the star in a number of concentric, conical beams. Some idea of the complexity involved may be seen in the case of PSR1237+25, where the outermost conical beams must be envisaged as subdivided into a number of yet smaller beams in steady rotation about the magnetic axis. This could explain the drifting sub-pulses. In a few cases, including the Crab pulsar, two pulses are observed for each single rotation

of the star. On the magnetic alignment theory this implies that the magnetic axis can sometimes be tilted nearly at right angles to the spin axis so that the beam from each magnetic pole can intersect the Earth.

It is likely that pulsars have extremely intense magnetic fields because many stars are known to possess magnetization and the final collapse to the neutron star diameter will intensify any pre-existing field by roughly 10,000 million times. In combination with rapid rotation there will be a dynamo effect generating huge voltages between different points on the star so that electrons and ions may be torn from the surface. Once ejected the motion of the charged particles will be controlled by electromagnetic forces and they will form an extended magnetosphere which is swept into rotation with the star. Beyond a certain distance, depending upon the rate of spin, the magnetosphere must either end or stream outward as a stellar wind; it cannot spin with the star as that would demand a speed exceeding the velocity of light, which is forbidden by Einstein's theory of relativity. The precise details of the magnetosphere involve theoretical problems which have yet to be solved, so the processes which cause radiation are currently unknown.

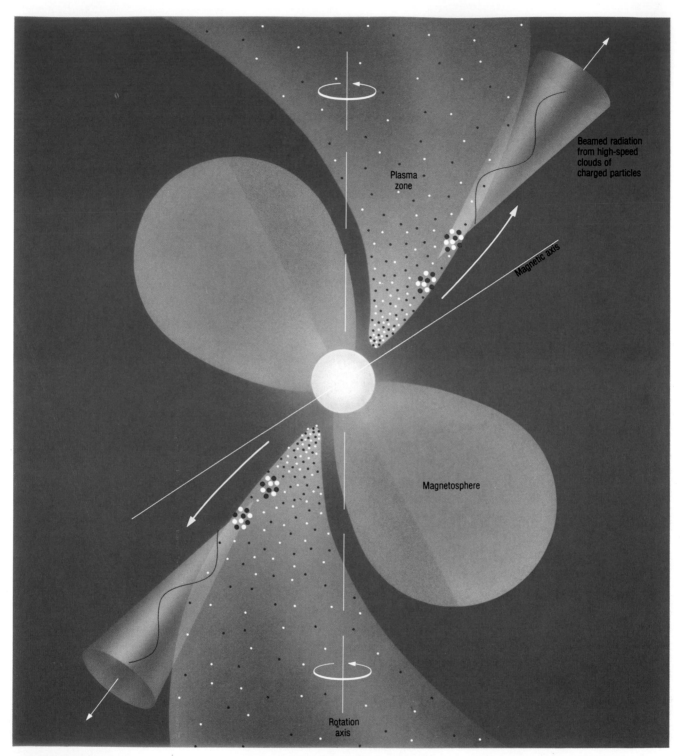

Plasma
zone

Beamed radiation
from high-speed
clouds of
charged particles

Magnetic axis

Magnetosphere

Rotation
axis

It is not yet known why pulsars emit beamed radiation. One theory, *above*, assumes that bunches of positive or negative charge are accelerated along the direction of the field near the magnetic poles of the neutron star. If the bunches travel at nearly the speed of light, radiation from them can only be emitted along their direction of motion. This produces conical beams which are swept around the rotation axis to produce the "lighthouse" effect.

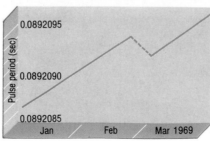

Accurate timing of pulses, *left,* shows that the steady slowing down, which increases the period between pulses, is occasionally interrupted by sudden "glitches" when the period decreases. This corresponds to a small increase in the speed of rotation by about one part in a million. This is probably caused by a rearrangement of a vortex structure in the liquid interior of the star and accords with the superfluid theory of the neutron state.

To generate radio waves of the required strength demands a well-organized motion of small clouds of positive or negative charge. One theory is that outward-moving clouds of charge might be formed above the magnetic poles and then guided by the curved magnetic field lines like beads on a wire. For speeds close to that of light this would both generate radiation and cause it to be beamed along the direction of motion. Such a mechanism qualitatively explains several observed features, for example the high linear polarization often found and the concentration into a narrow beam, but many theoretical difficulties remain to be solved.

Using pulsars as clocks

Soon after the discovery of pulsars it was realized that they provided, in effect, astronomical clocks of high accuracy which might be of great value in testing physical theories. It turned out that many pulsars were not perfect time keepers. Apart from the steady slowing down, which can be allowed for, some of them showed small irregular variations. This was particularly evident for the Vela pulsar, which occasionally *increased* its spin

rate abruptly by about one part in a million. The effects can be explained by small changes in the internal structure of neutron stars which cause a transfer of angular momentum between the superfluid component and the rigid shell. One pulsar was discovered in 1975, however, which did not show any detectable irregularities of spin. Moreover it was in orbit about another neutron star with a binary orbital period of 7 hours 45 minutes. Very accurate timing of the orbital period, by means of the Doppler effect and using the pulsar as a clock, has shown that the two neutron stars are losing energy of orbital motion and drawing slowly together. The only satisfactory explanation is that energy is being lost in the form of gravitational waves. This is the first verification of a prediction of Einstein's general theory of relativity and it has greatly stimulated other experiments aimed at the direct detection of gravitational waves reaching the Earth from astronomical sources.

The most recent surprise was the discovery, in 1982, of a pulsar with a period of 1.6 milliseconds. The idea of a star as massive as the Sun spinning at more than 600 revolutions per second boggles the imagination. Even

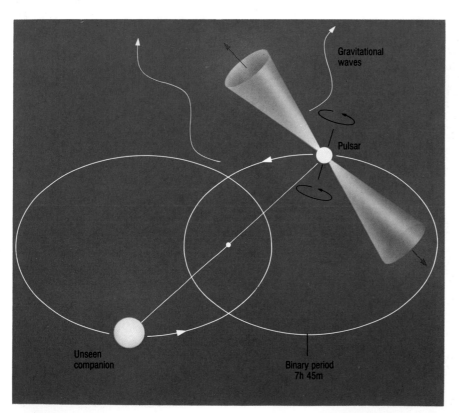

Neutron stars can occur as binaries, *left*, although this will be a rare phenomenon because the supernova explosions which give birth to the neutron stars are likely to disrupt the binary system. The pulsar PSR 1913 + 16 is one member of a neutron star binary which has an orbital period of 7 hours 45 minutes. The companion neutron star is undetectable, possibly because its "lighthouse" beam never points in our direction.

This binary pulsar provides a supreme check of Einstein's Theory of General Relativity because it can be used as an accurate clock to measure the effect on the orbit of spacetime curvature. One prediction of the theory is that gravity waves should be radiated from the orbiting masses. These are tidal forces which propagate at the speed of light, tending to stretch and compress alternately any object in their path. Distortions of this kind are extremely small and have not yet been detected, but careful timing of the binary pulsar shows that the system is radiating unseen energy, probably in the form of gravity waves.

Labels in figure: Gravitational waves · Pulsar · Unseen companion · Binary period 7h 45m

The pulsar PSR 1937 + 21 emits roughly 600 pulses per second and is the fastest pulsar known, *right*. It has the most steady pulse-rate so far found and provides an astronomical clock which may exceed the accuracy of laboratory atomic clocks. It will be necessary to check the stability of other pulsars similar to PSR 1937 + 21 before using them as standards of time in order to compare one against the other.

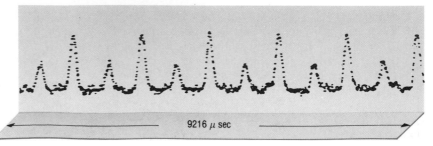

9216 μ sec

the intense gravitational attraction of a neutron star is stretched near the limit and this star must be close to disrupting like an exploding flywheel. Since all pulsars are observed to be slowing down, a very high rate of spin is what would be expected for a recently created neutron star such as the Crab pulsar. There is, however, no trace of a supernova remnant associated with the millisecond pulsar and the strength of the radiation is weak, suggesting that the neutron star is old enough for the magnetic field to have decayed to a low value. One possibility is that this pulsar was originally in orbit with a companion star and acquired a high rate of spin by dragging matter from the companion gravitationally. If this is so the companion must have exploded as a supernova long ago since no trace of it can be observed today.

Until more of the problems have been solved, our present understanding of neutron stars can only be regarded as fragmentary. Undoubtedly their discovery opened an exciting new era in astrophysics. In particular it focused attention upon the dramatic way in which midget stars can be reborn amidst the ashes of nuclear fusion in supernova explosions.

The millisecond pulsar, PSR 1937 + 21, is spinning so fast that it seems to have been recently created, but the strength of its radiation implies that the magnetic field is relatively weak. This suggests that it must be an old neutron star, in which the magnetisation has decayed, but which has acquired a high rate of spin due to material from a companion star being gravitationally attracted to its surface, *below*. The whirlpool effect of the in-falling gas then provides rotational acceleration. The companion star later became a supernova, and the two neutron stars then separated.

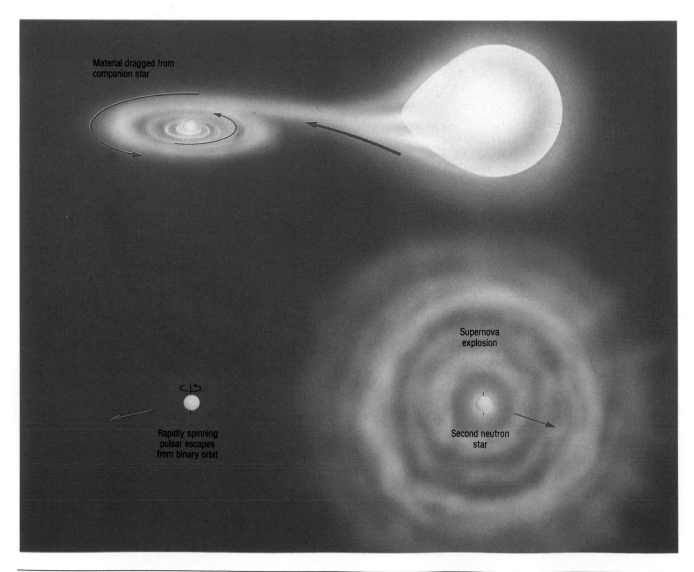

Material dragged from companion star

Rapidly spinning pulsar escapes from binary orbit

Supernova explosion

Second neutron star

650 million years old. It is slightly richer in metals than the Sun. The distance to Præsepe is about 160pc or 520 light-years. *GW*.

Præsepe: an open stellar cluster

Precession Properly, the precession of the EQUINOXES: the westward motion of the equinoxes, and the associated circular motion of the NORTH

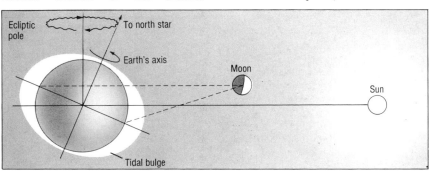

Precession: the shift of the equinoxes

CELESTIAL POLE (NCP) around the pole of the ECLIPTIC. Discovered by HIPPARCHUS about 150BC; the westward motion is 50".26 per year, giving a complete rotation in about 25,800 years. It is largely caused by the gravitational pull of the Sun and Moon on the Earth's equatorial bulge. The Earth acts like a spinning top because of its rotation, and the pull on the bulge produces an angular momentum in a perpendicular direction, according to the laws of dynamics. The NCP therefore traces a circle of radius equal to the inclination of the Earth's axis (23° 27'); it is presently near to POLARIS, and will be only 28' away in 2100. The "pole star" in 2500BC was Thuban (Alpha Draconis); in AD 4000 it will be Alrai (Gamma Cephei).

Since positional observations are made relative to the celestial poles at the epoch of observation, different observations are usually reduced to standard epochs, eg, 1950.0 and 2000.0. Precession comprises three compo-

nents: the luni-solar precession is the main component, amounting to 50".40 per year. There is a small similar component caused by the gravitational effects of the other planets, the planetary precession, amounting to 0".12 per year, but in the opposite direction. Relativity theory predicts another small effect called geodetic precession, of −0".02 per year. The net effect of these, called the general precession, is 50".26 per year.

Precession has moved the equinoxes about 1h 45m in R.A. since Hipparchus discovered it, so that each equinox has shifted into the adjoining constellation since then. Precession is also responsible for making the TROPICAL YEAR, measured with respect to the equinoxes, about 20' shorter than the SIDEREAL YEAR, which is measured with respect to the stars. Another effect is that 13,000 years from now ORION will be a summer constellation (in the northern hemisphere). *TJCAM*.

Priamus Trojan asteroid, 884, discovered in 1917. It never rises above magnitude 16.

Pribrăm fireball Fireball associated with the meteorite fall at Pribrăm, Czechoslovakia on April 7 1959. It was photographed from two stations and from them its orbit was calculated.

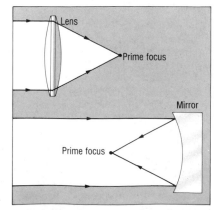

Prime focus: refractor and reflector

Prime focus The point at which the objective or primary mirror brings starlight directly to a focus (without the intervention of any additional lenses or secondary mirrors).

Primeval fireball A term sometimes applied to the situation at the time of the original BIG BANG.

Principia Common name for NEWTON's great work, published in 1687.

Prism A solid straight cylinder of glass (or other material) whose uniform cross-section is a simple polygonal shape (typically triangular). Triangular prisms are used in astronomical SPECTROGRAPHS to disperse light into its constituent colours; each colour is refracted by a different angle into a spectrum. For maximum dispersion, equiangular triangular prisms (60° angles) can be used in series to multiply the effect. Prisms are now less commonly used in spectrographs than reflecting or transmission gratings. Objective prisms are prisms of very small angles (1° or so — scarcely different in appearance from a window pane). These are used in front of telescope objectives to create very small spectra from star images. As thousands of images may be photographed at once this is an effective way to mass-survey a field of stars or galaxies. *PGM*.

P

Porro prism: correcting an image

Proclus Brilliant lunar crater, 18 miles (29km) in diameter, just west of Mare Chrisium; 16°N, 47°E. It is a ray-centre, but the system is asymmetrical.

Proctor, Richard Anthony (1837–88). English astronomer famous for his numerous writings on the subject, including *Saturn and Its System* which was published in 1865. He founded the popular scientific magazine *Knowledge* in 1881, the year he emigrated to the United States.

P

Richard Proctor: prolific writer

Procyon The star Alpha Canis Minoris, magnitude 0.38. It has a white dwarf companion of magnitude 13; the period is 40 years. The companion is a difficult object, because of the primary's glare.

Prognoz satellites Eight Soviet satellites designed to monitor solar activity and the interaction of the SOLAR WIND with the Earth.

Project Mercury First United States manned spaceflight programme. *See* MERCURY PROJECT.

Prometheus Second satellite in order of distance from Saturn, discovered from Voyager photographs. It measures 87 × 62 × 50 miles (140 × 100 × 80km), and is the inner SHEPHERD SATELLITE of the F ring.

Prometheus Active volcano on Jupiter's satellite IO (Plume 3): 3°S, 150°W. It was erupting during both the Voyager passes, and the activity was actually greater as seen from Voyager 2.

Propane (C_3H_8). A hydrocarbon which is a minor constituent of Saturn's atmosphere.

Proper motion See STELLAR PROPER MOTION.

Protonilus Mensa Martian mesa; 49–38°S, 325–303°W; diameter 641 miles (1,032km).

Proton launchers Large Soviet rockets used to place heavy satellites and the SALYUT space stations in orbit. Also used for launching lunar and planetary probes.

Proton-proton chain The proton-proton (p-p) chain is a set of nuclear reactions that results in the conversion of hydrogen into helium with the release of energy. The principal reactions are given in the article on NEUTRINOS.

The overall result is the conversion of four nuclei of hydrogen into one nucleus of helium. The p-p chain is the dominant source of nuclear energy in MAIN SEQUENCE stars with masses equal to or less than the Sun. *BW. See also* THERMONUCLEAR REACTIONS.

Protons The proton is the nucleus of the hydrogen atom, the most abundant element in the universe. It was first identified as a fundamental particle by J.J.Thomson in 1906. Protons and neutrons, collectively known as nucleons, attract each other with a nuclear force that is effective only at very short ranges, which results in formation of the nuclei of the elements. The proton has a positive electrostatic charge exactly equal in magnitude to the negative charge on the electron. The charge on a nucleus, or its atomic number, is thus equal to the number of protons that it contains.

The mass of a proton is 1.672×10^{-24} gram, or 1,836 times heavier than the electron. A proton has internal structure consisting of three QUARKS, but for nuclear reactions at the energies met with in astrophysical environments it behaves like a particle with radius about 10^{-13}cm, ie, about 10^{-5} the size of a hydrogen atom. At distances comparable with the radius of a proton the attractive nuclear force is stronger than the repulsive electrostatic force between protons.

Protons have intrinsic spin and an associated magnetic moment. It is the interaction between the proton and electron magnetic moments that produces the doublet structure of the lowest level of the hydrogen atom, with its associated 21cm radiation.

Modern theories of particle physics require that a proton should eventually decay into lighter elementary particles. The lifetime of a proton is calculated to be about 10^{33} years. Although much greater than the age of the universe, this rate of decay nevertheless would give one proton disintegration each year in a 1,000-tonne mass of water. Experiments are in progress to attempt to detect the proton decay. If successful these will connect together the internal structure of the proton and the ultimate fate of the universe. *BW.*

Proton satellites Very heavy Soviet satellites (12–17 tonnes), used for monitoring COSMIC and GAMMA RAYS and the development of the SALYUT space station.

Protoplanet A protoplanet is a planet in the making. As a consequence of the astronomical processes that led to the formation of the Sun, a disk of material was left moving in a flat broad ring around the Sun, with the ring extending at its inner boundary from the radius of the orbit of the innermost planet, Mercury, to the radius of the outermost planet, Neptune, at its outer boundary. Most of the material in the inner regions of the ring consisted of a swarm of small solid bodies usually referred to as planetesimals, with very little gaseous material surrounding them. Owing to their mutual gravitational attractions, the planetesimals could aggregate into larger and larger bodies until objects of about the size of the Moon were formed. Further aggregation could not arise by gravitation alone, however, and would have to depend on a degree of randomization arising in the orbits of the lunar-sized bodies, causing the orbits to interlace each other so that collisions between the bodies could occur. Collisions break up some bodies while allowing others to grow in a complex *mélée,* the eventual outcome being that a few of the bodies, those that can be designated protoplanets, grew toward becoming full-blown planets. As it turned out in the Solar System, four such protoplanets became planets in the inner regions of the Solar System — Mercury, Venus, Earth and Mars. The whole process of aggregation for the latter took about 10 million years.

A similar situation occurred toward the outer part of the ring, where, as it turned out, only two bodies grew to full-blown planets, Uranus and Neptune, but with a required timescale much longer than for the inner planets, a time scale for the outer regions of about 500 million years.

The middle part of the ring appears to have differed from the inner- and outermost parts in an important respect. Much gas, mostly hydrogen and helium, accompanied the solid planetesimals. When they had aggregated into protoplanets, also in a complex collisional *mélée,* the resulting bodies were able to begin acquiring gas through the growing strengths of their gravitational fields. Two protoplanets became full-blown planets in the middle part of the ring, Jupiter and Saturn, and they differ from all others in the great quantities of hydrogen and helium which they contain. The process in the middle of the ring probably took about 100 million years to complete. *FH*.

Protostars The earliest phases of STELLAR EVOLUTION. As dense clumps within molecular clouds collapse to form stars, several stages are distinguished. "Protostars" are the earliest — optically invisible — still accreting from the parent cloud; pre-MAIN SEQUENCE evolution follows the end of accretion and extends to the onset of hydrogen fusion in the core; the "main sequence" is the ensemble of mature stars stably burning hydrogen. The younger a star, the more rapid its evolution. The Sun's protostellar stage took about 0.1–1 million years; its pre-main sequence phase several hundred million; and its main sequence lifetime will be about 10,000 million years. (Post-main sequence evolution is also relatively rapid.) *MC*.

Proxima Centauri Discovered in 1915 by R.T.A.INNES, Proxima Centauri is a 13.1 magnitude star 2° away from ALPHA CENTAURI. It has a measured parallax of 0".764, or a distance of 1.31 parsecs, which makes it slightly closer than Alpha Centauri and is consequently the nearest known star to the Sun. Proxima and Alpha Centauri (itself a double star) revolve in orbit around each other with a period estimated to be about one million years.

Proxima is a dwarf star of spectral type M5 and an absolute magnitude +15, which makes it one of the least luminous stars known. It is a FLARE STAR, brightening to twelfth magnitude, and has been detected by satellite-borne X-ray telescopes. *BW*.

Psyche (Asteroid 16). Discovered in 1851; diameter 155 miles (250km). It is the largest known asteroid of the metallic iron type.

Ptolemæus Lunar walled plain 92 miles (148km) in diameter; 14°S, 3°W — one of a chain with Alphonsus and Arzachel. It has a darkish, relatively smooth floor.

Ptolemaic system The system of the universe proposed by PTOLEMY in his *Almagest*. It was geocentric, having the Earth at the centre of a spherically-shaped universe, with Sun, Moon and

planets orbiting round it. The system was accepted in Western and Arab countries for over 1,300 years.

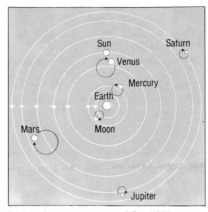
Ptolemaic system: accepted for 1,300 years

Ptolemy (*c*.AD100–*c*.170). Greek astronomer and mathematician, famous for writing the *Almagest*.

Little is known about Ptolemy's life. His dates are deduced primarily from his astronomical observations, while

Ptolemy: greatest astronomer of antiquity

even his name is uncertain. "Ptolemy" merely indicates that he had Greek or Greek-naturalized ancestors and lived in Egypt, which was then ruled by the Ptolemaic dynasty. The story that he was born in the city of Ptolemais Hermiou is a 13th-century suggestion and appears to have no independent support. Sometimes referred to as Claudius, this probably only means that Ptolemy's Roman citizenship goes back to the emperor Claudius I. It seems likely, though, that he lived and worked at the famous Library and Museum in Alexandria.

Ptolemy was a prolific author, but his earliest large work, *Almagest*, was

his greatest. Its original title was "matheematikee suntaxis" or "mathematical compilation", the word mathematical here referring to mathematical astronomy. Later it was often referred to as "the greatest compilation", probably to distinguish it from earlier and simpler works often called "the small astronomical compilation". When the book became translated into Arabic the Greek "majisti" meaning "greatest" gave rise to the title "Al" (the) "magest" by which it is now known.

The *Almagest* is a *tour de force* of Greek astronomy, and was the greatest astronomical work of antiquity. Divided into 13 sections or "books", it is in essence a basic textbook, expecting its reader to know only the fundamentals of Greek geometry and some familiar astronomical terms; the rest Ptolemy explains.

The first two books form an introduction; they contain a description of the spherical universe with its spheres carrying the planets, and excellent arguments proving the Earth is stationary at its centre. The basic mathematics making use of chords — the Greeks had not developed the sines and tangents of true trigonometry — is described. None of this was new, but all of it basic to what was to follow.

Book III discusses the apparent motion of the Sun, using primarily observations by HIPPARCHUS, for whose work he shows so great a reverence that he seems to have selected only those of his own observations which agreed with them. A table of the Sun's motion is then constructed.

The motion of the Moon is the subject of Book IV and here Ptolemy also follows closely the work of Hipparchus. Based on eclipses, it describes well the Moon's motion at conjunction (New Moon) and at opposition (Full Moon). Yet because of the way Hipparchus derived it, the theory was inadequate at other positions. In the next section, Book V, Ptolemy developed his own lunar theory.

Based on the device of the EPICYCLE and DEFERENT it was extremely ingenious. His observations showed him that the Moon changed in apparent size, and he took account of this by a kind of crank mechanism operating on the centre of the Moon's epicycle. His theory now accounted for the Moon at all positions of its orbit, yet it did not disturb the model of Hipparchus where this was successful.

Having dealt with the motions of both Sun and Moon, Ptolemy next moved on to consider eclipses. These

P

form the subject of Book VI, and he explains how to calculate every detail about them.

Books VII and VIII are about the "fixed" stars. Most of the space here is devoted to tables giving their positions and brightness, though precession is mentioned.

The final books, IX to XIII, are concerned with the thorny problem of planetary motion. There were fewer observations for Ptolemy to use here, and very little past theoretical work to help him. The epicycle and deferent were not totally satisfactory in the planetary context, because on their own they could not account for variations in the apparent backward or retrograde movement of a planet. Nor could they show why such retrograde motion occurred at seemingly irregular intervals. Ptolemy was faced with a daunting task when it came to this.

Typically he tackled his problems in a truly scientific way. Using his own observations as well as those of Hipparchus and the Babylonians, he discovered that each planet had two irregularities or "anomalies". One depended on the planet's apparent position east or west of the Sun — its "elongation" — the other on the planet's position along the ecliptic (the Sun's apparent orbit in the sky). With this evidence Ptolemy had now to evolve suitable theoretical mechanisms still based on the Greek geocentric universe.

Careful study showed him that the epicycle and deferent would account for the first anomaly. For the second he found it best to make use of the "eccentric". Here the centre of the planet's epicycle moved around a point M which was distant from the Earth by the planet's apparent eccentricity. Yet this was not all. It was another basic condition of Greek planetary motion that each planet moved at an unvarying rate about the centre of the universe though, of course, observation showed that not one of them in fact did this. Ptolemy satisfied this condition by using another point, E, on the opposite side of the Earth to M, but lying equally distant. Uniform motion took place about this second point. This was a brilliant solution which, with an additional minor modification, explained the motions of all the planets then known.

In recent years Ptolemy has been accused of faking his results in the *Almagest* in order to square with his theories. This seems quite mistaken. He may (and did) set too much store by

those of Hipparchus, and he also had doubts about some of his own observations — he was well aware of deficiencies in the accuracy of his equipment. He was bound, therefore, to make some selection of his data. After all, his aim was to write a textbook giving mathematical theories which would account in general for the observed phenomena of the heavens. This he achieved with such success, that his work was not seriously challenged until COPERNICUS did so almost 1,400 years later.

The *Almagest* was not Ptolemy's only great text. He wrote the *Tetrabiblos* on astrology, much of it on natural astrology — the physical effects of the Sun and Moon, for instance. Then he produced *Planetary Tables*, which were extracted from the *Almagest*, and a popular abridgement called *Planetary Hypotheses* which, however, extends some of his theoretical ideas and in particular his measurements of the distances and sizes of the Sun and Moon. There was also his *Phases of the Fixed Stars*, which went into more detail about rising and setting of the stars just before dawn and just after sunset, and the *Analemma*, a book on constructing sundials.

In additional works, Ptolemy wrote on music, mechanics, geometry, and also what seems to have been an excellent book on optics, though the original is now lost and is available only in an Arabic translation. Lastly, there is his magnificent *Geography*, which lists the longitude and latitude of a great number of places, and gives brief descriptions of various land areas. In addition it contains instructions on how to prepare maps. In spite of using some unreliable data based on the reports of others, this book is nevertheless a great scientific achievement which was used by many later generations in the West. *CAR.*

Puiseux, Pierre (1855–1928). French astronomer, who collaborated with LOEWY in a major lunar atlas.

Pulkovo Observatory One of the oldest observatories in Russia; it lies near Leningrad (59°46′N, 30° 20′ E). It was completely destroyed by the Germans during the siege of Leningrad in World War II, but was subsequently rebuilt. However, the observational conditions there are poor, and the main Soviet telescopes and other instruments have been set up at more favourable sites.

Pulsars *See* colour essay.

Pulsating stars Stars which swell and shrink, such as CEPHEID and RR LYRÆ VARIABLES.

Puppis (The Poop — part of the dismantled Argo Navis). It is a southern constellation, most of which is invisible from Britain or the northern United States.

Brightest stars

Name	Visual Mag.	Abs. Mag.	Spec.	Distance (light-yrs)
ζ (Suhail Hadar)	2.25	−7.1	05.8	2400
π	2.70	−0.3	K5	130
ρ (Turais)	2.81	−2.0	F6	300
τ	2.93	0.2	K0	82
ν	3.17	−1.2	B8	245
σ	3.25	−0.5	K5	166
ξ (Asmidiske)	3.34	−4.5	G3	750
L²	3.4 max	−3.1v	A0p	75

Also above magnitude 4: c (3.59), a (3.73) and 3 (3.96). L² is semi-regular, with a range of from 3.4 to 6.2 and a period of around 140 days.

Puppis is rich, with several bright open clusters such as M46, 47 and 93, and NGC2423 and 2477. There is no well-marked pattern, but the constellation is easy to identify. *PM.*

Purbach, Georg von (1423–61). Austrian astronomer and student of Ptolemaic theory whose table of lunar eclipses, published in 1459, was still in use two centuries later.

Purbach Lunar crater, 75 miles (120km) in diameter; a trio with Regiomontanus and Walter; 25°S, 2°W.

Purple Mountain Observatory Also named Zijinshan Observatory or Nanjing Observatory. A multiple-research-ing astronomical institute under the leadership of The Academy of Sciences, China. Located at 32°04′N, 118°49′E, 876ft (267m) above sea level, on the third peak of The Purple Mountain, north-east of Nanjing.

Founded on September 1 1934, the observatory was seriously damaged during the Sino-Japanese war of 1937–45. It has forged rapidly ahead since the 1950s. Today, 11 departments are in being: planets and comets; almanac; practical astronomy; stellar physics; solar physics; movement of artificial satellite; radio astronomy; space astronomy; theoretical astronomy; astronomical instruments; and computation. There is also a section on the history of Chinese astronomy. The main instruments it is equipped with include: a 24in (60cm) reflector; a 16in (40cm) double refractor; a 17–24in (43–60cm) Schmidt telescope; a 4in (10cm) photoelectric transit instrument; a 5½in (14cm) Lyot-type chromospheric telescope, a 16in (40cm) horizontal helioscope and a millimetre wave radio telescope. A new site for the millimetre wave radio telescope is under construction at Delingha, Qinghai Province.

The observatory publishes a non-periodic journal *Reports on Scientific Researches* and contains the editorial offices of *Acta Astronomica Sinica* and the *Astronomical Bulletin* of The Chinese Astronomical Society.

The average number of astronomical clear nights at Nanjing is less than 100 a year, and observations are being affected more and more by the city lighting. *DPB.*

Pyrolytic Release Experiment Viking Lander experiment which sought evidence of carbon assimilation as a function of photosynthesis on Mars. The observed synthesis was considered non-biological.

Pythagoras (*c*580–*c*500BC). Greek mathematician, philosopher and religious leader. Besides his theorem on right-angled triangles, he taught that the Earth is a sphere, and that the planets orbit in circles.

Pythagoras: geocentric theorist

Pythagoras Lunar crater, 70 miles (113km) in diameter; 65°N, 65°E. It has high walls and central peak, but as seen from Earth is very foreshortened.

Pytheas (*fl.* 300BC). Astronomer, navigator and geographer, who voyaged round Spain, visited Britain and penetrated to the Arctic Circle, possibly reaching Iceland.

Pyxis (the Compass). Originally part of Argo. It adjoins Carina, and has no star above magnitude 3.6, but there is one object of special interest, the recurrent nova T. Pyxidis (1890, 1902, 1920, 1944 and 1966).

Q Cygni Bright nova of 1876, discovered by J.Schmidt. It reached magnitude 3.

QSO Any *q*uasi-*s*tellar *o*bject, including those now generally known as QUASARS.

Quadrans Rejected constellation, now included in Boötes; however, it is remembered in the name of the Quadrantid meteor shower.

Quadrantids. A major annual meteor shower which peaks on about January 3 or 4 every year. The name comes from the obsolete constellation Quadrans Muralis (the Mural Quadrant); today the radiant point of the Quadrantids lies within the constellation of Boötes, at R.A. 15h 28m, Dec. +50°.

The Quadrantids do not follow the orbit of any comet currently observable. The annual occurrence of the shower was first noticed in about January 1835 and data on the shower are available almost continuously from the 1860s. The Quadrantids certainly seem to be a fairly young meteor stream, having a large proportion of small particles concentrated within a narrow region. The stream's cross-sectional diameter is only about 1,250,000 miles (2 million km). Because the stream of meteoroids is so narrow, the activity exceeds one-quarter of the peak rate for only 19 hours. Probably only once in every ten years can the Quadrantid maximum be observed under ideal conditions — when the peak of the 19-hour period occurs between 0200 and 0600 hours, and there is no interference from moonlight.

Some spectacular Quadrantid showers have been observed, with peak rates exceeding 150 meteors per hour (m/h), but records show that there is considerable fluctuation in Quadrantid activity from year to year. The stream has its ascending node close to the orbit of the giant planet Jupiter, and so it suffers severe perturbations which can shift the orbits of the stream particles from side to side. These Jovian perturbations are probably the main influences governing the strength of any given Quadrantid display, as the Earth will sample a different part of the meteor stream each year. For some displays, maximum rates are between 40m/h and 80m/h, but on one occasion in 1977, when peak rates were estimated as 200m/h, twelve meteors were noted in one minute! Many of the Quadrantid meteors are faint, and they are often described as having thin silvery or even bluish trails.

For the faintest meteor trails observed telescopically, the Quadrantid peak occurs almost seven hours before visual maximum. In every case, including the radio meteoroids, the maximum for faint meteors occurs earlier than for bright. Over the brightness range between magnitudes +1 and +8, the Quadrantid peak occurs about 1¼ hours earlier for every magnitude step toward the fainter particles. It has also been found that the large visual meteoroids have longer orbital periods (5.2 years) than the fainter radio and telescopic ones (4.7 years). This dispersion of the masses by radiational perturbations (e.g. the Poynting-Robertson effect) must have occurred when the Quadrantids had a much lower orbital inclination than the value of 71°.4 which we find today.

Recent studies have shown that only 1,500 years ago, the orbital inclination was 12°. The mass dispersion took place at this time. Furthermore, the perihelion distance has steadily increased from only 8 million miles (13 million km), to the present-day value of 91 million miles (146 million km). As the perihelion distance of the stream continues to increase, it will again become impossible for the Earth to intersect it, and the Quadrantid meteor shower may cease to be with us about 500 years from now. The Quadrantids therefore present us with a golden opportunity to monitor the evolution of a meteor stream over a fairly short period of time. *JM.*

Quadrature The position of the Moon or a planet when at right angles to the Sun as seen from Earth. Thus the Moon is at quadrature when at half-phase.

Quantum gravity A unified theory of gravity, taking into account quantum

Q

Q

effects, currently sought by cosmologists.

The current "best buy" theory of gravity, General RELATIVITY is a geometrical theory in which free-fall under gravity is interpreted as motions along straight lines (geodesics) in a curved four-dimensional space–time universe. This is rather in the spirit of NEWTON's third law that matter continues in its state of rest or uniform motion unless acted on by a force. Our understanding of the other three physical forces is quite different: the electromagnetic and nuclear forces work by exchanging particles carrying momentum (and other properties) between the particles experiencing the force. Attempts to unify these theories have taken two generally distinct forms.

First, there is the application of geometrical theories to explain the non-gravitational forces. Either 10- or 26-dimensional spaces offer interesting possibilities to explain all four forces. One curious feature is the status of the extra dimensions over and above the three familiar dimensions of space and one of time. These theories postulate that space is tightly curved in the extra dimensions and closes on itself on imperceptibly small scales.

The second idea follows the traditions of particle physics. However, this approach requires particle theories to double the number of types of "elementary" particle. A new family of undetected particles is created by the theory and this seems philosophically unsatisfactory.

Whatever the outcome of these fascinating speculations, their predictions matter only when the universe was very young and much smaller than it is now (ie, during the first trillionth of a microsecond of the life of the universe). Thus they are not easily accessible to direct astronomical tests and their appeal relies on predictions such as the existence of matter, the overall homogeneity of the Universe and the existence of galaxies. *MVP. See also* GRAND UNIFIED THEORY.

Quantum mechanics *See* WAVE MECHANICS.

Quarks Hypothetical sub-nuclear particles which are believed to be the fundamental building block of hadrons (baryons and mesons). They have fractional electrical charges ($+\frac{2}{3}$ or $-\frac{1}{3}$, compared with the unit charge carried by a proton or an electron), and have spin values of $\frac{1}{2}$.

There are believed to be six types, or "flavours", of quark, designated "up", "down", "strange", "charm", "top" and "bottom". The up, charmed and top quark have charges of $+\frac{2}{3}$, while the down, strange and bottom quark have charges of $-\frac{1}{3}$. Up and down quarks have masses of 0.3MeV, the others are heavier. Each flavour of quark has an antiquark with opposite properties.

Each baryon is thought to consist of a cluster of three quarks (or antiquarks), while mesons consist of quark-anti-quark pairs. For example, a proton consists of two up quarks and one down quark ($u + u + d = +\frac{2}{3} +\frac{2}{3} -\frac{1}{3} =$ net charge of 1), a neutron consists of one up quark and two down quarks ($u + d + d = +\frac{2}{3} -\frac{1}{3} -\frac{1}{3} =$ net charge of zero), and a positive pi-meson consists of an up and an anti-down quark ($+\frac{2}{3} +\frac{1}{3} =$ net charge of 1).

Quarks also have a property called "colour", which is analogous to electrical charge and has nothing to do with colour in the conventional sense. There are three different colours (and three equivalent anticolours for antiquarks). Hadrons have zero net colour, and this can be produced either by combining three quarks, each of different colour, or by combining two quarks, one with the anticolour of the other. The requirement for zero net colour explains why baryons are composed of three quarks and mesons of two.

The interquark force is known as the colour force, and the interaction between quarks is carried by massless particles known as GLUONS. The strong nuclear interaction between hadrons themselves is believed to be a remnant of the interquark force. The colour force is weak when quarks are close together, but becomes extremely powerful as quarks move apart and prevents quarks from escaping from within hadrons. Almost certainly, therefore, quarks cannot exist outside hadrons as separate free particles. *IKN.*

Quarter Moon The Moon when at half phase — because we then see one-quarter of the entire surface.

Quartz-crystal clock *See* ASTRONOMICAL CLOCKS.

Quasars Quasi-stellar radio sources. The most luminous objects in the universe.

The discovery of quasars in 1963 came about when radio astronomers located a number of places in the sky from which intense radio emission emanated. A major thrust of optical astronomy around that time was to identify the corresponding visible object in the hope that standard optical techniques would reveal the origin and nature of the radio emission. The first optical observation of a quasar was made in 1960, but the spectrum was puzzling, and put in the "too hard" bin because of the uncertainty in the radio position of the relevant source (3C 48), which did not allow its firm identification with the optical object. In 1963, however, the Moon passed in front of source 273 in the third Cambridge catalogue of radio sources. Observations of this event were possible from the PARKES radio telescope in Australia (with suitable modification to permit the dish to point closer to the horizon than it was designed to do).

By timing exactly when the signal was blacked out and reappeared, a team led by Dr Cyril Hazard pinpointed the radio source precisely. A starlike object of twelfth magnitude coincided with the position, and its spectrum caused much upset. Far from being a star, the object had a huge REDSHIFT which, if interpreted conventionally, implied an enormous distance. Whole galaxies at that distance were known to be hundreds of times fainter. What, then, was this object, smaller than a galaxy, that put out so much radiation and was by far the most luminous known object in the universe? For 3C 273, and subsequent examples (including 3C 48), the term quasar has been introduced. It derives from quasi-stellar radio source, because the object resembles a star on optical photographs (quasi: Latin "as if"). We now know that not all such objects are radio sources. Purists like to call them quasi-stellar objects, or QSOs.

At a redshift of 0.158, 3C 273 is neither the nearest nor the most luminous quasar. In general, if a type of object is rare in the universe, it is unlikely to be found close to us. The most luminous quasars lie at very great distances, and hence appear extremely faint. To today's quasar astronomer, both the redshift and the luminosity of 3C 273 are small. The hunt is on for quasars of high redshift and at the time of writing the record just exceeds 4. When we assess the luminosities of these very distant quasars, we find values one thousand times that of a galaxy.

More striking than the luminosity itself is the small space from which it emanates. The radio output of quasars varies rapidly, sometimes changing in a matter of days. Closely related to quasars, the BL LACERTÆ OBJECTS (also called blazars, in a singularly ugly portmanteau) vary even faster. Now, visualize a spherical object of large radius. Let it brighten simultaneously over its entire surface in a short space of time. An observer will see the brightening protracted because of the longer time it takes light from the edges to reach him than light from the frontmost portion. The brightening will appear sudden only if it occurs first at the most distant parts and spreads toward the observer at the speed of light, a most unlikely situation. Hence, if a quasar varies in a day, its radius must be less than the distance light travels in a day, one light-day or 170 AU. This is truly minuscule compared to a galaxy, which is typically 100,000 light-years across.

Highly detailed radio imagery of some quasars has shown them to comprise two or more components which are separating with time. If quasars have the distances we infer from their redshifts, then the components appear to be separating at many times the speed of light, apparently in contradiction of Einstein's Theory of RELATIVITY. The effect is called superluminal motion; together with the impressively high luminosity it has encouraged some astronomers to doubt the distances inferred. They attribute the redshift to some other cause, or claim that quasars have been ejected from galaxies so as to have high velocity but ordinary distance.

The strongest evidence in favour of the distances inferred from the redshift is the discovery that apparently ordinary galaxies, themselves at great distances, lie between us and some quasars. These have been found in the case of GRAVITATIONAL LENSES, but more commonly as absorption lines in the quasar's spectrum due to light passing through the gas clouds of galaxies. Quasars, in fact, provide a means of studying the galaxy population of the universe at such enormous distances that the galaxies themselves are too faint to detect by their emitted light, a field of research which is becoming very important. For the nearest quasars a surrounding galaxy has also been seen directly, almost lost in the glare of its quasar, and the redshift of the galaxy has been shown to be the same as that of the quasar. Quasars, it appears, inhabit the centres of galaxies. At such, they appear to be the extreme end of the range of activity and luminosity we see more feebly in the nearby SEYFERT GALAXIES.

The best existing explanation for the luminosity of quasars postulates at the centre of the host galaxy a BLACK HOLE. If such an object exists, the core of a galaxy is a good bet for where to find it. Moreover, there it can grow by swallowing gas clouds and stars. Any material swallowed is likely to be decomposed into a whirling disk of gas

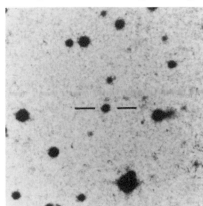

Quasars: universe's most luminous objects

as it falls in; this is called an accretion disk.

The energy available when gas is swallowed by a black hole is close to the maximum available according to Einstein's famous equation $E = mc^2$, or more than one thousand times the energy liberated in converting an equivalent mass of hydrogen to helium. A black hole accreting material steadily can be more luminous than a star for two reasons. First because of the factor of 1,000, and second because it can consume stars in vast numbers. The fastest that a black hole can swallow material is set by the EDDINGTON luminosity, which depends on its mass. The most luminous quasars would need black holes as massive as 100 million suns, or about 0.1 per cent of the mass of the host galaxy. They must swallow stars at a rate of about one every year, which in the crowded cores of galaxies is quite credible. Every star swallowed increases the mass by one star's worth; projecting back in time the black hole could have started quite small, perhaps at the death of a single massive star early in the galaxy's life.

Evidence in favour of this picture comes from optical and ultraviolet observations. Quasars are bright ultraviolet sources (which has provided one means of finding them), and it now appears that this hot radiation comes from the swirling accretion disk as it spirals to destruction. Also associated with the inner regions of quasar galaxies are emission lines of hydrogen,

helium and iron that show enormous spreads of velocity, indicating motions as rapid as 3,000 miles/s (5,000km/s), as might be expected near a massive black hole.

The superluminal expansion can also be explained. Here we must draw a parallel between quasars and RADIO GALAXIES, which may have similar central objects experiencing different environments. Radio galaxies emit jets of particles which cause outlying regions of radio emission. Quasars apparently also generate jets which produce additional components to the radio signal. The jets probably aim at right angles to the disk of accreting gas, and travel virtually at the speed of light. If we happen to look almost straight down a jet, the complex mathematics of relativity theory actually predicts the apparent superluminal separation of the end points. It also predicts that the central object will appear enormously brighter at radio wavelengths than one seen more nearly side on. Hence we can explain both why a surprising proportion of radio-emitting quasars have superluminal expansion and why only one quasar in 200 is a radio source. Those seen nearly end on to the jet will be the quasars with radio emission and also the ones for which superluminal motion is possible. Remarkable, however, is that the brightest quasar in the sky, 3C 273, should be one of these.

Because quasars are the most luminous objects in the universe, they can be seen to greater distances, and hence farther back in time, than anything else. Therefore they have afforded a means of examining the universe's youth. At redshifts near 4 we are seeing the universe as it was at about one-tenth of its present age. Despite many intensive searches for quasars of very high redshift, examples at redshifts greater than 3.5 are very rare. This appears to suggest that quasars were truly rare at those distant epochs. That may also indicate that galaxies were rare too — that we have found the time in the universe's evolution when galaxies began to form. Or it may simply mean that quasars did not afflict galaxies when they were very young. There is some evidence that quasar-like activity is possible within every galaxy (in the gas-free elliptical galaxies it may appear as a radio galaxy or BL Lacertæ object), and that at any given time the black hole is accreting matter in only a small proportion of galaxies. On the other hand, in some galaxies (including our own) the black hole, if it exists at all, may be so small that even at its most active it is completely swamped by the light of surrounding stars. *DAA.*

Quénisset, Ferdinand Jules (1872–1951). French astronomer. From 1906 until his death he worked at the Observatoire Camille Flammarion in Juvisy-sur-Orge, Essonne. He was best known for his photographic work and in 1934 he carried out some early experiments aimed at photographing the whole sky on one plate, a technique now known as all-sky photography.

Quetelet, Jacques Adolf (1796–1874) Belgian astronomer. In 1833, he was appointed Director of the Brussels Observatory. He is best remembered for his important work on meteor showers and their radiants. In 1839, he published a Catalogue des Principales Apparitions d'Étoiles Filantes, which listed 315 meteoric displays and for the first time called attention to the occurrence of the Perseid meteors on about August 9 every year. His catalogue and work formed an important basis for later workers in the field.

Quetzalcoatl Amor (Mars-crossing) asteroid 1915 with a librating orbit, its period being one-third that of Jupiter. A close approach to Mars occurs about every million years.

Radar astronomy A category of research in which astronomical bodies are investigated by radar techniques. Continuous wave or pulsed transmissions are made either from Earth or from space probes moving in close proximity to the object. Although paraboloidal or other forms of radio telescope are used, radar astronomy is distinguished from radio astronomy, which involves only the reception of radio waves having a natural origin in the universe. Because of the limits to transmitter power, telescope size and receiver sensitivity, investigations by radar are limited to the bodies of the Solar System.

The technique is almost entirely a post-World War II development, although for radar studies of meteors pre-war ionospheric workers can claim a degree of priority. The development

of radar during the war laid the foundations for its use as a research tool in astronomy. In September 1944, when Germany began the bombardment of

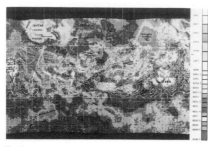

Radar astronomy: Orbiter's view of Venus

London by V2 rockets, J.S.Hey of the Army Operational Research Group (AORG) was asked to modify an anti-aircraft gun-laying radar in order to give early warning of the approach of a V2. By re-directing the aerial beam to an elevation of 60° radar echoes were obtained from the rockets, but many short-lived radar echoes were observed that had no connection with the rockets. Hey concluded that these transient echoes were associated with the entry of meteors into the upper atmosphere. After the cessation of hostilities military radar equipment was modified in order to study the meteoric phenomena by Hey's group at AORG, and by Lovell's group at Jodrell Bank. Similar techniques were soon in use in Canada and the United States. The radar waves are reflected from the long, thin columns of ionization formed at altitudes of about 60 miles (100km) when the meteoric body evaporates as it plunges into the Earth's atmosphere.

The technique enabled meteors to be observed without the hindrance of cloud or daylight and an important result was the discovery that major meteor showers were active in the summer daytime. Methods of measuring the velocity of the entry of meteors into the atmosphere were developed and a long-standing controversy about the origin of the sporadic meteors was settled. That is, they are moving in closed (elliptical) orbits around the Sun and not open (hyperbolic) orbits indicative of an origin in interstellar space, as many astronomers had maintained. Studies of the fluctuation and fading of the radar echoes from the meteor trails led to methods of measuring the wind systems in the 100km region of the atmosphere. As well as pulsed radar observations, the phenomena can be studied by using continuous wave transmissions. The reflections from the ionized meteor trails can be observed as

a "whistler" in a receiver tuned to a distant transmitting station. The whistler is an interference beat between the ground wave from the transmitter and the wave reflected from the ionized meteor trail. Many amateur astronomers studying meteors make use of commercial transmitters in the 10 to 20MHz band for these observations.

Whereas the radar studies of meteors can be made by using transmitters of low power, other work in radio astronomy requires powerful transmitters and high-gain aerial systems. After the end of World War II attempts were made to obtain radar echoes from the Moon. Success was first achieved by John H. De Witt of the US Army Signal Corps on January 10 1946 using modified army radar equipment. On February 6 1946 Z.Bay in Hungary also succeeded, again by using modified army radar equipment. Bay used an unconventional display system in which the radar waves scattered from the Moon were integrated for 30 minutes in a battery of water voltameters. Neither De Witt nor Bay achieved any significant scientific results with their Moon radar contacts, but in November 1947 F.J.Kerr and C.A.Shain of the Commonwealth Scientific and Industrial Research Organisation (CSIRO) in Australia investigated the erratic changes in the strength of the radar echoes scattered from the Moon. They made use of the Australian Government's shortwave transmitter normally used for broadcasting to North America and were able to distinguish two types of fading of the lunar echoes. A rapid fading with a period of seconds was superimposed on a longer period fading of about 30 minutes. They suggested that the rapid fading was associated with the libration of the Moon and that the long-period fading might have an ionospheric origin. These lunar radar phenomena were substantially clarified by scientists at Jodrell Bank, where lunar radar echoes were first obtained in 1949. Confirmation was obtained that the short-period fading was associated with the libration of the Moon and it was discovered that the long-period fading was caused by the rotation of the plane of polarization of the radar waves as they passed through the Earth's ionosphere (Faraday rotation). Further researches showed that the radar waves were not reflected uniformly from the whole disk of the Moon, but from a region at the centre of the visible disk which had a radius of only one-third of the lunar radius. This led to demonstrations of the possibility of a lunar voice-communication circuit. Another significant result of the Moon radar work has been the systematic measurement of the total electron content between the Earth and the Moon derived from an analysis of the longer period fading of the lunar echoes.

Using the high gain of a large steerable aerial, lunar radar echoes can be observed with relatively low-power transmitters. The attainment of echoes from the planets presents a far more formidable problem. At the closest approaches to Earth, Venus requires an overall improvement in sensitivity of a few million times, and about 100 million times for Mars. However, the scientific results to be obtained by a radar contact with Venus were so important that development of the necessary equipment was undertaken in the United States, the USSR and at Jodrell Bank. In particular, a radar measurement of the distance of the planet settled the ambiguities in the value of the solar parallax (hitherto known only to 0.1 per cent) and also the problem of the rate of rotation of the planet. Success was first achieved by NASA equipment at Goldstone on March 10 1961 using a continuous wave system. Pulsed radar contact was achieved at Jodrell Bank on April 8 soon followed by other successes in the United States and the USSR. From these distance measurements a definitive value of the solar parallax was agreed. The spin of the planet affects the spectrum of the scattered radar waves, and in 1962 and 1963 measurements in the USSR and the United States first established that the planet was in retrograde rotation with a period of 243 days.

In the 1960s the United States and the USSR developed powerful radars for the detection of missiles, and large radio telescopes with advanced computing techniques for the control and analysis of space probe data gave advantages for the pursuit of radar astronomy that could not be paralleled elsewhere. In 1962 the USSR achieved radar contact with Mercury; followed by the Jet Propulsion Laboratory in the United States in May 1963. In February 1963 radar contact was established with Mars both from the United States and the USSR. Procedures were developed for radar mapping of the planetary surfaces. In the case of Mercury and Mars, optical imaging from space probes soon provided superior data, but the radar mapping of the per-

petually cloud-covered surface of Venus has been of outstanding importance and radar maps obtained of the planet's surface have been made with a resolution of a few kilometres. To obtain this definition with the beam of a single telescope would require an aperture of several hundred kilometres even at the close approach of Venus to Earth. Equivalent apertures have been synthesized by measuring the amplitude and phase of the signal as the Earth moves the receiving antenna through the radiation scattered from the planet. Angular resolutions of a few milliarc-seconds have been obtained in one dimension by this means. Resolution in the second dimension is provided by measurement of the radar echo delay and ambiguities in the data have been resolved by using an interferometric method. Ground-based radar can map about one-quarter of the surface of the planet by this means.

In 1979 a computer analysis of several months of the data from the radar altimeter carried in the American Pioneer Venus Orbiter produced a topographical map of more than 90 per cent of the surface of Venus. The Soviet Venera 15 and 16 spacecraft placed in orbit around Venus in 1983 carried an APERTURE SYNTHESIS radar system that mapped the whole planetary surface above 30° north latitude with a resolution of about a mile (1–2km).

Other objects in the Solar System detected by radar include a number of asteroids, the four Galilean satellites of Jupiter, and Saturn's rings. Radar echoes were obtained from Comet Iras-Araki-Alcock by the JPL (Goldstone) and ARECIBO radars when it made a close approach to Earth (5 million miles/8 million km) on May 11 1983. This planetary radar work has been carried out by using radars operating in the centimetre waveband. Long wavelength radar (11.7m) was used at Stanford University in 1959 to obtain radar echoes from the Sun and similar work was carried out in the early 1960s by scientists of the MIT Lincoln Laboratory using a radar on a wavelength of 8m in Texas.

In 1964 I.I.Shapiro showed that radar astronomy could be used to make a fourth test of the General Theory of RELATIVITY. One of the classical tests was to measure the deflection of a ray of light from a star when passing close to the Sun. Shapiro pointed out that the General Theory also predicted that an electromagnetic wave should be retarded when grazing the solar disk because the speed of pro-

pagation depends on the gravitational potential along the path. For a radar pulse passing near to the Sun, and reflected from a planet, the time delay should be 2×10^{-4}s, equivalent to 40 miles (60km) difference in distance of the planet. The measurement was first made using the Haystack antenna of the Lincoln Laboratory in 1967 and more accurate results in accordance with the prediction of general relativity were subsequently obtained at the Lincoln Laboratory using radar pulses reflected from the planets Mars, Venus and Mercury and by the Arecibo radar using pulses reflected from Mercury and Venus. *BL.*

Radial motion *See* STELLAR MOTION.

Radial velocity Velocity directly toward (negative) or away from (positive) the observer. It may be determined from the DOPPLER EFFECT on radiation from an approaching or receding source.

Radian The angle subtended at the centre of a circle by an arc equal in length to the radius. It equals 57.296°, or 3438′, or 206265″. 2π radians $= 360°$.

Radiant A point on the CELESTIAL SPHERE, usually expressed by its coordinates in RIGHT ASCENSION and DECLINATION, from which a meteor appears to originate. All the members of a meteor shower seem to emanate from one small region of the sky — the shower radiant. Meteors will often start some distance from the radiant. Although the meteors in a particular shower are all actually travelling on nearly parallel paths, if their trails are extrapolated backward they will all intersect at the radiant. It is an effect of perspective that they appear to radiate out from one point. A meteor seen exactly on the radiant will appear as a brief, starlike point of light, with no apparent motion. Meteors have their maximum rate of streaking at an elongation of 90° from the radiant, when they are seen "sideways-on".

The radiant of a meteor shower moves a little eastwards (roughly one degree) every day, as the Earth travels through the meteor stream, in its annual path around the Sun. Meteor showers are normally named after the constellation in which the radiant lies. For example, the radiant of the April Lyrids is in Lyra, the August Perseids in Perseus, etc. *JWM.*

Radiation The study of radiant energy is fundamental to our understanding of

R

stellar structure, since it enables us to deduce the temperatures and luminosities of stars as well as their chemical constitutions.

In the late 17th century it was demonstrated that if a beam of white light is passed through a prism it forms a spectrum of its constituent individual colours and Christiaan Huygens put forward a wave model to account for this behaviour and also many of the other observed properties of light. Later developments of this theory by James Clerk Maxwell led to the representation of a ray of light as a series of transverse electric displacements which in turn generate a series of magnetic vibrations, the combined results being an electromagnetic wave which is propagated through a vacuum with a velocity c of 3×10^{10}cm/s. The distance between successive crests of these waves, the wavelength λ, is an indication of colour as also is the number of waves produced per second, the frequency v. These quantities are related so that $c = v\lambda$. As the eye responds to these various waves we perceive the short-wave, high-frequency rays as blue and the longer wave, lower frequencies as red in colour.

It was soon recognized that other "colours" exist beyond the red and blue ends of the spectrum for which the eye is unresponsive. We now realize that there is a wide-ranging electromagnetic spectrum running from very short gamma waves ($\lambda = 10^{-8}$cm) through X-rays, ultraviolet, visible, infrared, microwave radio and long-wave radio ($\lambda = $ tens of kms), and that all these are essentially the same phenomenon differing from one another only in their wavelengths or frequencies.

Since we are particularly interested in the relationship between the temperature of the surface of a hot body and the way that surface emits energy in the various parts of the electromagnetic spectrum, we must first consider what effect the nature of the surface itself might have.

It is well known that not all surfaces raised to the same temperature radiate energy in the same way. Some reflect well, but others may be highly absorbing. A perfect reflector will appear white (for diffuse reflection — not specular), but a totally absorbing surface will appear black and is referred to as a black-body. Such a body will also be a perfect radiator because the more energy it absorbs per second the hotter it becomes and the more energy it must radiate away per second. It is found

that, unlike reflection, where the reflected rays have the same colours as the incident rays, re-radiated energy is spread out over all wavelengths in a particular way, according to the laws of black-body radiation.

The experimental results were first obtained using a totally enclosed and blackened furnace by analyzing the radiation that emerged from a very small hole in its wall. This device approximated to a perfect black body and the intensity of the radiation was measured at different wavelengths. Distributions were obtained for differing furnace temperatures and the results plotted as a set of curves.

It was noticed that for each temperature there is a single wavelength corresponding to maximum intensity and this is at progressively shorter wavelengths as the temperature increases. Max Wien was able to formalize the relationship using classical physics as the expression $\lambda_{max}T = C$, where C is a constant and T is the absolute temperature of the furnace. This formula is known as Wien's Displacement Law.

Everyday experience shows that as objects become hotter their colours change from deep red through orange and yellow to white. They would eventually become blue — but melting and vaporization interfere with the experiment.

It can also be seen from the intensity–wavelength curves that the total energy radiated over all wavelengths at a particular temperature, ie, the area between the curve and the wavelength axis, increases steeply with increasing temperature. This was investigated experimentally by Stefan, and theoretically by Boltzmann, and led to the derivation of what is known as the Stefan-Boltzmann Law, $\varepsilon = \sigma T^4$. Here ε is the total energy radiated over all wavelengths and σ is the STEFAN CONSTANT. It should be noticed that this constant applies to unit area of radiating surface — a fact that is important in estimating the total surface areas of some stars.

Classical theory was totally unable to account for the shapes of the intensity–wavelength curves until, at the beginning of the present century, Max Planck introduced the ideas of the quantum theory that radiant energy behaves as if it is made of a beam of particles of energy, photons, rather than trains of waves. The relationship between the colours or wavelengths of the wave theory, and the sizes of the photons of the quantum theory, could be expressed as $E = h\lambda = hc/\lambda$, where E is the energy of the photon and h is a

universal constant known as PLANCK'S CONSTANT.

The final form of Planck's radiation law relates the intensity of emitted radiation at wavelength λ and absolute temperature T, $\varphi \lambda T$, to the fundamental constants c, h, k and e in the following expression: (k is Boltzmann's constant and e is the base of natural logarithms).

$$\varphi_{\lambda T} = \frac{2hc^2}{\lambda^5} \frac{1}{(e^{hc/k\lambda T} - 1)}$$

This rather complicated formula gives the amount of energy in each colour emitted per second from one square centimetre of a black body at T degrees above absolute zero.

The Wien and Stefan-Boltzmann laws can be applied, for example, to the HERTZSPRUNG-RUSSELL DIAGRAM to great advantage.

The H-R diagram is a plot of absolute magnitude of a star, or its luminosity, against its colour index and these quantities represent the total energy output of the star through its entire surface area, and the effective temperature of that surface. If, for instance, stars occur on the diagram at one particular colour but at several different luminosities, the laws tell us that this must be due entirely to differences in the surface areas and therefore diameters of these stars. The brilliant stars are literally giants and the fainter ones dwarfs.

All the radiation considered so far has been thermal radiation having a continuous spectrum and being emitted by hot incandescent solids or excited high-pressure gases. Under conditions of much lower pressures, where atoms behave as individuals and are relatively uninfluenced by their neighbours, emission can occur at discrete wavelengths giving rise to bright line spectra and, although the energy changes are also temperature dependent, the effects of the quantization of this energy are more obvious. In the visible region of the spectrum the colours emitted are determined by the sizes of the electron energy level jumps as described by the Planck relationship $E = hc/\lambda$ and the Bohr theory of the structure of the atom. As one moves from blue light toward ultraviolet and into X-rays the energy jumps become bigger over several energy levels corresponding to more energetic photons. In the gamma ray region the energy required is so great that the processes producing them take place inside the atomic nuclei and involve the reactions associated with radioactivity. It seems paradoxical that rays having the greater energy are the result of changes on the smaller scale.

Beyond the visible red into the infrared we have the weaker processes associated with the thermal vibrations of entire atoms and molecules and, to produce radio waves, we must deal with large-scale movements of electrical charges in magnetic fields or along wires. These non-thermal radiations tell us about the chemical nature of materials and the large-scale structures of stars and galaxies. *RCM.*

Radiation belts Regions of the magnetosphere of a planet that contain the energetic electrons and ionized nuclei trapped by the planet's magnetic field. These belts are generally doughnut-shaped regions. The particles which reside in these belts spiral along the magnetic field lines and travel backward and forward between the reflections that occur as they approach the magnetic poles. This motion produces the SYNCHROTRON RADIATION from the particles which is detected from the individual planetary systems. The particles are captured by the SOLAR WIND or they are formed by collisions between the COSMIC RAYS or ions in the planet's outer atmosphere.

The Earth, Jupiter, Saturn and Uranus are all known to possess radiation belts, while the magnetic field of Mercury is considered too weak to sustain

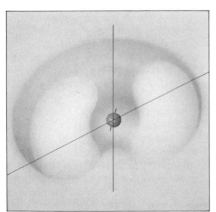

Radiation belts: Jupiter's are huge

this phenomenon. The radiation belts of Jupiter are huge in their spatial scale and about 10,000 times more intense than the belts that surround the Earth. They present a formidable obstacle for spacecraft venturing through this region, since the probe would receive an integrated dose of about 200,000 rads from the electrons and 50,000 rads from the protons in the region of the most intense radiation.

This may be compared with a dose of 500 rads, which is sufficient to kill a man. The intense bombardment of a spacecraft can saturate the sensitive instruments, upset the on-board computer systems and interfere with its communications with the Earth. These belts are a major hazard for the forthcoming GALILEO mission. The radiation belts of Saturn were not discovered until the encounter by the Pioneer 11 spacecraft in 1979, since the extensive system of rings and the satellites residing inside the magnetosphere sweep away the charged particles. However, Saturn does emit radio waves which are thought to be generated by synchrotron radiation and therefore evidence for the presence of radiation belts. The radiation belts around Uranus also have a strong interaction with the ring and satellite system, particularly in the neighbourhood of Miranda. We may anticipate a similar situation at Neptune. *GEH.*

Radiation era A hot period shortly after the origin of the universe when most of the energy existed as radiation rather than matter.

Radiation pressure Pressure exerted on a surface by incident ELECTROMAGNETIC RADIATION. This pressure is because each incident photon transfers a tiny quantity of momentum to the surface. One well-known observed effect concerns the gas or ion tails of comets, which are affected by the Sun's radiation pressure and therefore always point more or less away from the Sun — so that when moving outward, after perihelion, the ion tail of a comet more or less precedes the nucleus. For a fuller explanation of this phenomenon, *see* COMETS.

Radiation zones Regions or zones surrounding a planetary magnetosphere containing a population of trapped charged particles (*see* RADIATION BELTS, VAN ALLEN BELTS). These regions are found around the planets Mercury, Earth, Jupiter, Saturn, Uranus and Neptune.

Radiative diffusion The most important process by which heat energy is transported from a star's hot interior to its surface. It is known as radiative transport. In radiative diffusion, high-energy photons lose energy as they travel outward through the hot plasma. This loss of energy occurs as the photons are scattered, mainly by free electrons. The photons can also lose

energy if they are absorbed by an ion, and photoionization occurs. These processes take place in the radiative zones of the stars.

Radioactivity Some atoms of certain elements, which seem chemically the same, have a different mass. These atoms are called isotopes of the element. Many isotopes have unstable nuclei. This means that they eject energetic particles as the unstable nucleus spontaneously disintegrates into a more stable form. The process is called radioactive decay or radioactivity, and the energetic particles emitted are termed radiation. Isotopes which decay by this means are known as radioisotopes. Their rates of decay are unaffected by chemical changes, pressure, temperature or electro-magnetic fields. Each nuclide (nucleus of a particular isotope) has a characteristic rate of decay, called its half-life. This is the time taken for half the atoms in the isotope to break up. If the half-life is very long, then the isotope changes slowly. An example is Uranium 238 which has a half-life of 4,500 million years. If the half-life is short, then the decay of the radio-isotope is rapid, which means that it is very unstable. An example is Radon 222 which has a half-life of only 3.8 days. In 1902, the physicists Rutherford and Soddy suggested that a radioactive nuclide decays to a further radioactive nuclide, a series of transformations taking place, which ends with the formation of a stable "daughter" nucleus. For radioactive elements having high atomic mass number (A), three decay series exist. These are the uranium, actinium and thorium series.

Some radioisotopes such as uranium emit energetic particles all the time. It was a piece of material containing uranium which led Henri Becquerel to discover natural radioactivity in 1896. Three different types of radiation are given out by radioisotopes. The three types are called alpha particles, beta particles and gamma rays. Each of these forms of radiation is able to penetrate solid objects to a greater or lesser extent. Alpha particles are positively charged and consist of two protons and two neutrons (a helium nucleus). They are given off by radioisotopes which have too many protons to be stable. They travel at about one-twentieth the speed of light. Alpha particles are stopped by a thickness of paper or a thin piece of aluminium foil. Beta particles have a negative charge. They are very fast moving electrons, which may

R

be thrown out from the nucleus with a speed approaching that of light. They can pass through about 2mm of aluminium. Beta particles are given off by elements which have too many neutrons to be stable. Gamma rays are often emitted as well as alpha and beta particles. They travel at the speed of light, carry no charge, and have great penetrating power. A strong source of gamma rays such as a nuclear reactor needs a lead screen at least 150mm thick for good protection. *JWM*.

Radio astronomy The words "radio astronomy" were first used in the late 1940s to describe the new method of studying the universe by means of the reception and interpretation of the radio waves reaching Earth from space. The initial discovery that radio waves from outside the Solar System were detectable on Earth was made in 1931 by Karl JANSKY. This discovery was not thought to be of great significance to the development of astronomy until after World War II. Then, when the new techniques that had been developed during the war were used to study these radio emissions, it was appreciated that non-thermal processes must be operating in the Sun and in more distant regions of the universe. The scope of the observations and their theoretical interpretation grew with great rapidity when it was realized that significant information about the universe could be obtained in a region of the spectrum other than that available through the use of conventional optical telescopes observing through the relatively narrow spectral region from the violet to the red.

In the early years of the development of radio astronomy equipment in the metre waveband was most frequently used. With the availability of receiving equipment sensitive at shorter wavelengths, the observations were extended into the centimetre waveband. Ultimately observations in the millimetre waveband became possible. At sea-level sites observations down to wavelengths of 2cm or less can be made without serious interference from atmospheric conditions. At shorter wavelengths it is necessary to observe from high mountain sites in order to avoid atmospheric absorption — mainly by water vapour. At the long-wave end difficulties arise through absorption and scattering of the radio waves in the ionosphere. These ionospheric hindrances may occur in the 10-20m waveband region, depending on the ionospheric conditions and the

phase of the sunspot cycle. Between these extreme limits, set by the atmosphere in the low centimetre region and by the ionosphere in the long-wave metre waveband, the radio emissions from the universe are extensively observed. Given the appropriate receiving equipment, the main difficulty arises from man-made interference because of the widespread use of the radio waveband for communication and other purposes. International agreements are in force which restrict certain wavebands for the sole use of radio astronomy.

The receiving aerial is known as a RADIO TELESCOPE. The most conventional form is the steerable paraboloid. A cardinal item in the development of radio astronomy has been that of resolving power. Because wavelengths of reception in the metre waveband are of the order of a million times greater than that of light, it is difficult to achieve the resolution comparable to

Radio astronomy: Owens Valley, California

that of the human eye with a single radio telescope. For example, the 250ft (76.2m) Mark I paraboloid at Jodrell Bank, working on a wavelength of one metre, has a resolving power of a little under one degree of arc, which is about twenty times inferior to the few minutes of arc resolving power of the human eye. However, when the paraboloidal surface is accurate enough (deviations not greater than one-twentieth of a wavelength) then telescopes of this size operating in the low-centimetre waveband can achieve resolutions equivalent to that of the unaided eye.

Even that resolving power is some 200 times inferior to that of the large optical telescopes used under good see-

ing conditions, and the attainment of comparable resolution in the radio waveband stimulated the development of special techniques. Initially these consisted of two separate aerials or radio telescopes connected by cable to a common receiver to form a RADIO INTERFEROMETER. In one dimension a resolving power equivalent to that of a single telescope with an aperture equal to the distance between the aerials can be achieved. Using radio links between the sites with appropriate phase control, resolving power equivalent to that of a single aperture of a hundred or more kilometres can be obtained. Linkage of telescopes over distances of thousands of kilometres involving tape recording and rubidium frequency standards at the separate sites was first achieved by Canadian scientists in 1967.

The culmination of this interferometric technique was the development of APERTURE SYNTHESIS using two or more spaced radio telescopes. If the phase of the signals is preserved, then appropriate computer analysis can be used to combine the individual sets of observations to synthesize a two-dimensional radio image equivalent to that which would be obtained by a radio telescope having an aperture equal to the separation of the telescopes. This principle of aperture synthesis was first demonstrated by Sir Martin Ryle and his colleagues in Cambridge, England, in the 1950s.

The method was developed in Cambridge and elsewhere to produce two-dimensional radio maps with arc second resolution. Modern versions of the aperture synthesis system, represented, for example, by the MERLIN array at Jodrell Bank and the VERY LARGE ARRAY in New Mexico, are used to produce radio maps with a resolution of a few hundredths of seconds of arc. Thus within a period of about 30 years from 1950 to 1980 the resolving power of telescopes working in the radio waveband was transformed from those having a resolving power much inferior to that of the human eye to systems giving a resolution from ten to a hundred times superior to that of the large optical telescopes used under the best possible seeing conditions.

These developments in radio telescopes have been accompanied by advances in receiver design and data analysis giving very large improvements in sensitivity. For example, in the immediate post-war work sensitivity levels of several Jansky (1 Jansky = 10^{-26} Wm^{-2} hz^{-1}) were considered reasonable. By the 1980s sensitivity limits of 10 to 20 micro-Jansky were achieved — an improvement of the order of 100,000 times.

These developments have been associated with, and have been stimulated by, a sequence of remarkable discoveries about the nature of the universe. The pre-war measurements of Jansky and of Grote Reber led to the conclusion that the radio waves originated from FREE-FREE TRANSITIONS in the interstellar space of the Milky Way (the motion of an electron close to a proton without capture). The immediate post-war observations showed that the spectrum of the emission was incompatible with this process, and the discovery that there were localized (discrete) sources of radio emission led to the belief that a hitherto unknown population of dark radio-emitting stars existed in the Milky Way with a distribution similar to that of the visible stars. During the 1950–60 era this belief was overtaken by two discoveries of great significance. One was the development by Soviet physicists of the theory of synchrotron emission in the Milky Way. It was shown that the motion of relativistic electrons in the galactic magnetic field would produce radio emission such that the difficulty in the observed spectrum was removed. The other discovery concerned the nature of the localized sources of emission. Although localized sources existed in the Milky Way (particularly supernova remnants such as the Crab Nebula) collaboration with the astronomers using the 200in (5m) optical telescope on Palomar led to the surprising conclusion that the majority of the sources were distant extragalactic objects. These became known as RADIO GALAXIES — objects in which the strength of the radio emission exceeded by a large factor the integrated radio emission of that of a normal spiral galaxy such as the Milky Way or M31.

Since the radio galaxies were strong radio emitters, but faint optical objects, radio astronomy became an important avenue through which to extend astronomical observations to more distant regions of the universe. An important stage was reached in 1960 when, using the 200in telescope, Minkowski identified one of these radio galaxies with a galaxy having a REDSHIFT of 0.46 — the most distant object then known, then assessed to be 4.5×10^9 light-years away with a recessional velocity of 40 per cent of the velocity of light. This discovery enhanced the belief that the radio astronomical obser-

vations were of cosmological significance and in this connection researches of great importance followed in the 1960s. The attempts to find more distant radio galaxies led late in 1960 to the identification with the 200in Palomar telescope of hitherto unrecognized types of object. Initially these were believed to be stars in the Milky Way with peculiar properties. In 1963 when the redshifts were determined they were found, on the contrary, to be extragalactic objects more distant than any radio galaxy. These are known as QUASARS and by 1986 quasars with redshifts of $z = 4.01$ had been identified. The impact on astronomy of the discovery of radio galaxies and quasars is indicated by the statistic that there are 50,000 extragalactic objects in the optical catalogues, and 20,000 in the radio catalogues, of which only one half have so far been identified optically.

The detailed mapping of the radio galaxies and quasars with the milliarc-second resolution of the modern synthesis arrays has revealed the complexity of the processes. In many of the cases where a visible object has been identified, the radio emission arises from "lobes" separated from the object. In others the emission is core-dominated and in many cases jets of radio emission appear to be projected into space. Immense energies are involved in the central sources and jets, in some cases corresponding to the energy of the total annihilation of a million solar masses. The relatively small volume of space in which this energy is produced has led to speculations that BLACK HOLES may exist in these objects.

The realization that the majority of the localized sources observed by the radio telescopes were distant extragalactic objects led the Cambridge radio astronomers to study the relation between the number of such objects and their flux density. These number counts were decisively in favour of an evolutionary universe and could not be reconciled with the STEADY STATE cosmology then in vogue. In 1965 an even more decisive argument emerged when A.A.Penzias and R.W.Wilson announced that they had discovered radiation in the centimetre waveband isotropically distributed over the sky (*see* MICROWAVE BACKGROUND). The interpretation that this 3K microwave radiation is the red-shifted emission from an initial high-temperature phase of the universe has not been seriously challenged.

The radio emissions from the universe so far mentioned are continuum emissions, that is, given the appropriate equipment, the radio emission can be measured over a wide range of the spectrum, and indeed the nature of the spectrum provides important information about the origin of the radiation. These emissions may, in some cases, show variations in strength with time, but have a short-term continuity. It was, therefore, a matter of great surprise when, early in 1968, the Cambridge radio astronomers announced that they had discovered pulsating radio sources. The initial discovery was of a source emitting a pulse of radio waves every 1.337 seconds with a precision better than one part in 10^7. The investigation of these objects, known as PULSARS, developed rapidly at many radio observatories. By 1986 nearly 400 had been discovered, and although there are very few optical identifications it is widely accepted that they are NEUTRON STARS — the collapsed very high-density remnants of stars that have undergone a supernova explosion. The extensive study of pulsars is directed to the attempts to understand the mechanism of emission. Also by measurement of the dispersion of the radio waves, the electron density in the interstellar medium has been determined.

In 1951 the prediction made by Van de Hulst in 1945 that it should be possible to observe a spectral line from the neutral hydrogen in interstellar space was fulfilled by the discovery of a spectral line on a wavelength of 21cm. This corresponds to a transition between two closely-spaced energy levels in the hyperfine structure of the ground state of neutral hydrogen. Since the frequency of emission is precisely determined, measurement of the received frequency has provided a most powerful method for the study of the motion of neutral hydrogen clouds in the Milky Way — and indeed gave the first unambiguous determination of the spiral structure of the Galaxy.

Neutral hydrogen was already recognized as an important constituent of the Galaxy before the detection of the 21cm spectral line emission. The first interstellar molecule to be discovered at radio wavelengths was the hydroxyl radical (OH) in 1965 at a wavelength of 18cm. There are four components of this line (at frequencies of 1612, 1665, 1667 and 1720 MHz). Although in some dark galactic clouds these lines can be observed with the expected intensities, in other regions

R

R

one of the components may be a million times enhanced in strength. This amplification is considered to occur by means of a MASER process. The study of the complex behaviour of the lines from the OH radical is particularly important. There appears to be two main types of OH maser sources, those associated with late-type stars and those associated with regions of ionized hydrogen. Altogether if isotopes are included, spectral lines have been observed in the radio spectrum from over 100 molecules. It is the study of these molecular lines (the chemistry of space) that has stimulated the extension of the radio observations into the centimetre and millimetre spectral band.

Radio astronomy has revealed objects and features of the universe unknown through optical observations. It has also made significant contributions to well-known optical objects, such as the Sun and nearby stars, and has revealed the complex chemical composition of the interstellar clouds formerly thought to be composed solely of hydrogen. Much has been accomplished in the four decades since the end of World War II, but no one knows whether the sequence of spectacular discoveries will continue. The possibility of combining a radio telescope in Earth orbit with terrestrial telescopes (the Quasat project) would overcome the terrestrial limitations on the resolving power of the synthesis arrays and this is, perhaps, the major advance in the future that might again lead to further revolutionary discoveries. *BL. See also* colour essay on INTERSTELLAR MATTER.

Radio galaxies Radio galaxies are galaxies that emit radio radiation at rates around a million times stronger than the weak emissions from galaxies such as our own. Many have been found to have optical counterparts, which are almost invariably of the elliptical type. In our own Galaxy, the Milky Way, the radio radiation is believed to come mainly from supernovæ remnants and star formation zones. These remnants and zones occupy the same regions as the visible stars, and so the radio emission is generally found to be contained within the optical disk of the galaxy.

In contrast, the emission from a typical radio galaxy is concentrated in two huge radio "lobes" lying well outside the galaxy and frequently located symmetrically about it. In appearance these lobes look as though they have been explosively ejected from the central galaxy; and such may well be the case.

Lobe sizes for the biggest radio galaxies are around 5,000 kiloparsecs, comparable with the size of a typical cluster of galaxies. Such giant galaxies are the largest known objects in the universe. By way of comparison, our own Galaxy is a mere 20 kiloparsecs across.

Several radio surveys have been made to detect emission from "ordinary" galaxies similar to our own. They have found that very few galaxies emit detectable levels of radio energy: those that do are predominantly close-by bright objects. Furthermore they are almost invariably large galaxies of the spiral type. It therefore comes as a surprise to find dim, ELLIPTICAL GALAXIES listed as some of the strongest and most frequently occurring sources in radio catalogues.

The study of radio galaxies started in the 1950s with the measurement of an accurate position for the radio source

Radio galaxies: Cygnus A

CYGNUS A and its subsequent identification with a sixteenth magnitude peculiar galaxy (REDSHIFT 0.057). At first this galaxy was thought to be actually two galaxies in collision, but it was later realized that its double appearance was caused by a dark dust lane of absorbtion across the galaxy's centre. In fact, it was many years later before truly colliding — or "interacting" — galaxies were discovered. When they were, they were found to be only weak radio sources.

Following shortly after the identification of Cygnus A, the strong southern hemisphere radio source CENTAURUS A was identified with the bright, dusty, elliptical galaxy NGC5128. This object lies only about

4Mpc from our Galaxy, much closer than Cygnus A, and is one of the best studied of all radio galaxies.

Closer inspection of the structure of radio galaxies has been made possible by observations with APERTURE SYNTHESIS arrays. Radio astronomers have found that the extended radio lobes are almost invariably connected to the central galaxy by jets which seem to originate from the galactic nucleus. Yet again, high-frequency, Very long baseline interferometry (VLBI) observations have revealed strong, point-like, active radio sources in the nucleus of many objects.

We do not yet understand why some otherwise ordinary elliptical galaxies should emit such enormous quantities of energy in the radio part of the spectrum: so much, in fact, that it often dominates the energy we see in the visible portion of the spectrum. It seems clear that the basic "engine" that drives the explosion lies in the galaxy's nucleus. And in this regard, the radio galaxies are similar to the yet-more-energetic radio QUASARS and BL LACERTÆ OBJECTS.

Nor do we understand what this "central engine" is that produces the radio lobes. Certainly it must be very powerful: a conservative estimate suggests a total lobe energy of greater than 10^{60} ergs (10^{53} Joules) in the more violent examples. To produce this sort of energy would require the complete destruction of hundreds of thousands of stars with masses similar to that of the Sun. Furthermore it is possible that such explosions may occur more than once during the life of the galaxy.

We do know, however, that radio galaxies emit radio energy by the synchrotron process — so named after a similar process that occurs in terrestrial accelerators. The central "engine" in the galaxy's nucleus produces vast quantities of energetic electrons travelling at speeds very near to the speed of light (300,000km/s). When these electrons encounter a magnetic field, they spiral around and, in so doing, lose energy. It is this energy that we see as radio waves.

Although each spiralling electron emits energy over a range of radio frequencies, the faster, more energetic electrons emit preferentially at the higher frequencies. But the more energetic particles lose energy very much faster than their slower counterparts. Thus we find that there are far more low-energy electrons than high-energy electrons in the galaxy at any one time. This means that there is an excess of low-frequency radio emission compared to high-frequency emission.

The radio astronomer's measure of radiation from a radio source at any frequency is its flux density, normally given the symbol S. The unit of flux density is the Jansky — so-named after the first radio astronomer, Karl JANSKY. One Jansky corresponds to a minute amount of energy: 10^{-26} watts per square metre per Hertz of receiver bandwidth.

A typical radio galaxy would have a flux density of around 1 Jansky at a frequency of 1,400 MHz. On the other hand, the strongest objects, such as Cygnus A, range up to several thousand Janskys at the lower frequencies.

The way in which the flux density of a radio source varies with frequency nu (v) can be approximated by defining a spectral index alpha (α) so that:
$$S \propto v^{\alpha}$$
(Some early writers wrote minus alpha, but this practice is dying out.)

Radio galaxies have spectral indices which range from about $-\frac{1}{2}$ to about -3. In general, because of what we said earlier, we expect the older galaxies to have the more negative, or "steeper", spectral indices. By way of contrast, many quasars have values of alpha well above $-\frac{1}{2}$, showing them to be younger, more active objects.

Because of their spectral shape, with more flux at lower frequencies, radio galaxies have been found preferentially in low-frequency radio surveys. For example, in the famous Third Cambridge (3C) survey, about 70 per cent of sources are found to be radio galaxies. On the other hand, in surveys made at higher frequency, such as the Parkes 2,700 MHz survey, less than 50 per cent are believed to be radio galaxies.

Radio galaxies pose many as yet unanswered problems: why is the strong double-lobed structure only found in elliptical and not spiral galaxies? Why are the strongest radio ellipticals very often the brightest, central galaxies in clusters? A prominent example is the well-known Virgo A. What is the relation of the radio galaxies — if any — to the equally strong radio quasars and BL Lacertæ objects? What causes some otherwise ordinary elliptical galaxies to suffer a radio outburst in the first place? Do all galaxies pass through a strongly radio-emitting period some time in their evolution or is there an unknown "X-factor" that operates only in some objects? *AEW*[2].

Radio interferometer Name given to two (or a few) radio dishes separated by a distance comparable to, or greater than, their diameters and whose signals are combined electronically. Radio interferometers were developed in order to permit radio telescopes to attain much higher resolution than was possible with a single dish. Single-dish telescopes typically have beamwidths of around one minute of arc. This means that they can only resolve detail in a radio source on angular sizes above this value and measure positions to a few tens of seconds of arc. For investigating detail in radio sources on angular scales of a second of arc, and measuring similarly accurate positions, single-dish telescopes are inadequate.

The basic principles of the radio interferometer are quite easy to grasp: two radio dishes, separated perhaps by a kilometre or so will, in general, receive radio waves from a radio source at different times. This follows because the waves have to travel farther to one of the telescopes than the other. The important principle of the interferometer is that this delay is slightly different for different, but adjacent, points on the sky.

If the signals from the two separate dishes are combined correctly, they will either reinforce or cancel, depending on the amount of the delay: when the signals are exactly in phase (that is either have no delay or a delay corresponding to an exact whole-number of wavelengths) then a maximum combined signal response will be obtained. When the signals differ by an odd half-wavelength then they cancel, producing a minimum response. In principle, the size of the combined signal can be used to pinpoint the source on the sky.

It is sometimes helpful to think of the two-element radio interferometer as having a set of beams, or "lobes", arrayed across the sky and fanned out rather like the pages of an open book. When a lobe points at a radio source, the combined signal is a maximum. When the source lies between two lobes, the signal is a minimum. If the two dishes of an interferometer are arranged along an east-west line, a radio source moving across the sky will produce a series of alternating maxima and minima signals.

In practice, however, it is often difficult to decide whether the delay between the signals received at the two dishes is zero or a whole number of wavelengths. Alternatively, we can say we do not know which lobe our radio source is in. But these "lobe-ambiguities" as they are called can often be removed by observing with the dishes at different separations. For this reason many radio interferometer dishes used in the past were mounted on railway track to permit changing the spacings.

The main practical problem in constructing a successful interferometer is to ensure that the individual signals received remain exactly at the delay — or phase — differences they have when received at the aerials: any errors caused by electrical effects in the joining cables will produce errors of position on the sky.

Very long baseline (recording) interferometry (VBLI) uses no connecting cables, but a similar problem still exists. Signals from the separate dishes are recorded on high-quality video tape recorders. To combine the signals, separate tape recorders have to be synchronized at playback time to a very high accuracy. This requires very accurate time signals to be recorded on a time-track along with the sky signals. The "clocks" used are frequently hydrogen maser frequency standards and have an accuracy of better than a few seconds per century.

Even if the interferometer cannot be made "phase-stable" in the above sense, it is still possible to obtain valuable information about the radio sources being studied. If a source is of large angular extent — perhaps a galaxy or HII REGION — then different parts of it will arrive at the telescope at different times. This will smooth out the maxima and minima in the combined signal and eventually — if the source is large enough — produce no variations as the source passes through the interferometer response pattern. The source is said to have been resolved by the interferometer. Thus the smoothness of response — as compared with that to a point source — provides information about the angular size of the source, even if it is not possible to measure an accurate position for it.

If the spacing — or baseline — between the dishes can be changed, then the source may well be fully resolved on the longer baselines but not at all on the shortest baseline. When the maximum combined signal is plotted against baseline spacing, we obtain what radio astronomers call a "visibility function". From the visibility function it is often possible to make a fair estimate of the structure of a radio source. However, the results are rarely as good as can be obtained from a proper map made with an Earth-rotation APERTURE SYNTHESIS array. Using the visibility-function technique,

R

R

many peculiar stars which radiate abnormally at radio frequencies have been shown to be surrounded by extended clouds of gas.

In the normal type of two-dish radio interferometer — also called the Michelson interferometer — the voltage levels received at each dish are not added but instead multiplied. In this case, the power output of the combined signal is proportional to the product of the individual voltages and also to the product of the individual dish diameters. It follows that a small dish can be very effective as an interferometer when working with a second dish of large collecting area.

The first radio interferometers were built in the early 1950s and were used to measure the sizes and positions for the strong sources then known. Such

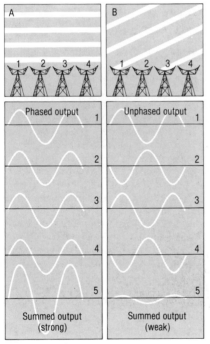

Radio interferometer: signal sizes

objects as the peculiar galaxy CYGNUS A were shown to be emitting vast quantities of radio energy from small regions. The identification of the strong radio source Taurus A with the CRAB NEBULA supernova remnant was only possible after an accurate position had been measured with the Dover Heights Sea Interferometer near Sydney in Australia. This instrument, in fact, used only *one* aerial perched on the top of a cliff overlooking the sea. The second aerial was formed by the reflection of the first in the water!

During the last 10 years or so, interferometers using two or three dishes have been superseded to a large extent

by the multi-dish, Earth-rotation, aperture-synthesis arrays, such as the VERY LARGE ARRAY located in New Mexico. These arrays are now used almost exclusively for accurate position measurement and mapping of radio sources down to scale sizes of an arc second or so. However, only very long-baseline interferometry can provide maps with structure on angular scales of a milliarcsecond or so. In the future most radio interferometry will probably be performed over distances spanning continents or even — as proposed — between Earth-based and space telescopes. Furthermore, VLBI will probably be used for astrometric work in the next few years. *AEW².*

Radio scintillation Irregular rapid changes in the apparent intensity of radio sources, caused by variations in the electron density in the ionosphere (mainly the F2 layer) and in the interplanetary gas. It occurs only with sources of less than 1″ diameter. *See also* SCINTARS.

Radio telescopes Instruments used to collect and measure ELECTROMAGNETIC RADIATION in the radio region of the spectrum emitted by astronomical bodies. Typical frequencies of operation range from about 10 megahertz (MHz) to 100,000 MHz or, equivalently, from wavelengths of several tens of metres down to a few millimetres.

Almost all the types of object studied with optical telescopes have also been observed with radio telescopes. These include the Sun, planets, stars, gaseous nebulæ and galaxies. Furthermore RADIO ASTRONOMY has been responsible for the discovery of several new and unsuspected types of astronomical phenomena, such as the QUASARS, PULSARS and the cosmic MICROWAVE BACKGROUND.

The first radio telescope was that built by Karl JANSKY in the early 1930s. With it he discovered radio noise coming from the Milky Way and the science of radio astronomy was born. However, lack of interest and the intervention of World War II caused Jansky's discoveries to be largely ignored at the time.

Following the war there was a glut of cheap (or free) radar and communications equipment. Using this equipment, experimental telescopes were built, mainly in Australia and England. Unexpectedly, many astronomical bodies were found to emit large amounts of natural radio energy. Since then increasingly sophisticated radio

telescopes have been constructed in order to study these radiations.

The basic radio instrument is the large single "dish" or parabolic reflecting telescope which can be directed at any part of the sky. It is the general-purpose work-horse of radio astronomy, capable of operating at a wide range of frequencies. Two or more such dishes can be linked together to form a RADIO INTERFEROMETER, or the energy collected by many dishes can be combined to form an Earth-rotation APERTURE SYNTHESIS array.

The single parabolic radio reflecting telescope is similar in many ways to its optical astronomy counterpart. Radio waves arriving at all parts of the dish are reflected in phase and (ideally) to a single focus. The smoothness and accuracy of the reflecting surface limit the highest frequencies (shortest wavelengths) at which the telescope can be used. Typically the dish is useless when the irregularities are a moderate fraction of a wavelength. However, this means that for the longer wavelengths — say around 21cm — it is perfectly adequate to use open wire mesh as the reflecting material rather than polished sheets of metal. Furthermore, the use of mesh has the advantage of reducing wind-loading on the dish.

It is important to realize that a single-dish telescope does not produce a "radio photograph" of a source — at least, not directly. The dish captures radio energy only from directions lying in a narrow "beam" along the dish axis. To build up a picture of a radio source, the dish must be scanned backward and forward and the resulting signals processed in a computer.

The twin goals that must be attained by any good radio telescope are high sensitivity and high resolution. Sensitivity is the dish's ability to collect the maximum possible energy in a given time, while resolving power is its ability to see fine detail in a radio source, that is, produce a narrow beam.

The largest existing single dishes are already very sensitive. The Bonn 328ft (100m) telescope, when looking at an averagely strong source and at a typical frequency, collects radio energy at a rate of only one hundred-million-millionth of a watt. Even added up over many years, this is still less than the energy expended by a flea making a single jump!

When the radio waves arrive at the focal plane, they are collected by a feed horn, which guides them to a dipole element, where they are converted into electrical signals. In a typical telescope, the signals are then amplified by a front-end receiver before being passed along cables to the main control room of the telescope where there is further amplification and, usually, analysis by computer.

The control system of a single-dish telescope has the job of accurately directing a dish, maybe weighing 1,000 tonnes, to any part of the sky. Radio telescopes are almost always exposed to the elements and typical specifications dictate that this enormous weight of steel must be kept accurately pointed at a radio source even under strong wind conditions. The accuracy required is often better than 10 seconds of arc which becomes millimetres when translated into dish movements.

Radio telescopes are normally built with altitude-azimuth mountings (see ALTAZIMUTH MOUNTINGS). That is, their basic movements are up-down and around parallel to the horizon. Occasionally, radio dishes have been built on EQUATORIAL MOUNTINGS, similar to those used for most optical telescopes, but these have not proved popular. The alt-az mount, as it is known, presents many engineering advantages, particularly with the advent of powerful computer control systems. So much so, in fact, that many new optical telescopes are being built with alt-az mounts.

Once the telescope is accurately pointing at the radio source, the control system has to track it to offset the effects of the Earth's rotation. At the heart of the control system of modern radio telescopes are sophisticated electronic computers. Large telescopes deform under their own weight and it is the function of the control system with its computers to model these imperfections and remove their effects, presenting a "near-perfect" telescope to the astronomer.

The second of the goals mentioned above — better resolution — can be achieved by building a bigger telescope. However, the world's biggest fully-steerable dishes, such as the Bonn 328ft (100m) dish, the JODRELL BANK 246ft (75m) dish or the PARKES 210ft (64m) telescope, have typical resolving powers of 100 seconds of arc or so at typical observing frequencies. This is too low for many astronomical investigations.

The angular resolution of any telescope is determined by the wavelength of the radiation divided by the diameter of the surface from which waves are reflected. This follows from the basic laws of physics. To get better (smaller) resolving power at a given

wavelength or frequency, we need a bigger telescope.

The Earth's atmosphere limits the resolution of optical telescopes typically to a second of arc or so. A single radio telescope having a similar resolving power would have to be many kilometres in size. To build such a telescope — at least if it is to be fully steerable — is quite impossible. Nevertheless, radio astronomers have overcome these problems by most ingenious means: higher resolution can be obtained by linking together separate radio dishes as an interferometer. Two small dishes separated by one kilometre, say, can have the same resolving power as a single dish 1km in diameter. Of course, the sensitivity remains just that of the two dishes added together.

Most of the new radio instruments either planned or under construction are of the aperture synthesis array type. Nevertheless, the large single-dish telescopes clearly have a role to play in the future of radio astronomy. They will provide complementary observations to the measurements made with the Earth-rotation aperture-synthesis arrays, and they will continue to be used for mapping large areas of the sky and for making observations of rapidly-varying sources, such as pulsars. Meanwhile very long baseline interferometers consisting of a few widely-spaced dishes will be needed to obtain the ultimate resolution presently feasible (less than a few thousandths of a second of arc) on ultra-compact sources. *AEW*[2].

Radio waves Electromagnetic radiation with wavelengths ranging from millimetres to hundreds of kilometres. They were first generated artificially by Heinrich Hertz in 1888.

Radio waves from Jupiter The first detection of radio emissions from Jupiter was made in 1955 by Burke and Franklin. These emissions are concentrated in wavelengths of tens of metres (decametric) and tenths of metres (decimetric). The decametric emission is due to electrical discharges along the Jovian field lines when IO crosses them.

Radio window The range of wavelengths from a few millimetres to about 20m to which the terrestrial atmosphere is transparent. *See also* ELECTROMAGNETIC SPECTRUM.

Ramsden Lunar crater 15 miles (24km) in diameter; 33°S, 32°W. A system of clefts lies nearby.

Ranger programme The first of NASA's programmes designed to investigate another Solar System body — in this case, the Moon. Originally intended to hard-land a seismometer on the lunar surface, early failures brought about a complete redesign of the spacecraft so that nesting television images of the surface at ever-increasing resolutions could be received from the spacecraft as they approached predetermined impact points. The first missions took place in 1961, but it was not until July 1964 that the first success was secured with Ranger 7. This was targeted to impact at a point in a small, isolated patch of mare material, subsequently named Mare Cognitum — "the sea that has become known". A mare area was selected because of its relatively level and uncratered surface, two essential criteria for the then future APOLLO PROGRAMME manned landings.

More than 400 images were received from six television cameras operating sequentially. The final wide-angle frame covered an area about one mile square, showing craters down to about 30ft (9m) diameter. The best-resolution images showed craters as small as 3ft (1m) across in an area about 100ft by 160ft (30m by 49m).

Ranger 8 was targeted to a point in south-west Mare Tranquillitatis, again in general support of the Apollo programme, but Ranger 9 was reserved for a more scientific target, a point on the floor of the crater Alphonsus that lay between the central peaks and the rilles near the east wall. Both missions were completely successful.

Besides showing that the lunar mare areas were free from rim-to-rim small deep craters, bottomless chasms, and impenetrable rock fields, and would support the weight of a spacecraft, the Rangers provided accurate values of the Moon's radius, and mass ratio with respect to the Earth. *EAW*.

Rapid burster Discovered by Lewin (MIT) in 1976, this unique source emits a string of bursts whose interval varies between 10s and 1,000s. The longest intervals follow the largest bursts. The rapid bursts are not caused by thermonuclear explosions, but are due to instabilities in the accretion of material from the companion star.

Rasalgethi Alpha Herculis, a red star varying between magnitude 3.1 and 3.9 with a semiregular period of 90 days. Its Arabic name means *Head of the Kneeler*, the "kneeler" being the constellation figure Hercules. The

R

R

spectral type of the star is M5II; it is a red giant. It is one of the few stars whose diameters were measured by the Michelson-Morely interferometer on the 100in (254cm) Mount Wilson telescope. At 0.030 arc sec, the star has the same apparent diameter as a 1cm coin held at a distance of 40 miles (70km). The true distance of Rasalgethi is uncertain but if it is at 490 light-years, the star has a diameter of nearly 500 times that of the Sun — it is about the same diameter as the Earth's orbit round the Sun.

Rasalgethi has a companion star of spectral type G5III, of magnitude 5.4; the two are separated by 700AU and orbit each other with a period of about 4,000 years. The companion is itself a binary with period 51 days, so this is a triple system. The spectra of all the stars show evidence of an expanding gaseous shell, which presumably originated from Rasalgethi and is large enough to enclose all the stars in the triple system. The rate of mass loss from the red giant is about one millionth of its mass in 1,000 years. *PGM*.

Rasalhague The brightest star in Ophiuchus (Alpha Ophiuchi); magnitude 2.08, type A5. It is rather isolated from the rest of Ophiuchus, and lies close to the red supergiant RASALGETHI or Alpha Herculis, which is variable between magnitudes 3 and 4.

Ra-Shalom Aten asteroid 2100, period 0.76 year, discovered by Eleanor Helin in 1978 from Mount Palomar using the 18in (46cm) Schmidt camera. Only three other Atens are known.

Rayet, George Antoine Pons (1839–1906). French astronomer who, with

George Rayet: found new class of stars

Charles Wolf, discovered a new class of high-temperature star (*see* WOLF-RAYET STARS). He was first director of the observatory at Floriac.

R Coronæ Borealis variables Stars that have eruptions in reverse: instead of outbursts, they are subject to sudden, unpredictable fadings. Normally they spend most of their time at maxima pulsating slowly. They are high-luminosity stars of spectral types Bpe-R, carbon- and helium-rich, but hydrogen deficient. The fadings can be anywhere between one and nine magnitudes; the resultant deep minima can last for a few weeks, months or for more than a year. These fadings occur completely at random, without the least sign of any periodicity, and once a decline has commenced it is usually rapid, especially if the fading is of several magnitudes. The subsequent rise back to maximum is very much slower, and there are many fluctuations in brightness as the star rises. R. Coronæ stars spend most of their time at maximum, some then pulsating about $0^{m}.5$ in a semi-periodic manner. Several show periods of around 35 to 50 days for such slow pulsations.

The stars are a small class of supergiants, numbering about three dozen. The type star, R Coronæ Borealis, was first found to be a variable in 1795. It normally is at sixth magnitude, fluctuating slowly, and may remain at this maximum brightness for as long as ten years. Then, suddenly, it will commence to fade. The minimum may be anywhere between seventh and fifteenth magnitude — there is no way of determining just how far it will fade. It will remain at minimum for a few weeks or several years: again, there is no way of knowing for how long. Other stars in this group behave in a similar manner, although they are not as bright at maximum as R. Coronæ Borealis. RY Sagittarii when at maximum is about half a magnitude fainter than R Coronæ Borealis, and the remaining members of the group are very much fainter.

A few of these stars have also been found in the Large MAGELLANIC CLOUD. W Mensæ is the brightest of these and has maxima around $13^{m}.8$. Three — RY Sagittarii, S Apodis and UW Centauri — show semi-periodic variations at maxima with amplitudes of around half a magnitude and periods of around 40 days. These small light variations are due to pulsations.

One estimate of the chemical abundances in these stars for which

adequate spectra are available is that they are 67 per cent carbon, 27 per cent hydrogen and six per cent light metals. Naturally, the greatest attention spectroscopically has been given to the two brightest members, RCrB and RY Sgr. All investigators agree that both stars are much more abundant in carbon than hydrogen and also that lithium is overabundant in both, with the C_2 absorption bands only weakly present. Other RCrB stars have strong C_2 absorption bands and are presumably cooler. In fact, their spectral types range from late R type to B and all are carbon-rich and hydrogen-poor. This implies that any model to account for their behaviour must provide for a range of temperatures.

Herbig, Payne-Gaposchkin and others have studied the spectra of RCrB and RY Sgr during their declines to minima. Both stars have rapid declines of from four to seven magnitudes in a month and during the initial decline the normal ABSORPTION spectrum is rapidly replaced by a rich EMISSION SPECTRUM. Payne-Gaposchkin suggested that, because of the lower temperature and density, the emission be regarded as chromospheric. At this stage radiation from the surface is decreased or obscured. This emission spectrum gradually fades and changes, appearing to proceed in an orderly manner with no sudden reversal on the rise. Evidence for some form of obscuration of the photosphere comes from the appearance of shell absorption lines in the spectra, both during some phases of minima and at some stages on the rise. All RCrB variables were shown by Feast and Glass to have an infrared excess. Variations in the infrared do not occur at the same time as changes in the visual and changes in either can be unrelated to changes in the other.

A number of models have been proposed to account for the behaviour of RCrB variables. O'Keefe was the first to suggest that a gas cloud forms above the photosphere and would be moving outward. As it does so it cools and condenses, forming small particles of graphite. These would be very efficient absorbers of light and so would veil the star and account for the sudden declines in brightness. The cloud would disperse, becoming at first patchy and transparent in places. Later, the temperature and pressure would become so low that the carbon particles would condense into soot, so that finally there was no veiling of the stellar light.

Payne-Gaposchkin considered that particles were formed in the upper photosphere. These would cut off the chromosphere from its source of excitation so that it gradually decays. A third theory suggests that a cloud is ejected from the star. The chromospheric spectrum then results from an eclipse by the dense cloud in the observer's line of sight. This spectrum is in many ways similar to that of the solar chromosphere during a total eclipse of the Sun. At first the cloud might be much smaller than the star, but as it moved away it would grow in size and, if it were centred over the star, cause an eclipse which would result in the deep declines that are observed.

According to Feast, the RCrB stars in the Large Magellanic Cloud show little variation in visual absolute magnitude with temperature; the cooler stars will have the larger diameters. This supposes that the rate of growth of the cloud size does not depend on temperature, which means that the time of decline would be greater for the cooler stars. Payne-Gaposchkin pointed out that the average time of decline, which ranges from 30 to 300 days, increased with decreasing stellar temperature. Infrared observations by Forrest and others suggest that clouds are formed and move outward away from the star, cooling and expanding as they do so. Other theories by Humphreys and Ney, and also by Wing and others, suggest particle clouds associated with a supposed companion star, or blotches in orbit around the star. Both theories appear to be open to objections.

Feast and Glass have drawn attention to similarity in chemical abundances between RCrB stars and the hot helium and cool HdC stars, implying that there is some relationship. These stars are thought to have ejected a hydrogen-rich shell and probably most belong to the old disk population. This has prompted several workers to suggest a link with the PLANETARY NEBULÆ. Herbig suggested that the presence of emission in RCrB stars near minima indicated a surrounding nebulosity. He also considered that the peculiar star V348 Sgr, which had many of the features of RCrB stars, was involved in nebulosity. Several workers have also drawn attention to the fact that there are some features in RCrB stars that are similar to those in old novae. The exact nature of these relationships of RCrB variables with other stars is unclear and hence there is still much argument as to how they have evolved.

The RCB variables provide a very good example of why professional astronomers welcome the assistance of well-organized amateurs. The professionals are interested in certain parts of the light curves of these stars, especially as they commence their deep declines. The cost of modern instruments and the demands on them mean that they cannot be used night after night, often over periods of years, in the hope that a decline of an RCrB star will be observed. The professionals therefore rely on amateurs to monitor these stars continuously and to advise them whenever a fading is detected. In this way the amateurs also produce the data needed to determine how they pulsate at maxima and the period in which they do so. *FMB*.

R Cygni A MIRA VARIABLE, discovered in 1852 by Pogson. Its extreme range is 6.1 to 14.2(v), though the average is 7.5 to 13.9. The mean period is 426.44 days, and the spectrum is S3,9e – S6,8e. The rise usually takes about 35 per cent of the total period. The co-ordinates are R.A.19h 36ʹ.8, Dec. + 50° 12ʹ (2000.0).

Reciprocity failure A problem encountered by astrophotographers in which the effective sensitivity or speed of a photographic emulsion diminishes in a non-linear manner with longer than optimum exposures.

Recombination The capture of an electron by a positive ion (the opposite process to IONIZATION).

Recurrent novæ Binary star systems in which the primary is a late type (G, K or M) giant filling its RÔCHE LOBE and transferring material to the WHITE DWARF secondary. At intervals of years to tens of years this accreted material on the surface of the white dwarf explodes in a THERMONUCLEAR REACTION, causing the brightness of the star to increase by seven to ten magnitudes for about 100 days in the visible and for double that at other wavelengths. The more time there is between outbursts, the brighter these are. *ENW*.

Red dwarfs Stars at the lower (cool) end of the MAIN SEQUENCE, spectral types K or M, surface temperatures between 2,500K and 5,000K and masses in the range 0.8 to 0.08 times that of the Sun.

Red giants Giant stars of spectral type K and M (and also the abnormal spectral types R, N and S), having surface temperatures cooler than 4,700K and diameters 10 to 100 times that of the Sun.

Redman, Roderick Oliver (1905–75). British astronomer who contributed to stellar spectroscopy and solar physics, and made a detailed analysis of the spectrum and temperature structure of the SOLAR CHROMOSPHERE.

Redshift The lengthening of the wavelength of ELECTROMAGNETIC RADIATION caused by either the recession of the source from the observer, or by the expansion of the universe.

The spectra of astronomical objects contain important information about the objects' motion. An approaching star or planet has its spectral lines shifted toward the blue side of the spectrum (thereby exhibiting a "blueshift"), whereas the spectral lines of a receding object are shifted toward the red (a "redshift"). The greater the speed along the line-of-sight between the source and the observer, the greater is the wavelength shift.

This phenomenon was first accurately described in 1842 by Christian DOPPLER, a professor of mathematics at Prague, and is today known as the Doppler effect. It is familiar to anyone who has heard an ambulance racing down a street with its siren wailing. While the sound-source is approaching you, you hear a higher-than-normal pitch. But as the source passes by you, you notice a distinct drop in pitch to a lower-than-usual frequency.

Suppose that λ_0 is the unshifted ("at rest") wavelength of a particular spectral line, as determined from laboratory experiments. (Such wavelengths are tabulated in voluminous reference books.) Suppose that you find this same spectral line at a wavelength λ in the spectrum of an astronomical object. You have therefore detected a wavelength shift of $\Delta\lambda = \lambda - \lambda_0$. Astronomers find it convenient to define a quantity z as

$$z = \frac{\Delta\lambda}{\lambda} = \frac{(\lambda - \lambda_0)}{\lambda}$$

which is the fractional change in wavelength. For example, z = 0.1 means that all the wavelengths of all the spectral lines of a source are lengthened by 10 per cent, for z = 0.2 they are lengthened by 20 per cent, and so forth. Christian Doppler proved that

$$z = \frac{v}{c}$$

and where v is the radial velocity between the source and the observer and c is the speed of light. This relationship is valid only for values of v very much smaller than the velocity of light. With this formula, astronomers convert observed wavelength shifts into

R

actual speeds in kilometres per second. This one formula provides crucial information about a wide range of circumstances, from the orbits of double stars to the rotation of entire galaxies.

In 1914, Veso M. SLIPHER, working at the Lowell Observatory in Arizona, began studying the spectra of objects then called "spiral nebulæ", but today known to be galaxies. He found that 11 of the 15 "spiral nebulæ" he examined had redshifts, thereby indicating a pre-

Redshift: shifting spectral lines

ference for recessional motion. Then years later, Edwin HUBBLE at the Mount Wilson Observatory in California proved that these objects are distant galaxies, each like our Milky Way. The fact so many galaxies exhibit redshifts piqued Hubble's interest and he began a systematic study of their spectra. By 1929, Hubble succeeded in proving that the redshift of a galaxy is proportional to its distance from Earth. This discovery, today called the Hubble law, can be written as

$$v = H_0 d$$

where d is the distance to a galaxy, v is its recessional velocity as deduced from its redshift, and H_0 is a number, called the "Hubble constant", roughly equal to 75km/s/Mpc. In other words, a galaxy has 47 miles/s (75km/s) of recessional velocity for each mega-parsec from Earth.

The Hubble law ranks among the most important astronomical discoveries of the 20th century. Using Hubble's formula, astronomers can deduce the distance to a remote galaxy just by measuring its redshift. For example, the galaxy in the Corona Borealis cluster shown in the accompanying illustration has a redshift of $z = 0.073$ corresponding to a

recessional velocity of 13,600 miles/s (22,000 km/s). To estimate the distance to this galaxy, we use Hubble law as follows:

$$d = \frac{v}{H_0} = \frac{22,000}{75} = 290 \text{Mpc}$$
$$= 950 \text{ million light-years}$$

The most distant discernible galaxies have redshifts of roughly $z = 1$, and some QUASARS have been discovered with redshifts as high as $z = 4$.

The Hubble law tells us that the universe is expanding. Indeed, a linear relationship between distance and speed is exactly what is meant by "uniform expansion." By way of example, consider the "expanding balloon analogy" shown in the accompanying sketch. Coins representing galaxies are glued to the surface of a balloon. As the balloon expands, all the coins recede from each other much in the same way that widely separated galaxies drift apart as the universe expands. Indeed, the speeds with which the coins recede from each other is directly proportional to their separation, just as in the case of galaxies obeying the Hubble law.

The expansion of the universe, which is manifested in the redshifts of galaxies, is properly described by the General Theory of RELATIVITY, formulated by Albert Einstein in 1915.

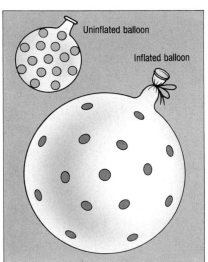

Redshift: a uniformly expanding universe

According to this description, space is not rigid and static, but rather expands with time. Quite literally, more and more space appears between widely separated points as time goes on.

It is important to realize that redshifts caused by the expansion of the universe are *not* due to the Doppler effect. To emphasize this distinction, the redshift exhibited by a remote

galaxy is called a "cosmological redshift." The Doppler effect results from *motion through space* whereas cosmological redshifts are caused by the *expansion of space*. Even though a distant quasar might have a redshift corresponding to a speed 90 per cent the speed of light, that quasar is not moving rapidly relative to its local surroundings. The quasar redshift that we observe is actually caused by the expansion of space, which literally stretches the wavelengths of the photons as they travel toward us.

For speeds greater than about one-third of the velocity of light, the simple relation between z and velocity must be modified according to the Special Theory of Relativity. Though it is possible to observe high red-shifts, it is not possible to travel at the speed of light. So, for example, if an object has a redshift of $z = 2$, this does not mean it is receding from us at twice the speed of light. Instead, the redshift is given by

$$z = \sqrt{\frac{1 + \frac{v}{c}}{1 - \frac{v}{c}}} - 1$$

According to this equation, a redshift of $z = 2.0$ corresponds to a speed of 80 per cent of the velocity of light. Ths most remote quasars seen so far have a redshift of 4. This corresponds to a speed of 92 per cent of the velocity of light. According to the Hubble Law, this implies that their distance is about 12,000 million light-years. Some people believe that the Hubble constant may be as low as 55km/s/Mpc. If this is so, these quasars would be about 16,500 million light-years away.

The Hubble law follows from the simple fact that the farther away a galaxy is, the longer is the travel time for its photons to get to Earth. The longer a photon's travel time, the more its wavelength is stretched, and thus the greater is its redshift.

The most highly redshifted photons are those in the 3° MICROWAVE BACKGROUND. These photons have been travelling toward us since matter and radiation decoupled from each other about 700,000 years after the BIG BANG. These photons exhibit a redshift of $z = 1,000$, meaning that their wavelengths have been stretched a thousand-fold during their 20-thousand-million-year journey to Earth.

Photons of the microwave background are the most highly redshifted photons that we can possibly observe. A redshift of $z = 1,000$ is an observational upper limit because the universe was optically opaque at times earlier than 700,000 years after the Big Bang. Since the most remote quasars have redshifts of roughly 4, there is a vast region of space (corresponding to the redshift range $4 < z < 1,000$) throughout which astronomers have never seen any objects. Exploring this enormous "wasteland" will be one of the major tasks of the HUBBLE SPACE TELESCOPE.

Accurate measurements of the distances and redshifts of remote galaxies by the Hubble Space Telescope will also yield fundamental information about the overall structure and evolution of the universe. Such data will show how fast the expansion of the universe is slowing down, thereby revealing whether or not the universe will expand forever. Thus redshifts provide clues about the ultimate fate of the entire universe. *WJK.*

Reference stars Stars whose co-ordinates (DECLINATION and RIGHT ASCENSION) are accurately known. They can thus be used to measure the relative positions of other celestial objects. The most accurately positioned are known as FUNDAMENTAL STARS.

Reference systems *See* CELESTIAL SPHERE and GALACTIC CO-ORDINATES.

Reflecting telescopes Although there was an unsuccessful attempt by James Gregory in 1663 to make an all-mirror telescope, the first workable reflecting telescope is attributed to Sir Isaac NEWTON in 1668. His initial attempt was regarded as a scientific toy, but his second telescope, built in the autumn of 1671, was presented to the Royal Society on January 11 1672 and examined by, among others, King Charles II and Christopher Wren.

The basic principle was to replace refracting lenses (*see* REFRACTING TELESCOPES) with a reflecting surface; as this is on the face of the mirror, there is no need for the material of which the mirror is made to be transparent and materials other than glass can be used.

The first telescopes experimented with metal mirrors, as these were easier to manufacture. A mixture of 68 parts of copper and 32 parts of tin, called "speculum metal", was used. It was not until two centuries after Newton that FOUCAULT made the first silver-on-glass mirror.

The purpose of an astronomical telescope is not, as is generally supposed, to give great manifying power, but is to collect as much light as possible. The NEWTONIAN TELESCOPE uses a parabolic mirror surface to reflect the light falling upon it back to a single point known as the prime focus. By introducing a small flat mirror at 45° just before this focus, the cone of light is guided outside, at right angles to the telescope, where it is passed through an EYEPIECE.

All telescopes have a FOCAL RATIO (or "f" number), derived from dividing the main mirror diameter into the focal length (eg, a 50cm diameter mirror and a 3m focal length $= 300/50 = $ f6). From this the magnification of the telescope is derived by multiplying the inverse ratio of the eyepiece focal length by the focal ratio of the telescope (eg, the focal length of a 25mm eyepiece equals $1/40$ of a cm. With an f6 telescope the magnification 1 equals $1/40 \times 6 = 240$ times).

After the 1600s other mirror arrangements were devised, though some, like the GREGORIAN TELESCOPE, fell out of common use because the telescope had to be too long.

In 1672 it was reported that a Frenchman named Cassegrain had invented a new reflecting telescope by placing a convex mirror inside the focal length, thus reflecting the light back down the tube and through a central hole in the main mirror (*see* CASSE-GRAIN TELESCOPE). Cassegrain was never identified and it was not until the mid-1700s that his system was fully developed and in general use.

In order not to have to perforate the main mirror, a small flat mirror can be introduced into the optical system to direct the light train outside the tube (sometimes along the declination axis). This system is known as "Cassegrain/coudé" (*coudé* is French for "elbow").

Nearly all of today's large telescopes are adaptable to prime-focus viewing, as well as the Cassegrain system, which produces a much longer focal ratio, permitting a large range of magnification for various astronomical uses.

Very large mirrors suffer from distortion due to expansion and contraction as the temperature changes. This problem was reduced in the 1930s by the development of Pyrex, which has a very low coefficient of linear expansion. The recent introduction of quartz and ceramics means that modern mirrors virtually do not expand or contract at all.

In the 1930s first SCHMIDT and then MAKSUTOV produced the first corrector-plate telescopes. They were initially intended for photographic use, but provide a portable instrument for direct observational use. All the various systems incorporate a corrector plate in front of the main mirror. There are several optical designs for this type of telescope, but each arrangement gives a wide field at prime focus which enables photographs to be taken of large areas of the night sky.

So that the astronomical telescope can point to every position of the sky, it must move in two axes at 90° to each other. When one of these axes is mounted to align with the rotational axis of the Earth, then, to follow a star, the telecope only has to move in the other axis. This aligned axis is called the "polar axis" and is pointed at the CELESTIAL POLE. It thus differs in its altitude with the latitude of the observer, and is vertical at the Pole and horizontal at the equator. The other axis is called the "declination axis" and is used to turn the telescope up and down the sky away from the polar orientation (*see* ALTAZIMUTH MOUNTINGS and EQUATORIAL MOUNTINGS).

Modern telescopes, computer controlled, use a far less costly type of altazimuth mounting in which the polar axis is vertical at any altitude. This, however, means that the telescope must be driven in two axes at the same time in order to remain fixed in relation to the night sky.

The most important and delicate parts of any reflecting telescope are the reflecting surfaces and over the years many different materials have been used to give a durable coating that will reflect back the maximum amount of light. For many years silver was used, as a newly-applied burnished finish will reflect 95 per cent of visible light. Silver's tendency to oxidize and tarnish renders it less durable than other coats, but it can readily be renewed.

In the 1930s aluminium was experimented with, as it reflects back 87 per cent of visible light, but its application requires that the mirror be placed in vacuum during deposition. ALUMINIZING is thus a job for professionals.

Other metals have been experimented with but due to their hardness require the use of abrasives for their removal. Rhodium is least affected by chemical and physical ageing but is very expensive and has less reflectivity than silver and aluminium. Overcoating of aluminium coated mirrors has been experimented with to try to lessen oxidization and has proved successful in extending the life of the mirror surface. The length of time between aluminizations, however, will always depend on the care taken in looking after the telescope, the amount of use it gets and air pollution at the telescope site. *RFT.*

R

Reflection grating A diffraction grating that has the form of closely-spaced parallel grooves which reflect light. A typical spacing density for the grooves in an optical reflection grating for astronomy is about 1,000 grooves per millimetre, and the size of the separation (about 1 micron) is not much bigger than the wavelength of light. Light of different wavelengths therefore reflects from the grating in different directions (diffraction), and this is how the reflection grating in an astronomical spectrograph creates a spectrum. *PGM.*

Reflection nebulæ *See* NEBULÆ

R

Refracting telescopes Telescopes that utilize the refraction of light through lenses to form images of distant objects. In its simplest form, a refracting telescope — or refractor — consists of two lenses, an OBJECTIVE and an EYEPIECE. In practice, both objective and eyepiece are compound lenses, consisting of two or more components. The objective, or object glass, is a lens of large aperture and long focal length which forms an image of a remote object at its focal plane. The eyepiece is a smaller lens of short focal length which magnifies the image. The magnifying power is the ratio of the apparent size of the image to the apparent size of the object seen without the telescope, and is equal to the ratio of the focal length of the objective to the focal length of the eyepiece.

The invention of the refracting telescope is usually credited to the Dutch optician Hans LIPPERSHEY, who is reputed to have discovered its principle purely by chance in 1608 when he noticed that magnified images of distant objects were produced when one lens was placed in front of another. It is surprising that the refractor was not invented earlier, for the properties of simple lenses had been investigated during, and spectacles had been invented by the end of, the 13th century. Although there are suggestions that the refractor had been invented in the late 16th century, there is no conclusive evidence to support these claims.

The first astronomer to make serious regular telescopic observations was GALILEO Galilei who, in 1610, with telescopes of his own design and construction, made a remarkable series of observations which revolutionized mankind's view of the universe. The Galilean refractor had a convex ("positive") objective and a concave ("negative") eyepiece which was placed in front of the focal point of the objective. As a result it produced bright, erect, images, but the field of view was small and the instruments were difficult to use. In 1611 Johannes KEPLER published an improved design which employed a convex eyepiece located behind the focal plane of the objective (because of this, it is possible to focus the eyepiece simultaneously on the image formed by the objective and on micrometer wires placed at its focal

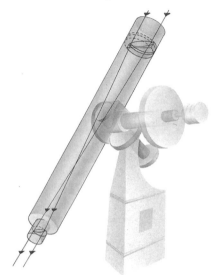

Refracting telescopes: date from 1608

plane (*see* MICROMETERS)). The Keplerian, or "astronomical", reflector gives a wider field of view and an inverted image.

Simple lenses and telescope systems suffer from a range of optical defects, or aberrations, notably chromatic aberration and spherical aberration. Chromatic aberration is the inability of a simple lens to bring all wavelengths of light to the same focus. The shorter the wavelength, the greater the amount of refraction; therefore the various wavelengths, or colours, are focused at a range of points along the axis of the lens with the shortest wavelengths (violet) focused nearest the lens and the longest wavelengths (red) farthest away. A star image formed by a simple lens will always be surrounded by a fringe of out-of-focus colours which impairs the observer's perception of its true colour and, by blurring the images, reduces the resolving power.

Spherical aberration is the inability of a simple lens (with surfaces which are portions of a sphere) to bring all rays to the same focus. Rays passing through the central regions of the lens are focused farther away than those which pass nearer to the perimeter of the lens. While spherical aberration can be overcome by giving the lens faces non-spherical curves, chromatic aberration cannot be eliminated from a single lens.

Seventeenth-century astronomers realized that spherical aberration and, to a lesser extent, chromatic aberration, could be reduced by making gently-curved lenses of very long focal length. This approach resulted in some extraordinary instruments. For example, Johannes HEVELIUS constructed several "long-tube" telescopes (with wooden lattice tubes which flexed badly) up to 150ft (46m) in length. Christiaan Huygens developed an alternative design — the aerial telescope — with the objective fitted in a short tube at the top of a mast, and connected only by a cord to an eyepiece at ground level.

In 1729 the English lawyer and amateur optician Chester Moor Hall realized that the effects of chromatic aberration could be substantially reduced by combining a positive lens made from ordinary "crown" glass with a negative lens made from "flint" glass, a harder glass which contained lead, and had a higher refractive index (a measure of the amount of refraction produced by a material) and a higher dispersion (a measure of the extent to which the different wavelengths are separated out by refraction). By careful selection of lens shapes, the chromatic aberration introduced by one lens could to a large extent be "cancelled out" by the other. A lens of this type is called "achromatic". The first achromatic lens was built for Hall by George Bass in 1733.

An achromatic doublet (two-lens objective) can bring only two wavelengths to a focus at exactly the same point, but the residual spread of focal points for the other wavelengths — the "secondary spectrum", or "longitudinal chromatic aberration" — can be reduced by a factor of 30 or so compared with a single lens objective. An apochromat is an objective — normally a triplet made from three types of glass — which brings three wavelengths to the same focus and reduces the secondary spectrum even further. In practice, very few astronomical refractors have anything other than two-element objectives. Eyepieces, too, must be designed to minimize optical aberrations.

With continuing improvements in materials, skills and optical technology, the 19th century became the heyday of the refractor. Although most refractors were conventional instruments, specialized designs were produced for specific purposes, eg, TRANSIT INSTRUMENTS and HELIOMETERS. Coudé refractors used mirrors placed after the objective to reflect light to a fixed observing position. Another variant involved a fixed refractor into which light was reflected by a siderostat or CŒLOSTAT — a flat, steerable mirror which could be driven to track objects across the sky.

The building of big refractors finally culminated in the 36in (0.91m) instrument at Lick Observatory in California, and the 40in (1.01m) refractor of the Yerkes Observatory, Wisconsin. Both instruments were funded by American millionaires — James Lick and Charles Yerkes — and both have objectives made by the optical company Alvan Clark & Sons. They were installed in 1888 and 1897, respectively.

The Yerkes instrument is highly unlikely ever to be surpassed. Compared with modern reflectors, large refractors have many disadvantages: chromatic aberration cannot wholly be eliminated. As lenses become larger and thicker a significant amount of light is absorbed in them. Since a lens is supported only round the edge (whereas a mirror is also supported across the back) a large lens tends to flex under its own weight; the long tube itself tends to flex and needs to be housed in a large, expensive building. An achromatic doublet has four surfaces to be accurately shaped whereas a mirror has only one and it follows, therefore, that a refractor is very much more expensive than a reflector of equal aperture.

Although all large modern telescopes are reflectors, smaller refractors are still widely used as guide instruments and refractors still have a number of attractions for visual observers. Their optical components are less likely to go out of alignment than those of a reflector. Refractors often seem to be less affected by temperature changes than reflectors, and the lack of obstructions in the light path (such as secondary mirrors and their supports) together with the wholly enclosed tube (which cuts down air currents in the light path) is reckoned by devotees to contribute to better, steadier images on average, than those produced by reflectors of equal aperture. *IKN. See also* LENS, REFLECTING TELESCOPES.

Refraction of light The change in direction of electromagnetic waves when passing from one medium to another in which they have a different velocity. For light, refraction is associated with a change in the optical density of the medium. When light passes into a denser medium, the wave path is bent towards the normal (the line perpendicular to the surface at the point of incidence). In this case, the ratio between the sine of the angle of incidence to that of the angle of refraction is called the refractive index of the medium. The refractive index varies with wavelength, with light of long wavelength (red) being bent the least and light of short wavelength (violet) being bent the most.

Regiomontanus Distorted lunar walled plain, 80 × 65 miles (129 × 105km) in diameter, between Purbach and Walter; 30°S, 48°E.

Regiomontanus, Johannes (Johann Müller) (1436–76). German astronomer and mathematician who completed the important *Epitoma Almagesti Ptolemai* (Epitome of Ptolemy's *Almagest*) began by his tutor Georg Peuerbach.

Regolith The 30–80ft (10–25m) thick layer of rock fragments covering the surface of the Moon, and similar atmosphereless bodies, produced from the underlying solid crust by the explosive impact of large and small meteoroids.

Regression of the nodes The slow westward motion of the NODES of the Moon's orbit, due to the gravitational pull of the Sun. A full circuit takes 18.6 years.

Regulus Alpha Leonis, sometimes known as the "Royal Star". Its magnitude is 1.35.

Reichenbach Valley Lunar feature in the southern uplands; it is not a valley, but a crater-chain. Reichenbach itself is a crater 30 miles (48km) in diameter; 30°S, 48°E.

Reinmuth, Karl (1892–1979). He worked at the Konigstuhl Observatory, Heidelberg, on the astrometry of asteroids. He discovered many asteroids and two short-period comets. In 1953 he published a catalogue of 6,500 accurate photographic asteroid positions.

Relativity Any point, P, on a plane — say, on a graph, the value inflation reached in 1980 — can be described in terms of the graph's co-ordinates, the x and y (inflation and time) axes, which are perpendicular to the origin, O. Similarly, any point on the surface of a sphere can be described by using the spherical polar co-ordinates r and θ. A continuous curve in a plane or on a sphere can be specified by a functional relationship between x and y — say, $y = f(x)$ or $r = \gamma(\theta)$. It makes no difference to a particular curve which system of co-ordinates is used. That implies some form of relationship between $y = f(x)$ and $r = \gamma(\theta)$ and the fact is that one is transformable into the other. This is expressed by saying that the curve is invariant with respect to the system of co-ordinates chosen.

There is no difficulty in extending this simple state of affairs from Euclidean geometry in three-dimensional space. Once again a fixed origin, O, is chosen. By taking three mutually perpendicular lines, the position of any point P can be specified by the three co-ordinates x, y, z, these being the shortest distances of P from each of the three lines in turn. Or the position of P may be specified by its distance r from O, together with two angles, θ, φ say, that can be chosen conveniently in a number of ways. Once again, the points lying on a surface can be specified by a functional relationship, $z = f(x, y)$ or $r = \gamma(\theta, \varphi)$, with similar transformable properties between f and γ when x, y, z are related appropriately to r, θ, φ.

This is geometry, not physics. In physics, the events which make up the world are each taken to have a position in three-dimensional space, however. So co-ordinates z, y, z or r, θ, φ can be used to describe the spatial position of an event, just as in geometry. But something more is also needed, because as well as knowing in what trajectory a particle moves, we want to know the time when the particle is at each point of its path. Hence to describe an event in physics we need the time, t, say, as well as the spatial co-ordinates, x, y, z say, with t being measured with the aid of a "clock", by a procedure that must be precisely specified.

Physics can be defined as the science of relating an event at x, y, z, t to other events at other co-ordinate values, for instance relating the position of a particle at time t to its positions at other times and to events not directly involving the particle itself, but often involving other particles, which exert forces on the particle in question, and to "fields" like the familiar electric field. The manner in which one event depends on others is always described

R

in physics by mathematical equations, just as curves and surfaces are in geometry, except that the equations in physics involve the time t. Also, as in geometry, the mathematical equations should have an invariance with respect to the particular way we have set up our co-ordinate system.

This last point was recognized in Newtonian physics, so far as changes in the co-ordinate system effected by a particular observer were concerned. It should not matter whether an observer elects to use x, y, z, or r, θ, φ, or elects to use either the spin of the Earth or the vibrations of an atom or the oscillations of a pendulum as a clock to obtain t. The equations of Newtonian physics had to be, and were, invariant with respect to the choices made by a particular observer.

Toward the end of the 19th century, physicists began to think more broadly about co-ordinates. Instead of merely permitting choices to a particular observer, suppose observers in motion with respect to each other use just the same prescriptions for measuring the co-ordinate values of events? Take as a simple case two observers in uniform motion with respect to each other, one measuring co-ordinates x, y, z, t and the other x^1, y^1, z^1, t^1, for the same event and for a similar agreed procedure. The first question that could be asked was: how do the values of x, y, z, t obtained by the first observer compare with the values of x^1, y^1, z^1, t^1 obtained by the second? The further question was whether the mathematical equations of Newtonian physics had the necessary invariance as one passed from x, y, z, t values to x^1, y^1, z^1, t^1 values for the same events? When the co-ordinates of the one observer were related to those of the other by what seemed intuitive commonsense considerations, nothing worked as it should have done, from which it was seen that something had to be terribly wrong. The question which tormented physicists in the closing years of the 19th century was, what?

The difficulty was compounded by the immense prestige that Newton's laws for determining the motions of particles had acquired over the centuries. Had physicists elected instead to believe more firmly in Maxwell's recently-discovered mathematical equations for the propagation of light, the problem would have not been so difficult. Nevertheless, a breakthrough was made in 1892 by the Irish physicist George Francis Fitzgerald,

who showed that in order to have any hope of a satisfactory outcome it would be necessary to abandon the supposed commonsense relation between the co-ordinate systems x, y, z, t and x^1, y^1, z^1, t^1. Then in a series of papers culminating in 1904, the Dutch physicist Hendrick Antoon Lorentz at last obtained the correct relation between the two co-ordinate systems, the so-called Lorentz transformation. Lorentz also showed in 1904 that Newton's laws would have to be modified so as to include a variation of the mass of a particle with its velocity, a discovery which became enshrined in later years in the cliché "$E = mc^2$".

Still there remained a general feeling of logical inconsistency, a feeling that was finally dispelled by Albert Einstein in 1905. Einstein showed that, with x, y, z, t and x^1, y^1, z^1, t^1, connected by the Lorentz transformation, Maxwell's equations for the propagation of light possessed the required property of invariance, and it became apparent that, had Maxwell's equations been used as the starting point, the Lorentz transformation might have been inferred with comparative ease. To complete this first development of what became known as the Special Theory of Relativity, in 1908 it was shown by Hermann Minkowski that the Lorentz transformation implied that the co-ordinates x, y, z, t belong to an extension of Euclidean geometry, an extension which became known as Minkowski spacetime, a mathematical system in four dimensions with respect to which both Maxwell's equations and a suitable modification of Newton's laws took elegant forms.

Ironically, the elegant success of Minkowski spacetime acted as an impediment to the next physical step, which was to obtain a similar invariance of the mathematical equations for co-ordinates x, y, z, t and x^1, y^1, z^1, t^1 measured by observers in non-uniform motion with respect to each other. Unless this further problem could be solved, physics was doomed to remain a prisoner of Newton's assumption of a so-called inertial system, which is to say an absolute system of spacetime. The more general problem amounted to showing that the mathematical equations representing the laws of physics have invariance with respect to general transformations of the kind $x^1 = X(x, y, z, t)$, $y^1 = Y(x, y, z, t)$, $z^1 = Z(x, y, z, t)$, $t^1 = T(x, y, z, t)$, where X, Y, Z, T are any suitably smooth transformation functions.

It was shown by Einstein in 1916 that to solve this general problem it was necessary to abandon Minkowski spacetime in favour of a more general geometry, first investigated in 1854 by Georg Friedrich Bernhard Riemann: Riemannian spacetime. Einstein's remarkable idea was to regard the difference between Riemannian spacetime and Minkowski spacetime as the true meaning of the phenomenon of gravitation. To this end he modified Newton's equations of motion so as to form a comprehensive scheme for calculating not just the motions of particles in a prescribed spacetime like that of Minkowski, but a determination of what the more complex Riemannian spacetime had to be. These "field equations" of Einstein are mathematically so difficult, however, that they have only been solved exactly for two very simple problems. Luckily, one of the exact solutions, that for a point mass, had consequences in the Solar System which could be tested by observation. Indeed, Einstein already showed in 1916 that his field equations could be solved approximately so as to clear up a long-standing difficulty connected with the orbit of the planet Mercury, namely on excess rotation of about 43 arc seconds per century of the major axis of the orbit which cannot be explained by Newton's theory.

Also in 1916, Karl Schwarzschild obtained the exact solution for a point mass, from which a spectacular prediction could be made. At any moment there is a background of stars in the direction of the Sun which can only be seen during a total eclipse of the Sun. Because the Sun moves a complete circuit during the year with respect to the stellar background, those stars which form its immediate background at a total eclipse will not do so some months later, say six months later. What was predicted was that the background at an eclipse would be slightly distorted from its form six months later, by a calculated amount arising from starlight passing close to the Sun. With World War I ending in 1918, arrangements were made to test this spectacular prediction immediately, at an eclipse that was to occur in 1919. The result agreed with the prediction to within a margin of accuracy of about 25 per cent, a margin that has been substantially reduced in modern times.

These developments became known as the "theory of relativity", for the historical reason that they came from a consideration of observers in relative motion with respect to each other.

Possibly a better way to describe the theory would be to say that it demands invariance of the physical laws with respect to all ways in which we care to choose co-ordinates x, y, z, t in order to describe the position of events in spacetime.

The theory became celebrated because it showed that a demand for logical consistency in a body of somewhat disordered physical knowledge could lead to remarkable new correspondences with observation that had not been expected at the outset, so that the human mind allied to an initial body of knowledge can itself perceive what must be true about the world. The early spectacular successes of the theory led many physicists to expect a whole sequence of continuing success, but unfortunately these have not been forthcoming in subsequent years. Instead of the theory being used to predict with success later observational discoveries, the theory has become employed more as a tool of research, especially in problems concerning the structure of the universe, and in recent years problems concerning extremely dense aggregates of matter. *FH*.

Réseau A reference grid superimposed photographically onto a photographic plate of a star field.

Resolving power The ability of a telescope to separate objects that are apparently close together. It depends on aperture, and the wavelength being studied. For an optical telescope it is $4\rlap{.}{''}56$ divided by the aperture in inches ($11\rlap{.}{''}48$ divided by the aperture in cm). The limit is related to the DIFFRACTION of light in the telescope, which causes point sources of light to appear as spurious disks (*see* AIRY DISKS), whose diameters are proportional to the focal length and the wavelength, and inversely proportional to the aperture. The resolving power is the distance between the centres of the spurious disks when the centre of one disk lies on the edge of the other. *TJCAM*

Resonance A gravitational effect which occurs when the orbital period of one object is an exact fraction of that of a larger neighbour. This causes systematic tugs on the smaller object, gradually changing its orbit. The KIRKWOOD GAPS in the asteroid belt are due to resonance effects between individual asteroids and Jupiter.

Rest mass The mass of a body when it is stationary. (Strictly speaking, it is the mass of a body measured when it is at rest in an inertial frame of reference.)

According to the theory of RELATIVITY, if a body has a finite rest mass, its mass increases as its speed increases and would become infinite if it could be made to travel at the speed of light. Photons (which travel at the speed of light) have zero rest mass.

Retardation The difference in the time of moonrise on successive nights. It may exceed an hour, or can be as little as 15 minutes at the time of HARVEST MOON. The variation is due to the Moon's northward or southward motion along the ECLIPTIC.

Reticulum (the Net). A small but quite distinctive far-southern constellation. Its leading star is Alpha (3.35; absolute magnitude -2.1, spectrum G0, distance 390 light-years). Alpha makes up a compact little pattern with Beta (3.85) and Gamma, Delta and Epsilon (about 4.5).

Retrograde motion Opposite of direct motion. Seen from "above the north pole", the Earth moves in a direct or anti clockwise sense round the Sun; HALLEY'S COMET is retrograde. Also the temporary apparent east-to-west movement of a superior planet in the sky before and after opposition.

Retrograde motion: clockwise round the Sun

Reversing layer A thin layer above the solar PHOTOSPHERE where the absorption lines of the solar spectrum are produced. The term lower CHROMOSPHERE is now preferred.

Revolution The orbital period of one body with respect to another.

Rhæticus, Georg Joachim (1514–76). German astronomer who visited COPERNICUS and persuaded him to send his great book for publication.

Rhea The second largest satellite of Saturn, with a diameter of 950 miles (1,530km) and a density of 1.1 gm/cm³. It must be composed chiefly of ices. Located in a nearly circular orbit at a mean distance of 327,000 miles (526,700km), the satellite is synchronously rotating with respect to

Rhea: composed chiefly of ices

Saturn and has a period of revolution of 4.5 days. The leading hemisphere is cratered and brighter than the following hemisphere, which contains bright wispy markings. *RFB. See also* colour essay on MOONS.

Rhea Mons One of two suspected shield volcanoes situated in BETA REGIO, on Venus. The radar-bright mountain rises to about 3 miles (4.5km) above the adjacent basaltic plains.

Rheita Valley The widest (24km) and deepest (2.5km) part of this fault-controlled lunar valley is 210 miles (340km) long and lies between Mallet C and Rheita, a crater 40 miles (70km) in diameter at 37°S, 47°E. The fault transects Young — another 40-mile (70km) crater — and downdrops part of Young in a graben 12 miles (20km) wide. Elsewhere the graben has crateriform edges or is crossed obliquely by north-to-south-trending ridges. At Mallet C the valley bends through 10-15° and becomes smaller: depth 3,900ft (1.5km), width 9 miles (14km), additional length 124 miles (200km).

Both parts of the valley point toward Mare Nectaris. These, and other, "selenofaults" radiate from the area covered by Mare Nectaris and may be cognate with it. *GF*.

Rho Cassiopeiæ A puzzling variable star near Beta Cassiopeiæ. Usually it is around magnitude 5, with slight fluctuations, but on rare occasions has fallen to below 6. Its type is unknown; the spectrum is F8.

R

Rhodium coating A metallic mirror coating formed by evaporating rhodium metal onto a glass substrate. Exceptionally hard and durable, but low reflectivity, 76 per cent. Difficult to remove.

Rho Ophiuchi Dark Cloud Despite its forbidding name, this nearby molecular cloud contains one of the most amazing collections of nebulæ in the sky. Rho Oph itself and other, similar, hot stars are embedded in dust and produce strongly blue reflection colours while the cool, red supergiant Antares is enveloped in an orange-yellow nebula of its own making. Smaller, yellowish nebulæ hint that a new generation of stars is forming deep inside the dark cloud, which can be traced over about 1,000 square degrees of sky.

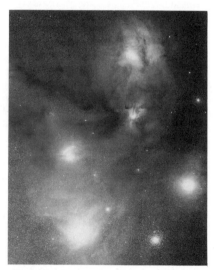

Rho Ophiuchi Dark Cloud: star-forming

R Hydræ A MIRA-type variable with a range of from 4.0 to 10.0. It lies 2°6 east of Gamma Hydræ. Its present period is about 386 days, but this has shortened from nearly 500 days in the early 18th century. Its variability was discovered by Maraldi in 1704.

Riccioli Lunar walled plain adjoining Grimaldi; 3°S, 75°W. Its maximum length is 100 miles (160km). It has very dark patches, but no large craters.

Riccioli, Joannes Baptista (1598–1671). Jesuit astronomer, of Bologna, who drew a lunar map in 1651 in which he gave names to the main craters; most of these names are still in use.

Richat Dome A circular area, in Mauritania, 30 miles (50km) wide, of disturbed and fractured rocks, with central dome, on a fault with smaller craters; probably volcanic.

Rich field adaptor Essentially a form of the telecompressor giving a wider field of view and a faster f ratio than a basic system — but for visual work only.

Richter scale The measure of earthquake magnitude, introduced by C.F. Richter. This is based on the amplitude of seismic waves as recorded by seismographs.

Riemann, Georg Friedrich Bernhard (1826–66). German mathematician who devised a geometry which contained no parallel lines and within which, for example, the angles of a triangle add up to more than 180 degrees (as on the surface of a sphere). Riemannian geometry is used in relativity theory.

Rigel Beta Orionis, the seventh brightest star in the sky. Its magnitude is 0.12 (very slightly variable); the spectral type is B8. The distance has been given as 900 light-years, in which case Rigel is some 60,000 times more luminous than the Sun. There is a seventh-magnitude companion at a separation of 9″2; no relative movement has been observed. Though lettered Beta, Rigel is the brightest star in Orion. It was 19 Orionis in Flamsteed's catalogue.

Right ascension (R.A.): A celestial co-ordinate, measured eastward from the VERNAL EQUINOX to the point where the HOUR CIRCLE of an astronomical body meets the CELESTIAL EQUATOR. It is occasionally measured in degrees, but more usually in SIDEREAL TIME and is the interval between the transit or CULMINATION of the true vernal equinox and that of the object. One hour equals 15°, one degree equals 4 minutes of time, and a full circle is obviously 24 hours. For example, RIGEL culminates 5h 14.5m after the vernal equinox, which is therefore its R.A. *TJCAM. See also* CELESTIAL SPHERE.

Ring galaxies Galaxies clearly distinct from spirals and ellipticals in which a ring of stars surrounds the nucleus like the rim of a wheel. Thought to arise as a ripple when a small galaxy passes through a larger one.

Ring micrometer A thin metallic ring, exactly circular, which is placed at the focus of the telescope objective, with its plane at right angles to the optical axis.

With the telescope remaining stationary throughout, the time at which stars appear and disappear behind the ring is noted. By this method accurate relative positions can be derived for stars, planets or comets.

It is necessary to know the radius of both edges of the ring in terms of ANGULAR MEASURE. This can be done by observing the TRANSITS of two stars whose DECLINATIONS are known accurately and whose separation is not much less than the diameter of the ring; for example, members of the Pleiades. Using an accurate stopwatch, the reappearance of each star behind the inner edge of the ring and the subsequent disappearance of the stars at the opposite edge are timed. This is repeated for the outer edge of the ring and the whole procedure should be carried out several times. When reducing the results the effects of differential refraction between the two stars must be considered.

To measure the difference in position of two objects, the telescope is directed so that DIURNAL MOTION takes them across the ring but as far from the centre as possible. Appearances and disappearances at both edges of the ring for each object are timed for maximum accuracy.

For objects that are moving relative to the stars, such as comets, minor planets or planets, a further correction is needed to allow for the movement of the object in question during the time taken to make the observation.

The main advantage of the ring micrometer is that it does not need an equatorial telescope or field illumination like the FILAR MICROMETER. However, observations are fairly lengthy, reduction is relatively complicated and the method is limited to wide pairs of stars. *RWA*[2].

Ring Nebula (M57, NGC6720). A PLANETARY NEBULA in Lyra, which was discovered by Antoine Darquier in 1779. It was he who first used the comparison between this nebula and a planet, leading William HERSCHEL to give currency to the term "planetary nebula".

The nebula has the form of a hollow ellipse centred on a blue star. The elliptical shape indicates that the three-dimensional shape is a toroid, or circular ring doughnut, seen at an inclined angle. The nebula is expanding at 12 mile/s (19km/s), and in fact over the 40 years in which large telescopes have been recording it, its image has grown larger overall by about 0.5 arc sec. This is the same expansion rate as a tree at a distance of 20 miles (35km) which is adding tree rings 1mm thick each year. The nebula is about 5,500 years old. *PGM*.

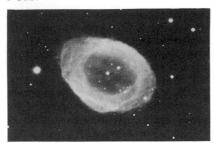

Ring Nebula: NGC6720, in Lyra

Riphæan Mountains Short but fairly conspicuous lunar mountain range in the Mare Nubium.

Ritchey-Chrétien telescope The Ritchey-Chrétien optical system is a modification of the CASSEGRAIN construction, having a hyperbolic primary mirror and an elliptical convex secondary. It is free of spherical aberration and coma over a wide field.

Ritchey, George Willis (1864–1945). American optical worker who figured the mirrors of the 60in (152cm) and 100in (254cm) reflectors at Mount Wilson.

R Leonis A long-period variable star, discovered by Koch in 1782; one of the largest stars known. Range 4.4 to 11.3 (visual); average period 312.43 days. The rise usually takes 43 per cent of the total period. Spectrum: M8IIIe. Average range 5.8 to 10.0(v). R.A. 9h 47.6m; Dec. +11° 26′ (2000.0).

R Leporis A MIRA-type variable in Lepus. It has a range of from 5.9 to 10.5, so that it is never a difficult object; the period is 432 days. It is of type N, and so red that it has been called "the Crimson Star".

R Monocerotis A star which has rarely, if ever, been glimpsed at optical wavelengths. The designation applies to the bright tip of a triangular nebula in Monocerotis, discovered by HERSCHEL and catalogued as NGC2261. The nebula is also known as Hubble's variable nebula and it is the association of the hidden star and the changing nebula which is so intriguing. The star is probably very young and still embraced in the dusty cloud from which it formed. The star itself is irregularly variable and its light is absorbed by the patchy dust in its vicinity, both effects which change the apparent brightness of the nebula. As is often the case with stars embedded in dust, R Mon and its nebula emit strongly in the infrared part of the spectrum. *DM*.

Roche One of the main craters on PHOBOS, the larger satellite of Mars.

Roche, Édouard Albert (1820–83). French astronomer and mathematician who studied the shapes of rotating fluid masses and who derived the ROCHE LIMIT for planetary satellites.

Roche limit Minimum distance that a satellite may have from its planet.

It was noted in 1848 by Édouard Roche, of Montpellier, that a satellite orbiting close to its planet is subjected to great stresses as the nearer portions try to orbit faster than the more distant parts. If close enough, these stresses exceed the strength of the rock, and the satellite will be ripped apart. Alternatively, it will never be able to form by ACCRETION. The Roche limit is the critical distance from the planet, usually quoted as 2.5 times the planet's radius, though the exact number depends on the composition of both bodies. The rings of Saturn lie within its Roche limit. *DAA*.

Roche lobes Surface defining the maximum sizes of stars in a binary system relative to their separation. If the star's surface gets closer than this, the companion's gravity will pull matter off it.

Rocket astronomy From 3,000Å – 10,000Å the Earth's atmosphere is fairly transparent and astronomical observations can be made from mountain tops or highly-flying aircraft. At shorter wavelengths, atmospheric absorption rapidly increases so that between 1,000Å and 10Å it is necessary to go to an altitude of 60 miles (100km) if 50 per cent of the radiation from a star or the Sun is to be detectable. At longer wavelengths than 1μm, the presence in the atmosphere of molecular species — H_2O and OH particularly — results in rapidly varying absorption with wavelength, until a wavelength of 1cm is reached. From here to about 20m, the atmosphere once again allows most of the incident radiation to pass through it, but at still longer wavelengths incoming rays are reflected by the charged particles of the ionosphere. Nearly all the radiative energy of the Sun passes through the atmosphere to low altitude, but only a fraction of the wealth of diagnostic information about its structure that is contained in its whole spectrum, from gamma rays to radio wavelengths, is allowed to pass. The same is true, perhaps *a fortiori*, for other stars and for external galaxies.

As soon as it became technically feasible to do so with the development of the high-altitude rocket, the quest began to find suitable experimental techniques to enable the solar spectrum to be measured over the widest possible range. The next step was to carry out sky surveys to bring about a similar broadening of the knowledge of stellar spectra, and at the same time to search for new objects in these previously unexplored wavebands. The result has been the development of "new astronomies" in broad wavebands from gamma rays to the infrared, though for this to come about, the availability of high-altitude rockets was primarily important in connection with the X-ray and ultraviolet parts of the spectrum. The increasing availability of orbiting spacecraft since the 1960s has resulted in most of this work now being done on satellites, to get longer observation times, though there is a limited range of solar UV observations still carried out by rocket, and they are still used for instrument development.

High-altitude rockets were first used in a programme of scientific research in 1946 when, at WHITE SANDS Proving Ground in the USA, captured German V2 rockets originally intended as weapons were modified for scientific use. Naturally enough, since the Earth's upper atmosphere was for the first time opened up to direct observation, this was a prime topic of the new research. But it was recognized from the start that an effective study of the upper atmosphere demanded a far better knowledge than was then available of the solar spectrum and its variations with time. The ultraviolet light and X-rays from the Sun which are absorbed in the atmosphere profoundly modify its composition, producing the charged particle layers of the IONOSPHERE, and these radiations also determine its temperature, density structure and dynamics.

The first rocket of the new programme was launched on April 16 1946. It was a dismal failure, but people have tried to do better ever since, with varying success.

In 1946, it was known that the solar spectrum in the visible approximates to that of a black body at 5,800K. It was realized that the ultraviolet component

R

R

in the tail of such a spectrum was insufficient to produce the ionosphere and would not explain its remarkable temporal variations. It was also recognized that evidence existed for the presence of a very hot SOLAR CORONA — this evidence included the white light corona observed at eclipse, the forbidden lines of highly excited atoms (the "coronal" lines which appear only in a high-temperature gas) and the Doppler widths of these lines. Moreover, it was understood that a corona at a temperature of a few million degrees K would very strongly enhance the X-ray spectrum of the Sun. But there was great reluctance to accept as real the existence of this layer in view of the difficulty in understanding the mechanism by which it is heated from the much cooler layers of the SOLAR PHOTOSPHERE (a problem which is still not resolved but which certainly involves wave motions generated in the convective layer below the photosphere).

An ultraviolet spectrograph launched on October 19 1946 by the Naval Research Laboratory on a V2 rocket showed clearly the extension of the spectrum from 3,000Å to 1,800Å as the ozone layer of the atmosphere was penetrated. Because of the falling intensity in the solar spectrum, observations at shorter wavelengths raised difficult experimental problems. These were progressively overcome and the solar spectrum has now been well explored to above 1MeV quantum energy (12Åm wavelength), with the aid of satellites as well as rockets.

At wavelengths less than 1,800Å, the familiar Fraunhofer absorption spectrum (see FRAUNHOFER lines), gives place to an EMISSION SPECTRUM of increasingly higher ionization states as the wavelength is reduced, indicating that the solar photosphere is overlaid with optically-thin layers of material at a range of temperatures up to about 5.10^6K. Observations of the ionospheric E-layer during a SOLAR ECLIPSE and subsequently direct solar images in X-rays each showed that these originated above the photosphere in the corona, confirming the high temperature of this region.

From 1962 onwards, spacecraft were increasingly employed to obtain solar spectra or spectroheliograms (see SPECTROHELIOGRAPH) of high angular and spectral resolution, because of the greater observing time available on a satellite than on a rocket. However, some special studies — such as the identification of spectral lines in very high resolution spectra — have continued to be done from rockets.

These studies of the Sun, and the new understanding of solar behaviour which came from them, naturally generated interest in the spectrum of other stars at short wavelengths. Instruments, usually detectors with filters giving them some sort of bandpass in the ultraviolet, were flown from the early 1960s onward. Thus, by 1961, observations had been made by the group working at GODDARD SPACE FLIGHT CENTER of about a dozen stars, either by filter photometers or objective grating spectrographs. A partial sky survey had also been made by a filter PHOTOMETER flown on a Skylark rocket from Woomera. However, the problems of instrument sensitivity are evidently far more severe than for the Sun, and satellites rapidly made rocket instruments obsolete, starting with the launch of the three Orbiting Astronomical Observatory satellites from 1966 onward, and culminating in the launch of the INTERNATIONAL ULTRAVIOLET EXPLORER in 1978.

As noted above, the existence of X-ray emission from the Sun was demonstrated in 1948. Only about 1 part in 10^6 of the solar luminosity is in this wavelength range, so that no ordinary star with similarly weak X-ray emission is detectable by the methods available to rocket astronomy. Nonetheless, the search for stellar X-ray emission began and eventually a rocket flight in 1962 by Giacconi, Gursky, Paolini and Rossi achieved sufficient sensitivity to detect a cosmic X-ray source, named Scorpius X-1. Soon after, Friedman detected a second source, Taurus X-1, which was later shown by the NRL group by an eclipse observation to be the X-ray counterpart of the CRAB NEBULA. Two years later, in 1966, the counterpart to Scorpius X-1 was found to be an inconspicuous blue star for which the X-ray luminosity is 10^4 *greater* than the optical luminosity. Thus the existence of a new class of astronomical object, whose emission was primarily at X-ray wavelengths, was demonstrated. Over the next 10 years, rocket surveys revealed about 30 sources, two of which were extragalactic (the active galaxy M87 and the nearest quasar 3C 273). *APW*.

Rockets A rocket is propelled by the reaction of escaping exhaust products, according to Newton's Third Law: "To every action there is an equal and opposite reaction." The amount of thrust produced is determined by the energy inherent in the propellant, the rate at which propellant is consumed and the exhaust velocity.

The big advantage of liquids is that the rocket engine can be throttled and stopped and re-started simply by closing and opening valves. Most liquid propellants have two components, a fuel (eg kerosene) and an oxidant (eg liquid oxygen). Some liquids like unsymmetrical dimethyl hydrazine and nitrogen tetroxide are hypergolic: they ignite spontaneously on contact. Contained in separate tanks, the fuel and

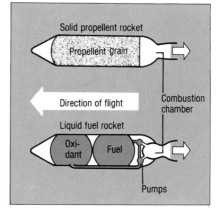

Rockets: reaction motors for space

oxidant are fed to the combustion chamber by high-pressure gases or turbopumps. There are also monopropellants in which a single liquid component is decomposed to produce the rocket efflux. A solid propellant, like liquid monopropellants, is one in which fuel and oxidant are combined, usually as a powdery or rubbery mixture. The mixture, called the "propellent grain", is packaged in the rocket casing, which serves both as storage and combustion chamber. Such rockets are simple to construct, requiring neither pumps nor valves, and they can be stored for long periods — but combustion is difficult to control or stop.

Solids find their greatest utility as boosters, which assist the launch of larger, liquid-propellant vehicles like Titan IIID and the SPACE SHUTTLE. Another form of propulsion is ion-drive, a low-thrust system for the attitude control and orbit positioning (station keeping) of satellites. In this concept, ions (atoms unbalanced by the removal of one or more electrons) are formed by passing a propellant such as cæsium through an ionizing device and accelerated to high speeds by electric fields. The ions can derive their propulsive energy from solar batteries. Larger systems, powered by a nuclear reactor and running for long periods, may propel future spaceships. *KWG*.

Rømer, Ole (1644–1710). Danish astronomer. In 1675 he determined the velocity of light (from observations of Jupiter's satellites) and later devised many instruments including the first TRANSIT instrument.

Rook Mountains Lunar mountains associated with Mare Orientale. Recent analysis has shown that they are composed of ANORTHOSITE.

Roque de los Muchachos Observatory
Spanish observatory on the "Mountain of the Boys" on La Palma, Canary Islands, containing Anglo-Dutch, Swedish, Anglo-Danish and Nordic optical telescopes.

The Canary Islands were discovered, astronomically, in 1856 by the Astronomer Royal for Scotland, Charles Piazzi Smyth. In his *Optics* (1730), Isaac NEWTON had written: "[Telescopes] cannot be formed as to take away that confusion of rays which arises from the tremors of the atmosphere. The only remedy is a most serene and quiet air, such as may perhaps be found on the tops of the highest mountains above the grosser clouds." Following this precept, Smyth site-tested on peaks on Tenerife, particularly Guajara (8,900ft/2,717m). Tenerife was the base for several astronomical expeditions in the following hundred years, including the solar eclipse expeditions of 1959, but no permanent observatory was set up until 1970. The following decade saw the Canary Islands established as a major astronomical centre, with observatories on two islands.

The foremost of these, in terms of its telescopes and the quality of the site for night-time observations, is Roque de los Muchachos Observatory, at a height of 7,870ft (2,400m) on the island of La Palma. It belongs to the Instituto de Astrofisica de Canarias (IAC) and is the result of international agreements between Spain, the United Kingdom, Denmark and Sweden. The observatory was inaugurated in 1985, ten years after the first meetings between the international authorities which led to international agreements in 1979. The agreements established an International Scientific Committee (known by its Spanish initials as the CCI), which governs the observatory. The telescopes remain, however, the property of the nations that built them.

In partnership with the Netherlands and the Republic of Ireland, the United Kingdom has established on La Palma the 165in (4.2m) Herschel Telescope,

the 100in (2.5m) Isaac Newton Telescope and the 39in (1m) Kapteyn Telescope. In a joint effort, the United Kingdom and Denmark have built and now operate the Carlsberg Automatic Meridian Circle. Sweden has built a 24in (60cm) solar telescope and a 24in (61cm) photometric telescope; and together with the other Nordic countries (Norway, Denmark and Finland) will have completed by 1988 a 2.5m telescope.

Spain provides the infrastructure for the observatory, including roads, power, communications, etc, and receives 20 per cent of the observing time on each instrument. A further five per cent of the observing time is available for collaborative projects for the countries that are attached to the international observatories. The remaining time is divided between the countries which participate in the operation of each telescope, according to numerous separate agreements; for example the United Kingdom and the Netherlands

Roque de los Muchachos Observatory

share the 75 per cent remaining time on the Herschel Telescope in the ratio of 4:1. The Anglo-Dutch telescopes are "common user" telescopes, operated for any astronomer from the appropriate countries who can propose challenging observations to make with them. About 400 astronomers visit La Palma each year to use the telescopes there.

The British interest in La Palma stemmed from its desire to have access to an excellent northern-hemisphere observatory to complement British astronomers' excellent facilities in the south, namely in Australia (154in/3.9m Anglo-Australian Telescope) and in South Africa (the telescopes of the South African Astronomical Observatory at Sutherland). After a year-long programme of site-testing in 1974-75, during which six sites in the Mediterra-

nean, North Atlantic and Hawaii were tested by modern meteorological and astronomical observations, the United Kingdom chose to negotiate for La Palma as the site for its telescopes. The island dips steeply into the sea and opportunity for large-scale tourist development (as in the major islands of Tenerife and Grand Canary) is limited. The vertical sides of the central caldera rise to the Roque de los Muchachos itself and the telescopes are situated near this summit. The site-testing of 1974-75, confirmed by the first years of telescope operation, showed that 55 per cent of the nights on the Roque were totally clear and 26 per cent were otherwise usable (with some cloud). The smooth contour of the mountain is presented to the northerly oceanic wind which flows laminarly over the hillside, and 42 per cent of the nights have seeing which is better than 1.0 arc second. The chosen site for the observatory is an excellent compromise between communication, infrastructure and technical base (all of which imply proximity to a sufficiently large population) and height, seeing, dark sky and freedom from pollution (all of which imply distance from people). In fact the geographical position of La Palma makes it competitive with the best sites in the world for optical astronomy independently of the base of infrastructure originally existing on the island. The weather of the western Canaries is oceanic, dominated by the cold Canary Current, flowing southward in the continuation of the clockwise circulation of the Gulf Stream in the North Atlantic Ocean. As the weather systems drift from the west the cold current produces a contraction of the convection at the ocean surface, and the inversion layer below which the cloud is contained lies at approximately 5,000ft (1,500m). Thus the highest points on La Palma protrude into clear air.

The sky over the island of La Palma is recognized by Spain as a natural resource of the Canary Islands, and Spain passed in 1986 the most comprehensive laws anywhere to protect it for astronomical observations. The law forbids lights that shine above the horizontal, permits street lights only of approved types, limits radio transmissions to a low output and restricts industrial development such as quarrying above 1,500m, the height of the inversion layer. This protection to avoid compromising the heights of the island is intended to ensure the observatory's life well into the 21st century. *PGM.*

R

Rosette Nebula This aptly-named nebula is almost two degrees in diameter and surrounds NGC2244, an open cluster of hot, bright stars in the constellation of Monoceros. The stars are very young; it is probably less than 500,000 years since they formed from the nebula which they illuminate. During

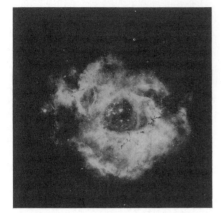

Rosette Nebula: 4,500 light-years away

this time radiation pressure from the stars has blown away material from the central part of the nebula, creating a cavity about 4pc (12 light-years) across. This expanding cavity will rapidly move through the dusty cloud and disperse it in a few million years, leaving NGC2244 as an undistinguished cluster of young stars on the outskirts of the Milky Way. The Rosette Nebula and its retinue of stars are at a distance of 1,500pc (about 4,500 light-years). *DM*.

Rosse, 3rd Earl of (1800-67). William Parsons, the 3rd Earl of Rosse, will be remembered as being the man who, unaided except by labourers whom he trained as assistants, built what was

3rd Earl of Rosse: built huge telescope

then much the world's largest telescope, and used it to make fundamental discoveries. *See also* BIRR CASTLE ASTRONOMY.

Rosse, 4th Earl of (1840-1908). Son of the 3rd Earl. He continued his father's work at Birr, though he concentrated upon making the first successful measurements of the heat received from the Moon. *See also* BIRR CASTLE ASTRONOMY.

Rotating stars All stars probably rotate, but their rates of rotation show a wide range. The Sun rotates on its axis with a period of about 28 days, as is shown by the motion of sunspots across its face. This is typical of the cooler dwarf stars, but the hot (O, B and A) stars generally rotate much faster. The fastest rotation among dwarfs occurs in B stars where periods of only a few hours can occur. The equators of such stars have velocities of 250 miles/s (450km/s), compared with only 2km/s on the Sun. Giant stars of all spectral types rotate fairly slowly and the largest supergiants take over a year to turn around.

Compact stars can and often do rotate much faster than dwarfs. Some noteworthy examples are the white dwarf in the nova remnant DQ HERCULIS, which rotates once every 71 seconds, and the millisecond PULSARS, which are NEUTRON STARS rotating nearly one thousand times a second. The immense gravitational fields of these stars prevents them from flying apart at such extreme rates of rotation.

The Sun does not rotate like a solid body: at the surface the polar regions rotate more slowly than the equator, and recent studies of SOLAR OSCILLATIONS suggest that the interior of the Sun rotates at about twice the rate of the surface. With the exception of A stars and stars with very strong magnetic fields, such differential rotation is expected to be common. Although most stars may be rotating quite rapidly when they are formed, slow mass loss through STELLAR WINDS, in combination with the normal magnetic fields of stars, produces magnetic braking, which slows the rotation. Old stars therefore rotate slower than stars recently formed.

Rotation in stars is measured in several ways: from variations in brightness of the star as a whole, from variations in the strengths of its emission lines (associated, as in the Sun, with regions of spot activity), from periodic variations in magnetic field, from broadening of the spectrum lines by the different Doppler shifts over the surface of the star, and from variations in radio or X-ray emission. *BW*.

Rotation The spinning motion of a body around an axis. Some degree of rotation seems to be a general property of all classes of celestial body. *See also* REVOLUTION.

Rowland, Henry Augustus (1848–1901). American physicist who devised a method of ruling DIFFRACTION GRATINGS more finely than previously (he achieved nearly 6,000 lines per centimetre). He also made a detailed map of the solar spectrum.

Rowton meteorite The only British iron, 7lb 11oz (3.5kg); it fell in 1876.

Royal Astronomical Society Founded in 1822 as the Astronomical Society of London; the main professional organization in Britain. Its headquarters is at Burlington House, Piccadilly.

Royal Greenwich Observatory *See* GREENWICH OBSERVATORY.

Royal Observatory, Cape of Good Hope Proposed in 1821 and built 1825-28 near Cape Town to serve as a southern hemisphere equivalent of the Royal GREENWICH OBSERVATORY. In the 19th century, under the successive Directors FALLOWS, HENDERSON Thomas Maclear and E.J.Stone, the observatory produced the bulk of accurate star positions in the southern sky. Under GILL the work of the observatory expanded to include construction of star catalogues from photographic surveys of the sky, measurement of stellar parallaxes and observations of spectra.

In the 20th century, particularly under the Directorship (1923-33) of Sir Harold Spencer Jones, large numbers of parallaxes and proper motions were measured photographically. In the 1950s and 1960s extensive measurements of stellar magnitudes and spectra were made.

At the end of 1971 the observatory, which had been run for 140 years by the British Admiralty (and later the Scientific Civil Service), was renamed the SOUTH AFRICAN ASTRONOMICAL OBSERVATORY. *BW*.

RR Lyræ variables are often called short-period CEPHEIDS or cluster-type variables. The latter name originated when S.I.Bailey, in 1895, discovered stars with Cepheid-like variations but extremely short periods of a fraction of a day. All his discoveries were in GLOBULAR CLUSTERS and were revealed by an examination of plates taken of these clusters at the Boyden Station in Peru. It was soon found that some of the brightest globular clusters contained many of these stars and within a few years hundreds had been found.

These rapidly-pulsating stars have light curves that differ from the classical Cepheids. Most rise very quickly to maximum in only a tenth or less of their total period. Their minima are comparatively prolonged, so that for a few hours their light remains constant. Their periods range from $0^{d}.2$ to $1^{d}.2$, and their amplitudes are from $0^{m}.2$ to two magnitudes. In 1900 the first short-period Cepheid was found outside of a globular cluster; it was discovered by W.P.Fleming, at Harvard, and was RR Lyræ, the type star of these variables. At first it was thought that this star must have escaped from a globular cluster, but as more and more such stars were discovered away from globular clusters this idea was dropped. There are now more than 2,000 of these stars known, and roughly half of them are found in clusters. Most have periods between 9 and 17 hours with a clustering around 13 hours. In the Small MAGELLANIC CLOUD some stars have periods as long as two days, but apart from these there is a fairly sharp cut-off in periods after $1^{d}.2$ days.

The spectral types of RR Lyræ variables are A – F. They are giants; those in our Galaxy belong to the spherical component. Some are known to have variable light curve shapes as well as variable periods. If these changes are periodic, they indicate what is called the BLASHKO EFFECT. Their maximum expansion velocity of their surface layers almost coincides with maximum light: in this respect they resemble the classical Cepheids. Some stars show two simultaneous operating pulsating modes, the fundamental and the first overtone; such stars are placed in a sub-type of RR Lyræ variables. Another sub-type has stars with asymetric light curves, with steep rises and periods from $0^{d}.3$ to $1^{d}.2$. Their amplitudes range from $0^{m}.5$ to two magnitudes. The type star, RR Lyræ, belongs to this type. Stars of another sub-type have asymmetric light curves with periods of $0^{d}.2$ to $0^{d}.5$ and amplitudes that do not exceed $0^{m}.8$. The absolute magnitude of RR Lyræ variables is about $+0.5$. They are too faint to be

seen in any but the nearest EXTERNAL GALAXIES, such as the dwarf system in Sculptor. Those found in the Magellanic Clouds are at about the observable limit.

RR Lyræ variables can be used as distance indicators out to about 200kpc. The method used depends on the star's motion in space. It is assumed that all the stars in a globular cluster are at about the same distance and therefore have the same velocities in space. The RADIAL VELOCITY is one-third of the total space velocity. With these assumptions the approximate distance to a globular cluster is found and this, in turn, gives its absolute magnitude. From the absolute magnitudes of clusters the size of the Galaxy can be calculated, since the globular clusters form a halo around it. It is then possible to extend these measurements to the nearest external galaxies. *FMB.*

RS CVn stars RS Canum Venaticorum is the type star for a distinctive set of binaries in which one star stirs up the other to produce giant star spots. They produce flaring radio emission.

R Serpentis A MIRA-type star in the head of Serpens (Caput). The range is from 5.7 to 14.4, but the period is 357 days, and as it comes to maximum only about a week earlier every year there are spells when maxima are difficult to observe at all — when the Sun is too close in the sky.

R stars The R spectral type was introduced as part of the original Harvard scheme of spectral classification and described stars with spectra identical in all ways with those of G and K stars but with the addition of absorption bands of molecular carbon. It is now known that the R stars are almost all giant stars and are the hotter members of the class known as CARBON STARS. Their atmospheres are rich in carbon mixed from the material processed through nuclear reactions near the centre of the star.

The R spectral type was divided into ten subclasses R0–R9, which approximate a temperature sequence analogous to the G and K giants. This classification scheme is often replaced now with the later comprehensive classification of carbon stars used by Keenan and Morgan (*see* MORGAN-KEENAN CLASSIFICATION) in 1941, which has types C0–C9, which accurately form a temperature sequence. The R0 to R3 or R4 classes correspond to the C0–C3 types, but the C4–C9 types

embrace both the R5–R9 and all of the N-type spectra. Strengths of the carbon features in the spectra are indicated by addition of a further number, eg, C2.3 or C6.5. A number of revisions to the classification scheme have been made, but the basic objectives remain — that of providing a system of spectral types which give both temperature and abundance indices.

Among the R stars there is a small fraction of peculiar stars. The hydrogen-deficient carbon stars show bands of molecular carbon and carbon compounds but have weak or absent hydrogen lines or CH bands. The R CORONÆ BOREALIS VARIABLES fall in this class, some members of which are hotter than the usual G–K equivalent of the R stars, and they are considerably more luminous than the normal R stars. *BW.*

Rümker Semi-ruined lunar plateau on the Sinus Roris; diameter 30 miles (49km); 41°N, 58°W.

Rupes Ridges or cliffs on planetary surfaces.

Russell, Henry Norris (1877–1957). American astronomer who, in 1913, published a diagram which plotted the absolute magnitudes of stars versus spectral type which, independently of Hertzsprung's work, revealed the division between MAIN SEQUENCE stars and giants. He first suggested (correctly) that the Sun and stars are composed mainly of hydrogen and proposed (incorrectly) that stars progress along the main sequence as they evolve.

Rutherfurd, Lewis Morris (1816–92). A highly-regarded American astrophoto-

Lewis Rutherfurd: astrophotographer

R

grapher of the Moon, Sun, planets and stars who subsequently turned to spectroscopy using improved DIFFRACTION GRATINGS of his own design.

Rutilicus Zeta Herculis, a fine binary with a period of 34 years.

RV Tauri variables A small group of highly luminous, pulsating variable stars. They are mainly yellow or orange supergiant stars in spectral classes G and K, with some F stars. Examples are RV Tauri, R Sagittae and R Scuti. They have enormous, very extended atmospheres of gas that emit infra-red radiation. The light curves of the RV Tauri stars are very characteristic, with alternate deep and shallow minima. Their periods range from about 30 to 145 days. Sometimes the light variations can become rather irregular, particularly for the stars having the longest periods. For this reason the RV Tauri stars are classified as semi-regular variables. They also have a variation in their colour index which looks like the light curve, but goes through its maximum shortly before the star reaches minimum brightness. The colour index is an indication of the temperature of the star. A small nunber of RV Tauri stars have double periods, with a rapid oscillation being superimposed on a much slower oscillation, having a greater amplitude. *JWM.*

R

RW Aurigae variables A sub-class of the group of very young stars known as nebular variables or T TAURI STARS. The RW Aurigae stars have quite large amplitudes and are almost all G-type dwarfs. Very few of them are brighter than tenth magnitude at maximum. They show rapid and extremely irregular variations in their light curves. With some stars there does however appear to be a pseudo-periodic wave superimposed upon the primary light variations. A large number of RW Aurigae stars are found associated with the T Orionis stars.

Ryle, Sir Martin (1918–84). British pioneer of radio astronomy; Astronomer Royal from 1972 to 1982. He made valuable contributions to COSMOLOGY and was awarded a Nobel Prize in 1974 for his work on PULSARS.

Ryle began work at the Cavendish Laboratory, Cambridge, after World War II. In order to study solar emission at 1m wavelengths he chose to build an interferometer, thus introducing the kind of equipment for which Cambridge is now famous.

Sabæa Terra Dark albedo feature on Mars, around 1°S, 325°W; formerly known as Sinus Sabæus.

Sabine Lunar crater on Mare Tranquillitatis; diameter 19 miles (31km); 2°N, 20°E. It has an almost identical twin close by, Ritter.

Sagitta (the Arrow). A small but easily found constellation in the Aquila area. Its brightest star is Gamma (3.47; absolute magnitude −0.3, spectrum K5, distance 166 light-years). Then comes Delta (3.82). Both Alpha and Beta, which complete the pattern, are of magnitude 4.4.

Sagittarius (the Archer). The southernmost of the Zodiacal constellations, and notable as containing the lovely star-clouds which mark the direction of the centre of the Galaxy. There is no really well-marked pattern, though the outline is sometimes likened to that of a teapot!

Brightest stars

Name	Visual Mag.	Abs. Mag.	Spec.	Distance (light-yrs)
ε (Kaus Australis)	1.85	−0.3	B9	85
σ (Nunki)	2.02	−2.0	B3	209
ζ (Ascella)	2.59	0.6	A2	78
δ (Kaus Meridionalis)	2.70	−0.1	K2	82
λ (Kaus Borealis)	2.81	−0.1	K2	98
π (Albaldah)	2.89	−2.0	F2	310
γ (Alnasr)	2.99	0.2	K0	117
η	3.11	−2.4	M3	430
φ	3.17	−1.2	B8	245
τ	3.32	0.0	K1	130

Also above magnitude 4: Xi² (3.51), Omicron (3.77), Mu (3.86), Rho¹ (3.93), Beta¹ (3.93) and Alpha (3.97).

There are plenty of variable stars, including the Cepheids X and W, which range between magnitudes 4 and 5; the Milky Way is at its richest, and there are no less than 15 Messier objects — globular clusters, open clusters and gaseous nebulæ, including the famous LAGOON NEBULA, of magnitude 6. Curiously, both Alpha (Rukbat) and Beta (Arkab) are relatively faint. *PM.*

Sagittarius A The very centre of our GALAXY.

The centre of our Galaxy has been likened to a dustbin. Gravity is so effective at pulling things there that almost any type of object might have found its way into the inner few light-years. We lack the ability to look into that dustbin, because of intervening banks of opaque dust, and must explore it by the techniques of INFRARED and RADIO ASTRONOMY, and to a lesser degree using X-ray satellites. In so cluttered a region, our view is confused and our interpretation changes rapidly. The description given here corresponds to the 1986 view.

Sagittarius A (Sgr A) is the name given to the most intense part of the

Sagittarius A: galactic centre

Milky Way's radio emission. Detailed maps, mostly made by the VERY LARGE ARRAY in New Mexico, show remarkable structures, including a series of narrow parallel streaks at right angles to the plane of the Milky Way, which radiate by the SYNCHROTRON mechanism but whose nature and origin remain a puzzle. Within 3 light years of the centre is a theta-shaped pattern. The ring of the theta is made of gas that probably has piled up there as a result of some outflow from the centre; alternatively it may be the tattered remnant of one or more nebulæ that were shredded by moving in the gravitational pull of the central stars. The bar of the theta, also gaseous, is surprisingly hot and apparently had a different origin.

Close to the bar is an intense, variable radio source known as Sgr A*, which has extremely small size, less than 20AU across. Probably coincident with this is a variable X-ray source as well as an object emitting positrons (anti-electrons) at a variable rate. The origin of these bizarre emissions is believed to be a BLACK HOLE. Some estimates place the hole's mass at five million solar masses, but recent evidence from infrared observations suggests a mass only one hundred times that of the Sun. A small black hole at the Galaxy's heart can be thought of as a very miniature QUASAR.

The black hole does not provide much of the light from the Galaxy's central regions, which instead arises from an infrared source called IRS 16, comprising several compact groups of young, hot stars that probably formed a few million years ago. These groups intermingle with the older and dimmer stars that cluster together as densely as in a GLOBULAR CLUSTER. Whatever else lies in this region, the densest collection of these old stars, when eventually it can be exactly located, defines the true Galactic centre. Whether this coincides with Sgr A* remains to be determined. *DAA.*

Sagittarius B2 Dark cloud of gas and dust lying near the centre of our GALAXY, found by radio astronomy.

Saha, Meghnad (1894–1956). Indian astrophysicist who derived the formula that links the degree of IONIZATION in a gas to temperature and electron pressure and which is vital to the interpretation of stellar spectra.

Saint Michel Observatory The "Observatoire de Haute Provence" (OHP) was established for astrophysics at St. Michel, France, in 1946, with a 76in (193cm) telescope combined with a multi-gratings spectrograph, a 60in (152cm) spectrographic telescope, a 47in (120cm) reflector, a 16in (40cm) prism-objective refractor, a SCHMIDT TELESCOPE and several dedicated instruments. Major results have been obtained on stellar and galactic spectroscopy, radial velocities and comets.

Saint Michel Observatory: in Provence

Saiph The star Kappa Orionis, magnitude 2.06. It is very luminous and remote.

Sakigake ("Pioneer"/MS-T5). Probe launched in January 1985 by Japan's Institute of Space and Astronautical Science (ISAS) to flyby HALLEY'S COMET on March 11 1986. It was a spin-stabilized spacecraft of cylindrical shape with a diameter of 460ft (140m) and a height of 27in (70cm), and weighed 304lb (138.1kg). Flyby distance was 4.3 million miles (6.9 million km) on the sunward side and the principal investigation was of solar wind/cometary interaction at large distances from a comet.

Salpeter process Usually called "triple-alpha", this is the process of conversion of helium ultimately into carbon. It occurs in sufficiently massive stars after the exhaustion of hydrogen in the interior.

Salyut vehicles Soviet space stations, about a quarter the size of SKYLAB and weighing 18.5 tonnes. Crews were ferried up to Salyut by the SOYUZ ferry craft. Salyut 1 was launched in 1971 and by 1982 seven had been placed in

Salyut vehicles: Salyut 1, 1971

orbit. Commencing with the Salyut 6 flight, automatic ferry flights (called Progress) have carried supplies and fuel to Salyut, this enabling crews to endure very long periods in space. Although Salyut 7 is still in orbit (1987), it appears that it is now being used in conjunction with the new MIR space station. *HGM.*

S Andromedæ The first extragalactic SUPERNOVA to be detected. Discovered near the centre of the Andromeda Nebula (M31) in 1885, the star attained an apparent magnitude of +6.5. Supernovæ were then not recognized as a distinct class of objects. The assumption that the star was an ordinary NOVA led to a distance estimate for M31 of about 8,000 light-years, ie, well within our

own Galaxy. Later it was realized that M31 was a galaxy similar to our own and that S Andromedæ, shining temporarily with a sixth of the total light of the galaxy, was vastly more brilliant than a typical nova. *FRS.*

San Marco Italian platform in the Indian Ocean off the coast of Kenya from which rockets and satellites have been launched.

Sappho. Feature on Venus: 13°N, 27°W. It seems to be a crater 125 to 185 miles (200 to 300km) across, on an elevated region with linear features radiating from it; presumably it is volcanic in origin.

Sargas The star Theta Scorpii, magnitude 1.87 — the southernmost of the bright stars in Scorpius.

Saros A cycle of 18 years 11.3 days after which the Sun, Moon, and the NODES of the Moon's orbit return to almost the same relative positions. It is due to the REGRESSION OF THE NODES. The Saros was known to the ancient Babylonian astronomers, and was used to predict eclipses, since any eclipse is usually followed by a similar one 18 years 11.3 days later. Slight variations from cycle to cycle mean that the eclipses are not identical. For example, the total solar eclipse of July 10 1972 was visible in parts of Alaska and Canada: the next one in the cycle, on July 22 1990, will be visible in parts of Finland and Russia. *TJCAM. See also* SOLAR ECLIPSES.

SAS II *See* GAMMA RAY ASTRONOMY.

Satellites *See* ARTIFICIAL SATELLITES, essay on MOONS AND MINOR PLANETS, and PLANETARY SATELLITES.

Saturn The sixth planet in the Solar System and the most distant object known to ancient man who observed the heavens before the development of the telescope. It is the second largest planet with an equatorial diameter of 74,500 miles (120,000km) and orbits the Sun at a distance of 9–10.1AU every 29.5 years. The rotational period of the planet, which corresponds to the length of the Saturn day, is 10 hr 39.4 min. The polar diameter of Saturn is only 67,000 miles (108,000km) so that it is the most oblate planet in the Solar System. The density of Saturn is only 0.7g/cm³, which is less than that of water and is lower than any other known planet. The planetary albedo is

S

S

32.8 per cent and its brightness temperature 93.6K, which is greater than would be expected from a black body at this distance. Consequently, Saturn behaves like a failed star and emits more than 1.8 times the energy it receives from the Sun.

The telescopic appearance of the planet is dominated by the majestic system of rings, which were probably first seen by Galileo in 1610, even though he did not recognize their nature (he believed Saturn to be a triple planet). The rings lie in the plane of the planet's equator and are tilted by 27° with respect to its orbit. Consequently, the faces of the rings will be alternately inclined towards the Sun and then the Earth by up to 27°. At intervals of approximately 15 years, the rings become edge on to the Earth and are virtually invisible to the observer. This situation last occurred in 1979–80. The rings are extremely reflective, so that they can add significantly to the total brightness of the planet. The distance from one edge of the rings to the other is more than 173,000 miles (279,000km), which is nearly three-quarters the distance of the Earth to the Moon. Saturn is now known to have at least 23 satellites.

Saturn is composed of hydrogen and helium, but unlike Jupiter, they are not in solar proportions. There is a helium depletion in Saturn where the mass fraction is only 11 per cent. This significant difference, when compared with Jupiter, may be due to the differing internal structures and current stages of evolution of the two planets. It is thought that the internal temperatures are too low for helium to be uniformly mixed with hydrogen throughout the deep interior. Instead, the helium may be condensing at the top of this region, where the gravitational energy is then turned into heat. This process may have started 2,000 million years ago when the temperatures first dropped to the helium condensation point. For Jupiter, this situation can only have been reached recently. The other primary constituents of the atmosphere are ammonia, methane, acetylene, ethane, phosphine and water vapour.

Saturn has a similar visible appearance to Jupiter, with alternating light and dark cloud bands, known as zones and belts, respectively. However, Saturn's clouds are more subtle and yellowish and therefore less colourful than those found on Jupiter. Consequently, the cloud features and associated spots, although varied in nature and colour, are less prominent on

Saturn. Several stable ovals of various colours (white, brown and red) have been observed in the Saturn atmosphere. The white clouds are composed of ammonia particles and the other colours are generated through dynamical and photochemical actions and reactions in the atmosphere. Three brown spots, situated at 42°N during the recent Voyager encounters, were seen to behave in a similar fashion to the Jovian white ovals. The largest features seen are a reddish cloud 6,200 by 3,700 miles (10,000 by 6,000km) at 72°N and a red spot 3,100 by 1,800 miles (5,000 by 3,000km) at 55°S. These features demonstrate the non-uniqueness of the Jovian Great Red Spot (GRS) and other cloud ovals. A major

Saturn: a Voyager 1 photo taken in 1980

characteristic of the Saturnian mid-latitude weather systems is a JET STREAM, which produces alternate high- and low-pressure systems in the same way as the terrestrial phenomenon. Anti-cyclonically rotating cloud systems, like the Jovian GRS and trains of vortices, familiar in the Earth's atmosphere, are also seen. The Saturnian weather systems, like those of Jupiter, are strongly zonal. At the equator the cloud top winds reach 1,600ft/s (500m/s), which is equivalent to three-quarters of the speed of sound at this level. Although Saturn has a strong internal heat source, the weather systems of Saturn (and of Jupiter and the Earth) are driven by the transport of energy from small-scale features into the main zonal flow.

The interior in Saturn is thought to consist of an Earth-sized iron-rich core of ammonia, methane and water, which is enclosed by about 13,000 miles (21,000km) of liquid metallic hydrogen above which extends the liquid molecular hydrogen and the extensive cloud layers. It is in this metallic hydrogen region that the magnetic field is created

by the dynamo action from the rapidly-rotating planet. The first detection of the Saturnian magnetic field was made from the Pioneer 11 spacecraft in 1979. At the cloud tops the equatorial field has a strength of 0.22G compared with the Earth's value of 0.3G. The magnetic axis is within 1° of the axis of rotation, and is therefore the least tilted field in the Solar System.

The MAGNETOSPHERE of Saturn is intermediate between those of the Earth and Jupiter in terms of both extent and population of the trapped charged particles. The average distance of the BOW SHOCK is 1,100,000 miles (1,800,000km), while the magnetosphere itself lies much closer to the planet at 310,000 miles (500,000km). These distances are of course extremely variable, since their precise positions will depend on the temporal behaviour of the solar wind. Consequently, the largest satellite, Titan, is situated at the magnetosphere boundary and regularly crosses this division. The magnetosphere is divided into several definite regions. At about 250,000 miles (400,000km) there is a torus of ionized hydrogen and oxygen atoms and the plasma's ions and electrons spiral up and down the magnetic field lines, contributing to the local field. Beyond the inner torus, there is a region of plasma that extends out to 620,000 miles (1 million km) produced by material coming partly from Saturn's outer atmosphere and also from Titan. The magnetotail has a diameter of about 80R_s. There is a strong interaction between the charged particle environment and the embedded satellites and rings which surround Saturn. All these bodies absorb the charged particles and have the effect of sweeping a clear path through the region where they are located. There is also auroral activity in the polar regions where the charged particles cascade into the upper atmosphere. The Saturnian auroræ are about two to five times brighter than the equivalent terrestrial phenomena. Saturn is also a powerful radio source emitting broad band emissions in the range from about 20KHz to about 1MHz; The maximum intensity occurs between 100 and 500KHz with a period of 10hr 39.4min, which corresponds to the SYSTEM III rotation period. The ionosphere, which resides above the neutral mesosphere, contains mainly ionized hydrogen whose population is strongly controlled by the solar wind.

The interaction between the charged particle environment and the rings is unique at Saturn, where radial features or spokes have been seen in the B ring. These features are confined to the central portion of the B-ring in a region from 26,700 to 35,400 miles (43,000 to 57,000km) above the cloud tops and appear to correspond to a location where only small micron size particles are present. The ring particles become electrically charged and appear as a torus above the plane of the rings. Each spoke, which is generally wedge-shaped and about 3,700 miles (6,000km) long, has a lifetime of a few planetary rotations. The detection of electrostatic discharges (lightning) at radio wavelengths does suggest that this mechanism is related to the formation of the spokes. *GEH*.

Saturn Nebula (NGC7009). A planetary nebula with *ansæ* or handle-like protuberances extending from the long axis of its slightly oval shape, giving it the appearance of Saturn. The nebula is very small, but extremely bright, and reveals an internal structure which is probably a series of shells ejected by the moribund central star.

Saturn rockets Large American rockets designed for manned spaceflights. The largest, Saturn V, was used for the APOLLO mission to the Moon. *See also* essay on EXPLORING SPACE.

Saturn rockets: Apollo's Saturn V

Scale height The height over which a physical quantity (eg, atmospheric density or pressure) declines by a factor of e ($e = 2.71828$). For example, at an altitude of one scale height, the density of an atmosphere is

$$\frac{1}{e} (= 0.37)$$

of the value at its base.

Scarp A sharp break in slope between two surfaces, usually developed by erosion along a boundary between lithologies of differing resistance, or a fault plane.

Scattering The deflection of electromagnetic waves by particles. Where the particles are very much larger than the wavelength, scattering consists of a mixture of reflection and DIFFRACTION, and is largely independent of wavelength. Where the particle sizes are very much smaller than the wavelength, the amount of scattering is inversely proportional to the fourth power of wavelength (Rayleigh scattering) and so blue light is scattered about ten times more efficiently than red. The blueness of the sky is due to Rayleigh scattering of sunlight by atoms and molecules in the atmosphere. The setting sun appears red because far more blue than red light is scattered out of the line of sight and thus a greater proportion of red light penetrates to the observer. Scattering is described as elastic if no change in wavelength is produced, and inelastic if the wavelength is increased or decreased by the encounter. INTERSTELLAR DUST produces both scattering and ABSORPTION of starlight; the combined effect is known as extinction. The amount of extinction is inversely proportional to the wavelength. *IKN*.

Scheat Beta Pegasi, in the Square. It is a semi-regular variable with a period of about 35 days; but the period is by no means regular; and the magnitude range is small — about 2.4 to 2.8. It is an orange star.

Scheiner, Christopher (1575–1650). A German Jesuit astronomer, who was for some time professor of mathematics in Rome. He discovered sunspots independently of his contemporaries and was one of the first to make systematic observations of them. He wrote a book *Rosa Ursina* which contains observations of the Sun made between 1611 and 1625. Scheiner was unfriendly toward Galileo, and played a discreditable role in the events leading up to Galileo's trial and condemnation.

Schiaparelli, Giovanni Virginio (1835–1910). Italian astronomer. Assistant at Brera, Milan, became Director after the death of his chief Carlini (1862). He discovered the asteroid Hesperia (1861); observed numerous double stars, and later produced authoritative

studies of Biblical, Babylonian, Greek and medieval astronomy. Schiaparelli is best known for his explorations of the terrestrial planets, especially Mars; in 1877 he remarked upon the so-called "canali", and two years later their apparent duplication. His Mars work was epoch-making, and from LOWELL earned him the epithet "Columbus of a new planetary world." He made pioneer observations of Uranus (1883–84), and from certain markings on Mercury and Venus believed they had SYNCHRONOUS ROTATION. *RMB*.

Giovanni Schiaparelli: planetary observations

Schickard Major lunar walled plain; diameter 125 miles (202km); 44°S, 54°W. It has low walls, and a floor which is darkish in places. It is one of a group which includes the plateau Wargentin.

Schiller Lunar formation, measuring 112 × 60 miles (180 × 97km); 52°S, 39°W. It is made up of the fusion of two older formations.

Schmidt, Bernhard Voldemar (1879–1935). Estonian optician who devised the SCHMIDT TELESCOPE.

Schmidt telescope REFLECTING TELESCOPE which uses a specially-shaped glass correcting plate to achieve wide field of view combined with fast FOCAL RATIO, invented by Bernhard SCHMIDT.

The rather unassuming title of Schmidt's one and only scientific paper — "A rapid coma-free mirror system" — betrayed little of the enormous significance of the invention described within it. Schmidt's formal announcement to the world of his new telescope design appeared in the Communications of the Hamburg Observatory in 1932, two years after he had perfected a 14in (0.36m) prototype and dazzled his

S

colleagues at Hamburg with the quality of the images it produced. Within a few years, the new optical system had won wide acclaim among astronomers, particularly in the United States, and by the time of Schmidt's death in 1935 his immortality was assured.

To understand something of the excitement generated by Schmidt's "rapid coma-free mirror system", it is necessary to look at what it achieved against the background of the optical technology of its time. Large telescopes, then as now, were mostly reflecting telescopes, whose concave primary mirrors were not segments of spheres (which would produce SPHERICAL ABERRATION) but were shaped in cross-section like a PARABOLA. Such a "paraboloidal" mirror will form a perfect point image when parallel light (eg, from a star) is incident upon it — so long as the incoming light is exactly aligned with its axis. Departure from this condition causes steadily increasing degradation of the images due to aberrations — mostly COMA, which gives stars the appearance of half-open umbrellas or comets (hence the name) with their tails pointing away from the centre of the field of view. Coma is noticeable at "off-axis" angles as small as a few minutes of arc, and becomes more acute with the "fast" or "rapid" (ie, numerically small) focal ratios that are highly desirable for recording images of extended objects like nebulae and galaxies. Until Schmidt's time, there was no way of building a reflecting telescope with a focal ratio faster than about f3 and an appreciable field of view.

Schmidt was attracted by the possibility of somehow using a spherical mirror in a fast, wide-field telescope. He realized that if the incoming beam were limited in size by an aperture (or "stop") placed at the mirror's centre of curvature there would be no preferred axis — and therefore no coma, only spherical aberration, which would be the same for all angles. Schmidt's breakthrough was to place a thin glass correcting plate at the centre of curvature which would introduce into the incoming beam just enough spherical aberration to balance exactly that of the mirror. It was a brilliantly simple solution to the problem, and proved to be highly successful in practice. Schmidt's prototype had a focal ratio of f1.7 and gave perfect images over a field of view 12° in diameter — enormous by the standards of the time.

The thin glass plate in a Schmidt-type telescope — the "Schmidt correc-

tor" — is a flat piece of glass with a very shallow profile worked into one or both of its surfaces. The profile is complex in shape and has to be precisely computed to match the spherical aberration of the mirror. Because the plate is not a lens as such (it is the mirror that does the focusing), flexure in the glass presents little difficulty and Schmidt correctors can be made larger than the OBJECTIVES of refractors. For the most exacting work with very large Schmidt telescopes, the dispersion of light in the corrector becomes a problem, and an ACHROMATIC corrector becomes necessary.

Schmidt telescopes are usually used photographically, and are often referred to as "Schmidt cameras". Hybrid designs incorporating secondary mirrors (Schmidt-Cassegrains) or additional correctors (eg, BAKER-SCHMIDTS) are also possible.

The image in a Schmidt telescope is formed on a curved surface which is a segment of a sphere concentric with the mirror. To match this surface, the

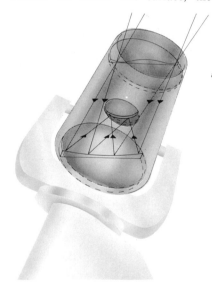

Schmidt telescope: a brilliant solution

photographic film or plate (made of very thin glass) must be deformed by clamping it in a suitably-shaped holder. The curved focal surface is one of the few drawbacks of the Schmidt system; another (which is important from the point of view of cost in very large instruments) is that the telescope tube must be made twice as long as its FOCAL LENGTH to support the corrector at the mirror's centre of curvature.

Notwithstanding these shortcomings, Schmidt telescopes have found great favour among astronomers because of their ability to record detailed images of very large areas of

sky. This makes them ideal for observing extensive objects such as comets or certain nebulae, or for recording large numbers of more compact sources — stars or distant galaxies — for statistical studies. (A single 14in (356mm) square photographic plate taken with the UK Schmidt telescope will typically record about a million objects!) Perhaps their most important rôle is as survey telescopes used for identifying unusual or specific kinds of object, which can then be investigated in more detail with larger, conventional (narrow-field) instruments.

To this end, major surveys of the entire sky have been undertaken by some of the world's largest Schmidt telescopes, producing the most complete atlases of the sky at optical wavelengths available to astronomers, and showing objects as faint as the 22nd magnitude. The most important of these carried out to date are the National Geographic Society/Palomar Observatory Sky Survey (undertaken in the 1950s with the Palomar Schmidt) and its more recent southern counterpart, the European Southern Observatory/Science and Engineering Research Council Southern Sky Survey (ESO and UK Schmidts). Other surveys are currently being undertaken, and all have produced many new discoveries, ranging from comets and asteroids in the Solar System to superclusters of galaxies and the large-scale structure of the universe.

What is the future of Schmidt telescopes? There is occasional discussion among astronomers of a new Schmidt, larger than any yet built, although the corrector plate would be unlikely to exceed 8ft (2.4m) in aperture because of the difficulty of casting a sufficiently homogeneous piece of glass. The possibility of a Schmidt telescope in Earth orbit — the "Space Schmidt" — has also been raised. However, perhaps the most significant pointer to the future is that astronomers are beginning to use the enormous information-gathering capability of Schmidt telescopes in completely new ways. Leading the field in this is the UK Schmidt telescope, whose FLAIR system (Fibre-Linked Array Image Reformatter) uses fibre optics to bring light from up to 80 celestial objects in the field of view for simultaneous analysis with spectrographs in the dome. The very large gains in observing efficiency achieved in this way can be directly attributed to the effectiveness of the telescope itself, and no doubt Bernhard Schmidt would have been delighted at this novel extension to his "rapid coma-free mirror system". *FGW*.

Largest Schmidt telescopes

Designation	Location	Aperture of corrector (metres)	Date of completion
Tautenburg	E.Germany	1.34	1960
Palomar*	USA	1.24	1948
United Kingdom*	Australia	1.24	1973
Kiso	Japan	1.05	1976
Byurakan	Armenia	1.00	1961
Kvistaberg	Sweden	1.00	1963
European Southern Obs.*	Chile	1.00	1972
Llano del Hato	Venezuela	1.00	1978
* Achromatic corrector			

Schrödinger Lunar walled plain, 75°S, 133°E, and therefore not visible from Earth. It is associated with the great Schrödinger Valley, which was first photographed from the Orbiter unmanned probes.

Schrödinger Valley Major valley on the Moon; associated with the crater Schrödinger (47°N, 133°E). It lies on the lunar far side, and is therefore unobservable from Earth.

Schröter effect When Venus is at dichotomy (exact half-phase) there is usually a difference of at least a day or two between theoretical and observed phase; during western (morning) elongations dichotomy is late, at eastern (evening) observations it is early. The phenomenon, due presumably to the effects of Venus' atmosphere, was first noted in the 1790s by J.H.SCHRÖTER, and the term "Schröter effect", first applied by Patrick Moore, has come into general use.

Schröter, Johann Hieronymus (1745–1816). German amateur astronomer; chief magistrate of Lilienthal, where he erected his observatory. He was the real founder of lunar observation, and also carried out planetary work. His observatory was destroyed in 1814 by invading French soldiers.

Schröter Valley The great winding valley extending from the lunar crater Herodotus.

Schubert Crater on Mercury, 100 miles (160km) in diameter. 47°S, 55°W.

Schwabe, Samuel Heinrich (1789–1875). German chemist and amateur astronomer who made daily sunspot records from 1925 onward and who, in 1843, announced the discovery of the sunspot cycle.

Schwarzschild, Karl (1873–1916). German astronomer and theoretician who computed the first exact solutions of Einstein's field equations of general RELATIVITY relating to the gravitational field of a point mass. This work provided the basis for the theory of BLACK HOLES. He also made important contributions to astronomical photometry and spectroscopy.

Karl Schwarzschild: German astronomer

Schwarzschild limit. The maximum permissible density for a self-supporting body. If the density of a body exceeds this limit, it will collapse to form a BLACK HOLE. The greater the mass of a body, the lower the value of the Schwarzschild limiting density (and the larger the value of the SCHWARZSCHILD RADIUS). Thus, while a body of one solar mass has a limiting density of about 10^{19}kg/m^3, the value for a 10 solar mass star is about 10^{17}kg/m^3, and for a hundred million solar mass body is only about 10^3kg/m^3, ie, ordinary water density. The ratio of the observed density of a body to the Schwarzschild limiting density is called the "filling factor". The filling factor for the Sun is about 3×10^{-6}, while for a neutron star it is, typically, about 0.25. *IKN*.

Schwarzschild radius The critical radius (Rs) around a collapsed mass (M) at which an emitted photon would be unable to escape and would remain stationary. It is the radius at which a collapsing mass (strictly speaking, an uncharged, non-rotating mass) becomes a BLACK HOLE. Its value was calculated by K.SCHWARZSCHILD, using the general theory of RELATIVITY, in 1916, and is given by: Rs = $2GM/c^2$, where G = the gravitational constant and c = the speed of light. Rs for the Sun is about 10,000ft (3km) and for the Earth is about $\frac{3}{8}$in (0.9cm). *IKN*.

Schwassmann-Wachmann 1 Comet Comet with a 15.0-year period. It is unusual for a near circular orbit between Jupiter and Saturn and for sudden brightenings.

Scintars Radio sources that show the effects of interplanetary SCINTILLATION and which therefore are point-like rather than extended. This property distinguishes QUASARS and radio stars from RADIO GALAXIES and NEBULÆ. The principle by which the two kinds of radio sources are separated is the same as the principle by which atmospheric scintillation distinguishes stars and planets. The twinkling of stars is caused by starlight passing through moving bubbles of air at different densities. Each bubble deflects a portion of the starlight by a small angle and a given bubble through which we perceive the star may diminish or intensify its brightness momentarily. However, planets have disks, each point of which is subject to the effect of different bubbles. The perceived light from a planet is averaged over many bubbles, and is relatively constant. Thus stars twinkle, but planets shine steadily.

Radio waves are refracted by the plasma clouds which move outwards from the Sun through the Solar System. Star-like radio sources are affected by the clouds, just as stars are affected by air bubbles, and such radio sources scintillate. The point-like sources are either intrinsically very small, like true stars which emit radio waves, or sources which are so distant that they subtend a small angle, like quasars.

The scintillation method was used to discover quasars by Antony Hewish and Jocelyn Bell Burnell in 1967 with the 4.5-acre radio telescope in Cambridge. To see the twinkling it was necessary to make the radio receiver respond very quickly to changes in radio intensity. This made it possible for Hewish and Bell Burnell to discover PULSARS during the course of the scintillation measurements. *PGM*.

Scintillation The twinkling of the stars and, to a lesser extent, of planets, etc, due to the uneven refraction of light in

S

S

areas of different density in the Earth's atmosphere. To the naked eye, it appears as a change in brightness and colour, while in the telescope it may also make the star appear to make rapid slight movements. It is greatest at low altitudes, when the object's light shines through a greater amount of the atmosphere. Most of the effect is caused below 30,000ft (9,000m).

The planets twinkle less than stars as they present small disks rather than points of light and the whole disk is unlikely to be affected simultaneously. In a telescope, however, the disk will appear blurred, usually referred to as bad SEEING.

It depends on the weather, and is usually greatest about noon. Modern observatories are sited where the effect occurs least. *TJCAM. See also* RADIO SCINTILLATION, SCINTARS.

Scorpius (the Scorpion — often, less correctly, called Scorpio). In mythology it was the insect that caused the untimely death of Orion, and with its long line of bright stars it really does conjure up an impression of a scorpion, with the red "heart" (Antares) and the "sting".

Brightest stars

Name	Visual Mag.	Abs. Mag.	Spec.	Distance (light-yrs)
α (Antares)	0.96	−4.7	M1	330
λ (Shaula)	1.63	−3.0	B2	360
θ (Sargas)	1.87	−5.6	F0	900
ε (Wei)	2.29	−0.1	K2	65
δ (Dschubba)	2.32	−4.1	B0	550
κ (Girtab)	2.41	−3.0	B2	390
β (Graffias)	2.64	−3.7	B0.5 + B2	815
u (Lesath)	2.69	−5.7	B3	1570
τ	2.82	−4.1	B0	780
σ (Alniyat)	2.89	−4.4	B1	590
π	2.89	−3.5	B1	620
i¹	3.03	−8.4	F2	5500
μ¹	3.04	−3.0	B1.5	520
G	3.21	−0.1	K2	150
η	3.33	0.6	F2	68

Also above magnitude 4: Mu² (3.57), Zeta² (3.62), Rho (3.88) and Omega¹ (3.96). Nu is of exactly magnitude 4.00. Both Mu and Zeta masquerade as doubles, but are optical pairs.

In addition to being immersed in a very rich part of the Milky Way, Scorpius contains some magnificent clusters, notably M4 (near Antares), an easy binocular object, and M6 and M7, visible with the naked eye. The "sting" is made up of Shaula, Lesath, Girtab, and Q (4.29). The obscure-looking Iota¹ is exceptionally luminous and may equal about 200,000 Suns! *PM.*

Scorpius X-1 First cosmic X-ray source to be discovered and the brightest known apart from certain transients. Given its brightness, it received much attention in the early days of the subject. It is now thought to be one of the low-mass X-ray binary class with an orbital period of 19.2 hours and an X-ray luminosity of about 10^{38} erg/s. The companion star cannot be classified, but exhibits a UV excess and optical flaring. X-rays are thought to originate as black body emission from the associated NEUTRON STAR and as Compton scattered emission from a thick ACCRETION disk. *JLC.*

Sculptor (the Sculptor). A very barren constellation adjoining Phœnix. It contains no star above magnitude 4.3, but is rich in faint galaxies.

Sculptor galaxy Dwarf elliptical galaxy in the southern constellation Sculptor, discovered in 1976 as a new member of the LOCAL GROUP.

Scutum (the Shield; originally Scutum Sobieskii, Sobieski's Shield). A very small constellation adjoining Aquila. It has only one star above the fourth magnitude (Alpha, 3.85), but it contains the splendid open cluster M11, nicknamed the Wild Duck, and is crossed by the Milky Way.

S Doradûs variables *see* ERUPTIVE VARIABLES.

Seasons Due to the OBLIQUITY OF THE ECLIPTIC the Earth's Equator is inclined at 23° 27′ to its orbit, tilting the northern and southern hemispheres alternately towards the Sun, as the Earth revolves around it yearly, giving summer and winter, with spring and autumn in between.

Secchi, Pietro Angelo (1818–78). Italian Jesuit Father, director of the Collegio Romano Observatory. More interested in the physical composition of celestial bodies than in their position, he became a pioneer in astrophysics with his spectroscopic observations of the Sun and of the stars. He was the first to describe the SOLAR SPICULES and to classify PROMINENCES as quiescent and eruptive. Following his systematic observations of stellar spectra, he introduced the classification of stars into four classes based on the colour of the stars and on the characteristics of their spectra, a classification which was later superseded by the Harvard classification. *VB.*

Angelo Secchi: studied solar phenomena

Secular acceleration A continuous, non-periodic rate of change of the rotational or orbital motion of a body or system of bodies. For example, in the Earth-Moon system, frictional effects due to the tides are gradually slowing the Earth's axial rotation, while the gravitational interaction between the tidal bulge and the Moon transfers angular momentum to the Moon and causes it to recede; as a result, the orbital period of the Moon slowly increases and the mean angular motion of the Moon decreases. The Earth's rotation period is increasing by 0.000015 seconds a year, and the Moon is receding at about 1¼in (3.5cm) a year. *IKN.*

Secular variables Stars that are suspected of having increased or decreased markedly and presumably permanently since ancient times. Thus PTOLEMY and other early observers ranked MEGREZ (Delta Ursæ Majoris) as the equal of the other stars in the Plough pattern, whereas it is now obviously fainter; DENEBOLA, in Leo, was ranked of the first magnitude, but is now below the second, and so on. On the other hand RASALHAGUE (Alpha Ophiuchi) was given as magnitude 3, and is now 2. On the whole it seems very unlikely that any of these changes are real, and more probably we are dealing with errors of observation or – more plausibly – translation or interpretation. *PM.*

Seeing The deterioration in telescopic image quality caused by the atmosphere. The term is also used to denote the atmospheric conditions which cause image deterioration.

Rapid changes in image detail arise from air turbulence within the telescope tube. Steady but blurred images are caused by similar turbulent cells in high-speed air streams some distance away. Sharp images with lateral motion up to 10 arc seconds result from larger-scale turbulence at high altitudes.

Visual observation is usually little impaired by lateral motion alone, but all three forms may be present simultaneously and any one vitiates photography. *LMD.*

Seismology The study of the various types of earthquake or seismic waves. The strength of such waves are recorded on sensitive instruments called seismographs. The size of an earthquake can be obtained from the trace and is commonly interpreted on the Richter scale.

Selected areas A set of 262 small, uniformly distributed regions of sky in which the magnitudes, spectral and luminosity classes of stars have been accurately measured to provide standard comparison data and statistics on the distribution of stars.

Selenography The study of the features of the surface of the Moon.

Selenology The study of the nature and history of the Moon.

Seleucus Crater on the lunar Oceanus Procellarum; 28 miles (45km) in diameter; 21°N, 66°W. It has terraced walls.

Semi-regular variables Pulsating giants and supergiants of late spectral type with periods 20 to 2,000 days or more and amplitudes from a few hundredths to several magnitudes. Some show a definite periodicity, interrupted at times by irregularities. Others differ little from MIRA stars except in having amplitudes less than 2^{m}5, but the shapes of their light curves vary. A small sub-division of these stars includes supergiants of late spectral type with amplitudes of no more than one magnitude and periods from 30 to several thousand days. There are also stars belonging to the semi-regulars with spectral types F, G and K, sometimes with emission lines.

Serpens (the Serpent). A constellation in two parts, Caput (the Head) and Cauda (the Body), separated by Ophiuchus.

Brightest stars

Name	Visual Mag.	Abs. Mag.	Spec.	Distance (light-yrs)
α (Unukalhai)	2.65	−0.1	K2	85 (Caput)
η	3.26	1.7	K0	52 (Cauda)
θ (Alya)	3.4	2.1 2.1	A5 + A5	101 (Cauda)

Next come Xi (3.54) in Cauda, and Beta (3.67), Epsilon (3.71), Delta (3.8) and Gamma (3.85) in Caput.

Alya is a wide, easy pair, with components of magnitudes 4.1 and 5.0, separation 22″.6. Delta is also an easy telescopic double, with components of magnitudes 4.2 and 5.2, separation 3″.9. The MIRA-type variable R Serpentis can reach magnitude 5.7 (minimum 14.1, period 367 days) and there are two bright Messier objects, M5 (globular cluster) and M16 (nebula and embedded cluster). *PM.*

Serpentarius Obsolete name for OPHIUCHUS (the Serpent-bearer).

Service module *see* APOLLO PROGRAMME.

SETI Search for Extra-Terrestial Intelligence. *See* LIFE IN THE UNIVERSE.

Seven Sisters *see* PLEIADES.

Sextans (the Sextant). A very obscure constellation between Leo and Hydra. It has no star above magnitude 4.4.

Seyfert, Carl Keenan (1911–60). American astrophysicist best known for his study, begun in 1943, of what are now known as SEYFERT GALAXIES.

Seyfert galaxies Active galaxies — usually spiral or barred spiral in form — with highly luminous compact nuclei whose spectra display broad EMIS-

Seyfert galaxies: NGC1566; an AAT photo

SION LINES. Many Seyfert nuclei are too small to be resolved and look "starlike" in appearance. Seyferts emit strongly at ultraviolet and infrared wavelengths, and exhibit a degree of short-term variability. Some are strong X-ray sources, but few are particularly strong radio emitters. Seyfert nuclei closely resemble QUASARS in many of the above properties. The broad emission lines imply the presence of gas clouds moving at speeds of several thousand kilometres per second. These high velocities many be due to the gravitational influence of a massive compact object in the nucleus — possibly a BLACK HOLE, which acts as the underlying energy source for these highly energetic objects. Named after Carl SEYFERT, Seyfert galaxies comprise a few per cent of all bright galaxies. *IKN.*

Shadow bands Occurring just before and just after total eclipse this, not yet fully understood, phenomenon is seen as faint, slowly moving lines of shadow. Due to their very low contrast the lines are difficult to photograph successfully.

Shapley, Harlow (1885–1972). American astronomer who spent part of his career at Mount Wilson. It was he who first measured the size of our Galaxy, from observations of variable stars in

Harlow Shapley: the size of the Galaxy

globular clusters, and he made many other major contributions. He was also a pioneer of international collaboration in astronomy.

Shedir The star Alpha Cassiopeiæ, in the W. Its magnitude is 2.2, but it is suspected of slight variability. It is orange, of type K.

Sheliak *See* BETA LYRÆ.

S

Shell galaxies Isolated elliptical galaxies around which lie extremely faint, incomplete rings of light. The first was discovered by David Malin in 1979. Research has shown them to be stars lying in thin shells. The best explanation for their origin is that a spiral galaxy merged into the elliptical, and their dense centres oscillated back and forth as they merged, throwing off stars into the shells. As many as 28 shells have been counted around a single galaxy, each corresponding to an oscillation of the centres.

Shell galaxies differ from RING GALAXIES in that the shells lie much farther out, are very faint, and comprise old stars. *DAA.*

Shell stars Circumstellar gaseous envelopes are found in several classes of O-, B- and A-type emission stars, including interacting binaries. "Shell star" is something of a misnomer, since shells are frequently equatorial disks episodically thrown out by rapid rotation. Strong shell spectra are seen in B0e-B3e giants and dwarfs. Stars such as Gamma Cassiopeiæ, Pleione and 48 Libræ are classical examples.

Typical strong Be/shell spectra show a rapidly rotating underlying B star with broad ABSORPTION LINES superposed with EMISSION LINES and additional sharp absorption features from the disk. The latter appear at various shortward-displaced wavelengths, indicating variable mass outflow at different velocities. Far-ultraviolet Be spectra reveal persistent STELLAR WINDS in high-temperature ionic lines, reminiscent of supergiants, and coronal X-rays are observed possibly originating from above and below the disk. *ADA.*

Shen Gua (1031–95). Chinese astronomer and polymath.

A noted civil and military administrator, Shen Gua was a man of immense technical ability and an able mathematician, who took a great interest in the natural world. Like his contemporaries, he was also a believer in forms of divination, and this as well as his scientific curiosity led to his interest in astronomy, for in ancient China celestial portents were considered important in indicating the quality of an emperor's administration.

Shen was one of the few Chinese astronomers to try to visualize planetary motion. This stimulated him to evolve mathematical techniques so that the apparent retrograde movements could be accounted for better; in turn this led to improvements in Chinese lunar theory. This work was probably why he was appointed head of the government Astronomical Bureau in 1067, an appointment of importance for Chinese astronomy.

In the first place it led to the adoption of Shen's re-designs of many astronomical instruments. There was an improved GNOMON — a special vertical rod — then still used for measuring the Sun's noonday shadow and its altitude at a solstice. Secondly, his superior ARMILLARY SPHERE — a sphere with metal rings representing celestial coordinates — was used for measurements of celestial positions. Shen's innovation was to improve the sighting tube of the instrument.

Thirdly, Shen Gua improved the clepsydra or water-clock by a new approach to its calibration which led to greater precision in its timekeeping.

Shen's second important course of action at the bureau was to institute a reform of the calendar. It was his wish to introduce a purely solar calendar instead of the luni-solar one then in use. This depended on having more frequent and regular observations of the Moon and planets, which he then made, thus presaging those of Tycho BRAHE five centuries later. *CAR.*

Shepard, Alan (1923–). American astronaut. He saw duty at sea in World War II and was later a naval pilot. He was selected as an astronaut in 1959, and made the first manned Mercury flight (suborbital) in 1961. Grounded for medical reasons in 1964–69, he commanded the Apollo-14 lunar landing, in 1971. He retired as a Rear Admiral, and is now a Houston businessman.

Alan Shephard: 1961 Mercury flight

Shepherd satellites Minor satellites in the outer Solar System that guard or share the orbit of the particles in a planetary ring system. At Saturn there are eight such satellites; located by the outer edge of Ring A; on either side of the F ring; two co-orbiting moons beyond the F ring and two satellites situated at the Lagrangian points of the satellite Tethys. There are two Shepherd satellites in the Uranian system, Cordelia and Ophelia, situated on either side of the Epsilon ring.

Shklovskii, Iosif Samuilovich (1916–85). Soviet astrophysicist who in 1953 proposed that the radio and X-ray emission from the CRAB NEBULA was SYNCHROTRON RADIATION and predicted that OH would radiate at microwave frequencies. He has also made prominent contributions to the debate on extra-terrestrial life.

Shocks Important processes in many branches of astronomy and space science. Though the simple shock produced by a supersonic aircraft is familiar, the processes involved in astrophysical shocks are less well understood. The general description of a shock is that there is some boundary, the shock surface, and that as matter crosses the boundary there is a rather sudden change in the state of the matter. This change is an irreversible change in thermodynamic terms. The usual form of shock is that there is conversion of kinetic or flow energy into random thermal energy. The temperature is higher on one side of the shock.

Examples of shocks are found in the SOLAR WIND including the BOW SHOCK caused by planetary MAGNETO-SPHERES. At the Sun, shocks are probably involved in SOLAR FLARES. In the interstellar medium, the clouds of gas and dust that form the region between the stars as well as the more spectacular nebulæ, shocks form important boundaries. They are thought to be involved in various early stages of star formation from these clouds and are certainly involved in the late stages of STELLAR EVOLUTION such as supernovæ. At a larger scale it is thought that shock processes occur in radio galaxies. *LJCW.*

Shooting Stars *see* METEORS.

Short, James (1710–1768). Optician, and son of William Short, a joiner in Edinburgh. James Short was the first to give speculum mirrors a true parabolic figure and also one of the first to construct telescopes to Cassegrain's design (c. 1740). Although Short favoured the Gregorian, his instruments were often supplied with an auxiliary convex mirror for use as an alternative to the concave secondary, increasing the angular magnification of the instrument. From many observations of the Venus transit, on June 6 1761 Short deduced a value for the solar parallax, long accepted as authoritative.

Shortt clocks The Shortt clock was developed by W.H.Shortt, a railway engineer, in collaboration with the Synchronome Clock Company, which made as its main product a clock system with multiple electrically-driven dials.

The Shortt clock's distinguishing feature was the use of separate pendulums to perform the two tasks of reproducing and counting seconds. The master pedulum, swinging in a temperature-controlled and partially-evacuated enclosure, was kept free of all disturbances except a highly reproducible driving impulse delivered every 30 seconds close to and symmetrically about dead centre. The slave clock, only slightly modified from the company's commercial units, controlled the regular resetting of the impulse mechanism and was synchronized by its release.

Shortt clocks were used by the main timekeeping observatories between the early 1920s and the mid-1940s, when their stability was surpassed by groups of quartz clocks. *JDHP.*

Sickle of Leo Curved pattern of stars in Leo, of which Regulus is the brightest.

Sidereal clocks Sidereal clocks show sidereal time, which is a direct representation of the Earth's angular position around its axis of rotation with respect to the stars; in contrast, normal clocks show mean solar time, which represents, approximately, the Earth's orientation relative to the Sun. Because of the Earth's orbital motion around the Sun, sidereal clocks must gain approximately 3m 56s per day, or one day per year, on clocks keeping mean solar time.

The celestial reference direction used in the definition of sidereal time, called the First Point of Aries, lies on the intersection of the planes of the ecliptic and the equator and is also the zero point of RIGHT ASCENSION. It moves slowly with respect to the stars because of PRECESSION. Sidereal and solar times measured from noon coincide near the spring equinox, when the Sun lies close to this direction. Objects cross the local meridian at a local sidereal time equal to their right ascension. *JDHP. See also* ASTRONOMICAL CLOCKS.

Sidereal day The time interval between two successive upper transits across an observer's meridian of the VERNAL EQUINOX, or a given star. It is the axial rotation period of the Earth relative to the background stars and is equal to 23 hours 56 minutes 04.1 seconds of mean time. *See also* DAY.

Sidereal month A month measured relative to the stars: 27.32166 days.

Sidereal period The real orbital period of a body relative to the stars.

Sidereal time Time measured relative to the stars.

Sidereal year A year measured relative to the stars: 365.25636 days.

Siderite Old name for an iron METEORITE.

Siderolite Old name for stony-iron METEORITE.

Siding Spring Observatory. Australia's major optical observatory.

Time has sandpapered the island-continent of Australia to a great degree. There are no towering mountains of the type sought by astronomers

Siding Spring Observatory

for their observatories, and the few peaks that top 6,500ft (2,000m) altitude are mere ripples in a rising swell of land that carries its weather aloft with it. When, therefore, the suburbs of burgeoning Canberra grew near to MOUNT STROMLO OBSERVATORY, and the city lights polluted its skies, few alternative sites were available. Testing was undertaken at several, and two proved favourable. Of these the one chosen offered much the better access and facilities. It lay on a shapely peak known to the Aborigines as Woorut, and named by white settlers Siding Spring Mountain. This peak, one of half a dozen that top 3,200ft (1,000m) elevation, lies in the Warrumbungle range, 17 miles (27km) from the small but pleasant town of Coonabarabran. The town name is reported to be Aboriginal for "inquisitive man".

Most of the Warrumbungles form a national park that preserves some splendid scenery, including spires and plugs of volcanoes that were active 13 million years ago. Siding Spring Mountain is also a volcanic peak, but less vertiginous, so that a good road leads easily to its large summit, where are to be found a collection of buildings including seven telescopes, extensive workshops, a visitors' centre, a motel-style lodge for observers and nearly a dozen houses for employees.

Initially, Siding Spring Observatory was built as an outstation for the Australian National University, which also operates Mount Stromlo Observatory. In 1962 construction of three small telescopes commenced, with apertures of 16, 24 and 40in (0.4, 0.6 and 1.0m). The smallest of these is used mostly for photometry. The 0.6m was designed for accurate polarimetry, and was constructed so that it could be rotated about its own axis: two measures of polarization, made with the telescope in opposite orientations, provide a more reliable determination than one. The largest of the telescopes has at times carried a range of instruments.

The Australian National University has expanded its holding by constructing a 90in (2.3m) telescope. By late 1986 the 2.3m was nearing the end of its engineering work, and had additionally been used in intermittent astronomical research for more than a year. It incorporates much of the cost-saving technology of recent telescope design, including a thin primary mirror, an altazimuth mount and a rotating box-shaped building.

The remaining telescopes on Siding Spring Mountain are owned by other institutions. A Schmidt telescope of clear aperture 20in (0.5m) is operated by UPPSALA OBSERVATORY, Sweden, and was transported from Mount Stromlo in 1981. It was employed extensively for the visit of Halley's Comet, and has been used for many other research projects.

A larger Schmidt telescope, of 48in (1.2m) clear aperture, is operated by

S

S

the Royal Observatory, Edinburgh. Known as the United Kingdom Schmidt Telescope (UKST), it has achieved world acclaim for the quality of its photographic work. The UKST was opened in 1973, and its initial task was to undertake a full survey of the southern sky in blue light. The parallel survey in red light is being made by a smaller Schmidt telescope at the EUROPEAN SOUTHERN OBSERVATORY, which also produced an initial "quick look" blue survey. The UKST blue survey set new standards in quality, and in particular detected stars, galaxies and nebulæ about two magnitudes fainter than the older Palomar survey of the northern sky. The major cause of this advance was the development of new emulsions, but these in turn forced much more stringent techniques. Having completed its first survey, the UKST is now continuing other photographic surveys, both in the near infrared, and using OBJECTIVE PRISMS for spectroscopic surveys. Much of the telescope's time is now used for research projects. Examples of this include searches for QUASARS, VARIABLE STARS, optical counterparts of radio, X-ray or infrared sources, stars of high proper motion, or those with unusual colours. A recent innovation has been the use of optical fibres to bring the light from as many as 60 objects at the focus of the telescope to a spectrograph on the dome floor. This technique promises to be extremely valuable.

The last telescope on Siding Spring Mountain is the Anglo-Australian Telescope, of 154in (3.9m) aperture. A binational facility, it is owned and funded equally by Australia and the United Kingdom. This telescope, too, has won international acclaim for its performance, the instruments that are used on it and the science it has yielded. Constructed in the early 1970s, it was opened by HRH The Prince of Wales in October 1974, and scheduled observing began in mid-1975. The AAT was the first telescope to be fully controlled by a computer, and quickly set new standards of accuracy for setting on and tracking astronomical objects. It was the AAT's demonstration of the power of computer control that paved the way for large altazimuth telescopes like the neighbouring 2.3m.

The AAT is a general-purpose telescope, equipped with a large suite of state-of-the-art instruments which include photographic facilities, CHARGE-COUPLED DEVICES, spectrographs and infrared equipment.

Among the scientific results from the AAT are the discovery of optical flashes from the VELA PULSAR (then the faintest star ever studied), the identification of SS433, the study of the extremely faint SHELL GALAXIES, the discovery of patterned clouds on the dark side of Venus, the demonstration that organic material is shed by HALLEY'S COMET, and several detailed studies of SAGITTARIUS A, the very central region of our Galaxy.

The astronomical climate of Siding Spring Observatory does not place it in the top league. About 60 per cent of nights are usable for some form of observing, a lower percentage than at some observatories. The sky is, however, extremely dark and likely to remain so for a very long time. Moreover Coonabarabran, population 3,000, offers a pleasant place to live for the skilled technical staff required to keep today's highly sophisticated instruments in top working order. The telescopes on Siding Spring Mountain have proved beyond doubt that the best science comes from having good equipment ready to make use of every minute of clear sky, and this is not possible at a remote and inhospitable observatory. As such, Siding Spring Observatory has forced a rethink of what astronomers really seek when selecting an observatory site. *DAA.*

Sigma Octantis The south polar star, within 7° of the pole. It is only of magnitude 5.46. By AD2000 the distance from the pole will have increased to 1°.

Sikhote-Alin meteorite Largest observed fall; iron; East Siberia 1947.

Silicon: The eighth most abundant element in the universe and the second most abundant in the Earth's crust. It is a major component of rock-forming materials. Chemical symbol Si, atomic number 14, atomic mass 28.086.

Silicon stars One of the groups of A-type stars with peculiar spectra. The exceptionally strong lines in their spectra arise from large overabundances of silicon. Many of the rare earth elements are also overabundant in these stars.

Singularity A point in spacetime at which the laws of physics break down. For example, the singularity at the centre of a Schwarzschild BLACK HOLE is a point at which conventional theories predict infinite gravitational forces and the compression to infinite density of infalling matter.

Sinope The ninth satellite of Jupiter, discovered in 1914; diameter 22 miles (35km). It has retrograde motion, and may be asteroidal.

Sinus Æstuum (The Bay of Heats). Lunar feature east of the crater COPERNICUS, bordered by the south end of the Apennine range. Its dark floor contains no large craters.

Sinus Iridum A lava-flooded lunar bay, 45°N, 31°W, walled only partly by the Jura Mountains; themselves more than two miles (3.5km) high in places and forming an arc (diameter some 160 miles/255km) terminating in Promontories Laplace (to the east) and Heraclides (to the west).

Sinus Margaritifer. Prominent telescopic dark marking of Mars. VIKING imagery reveals it to be a region of chaotic terrain into which a number of prominent channels drain.

Sinus Medii (the Central Bay). A small dark lunar feature near the apparent centre of the Moon's disk as seen from Earth.

Sinus Roris (the Bay of Dews). Lunar mare area, joining Oceanus Procellarum to Mare Frigoris.

Sinus Sabæus Prominent telescopic dark marking on the equator of Mars. Viking imagery reveals it to be an area of densely-cratered terrain.

Sirenum Terra (formerly known as Mare Sirenum). Dark feature on Mars; 50°S, 150°W at its centre. It is one of the most conspicuous of the albedo features.

Sirius The Dog Star, Alpha Canis Majoris, the brightest star in the sky. Sirius, with its companion SIRIUS B, forms the sixth-nearest star system,

Sirius: brightest star in the sky

only 8.7 light-years away. Its nearness accounts for its brightness and its large proper motion: in 1,350 years Sirius moves southwest by a distance equal to the diameter of the Full Moon.

Sirius was classified by PTOLEMY in AD144 as a red star, but is clearly a brilliant white at the present time. Astronomers have enjoyed inventing reasons for this, based on rapid changes in Sirius itself, but most prefer to think that Ptolemy's record has been mistranscribed. *PGM*.

Sirius B Faint white dwarf companion star to SIRIUS. In 1834 F.W.BESSEL noticed that in its proper motion Sirius deviated from the straight by two arc seconds with a period of 50.09 years, pulled from side to side by an invisible companion star. This fainter star, Sirius B, was detected in 1862 by the son of telescope-maker Alvan Clark during the test of a new 18.5in (47cm) refractor. As he waited for Sirius to appear from behind the wall of a building which obstructed his view, he saw a star 10 magnitudes fainter than Sirius come into view before Sirius itself. The separation of Sirius A and B ranges up to a maximum 11 arc seconds, which last occurred in 1973. The stars' masses are 2.2 and 0.94 times solar mass respectively, and Sirius B has a density of one-sixth of a tonne per cubic centimetre. The realization that Sirius B was such a small, dense star led to the recognition of the star as the first of a newly discovered kind, the WHITE DWARF stars. *PGM*.

Sirsalis Lunar double crater, 20 miles (32km) in diameter, just outside Oceanus Procellarum; 13°S, 60°W.

Sisyphus Apollo asteroid 1866, with a comet-like orbit.

61 Cygni A fifth-magnitude binary in Cygnus which was the first star to have its parallax measured (by F.W.BESSEL in 1838). Parallax: 0.294 arc sec; distance: 11.1 light-years.

Skelnaté Pleso Observatory (Czechoslovakia). Known mainly for an internationally-used star atlas.

Skjellerup-Maristany Comet (1927 IX). Brilliant comet, reaching magnitude −6 in December. Its period is about 36,500 years.

Skylab The first American space station, weighing 75 tonnes and developed from the third stage of the Saturn V rocket. It was launched into a near circular orbit 270 miles (433km) above the

Earth's surface on May 14 1973.

Skylab consisted of four sections, the largest being the orbital workshop, 48ft (14.7m) long and 22ft (6.6m) in diameter. This section also contained the living quarters. An airlock module contained equipment for the operational control of the station and also the hatch for space walks. The module was designed to allow the crew to leave the station without having first to depressurize the whole spacecraft. A multiple docking facility contained an Apollo docking port at one end and a reserve and rescue port on one side. In this section were also the controls for the Earth-resources scanners, a vacuum chamber, and a furnace for carrying

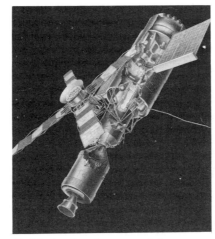

Skylab: first American space station

out experiments in material processing. The controls and display console for the telescope mount were also located in this section. The telescope mount contained six telescopes for monitoring the Sun. They were powered by a windmill-shaped array of four solar panels.

Just after launch, the micrometeorite shield deployed prematurely and broke away, and in doing so destroyed one solar panel and damaged another. After being briefed on the damage that had been done to the satellite, the first crew, originally scheduled to join the station just 24 hours after its launch, were launched on May 25 and after some difficulties docked with the parent craft the next day. They erected a sunshield and during a space walk cut free the jammed solar panel. After a record stay of just over 28 days, the crew of three, Charles Conrad, Joseph Kerwin and Paul Weitz, returned to Earth on June 22. In spite of all the technical problems, this first mission was highly successful. The Apollo telescope mount was operated for 88 per cent of the planned time, making more

than 30,000 exposures. More than 16,000 exposures were made with the three cameras of the Earth Resources Experimental Package (EREP), and the three sensors forming part of the EREP used 46,000ft (14,000m) of tape.

On July 28, a second crew (Alan Bean, Owen Garriott and Jack Lousma) were launched and docked with the Space Station the same day. They installed a new and more efficient sunshield and also installed a micrometeorite detector on the telescope mount truss. During the mission they took 77,000 photographs of the Sun and studied in detail more than 100 flares, including a major disturbance on August 21. In the Earth-scan experiments the crew took 16,800 pictures of the Earth for use in crop surveys, land-use planning and searches for natural resources. Space-welding techniques revealed that strong new alloys could be formed under weightless conditions. On their return to the Earth on September 25 the astronauts brought back nearly 18 miles (30km) of magnetic tape recorded during their 59½ days spent in space, then a record.

The third crew (Gerald Carr, Edward Gibson and William Pogue) joined Skylab on November 16. With them in the Saturn rocket was nearly a tonne of supplies, including film, recording tape and food. They spent much of their time repairing and replacing failed equipment, which required a seven-hour EVA. Nevertheless, they spent nearly 2½ times longer than planned observing the Sun and Kohoutek's Comet and more than three times the alotted time on materials and space manufacturing experiments. Before returning to Earth the crew spent 5 hours 19 minutes outside Skylab retrieving film and collecting a piece of the spacecraft's metal skin for analysis of the effects of exposure to extra-terrestrial conditions. A successful return to Earth was achieved on February 8 1974 after a record 84-day mission. Although film and other equipment were left on board the spacecraft, no more visits were planned, but just prior to undocking Skylab was boosted into a slightly higher orbit to extend its lifetime, just in case plans were altered. During the three missions over 200,000 photographs of the Sun were taken and 45 miles (72km) of magnetic tape were used in recording data from onboard experiments. However, later in 1974 the project was officially cancelled.

Much concern was expressed about the possible damage that the 75-tonne

S

spacecraft could do when it re-entered the atmosphere, with possibly large pieces reaching the Earth's surface. After the orbit had been modified on several occasions, the spacecraft re-entered the atmosphere, over the Indian Ocean, scattering débris over the Ocean and also Australia. Fortunately no one was injured, but a spectacular fireball was observed over Western Australia.

One of the most important results of the Skylab missions was to show that man could exist and work for long periods in the weightless condition. It also proved beyond doubt that future developments for scientific research in space would be carried out in manned space stations. *HGM.*

S

Skylark First British scientific sounding rocket.

Slipher, Earl (1883–1964). Younger brother of V.M. SLIPHER. He worked at the Lowell Observatory and was noted for his planetary photography.

Slipher, Vesto Melvin (1875–1969). American astronomer, who worked at the Lowell Observatory and was Director from 1916 to 1952. He initiated the successful search for Pluto, and earlier had been the first to recognize the REDSHIFTS in the spectra of the objects we now know to be galaxies — a fundamental step in the development of modern cosmology.

V.M.Slipher: began the search for Pluto

Slow motions Telescopes without electric or clockwork drives require slow motions to enable a star or planet to be followed across the sky.

For an equatorially-mounted telescope only the right ascension axis requires a slow motion control; if the telescope is on an altazimuth mounting then slow motion controls are required on both axes.

The controls allow the telescope to be moved smoothly by means of a handwheel or fine screw. The right ascension control moves the telescope horizontally and the declination control vertically. The combination of these two motions allows smooth accurate tracking of the celestial object. *See also* ALTAZIMUTH MOUNTING, EQUATORIAL MOUNTING.

Smithsonian Astrophysical Observatory Important American observatory at Cambridge, Massachusetts, which in 1973 officially joined with the Harvard College Observatory to become the Center for Astrophysics. At Harvard, pioneer astronomical photography was carried out, and at the outstation at Arequipa, in Peru, in the early 20th century, photographic surveys were carried out, one of which enabled Henrietta LEAVITT to establish the period-luminosity law of CEPHEID VARIABLES. The DRAPER CATALOGUE of stars was also drawn up at Harvard. Today the centre controls several major outstations, including the WHIPPLE OBSERVATORY on Mount Hopkins (Arizona) where the main instrument is the MMT or MULTIPLE-MIRROR TELESCOPE. *PM.*

Smyth, William Henry (1788–1865). English naval officer; hydrographer; and a founder of the Royal Geographical Society. He built an observatory in Bedford; and wrote *Cycle of Celestial Objects* and *The Bedford Catalogue* (1844).

Snow telescope The horizontal solar telescope of Mount Wilson. It consists of a CŒLOSTAT and of a concave mirror of 24in (61cm) aperture and a focal length of 59ft (18m).

Sobkou Planitia Plain on Mercury: 39°N, 128°W. It lies east of the Caloris Basin.

Sol The modern term for the rotation period of Mars: 24h 37m 22.6s.

Solar apex The point on the CELESTIAL SPHERE toward which — relative to the local stars — the Sun and Solar System are moving at about 12 miles/s (19.5km/s). It lies in Hercules at R.A. 18 hours, Dec. +34°.

Solar chromosphere The layer of the Sun's atmosphere above the PHOTO-SPHERE. It consists mostly of hydrogen, helium and calcium and envelops the whole of the photosphere, but because of its low density it cannot be seen except for a very brief time during a total SOLAR ECLIPSE. It appears at the limb of the Sun, just before the beginning and at the end of totality when the brightness of the photosphere is eclipsed by the disk of the Moon. The chromosphere is pinkish in colour and the upper boundary appears irregular with a spiky structure.

Nowadays the SPECTROHELIOGRAPH or birefringent filters enable astronomers to study the chromosphere at any time and not only at the limb of the Sun but in its whole extent projected on the disk of the Sun. Spectroheliograms and filtergrams are monochromatic images generally taken in the spectral line of hydrogen (Hα) or of ionized calcium (H and K). They show large bright patches and irregular dark filaments which are the projections of SOLAR PROMINENCES on the disk, together with a more general network which appears to be associated with magnetic fields. Studies of the structure of the chromosphere have shown that it extends to about 6,000 miles (10,000km) above the photosphere and that it merges into the SOLAR CORONA. The chromosphere is generally divided into low, middle and high. Its density increases with height while its temperature increases rapidly, reaching about 100,000K at the limit of the upper chromosphere. Filtergrams, taken in the UV by space vehicles using the Lyman alpha line of hydrogen (Lyα), show a great similarity with those taken in CaII from Earth, Intensive studies of chromospheric phenomena (*see* SOLAR SPICULES, PROMINENCES and FLARES) have increased our knowledge of the structure and behaviour of the chromosphere. In astrophysics the term chromosphere is also used in stellar models to represent the layer above the photosphere of a star. *VB.*

Solar constant The amount of solar energy received per square metre per second at a distance of 1AU from the Sun. Its mean value is 1,370 w/m²/s, but it varies slightly.

Solar corona The region of the solar atmosphere which is normally only visible during a total eclipse of the Sun. By means of filters with a radially varying transmission which compensates for the decrease in light intensity with distance from the PHOTOSPHERE, the corona can be seen to extend for a few solar radii above the photosphere. A coronagraph permits routine observations of the corona at times other than during eclipses. In principle a coronagraph is a fairly simple solar telescope which has a disk to occult or artificially eclipse the photosphere, but in practice, because the photosphere is about a million times brighter than the corona, much care and a complex design are required to remove scattered photospheric light. Lenses of very high quality with no blemishes are needed. Similarly, coronagraphs are usually sited at mountain peaks to avoid the effects of dust scattering in the Earth's atmosphere.

The corona consists of a plasma or ionized gas and typical values for the coronal temperature are greater than 1,000,000K. Much of the interest in coronal physics concerns the processes involved in heating and heat flow within the corona and into the SOLAR WIND and also with the highly variable structure of the corona. It appears different at each eclipse, with regions where it seems to be almost absent, CORONAL HOLES, existing on many occasions.

The most obvious feature seen when viewing the corona is that there appear to be filaments or streamers directed more or less away from the Sun. Near solar minimum they appear in a clear pattern which is akin to the field lines from a dipole (bar) magnet with plumes from the polar regions. Near solar maximum the corona is a few times brighter and the pattern is less regular with coronal holes and a variety of loop, arch and prominence structures. These latter are associated with SOLAR FLARES.

When viewed during an eclipse the corona appears to be white and this continuum radiation is light from the photosphere which is weakly scattered from the free electrons in the corona. The coronal light is polarized. In addition to the continuum radiation there are EMISSION LINES in the coronal spectrum which were a major puzzle. These lines did not seem to be at the wavelengths of any laboratory emission lines due to known elements. CORONIUM was suggested as a mysterious new element with these lines in its spectrum. In 1940 the Swedish physicist Edlén showed that multiply-ionized iron and calcium atoms, ie, atoms with more than one electron removed, would emit light at the coronium wavelengths. The different ionization states, or numbers of electrons per atom removed, tend to occur at different temperatures. Thus, by viewing the corona at different wavelengths, different temperature regions are observed.

Apart from the emission lines there are ABSORPTION LINES in the spectrum of light from the outer regions of the solar corona. This is mainly due to the scattering of a Fraunhofer spectrum by interplanetary dust grains. It is important to realize that the solar corona is optically thin at visible wavelengths and thus is largely transparent to the light from the photosphere.

Extreme ultraviolet (EUV), and X-ray observations are not confused by large fluxes from the "low" temperature photosphere and thus X-ray astronomy is a powerful technique for studying the corona. The X-rays are emitted essentially only from those regions with temperatures greater than 10^6K and thus coronal structures on the disk can be observed. Throughout most of the corona the temperature increases with radial distance from the surface and so images at EUV and X-ray wavelengths produced by atomic transitions associated with higher temperatures correspond to images of higher regions.

The problem of heating the corona has been one of the central issues of solar physics for many years. The photospheric temperature of around 5,800K is cooler than the coronal temperature and so by the laws of thermodynamics the net flow of heat must be downward toward the photosphere. Thus solar radiation cannot be the source of the heating which causes the coronal temperatures. The high temperatures have to be the result of energy being transported into the corona by other means than radiation and then dissipated within it. Various processes have been suggested, including the conversion of gravitational potential energy from cometary dust grains falling toward the Sun.

A mechanism which has been favoured since the work of Biermann and, independently, SCHWARZSCHILD in 1946, involves the ion acoustic waves generated by the turbulent motions in the granulation of the photosphere. Sound waves cannot propagate in the vacuum, ie, in the absence of collisions between the gas molecules, but ion acoustic waves are plasma phenomena which are analogous to sound waves. The dissipation of the ion acoustic waves can take place by non-linear processes, especially by the formation of shock waves. Although this theory has some attractions, there are certain difficulties. Observations which should show the motion associated with the waves fail to do so. The process seems to be efficient at heating the chromosphere, so efficient that there is not sufficient energy to heat the corona. In particular, it is very difficult to use ion acoustic wave energy to provide the heat associated with coronal loops and flares. These regions are dominated by their magnetic field structures and magnetic processes are now thought to be responsible for their heating. It is still possible that wave heating from the photosphere is taking place in the corona to some extent. The magnetic processes may also involve plasma waves.

High spatial resolution images of the corona show that the most prevalent structures are loops; some are "closed" in the form of arches and some are "open" and extend to the high corona. These loops are anchored in footprints at the solar photosphere and are caused by the processes responsible for SUNSPOTS and the SOLAR CYCLE. The two footprints associated with each loop have opposite magnetic polarity as can be seen in a magnetogram, an image which is sensitive to regions of one magnetic polarity. Plasma is confined by these magnetic loops and heating may be caused by electric currents associated with twists in the loops. Other mechanisms for heating in coronal loops include magnetic reconnection, a process in which the magnetic field becomes reconfigured, and heating by magnetic shock waves. It is likely that all of these processes are all important at different times. For example, magnetic reconnection is probably important in solar flares. *LJCW.*

Solar cycle The period, about 11 years, of the rise and decay of solar optical activity. At the beginning of the cycle sunspots appear in latitudes about 35° north and south of the equator. Their number increases and their lateral distribution moves towards the equator throughout the cycle. All other forms of solar activity also follow the cycle.

Solar day The mean time interval between two successive noons. *See also* DAY.

Solar eclipse An eclipse of the Sun occurs when the Moon's shadow falls on Earth; for this to happen the Moon must be in line between Sun and Earth. In fact, it is an OCCULTATION of the

S

S

Sun by the Moon and the purist will call it an eclipse of the Earth, since an eclipse is by definition the passage of a celestial body through the shadow of another.

From the above it will be seen that a solar eclipse can occur only at New Moon (when the Moon is in conjunction with the Sun), but not at every New Moon because of the approximately 5° inclination of the Moon's orbit to the plane of the ecliptic. An eclipse will result only when the New Moon roughly coincides with a node (the intersection between the orbit of the Moon and the Earth). This coincidence need not be exact, an eclipse can occur up to $18\frac{3}{4}$ days before or after the alignment, thus creating the "eclipse season". Two eclipses can occur in every eclipse season because the synodic month ($29\frac{1}{2}$ days) is less than the eclipse season ($37\frac{1}{2}$ days).

The nodes shift gradually westward along the ecliptic, so the Moon reaches the opposite node less than six months later and realignment with the original node takes place after 346.6 days,

($29.5306 \times 223 = 6{,}585.32$ days).

This cycle closely corresponds with 19 eclipse years ($346.62 \times 19 = 6{,}585.78$ days), hence eclipses recur after such cycles and form series. The added 0.32 of a day of the Saros is responsible for the westward shift of subsequent eclipses by one-third of the Earth's circumference (120° longitude). They also shift 2° to 3° north or south due to the 0.46 day difference between 19 eclipse years and the Saros. This eventually causes the series to end by passing one or the other pole. Each series comprises some 70 eclipses over a period of about 1,200 years.

Types of eclipses By some strange coincidence the apparent sizes of Sun and Moon as seen from Earth are very similar, but subject to variation due to their elliptical orbits.

A total eclipse occurs when the Moon appears larger than the Sun and the shadow cone reaches the Earth. When the Moon is at its largest (at perigee) and the Sun at its smallest (at aphelion) a long total eclipse occurs; the maximum duration is $7\frac{1}{2}$ minutes

eclipse can be seen only on the narrow path caused by the shadow-cone sweeping over the Earth's surface from west to east with a velocity of some 2,000mph (3,200kph), the maximum width of this path is 170 miles (270km). An observer situated outside this path will see the partial phase.

Phenomena at a total eclipse The mystery of a total eclipse is enhanced because the uninitiated have no warning of the impending spectacle as the Moon cannot be seen approaching the Sun: the partial phase passes unnoticed unless specifically looked for.

It is dangerous to look at the Sun at any time, especially with optical aids. The only safe way to observe the partial phase is to project the image of the Sun onto a white surface. Only during totality is it perfectly safe to look directly at the occulted Sun and the corona.

The eclipse begins when the east limb of the Moon appears in the same line of sight as the opposite limb of the Sun and seems to encroach upon the Sun. This is the first contact. This moment goes by quite unnoticed.

The projected image of the Sun will show a small notch some 10 seconds after the contact. This notch increases in size as the Moon travels across the face of the Sun during the next hour or so. The light reduction is imperceptible at first and the temperature drops very little until the last five minutes of the partial phase.

Then the real drama begins: the sky becomes darker, often with an eerie greenish tinge which is quite indescribable and quite unlike the darkening caused by clouds. Far on the western horizon a cloud-like darkening appears to be increasing in size; it is the approaching shadow of the Moon. At the same time curious moving ripples of dark and light bands appear on any white smooth surface — a strange atmospheric phenomenon known as "shadow bands". During the last few seconds of the partial phase light fails rapidly, it becomes noticeably cooler, birds settle down to roost, some flower petals close, and the wind tends to drop. As the last sunrays fade, a dramatic change of the scene occurs: darkness descends on the countryside. The last sliver of the Sun is broken up by the mountains of the Moon forming BAILY'S BEADS and as the last bead disappears second contact has occurred: totality has begun. Earth is engulfed in darkness only illuminated by the beautiful pearly white corona surrounding the pitch black Moon. The resulting

Solar eclipse: area of partial eclipse and path of totality

representing the "eclipse year".

Thus four eclipses can occur in one year, but since the calendar year is greater than the eclipse year a fifth eclipse is possible in one calendar year on rare occasions, and then only in January or December. The maximum of five solar eclipses occurred in 1935 and will happen again in 2206. At least two solar eclipses of some kind must occur every year. An eclipse of the Moon precedes or follows a solar eclipse by about two weeks because the same conditions prevail for the Moon before or after that interval.

Recurrences of eclipses It has been known since Babylonian times that the nodes regain their original positions after 18 years and $10\frac{1}{2}$ days, the SAROS period. It lasts 223 synodic months

and this can happen only if the shadow cone reaches the Earth near the equator around local noon.

An annular eclipse results in the opposite situation when the shadow cone fails to reach the Earth. A ring of Sun will surround the Moon at mid-eclipse.

An annular-total eclipse occurs if the apparent size of Sun and Moon are the same; it will be annular along most of the path, but in the middle of the eclipse path the shadow cone will just reach the Earth and in this location a very short totality is seen.

A partial eclipse results if only part of the Sun is occulted by the Moon. A partial eclipse is seen over a wide area on Earth where the Moon's penumbra reaches the Earth's surface. A total brightness on the Earth's surface varies from eclipse to eclipse; it is comparable with that of the Full Moon. In contrast to the slow progress of the partial phase, events around second contact progress with incredible rapidity leaving the observer completely overtaken by surprise, awe and wonderment.

For a few seconds after the disappearance of the last Baily bead the pink chromosphere becomes visible, only to be covered quickly as the Moon advances. The corona is the most striking feature of the total eclipse. The bright inner corona contains elegantly shaped arches, loops and helmet-like structures tapering off into the fainter streamers of the outer corona for a distance of several solar diameters. These various forms are created by the solar magnetic field. The shape of the corona varies with the 11-year solar cycle: at minimum, the corona is of equatorial disposition with long beautifully-shaped streamers extending to enormous distances east and west whilst the poles are studded with shorter plume-like jets. At solar maximum, the corona surrounds the Sun more evenly, the whole circumference of the Sun being surrounded by medium-sized streamers of intricate structure.

Prominences of various shapes and sizes are seen during totality as pink flame-like projections. Large prominences remain visible throughout totality while smaller ones appear and disappear as the advancing Moon uncovers or covers them.

It pays to take the eye off the features surrounding the Moon and look at the sky, where planets and bright stars can be seen with the now dark-adapted eyes. The surrounding landscape shows a 360° orange glow bordering the shadow of the Moon. There is so much to be seen, measured and photographed, that time passes with incredible speed and the observer is invariably taken by surprise when the west limb brightens and the first ray of sunlight appears, indicating the end of totality: third contact has occurred. The first sunbeam shining through a lunar valley gives rise to the famous DIAMOND RING EFFECT. The corona and the brightest planets may be discernible to the still dark-adapted eyes for some 10–20 seconds, but the main spectacle is almost over. Events now happen in reverse: as the sky brightens one can see the receding shadow of the Moon toward the eastern horizon, shadow bands reappear and may last for another four to five minutes; the temperature gradually rises, cocks

crow as in the early morning, and day-time activity resumes again after the short interruption. The projected image of the Sun shows the gradual uncovering of the solar disk by the advancing Moon. The partial phase lasts another hour or so until the last notch on the solar disk dwindles and finally disappears: the Moon has parted from the Sun: fourth contact has occurred and the show is over.

Information gained by a total eclipse
The corona, the outer atmosphere of the Sun, can be studied both visually and spectroscopically only during a total or annular eclipse. However, LYOT's coronagraph creates an artificial eclipse and allows limited study of the inner corona from Earth. It is used in space with great advantage.

Prominences, also first seen during an eclipse, can now be studied in more detail with the spectroscope or the interference filter.

Timing the four contacts is still important today. The results may show

Solar eclipse: total eclipse of 1976

a slight discrepancy between recorded and predicted times and this may possibly be due to an alteration of the solar diameter — a highly controversial issue, ever since Eddy proposed the idea of the shrinking Sun. Timing may also help to identify perturbations of the lunar orbit and irregularities of Earth's rotation.

The FLASH SPECTRUM can be photographed at second and third contacts when the dark absorption bands of the photosphere change to the bright EMISSION LINES of the lower chromosphere, thus giving information on the intensities of the various spectral lines.

Shadow bands are atmospheric phenomena, ill understood; they are difficult to record photographically; they deserve further study.

Einstein's theory of general RELATIVITY was first tested during the total

solar eclipse of May 29 1919. Starlight was proved to be deflected by the Sun's gravitational field. Hitherto undetected comets may be found in the vicinity of the Sun during totality and ionospheric studies found cut-off of the ultraviolet radiation during totality; this resulted in an alteration of the conductivity of the upper atmosphere. *EHS.*

Solar faculae The bright surface of the Sun (the PHOTOSPHERE) frequently shows areas of even greater brightness, best seen in the regions near the limb where the limb darkening increases the contrast.

These faculae take the shape of irregular patches, streaks or networks and occur typically in the vicinity of sunspots. They are clouds of incandescent gases in the upper regions of the photosphere. They denote increased solar activity and magnetic disturbances. They surround sunspot groups over a large area, but often precede the appearance of a sunspot by a few days and persist in the location of a sunspot group after it has disappeared. Faculae occur predominantly in sunspot zones and their frequency corresponds with the SOLAR CYCLE.

Chromospheric faculae (SOLAR FLOCCULI) are similar areas seen in the light of hydrogen or calcium, and are also indicative of increased activity around sunspot groups; they can be clearly seen at the centre of the solar disk as well as near the limb.

Different faculae are found in the polar regions. Polar faculae are smaller, elliptical or circular; they can only be observed in good seeing conditions, when SOLAR GRANULATIONS can also be seen. They are most numerous near sunspot minimum, especially from four years prior to minimum to one year after minimum. They occur in latitudes higher than 65° and each facula has a lifetime which ranges from 12 minutes to a few days. They are best observed if the Sun is projected on a large piece of white paper to a diameter of about 25in (64cm). It is thought that polar faculae may be associated with coronal plumes, which are so evident at a total SOLAR ECLIPSE that occurs near minimum. Both polar faculae and plumes are determined by magnetic fields. *EHS.*

Solar flares Sudden violent releases of energy which occur above complex active regions on the solar surface. They eject charged particles and emit radiation ranging from hard X-rays (and, in some cases, gamma rays) of

S

S

wavelengths less than 0.1nm to radio wavelengths of metres or even kilometres. The greatest output of energy is in the form of soft X-rays and extreme ultraviolet (*see* ELECTROMAGNETIC SPECTRUM). Shock waves spread out through the SOLAR CORONA and across the SOLAR CHROMOSPHERE, and large masses of plasma may be ejected into interplanetary space.

Flares last from a few minutes to a few hours, with a typical duration of about 20 minutes, reaching peak brightness after a few minutes and declining more slowly. They occur in the low corona and chromosphere usually, but not always, close to the twisted neutral line which separates areas of opposite magnetic polarity in complex active (sunspot) regions.

The radiant energy released in flares ranges between about 10^{21} and 10^{25} Joules (10^{25} Joules is equivalent to the energy released by 10 billion one-megaton nuclear bombs!). The emitted particles are mainly ELECTRONS and PROTONS. Some of the electrons are accelerated to speeds as great as half the speed of light. Smaller numbers of neutrons and atomic nuclei may also be ejected. Energetic protons and nuclei are known as solar COSMIC RAYS.

Bulk material ejected with speeds in excess of the solar escape velocity of 380 miles/s (618 km/s) at the surface can escape into interplanetary space. These ejections take the form of flare surges and eruptive SOLAR PROMINENCES visible in hydrogen light, plasmoids — clouds of plasma containing magnetic fields — and coronal transients — loop-like shocks which surge out through the corona, expelling up to 10^{13} kg of matter into space.

Flares trigger a wide range of radio emissions. Radio emissions with wavelengths of around a metre have been classified into five major types, designated types I to V. Type I events, "noise storms", which can last from hours to days, consist of spikes of radio emission, and are not specifically flare-related. Type II and III events are known, respectively, as "slow-drift" and "fast-drift" bursts because of the way in which their frequencies decline with time. Both are thought to be due to plasma oscillations (radiation emitted when electrons in the coronal plasma oscillate to and fro relative to the local protons). Type IV and V bursts are basically SYNCHROTRON RADIATION produced by electrons in magnetic fields.

The first flare ever to be seen was observed in 1859 by Richard Carrington. It was visible in white light, but it is only relatively rarely that a "white-light flare" can be seen visually against the brilliance of the PHOTOSPHERE. Traditionally, flares have been studied in hydrogen-alpha (Hα) light (Hα is one of the lines at which hydrogen emits or absorbs; it is in the red part of the spectrum) and have been classified according to their surface area as seen at this wavelength. However, the X-ray classification scheme gives a better idea of the energy output and likely terrestrial effects of a flare.

Classification of flares – Hα

Area		Class
Millionths of Hemisphere	Square °	
< 100	< 2.0	s
100–250	2.0–5.1	1
250–600	5.2–12.4	2
600–1,200	12.5–24.7	3
> 1,200	> 24.7	4

The maximum brightness of a flare in Hα is added to the importance class: F = faint, N = normal, and B = bright. The least important flare is 1F and the most important 4B.

Classification of flares – soft X-rays

Peak flux, φ, in the range 0.1 to 0.8nm	Class
$\varphi < 10^{-5}$ W m^{-2}	C
$10^{-5} \leqslant \varphi < 10^{-4}$	M
$\varphi > 10^{-4}$	X

A one-digit number from 1 to 9 is added as a multiplier. Thus a C5 event has a flux of 5×10^{-6} W m^{-2} and X7 corresponds to 7×10^{-4} W m^{-2}. The class C0 is used for events below 10^{-6} W m^{-2}. This classification dates from January 1 1969.

A typical flare will pass through the following phases:

(i) A preflare stage which involves a gradual brightening in Hα, soft X-ray and centimetre-wave radiation. Changes may occur in the structure of the magnetic field, and if there is a filament (prominence) in the area it may be triggered into activity.

(ii) The impulsive (flash) phase, which lasts about five minutes or less, during which the most rapid release of energy occurs and the highest-energy photons are released. Within a hundred seconds about 10^{36} electrons are accelerated typically to about one-third of the speed of light (energies of 10–100keV). Bursts of hard X-rays (photon energies greater than 20keV) occur simultaneously with centimetre-wave and Type III bursts, and small knots of Hα emission are seen. These knots, or flare "kernels", appear either side of the neutral line and spread rapidly, often merging to form two ribbons of emission which drift apart. The Hα intensity rises to a maximum in about 300 seconds.

Soft X-ray and EUV radiation is often seen to occur at the top of a magnetic arch, or an arcade (row of arches), in the low corona. The Hα emission develops lower down, in the chromosphere. The hard X-ray bursts seem to occur at the bases or "footprints" of magnetic arches. Gamma ray bursts, when they occur, are virtually simultaneous with the hard X-ray bursts. Line emission at gamma ray wavelengths shows that nuclear reactions occur in some flares.

(iii) The main phase, which lasts for typically 3,000 seconds. The flare reaches its maximum area of some 10^9 to 10^{10} square kilometres and the energy output slowly decays.

X-ray and ultraviolet radiation from a flare reaches the Earth in about eight minutes, causing extra ionization in the ionosphere and leading to geomagnetic fluctuations and short-wave radio fade-outs. Charged particles (ions and electrons) arrive 20–40 hours later. They squeeze the Earth's MAGNETOSPHERE on the sunward side to cause magnetic storms, and shake particles out of the magnetosphere into the upper atmosphere to generate auroræ. Solar cosmic rays arrive about an hour after.

Flares are believed to be caused by the sudden release of magnetic energy built up and stored in the magnetic fields of complex active regions. The basic mechanism is thought to be magnetic reconnection — the joining together of oppositely-directed lines of magnetic force. This results in the "annihilation" of a proportion of the field and the conversion of this magnetic energy to heat and kinetic energy. There are various possible ways in which suitable magnetic configurations might arise. For example, above a closed magnetic loop, oppositely-directed lines of force stream away into interplanetary space. A slight disturbance can be enough to reconnect these lines at a "neutral point" immediately above the loop. Another possibility is for flaring to occur where a new loop of magnetic flux emerges close to an established one.

An alternative theory is that a
(Continued on page 401)

Superclusters

Gérard de Vaucouleurs
Professor of Astronomy
University of Texas

Man has acquired his knowledge of the universe
in much the same way that a child learns about the society
of which he is a part. Just as a growing child progressively
becomes aware of larger units of human organization —
family, neighbourhood, city and so on — astronomers have
slowly come to recognize the hierarchical arrangement
of the heavens. Until the mid-20th century
we had the following picture of the universe as

This cluster of galaxies in the southern constellation of Fornax is the densest part of the "Southern Supergalaxy", the supercluster nearest to the Local Supercluster. Because of the large angular size of nearby superclusters it is not possible to obtain direct photographs of them with the large telescopes needed to record their member galaxies. Wide-angle photographs of the sky with small cameras show only the galactic stars in the foreground. Hence illustrations of superclusters must be hand- or computer-drawn from careful counts of galaxies, excluding all stars.

a whole: our Sun was a member of a galaxy of stars beyond which other galaxies were scattered without apparent organization. Since then we have gathered evidence that our Galaxy and those relatively near it form a distinct galaxy of galaxies, a system variously described as a galaxy of the second order, a supergalaxy, and more recently a supercluster of galaxies. Beyond our Local Supergalaxy, many others have been identified in recent years and their shapes, composition and, to some extent, their dynamics have been explored. Their irregular filamentary structures seem to form the basic skein of the universe, although still larger complexes – third-order clusters or "super-superclusters" – are beginning to be dimly perceived. This is a very different view of the universe from that which prevailed just half a century ago.

The world of galaxies

The great distances of the spiral nebulae, placing them well beyond the confines of our Galaxy, were established between 1918 and 1924. From a universe of stars the "system of the world" expanded to a universe of galaxies. Except, perhaps, for an occasional group or cluster, the extragalactic universe was taken to be statistically homogeneous (on the average) when sampled on a "large enough" scale.

During the 1930s the Swiss astronomer Fritz Zwicky, working at Palomar Observatory, demonstrated that clusters of galaxies, millions of light-years in diameter,

are the rule rather than the exception, and Harlow Shapley and collaborators at Harvard Observatory discovered vast galaxy *clouds* stretching over tens and perhaps hundreds of millions of light years across space. The universe, they found, is not uniformly populated by galaxies, not only on the "small" scale of groups and clusters, but also on the much larger scale of galaxy clouds. In 1938 Zwicky advanced the idea that the basic building blocks of the universe were not the galaxies themselves, but the clusters of galaxies which, he thought, were distributed at random throughout space. He visualized space as being divided into unequal "cluster cells" which, he wrote, fill the universe "as suds divide a volume of suds". Each cell was thought of as the "sphere of influence" of the cluster in its centre. This, however, was only a first step toward our present concept of an inhomogeneous universe.

A clumpy universe

The notion of a clumpy universe was far from new, even then. During the 19th century the concept of an infinite universe had been abandoned because, it was argued, the total attraction of such a universe acting on any of its parts would be infinite, and the luminosity of the sky would have to be comparable to that of the Sun day and night since in all directions the line of sight would eventually meet the surface of a star (*see* OLBERS' PARADOX). Between 1908 and 1922 the Swedish mathematician and astronomer C.V.L.Charlier showed that this

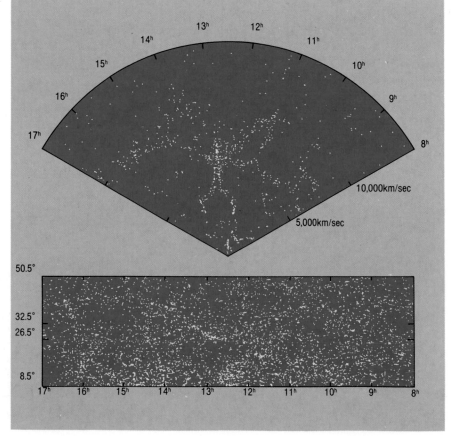

A slice of the universe. The fan-shaped diagram, *right*, shows the redshifts (in km/s) of galaxies brighter than magnitude 15.5 in a narrow wedge of space between north declinations +25.5° and +32.5° and right ascensions 8 to 17 hours. The map, *below*, shows the apparent distribution on the sky of these same galaxies and in a wider range of declinations The heavy concentration of galaxies between 12 and 13 hours near the lower edge of the map marks the Virgo cluster; the narrow, elongated structure near the centre of the map is the Coma supercluster which is seen in the upper diagram to extend over a large range of right ascensions in a redshift range of 6,000 to 8,000 km/s. The concentration of galaxies near 16 hours with redshifts of 10,000 to 12,000 km/s is part of the great Hercules supercluster. Note the near empty bubble-like regions or *voids* separating the filamentary concentrations of galaxies.

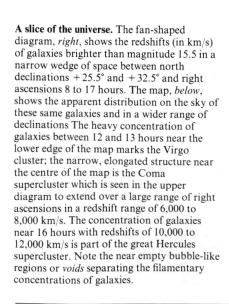

The apparent distribution of one million galaxies brighter than the 19th magnitude in the north galactic hemisphere is depicted on this map, *right*, of the Lick Observatory galaxy counts of C.D.Shane and C.Wirtanen. The north galactic pole is at the centre of the map where the dense clump is the Coma cluster; the galactic equator is at the periphery where extinction of light by interstellar dust obscures the galaxies. The vacant sector in the lower part of the map is a part of the southern (celestial) hemisphere not covered by the Lick survey because it is too low above the southern horizon of the observatory. The large concentration of galaxies left of centre is the great Hercules supercluster at an estimated distance of 350 million light-years. Note the prevalent filamentary structure of the galaxy distribution and the dark empty patches or *voids* between them. On this computer-generated map produced by James Peebles and collaborators at Princeton University the dots do not represent individual galaxies but the numbers of galaxies in each of the 10 × 10 arc-minutes squares used to make the counts.

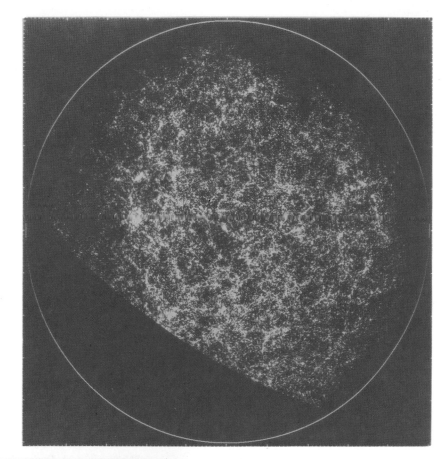

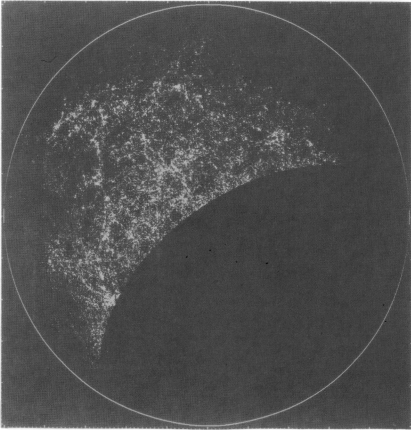

The distribution of galaxies brighter than the 19th magnitude in a part of the southern galactic hemisphere accessible from Lick Observatory, *left*, mapped as above, shows also some striking filamentary structures and great voids. The south galactic pole is at the centre of the map, the galactic equator at the periphery where the belt or galactic obscuration is also in evidence.

was not necessarily so; he demonstrated "how an infinite universe may be built up" and escape the Olbers' paradox, if it possesses a *hierarchical* structure that obeys some simple inequalities relating the masses and luminosities to the radii of its units and sub-units of successive orders. He suggested that if the "nebulæ" were in fact "island-universes" they could be part of a "nebula of the second order", which he attempted to detect by counting the nebulæ in different zones of the sky. The idea did not attract much attention outside Sweden at the time and for several decades, but it was kept alive by Charlier's successors, mainly Knut Lundmark and his students Erik Holmberg and Anders Reiz who, in theses published in 1937 and 1941, discussed the distribution of galaxies and concluded that our Galaxy was situated within a large "metagalactic cloud or supercluster" and "somewhat apart from the centre of it" which they located in the general direction of the north galactic pole. The far edge of the cloud was at an estimated distance of the order of 100 million light-years (on the current revised scale of extragalactic distances) in the general direction of the north galactic pole. These results were ignored by most astronomers at the time.

This is the more curious since there was already abundant evidence that galaxies and groups of galaxies are not distributed at random in our neighbourhood. In the 1920s and 1930s. Lundmark and also John H.Reynolds in England had on several occasions called attention to the curious distribution of the brighter spiral nebulæ along a great circle of the celestial sphere almost perpendicular to the galactic plane, but neither offered an interpretation. Then in 1932, Harlow Shapley and his assistant, Adelaide Ames, at Harvard Observatory published a famous catalogue of the 1,250 brighter galaxies based on a uniform photographic survey of the whole sky. Maps of the apparent distribution of the "Shapley-Ames galaxies" strikingly confirm the phenomenon discovered by Reynolds and Lundmark. In 1938 Zwicky pointed out that, considering the large sizes of many of the clusters he had discovered and the great apparent diameter of the galaxy cloud complex centred in Virgo depicted by the Shapley-Ames maps, the outskirts of this extended Virgo cluster might well reach almost as far out as the Local Group of galaxies, including our own. However, he did not suggest a local supercluster and to the end of his life he vigorously opposed the concept.

Superclustering and the Local "Supergalaxy"
The modern studies on superclustering and the Local "Supergalaxy" began in 1951-52 when the author was at Mount Stromlo Observatory, Australia, working on the long-neglected southern galaxies and preparing a

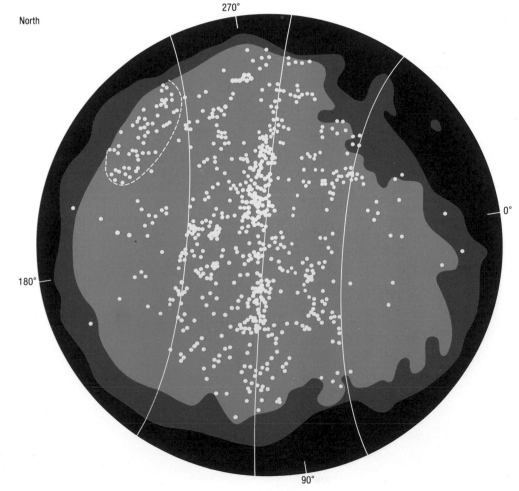

The apparent distribution of the Shapley-Ames galaxies brighter than the 13th magnitude in the north galactic hemisphere, *right,* demonstrates clearly the concentration of nearby galaxies along a great circle of the sphere or "supergalactic equator" (solid vertical line) with lesser concentration between the +30° and −30° parallels of supergalactic latitude (lateral solid lines). The galactic pole is at the centre of the map, the galactic equator at the periphery. The zones of total and partial obscuration by interstellar dust in the Milky Way — where very few galaxies are visible — are indicated by the hatchings. By a fortunate accident, the galactic and supergalactic planes are nearly perpendicular to each other. Had these planes been coincident, very few nearby galaxies would be visible and discovery of the Local Supercluster would have been difficult or impossible.

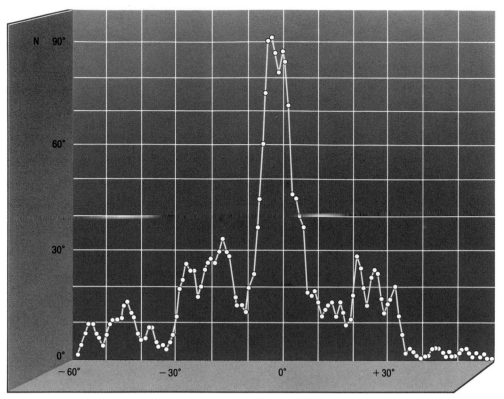

The apparent distribution of bright (Shapley-Ames) galaxies in the northern galactic hemisphere as a function of supergalactic latitude (*right*) demonstrates their strong concentration within ±30° and particularly toward the equator. The secondary maximum near −45° latitude is due to the Hydra cloud, and is not part of the Local Supercluster. The low counts between +30° and +60° are not caused by galactic obscuration, but reflect a real scarcity of nearby galaxies in the polar regions of the flattened supersystem.

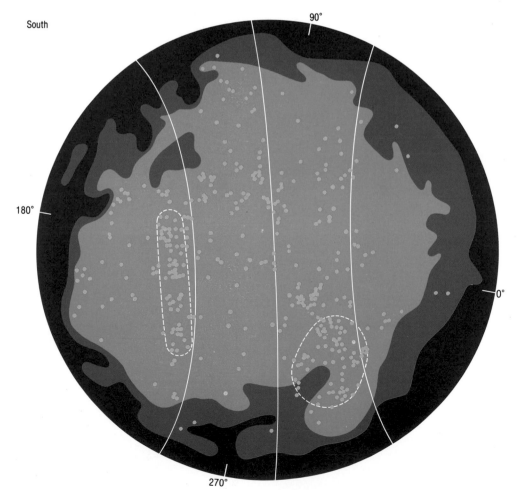

South

The distribution of the Shapley-Ames galaxies in the south galactic hemisphere, *left*, shows little concentration to the supergalactic equator, indicating that our Local Group of galaxies is near the southern edge of the Local Supercluster. Three of the nearer external superclusters are outlined (dashed lines), the Hydra cloud in the northern galactic hemisphere (opposite), the Pavo-Indus cloud (lower right) and the elongated "Southern Supergalaxy", centred at the Dorado and Fornax clusters (left of centre) in the southern hemisphere. The more distant Perseus supercluster is not visible on this chart. These maps of the Shapley-Ames galaxies, originally prepared in the 1950s, are in the old galactic co-ordinate system in which the Galactic centre was at longitude 327°.

revision of the Shapley-Ames catalogue. The remarkable concentration of the brighter galaxies toward and along a great circle of the sphere once again attracted attention. This "Milky Way" of galaxies could be traced all around the sky in both hemispheres, although it is most conspicuous in the northern galactic hemisphere. It seemed obvious that unless improbable chance encounters had brought together a number of independent clusters and clouds of galaxies, this was evidence for a flattened supersystem of galaxies — a "Supergalaxy" — within which the Local Group (and our own Galaxy) were situated, rather far from the centre. Counts of galaxies along the "supergalactic" equatorial belt placed the centre in the direction of the Virgo cluster, which thus appeared to act as a nucleus for the supersystem.

It was clear from the beginning that ours was in no way the only supergalaxy in the universe and several others were soon identified. A nearby, elongated "stream" of galaxies, dubbed the "Southern Supergalaxy", centred in the Fornax cluster, was the first example and several others were noted in Hydra, Pavo-Indus, Perseus and elsewhere. The statistical evidence that superclustering is a general phenomenon, not a local accident, was accumulating rapidly. In 1958 George Abell in a massive dissertation project at the California Institute of Technology catalogued and discussed the distribution of several thousand large clusters of galaxies. He concluded that the clusters are not distributed at random (as Zwicky insisted) but tended to associate more than chance would permit, indicating second-order clustering. These conclusions have been abundantly confirmed during the past 20 years, particularly through the work of James Peebles and collaborators at Princeton who in the 1970s analyzed galaxy and cluster distributions in terms of a "co-variance function" measuring the degree of association of density fluctuations in excess of chance expectation.

Nevertheless, it was not until 1976 that the reality of the superclustering phenomenon in general, and of the Local Supercluster in particular, became generally accepted. The enormous concentration of the nearby galaxies to the supergalactic plane was strikingly illustrated by Brent Tully and Richard Fisher of the National Radio Astronomy Observatory, who produced computer-generated stereoscopic views of the Local Supercluster and even movies of a "space-flight" around it.

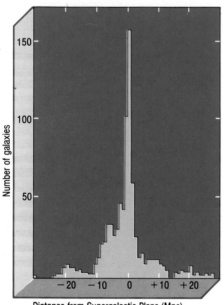

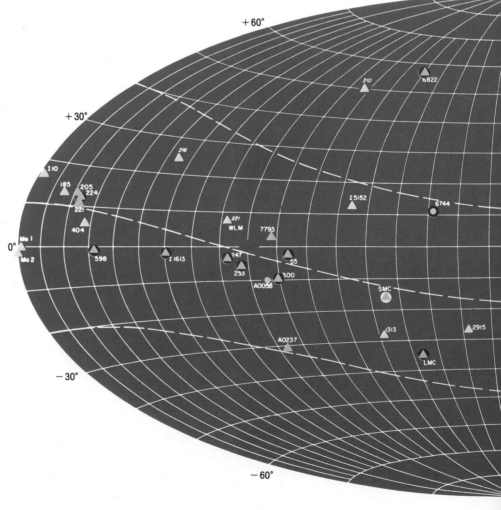

The space density distribution, *above,* of bright galaxies as a function of distance from the supergalactic plane in Megaparsecs (30 Megaparsecs = 100 million light-years) demonstrates even more vividly than the apparent distribution their enormous concentration to the plane of the Local Supercluster. There are very few galaxies more than 10Mpc (30 million light-years) from the plane of the system. Note the curious — and so far unexplained — asymmetry of the distribution marked by a relative excess south of the plane (near −5Mpc) compared with the north side (near +5Mpc).

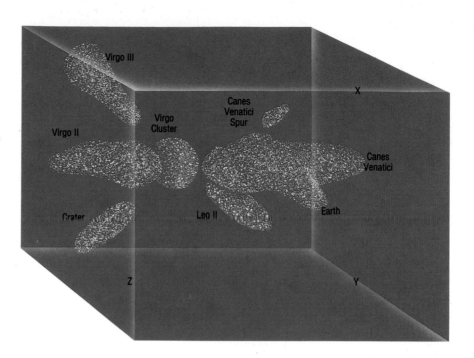

An artist's concept of the Local Supercluster in space. In this highly schematic representation, *right*, the major galaxy clouds within a cube 100 million light-years on a side, encompassing the Local Supercluster, are pictured as shadowed surfaces. Note the location of the Galaxy and the Local Group on a "spur" of the Ursa Major-Canes Venatici cloud. The outlying location of our Galaxy in the supersystem is evident and accounts for the asymmetry between the north and south galactic hemispheres.

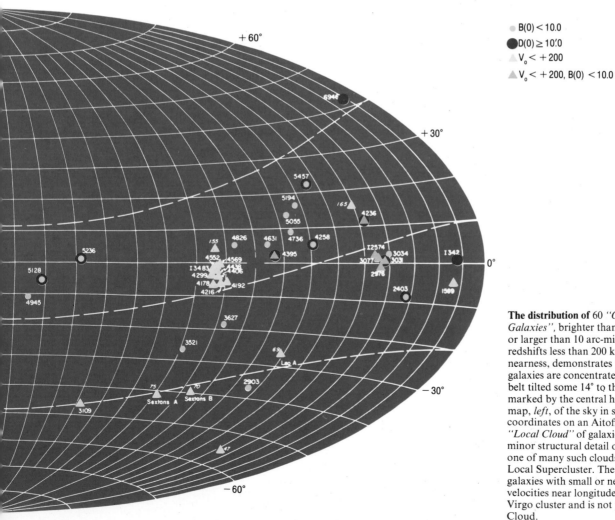

The distribution of 60 *"Outstanding Galaxies"*, brighter than the 10th magnitude, or larger than 10 arc-minutes, or having redshifts less than 200 km/s, all indicators of nearness, demonstrates that the nearest galaxies are concentrated within a 60° wide belt tilted some 14° to the supergalactic plane, marked by the central horizontal line on this map, *left*, of the sky in supergalactic coordinates on an Aitoff equal area grid. This *"Local Cloud"* of galaxies is a relatively minor structural detail of the Supergalaxy, one of many such clouds making up the Local Supercluster. The concentration of galaxies with small or negative radial velocities near longitude 100° marks the Virgo cluster and is not part of the Local Cloud.

Kinematics and dynamics of the Local Supercluster

In an initially unrelated development George Gamow, the noted atomic physicist, had remarked in a letter to *Nature* in 1946 that rotation is a universal phenomenon among atoms (and sub-atomic particles), molecules, planets, stars and galaxies, and he suggested that astronomers search for differential rotation effects in the radial velocities of galaxies (much as the motions of nearby stars reflect differential rotation of the Galaxy). Such effects should be present, he thought, if the galaxies were part of a vast universal supersystem (he did not think of intermediate-scale systems, such as superclusters). In 1951 Vera Cooper-Rubin made at Cornell University the first such analysis, seeking "evidence for a rotating universe". A similar, independent study was made the next year by K.F. Ogorodnikov in the Soviet Union with very different results as to the directions of the presumed centre of rotation and of the pole of the system. The disagreement was due in part to the near indeterminacy of the solutions based on the very small sample of only 100 galaxies with known redshifts available at the time.

In 1956, however, a large list of more than 800 galaxy redshifts was published by Milton L.Humason, Nicholas U.Mayall and Allan Sandage summarizing two decades of spectral observations at the Mount Wilson, Palomar and Lick observatories. From a detailed analysis of these velocities the author showed in 1958 that the expansion of the universe is neither linear nor isotropic in our vicinity, particularly within the Local Supercluster. In other words, conspicuous systematic departures from the idealized Hubble law of uniform expansion were indicated varying with distance and direction. These departures could be explained quantitatively within the framework of a simple model of the supergalaxy visualized as a flat rotating disk in differential expansion around its centre. The basic idea was that in regions where the space density of matter is high — such as groups, clusters and superclusters — gravitational attraction should stop or at least reduce the general expansion of space. This general principle has been repeatedly confirmed since, although it too met much opposition at the time, in part because local distortions of the velocity field complicate the determination of the Hubble constant measuring the expansion rate of the universe.

The Local Supercluster and the Hubble constant

In an irregularly clumpy universe the local value of the Hubble ratio (redshift/distance) must fluctuate with the average local density of matter all the way from zero (no

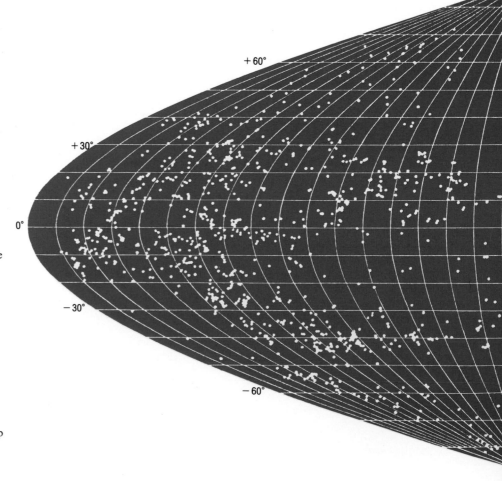

This map of the sky in supergalactic coordinates on a Flamsteed equal area grid, *right*, shows the distribution of over 4,000 galaxies having an apparent diameter greater than one arc-minute. The concentration of galaxies along the supergalactic equator, particularly in the north galactic hemisphere (right half of map), is in evidence, as well as the fact that galaxies are clustered on many different scales from small groups, such as the Local Group, to large clusters, such as the Virgo cluster (in evidence near longitude 100°–110°, slightly south of the equator), and clouds, such as the Ursa Major – Canes Venatici cloud (in evidence between longitudes 30° and 90°, slightly north of the equator) and the *"Southern Supergalaxy"* (distinguishable along the −30° parallel left of centre). The Perseus-Pisces supercluster is faintly marked by its brighter and larger members near the left edge of the map. Note also the many vacant regions separating galaxy clouds. On this map, prepared by H.G.Corwin, University of Texas, the Milky Way is along the curved edges and straight up across the vertical diameter where few galaxies are observed because of interstellar obscuration.

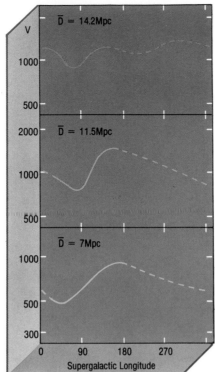

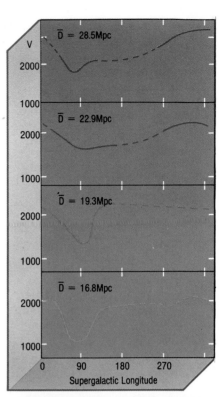

The variations with direction (longitude) near the supergalactic plane of the mean radial velocities (redshifts) of galaxies in different distance intervals D (in Megaparsecs), *right*, demonstrate the anisotrophy and departures from linearity (ie, Hubble's law) within the Local Supercluster and in its vicinity. Note that the deep minima between longitudes 60° and 90° are not in the direction of the Virgo cluster (longitude 104°). The dashed sections of the curves represent incompletely observed parts of the sky, mainly in the southern hemisphere.

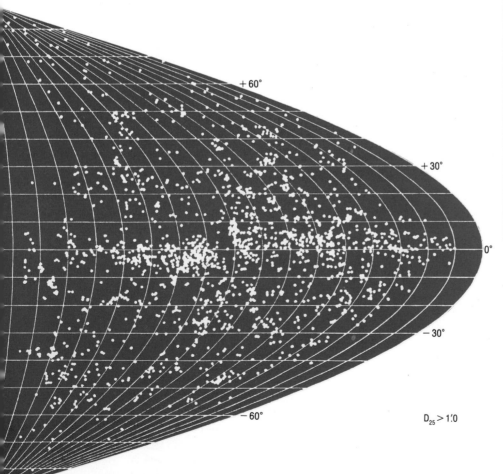

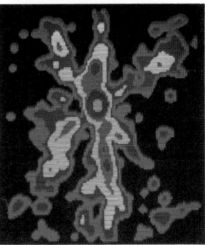

The computer-graphics display, *above*, of the Local Supercluster is a smoothed, colour-coded representation of the surface of density of galaxies projected onto the X, Z plane perpendicular to the direction of the north galactic pole. It approximates a cross-section of the system perpendicular to its own plane. Note that in addition to the major galaxy clouds in its plane, outlying elongated clouds are pointing toward (or away from) the peak density in the Virgo cluster. Successive contours represent surface densities of 0.5, 1, 2, 4, 8, 16, and 32 galaxies per square Megaparsecs (about 10 million square light-years).

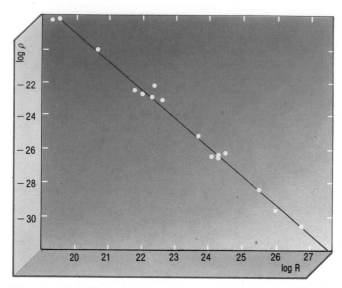

The universal density-radius relation between the logarithms of the radii R (in cm) and the logarithms of the mean densities (in grams per cubic centimetre) of the *densest* known systems of stars and galaxies of any given size, *above*. The relation, valid from the densest nuclei of compact elliptical galaxies (upper left) to the volume probed by the Lick galaxy counts (lower right) is consistent with a Charlier-type of hierarchical clustering in an inhomogenous world. If large-scale homogeneity prevails, it must be on scales larger than one thousand million light-years.

The spectrum of clustering, *below*. The upper panel shows, after I.D.Karachentsev, how the logarithm of the velocity dispersion (in centimetres per second) increases with the logarithm of the characteristic radius R (in centimetres) of systems of galaxies, from pairs (P) to triplets (T), groups (G), small and large clusters (C_1, C_2) and superclusters (SC). The lower panel shows, after L.M.Ozernoy, how the relative density contrast (on a log scale) between a system and the system of the next larger order varies with its size. Indications are that the contrast becomes negligible on scales larger (but not much larger) than superclusters.

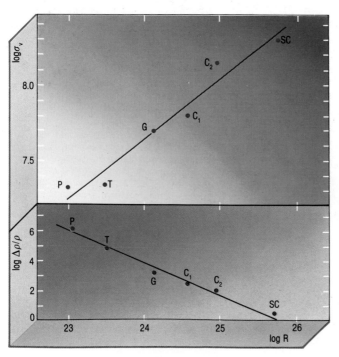

expansion) in regions of high density, such as galaxies, to a maximum or asymptotic value H_0 (the Hubble constant) in regions of very low density, such as between independent superclusters and in the universe at large. Within a supercluster such as ours the expansion rate is somewhat reduced and varies with direction in a complicated way which is only dimly perceived and still less understood as yet.

The mapping of the velocity field within the Local Supercluster has received a great deal of attention in recent years, but with still somewhat conflicting results. Many astronomers have attempted to fit some greatly simplified models of the supercluster (assuming spherical symmetry, a mean radial density distribution, and no rotation) to various data sets: the results of such studies reflect more the properties of the model than of the real world. Others (including the author) have attempted to derive in a purely empirical manner the local velocity field and the space motion of the Local Group. Peculiar motions of galaxy groups and clouds within and without the Local Supercluster may be substantial and further distort the picture. Nevertheless, the space motion of the Local Group, relative to the Virgo cluster, about 150 miles/s (250km/s), seems now fairly well determined, but it is not pointing directly to Virgo, but toward the centre of mass of the nearby galaxies some 20–30° away. Much work remains to be done to map the departures from the ideal Hubble law within the Local Supercluster and its nearest neighbours.

Superclusters and voids

Several of the nearer superclusters, including the Local one, appear on the sky as elongated structures. Typical examples are the "Southern Supergalaxy" and the Perseus supercluster. As more examples were discovered the initial assumption that they were flat disks seen edge-on became untenable. Many superclusters must be elongated string-like or sheet-like structures. Jan Einasto and collaborators at Tartu Observatory, Estonia, have presented compelling evidence that clusters and superclusters are found mainly in the faces and, especially, along the intersections of the faces of polyhedral cell structures, with very few galaxies within the cells which, consequently, appear as nearly empty alveoles. This bizarre and unexpected picture was quickly confirmed by supporting observations reported by several teams of United States astronomers, particularly at the Harvard-Smithsonian Center for Astrophysics, Cambridge, Massachusetts, where a massive programme of redshift measurements by John Huchra and colleagues has yielded detailed information on the space distribution of galaxies within several hundred million light-years. By plotting redshift versus direction in the sky one obtains a crude map of the space distribution of the galaxies in a fan-shaped sector centred at the Galaxy. Such diagrams show in a spectacular fashion the extreme irregularity of the space distribution of galaxies and their concentration in long filamentary structures separated by great voids where few or no galaxies are found. The same filamentary structure in the large-scale distribution of galaxies

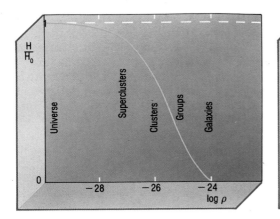

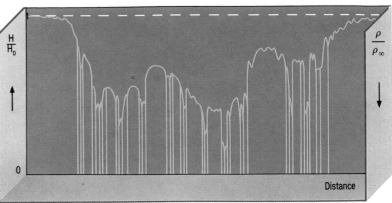

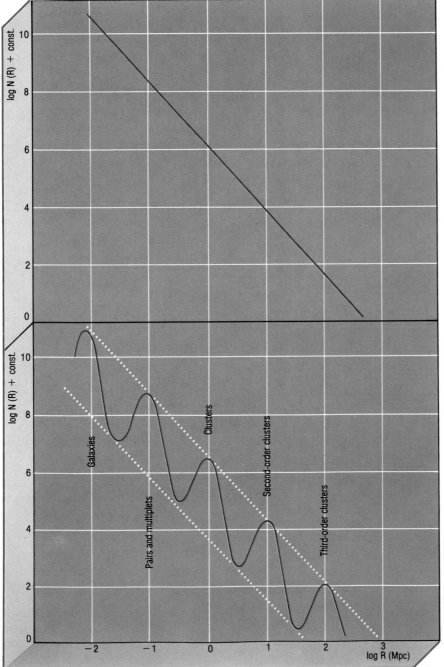

Effect of hierachical clustering on the local expansion rate of the universe, *above*. At *left:* schematic relation between Hubble *ratio* H (as a fraction of the Hubble *constant* H_0) and logarithm of mean space density ρ (in grams per cubic centimetre). Note how expansion is slowed down in systems of increasing densities (superclusters, clusters, groups) and stopped completely in still denser galaxies. At *right:* as a consequence of the dependence of expansion rate on density, the Hubble ratio H fluctuates along a diameter of a supercluster as it crosses galaxies, groups, clusters and clouds as schematically indicated by the line. Only outside the supercluster is H approaching H_0.

The spectrum of clustering. The two panels, *left*, illustrate two possible concepts of hierarchical clustering. The curves show on a logarithmic scale the numbers N(R) (per unit volume of space) of systems of different sizes. In the upper panel there is a smooth distribution (a "continuous spectrum") indicating no preferred scale of clustering. In the lower panel there are some preferred sizes for the systems of different orders, from galaxies which have characteristic radii of order 10 kiloparsecs (log R = −2) to pairs and multiplets, and clusters of first-, second-, and third- orders which have typical scales of 0.1, 1, 10 and 100 Megaparsecs. Whether the real universe follows more closely one or the other scheme (or some intermediate situation) is not yet known.

had already been noticed by James Peebles and co-workers at Princeton University on the maps of the Lick Observatory counts of faint galaxies. Thus the latest view of the universe is almost the opposite of that advocated by Fritz Zwicky in 1938: it can still be likened to a "volume of suds", but instead of finding clusters and superclusters in the centres of the "cluster cells" we now find them populating the walls and intersections of the bubbles, much as the soapy water in the suds!

How this strange structure came about is not known, although suggestions abound. A theory of the formation of large-scale structures in the early universe proposed by Ya.B.Zeldovich and colleagues in the Soviet Union invokes the growth of small initial density perturbations in an expanding space; it does give rise to a cell-like

Evidence for third-order clustering, *below.* The correlation coefficient between the space densities of large clusters of galaxies in the Abell Catalogue in cubes of either 50Mpc (dots) or 100Mpc (crosses) on a side are plotted versus their separation S (in Megaparsecs). Note that positive correlation is still indicated, particularly for the 100Mpc cubes, at separations in excess of 300Mpc (1,000 million light-years), indicating clustering tendencies on scales much larger than a typical supercluster (30Mpc = 100 million light-years.

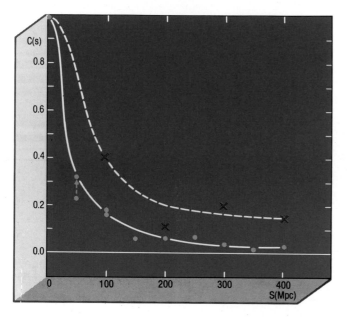

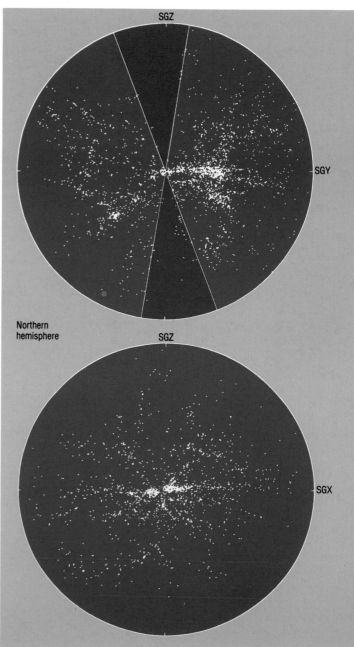

Projections of 2,715 nearby galaxies on different planes, *right* and *opposite.* Upper right: projection onto the equatorial plane (X, Y) of the Local Supercluster. The densest concentration is the Virgo cluster (right of centre); the north galactic hemisphere is on the right of the vacant sectors obscured by dust in our Galaxy. Upper left: projection on the Y, Z plane perpendicular to the supergalactic plane gives an edge-on view of the supercluster. Note outlying location of our Galaxy and Local Group at the apex of the obscured sectors and deficiency of galaxies beyond the Virgo cluster. Lower left: another edge-on view of the supercluster projected on the X, Z plane as it appears to an observer looking toward the north galactic pole. Note extreme flatness of the densest part of the system and the tendency of outlying galaxy clouds to form filaments pointing toward the Virgo cluster. Lower right: an observer looking in the opposite direction toward the south galactic pole and away from the dense central regions of the Local Supercluster sees little or no concentration toward the supergalactic plane.

structure and to the formation of flat systems or "pancakes" of galaxies. More detailed numerical simulations with super-computers predict the formation of elongated, straight or twisted galaxy clouds separated by roundish voids, which look very similar to the actually observed distribution of galaxies. The details are sensitive, however, to the nature of the "particles" carrying most of the mass, whether ordinary baryonic matter in stars and galaxies, or some exotic (and so far hypothetical) form of "dark matter" (such as heavy neutrinos) whose distribution may or may not be traced by that of visible matter. Clearly much work remains before we can claim to understand the curious space distribution of galaxies revealed by modern studies, particularly as clustering on still larger scales may be significant.

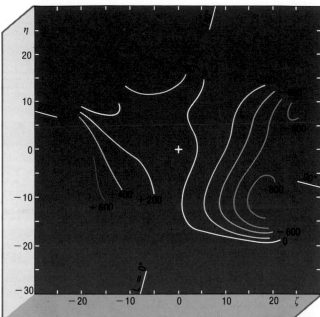

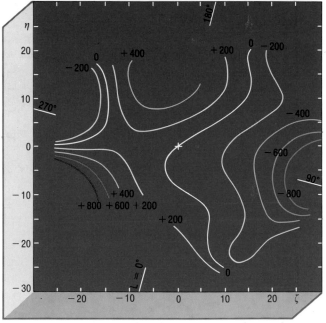

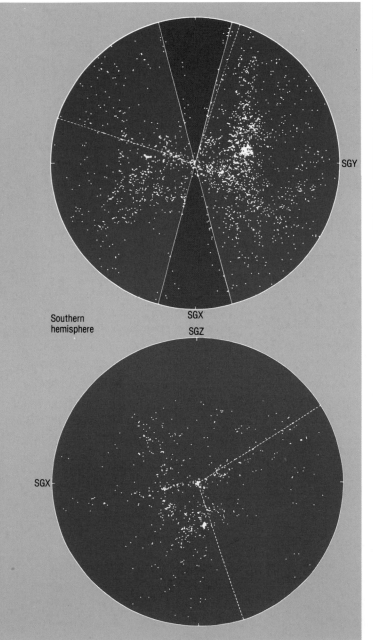

Departures of velocity field near the plane of the Local Supercluster from a uniform, isotropic expansion, *top*, for galaxies within 5Mpc, *below*, for galaxies within 10Mpc from the supergalactic plane and within a radius of about 30Mpc from the Local Group (marked by a cross at the centre of the map). The contour lines show the departures (in km/s) of the observed velocities from those expected for a uniform, isotropic expansion obeying Hubble's law. Velocities are sub-normal in the general direction of the centre of the Local Supercluster, near longitude 90° (right of centre), and excessive in the opposite (anti-centre) sector. This asymmetry is, at least partly, the result of systematic motions within the inhomogenous Local Supercluster.

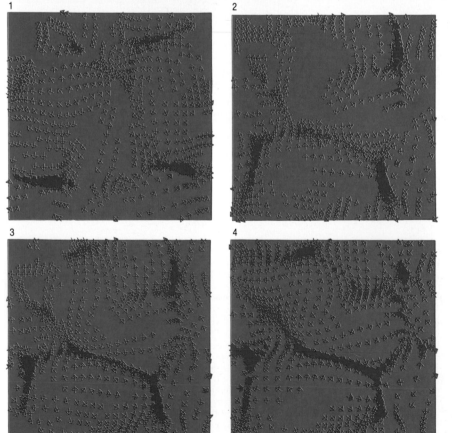

Three dimensional computer simulation, *above*, of superclusters (left) and voids (right) in a neutrino-dominated universe. A million test particles were followed in an expanding volume of space about 300 million light-years on a side at the present time. At left, the contours enclose all regions (filaments and pancakes) within which the density is more than half the present mean density. At right, the contours enclose the regions (voids) where the density is less than half the mean density.

Computer model, *left,* illustrates the type of clustering that could develop from small initial perturbations in a hypothetical expanding universe dominated by collisionless particles, such as heavy neutrinos. These four typical cross sections of a cube, perhaps 400 million light-years on a side, are spaced some 15 million light-years apart to show the cell structure arising from self-gravitation. Note the continuity from one section to the next of the zones of concentration, indicating that these are not isolated "strings", but extended "caustic" surfaces forming the walls of polyhedral alveoles or bubbles. It is speculated that ordinary baryonic matter (mainly hydrogen and helium atoms initially) would form galaxies in the denser regions only, leaving substantially empty voids in between. The arrows indicate the mean mass motion of the neutrinos, possibly carrying along atoms and, later, galaxies.

Higher-order clustering?

The original Charlier scheme envisioned an indefinite, open-ended hierarchy of systems of galaxies of ever-increasing mass and size, but with enough intervening voids that the average density of larger and larger volumes of space could be arbitrarily small. The search for structures on scales much larger than a typical super-cluster (about 100 million light-years in its long dimension) has been pursued since the 1960s by a few astronomers. In 1967 T.Kiang, a Chinese astronomer working in Ireland, found that the space distribution of the galaxy clusters catalogued by Abell was still showing signs of positive correlation over distances far in excess of typical supercluster sizes. He confirmed this result more precisely in 1969, in collaboration with British astronomer W.C.Saslaw, but few astronomers paid much attention to their results. A Bulgarian astronomer, M.Kalinkov, reported similar results at several symposia of the International Astronomical Union in the 1970s. New studies of the large-scale distribution of Abell clusters at Princeton University by Neta Bahcall have recently confirmed these previous findings.

The question arises, then, of how far and how long the quest for higher-order clustering will continue. The answer is: probably not beyond third order — say, a thousand million light-years. This, at least, is suggested by studies of the clustering "spectrum".

The Carpenter density-radius relation and the spectrum of clustering

During the 1930s Edwin Carpenter, at the Steward Observatory of the University of Arizona in Tucson, made an original study of the sizes and populations of 25 galaxy clusters for which galaxy counts had been published by Shapley at Harvard. He found that the larger the cluster, the lower its mean density. He perceived this to be the effect of some balance between the maximum mass and kinetic energy content that Nature can pack in a given volume of space. At the time no one paid much attention to this profound finding. In 1960 the present author re-discovered this relation as a result of the studies of the masses of groups and clusters of galaxies derived from the "virial theorem" relating the total kinetic and potential energies of systems of stars or galaxies in stable statistical equilibrium under their own aggregate gravitation field. Curiously, the Carpenter relation (actually an upper envelope) was soon found to extend to much larger and smaller systems, all the way from the dense cores of compact elliptical galaxies, such as M32, and the nuclei of spirals, such as M31, to whole galaxies, then through groups and rich clusters, such as the Coma cluster, even to the Local Supercluster and, perhaps, the whole region of the universe probed by the Lick Observatory galaxy counts. This universal density–radius relation remains unexplained, but clearly represents a fundamental datum for cosmology which has been ignored for far too long.

If this relation would hold for all sizes, no matter how large, the universe would be a realization of the Charlier concept of indefinite clustering. However, this is

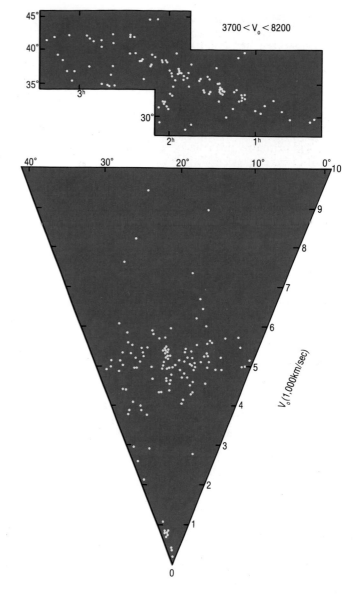

A typical wedge diagram of the Perseus supercluster, *above*, shows the distribution of radial velocities of galaxies (scale on the right side) versus angular distance in degrees (scale at the top) along the supercluster. Beyond the few nearby galaxies with redshifts less than 1,000km/s in the foreground, most galaxies in this region have velocities close to 5,000km/s, indicating that the supercluster is not merely an accidental grouping of galaxies in projection on the sky, but is really a concentration in space at an average distance of some 150 million light-years from us. The apparent great depth of the galaxy distribution, extending from 4,000 to 6,000km/s, is mainly an artefact reflecting the velocity dispersion of the galaxies about the mean. The true depth of the supercluster along the line of sight is probably not much more than its apparent width on the sky.

probably not the case. In 1967 and 1969 the Russian astronomers I.D.Karachentsev and L.M.Ozernoy demonstrated that the *contrast* between systems of different orders decreases steadily as the size (order) of the system increases. Thus while stars have enormously greater densities than the galaxies to which they belong, and galaxies are much denser than the groups in which they congregate, groups and clusters are only moderately denser than the superclusters they form, and superclusters are probably not much denser than the third-order systems, as evidenced by the difficulty of even detecting such third-order clustering. Finally, counts of faint radio sources which allow us to pinpoint extremely distant active galaxies and quasars (most optically too faint to

be photographed) have so far failed to detect any indication of clustering on scales of thousands of millions of light-years, where extrapolation of the relation would in any case predict negligible contrast between systems of order greater than the third. It seems, therefore, plausible that while clustering and superclustering of the second and probably third order are dominant properties of the distribution of galaxies on scales of millions, tens of millions and hundreds of millions of light-years, the hierarchical structure dies down on scales in excess of a thousand million light-years; that is, on *very* large scales the universe appears to be uniform and isotropic, much the same in the mean everywhere and in all directions.

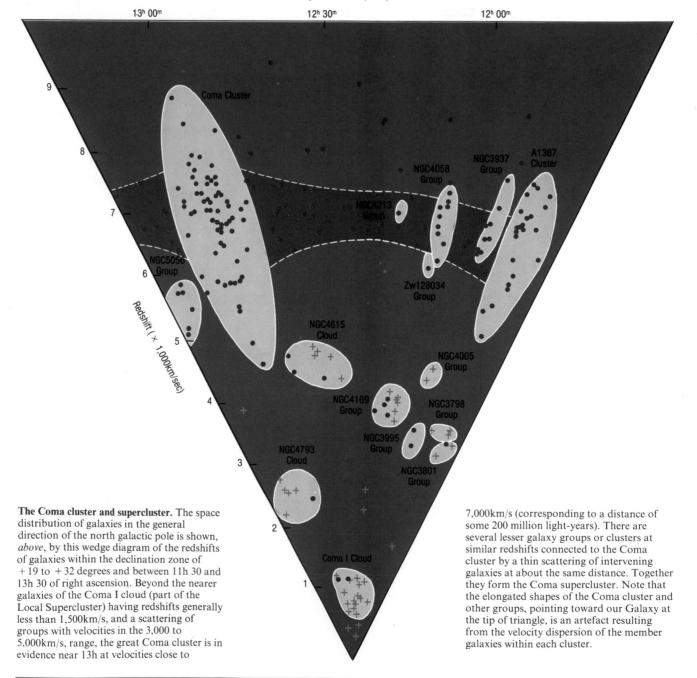

The Coma cluster and supercluster. The space distribution of galaxies in the general direction of the north galactic pole is shown, *above*, by this wedge diagram of the redshifts of galaxies within the declination zone of +19 to +32 degrees and between 11h 30 and 13h 30 of right ascension. Beyond the nearer galaxies of the Coma I cloud (part of the Local Supercluster) having redshifts generally less than 1,500km/s, and a scattering of groups with velocities in the 3,000 to 5,000km/s, range, the great Coma cluster is in evidence near 13h at velocities close to

7,000km/s (corresponding to a distance of some 200 million light-years). There are several lesser galaxy groups or clusters at similar redshifts connected to the Coma cluster by a thin scattering of intervening galaxies at about the same distance. Together they form the Coma supercluster. Note that the elongated shapes of the Coma cluster and other groups, pointing toward our Galaxy at the tip of triangle, is an artefact resulting from the velocity dispersion of the member galaxies within each cluster.

process known as the "tearing mode instability" breaks up magnetic arches into numerous "magnetic islands" separated by neutral points, so that flaring can occur at many points along the arch. This view is supported by soft X-ray and EUV observations which show that flaring tends to begin at the tops of arches.

The reconnection process heats the plasma locally and accelerates charged particles (especially electrons) to substantial fractions of the speed of light. High-speed electrons are channelled down the magnetic arches into the chromosphere where they produce

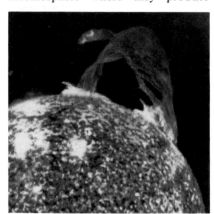

Solar flares: the great flare of 1974

hard X-ray bursts at the footprints of the arches. This radiation is produced by the BREMMSSTRAHLUNG (free-free) mechanism when the electrons make close encounters or collisions with ions. The upward-moving, high-speed electron streams give rise to Type III radio bursts. The downward conduction of heat, together with the excitation and ionization of hydrogen atoms by electrons, produces Hα emission. Hot plasma is squeezed and ejected upward at around 1,000km/s. As it moves into less dense regions of the corona it sets up a shock wave which triggers Type II radio emission. In some cases, a plasmoid with its own self-contained magnetic field is expelled, and becomes a source of a moving Type IV burst.

Despite considerable progress, many aspects of the theory of flares remain poorly understood. *IKN.*

Solar flocculi The bright and dark markings in spectroheliograms (*see* SPECTROHELIOGRAPH).

Solar granulation The mottled appearance of the solar PHOTOSPHERE caused by bright convection cells of hot gas, typically 620 miles (1,000km) in diameter, rising to the solar surface.

Solar Maximum Mission A satellite launched in 1980 to monitor the Sun during the peak years of activity. The craft operated perfectly for 9½ months, providing much data on the nature and origins of flares, but then the attitude control system failed. It was the first satellite designed for retrieval by the SPACE SHUTTLE and in the spring of 1984, after many difficulties, the SMM was brought into the shuttle's cargo bay and repaired. It was then placed back in orbit and resumed monitoring the Sun. One of the important results from SMM was the discovery of the variability of the SOLAR CONSTANT. *HGM.*

Solar nebula A cloud of material associated with the Sun in its early stages. The planets, including the Earth, are believed to have been formed by ACCRETION from the solar nebula, which was in rotation and therefore became disk-shaped.

Solar oscillations Vibrations of the Sun with periods ranging from five minutes to hours. The five-minute oscillation seems to be a surface phenomenon affecting only the outer layers, but longer-period phenomena, such as the 50-minute oscillation, represent global pulsations of the Sun as a whole and in principle these can yield information about the internal structure of the Sun. The amplitudes of the oscillations (the amounts by which the solar surface rises and falls) are only a few kilometres. There is also some evidence to suggest that the Sun undergoes larger pulsations over decades or centuries and these may be linked to variations in the sunspot cycle and the SOLAR CONSTANT. *IKN.*

Solar particles The Sun's energy is believed to be generated by THERMONUCLEAR REACTIONS occurring deep in its core. Here, at temperatures estimated to be of the order of 15×10^6K, the most important reactions involve the changing of hydrogen into helium. This process can be realized in several ways and, in the course of some of the steps of the pertinent reactions, NEUTRINOS are produced. Particles of this kind were first hypothesized to exist on theoretical grounds (in an atomic physics context) by Wolfgang Pauli (1930), who suggested that they would be without mass and possess no electromagnetic properties (eg, charge, magnetic moment, etc). In the case where a neutrino is produced deep inside the Sun, it can usually escape

into space directly because of its negligible interaction with matter and, in consequence, this radiation can provide a terrestrial observer with information concerning the physical conditions pertaining at the solar centre.

In the mid-1960s, Raymond Davis designed an experiment to detect solar neutrinos based on the fact that, very occasionally, such a particle will interact with the nucleus of an atom of chlorine and transform it into an isotope of radioactive argon. The frequency of such interactions is so low that, even in the volume of 400,000 litres of C_7Cl_4 (cleaning fluid), which forms the Davis detector, only of the order of single neutrino–chlorine interactions are expected to occur in a day. Periodically, the argon produced is removed (using a chemical method) from the tank and the number of individual atoms of argon present in the sample determined. Results obtained to date based on these data indicate that the number of neutrinos actually emitted by the Sun is substantially smaller than the number predicted to be produced theoretically. This poses a serious problem in astronomy since it places in doubt our existing models of stellar processes. At the present time the Davis experiments continue, using refined methodology, and the recently developed branch of solar physics known as helioseismology (whereby detailed observations of solar oscillations are used to infer information concerning the structure and dynamics of the solar interior).

On the basis of a variety of observations (including the anti-solar alignment of ionized comet tails and spectroscopic determinations of the temperatures, about 2×10^6K, pertaining in the outer corona), Eugene Parker anticipated in 1958 that, at a critical distance of about six solar radii, the thermal energy of the solar plasma overcomes the gravitational attraction of the Sun, allowing coronal gas to expand supersonically into interplanetary space. This assumed continuous outflux of plasma he called the SOLAR WIND and confirmation of its existence was provided by experiments performed aboard the Mariner 2 spacecraft in 1962. At the distance of the Earth from the Sun (one astronomical unit or 1AU), the solar wind has a typical density of 3–10 ions/cm³ and it is made up of approximately equal numbers of ELECTRONS and PROTONS, with a small percentage of alpha particles and heavy nuclei. This expanding plasma couples tightly to lines of

S

magnetic force and, in consequence, it carries a part of the solar magnetic field along with it. Since its flow speed exceeds the Alfvén speed (a characteristic speed for the propagation of magnetic disturbances in a plasma), the solar wind may be described as a supersonic, super-Alfvénic, strongly ionized flow that transports plasma, energy, angular momentum and magnetic field past the planets of the Solar System. It is finally decelerated to subsonic speeds by its interaction with INTERSTELLAR MATTER at a distance from the Sun of approximately 50–100AU. The heliosphere is defined to comprise that region extending from the Sun to the boundary where the pressure of the solar wind is balanced by the pressure of the interstellar magnetic field. This interface also defines the extent of the domain of the Solar System.

Observations from space during the past two sunspot cycles indicate that the plasma outflow comprising the solar wind is not spherically symmetric, as described by the Parker model, but rather comprises outward convecting streams of different flow velocities. High-speed solar wind streams greater than 250 miles/s (400km/s) can normally be traced back to long-lived CORONAL HOLES (ie, areas in which the coronal luminous intensity and temperature are abnormally low but which are characterized by open magnetic field lines along which the plasma preferentially escapes). It is presently deemed likely that slow and fast wind streams are accelerated by different (and as yet unidentified) mechanisms in well separated regions of the corona. There is evidence to suggest that the slow solar wind (which carries most of the angular momentum) may become supersonic at 30–40 solar radii (R_o) while the fast streams correspondingly escape at 10–20R_o. In support of this, the helium content of the slow component is only 2.5 per cent and highly variable, while the fast component maintains a helium content which remains relatively stable at 3.6 per cent (a compositional difference which corresponds with the known depletion of heavier ions with height in the solar atmosphere). Although the particulate constituents of the solar wind move out almost radially, the rotation of the Sun gives to the magnetic field a spiral configuration such that, at the distance of the Earth, the magnetic field lines make an angle of about 45° with the radial direction. Close to the Sun, where the magnetic field is still not substantially curved, the fast and slow solar wind

streams "slip" past each other. However, farther out, where the spiral angle is tighter, fast streams plough into the slow steams ahead of them. The discontinuities thus formed create individual interaction regions in which plasma is compressed to a high density and forward and reverse SHOCKS associatively generated. The region between the shocks contains highly turbulent plasma and, in this location, solar particles may be accelerated to high energies. Such so-called "Co-rotating Interaction Regions" (CIRs) are observed to become fully developed in the ecliptic plane at distances from the Sun of the order of a few astronomical units and they persist to distances of approximately 30AU (and perhaps beyond). The CIRs are responsible for producing a general smoothing of the solar wind at large radial distances from the Sun.

Major disturbances in the "quiet" solar wind are occasioned by the spectacular (and frequent) mass ejections styled coronal transients. Though truly immense (sometimes exceeding in white light the size of the visible solar disk), coronal transients contain very little matter and carry away only a negligible part of the Sun's total mass. The faster events (speeds greater than or equal to 620 miles/s, 1,000km/s) are accompanied by SOLAR FLARES and by nonthermal radio bursts, these latter allowing the magnetic field strength to be derived. In all cases to date where the magnetic field strength could be deduced, its energy was found to be larger than the thermal, potential or kinetic energy of the ejected matter. It is therefore supposed that magnetic forces drive transient material away from the Sun.

Major flares are almost always accompanied by coronal transients, although the causal physical mechanisms linking these two phenomena are not presently well understood. Other kinds of mass ejection associated with some flares include surges, eruptive SOLAR PROMINENCES and sprays. Surges follow straight or curved paths upward and return to the Sun along the same trajectories. They probably outline magnetic flux tubes and have velocities in the range 60–120 miles/s (100–200km/s). In most cases they have a small flare or chromospheric brightening at their base and are located near sunspot penumbræ or near polarity reversals associated with satellite sunspots. Eruptive prominences can occur either in active regions or involve quiescent prominences located far from

active regions with sunspots. In the latter instance, during a "disparition brusque" (sudden disappearance), a previously quiescent prominence erupts and ascends, slowly at first, then with increasing velocity to finally escape the constraint of solar gravity. Sprays are extremely violent ejections which are always associated with flares. In narrow band observations matter appears not to be constrained by the magnetic field but to fly outward in fragments. However, broadband observations indicate that most of the material is, in fact, as is the case with eruptive prominences, entrained on expanding magnetic loops.

The enormous variety of energetic flare associated particle emissions was only fully recognized when direct measurements were made in the interplanetary medium using satellites and spacecraft. These experiments provide information concerning fluxes and energy spectra and allow different particle species to be distinguished (protons, alpha particles, heavier nuclei and electrons). In certain rare instances, solar flares may eject into space protons with energies in the BeV range. Such radiation poses a serious hazard to living organisms not protected by the terrestrial atmosphere and passengers in supersonic aircraft flying over the poles as well as astronauts on space missions may be at risk. Such protons can arrive at the Earth within less than 30 minutes, followed, after some hours, by lower-energy particles in the MeV range. Finally, one or two days after such a flare, a geomagnetic storm may occur, indicating the arrival at the Earth of particles with energies of the order of 10^3–10^5 eV.

If the solar disk co-ordinates of high-energy (COSMIC RAY) flares, associated with the generation of greater than 500 MeV protons recorded at the Earth in less than or equal to 30^m, are plotted, they show a remarkable longitudinal asymmetry in that they cluster predominantly to the west of the Sun's central meridian, with many events occurring close to the west limb. It is inferred that (under suitable circumstances) high-energy solar particles produced in flares occurring within a broad longitude band extending from about 20°–90° west have nearly immediate access to the magnetic field lines extending from the Sun to the Earth in the ecliptic plane. There is a general tendency for events occurring near the west limb to produce particles having the shortest Sun–Earth transit times and this is related to the distortion westward of the magnetic field lines produced by solar rotation.

High-energy particles originating from flares in eastern longitudes can, although perhaps delayed by up to several tens of hours, arrive, under favourable circumstances, at the Earth by a process which probably involves diffuse transport within the corona to locations westward of the flare site, followed by escape from the Sun and interplanetary propagation earthward. The details of the method whereby transfer is achieved is a matter of controversy and multispacecraft observations of the spatial and temporal evolution of flare particles in different energy regimes (as sampled at various heliographic longitudes) are currently being studied to investigate azimuthal propagation in solar longitude: coronal trapping and storage and interplanetary transport mechanisms.

Flare particles with energies greater than 15 MeV/nucleon are observed to exhibit the same elemental abundances as those identified in the corona. At lower energies however there seems to be an "overabundance" of elements greater than or equal to Ne and, in the so-called ^{3}He-Fe rich events, there is a dramatic overabundance of ^{3}He as compared with ^{4}He, Ne and heavier elements. There is in addition a large variation in elemental composition observed from one event to another. At the present time, a major interpretational effort is required in order to disentangle from these data the influence on particular elemental ratios of interplanetary propagation effects so that information on compositional variations which are "truly" solar in origin may be obtained.

Up until now, observations of the interplanetary medium have only been made from spacecraft situated within, or close to, the ecliptic plane. Nevertheless, a picture is emerging of a heliosphere with a three-dimensional structure within which the ecliptic occupies a rather special position, ie, perpendicular to the solar rotation axis and close to the equatorial plane of the large-scale solar magnetic field. It is now accepted that, except during the inversion phase of the solar magnetic cycle, the heliospheric magnetic field resembles a DIPOLE deformed by the expansion of the solar wind. Fields from the unipolar coronal holes are dragged out and separated by a vast neutral sheet lying more or less in the equatorial plane. This current sheet is slightly warped so that it extends up to 15°–20° above and below the equatorial

plane and it is thus often picturesquely referred to as the "ballerina skirt" of the Sun. To date the only significant excursion by any spacecraft out of the ecliptic plane was that of Pioneer 11, which reached heliographic latitude 17°. Thus, solar particles and fields have hitherto only been sampled at the Earth and from space over a narrow, and indeed very non-representative, range of latitudes. It is recalled that most solar activity originates in the sunspot belts, which are located between 10°–40° in both solar hemispheres, with a superimposed variation in the latitude distribution of sunspots which is related to the solar cycle. However, the characteristic features of the solar wind and of the magnetic field above this activity zone have not yet been investigated. Also, at high solar latitudes, where the magnetic field lines and the solar wind flow are predicted to become parallel, no observations are available.

It is clear that in situ measurements in the hitherto uncharted third dimension of the heliosphere are required to provide the next major step in understanding solar particle emissions. NASA's International Solar Polar Mission (spacecraft ULYSSES) has already been designed to exploit "Jupiter gravity assist" to achieve an out-of-the-ecliptic trajectory for a payload of sophisticated solar instruments dedicated to studying the heliolatitude dependence of a variety of solar phenomena. Ulysses was scheduled to be launched in May 1986 using the Space Shuttle "Challenger". Sadly, however, an explosion over Cape Canaveral in the previous January has seriously delayed the realization of this important mission, the results of which are ultimately expected to provide the resolution of many outstanding problems in solar particle physics as well as bring new and as yet unanticipated phenomena to light. *SMcKL.*

Solar penumbra Grey area surrounding the dark umbra of a sunspot. It consists of numerous delicate fibrils which seem to emanate from the umbra. Its outline is fairly sharp but jagged and corresponds approximately with the shape of the umbra. The fibrils are thought to represent convection channels in the magnetic field of the sunspot.

Solar plages Bright, active regions in the solar atmosphere occurring especially in the vicinity of sunspots. They are best seen in monochromatic light, such as hydrogen or calcium, when they are

also described as SOLAR FLOCCULI (little clouds). Their brightest features are also seen in white light as SOLAR FACULÆ.

Solar prominences Large clouds of matter extending above the SOLAR CHROMOSPHERE. They are of the same colour as the chromosphere and can be seen at the limb of the Sun during a total eclipse. With the aid of special instruments such as the coronagraph, the SPECTROHELIOSCOPE or birefringent filters, prominences can be observed at any time, not only at the limb of the Sun but also as dark filaments projected on the solar disk, so that their structure, motions and the development of their shape can be studied in detail.

Prominences are divided into two main classes: quiescent, which show no large motions, may last for several solar rotations and dissolve slowly; active, also known as eruptive or metallic, which rise very rapidly, have complex forms and may disappear in less than one hour. Active prominences

Solar prominences: March 31, 1971

may reach great heights like the one observed on June 4 1946, which, in one hour, reached a height of 435,000 miles (700,000km). Prominences show a great variety of shapes, which, from their appearance, have been described as loops, filaments, jets, trees and fountains. Loop prominences may be connected with SOLAR FLARES (flare loops) or with sunspots (single loops); tree and hedgerow prominences are typical quiescent prominences.

Various classifications have been proposed since Secchi introduced the first classification in 1875. The criteria used in these classifications are the shape, the activity of prominences, their origin, whether in the corona or chromosphere, and whether or not they are associated with sunspots. The spectrum of prominences show bright emission lines of H, He and CaII; occasionally lines of Fe, Mg and Na are also present. Recent theories link

S

the existence and structure of prominences with the magnetic fields of the Sun. The material in the corona which condenses into prominences probably originated in the SPICULES. *VD*.

Solar spectrum The Sun's light analyzed colour by colour into a spectrum of type G_2 V.

Solar spicules Fast-moving, short-lived jets of gases of a spiky appearance in the upper chromosphere of the Sun. Probably they supply matter to the corona.

Solar System The system of the Sun, consisting of the Sun itself, the planets, the planetary satellites, comets, asteroids, meteoroids and a great deal of tenuous interplanetary material.

Solar System dust rings As well as the expected dust throughout the Solar System (*see* ZODIACAL LIGHT), IRAS found three narrow dust bands near the ecliptic. The bands lie at the distance of the asteroid belt and represent collisional debris from three prominent asteroid families.

Solar-terrestrial physics The study of the complex interaction between the Sun and the Earth. It is concerned mainly with the response of the fluids of the atmosphere and the oceans and the plasma of the ionosphere and magnetosphere to the energy of the Sun. This influence is exerted by three routes. The main energy transfer is by solar ELECTROMAGNETIC RADIATIONS, light, ultraviolet, X-rays, etc, which reach the Earth approximately eight minutes after leaving the Sun. Reaching the Earth more slowly are the particles which form the SOLAR WIND — taking between typically four and six days to reach the Earth. The solar wind carries far less energy from the Sun than the electromagnetic radiations. Even lower energies are associated with the third mechanism — that of the magnetic field of the Sun. The response of the fluids surrounding the Earth to these stimuli can be extremely complex. They are frequently non-linear and can involve triggering to give a large response to a small change in the controlling parameter. An understanding of these complicated processes has practical importance to radio communications, satellite engineering and meteorology. *LJCW*.

Solar tower Vertical fixed telescope for the study of the Sun, first constructed

by G.E. Hale, consisting of a CŒLOSTAT and an OBJECTIVE of great focal length.

Solar wedge Prism-shaped wedge inserted into optical path of a telescope when observing the Sun to reduce the image brightness to safe levels. It should be used with great caution.

Solar wind The stream of PLASMA flowing from the Sun throughout the Solar System. First predicted by Biermann in 1951 from the study of cometary tails, it has been measured by many spacecraft. The plasma is composed mainly of PROTONS and alpha particles (helium nuclei) with ELECTRONS, in approximately the same abundances as in the SOLAR CORONA, and typically between one and ten particles per cubic centimetre when measured in the vicinity of the Earth. The velocity is between 125 and 560 miles/s (200 and 900km/s) away from the Sun with low-speed streams coming from coronal loops and high-speed streams from CORONAL HOLES. The plasma flow drags the solar magnetic field radially, and this combined with the solar rotation causes the interplanetary magnetic field to spiral outward from the Sun. *LJCW*.

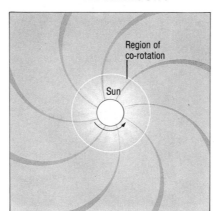

Solar wind: streams of plasma

Solis Planum (formerly known as Solis Lacus). Dark feature on Mars: 20 to 30°S, 098 to 088°W. It is decidedly variable in darkness and extent.

Solstice The extreme northern and southern positions of the Sun in its apparent annual path among the stars, denoting midsummer's and midwinter's day (usually, in the northern hemisphere, June 22 and December 22).

Solstitial colure The HOUR ANGLE passing through the SOLSTICES.

Sombrero Hat galaxy (M104, NGC4594). Galaxy in Virgo, seen edge-on. The galaxy is crossed by a thick dark lane of material; it has an unusually large central bulge containing numerous globular clusters of stars.

Sombrero Hat Galaxy: NGC4594, in Virgo

Sosigenes Greek astronomer, who devised the Julian calendar in 46BC.

South African Astronomical Observatory Founded in 1972 as an observatory run jointly by the British Science Research Council and the South African Council for Scientific and Industrial Research. The headquarters are at the site of the old ROYAL OBSERVATORY, CAPE OF GOOD HOPE, but the principal telescopes are installed near Sutherland in the Karoo semi-desert about 200 miles (320km) northeast of Cape Town. The site is at 32°23′S, 01°23′.2E at an elevation of 5,775ft (1,760m).

The main instruments are a 74in (188cm) reflector with a coudé spectrograph, moved from the Radcliffe Observatory, Pretoria, in 1976, and a 40in (102cm) reflector, originally erected at the Cape Observatory in 1964 and known as the Queen Elizabeth Telescope. There are also 30in (76cm) and 20in (51cm) reflecting telescopes used principally for optical and infrared photometry.

The 26½in (67cm) refractor of the Union (later Republic) Observatory in Johannesburg also comes under the management of the South African Astronomical Observatory. *BW*.

South African Astronomical Observatory

South, Sir James (1785–1867). British astronomer; a pioneer of the observation of DOUBLE STARS.

South tropical disturbance Major feature on Jupiter, in the same latitude as the GREAT RED SPOT, which it periodically passed; its mean rotation period was 9h 55m 27.6s. It was seen from 1901 to 1940, but has not subsequently been observed, and has presumably disappeared permanently.

Southern Cross *See* CRUX.

Southern lights *See* AURORÆ.

Soviet Mountains Range reported on the Moon's far side from the Lunik 3 pictures in 1959. They do not, however, exist!

Soyuz programme A continuing series of Soviet manned spacecraft, the first of which — Soyuz 1 — was launched on April 23 1967. The word "Soyuz" means "Union". The spacecraft con-

Soyuz programme: Soviet spacecraft

sists of three principal sections, the orbital module (OM), the descent vehicle (DV) — the only part to return to Earth — and the Instrument-Assembly Module (IAM), which includes the orbital propulsion system. The OM is used as an experimental, rest and sleeping area for the crew while in orbit and the DV provides accommodation for up to three cosmonauts during launch into and descent from orbit. A more sophisticated version, Soyuz-T, was introduced in 1981 and a further upgraded version, Soyuz-TM, made its first flight in 1986. The overall length of the complete Soyuz-T spacecraft is 23ft (6.98m) and the weight 6.85 tonnes.

Soyuz 1 crashed when its parachute lines became tangled during re-entry, and the pilot, Vladimir Komarov, was killed. The early phase of the programme included the first docking and crew transfer between two spacecraft

(Soyuz 4 and 5) in 1969, and the first transfer of crew to an orbital space station (Salyut 1) from Soyuz 11 in 1971. The Soyuz 11 flight ended tragically when the spacecraft depressurized during re-entry, killing all three of the crew. The subsequent safety record of Soyuz has been good.

The Soyuz spacecraft is the workhorse of the Soviet manned space programme, its principal role being the transfer of cosmonauts to and from the SALYUT series of orbital space stations. The usual procedure for a long-duration Salyut mission is as follows. The crew are taken to the space station in one Soyuz vehicle. At a later date, a visiting crew docks with Salyut and returns to Earth in the original Soyuz craft, leaving their freshly-fuelled Soyuz docked for subsequent use by the returning long-stay cosmonauts. An unmanned version of Soyuz — the "Progress" craft — is used to ferry fuel and supplies to, and return "cargo" from, Salyut. *IKN.*

Space docking The joining in space of two independently-launched vehicles. The first was made between GEMINI 8 and an Agena target, but it is now routine.

Space docking: Apollo and Soyuz

Spacelab A manned laboratory developed by the EUROPEAN SPACE AGENCY to fly scientists and experiments into space in the cargo bay of the SPACE SHUTTLE. There are two basic elements — a pressurized laboratory, within

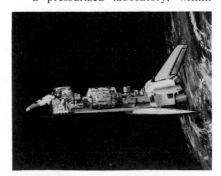

Spacelab: ESA's manned laboratory

which scientists work in a "shirt-sleeve" environment, and an external pallet, or pallets, which carry instruments for direct exposure to space. Various combinations of elements are possible. The first Spacelab mission was accomplished in 1983.

Space probe A space vehicle launched to investigate the Moon, the planets and their satellites, or to study interplanetary space.

Space research Defined as those areas of science in which new information can be obtained by means of space vehicles, whether sounding rockets, satellites, lunar or planetary probes. These can be either manned or unmanned and experience has shown that there is an important role for both. Space science has had the remarkable effect of bringing together many branches of science which hitherto had been more or less considered as separate subjects.

Although sounding rockets containing scientific experiments had been launched since the late 1940s, it was the launching of the first artificial satellite by the Soviet Union in 1957 which created a worldwide interest in space research. Both the United States and the Soviet Union had realized the importance, both scientifically and politically, of putting satellites in Earth orbit, and both had published details of their plans, although reports from the Soviet Union were either not read or just ignored.

There are two main reasons why astronomers require an observing platform above the atmosphere. Earthbound observers had obtained a very distorted picture of the universe due to the fact that the atmosphere is transparent to radiation from space in only two very narrow regions, known as the optical and radio windows. With the exception of a range of wavelengths in the infrared, where it is possible to carry out limited observations from the tops of high mountains, it was impossible to learn anything about radiations outside these windows. Above the atmosphere no such restrictions exist. Even in the transparent windows the observing conditions are far from ideal. Such problems as turbulence degrade the quality of an observation quite considerably. Once again no such difficulty exists from a position above the atmosphere. The net result of having orbiting observatories has been the rapid development of the so-called new astronomies, the ULTRAVIOLET, X-RAY and GAMMA-RAY ASTRONOMIES as well as extended INFRARED and RADIO studies. These studies have revealed a large range of new types of object and they have revolutionized our ideas of the structure of the universe.

The near-Earth regions lying just above the atmosphere have been the target for a very large number of satellites. The first successful American satellite, Explorer 1, provided data for the discovery and identification of the VAN ALLEN radiation belts. Subsequent probes have mapped these zones in great detail and have shown the close relationship between the ionosphere and solar activity. Other probes have mapped in detail the Earth's magnetic field and have shown the existence of a long magnetic tail reaching out well beyond the orbit of the Moon.

One important aspect of space research does not, however, require instruments on board the satellite or orbiting rocket casing. For a long time it has been known that the Earth is not a perfect sphere but an oblate spheroid, but the degree of flattening was known only approximately. By accurately tracking the behaviour of an orbiting object, an accurate value for the flattening was soon obtained. Subsequent analysis has shown that the northern and southern hemispheres were of different shapes, the latter being more flattened than the former. Analysis of orbital behaviour over the last quarter century has provided a very accurate figure for the shape of the Earth revealing high and low regions, results which would have been virtually impossible to obtain from Earth-based data. Analysis of orbital behaviour has also provided much information on the properties of the upper atmosphere, a region more or less barred to the Earth-bound investigator.

One of the more sensational aspects of space research has been the sending of probes to the Moon and planets. Both the Americans and the Russians have studied the Moon in great detail, but the highlight must be the American APOLLO project, that of sending a man to the Moon and bringing him back.

Venus and Mars have been explored in detail, revealing features totally unexpected. American probes have now flown past all the major planets except Neptune, although this is scheduled to occur in 1989. Surface details of the planets, their ring systems, and the surfaces of their many satellites have provided a mass of facts which will take years to unravel. The monitoring of the Earth's surface from space has provided scientists with a tremendous amount of information. Meteorological studies have benefited considerably. The use of cloud photographs taken from space are shown on television as a matter of routine. The monitoring of hurricanes is just one aspect of the way weather forecasting has improved.

Possibly the biggest effect that space research has had on us has been the revolution in communications, both telephone and worldwide television broadcasting. This has been achieved because of the necessity of the space scientist to miniaturize all equipment. The miniaturizing of domestic equipment has brought a revolution in the home. Closely linked with the advances in electronics have been the worldwide navigational systems using satellites, now used extensively by ships and aircraft. Space research has also involved the study of the way man would react to prolonged weightlessness. Much has been learned about human physiology and life processes, this knowledge now finding its way into routine medicine. The cost of all this research has been very high, both financially and in human life. Both American and Soviet astronauts have been killed, the worst from a numerical point of view being the destruction of the SPACE SHUTTLE "Challenger". Catastrophes such as these drive home the point that in space activities, man is constantly pushing into the unknown with all its attendant problems and that one small slip can have devastating effects.

Mention has so far been made only of civil experiments and applications. However, military aspects have been a major component of space research. For example, surveillance satellites are now a matter of routine.

In the early days of the space age most of the activity was carried out by the Soviet Union and the United States, but now many nations have their own satellites in orbit and are setting up space agencies. Currently a rival to the two main organizations is the EUROPEAN SPACE AGENCY (ESA) with its orbiting research laboratory SPACELAB. *HGM. See also* ARTIFICIAL SATELLITES, HUBBLE SPACE TELESCOPE, ROCKETS, SOYUZ PROGRAMME, SPACE STATIONS, VOYAGER PROBES, and entries on specific probes, programmes and vehicles.

Space Shuttle Generic term for NASA programme of a large, winged, reusable manned space vehicle.

Conceived during the APOLLO era as a follow-on spacecraft to transport cargo to and from orbit, and to service planned space stations, its design was modified to act primarily as an economic space cargo carrier. Key features included delta wings (versus more stable but lower lift straight wings), a large disposable external propellant tank, and high-thrust solid boosters for the early launch phase. The required technology advances included a digital flight control system, an advanced thermal protection system (the tiles), fuel cells for electrical power, and hydraulic power units to drive aero-surfaces and engine gimbelling.

Space Shuttle: "Challenger" lifts off, 1984

To be economic, the system had to monopolize US space launchings; this became national policy, imposed by decree on the Defense Department and commercial customers.

The first vehicle, 101 ("Enterprise"), was used for air drop tests in 1977, and later for pad fit checks in Florida and California. The first orbital vehicle, 102 ("Columbia"), was launched on April 12 1981; subsequent vehicles were 99 ("Challenger") flown in 1983, 103 ("Discovery") in 1984, and 104 ("Atlantis") in 1985; the loss of "Challenger" in 1986 led to the construction of 105, due for delivery in 1992.

The crew compartment (at 15psi Earth-normal air pressure) consists of two decks, with a commander and pilot, plus two flight engineers, on the flight deck and additional seating for engineers and scientists on the middeck (up to eight people have flown on one mission); an airlock leads into the payload bay capable of holding 65,000lb (29,500kg) of cargo (deployables, or Spacelab modules, or other rack-mounted instruments). Auxiliary equipment includes a "Remote Manipulator System" (RMS) for moving pay-loads and a "Manned Manoeuvring Unit" (MMU) for free-flight of astronauts.

On 24 flights between 1981 and 1986, the shuttles carried commercial, scientific, and military payloads, plus laboratory modules, and also repaired and recovered several broken satellites; the maximum mission was ten days.

The shuttles were launched from the former Apollo Moon rocket facilities at KENNEDY SPACE CENTER's Pad 39 on Merritt Island (north of Cape Cana-veral); plans for manned launchings into polar orbit from Vandenberg Air Force Base have been indefinitely postponed.

The flight profile involves a nine-minute ascent to orbit with three hydrogen-fuelled main engines and initial assist from two solid rocket boosters; tank jettison and insertion into orbit using smaller engines of the orbital manoeuvring system; payload bay door opening to expose high-gain antenna, thermal radiators, and pay-loads; on-orbit operations; de-orbit burn and atmospheric entry to a runway landing. Besides four-person flight crews, the shuttles carried specialists (eg, scientists, commercial researchers, oceanographers); payload customer representatives (civilian and military, plus foreign nationals such as Saudi Arabian, Dutch, German, and Mexican); diplomatic passengers (Canadian and French); congressional observers; and "citizen in space observers" (the teacher-in-space was among those killed on the "Chal-lenger" catastrophe on January 28 1986). That disaster was due to a combination of adverse weather and design weaknesses, which led to a burn-through of the solid booster at a joint, with subsequent separation and struc-tural failure of the tank, disintegration of the orbiter from aerodynamic forces, incapacitation of the crew members following cabin depressurization, and their deaths when the cabin section hit the ocean. Redesigned boosters are planned for future flights, scheduled to begin in 1988. *JEO.*

Space station A large orbiting structure with substantial living and working accommodation which is designed to be permanently or intermittently manned. The potential roles of space stations include experimental and observational work in the pure sciences (astronomy, physics, the life sciences), environmental monitoring, research and development work in applied sciences, industrial activity utilizing the space environment (virtual zero gra-vity, ultra-high vacuum) for materials processing, providing a servicing base for satellites and free-flying structures and a base for further constructional work in space. The potential military applications include surveillance, ser-vicing of satellites and anti-missile activity. Space stations will also

Space station: the Soviet Union's Mir

provide a base for the launching and return of interplanetary exploration missions, manned or unmanned.

The first prototype space stations actually to be placed in orbit were the Soviet SALYUT series — which com-menced with Salyut 1 in 1971 — and the American SKYLAB, a one-off ven-ture, launched in 1973.

The next generation of Soviet space stations is likely to be assembled from Salyut-sized units joined together in orbit. The first experimental step in this direction was the linking of the module, COSMOS 1267, to Salyut 6 in 1981. In February 1986 the Soviet Union launched its MIR ("Peace") station. Measuring 50ft (15m) long by about 14ft (4.2m) wide, and weighing some 20 tonnes, Mir is similar in size to Salyut. Like Salyut, it has front and rear dock-ing ports, but it has four additional radial ports to which further units may be attached. Mir would provide the liv-ing quarters for an expanded station and is kitted out with individual cubicles to provide a modicum of crew privacy.

The United States, with European, Canadian and Japanese participation, intends to have a permanently-manned station in orbit by the mid-1990s. As presently envisaged the basic configu-ration consists of two vertical "keels" 360ft (110m) long connected top and bottom by 148ft (45m) booms, and the structure will measure 500ft (153m) from tip to tip of the transverse boom on which the solar power modules are mounted. Fitted to this framework will be two modules 45ft (13.6m) long by 14ft (4.2m) wide, one to serve as a laboratory, and the other to provide crew quarters for up to eight astro-nauts. An additional logistics module will be interchanged on a 90-day cycle with an identical module which has been replenished on the ground. Addi-tional payloads may be attached at several points on the structure and the station will include facilities for ser-vicing free-flying spacecraft. About 14 shuttle flights will be needed to assemble the basic station. *IKN.*

Space Telescope *See* HUBBLE SPACE TELESCOPE

Space travel *See* colour essay on EXPLORING SPACE.

Space walk Human activity carried out in space outside and beyond a space-craft. Otherwise known as "extra-vehicular activity" (EVA). The first spacewalk was accomplished by Alexei Leonov from the Voskhod 2 spacecraft on March 18 1965.

Space walk: Bruce McCandless, 1984

Spallation High-energy nuclear reaction where a nucleus is struck by a particle of energy greater than 50MeV. Also, the ejection of debris from the back of a surface hit at high speed, but not penetrated, by an object, eg, with meteoroid impacts on spacecraft.

Speckle interferometry A technique first pointed out in 1970 by the French astronomer Antoine Labeyrie, and then developed by him, for obtaining diffraction-limited angular resolution of astronomical objects with a large telescope despite the disturbing effect of the atmosphere through which it has to view them. Speckle interferometry is the most efficient interferometric tech-nique for normal telescopes. While a two-beam (Michelson) OPTICAL INTER-FEROMETER is restricted to the light

S

S

from individual portions of aperture, or from a pair of relatively small telescopes, speckle interferometry can take advantage of all the light collected by a telescope of the largest available aperture. Conventional interferometers can, however, achieve greater angular resolution because the elements can be more widely separated than the dimensions of the largest telescope mirrors.

The English astronomer Sir George AIRY described the pattern made up of a prominent bright disk with a system of concentric rings that sets the limit to the angular resolution of optical observations with small telescopes. This pattern (termed a diffraction pattern) results from optical interference effects over the open aperture of the telescope and is the appearance presented by the image of a distant (unresolved) star which otherwise would show as a point. The scale of the pattern, for example the diameter of the bright disk, depends directly on the wavelength of the light and inversely on the diameter of the telescope. For a given band of wavelengths, the larger the telescope the smaller is the diameter of the Airy disk and the greater is the angular resolving power, if the optical quality of the telescope is adequate. But it has long been recognized by observers that the pure Airy pattern is rarely seen with telescopes of aperture larger than about 10in (25cm). This is because of the disturbing effect of the turbulent atmosphere which causes the pattern to break up into a dancing mass of bright granules, known as speckles, having a random appearance. This effect, too, is dependent on wavelength so speckles are best seen in filtered light.

To understand how a speckled image is formed, we can consider the Earth's atmosphere to be a collection of closely packed cells, each uniform within itself and all about the same size which, at different times, range from only a few centimetres to a few tens of centimetres in diameter. For a distant star the radiation ultimately reaching a large telescope can be thought of as approaching the atmosphere with a plane wavefront which then is intercepted by a large number of these atmospheric cells. The cells break up the wavefront into a number of patches, with the result that the radiation continuing to the telescope is effectively burned into a large bundle of individual beams. We first note that the focused image of the star due to each cell has a relatively large Airy diffraction pattern, appropriate to its small size, and the patterns from all cells are

more or less superimposed. This composite pattern defines the overall scale of the time-averaged "seeing" disk (when the speckles are smeared) corresponding to the ambient atmospheric conditions. When the seeing is said to be bad, it is because the atmospheric cell size happens to be small and the telescope delivers a best-focused image of perhaps several arcseconds in diameter; in very good seeing at the best sites, images smaller than half an arcsecond can be observed. This is still more than ten times worse than the potential intrinsic performance of large telescopes with the apertures now available. Now to the formation of the speckles: the individual beams emanating from the atmospheric cells are angled in slightly different directions one to another (the cells can be regarded as being randomly distributed sections of very shallow-angled prisms) and in a given direction mutually can constructively interfere in the telescope focal plane (when wavecrests forming different beams are superimposed) or in another direction destructively interfere (when crests are superimposed on troughs). This results in a series of bright and dark regions along any line through the extent of the focused but seeing-limited stellar image, giving an overall highly contrasted mottled or grainy appearance. The bright regions are called speckles. Because the overall interference pattern will involve light from all over the telescope aperture, an individual speckle has a dimension close to that of the Airy disk corresponding to a perfect telescope of the same diameter operating in a perfect atmosphere, and when viewing a distant star. For the Palomar 200in (5m) telescope the speckle size is about 0.02 arcsecond in yellow light and thus is much smaller than the size of the seeing disk. Several hundred speckles may then be seen. Because the atmosphere is turbulent, this pattern changes quite rapidly with time so that in photographic images of stars taken over exposures of about a second or longer the speckle pattern becomes smeared into the familiar smooth overall seeing profile. For short enough time-exposures, of about one-fiftieth of a second, the atmospheric motion is "frozen" and the full speckle pattern is revealed if the image scale is made large enough.

A speckle interferometer is a camera set at the focal plane of the telescope and incorporating optics that extend the effective focal length of the telescope, filters, an atmospheric dispersion compensator, and a recording

camera capable of repeated short exposures preceded by an IMAGE INTENSIFIER. Focal extension is needed to adapt the speckle size to the spatial sampling characteristics of the camera, where several image elements are required per speckle for adequate sampling; focal ratios from f200 to f400 are generally adopted. Both film cameras and television cameras are used, but for the faintest sources a fully digital IMAGE PHOTON-COUNTING SYSTEM takes the place of the intensified camera. The filter bandwidth usually is not greater than about 300Å. In Labeyrie's original design a concave holographic grating located in an intermediate image plane served both as a tunable filter and an atmospheric dispersion compensator.

Speckle patterns can also be observed visually at a telescope eyepiece of suitable magnification. It is interesting to note that the impressive performance of naked-eye observers of binary stars, who routinely measured angular separations below the seeing limited resolution, probably relied upon the mental processing of speckle phenomena.

For stars whose angular size appears much smaller than the speckle size, the speckles closely approximate to the Airy disk appropriate for the telescope aperture. The star then is said to be unresolved. For stars which can be resolved, each speckle is a convolution of the stellar disk and the Airy disk. A direct method of obtaining a crude image of a resolved star, then, is simply to superimpose the records of the brightest speckles, which can be compared with the similarly treated speckles of unresolved stars. A more satisfactory method frequently used to obtain simple information about the object is to autocorrelate the whole speckle pattern; this is particularly useful for the measurement of stellar diameters or of separations of close binary stars. Auto correlation, however, does not yield true images. On the other hand, true images can be extracted if within the available small field an additional point source is present, with the aid of which the image can be deconvolved. Other, computation-intensive, methods are available which do not require an additional point source to produce an image. *AB.*

Spectral classification From the middle of the 19th century, when it became possible to observe the brighter stars with visual SPECTROSCOPES, it was recognized that their spectra have a variety of appearances and some system of description or classification is necessary. Before the advance of physics provided understanding of the reasons for the different kinds of spectra, a simple descriptive type was all that could be achieved. Of several schemes, that by Father Angelo Sechi was the most widely adopted until the 1890s. It contained five types, based as much on the overall colour of the star as on details within the spectrum: Type I contained blue-white stars with strong hydrogen absorption lines, Type II referred to yellow and orange stars with numerous spectrum lines, Types III and IV included spectra with broad dark bands, occurring in red stars, and Type V was reserved for stars with bright spectrum lines.

As spectroscopes improved, more details could be seen in the spectra and a more comprehensive description was needed. This was provided by E.C. Pickering at Harvard. His original scheme used the letters A to Q for the various defined spectral types and the first catalogues made from photographs of the spectra of the brightest stars were published in 1890. Further considerations of the relationships between the spectral types suggested regrouping and reordering so that by the time the Henry Draper (HD) Catalogue of spectral types began to be published in 1918, the Harvard system of classification consisted of the sequence O, B, A, F, G, K and M. As it was possible to classify the spectra of stars to even finer divisions than the seven classes of the Harvard sequence, decimal subdivisions were introduced — with G5 indicating a star midway in type between G0 and K0. Different subdivisions were used for O and M stars: Oa to Oe and Ma to Md.

From the relationship of the spectral types to the colours of the stars — O and B stars are blue-white, M stars are red — it was realized that the O, B . . . M series is a temperature sequence. Three new spectral types had to be added when it was found that some cool stars had strong absorption bands not usually seen in other stars of the same colour: classes R and N with strong bands of molecular carbon, and class S with bands of zirconium oxide.

A number of comments could be added to the basic spectral type to describe the spectrum more fully. When it became possible to distinguish dwarfs and giants from their spectra, types such as dG2, and gF5 were adopted. Other information was given as suffixes: e for emission lines, s for sharp spectrum lines, n for broad spectrum lines, k for interstellar absorption lines present in the spectrum of the star. Examples of spectral types that use these are B5ne, Blnk, dMe.

With careful use of the spectral types came the realization that the Harvard system was not adequate to deal with the differences between dwarfs, giants and supergiants, ie, stars of different luminosities at a given temperature. In 1943 Morgan, Keenan and Kellman redefined the spectral types and introduced a luminosity classification scheme. They retained the decimally subdivided O, B . . . M sequence, giving precise details of which ratios of spectrum line strengths are to be used to determine the spectral type, and provided a list of bright stars that act as standards for the spectral types. This is now known as the MK system and is used universally for the classification of the spectra of normal stars.

Harvard and MK systems

Star type	General spectral characteristics
O	Lines of ionized helium. Lines of neutral helium and often weak hydrogen lines are also visible.
B	Neutral helium lines, with hydrogen lines strengthening in the later (B6, B7, B8, B9) subtypes.
A	Very strong hydrogen lines at A0, decreasing towards A9. Lines of ionized calcium increase in strength from A0 to A9.
F	Ionized calcium continuing to increase in strength and hydrogen weakening. Lines of other elements begin to strengthen.
G	Ionized calcium very strong, hydrogen weaker. Metal lines, particularly of iron, become prominent.
K	Strong metallic lines; molecular bands of CH and CN become prominent.
M	Red stars with strong absorption bands of titanium oxide and large numbers of metallic lines.

MK system luminosity classes

Class	Star type
0	Extremely luminous "super" supergiants, present in only small numbers in the Magellanic Clouds and the Galaxy.
Ia	Luminous supergiants.
Ib	Supergiants of lower luminosity.
II	Bright giants.
III	Ordinary giants.
IV	Subgiants.
V	Dwarfs (Main Sequence stars).
VI	Subdwarfs.

In the most accurate work, even finer subdivisions may be necessary, such as Iab and IIa or IIb. An example of the information contained in a spectral type is given by O9.5 IV-V, which means a star whose spectral type (and therefore temperature) is midway between that of an O9 and B0 star and whose luminosity is midway between that of a dwarf and a subgiant at that spectral type.

The MK system is applicable only to stars of normal chemical composition; fortunately this includes about 95 per cent of all stars. The various types of peculiar stars are given their own special classification schemes. The Harvard R and N types are now combined into one CARBON STAR class with spectral types such as C2,4 which include a temperature type and a carbon band strength. The S STARS similarly use a binary type: S1,4. White dwarfs have usually been classified on a Harvard-type scheme, with D preceding the type: DO, DB . . ., with the addition of DC for continuous spectra, but a more complex scheme is now available.

For stars that contain minor spectral abnormalities, an MK type may be assigned together with an indication of the strength of the peculiarity, eg, K0III-CN3 shows that the star has anomalously strong bands of CN, or K2II-Ba5 indicates that the K2 giant is an extreme BARIUM STAR.

To be classified on the MK system the spectrum of a star must be observed with a spectrograph of prescribed type. For example, the presence of ionized helium lines only labels the star as being of type O if the observation is made at "classification dispersion" — at high spectral resolution ionized helium lines can be seen in the hottest B stars. However, systems of classification approximating the MK or Harvard system can be achieved at low spectral resolution. This may be used with objective prism spectroscopy, in which spectra of a great many stars in a region of sky can be photographed simultaneously. In this way a search may be made for all stars of a particular spectral type. Such a study is helpful, for example, in finding distant B stars that can aid studies of the structure of the Galaxy. Nearby B stars are easily recognized from their blue-white colour, but distant ones are reddened by interstellar dust. The spectral type of a star, however, being determined by ratios of strengths of neighbouring lines in the spectrum, is independent of distance or reddening. Furthermore, as

S

the MK types have been calibrated against temperature, colour and absolute magnitude, the intrinsic colour of a reddened star may be found, and hence the amount of reddening and interstellar absorption. *BW.*

Spectral types Classification of stars according to their temperature. The system evolved from typing of stars by their colours. The types are O, B, A, F, G, K, M, R, N and S, each being subdivided numerically, eg, the Sun being G2. *See also* SPECTRAL CLASSIFICATION.

Spectrograph *See* SPECTROSCOPE.

S

Spectroheliograph An instrument, adjustable in wavelength, which presents a monochromatic image of all or part of the solar disk and its prominences to a photographic emulsion. In principle it is a SPECTROHELIOSCOPE with a photographic surface placed very closely behind the exit slit and an exposure is made by a single scan, at constant speed, of the two slits across the primary and final images. Increasingly finer detail is resolved as the slit widths are reduced, but this lessens the final image brightness so that, to maintain the required photographic exposure, the speed of the scan is reduced in approximate proportion to the slit widths.

Small observatory and amateur practice for occasional spectroheliograph work is to replace the eyepiece of the spectrohelioscope with a 35mm camera focused on the final image plane. The principal limitation of this technique is the non-uniform scanning speed of the spectrohelioscope slits. This gives effective exposures which increase from the middle toward the ends of the scan.

In 1892 HALE, in California, developed the first spectroheliograph, which he used in a Doppler survey of radial velocities of solar gases. About the same date DESLANDRES, in France, built a similar instrument for the same purpose. Tangential velocities were deduced from measures of radial velocities. Both workers extended their observations into ultraviolet light. Using a modified spectroheliograph, a magnetograph, Hale also showed that some of the observed line broadening and splitting is ZEEMAN EFFECT caused by strong magnetic fields in sunspots. Over the 80 years since this pioneer work, monochromatic studies of solar gas motions and magnetic fields have been refined and extended to many

wavelengths, using various sensors and data recorders, both terrestrially and in space vehicles. The term spectroheliograph is today applied loosely to denote any of the modern instruments used provided a dispersive element is included. *LMD.*

Spectrohelioscope A high-dispersion SPECTROSCOPE arranged to scan all or part of the solar disk and its prominences with a repetition rate exceeding ten per second so that, by persistence of vision, the observer sees a stationary monochromatic image. By setting an exit slit in the final image plane at a specific position within a chosen Fraunhofer line the observer can select the element and the depth in the solar atmosphere that he wishes to view. The wide range of possible settings and, additionally, its suitability for the measurement of radial velocity of violent solar disturbances make it a more versatile monochromatic instrument than optical filters.

Scanning is commonly by electro-mechanical oscillation of the slits but some instruments have fixed slits which are scanned optically.

This apparatus is usually so bulky that it is mounted with its entrance slit at the focus of a fixed telescope served by a HELIOSTAT or CŒLOSTAT and second mirror. In small observatory practice the focal length will be between 78 and 196in (2 and 5m) giving a disk image diameter of 0.8in to 2in (20 to 50mm). Large installations are up to ten times bigger. The spectroscope focal ratio must be the same as the fixed telescope so that, although the spectroscope optics are invariably folded to put the entrance and exit slits adjacent to each other, the complete system is longer than twice the focal length of the fixed telescope.

The detail seen increases as the slit width is reduced. A slit passing 0.2 to 0.4 Ångstroms is desirable, but this necessitates a high scanning rate to ensure adequate image brightness. Wider slit widths show limb prominences only, but disk detail is not visible.

Radial velocity is indicated by a change of contrast of a solar feature on a time scale of the order of tens of seconds. This is a Doppler effect. Contrast may be restored by rotation, which is proportional to radial velocity, of a line shifter, a thin parallel-sided glass plate, sited immediately behind the exit slit. To a close approximation the velocity is proportional to the rotation. *LMD.*

Spectroscope An instrument that splits radiation of a range of wavelengths, eg, white light, into a series of individual wavelengths or colours. The different wavelengths can then be viewed using a small telescope (a spectrometer), or recorded for future study (a spectrograph).

A traditional spectrometer uses a prism to split light into its component wavelengths. Each wavelength is refracted through a slightly different angle as it passes through the prism. A movable telescope, which swings around the prism on an accurately marked base, is used to measure the angle through which each wavelength has been bent. In a spectrograph the eyepiece of the telescope is replaced by a photographic plate which records all the different wavelengths simultaneously. The plate can then be examined later using powerful microscopes. Modern spectrographs record the spectrum produced on electronic detectors like CHARGE-COUPLED DEVICES (more sensitive than photographic film) or BOLOMETERS (sensitive to other wavelengths).

Spectrographs can be fitted to telescopes and used to examine astronomical objects. A slit is usually positioned in front of the prism so only an image of the slit, rather than the star image, which will be blurred by atmospheric effects and imperfections in the telescope, falls on the detector. This prevents adjacent wavelengths being smeared out across the detector, and keeps the spectrum sharp. When observing objects which emit most of their light in a small number of emission lines, eg, nebulæ, the slit is often omitted because, unless the lines are very close together, overlapping of adjacent wavelengths on the detector is not usually important.

Some spectrographs use a DIFFRACTION GRATING, a piece of reflective material ruled with thousands of parallel lines, to split up the light. This avoids absorption by the prism and allows the study of wavelengths that cannot pass through glass. *JKD.*

Spectroscopic binaries BINARY STARS which are too close to be seen in any telescope. Orbital motion is deduced from periodic shifts in the spectral lines indicating variable RADIAL VELOCITY.

Spectroscopic parallax A method of obtaining the distances of stars from their spectral classes (*see* SPECTRAL CLASSIFICATION) and apparent magnitudes. Analysis of the spectrum reveals the spectral class. Knowing the absolute magnitude of a star of that type, the star's distance can be calculated by comparing its absolute and apparent magnitudes (*see* DISTANCE MODULUS). This is the most common method of determining stellar distances and is usually assumed to be accurate to within about 10 per cent. However, the calibration of spectral class versus absolute magnitude rests on the accuracy with which the distances of nearby stars can be measured by TRIGONO-METRICAL PARALLAX. *IKN.*

Spectroscopy *See* ASTRONOMICAL SPECTROSCOPY.

Spectrum The distribution of intensity of ELECTROMAGNETIC RADIATION with wavelength; a "map" of brightness plotted against wavelength. A continuous spectrum is an unbroken distribution over a broad range of wavelengths (at visible wavelengths, for example, white light may be split into a continuous band of colours ranging from red to violet). An EMISSION LINE spectrum comprises light emitted at particular wavelengths only, and an ABSORPTION LINE spectrum consists of a series of dark lines superimposed on a continuous spectrum. The spectrum of a star normally consists of a continuous spectrum together with dark absorption lines and, in some cases, emission lines. Emission line spectra are typical of luminous nebulæ. The term "spectrum" is also applied to a distribution of particle energies; ie, an energy spectrum is a plot of the numbers of particles with particular energies against the range of possible energies. *IKN.*

Spectrum variables The first variable stars to be discovered were, by definition, variable in brightness. As early as 1906, however, it was realized that some stars, although not changing noticeably in luminosity, have spectra with ABSORPTION LINES that vary in strength by large amounts, some in a clearly periodic manner. This class of star was given the name spectrum variable and included stars of spectral type A and early F: because of the spectral peculiarities these formed part of the Peculiar A (Ap) and Fp stars.

With photometry of improved accuracy it was discovered in 1950 that the Ap-type spectrum variables do in fact possess very low amplitude brightness variations (of one or two per cent of the brightness of the star). Although the term spectrum variable is still retained

for these stars, and is often reserved solely for the Ap and Fp variables, there has in recent years been a tendency to broaden the definition to include other types of stars with very low amplitudes of brightness but clearly variable spectra.

To illustrate the remarkable behaviour shown by the classical spectrum variables we will describe some of the variations that occur in the brightest and best-studied member of the class: Alpha² Canum Venaticorum. In this star the variations in spectrum, brightness and measured magnetic field are all cyclical with a period of 5.46939 days, the rotation period of the star.

The spectrum lines that change in strength are those that are in any case anomalously strong and give the star its Ap classification. Lines from the rare earth elements europium, gadolinium and dysprosium vary in strength by a factor of five or more over the 5.4-day cycle. Lines of the iron group elements titanium, chromium and iron also vary by a factor of about five, but reach maximum strength at a different phase of the 5.4-day cycle than the rare earths. Furthermore, lines from these iron group elements are often split into two or three components, each of which reaches maximum strength at different phases. Finally, measurements of the RADIAL VELOCITIES of all the lines show that the velocities of the rare earths reach maximum at one phase of the cycle, but the iron group elements show four maxima. In all cases maximum line strength occurs when the radial velocity of the line is passing through its average value.

The explanation of this complex behaviour, deduced from these observations, is actually relatively simple, but it demonstrates the amazing structure of the atmospheres of the Ap-type spectrum variables. If a chemical element is confined to a particular patch on the surface of a star, then as the star rotates and brings the patch into view around the limb of the star, lines of that element will show an approaching Doppler shift, ie, a negative radial velocity. As the patch crosses the centre of the star's disk it is viewed in its most favourable position from the Earth and we see the spectrum lines at their strongest. As the star rotates further and the patch recedes behind the limb again, the lines show a positive radial velocity. Clearly maximum line strength corresponds to the phase of average velocity.

Applying the reasoning to Alpha²CVn shows that the rare earths are gathered into two patches, one con-

taining about five times the concentration of the other. The iron group elements, on the other hand, are concentrated in four patches, distinct from the rare earth patches. From a separate analysis of the variation in the magnetic field of Alpha²CVn it is possible to show that, similarly to what occurs on Earth, the magnetic field has a distinct north and south pole which are displaced from the poles of rotation. It turns out that the two rare earth patches coincide with the magnetic poles of Alpha²CVn and the four iron group concentrations are spaced along the magnetic equator.

A wide range of different distributions of element concentrations is shown among the other Ap-type spectrum variables. A complete explanation of this variety is not yet available, but the principal mechanism at work is that of diffusion (*see* A STARS). Not all of the Ap spectrum variables are as regular in their behaviour as Alpha²CVn. Some of them have non-repetitive, irregular variations; all of these have quite small magnetic fields. Of the Ap stars as a group, about half are spectrum variables.

The low amplitude variations of brightness are a direct result of the periodic changes in spectrum, but not from the variation in strengths of the lines in the visible part of the spectrum — the variable amount of light that they absorb is not enough to produce a one per cent change in luminosity. Rather it is an effect of the very much larger range of brightness that is observed by satellites in the far ultraviolet. This is caused by the increase in continuous absorption (as opposed to line absorption) as a metal-rich patch passes across the disk. The ultraviolet radiation that is prevented from escaping through the patch emerges at longer wavelengths, ie, in the visible, creating an increase in brightness there.

Just as the peculiarities that occur among the A stars extend into the B STARS, so do the spectrum variables occur among earlier spectral types. The helium-rich B stars show correlated variations in brightness, radial velocity and helium line strength. The helium-weak B stars behave similarly; analysis of the observations of one star point to a silicon-rich, helium-poor patch near one magnetic pole, with the reverse situation at the other pole. *BW.*

Speculum metal Alloy of copper and tin, used for the mirrors of REFLECTING TELESCOPES before glass could be worked with sufficient precision.

S

Spherical aberration Deficiencies in an optical image caused by the way that different parts of spherical mirrors or lenses cause light from the image to focus at different places.

Spica The bright star Alpha Virginis. It is a very close binary.

Spiral galaxy Type of galaxy in which a central bulge of cool stars is surrounded by a spiral pattern of hotter stars in a flattened disk.

Spiral galaxy: NGC5194, in Canes Venatici

Spörer, G.F.W. (1822–95). Spörer studied mathematics and astronomy at Berlin University, worked with Encke at Potsdam and investigated the distribution of sunspots. The result later

Gustav Spörer: investigated sunspots

became known as SPÖRER'S LAW. He took part in an expedition to India to see the total eclipse of the Sun in 1868. His study of historical records revealed a scarcity of sunspot sightings between 1645 and 1715, a period now known as the Maunder minimum.

Spörer's Law Describes the distribution of sunspots during the solar cycle. A new cycle starts with high-latitude sunspots (lat. 30°–40°). As the cycle progresses the spots appear at decreasing latitudes until they occur around the 5° mark, north or south of the equator at solar minimum. Before minimum has been reached, high-latitude spots of the new cycle appear, thus causing an overlap of the cycles.

SPOT (*Système Probatoire d'Observation de la Terre*). High resolution French operational/commercial remote sensing satellite system — using CHARGE-COUPLED DEVICE linear arrays — first vehicle of which was launched in February 1986.

Spring tides *See* TIDES.

s process A process by which heavy nuclei are built up from iron group elements by the successive slow capture of NEUTRONS. It is believed to occur within some second-generation RED GIANTS.

Spurious disk *See* RESOLVING POWER.

Spurr, Josiah Edward (1870–1950). American geologist who made careful studies of the Moon and perfected a theory that the craters are of internal rather than impact origin.

Sputnik Name given to the first ten satellites launched by the Soviet Union. Sputnik 1, launched on October 4 1957 was the first ARTIFICIAL SATELLITE. It weighed 186lb (84.6kg) and orbited the Earth in 96 minutes. Sputnik 2 (November 3 1957) carried the dog Laika. Sputnik 3 (May 15 1958) was a highly successful scientific satellite. Sputniks 4, 5 and 6 tested the VOSTOK

Sputnik: first artificial satellite

re-entry capsules, Sputnik 5 making the successful return from space of the dogs Belka and Strelka. Sputniks 7 and 8 were associated with the Venus space

probe. Sputniks 9 and 10 were also Vostok test flights. Subsequently the name COSMOS was used. *HGM*.

SS Cygni stars *See* U GEMINORUM STARS.

SS433 Attention was first drawn to the bizarre star SS433 in June 1978, as the result of peculiarities in its spectrum. The star gained its rather unusual name from being object number 433 in a

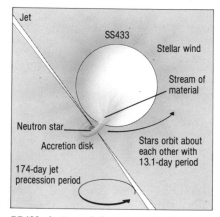

SS433: the "cosmic lawn-sprinkler"

catalogue of emission-line stars published in 1977 by Bruce Stephenson and Nicholas Sanduleak of Case Western Reserve University.

The unique nature of SS433, which lies at the centre of a SUPERNOVA remnant called W50, comes from EMISSION LINES in its spectrum that drift in wavelength in a periodic fashion. Emission lines in stellar spectra are reasonably commonplace. Indeed, the shift of an emission line from the wavelength that would be measured in the laboratory is in itself not surprising, and can be ascribed to the DOPPLER EFFECT whereby light from a source moving away from an observer is "red-shifted" (that is, the wavelength is increased), while light from an approaching source is "blue-shifted" (that is, the wavelength is decreased). But what Bruce Margon, then at the University of California, Los Angeles, discovered was that in addition to strong "stationary" emission lines indicating the presence of hydrogen and helium, the spectrum of SS433 also showed these lines dramatically red-shifted *and* blue-shifted. Perhaps most unexpected of all, the Doppler-shifted features drifted backward and forward in wavelength by an enormous amount, over an interval of 164 days. In addition, the so-called "stationary" emission features showed a much more modest periodic Doppler shift, over an interval of 13.1 days.

Martin Rees and Andrew Fabian of Cambridge University, and independently Mordehai Milgrom of Israel's Weizman Institute of Science, proposed that the spectacular Doppler-shifted features could be explained by the presence of JETS of material travelling at enormous speed — about a quarter of the speed of light. While such high-speed jets were believed to be present in QUASARS, they had never previously been detected in stars.

Although there is by no means universal agreement on the true nature of SS433, few astronomers would now object to the model depicted in the accompanying diagram. This shows a BINARY STAR system made up of a hot massive star (some 10 to 20 times the mass of the Sun) and a NEUTRON STAR, orbiting each other every 13.1 days. The neutron star is presumably the stellar remnant from the supernova that produced W50. A stellar wind from the hot star produces the "stationary" emission features. Matter is transferred from the massive star to the neutron star via a so-called accretion disk. However, not all the material reaches the surface of the neutron star; some is ejected at high speed (probably by radiation pressure), via two finely-collimated "jets". The "jets" sweep around the sky every 164 days, producing the peculiar drifting spectral features that appear to make SS433 unique in astronomy. *DHC.*

S stars The S stars are giant stars with the same temperature range as M stars, but their spectra contain absorption bands of zirconium oxide (ZrO) as well as the titanium oxide characteristic of the M stars. The over-abundance of zirconium, as well as carbon and many heavy elements, is a result of convective mixing which brings products of nuclear reactions in the interiors of stars up to their surfaces. As a result, many S stars show spectral lines of the radio-active element technetium.

M stars with very weak ZrO bands are assigned the intermediate spectral type MS. S stars often show hydrogen emission lines in their spectra and are MIRA VARIABLES. *BW.*

Stadius Lunar "ghost crater", 44 miles (70km) in diameter; 11°N, 14°W. It lies west of Copernicus. The walls are now extremely low, and the whole area is pitted with small craterlets and craterlet-chains.

Standard time Legal time in any community, based on TIME ZONES. For example, standard time in Great Britain is UNIVERSAL TIME in winter, UT + 1h in summer.

Starburst galaxies Galaxies that appear much brighter at infrared wavelengths than would be expected from the amount of visible light which they emit.

When the INFRARED ASTRONOMICAL SATELLITE (IRAS) surveyed the sky in 1983 it detected thousands of galaxies outside the Milky Way. Many of these were normal spiral galaxies, but a small fraction were found to be unusually bright in the infrared, most noticeably at wavelengths of about 60 microns. These unusual galaxies seem to emit up to 50 times more energy as infrared radiation than they do as visible light. For comparison the nearby Andromeda Spiral, a normal galaxy, emits only about 4 per cent of its total energy as infrared radiation. The extra infrared emission from these objects is probably caused by an unusually high rate of star formation, and they have been named starburst galaxies.

Why galaxies should undergo a sudden burst of star formation is not known. Some starburst galaxies have other galaxies nearby and it may be that the gravitational influence of a neighbour is responsible for triggering the starburst. However, many starburst galaxies have no close companions, so interaction between galaxies cannot be the only explanation. The starburst galaxies may be another type of active galaxy, broadly similar, but not identical, to the MARKARIAN and SEYFERT GALAXIES. *JKD.*

Star catalogues The stars are so numerous that no naming system can accommodate them all. Many catalogues list stars of certain types. Some of the brighter or otherwise unusual stars feature in several catalogues. The first system used the constellation boundaries, with Greek letters (Bayer, 1603) and numerals (Flamsteed, 1725) to distinguish stars within each. Today only variable stars follow the constellation-based schemes, other catalogues covering the entire sky. Commonly referred-to catalogues are:

BS = HR (9,110 stars) — the bright star catalogue of basic data.

BD, CoD, CPD (over 2 million) — Bonn, Cordoba and Cape Photographic Durchmusterung, positions.

CG (33,342) — Boss's catalogue of stars of known proper motion.

SAO (258,997) — Smithsonian Astrophysical Observatory, positions.

HD (359,083) — Henry Draper, spectral types.

GCVS (28,450) — General catalogue of variable stars.

LTT (18,546) — Luyten's two-tenth's arc second catalogue, stars with high proper motion, generally nearby. *DAA.*

Star clouds In the clear, dark skies of the southern hemisphere the brightest parts of the Milky Way appear overhead as thin wisps of cloud caught in the light of a quarter moon. This connection with clouds is not dispelled even when the stars are resolved as individuals with the aid of a telescope. Toward the centre of our Galaxy, in the direction of Sagittarius, we see a portion of the nuclear bulge of our own Galaxy. This homogeneous collection of billions of old, cool stars forms the brightest part of the Milky Way. Much of it is obscured by dust in the plane of the Galaxy, giving the star clouds an irregular appearance. The intrinsic colour of the stars and their association with dust gives them a yellow appearance on long-exposure photographs. Farther along the galactic plane, in Centaurus and Crux, the population of bulge stars is replaced by a more mixed collection with fewer, brighter stars and less obscuration which gives the star clouds a more uniform appearance to the eye and a more nearly white hue on deep photographs. *DM.*

Star clusters *See* STELLAR CLUSTERS.

Star diagonal Attachment enabling an object to be viewed at right angles to the direction in which a telescope is pointing. The image is reversed left-to-right, but is the correct way up.

Starlink A network of powerful and similar minicomputers which provides United Kingdom astronomers with data analysis, interactive image-processing, and access to data bases.

The system was set up in 1979 in response to a "data explosion" in United Kingdom astronomy. Observing installations for United Kingdom astronomers now include the optical telescopes at SIDING SPRING OBSERVATORY in Australia, the optical telescopes at Observatorio del ROQUE DE LOS MUCHACHOS in the Canary Islands, the millimetre-wave and infrared telescopes on MAUNA KEA, Hawaii, and various satellites. Each of these facilities returns magnetic tapes full of digital data to astronomers every 24 hours. To analyze these data efficiently, inter-

S

active image-processing is essential so that the astronomer can assess the data and make rapid decisions on how to process it. Similar computers for all astronomers allow the sharing of software and analysis expertise, preventing overlapping efforts in the development of computer programs. Linking the computers enables communication of both software and data around the network, and permits access to large banks of astronomy data — "data bases" — which may be held at a single location. Best of all, perhaps, the linking provides rapid communication between collaborators at different institutions. The Starlink network can now access international computer links, and thus it is a basis for the development of remote observing with ground-based telescopes.

A typical Starlink node consists of a Digital Equipment Corporation VAX 32-bit minicomputer with 4 megabytes of memory, disk-drive capacity of 1,000 megabytes, line-printer and laser printer/plotter, fast tape drive input, one or two colour graphics display systems, and perhaps 15 terminals, several of which have graphics capability. It is linked to the network by a dedicated telephone line with a speed of 9,600 bits per second. The nodes and network are funded by the Science and Engineering Research Council; hardware and software development is co-ordinated at Rutherford Appleton Laboratory. *JVW*.

Star of Bethlehem The biblical "Star in the East" which guided the Wise Men. It is mentioned only by St. Matthew, so that our information is limited. Various explanations have been given — a comet, a nova, a planetary conjunction, even an aurora — but all these seem most unsatisfactory, and we have to admit that there is no scientific explanation. It is always possible that St. Matthew's account was not meant to be taken literally.

Stars Gravitation is the dominant force in the universe for all objects much larger than molecules. It pulls material together, and thus counters the general expansion of the universe. The balance of these two effects is such that in regions the size of galaxies or clusters of galaxies gas tends to condense, while the galaxies and clusters themselves spread apart from one another.

Because the gravitational force falls as the square of separation, the infall of gas from distances comparable to the size of galaxies is extraordinarily slow.

Locally denser portions tend to separate out during the gradual collapse, and themselves more rapidly condense. These local condensations form stars, which are therefore the fundamental building blocks of galaxies.

A star is the natural result when a cloud of gas collapses. Stars are, in fact, a temporary halting point in the process of collapse. The halt is caused by the introduction of a balancing force — the outward flow of radiation generated deep within, and a star is formed when that radiation switches on. The source of the radiation is nuclear fusion (*see* NUCLEAR REACTIONS and CARBON-NITROGEN CYCLE) — the combination of hydrogen atoms to make helium, a process which is triggered once the temperature exceeds a critical value. The temperature rises within a collapsing cloud because the energy of infalling atoms as they collide is liberated as heat, and it rises highest in the centre of the star. The "burning" of hydrogen to make helium occurs

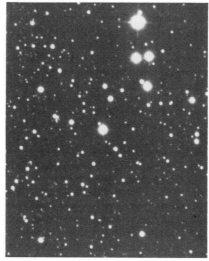

Stars: at the centre, SS433

fastest in the hottest regions, so it too is concentrated at the centre. The energy released has to force its way out through opaque gas, and at every point within the star the gravitational pull of the parts within exactly balances the outward thrust. The structure of a star, worked out in the 1920s by the brilliant English astrophysicist Sir Arthur EDDINGTON, is a mathematically defined function called a polytrope, and is the same in all normal stars.

What causes the differences from star to star is the amount of material involved. Massive stars come out very different from tiny ones. The range of sizes is quite large. The most massive stars are about 100 times heavier than

the Sun. Above this mass the star is not very stable, and is likely to break up into two portions. The smallest stars contain less than one-twentieth the mass of the Sun: below that mass the star does not grow hot enough for nuclear reactions to take place, and it becomes instead a BROWN DWARF. The very smallest stars are so faint that they cannot be found easily, so we have very incomplete knowledge of them.

After settling down from the trauma of nuclear fusion, stars have temperatures and luminosities prescribed precisely by their masses, and these define the well-known MAIN SEQUENCE. Large stars are very luminous and hot, and therefore blue. Stars in the middle range are yellow (the Sun being our best-loved example) while small stars simmer a dull red. The colours (and hence temperatures) gave rise to a classification scheme. At the top of the main sequence we find the blue-white O stars, and below them the sequence continues B, A, F, G, K, M. Main sequence stars of types G, K and M are also known as DWARFS. Examples of O stars are the components of THETA ORIONIS. Typical B stars are the brightest, blue members of Orion. SIRIUS is an A star and PROCYON of type F. We need look no further than the Sun for a G star. K and M dwarfs have very low luminosities, and thus are not prominent, though they are the most numerous type of star. The brightest K dwarf is Tau Ceti, and there is no M dwarf visible to the naked eye. BARNARD'S STAR is an example of an M dwarf.

A star lives most of its life "on the main sequence", by which we mean that its luminosity and temperature are defined by its mass (with a very slight variation caused by the chemical mix). However, at the start of its life, and in its later stages, a star behaves very differently. In order to describe a star more fully, therefore, we must specify both its mass and age.

Around the time of formation it is called a PROTOSTAR. By the time it becomes visible optically it has already begun to settle down within, and is called a pre-main-sequence star. It then has the temperature it will end up with, but is more luminous. The T TAURI STARS are seen at this stage.

The changes in old age are more dramatic. As hydrogen is consumed, a core of hot but inert helium grows within, and the hydrogen burning proceeds on the outside of this core. The core grows hotter to a point at which helium too can partake in nuclear fusion, to produce carbon and oxygen.

The change in the central regions is sudden, and is called the HELIUM FLASH. At its onset the core shrinks to a density quite unknown on Earth, and the outer parts correspondingly expand. The luminosity rises somewhat (more for smaller stars than for larger ones) and the temperature falls rapidly. The star becomes a GIANT, perhaps 100 times its original diameter, and sufficiently cool that it has a spectral type of K or M. ARCTURUS is a K giant.

The giants change slowly, growing brighter and cooler, but have a chequered history as further fusion reactions within cause additional fluctuations. At length they become MIRA VARIABLES. At this stage the internal chemistry can change, or gas of a different chemical mix deep within can rise to the surface by convection. They may then appear as CARBON STARS (C stars), for which the spectral sequence is extended by classes R and N, or S STARS (spectral class S). Intermediate types are also known: SC, CS or MS stars. Most of these objects are extremely red and thus rather faint; a carbon star visible to the naked eye is 19 Piscium.

Eventually the giants shed their outer layers, revealing the hot core of extreme density. In so doing they become PLANETARY NEBULÆ, after which only a WHITE DWARF remains. It is believed that white dwarfs cool for ever, becoming BLACK DWARFS, mere cinders of stars.

This sketch of the later life of a star is correct only for those less than a few times the mass of the Sun. If the core exceeds 1.4 solar masses, the CHANDRASEKHAR LIMIT, a different fate is in store. Massive stars exhaust their hydrogen quickly, and in progressing toward spectral type M they may be caught up in a more catastrophic internal convulsion. If they do make it to type M, they will be even larger and more luminous than the M giants, and are therefore called supergiants. BETELGEUX, in Orion, is an M supergiant, as is ANTARES in Scorpius. Because M supergiants are so very luminous we can easily see them right across our Galaxy and in other galaxies. At some point, either before or after the M supergiant stage, the central nuclear reactions become so violent as to be explosive. The star is ripped apart as a SUPERNOVA.

A star which has passed through the supernova stage at first sight appears not to be a star any longer. However, the explosive events take place not at the centre, but on the exterior of a dense white-dwarf core. As Newton's first law of motion predicts, the explosion that throws off several solar masses of gas at speeds of many thousands of km/s must also push inward on the core with unimaginable impact. The effect of this implosion is to crush the already extraordinarily dense white dwarf yet further. At the very least, the protons and electrons are forced together to create a NEUTRON STAR, and it may be that some supernovæ produce BLACK HOLES.

There are various configurations of luminosity and temperature (one may also categorize the latter by radius because the three are linked by Stefan's law: see RADIATION) at which the star is prone to oscillate in size. The balance between inward and outward forces does not itself guarantee a fixed size, just as a weight hanging on a spring may be static or may oscillate up and down according to its circumstances. On the main sequence, stars resist oscillations — in the weight and spring analogy we may picture the entire system submerged in highly viscous syrup. When the changes in later life take place, the oscillations can grow and continue for a long time. Mira variables, noted above, are one such phase. Others include RR LYRÆ STARS and CEPHEID VARIABLES. The peculiar star ETA CARINÆ, possibly the most massive star in our Galaxy, is also at an unstable stage, though its variations are ponderous and extreme.

Most stars are born not as only children but in groups or pairs. If the component stars of a pair are widely separated, they will evolve independently. Closer pairs may behave very differently, however. As a simple case, consider two stars like the Sun, with slightly different masses. The heavier first evolves to become a red giant, but in so doing its outer surface can enter the region where the gravitational pull of its companion exceeds that of the giant. Gas will begin to pour off the giant and onto its companion, being heated as it falls. We recognize the resulting system as a SYMBIOTIC STAR. Other configurations in which gas is transferred are possible. If one star has become a white dwarf, we may have a CATACLYSMIC VARIABLE, a POLAR variable, or a different kind of symbiotic star. If the companion has become a neutron star or even a black hole, we will find an X-RAY BINARY. In all these cases the transfer of material also changes the smaller star, hastening its evolution or, if it is a white dwarf, turning it into a NOVA.

When we look up at the twinkling stars we incline to think of them as distant suns. Research has found a much greater variety among them, and will continue to do so. *DAA.*

Star streaming In their passage around the Galaxy, stars move in parallel groups called star streams, whose paths cross so that the stars intermingle like marching bandsmen in a military display. In the neighbourhood of the Sun, the major star stream is associated with the HYADES star cluster.

Steady State Theory A cosmology, due to Bondi, Gold and Hoyle (1984), designed to satisfy the PERFECT COSMOLOGICAL PRINCIPLE, which states that the local properties of the universe (when averaged over some suitable distance) are the same when determined from any point in space and time. It avoids the problem of an instrument of creation which is associated with BIG BANG theories, but does so at the expense of postulating continuous creation of matter at all places and times to maintain a constant density of matter in an expanding universe. The discovery of the cosmic MICROWAVE BACKGROUND in 1965, which cannot naturally be explained by this theory, led to its demise. *APW.*

Stebbins, Joel (1878–1966). American astronomer who pioneered photoelectric photometry in astronomical research.

Stefan constant (σ) Value of 5.669×10^{-8} Wm^{-2} K^{-4}. *See* RADIATION.

Stellar associations Groups of stars which have formed together, but are more loosely linked than clusters. Many are young and define portions of the spiral arms of our Galaxy.

Stellar classification *See* SPECTRAL CLASSIFICATION.

Stellar clusters Localized regions of enhanced star density, as compared to the immediate surroundings. A few have been known since antiquity, eg, the HYADES, the PLEIADES, PRÆSEPE, and COMA BERENICES. Several star clusters were included in MESSIER's 18th-century catalogue, and still more are found in the NGC. On a system adopted by the IAU, clusters are designed "Chhmm±ddd" where C is for cluster, hhmm is the RIGHT ASCENSION in hours and minutes, and ±ddd is the DECLINATION in tenths of degrees

(1950 co-ordinates). Lists of selected clusters appear yearly in *The Astronomical Almanac*, giving positions, distances, and other pertinent data.

The stars belonging to a cluster not only lie close together in space. They also share a common origin, and generally travel together through space, although they also to some extent move within the cluster. A sometimes — especially in star-rich regions — severe problem in studies of stellar clusters is the separation of true cluster members from field stars in the same region of the sky.

The stellar clusters are divided into two classes, globular clusters and open clusters. Related to the open clusters are associations, loose groupings of stars of common origin, which, however, are not dense enough to be readily apparent against the surrounding distribution of stars. Often one or more open clusters are found in the central parts of associations.

The globular clusters contain great numbers of stars (10^5 to 10^7) within sometimes slightly flattened spheroidal volumes of typical radii 10–25pc. The stars are strongly concentrated toward the centres of the clusters, with steep density gradients outward.

All the 140-odd known globular clusters belonging to our Galaxy are extremely old systems, with ages presumably greater than ten thousand million years. They are preferentially found in the GALACTIC HALO, with a strong concentration toward the galactic centre. Probably about the same number of globulars remain undiscovered in the Galaxy, hidden behind the vast dark clouds near the galactic plane.

Globular clusters may be classified after their apparent degree of concentration, or according to the intregrated SPECTRAL CLASSIFICATION; F-type clusters are very poor in heavy elements ("metals") and are found throughout the halo, whereas G-type clusters are only moderately metal-poor, and lie closer to the nuclear bulge of the Galaxy. These are also generally smaller and less concentrated.

Through studies of the distribution of globular clusters, H. SHAPLEY could show that the centre of our Galaxy lies far from the Sun in the direction of Sagittarius.

Globular cluster distances are most reliably obtained through studies of variable stars, primarily of the RR LYRÆ type, which have fairly well-determined mean absolute magnitudes. For clusters lacking known variable

stars the apparent magnitudes of MAIN SEQUENCE and horizontal branch stars are used. Alternate, but less reliable, methods are measurements of cluster angular diameters or integrated magnitudes. In this way globular clusters may also serve as distance indicators to external galaxies.

The compactness and large total masses of the globular clusters make them dynamically very stable systems, that may remain largely unchanged over times of many thousands of millions of years. There is, however, a slow but steady loss of primarily low-

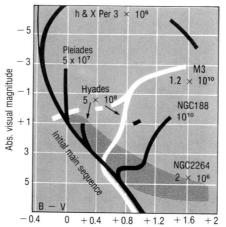

Stellar clusters: stars of common origin

mass stars, that through encounters with other cluster stars gain enough kinetic energy to leave the clusters (cluster evaporation).

A few globular clusters have been found to contain X-RAY SOURCES. One proposed interpretation of this involves BLACK HOLES in the centres of the clusters, which are formed through the collapse of massive stars and give rise to the (sometimes variable) X-ray emission by devouring remnant gas.

Whereas the globular clusters in the Galaxy are all very old, a number of relatively young ones have been found in the MAGELLANIC CLOUDS. These differ from the older ones by being much bluer in colour (reflecting their content of young, bright blue stars), and by having a less concentrated distribution of stars.

Open clusters are much less rich in stars than the globulars. They sometimes contain no more than a few tens of stars, and seldom more than a few thousands. Their radii are generally of the order of a few parsecs. The distribution of stars may vary considerably from cluster to cluster, but it is never as concentrated toward the centre as in globular clusters.

Open clusters are susceptible to dis-

ruption, not only through evaporation, but also through tidal forces from the Galaxy or encounters with interstellar clouds. Only the richest may survive more than 10^9 years, while the smallest and least tightly bound do not last more than a few million years. Their hazards are strengthened by their being found only near the galactic plane, which is the reason for the older but unsuitable designation "galactic clusters". The number of known open clusters in the Galaxy is nearly 1,200, but this is certainly only a very small fraction of the total number.

As all stars in a cluster take part in the general motion of the cluster, it is possible to derive the distances of nearby clusters by the "moving cluster" method (*see* HYADES).

All stars are thought to be born in clusters, most of which, however, are gravitationally unbound, and thus rapidly disrupt and spread their contents throughout the general field. In still recognizable clusters all members may thus be considered equally old (barring the youngest clusters and associations, where the ages are comparable to the spread in star formation time, of order 10^6 years). Studies of star clusters therefore yield good tests for theories of STELLAR EVOLUTION. As evolution proceeds more rapidly for the more massive stars, these leave the MAIN SEQUENCE stage before the less massive ones. Accordingly the position of the brightest main sequence stars in a COLOUR-MAGNITUDE DIAGRAM gives an estimate of the cluster age. For extremely young clusters it is instead the faintest main sequence stars that yield the age, as no stars have yet left the main sequence, but the least massive ones have not yet got there. Such clusters are often still embedded in their parent clouds of gas and dust. *GW.*

Stellar coronæ Observations with the EINSTEIN OBSERVATORY have shown that almost all kinds of normal stars emit X-rays over a range of luminosity (L_x) which extends from 10^{26}ergs^{-1} to 10^{34}ergs^{-1}. Stars earlier than spectral type B5 have L_x proportional to overall luminosity. X-ray production is not well understood but may be due to shock heating in the high-speed gas streams or winds that flow out from these stars. For stars later than A5, L_x is proportional to the stellar rotation velocity. Hence these stars probably have high-temperature coronæ, which, like the Sun, are heated by magnetic dynamo action. *JLC.*

Stellar evolution Stars shine by radiating energy generated in their interiors by nuclear fusion. Eventually, the nuclear fuel must become exhausted, causing the star to evolve to a different state. Timescales for evolution lie between hundreds of thousands, and thousands of millions of years; hence the evolution of a particular star cannot be directly observed. The theory of stellar evolution is based on physical models of stellar interiors. It gives a fairly detailed account of the evolution of any particular star, and can be checked by observation of stars at each of the predicted phases.

Stars form from clouds of dust and gas. A large cloud is able to collapse under its own gravity. Subsequently it will fragment into several hundred smaller clouds, each of which contracts further, and heats up to become a PROTOSTAR. Protostars are very red, and can be observed at infrared wavelengths; they are found in regions of our Galaxy where there is an abundance of gas and dust. A protostar will continue to collapse under gravity until its centre is hot enough (about 15 million degrees K) for nuclear fusion reactions to commence. As this is occurring, its outer layers will also fall in to take up a normal starlike form and become bluer in colour. T-TAURI STARS are low-mass stars in this final phase.

Once nuclear fusion has commenced in the core of a star, the star adopts a stable structure, with its gravity being balanced by the heat from its centre. The controlled nuclear fusion reaction is the fusion of hydrogen atom nuclei (protons) to form helium nuclei. This is known as the MAIN SEQUENCE phase of a star's life; it is the longest phase in the life of a star, and stars are most commonly of main sequence type. The Sun has a main sequence lifetime of some ten thousand million years, of which only a half has expired. Less massive stars are redder and fainter than the Sun, and stay on the main sequence longer; more massive stars are bluer and brighter and have a shorter lifetime. As a star ages on the main sequence, the change of composition of its core from mainly hydrogen to mainly helium causes it to become very slightly brighter and bluer.

The internal structure of a main sequence star depends on its mass. Massive stars have a convective core and a radiative mantle (outer layer). Low-mass stars, including the Sun, have a radiative core and a convective mantle. The stirring which occurs in the core of a massive star causes it to have a uniform composition within that core, and affects the details of the next phase of evolution.

Eventually, the hydrogen supply in the core of the star runs out. With the central energy source removed, the core will collapse under gravity, and heat itself up further, until hydrogen fusion is able to take place in a spherical shell surrounding the core (hydrogen shell burning). As this change occurs, the outer layers of the star expand considerably, and the star becomes a red giant, or in the case of the most massive stars, a red supergiant. The speed of evolution to the red giant phase depends on the mass. A low-mass star changes gradually to become a red giant, as the hydrogen exhaustion spreads outward from its centre. A high-mass star evolves quickly to become a red giant, because its whole convective core runs out of hydrogen at the same time. Red giant stars can be seen at great distances owing to their high luminosity, and many can be observed. Whilst it is a red giant, the central temperature will reach 100 million degrees K, and the fusion of helium to carbon will commence in the core. In the case of low-mass stars, the onset of helium fusion is sudden (the HELIUM FLASH); the star reduces its radius to become a bluer but fainter "horizontal branch" star as observed in globular clusters; it may subsequently return to being a red giant. In high-mass stars, the onset of helium fusion occurs more gradually, and the star remains a red giant.

The evolution of a star beyond the helium fusion phase is more difficult to calculate with certainty. When the helium fuel in the core has been converted into carbon, the core of the star will again collapse and heat up. In the case of a low-mass star, the central temperature will not rise high enough to initiate carbon fusion, and the outer layers of the star will contract, then gradually cool down, so that the star becomes a WHITE DWARF. It is likely that before a white dwarf cools appreciably, it throws off its outer layers in the form of a PLANETARY NEBULA. White dwarf stars are plentiful, but they are difficult to observe owing to their faintness. In the case of a high-mass star, the contraction of the carbon core will lead to further episodes of nuclear fusion, during which the star will remain very bright. When the core is composed of elements close to iron in the periodic table, no further fusion is possible. At this point, it is most likely that the core will collapse explosively, and the star will throw off its outer layers in a SUPERNOVA explosion, becoming for a few weeks bright enough to outshine the galaxy in which it is situated; the dead core of the massive star will remain as a NEUTRON STAR or BLACK HOLE. *DJA.*

Stellar populations Groups of celestial objects classified according to their age and location in the Galaxy. There are two principal populations:

Population I consists of the relatively young celestial objects located in the plane of the Galaxy, in its spiral arms;

Population II consists of relatively old celestial objects dispersed throughout the entire Galaxy and prominently visible in its centre and its distant halo.

Astronomers also occasionally refer to a conjectural *Population III* of celestial objects, which were created when the Galaxy was first formed and which have now disappeared.

Across these broad divisions there is a gradation of population types; for example, astronomers refer to "disk Population II", which is intermediate in age and which is located in the galactic disk enclosing the spiral arms but lying within the halo.

Because the youngest objects are confined to the thin plane of the Galaxy, while older objects are more dispersed outside the plane, there is a correlation between the average height above the galactic plane of objects and their age; this can be used to classify celestial objects into populations.

The observational basis for the idea of stellar populations originated with Walter BAADE. Baade obtained photographs of galaxies, exposing blue- and red-sensitive emulsions in order to emphasize the contributions of different-coloured light. He discovered that the central regions of spiral galaxies were smooth distributions of red light, from innumerable faint stars, whereas the spiral arms were patchy distributions of bright blue stars and nebulæ. Elliptical galaxies had the appearance of the central regions of spirals. These two contrasting kinds of stellar material were the original Populations I and II.

The distinction between the two populations was explained in terms of stellar evolution, and given its modern expression, in 1957 by, among others, Alan Sandage.

When stars are first formed in clusters from the interstellar material of dust and gas in a spiral arm, they consist of the MAIN SEQUENCE of bright blue dwarf stars and faint red dwarfs.

S

Although the faint stars are much more numerous than the bright ones, the numbers of each kind (luminosity function) are such that the few bright lighthouses still outshine the numerous faint glow worms. Thus the initial Population I of stars (the Zero Age Main Sequence — ZAMS — of dwarf stars) is dominated by small numbers of bright blue stars in a patchy distri-

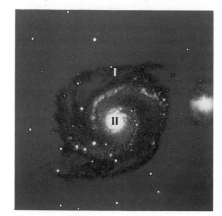

Stellar populations: principal groups

bution of dust and gas, illuminated as reflection and diffuse nebulæ by the blue stars.

After a long time (say 100 million years), the bright stars have evolved to WHITE DWARFS, NEUTRON STARS, BLACK HOLES, etc, and effectively have disappeared. The faint red stars are still in sedate middle age, not much different from their appearance on the ZAMS. Some of the intermediate aged stars have evolved off the main sequence and are now red giant stars — bright, but not as bright as the blue ZAMS stars were. The luminosity function of the second population of stars is therefore some fairly bright red giants and numerous faint red dwarfs — the total contribution of light from the giants and dwarfs is more or less uniform and red stars of all kinds form this population. Because the dust and gas is used up, and because the clusters of stars which first formed have settled into equilibrium, and because there are so many stars contributing to its appearance, Population II no longer has an irregular patchy look, but is smooth and uniform. As there are few blue stars, any gas and dust left over is seen not as bright nebulæ but as sparse DARK NEBULÆ.

The theory of stellar evolution explains the appearance of the different populations, just as the theory of human ageing explains why a town of old people looks different from a town dominated by young, but it does not explain why the populations are separated in different parts of the Galaxy. For this is needed a theory of stellar sociology, just as a theory of human sociology could explain the town of old people as a retirement area and a town of young people as a development centre. The distribution of populations in our Galaxy was explained by O. Eggen, D. Lynden-Bell and A. Sandage, in terms of the collapse of the Galaxy during its formation. The oldest population of stars was made during the infall, and it is this fact which leaves Population II its distribution throughout the galactic halo and central regions of the Galaxy. These stars, including those in globular clusters, still "remember" their infalling motion and continue to have elliptical orbits around the Galaxy. After the Galaxy had developed its flat rotation plane and spiral arms, the continuing formation of stars in these regions now represents Population I, journeying in circular orbits. Between these two extremes lie intermediate populations showing progressive flatter and flatter distributions, representing successive stages in the collapse of the Galaxy.

Because the chemical elements are generated in stars and recycled back into the interstellar material by stellar winds, supernova explosions, etc, Population I stars have more abundant concentration of "metals" (chemicals) than Population II. This makes it possible to distinguish the populations even of stars which are slow to evolve and which occupy parts of the Galaxy where populations overlap. For example, the Sun is typical in evolutionary state of yellow dwarfs of both Populations I and II, and is near to the galactic plane, so it could belong to a spiral arm or the disk or halo of the Galaxy. Its content of chemical elements puts it neither among the oldest metal-poor, nor the youngest metal-rich stars and it is thus in the intermediate disk population. This accords with its age of 5,000 million years.

It is the chemical history of the Galaxy which suggests the existence of the conjectural Population III. There are too few stars found in which there is a minimum of metals. Astronomers hypothesize that at the very formation of the Galaxy, there was a wave of formation of very bright stars of near-zero metal content. These stars would have manufactured the metal content now present in Population II and then disappeared, forming invisible compact stars like neutron stars. *PGM.*

Stellar populations

	Population I		Population II	
	Extreme Pop. I	Older Pop. I	Old disk	Halo
Occurrence	Spiral (& irregular) galaxies		Elliptical and central parts of spiral galaxies	
Objects	Interstellar gas	Galactic disk	Planetary nebulæ	Globular clusters
	Dust		Galactic nuclear regions	Elliptical galaxies
	Diffuse nebulæ			
	Reflection nebulæ			
	Open clusters			
	Spiral arms			
Stars	Supergiants	Yellow giants	Red giants and dwarf stars	
	Cepheids		Short-period RR Lyræs	Long-period RR Lyræs
	Blue giants	The Sun and main sequence stars nearby		
	T Tauri stars		Mira variables, carbon stars	
	Type II supernovæ	Type I & II supernovæ	Type I supernovæ	Type I supernovæ
	Wolf-Rayet stars			
Mean height above galactic plane	400	500	1,200	6,000 light-years
Distribution	Very patchy	Patchy	Smooth	Smooth
Age	Less than 0.1	1	3	Greater than 6 thousand million years
Composition	Metal-rich	Metal-rich	Metal-poor	Extremely metal-poor
Orbits	Circular	Almost circular	Very elliptical	Almost radial

S

Stellar proper motion Stars in our Galaxy move relative to each other and relative to the Sun, but because of their great distances their apparent movements in the sky are very small. The apparent angular displacement of a star due to its motion over a year is called its proper motion. Proper motion is usually denoted by the Greek letter Mu (μ) and is measured in seconds of arc per year ("/year). Barnard's star in the constellation Ophiuchus has the largest known proper motion of 10".27/year. About 300 known stars have proper motions larger than 1"/year; but most proper motions are smaller than 0".1/year.

Proper motions are measured by differencing the accurate positions of stars obtained at two or more epochs. This gives the components of the proper motion in right ascension, μ_α, measured in seconds of time per year, and in declination, μ_δ, measured in seconds of arc per year. The combined proper motion, μ, measured in seconds of arc per year, is given by

$$\mu = (225\, \mu_\alpha{}^2 \cos^2\delta + \mu_\delta{}^2)^{\frac{1}{2}}$$

The transverse velocity in space of a star relative to the Sun, V_T (km/s), is related to its observed proper motion, μ ("/year) by

$$V_T = 4.74 \frac{\mu}{\pi},$$

where Pi (π) is the PARALLAX of the star. For example, the parallax of Barnard's star is 0".545, so its transverse velocity is 89.3 km/s.

The type of telescopes used for measuring the positions of stars from which the proper motions are derived, are Transit or Meridian Circles. They measure the RIGHT ASCENSION and DECLINATION of the bright stars as they transit the north–south meridian on which the telescope is sited. The detailed analysis of many observations made with these telescopes over the past 100 years or so has led to the creation of the Fourth FUNDAMENTAL CATALOGUE (FK4), with positions and proper motions of 1,535 stars brighter than about seventh magnitude. This catalogue defines the basic reference frame to which the kinematics of all stars in our Galaxy are ultimately referred.

The most reliable extension of the FK4 system to fainter stars is to be found in the catalogues AGK3R and SRS, which each contain about 20,000 reference stars in the northern and southern hemispheres, respectively. These reference stars have been used to derive the positions of about a quarter of a million stars using photographic plates. The reference stars define the scale and orientation of the plates.

The proper motions in the AGK3, Yale and Cape catalogues (*see* accompanying table) were obtained by differencing the positions from plates taken about 30 years apart.

Surveys of stars with large proper motions have been carried out mainly by Luyten using the "blink microscope" (*see* BLINK COMPARATOR), which compares alternately two photographic plates of the same star field taken several years apart. When the plates are aligned, the stars with large proper motion appear to jump back and forth when alternate plates are viewed in rapid succession. By this method, about 3,500 stars have been found with proper motions greater than 0".5/year and about 50,000 greater than 0".2/year.

Proper motions of stars have been measured with respect to galaxies. The proper motions of galaxies are negligible, so they provide a fixed frame of reference against which to measure stellar proper motions. This method is therefore independent of the positions and proper motions of the bright reference stars measured by meridian circles. The reference system defined by the bright stars is known to contain systematic errors which propagate through all the catalogues listed in the table, except the Lick and Pulkovo proper motions, which are measured with respect to galaxies. However, it has been shown that even these proper motions are not entirely free of systematic errors, especially in right ascension. There is also the problem of completeness because faint galaxies are obscured near the plane of our Galaxy, and thus proper motions cannot be derived in the zone bordering the galactic plane.

Using the VERY LARGE ARRAY telescope, faint radio emission can be detected from stars in our Galaxy. The radio positions of these stars can be measured with an accuracy of about 0".03 (in 1986) with respect to galaxies whose radio positions are known through VLBI measurement to the order of 0".001. After the elapse of about ten

Proper motion catalogues

	Declination range	No. of stars (approx)	Average error ("/year)
General catalogues			
Fourth Fundamental Catalogue (FK4)	−90 to +90	1,500	0.002
Fifth Fundamental Catalogue (FK5) (in preparation 1986)	−90 to +90	5,000	0.002
General Catalogue (GC)	−90 to +90	33,300	0.010
Reference Stars for Third Astronomische Gesellschaft Catalogue (AGK3R)	−5 to +90	20,000	0.004
Southern Reference Stars (SRS)	−0 to −90	20,000	0.004
Zone catalogues			
Third Astronomische Gesellschaft Catalogue (AGK3)	−2.5 to +90	183,000	0.009
Yale Catalogues	−90 to +30 +50 to +90	150,000	0.02
(First) Cape Photographic Catalogue (CPC)	−35 to −90	61,000	0.004 to 0.02
Cape Zone Catalogue (CZ)	−40 to −52	20,800	0.01
Second Cape Photographic Catalogue (CPC2) (in preparation 1986)	−0 to −90	about 200,000	about 0.01
Stastical surveys in selected areas			
Kapteyn Selected Areas (Radcliffe and Pulkovo)	northern	50,000	0.007
Parallax Stars (McCormick)	northern	29,000	0.009
Lick and Pulkova proper motions with respect to galaxies	−30 to +90	about 300,000	0.004 to 0.007
Surveys of large proper motions			
Luyten Large Proper Motion Surveys (LHS): greater than 0".5/year	−90 to +90	3,500	0.014
(NLTT): greater than 0".2/year	−90 to +90	50,000	0.02
Luyten Bruce Proper Motion Survey	−90 to −0	98,000	0.02

S

S

years, this will give proper motions of these stars with respect to galaxies with an accuracy of about 0".004/year. This should provide a useful link between the proper motion system derived from bright reference stars and the proper motion system derived from the fixed frame of the galaxies.

Probably the greatest impact on the number and quality of proper motion determinations will be made by the astrometric satellite, HIPPARCOS, to be launched by ESA in 1988. In its 2½-year mission, Hipparcos will measure the proper motions of about 100,000 selected stars with a mean accuracy of 0".002/year.

Besides the direct application of predicting the positions of stars in the future, proper motions are of central importance in the investigation of the structure, kinematics and dynamics of our Galaxy.

The local motion of the Sun relative to stars within a 20-parsec radius can be derived from proper motions. The average proper motion of the local stars depends on the angle between the direction of the Sun's motion and the

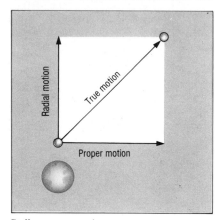

Stellar proper motion: star movement

direction to the stars. If the stars lie in the direction of the apex (or antapex) of the Sun's motion, the average of their proper motions will be zero. The calculated position of the apex and the solar velocity toward it depends to some extent on the number and type of stars used in the analysis. The average result is that relative to the nearby stars the Sun is travelling about 12 miles/s (20km/s) in the direction of the constellation Hercules. When the solar motion is subtracted from the observed proper motion of a star we obtain its peculiar proper motion. The apparent displacement of a star arising from the solar motion is called its parallactic motion. By measuring the parallactic motion of a homogeneous group of stars, the so-

called secular parallax of the group is obtained. This method has been applied to estimate the average distances of groups of variable stars, such as RR LYRÆ and CEPHEIDS, which are beyond the range of TRIGONOMETRICAL PARALLAX.

Another important application of proper motions is in determining the distances of moving clusters. Since stars in a cluster have a common space motion, their proper motions appear to converge to a point on the celestial sphere. If θ is the angle between the direction to the convergent point and the direction to the star, and V_R is the star's RADIAL VELOCITY in km/s, then its parallax, and thus the parallax of the group, is given by

$$\pi = \frac{4.74\mu}{V_R \tan \theta}$$

Applying this method to Delta Tauri in the HYADES cluster for which $\mu = 0".115/year$, $\theta = 29°.1$ and $V_R = 38.6$ km/sec, we find $\pi = 0".025$; this is a distance of 40pc. A systematic error in the proper motions will lead directly to a systematic error in the distance to the Hyades. Great care has to be taken to eliminate these errors.

Proper motion surveys have an important application in investigating the distribution and luminosity of stars in the solar neighbourhood. Stars close to the Sun tend to have larger values of proper motion than more distant stars, so by selecting stars from proper motion surveys which are complete to some apparent magnitude, a statistical sample of the stars within a specified volume around the Sun is obtained. By combining this sample with the known parallaxes of some of the stars, and making allowances for the incompleteness of the sample, one can arrive at a statistical estimate of the distribution and luminosity of the stars in the vicinity of the Sun. This shows that most stars in the solar neighbourhood are intrinsically faint.

Proper motions have another important application in the subject of galactic rotation. Proper motions provide the only direct method of measuring Oort's constant B, which essentially describes how the rate of change of the angular velocity of rotation changes with distance from the centre of our Galaxy. If l is the galactic longitude of a star and μ_l ("/year) its component of proper motion parallel to the galactic plane, then

B = 4740 μ_l − A cos 2l,

where A is the other Oort constant.

Using this method B is found to be about − 11 km/sec/kpc. *LVM.*

Stellar structure The physical state within stars. The gas in stars must lie in equilibrium everywhere, balancing the pressure exerted by the outer layers with the internal pressure of hot atoms. The theory was published by EDDINGTON in 1926.

Stellar wind A stream of charged particles, mostly PROTONS and ELECTRONS, from the surface of a star. The strength of the wind depends on the type of star, but the wind velocity is often quite high, ranging from a few hundred to several thousand km/sec.

Young stars evolving toward the MAIN SEQUENCE have powerful stellar winds, sometimes up to a thousand times stronger than the SOLAR WIND. These winds crash into surrounding gas clouds and ionize them, producing expanding shock waves in the interstellar medium. Old stars evolving off the main sequence to become red giants also have strong stellar winds. *JKD.*

Stephan's Quintet A small cluster of galaxies discovered in 1877 by the French astronomer M.E.Stephan. Of these, four have nearly the same REDSHIFT (z = 0.02), but in 1961 G. and M.Burbidge found that the other, NGC7320, has a much lower redshift (0.003), so that either the redshifts are unreliable as distance indicators or else NGC7320 merely happens to lie in the same line of sight. Some astronomers now refer to the group as Stephan's Quartet.

Step-rocket A rocket made up of separate propulsive sections or stages which are jettisoned progressively as propellant is used up; avoids accelerating "deadweight".

Steward Observatory Observatory of the University of Arizona, Tucson, with its main telescopes (36 and 90in/91 and 229cm reflectors) sited on KITT PEAK.

Stjerneborg The second of Tycho BRAHE's observatories on Hven; it was set up close to Uraniborg.

Stockholm Observatory Founded in 1748 (the old building is still in use by amateurs), and moved to suburban Saltsjöbaden in 1931. Main instruments are a large double refractor, a 16in (40cm) astrograph, and a 39in (1m) reflector. Earlier, the observatory had a solar station on Capri, Italy.

Stöfler Lunar walled plain south-east of Walter; diameter 90 miles (145km); 41°S, 6°E. The floor is darkish, and the wall is broken by a smaller but deeper crater, Faraday.

Stony-iron meteorites Meteorites with roughly equal proportions of metal to stony material by weight. There are two major groups, plus eleven oddities. Mesosiderites (25 known) are mixtures of chunks of silicate, related to eucrites and howardites (achondrites), with slugs of metal, related to Group IIIAB irons. Mixing was associated with violent impact, near an asteroidal surface. Pallasites (36 in the "main group") comprise gently mixed molten metal and olivine, possibly at the interface between an asteroidal core and silicate mantle. The Bencubbin (Western Australia) and Weatherford (Oklahoma) meteorites are complex mixtures of metal with aubrite achondrite and two types of CHONDRITE. *RH.*

Størmer, Carl Fredrik Mülertz, (1874–1957). Norwegian physicist who made a study of auroral particle trajectories and stereo photo height measurements.

Straight Range A short but remarkably regular lunar range in the MARE IMBRIUM, not far from Plato.

Straight Wall The most famous fault on the Moon, in the MARE NUBIUM, west of Thebit. It runs roughly SSE, from 20°S to 23°S. Before full moon it appears as a dark line, because of its shadow; after full it shows as a bright line. The actual angle of the slope is no more than 40°.

Strathmore meteorite Largest Scottish fall, chondrite, 1917.

Stratosphere A gaseous layer in the Earth's atmosphere extending above the tropopause (the upper limit of the troposphere) as far as the mesosphere. From the temperature minimum which exists at the tropopause, the temperature at first remains steady with increasing height. It then rises to a maximum of about 0°C at an altitude of 30 miles (50km), which marks the stratopause, the top of the stratosphere. The ozone layer or ozonesphere lies between about nine and 30 miles (15 and 50km) altitude and is essentially identical with the stratosphere. The highest concentration of ozone occurs at a height of about 15 miles (25km). *JWM.*

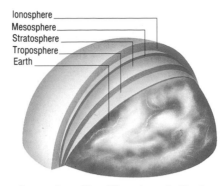

Ionosphere
Mesosphere
Stratosphere
Troposphere
Earth

Stratosphere: 25 to 50km above the Earth

Struve, Frederich Georg Wilhelm (1793–1864). The first in a long line of astronomers from the same family, F.G.W. Struve was a pioneer in the observation of double stars. His career began in 1813 at the Dorpat Observatory in Estonia, and in 1824 he set up the Fraunhofer 9in (23cm) refractor using it to discover 2,343 new double stars. His results appeared in 1837 in the "Mensuræ Micrometricæ" and he was awarded the Gold Medal of the Royal Astronomical Society for this work. He later supervised the building of the Pulkovo Observatory and was Director from 1839 to 1861. *RWA.*

Sualocin The star Alpha Delphini. Beta Delphini is "Rotanev"; the names were given by Nicolaus Venator!

Subduction zones Zones of the Earth's crust along which lithospheric plates converge. Such zones are characterized by volcanicity, seismicity and higher-than-average heat flow due to the downthrusting of one plate beneath the other. The zone of violent volcanic and earthquake activity that surrounds the Pacific from New Zealand to southern Chile is a modern manifestation of this ancient, continuing process.

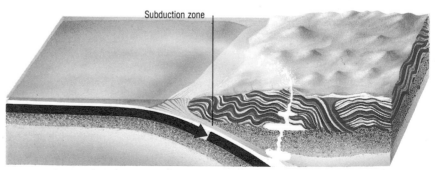

Subduction zone

Subduction zones: areas of volcanic activity in the Earth's crust

Subdwarfs Stars that are less luminous by 1 to 2 magnitudes than MAIN SEQUENCE stars of the same spectral type. Denoted by the symbol sd.

Subgiants At the end of their lives as MAIN SEQUENCE dwarfs, stars expand, becoming subgiants and then giants.

Subluminous stars Stars less luminous than MAIN SEQUENCE stars of the same temperature. They are usually old evolved (ie, Population II) stars. Examples are WHITE DWARFS, hot SUBDWARFS, NOVA remnants.

Suisei Second of two Japanese probes to HALLEY'S COMET which was launched in August 1985 and flew past the comet at a distance of 94,000 miles (151,000km) on the sunward side on March 8 1986. Suisei was a spin-stabilized spacecraft with a diameter of 4½ft

Suisei: Japanese probe to Halley's Comet

(1.4m) and a height of 28in (70cm), weighing 307lb (139.5kg). Its major investigations were of the growth and decay of the comet's hydrogen corona and the interaction of the solar wind with the cometary ionosphere.

Sulphur An element — chemical symbol S, atomic number 16, atomic mass 32.06 — which is tenth in descending order of cosmic abundance.

S

Suisei Planitia Plain on Mercury: 59°N, 157°W, north-east of the CALORIS BASIN.

Sulphur dioxide A compound, of sulphur and oxygen — chemical symbol SO_2.

Sulphuric acid on Venus Sulphuric acid droplets make up the main composition of the clouds of Venus. The droplets are formed from the reaction between H_2O and SO_2 high in the atmosphere and aided by the solar UV radiation. The sulphuric acid clouds form a layer extending from 25 to 50 miles (40 to 80km) of varying concentration with rain occurring in the lower layers.

Sulpicius Gallus Very bright crater, 8 miles (13km) in diameter, near the Hæmus Mountains bordering the lunar MARE SERENITATIS: 20°N, 12°E.

Summer triangle Unofficial nickname given to the arrangement of Vega, Altair and Deneb. It was given by Patrick Moore in a television "Sky at Night" broadcast around 1958, and has come into common use, though it is unofficial and does not apply to the southern hemisphere!

Sun A star of the MAIN SEQUENCE of class G2V. Life on Earth depends on it. To astronomers this star is the central body of the Solar System and the only star which, as seen from the Earth, shows a disk with surface details which can be observed and studied. Astrophysicists build models based on these observations which help them to study STELLAR STRUCTURE and EVOLUTION.

Solar data

Mean radius (photosphere)	16.0 arc min = 432,000 miles (696,000km)
Solar parallax	8.794 arc sec
Mean distance from the Earth (AU)	92,955,630 miles (149,597,870km)
Solar mass ($M_\odot$)	$1,989 \times 10^{33}$ g
Solar constant	1.39×10^6 erg cm^{-2} sec^{-1}
Apparent magnitude (m_v)	−26.7
Absolute magnitude (M_v)	+4.79
Effective temperature	5,800K
Mean period rotation	27 days
Period revolution (around centre Galaxy)	2.2×10^8 years

Structure of the Sun Like all stars the Sun consists of a mass of gas. The sharply defined disk that we can see through thin clouds, or at sunset, is called the PHOTOSPHERE, from which is emitted most of the sunlight that we see. As we proceed deeper into the Sun the opacity of the gases increases. The photosphere represents the boundary between transparent and opaque gases. In the interior of the Sun there is a core, where the nuclear reactions take place which provide the energy of the Sun (*see* PROTON-PROTON CHAIN). The core occupies about 25 per cent of the radius of the Sun. The temperature at the centre is of the order of 15 million degrees K̊ decreasing to about 13 million at the outer limit of the core. Beyond the core, extending to about 70 per cent of the radius, is the radiative zone, where energy is diffused only by radiation processes (*see* RADIATIVE DIFFUSION). Surrounding the radiative zone and extending to the photosphere is the convection zone, a turbulent region where high opacity begins to be evident.

Next to the photosphere is the layer called the SOLAR CHROMOSPHERE, so named because of its red colour, which can be seen during a SOLAR ECLIPSE. The chromosphere is a thin and nearly transparent layer which merges into the SOLAR CORONA, and its radiation is less intense than that of the photosphere. Between the chromosphere and the corona is the transition region, where the temperature rises to reach the high temperature of the corona. Finally we reach the corona, which can be seen during an eclipse of the Sun and is considered to be its outermost layer.

Modern instrumentation allows observation of the outer layers of the Sun, but because of the opacity below the photosphere, it is not possible to observe directly the interior of the Sun. Indirect methods, however, have been used. Attempts have been made to measure the neutrino flux which originates in the core as a result of the nuclear reactions (*see* NEUTRINOS). Another method is based on the study of pressure waves which travel through the convection zone and are reflected back from the photosphere (helioseismology) (*see* SOLAR OSCILLATIONS). The radiation emitted by the Sun is distributed throughout the whole of the electromagnetic spectrum, from the very short wavelengths of X-rays to the longer radio wavelengths. About 41 per cent of the total radiation is in the optical or visible range of the spectrum and because the photosphere can be easily observed in white light, the photospheric phenomena (*see* SOLAR FACULÆ, GRANULATION and SUNSPOTS) were the first to be studied. In the case of sunspots we have records covering a period of more than 300 years. Systematic observations of sunspots have revealed the differential rotation of the Sun and the existence of the SOLAR CYCLE. The chromosphere, on the other hand, can only be observed in monochromatic light. The study of chromospheric phenomena (*see* SOLAR FLARES, PLAGES, SPICULES and PROMINENCES) has developed rapidly this century with the introduction of new instruments and has added to our knowledge of the physical constitution of the Sun.

To study the Sun at X-ray wavelengths it is necessary to observe from outside the Earth's atmosphere. Photographs of the Sun, taken at these very short wavelengths by special instruments aboard orbiting space vehicles, have shown the structure of the corona as seen from above and revealed the existence of dark, cool regions (CORONAL HOLES) and bright, active, short-lived points (coronal bright points). In the radio waves region of the ELECTROMAGNETIC SPECTRUM, the Sun can be observed from Earth by means of RADIO TELESCOPES. Measurements made at various wavelengths show a

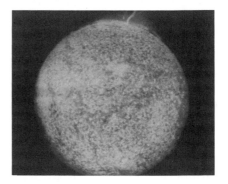

Sun: central body of the Solar System

different aspect of the Sun (radio Sun) with an outline which is elliptical and which has a marked limb brightening. It was established that the radiation from the quiet Sun is different from that of the active Sun when bursts and outbursts of radiation occur.

Solar magnetism Magnetic fields of the order of 3,000 gauss exist in sunspots, while the general magnetic field of the Sun is very weak, about 1 gauss, and hence more difficult to measure. There are indications that this general field is subject to reversal of polarity. Daily recordings of the distribution of magnetic fields over the solar disk (magnetic active regions) are obtained by means of magnetographs. It appears that groups of sunspots are bipolar with a reversal of polarity at the beginning of every solar cycle. If magnetic polarity is taken into account, then the period of the solar cycle is of 22 years against the 11 years of the optical cycle. Solar magnetism plays an important part in many solar phenomena.

Solar terrestrial effects At the beginning of the century Maunder suggested the existence of a link between sunspots and MAGNETIC STORMS on the Earth. A correlation between solar activity, terrestrial magnetism, frequency of AUR ORÆ and ionospheric disturbances has now been established. The sudden enhanced radiation from the Sun, particularly when a large flare is present, has the effect of altering the composition and reflectivity of the D, E and F layers of the IONOSPHERE. As a result of this we observe freak receptions in radio communications at very short wavelengths, fade-out in the short waves range, enhancement of long-wavelength signals and sudden ionospheric disturbances (S.I.D.).

Sun and weather The energy required to maintain the weather system of the Earth is supplied by the Sun. Attempts have been made, so far unsuccessfully with the exception perhaps in the case of MAUNDER'S MINIMUM, to establish a correlation between solar activity and the Earth's weather. A correlation probably exists, but it is difficult to determine because many factors are involved, such as: variations of the SOLAR CONSTANT; oscillations in the size of the Sun's diameter; changes in the ultraviolet radiation from the Sun; and conditions of the ozone layer. In addition we have to take into account the possible interactions between the SOLAR WIND and the magnetic field of the Earth and the ensuing exchange of energy. The determination of a correlation between solar activity and the weather is further complicated by phenomena of terrestrial origin. Violent volcanic eruptions inject large amounts of dust and ashes into the circulation of the upper atmosphere, blocking part of the solar radiation. On the other hand, an increase in the carbon dioxide content of the atmosphere, produced by man's activities on Earth, generates a greenhouse effect and hence an increase in the average temperature of the Earth, which would have a bearing on its general climate. *VB.*

Sundogs Patches of shimmering light, also called parhelia, which are images of the Sun refracted by thin clouds of ice crystals in the lower atmosphere of the Earth.

Sunspots Dark patches on the solar surface. They can be as small as 600 miles (1,000km) in diameter or larger than 25,000 miles (40,000km). Most spots have a central dark umbra and a grey surrounding region, the penumbra. Small spots may last for a few hours, large spots may survive for weeks and months. They are some 1,500K cooler than the PHOTOSPHERE. Spots are produced by strong magnetic fields which cause cooling of the region by suppression of the normal convection of energy from lower levels.

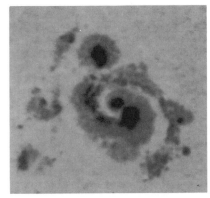

Sunspots: produced by magnetic fields

Superclusters *See* colour essay.

Supergiants The most luminous stars known. Supergiants occur with spectral types from O to M. Red (M type) supergiants have the largest radius, of the order of 1,000 times that of the Sun, the mean radius of which is 432,000 miles (700,000 km).

Superior conjunction The passage of an INFERIOR PLANET on the far side of the Sun, when they have the same RIGHT ASCENSION.

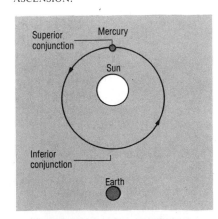

Superior conjunction: planetary position

Superior planets All planets whose orbits around the Sun lie beyond that of the Earth.

Supernova A stellar explosion involving the disruption of virtually an entire star. Also, supernova remnant (SNR): the remains of a supernova explosion, either gaseous like the CRAB NEBULA or, less commonly, stellar (eg, a PULSAR or BLACK HOLE).

Stable stars are a balance between their internal pressure and their own force of gravity, and the failure of this balance is the underlying cause of a supernova. If the pressure, which is supplied from the heat of the internal material, suddenly greatly increases, then the star will rapidly expand. The increase of the star's surface area and the release of the internal energy contribute to a sudden brightening of the star. At expansion speeds of 6,000 miles/s (10,000km/s), a star will grow from Sun-sized to the size of the Solar System in only a day, and its surface area and therefore brightness in this short period will increase many millions of times. It is the sudden immense brightening of an otherwise unnoticed star in such an explosion which is observed as a supernova.

For a period of a week or so, a supernova may outshine all the other stars in its parent galaxy. The luminosity of a supernova (absolute magnitude −19.5) is some 25 magnitudes brighter than the Sun (a factor of 10,000 million),

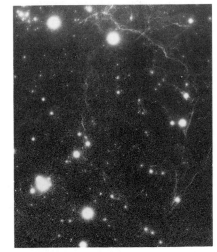

Supernova: stellar explosion

and the energy released in the explosion is equal to the energy released over the entire life of the star beforehand.

Supernovæ are discovered in external galaxies by repeated photography of the sky, and by comparing images of galaxies to spot the "new" star. Super-

S

novae are named with the letters SN, followed by the year in which they were at maximum brightness, followed by a lower case letter. For example the brightest extragalactic supernova was SN1937c in the galaxy IC4182. F. ZWICKY and C. Kowal are the two names most associated with the discovery of supernovæ. each being responsible for over 100 discoveries using SCHMIDT TELESCOPES at Mt Palomar.

Supernovæ are relatively rare, discovered at the rate of one about every century in an average galaxy. They are discovered relatively infrequently in edge-on spiral galaxies because their light is dimmed by dust. There is probably about one supernova every 30 years in a galaxy like our own Milky Way Galaxy, but only five have been discovered in the last millennium.

Supernovæ

Year	Constellation	Duration	Mag.*
1006	Lupus	Two years	Much brighter than Venus
1054	Taurus	24 months	−3.5
1181	Cassiopeia	6 months	−1
1572	Cassiopeia	483 days	−4.0
1604	Ophiuchus	365 days	−2.6
*Maximum brightness			

The supernova of 1054 was identified by Edwin HUBBLE as the progenitor of the Crab Nebula, and like the supernovae of 1006 and 1181 was recorded by oriental astronomers as they scanned the sky for celestial portents. Chinese, Korean, Japanese, Arabic and European astronomers contributed to records of these supernovæ: the 1054 supernova was probably depicted in Pueblo Indian art.

The supernova of 1572 was carefully observed by Tycho BRAHE. He recorded data on its unchanging position and its stellar magnitude as it faded day by day. Brahe demonstrated that the supernova, which was circumpolar from Denmark, had no PARALLAX as the Earth rotated. This placed the star well beyond the Moon. Its lack of motion over the 18 months that it was visible meant that it was beyond Saturn, the most distant planet then known. This placed the supernova among the "fixed stars" and proved that they were subject to the same laws of change as terrestrial phenomena.

The supernova of 1604 is known as Kepler's Supernova, although Kepler was not the first to see it, in October of that year.

Both these supernovæ inspired a wealth of comment by many writers of the 16th and 17th centuries, including Richard Corbet, Henry More, John Donne, Edmund Spenser and John Dryden.

Supernovæ were recognized by R. MINKOWSKI as being of two main types, called Type I and Type II. Type II supernovæ have hydrogen lines in their spectra, and those of Type I do not. The two types have distinctive light curves. Enough is known of the light curves of SN1572 and SN1604 to know that both were Type I. Type I supernovæ have been found in all kinds of galaxies, but Type IIs have never been found in elliptical galaxies. Since elliptical galaxies are different from others in that they contain no bright massive stars, astronomers infer that massive stars give rise to Type II supernovæ, and regard a supernova explosion as the end-point of the evolution of stars of original mass larger than eight solar masses. At the end of its life such a large star consists of a series of concentric shells of different composition. The outer shell is hydrogen, which is burning to make helium. Inside this a shell of recently formed helium is burning to make carbon, oxygen and neon. Inside this are shells of magnesium, silicon, sulphur and at the centre an iron-nickel core. When the nuclear fuel gives out, the core collapses to form a NEUTRON STAR. The collapse of the interior of the star releases energy which is picked up by the outer layers which are ejected into space. These expanding layers form the supernova.

Type I supernovæ are believed to be formed in an explosion on the WHITE DWARF component of a BINARY STAR. Hydrogen from the companion leaks onto the white dwarf and drives it over the critical mass of 1.4 solar masses: it explodes.

After a couple of years the supernova has expanded so much that it becomes thin and transparent. But then for hundreds or thousands of years the ejected material is visible as a NEBULA. The ejected layers of an exploding supernova produce a gaseous supernova remnant by collision with the surrounding interstellar gas. The collisional energy heats the gas as if a piston is being driven into it. Its temperature rises to perhaps a million degrees and the gas emits X-rays. As the gas cools to 10,000 degrees, it emits numerous distinctive optical spectral lines. Electrons in the gas interact with the magnetic fields in the gas compressed by the piston, and emit radio waves from the synchrotron process. Thus supernova remnants are detected by optical, by radio and by X-ray telescopes such as the EINSTEIN OBSERVATORY. One hundred and thirty-five radio supernova remnants are known, of which 40 have been seen optically and 33 by X-ray, including the most intense radio source in the sky, Cassiopeia A. This object shows a rate of outward motion which, if uniform, puts its explosion date within three years of 1658. It may be identified with the missing sixth magnitude star 3 Cassiopeiæ observed by John FLAMSTEED in 1680, if the explosion has been slowed by collision with interstellar gas.

Supernova remnants (SNR) show usually as hollow shells of outflowing gas. Tycho's and Kepler's SNR are of this type, as well as Cas A. Six supernova remnants show as filled balls of radio and X-ray emission, including the Crab Nebula and 3C58. Such SNR are called "plerionic" (filled). This implies that they contain relativistic electrons produced by an active pulsar, the neutron star formed by the collapse of the iron-nickel core in the supernova explosion.

When one member of a binary star explodes as a supernova, the other member recoils and is flung off like a sling-shot. Such stars are called runaway stars. 53 Arietis, AE Aurigae and Mu Columbæ are three runaway stars with a common origin point in Orion and may result from a supernova in what was a quadruple star system approximately three million years ago.

Supernovæ are responsible for the production of COSMIC RAYS. The explosion of a supernova within 10 parsecs of the Sun produces a ten to one-hundredfold increase in the cosmic ray background at the Earth, as the ejecta of the supernova sweeps past the Solar System. This will accelerate radiation exposure for living beings. About 100 such events have taken place in the lifetime of the Earth. I. Shklovsky has speculated that supernova explosions contribute to terrestrial biological evolution, eg, killing the dinosaurs at the end of the Cretaceous Era. *PGM*.

Supernova SN 1987A A bright supernova which flared up in the Large Cloud of Magellan in February 1987, reaching naked-eye visibility — the first supernova to do so since Kepler's Star of 1604.

Surface gravity The acceleration due to gravity experienced by a freely-falling object close to the surface of a massive body. Denoted by the symbol g.

Surt Active volcano on Jupiter's satellite IO: 46°N, 336°W.

Surveyor programme A series of seven soft-landing lunar spacecraft, five being highly successful, launched between May 1966 and January 1968. Each Surveyor had a steerable television camera with filters, while some had a combined surface-sampler/trench-digger and/or a simple major-element analyzer A wealth of data was returned, including the appearance of the surface (panoramas and close-ups); its bearing strength, optical, and thermal properties; its content of magnetic material; and its major-element chemistry, later confirmed by Apollo samples analysis.

Su Song (Su Sung) (1020–1101). Chinese astronomer, noted for his clock-driven astronomical instruments.

In 1090 Su Song finished his two-storey clock tower constructed at the then capital, Kaifeng. Inside was a clock-driven celestial globe, while moving figures or "jacks," indicating time, were visible from the outside. Surmounting the tower was an ARMILLARY SPHERE (a framework with rings to represent celestial co-ordinates), fitted with a sighting tube. Globe and sphere automatically followed the rotation of the skies, and were the first astronomical instruments to do so, the sphere anticipating Western automatic clock-driven instruments by six centuries. *CAR.*

Sven Hedin (or Hedin). Broken lunar formation, 60 miles (98km) across, west of Hevel; 5°N, 75°W. The name is omitted from some recent maps, but lunar observers use it.

Swift, Lewis (1820–1913). American astronomer who specialized in hunting for comets and nebulæ. He found 13 comets including P/Tempel-Swift (period 5.7 years), P/Di Vico-Swift (period 6.3 years) and two periodic comets P/Swift 1 (period 8.9 years) and P/Swift 2 (period 7.2 years). He also jointly discovered the Great Comet (P/Swift-Tuttle) of 1862, which is the parent of the annual Perseid meteor stream.

Swift's Comet Also known as the Great Comet of 1862, this periodical comet is correctly designated P/Swift-Tuttle

1862 III. It was discovered by Lewis SWIFT in Camelopardus on July 16 1862, and was independently discovered three days later by Horace Tuttle (Harvard Observatory, Massechusetts). The comet became a naked-eye object, attaining magnitude 2 in the first days of September, when the tail length was estimated as about 30°. Thereafter, the comet rapidly faded and was last seen on October 31 1862. After debate about its period, it was predicted to return in 1982. It was not seen, and Brian Marsden of the Smithsonian Astrophysical Observatory has proposed that Comet Kegler, seen in 1737, may be an earlier return of P/Swift-Tuttle. If this is so, then the period may be nearer 130 years, and it might be back in late 1992. *JWM.*

SX Phœnicis Dwarf Cepheid, 6°5 west of Alpha Phœnicis, with a period of only 79 minutes. The visual range is from magnitude 7.1 to 7.5. Distance, about 140 light-years.

Symbiotic stars Stars that combine two grossly different temperature regimes. Absorption bands found only below 4,000K, and gas heated by a source of about 200,000K, are seen. They are believed to be binaries in which gas from the cool star falls onto, and heats, a smaller, invisible companion.

Synchronous orbit An orbit in which a satellite revolves with the same period as its primary's "day". For Earth, the radius of the orbit is 26,000 miles (42,000km). The satellite "hovers" over one point, neither rising nor setting.

Synchronous rotation Occurs when a satellite spins with the same period as its orbital period. Tidal friction has "locked" it into this condition and it always presents a constant face to its primary (*cf* the Moon).

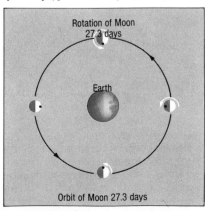

Rotation of Moon 27.3 days

Earth

Orbit of Moon 27.3 days

Synchronous rotation: Earth and Moon

Synchrotron radiation ELECTROMAGNETIC RADIATION emitted by charged particles (usually electrons) moving in magnetic fields at large fractions of the speed of light. The charged particles follow helical paths along magnetic lines of force and emit radiation because of the accelerations to which they are subjected. The higher the energy of the electrons, the shorter the wavelength of the emitted radiation. Synchrotron radiation is polarized.

Syncom satellites The first communications satellites to be placed in geostationary orbits. Now replaced by INTELSAT.

Synodic period The interval between successive OPPOSITIONS of a SUPERIOR PLANET.

Syrtis Major Planitia The most conspicuous feature on Mars; it is dark, triangular, and an easy telescopic object. It was formerly known as the Syrtis Major, and was drawn by Huygens as long ago as 1659. It is now known to be a plateau rather than a vegetation-filled depression, as was once thought. It extends from 20°N to 1°S, 283° to 298°W.

Systems I and II Grouping by rotation period of the clouds of Jupiter, which do not rotate uniformly. The groups are:
I 9^{h}50^{m}30^s equatorial regions
II 9^{h}55^{m}41^s above latitude 10° N or S
A third period, defined by radio outbursts, is 9^{h}55^{m}29^{s}37. *See also* JUPITER.

Syzygy CONJUNCTION or OPPOSITION, or NEW or FULL.

Tacitus Lunar crater 25 miles (40km) in diameter; 16°S, 19°E. It is polygonal in outline.

TAI *See* INTERNATIONAL ATOMIC TIME.

Tarantula Nebula (NGC2070, 30 Doradûs; 2000 co-ordinates 5^{h}38^{m}3, −69°03′). Nebula faintly visible to the naked eye on the south-east edge of the Large MAGELLANIC CLOUD, 160,000 light-years away. In actual size, it is bigger and brighter than any nebula in

T

the Milky Way or other nearby galaxy, being 1,000 light-years across and 500,000 times the Sun's mass. It has a complex filamentary structure with a cluster of stars at the centre. It glows in the ultraviolet light from a source in the central cluster which is producing 50 million times the Sun's energy. It is not known whether the source, called R136al, is a very compact cluster of super-luminous stars or a single star more massive than any other. *ACG*.

Tarantula Nebula: 1,000 light-yrs across

Tarazed The star Gamma Aquilæ, near Altair. It is of magnitude 2.7, and clearly orange. To the far side of Altair is Alshain or Beta Aquilæ (3.7).

Taruntius Concentric lunar crater 37 miles (60km) in diameter; 6°N, 48°E, on Mare Fœcunditatis. Its walls are rather low.

T-associations Regions of recent and active star formation in which the dominant visible population is low-mass, roughly solar, stars. These stars, in their pre-main sequence phase of life, are styled T TAURI STARS after the prototype in the Taurus-Auriga complex of dark clouds which, at 500 light-years' distance, is one of the nearest stellar nurseries to the Sun.

Low-mass stars may form throughout the dark clouds of T-associations but can only be seen optically in the surface layers of these clouds. Embedded within the complexes there may be hundreds of other young solar-type stars, too heavily obscured by dust to detect at visible wavelengths. Occasionally high-mass young stars are also found in T-associations, but only within the darkest, densest cores of clouds.

The importance of T-associations is that their members afford a series of snapshots of our young Sun through its

infancy. If we could order these different stars correctly, we would have constructed an evolutionary sequence for our Sun. It might also be clear how and when the planets formed, and over what period.

The T Tauri stars themselves are intriguing, irregular optical variables. In the nearby Taurus-Auriga T-association they are of the eleventh to nineteenth visual magnitude. Infrared brightness characterizes the stars of a T-association, since they are still surrounded by dusty obscuring circumstellar material whose fate is eventually to be accreted by the central star. Of topical interest is the manner in which the already-formed stars of a T-association interact with their parent gas clouds. Vigorous winds terminate the protostellar accretion phase of T Tauri stars (perhaps after only a million years) and stir up the ambient medium of the T-association through bipolar flows. Such flows are likely to act against the ready formation of new low-mass stars in their immediate vicinities. *MC*.

Tau Ceti. One of the nearest stars sufficiently like the Sun to be regarded as a possible centre of a planetary system. It is 11.9 light-years away, and of type K0, with an apparent magnitude of 3.5 and an absolute magnitude of 5.7.

Taurid meteors From about October 20 until November 30 every year, meteors can be seen emanating from Taurus, the Bull. The maximum of the shower, which occurs between about November 3 and 10, is extremely broad, and the activity is low — not more than about 12 meteors per hour for a visual observer. The Taurid stream has both northern and southern branches, with radiants at R.A. 3h 44m, Dec. +14°, and R.A. 3h 44m, Dec. +22°, near maximum. In 1940, Whipple showed that the Taurid meteors moved in unusually short-period orbits, and that the stream was associated with comet P/Encke. Later it was shown that the night-time autumn Taurid shower is the return of the Beta Taurid stream which peaks on about June 30, during daylight hours. *JWM*.

Taurus (the Bull). In mythology, the bull into which Jupiter changed himself when he wished to carry off the daughter of the King of Crete.

Taurus is distinguished by the presence of the two most famous open clusters in the sky, the PLEIADES and the HYADES, and by the CRAB NEBULA,

M1, near Zeta; there is also the prototype T Tauri variable, associated with nebulosity. *PM*.

Brightest stars

Name	Visual Mag.	Abs. Mag.	Spec.	Distance (light-yrs)
α (Aldebaran)	0.85	−0.3	K5	68
β (Al Nath)	1.65	−1.6	B7	130
η (Alcyone)	2.87	−1.6	B7	240 (Pleiades)
ζ (Alheka)	3.00	−3.0	B2	490
λ	3.3(max)	−1.7	B3	325
θ²	3.42	0.5	A7	125 (Hyades)

Then come Epsilon (3.54), Omicron (3.60), 27 (3.63), Gamma (also 3.63), 17 (3.70), Delta¹ (3.76), Theta¹ (3.85), 20 (3.88), Xi (3.74), and Nu (3.91). Al Nath was formerly included in Auriga, as Gamma Aurigæ. Lambda is an eclipsing binary; range 3.3 to 4.2, period 3.95 days.

Taurus dark clouds Extensive volume of dust-rich gas clouds about 500 light-years distant in the constellation Taurus. Site of continuing star formation and of many T TAURI STARS.

Taygete The star 19 Tauri, in the Pleiades; magnitude 4.30.

Taylor column A relatively stagnant column of fluid which is formed over an obstacle in a rotating fluid. Such columns were first studied in the laboratory by Sir Geoffrey Taylor, and in the 1960s a Taylor column was suggested as the explanation of the Jovian GREAT RED SPOT. This theory has been subsequently discounted.

T Coronæ Borealis A recurrent nova which erupted in 1866 and 1946. Binary period about 227ᵈ5, due to M3 giant and white dwarf. Possible second period aobut 53ᵈ69, due to third component.

Tebbutt's Comet Few comets created greater sensation in the 19th century than the Great Comet, 1861 II. It was discovered by an Australian astronomer, John Tebbutt, on May 13 1861, before its perihelion passage. Moving from the southern hemisphere into the northern in the second half of June, it became a brilliant object. On June 29, John Herschel observed it to have a head far brighter than any star or planet except Venus, and a tail over 30° in length. On June 30, the Earth passed through the tail of the comet, which was then 11 million miles (18 million km) inside the Earth's orbit.

Technetium Unstable radioactive element which does not occur naturally on Earth but was the first element to be created artificially. It is present in certain types of variable star, within which it is continuously created. Chemical symbol Tc, atomic number 43, atomic mass 98.906.

Tektites Objects of natural glass, unrelated to vulcanism, occurring over four main areas ("strewnfields"). (1) Australasia: Tasmania, across Australia to Sumatra and the Philippines and south-east Asian mainland. Australian tektites (australites) are 830,000 years old, compared with 690,000 for the remainder of the strewnfield, so two groups may be represented. (2) Ivory Coast tektites: 1.3 million years old, related to the Lake Bosumptwi, Ghana, impact crater of similar age. (3) Moldavites: from Czechoslovakia, 14.7 million years old, linking them with the Ries impact structure, south Germany. (4) North American tektites: Texas and Georgia (one only from Martha's Vineyard, Massachusetts), 34 million years old. Numbers 1 and 4 are not linked to known impacts. Microtektites related to 1, 2 and 4 occur in marine sediments.

Tektites often have shapes (teardrop, disk, dumbbell) suggesting they were once fluid, but they may be corroded. The largest are up to 33lb (15kg), from south-east Asia, and have

Tektites: formed by meteorites

laminar flow structures (Muong Nong type). Many from south-east Australia have features indicative of high-velocity atmospheric flight, such as a "button" shape, atmospheric ablation having melted the front part of an original sphere, causing melt to flow backward to form a flange.

Ivory Coast and Australasian tektites are opaque, unless very thin, Moldavites are green and North American tektites brown. All are silica-rich and contain inclusions of lechatelierite (pure silica glass) and other resi-

dual minerals. A lunar origin was suggested, but the return of lunar samples reinforced the terrestrial chemical and isotopic signatures. With silica glass from the Libyan desert, Egypt, they are interpreted as terrestrial materials subjected to meteoritic impacts big enough to blow holes through the atmosphere. Molten or vaporized ejected material coagulated and re-entered the atmosphere, to land as tektites. *RH*.

Tele-compressor A positive lens inserted into the condensing cone of a telescope which effectively gives a wider field of view and a faster focal ratio for photographic purposes.

Tele-extender A projection tube used in conjunction with an ocular to increase the effective focal length of a telescope system.

Telescopes Optical scientific instruments which, in an astronomical application, allow the user to obtain an enhanced and magnified view of the heavens over the visual part of the electromagnetic spectrum.

At optical wavelengths the telescope consists of a light collector in the form of a primary mirror or an objective lens or a combination of these, together with an eyepiece which enables the observer to examine the image produced by the collector. Various optical and mechanical arrangements of mirrors and lenses may be employed and these are discussed below. A photographic camera, SPECTROSCOPE or other instrument may be used in place of the eye. The "size" of a telescope is usually quoted as the diameter or aperture of the primary mirror or lens.

Whatever the type of telescope, it is necessary for it to be mounted on an appropriate stand since even a small-aperture instrument will have a magnification which will make hand-held operation impossible. The mounting should be as massive and stable as possible, and ideally should be of the equatorial type.

A common form of astronomical telescope is the refractor. This consists of an objective lens, usually an achromatic doublet to minimize CHROMATIC ABERRATION, contained in a sealed tube, with an eyelens, commonly called an eyepiece, mounted in another tube coaxial with the first at a distance just in excess of the focal plane of the objective lens. This eyepiece has a considerably shorter focal length than the

objective lens, and is so mounted that its position is variable in order to allow the eyepiece to focus exactly on the primary image. The refractor is a deservedly popular telescope which gives a good all-round performance, is easy to maintain and retains its optical alignment. Whilst small (3in/75mm aperture) refractors are inexpensive, the larger-aperture telescopes become costly, far from portable and require a substantial mounting.

Another commonly used form of astronomical telescope is the reflector. The three principal forms of reflecting telescope are the Newtonian, Cassegrain and the Schmidt-Cassegrain. Each form uses a primary mirror as a light collector and a secondary mirror to direct the image formed by the primary mirror to a place where it may be examined by the eyepiece. Where these forms differ is in the shape of the curves employed in the mirrors and in the optical and mechanical configuration.

The Newtonian reflector employs a parabolic primary mirror which directs its light onto a plane secondary mirror sometimes called the "flat", mounted on the optical axis but at an angle of 45° in order to redirect the final image by 90° to a point outside the main tube assembly. Here, the eyepiece, mounted in a sliding tube, is able to focus on the primary image. The Newtonian format is characterized by its convenience, relative simplicity and good on-axis images. It is also relatively cheap. The disadvantages are that the flat field of view is comparatively small and off-axis images can be rather poor.

The Cassegrain reflector also uses a parabolic primary mirror, but in this case the mirror is pierced by a central hole and the secondary mirror reflects the primary image back along the optical axis through the hole to a point just behind the primary mirror, where the eyepiece is again mounted in a sliding tube. The secondary mirror is figured with a hyperbolic curve, and the combination of the parabolic primary and hyperbolic secondary allows a relatively long focal length to be obtained in a short tube because of the inherent folding of the optical path that is achieved. Comparatively large (12in/30cm) aperture Cassegrains are fairly small and portable and convenient to use; they give good on-axis images, although with a typical focal ratio of f14 the field of view is rather small.

The Schmidt-Cassegrain telescope, which has become extremely popular

T

during the last few years, is strictly a catadioptric telescope employing both lenses and mirrors. The primary light collector is a spherical or paraboloidal mirror depending on the system employed, pierced with a central on-axis hole. The secondary mirror has a matching curvature and reflects the image back through the primary hole to behind the main mirror as in a standard Cassegrain. The weak correcting lens is placed just outside the secondary mirror and this mirror-lens combination results in a sealed, compact, reliable, long-focal-length, virtually achromatic telescope which offers a good all-round performance at a reasonable price.

There are a number of different reflecting and refracting telescopes together with various catadioptric systems which have been designed for special purposes such as astronomical cameras, rich field telescopes, astronomical astrometry, etc.

The main qualities which are looked for in an astronomical telescope are light-gathering power, resolution and finally magnification. In all the above telescopes the light-gathering power

Telescopes: the Great Melbourne Telescope

depends on the area of the objective lens or primary mirror. To compare with the human eye, assume the diameter of the dark adapted pupil to be 7mm. The area will be about 38sq mm. Assume the telescope's objective lens to have a diameter of 75mm; its area will be about 4,418sq mm. The ratio is therefore 4,418/38 = 116.26, which corresponds to just over 5 stellar magnitudes. Clearly, the larger the telescope objective lens or mirror, the better its light-gathering power.

The resolution of a telescope also depends on the size of the objective lens

or primary mirror. Resolution may be defined as the ability of the telescope to differentiate or separate two close astronomical objects such as components of a double star or two close craters on the moon. Resolution is given in the angular measurement of arc seconds, ie, 1/3600 of a degree. it may be calculated from

$$\text{Resolution} = \frac{1.22\lambda \times 206265}{d} \text{ arc sec}$$

where λ is the wavelength of the incoming light in nanometers and d is the diameter of the objective lens or mirror in metres. For example: the resolution of a 6in (15cm) diameter reflecting telescope at 550 nanometers is given by

$$\frac{1.22 \times 550 \times 10^{-9} \times 206265}{6 \times 2.54 \times .01}$$
$$= 0.9 \text{ arc seconds}$$

The magnifying power is determined by the ratio of the focal length of the objective lens or primary mirror to the focal length of the eyepiece being used. That is, virtually any magnification may be used on any telescope given the appropriate eyepiece. An example would be: assume an f8, 6in (15cm) primary mirror. The focal length is $8 \times 15 = 120$cm = 1,200mm. Insert a 25mm eyepiece and the effective linear magnification would be 1200/25 = 48×. Note that if a 4mm eyepiece was used the magnification would increase to 1200/4 = 300 times, generally well beyond the capability of the telescope.

Telescopes are also used at wavelengths other than that of light, eg, RADIO TELESCOPES. Here the collecting element may be a dish, the analogue of the mirror in the optical telescope, but equally may be of some other design depending on the wavelength of operation. And the detector is frequently a solid state electronic device. This also applies to infrared telescopes, X-ray telescopes and telescopes operating from gamma rays to millimetre waves. *JCDM. See also* ALTAZIMUTH MOUNTING, EQUATORIAL MOUNTING, EYEPIECES, LENS, OBJECTIVE, REFLECTING TELESCOPES, REFRACTING TELESCOPES.

Telescopic drives The mechanisms by which telescopes are moved quickly from one part of the sky to another (slewed) or moved slowly to follow stars on their sidereal motion (tracked).

In its simplest form, a telescope is slewed by being pushed. The telescope reaches the appropriate direction, toward the star or planet of interest. The telescope is then clamped by friction to the mounting and the star or

planet drifts slowly through the telescope's field of view. As it does so the telescope is tracked manually to reacquire the object.

All telescope drives are sophisticated and accurate versions of this process.

The most essential item to add to the simple process is a tracking motor. This motor turns the telescope about the polar axis of an EQUATORIAL MOUNTING. It drives at a constant speed, geared down to the equivalent of one revolution in 23hr 56m (once per sidereal day) or 15 arc seconds per sidereal second. Early forms of the motor used clockwork to provide both the power and the speed control.

As telescopes got heavier there was not enough power in a clockwork mechanism and falling weights were used — the weight was cranked by hand at the beginning of the night to the top of its fall. The rate of fall was initially controlled by a governor. This regulating device is a pair of balls pivoted on arms so that they can rise and fall. The arms are rotated about a vertical axis. If the rotation speed increases above nominal, centrifugal force flings the balls outward and they rise, rotating the arms about the pivots and thereby working a frictional device which slows the rotation speed back to nominal. The governor has to be set to regulate the rotational speed to the nominal rate. Fraunhöfer invented a friction regulator of this kind for the telescopes which he made. The rough regulation of a governor was improved by a pendulum clock. It is not uncommon still to find such mechanisms in use on old but working telescopes.

The telescope position has been traditionally read from large scales like protractors attached to the telescope axes. These "circles" measure the HOUR ANGLE and DECLINATION to which the telescope points. In order to convert hour angle to RIGHT ASCENSION, the telescope is pointed at the beginning of the night to a star whose position is accurately known (a "clock star") and the hour angle circle rotated until it reads the correct right ascension. The right ascension circle is then fixed through a clutch mechanism to the telescope tracking motor, in order to advance it at the sidereal rate.

At the turn of the 20th century electric motors replaced the falling weight as a source of power for the drive motion, but pendulum clocks were still used as regulators for the speed control. Tuning forks and valve-type oscillators enjoyed a brief supremacy, until replaced by the present state-of-the-art quartz crystal oscillators. These represent the ultimate requirement in clocks used for telescope drives, since a telescope regulated by one can in principle follow a star unwaveringly from horizon to horizon. If the mains AC power supply available at the telescope is well regulated by such a clock, then synchronous motors provide almost as good a telescope drive, although the frequency of mains AC is liable to fluctuate as the power company sheds load at times of high demand. The ease of use of such motors and their relative accuracy makes them very suitable for moderate-sized amateur telescope drives.

Whilst the clock is the ultimate limit to the accuracy of a telescope drive, the clock motion is communicated to the telescope by mechanical connections, and the telescope motion is only as smooth as these connections are accurate. A typical arrangement is to have the motor drive a worm, that is to say a spiral cut on the surface of a cylindrical rotating shaft. The worm engages the teeth of a large gear attached to the polar axis. The gear teeth must be positioned relative to one another at the accuracy with which it is desired to track the stars, in the ultimate to a fraction of the size of the seeing disk of the stars (about 1 arc second). The gear teeth and worm are matched together and ground to suit each other. The gear must be preloaded to stop the telescope "floating" back and forth as first one and then the next tooth touches the worm. To represent the mid-20th-century state of the art, the Palomar 200in (5m) telescope built between about 1930 and 1947 used worm wheels 14ft (4.3m) in diameter driven by synchronous motors controlled by a quartz crystal clock.

The residual errors of the drive mechanism have to be eliminated by the astronomer, who views a star in an EYEPIECE and advances or retards the telescope by pressing buttons on a "guide paddle". These last guiding motions are necessary to compensate for the flexure of the telescope and its mounting as its position changes, for atmospheric refraction (which raises the position of stars in the sky by an amount which depends on their direction) and for errors in the gear construction.

In fact the imperfections of the telescope can be discovered empirically in a "pointing test" by observing numerous stars well distributed over the sky and finding by how much the telescope misses them. The most

advanced telescope drives available at the close of the 20th century use computers to control the telescopes and can take account of the telescope imperfections, representing them by carefully chosen formulæ. The first widely acclaimed successful computer-controlled telescope was the Anglo-Australian Telescope in 1974. The fact that such computer systems can be and have been added to older telescopes like the Palomar telescope is a tribute to the original standard of their mechanical construction and of their drive systems.

In a computer-controlled telescope drive, the computer is set to calculate the desired position of the telescope at a rapid rate. In the UK-NL telescopes of the ROQUE DE LOS MUCHACHOS OBSERVATORY, the control computers read quartz clocks ten times each second. The computer then determines for that time the hour angle and declination of the star which it has been instructed to follow, modifying them for atmospheric refraction and for the telescope imperfections. The computer compares the position of the telescope which it has calculated with the actual position. It reads the telescope position from encoders, of the type used to control machine tools in automated workshops. The encoders are attached to the two telescope axes, and replace the traditional "circles". If they do not read correctly, the telescope is accelerated or decelerated accordingly.

It is possible in principle to attain an accuracy of 1 arc second in the pointing of a telescope by these methods, so that the image of the star of interest always falls on the fiducial mark in the telescope: however, this standard has not quite been attained in practice (as of 1986). The final adjustment of the telescope position is made by inspection of the star field by eye as before, or by measurement of the position of a star in the field, using image dissectors. These devices, called fine error sensors in space technology, photoelectrically sense the star's position and, if it drifts off centre, create error signals which are fed to the telescope control motors. In this way the drive is always adjusted to track the star of interest. *PGM*.

Telescopium (the Telescope). A faint constellation near Ara, with no star above magnitude 3.5; it contains little of interest.

Telesto Small satellite of Saturn; 21 × 17 × 16 miles (34 × 28 × 26km) in size. It is co-orbital with Tethys.

Telstar Series of two American communication satellites launched in 1962–63. They were placed in elliptical orbits with a minimum height of about 600 miles (1,000km). Because of the fairly low height, the ground station antennæ had to track the satellite as it crossed the sky, thus requiring a series of satellites to give a continuous communications signal. It could carry 600 telephone circuits or one television channel. Telstar 1 carried the first live transatlantic television broadcast linking the United States with the United Kingdom and France. It was replaced by geostationary satellites.

A series of geostationary satellites under the name Telstar was begun in 1983. *HGM*.

Telstar: first satellite link

Tempe Martian region east of Acidalia Planitia, associated with the 730-mile (1,180km) "ditch" Tempe Fossa (35° to 50° S, 82° to 62°W).

Tempel 1 Comet A short-period comet with a period of only 5.49 years, discovered in 1867 by the German astronomer E.W. Tempel. It is always faint and was lost from 1879 until 1967. The perihelion distance is 1.49AU, aphelion distance 4.73AU, orbital eccentricity 0.52 and inclination 10°.55. It was studied from the IRAS satellite in 1983, but no dust-tail was found.

Tempel-Tuttle Comet Periodic comet discovered by German astronomer Ernst W. Tempel on December 19 1865. It came to perihelion in January 1866, and was known as comet 1866 I. It was independently discovered by Horace Tuttle. The comet has a period of 32.9 years, and the elliptical orbit is retrograde with an orbital inclination of 162°.7, an eccentricity of 0.904, and a perihelion distance of 0.982AU. It last

T

returned to perihelion in 1965 and is due back in 1998. The comet is associated with the Leonid meteors: spectacular Leonid meteor storms have occurred near to the times of the comet's perihelion passage. The last such storm was on November 17 1966.

Tempel-2 A short-period comet, discovered in 1873. Observations from IRAS in 1983 showed that Tempel-2 had a hitherto undiscovered dust tail, more than 60,000,000 miles (100,000,000km) long.

Temperature scales The two principal scientific temperature scales are the absolute (or Kelvin) scale and the Celsius, or Centigrade, scale. The unit of temperature on the absolute scale, the Kelvin (K), is equal in magnitude to one degree on the Celsius scale (°C). Zero Kelvin corresponds to −273.16°C (*see* ABSOLUTE ZERO). To convert from Kelvin to Celsius, substract 273.16; to convert from Celsius to Kelvin, add 273.16.

Tereshkova, Valentina Vladimirovna (1937–). Soviet cosmonaut and the first woman to make a space flight. Between June 16 and 19 1963 she made 48 orbits of the Earth aboard the Soviet Union spacecraft Vostok 6.

Valentina Tereshkova: 1st woman in space

Terrestrial planets Mercury, Venus, the Earth and Mars; Pluto is not now generally included in this group.

Tethys A satellite of Saturn. At its mean distance of 183,100 miles (294,700km) from the planet, it was the third in order of the satellites known before the Space Age; its orbit lies between those

of ENCELADUS and DIONE. Its revolution period is 1.89 days.

Tethys is 650 miles (1,050km) in diameter, slightly smaller and considerably less massive than Dione; its density is only 1.1 times that of water, so that it must consist largely of ice. The surface is cratered; one crater, Odysseus, is 250 miles (400km) in diameter. The main feature is, however, Ithaca Chasma, a trench 1,250 miles (2,000km) long, running from the north pole across the equator through to the south polar region. Its average width is 60 miles (100km) and it is 2.5 to 3 miles (4 to 5km) deep. Tethys has two small co-orbital satellites, CALYPSO and TELESTO. *PM.*

Thales of Miletus (*c.* 625–547BC). Greek polymath, famous for his teaching that water was the fundamental material of the universe.

Tharsis ridge The main volcanic region of Mars. On it stand Arsia Mons, Pavonis Mons and Ascræus Mons.

Tharsis Ridge: Martian volcanic region

Thaumasia Martian region south of Solis Planum. It is associated with the 500-mile (802km) "ditch" Thaumasia Fossa (33° to 45°S, 100° to 85°W).

Theætetus Lunar crater on ·Palus Nebularum 16 miles (26km) in diameter; 37°N, 6°E. It has a low central peak.

Thebe The 15th satellite of Jupiter, discovered in 1980. Its orbit lies between those of Amalthea and Io. Thebe is no more than 47 miles (75km) in diameter, and is of magnitude 20.

Theia Mons (Venus). One of two radar-bright mountains situated in Beta Regio. It rises to about 3 miles (4.5km) above datum and is considered to be a volcanic shield.

Themis (Asteroid 24). Discovered in

1853. The name was also given to a satellite of Saturn reported in 1904 by W.H. Pickering, which, however, has never been confirmed and certainly does not exist.

The Norc (Asteroid 1625). Named after the Naval Ordnance Research Calculator at Dahlgren, Virginia. Its magnitude is 14.

Theon Theon Junior and Theon Senior are two lunar craters, 10 miles (16km) in diameter, near Delambre; both have bright walls.

Theophilius Magnificent lunar crater, 63 miles (101km) in diameter; 12°S, 26°E; it forms a chain with Cyrillus and Catharina. It has high, massive walls and a complex central mountain group.

Theophilus: part of a lunar crater chain

Thermal equilibrium The state in which all parts of a system have the same temperature.

Thermonuclear reactions NUCLEAR REACTIONS in which nuclei of light elements fuse together to build heavier elements, releasing energy in the process. Nuclei are very tightly bound and highly charged, so that very high velocities of collision are necessary for them to fuse. This in turn implies extremely high temperatures of the order of ten million degrees. There are two regimes where appropriate conditions exist or existed: the interiors of stars, and the universe itself, seconds to minutes after the BIG BANG, when the whole of the universe was a fusion reactor. Appropriate conditions have been created by man, in the hydrogen (fusion) bomb, and in successful fusion-reactor experiments exploring the peaceful generation of thermonuclear energy.

A star has two sources of energy. It can convert its gravitational energy into heat by contraction, the outer layers falling in to compress and warm the centre, or it can burn its own fuel. Geological records indicate that the Sun has shone with its present output for at least 4.5×10^9 years. Kelvin and Helmholtz demonstrated in the 19th century that the thermal energy which the Sun could generate from its gravitational energy would suffice for less than one per cent of this. If the entire Sun were burning in any *chemical* reaction the energy produced would fall short by a factor of one million or more. Such estimates led Perrin in 1919 and Eddington in 1920 to propose that the Sun burnt its own atomic fuel by nuclear reactions. The prodigious energy release is more than adequate: if the entirety of the Sun were hydrogen converting to helium at a rate set by the present radiation, the total lifetime would be 10^{11} years.

From the point of view of energy production (starlight; sunlight), the most important thermonuclear processes fuse hydrogen to helium. It was 1938 before the details of the reactions were discovered by Bethe, von Weizsäcker and Critchfield. There are two basic processes; the PP cycle (proton-proton reaction; *see* PROTON-PROTON CHAIN), and the CNO (carbon-nitrogen-oxgen) cycle. In the former, the nuclei of four hydrogen atoms, ie, four protons, fuse together (in any one of three building processes) to form a helium nucleus (two protons and two neutrons, an ALPHA PARTICLE), with the release of energy in the form of PHOTONS and NEUTRINOS. In the second process, the CNO cycle, hydrogen nuclei (mass number 1) combine initially with carbon nuclei (mass number 12) to build up ISOTOPES of nitrogen and oxygen, until in the final component of the cycle, hydrogen combines with a nitrogen isotope of mass number 15 to produce a helium nucleus (mass number 4) and the original carbon isotope (mass number 12). Other thermonuclear reactions in stellar cores are vital in synthesizing the heavier elements up to iron, as elucidated by Burbidge, Burbidge, Fowler and Hoyle in 1957. These reactions contribute negligibly to the energy budget of a star except during the short, final and explosive stages of its evolution.

In a typical star such as the Sun, only about 12 per cent of the mass — the core region — is involved in nuclear reactions. The remaining mass stabilizes the reaction rate so that the star does not self-destruct. Slight contractions of the core heat it, increasing the reaction rate and energy output, caus-ing increased pressure which opposes the original contraction; and vice versa for slight expansions. The energy radiated at the surface is equal to that produced in the core; if the core is not producing enough, slight contraction takes place in which gravitational energy from the (non-reacting) mass of the star heats the core, increasing the reaction rate to make up the deficit. In this way a star is stable during the hydrogen-burning in its core, and it lives perhaps 80 per cent of its life in this way — the MAIN SEQUENCE phase of STELLAR EVOLUTION. The PP cycle is the dominant mechanism for most stars, while on the main sequence — it accounts for 99 per cent of the Sun's output. The CNO cycle comes into play for massive hot stars on the main sequence, and for stars of normal mass when they have completed burning the core hydrogen and begin to burn the core helium, evolving from the main sequence into the RED GIANT phase. The shell around the helium-burning core continues to burn hydrogen, but now at such a high temperature as to favour the CNO cycle. Most of what we see in the universe, however, stars, clusters of stars, galaxies, we see by means of starlight generated in the PP reaction.

Thermonuclear reactions in the early universe took a somewhat different course. Immense temperatures and densities were present immediately after the Big Bang, but nuclear reactions could not commence until neutrons, stable particles under the initial extreme conditions, began to disintegrate to create quantities of protons. This happened after about 60 seconds, and in the next two minutes protons and neutrons fused to deuterium (mass number 2) nuclei, which subsequently fused with each other to create helium (mass number 4), together with small leftover quantities of lithium (mass number 7), the lighter helium isotope (mass number 3), and some deuterium. The universe continued to expand; falling temperature and density stopped the reaction after about three minutes. Within this time the universe operated as a fusion reactor burning 25 per cent of its mass into helium.

Measurements of helium abundance in all types of object consistently determine mass fractions of 20 to 35 per cent; nothing showing a significantly lower value has been found. Stars cannot synthesize more than a percent or two during the lifetime of the universe, and they cannot destroy helium. It is thus a major triumph of Big Bang cosmology (and of thermonuclear reaction theory) that a primordial helium abundance in the range 20 to 30 per cent is predicted. In recent years the calculations have been refined. The precise amount of primordial helium depends on the density of nuclear matter (the baryonic density), the velocity of expansion of the universe, and the number of neutrino types. The best measurements of primordial helium come from primitive (blue irregular) galaxies, currently giving a fraction by mass of 0.24. With present estimates of expansion velocity this yields an estimate of the density of baryonic matter as one-tenth of the critical density required to close the universe. Such a density is below that required for gravitational stability of clusters of galaxies. The implication is that much of the matter of the universe is non-baryonic and invisible (dark matter). The present estimate of primordial helium fraction also limits the number of neutrino types to three. Three are already known; the discovery of more neutrino types will endanger the currently accepted Big Bang model.

Thermonuclear reactions thus provide us with a visible universe, and with theoretical and observational physics to determine the kind of universe in which we live. *JVW.*

Thermosphere The layer in the Earth's atmosphere above the mesopause (height 53 miles/85km). The temperature rises with height as the Sun's far-ultraviolet radiation is absorbed by the oxygen and nitrogen of the thin air. This process produces ionized atoms and molecules. It also causes the formation of the layers in the ionosphere between altitudes of 37 and 310 miles (60 and 500km). This is also the region in which auroræ and meteors occur. Within the thermosphere, the temperature climbs to 500°C at 90 miles (150km) and about 1,300°C at 500km. However, the atmospheric density is extremely low, decreasing from seven-millionths of its sea-level value at the mesopause to only about one-million-millionth at 500km altitude, although the density does vary with solar activity. Above this height lies the exosphere. *JWM.*

Theta Carinæ B-type star in Carina, magnitude 2.76, associated with a bright open cluster.

Theta Orionis Multiple star in the ORION NEBULA (M42). Its four main components are so arranged that it has been nicknamed the Trapezium. The stars are very hot, and are responsible for the luminosity of the nebula.

Three-body problem The problem of computing the motion of three bodies moving under the action of their mutual gravitational attractions. This problem has no exact analytical solution, but the motion of the three bodies can be computed numerically to the required level of accuracy.

3C147 (0538 + 49). Strong, compact radio quasar; $z = 0.545$.

3C295 (1409 + 52). Strong radio galaxy; $z = 0.5$.

3C273 (1226 + 02). First-identified and brightest quasar; $z = 0.158$.

3 kiloparsec arm The spiral arm lying about 3.5kpc from the galactic centre, and visible as 21cm radio emission within about 20° longitude from it. It is expanding outward at about 30 miles/s (50km/s).

Thuban The star Alpha Draconis, magnitude 3.65. In Egyptian times it was the north pole star.

Thyle Chasma Linear chain on Mars; 69° to 73°S, 230° to 235°W, length 146 miles (235km).

Tides The rise and fall of the sea due to the attractions of the Moon and, to a lesser extent, the Sun. The pull of the Moon causes a rise in the waters

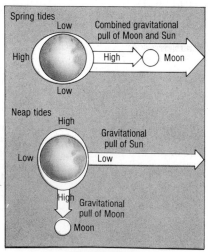

Tides: effects of Sun and Moon

beneath it and a corresponding one on the far side of the Earth, where the Moon's pull is weakest. As the Earth rotates, these bulges sweep around it

causing two high tides and two low tides in between, in just over a day. The bulges are not directly in line with the Moon due to the frictional effects of the land masses.

When the pulls of the Sun and Moon are aligned at new and full moon the combination magnifies the effect to give higher amplitude *spring* tides: at QUADRATURE, the effects partially cancel each other out, giving lower amplitude *neap* tides (actually, these phenomena occur just after SYZYGY and Quadrature).

Very small tidal effects have been measured in the Earth's crust. Similar effects occur in the Moon, other satellites, close binaries, etc. *TJCAM.*

Tikhonravov, Mikhail Klavdievich (1900–74). Soviet rocket engineer who, with Sergei Korolev, launched the first Soviet liquid-fuelled rocket on August 17 1933.

Timaeus Bright lunar crater, 21 miles (34km) in diameter, on the edge of Mare Frigoris; 63°N, 1°W.

Time dilation effect According to the theory of RELATIVITY, time recorded on a clock moving at velocity v relative to an observer will pass more slowly than time recorded on a clock which is stationary (relative to that observer) by a factor of $\sqrt{(1 - v^2/c^2)}$, where c denotes the speed of light. The closer the moving clock approaches the speed of light, the slower time passes. For example, during one hour recorded on the stationary observer's clock, 0.44 hours would pass on a clock moving at $0.9c$, while only 0.04 hours would elapse on a clock moving at $0.999c$. Thus if an astronaut were to travel to and from a star 12.5 light-years distant at a constant speed of $0.999c$ (ignoring acceleration and deceleration times), 25 years would have elapsed on Earth, but the astronaut would have aged by only one year. Relativity also predicts a gravitational time dilation, whereby time passes more slowly in a strong gravitational field than in a weak one. Both kinds of time dilation have been confirmed experimentally. *IKN.*

Time-scales A time-scale is a scheme for associating with each instant of time a corresponding number, later instants being assigned larger numbers. Most time-scales used in astronomy are based either on the rotation or orbital motion of massive bodies or on a continuous count of units of time interval defined by atomic resonators. Such

time-scales are almost uniform, in the sense that equivalent physical processes occur in identical intervals on the scale; other scales, based for example on radioactive decay or geological processes, may be highly non-uniform.

Important qualities of a time-scale are the frame of reference in which it is defined, and its uniformity, continuity, accessibility, precision and accuracy.

It is now often more appropriate to use INTERNATIONAL ATOMIC TIME (TAI) as a fundamental time-scale for use on or near the Earth and to think of UNIVERSAL TIME, for example, as a measure of the Earth's rotation rather than a time-scale in its own right. In precise astronomical work it may be necessary to take account of non-uniformities of up to 1.7ms produced in TAI by the varying gravitational potential of the Earth (and the atomic clocks) with respect to the BARYCENTRE of the Solar System. *JDHP.*

Time zones Division of the world into 24 zones, each 15° (or 1h) of longitude wide. Since the 1880s, used as the basis of the STANDARD TIME system. *See also* GREENWICH MEAN TIME.

Timocharis Lunar crater 22 miles (35km) in diameter, on the Mare Imbrium; 170°N, 13°W. It is a minor ray-centre.

Tiros satellites The first purpose-built weather satellites. They monitored heat flow from the Earth into space and photographed cloud cover.

Tir Planitia Plain on Mercury; 3°N, 177°W. It lies south of the Caloris Basin.

Titan The largest satellite of the Saturn system, discovered by Christiaan Huygens on March 25 1655. It has a density of 1.88g/cm³ and a diameter of 3,200 miles (5,150km) and is therefore just slightly smaller than Ganymede of the Jovian system. It orbits Saturn at a mean distance of 760,000 miles (1,222,000km) and is situated at the edge of the planet's magnetosphere so that its position is sometimes inside and then occasionally outside this protective region.

Titan has a synchronous period of 15.95 days and an escape velocity of 2.5km/s. The satellite appears reddish-orange in colour with a small but noticeable difference between the hemispheres. It has an extensive atmosphere of mainly nitrogen and traces of methane, ethane, acetylene, propane, diacetylene, methylacetylene, hydrogen cyanide, cyanoacetylene, carbon dioxide and carbon monoxide. Titan and the Earth have the only currently known atmospheres in the Solar System to be composed primarily of nitrogen. The photodissociation of the atmospheric constituents produces hydrogen, which then escapes into space so that Titan is surrounded by a doughnut torus of hydrogen atoms which extends from 8 to 25 R_s. The outer edge of the magnetosphere prevents this gas cloud extending any farther beyond the Titan orbit.

The surface pressure is 1.6 atmospheres and the surface temperature is 92K, so that it is warmed by a small greenhouse effect. The coloration of the atmosphere is due to photochemical reactions between the gaseous, fog and aerosol particles which extend from near the surface to an altitude of about 190 miles (300km). The atmosphere is opaque to all radiation except

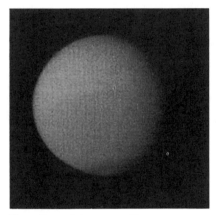

Titan: Saturn's largest satellite

those at radar wavelengths and therefore obscures the surface, which may be covered in methane ices and oceans of ethane.

Titan does not appear to possess an intrinsic magnetic field and may not, therefore, have an electrically conducting core. The extraordinary satellite is the prime objective for the Cassini mission, which will drop an instrumented probe to examine the atmosphere and surface. *GEH. See also* colour essay on MOONS.

Titania The largest satellite of Uranus; diameter 982 miles (1,580km). It has an icy; cratered surface, as was shown by Voyager 2 in 1986; there are also spectacular linear troughs or fault valleys hundreds of miles long.

There is some evidence of the emplacement of smoother material from the interior in association with some of these fractures.

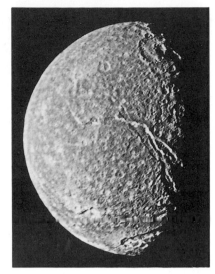

Titania: Uranus' largest satellite

Tithonia Catena Crater-chain on Mars; 5° to 6°S, 80° to 87°W. It extends from Valles Marineris. Nearby is the Tithonium Chasma, 550 miles (880km) wide.

Titius – Bode Law. *See* BODE'S LAW.

Titov, Gherman Stepanovich (1935–). Soviet cosmonaut who piloted the Vostok 2 spacecraft in August 1961, so becoming the second man to orbit the Earth.

Tokyo Astronomical Observatory Japan's national observatory, founded in 1888 as an institute of the University of Tokyo, the country's first national university, which was founded 11 years earlier. It is still a research institute of the university with the duties of the national observatory. Its main office is at Mitaka, Tokyo, 12 miles (20km) west of the centre of the city; it was moved there in 1920 from Azabu, Tokyo, where the Japanese datum point remains. There are about 70 research staff at the observatory and they are also responsible for teaching graduate students of the university.

After the observatory was moved to Mitaka, new instruments such as a meridian circle, transit instruments, a 26in (65cm) Zeiss refractor and a solar tower telescope, as well as other solar observing facilities, were installed there in 1920-30. Some of them are still in use, but most of the modern instruments were installed not at Mitaka but, as astronomical observing conditions at Mitaka became worse, particularly after 1950, at several branch observatories far from Tokyo.

As the national observatory, Tokyo is responsible for producing astronomical ephemerides and time-keeping, which is undertaken at Mitaka, where solar monitoring and bright star observations are also routinely made with the 65cm refractor, a 4in (10cm) photoelectric meridian circle, a photographic zenith tube and a 5½in (14cm) monochromatic heliograph as well as other solar observing devices. There is also a high-speed computer and other instruments for handling data.

The first telescope installed outside Mitaka was a 4in (10cm) coronagraph made in 1950. It is located on Mount Norikura, 9,435ft (2,876m) high and with a 10in (25cm) coudé-type coronagraph made in 1972 has been used for solar observations.

The largest optical telescope the observatory has is a 74in (188cm) reflector, made by Grubb Parsons in 1960, at Okayama Astrophysical Observatory. Together with a 36in (91cm) reflector and a 24in (60cm) solar telescope it has been used by not only the staff of the observatory but also by astronomers of other universities and institutes. At Kiso Observatory a 41in (105cm) Schmidt telescope made

T

Tokyo Astronomical Observatory

in 1974 is used by many Japanese astronomers. Airglow observations are made routinely at Kiso.

At Dodaira Observatory a 36in (91cm) reflector, a 20in (50cm) Schmidt telescope and satellite tracking facilities have been operating since 1962.

Radio observations of the Sun began at Mitaka in 1949, but were later moved to Nobeyama, 93 miles (150km) north-west of Mitaka. There, a 17GHz multi-correlator interferometer and a

160MHz interferometer as well as a 17GHz polarimeter and others have been in use.

A new observatory for cosmic radio observations was inaugurated at Nobeyama in 1982 with a 148ft (45m) millimetre wavelength telescope and a 5-element, 33ft (10m) dish interferometer, the surface accuracy of the 45m dish being as high as 0.15 millimetres in rms. The facilities are open to all Japanese astronomers and many others have used them to observe the interstellar medium, the centre of the Galaxy, dark nebulæ and so on.

Besides conducting ground-based observations, several staff members of the observatory have engaged in space science activities by joining projects of the Institute of Space and Astronautical Science involving satellites, rockets and balloons. *YK*.

Tolstoy Crater on Mercury, 250 miles (400km) in diameter; 15°S, 165°W. Apart from Beethoven, it is the largest named crater on the planet.

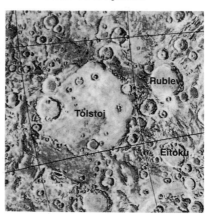

Tolstoy: second largest crater on Mercury

T Orionis variables Irregular eruptive stars with abrupt Algol-like fadings. Like other Orion variables, they are found connected to bright or dark diffuse nebulæ.

Toro (Asteroid 1685). An Apollo, discovered by A. Wirtanen at Lick Observatory on July 17 1948. Its orbit is remarkable for a resonance involving both Earth and Venus. Physical observations have shown it to be an S-type asteroid of 3 miles (5km) diameter, rotating in 10.2 hours, and having a large light curve amplitude of 0.8 magnitude.

Torricelli Irregular, compound formation on the lunar Mare Nectaris; 5°S, 29°E. It is no more than 12 miles (19km) in diameter.

Transfer orbit The path followed by a satellite or spacecraft in moving from one orbit to another, for example from low Earth orbit to a higher one, or from the Earth to another planet.

Transient Lunar Phenomena (TLP). Localized, short-lived glows or obscurations on the surface of the Moon. They have been reported by most experienced lunar observers, both professional and amateur; they are most frequent near the time of lunar perigee, and in areas associated with clefts or rilles. The area which has shown the most TLP is that of the brilliant crater ARISTARCHUS. The first TLP spectrum, of an event in Alphonsus, was secured by N.A. Kozyrev at the Crimean Astrophysical Observatory in 1958. It is generally thought that TLP are due to gaseous emissions from below the visible surface of the Moon.

Transient Lunar Phenomena: gas emissions

Transit The passage of a celestial body across an observer's meridian. The term also describes the passage of a smaller body across the visible disk of a larger one (eg, the passage of Mercury or Venus across the face of the Sun).

Transit instrument Telescope specially mounted for observing the precise moment a body crosses (transits) the meridian.

The simple transit instrument consists of a telescope fixed at right angles to a horizontal axis and free to rotate upon pivots on two fixed piers. The telescope can be moved up and down, but not from side to side, so that its optical axis will always be in the plane of the observer's meridian. The transit telescope has several horizontal and vertical "wires" — actually they are usually spiders' threads set at the telescope focus — which can be seen in the field of view.

Most transit instruments are fitted with a setting circle or semicircle, and have some means of illuminating the

crosswires at night. A few have some apparatus for counterbalancing the moving parts so that the weight pressing down on the pivots, and the resulting friction, is kept to a minimum.

The invention of the transit instrument is ascribed to Ole RØMER, who first used such an instrument in Copenhagen in 1689. The first to be used in England was set up by Edmond HALLEY at Greenwich in 1721, where it can still be seen.

For positional astronomy, the transit instrument was used with a clock to measure RIGHT ASCENSION, and it was necessary to have a separate meridian instrument to measure ZENITH DISTANCE to obtain DECLINATION. The development early in the 19th century of graduated instruments with a large-radius full circle (as opposed to, say, a quadrant) made it possible to measure both co-ordinates with one instrument, called the transit circle or meridian circle. Thereafter the simple transit instrument was generally used only for time determination.

The Greenwich instrument (Airy's transit circle), which since 1884 has defined the world's prime meridian, was brought into use in 1851. *HDH*.

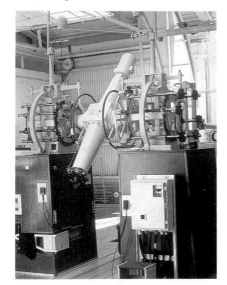

Transit instrument: a 6in version

Transits of inferior planets Since their orbits are slightly inclined to the ecliptic, Mercury and Venus usually pass north or south of the Sun at inferior conjunction; but if conjunction occurs when the inferior planet is near its node, then from Earth it is seen as a small, dark spot moving from east to west across the Sun's disk along a path sensibly parallel to the ecliptic. This is known as a transit. There are two phases at the start (ingress), and two at the close (egress), making four in all. These are: (i) first exterior contact, when the planet first appears to touch the Sun's edge or limb; (ii) first internal contact, the planet is fully upon the Sun but still contiguous with its inner edge; (iii) second internal contact, when the planet, its traverse complete, first touches the inside edge of the opposite limb of the Sun; and (iv) second external contact, the moment when the planet finally disappears. In the past, transit observation was a method of determining the ASTRONOMICAL UNIT.

Transits of Mercury occur around May 7 and November 9. At May transits the planet is near its aphelion, and much nearer the Earth than normal, hence the transit limit is less than for a November event, when both the Earth and planet are close to the Sun; accordingly November transits are twice as numerous. The interval between successive events varies from three to·13 years; transits are due in 1993 and 1999. Kepler first drew attention to Mercury transits in 1627, and predicted that of November 1631.

Unlike those of Mercury, Venus transits are rare, and visible to the naked eye. They occur a day or so either side of June 7 and December 8. The next are due in 2004 and 2012. *RMB.*

Transmission grating A diffraction grating made from transparent material, on the surface of which are closely spaced, parallel grooves. As light is transmitted through the grating it diffracts into separated colours. *See also* REFLECTION GRATING.

Transverse velocity Velocity perpendicular to the observer's line of sight.

Transverse waves Waves which vibrate at right angles to their direction of propagation, eg, water waves, light waves.

Trapezium *see* THETA ORIONIS.

Treptow Observatory East German observatory now known officially as the Archenhold Observatory in honour of its first Director, F.S.Archenhold, who held office from 1891 to 1931. It lies not far from Berlin, and is mainly educational; it includes an unusual refracting telescope with a 27in (70cm) objective and a focal length of 69ft (21m). Visitors annually exceed 40,000.

Triangulum (the Triangle). A small but easily-found constellation between Andromeda and Aries.

Brightest stars

Name	Visual Mag.	Abs. Mag.	Spec.	Distance (light-yrs)
β	3.00	0.3	A5	114
α (Rasalmothallah)	3.41	2.2	F6	59

These form a well-marked triangle with Gamma (4.01). The main feature of interest is the spiral galaxy M33, not hard to find with binoculars but apt to be elusive with a small telescope. *PM.*

Triangulum: between Andromeda and Aries

Triangulum Australe (the Southern Triangle). A well-marked constellation near Alpha Centauri; the three leading stars do indeed make up a triangle.

Brightest stars

Name	Visual Mag.	Abs. Mag.	Spec.	Distance (light-yrs)
α	1.92	−0.1	K2	55
β	2.85	3.0	F5	33
γ	2.89	0.6	A0	91

Next comes Delta (3.85). The open cluster NGC6025 is visible with binoculars.

Triesnecker Lunar crater, 14 miles (23km) in diameter; 4°N, 4°E, in the Mare Vaporum area.

Trifid Nebula An emission nebula in the constellation of Sagittarius which is apparently divided into three sectors by dust lanes.

Trifid Nebula: divided by dust lanes

Trigonometrical parallax Determinant of stellar distance. As the Earth revolves around the Sun the directions to nearby stars change relative to the distant, background stars. The angular displacement of a nearby star as seen from opposite ends of the diameter of the Earth's orbit at right angles to the direction from the Sun to the star, is a direct measure of the star's parallax. The parallax (π) is defined as the angle subtended at the star by the radius of the Earth's orbit, which is 1 Astronomical Unit (AU). Hence, in the diagram on the next page,

$$\tan \pi = \frac{1 \text{AU}}{d},$$

where d is the distance between the Sun and the star. Since π is always a small angle and is usually measured in seconds of arc ("), we may write

$$\frac{\pi}{206265} = \frac{1 \text{AU}}{d}.$$

A convenient unit of distance is the parsec (pc), which is the distance at which a star would have a parallax of 1″. In this case, 1 pc = 206,265 AU = 3.086×10^{13} km, Hence,

$$d = \frac{1}{\pi} \text{pc}.$$

The star Alpha Centauri has the largest known parallax of 0″.75, which is equivalent to a distance of 1.3pc.

In practice, the parallax of a star is measured from several photographic plates taken about six months apart. Allowance is made for the proper motion of the star (*see* STELLAR PROPER MOTION) and the possibility that some of the background stars may have significant parallaxes. Parallax is "absolute" when it is corrected for the parallaxes of the background stars.

Typical errors in measuring parallax photographically are about ± 0″.01, but by using a large number of plates and modern techniques it is possible to reduce this error to around ± 0″.004. With CHARGE-COUPLED DEVICES (CCD) fitted to large-aperture telescopes it is possible to attain an accuracy of ± 0″.002 for stars as faint as the

Parallax catalogues

Catalogue	Author	Number of stars
Yale Parallax Catalog (new edition nearing completion 1987)	Van Altena et al.	~6,000
Catalogue of Nearby Stars (within 20pc)	Gleise (1969) Veröff Rechen-Inst. Heidelberg, 3, No. 22	900
Catalogue of Stars within 25pc of Sun	Woolley et al. (1970) Royal Gr. Obs. Ann. No. 5	1,700

T

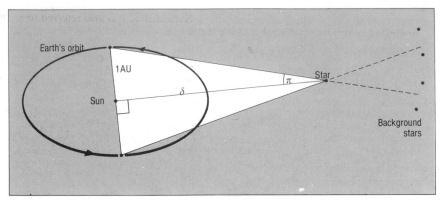

Trigonometrical parallax: star distances

twentieth magnitude. The HIPPARCOS satellite, to be launched by ESA in 1988, will measure the parallaxes of about 100,000 stars to $\pm 0.''002$.

The parallax of a star at 25pc is $0.''04$; so with the best photographic determinations (which account for the great majority) the uncertainty is around 10 per cent. Stars within this distance-limit have been collected into special catalogues. *LVM.*

Triple-alpha process A nuclear reaction that converts helium to carbon by the successive combination of three alpha particles (helium nuclei). In each reaction two helium nuclei combine to form a beryllium nucleus which, in turn, captures a further helium nucleus to form a carbon nucleus. Additional helium captures can produce oxygen, neon, and a number of heavier elements. The reaction takes place at temperatures in excess of 100 million K. When a star exhausts the hydrogen fuel in its core, the core contracts and its temperature rises until the triple-alpha reaction is initiated. The reaction is believed to be the dominant energy-generating process in RED GIANTS. *IKN.*

Triple point The temperature and pressure at which a substance can exist as a solid, liquid and gas.

Triton The larger satellite of Neptune. Its diameter is uncertain, but it is probably larger than our Moon, and may have a tenuous atmosphere; it has been suggested that its ocean may include seas of liquid nitrogen. It is unique among large planetary satellites in having retrograde motion.

Trojan asteroids Asteroids that have orbits very similar to Jupiter and are positioned around the leading and following LAGRANGIAN POINTS, 60° in front of and behind the planet. They are named after the heroes of Homer's

Iliad, the leading set being the Greek warriors and the followers the defenders of Troy. Over 220 are known, there being 3.5 times more at the leading point than at the following one. Their mass distribution is similar to main-belt asteroids, indicating that they have been at their present location since condensation from the solar nebula. Saturnian "Trojans" have not been found and have probably been removed by Jovian perturbation.

Tropics Two parallels of latitude at 23° 27' North (Tropic of Cancer) and South (Tropic of Capricorn) of the equator (equals the OBLIQUITY OF ECLIPTIC), marking the extremes of the overhead Sun.

Tropopause The name for the temperature minimum which marks the boundary between the troposphere and the stratosphere. The temperature is about $-55°C$ at the polar tropopause and maybe $-80°C$ at the equatorial tropopause. The altitude of the tropopause varies from 5 miles (8km) at the poles to 11 miles (18km) at the equator.

Troposphere The lowest layer in the Earth's atmosphere. It extends from sea level to the tropopause. The troposphere contains three-quarters of the atmosphere by mass, and is the layer of clouds and weather systems. It is the region which is heated by infrared radiation and convection from the ground. Within the troposphere, the temperature falls with increasing height, reaching a minimum at the tropopause. At the same time the atmospheric density decreases to one-quarter of its sea-level value.

Tsiolkovskii Lunar formation on the Moon's far side; it has a dark floor and a major central peak, so that it seems to be intermediate in type between a mare and a crater.

Tsiolkovskii, Konstantin Eduardovich (1857–1935). Russian rocket pioneer; he was born at Izhevsk of a Polish father and Russian mother. A boyhood illness impaired his hearing, but he educated himself and became a teacher of mathematics and physics. He studied the theory of dirigibles, and investigated the possibilities of rocket flight; between 1903 and 1926 he published papers which contained some modern-sounding ideas, such as the use of liquid propellents and step-vehicles. He also proposed other ideas which are now standard practice. For years his work, published in obscure Russian journals, remained unknown outside his own country, but although he never carried out any practical rocket experiments he is justly regarded today as the "father of astronautics". *PM.*

Konstantin Tsiolkovskii: rocket pioneer

T Tauri stars Very young stars still settling onto the MAIN SEQUENCE, and typified by the variable star T Tauri. Characteristics of the T Tauri stars are emission lines in the spectrum, due to an extended atmosphere of gas, high velocities in outflowing or infalling gas as the stars adjust to the recent onset of nucleosynthesis, association with dark clouds where star formation occurs, and a high proportion of lithium, which is radioactive and quickly becomes depleted in most stars.

Other characteristics include the proximity of HERBIG-HARO OBJECTS, the presence of jets of gas, or an even younger companion star. T Tauri stars are usually less massive than the Sun. Heavier stars either pass through the T Tauri phase while still obscured in their natal clouds, or have a different appearance at the same stage of life. *DAA. See also* T-ASSOCIATIONS.

Tucana (the Toucan). The faintest of the Southern Birds, but redeemed by the presence of part of the Small Cloud of Magellan and also by two superb globular clusters — 47 TUCANÆ or NGC104, inferior only to Omega Centauri, and NGC362, which is on the fringe of naked-eye visibility. The brightest star in Tucana is Alpha (2.86; absolute magnitude −0.2, spectrum K3, distance 114 light-years).

Tunguska Meteorite Early on June 30 1908, there was an explosion in central Siberia, whose location was identified by L.A.Kulik in 1927 by an area 25 miles (40km) across where the forest was dead. In the central 9 miles (15km) the trees stood, stripped of branches, but farther out the trunks had been felled and radially aligned. Swampy depressions are unrelated to the event, but soils and sediments have a micrometeorite component. Although anti-matter or a black hole have been suggested as the cause, a fragment of ENCKE'S COMET probably arrived, tail first, and detonated in the atmosphere. *RH*.

Turbulence The name given to a type of irregular flow of fluids in which the motion of the fluid at any point varies rapidly in both magnitude and direction. Most natural fluid motion is turbulent. A value called the Reynolds Number determines whether fluid flow is smooth and well defined (laminar), or is turbulent.

Turner, Herbert Hall (1861–1930). English astronomer. Second Wrangler, Cambridge, 1882; Chief Assistant, Greenwich, 1884-93; Savilian Professor

Herbert Turner: organized expeditions

of Astronomy, Oxford, 1893-1930. He organized many international projects and eclipse expeditions.

Tuttle's Comet Originally found by Méchain in 1790 and rediscovered by Tuttle in 1858, period 13.7 years. Seen at 10 returns to 1980.

Twinkling *See* SCINTILLATION.

Tycho Perhaps the best-known lunar ray crater, diameter 54 miles (87km) and depth 3 miles (4.5km). Central mountains rise to 7,500ft (2.3km) above the rough floor, which features volcanic vents and tumuli, and major cracks. The walls have an average slope of 34° inward and 26° outward; but the inner rim is terraced. Lava flows with strongly developed flow ridging occur here. Major flows with well-developed levées and fronts, together with lava lakes occupying declivities, can be found in the outer rim. The floor lavas and wall lakes are much younger than the outer rim unit of the Tycho caldera. *GF*.

Tycho: best-known lunar ray crater

Tycho's Star Appearing in Cassiopeia in 1572, this was the first supernova whose position and changing brightness were accurately estimated. The star was visible for 18 months.

Tyrrhena Terra Prominent albedo feature in the southern hemisphere of Mars.

Tyuratam Soviet launch site used for manned space missions. Located near Tyuratam in the Soviet province of

Kazakhstan, it is also referred to as the Baikonur Cosmodrome, though, in fact, it is some 230 miles (370km) south-west of the town of Baikonur.

UBV System A system of colour indices using differences of magnitudes at three selected spectral bands. The bands are (U) ultraviolet (3,700Å), (B) blue (4,000Å) and (V) visual (5,550Å).

U Cephei An interesting Algol-type eclipsing variable star, which varies between magnitudes 6.6 and 9.8. The primary minimum lasts just under 10 hours and occurs every 2.493 days. Measurements have shown that the period of the star is slowly increasing, and also shows some irregular, spontaneous variations. The system consists of two stars moving in a nearly circular orbit. The smaller of the two stars is the brightest and more massive. It is of spectral type B8V and has a mass 4.7 times that of the Sun. The larger component is a G-type giant star with a mass of only 1.9 times that of the Sun. A stream of gas is flowing off from the larger, less dense star. This slows up the rotation of the large star and speeds up the smaller, more massive companion. The surfaces of the two stars are only about 6 million miles (10 million km) apart, just about the same as the diameter of the giant star. *JWM*.

U Cygni A long-period variable, range 5.9 to 12.1(v), period 462.4 days. The rise and fall are almost equal. The spectrum is Npe (C7,2e – C9,2e). R.A. 20h 19.6m, Dec. +47° 54′ (2000.0).

U Geminorum stars Variable stars, also known as SS Cygni Stars after the brightest member, and, along with the Z CAMELOPARDALIS stars, as Dwarf Novae. Named after the prototype, discovered by Hind in 1855; almost 200 are now known. They are characterized by rapid outbursts of between 2 to 6 magnitudes, followed by a slower return to minimum, where they remain until the next outburst.

Periods range from 10 days to several years, and are fairly constant when averaged out over many cycles, but there may be considerable variations in both period and brightness

U

U

from one cycle to the next. The SU Ursae Majoris sub-class have both normal maxima and "supermaxima", which last about five times as long and are twice as bright. Another group, with periods between 50 and 80 days, sometimes show "anomalous" maxima

Brightest U Geminorum stars

Name	Range	Period (d)
VZ Aqr	11.8–15.0	49
UU Aql	10.6–16.2	56
SS Aur	10.5–14.5	54.1
WW Cet	9.3–16.8p	31.2
WX Cet	9.5–18.0p	450
SS Cyg	8.2–12.4	50.1
EY Cyg	10.9–15.1	40.8
U Gem	8.2–14.9	102.9
IR Gem	10.8–13.1p	?
VW Hyi	8.4–14.4	27.8
X Leo	12.0–15.1	22
AY Lyr	12.1–15.5	23
CZ Ori	11.4–15.8	38
RU Peg	9.0–13.1	68
UV Per	12.3–15.2	300
FQ Sco	12.4–15.4	?
UZ Ser	12.0–16.6	40
SU UMa	11.1–14.5	16
SW UMa	10.8–16.2	1000?
p = photographic magnitude		

in which the rise takes several days, instead of the usual day or so.

Almost all members show rapid irregular flickering of around 0.5m at minimum. As they are all binaries, some also show eclipses at minimum, such as that of U Geminorum, which is of 0.5m.

All are close binaries, comprising a subgiant or dwarf star of type K to M, which has filled its ROCHE LOBE, and a WHITE DWARF, surrounded by an ACCRETION disk of infalling matter. Variations in the flow of matter and/or variations in the atmospheres are

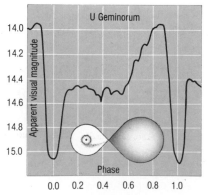

U Geminorum stars: light curve

thought to cause the outbursts. The longer the period of the star, the greater the variation, which suggests a

relationship with the RECURRENT NOVAE, though the physical basis for this, if any, is unclear.

There have been suggestions that they evolve from W URSAE MAJORIS variables, and the Z Camelopardalis group are obviously closely related. At minimum the spectra show wide emission lines, which during the brightening merge gradually into the continuous background, reappearing at maximum as absorption lines. At least two possible U Geminorum stars have been found in GLOBULAR CLUSTERS (M5 and M30). *TJCAM.*

Uhuru satellite A NASA satellite launched on December 12 1970. Completed the first X-ray sky survey. Discovered NEUTRON STARS in binary systems and extended sources in clusters of galaxies. *See also* X-RAY ASTRONOMY.

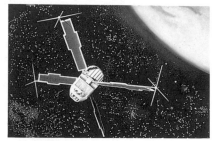

Uhuru satellite: X-ray sky survey

UKIRT *See* MAUNA KEA OBSERVATORY.

Ultraviolet astronomy The study of astronomical objects using their ultraviolet emission between wavelengths of 900 and 3,000 Ångströms. The atmosphere is opaque to the ultraviolet, and these observations cannot be carried out from the ground.

There is not much difference in the transmission of the atmosphere between sea level and the best mountain sites because the ozone responsible for absorption near 3,000Å lies high in the stratosphere. However, scientific balloons can reach an altitude of 28 miles (45km), where there is appreciable penetration of near ultraviolet radiation. Thus some ultraviolet observations down to about 2,000Å have been performed from balloon platforms. On the other hand, most ultraviolet data have come initially from rockets and now from satellites.

The first experiments undertook studies of the ultraviolet spectrum of the Sun, culminating in an American series of satellites, the Orbiting Solar Observatories. Later attention has switched to "cosmic" (ie, non-solar) sources. Various early satellites, in par-

ticular the American Copernicus, OAO-2, the European TD-1 and the Dutch ANS, laid the groundwork of the subject. However, ultraviolet astronomy has come to maturity with the wide availability to non-specialists of the INTERNATIONAL ULTRAVIOLET EXPLORER satellite, which was launched in 1978. Further major steps are anticipated following the eventual launch of the HUBBLE SPACE TELESCOPE.

These satellites have given astronomers access to ultraviolet wavelengths down to about 1,150Å. Normal telescopes have severe problems below this because of the difficulty of making highly reflecting mirrors for shorter wavelengths. If aluminium is used on the mirrors, it soon oxidizes on the ground (or from residual oxygen in orbit) and becomes covered with a layer of aluminium oxide, opaque below about 1,600Å. Overcoating the aluminium with magnesium fluoride works well down to 1,200Å but the overcoat is opaque below that. Copernicus used an overcoat of lithium fluoride which extends this coverage to 1,050Å but is hygroscopic, a great inconvenience prior to launch in the humidity of Cape Canaveral.

The interstellar medium itself does not transmit shortward of 912Å because of photoelectric absorption of hydrogen, but a substantial and interesting wavelength range immediately above this remains almost unexplored. In addition space becomes transparent again in the extreme ultraviolet toward 100Å. At the time of writing (1987) only one source outside the Solar System, a very hot white dwarf HZ43, is known at this wavelength, but detection of sources is expected soon by the American Extreme Ultraviolet Explorer and a British Wide Field Camera experiment on the German ROSAT satellite.

The European Space Agency is currently engaged on a study of a new satellite project called LYMAN which will undertake spectroscopy between 100 and 2,000 Ångströms. It overcomes the lack of efficient reflectors by using a grazing incidence telescope. The earliest possible date for launch is 1994.

The character of ultraviolet astronomy is different from other "new" astronomies like radio, infrared and X-ray astronomy. The discovery of first radio and then X-ray sources led to the discovery of hitherto unanticipated types of astronomical objects — RADIO GALAXIES and QUASARS and X-RAY

BINARIES containing compact objects. Infrared observations also discovered very cool PROTOSTARS with unexpected properties. Perhaps the major contribution of these wavelengths has been the discovery of new types of astronomical objects, the elucidation of their nature often requiring optical observations. On the contrary, few new sources have been discovered in the ultraviolet; those which have been found in this way have turned out to be unusually hot, but normal stars.

As a result, ultraviolet astronomy has been concentrated on ASTRONOMICAL SPECTROSCOPY. In the optical region this is already a powerful way of determining the physical state (temperature and density), composition and radial velocity of astronomical objects. The ultraviolet region is even better than the optical because it is here that many of the abundant atoms and ions have their strongest spectral lines. Moreover, the same data analysis methods apply in the ultraviolet as in the optical, making the subject very accessible to optical astronomers.

Ultraviolet observations contribute to all branches of astronomy, but they have been particularly relevant to studies of the interstellar medium and hot stars. Most atoms and ions found between the stars have their strongest reasonance absorption lines in the ultraviolet. Thus observations in this wavelength region have been most significant for studies of the composition and motions of the interstellar gas. Discoveries include the way some elements are "depleted" from the gas by ACCRETION onto interstellar grains, the existence of highly ionized regions in the interstellar gas which may be "fossil" supernova remnants and an extensive ionized halo to our Galaxy.

Ultraviolet data have also been crucial in extending our understanding of the gas flows around hot stars. Many hot stars show strong STELLAR WINDS, and BINARY STARS show evidence of transfer of material from one star to the other. In both cases emission lines provide information on the nature of the flows and their effect on the evolution of the stars concerned. In addition, the emission spectra of gas nebulæ excited by the radiation from hot stars pose interesting spectroscopic problems and lead to accurate measurements of abundances of materials, like carbon, which have few strong spectral lines at optical wavelengths.

For Solar System objects, opening a new wavelength region has led to the discovery of new constituents of planetary atmospheres. However, it is with comets that some of the most striking results have been obtained. Ultraviolet spectroscopy has shown the way in which the gases spreading out from the cometary nucleus are gradually dissociated and ionized by solar radiations. These studies also

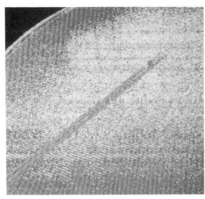

Ultraviolet astronomy: NGC6752

reveal details of the composition of the cometary material itself. This is most interesting because comets may be samples of material unchanged since the formation of the Solar System.

Ultraviolet emission comes most strongly from hot plasmas, but there are very hot regions in the outer parts of the atmospheres of cool stars. These are analogues of the chromosphere, transition region and corona of the Sun. Ultraviolet observations of cool stars have shown that hot regions are more extensive in double stars of the RS Canum Venaticorum type than in normal stars. Among single stars it has been found that coronæ and transition regions are absent for stars cooler than a certain limit: this is thought to reflect a difference in the details of the stellar wind flows from these stars.

Lastly, ultraviolet astronomy has made important contributions to extragalactic studies. Firstly, it was shown that in many ways the ultraviolet spectra of nearby active galaxies resemble very closely those of the quasars — whose ultraviolet emissions are seen from the ground by virtue of the REDSHIFT. Secondly, the stellar content of galaxies has been studied in the ultraviolet where the galaxy spectrum is sensitive to the presence of hot stars in the galaxy. But perhaps one of the most spectacular results in this field is the work on the variations of the ultraviolet spectrum of the active galaxy NGC4151, which led to an estimate of 1,000 million solar masses for the weight of the BLACK HOLE in the nucleus of the galaxy. *MVP*.

Ultraviolet radiation Electromagnetic radiation with a shorter wavelength than visible violet light. It cannot be seen with the human eye. The wavelength range extends from about 2 to 380nm. Ultraviolet (UV) radiation having wavelengths near to that of visible light is called near-UV. The term far-UV applies to the shorter wavelengths. UV radiation constitutes five per cent of the energy radiated by the Sun. However, most UV which falls upon the Earth is filtered out by atmospheric oxygen and ozone. Near-UV radiation with wavelengths above 310nm can penetrate to the Earth's surface. UV is also strongly absorbed by glass, so lenses and prisms have to be made of quartz or fluorite. It can be detected by photographic film and with a fluorescent screen. *JWM*.

Ulugh Beigh (or Beg) (1394–1449). Islamic astronomer, whose name means "great prince". A grandson of Tamerlane, his real name was Muhammad Taragay. Ruler of Maverannakhr (now Uzbekistan); he built a famous observatory in its chief city, Samarkand. The observatory was noted for its important astronomical tables.

Ulysses Shuttle-launched ESA spacecraft to study the poles of the Sun out of the ecliptic plane — the first to do so. This vehicle is the remaining element of an originally much more ambitious international project which was further seriously delayed by the SPACE SHUTTLE "Challenger" tragedy in January 1986.

Umbra The dark central cone of the shadow cast by a planet or satellite. The term also describes the darker and cooler central region of a sunspot.

Umbriel Satellite of Uranus: with its diameter of 730 miles (1,174km) it comes third in order of size. Voyager 2 surveyed it in 1986; showing that it has a relatively dark surface with only a few bright albedo features; it shows much less sign of past activity than the other major satellites.

Undina (Asteroid 92). Discovered in 1867. It is 155 miles (250km) in diameter, and therefore one of the largest members of the swarm.

United States Naval Observatory Major observatory at Flagstaff, Arizona, though not associated with the LOWELL OBSERVATORY. The main telescope is the 61in (155cm) reflector. It was this

U

United States Naval Observatory: at Flagstaff, in Arizona

U

telescope which was used to take the photographs leading to the discovery of CHARON, the satellite of Pluto.

Universal time Mean solar time on the prime meridian, measured from midnight.

Uppsala Observatory Swedish observatory with a long tradition. The present building was built around 1850, replacing the old Celsius observatory. Its main instrument, a double refractor from 1892, is now out of scientific use. It has one of the finest collections of old astronomical literature in the world. Branch stations are Kvistaberg Observatory, 30 miles (50km) south of Uppsala, with a large Schmidt telescope and a small reflector, and Uppsala Southern Station, at SIDING SPRING OBSERVATORY, Australia, with a small Schmidt. *GW.*

Uranienborg Castle on the island of Ven in Denmark, built by the Danish astronomer Tycho BRAHE and completed in 1580. Brahe had received the island as a gift from the Danish king Frederik II after having become famous throughout Europe for his work *De Stella Nova* (1573), which was a scientific report on Brahe's observations in 1572 of a supernova in the constellation of Cassiopeia.

The castle was built in a Dutch-Italian renaissance style and was situated on a plateau in the middle of the island, approximately 130ft (40m) above sea level. It was built as a square with each side 50ft (15m) long and with two entrances, one facing due west, the other facing due east. The northern and southern sides of the castle each had a round tower, which served as astronomical observatories. These were very well equipped, and for two decades Brahe made very accurate observations of the positions and movements of the heavenly bodies. His observations were

to become very important for his future pupil, Johannes KEPLER in Prague, as they later enabled Kepler to discover the laws of planetary movements.

By improving his instruments and methods, Brahe pioneered the art of accurate observation. After four years, the Uranienborg observatories became too cramped for his instruments, and he built a sister observatory, Stjerneborg, south of the castle. This was one of the most original observatories ever built as it consisted of five subterranean rooms or crypts; in the middle was Brahe's study, and the four other rooms each contained one large instrument. The reason for building an underground observatory was that Brahe wanted to increase the stability of his measuring instruments. The roofs of the crypts consisted of sliding or removable cupolas. Brahe's observatories were unsurpassed in their time; today only ruins remain. *LH.*

Uranienborg: castle built by Tycho Brahe

Uranium The heaviest element known to occur naturally; atomic number 92.

Uranium decay A series of radioactive decays whereby uranium-238 emits a succession of alpha and beta particles to become the stable isotope lead-206.

Uranius Patera. Martian volcanic structure; 26°N, 93°W.

Uranius Tholus Minor Martian volcano: 26°N, 98°W, with a 40-mile (65km) base. It lies near the volcanic structure Uranius Patera.

Uranus The seventh planet in the Solar System and the first to be discovered using telescopic observations when, in 1781, William HERSCHEL, observing from Bath, detected the body during a routine star search of the northern sky. It is the third of the major planets in order from the sun with a radius of 16,000 miles (26,200km) and is more than four times the size of the Earth. However, Uranus has 15 per cent more mass than the Earth and only 5 per cent the mass of Jupiter.

The density of Uranus is $1.2g/cm^3$ and it is therefore denser than the larger gas giants, Jupiter and Saturn. It is composed primarily of hydrogen and

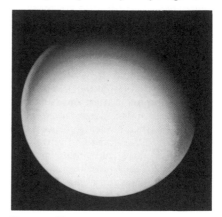

Uranus: a Voyager 2 image, 1986

helium, in approximately solar proportions, with small amounts of methane, ammonia and their photochemical products. Uranus is also thought to contain heavier materials such as oxygen, nitrogen, carbon, silicon and iron. Beneath the extensive layers of clouds, there is thought to be an ocean of super-heated water in the region of the mantle overlying the Earth-sized rocky core. It has been suggested that this extensive ocean may originate from collisions with billions of comets, which are primarily composed of water substances and reside in part of the planetary system, during the early formation of the Solar System. The numerous collisions and high pressure of the region where the water is currently located, accounts for its super-heated state.

Uranus orbits the sun every 84 years at a distance which varies between 18.3 and 20.1AU. The planet has a brightness temperature of about 55K, a greenish visual appearance and a period of rotation of 17.24 hours. The visible colour of the planet is due to the absorption of the sunlight in the blue/green portions of the spectrum, by the atmospheric methane gas.

Uranus is an extraordinary planet, whose axis of rotation is tilted by 98° with respect to its orbit. Consequently, it is unique in our Solar System since its axis of rotation is close to its orbital plane. In its motion around the Sun, each pole is presented toward the Sun and the Earth. During the encounter by the Voyager 2 spacecraft in January 1986, the south pole of the planet was directed toward the oncoming probe and the Earth. It is possible that a major collision between Uranus and some other object(s) occurred before the planet and its satellite system were formed. A possible candidate for these collisions is a swarm of comets, an event which would also be consistent with the deep oceans of super-heated water in the interior of Uranus. This unusual interior structure may also account for Uranus not possessing a significant internal heat source. This is quite dissimilar to the other major planets, Jupiter, Saturn and Neptune.

Since Uranus is so distant from the Earth, the recent Voyager 2 spacecraft encounter has provided a dramatic increase in the understanding of it. The planetary magnetic field is also extra-ordinary, since it is inclined to 60° to the rotational axis of the planet. The dipole field has a strength of 0.25 Gauss, which compares with field strengths of 0.21G at Saturn and 0.31G at the Earth. This is the most inclined magnetic field in the Solar System, with a correspondingly large difference from the planetary rotational axis. It is possible that the magnetic field may be in some stage of reversing its polarity, which is currently south-seeking toward the Earth. The planet has an extensive magnetosphere which stretches for more than $18R_u$ and contains a hot plasma environment with temperatures reaching more than 10,000K. There is a complicated inter-action between this hostile region con-taining the trapped charged particles and the embedded rings, satellites and the upper atmosphere of the planet. It sweeps away the tiny particles from the rings and affects the chemistry of the satellites by heating their surfaces by particle bombardment. On the sunlit side of the planet there is an intense electroglow, while on the dark side auroral activity is detected through these magnetospheric/atmospheric interactions, which are also related to a

huge glow in the atmosphere which spreads outwards for about 31,000 miles (50,000km) $(2R_u)$.

Uranus, like Jupiter and Saturn, has a visible banded appearance and the cloud motions reveal a predominantly zonal circulation where the winds are blowing in a east-west direction rather than from north to south. This circula-tion resembles the flow on Jupiter and Saturn and, to a lesser extent, the motions on every planet in the Solar System. The motions are zonal in spite of the substantial differences in their solar heating distributions so that this confirms that it is the rotation of the planet that organizes the weather systems. However, unlike all the other planets, the polar regions are slightly warmer than the equatorial regions of the planet. Extensive haze layers, created photochemically and com-posed of acetylene and ethane particles, obscure the major weather systems on Uranus. During the Voyager en-counter, several discrete clouds were seen in the southern mid-latitudes, at the 1–2 bar pressure level and com-posed of methane particles. Some of these clouds resemble convective systems seen on the Earth. The temper-ature at these cloud-top levels is approximately 64K. There is a small seasonal variation of about 5K in these cloud-top temperatures, but this will have a negligible effect on the meteoro-logical systems since the Uranus atmosphere responds sluggishly to solar radiation, like a very deep ocean. Although the upper clouds are com-posed of methane, layers of ammonia, ammonia polymers and water clouds may exist deeper in the atmosphere.

Uranus is now known to possess 15 satellites. The five main satellites, Miranda, Ariel, Umbriel, Titania and Oberon, were discovered from Earth-based telescopic observations. These satellites have densities in the range 1.26–1.5g/cm³ and are therefore com-posed mainly of water ice and rock with small amounts of methane clath-rate. However, they do appear to have varied surfaces. Oberon, Titania and Umbriel are the least geologically active of the satellite family, with Umbriel the darkest and most inactive of this set. The tectonic activity increases from Ariel to Miranda, which both show evidence of surprising sur-faces with fracture patterns, scarps, valleys and layered terrains. Miranda has the appearance of a reassembled body following a major collision at some earlier stage in its life, with the grooves of Ganymede, the canyons of

Mars and the compressional faults of Mercury, all on this tiny body. The 10 new satellites all reside in the region between Miranda and the rings, with two of them acting as shepherds for the outer ε ring. These satellites range in size from about 9 to 100 miles (15 to 170km) in diameter. They are all sur-prisingly dark with albedos of only about 5 per cent. It is possible that the darkness of all the satellites and the rings, too, may be due to the bombard-ment of the charged particles in the magnetosphere altering their surface heating and chemistry.

The stellar occultation observations in 1977 provided the first evidence of a system of nine rings around Uranus. Two further rings have been found during the Voyager encounter. This system of 11 rings now extends from 23,000 to 32,000 miles (37,000 to 51,160km) from the planet, although there appear to be hundreds of very thin rings and ring arcs around the planet. The principal arcs range in width from about 60 miles (100km) at the widest part of the ε ring to a few kilometres for most of the others. The ε ring seems to be composed of 3ft (1m) boulders rather than the micron-size particles that are typical of the rings of Jupiter and Saturn. It is perhaps sur-prising that only one pair of shepherd-ing satellites have been found so far, namely, those bodies adjacent to the ε ring. It is possible that smaller satellites do exist, shepherding the other rings which are currently below the level of detection. *GEH*.

Ursa Major (the Great Bear). Constel-lation whose main pattern is known as the Plough or the Big Dipper.

Brightest stars

Name	Visual Mag.	Abs. Mag.	Spec.	Distance (light-yrs)
ε (Alioth)	1.77	0.2	A0	62
α (Dubbe)	1.79	0.2	K0	75
η (Alkaid)	1.86	−1.7	B3	108
ζ (Mizar)	2.09	0.4,2.1	A2 + A6	59
β (Merak)	2.37	1.2	A1	62
γ (Phad)	2.44	0.6	A0	75
ψ	3.01	0.0	K1	120
μ (Tania Australis)	3.05	−0.4	M0	155
ι (Talita)	3.14	2.4	A7	49
θ	3.17	2.2	F6	46
δ (Megrez)	3.31	1.7	A3	65
ο (Muscida)	3.36	−0.9	G4	230
λ (Tania Borealis)	3.45	0.6	A2	120
ν (Alula Borealis)	3.48	−0.2	K3	150

U

Next come Kappa (3.60), h (3.67, Chi (3.71), Xi (3.79) and Upsilon (3.80). Mizar is the famous double, and makes a naked-eye pair with Alcor (magnitude 4.01). Dubhe and Merak are the Pointers to the Pole Star. Five of the "Plough" stars make up a moving cluster, the exceptions being Dubhe and Alkaid.

There are many galaxies in Ursa Major, including several Messier objects (81, 82, 101, 108 and 109), as well as M97, the "Owl Nebula", one of the most famous of all planetary nebulae even though it is a very faint object with small telescopes. *PM*.

Ursa Minor (the Little Bear). The north polar constellation.

Brightest stars

Name	Visual Mag.	Abs. Mag.	Spec.	Distance (light-yrs)
α (Polaris)	1.99	−4.6	F8	680
β (Kocab)	2.08	−0.3	K4	95
γ (Pherkad Major)	3.05	−1.1	A3	225

There are no other stars above magnitude 4, but the pattern which also includes Delta (4.4), Epsilon (4.2), Zeta (4.3) and Eta (4.9) gives a slight impression of a faint and distorted Plough. *PM*.

Ursid meteors A minor meteor stream that peaks on about December 22 or 23 every year. The radiant is at R.A. 14h 28m, Dec. +78°, not far from the star Beta Ursæ Minoris. Ursid meteors may be seen between about December 18 and 25. On December 22 1945, a strong Ursid meteor shower occurred, with peak rates of around 100 meteors per hour (m/h). Since that time, Ursid rates have remained low, with various estimates of between 5 and 15m/h at peak. The stream may be rich in rather faint meteors. The Ursids are associated with periodic comet Tuttle 1790 II, which has a period of 13.7 years. It last returned in 1980. The strong 1945 display was due to a concentration of dust on the opposite side of the orbit to the comet. *JWM*.

Ussher, Archbishop James (1581–1656). Archbishop of Armagh, who, by adding up the ages of the patriarchs, dated the Earth's creation as precisely 10 o'clock in the morning of October 26 4004 BC. He was in England when the Irish rebellion broke out in 1641 and he never returned to Ireland. His learning was much admired and on the orders of Oliver Cromwell he was given a state funeral in Westminster Abbey.

James Ussher: dated the creation

Utopia Planitia Martian plain, 35° to 50°N, 195° to 310° W — the landing-site of Viking 2 in 1976. The most prominent crater in Utopia is Mie.

UV Ceti The active red-dwarf FLARE STAR, UV Ceti, discovered by Luyten and Carpenter. This nearby twelfth-magnitude binary at 2.7 parsecs, with dM6e and dM4e components, exhibits large flares every few hours.

UX Ursæ Majoris stars A small group of eclipsing variable stars with Algol-like characteristics but extremely short periods. Unlike normal Algol stars, there are rapid and completely random brightness variations outside of eclipse, up to as much as $1\frac{1}{2}$ magnitudes in the case of RW Trianguli. During eclipse this star has an asymmetric light curve often with "shoulders" before and after minimum. The maximum brightness of UX Ursæ Majoris is subject to variations of amplitude 0.01 to 0.2 magnitude over periods of one to 20 minutes. The average maximum and minimum brightnesses also vary, and there are oscillations in the period. These stars have a bright gaseous cloud surrounding the smaller, brighter component of the system, which ejects gas erratically. The larger, less luminous companion may lose mass to the smaller star. *JWM*.

Valhalla The largest ringed basin on Jupiter's satellite Callisto; there is a

bright circular region 370 miles (600km) in diameter, surrounded by concentric rings extending out to 930 miles (1,500km) from the centre of the structure. The positon of the centre is 10°N, 55°W.

Vallis Alpes *See* ALPINE VALLEY.

Vallis Rheita *See* RHEITA VALLEY.

Vallis Marineris *See* MARINER VALLEY.

Van Allen belts The regions in the magnetosphere where charged particles become trapped and oscillate forward and backward between the magnetic poles as they spiral around the magnetic field lines. These regions of the terrestrial magnetosphere were discovered by James Van Allen from the analysis of observations made from the Explorer 1 satellite in 1958.

The terrestrial Van Allen belts consist of a lower region located between 600 and 3,000 miles (1,000 and 5,000km) above the equator, which contains protons and electrons either

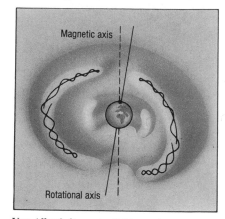

Van Allen belts: potential hazard

captured from the solar wind or orientating from collisions between upper atmosphere atoms and high-energy cosmic rays. The second region is located between 9,000 and 15,000 miles (15,000 and 25,000km) above the equator but curving downward toward the magnetic poles. This region contains mainly electrons from the solar wind.

These regions are a potential hazard to Earth-orbiting spacecraft, since an interaction with these regions of charged particles will have an adverse effect on the electronic sub-systems and on-board instrumentation. The regions around the planetary systems are much more intense than their terrestrial counterparts and therefore present formidable hazards for space missions such as GALILEO to Jupiter.

Similar regions of charged particles have been discovered around Mercury, Jupiter, Saturn and Uranus and they are expected to be discovered around Neptune by Voyager 2 in 1989. The charged-particle environment around these major planets has unique interactions with their individual systems of satellites and rings. All of these bodies embedded in the local Van Allen belts sweep their orbital path clear of the trapped particles. This interaction with the rings and satellites causes a local heating and alters their surface chemistry. *GEH*.

Van Biesbroeck, George (1880–1974). Belgian-born astronomer who worked at Yerkes Observatory. He discovered three comets and numerous double stars, including the eighteenth-magnitude companion to BD + 4°4048, until recently the intrinsically faintest star known.

George Van Biesbroeck: found double stars

Van de Graaf A deep basin, 62 miles (100km) in diameter, on the far side of the Moon at latitude 27°S and longitude 173°E, probably formed by the impact of a large body. Unlike such basins on the near side, the circular maria, those on the far side have not been flooded with lava.

Vandenberg Air Force Base American launch facility near Lompoc, in southwest California, used for launching military satellites, and which will be used for military Shuttle missions. It covers 3,456 acres (1,399 hectares). In the illustration at the top of this page, technicians at the base are checking the IRAS.

Van Maanen, Adriaan (1884–1946). Dutch astronomer who spent much of

Vandenberg Air Force Base: IRAS

his career in America, at Yerkes and Mount Wilson. He carried out excellent work in connection with stellar movements and spectroscopy, though he was mistaken in believing that he had detected individual motions in the outer parts of spiral galaxies.

Variable stars Stars whose brightness changes with time. These variations in light may be due to some inherent feature in a star or its atmosphere, or to some geometrical alignment. The former are termed intrinsic variables; the latter extrinsic. There are about 29,000 known variable stars and a further 14,810 objects which are suspected to vary in light, but these have not yet been confirmed. These numbers do not include variables in GLOBULAR CLUSTERS or those in other galaxies. A few variable stars are known by their Greek letters or proper names, ie, Alpha Orionis (Betelgeux). The system of naming variable stars is to assign the letter R to the first discovered in a constellation; the second becomes S and so on down to Z. Then comes RR to RZ; SS to SZ. This sequence ends at ZZ, after which comes AA to AZ; BB to BZ and so on down to QZ, but the letter J is never used. After QZ the next variable in any constellaton would be the 335th: it becomes V335, and each subsequent variable in that constellation gets a higher number. The letters and numbers are followed by the name of the constellation.

Early discoveries of variables were made visually and a few are still discovered by this method. However, most discoveries result from photography. Two plates taken of the same region of the sky can be placed in a blink machine, which enables each plate to be quickly and alternately scanned so that any difference in a star is soon noticed. Other methods use positive and negative films superimposed. The plates, or films, must be taken some time apart and with the same instrument and under the same sky conditions if possible.

Variable stars work is a field in which the amateur astronomer makes a large contribution provided he works in co-operation with some central organization such as the American Association of Variable Star Observers, the British Astronomical Association, or the Royal Astronomical Society of New Zealand. The reason for this is that there are so many variable stars that the professionals can only observe a few. They wish to select those stars that pose special problems, or to observe them at certain phases of their light variations. As a result, they rely on amateurs to provide data that will enable them to select the stars that interest them and to provide light curves and predictions for the more regular variables so that they can select the phases that they wish to observe. The amateur astronomers are thus able to make a very valuable contribution by consistently following these stars. An observer working alone in this field must pool his results at one of the recognized international centres for amateurs so that his results and those of many other amateurs can then give sufficient data to provide complete and reliable light curves which are of great use to the professionals.

The intrinsic variables are divided into many classes, each of which has a number of divisions depending on the way in which the stars change in brightness or because of some other property inherent in them. Broadly speaking there are seven classes. The first are called ERUPTIVE VARIABLES because of violent processes taking place in their chromospheres and coronæ. The resultant flares are often accompanied by shell events due to matter being carried off by stellar winds and by interaction with the interstellar matter that surrounds the star. Typical examples are the various types of Orion variables. The second class are the pulsating variables, which have more or less regular expansion and contraction of their surface layers. MIRA VARIABLES and the many types of CEPHEIDS belong to this class. The third class consists of ROTATING STARS, with light variations due to their axial rotation or to star spots or some feature of their atmospheres caused by a magnetic field. So this class includes both intrinsic stars and extrincis variables. The fourth class are the explosive or cataclysmic stars such as NOVÆ, SUPERNOVÆ and DWARF NOVÆ. The fifth class are the ECLIPSING BINARY systems such as the Algols, which are really extrinsic variables. The sixth class consists of X-RAY SOURCES in which the variability is in the X-ray radiation. Finally, there is a seventh class of unique objects that cannot be assigned to any other type. Included in this class are the BL LACERTÆ and optically varying QUASARS, which are, of course, extragalactic objects.

The observation of many of the foregoing stars has undergone a change in recent years. Previously it was fairly useless for amateurs to observe most of the short-period variables simply because they are not suitable for visual observation. However, the availability of photoelectric photometers and their associated equipment has placed these within the reach of many amateurs as their prices are now reasonable. This has resulted in many advanced amateurs doing what was formerly the exclusive field of the professional by obtaining precise three-colour data on a wide range of stars. Such observations have been welcomed by the professionals simply because if all their telescopes were used to observe variable stars — and that of course is impossible and undesirable — they still could not cope with all the stars to be observed. This can be illustrated when the growth in the number of variable stars is considered. In 1786 there were only 12 known variable stars. By 1866 this had grown to 119. By 1907 the number had increased to 1,425 and the advent of photographic methods of discovery saw numbers grow rapidly thereafter until by 1941 there were 8,445 known variable stars. Since then the number has continued to increase rapidly so that today there are tens of thousands of these stars. Even with the aid of thousands of amateurs throughout the world, only a limited number of these can be adequately studied. Many have been entirely neglected since their original discovery and it is believed that hidden among these may be a number of very interesting objects that will repay study because they would pose some interesting astrophysical problem. The trouble is to find these among the host of neglected stars since so often the scanty data made at the time of the original discovery provide few clues as to the true nature of the stars concerned.

In order to observe variable stars both professionals and amateurs need charts, which are detailed maps of small areas of the sky in which the variable is situated and on which both the variable and the surrounding stars are clearly identified, preferably with a sequence of comparison stars of constant brightness with which the variable can be compared. This is often impossible for the simple reason that the magnitudes of the surrounding stars have not been accurately determined. This can be overcome in several ways, one of which is to select what appear to be suitable comparison stars from photographic plates and to assign letters to them as symbols. This then enables visual observations to be made pending the determination of accurate magnitudes later. For making a visual estimate of a variable star, the observer compares it with two stars of known, unchanging brightness. There are two methods. The first is Pogson's step method, in which the observer trains himself to measure changes of a tenth of magnitude. Suppose that he has two comparison stars, A (mag. 7.0) and B (7.6) and finds that the variable is midway between them; its magnitude will then be 7.3. If he estimates as 0.1 mag. below A and 0.5 above B, then its magnitude will be 7.1, and so on.

In the fractional method, two stars must be used (whereas with the Pogson, one comparison will serve if need be). The fractional method is to take the comparisons and mentally divide the interval between them into ten stages. The variable is then placed in its correct position to the sequence. Thus if it is one-quarter of the way from A to B (and hence threequarters of the way between B to A) the record will read A1V3B. Knowing the magnitudes of the comparison stars, that of the variable can be worked out. If the brightness of the comparison stars are unknown, the observations can either await their accurate determination or the series of observations can be evaluated by a step method and a light curve still produced.

The main sources of the necessary charts are the Variable Star Sections of the British Astronomical Association and the Royal Astronomical Society of New Zealand, and the American Association of Variable Star Observers. The importance of variable stars is shown by the fact that about one-third of the astronomical literature is concerned with them in some way. *FMB.*

Variation The term has several meanings in different contexts:

Lunar variation An inequality in the motion of the Moon resulting from the changing perturbing force exerted by the Sun as the Moon moves round the Earth.

Annual variation The annual rate of change of the co-ordinates of a star due to the combined effects of precession and proper motion.

Secular variation The rate of change per century of the annual precessional change in a star's co-ordinates.

Magnetic variation The angle between the directions of magnetic north and true north at any particular location on the Earth's surface; its value changes with time. *IKN.*

Vastitas Borealis The northern circumpolar plain of Mars; it extends from latitude 55° to 70°N.

Vega Alpha Lyræ, fifth brightest visual star, visual magnitude 0.0, distance 8pc, A0 dwarf. Standard star for magnitude scale zero point. IRAS found heat emission from large (greater than 1mm) dust particles at a temperature of −190°C, and extending out to 85AU from the star. The mass involved is similar to that of our own planetary system. This may be a planetary system in the process of formation around Vega. Similar dust has been seen by IRAS in other A-type dwarfs (*see* FOMALHAUT, BETA PICTORIS), indicating that such systems may be common. *See also* EPSILON ERIDANI.

Vega: standard star for magnitude scale

Vega 1 and 2 Two Interkosmos probes launched in December 1984 by the Soviet Union to drop balloons in the Venus atmosphere in June 1985 and to encounter HALLEY'S COMET on March 6 and 9 1986. The name Vega is derived from Venera + Galley (the Russian pronunciation of Halley).

The Vegas were three-axis stabilized spacecraft measuring 36ft (11m) overall, including the four large solar cell panels, and weighed about four tonnes at launch. Fly-by distance was 5,520 and 4,990 miles (8,890 and 8,030km) on the sunward side of the nucleus. The probes carried identical payloads of two cameras, infrared spectrometers, gas and dust mass spectrometers, dust impact detectors and plasma analyzers. *RR.*

Vega 1 and 2: Venus and Halley's Comet

Veil Nebula Part of the Cygnus Loop, a supernova remnant 3° in diameter. The strands of nebulosity are the remains of a massive star which exploded about 30,000 years ago. The nebula is at a distance of about 800pc (2,500 light-years) and is expanding at more than 60 miles/s (100km/s).

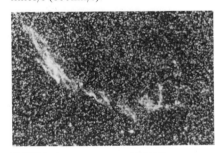

Veil Nebula: a supernova remnant

Vela The Sails of the dismembered Argo Navis. It is a rich region; Delta and Kappa make up the "False Cross" with Iota and Epilson Carinæ.

Brightest stars

Name	Visual Mag.	Abs. Mag.	Spec.	Distance (light-yrs)
γ (Regor)	1.78	−4.1	WC7	520
δ	1.96	0.6	A0	68
λ	2.21	−4.4	K5	490
κ (Markeb)	2.50	−3.0	B2	390
μ	2.69	0.3	G5	98
N	3.13	−0.3	K5	147

Next come Phi (3.54), Psi (3.60), Omicron (3.62), c (3.75), p (3.84), b (3.84), q (3.85) and a (3.91). *PM.*

Vela pulsar 0833-45, a strong, short-period (89 milliseconds) radio pulsar discovered in 1968, and identified with a supernova remnant. The optical counterpart is a twenty-fourth magnitude star discovered in 1977.

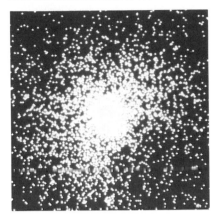

Vela pulsar: lies inside an SNR

Velocity Rate of change of position in a specified direction. Velocity has both magnitude ("speed") and direction, and so is a vector quantity.

Velocity of light The velocity of light in a vacuum is 299,792.458km/s, or 186,282 miles/s.

Vendelinus Lunar walled plain; one of the Langrenus chain; 16°S, 62°E. It is 103 miles (165km) in diameter, but is irregular and broken, and much less perfect than Langrenus or Petavius.

Venera probes Series of Soviet Venus probes. The first successful landing (Venera 7) was accomplished in 1970, since when Veneras 8 through 16 have been deployed successfully. Experiments on board the craft allowed for

Venera probes: explored Venus

the collection of atmospheric data during descent mode and, on later missions, included a gamma ray detector and instruments to measure electrical discharges. On landing, each probe returned television pictures to Earth. Veneras 13 through 16 also scooped up samples of the local regolith and have provided scientists with the first analyses of the planet's crust. *PJC.*

Venus The second planet in the Solar System from the Sun and, apart from the Moon and the Sun, the brightest object in the sky. Since its orbit lies within that of the Earth, it passes through inferior conjunctions rather than oppositions. Furthermore, GALILEO's observations of the phases of Venus in the 17th century helped him to confirm the Copernican theory of planetary motions. Venus orbits the Sun every 225 days at a distance which varies between 0.72 and 0.73AU, so that its orbit is very nearly circular. However, it rotates in a retrograde manner with a period of 243 days. It is the nearest planet to the Earth, with a radius of 3,760 miles (6,051km) and a density of 5.2g/cm³ so that for centuries it was thought to be the sister of our planet. The planet has been visited by numerous Soviet and some American spacecraft missions involving fly-bys, atmospheric probes and the Soviet/French floating balloons from the VEGA mission *en route* to the Comet Halley encounter in March 1986. The modern information has shown that the current atmospheric composition, meteorology and the surface conditions of Venus are quite different from their terrestrial counterparts.

Venus does not have a magnetic field in spite of possessing a large nickel-iron core. This is due to the very slow rotation of the planet, which is therefore unable to generate a field by a dynamo action. However, a very weak magnetic field is induced in the planet's ionosphere through the interaction of the SOLAR WIND and the resulting BOW SHOCK region acts as a buffer between the interplanetary medium and the atmosphere of Venus.

The visible appearance of Venus is a uniform, unbroken layer(s) of yellowish cloud whose top may be at an altitude of about 60 miles (100km) above the surface of the planet. The cloud layer reflects 79 per cent of the incident sunlight and this value is the highest known planetary albedo in the Solar System. This is in complete contrast with the Earth, which, on average, has only 50 per cent cloud cover and an albedo of about 30 per cent. The Venus atmosphere is huge, more than 90 times more massive than the Earth and composed primarily of carbon dioxide

(CO$_2$). Traces of hydrochloric and hydrofluoric acids, carbon monoxide, water vapour, hydrogen sulphide, carbonyl sulphide, sulphur dioxide, argon, krypton and xenon have also been detected in the atmosphere. The Venus atmosphere extends to an altitude of 150 miles (250km) above the surface, but possesses only a thermosphere and troposphere in its layered structure. On the day side of Venus, there is a terrestrial-type thermosphere with temperatures increasing from about 180K at 60 miles (100km) to about 300K in the exosphere. The thermosphere does not exist on the night side of the planet, where the temperature falls from about 180K at 60 miles (100km) to 100K at 90 miles (150km). The transition from day-side to night-side temperatures across the terminator is very abrupt.

The atmosphere is extremely variable in the neighbourhood of the cloud tops in the altitude range of 45–60 miles (75–100km) and diurnal fluctuations of as much as 25K have been observed at the 59 mile (95km) level. Below these variable haze layers are the ubiquitous clouds, which occupy a substantial portion of the troposphere. There are three distinct cloud layers in the region of 28–37 miles (45–60km) with differing particle sizes and concentrations. The particles are composed of varying concentrations of sulphuric acid droplets and solid and liquid sulphur particles. The temperature increases steadily from the cloud tops at about 300K to the surface at 737K, where there is virtually no wind.

Beneath the clouds and on the surface of Venus, the temperatures are the same everywhere. About 90 per cent of the volume of the entire atmosphere lies between the surface and an altitude of 17 miles (28km) and at this level the atmosphere resembles a massive ocean; dense and sluggish in its response to the very weak solar heating. Only 2 per cent of the incident sunlight actually reaches the surface of the planet. The Venus surface pressure is 90 times greater than the Earth's value and the surface temperature is the highest known value in the Solar System and more than 2.5 times the highest known value on the surface of the Earth. This crushing CO$_2$ environment at the surface of Venus is the result of a runaway greenhouse effect. The basic make-up of the Earth and Venus is very similar, although there is now as much CO$_2$ in the atmosphere of Venus as we find in the limestone rocks of the Earth. However, the orbit of Venus is nearer to the Sun so that it receives twice the

sunlight that is incident on the Earth, and thus the surface of Venus has rapidly heated by the greenhouse mechanism to its current state. This effect cannot be achieved by the CO$_2$ alone. The small traces of H$_2$O and SO$_2$ are essential for the efficient greenhouse effect. The surface temperature will not increase further since the atmosphere and surface are in chemical equilibrium. However, this situation is a frightening lesson in uncontrolled atmospheric pollution.

The weather systems of Venus, at the level of the cloud tops, are strange too. Although the planet itself is rotating very slowly, the equatorial clouds have a rotation period of 4 days, which is equivalent to winds of 330ft/s (100 m/s). Consequently, the cloud tops are moving 60 times faster than the surface of the planet and in the opposite direction to the Sun. Almost all the solar energy is absorbed in the cloud tops and this provides the main driving mechanism for the super-rotation of the atmosphere. The cloud tops of TITAN and the upper atmosphere of the Earth are the only other regions known to super-rotate.

The surface of Venus is surprisingly varied and may suggest that the initial geological developments took place before the massive atmosphere evolved into its current state. The images from

Venus: a planet of mystery

the Soviet Venera landers showed a stony desert landscape with outcrops and patches of dark material which suggest some chemical erosion. The subsequent radioactive analyses of the Venus soil suggest the composition is similar to basalt, but with an unusually high concentration of potassium. Some of the basaltic materials are similar to those found on the terrestrial sea bed. About 70 per cent of the surface is covered by huge rolling plains, 20 per cent by depressional regions and the

remaining 10 per cent by highlands which are concentrated in two main areas. Basins are comparatively rare on Venus, but they do exist. The most extensive is Atalanta Planitia in the northern hemisphere, which is about the size of the Gulf of Mexico and whose depth is about 4,500ft (1.4km) below the mean radius of the planet. There are some craters about 15 to 30 miles (25–50km) in diameter which suggest the impacts occurred before the atmosphere reached its current opacity. The highland areas include the volcanic areas which are therefore important in the evolution of the planet and its atmosphere. The two main areas are Ishtar Terra in the north and Aphrodite Terra, which crosses the equator. Ishtar Terra is a flattish plateau, about the size of Australia, whose average altitude is 9,800ft (3km) and which carries several mountainous areas. On the eastern side is Maxwell Montes, which is the highest peak on the surface of Venus, rising to 7 miles (11km) above the mean surface of the planet. The Aphrodite Terra region is about the size of the continent of Africa and is made up of mountain areas at the west and east separated by a lower region. There is also a circular feature which is thought to be a volcanic caldera. The third upland area, Beta Regio, contains two large shield volcanoes, Rhea Mons and Theia Mons, which rise some 13,000ft (4km) above the surface on a fault line which extends in a north-south direction. They are similar to the Hawaiian volcanoes and may well be active. Indeed, there has been a noticeable reduction in the measured amount of atmospheric SO$_2$ during the past 10 years which could be explained by a past volcanic eruption and the associated atmospheric adjustment. Furthermore, there have been suggestions of lightning beneath the clouds from measurements made during the descent of the Venera 11 and 12 probes, which could be caused by volcanic activity. Alpha Regio is a smaller area which is similar to the Tharsis region of Mars. A huge rift valley, Diana Chasma, with a depth of 6,500ft (2km) below the surface and a width of nearly 180 miles (300km) is found in the Aphrodite region. This valley is much larger than any feature on the Earth and may be compared with Valles Marineris on Mars. It is probably of tectonic origin. There is little doubt that Venus is a planet of mystery. *GEH*.

Venus tablet Assyrian astrological text found by Layard at Nineveh (1849–50), published in 1870 and deciphered by Kugler (1911). It records early Babylonian sightings of Venus (c. 1702–1681BC).

Vernal equinox *See* EQUINOX.

Verne, Jules Ray-crater on the far side of the Moon; first shown on the 1959 photographs from Lunik 3.

Very Large Array The most complex single-site radio telescope in the world. Twenty-seven 82ft (25m) dishes lie along the arms of a giant "Y" at 7,000ft (2,000m) in the plains of New Mexico, USA. Each arm can be up to 13 miles (21km) and four different configurations are used, from compact to far-flung. Every pair of dishes acts as

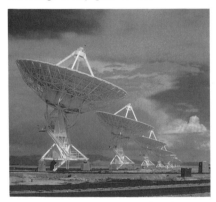

Very Large Array: complex radio telescope

an interferometer when constructive and destructive interference occurs as radio wavefronts encounter different antennæ at different times.

The VLA serves as 351 independent interferometers, sprinkled across the surface of a virtual dish some 21 miles (34km) in diameter! As the Earth rotates it sweeps the interferometers across the sky, providing more information from different separations of antennae. The dishes each contain their own separate, remotely commanded, receivers for different observing frequencies. The angular resolution achieved is that of a vast dish, equivalent in scale to millions of radio wavelengths. Currently the VLA's commonly-used continuum wavelengths are 1.3, 2, 6 and 20cm and, at 2cm, it can resolve structures smaller than 0.2 seconds of arc (3cm at 25km). To analyze and decipher the huge volumes of data requires hours of labour assisted by sizeable computers, before the end-product, a detailed map of a small piece of sky, is attained. Typical observing programmes may be awarded from one to 20 hours, to

observe one to 50 sources, at up to several frequencies.

The VLA can also observe radio spectral lines by isolating very small slices of spectrum to make a series of maps at closely spaced frequencies. This yields velocity information on a region through the Doppler shift whereby line-emitting objects approaching us have their line frequency shifted to shorter wavelengths "blue" if visible radiation) and objects receding from us are "red-shifted" (to longer wavelengths). *MC.*

Vesta (Asteroid 4). Discovered by Olbers in 1807. It is the brightest of all the asteroids; and at its best can just attain naked-eye visibility. It is classed as of the eucritic type, with an albedo of 0.23; estimated mass 2.4×10^{20}kg.

Vignetting Shadowing of a two-dimensional image caused by components of the optical system, such as an obstruction in the light path, or, more subtly, optical components that are too small to pass all of the light.

Viking missions The Viking missions to Mars incorporated two orbiter-lander spacecraft. Viking 1 was launched by a Titan-Centaur booster on September 9 1975, and injected into Mars orbit on June 19 1976. The lander spacecraft was separated from the orbiter and landed on the surface of Mars on July 20 1976. Viking 2 was launched on August 20 1975, injected into Mars orbit on August 7 1976, and its lander spacecraft reached the surface of Mars on September 3, 1976. The first month

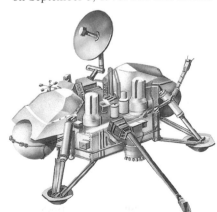

Viking missions: Viking 1, Mars lander

of orbiter operations consisted of locating and verifying a safe landing site for the lander spacecraft.

Because Mars is the most hospitable of the planets, and because of the long history of debate regarding the possi-

bility of life on Mars, great emphasis was placed on lander experiments designed to search for life. All of the experiments operated successfully, and no indications of life were found at the landing sites. However, this does not rule out life at other locations. Meteorological instruments on the landers regularly reported the weather at their locations.

Each orbiter carried two vidicon cameras and each lander carried two facsimile cameras. The orbiters mapped the entire surface of Mars at 490–980ft (150–300m) resolution and selected areas at resolutions down to 26ft (8m). In all, the orbiters returned more than 55,000 images, including many high-resolution pictures of the satellites Phobos and Deimos. The landers acquired thousands of pictures and monitored surface changes and seasonal effects throughout the Martian year. *MED.*

Virgo (the Virgin). A very large Zodiacal constellation crossed by the equator.

Brightest stars

Name	Visual Mag.	Abs. Mag.	Spec.	Distance (light-yrs)
α (Spica)	0.98	−3.5	B1	260
γ (Arich)	2.75	2.6	F0 + F0	36
ε (Vindemiatrix)	2.83	0.2	G9	104
ζ (Heze)	3.37	−1.7	A3	1100
δ (Minelauva)	3.38	−0.5	M3	147

Next come Beta or Zavijava (3.61), 109 (3.72), Mu (3.88) and Eta or Zaniah (3.89). Arich is a splendid binary with equal components; it is closing, and will be a difficult object by the end of the century.

Virgo is one of the richest of all areas for faint galaxies, including the Messier objects 58, 61, 49, 84, 86, 87 (the giant elliptical), 89, 90, 59, 60 and 104. However, all these are below the eighth magnitude, and are not easy to see with small telescopes. *PM.*

Virgo cluster Rich concentration of galaxies seen in the direction of the constellation Virgo. This example is typical of more distant rich clusters, but its relative proximity of 60 million light-years makes it prominent in our skies. MESSIER catalogued nearly 20 members of the Virgo cluster, and hundreds more are known.

The Virgo cluster, the centre of our local supercluster, is so massive that it influences the expansion of the universe around it. Our Galaxy and the LOCAL GROUP are separating from the Virgo cluster at about 500 miles/s (800km/s) instead of the 750 miles/s (1,200km/s) that would be expected. *DAA.*

Virgo cluster: concentration of galaxies

Virtual particles Short-lived particles that cannot be detected directly but whose effects can be measured. The Uncertainty Principle of quantum mechanics allows them to come briefly into existence, but dictates that the higher their energies, the shorter their lifetimes. They are believed to act as "messenger" particles to convey interactions between real particles. For example, the electromagnetic force is conveyed by the exchange of virtual photons, while the strong nuclear interaction is conveyed by mesons.

Visual magnitude The apparent brightness of an astronomical body as seen by the eye, whose maximum sensitivity is at a wavelength of 5,600Å. Such magnitudes are now determined photographically, using appropriate filters, and are called PHOTOVISUAL MAGNITUDES.

Vitello Concentric lunar crater, 24 miles (38km) in diameter in the Mare Humorum area; 30°S, 38°E. Orbiter pictures have shown inside it the track of a rock that has rolled down a slope.

Vitruvius Lunar crater, 19 miles (31km) in diameter, between Mare Serenitatis and Mare Tranquillitatis; 18°N, 31°E. It has low but rather bright walls.

Vlacq One of a group of six lunar walled plains south-west of Janssen; 53°S, 39°E; diameter 56 miles (90km).

VLBI *V*ery *L*ong *B*aseline *I*nterferometry (See RADIO INTERFEROMETER).

Volans (the Flying Fish). A small constellation intruding into Carina. There are four stars above the fourth magnitude; Gamma (double, combined magnitude 3.6), Beta (3.77), Zeta (3.95) and Delta (3.98).

von Braun, Wernher (1912–1977). German rocket pioneer, who was involved in making V2 weapons (to 1945) and then led the American team that launched Explorer 1.

Wernher von Braun: rocket pioneer

von Kármán, Theodor (1881–1963). Aeronautical engineer and "father of supersonic flight" born Budapest. In 1930 he became head of the aeronautics lab for California Institute of Technology, and later organized pioneering rocket engine research leading to the WWII JATO system.

Theodor von Kármán: rocket research

von Zach, Baron Franz Xavier (1754–1832). Hungarian astronomer (Director of the Seeberg Observatory). He undertook many mathematical researches in astronomy, played a major role in the hunt for the "missing

planet" between Mars and Jupiter, and was energetic in international collaboration among astronomers.

Baron von Zach: hunted "missing planet"

Voskhod First Soviet spacecraft to carry three men October 12-13 1964. From Voskhod 2, on March 18 1965, Alexei Leonov made the first space walk.

Vostok Soviet spacecraft in which Yuri GAGARIN became first man to orbit the Earth on April 12 1961. Valentina TERESHKOVA, first space woman, flew June 16-19 1963 in Vostok 6.

Vostok: Yuri Gagarin's vehicle

Voyager Two Voyager probes were launched from Cape Canaveral in 1977. Their targets were the outer planets, and both were extremely successful. The two probes were identical except for the more powerful Radioisotope Thermoelectric Generator (RTG) on Voyager 2, which was intended to rendezvous with Uranus and Neptune as well as Jupiter and Saturn. The launch weight of the vehicle and propulsion rocket was in each case 4,444lb (2.016kg), of which the Voyager itself accounted for 1,746lb (792kg). Ten instruments were carried on each vehicle.

Voyager 2 was launched first, on August 20 1977; Voyager 1 followed on September 5. However, Voyager 1 was travelling in a more economical orbit, and "caught up" with its twin during the crossing of the asteroid belt. On March 5 1979 Voyager 1 passed Jupiter at a distance of 217,500 miles (350,000km), and sent back excellent images as well as a mass of miscellaneous data; the results were far superior to those of the earlier Pioneers. Voyager 1 then went on to a rendezvous with Saturn on November 12 1980, at a minimum distance of 77,000 miles (124,200km); as well as the planet itself, good images were obtained of the satellites TITAN, RHEA, DIONE and MIMAS. The complex nature of the rings was fully revealed, and new data were obtained with regard to the magnetic field, radiation zones and other phenomena. Voyager 1 was not scheduled to rendezvous with Uranus or Neptune, and indeed could not do so, as the encounter with Titan meant that it continued in an orbit well out of the ecliptic. Had Titan not been satisfactorily imaged, then Voyager 2 would have had to carry out a survey, and would have been unable to rendezvous with either of the outer giant planets.

Voyager 2 passed Jupiter on July 9 1979 at 444,000 miles (714,000km). The results fully complemented those of its predecessor, and showed marked changes in the Jovian scene, notably with regard to the Red Spot and the Ionian volcanoes. Voyager 2 then went on to Saturn, by-passing that planet on August 25 1981 at 63,000 miles (101,300km), and also obtaining good images of the satellites IAPETUS, HYPERION, TETHYS and ENCELADUS. However, after closest approach to Saturn, the scan platform, carrying the camera, gave trouble, and apparently jammed, so that some vital information was lost. It was at first thought that the platform had been damaged by an icy particle in the region of Saturn's rings, but this proved not to be the case; the problem was due to lack of lubrication when the platform was being manoeuvred. The problem solved itself, but subsequently the platform was manoeuvred only at reduced rate.

After leaving the neighbourhood of Saturn, Voyager 2 was put into its "cruise mode" preparatory to the encounter with Uranus. This was achieved on January 24 1986, at a minimum distance of 67,000 miles (107,000km) (less than 51,000 miles/

82,000km above the Uranian cloud-tops). This part of the mission was particularly important, because comparatively little had been known about Uranus; even at this immense distance Voyager 2 functioned perfectly. This was all the more remarkable because it had been in space for more than eight years, and so could justifiably be regarded as an old vehicle. Many discoveries were made, including the small inner satellites and the revelations about Uranus' remarkable magnetic field.

The final target for Voyager 2 is Neptune, in August 1989; there seems no reason why this part of the mission should not be just as successful as the earlier encounters. After that Voyager 2 will move away from the Sun into interstellar space, as also will Voyager 1; it is hoped to keep contact with them until they reach the boundary of the HELIOSPHERE, and it has even been suggested that perturbations in their movements might give a key to the position of an unknown planet beyond Neptune and Pluto — if, of course, such a planet exists, which is by no means certain.

There is always the chance, admittedly remote, that in the future the Voyagers will be picked up by some alien race in another Solar System.

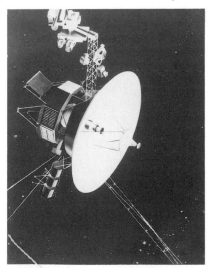

Voyager: Voyager 2, *en route* to Neptune

Each probe therefore carries a 12in (31cm) copper record called "Sounds of Earth", together with information upon how to play it, using the cartridge and needle which have been provided.

It is fair to say that most of our detailed knowledge of Jupiter, Saturn and Uranus has been drawn from the Voyager missions, and before the end of the decade the same may be true of

Neptune. Undoubtedly the Voyagers are among the most successful probes so far launched. *PM*.

V.603 Aquilæ The brilliant nova of 1918; it reached magnitude −1.1.

V2 German WWII rockets, built by the Peenemünde team led by Wernher von Braun and fired at England in the late stages of the war. After the German surrender in 1945, captured V2s were taken to America and used for testing. It may be said that the V2 is the direct ancestor of the modern spacecraft.

V2: forerunner of today's launch vehicles

Vulcan Hypothetical intra-Mercurial planet invented by LE VERRIER (1859) to explain anomaly in orbital theory of Mercury; destroyed at the stroke of a pen by EINSTEIN (1915).

Vulpecula (the Fox, originally Vulpecula et Anser, the Fox and Goose). A very dim constellation near Cygnus. It contains M27, the DUMBBELL NEBULA, probably the finest planetary nebula in the sky, but there is little else of interest here.

Wabar craters Two small Arabian impact craters produced by iron meteorites.

Wallops Flight Facility Part of NASA's Goddard Space Flight Center, which is located on Wallops Island, Virginia, and is used mainly for operating suborbital rockets and balloons.

Walter Lunar walled plain 80 miles (129km) across, 33°S, 1°E. It forms a chain with Regiomontanus and Purbach.

Wargentin The best example of a lunar plateau. It is 55 miles (89km) in diameter, and fairly regular in outline; position 50°S, 60°W, not far from Schickard and Phocylides. Wargentin is filled with lava to its brim, so that in places the wall does not rise above the interior; on the plateau there are ridges and some small craterlets. There are other plateaux on the Moon, but none is nearly so perfect as Wargentin. Unfortunately it is very foreshortened as seen from Earth. *PM*.

Wargentin: perfect lunar plateau

Wargentin, Pehr (1717–83). Swedish astronomer, and one of the foremost astronomers of his time. His outstanding tables of Jovian satellites were long used for longitude determinations.

Wave mechanics During the 19th century, there were two rival theories of the nature of light: either it consists of particles, as originally postulated by NEWTON, or it has a wave nature, as suggested by the interference experiments of Thomas Young and later developed theoretically by Maxwell. With the introduction of the Quantum Theory by Planck and EINSTEIN at the beginning of the 20th century, it became evident that light really has a dual nature: it can interact with matter either as a photon, exchanging energy and momentum, or as a wave, producing interference and diffraction phenomena. In 1924, de Broglie introduced the concept that all forms of matter possess the dual behaviour of particles and waves. The de Broglie wavelength λ of a particle is given by $\lambda p = h$, where p is the momentum of the particle and h is PLANCK'S CONSTANT. This was confirmed experimentally by Davisson and Germer, who showed that electrons interfere with each other in a wave-like manner. The mathematical equations that describe the wave structure of atomic particles, and the laws that govern their interactions, were formulated by de Broglie and Schrödinger and are known as wave mechanics. With the latter development of quantum mechanics, principally by Dirac and Heisenberg, wave mechanics is now seen as an alternative mathematical description, completely contained within the more general structure of quantum mechanics.

The wave mechanical approach leads to an entirely different picture of ATOMIC STRUCTURE from that of the "miniature Solar System" model of the Bohr atom. In the hydrogen atom, for example, the electron is no longer conceived as a particle in orbit around the nucleus: instead, the electron has a wave structure that extends throughout space. The amplitude of the wave is greatest in the vicinity of the classical Bohr electron orbit. In the quantum mechanical picture, this is understood as meaning that there is a finite probability that the electron can be anywhere in the universe, but the probability density (which in a simplified way is proportional to the square of the wave amplitude) is highest at the classical orbit.

The quantization of electron orbits emerges in a very straightforward manner in Schrödinger's wave formulation: the "orbit" corresponds to the region where there is an integral number of waves around the atom. That is, the position where a stationary interference pattern is established by the wave train that represents the electron. This three-dimensional pattern can also be thought of as the probability density of a cloud of negative charge (integrating to the charge on a single electron) around the atom. The nucleus of the atom, being constructed of individual protons and neutrons, is also described by wave mechanics. Its greater mass results in a more concentrated probability distribution than for an electron.

A Schrödinger equation can be written down for any atomic system. It will include the effects of the electrostatic and magnetic interactions of all the constituent particles. The solution of this equation gives an exact description of the energy states of the system. Only in the simplest of atoms (hydrogen and helium) can accurate solutions — called wave functions or eigenfunctions — of the Schrödinger equation be obtained.

As with the classical Bohr atom described in atomic structure, the various energy states of an atomic system are described by quantum numbers, but in wave mechanics these emerge naturally from the properties of the spherical harmonic functions (which are generalizations of simple sine waves to three dimensions). Thus, the principal quantum number, n, is related to the number of nodes (zero-crossing points) of the radial wave function and the quantum number l is related to the number of nodes in the azimuthal wave function. Only in the simplest atomic systems can the radial and angular wave functions be treated independently; because the wave functions extend to infinity there is overlap, which can lead to significant interference, or mixing, of the wave functions of one state with another.

For complex structures (atoms containing many electrons, or the atomic lattice structure of crystals) only approximate solutions to the Schrödinger equation are, in general, possible. However, these solutions are often required not only to investigate the internal structure of an atom or ion, but also to determine how such particles interact with each other and with photons. Thus solutions of the appropriate wave equations lead to estimates of collision cross sections, of spontaneous and stimulated transition probabilities and of ionization and dissociation energies, all of which may be of interest in the study of laboratory or astrophysical plasmas. Modern high-speed computers are capable of furnishing solutions of wave equations for quite complex systems. In the area of astrophysics, such solutions can give, for example, the energy levels and hence emission line wavelengths of highly ionized species or of complicated molecules not observable in the laboratory. In this way, identifications of, respectively, ion lines in the far ultraviolet spectrum of the solar corona and microwave radio emissions from interstellar molecular clouds are aided.

In another important branch of astrophysics, that of quantitative analysis of the shapes and strengths of ABSORPTION LINES in stellar spectra, the profiles of such lines are determined by various line-broadening mechanisms. The atomic parameters (Stark coefficients, van der Waals interaction constants, lifetimes of state) that are needed to evaluate these mechanisms must often be computed by the methods of wave mechanics.

The principal structures of atomic nuclei and the ways in which they interact are also governed by quantum and wave mechanics. Some of the NUCLEAR REACTIONS of interest in the construction of theoretical models of stars cannot be measured in the laboratory because they occur at too low a rate. For such reactions, calculations of rates must be made, which involve solutions of the wave equations for the two nuclei as they approach, collide and recede from one another.

At the highest energies obtained in nuclear accelerators, in which the deep structure of nuclei and the internal structures of protons and neutrons themselves are probed, it is found that the quantum mechanics which satisfactorily accounts for the lower-energy experiments on atoms and nuclei is incomplete. A theory known as Quantum Chromodynamics is currently being developed to interpret these high-energy phenomena.

The probabilistic nature of quantum mechanics, which is present in wave mechanics as the probability distribution for individual particles, provides an explanation for the ability of particles to "tunnel" from one state to another (of equal energy) without having to acquire the extra energy to pass over the potential barrier that separates the two states. Such tunnelling, for example, allows nuclear reactions to occur at lower energies (and hence lower temperatures) than they would otherwise require.

On a somewhat grander scale, attempts are currently being made to examine the properties of wavefunctions that represent the entire universe, to see if the Big Bang (*see* essay) could be the result of tunnelling from a different state. *BW*.

W Comæ Faint object in Coma Berenices, formerly regarded as a variable star but now as a QUASAR.

Webb, the Rev. Thomas William (1806–85). English amateur astronomer, author of the famous book *Celestial Objects for Common Telescopes*. He was also an excellent observer.

Wegener, Alfred Lothar (1880–1930). German meteorologist noted for his contention that Earth's continents changed their position through time. This hypothesis of "continental drift" was published in 1912, but was not taken seriously until after his death.

Werner Regular lunar crater, 41 miles (66km) in diameter; 28°S, 3°E; a pair with Aliacensis.

Westerbork Observatory Radio astronomy observatory in the northeast of the Netherlands, operated by the Netherlands Foundation for Radio Astronomy. Its major instrument, completed in 1970, consists of 12 82ft (25m) reflectors along an east-west baseline one mile (1.6km) long, with the two end dishes movable on rails.

Westerbork Observatory: 12 25m "dishes"

Westphal's Comet Naked-eye comet discovered in 1852; period 61.2 years. It faded during its return in 1913, and did not reappear as expected in 1974–5, so that it is probably defunct.

West's Comet A naked-eye comet seen in March 1976 as a conspicuous object. After perihelion it showed obvious signs of distintegration. Its period is very long, amounting to tens of thousands of years.

Wezea The star Delta Canis Majoris, magnitude 1.86. It is very luminous and remote.

Whirlpool Galaxy A well-defined type Sc spiral galaxy, which has a nucleus that is not very obvious, but very open and prominent spiral arms. The galaxy

Whirlpool Galaxy: prominent spiral arms

appears almost "face-on" to us. It was given the number M51 in Charles Messier's catalogue and NGC5194 in

J.L.E.Dreyer's New General Catalogue.

The Whirlpool lies in the constellation of Canes Venatici, not far from the end of the tail of Ursa Major, the Great Bear. A small companion galaxy NGC5195 appears to be connected to it by an extension of one of the spiral arms. The galaxy is estimated to be about 37 million light-years from us. It has a total mass less than half that of our own Milky Way Galaxy.

Whistler An effect that occurs when a plasma disturbance, caused by a lightning stroke, travels out along lines of magnetic force of the Earth's magnetic field, and is reflected back to its point of origin. The disturbance may be picked up electromagnetically and converted directly to sound. The characteristic drawn-out descending pitch of the whistler is a dispersion effect caused by the greater velocity of the higher-frequency components of the disturbance.

White dwarfs Stars with masses similar to the Sun, but whose radii are only about as large as that of the Earth. The density of the matter of which they are made is in consequence enormously greater than that of any terrestrial material: one cubic centimetre of white dwarf matter weighs approximately one tonne. The first white dwarf to be discovered was the companion of Sirius, now called Sirius B. In 1844 Friedrich BESSEL discovered irregularities in the motion of Sirius, and concluded that it must have a companion, the pair forming a binary system with a period of about 50 years. In order for its gravity to swing Sirius around in its orbit, it was clear that the companion must have a comparable mass (quantitative analyses using KEPLER'S third law for the binary orbit giving a mass between 75 per cent and 95 per cent of the Sun's mass), and yet it must be very faint to have remained undetected by direct observation. In 1862 Alvan Clark discovered a very faint star close to Sirius, which was subsequently identified as the elusive companion. As the distance to the system was known, this allowed the companion's luminosity to be estimated as only about 1/360 of that of the Sun. In 1914 W.S.Adams made the surprising discovery that Sirius B had the spectrum of a "white" star with a surface temperature about two and a half to three times that of the Sun. The total radiation from the surface of a hot body rises as the fourth power of its temperature, so that each square centimetre of Sirius B's surface should radiate between thirty and eighty times as much as the Sun. The only way to reconcile this with the very low total luminosity inferred above was to conclude that the radius of Sirius B was far smaller than that of any star known at the time. The result remained uncertain until in 1925 Adams succeeded in detecting the gravitational redshifts of several absorption lines in the spectrum of Sirius B. This general relativistic effect depends on the ratio of the star's mass to its radius. As the mass was already known from the properties of the orbit, this allowed Adams to find the radius of the star in an independent way, giving an estimate consistent with his earlier result.

It was very quickly realized that their enormous matter densities make white dwarfs fundamentally different in nature from "normal" stars. In a star like the Sun, it is the thermal pressure of the gas and radiation of the star which prevents it simply collapsing under its own weight. Already in 1926, R.H.Fowler showed that this cannot be the case for white dwarfs. Instead, the pressure required to hold up these stars against their own gravity is provided by a fundamental quantum-mechanical effect which had been discovered only months before by Fermi and Dirac. As in most stars, the matter inside a white dwarf is almost completely ionized, that is, the electrons have all been ripped off their parent atoms (here because of the frequent collisions between atoms) leaving the matter as a mixture of free electrons and nuclei. What Fermi and Dirac had discovered is that there is a fundamental limit to how close together two electrons can be pushed. Electrons resist being in the same place with the same speeds: push them closer together and they will react by moving faster. This motion amounts to a pressure, called degeneracy pressure: it differs from ordinary thermal pressure in depending only on the density of the constituent particles rather than on the product of the density and the gas temperature. The larger the mass of a star supported by degeneracy pressure, the faster the electrons inside it must move, until ultimately, for stars of masses only slightly greater than that of the Sun, they are moving with speeds close to that of light. The theory of RELATIVITY forbids motions faster than light, so that relativistic corrections must be made to the degeneracy pressure for such stars. The pressure is then found to increase rather less rapidly with the density.

In 1931 S.Chandrasekhar computed the structure of a white dwarf with these relativistic effects included, and made the momentous discovery that a white dwarf cannot have a mass more than about 1.4 times the Sun's mass. For larger masses gravity would always overwhelm the pressure forces and the star would collapse under its own weight. Subsequent observation has amply verified this prediction. Its tremendous importance comes from the fact that white dwarfs must represent one possible endpoint to the evolution of normal stars. All stars lose energy into space by radiation. In response to this constant drain of energy, the central parts of the star gently sink closer together during huge expanses of time, becoming denser and more tightly bound by their mutual gravitational attraction. If the star is supported in the normal way by thermal pressure, it has to become hotter in order to supply the increased pressure demanded by the stronger gravity. Thus in response to the loss of energy due to radiation from its surface, the star has actually had to become hotter. This means that radiation losses from its surface will continue, driving the evolution to more tightly bound and hotter configurations ever onward: the star cannot simply cool down and stop evolving. In the white dwarf state, however, we have seen that the supporting degeneracy pressure is independent of the temperature of the stellar material: the star can now radiate away all its thermal energy into space without any consequences for its internal structure, and the evolution of the star ends with it quietly cooling down to a cold, dark, inert configuration. In practice the star will by then consist entirely of helium and heavier elements, its original hydrogen having been transmuted in the course of its ealier evolution. The existence of the Chandrasekhar limiting mass means that this endpoint is not available to all stars, but only to those whose mass is below the limit when they reach the white dwarf state. We now know that more massive stars must end their evolution either as NEUTRON STARS or BLACK HOLES. Neutron stars are also supported by degeneracy pressure, but of neutrons rather than electrons; for them, too, there is a maximum mass, although its precise value is less certain than for white dwarfs.

Many white dwarfs are now known, and their observed properties are in good accord with theoretical predictions. Much interest now centres on white dwarfs in binary systems which are close enough that gas is pulled off the companion star onto the white dwarf. Such systems include the NOVAE and the DWARF NOVAE. *ARK.*

White, Edward Higgins II (1930–67). First American astronaut to make a spacewalk (from Gemini 4, in 1965). He died in a fire during a ground test of an Apollo spacecraft.

White hole Time reversal of a BLACK HOLE.

White Sands American rocket testing ground in New Mexico; formerly the main site, but now used chiefly for sounding rockets and short-range vertical firings.

Widmanstätten pattern Four parallel sets of plates of nickel-poor kamacite, set in taenite; present in most iron meteorites (octahedrites); revealed by polishing and etching with acid.

Wien's law The wavelength at which a BLACK BODY emits its maximum amount of radiation is inversely proportional to its absolute temperature. The law was derived by Wilhelm Wien (1864–1928), a German, who in 1911 was awarded the Nobel Prize.

Wild Duck Cluster Nickname for the open cluster M11, in Scutum.

Wilhelm Förster Observatory The main observatory in West Berlin. Research is carried out there, but the observatory is largely concerned with educational projects.

Wilhelm Humboldt Vast lunar formation 120 miles (193km) in diameter; 27°S, 81°E. It is so near the rim that it cannot be well seen from Earth, but it has much floor detail, including systems of clefts or rilles.

Willamette Meteorite A 14-tonne meteorite found in Oregon, USA.

William Herschel Telescope The 14ft (4.2m) Anglo-Dutch telescope of the ROQUE DE LOS MUCHACHOS OBSERVATORY, La Palma. Built in 1986 at a cost of £15 million, the WHT is the world's third largest single-mirror telescope. It is carried in a Nasmyth altazimuth mount. The conventional Cassegrain optical design features a third switchable mirror to direct the light beam to any of three foci. This, together with an integrated data collection and instrument control system called "4MS", makes it the first large telescope designed at the outset for remote operation from overseas.

William Herschel Telescope: at La Palma

Wilson, Alexander (1714–86). In his early days Wilson was apprenticed to a surgeon-apothecary in St. Andrews and London. His career changed and he set up a type foundry in St. Andrews. He was always interested in astronomy and was appointed Professor of Practical Astronomy in Glasgow in 1760. He carefully observed sunspots and found that they are often depressions in the solar surface. He published his work in 1774 (*see* WILSON EFFECT).

Alexander Wilson: studied sunspots

Wilson effect A roughly circular sunspot appears elliptical when seen near the limb due to foreshortening. In 1769 Alexander WILSON observed that the penumbra farthest from the limb appears compressed and the penumbra nearest to the limb appears relatively wider (the "Wilson effect"): this is due to the perspective effect of the funnel-shaped depression of the sunspot.

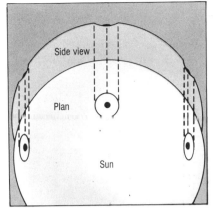

Wilson effect: sunspots in perspective

Wilson-Harrington Comet A comet discovered in 1949. It had an estimated period of 2.3 years — the shortest known — but has not been seen again.

Wolf Creek Crater Western Australia, 900 yards (820m) wide, caused by the impact of an iron meteorite.

Wolf, Maximilian Franz Joseph (1863–1932). German astronomer always known as Max Wolf, noted for his photographic and spectroscopic work. He discovered 232 asteroids, including the first to be found photographically (323 Brucia, in 1891).

Wolf number (or Zürich number). A count of sunspot activity. The Wolf number $Z = k(f + 10g)$, where g is the number of groups, f the total number of individual spots, and k a constant depending on the calibre of the observer (usually about 1).

Wolf-Rayet stars The existence of stars whose spectra contain entirely bright instead of dark lines was first recognized by the French astronomers C. Wolf and G. Rayet in 1867. These Wolf-Rayet (W–R) stars are divided into two kinds: the WN type, in which emission lines from nitrogen ions dominate the spectrum, and the WC type, in which emission lines of carbon and oxygen ions predominate. Both types have strong lines of helium and a few have moderate or weak lines of hydrogen as well. All of the emission lines are broad.

The W–R stars have surface temperatures between 25,000K and 50,000K,

luminosities between 100,000 and one million times that of the Sun, and masses from 10 to 50 times solar. The very high luminosities generate radiation pressure that drives a wind of gas leaving W–R stars at a rate of 3 solar masses in a million years. Many central stars of PLANETARY NEBULÆ are W–R stars, which shows that even higher rates of mass ejection can occur (but probably immediately prior to the W–R phase of evolution).

W–R stars demonstrate quite dramatically the effect that mass loss can have on massive stars. Their original hydrogen-rich envelopes have been stripped away to expose regions in which the products of NUCLEAR REACTIONS are present: elements made in the CNO cycle in the case of WN stars, and elements from the helium-burning which formed a carbon–oxygen core in the WC stars. The type of W–R star that emerges after the mass loss depends on how far evolution had progressed before the hydrogen envelope was lost.

Despite their rarity, one W–R star is easily visible to the unaided eye: the second-magnitude southern-hemisphere star Gamma Velorum. It is a WC star with a spectroscopic binary companion of spectral type 07. *BW.*

Wollaston, William Hyde (1766–1828). English doctor; he invented the camera lucida. In 1802 he observed dark lines in the solar spectrum, but took them merely for the boundaries between different colours.

William Wollaston: invented camera lucida

Woolley, Sir Richard van der Riet (1906–86). The tenth Astronomer Royal, who held office from 1956 to 1971; he also held office at various times at Mount

Wilson, Cambridge and Canberra. While in Canberra he developed the MOUNT STROMLO OBSERVATORY into a centre of vital importance in astronomy. After retiring from the Royal Observatory Greenwich in 1971 he became the first Director of the South African National Observatories, retiring in 1976. Woolley made great contributions to astrophysical theory, and also wrote a classic book about eclipses jointly with Sir Frank DYSON, (1868–1939), his predecessor as Astronomer Royal.

Richard Woolley: tenth Astronomer Royal

Woomera Rocket Range British proving ground in Australia — now virtually disused.

Wright of Durham, Thomas (1711–86). English teacher of astronomy. In *An original theory of the universe* (1750) he explained the Milky Way as the optical

Thomas Wright: explained the Milky Way

effect of the Sun's immersion in either a thin hollow shell of stars, or a flat hollow disk of stars, the stars in either case surrounding a supernatural centre.

W Serpentis An interesting eclipsing variable star with extremely unusual light variations, between magnitudes of about 8.9 and 10.3. The generally accepted period is 14.153 days. There appears to be an overlying longer period of about 270 days. In addition to the deep minimum, there are also two shallow minima, and three maxima. The system consists of two stars of roughly the same size, but of different brightness, rotating inside a common envelope. From one of the components (of spectral class A or F) ejection of material is taking place. The system evolves rapidly and the period is increasing. Maybe one of the two stars is itself variable.

W stars *See* WOLF-RAYET STARS.

W Ursæ Majoris stars. Pairs of stars almost in contact as they orbit one another. They differ from BETA LYRAE stars in being smaller, less luminous stars of nearly identical brightness.

W Virginis stars Pulsating variable stars superficially similar to Cepheids. On the HR diagram they typically lie in the spectral type range G0 to M0 and the absolute luminosity range $M_v = -1$ to -4 and thus are giant stars. Their masses could be as low as 0.5 solar mass, suggesting that they have evolved from low-mass MAIN SEQUENCE stars. They lie on the INSTABILITY STRIP on the HR diagram which goes from Delta Scuti stars on the main sequence through the RR Lyrae stars and Cepheids to the irregular variables and Mira stars and therefore the cause of their variation is thought to be the same, ie, a layer of ionized helium within the star.

Their light curves can be distinguished from those of the classical Cepheids by the less-regular shape of their light curves and the double-peaked nature of their maxima. Their periods range from approx. 1^d to 100^d and their period–luminosity relationship is distinctly different from that of the Cepheids. They do not form a homogeneous group as some belong to Population I and some to Population II. Occasionally W Vir stars show small period changes, but in the remarkable case of one star, RU Cam, its variations suddenly stopped in 1964 for about three years. The spectral type of

RU Cam (K0 to R2 with excess carbon) is later than usual for W Vir stars, but the star is otherwise unremarkable. It is not understood how pulsations can stop, or start, in such a short time.

Because these stars are brighter than the RR Lyraes, but fainter than the Cepheids, and do have a well-defined period–luminosity relationship, they can be used as distance estimators for both galactic and extragalactic objects. Once their periods are known, their absolute magnitudes are known, and hence their distances can be calculated from the difference between their real and apparent magnitudes. *ENW. See also* CEPHEID VARIABLES, DISTANCE MODULUS.

WZ Sagittæ Originally classified as a recurrent nova, this star has many similarities with the U GEMINORUM class of dwarf novæ. It may be a link between the two classes. Three outbursts have been observed, in 1913, 1946 and 1978. It is also an eclipsing system with a period of 81.5 minutes. The light curve at eclipse resembles that of a W Ursæ Majoris star. WZ Sagittæ is normally around magnitude 15.5 at minimum, but has reached 7.0 during an outburst. Possible masses of the components are 0.59 and 0.03 times the Sun's mass.

Xenon A noble (inert) gas. Chemical symbol Xe, atomic number 54, atomic weight 131.30.

Xenon abundance Xenon comprises, by volume, about 0.08 parts per million of the atmospheres of Earth and Mars. The relative deficiency of noble gases in those atmospheres, compared with Venus, suggests that during the formation of the Solar System the material which eventually formed Venus absorbed a large proportion of the noble gas content of the SOLAR WIND.

Xenophanes High-walled lunar crater 67 miles (108km) in diameter; 57°N, 77°W, in the Roris area.

X Persei Massive X-RAY BINARY. X-ray luminosity 10^{34} erg/s. Pulsation period 835s. Optical companion is a Be star, magnitude 6.0–6.7. There is a possible 580-day binary period.

X-ray astronomy Branch of astronomy which began in 1949 with the discovery, by Burnight and co-workers of the United States Naval Research Laboratory (NRL), that the Sun emitted X-rays. The radiation is completely absorbed in the Earth's atmosphere and so detectors and later X-RAY TELESCOPES had to be carried above the surface of the Earth to a height of more than 75 miles (120km). Much of the early work was undertaken by Friedman and co-workers at NRL who used the V2 and other early sounding rockets to establish the general properties of solar X-ray emission. The use of X-ray proportional counter detectors on the United Kingdom Ariel I satellite, built and launched by NASA in 1962, established that high-temperature (about 10^7K) plasma was created in the corona during SOLAR FLARES. Solar X-ray studies continued through the 1960s and 1970s with the highly successful NASA Orbiting Solar Observatory series and with the flight of the Apollo Telescope Mount on the Skylab mission. Solar observations are continuing with NASA's SOLAR MAXIMUM MISSION, which was the first satellite to be successfully repaired in space.

Based on a knowledge of the solar X-ray luminosity, it seemed clear in the late 1950s that the flux at Earth from the star of similar brightness at a distance of only a few light-years would be quite undetectable by the then available detectors. It was therefore a major surprise when Giacconi and his colleagues discovered a strong X-ray source in the constellation of Scorpius. This observation, made from a sounding rocket in 1962, was followed by the discovery of many other X-ray sources by groups of workers at NRL, Lockheed and the Lawrence Livermore Laboratory. By 1970, after some dozens of sounding rocket flights which gave a total exposure of only a few hours, about 50 individual X-ray sources had been discovered, including the CRAB NEBULA and several other supernova remnants, the peculiar galaxies NGC1275 and M87, the nucleus of the giant radio galaxy CENTAURUS A, the nearby QUASAR 3C273 and a diffuse X-ray emission component (the X-ray background) which appeared to be isotropic at energies above 2keV.

The launching of the Uhuru satellite by NASA in December 1970 marked the beginning of a new era for cosmic X-ray astronomy. Instrumented by Giacconi and his colleagues at American Science and Engineering Inc., Uhuru carried out a complete sky survey which resulted in the discovery of more than 400 discrete X-ray sources. The optical identification of a number of bright X-ray sources in the galaxy together with observations of periodic X-ray variability, established that NEUTRON STARS were present in binary systems. By accreting matter from its normal stellar companion, the neutron star heats the gas that falls onto its surface to temperatures in excess of 10^9K.

In the case of one of these systems, CYGNUS X-1, mass estimates based on optical observations strongly suggested that the compact X-ray source in the system was about ten times more mas-

X-ray astronomy: Centaurus A in X-ray

sive than the Sun. Since extreme X-ray variability required that the source be about 6 miles (10km) in diameter, and since it is believed that neutron stars cannot exceed three solar masses, it was concluded that the Cygnus X-1 system includes a BLACK HOLE. Uhuru observations also showed that clusters of galaxies always included extended sources of X-ray emission which were unrelated to the individual cluster galaxies but seemed to originate from an intra-cluster medium.

In the decade that followed the launching of Uhuru, several satellites were flown that carried X-ray detectors. Prominent among these was the UK Ariel V. An instrument provided by University College London's Mullard Space Science Laboratory (MSSL) detected emission lines characteristic of highly ionized iron in the X-ray spectra of the extended cluster sources. This demonstrated beyond doubt that the space between the member galaxies was filled with plasma at temperatures of around 10^8K. This instrument also discovered the class of slowly rotating neutron stars with periods of up to several tens of minutes, while a sky survey instru-

ment provided by the University of Leicester established the importance of active galaxies (Seyferts, quasars and the like) as highly luminous X-ray sources. The Astronomical Netherlands Satellite discovered burst sources which were studied in great detail by the NASA Small Astronomy Satellite 3, whose instruments were provided by the Massachusetts Institute of Technology. The small MSSL X-ray telescopes on the NASA Copernicus satellite made the first detailed maps of the emission from supernova remnants and, together with Ariel V spectral observations, confirmed that the radiation came from shock-heated interstellar gas. Instruments on the NASA Orbiting Solar Observatory–8 and High Energy Astronomy Observatory–1 from the Goddard Space Flight Center provided much detailed information on the spectra of extragalactic X-ray sources and increased the number of known X-ray objects.

With a few exceptions, the observations carried out by Uhuru and its successors were done with detectors whose angular resolution was limited to a few arc minutes or significantly worse for faint sources. The NASA EINSTEIN OBSERVATORY launched in November 1978 carried an X-ray telescope with few arc second angular resolution. This instrument ushered in the next era in X-ray astronomy. It consisted of a large nested grazing incidence telescope of Wolter I design. It had a focal length of 10ft (3m) and responded to photons in the 0.1 to 4kcV energy range. Four instruments were available for use at the telescope focus — two for imaging which had fields of view of around 1^0 by 1^0 together with a high-resolution Bragg crystal spectrometer and a moderate spectral resolution grating spectrometer and a solid state detector. The combination of good angular resolution and high sensitivity led to the discovery of several thousand new X-ray sources during the 2.5-year operational life of the satellite. Much new information was obtained on the structures of extended sources such as supernova remnants, clusters of galaxies and the jet features at the centres of a number of radio galaxies. Objects in nearby normal galaxies like M31 and the MAGELLANIC CLOUDS were resolved and studied for the first time. However, the most striking result was the discovery that almost all types of ordinary stars emit X-rays. Thus the nature of X-ray astronomy was fundamentally changed. Instead of being concerned

X

with the study of a small number of exotic high-energy or high-temperature objects, its range now extends throughout the entire subject along with radio, infrared, optical and ultraviolet astronomy. The European Space Agency's EXOSAT satellite, launched in May 1983, carried two small imaging X-ray telescopes which together had about one-third the collecting area of the Einstein telescope. This satellite was placed in a highly eccentric orbit with an apogee of almost 12,500 miles (200,000km) and was therefore able to observe individual sources for periods of up to 90 hours. Thus in spite of its lower sensitivity, it has brought about significant advances in the study of binary systems and many other variable X-ray sources.

Given that X-ray astronomy is now a mature and wide-ranging branch of the subject, there is a need for permanent access to X-ray observatories. Both NASA and ESA have formulated plans for such facilities. NASA has begun the design of a major spacecraft known as the Advanced X-ray Astrophysics Facility (AXAF). It will include a 4ft (1.2m) diameter X-ray telescope which will allow a sensitivity ten times greater than was available with the Einstein mission. Equipped with powerful focal plane instruments for imaging and spectroscopy, it will allow the essential next steps to be taken in this rapidly developing subject.

ESA has designated as one of its major future projects the provision of a large observatory spacecraft for high-throughput X-ray spectroscopy. Known as the X-ray Multi-mirror Mission (XMM), it will have a somewhat lower angular resolution than AXAF but, with a collecting area about 20 times greater at 8keV, it will concentrate on detailed studies of X-ray spectra. It is hoped that these two large observatories will be available to undertake a wide range of complementary investigations in the second half of the next decade. *JLC.*

X-ray binaries Luminous X-ray sources first discovered by UHURU satellite. Usually consist of an ordinary star in a binary pair with a compact degenerate object; most commonly a NEUTRON STAR, although in a few cases (notably CYGNUS X-1) the compact X-ray emitter may be a BLACK HOLE. Gas flows from the atmosphere of the ordinary star onto the compact object. Since the latter will typically be 1–3 times the mass of the Sun, but will have a diameter of only 6 miles (10km) it

provides a very strong gravitational field. Gas falling through this field will be heated to about 10^9K and will thus emit X-rays. *JLC.*

X-ray galaxies A number of galaxies and many QUASARS are powerful emitters of X-rays and are also often strong radio sources. Those spiral galaxies which emit X-rays seem to have bright nuclei containing much ionized gas moving at a high velocity and exhibiting a rich and distinctive pattern of optical emission lines. These objects are usually classified as SEYFERT GALAXIES. Similar to, but much more active than, Seyferts are quasars whose X-ray luminosity may be 1,000 times greater. Galaxies may also emit X-rays as a consequence of merging (CENTAURUS A, NGC5128 for example) or as a result of rapid movement through an extremely hot intergalactic plasma (M87, NGC4486). The production of X-rays seems to reflect the interaction of ionized matter under extremes of temperature or velocity which cannot be reproduced on Earth. *DM.*

X-ray novæ Soft X-ray transients. X-ray binaries with low-mass optical companions that are sometimes observed to brighten rapidly by factors of 10^2–10^6 and to decline in weeks to months.

X-ray sources Following the discovery of Scorpius X-1 by Giacconi, in 1962, many thousands of X-ray sources have been found throughout the universe. *See* X-RAY ASTRONOMY.

X-ray stars The detection of X-rays from a star usually indicates the presence of a close binary system where matter is transferred from the extended atmosphere of one star to the surface of a smaller, denser companion, which may be a NEUTRON STAR or a BLACK HOLE.

X-ray telescopes In X-RAY ASTRONOMY, most interest centres on the photon wavelength range 10 to 1,000nm. These X-rays are easily absorbed by quite small amounts of matter and thus cannot be handled by refracting optics. In addition, because the refractive index for X-rays at a vacuum–material boundary is typically $\mu = 0.9995$, the rays cannot be reflected at anywhere near normal incidence. However, for angles of grazing incidence less than about 2°, the rays can undergo total external reflection and grazing incidence reflecting systems can be constructed. X-ray reflectivity is shown plotted against angle of grazing incidence in the accompanying diagram.

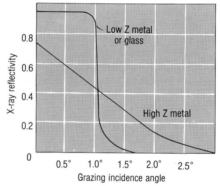

X-ray telescopes: reflectivity

For a glass or a metal of low atomic number, reflectivity cuts off sharply with angle. For metals of high atomic number (eg, gold, nickel), the cut-off is more gradual. A number of possible telescope geometries are also illustrated. They were originally proposed by Wolter as possible X-ray microscope elements and were adapted for use in X-ray telescopes following a suggestion of Giacconi and Rossi. The concentric nesting of reflectors in an attempt to increase collecting area can be used for all three configurations. The Wolter Type 0 configuration

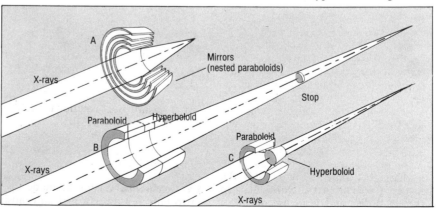

X-ray telescopes: instrument types

cannot form an image due to severe comatic aberration. The Wolter 1 and 2 configurations can image by virtue of providing two reflections for the incoming radiation, thus fulfilling the Abbe sine condition for the elimination of coma. Because of the short wavelength of X-rays, the reflecting surfaces must be figured and polished with submicron precision. The first large astronomical X-ray telescope was placed in orbit in 1978 when NASA launched the EINSTEIN OBSERVATORY. The future of the subject requires the launching of large X-ray observatories and both NASA and ESA plan to do so in the next decade. *JLC.*

Year The time taken by the Earth to complete one revolution around the Sun. There are several different values, according to the reference system. (1) **Sidereal**: one revolution relative to the fixed stars (365.25636 mean solar days). (2) **Tropical** or solar: one revolution relative to the equinoxes (365.24219 days). (3) **Anomalistic**: one revolution relative to perihelion: as the perihelion point moves eastward, this is about 4m 43.5s longer than a sidereal year, or 365.25964 days. (4) **Civil** or **Calendar**: obviously a whole number of days, usually 365, but 366 in a leap year (*see* CALENDAR): it averages 365.2425 days. (5) **Lunar**: 12 synodic months, or about 354 days. (6) **Eclipse**: one revolution relative to the same node of the Moon's orbit (346.62003 days). Nineteen eclipse years are 6585.78 days, almost exactly the same as the SAROS.

Relative to the stars, the Sun completes: in a sidereal year, 360°; in a tropical year, 360° − 50″.26; in an anomalistic year, 360° + 11″.25. *TJCAM.*

Yerkes Observatory Major American observatory at Williams Bay, in the Chicago area. Unlike many observatories it is readily accessible, and actually backs onto a golf course! It was founded in 1897, due to the persuasion of George Ellery HALE, who persuaded the millionaire Charles T. Yerkes to finance the main telescope — a 40in (101.6cm) refractor, still the largest refractor in the world. It is in perfect order, and is in use on every clear night. The observatory also includes a 41in (104cm) reflector and a 24in (61cm) reflector. *PM.*

Yerkes Observatory: largest refractor

"Y-feature" (Venus). In 1957, the French amateur astronomer Charles Boyer recorded sequences of ultraviolet images of Venus and noted a dark horizontal Y feature recurrent every four days. He advocated a retrograde atmospheric circulation in four days. This was received with scepticism outside France, but in 1974 the spacecraft Mariner 10 passed Venus and produced an ultraviolet film which confirmed the fact. In 1985, the Soviet Union placed two balloons in the atmosphere of Venus which were carried by the current.

Young, Charles Augustus (1834–1908). American astronomer who made some of the first spectroscopic studies of the SOLAR CORONA and who discovered the flash spectrum of the SOLAR CHROMOSPHERE during a total eclipse in 1870. He also compiled a catalogue of bright spectral lines in the Sun and used these to measure its rotational velocity. Young wrote a number of astronomy textbooks which became bestsellers. Most of his research was carried out at Dartmouth College.

YY Geminorum Castor C, made up of two faint red dwarfs: it is an eclipsing binary.

Z Andromedæ Prototype of a class of peculiar variable stars, otherwise known as symbiotic stars; such a system is made up of a hot dwarf together with a red giant. The components are close together, and there are spectral changes hard to interpret. Z Andromedæ itself has a range of from 8.3 to 12.4. Other examples of symbiotic variables are V Sagittæ and R Aquarii.

Zaniah The star Eta Virginis, now of magnitude 3.9. It was ranked as magnitude 3 in ancient times, but it is unlikely that any real change in magnitude has taken place.

Z

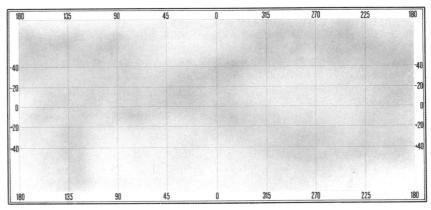

"Y-feature": a dark, horizontal feature that appears every four days

Zanstra, Herman (1894–1972). Dutch theoretical astrophysicist whose main work was on spectra of diffuse and planetary nebulæ and on the theory of thermal ionization of stellar atmospheres.

Zap crater/pit Minute micrometeorite crater in rock.

Z Camelopardalis stars A small subgroup of the class of eruptive variable stars known as dwarf novæ. Most dwarf novæ are classified as U GEMINORUM STARS, of which the brightest member is SS Cygni. This sub-group has a fairly smooth decline in brightness from the maximum. The Z Camelopardalis stars are different in that they experience occasional "standstills", remaining more or less constant in brightness for a long period. These standstills seem always to begin during

Z

a decline from maximum. When the standstill ends, the star drops to minimum brightness, and then resumes its "normal" behaviour. Both the occurrence of the standstills and their duration, ranging from a few days to many months, are quite unpredictable.

Z Camelopardalis itself is the brightest member of the class. It can reach magnitude 10.2 at brightest and falls to about 14.5 at minimum. The "normal" interval between outbursts is roughly 22 days. *JWM*.

Zeehæn Rupes Major ridge on Mercury; 50°N, 158°W, east of the CALORIS BASIN.

Zeeman effect The splitting of a spectral line into several components by the presence of a magnetic field. Where the components are unresolved, a broadened line is seen. The effect allows magnetic fields on the Sun and stars to be measured.

Zelenchukskaya Astrophysical Observatory Major Russian observatory at Mount Semirodriki, in the Caucasus:

Zelenchukskaya Astronomical Observatory

43° 49′ 32″.N, 41° 35′.4E. Its altitude above sea-level is 3,000ft (973m). The main instrument is the 236in (6m) reflector, the largest telescope in the world, which was completed in 1976. Though its aperture is almost a metre larger than its nearest rival, the Hale reflector at Palomar, it cannot be said that the telescope has been a real success. The mounting, of the altazimuth type and computer-controlled, is satisfactory enough, but there have been problems with the optics which have yet to be fully resolved. However, modifications are being put in hand,

and it is hoped that the reflector will eventually fulfil its undoubted potential. *PM*.

Zenith The overhead point, 90° from the horizon (Astronomical Zenith). Since the Earth is not a sphere, Geocentric Zenith is distinguished as a line joining the centre of the Earth to the observer, extended to the CELESTIAL SPHERE.

Zenithal hourly rate Often abbreviated to ZHR. This is the probable number of meteors observed per hour from a meteor shower, if its radiant were at the observer's zenith (ie, directly overhead) under ideal seeing conditions and in the absence of moonlight. Shower meteor rates appear to vary with the altitude (a) of the radiant. To obtain the normalized ZHR, the observed rate is usually multiplied by a factor (F), which is given by F = cosec a or F = cosec (a + 6°).

Zenith distance Angular distance from the ZENITH to a celestial body. Usually topocentric, ie, measured from the observer's position, but sometimes geocentric, ie, measured from the centre of the Earth.

Zero-Age Main Sequence Sometimes abbreviated to ZAMS. This is the MAIN SEQUENCE as defined by stars of zero-age, that is before they have undergone any substantial evolution.

Zero gravity The apparent absence of gravitational forces within a system which is in a state of free fall; a body in a "zero gravity" or free-fall state experiences no sensation of weight.

An observer standing on the Earth's surface feels the sensation of weight because the ground prevents him from accelerating toward the Earth's centre under the action of the Earth's gravity by exerting an equal and opposite upward force against his feet. If the observer were to be placed in a freely-falling container both he and the floor of the container would accelerate downward at exactly the same rate; there would be no force between his feet and the floor, and no sensation of weight. The observer would be unable to tell whether he was falling freely in a gravitational field or floating in space far removed from any gravitating mass.

The term "zero gravity" does not imply a total absence of gravity, rather it refers to an absence of any detectable gravitational forces. Although gravity becomes very weak at large distances from massive bodies, it nowhere

declines absolutely to zero. *IKN*.

Zeta Aurigae Sometimes known by its old proper name of Sadatoni, Zeta Aurigae is the faintest of the three Hædi or "Kids" near Capella. It is an eclipsing binary with a period of 972 days; the visual range is from magnitude 3.7 to 4.2. The system consists of a hot B-type star and a supergiant companion of type K; the distance is over 500 light-years. During the partial phase of the eclipse of the hot star, its light shines through the outer, rarefied layers of the supergiant, and there are complicated and informative spectroscopic effects. The eclipse of the B-type star is total for 38 days; this is preceded and followed by partial stages lasting for 32 days each. *PM*.

Zeta Herculis A fine binary, with a period of 34 years; the components are of magnitudes 3.1 and 5.6. The separation is always below 2″. The old proper name of "Rutilicus" is still occasionally used.

Zeta Ophihchi (Han). 09.5-type star; magnitude 2.56; it is 5,000 times as luminous as the Sun.

Zeta Orionis (Alnitak). The southernmost star of Orion's Belt; type 09.5; distance 1,100 light-years; luminosity 19,000 times that of the Sun. It has a physically-associated companion of magnitude 4.2; the present separation is 2″.4.

Zeta Tauri Third-magnitude star in Taurus, close to the Crab Nebula. Its old name is Heka.

Zodiac A belt about 8°–9° on either side of the ECLIPTIC, in which the Sun, Moon and planets (except Pluto) appear to move. It passes through the 12 Zodiacal constellations or signs: Aries, Taurus, Gemini, Cancer, Leo, Virgo, Libra, Scorpio, Sagittarius, Capricornus, Aquarius and Pisces. The signs are each 30° long, but due to PRECESSION no longer coincide with the constellations of the same name. The ecliptic also passes through Ophiuchus, and the Zodiac also includes parts of Cetus, Orion and Sextans.

Zodiacal Band A very faint band of light extending along the ECLIPTIC which joins the ZODIACAL LIGHT with the GEGENSCHEIN at the anti-solar point. The Zodiacal Band is caused by the scattering of sunlight toward the Earth by dust particles in interplanetary space which form a cloud surrounding the Sun. It can be observed only in conditions of extreme clarity under moonless conditions and even in the tropics is a difficult object. The band is parallel-sided and fades in brightness on either side of the median line. The band is faintest at some 135° from the Sun and brightens toward the cone of the Zodiacal Light and toward the Gegenschein. The brightness is variable. *RJL.*

Zodiacal dust Interplanetary dust orbiting the Sun in a disk roughly symmetrical about the ECLIPTIC plane. Sunlight scattered from the dust produces the ZODIACAL LIGHT. The smallest grains probably weigh about one million-millionth of a gram. Since such small particles eventually fall into the Sun, or are blown out of the Solar System entirely, the dust must be continuously replaced by new material.

Observations from IRAS showed that most of the dust lies in the asteroid belt and that it has a banded structure. Since these bands are unstable and can only last a few thousand years, new dust is probably formed during collisions between ASTEROIDS. Each collision produces a short-lived ZODIACAL BAND which eventually spreads out and merges with the general background of interplanetary dust. Additional material is doubtless released from comets and spread throughout the Solar System. *JKD.*

Zodiacal Light A faint conical beam of light rising from the western horizon after sunset and the eastern horizon before sunrise. It is best observed in the tropics, where the axis of the cone makes the steepest angle with respect to the horizon. In the northern hemisphere it is best visible in the evening at the spring equinox and alternatively in the morning at the autumnal equinox, and vice versa in the southern hemisphere. The beam is brightest at the centre and fades slowly toward the diffuse edges. The base is about 20° to 30° in width and the cone has a height to width ratio of about 3. The apex of the cone may be 60° to 90° from the Sun in the high latitudes and 100° to 110° in the tropics, where the light is brighter, resembling the densest part of the Milky Way in the field of Sagittarius.

Were the Zodiacal Light to be seen in full it would comprise an ellipse with the Sun at the centre. The major axis would lie at an angle of about 7° to the ecliptic and coincident with the solar equator. The spectrum is similar to the Sun and yellower than the Milky Way. Some Doppler broadening together with partial polarization both indicate that the light is mainly due to particle scattering rather than to electron scattering. The Zodiacal Light is therefore sunlight reflected from dust probably some 1 to 300 microns in particle size

Zodiacal Light: reflected sunlight

comprising a tenuous cloud surrounding the Sun and, as Blackwell (1960) says, the outer F corona.

The Zodiacal Light was discovered by Cassini in 1683, who noted it as a celestial light in the ZODIAC and predicted the presence of an aura of material surrounding the Sun. Weill (1966), Dufay (1966) and Robley (1980) have researched variations in the brightness of the Zodiacal Light in relation to the sunspot cycle and the variations in the SOLAR WIND which may provoke a fluorescent response.

Variations in brightness have been said to rise by a factor of two after SOLAR FLARES. Legrand (1985) has linked maximum brightness with the recurrent auroræ. *RJL.*

Zöllner, Johann Karl Friedrich (1834–82). German astronomer; inventor of the polarizing PHOTOMETER. He determined the albedos of the Moon and planets, and proposed a theory of the constitution of the Sun.

Zöllner photometer A visual photometer which uses a fixed and a rotating polarizing element to vary the apparent brightness of an artificial star until it is the same as a real star seen in the same field of view. The amount of rotation of the polarizer can be calibrated to give the apparent magnitude of the real star.

Zond probes Unmanned Russian space-

craft; eight were launched (1964 to 1970). Zonds 5, 6 and 7 went round the Moon and returned to Earth, making controlled landings.

Zone of avoidance A region of the sky close to the plane of the Milky Way where almost no external galaxies can be seen because of extinction by interstellar dust in or near the main plane of our own Galaxy.

Zucchius Lunar crater, 39 miles (63km) in diameter; 61°S, 50°W; in the Schiller area. It makes a pair with Segner, and is well formed.

Zupus Low-walled irregular lunar formation, distinguished by its very dark floor, 16 miles (26km) in diameter; 17°S, 52°W, south of the even darker-floored Billy.

Zürich number *See* WOLF NUMBER.

Zwicky, Fritz (1898–1974). Swiss astronomer who spent much of his career in the United States. He undertook much valuable research, but is probably best remembered for his discoveries and measurements of SUPERNOVÆ in external galaxies.

Z

Tables

The Messier Objects

*Identification uncertain

M	NGC	RA (2000.00) h m	Declination ° '	Mag. (vis.)	Size (mins of arc)	Distance	Constellation	Description
1	1952	05 34.5	+ 22 01	8.4	6 × 4	1050pc	Tau	Crab nebula
2	7089	21 33.5	− 00 49	6.3	12	16kpc	Aqr	Globular cluster
3	5272	13 42.2	+ 28 23	6.4	19	14kpc	CVn	Globular cluster
4	6121	16 23.6	− 26 32	6.4	23	2.3kpc	Sco	Globular cluster
5	5904	15 18.6	+ 02 05	6.2	20	8.3kpc	Ser	Globular cluster
6	6405	17 40.1	− 32 13	5.3	26	630pc	Sco	Galactic cluster
7	6475	17 53.9	− 34 49	4:	50	250pc	Sco	Galactic cluster
8	6523	18 03.8	− 24 23	6:	90 × 40	1.5kpc	Sgr	Lagoon nebula
9	6333	17 19.2	− 18 31	7.3	6	7.9kpc	Oph	Globular cluster
10	6254	16 57.1	− 04 06	6.7	12	5.0kpc	Oph	Globular cluster
11	6705	18 51.1	− 06 16	6.3	12	1.7kpc	Sct	Galactic cluster
12	6218	16 47.2	− 01 57	6.6	12	5.8kpc	Oph	Globular cluster
13	6205	16 41.7	+ 36 28	5.7	23	6.9kpc	Her	Globular cluster
14	6402	17 37.6	− 03 15	7.7	7	7.2kpc	Oph	Globular cluster
15	7078	21 30.0	+ 12 10	6.0	12	15kpc	Peg	Globular cluster
16	6611	18 18.8	− 13 47	6.4	8	1.8kpc	Ser	Gaseous nebula
17	6618	18 20.8	− 16 11	7.0	46 × 37	1.8kpc	Sgr	Omega nebula
18	6613	18 19.9	− 17 08	7.5	7	1.5kpc	Sgr	Galactic cluster
19	6273	17 02.6	− 26 16	6.6	5	6.9kpc	Oph	Globular cluster
20	6514	18 02.6	− 23 02	9:	29 × 27	1.6kpc	Sgr	Trifid nebula
21	6531	18 04.6	− 22 30	6.5	12	1.3kpc	Sgr	Galactic cluster
22	6656	18 36.4	− 23 54	5.9	17	3.0kpc	Sgr	Globular cluster
23	6494	17 56.8	− 19 01	6.9	27	660kpc	Sgr	Galactic cluster
24	6603	18 16.9	− 18 29	4.6	4	5.0kpc	Sgr	Galactic cluster
25	IC4725	18 31.6	− 19 15	6.5	35	600pc	Sgr	Galactic cluster
26	6694	18 45.2	− 09 24	9.3	9	1.5kpc	Sct	Galactic cluster
27	6853	19 59.6	+ 22 43	7.6	8 × 4	200pc	Vul	Dumbbell
28	6626	18 24.5	− 24 52	7.3	15	4.6kpc	Sgr	Globular cluster
29	6913	20 23.9	+ 38 32	7.1	7	1.2kpc	Cyg	Galactic cluster
30	7099	21 40.4	− 23 11	8.4	9	13kpc	Cap	Globular cluster
31	224	00 42.7	+ 41 16	4.8	160 × 40	700kpc	And	Galaxy
32	221	00 42.7	+ 40 52	8.7	3 × 2	700kpc	And	Galaxy
33	598	01 33.9	+ 30 39	6.7	60 × 40	700kpc	Tri	Galaxy
34	1039	02 42.0	+ 42 47	5.5	30	440pc	Per	Galactic cluster
35	2168	06 08.9	+ 24 20	5.3	29	870pc	Gem	Galactic cluster
36	1960	05 36.1	+ 34 08	6.3	16	1.3kpc	Aur	Galactic cluster
37	2099	05 52.4	+ 32 33	6.2	24	1.3kpc	Aur	Galactic cluster
38	1912	05 28.7	+ 35 50	7.4	18	1.3kpc	Aur	Galactic cluster
39	7092	21 32.2	+ 48 26	5.2	32	250pc	Cyg	Galactic cluster
40*		12 22.4	+ 58 05	9.0, 9.3	−		UMa	Double star
41	2287	06 47.0	− 20 44	4.6	32	670pc	CMa	Galactic cluster
42	1976	05 35.4	− 05 27	4:	66 × 60	460pc	Ori	Orion nebula
43	1982	05 35.6	− 05 16	9:		460pc	Ori	Orion nebula
44	2632	08 40.1	+ 19 59	3.7	90	158pc	Cnc	Praesepe
45	−	03 47.0	+ 24 67	1.6	120	126pc	Tau	Pleiades
46	2437	07 41.8	− 14 49	6.0	27	1.8kpc	Pup	Galactic cluster
47*	2422	07 36.6	− 14 30	5.2	25	548pc	Pup	Galactic cluster
48*	2548	08 13.8	− 05 48	5.5	35	480pc	Hya	Galactic cluster
49	4472	12 29.8	+ 08 00	8.6	4 × 4	11Mpc	Vir	Galaxy
50	2323	07 03.2	− 08 20	6.3	16	910pc	Mon	Galactic cluster
51	5194−5	13 29.9	+ 47 12	8.1	12 × 6	2Mpc	CVn	Whirlpool
52	7654	23 24.2	+ 61 35	7.3	13	2.1kpc	Cas	Galactic cluster
53	5024	13 12.9	+ 18 10	7.6	14	20kpc	Com	Globular cluster
54	6715	18 55.1	− 30 29	7.3	6	15kpc	Sgr	Globular cluster
55	6809	19 40.0	− 30 58	7.6	15	5.8kpc	Sgr	Globular cluster
56	6779	19 16.6	+ 30 11	8.2	5	14kkpc	Lyr	Globular cluster
57	6720	18 53.6	+ 33 02	9.3	1 × 1	550kpc	Lyr	Planetary nebula
58	4579	12 37.7	+ 11 49	8.2	4 × 3	11Mpc	Vir	Galaxy
59	4621	12 42.0	+ 11 39	9.3	3 × 2	11Mpc	Vir	Galaxy
60	4649	12 43.7	+ 11 33	9.2	4 × 3	11Mpc	Vir	Galaxy
61	4303	12 21.9	+ 04 28	9.6	6	11Mpc	Vir	Galaxy
62	6266	17 01.2	− 30 07	8.9	6	6.9kpc	Oph	Globular cluster
63	5055	13 15.8	+ 42 02	10.1	8 × 3	4Mpc	CVn	Galaxy
64	4826	12 56.7	+ 21 41	6.6	8 × 4	6Mpc	Com	Galaxy
65	3623	11 18.9	+ 13 05	9.5	8 × 2		Leo	Galaxy
66	3627	11 20.2	+ 12 59	8.8	8 × 2		Leo	Galaxy
67	2682	08 50.4	+ 11 49	6.1	18	830pc	Cnc	Galactic cluster
68	4590	12 39.5	− 26 45	9:	9	12kpc	Hya	Globular cluster
69	6637	18 31.4	− 32 21	8.9	4	7.2kpc	Sgr	Globular cluster
70	6681	18 43.2	− 32 18	9.6	4	20kpc	Sgr	Globular cluster
71	6838	19 53.8	+ 18 47	9:	6	5.5kpc	Sgr	Globular cluster
72	6981	20 53.5	− 12 32	9.8	5	18kpc	Aqr	Globular cluster
73	6994	20 58.9	− 12 32	9.0	3		Aqr	Star group
74	628	01 36.7	+ 15 47	10.2	8	8Mpc	Psc	Galaxy
75	6864	20 06.1	− 21 55	8.0	5	24kpc	Sgr	Globular cluster
76	650	01 42.4	+ 51 34	12.2	2 × 1	2.5kpc	Per	Planetary nebula
77	1068	02 42.7	− 00 01	8.9	2	16Mpc	Cet	Galaxy
78	2068	05 46.7	+ 00 03	8.3	8 × 6	500pc	Ori	Gaseous nebula
79	1904	05 24.5	− 24 33	7.9	8	13kpc	Lep	Globular cluster
80	6093	16 17.0	− 22 59	7.7	5	11kpc	Sco	Globular cluster
81	3031	09 55.6	+ 69 04	7.9	16 × 10	3Mpc	UMa	Galaxy
82	3034	09 55.8	+ 69 41	8.8	7 × 2	3Mpc	UMa	Galaxy
83	5236	13 37.0	− 29 52	10.1	10 × 8	4Mpc	Hya	Galaxy
84	4374	12 25.1	+ 12 53	9.3	3	11Mpc	Vir	Galaxy

M	NGC	RA (2000.00) h m	Declination ° '	Mag. (vis.)	Size (mins of arc)	Distance	Constellation	Description
85	4382	12 25.4	+18 11	9.3	4 × 2	11Mpc	Com	Galaxy
86	4406	12 26.2	+12 57	9.7	4 × 3	11Mpc	Vir	Galaxy
87	4486	12 30.8	+12 24	9.2	3	11Mpc	Vir	Galaxy
88	4501	12 32.0	+14 25	10.2	6 × 3	11Mpc	Com	Galaxy
89	4552	12 35.7	+12 33	9.5	2	11Mpc	Vir	Galaxy
90	4569	12 36.8	+13 10	10.0	6 × 3	11Mpc	Vir	Galaxy
91*	4571	12 35.4	+14 30	0.2	–	–	Vir	Planetary nebula
92	6341	17 17.1	+43 08	6.1	12	11kpc	Her	Globular cluster
93	2447	07 44.6	−23 52	6.0	18	1.1kpc	Pup	Galactic cluster
94	4736	12 50.9	+41 07	7.9	5 × 4	6Mpc	CVn	Galaxy
95	3351	10 44.0	+11 42	10.4	3	9Mpc	Leo	Galaxy
96	3368	10 46.8	+11 49	9.1	7 × 4	9Mpc	Leo	Galaxy
97	3507	11 14.8	+55 01	12.0	3	800pc	UMa	Owl nebula
98	4192	12 13.8	+14 54	10.7	8 × 2	11Mpc	Com	Galaxy
99	4254	12 18.8	+14 25	10.1	4	11Mpc	Com	Galaxy
100	4321	12 22.9	+15 49	10.6	5	11Mpc	Com	Galaxy
101	5457	14 03.2	+54 21	9.6	22	3Mpc	UMa	Galaxy
102*			= =		=		–	–
103	581	01 33.2	+60 42	7.4	6	2.6kpc	Cas	Galactic cluster
104	4594	12 40.0	−11 37	8.7	7 × 2	4.4Mpc	Vir	Sombrero nebula
105	3379	10 47.8	+12 35	9.2	2 × 2		Leo	Galaxy
106	4258	12 19.0	+47 18	8.6	20 × 6		UMa	Galaxy
107	6171	16 32.5	−13 03	9.2	8		Oph	Globular cluster
108	3556	11 11.5	+55 40	10.7	8 × 2		UMa	Galaxy
109	3992	11 57.6	+53 23	10.8	7	–	UMa	Galaxy

Galactic Clusters not in Messier's Catalogue

With Integrated Magnitude above 7.0
(Class: c = v. loose, d = loose, poor, a = moderately rich, f = fairly rich, g = rich and condensed)

NGC	R.A. h m	Declination ° '	Diameter	Magnitude	Number of stars	Class	Constellation
752	01 57.8	+37 41	45	7.0	70	d	Andromeda
869	02 19.0	+57 09	36	4.4	350	f	Perseus (h Persei)
884	02 22.4	+57 07	36	4.7	300	e	Perseus (Chi)
1245	03 14.7	+47 15	30	6.9	40	e	Perseus
1444	03 49.4	+52 40	4	6.4	15		Perseus
1502	04 07.7	+62 20	7	5.3	15	e	Camelopardalis
1528	04 15.4	+51 14	25	6.2	80	e	Perseus
Mel.2	04 27	+16	330	0.8	40	c	Taurus (Hyades)
1647	04 46.0	+19 04	40	6.3	30	c	Taurus
1746	05 03.6	+23 49	45	6.0	60	e	Taurus
2169	06 08.4	+13 57	5	6.4	18	d	Orion
2175	06 09.8	+20 19	18	6.7	15		Orion
2244	06 32.4	+04 52	40	6.2	16	c	Monoceros
2264	06 41.1	+09 53	30	4.7	23		Monoceros
2281	06 49.3	+41 04	17	6.7	30	e	Auriga
2301	06 51.8	+00 28	15	5.8	60	d	Monoceros
2353	07 14.6	−10 18	20	5.3	25	d	Monoceros
2422	07 36.6	−14 30	25	4.5	50	d	Puppis
2423	07 37.1	−13 52	20	6.9	60	d	Puppis
2451	07 45.4	−37 58	45	3.6	50		Puppis
2477	07 52.3	−38 33	25	5.7	300	g	Puppis
2516	07 58.3	−60 52	60	3.0	80	g	Carina
2547	08 10.7	−49 16	15	5.1	50	d	Vela
2546	08 12.4	−37 38	40	4.6	50		Puppis
IC2931	08 40.2	−53 04	40	2.6	20	c	O Velorum
IC2395	08 41.1	−48 12	10	4.6	16	e	Vela
H.3	08 44.8	−45 59	7	6.2	35	e	Vela
3114	10 02.7	−60 07	30	4.4	100	e	Carina
3228	10 21.8	−51 43	30	6.5	12	f	Vela
IC2581	10 27.4	−57 38	5	5.2	35	f	Carina
IC2602	10 43.2	−64 24	70	1.6	32	c	Carina
3532	11 06.4	−58 40	60	3.3	130	f	Carina
3766	11 36.1	−61 37	10	5.1	60	g	Centaurus
Mel.111	12 25	+26	275	2.7	30	c	Coma Berenices
4755	12 53.6	−60 20	10	5.2	50	g	Crux
5460	14 07.6	−48 19	30	6.3	25	d	Centaurus
6025	16 03.7	−60 30	10	5.8	30	d	Triangulum Australe
6067	16 13.2	−54 13	15	6.7	120	f	Norma
6087	16 18.9	−57 54	20	6.0	35	d	Norma
6124	16 25.6	−40 40	25	6.3	120	e	Scorpius
6167	16 34.4	−49 36	18	6.4	110		Ara
6193	16 41.3	−48 46	20	5.0	30	e	Ara
6383	17 34.8	−32 24	6	5.5	12	e	Scorpius
6530	18 04.8	−24 20	10	6.3	25	e	Sagittarius
6633	18 27.7	+06 34	20	4.9	65	d	Ophiuchus
IC4756	18 39.0	+05 27	70	5.1	80	d	Serpens
6716	18 54.6	−19 53	7	6.9	20		Sagittarius
6871	20 05.9	+35 47	37	5.6	60		27 Cygnus
6910	20 23.1	+40 47	8	6.7	40	d	Cygnus
IC1396	21 39.1	+57 30	50	5.1	30		Cepheus
7160	21 53.7	+62 36	7	6.6	25		Cepheus

Galaxies not in Messier's Catalogue

With Integrated Magnitude above 10.0

NGC	R.A. h m	Decl. ° '	Photo Mag.	Diameter	Type	Constellation
55	00 14.9	−39 11	7.8	25.0 × 3.0	S	Sculptor
205	00 40.4	+41 41	8.9	10.0 × 4.5	E6	Andromeda
247	00 47.1	−20 46	9.5	18.2 × 4.5	S	Cetus
253	00 47.6	−25 17	7.0	24.6 × 4.5	Sc	Sculptor
–	00 52.7	−72 50	1.5	216 × 216	1	SMC (Tucana)
–	05 23.6	−69 45	0.5	432 × 432	1	LMC (Dorado)
2403	07 36.9	+65 36	8.8	16.8 × 10.0	Sc	Camelopardalis
2903	09 32.2	+21 30	9.5	11.0 × 4.6	Sb	Leo
4449	12 28.2	+44 06	9.9	4.1 × 3.4	1	Canes Venatici
4631	12 42.1	+32 32	9.7	15.1 × 3.3	Sc	Canes Venatici
4945	13 05.4	−49 28	9.2	11.5 × 2.0	S	Centaurus
5128	13 25.5	−43 01	7.2	10.0 × 8.0	1	Centaurus A
6822	19 44.9	−14 48	9.2	16.2 × 11.2	1	Sagittarius
6946	20 34.8	+60 00	9.7	11.0 × 9.8	Sc	Cepheus
7793	23 57.8	−32 35	9.7	6.0 × 4.0	S	Sculptor

Globular Clusters not in Messier's Catalogue

With Integrated Magnitude above 7.5

NGC	R.A. h m	Decl. ° '	Dia	Magnitude (v – visual) (p – photographic)	Constellation
104	00 24.1	−72 05	23.0	3.0p	47 Tucanæ
288	00 52.8	−26 35	10.0	7.2p	Sculptor
362	01 03.2	−70 51	5.3	6.8p	Tucana
1851	05 14.1	−40 03	5.3	8.1v	Columba
2808	09 12.0	−64 52	6.3	5.7v	Carina
3201	10 17.6	−46 25	7.7	7.4v	Vela
4833	12 59.6	−70 53	4.7	6.8v	Musca
5139	13 26.8	−47 29	23.0	3.7v	Omega Centauri
5897	15 17.4	−21 01	7.3	7.3v	Libra
6362	17 31.9	−67 03	6.7	7.1p	Ara
6388	17 36.3	−44 44	3.4	7.1p	Scorpius
6397	17 40.7	−53 40	19.0	4.7p	Ara
6541	18 08.0	−43 42	6.3	5.8p	Corona Australis
6723	18 59.6	−36 38	5.8	6.0p	Sagittarius
6752	19 10.9	−59 59	13.3	4.6p	Pavo

Planetary Nebulæ not in Messier's Catalogue

NGC	R.A. h m	Decl. ° '	Diameter (secs of arc)	Magnitude	Constellation
246	00 47.0	−11 53	240 × 210	8.5	Cetus
2392	07 29.2	+20 55	47 × 43	8.3	Gemini
3132	10 07.7	−40 26	84 × 53	8.2	Antlia
3918	11 50.3	−57 11	32 × 13	8.4	Centaurus
7009	21 04.2	−11 22	44 × 26	8.4	Aquarius
7293	22 29.6	−20 48	900 × 720	6.5	Aquarius

Bright Diffuse Nebulæ not in Messier's Catalogue

	R.A.		Declination		Diameter	Magnitude	Constellation	Description
	h	m	°	'	(secs. of arc)			
NGC1499	04	00.7	+ 36	37	145 × 40	4.0	Perseus	California Nebula
NGC1554	04	21.8	+ 19	32	variable	var.	Taurus	Hind's Nebula (T Tauri Neb)
NGC2070	05	38.7	− 69	06	20 × 20		Dorado	Tarantula Nebula
2237/9	06	32.3	+ 05	03	64 × 61		Monoceros	Rosette Nebula
2261	06	39.2	+ 08	44	variable	var.	Monoceros	Hubble's Nebula (R. Monoc.)
2264	06	40.9	+ 09	54	60 × 30	4.7	Monoceros	Cone Nebula (S. Monocerotis)
IC1318	20	27.9	+ 40	00	40 × 20		Cygnus	Cirrus Nebula
6960	20	45.7	+ 30	43	70 × 6		52 Cygnus	Cirrus Nebula
IC5068	20	50.8	+ 42	31	85 × 75	1.3	Cygnus	Pelican Nebula (nr Deneb)
6992/5	20	56.4	+ 31	43	78 × 8		Cygnus	Cirrus Nebula
7000	20	58.8	+ 44	20	120 × 100	1.3	Cygnus	N. America Nebula

The Brightest Stars
List includes all the stars above magnitude 2.00 (Epoch 2000.00)

Star		R.A.			Declination			Magnitude (apparent)	Magnitude (absolute)	Spectrum
		h	m	s	°	'	"			
Sirius	Alpha Canis Majoris	06	45	09	− 16	42	58	−1.46	+ 1.5	A0
Canopus	Alpha Carinæ	06	23	57	− 52	41	44	−0.72	− 8.5	F0
	Alpha Centauri	14	39	37	− 60	50	02	−0.27	+ 4.4	G2 + K1
Arcturus	Alpha Boötis	14	55	40	+ 19	10	57	−0.04	− 0.2	K0
Vega	Alpha Lyræ	18	36	56	+ 38	47	01	0.03	+ 0.5	A0
Capella	Alpha Aurigæ	05	16	41	+ 45	59	53	0.08	0.3	G8
Rigel	Beta Orionis	05	14	32	− 08	12	06	0.12	− 7.1	B8p
Procyon	Alpha Canis Minoris	07	39	18	+ 05	13	30	0.38	+ 2.6	F5
Achernar	Alpha Eridani	01	37	43	− 57	14	12	0.46	− 1.6	B5
Betelgeux	Alpha Orionis	05	55	10	+ 07	24	26	0.5v	− 5.6v	M2
Agena	Beta Centauri	14	03	49	− 60	22	22	0.61	− 5.1	B1
Altair	Alpha Aquilæ	19	50	47	+ 08	52	06	0.77	+ 2.2	A7
Aldebaran	Alpha Tauri	04	35	55	+ 16	30	33	0.85	− 0.3	K5
Acrux	Alpha Crucis	12	26	36	− 63	05	57	0.83	−3.9, − 3.4	B1 + B3
Antares	Alpha Scorpii	16	29	24	− 26	25	55	0.96v	− 4.7	M1
Spica	Alpha Virginis	13	25	11	− 11	09	41	0.98v	− 3.5	B1
Fomalhaut	Alpha Piscis Austrini	22	57	39	− 29	37	20	1.16	+ 2.0	A3
Pollux	Beta Geminorum	07	45	19	+ 28	01	34	1.14	+ 0.2	K0
Deneb	Alpha Cygni	20	41	26	+ 45	16	49	1.25	− 7.5	A2p
	Beta Crucis	12	47	43	− 59	41	19	1.25	− 5.0	B0
Regulus	Alpha Leonis	10	08	22	+ 11	58	02	1.35	− 0.6	B7
Castor	Alpha Geminorum	07	34	36	+ 31	53	18	1.58	+ 1.2	A0 + A2
	Gamma Crucis	12	31	10	− 57	06	47	1.63	− 0.5	M3
Adhara	Epsilon Canis Majoris	06	58	38	− 28	58	20	1.50	− 4.4	B1
Alioth	Epsilon Ursæ Majoris	12	54	02	+ 55	57	35	1.77	+ 0.2	A0
Bellatrix	Gamma Orionis	05	25	08	+ 06	20	59	1.64	− 3.6	B2
Shaula	Lambda Scorpii	17	33	36	− 37	06	14	1.63	− 3.0	B2
Avoir	Epsilon Carinæ	08	22	31	− 59	30	34	1.86	− 2.1	K0
Alnair	Alpha Gruis	22	08	14	− 46	57	40	1.74	− 1.1	B5
Alnilam	Epsilon Orionis	05	36	13	− 01	12	07	1.70	− 6.2	B0
Al Nath	Beta Tauri	05	26	17	+ 38	36	27	1.65	− 1.6	B7
Miaplacidus	Beta Carinæ	09	13	12	− 69	43	02	1.68	− 0.6	A0
Sargas	Theta Scorpii	17	37	19	− 42	59	52	1.87	− 5.6	F0
	Alpha Trianguli Australe	16	48	40	− 69	01	39	1.92	− 0.1	K2
Menkarlina	Beta Aurigæ	05	59	32	+ 44	56	51	1.90	+ 0.6	A2
Mirphak	Alpha Persei	03	24	19	+ 49	51	40	1.80	− 4.6	F5
Alkaid	Eta Ursæ Majoris	13	47	32	+ 49	18	48	1.86	− 1.7	B3
Alnitak	Zeta Orionis	05	40	46	− 01	56	34	1.77	− 5.9	09.5
Regor	Gamma Velorum	08	09	32	− 47	20	12	1.78	− 4.1	WC7
Alhena	Gamma Geminorum	06	37	43	+ 16	23	57	1.93	− 0.1	A0
Dubhe	Alpha Ursæ Majoris	11	03	44	+ 61	45	03	1.79	+ 0.2	K0
Kaus Australis	Epsilon Sagittarii	18	24	10	− 34	23	05	1.85	− 0.3	B9
Wezea	Delta Canis Majoris	07	08	23	− 26	23	36	1.86	− 8.0	F8p
	Alpha Pavonis	20	25	39	− 56	44	06	1.94	− 2.3	B3
Mirzam	Beta Canis Majoris	06	22	42	− 17	57	22	1.98	− 4.8	B1
Alphard	Alpha Hydræ	09	27	35	− 08	39	31	1.98	− 0.2	K3
Polaris	Alpha Ursæ Minoris	02	31	50	+ 89	15	51	1.99v	− 4.6	F8
Algieba	Gamma Leonis	10	19	58	+ 19	50	30	1.99	+ 0.2	K0 + G7

Catalogue of Variable Stars

Type: M = Mira type. EA = Eclipsing binary, Algol type. EB = Eclipsing binary, Beta Lyræ type. C = Cepheid. CW = W Virgubus type. RR = Lyrae type. SR = Semi-regular. I = Irregular. RV = RV Tauri type. RCrB = R Coronae Borealis type. RN = Recurrent nova. N = Nova. Symb = Symbiotic. Z And = Z Andromedae type. UG = Geminorum (SS Cygni type).

Variable	R.A.		Decl.		Mag.		Period	Type
	h	m	°	'	max	min	(days)	
W Ceti	00	02.1	− 14	41	7.1	14.8	351.3	M
T Andromedae	00	22.4	+ 27	00	7.7	14.5	280.8	M
T Cassiopeiæ	00	23.2	+ 55	48	6.9	13.0	445	M
R Andromedæ	00	24.0	+ 38	35	5.8	14.9	409.3	M
Alpha Cas.	00	40.5	+ 56	32	2.2	2.5?	−	Suspected
Gamma Cas.	00	56.7	+ 60	43	1.6	3.3	−	I
R Piscium	01	30.6	+ 02	53	7.1	14.8	344.0	M
UV Ceti	01	38.8	− 17	58	6.8	13.0	−	Flare star
R Arietis	02	16.1	+ 25	03	7.4	13.7	186.8	M
Omicron Ceti	02	19.3	− 02	59	1.7	10.1	332.0	M
S Persei	02	22.9	+ 58	35	7.9	11.5	Long	SR
R Ceti	02	26.0	− 00	11	7.2	14.0	166.2	M
U Ceti	02	33.7	− 13	09	6.8	13.4	234.8	M
R Trianguli	02	37.0	+ 34	16	5.4	12.6	226.5	M
T Arietis	02	48.3	+ 17	31	7.5	11.3	317	SR
R Horologii	02	53.9	− 49	53	4.7	14.3	404.0	M
T Horologii	03	00.9	− 50	39	7.2	13.7	217.7	M
Rho Persei	03	05.2	+ 38	50	3.3	4.0	50	SR
Beta Persei	03	08.2	+ 40	57	2.2	3.4	2.87	EA
R Persei	03	30.1	+ 35	40	8.1	14.8	210.0	M
T Eridani	03	55.2	− 24	02	7.4	13.2	252.2	M
Lambda Tauri	04	00.7	+ 12	29	3.3	3.8	3.9	EA
R Reticuli	04	33.5	− 63	02	6.5	14.0	278.2	M
R Pictoris	04	46.2	− 49	15	6.7	10.0	164.0	SR
R Leporis	04	59.6	− 14	48	5.5	11.7	432.1	M
Epsilon Aurigæ	05	02.0	+ 43	49	2.9	3.8	9892.2	Eclips.
Zeta Aurigæ	05	02.5	+ 41	05	3.7	4.1	972.1	Eclips.
T Leporis	05	04.8	− 21	54	7.4	13.5	368.1	M
W Orionis	05	05.4	+ 01	11	5.9	7.7	212	SR
R Aurigæ	05	17.3	+ 53	35	6.7	13.9	457.5	M
T Columbæ	05	19.3	− 33	42	6.6	12.7	225.9	M
R Octantis	05	26.1	− 86	23	6.4	13.2	405.6	M
S Orionis	05	29.0	− 04	42	7.5	13.5	419.2	M
Beta Doradûs	05	33.6	− 62	29	3.7	4.1	9.8	C
U Aurigæ	05	42.1	+ 32	02	7.5	15.5	408.1	M
Alpha Orionis	05	55.2	+ 07	24	0.1	0.9	2110	SR
U Orionis	05	55.8	+ 20	10	4.8	12.6	372.4	M
Eta Geminorum	06	14.9	+ 22	30	3.1	3.9	233	SR
V Monocerotis	06	22.7	− 02	12	6.0	13.7	333.8	M
T Monocerotis	06	25.2	+ 07	05	6.0	6.6	27.0	C
X Geminorum	06	47.1	+ 30	17	7.5	13.6	263.7	M
R Lyncis	07	01.3	+ 55	20	7.2	14.5	378.7	M
Zeta Geminorum	07	04.1	+ 20	34	3.7	4.1	10.2	C
R Geminorum	07	07.4	+ 22	42	6.0	14.0	369.8	M
R Canis Min.	07	08.7	+ 10	01	7.3	11.6	337.8	M
VZ Camelopardalis	07	31.1	+ 82	25	4.8	5.2	23.7	SR
Z Puppis	07	32.6	− 20	40	7.2	14.6	499.7	M
T Geminorum	07	49.3	+ 23	44	8.0	15.0	287.8	M
V Puppis	07	58.2	− 49	15	4.7	5.2	1.45	EB
R Cancri	08	16.6	+ 11	44	6.1	11.8	361.6	M
V Cancri	08	21.7	+ 17	17	7.5	13.9	272.1	M
V Carinæ	08	28.7	− 60	07	7.0	7.8	6.7	C
T Velorum	08	37.7	− 47	22	7.7	8.3	4.6	C
S Hydræ	08	53.6	+ 03	04	7.4	13.3	256.4	M
X Cancri	08	55.4	+ 17	14	5.6	7.5	195	SR
T Pyxidis [1]	09	04.7	− 32	23	6.3	14.0	−	RN
W Cancri	09	09.9	+ 25	15	7.4	14.4	393.2	M
R Carinæ	09	32.2	− 62	47	3.9	10.5	308.7	M
ZZ Carinæ	09	45.2	− 62	30	3.3	4.2	35.5	C
R Leonis Mi.	09	45.6	+ 34	31	6.3	13.2	371.9	M
R Leonis	09	47.6	+ 11	25	4.4	11.3	312.4	M
R Ursæ Maj.	10	44.6	+ 68	47	6.7	13.4	301.7	M
Eta Carinæ	10	45.1	− 59	41	− 0.8	7.9	−	I
VW Ursæ Maj.	10	59.0	+ 69	59	6.8	7.7	125	SR
X Centauri	11	49.2	− 41	45	7.0	13.8	315.1	M
Z Ursæ Maj.	11	56.5	+ 57	52	7.9	10.8	196	SR
R Corvi	12	19.6	− 19	15	6.7	14.4	317.0	M
RY Ursæ Maj.	12	20.5	+ 61	19	6.7	8.5	311.2	SR
R Crucis	12	23.6	− 61	38	6.4	7.2	5.8	C
U Centauri	12	33.5	− 54	40	7.0	14.0	220.3	M
T Ursæ Maj.	12	36.4	+ 59	29	6.6	13.4	256.5	M
R Virginis	12	38.5	+ 06	59	6.0	12.1	145.6	M

Variable	R.A. h m	Decl. ° '	Mag. max min	Period (days)	Type
S Ursæ Maj.	12 43.9	+61 06	7.0 12.3	226.2	M
RY Draconis	12 56.4	+66 00	6.5 8.0	172.5	SR
R Hydrae	13 29.7	−23 17	4.0 10.0	389	M
S Virginis	13 33.0	−07 12	6.3 13.2	377.4	M
T Centauri	13 43.4	−55 38	5.5 9.0	90.6	SR
Theta Apodis	14 05.3	−76 48	7.1 8.6	−	I
R Centauri	14 16.6	−59 55	5.3 11.8	546.2	M
R Camelopardalis	14 17.8	+83 50	7.0 14.4	270.2	M
S Boötis	14 22.9	+53 49	7.8 13.8	270.7	M
R Boötis	14 37.2	+26 44	6.2 13.1	223.4	M
Delta Libræ	15 01.0	−08 31	4.9 5.9	2.3	EA
S Coronæ Bor.	15 21.4	+31 22	5.8 14.1	360.3	M
R Normæ	15 36.0	−49 30	6.5 13.9	492.7	M
R Coronæ Bor.[2]	15 48.6	+28 09	5.7 15	−	RCrB
R Serpentis	15 50.7	+15 08	5.1 14.4	356.4	M
T Coronæ Bor.	15 59.5	+25 55	2.0 10.8	−	RN
R Herculis	16 06.2	+18 22	7.7 15.0	318.4	M
U Serpentis	16 07.3	+09 56	7.8 14.7	237.9	M
W Coronæ Bor.	16 15.4	+37 48	7.8 14.3	238.4	M
U Herculis	16 25.8	+18 54	6.5 13.4	406.0	M
R Draconis	16 32.7	+66 45	6.7 13.0	245.5	M
W Herculis	16 35.2	+37 21	7.6 14.4	280.4	M
S Herculis	16 51.9	+14 56	6.4 13.8	307.4	M
RS Scorpii	16 55.6	−45 06	6.2 13.0	320.0	M
RR Scorpii	16 56.6	−30 35	5.0 12.4	279.4	M
R Ophiuchi	17 07.8	−16 06	7.0 13.8	302.6	M
Alpha Herculis	17 14.6	+14 23	3.0 4.0	−	SR
T Draconis	17 56.4	+58 13	7.2 13.5	421.2	M
S Octantis	18 08.7	−86 48	7.3 14.0	258.9	M
R Pavonis	18 12.9	−63 37	7.5 13.8	229.8	M
W Lyræ	18 14.9	+36 40	7.3 13.0	196.5	M
T Lyræ	18 32.3	+37 00	7.8 9.6	−	I
R Scuti	18 47.5	−05 42	4.4 8.2	140	RV
Beta Lyræ	18 50.1	+33 22	3.3 4.3	12.9	EB
Kappa Pavonis	18 56.9	−67 14	3.9 4.7	9.1	C(W)
R Lyræ	18 55.3	+43 57	3.9 5.0	46	SR
V Aquilæ	19 04.4	−05 41	6.6 8.4	353	SR
R Sagittarii	19 16.7	−19 18	6.7 12.8	268.8	M
U Sagittæ	19 18.8	+19 37	6.6 9.2	3.4	EA
CH Cygni	19 24.5	+50 14	6.4 8.7	97	Z And
RR Lyræ	19 25.5	+42 47	7.1 8.1	0.6	RR
U Aquilæ	19 29.4	−07 03	6.1 6.9	7.0	C
R Cygni	19 36.8	+50 12	6.1 14.2	426.4	M
Chi Cygni	19 50.6	+32 55	3.3 14.2	406.9	M
T Pavonis	19 50.7	−71 46	7.0 14.0	244.0	M
Eta Aquilæ	19 52.5	+01 00	3.5 4.4	7.2	C
Z Cygni	20 01.4	+50 03	7.4 14.7	263.7	M
R Delphini	20 14.9	+09 05	7.6 13.8	284.9	M
P Cygni	20 17.8	+38 02	3 6	−	I
U Cygni	20 19.6	+47 54	5.9 12.1	462.4	M
T Microscopii	20 27.9	−28 16	7.7 9.6	344	SR
EU Delphini	20 37.9	+18 16	5.8 6.9	59	SR
HR Delphini Nova 1967	20 42.3	+19 10	3.6 12.4	−	N
U Delphini	20 45.5	+18 05	7.6 8.9	110	SR
W Aquarii	20 46.4	−04 05	8.7 14.9	381.1	M
T Aquarii	20 49.9	−05 09	7.2 14.2	202.1	M
R Vulpeculae	21 04.4	+23 49	7.0 14.3	136.4	M
T Cephei	21 09.5	+68 29	5.2 11.3	388.1	M
T Indi	21 20.2	−45 01	7.7 9.4	320	SR
S Cephei	21 35.2	+78 37	7.4 12.9	486.8	M
W Cygni	21 36.0	+45 22	6.8 8.9	126	SR
SS Cygni	21 42.7	+43 35	8.4 12.4	50	UG
Mu Cephei	21 43.5	+58 47	3.4 5.1	−	I
R Gruis	21 48.5	−46 55	7.4 14.9	331.9	M
Delta Cephei	22 29.2	+58 25	3.5 4.4	5.4	C
W Cephei	22 36.5	+58 26	7.0 9.2	Long	SR
AR Aquarii	22 51.6	+85 03	7.0 7.9	?	SR
S Aquarii	22 57.1	−20 21	7.6 15.0	297.3	M
Beta Pegasi	23 03.8	+28 05	2.3 2.7	38	SR
R Pegasi	23 06.6	+10 33	6.9 13.8	378.0	M
V Cassiopeiæ	23 11.7	+59 42	6.9 13.4	228.8	M
R Aquarii	23 43.8	−15 17	5.8 12.4	387.0	Symb.
Rho Cassiopeiæ	23 54.4	+58 30	4.1 6.2	−	I
R Phoenicis	23 56.5	−49 47	7.5 14.4	267.9	M
R Cassiopeiæ	23 58.4	+51 24	4.7 13.5	430.5	M

1 Outbursts 1920, 1944   2 Outbursts 1866, 1946

Catalogue of Double Stars

Star	R.A. h m	Decl. ° '	Mag.	P.A. °	Dist. "
Beta Tucanæ	00 31.5	−62 58	4.4, 4.8	169	27.1
Eta Cassiopeiae	00 49.1	+57 49	3.4, 7.5	293	10.1 Binary
Lambda Tucanæ	00 55.0	−69 31	5.3, 7.3	080	20.8 Optical
Beta Phœnicis	01 06.1	−46 43	4.0, 4.2	346	1.4
Zeta Piscium	01 13.7	+07 35	5.6, 6.5	063	23.0
Omega Andromedæ	01 27.7	+45 24	4.8, 12.0	103	2.4
Epsilon Sculptoris	01 45.6	−25 03	5.4, 8.6	034	4.7
Gamma Arietis	01 53.5	+19 18	4.8, 4.8	000	7.8
Alpha Ursæ Minoris	02 31.8	+89 16	2.0, 9.0	218	18.4
Lamba Arietis	01 57.9	+23 36	4.8, 7.4	046	37.4
Alpha Piscium	02 02.0	+02 46	4.2, 5.1	297	2.1
Gamma 2 Andromedæ	02 03.9	+42 20	2.3, 5.5	063	9.8 (B is double, 0".3)
Iota Trianguli	02 15.9	+33 21	5.4, 7.0	071	3.6
Omicron Ceti	02 19.3	−02 59	var, 9.5	005	0.0
Iota Cassiopeiæ	02 29.1	+67 24	4.6, 6.9	232	2.4
Omega Fornacis	02 33.8	−28 14	5.0, 7.7	244	10.8
Nu Ceti	02 35.9	+05 36	4.9, 9.5	083	8.1
Gamma Ceti	02 43.3	+03 14	3.5, 7.3	294	2.8
Theta Persei	02 44.2	+49 14	4.1, 9.9	215	19.8
Theta Eridani	02 58.3	−40 18	3.4, 4.5	088	8.2
Epsilon Arietis	02 59.2	+21 20	5.3, 5.6	191	1.5
Zeta Persei	03 54.1	+31 53	2.9, 9.5	208	12.9
Epsilon Persei	03 57.9	+40 01	2.9, 8.1	010	8.8
40 Eridani	04 15.2	−07 39	4.4, 9.5	104	83.4
Phi Tauri	04 20.4	+27 21	5.0, 8.4	250	52.1
Chi Tauri	04 22.6	+25 38	5.5, 7.6	024	19.4
Alpha Tauri	04 35.9	+16 31	0.9, 13.4	110	30.4
Iota Pictoris	04 50.9	−53 28	5.6, 6.4	058	12.3
Omega Aurigæ	04 59.3	+37 53	5.0, 8.0	359	5.4
Gamma Caeli	05 04.4	−35 29	4.6, 8.1	308	2.9
Rho Orionis	05 13.3	+02 52	4.5, 8.3	064	7.0
Kappa Leporis	05 13.2	−12 56	4.5, 7.4	358	2.6
Beta Orionis	05 14.5	−08 12	0.1, 6.8	202	9.5
Eta Orionis	05 25.5	−02 24	3.8, 4.8	080	1.5
Theta Pictoris	05 24.8	−52 19	6.9, 7.2	287	38.2 (A is double, 0".2)
Delta Orionis	05 32.0	−00 18	2.2v, 6.3	359	52.6
Lambda Orionis	05 35.1	+09 56	3.6, 5.5	043	4.4
Theta Orionis	05 35.3	−05 23	5.9, 6.8 / 6.8, 6.8		The Trapezium
Iota Orionis	05 35.4	−05 55	2.8, 6.9	141	11.3
Sigma Orionis	05 38.7	−02 36	4.0, 6.0 / 6.5,		7.5 Multiple
Alpha Columbæ	05 39.6	−34 04	2.6, 12.3	359	13.5
Zeta Orionis	05 40.8	−01 57	1.9, 4.0	162	2.4
Gamma Leporis	05 44.5	−22 27	3.7, 6.3	350	96.3
Theta Aurigae	05 59.7	+37 13	2.6, 7.1	320	3.6
Eta Geminorum	06 14.9	+22 30	var, 8.8	266	1.4
Epsilon Monocerotis	06 23.8	+04 36	4.5, 6.5	027	13.4
Beta Monocerotis	06 28.8	−07 02	3.7, 4.7	132	7.3
Alpha Canis Majoris	06 45.1	−16 43	−1.5, 8.5	var	< 9
Pi Canis Majoris	06 55.6	−20 08	4.7, 9.7	018	11.6
Mu Canis Majoris	06 56.1	−14 03	5.3, 8.6	340	3.0
Epsilon Canis Majoris	06 58.6	−28 58	1.5, 7.4	161	7.5
Gamma[2] Volantis	07 08.8	−70 30	4.0, 5.9	300	13.6
Lambda Geminorum	07 18.1	+16 32	3.6, 10.7	033	9.6
Tau Canis Majoris	07 18.7	−24 57	4.4, 10.5	090	8.2
Delta Geminorum	07 20.1	+21 59	3.5, 8.2	223	6.0
Sigma Puppis	07 29.2	−43 18	3.3, 9.4	074	22.3
Alpha Geminorum	07 34.6	+31 53	1.9, 2.9	083	2.5 Widening
Kappa Geminorum	07 44.4	+24 24	3.6, 8.1	240	7.1
Epsilon Volantis	08 07.9	−68 37	4.4, 8.0	024	6.1
Gamma Velorum	08 09.5	−47 20	1.9, 4.2	220	41.2
Zeta Cancri	08 12.2	+17 39	5.1, 6.0	088	5.7
Theta Volantis	08 39.1	−70 23	5.3, 10.3	108	45.0
Delta Velorum	08 44.7	−54 43	2.1, 5.1	153	2.6
Iota Cancri	08 46.7	+28 46	4.2, 6.6	307	30.5
Iota Ursae Majoris	08 59.2	+48 02	3.1, 10.8	016	4.6
Upsilon Carinae	09 47.1	−65 04	3.1, 6.1	127	5.0
Gamma Leonis	10 20.0	+19 51	2.2, 3.5	122	4.3
Xi Ursae Majoris	11 18.2	+31 32	4.3, 4.8	060	1.3
Nu Ursae Majoris	11 18.5	+33 06	3.5, 9.9	147	7.2
Gamma Crateris	11 24.9	−17 41	4.1, 9.6	096	5.2
Alpha Crucis	12 26.6	−63 06	1.4, 1.9	115	4.4
Gamma Crucis	12 31.2	−57 07	1.6, 6.7	031	110.6
Gamma Centauri	12 41.5	−48 58	2.9, 2.9	353	1.4
Gamma Virginis	12 41.7	−01 27	3.5, 3.5	287	3.0 Closing
Iota Crucis	12 45.6	−60 59	4.7, 9.5	022	26.9
Beta Muscae	12 46.3	−68 06	3.7, 4.0	014	1.4
Alpha Canum Venticorum	12 56.0	−38 19	2.9, 5.5	229	19.4
Zeta Ursae Majoris	13 23.9	+54 56	2.3, 4.0	152	14.4
Iota Bootis	14 16.2	+51 22	4.9, 7.5	033	38.5
Alpha Centauri	14 39.6	−60 50	0.0, 1.2	215	19.7 (epoch 1990)
Zeta Boötis	14 41.1	+13 44	4.5 4.6	303	1.0
Epsilon Boötis	14 45.0	+27 04	2.5, 4.9	339	2.8
Xi Boötis	14 51.4	+19 06	4.7, 7.0	326	7.0
Pi Lupi	15 05.1	−47 03	4.6, 4.7	073	1.4
Kappa Lupi	15 11.9	−48 44	4.0, 4.1	144	27.0
Eta Coronæ Borealis	15 23.2	+30 17	5.6, 5.9	024	1.0
Gamma Circini	15 23.4	−59 19	5.1, 5.5	049	0.9
Delta Serpentis	15 34.8	+10 32	4.2, 5.2	179	3.9
Zeta Coronæ Borealis	15 39.4	+36 38	5.1, 6.0	305	6.3
Gamma Coronæ Borealis	15 42.7	+26 18	4.2, 5.6	118	0.6
Eta Lupi	16 00.1	−38 24	3.6, 7.8	020	15.0
Beta Scorpii	16 05.4	−19 48	2.6, 4.9	021	13.6
Kappa Herculis	16 08.1	+17 03	5.3, 6.5	012	28.4
Nu Scorpii	16 12.0	−19 28	4.3, 6.4	337	41.1
Eta Draconis	16 24.0	+61 31	2.7, 8.7	142	5.2
Alpha Scorpii	16 29.4	−26 26	1.2, 5.4	275	2.9
Zeta Herculis	16 41.3	+31 36	2.9, 5.5	089	1.6 Binary
Alpha Herculis	17 14.6	+14 23	var, 5.4	107	4.7
Rho Herculis	17 23.7	+37 09	4.6, 5.6	316	4.1
Nu Draconis	17 32.2	+55 11	4.9, 4.9	312	61.9
Epsilon Lyræ	18 44.3	+39 40	4.7, 5.1	173	207.7
Epsilon 1	"	"	5.0, 6.1	357	2.6
Epsilon 2	"	"	5.2, 5.5	094	2.3
Zeta Lyræ	18 44.8	+37 36	4.3, 5.9	150	43.7
Beta Lyræ	18 49.8	+32 49	var, 8.6	149	45.7
Theta Serpentis	18 56.2	+04 12	4.5, 4.5	104	22.3
Zeta Sagittarii	19 02.6	−29 53	3.2, 3.4	327	0.2
Gamma Coronæ Australis	19 06.4	−37 04	4.8, 5.1	109	1.3
Beta Cygni	19 30.7	+27 58	3.1, 5.1	054	34.4
Delta Cygni	19 45.0	+10 46	2.9, 6.3	246	2.1
Alpha Capricorni	20 18.1	−12 33	3.6, 4.2	291	377.7
Alpha 2	20 18.1	−12 33	3.8, 10.6	158	7.1
Beta Delphini	20 37.5	+14 36	4.0, 4.9	187	0.3
Gamma Delphini	20 46.7	+16 07	4.5, 5.5	268	9.6
61 Cygni	21 06.9	+38 45	5.2, 6.0	148	30.0
Zeta Aquarii	22 28.8	−00 01	4.3, 4.6	200	2.0 Widening
Delta Cephei	22 29.2	+58 25	var, 7.5	191	41.0
Sigma Cassiopeiae	23 59.0	+55 45	5.0, 7.1	326	3.0

Acknowledgements

Photographic credits

Abbreviations used are; l, left; r, right; c, centre; t, top; b, bottom
AAT, Anglo-Australian Telescope Board;
AIP, American Institute of Physics;
Caltech, California Institute of Technology;
CFA, Harvard-Smithsonian Center for Astrophysics;
CTIO, Cerro Tololo Inter-American Observatory (La Serena, Chile);
ESA/ESTEC, European Space Agency-European Research & Technology Centre;
IPC, International Portrait Catalogue, Archenhold Sternwarte, Berlin, GDR;
JPL, Jet Propulsion Laboratory/California Institute of Technology;
KPNO, Kitt Peak National Observatory (Tucson, AZ);
Mansell, The Mansell Collection;
NASA, National Aeronautics and Space Administration;
NOAO, National Optical Astronomy Observation;
NOAA, National Oceanic and Atmospheric Administration;
NRAO/AUI, The National Radio Astronomy Observatory, operated by Associated Universities, Inc under contract with the National Science Foundation;
PM, Patrick Mocre Collection;
RAS, Royal Astronomical Society, London;
ROE, Royal Observatory, Edinburgh;
ROE/Malin, Photography by D.F.Malin of the Anglo-Australian Observatory from original negs by United Kingdom Schmidt telescope © Royal Observatory, Edinburgh;
SAO, Southern Astronomical Observatory;
USNO, US Naval Observatory;
Yerkes Observatory, Yerkes Observatory/University of Chicago.

Cover (background), UK and US editions: AAT © 1982, 1981. Cover (insets, l to r), UK edition: Lockheed Missiles and Space Co.; NOAO; NASA; Paul Doherty. US edition: Lockheed Missiles and Space Co.; NASA; Paul Doherty. Background to initial letters, AAT © 1984.

1, AAT © 1977; 2, AAT © 1981; 4, AAT; 7, ROE/Malin © 1982; 8, Joseph Sutorik/NOAA/US Department of Commerce; 10l, Thomas Stephenson/SAO; 10r, Dane Penland, from Paul Gorenstein and Dan Fabricant/CFA; 11tr, Dennis di Cicco; 11b, Erick T. Young/University of Arizona/Steward Observatory; 12t, Dr. C.G.T.Haslam, H.Stoffel, Dr. C.J.Salter, Dr. W.E.Wilson (100m Effelsberg; 250ft Jodrell Bank; 64m Parkes); 12b, T.N.Gautier/Caltech; 13, F.R.Harnden Jr./Harvard-Smithsonian Center for Astrophysics; 14l, Farhad Yusef-Zadeh, Mark R.Morris, Don R.Chance/NRAO/AUI; 14r, Department of Planetary Sciences, University of Arizona; 15, J.Duerst; 16l, R.Donald P.Schneider/School of Natural Sciences/Institute for Advanced Study, Princeton; 20, AIP/Niels Bohr Library; 22, RAS; 25tl, NASA; 26, RAS; 27, Royal Greenwich Observatory; 28l, RAS; 28r, AAT © 1977; 32, NASA; 34, The Arecibo Observatory is part of the National Astronomy and Ionosphere Center which is operated by Cornell University under contract with the National Science Foundation/Equinox (Oxford) Ltd.; 35, Palomar Observatory; 36, Centre National d'Etudes Spatiale; 37r, US Geological Survey/Photo: D.J.Roddy; 38l, Mansell; 38br, The Mary Evans Picture Library; 38tr, NASA/SPL; 43, Mansell; 46, Mansell; 50, RAS; 56tr, W.Dieckvoss, Sternwarte Hamburg/IPC; 56br, AAT, © 1986; 57l, AIP/Niels Bohr Library; 58, AAT, © 1980; 59tl, AAT, © 1976; 59bl, Dr. John F.Arens/SPL; 59r, IPC; 61, Ann Ronan Picture Library; 63, RAS; 66, Cubic Space Division, 1952, © M.C.Escher Heirs c/o Cordon Art, Baarn, Holland; 67l, AT & T Bell Laboratories; 68t, Circle Limit 1; © M.C.Escher Heirs c/o Cordon Art, Baarn, Holland; 68b, M. Seldner, B.L.Siebers, E.J.Groth, P.J.E.Peebles/Astronomical Journal 82, 249, 1977; 69t, AAT, © 1980; 71b, Rudolph Schild/SAO; 72b, Dane Penland, from Stephen Murray/CFA; 74t, Fermilab National Accelerator Laboratory; 76, Brookhaven National Laboratory; 77r, Popperfoto; 77l, NASA; 78l, AIP/Niels Bohr Library; 78r, Academie Royale des Sciences, des Lettres et des Beaux-Arts de Belgique; 78/9tc, Dr. Kurt Weiler/SPL; 79c, RAS; 79r, Popperfoto; 81, ROE, © 1978; 82, Kitt Peak National Observatory/AAT; 83, AAT, © 1986; 84, AAT, © 1986; 86l, Novosti Press Agency; 86r, NASA; 87, NASA; 89, AAT, © 1977; 91, National Maritime Museum, London; 94, AAT, © 1980; 95, NOAO/CTIO; 97, USNO; 99l, PM; 99r, RAS; 100bl, NASA; 100c, AAT © 1977; 101l, RAS; 105, NASA; 106l, RAS; 107, Harvard College Observatory/Hansen Planetarium; 110l, Novosti Press Agency; 111, NASA; 112tl, Novosti Press Agency; 112bl, The National Maritime Museum, London; 112c, RAS; 113l, AAT, © 1980; 113r, W.R.Dawes/IPC; 114, NASA; 115l, Mansell; 115r, Paul Doherty; 116l, RAS; 116c, 116r, NASA; 117t, University of Toronto; 117br, IPC; 118c, Palomar Observatory; 119l, Lick Observatory/RAS; 119c, AAT © 1986; 120l, RAS; 120r, NOAO/KPNO; 121, NASA; 123l, IPC; 123r, Max-Planck-Institut fuer Radioastronomie/Photo: G.Hutschenreiter; 124, NASA; 129, NASA; 130l, NASA; 130/1, NRAO/AUI; 131r, NASA; 132, NASA; 133t, 133bl, 133br, NASA; 134/5, NASA; 135t, NASA; 135b, Rutherford Appleton Laboratory/SERC; 136/7, NASA; 138t/b, TASS News Agency; 139t, JPL; 139b, USNO/National Air & Space Museum; 140/1, JPL; 142, NASA; 143l, NASA; 143r, NASA; 145c, Mansell; 145r, NASA; 147, Paul Doherty; 151, Royal Observatory, Cape, South Africa; 152l, ESA-ESTEC; 152r, NASA; 157l, Mansell; 159l, AIP/Niels Bohr Library; 159r, NASA; 160, Novosti Press Agency; 162, Mansell; 165l, AIP/Niels Bohr Library. Physics Today Collection from Center for History of Physics, NY; 165r, NASA; 168l, NASA; 168, NOAO/KPNO; 169, Moore & Chappell, Lick Observatory; 170, USNO; 171l/tr, 171br, NASA; 174l, NASA; 175l, PM; 176, RAS; 177l, Mansell; 177r, European Southern Observatory/PM; 179, AAT © 1978); 180bl, AAT © 1986; 181, Mansell; 186l, Brookhaven National Laboratory; 186c, AAT © 1984; 187l, Palomar Observatory; 187r, Space Telescope Science Institute, Baltimore, MD; 188, Lockheed Missiles and Space Co.; 191, NASA; 193, USNO; 194t, AAT © 1981; 195t, USNO; 195bl, NASA; 195br, Caltech; 196t, Rudolph Schild/SAO; 196b, ROE © 1980/Malin; 197t, ROE © 1985/Malin; 198b, ROE © 1980/Malin; 199t, ROE, © 1980; 199b, AAT © 1980; 200b, NOAO/KPNO; 202/3b, NASA; 204/5b, Dr. C.G.T.Haslam, H.Stoffel, Dr. C.J.Salter, Dr. W.E.Wilson (100m Effelsberg; 250ft Jodrell Bank; 64m Parkes); 203t, Dr. Carl Heiles/SPL; 205t, R.Allen/SPL; 207b, Dr. W.Kaufmann/Palomar Observatory; 208, AAT © 1980; 210, SERC; 213/4, NASA; 217tl, RAS; 217bl, AIP/Niels Bohr Library/Physics Today Collection. From: Center for History of Physics; 217c, © 1986 Max-Planck Institute fuer Aeronomie; 217r, AAT © 1977; 218, PM; 219/220, NASA; 221tr, Mansell; 221br, NASA; 222, PM; 223c, RAS; 223r, Sidney Wolff/NAOA/KPNO; 224, Ames Research Center/NASA; 225c, IPC; 226r, AIP/Niels Bohr Library/Shapley Collection/Center for History of Physics; 228, AAT © 1977; 229, IPC; 230, Lick Observatory/RAS; 232l, Novosti Press Agency; 232r, Berntzen/Hart/Seeley/IPC; 234, NASA; 236c, NASA; 236tr, AAT, © 1976); 236br, NASA; 237c, NASA; 237r, IPC; 239, ROE © 1978; 242, World Data Center A for Rockets and Satellites/Goddard Space Flight Center; 243, NASA; 244, NASA; 245, NASA; 246l, IPC; 246r, D.P.Cruikshank; 247, Mansell; 248l/r, NASA; 250, PM; 252, NASA; 253l, ESA-ESTEC; 255, Messerschmidt/ZEFA Picture Library, London; 256, NASA; 257–61, NASA; 262l NASA/SPL; 263–72, NASA; 274, Yerkes Observatory; 275, NASA; 276, Palomar Observatory; 277l, Dr. Wayne Orchiston; 277r, RAS; 279, Mullard Radio Astronomy Observatory/University of Cambridge; 280, Smithsonian Institution — University of Arizona; 281, NASA; 284, NOAO/CTIO; 290, Mansell; 291c, Royal Greenwich Observatory; 292, R.K.Pilsbury; 293, PM; 297, Hamburger Sternwarte/Universitaet Hamburg/IPC; 299, IPC; 300c/r, Dr. M.H.Carr/National Space Science Data Center; 301c, Yerkes Observatory; 301r, RAS; 304l, ROE/Malin © 1982; 304c, PM; 306t, Observatoire de Paris/Photo: J. Counil; 306b, PM; 307, NASA; 309tr, NASA; 312/3, IPC; 316, USNO; 318l, Mansell; 318tc, ROE/Malin © 1985; 318bc, Robin Scagell; 319, USNO; 320tc, Royal Greenwich Observatory; 320bc, Istituto ē Museo di Storia della Scienza, Firenze; 322tl, Cavendish Laboratories/University of Cambridge; 322bl, Philip Daly; 322–3, The Arecibo Observatory, part of the National Astronomy and Ionosphere Centre which is operated by Cornell University under contract with the National Science Foundation; 329b, Palomar Observatory; 330t, Harvard-Smithsonian Center for Astrophysics/SPL; 330c, Starlink/AAT/SPL; 330/lt, ROE © 1979/Malin; 337tl, Robin Scagell; 338/9b, 341, Mansell; 343/4, NASA; 348, NASA; 352, NOAO/KPNO; 354, Royal Greenwich Observatory; 361r, NASA; 362, ROE © 1980/Malin; 363, Palomar Observatory; 365, Royal Greenwich Observatory; 366t, NASA; 366b, Mansell; 367, IPC; 368r, Lick Observatory; 369l, Observatorie de Haute-Provence/Centre National de la Recherche Scientifique; 369c, Novosti Press Agency; 370, NASA; 371l, NASA; 371r, RAS; 373, IPC; 374, Yerkes Observatory; 375c, AAT © 1984; 375r, Yerkes Observatory; 376, NASA/SPL; 377, Robert Scagell; 378, Lick Observatory; 380, Lowell Observatory; 383, E.H.Strach; 385, ROE © 1984/Malin; 401, NASA; 403, Jay M.Passachoff © 1978/Hansen Planetarium; 404tr, USNO; 404br, Dr. E.I.Robson/SPL; 405l, 405ct, NASA; 405cb, NASA/SPL; 407l, NASA; 407c, Novosti Press Agency; 407r, NASA/SPL; 412tl, Palomar Observatory; 412bl, IPC; 412c, Novosti; 412r, AAT © 1985; 418, USNO/SPL; 412r, National Space Development Agency of Japan; 422, Naval Research Laboratory/Hansen Planetarium; 423tc, NOAO; 423r, ROE © 1979/Malin; 426, AAT © 1984; 427, PM; 428, PM; 429, British Telecom International; 430l, Novosti Press Agency; 430c, 430r, NASA; 433l, 433c, NASA; 433r, PM; 434l, US Geological Survey; 434c, RAS; 434r, USNO; 435t, Palomar Observatory; 435b, Palomar Observatory; 436r, Novosti Press Agency; 437l, RAS; 437r, NASA; 438r, NASA; 439, NASA; 440tl, USNO/Photo: G.Cleere; 440c, Mansell; 440r, NASA; 442l, Ann Ronan Picture Library; 443l, University of Arizona, Tucson; 443r, NASA; 444, Paul Doherty; 445tl, TASS News Agency; 445bl, Palomar Observatory; 445tc, NASA; 445bc, PM; 446, NASA; 447l, NRAO/AUI; 448l, ROE © 1982; 448tc, NASA; 448bc, AIP/Niels Bohr Library; 448tr, Philosophical Magazine, 1836, 3rd series, vol. 9/Photo: Eileen Tweedy; 448br, TASS News Agency; 449l, NASA; 449r, US Air Force/MARS; 450, Lick Observatory; 451t, Dr. A.Lazenby; 451b, Royal Greenwich Observatory; 453tl, Royal Greenwich Observatory; 453bl, University of Glasgow, University Observatory; 453r, Mansell; 454tl, RAS; 454bl, Ann Ronan Picture Library; 455, Dane Penland, from Leon Van Speybroeck, CFA; 457tc, Yerkes Observatory; 458, Novosti Press Agency; 459, ATT © 1985.

Artwork credits

All A–Z artwork not otherwise credited is by Paul Doherty.

9, 11, Kuo Kang Chen; 35, Max Rutherford; 65–7, Paul Doherty; 69b, Paul Doherty; 70–5, Paul Doherty; 80, Paul Doherty after M. Rees; 144, Paul Doherty; 200–7, David Mallott; 258–72, Mike Saunders; 321, Kuo Kang Chen; 324–36, Paul Doherty; 386, Kuo Kang Chen. 388–91, David Mallott; 392–95, Kuo Kang Chen; 396–99, David Mallott; 400, Mike Saunders.

Attributions and references

67tr, based on an illustration in *The Cambridge Atlas of Astronomy*, Cambridge University Press, 1985; 192 © Uitgeverij het Spectrum B.V., 1974; 204tl, after B.A.Twarog, 1980; 206bl, after A.Boksenberg and Professor W.L.W.Sargent; 386b, after V. de Lapparent, M. Geller and J.Huchra, Harvard-Smithsonian Center for Astrophysics, © American Astronomical Society and *The Astrophysical Journal*, 1986; 387t, after J.Peebles, © American Astronomical Society and *The Astrophysical Journal;* 387b, after V. de Lapparent, M.Geller and J.Huchra, Harvard-Smithsonian Center for Astrophysics, © American Astronomical Society and *The Astrophysical Journal*, 1986; 388b, 389t, 388b, Gérard de Vaucouleurs; 390bl, after R.Brent Tully, University of Hawaii. 390–91b, Gérard de Vaucouleurs; 391t, after Rob Hess and by permission of *Sky and Telescope*, Cambridge, Mass; 392–93b, from the *Third Reference Catalogue of Bright Galaxies*, by G.de. Vaucouleurs, A. de Vaucouleurs and H.G.Corwin, Austin; University of Texas Press, 1976; 393t, Gérard de Vaucouleurs; 393cr, after R.Brent Tully, University of Hawaii, © *Sky and Telescope*, Cambridge, Mass; 394, 395t, 395lbc, Gérard de Vaucouleurs; 396cl, after T. Kiang and W.C.Saslaw; Monthly Notices of the Royal Astronomical Society, 1969; 396–97b, courtesy R.Brent Tully, University of Hawaii; 397ct, after G. de Vaucouleurs and W.L.Peters, *The Astrophysical Journal*, 1985; 398t, after J.Centrella and A.Melott, © *Physics Today*, 1983; 398b, after P.R.Shapiro, C.Struck-Marcell and L.A.Melott; © *The Astrophysical Journal*, 1983; 399, after S.A.Gregory, L.A.Thompson and W.G.Tifft; © *The Astrophysical Journal*, 1981; 400, after S.A. Gregory and L.A.Thompson; © *The Astrophysical Journal*, 1978.